CONTENTS

PART II
COASTAL PROCESSES AND SEDIMENT TRANSPORT

PART III
COASTAL STRUCTURES AND RELATED PROBLEMS

PART IV
COASTAL, ESTUARINE AND ENVIRONMENTAL PROBLEMS

PART V
CASE STUDIES OF SOLUTIONS TO COASTAL ENGINEERING PROBLEMS

PART VI
SHIP MOTIONS

Costa de Benalmadena — Malaga

PART III

COASTAL STRUCTURES

Playa D'Aro — Gerona, costa Brava

CHAPTER 147

GENTLE SLOPE SEAWALLS COVERED WITH ARMOUR UNITS

by
Osamu Toyoshima

ABSTRACT

Since the early 1960's, many seawalls against the beach erosion have been constructed in Japan, most of which were of vertical type. As a result, some of the seawalls even encouraged the beach erosion due to the reflected waves on the steep front of the seawalls. The author then proposed seawalls of new types with gentle front slope(1 to 3) covered with armour units in 1981, and over one hundred fieldworks have successfully been carried out.

In 1985, the author proposed reforming the existing vertical type seawalls into the gentler front slope (1 to 5 or 6) seawalls. The laboratory test on the gentle slope seawall was made, and some experimental fieldworks were carried out. At this time, these new type gentler slope seawalls are successful.

I. INTRODUCTION

On the 19th International Conference on Coastal Engineering, the author presented a paper titled " **NEW TYPE BLOCKS FOR SEAWALL SLOPE PROTECTION** " and proposed a gentle slope (1 to 3) seawall, covered with new type block named "Lotus-Uni" for measures against beach erosion. After the author's proposal, over one hundred works of the new type seawall have been carried out on various erosive sandy beaches in Japan, and most of these works have proved successful.

In 1985, the author proposed reforming the existing vertical type seawalls into the gentle slope (1 to 5 or 6) seawalls covered with Lotus- Uni. The reason for reforming is that, the existing vertical type seawalls have been disliked by inhabitants near the coast because of spray of seawater and environmental disruption.

II. THE EXPERIMENTAL WORKS OF THE GENTLE SLOPE SEAWALL

The first experimental works of the gentle slope (1 to 3) seawall was carried out on Muroran Coast, Hokkaido Prefecture, facing the Pacific Ocean, in 1982.

Dr.Eng.,Professor, School of Marine Scie. & Tech., Tokai Univ. Orido 3-20-1, Shimizu, Shizuoka Pref. 424 Japan.

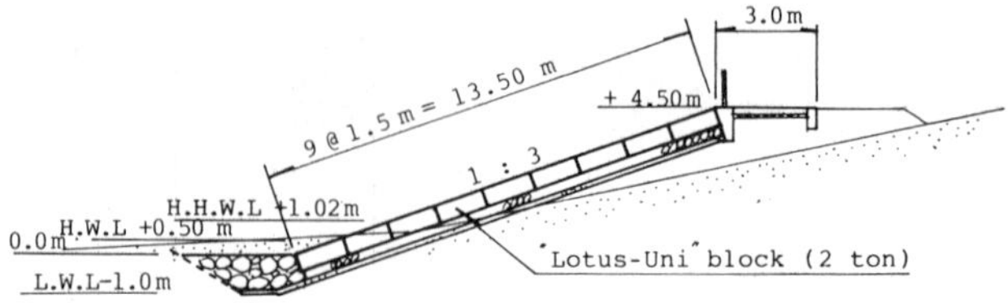

Fig.1. Cross section of the new type seawall on Muroran Coast.

Fig.1 shows the standard cross section of the new type gentle slope seawall covered with Lotus-Uni blocks. Nine blocks are arranged in each column, and no foundation nor shore protection works such as wave dissipation mounds were provided.

Photo 1. Experimental seawall immediately after the completion (1982).

As shown in Fig.1, the lowest blocks of the slope were set on the height of the mean water level, and there existed very narrow foreshore in front of the seawall. Uprush water mass with sands and gravels runs up on to the slope. Under the block facing, there lie a 50cm thick layer of cobbles and a 10cm layer of gravels. Part of the uprush flows into the holes of and the spaces between the blocks as shown in Photo 1. Thus the flow rate of backrush decreases, and wave reflection diminishes, the erosive beach tunning accumulative.

A later view of the beach is given in Photo 2. The shoreline has advanced remarkably, and the sand beach has grown extensively. Most part of the seawall is buried under the accumulated sand, only one block row being visible. The experimental seawall successfully change the erosive beach into a stable beach.

The result of this experimental works supported the author's confidence that a seawall against beach erosion should have a rough and permeable facing, and its front slope should be gentle enough to diminish wave reflection.

Photo 2. A later view of the experimental seawall.

III. GENTLER - SLOPE SEAWALL

When a gentle slope seawall is constructed on a narrow foreshore, the toe will be under the water. A gentler slope will yield a deeper toe, which may cause higher wave runup. Hence the determination of crown height as well as slope gentleness becomes an important problem.

3.1 Laboratory tests on wave runup on gentle slope

In general, wave runup on a gentle slope is lower than that on a steep slope. Fig.2 shows diagrams of the wave runup height based on laboratory measurements by Saville (1958), where H_o is the offshore wave height and H_o/L_o the offshore wave steepness. The laboratory test was conducted using periodic waves on smooth impermeable slopes. The slope of the model seabed was 1/10.

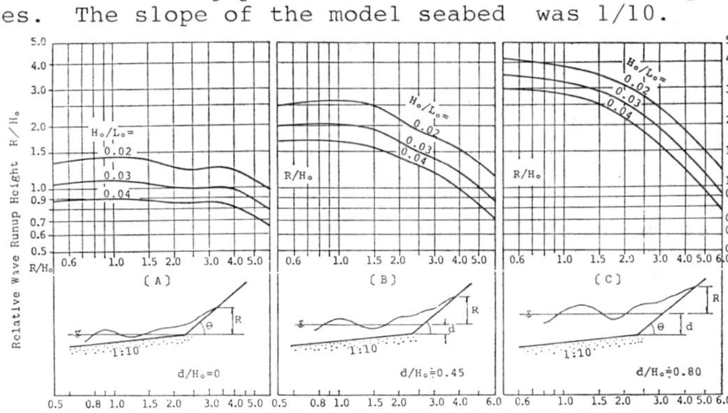

Fig.2 Wave runup height on smooth, impermeable slope faces on the seabed of 1/10 slope(Drawn after Saville,1958).

The diagram [A] is for cases of the relative toe depth $d/H_o=0$, [B] for $d/H_o=0.45$, and [C] for $d/H_o=0.80$.

The relative wave runup height R/H_o in[A] is comparatively small and varied little with the structure slope (1/cot), where as R/H_oin[B] and [C] shows a considerable increase with the relative toe depth d/H_o and with the slope. In the case of the structure slope of 1:5 or 1:6 ($\cot\theta$ = 5 or 6), however, the runup height is considerably small and its variation with the toe depth is rather insignificant. then, the slope of 1:5 and 1:6 are very favorable to wave runup in comparison with the slope of 1:3 or 1:4. This is why the author regards gentler-slope seawalls as the optimum measure for shore protection.

Simultaneously with the proposal of the reforming the existing vertical type seawalls into the gentler-slope (1:5 or 1:6) seawalls,the author made some laboratory tests on the gentler-slope seawalls in the Tokai University.

The length of the two dimensions water tank was 52 m, water depth was 0.70m, slope of the model seabed was 1:20, all model slopes were made as smooth and impermeable, and regular waves were used.

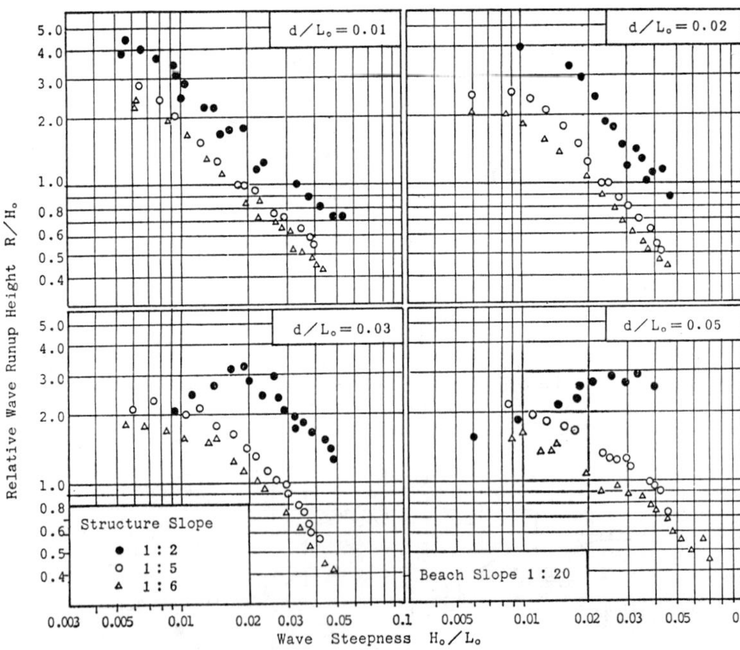

Fig.3 Wave runup height on smooth, impermeable slope faces on the seabed of 1:20 slope(by Toyoshima, 1985).

From the results of the laboratory test, as shown in Fig.3, it has become clear that,

(1) the wave runup height on 1:5 and 1:6 slope are considerably small in comparison with 1:2 slope.

(2) the depth at the toe of seawalls become deeper, gentler slope seawalls are advantageous on the runup heights, generally.

3.2 Experimental works of a reformed type seawall

A reformed type seawall with gentler slope, which toe was under the water, was experimentally constructed on Hongo Coast.

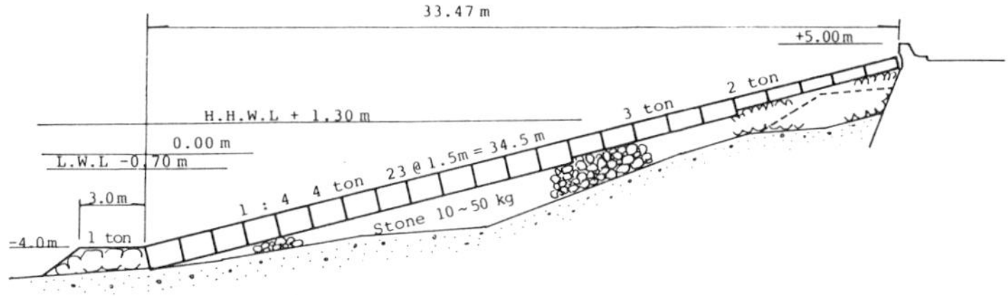

Fig.4 Reformed type experimental seawall on Hongo Coast.

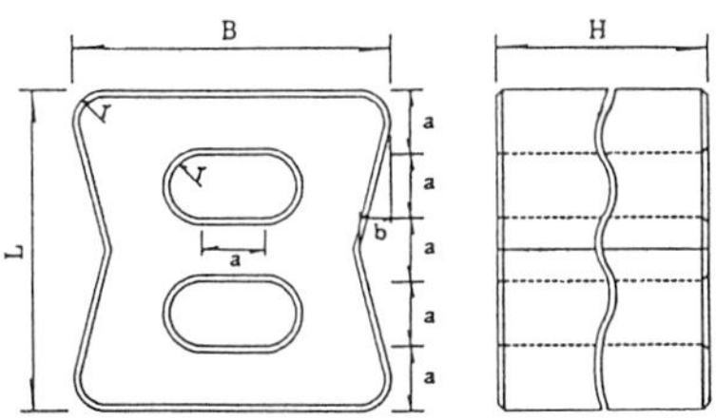

Model	B	L	H	Weight
50 H			50 cm	2 ton
75 H	1.5 m	1.5 m	75 cm	3 ton
100 H			100 cm	4 ton

Fig.5 Lotus-Uni block specification.

Fig.4 shows the standard cross section of the experimental works of reformed type seawall with gentler slope (1:4) on Hongo Coast. Three models of Lotus-Uni block were used as facing blocks without any special foundations or rubble stone works.

Photo 3. A former view of Hongo Coast (1977).

Photo 3 shows a former view of Hongo Coast in 1977 after the construction of the old seawall. A little sand beach were remaining.

Photo 4. Before the experimental works on Hongo Coast.

Photo 5. The experimental seawall under construction.

Photo 6. The experimental seawall completed.

Sand beach remaining in Photo 3 disappears in Photo 4 which was taken before the experimental works. In winter stormy weather, considerable amount of splash overtopped the seawall in spite of armour units placed in its front. Photo 5 shows the experimental seawall under construction. The depth at the toe of the slope was about 4 m.

The works was completed in August,1985,(Photo 6). Although this seawall has been attacked annually by severe winter waves, no collapse of the slopes or dispersion of the facing blocks has occured up to now.

IV. GENTLER SLOPE SEAWALL ON KUROBE COAST

4.1 Construction of the experimental seawall

The first, real, reformed type, gentler (1:5) slope seawall, which toe was under the water, was experimentally constructed on Kurobe Coast, in fiscal 1986 and 1987. The location of Kurobe Coast is shown in Fig.6, together with Muroran and Hongo Coasts. The Kurobe Coast is one of the most eroded coast in Japan. Fig.7 and Fig.8 show the location and area of the experimental seawall works in Kurobe Coast.The wave recorder located about 500 m offshore from the coast, and 5 km to the experimental works site. The water depth of the wave recorder site is 15 m,and the wave recorder is set at -14m depth.

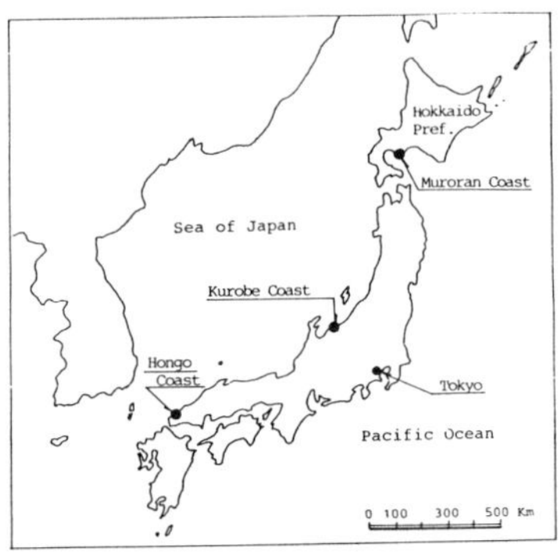

Fig.6 Location of Kurobe Coast and others.

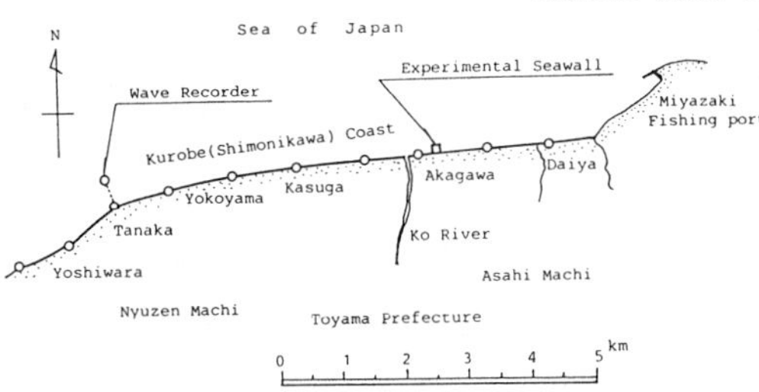

Fig.7 Location of the experimental seawall works.

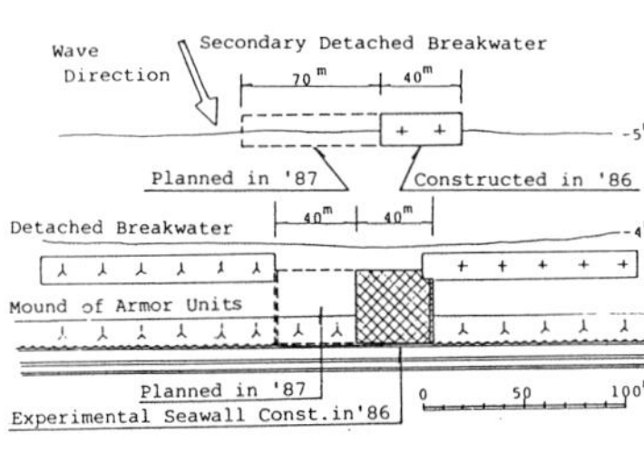

Fig.8 Areas of the experimental seawall works.

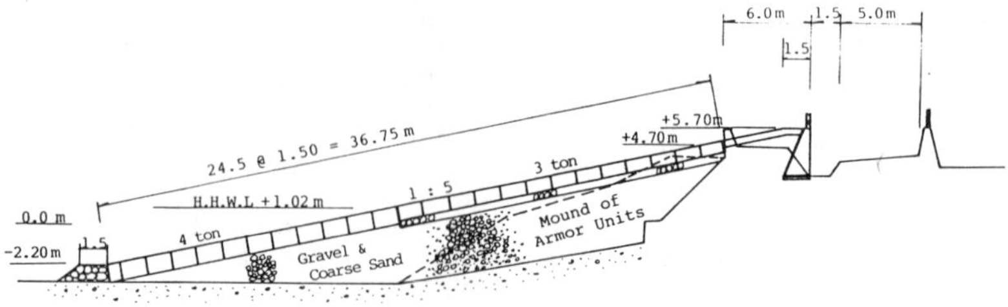

Fig.9 Standard cross section of the experimental seawall.

In fiscal 1986, a part of the experimental seawall and the secondary detached breakwater were constructed. The predominant direction of severe winter waves is NNW. The cross section of the experimental seawall is shown in Fig.9. The front slope is 1:5, and two models of Lotus-Uni, 4 ton and 3 ton, are used as the facing blocks without any special foundations. Removing the existing parapet wall, and reformation of the crown protection and retaining wall were carried out in fiscal 1987.

Photo 7 shows a latest view on the left-hand side of the experimental seawall construction site, and that is the same conditions before the experimental seawall works. There are vertical type seawalls, large mound of armour units for wave dissipation works and detached breakwaters. The weight of armour units for wave dissipation works and detached breakwaters are 8 ton and 20 ton respectively.

Photo 7. Left-hand side view of the experimental works.

First, as shown in Photo 8, the existing mound of armor units were removed.The existing seawalls are visible. Until the early 1960's, some narrow sandy beaches had been remaining in this area. However, immediately after the construction of this seawall, the beaches have vanished away. Then, large mound of armor units have been built, in order to protect the toe of seawalls from scouring and reduce the wave overtopping and splashing of sea water. However, these have been not enough to reduce the wave overtopping and splashing. And sometimes, the seawalls were damaged due to the leakage of the backfill sand.

Photo 8. First, mound of armour units were removed.

Photo 9. Filling of sand and gravel, which were diverted from the dredging of Miyazaki fishing port, located 4 km away from the site (cf. Fig.7).

Photo 10. Setting the Lotus-Uniblock on the rubble stone. The lowest block were set at -3.3 m water depth.

Photo 11. The works of the fiscal 1986 have completed in September 1986. The length of works is 40 m.

4.2 Observation of the wave runup on the seawall

After the completion of the experimental works, the first heavy wave struck the seawall in November 1986. On the 20th December 1986, pretty heavy waves attacked the Kurobe Coast, and overtopped the crown of the experimental seawall, as shown in Photo 12.

Photo 12. Heavy waves overtopped the crown of the experimental seawall, on the 20th December 1986.

The number of times of heavy wave attacks in the first winter season amounted to ten times. The wave runup heights on the experimental seawall were observed by eye measurement and video cameras in the field. Color lines were marked on the blocks for the observation of the wave runup height. These field observations have been done on each twenty minutes of the every O'clock in the day time on stormy weather days.

Table 1 shows the data of the waves which were obtained from the wave recorder, and of the calculated relative runup heights on the experimental seawall based on the field observation.

Table 1. Data of the waves and the calculated relative runup height on the experimental seawall based on the field observation.

Date	Time	① Nos. of Runp	② Nos. of Waves	③ Runup Count In	④ H_o ($H_{1/10}$) m	⑤ T ($T_{1/10}$) s	⑥ L_o ($L_1{}'_{10}$) m	⑦ Water Level +m	⑧ Water Depth dm	⑨ Runup Height Rm	⑩ H_o/L_o	⑪ d/L_o	⑫ R/H_o
'86. 11. 26	10	120	245	25	2. 28	7. 4	85. 4	0. 38	3. 58	3. 09	0. 027	0. 042	1. 36
	11	125	233	23	2. 16	7. 5	87. 8	0. 37	3. 57	3. 12	0. 025	0. 041	1. 44
	12	127	231	23	2. 12	7. 8	94. 9	0. 36	3. 56	2. 97	0. 022	0. 038	1. 40
	13	117	213	21	2. 58	8. 9	123. 6	0. 32	3. 52	2. 99	0. 021	0. 028	1. 16
	14	108	223	22	2. 25	8. 6	115. 4	0. 31	3. 51	3. 49	0. 020	0. 030	1. 55
	15	109	210	21	2. 25	9. 2	132. 0	0. 27	3. 47	2. 75	0. 017	0. 026	1. 22
11. 27	12	96	178	18	1. 70	9. 9	152. 9	0. 22	3. 42	4. 92	0. 011	0. 022	2. 89
	14	89	199	20	1. 32	9. 5	140. 8	0. 15	3. 35	4. 68	0. 009	0. 024	3. 55
12. 15	10	145	213	21	2. 18	7. 2	80. 9	0. 25	3. 45	2. 79	0. 027	0. 043	1. 28
	11	125	210	21	2. 38	6. 7	70. 0	0. 27	3. 47	3. 07	0. 034	0. 050	1. 29
	12	131	218	22	2. 73	6. 9	74. 3	0. 30	3. 50	3. 48	0. 037	0. 047	1. 27
	13	126	233	23	2. 61	6. 7	70. 0	0. 35	3. 55	3. 09	0. 037	0. 050	1. 18
	14	126	233	23	2. 45	6. 5	65. 9	0. 38	3. 58	3. 02	0. 037	0. 054	1. 23
	15	118	245	25	2. 33	6. 5	65. 9	0. 38	3. 58	3. 14	0. 035	0. 054	1. 35
12. 20	10	80	178	18	3. 34	9. 9	152. 9	0. 26	3. 46	5. 14	0. 022	0. 023	1. 54
	11	92	159	16	3. 66	9. 2	132. 0	0. 26	3. 46	5. 29	0. 028	0. 026	1. 44
	12	85	184	18	3. 09	9. 6	143. 8	0. 24	3. 44	5. 17	0. 021	0. 024	1. 67
	13	95	164	16	3. 45	10. 2	162. 3	0. 23	3. 43	5. 12	0. 021	0. 021	1. 48
	14	99	147	15	3. 70	10. 3	165. 5	0. 25	3. 45	5. 42	0. 022	0. 021	1. 46
	15	89	144	14	3. 22	10. 3	165. 5	0. 28	3. 48	4. 93	0. 019	0. 021	1. 53
'87. 1. 14	10	100	271	27	1. 05	8. 5	112. 7	0. 11	3. 31	3. 05	0. 009	0. 029	2. 90
	11	98	253	25	1. 36	8. 8	120. 8	0. 11	3. 31	3. 18	0. 011	0. 028	2. 34
	12	107	260	26	1. 16	7. 6	90. 1	0. 15	3. 35	3. 43	0. 013	0. 037	2. 96
	13	107	251	25	1. 14	9. 1	129. 2	0. 18	3. 38	3. 03	0. 009	0. 026	2. 66
	14	102	261	26	1. 10	9. 2	132. 0	0. 21	3. 41	2. 94	0. 008	0. 025	2. 67
	15	100	233	23	1. 29	9. 4	137. 8	0. 24	3. 44	3. 14	0. 009	0. 025	2. 43
2. 4	10	88	161	16	3. 33	10. 5	172. 0	0. 12	3. 32	5. 36	0. 019	0. 019	1. 61
	11	82	164	16	3. 06	9. 6	143. 8	0. 07	3. 27	5. 46	0. 021	0. 023	1. 78
	12	87	171	17	2. 73	9. 3	134. 9	0. 03	3. 23	5. 32	0. 020	0. 024	1. 95
	13	87	161	16	2. 59	9. 4	137. 8	0. 03	3. 23	5. 42	0. 019	0. 023	2. 09
	14	91	173	17	2. 25	9. 3	134. 9	0. 04	3. 24	5. 81	0. 017	0. 024	2. 58
	15	80	191	19	2. 46	9. 5	140. 8	0. 03	3. 23	5. 21	0. 017	0. 023	2. 12
2. 26	10	89	217	22	2. 68	8. 3	107. 5	0. 13	3. 33	4. 35	0. 025	0. 031	1. 62
	11	105	212	21	2. 79	8. 7	118. 1	0. 19	3. 39	4. 45	0. 024	0. 029	1. 59
	12	93	243	24	2. 40	9. 0	126. 4	0. 23	3. 43	4. 45	0. 019	0. 027	1. 85
	13	89	244	24	2. 36	7. 9	97. 4	0. 28	3. 48	4. 26	0. 024	0. 036	1. 80
	14	90	196	20	2. 58	9. 8	149. 8	0. 31	3. 51	4. 50	0. 017	0. 023	1. 74
	15	89	199	20	3. 00	10. 7	178. 6	0. 32	3. 52	4. 73	0. 017	0. 020	1. 58
2. 27	10	77	141	14	2. 82	12. 5	243. 8	0. 12	3. 32	5. 37	0. 012	0. 014	1. 90
	11	75	145	15	3. 03	12. 9	259. 6	0. 16	3. 36	5. 12	0. 012	0. 013	1. 69
	12	68	153	15	2. 58	13. 1	267. 7	0. 21	3. 41	4. 85	0. 010	0. 013	1. 88
	13	82	159	16	2. 59	12. 8	255. 6	0. 28	3. 48	4. 73	0. 010	0. 014	1. 83
	14	72	194	19	2. 29	12. 5	243. 8	0. 34	3. 54	4. 61	0. 009	0. 015	2. 01
	15	83	154	15	2. 65	11. 9	220. 9	0. 35	3. 55	4. 32	0. 012	0. 016	1. 63

Because of the great differences between number of recorded waves and observed runup waves to the experimental seawall, the author used the one-tenth highest wave and the same numbers of the highest wave runup, in the arrangement. For example, in the case of

the date the 20th December 1986,
time 14 o'clock

in the Table 1,

(1) the number of the observed runup waves to the experimental seawall in twenty minutes is 99.
(2) the number of the waves recorded in the wave recorder in twenty minutes is 147.
(3) calculated number of the one-tenth highest wave is 15. and the same number of highest wave runup are adopted.
(4) one-tenth highest wave height is 3.70 m.
(5) one-tenth highest wave period is 10.3 s.
(6) calculated offshore wave length L_o is 165.5 m.
(7) sea water level at that time is +0.25 m.
(8) water depth at the toe of the seawall is 3.45 m.
(9) the mean value of the highest 15 waves runup height, adopted in (3), measured from the sea water level is 5.42 m.
(10) wave steepness H_o/L_o (4)/(6)=3.70/165.5 = 0.022.
(11) relative water depth d/L_o(8)/(6)= 3.45/165.5 = 0.021.
(12) relative wave runup height R/H_o (9)/(4) = 1.46

The wave runup height is

5.42 m + 0.25 m = + 5.67 m ≒ + 5.70 m (Crown height)

Naturally, a number of runup waves overtopped the crown of the experimental seawalls, as shown in Photo 12.

4.3 Comparing the field observation data with laboratory test results

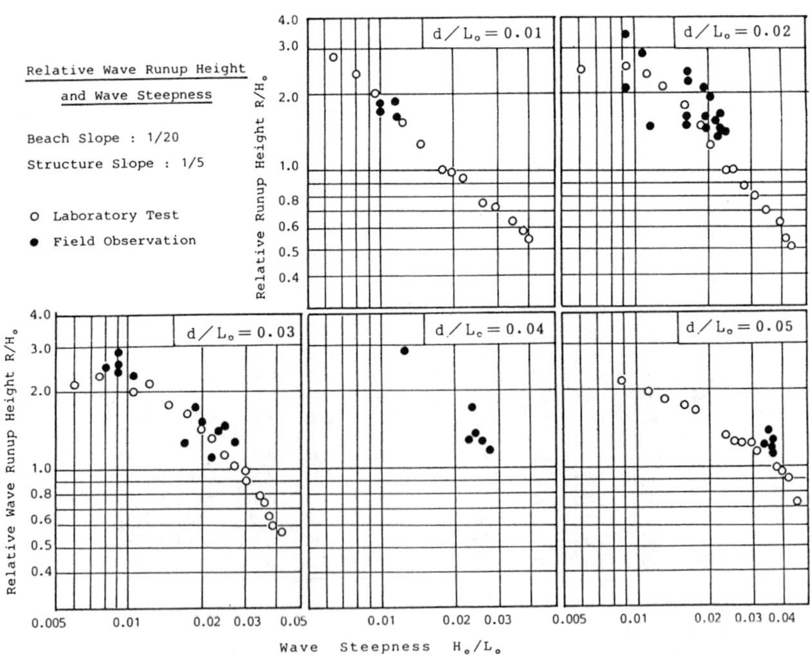

Fig.10. Comparison between the laboratory test and the field observation on the wave runup height on gentle slope seawalls.

The data of the field observation are ploted in Fig.10, together with the data of the laboratory test. Open circles are laboratory test and solid dots are field observation. At a grance, both of these data show a good agreement. The slope of the seawalls in the field are covered with armour units, but in the laboratory test all slope of model are smooth and impermeable. The waves in the laboratory test were regular waves. The wave recorder is 5 km away from the site of works, the runup waves are not the same waves as recorded by the wave recorder.

Although,there are many problems in these data and their caluculation processes, the author considers that these results on the gentle slope 1:5 seawalls are very useful for designing of the shore protection in future.

4.4 The works of fiscal 1987 was completed

Photo 13. The works of fiscal 1987 was completed. The left half are of fiscal 1986.

Photo 14. Heavy waves attacked the Kurobe Coast, and some of them overtopped the experimental seawalls.

On 6th November 1987, heavy winter waves attacked the Kurobe Coast, and some of them overtopped the experimental seawalls. However, the spray of sea water on the gentle slope seawalls were excessively small in comparison with that of existing vertical type seawalls with large mound of armour units.

Photo 15. The latest view of the experimental seawalls.

After the construction of the experimental works, one or two winter season are over, but there has not been any trouble, damage or collapse, on the body, slope, covering blocks, foot and toe of the experimental works. At this time, these new type gentler slope seawalls are successful.

V. CONCLUSION

Many of existing vertical type seawalls have lost their foreshore, and wave dissipation works are not effective enough to prevent the overtopping of the splash and sea water mass. On the basis of the successful results and experiences obtained through the experimental works, the author proposed to reform existing vertical type seawalls to gentler slope seawalls with armour unit facing.

In conclusion, in order to maintain better coastal environment, structures on an erosive coast are desired to satisfy the following conditions:

(1) the wave reflection from the front slope should be minimized,
(2) scouring at the structure toe should be prevented as well as erosion of the foreshore,
(3) the whole structure should not collapse even if partial breakdown took place,
(4) the crown of the structures should not be too high,
(5) repairs and reinforcements should be easy, and
(6) the structure should be rather simple and not too costly.

In this view, the gentler slope seawall with rough and permeable front slope is one of the most relevant countermeasures against the beach erosion.

REFERENCES

Saville,T,Jr.(1958): Wave runup on composite slopes. Proc. 6th Coastal Eng. Conf.

Toyoshima,O. (1984): New type blocks for seawall slope protection.Proc.19th Coastal Eng. Conf.

CHAPTER 148

ON BERM BREAKWATERS

by
Alf Tørum[1], Steinar Næss[2],
Arne Instanes[3] and Svein Vold[4]

ABSTRACT

Two and three dimensional laboratory studies have been carried out on the stability of a berm breakwater concept. The study has to some extent been general and to some extent been connected to a project study of the stability of a berm breakwater for the fishing port of Årviksand, Norway.

INTRODUCTION

The main feature of a berm breakwater is that it has a rather thick cover layer of stones, relatively much smaller than on a conventional breakwater with one or two layers of cover blocks. The berm breakwater has been adopted several places as an economic solution when large cover blocks of natural stones are not available. It might also be an economical solution even when large cover blocks for a conventional breakwater are available.

The berm breakwater concept has become of interest in Norway in connection with plans for an extension of a breakwater at the Årviksand fishing port. Fig. 1 shows the layout of the harbour. The breakwater will be extended about 90 m into a maximum water depth of approxi-

---

1 Head of Research/Professor. Norwegian Institute of Technology/ Norwegian Hydrotechnical Laboratory, 7034 Trondheim, Norway.
2 Chief Engineer. Norwegian Coast Directorate. P.O. Box 8158 - Dep 0033 OSLO 1, Norway.
3 Research Engineer, Division of Structural Engineering, Norwegian Institute of Technology, 7034 Trondheim, Norway.
4 Research Engineer, Norwegian Hydrotechnical Laboratory, 7034 Trondheim, Norway.

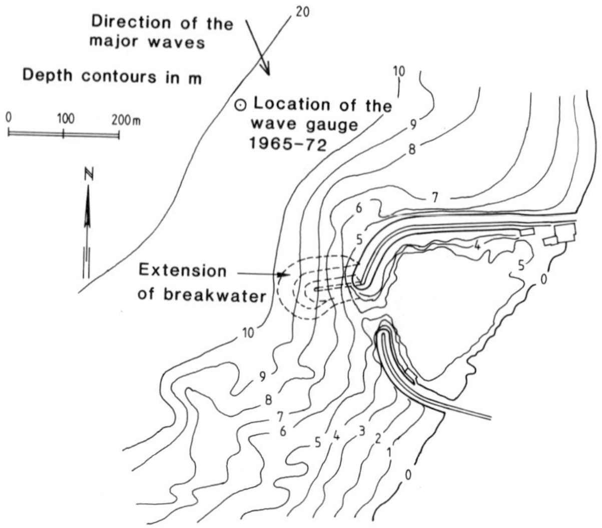

Fig. 1. Årviksand fishing port.

mately 13 m. The wave climate at the breakwater site has been evaluated from wave measurements during the time period 1965-1972 at 20 m water depth outside the harbour and from hindcast wave data in deep water from the time period 1955-1985 and refraction analysis. Based on the wave measurement and a Weibull distribution formulation the 50 and 100 year significant wave heights were estimated to be approximately 6.4 m and 7.2 m respectively. From the hindcast data and the refraction analysis the 100 year wave height was estimated to be approximately 6.2 m.

The differences in the estimated 100 year wave height reflects the uncertainty on the estimated wave climate which always is present at any harbour location on an open coast.

Based on a Weibull distribution fitting of the measured daily maxima, a simulation study has been performed based on a procedure described in [5]. The Weibull distribution which fits, by the method of moments, the data best has been used as the "parent" distribution. By a Monte-Carlo procedure daily maxima for seven-years of observation have been simulated one hundred times. Fig. 2 shows the probability density function for 50 year occurrence of significant wave height for water depth 20 m. The Weibull distribution parameter for the measured daily maxima were: Shape parameter 0.77, scaling parameter 0.660 m and location parameter 0.057 m. The shape parameter is rather low for this location and explains partly the large scatter of the estimated 50 year significant wave height through the simulation study.

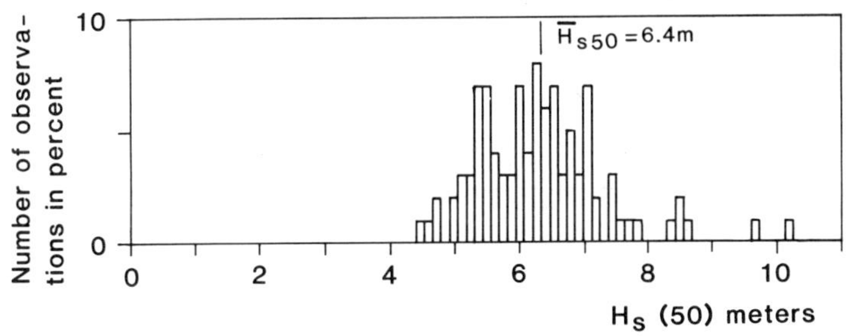

Fig. 2. Results of wave statistics simulation.

The uncertainty in the wave climate favours a breakwater design that is not too sensitive to the wave height with respect to stability. The berm breakwater is a concept that is of interest in this respect.

The stability results reported in this paper are partly from a student thesis work [1], to some extent related to Årviksand harbour, partly from a general investigation of the stability of berm breakwaters and partly from a project study for Årviksand harbour.

INTRODUCTORY TESTS ON THE STABILITY OF THE BERM BREAKWATER FOR ÅRVIKSAND HARBOUR

The existing north breakwater at Årviksand is built as a conventional breakwater with one layer of cover blocks of natural stones. The average block weight on the outer most exposed part is 10 tons. Some introductory tests showed that if the breakwater was extended to a maximum water depth of 12 m, a breakwater with 10 tons cover blocks will be stable for a significant wave height of 4.5 - 5.0 m. The estimated necessary block weight for a conventional breakwater to stand wave heights of Hs = 7.2 m would be 25 - 30 tons. Flume tests were then carried out on a berm breakwater design as shown in Fig. 3. One test was also carried out for a design shown on Fig. 4.

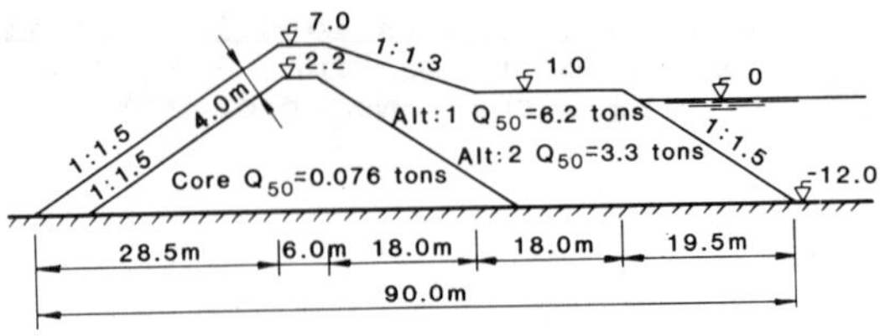

Fig. 3. Berm breakwater section - Introductory tests.

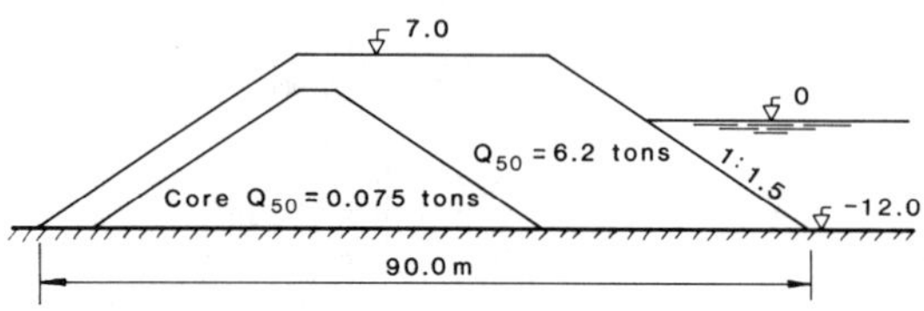

Fig. 4. Berm breakwater section - Introductory tests.

The model scale was 1:40. Two block weights in the berm were used: 1 average 3.3 tons, range 1 - 6 tons. 2 average 6.2 tons, range 1 - 14 tons.

Fig. 5 shows the flume with the breakwater model. The bottom configuration in front of the breakwater corresponded to the bottom configuration in front of the planned extension of the breakwater at Årviksand fishing port.

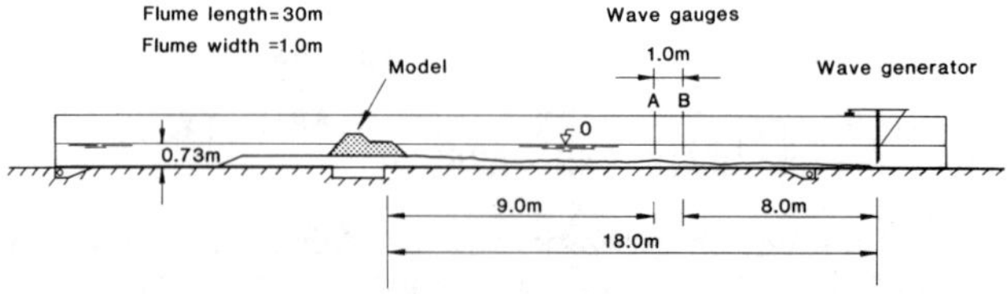

Fig. 5. Test flume.

During the tests two wave gauges were used. The water depth at wave gauge A corresponded to 20.8 m or approximately the water depth at the location of the wave gauge outside Årviksand in the time period 1965 - 1972.

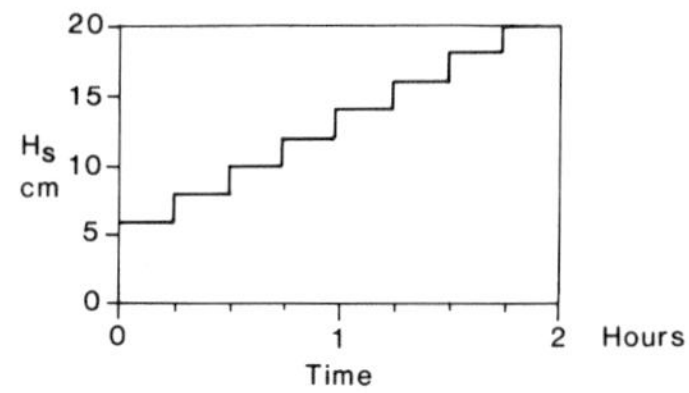

Fig. 6. Hs versus time.

The test programme used during the tests is shown in Fig. 6. The evolution with time of the significant wave height corresponds to a typical evolution during a heavy storm on the Norwegian coast. The wave spectrum was narrow with a peak period of 11.4 - 12 sec. The peak period was the same for all significant wave heights. The incoming and reflected waves were found by a procedure described by Goda and Suzuki [2]. It should be noted that for the highest significant waves the waves were non-Gausian. For exampel were the skewness 0.78 and the kurtosis 3.82 for a significant wave height of 8.3 m. The highest waves would then break before they came to the breakwater.

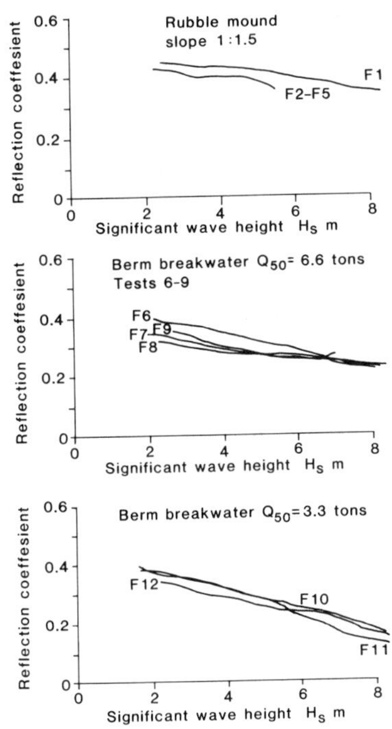

Fig. 7. Reflection coefficients.

The obtained reflection coefficients as a function of the significant wave height is shown in Fig. 7.

Run-up on the breakwater was also measured. Fig. 8 shows a run-up distribution. The run-up r is defined as the maximum run-up of each individual wave on the slope or within the breakwater.

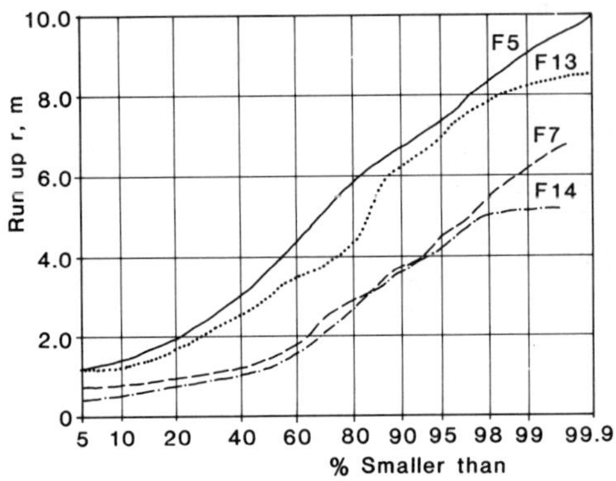

———	Hs = 5.3 m	Rubble mound	10.3 tons	Test F5
······	Hs = 5.3 m	Berm breakwater*	6.2 tons	Test F13
— — —	Hs = 5.3 m	Berm breakwater	6.2 tons	Test F7
—·—·—	Hs = 5.3 m	Berm breakwater	3.3 tons	Test F14

* "Full" berm, Fig. 4

Fig. 8. Run-up

The run-up on the berm breakwater is much less than on the conventional breakwater. It is also seen that the run-up on the "full" berm breakwater, Fig. 4, is not much less than on the conventional breakwater. Hence it is concluded that a low berm is very efficient from the run-up point of view.

Figs. 9 shows profiles after the tests for the maximum significant wave heights 8.60 m and 8.32 m respectively were completed.

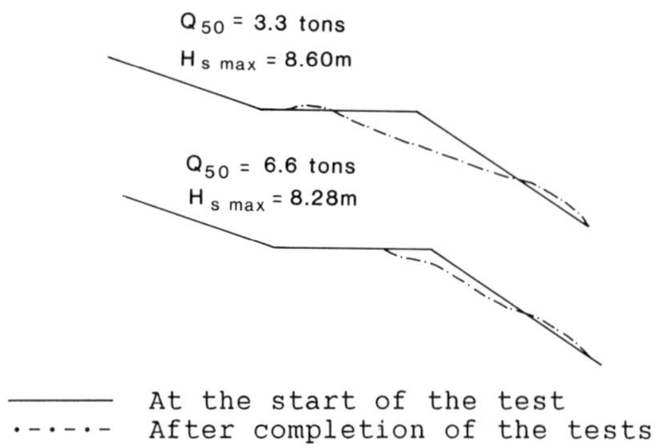

Fig. 9. Cross-section

The results of these introductory tests indicated that a berm breakwater could be build with an average block weight in the range 3.3 - 6.6 tons. The berm breakwater showed also less sensitivity to the variation in the significant wave height than the conventional breakwater. There was not a sudden collapse as normally is the case for a conventional breakwater. This is very useful for a location where there is a significant uncertainty on the wave climate. However, it was felt that there was a need for a more general flume study on the stability of the berm breakwater concept.

GENERAL TEST SERIES

The objective of these tests was to investigate more closely how the berm evolved with the wave height, wave spectrum and duration of the "storm". The general test series were conducted in the same flume as shown in Fig. 5 in scale 1:40. The tested cross-section is shown in Fig. 10. Two berm widths were tested: 30 and 45 cm (12 and 18 m).

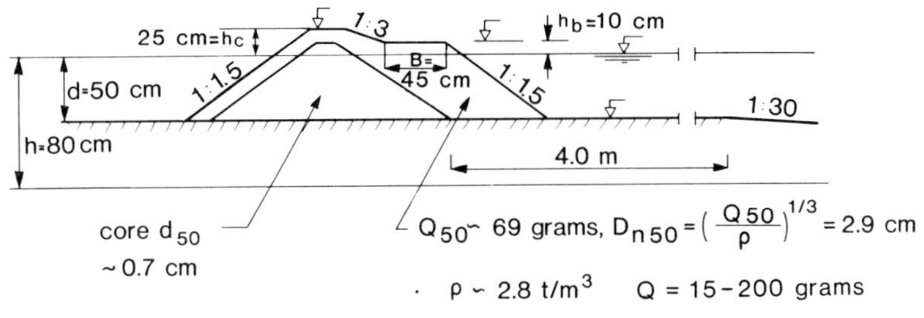

Fig. 10. Berm breakwater - General tests.

The "peak" period Tp corresponded to 12 and 16 sec. The spectrum corresponded to a JONSWAP spectrum. Two values of the peakedness parameter $\gamma = 1.0$ and $\gamma = 7.0$ were used. The wave height was varied in steps in different ways.

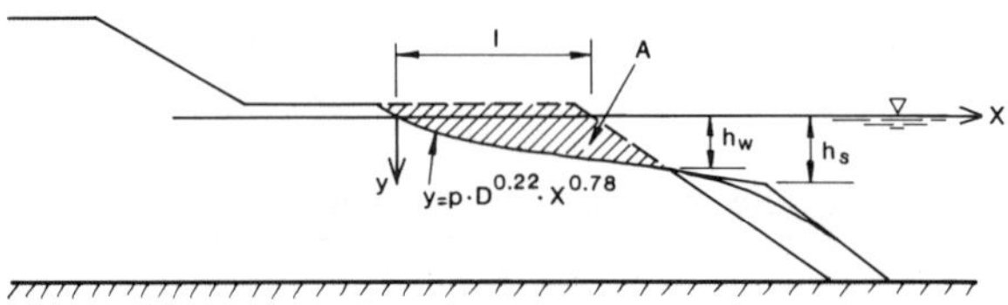

Fig. 11. Definitions.

The change of the berm is essential for the consideration of the stability of the breakwater. This change has been quantified by the number S introduced by van der Meer [3] as damage to a conventional breakwater. With reference to Fig. 11, S is defined as;

$$S = A/(D_{n50})^2$$

where:

A = cross sectional area removed from the berm

$$D_{n50} = [\frac{Q_{50}}{\rho_s}]^{1/3}$$

Q_{50} = 50 % of the stones has a weight larger than Q_{50}

ρ_s = spesific density of the stone material

The recession of the berm, l, see Fig. 11, has been non-dimensionalized as

$$l/D_{n50}$$

If $l/Dn_{50} > B/Dn_{50}$, where B = berm width, inacceptable damage may occur.

In conventional breakwater design there is a consideration on static-stability. This means that the requirement is that the stones should not move at all. Sand and gravel beaches may on the other hand be dynamically stable. That is the individual grains may move heavily, but the profile does not in the long term sence change. This is also true for a berm breakwater. The profile is then dynamically stable. This means that the individual stones may move up and down the slope, but the profile is stable in the long term sence. This applies especially to a two-dimensional profile, when the waves are coming normal to the breakwater. If the wave attack is oblique to the breakwater, stones can also move along the breakwater front. In this case the breakwater may not be dynamic stable.

The present program included basically two test series: One series, when the spectrum shape and peak period were varied, had a time evolution of the significant wave height, as shown in Fig. 12, corresponding to a typical evolution of an intense storm on the Norwegian open coast. In the other test series the duration of the tests for each variation of the wave height was much longer. The tests were carried out with no "repair" work on the breakwater after each wave height step had been completed. The program included one test, A, with a berm width of 30 cm (12 m). The remaining tests, B, were all for a berm width of 45 cm (18 m).

	Prototype		Model	
A	Tp = 12 sec	γ = 1.0	Tp = 1.9 sec	γ = 1.0
B 1-2	Tp = 12 sec	γ = 1.0	Tp = 1.9 sec	γ = 1.0
B 3	Tp = 12 sec	γ = 7.0	Tp = 1.9 sec	γ = 7.0
B 4	Tp = 16 sec	γ = 1.0	Tp = 2.53 sec	γ = 1.0
B 5	Tp = 12 sec	γ = 1.0	Tp = 1.9 sec	γ = 1.0

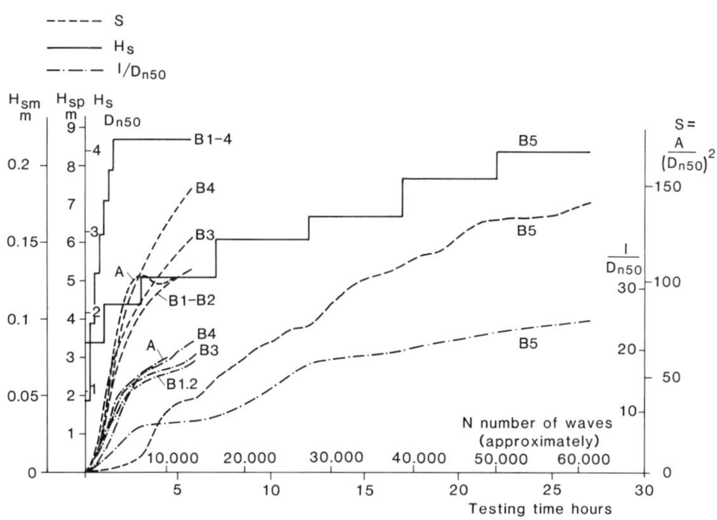

Fig. 12. General tests - Results.

The evolution with time and wave height of S and $1/Dn_{50}$ is also shown in Fig. 12.

It is seen that the value of S is depending on the peak period as well as on the width of the wave spectrum. It has also previously been found for exampel by Carstens et al. [4] that a narrow spectrum give a harder attack on a conventional rubble mound breakwater than a wide spectrum provided that the peak period is the same and provided that the significant wave height is the same. Van der Meer [3] found that the significant period was the average zero crossing period T_{02}. According to van der Meer [3] damage to a breakwater will be the same provided that the significant wave height is the same and the average zero upcrossing period is the same, irrespective of the spectrum shape.

However, on a conventional breakwater the damage is depending also on the Irribarren number

$$\xi = tg\alpha/(2\pi H_s/g\ T_{02}{}^2)$$

where α is the breakwater slope and $T_{02} = \sqrt{m_0/m_2}$, where mn is the nth moment of the spectrum. A berm breakwater has not a well defined slope. However, the results presented in Fig. 12 indicate that the "damage" is depending on the peak period Tp.

In the Figs. 13 and 14 are shown typical evolution of the profiles for two test series. B2 is for the shorter duration tests and B5 is for the longer duration tests.

From Fig. 13 and Fig. 14 is seen that there seems to be a significant point located at a distance hw as shown in Fig. 11. The intersection between the profile and the original profile seems to be more or less at this point, irrespective of the scour of the slope.

Van der Meer [3] has investigated many of the parameters of importance for the evolution of the slope. He also gives an equation for the profile below the water line as indicated in Fig. 11. With the local origo given as the intersection between the still water line and the profile the profile is given by

$$y = p \cdot D^{0.22} \cdot x^{0.78} \qquad (1)$$

It is not clearly stated by van der Meer which diameter, D, whould be used, but it is assumed that it is the diameter Dn_{50}. The coefficient p is primarily a function of the wave steepness, $2\pi Hs/(gT_{02}{}^2)$. For the present test parameter values van der Meer´s results give $p \approx 0.5$. With this value of p we have shown in Figs. 13 and 14 the profile as given by equation (1). For some reason there is an apparent difference between the profiles as obtained in the present study and the study by van der Meer [3]. Van der Meer carried out most of his tests in relatively deeper water than used in the present study. However, van der Meer also carried out some tests with shallower water and states that there should not be any significant effects of the water depth, h, if h/Hs > 2.2, which is the case in the present study.

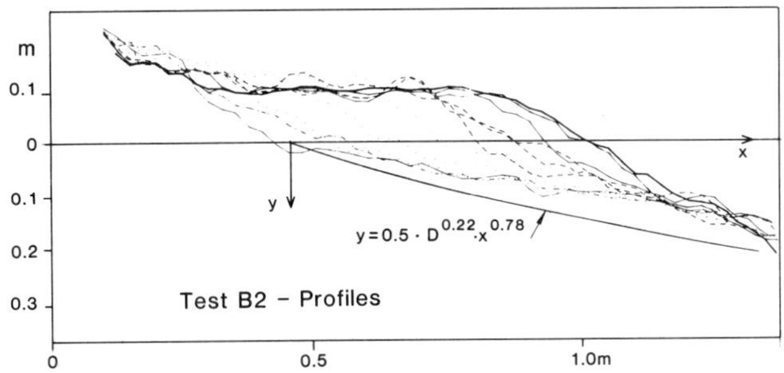

Fig. 13. Profile evolution.

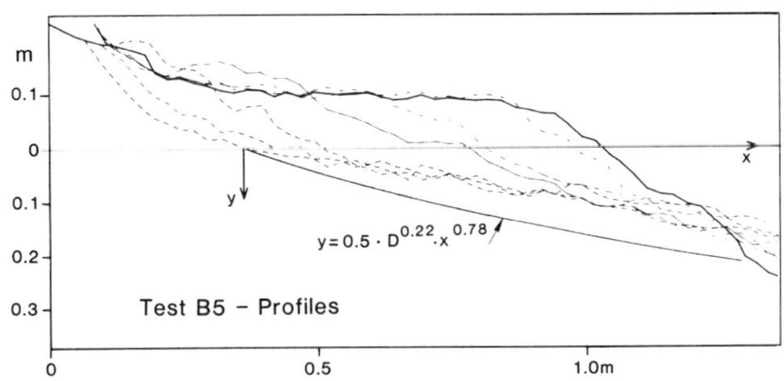

Fig. 14. Profile evolution.

There is obviously a time effect in the development of a dynamic stable profile. If we for example take the evolution with time of the wave height for the tests B1 - B4 (this evolution could be a typical evolution for the Norwegian coast) the value of l/Dn_{50} is equal to the berm width value B/Dn_{50} = 15.5 after approximately 6000 waves. The maximum significant wave height is then corresponding to $Hs/\Delta Dn_{50} \approx 4.0$, where $\Delta = \rho s/\rho w - 1$, ρw = specific density of the water.

For the long duration tests, B5, we have l/Dn_{50} = 15.5 after approximately 11 hours testing time or after approximately 25000 waves. The maximum significant wave height is then $Hs/\Delta Dn_{50} \approx 3.0$. This means that the design of the breakwater should be for $Hs/\Delta Dn_{50}$ somewhere between 3 and 4.

The tests with the 12 m wide berm (A-test) showed a berm

recession evolution about the same as for the tests with the wider berm (B-tests). However, due to the narrower berm, the core material finally became visible. This did not occur for any of the B-tests.

THREE-DIMENSIONAL STABILITY TESTS FOR THE BREAKWATER AT THE FISHING PORT OF ÅRVIKSAND

Based on the previous studies, the design shown in Fig. 15 was tested in a three-dimensional model of Årviksand fishing port. The model had a length scale of 1:60 and with a layout as shown in Fig. 1. The tests were run with high water + 3.0 m. The waves referred to were measured at a location corresponding to the location of the wave gauge during the field measurements in the time period 1965 - 1972. The peak period for the waves was Tp = 14 sec and the wave spectrum was approximately a JONSWAP spectrum with γ = 7.0. The time variation of the significant wave height was as shown in Fig. 16.

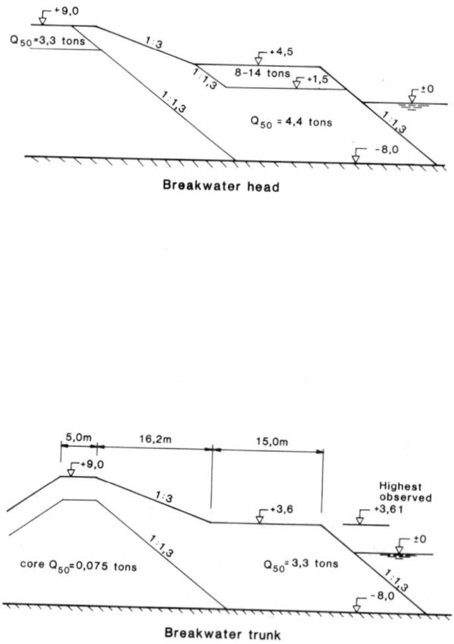

Fig. 15. Tested design for the Årviksand fishing port.

The first tests were carried out with a crest height + 9.0 m, or 6 m above the still water level.

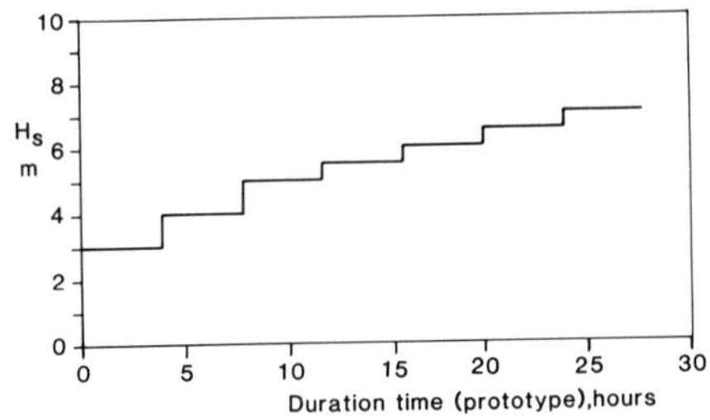

Fig. 16.

Fig. 17 shows the contours of the breakwater after 3.5 hours of testing time, or approximately 10.000 waves, with a maximum significant wave height of 7.0 m. It is seen that there is a damage on the back side of the breakwater due to overtopping at the middle part of the breakwater. The location of the damage is due to a consentration of the waves due to refraction towards the middel part of the breakwater.

The wave crest was then raised by 1.0 m to 10.0 m, or 7 m above the still water line. The breakwater was then tested for the same wave program, Fig. 16, as the previous design. Fig. 18 shows the contour map for the tested breakwater after 3.5 hours testing time. Although there were motions of stones up and down the breakwater slope as well as along the breakwater, mainly landwards, there was not any major damage to the breakwater. Some stones even moved over the breakwater crest.

Since storm duration is essential for consideration of the dynamic stability of the berm breakwater it is essential to consider statistically all the major storms the breakwater will encounter during its expected life time. The water level variation has also to be considered. The result of such a "fatique" study is not included in this paper.

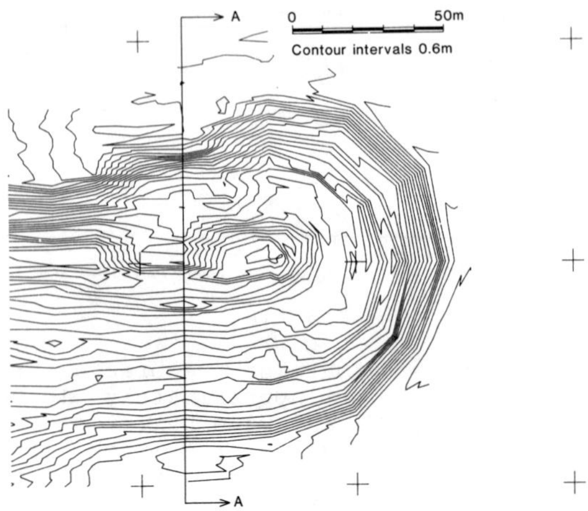

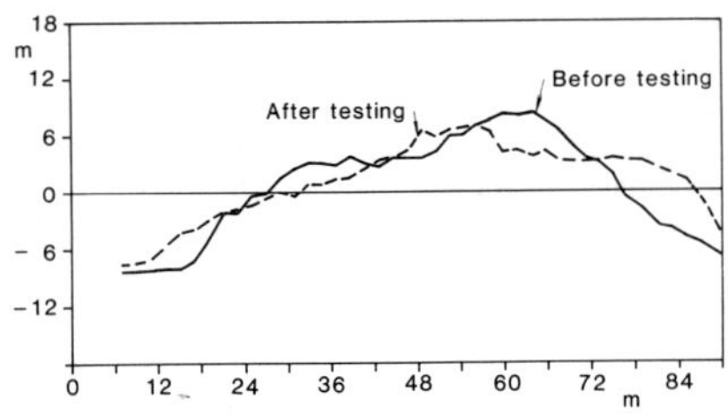

Fig. 17. Section A-A. Crest height + 9.0 m.

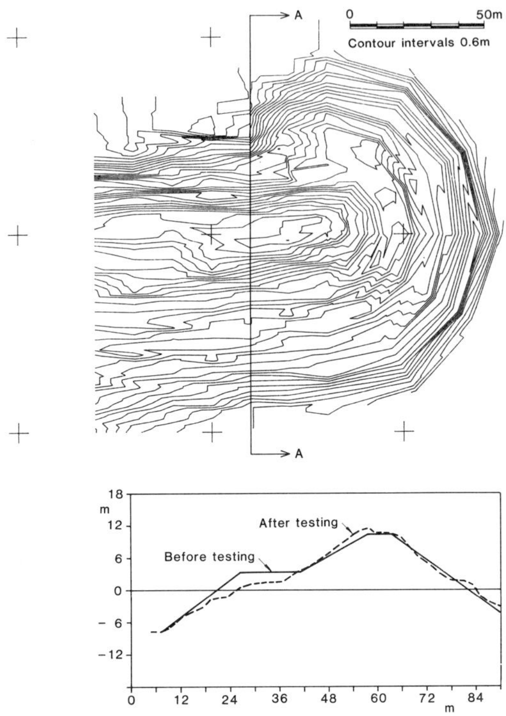

Fig. 18. Section A-A. Crest height + 10.0 m.

CONCLUSION

Based on the present study it is preliminary concluded that a berm breakwater should be designed for $Hs/\Delta Dn_{50}$ = 3 - 4, provided that the water depth ratio h/Hs > 2.5.

A berm breakwater should be subjected to a "fatique" analysis. It is essential that a three-dimensional study is conducted on the stability of berm breakwaters.

ACKNOWLEDGEMENT

We appreciate the finacial support to carry out this study from the Norwegian Coast Directorate and Selmer-Furuholmen Anlegg A/S.

REFERENCES

1. Instanes, A: Molostabilitet (Breakwater stability). M.Sc.-thesis, Norwegian Institute of Technology, December 1987 (In Norwegian).

2. Goda, Y. and Suzuki, Y: Estimations of incident and and reflected waves in random wave experiments. Proceedings Coastal Engineering Conference, 1976, pp 828 - 845.

3. Van der Meer, J: Rock Slopes and Gravel Beaches under Wave Attack. Thesis Delft Technical University 26 April 1988.

4. Carstens, T, Trætteberg, A. and Tørum, A: The stability of ruble mound breakwaters against irregular waves. Xth Conference on Coastal Engineering, Tokyo, 1966.

5. Environmental conditions on the Norwegian Continental Shelf. Barents Sea. Report - Phase 1 to ODAP. Norwegian Hydrotechnical Laboratory, Trondheim, 1 July 1988.

CHAPTER 149

NUMERICAL MODEL OF BREAKWATER WAVE FLOWS

Alex C Thompson*

ABSTRACT

A mathematical model of flow on a sloping breakwater face is described and results of calculations compared with some experimental results to show how the model can be calibrated. Flow above the surface of the slope is represented by the shallow water wave equations solved by a finite difference method. Flow within the breakwater is calculated by one of two methods. A solution of the linear seepage flow equations, again using finite differences or a simplified model of inflow can be used. Experimental results for runup and reflection coefficient are from tests performed at HRL Wallingford.

INTRODUCTION

The behaviour of waves on a breakwater particularly as far as they affect stability of the armour layer has been investigated extensively. Despite this the process is still poorly understood especially for slender concrete armour units, Burcharth 1987. Expensive physical model tests are undertaken for each new breakwater and the interpretation of these is uncertain. The writer has conducted experiments in simple flows to try and understand armour stability and scale effects Burcharth & Thompson 1983, 1984. If similar tests can be combined with a numerical model giving the velocities on a breakwater face the understanding of armour stability may be increased and the amount of physical model testing reduced. The calibration of the numerical model for a particular type of armour would be through simplified tests. This paper is concerned with a suitable numerical model for the task. One such model has ben shown by Hibberd & Peregrine 1979 to give good results for a wave breaking and running up a shallow slope. This model is based on the shallow water wave equations so it calculates only the depth averaged velocity at any point on the slope. The detailed velocities in the breaking part of the wave may not be necessary to predict runup reflection and some measures of stability however. This seems to be confirmed by the work of Kobayashi 1986.

* RNC Greenwich UK

He has applied a model based on Hibberd & Peregrine's work to a steeper slope typical of a rock armoured breakwater with some success. Two main areas require investigation if the model is to be applied to slopes covered with concrete armour. Firstly it must be shown that the model works with a steeper slope (upto 1 on 1.5 say).
Secondly the effects of a porous armour layer and core must be accounted for. The assumption of an impermeable slope as used by Kobayashi may be reasonable for a rock armoured slope, it must be less valid for more porous armour such as those tested by Allsop 1983, Burcharth 1979 and Timco 1984. To separate the effects of core permeability and armour permeability and so make calibration of the numerical model easier physical model tests with an impermeable core have been performed. These are compared with calculations for an impermeable slope, a thin permeable layer and a permeable core.

MATHEMATICAL MODEL

The motion of the water on the breakwater face is described in this model by the one dimensional equations for nearly horizontal flow.

$$\frac{d(hu)}{dt} + \frac{d(uhu)}{dx} = \frac{-ghdn}{dx} \quad \frac{-f/u/u}{2} \qquad (1)$$

$$\frac{dh}{dt} + \frac{d(hu)}{dx} = 0 \qquad (2)$$

Where h is the depth of water above the slope and u the mean velocity. Both are at a distance x from the toe of the slope and a time t after the start of motion. The water surface elevation above still water level is n (n=h-D). D is the still water depth and f the friction factor.

These equations are solved by writing them in finite difference form using the Lax Wendrof method. If equations (1) and (2) are combined and expressed in vector form this gives:

$$\frac{dU}{dt} + \frac{dF}{dx} + G = 0 \qquad (3)$$

Where $U = \begin{Bmatrix} hu \\ h \end{Bmatrix}$ $F = \begin{Bmatrix} huu+gh/2 \\ hu \end{Bmatrix}$ $G = \begin{Bmatrix} ghtan\alpha+f/u/u \\ 0 \end{Bmatrix}$

$\tan\alpha$ = tangent of slope angle.

A finite difference grid of constant space interval δx and time step δt is used. If the quantities at $x=j\delta x$ and time $t=n\delta t$ are given subscripts j and n then the finite difference equations are:

$$U(j,n+1) = U(j,n) - \delta t/\delta x[F(j+1,n)-F(j-1,n))/2+\delta x G(j,n)] + (\delta t/\delta x)^2[g(j,n)-g(j-1,n)-\delta x S(j,n)]/2 \quad (4)$$

$$g(j,n)=[A(j+1,n)+A(j,n)][F(j+1,n)-F(j,n)+dx(G(j+1,n)+G(j,n))/2]/2$$

$$A = \frac{dF}{dU} = \begin{Bmatrix} 2u & gh-uu \\ 1 & 0 \end{Bmatrix} \qquad S = \frac{dG}{dt} = \begin{Bmatrix} -g\tan\alpha d(hu)/dx \\ 0 \end{Bmatrix}$$

Dissipation is introduced to damp the short waves which appear in the solution when a steep wave front is formed. This is applied after every time step using a dissipation constant AD with the value 0.1 in the equation:

$$U(j,n) = AD*U(j+1,n)+(1-2*AD)*U(j,n)+AD*U(j-1,n) \quad (5)$$

The boundary conditions applied to produce a model of the flow on a slope were as follows. At the shore boundary the depth of flow becomes zero and the position of this point varies in the finite difference grid. Hibberd and Peregrine developed a comprehensive method of dealing with this condition which worked well on the thin runup layer formed on a smooth shallow slope. A simpler condition has been used here which is belived to be adequate for the steep rough slopes considered. Sufficient points are included in the calculation to cover the maximum runup. The depth h at each point is tested and if it is less than a minimum value hm then h is set equal to hm and the discharge hu is made zero. hm is made 0.004 times the depth at the toe of the slope.

At the seaward end of the calculation both the water depth h and the discharge hu must be given as a function of time. The surface elevation of an incident wave ni can be specified, as discussed below. The elevation of the reflected wave nr can be found as in Kobayashi's work. These two quantities are added to give the total surface elevation at the toe of the slope. The discharge can then be found from the quantity β found in the backward characteristic equation. This quantity is defined as:

$$\beta = u-2(gh)^{\frac{1}{2}} \quad (6)$$

The equation for β is:

$$\frac{d\beta}{dt} + (u-c)\frac{d\beta}{dx} = g\tan\alpha \quad (7)$$

Where $c = (gD)^{\frac{1}{2}}$ is the wave speed.

The value of β can therefore be calculated at the seaward boundary from equation (7) at the start of a new time step. Using the definition of depth,

$$h = D + ni + nr$$

And from linear long wave theory,

$$u = (c/D)(ni - nr)$$

Putting these in equation (6) and using the binomial theorem to expand the square root gives:

$$nr = (-c/g)(\beta/2 + c)$$

The incident wave ni is taken as a sine wave for some calculations but in most cases a suitable profile is taken from Dean's stream function theory. A surface profile is chosen from Dean's tables according to the period and height of the wave selected and the surface co-ordinates stored in a data file.

Typical values of the time step etc used in the model can be given for the example of modelling waves on a 1:2 slope in a depth of 0.3m. This is the slope tested in a wave flume at HRL. The grid size δx in the x direction is 0.0075m and 150 points are used in the calculation. The time step δt is 0.00375s requiring 667 steps to calculate one period of a 2.5s wave, the longest period wave investigated.

Permeable Slope Model

The model described above is used to calculate wave action on an impermeable slope which forms the largest part of the work described here. It can also be linked to two models of flow in a permeable slope and a smaller number of calculations are made in this mode.
The first model of a permeable slope assumes that flow takes place only in a thin layer just below the surface of the slope. This might approximate to the flow on a slope armoured with a very porous layer such as the experiments to be described. The water surface in this layer is assumed to rise at the same rate as the uprushing wave tip. If the thickness and porosity of the layer are specified the discharge into the layer from the water above the slope can be found. This discharge is taken to arise from a small number of points near the wave tip and to cause an equal velocity into the slope at each of these points. The water depth h and discharge hu at these tip points are then adjusted before the next time step of the Lax Wendrof computation is executed. The equations for this adjustment are:

$$h^1 = h - v\delta t \qquad hu^1 = hu(h^1/h) \tag{9}$$

Where h^1, hu^1 are the adjusted values of h, hu and v is the velocity into the slope. The downward velocity of water in the permeable layer is assumed to have a maximum value and outflow from the layer is constant once this is reached. The wave tip can therefore leave the water surface in the layer behind and inflow only resumes when the wave tip rises above the water in the layer again. Typical parameters for this model for the HRL test case are: layer depth 0.042m (.07mxporosity0.6), number of tip points 5, maximum downward velocity 0.01m/s.

The second model of a permeable slope allows for a large region of any shape to underlie the slope. This has a constant permeability at present but the method can easily allow for variations in permeability, see Thompson 1986. The water motion in this permeable region is assumed to follow Darcy's law, implying a linearisation of the actual turbulent flow relationship between head gradient and velocity. The equation for the total head H at any point is then:

$$\frac{d(kdH)}{dx(dx)} + \frac{d(kdH)}{dy(dy)} = 0 \qquad (10)$$

This equation is solved using a finite difference method as follows. The permeable region is covered by a grid with spacing Δx horizontally and Δy vertically. These spacings are equal and larger than the spacing δx in the grid for the Lax Wendrof solution. The finite difference version of equation (10) is the familiar five point formula solved by successive over-relaxation. This iterative method allows the solution for one time step to start from that for the previous step which will be quite close to the required solution. The same time step δt is used as for the Lax Wendrof solution of the flow above the slope. The free surface in the permeable region is moved using this time step and the boundary conditions on the slope taken from the next solution for water depth above the slope. Motion of the free surface is calculated by the formula:

$$\frac{dY}{dt} = -\left(\frac{dH}{dy} - \frac{dH}{dx}\tan\theta\right)\frac{k}{e} \qquad (11)$$

where Y is the free surface elevation in the permeable region, H the total head, $\tan\theta$ is dY/dx, k is the permeability and e the porosity.

The total head H is put equal to Y on the free surface. The location of this in the grid changes as the surface moves and this is allowed for by taking the nearest grid point to the surface on each vertical grid line as the point where H is specified. In a similar way the surface of the slope is represented by the nearest points in the grid. The head H at these points is taken as the water depth above the slope plus the height of the point above the toe of the slope. Water depths are found from the solution for flow above the slope at grid points in the Lax Wendrof routine which coincide with the interior grid. The remaining boundaries of the permeable region are taken as zero flow boundaries and the gradient of H is set to zero at these points. For example this condition is applied along a horizontal line level with the toe of the slope which is the bottom of the permeable zone in the present work. The seaward end of the free surface is forced to coincide with the tip of the wave on the slope.

The model of wave flow on the slope is linked with the interior flow model via the velocities into the slope. These are calculated from the head gradients at the slope surface found in the interior flow model. The discharge into the slope is calculated at each grid point on the slope in the permeable region this is then shared among the neighbouring grid points in the wave flow model and a velocity into the slope found for each of these. The velocities are then used to modify the wave flow solution as before from equation (10).

A typical application of the model is to the region shown in Fig 11. The permeable region under the slope has a maximum length of 1.17m and maximum height of 0.45m. This is covered by a grid of (40x16) points with spacing $\Delta x = \Delta y$ of 0.03m. The wave flow grid has spacing $\delta x = 0.0075$m ($\Delta x/4$) and 150 points. A permeability k of 1.0m/s is used and a porosity e of 0.4.

PHYSICAL MODEL TESTS

The calculations are compared with results of tests performed at HRL Wallingford and described more fully in a companion paper at this conference, Stephens et al 1988. These tests are made on an impermeable slope of 1:2 with a water depth of 0.3m at the toe. Regular waves with period of 1.0, 1.5 2.0 and 2.5s are generated in a long flume by a hydraulically actuated paddle and measurements made before reflections from the paddle build up. Incident and reflected wave heights are measures from two twin wire wave probes mounted offshore. A further 10 wave gauges are closely spaced above the slope and another gauge lies flat on the slope so that the wave motion can be followed in detail here. Video recordings of the wave on the slope are made for each test which can also be analysed to give details of the flow. Only the results on reflection coeffient will be discussed in this paper. Tests are made on a smooth slope and also on a slope covered with a single layer of SHED units 40mm deep.

CALCULATIONS MADE

The mathematical models described above are used to calculate the same situation as the physical model tests. That is regular waves are directed towards a plane slope of 1 on 2 in a water depth of 0.3m. The wave periods are 1.0, 1.5, 2.0 and 2.5s. A number of wave heights are produced at each period in the range .03 to .09m. The calculation produces values for the incident and reflected waveheight so that values of the reflection coefficient can be compared with the measured values. Also calculated are the runup of the waves on the slope and a stability number indicating the start of motion of rip rap or rock armour placed on the slope. The calculations are started with the water above the slope at rest and continued until the values of reflection coefficient etc repeat at each wave. Usually 4 to 5 wave periods are sufficient with the short waves and 2 to 3 with the long waves. For the bulk of the computations the wave profiles chosen were Dean's 7A, 6A, 5A and 4A for the 1.0, 1.5, 2.0 and 2.5s waves respectively. The reflection coeffient etc are plotted against a surf similarity parameter or Irribaren number Ir defined as $T\tan\alpha/(2\pi H/g)^{\frac{1}{2}}$.

Impermeable Slope Model

Fig 1 shows the reflection coefficient measured and calculated for a rough slope. The measured results from the impermeable slope covered with a single layer of model SHED units are shown as crosses. It will be seen that these fall onto separate lines, one for each wave period. This may be because at a given wave period Ir increases as H decreases. Thus for fixed Ir longer wave periods have higher waves. The ratio of wave height to slope roughness will be less for the longer waves and this may explain the higher reflection at a given Ir for these waves. The different wave profiles at each period may be another cause of this effect. The computed results shown by the continuous lines are for a friction factor of 1.0 and of course only the model for water above the slope employed. This friction factor value gives a better agreement with the experimental results than the value of f=0.3 used by Kobayashi except for T=2.5s. Results with f=0.3 can be seen in Fig 3.
Fig 2 shows the reflection coefficient for a smooth slope. The measurements again appear as separate symbols and this time there is much less separation into different groups for each wave period. In this case, the reflection coefficient for T=2.5s are slightly lower than for t=2.0s. The computed results use a friction factor f=0.01 and fit the experimental results quite well except for T=1.0s when the reflection is too high. It may be that the breaking process is not modelled well enough here or that a higher friction factor is appropriate for T=1.0s. The present model uses a constant friction factor for simplicity.

Permeable layer model:

Fig 4 shows the results from the permeable layer model linked to the model for flow above the slope and used to calculate reflection coefficient. The effective depth of the layer is .024m calculated from a depth of .04m and porosity of .60 for the SHEDs in the physical model. A maximum downward velocity in the layer of .10m/s is assumed. The number of points affected by inflow at the wave tip is 20 which is found to be better than 5. The friction factor f is 0.3 the value which Kobayashi found suitable for a comparable surface roughness but which is too small to model the SHED layer if no inflow is allowed. With inflow included as here the calculated results agree quite closely with the experimental findings. In particular the separation into distinct lines for each wave period is reproduced. The results for T=2.5s have been further enhanced by changing the wave profile from Dean's 4A to 7A as the wave height is reduced and Ir increases.

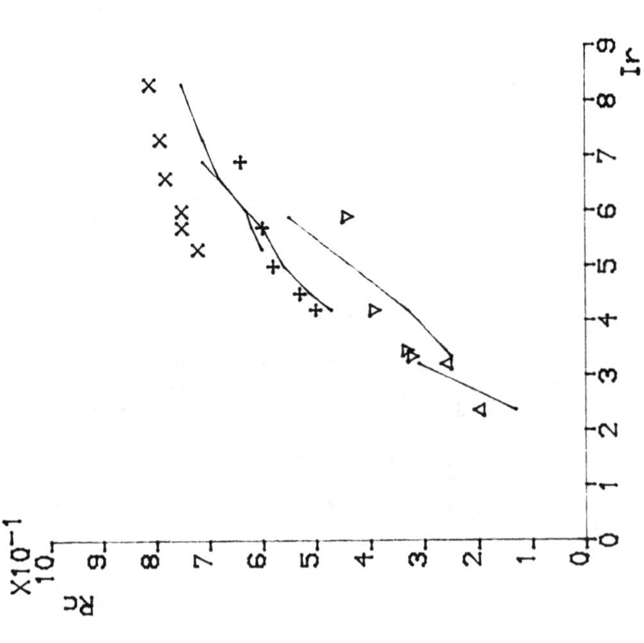

FIG 1 Reflection Coef. Rough Slope.
Δ, ∇, +, x measured T=1, 1.5, 2, 2.5s
- -calculated f=1.0 impermeable.

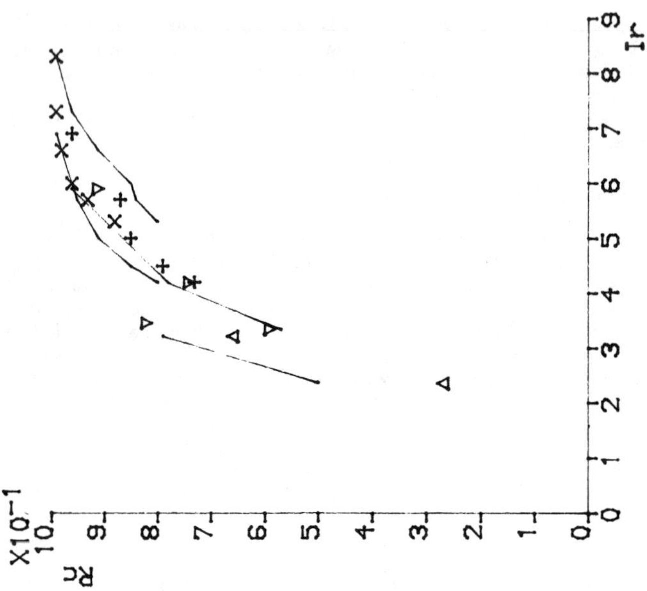

FIG 2 Reflection Coef. Smooth Slope.
Δ, ∇, +, x measured T=1, 1.5, 2, 2.5 s
- - calculated f=0.01 impermeable.

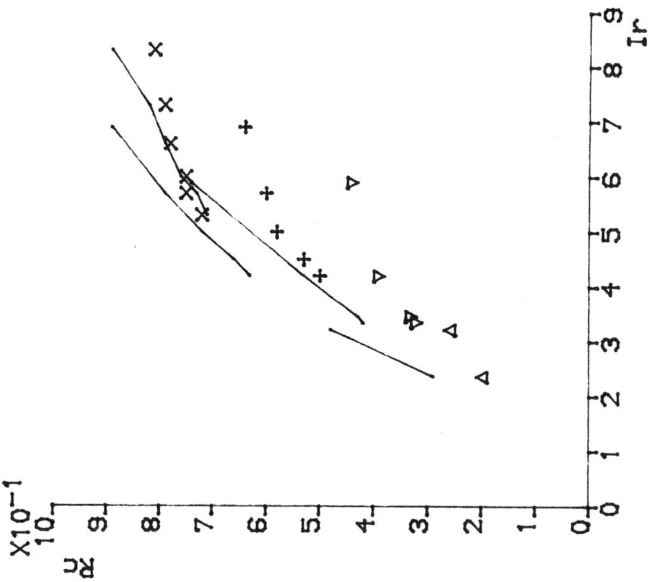

FIG 3 Reflection Coef. Rough Slope.
Δ, ∇, +, x measured T=1, 1.5, 2, 2.5 s
-- calculated f=0.3 impermeable.

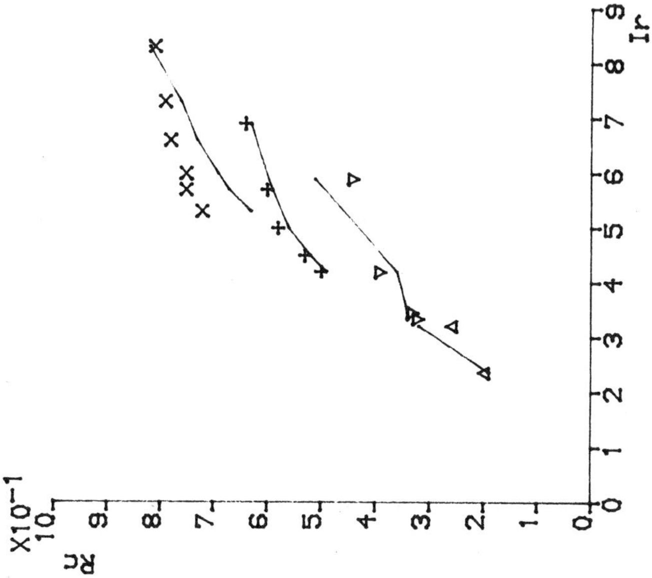

FIG 4 Reflection Coef. Rough Slope.
Δ, ∇, +, x measured T=1, 1.5, 2, 2.5 s
-- calculated f=0.3 permeable layer.

It is interesting that the wave profile assumed has a marked affect on many of the results of the computations. The run-up shows this especially as can be seen from Fig 5. Here the run-up divided by waveheight R/H is plotted against Irribaren number Ir. The full line shows results found with Dean's profiles and the crosses those with a sine profile. In both cases the impermeable slope model with f=1.0 was employed. Run-up/H calculated with a sine profile is more strongly grouped by wave period and reaches a maximum at around Ir=5. With Dean's profiles however, results from all wave periods are closer to a single line and R/H continues to increase with Ir.

Fig 6 shows a comparison of calculated run-up with some experimental results from other sources. The higher solid line is based on work reported by Ahrens 1983 for smooth impermeable slopes. Starting from Ir=0.4 it shows the line R/H=Ir up to R/H=1.86, the vlaue predicted by Miche's law for sine waves on a slope of 1:2. This is then joined to points of increased R/H calculated from the crest height/wave height ratio for the wave profiles used at each wave period as in Ahrens 1983. The results calculated for an impermeable slope for f=.01 (x) fall close to this line. With f=.3 (+) calculated points fall below this but above the lower full line. This line is the experimental result quoted by Kobayashi 1986 from work by Ahrens on slopes covered with rip rap. The calculations with the permeable layer model (Δ) fall close to this line.

Fig 7 shows the calculated results for stability number Ns and some experimental results for rip rap. Ns is defined as H/ΔD where D the diameter of rock which is just stable is calculated using a similar model to that of Kobayashi with the same constants. The full line is the result of van der Meer & Pilarczyk for a slope of 1:2 and the lowest damage level. The calculated results with the permeable layer mode (x) and the impermeable slope model (+) both fall close to this line for Ir greater than 3.

Permeable Core Model:

Results from the permeable core model are more preliminary than those above and have yet to be compared with experimental data. For these first computations the permeable region is shown in Fig 11. All of this region is assumed to have a permeability of 1.0m/s and a porosity of 0.4. The permeability value follows from the linearisation of the true friction law used by Harlow 1980, K=gd/kVav, with k=1.0, d=.01m, Vav=0.1m/s. Fig 8 shows the reflection coefficient calculated for 4 wave period/height combinations (+). The friction factor f is 0.3 as for the permeable layer model. The full line shows the reflection suggested by Seelig 1983 for breakwaters. This lies well below the results for an impermeable slope with only a single layer of armour units, for example Fig 4. The run-up found with a permeable core is given in Fig 9 (+). The full line is 0.5 times the smooth slope line from Fig 6 and lies just above calculated points. Stability numbers computed are given in Fig 10 (+). These lie well above the experimental value for an impermeable core shown as a full line. Finally Fig 11 shows a typical water surface profile above and

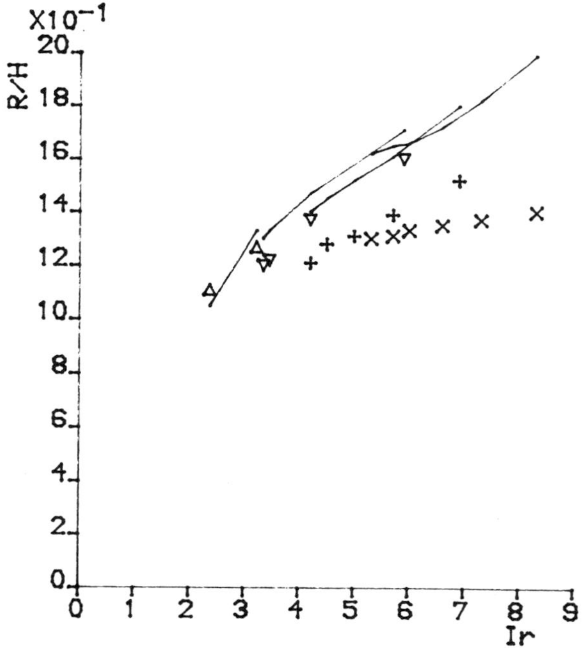

FIG 5 Runup/H Rough Slope.
-- calculated Dean wave profiles
Δ, ∇, +, x calculated sine waves.

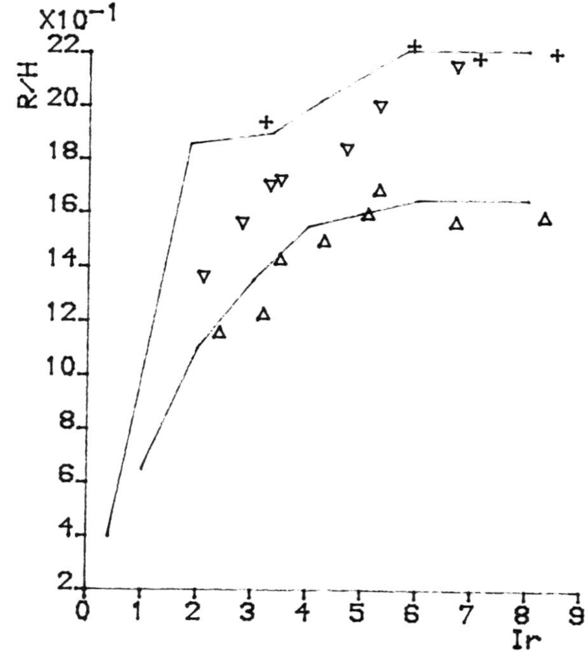

FIG 6 Runup/H.
Δ permmeable layer f=0.3, ∇, + impermeable f=0.3, 0.01
-- upper Ahrens, -- lower Kobayashi.

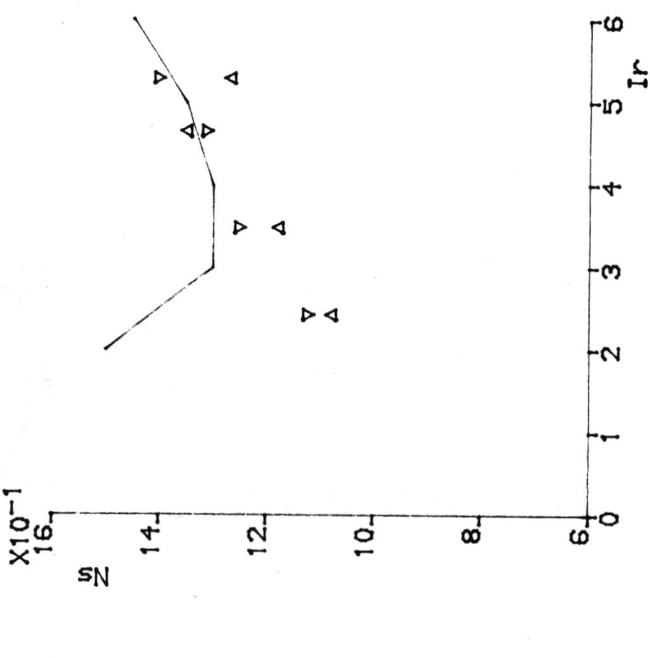

FIG 7 Stability Number.
Δ, ∇ calculated f=0.3 permeable, impermeable.
-- Pilarczyk.

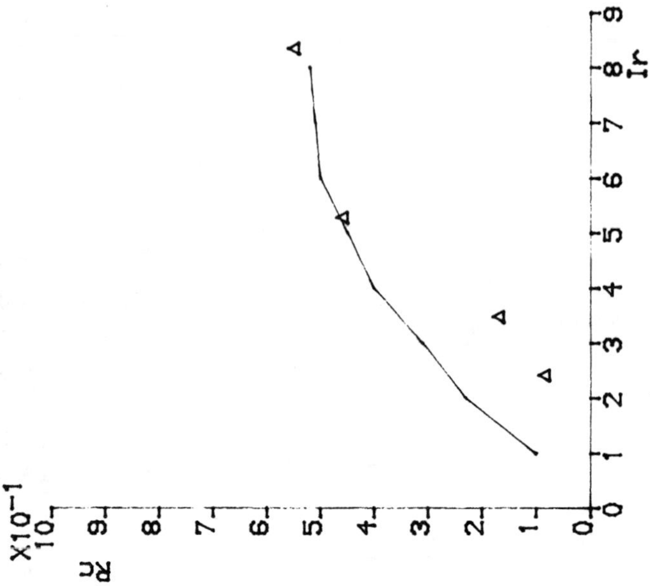

FIG 8 Reflection Coef. Permeable Core.
Δ calculated, -- Seelig.

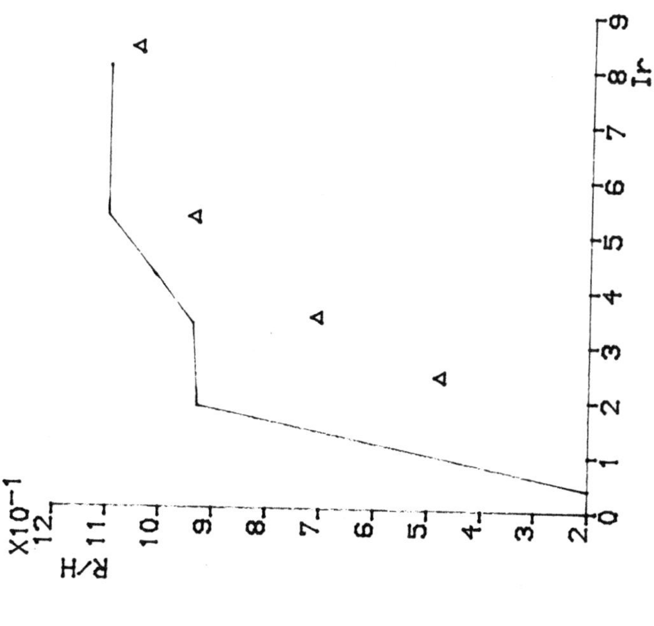

FIG 9 Runup/H Permeable Core.
Δ calculated, -- Ahrens/2.

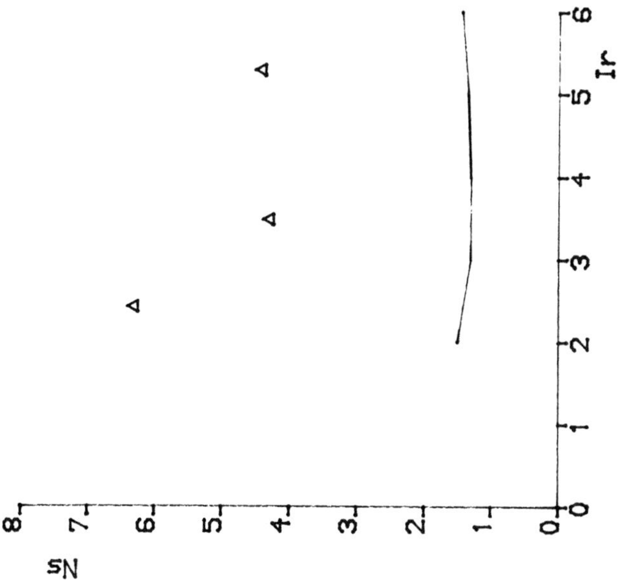

FIG 10 Stability Number Permeable Core.
Δ calculated, -- Pilarzcyk.

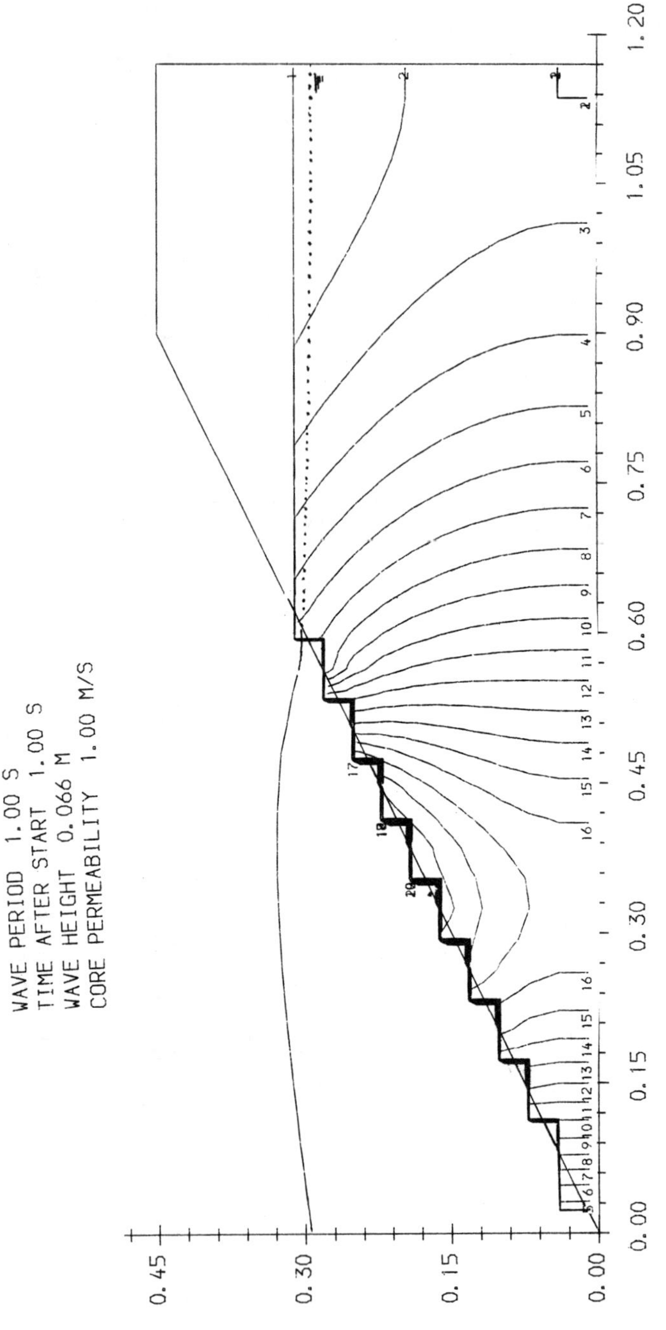

FIG 11 Head in core and on boundaries.

within the core and contours of Run-up/Waveheight R/H head within the core. Contour number 1 shows the location of the lowest head in the core and contour number 20 the highest head, the head values being the height of the water surface at these points.

CONCLUSIONS

The first model described above, based on the equations of nearly horizontal flow is able to calculate many of the important features of wave motion on a steep impermeable slope. It can be linked with the other models described which show promise in calculating the effects of permeable regions underlying the slope.

REFERENCES

AHRENS J P, "Wave run-up on idealised structures", Coastal structures '83, Washington 1983.

ALLSOP N W H & WILKINSON A R, "Hollow block breakwater armour units", Coastal structures '83, Washington 1983.

BURCHARTH H F, "The effect of wave grouping on onshore structures", Coastal Engineering, 2 (1979) 189-199.
BURCHARTH H F & THOMPSON A C, "Stability of armour units in oscillatory flow", Coastal structures '83, Washington 1983.
BURCHARTH H F, "Lessons from recent breakwater failures", World fed. of Eng, orgnaisations, Vancouver 1987.

HARLOW E, "Large rubble mound breakwater failures", J. ASCE vol 106 WW2 1980.

HIBBERD S & PEREGRINE D, "Surf and run-up on a beach of a uniform bore", JFM vol 95 pt 2 1979.

KOBAYASHI N, ROY I & OTTA A K, "Numerical simulation of wave run-up and armour stability", OTC 5088, Houston 1986.

SEELIG W N, "Wave reflection from coastal structures", Coastal structures '83 Washington 1983.

STEPHENS R V, SMALLMAN J V & ALLSOP N W H, "Development and application of a mathematical model of wave action on steep slopes", 21st Coastal Eng. Conf., Torremolinos, 1988.

THOMPSON A C & BURCHARTH H F, "Stability of armour units in flow through a layer", 19th Coastal Eng. Conf., Houston 1984.
THOMPSON A C, "BASIC Hydrodynamics", Butterworths Scientific Ltd, 1986.

TIMCO G W, MANSARD E P D & PLOEG J, "Stability of breakwaters with variations in core permeability", 19th Coastal Eng. Conf., Houston, 1984.

CHAPTER 150

STATISTICAL ANALYSIS OF DETACHED BREAKWATERS IN JAPAN

Takaaki UDA*

ABSTRACT

Construction of detached breakwaters is one of the main countermeasures against beach erosion in Japan. The number of breakwaters has been rapidly increasing, with about 2500 constructed as of the end of 1981. This study aims at the investigation of the effect and stability of detached breakwaters along Japan's coasts on the basis of survey of 1552 breakwaters constructed by the Ministry of Construction. The existing conditions of the dimensions of the detached breakwaters are statistically analyzed; the optimal dimensions for sand deposition behind a breakwater are proposed, and the critical conditions for advance of the shoreline facing an opening of breakwaters are investigated. Furthermore, the relations between the scattering rate of concrete blocks and various conditions such as bottom slope, the depth at the breakwater, the offshore distance of the breakwater and the weight of the blocks are studied. It is concluded that, in order to prevent scattering, the weight of the blocks should be at least 1.5 times heavier than that calculated from the Hudson formula.

I. INTRODUCTION

The coastal zone has been highly utilized in Japan because of the shortage of plains in the Japanese archipelago. Every effort has been made to protect the coastline from erosion due to sea waves. In 1950's, full-scale coastal protective measures were initiated in Japan. In this period, coastal dikes were mainly constructed to prevent coastal disasters brought by storm surges. In the late 1950's, coastal dikes and revetments of the vertical wall type were introduced against beach erosion. Thereafter, groins were constructed, but there were several cases with less effect. In the 1960's, the field test of detached breakwaters was conducted for the first time to confirm their effectiveness. Since then, detached breakwaters have been constructed extensively to preserve or revive the sandy foreshore, as the damage to coastal dikes and revetments, triggered indirectly by the disappearance of the foreshore, increased. At present, construction of detached breakwaters is one of the main countermeasures against beach erosion, and many of them have been constructed along Japan's coasts to dissipate wave energy and prevent beach erosion. The number of detached breakwaters has been rapidly increasing with about 2500 constructed as of 1981. However, there are still many problems to be solved for construction of detached breakwaters. The planning method to ensure the effect and

* Dr.Eng., Head, Coastal Eng. Div., Public Works Res. Inst., Ministry of Construction, Ibaraki 305, Japan.

stability of detached breakwaters has not yet been established. This study, based upon survey of 1552 breakwaters built by the Ministry of Construction, aims at the investigation of the effect and stability of detached breakwaters(Seiji et al., 1987).

II. METHOD OF INVESTIGATION

The survey was conducted in order to study the existing situations of detached breakwaters in Japan. Questionnaires were distributed to the prefectural government offices and to the work offices of the Ministry of Construction to collect the information on each of the 1552 detached breakwaters constructed by the Ministry or by the prefectural governments as subsidiary works of the Ministry by the end of 1982.

The contents of the questionnaire are summarized in Table 1, in which the principal items are general matters, natural conditions, dimensions of the breakwater, kinds of concrete blocks, information on the scattering of the blocks and the effect of the breakwater. Although the details of the items are summarized in Table 1 , there are some additional notes to be given. A serial number was attached to identify each breakwater. The difference between permeable and impermeable types was of interest, but all the breakwaters were of the permeable type. The water depth at the breakwater and the offshore distance to the breakwater

Table 1 Contents of the questionnaire.

Classification	Items
General Matters	name of the prefecture name of the coast number of the detached breakwater construction date
Design Conditions	design wave height design wave period tidal range bottom slope at the site water depth (reference: H.W.L.) offshore distance (reference: H.W.L.)
Dimensions of the Detached Breakwater	type (permeable or impermeable) placing (pellmell or uniform) crown height length elevation above the H.W.L. foundation
Kinds of Concrete Blocks	kind weight K_D value in the Hudson formula
Effect and Stability	subsidence of the blocks scattering of concrete blocks method of placing tombolo formation (yes or no) foundation (yes or no)

were measured above the H.W.L., whereas the crown height of the breakwater was measured above the standard mean sea level of Tokyo Bay.

In order to study the effect and stability of the detached breakwaters in general, the type of coasts was classified into five categories of A through E, as shown in Fig.1, based on the beach profile. The characteristic features of these coasts are as follows:

Type A: The coasts facing a bay or an inland sea, such as the Aomori Coast and the Toban Coast. The wave heights are low on these coasts compared with those on the coasts facing the open sea, and the critical depth for sand movement is small.

Type B: The coast with a fairly developed bar-trough topography, such as the Niigata Coast, the Ishikawa Coast and the Enshu Coast. The bottom slope in a region shallow enough for sand movement is mild, and the direction of mean incident waves is almost normal to the coastline.

Type C: The coast where the bottom slope in the shallow region is relatively steep without the formation of bar-trough topography, e.g., the Shimoniikawa Coast and the Suruga Coast.

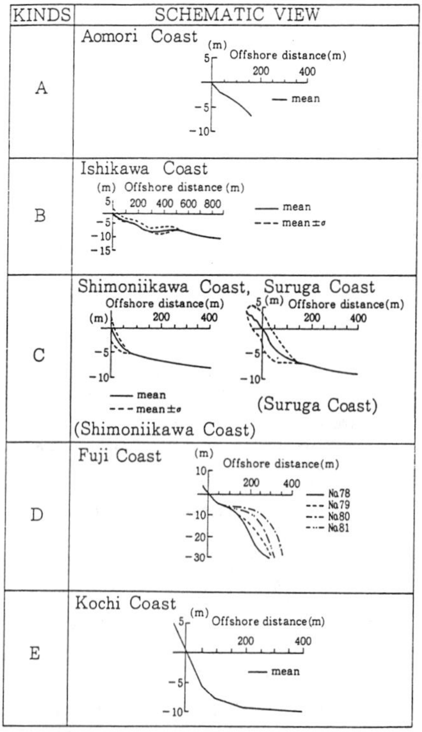

Fig.1 Type of the coasts classified into five categories of A through E.

Type D: The coast with an extremely steep slope of 1/3-1/10, e.g., the Fuji Coast in Suruga Bay.
Type E: The coast similar to Type C, but with a bar-trough topography sometimes observed in a far offshore zone, e.g., the Kochi Coast.

III. RESULTS OF SURVEY

3.1 Situations of detached breakwaters in Japan

The results of the survey about the dimensions of the detached breakwaters are shown in Figs.2 (a) to (h).

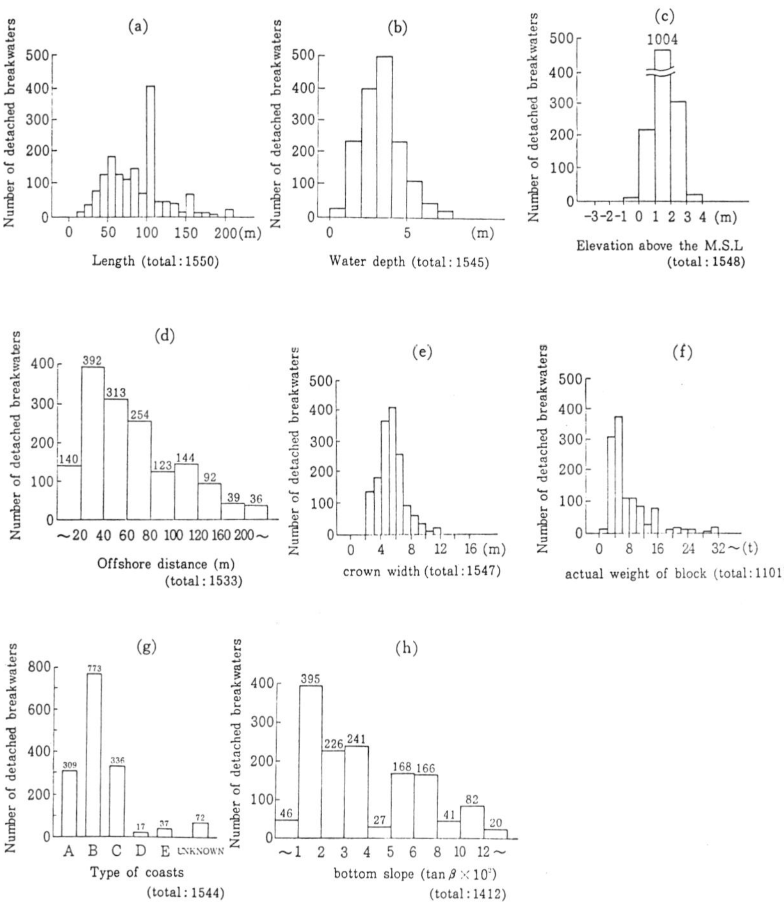

Fig.2 Frequency distribution of various dimensions of the detached breakwaters; (a) length, (b) water depth, (c) elevation above the M.S.L., (d) offshore distance, (e) crown height, (f) actual weight of block, (g) type of coasts and (h) bottom slope.

The maximum frequency of breakwater length, divided into 10m intervals, appears in the interval of 100-110m, amounting to 20%. The secondary peak is found in the interval of 50-60m. The maximum frequency of the water depths at the breakwaters is 3-4m, and it amounts to 30%. 90% of the breakwaters are located at a water depth less than 5m. The most frequent elevation of the breakwater above the M.S.L. is 1-2m and the frequency amounts to 65%. Regarding the distribution of the offshore distances, the number of breakwaters in each interval generally decreases with the increase in the offshore distance, except in the case less than 20m. The maximum number, 392, is included in the interval of 20-40m, amounting to almost 25%. The crown width varies between 2 and 12m, and the most frequent value is in the interval of 5-6m. The weight of the concrete blocks is mostly in the range from 2 to 6 tons, but there is an exceptional case of the Fuji Coast in Suruga Bay where 50-ton blocks were used (Kohno et al., 1986). Most breakwaters were built on the coasts of type B. All the data for type D, whose number is rather limited, belong to the Fuji Coast. The bottom slope ($\tan\beta$) at the breakwater was determined from the bottom topography before the construction of the breakwater, and its peak appears in the interval of 0.01-0.02. Few breakwaters were built on a coast with a slope milder than 0.01.

The sand deposition effect of the detached breakwaters was investigated statistically on the basis of tombolo formation. Figure 3 shows the percentage of tombolo formation. It is clearly seen that tombolos were formed in about 60% of all the cases. The frequency distribution of the maximum shoreline advance due to tombolo formation is shown in Fig.4. The shoreline advance ranges from 0 to 140m. The maximum frequency is found in the interval of 10-20m, and the cases with a shoreline advance of more than 100m is quite limited in number.

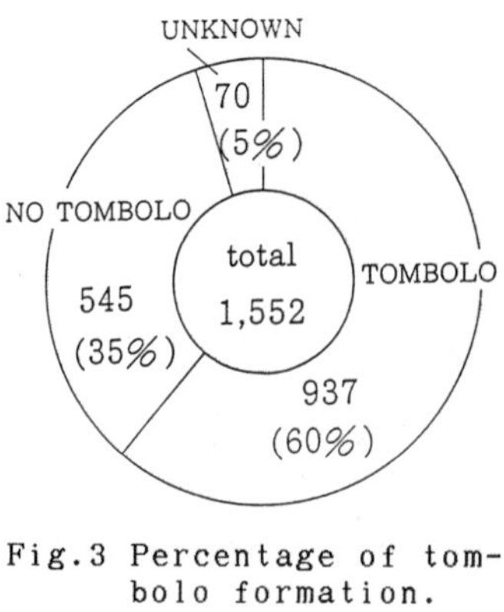

Fig.3 Percentage of tombolo formation.

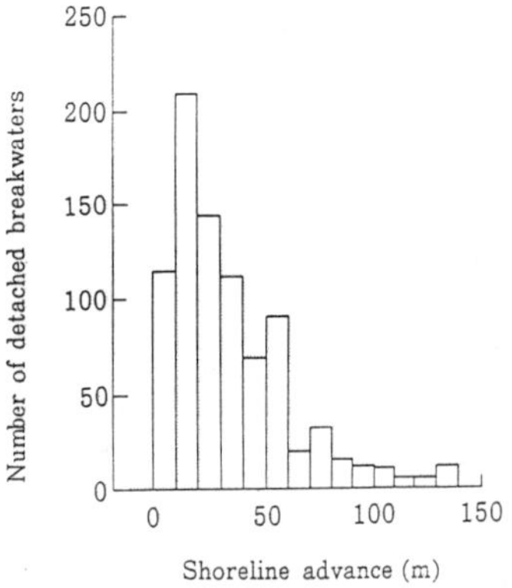

Fig.4 Frequency distribution of the maximum shoreline advance due to tombolo formation.

The number of the detached breakwaters with or without subsidence was investigated statistically as shown in Fig.5. It is found that the subsidence of the detached breakwater is observed in about half of the total number (1552). From this it is realized that the subsidence of the detached breakwater becomes an important problem regarding the stability of the breakwater. On the basis of 503 detached breakwaters whose subsided height is available, frequency distribution of the subsided height of detached breakwater is investigated as shown in Fig.6 . The subsided height ranges between 0 and 2.8m . The most frequent subsided height of the breakwater is 1.0-1.2m , and the number amounts to 158.

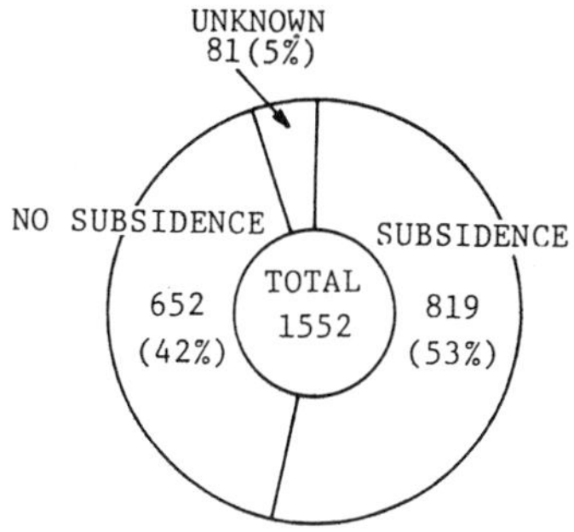

Fig.5 Percentage of the number of detached breakwater with or without subsidence of blocks.

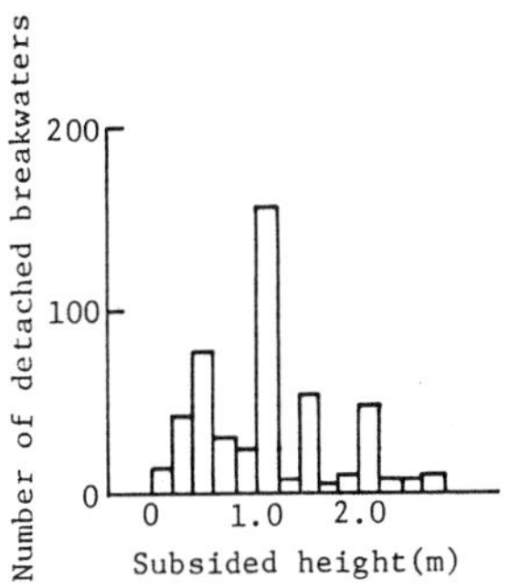

Fig.6 Frequency distribution of the subsided height of detached breakwater.

Table 2 shows the percentage of the number of the subsided detached breakwater corresponding to the kinds of the bed materials. The ratio of the number of the subsided breakwater relative to the total number accounts for 64% (sand), 43% (gravels) and 10% (rock), respectively. The reason of the subsidence of the breakwater for the rocky bottom materials may be due to the fact that the interlocking of the concrete blocks near the top of the

Table 2 Percentage of the number of the subsided detached breakwater corresponding to kinds of the bed materials.

Bottom materials \ Subsidence	Exist	Nil	Unknown	Total
rock	(10%) 19	(73%) 141	(17%) 34	(100%) 194
gravel	(43%) 139	(56%) 183	(1%) 5	(100%) 327
sand	(64%) 678	(33%) 347	(3%) 42	(100%) 1067
silt	(100%) 1	(0%) 0	(0%) 0	(100%) 1
Total	(53%) 837	(42%) 671	(5%) 81	(100%) 1589

breakwater may be loosened by wave action. Comparing both foundations of sand and gravels, the ratio of the subsidence in the latter case is small. However, the ratio of the subsidence of the blocks is large enough in both cases, and it is necessary to take effective measures against subsidence if detached breakwater is constructed on the foundation of sand or gravels.

Percentage of the number of detached breakwater with or without scattering of blocks is shown in Fig.7 . Scattering of the blocks can be observed in 14% of the total. In addition, frequency distribution of the scattered distance of blocks is shown in Fig.8 in the case whose scattered distance is available. Scattered distance ranges between 0 and 13m, and the most frequent case of scattering distance is 4-5m .

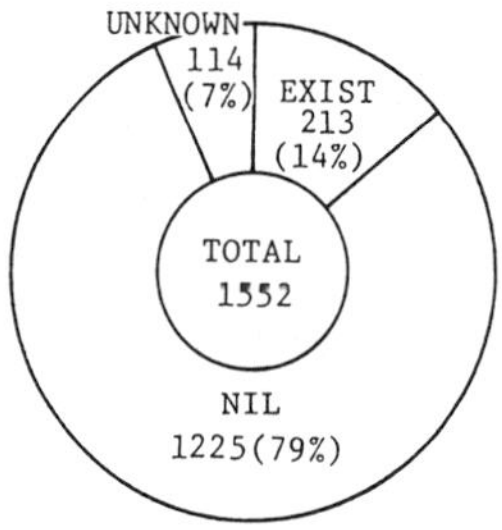

Fig.7 Percentage of the number of detached breakwater with or without scattering of blocks.

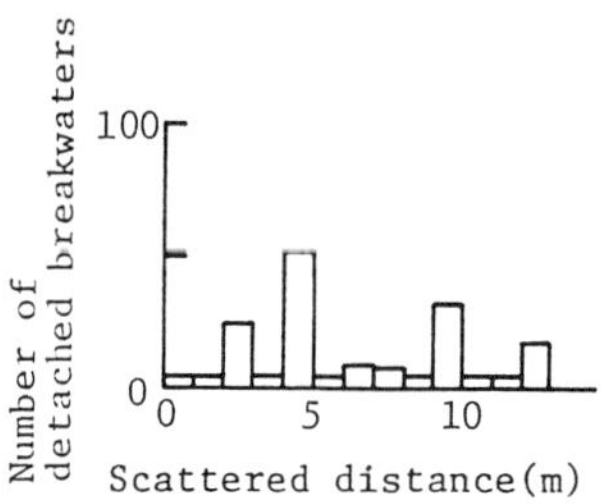

Fig.8 Frequency distribution of the scattered distance of the blocks.

3.2 Effect of detached breakwaters

The effects of detached breakwaters are mainly divided into two categories: the first is to cause the sand deposition behind the breakwater, and the second is to dissipate the incident wave. The former effect basically depends not only on the dimensions of the breakwater but on the rate of sand supply due to littoral drift. It is, however, difficult to study the influence of the littoral drift in general, because the conditions of sand supply differ to a large extent from place to place. Therefore, only the influence of the dimensions on sand deposition is discussed in this study.

The length l, the width of the opening l' , the offshore distance Y, and the water depth at the breakwater h' , are selected as the characteristic dimensions, as schematically shown in Fig.9. In order to indicate the effect of the detached breakwater on sand deposition behind the breakwater, an index T_a is introduced, which is defined by the ratio of the area of formed tombolo ($\approx 0.5\,L_t\bar{B}_t$) to the area of the original sea surface behind the breakwater ($= l \cdot Y$). Figure 10 shows the relations between the index T_a and the nondimensional parameters showing the breakwater

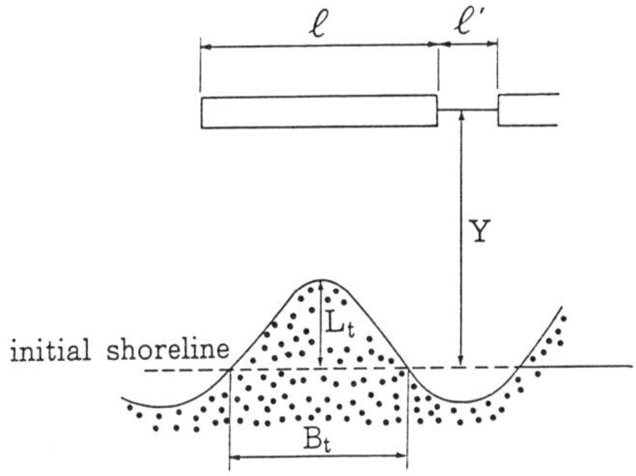

Fig.9 Characteristic dimensions of the detached breakwater (l: the breakwater length, l' : opening width, L_t : the tombolo length, B_t : the tombolo width, and Y: offshore distance.)

dimensions such as the ratios of the length to the offshore distance (l/Y) and of the relative water depth at the breakwater to the breaker depth (h'/h_b). The breaker depth was determined as follows. First, the data of the significant wave heights at each coast, measured in a typical year(1981), were ordered from the largest to the smallest. Secondly, the five largest data were selected and averaged to obtain a local reference wave height under relatively high wave conditions. This number of waves was selected since it seemed to be most significant in the discussion of stability and sand deposition under high wave conditions. Finally, the breaking depth was evaluated on the assumption that the ratio of the breaker height and the water depth at the breaking point is 0.78.

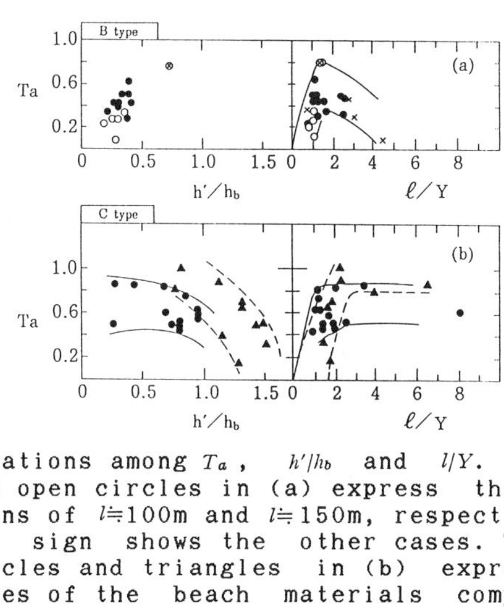

Fig.10 Relations among T_a , h'/h_b and l/Y. (The solid and open circles in (a) express the conditions of $l \fallingdotseq 100$m and $l \fallingdotseq 150$m, respectively. The 'x' sign shows the other cases. The solid circles and triangles in (b) express the cases of the beach materials composed of sand and gravels, respectively.)

Figure 10 shows the relations among T_a, h'/h_b and l/Y on the coasts of type B (e.g., the Ishikawa Coast and the Niigata Coast) and type C (e.g., the Shimoniikawa Coast and the Suruga Coast), with additional classifications on the length of breakwaters and the kind of the beach materials. The solid and broken lines in the figure show the upper and lower limits of the data. The coasts of these two types were selected as typical coasts in Japan. Although the data were limited to the case of $h'/h_b<1$ for the coasts of type B, T_a increase with h'/h_b under these conditions. The sand deposition effect due to the detached breakwaters improves with larger h'/h_b, and the relative scale of tombolo is larger when the location of the breakwater is closer to the breaking point. It was found that T_a takes the maximum value at $l/Y\approx1.6$; that is, the relative area of the tombolo is maximized when $l\approx1.6Y$.

In the case of type C, the data are scattered compared with the data for type B. The value of T_a tends to decrease if $h'/h_b<1$, which means that the relative area of tombolo decreases as the distance from the breaking point to the breakwater increases. As for the relation between T_a and l/Y, T_a appears to be maximum at $l/Y\approx2.0$, and thereafter it remains constant in contrast with the decrease in the case of type B.

As described above, a tombolo is formed due to the sand deposition effect of detached breakwaters, if various conditions are satisfied. The formation of a tombolo behind the breakwater is normally considered to be a favorable effect. On the contrary, the retreat of the shoreline facing openings can be caused by the construction of the detached breakwaters. This sometimes causes serious problems such as wave overtopping over the revetment behind a detached breakwater and/or the scouring at the foot of the revetment, when the original foreshore is narrow. Figure 11 shows the relations between the change in the shoreline facing the openings and the dimensions of detached breakwaters on the basis of the field data. Three dimensionless parameters l'/l, l'/Y and l/Y were selected to indicate the breakwater dimensions, where l', l and Y are the opening width, the breakwater length and the offshore distance, as illustrated in Fig.9. For the detached breakwaters built on the Japan's coasts, the ratios of the opening width to the length concentrate to certain values, so that the data are plotted separately for different l'/l values. In Figs.11 (a) to (c), the effect of each parameter is shown without regard to the other parameters. Let us cosider the case of $l'/l\approx0.3$ for example. In Figs.11 (b) and (c), l'/Y and l/Y are respectively taken in abscissa. Between these variable, the following relations hold:

$$l'/Y=l/Y\cdot l'/l\approx0.3l/Y \tag{1}$$

Consequently, Figs.11 (b) and (c) give the similar result. In the case of $l'/l\approx0.3$, shoreline facing the openings of detached breakwaters always advance regardless of l/Y. The tendency changes as l/Y approaches to 2; that is, if the offshore distance of the breakwater approaches to one half of the breakwater length.

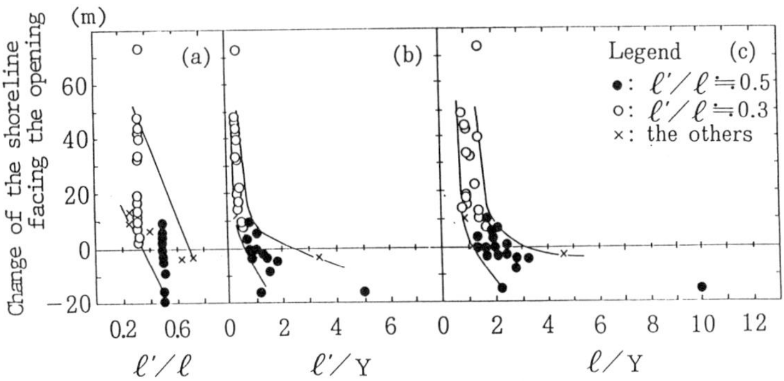

Fig.11 Relations between the change in the shoreline facing the openings and the nondimensional parameters indicating breakwater dimensions.

In the case of $l'/l \approx 0.5$, the possibility of the retreat of the shoreline facing the opening becomes high in contrast with the case of $l'/l \approx 0.3$. The critical condition for the shoreline advance may also be given by $l/Y \approx 2.0$. In other words, the shoreline retreats under the condition of breakwater length, if the offshore distance Y is less than $0.5l$. As described above, the shoreline facing the openings may retreat depending on the breakwater dimensions. This is due to the fact that the breakwater construction causes the increase in the wave height at a opening and, therefore, in the littoral drift from the opening toward the lee of the breakwater.

The second function of the detached breakwater is to dissipate the incident waves. This function can be discussed through the investigation of wave transmission coefficient. For the evaluation of the wave transmission coefficient K_T , the following formula has been proposed (Numata, 1975).

$$K_T=1/\{1+1.135(B_h/D)^{0.66}\cdot(H_i/L_i)^{0.5}\}^2 \tag{2}$$

where B_h is the mean width of the breakwater at the still water level, D is the height of a concrete block, H_i is the incident wave height and L_i is the incident wave length. Equation (2) holds only if the elevation of the crown of detached breakwater above the sea level is higher than the incident wave height.

The validity of the formula is examined by using the field data obtained through field investigations conducted on three coasts: namely, the Shimoniikawa Coast, the Niigata Coast and the Enshu Coast. As shown in Fig.12, the average of the field data can be predicted fairly well by use of Eq.(2) if $H_i/L_i>0.02$, whereas K_T is underestimated if the wave steepness is smaller than 0.02.

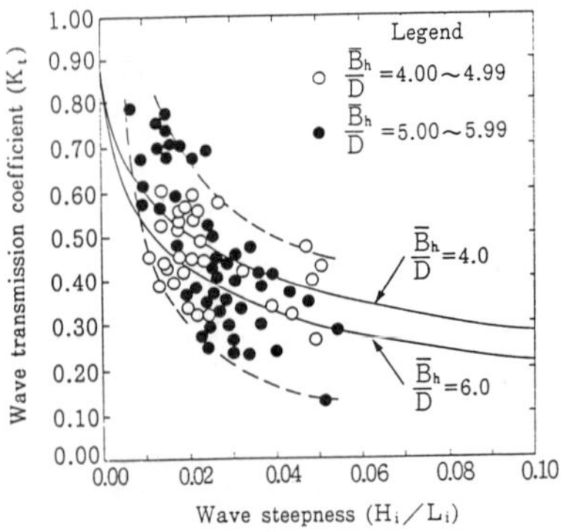

Fig.12 Comparison of the measured and predicted wave transmission coefficients.

3.3 Scattering of concrete blocks

Figure 13 shows the scattering rate, i.e., the ratio of the number of the breakwaters that suffered the scattering of concrete block to the total number of the breakwaters on the coasts of types A through E. The overall average scattering rates is 14.2%. The scattering rate of the coasts of type A is low compared with the others. This may be because the wave height is usually low in a bay and in an inland sea and consequently the wave forces acting on the concrete blocks and the changes in the beach topography caused by scouring around the blocks are small. On the other hand, the scattering rate for the coast of type D is as high as 52.9%. The scattering of concrete blocks on the coasts of this type was reported in the preceding paper (Kohno et al., 1986). Several reasons can be raised for this. First, on the Fuji Coast, incident waves tend to act on the concrete blocks without large attenuation, since the bottom slope is steep (1/3-1/10). The second reason is that strong wave actions cause the severe topographical changes around breakwaters.

The scattering rate of the concrete blocks is related to the various parameters characterizing breakwaters. Here,

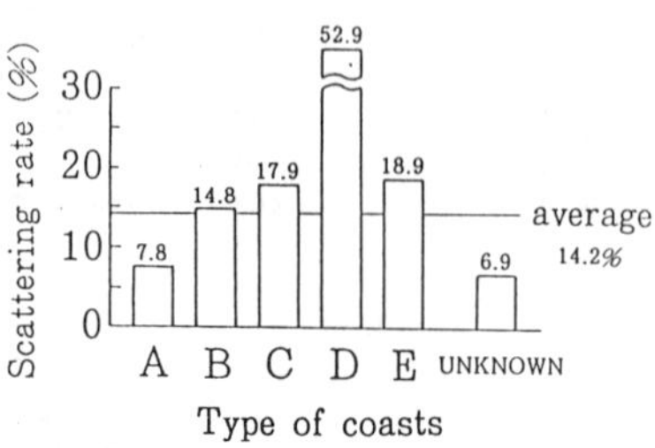

Fig.13 Scattering rate for the coasts of each type.

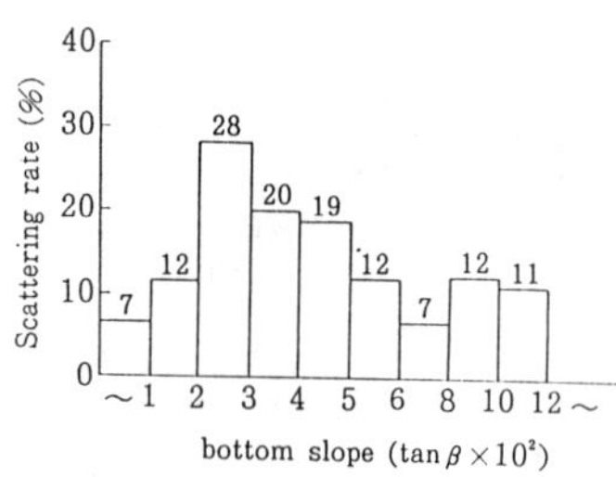

Fig.14 Scattering rate versus bottom slope.

three parameters are considered: the bottom slope, the water depth at the breakwater and the offshore distance. Figure 14 describes the relation between the scattering rate and bottom slope. The scattering rate has a maximum value of 28% for the bottom slope of 0.02-0.03, and it decreases for steeper beach slope, although it is expected that the steeper beach slope always causes the higher scatteing rate. On mildly sloping coasts, the beach material is usually fine sand, and the bar-trough topography is easily formed. The instability associated with the subsidence of the blocks and the scouring around the blocks can be the main reason of serious scattering.

The scattering rate depending on the water depth is shown in Fig.15. The scattering rate tends to increase with the water depth, fluctuating in the range from 6 to 27%. This is due to the fact that the concrete blocks are scattered by the action of breaking waves. The breaking point gets closer to the location of the breakwater under storm wave conditions. The high scattering rate at the depth of 6-7m is doubtful, because only 41 breakwaters fall in this range as shown in Fig.2 (b). In this view, there might be a tendency that the scattering rate decreases as the water depth exceeds 5m.

Similarly, the scattering rate depending on offshore distance is plotted in Fig.16. The offshore distance less than 120m are divided into 20m intervals, while those greater than 120m are divided into 40m intervals. The scattering rate increases with the offshore distance up to 100m, and then tends to decrease. High scattering rates were measured at offshore distance between 80 and 120m possibly because the locations of the detached breakwaters are close to the breaking points.

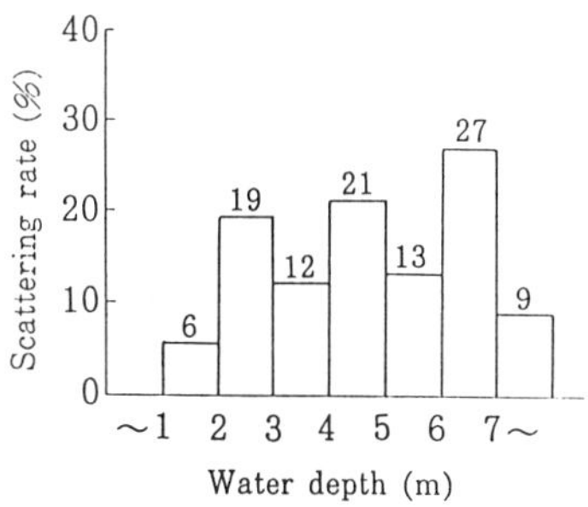

Fig.15 Scattering rate versus water depth.

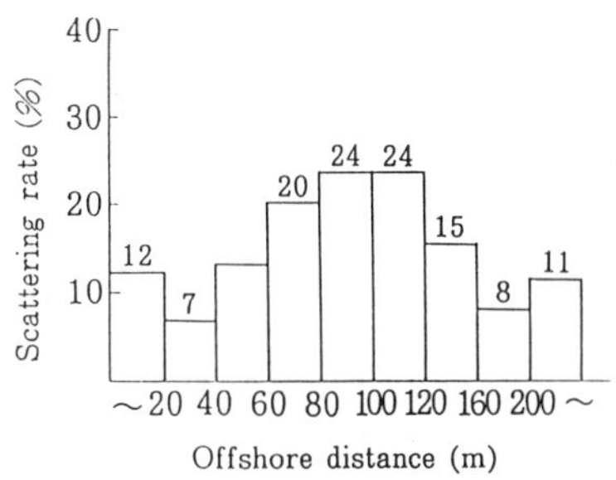

Fig.16 Scattering rate versus offshore distance.

The scattering rate was also investigated in relation with the weight of the concrete blocks. For this purpose, the ratio of the actual weight of the block to the weight calculated from the Hudson formula is introduced, where the design wave height is assumed to be equal to the water depth multiplied by 0.78, namely, the breaking wave height. The distribution of the number of the breakwaters that suffered the scattering and the distribution of the scattering rate are shown in Figs.17(a) and (b). The largest number of the detached breakwaters appears in the interval where actual weight-calculated weight ratio is 1.1-1.2. A high scattering rate is observed when the actual weight is smaller than the calculated weight multiplied by 1.1, and the scattering rate tends to decrease as the ratio of the actual to the calculated weight increases. The extent of scattering was relatively small for majority of the breakwaters with concrete blocks heavier than 1.5 times the calculated weight. If the scattering rate is re-investigated excluding the cases with small extent of scattering, the scattering rate reduces to 3.1% as shown in Fig.17(b). Effective preventive measures against scattering is the use of the concrete blocks whose weight is at least 1.5 times as large as that calculated from the Hudson formula. A firm foundation such as a rock mound is also effective.

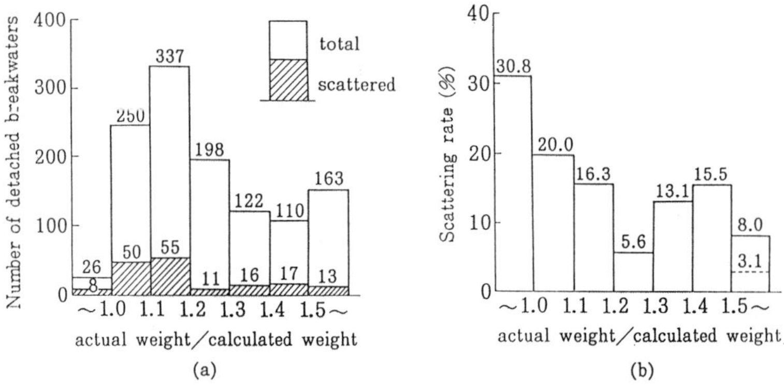

Fig.17 Number of scattered breakwaters and the scattering rate in relation with the weight of the concrete blocks.

Finally, the scattering rate of the concrete blocks is studied in relation with the kind of block placing, the foundation performance and the subsidence of the blocks. According to the preliminary investigation of the relation between the scattering rate and the kind of placing, the pellmell and uniform placings resulted in 19.5% and 11.0% scattering respectively. In studying the scattering of concrete blocks, it is important to consider the combined effects of the kind of placing and foundation performance as shown in Fig.18. The scattering rates in the cases of pellmell and uniform placings are almost the same if a foundation exists. Without foundations, however, the scattering rate in the case of pellmell placing becomes 18%, which is much larger than 8.5% in the case of uniform placing; the extent of scattering is greatly influenced by the kind of block placing. The influence of subsidence on the scattering of blocks is shown in Fig.19. The scattering rate for the breakwaters that subsided due to waves is 17.9%, which is nearly twice as large as the rate 9.4% for the breakwaters without subsidence.

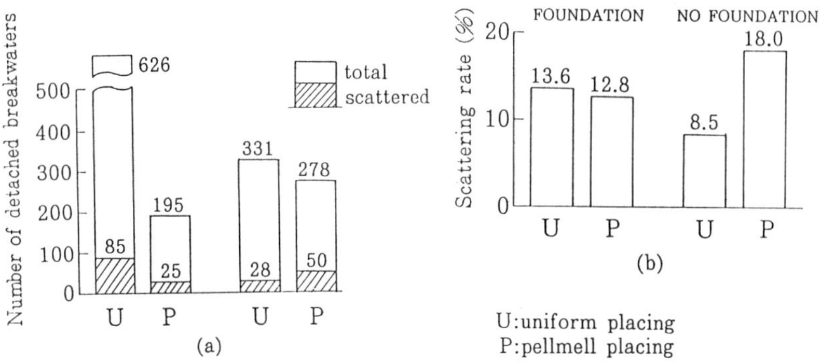

Fig.18 Combined effects of the kind of placing and the foundation performance on the scattering rate.

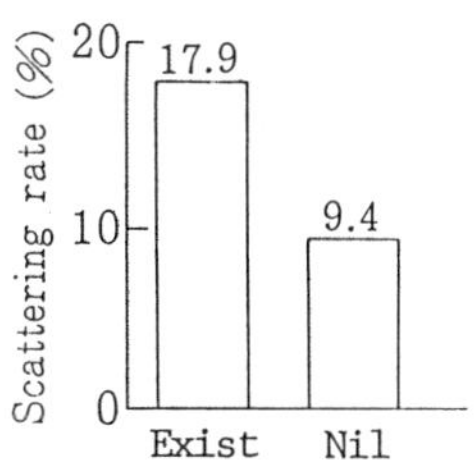

Fig.19 The scattering rate depending on the subsidence of concrete blocks.

IV. CONCLUSIONS

The dimensions of the detached breakwaters were satistically analyzed on the basis of the survey of 1552 breakwaters constructed by 1982. The main conclusions of this study are summarized as follows:

1) The optimal of a detached breakwater was evaluated for sand deposition behind the breakwater as well as the critical conditions for advance of the shoreline facing a breakwater opening.

2) The wave transmission coefficient to be used for prediction of the wave height distribution behind the breakwater can well be estimated by the formula given by Numata.

3) The scattering of concrete blocks was found in 13.7% of all the breakwaters surveyed.

4) The ratio of the number of the breakwaters that suffered scattering to the total number of the breakwaters on each coast takes the maximum value, when the bottom slope, the depth at the breakwater, and the offshore distance of the breakwater are equal to 0.02-0.03, 4-5m and 80-120m, respectively.

5) A high scattering rate appears when the actual weight of concrete blocks is smaller than 1.1 times the

weight calculated from the Hudson formula. The scattering rate reduces to 3.1% when concrete blocks heavier than 1.5 times the calculated weight are used.

6) The scattering rates of concrete blocks were 19.6% and 11.6% for the cases of pellmell and uniform placings. The scattering rate does not significantly depend on the kind of block placing, if a foundation is constructed. Otherwise, the scattering rate in the case of pellmell placing is more than twice as high as that in the case of the uniform placing. The extent of scattering is particularly influenced by the kinds of block placing in the case of detached breakwaters without foundation.

REFERENCES

Kohno,H. , T.Uda and Y.Yabusaki (1986): On the scattering of concrete armor units of detached breakwaters due to waves, Proc. 20th Inter. Conf. on Coastal Eng., ASCE, pp.1972-1986.

Numata,J. (1975) : Experimental study on wave dissipating effect of the breakwater composed by concrete armor units, Proc. 22nd Japan. Conf. on Coastal Eng., JSCE, pp.501-505.(in Japanese)

Seiji,M. , T.Uda , and S.Tanaka (1987) : Statistical study on the effect and stability of detached breakwaters, Coastal Eng. in Japan, Vol.30, No.1, pp.131-141.

CHAPTER 151

Design aspects of block revetments

K.W. Pilarczyk*

1. Introduction

The increasing shortage and costs of natural materials in certain geographical areas has resulted in recent years, inter alia, in the rapid development of artificial (concrete) block revetments.
In general, two main types of revetments can be distinguished: permeable (stone pitching, placed relatively open block-mats) and (relatively-) impermeable (closed blocks, concrete slabs). Regarding the shape and/or placing technique a distinction can be made between: a) free (mostly rectangular-) blocks and b) interlocking blocks of different design (tongue-and-groove connection, ship- lap, cabling, blocks connected to geotextile by pins etc.).
In all these cases the type of sublayer (permeable/impermeable) and the grade of permeability of the toplayer are very important factors in the stability of these revetments. The design also needs to be made (executed) and maintained. Both aspects must therefore already be taken along within the stadium of designing.

At the moment there is a large variety of types of revetment-blocks and other defence systems (i.e. block-mats), see Fig. 1. Until recently no objective design-criteria were available for most types/systems of blocks. The choice (type and size) of the revetments built sofar is only based on experience and on personal points of view, sometimes supported by small-scale model investigations.
In the light of new (stricter) rules regarding the safety of the Dutch dikes, as they have been drawn up by the Delta-Commission, the need for proper design-criteria for the revetments of dikes has evidently grown.

Because of the complexity of the problem no simply, generally valid mathematical model for the stability of the revetment are available yet. For restricted areas of application however, fairly reliable criteria (often supported by large-scale tests) have been developed in the Netherlands not only for the kind of revetment, but also for conditions of loads. This new approach is discussed in (Klein Breteler, 1988).
This paper presents a short state-of-the-art review of existing knowledge on the designing of different types of revetments and, where ever possible, the available stability criteria are mentioned. There is also given some comparison of the different types of revetments with their advantages and disadvantages and suggestions regarding their practical application.

2. Design requirements

In the design process of revetments, besides the specific functional requirements, the following technical aspects have to be taken under consideration:

*) Head Research & Development, Hydraulics and Geotechnics. Rijkswaterstaat, Road and Hydraulic Engineering Division, P.O. Box 5044, 2600 GA Delft, The Netherlands

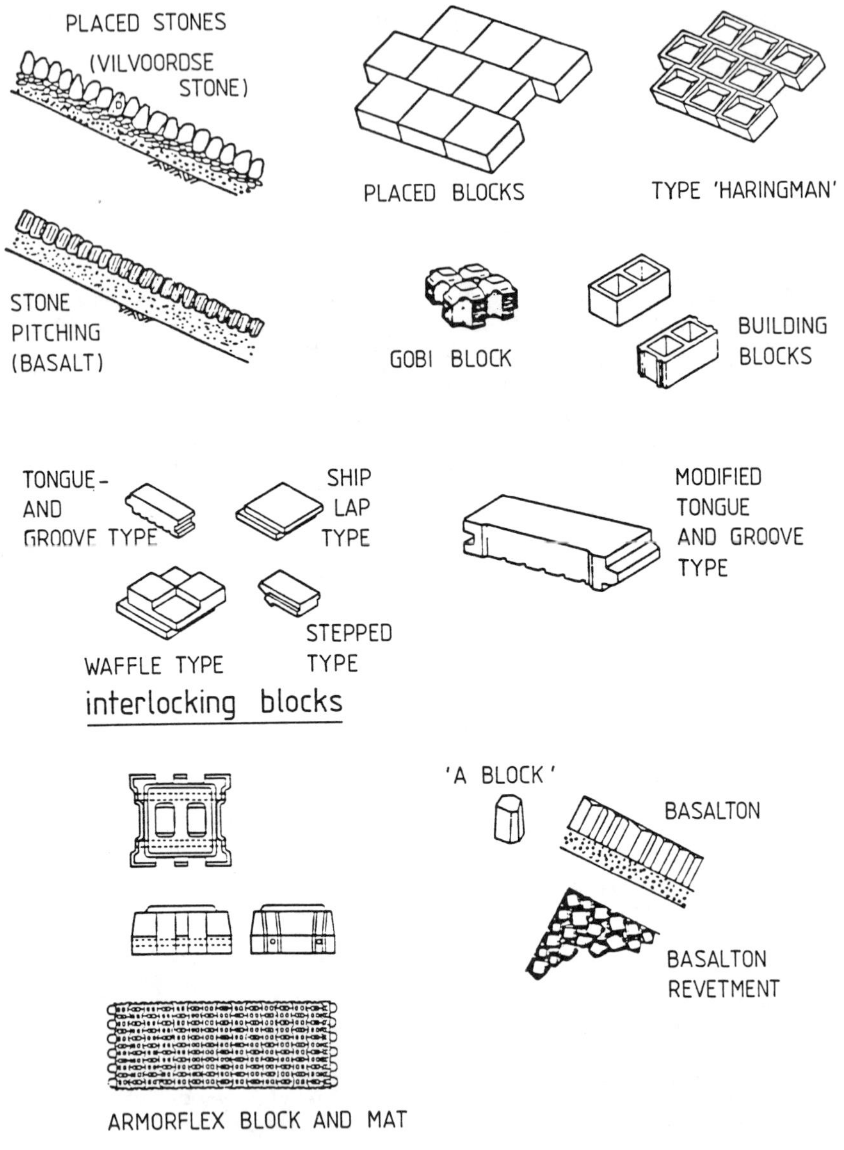

Figure 1. Examples of blocks tested on large scale

- stability (toplayer, sublayers, suboil)
- flexibility
- durability (toplayer/concrete, geotextile, cables etc.)
- possibility of inspection of failure (monitoring of damage)
- low cost (construction/maintenance)

The best revetment is one which combines all these functions.
Some of the aspects mentioned above will be discussed successively.

3. Stability criteria ("black-box" approach)

The total strength of a stable toplayer can be regarded as the sum of the contributions of different strength factors i.e. the static weight of the toplayer, the friction forces on the interface of adjoining elements, the jamming forces induced by sand or other granular materials in the voids/interspaceses between adjoining elements. The toplayer is instable when the hydraulic loads exceed the stabilizing forces. Hydraulic loads are created by waves inducing time dependent pressures on the toplayer, that are transmitted to the sublayers causing pressure differences across the toplayer, that tend to lift blocks from the revetment. These pressure gradients cause also watermotions in the sublayers that endanger the internal stability of these granular layers or the stability of the interface between successive sublayers, causing a gradular deformation which may affect the stability of the toplayer.
When the hydraulic loads (uplift pressures) exceed the stabilizing forces than the additional strength factors are mobilized to prevent damage to the toplayer i.e. inertia forces of the block and the water underneath, and pressure fall under the block to enable the inflow of water in the growing space between the moving element and the sublayers. For relatively smooth block revetments and for openings-percentage smaller than 10%, the drag forces are neglectable, and the (in-)stability of the block revetments is mainly influenced by the uplift forces. These uplift forces are directly dependent on the composition (hydraulic permeability) of the sublayers. That means also that the stability of the block revetments can not be treated separately from the sublayers and/or subsoils.

This last remark (conclusion) has very important practical consequences, namely:

- results from the small-scale model have limited value for practical application because these models are not able to reproduce the prototype sublayers and subsoils (the scaling rules are still not properly known); this explains also a large scatter in the results from the small-scale models,
- "black-box" approach (even based on a large-scale data) has limited value and should be applied with caution; in the black-box approach all the forces discussed above are integrated to one strength number - the changing of the composition of revetment (i.e. permeability of toplayer or sublayer) may strongly affect the total stability which is difficult to deduce from the black-box stability numbers,
- the problems mentioned above can be overbridged by applying the mathematical models as being recently developed in the Netherlands (Klein Breteler, 1988).

For the rough (first) approximation of the necessary dimensions of block revetments the following "black-box" model is being developed by Pilarczyk (1987) (PIANC, 1987) (see Fig. 2):

$$\frac{H_s}{\Delta D} = \psi \cdot \frac{\cos\alpha}{\sqrt{\xi_z}} \approx \psi \cdot \xi_z^{-0.5}$$

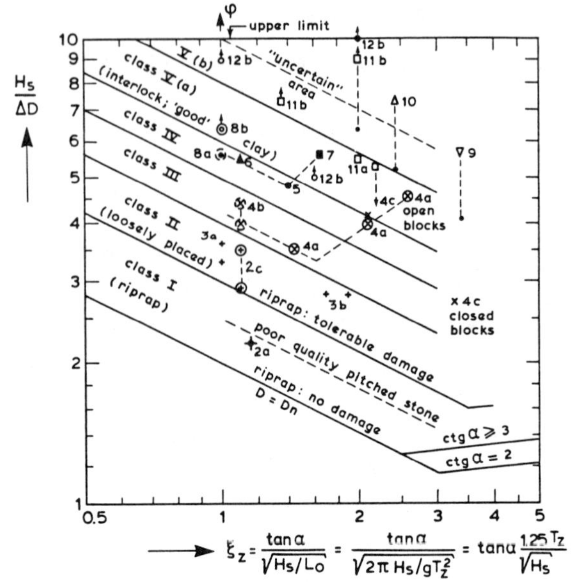

Note :

(arrow symbol) - no damage appeared

Δ^{10} - numbers refer to table

(dashed line) ← possible replacement due to translation to irregular waves

D - block thickness
D = D_n for riprap

$D_n = \left(\frac{M_{50\%}}{\rho_s}\right)^{1/3}$

$M_{50\%}$ = average mass of stone

ρ_s = density of stone

H_s - significant wave height

T_z - average wave period

$\tan\alpha$ - slope gradient; $\text{ctg}\,\alpha \geq 2$

φ - strength coef. defined at $\xi_z = 1$

ξ_z - breaker similarity parameter

$$\frac{H_s}{\Delta D} = \varphi \frac{\cos\alpha}{\sqrt{\xi_z}} \approx \frac{\varphi}{\sqrt{\xi_z}}$$ (assumed function)

$\xi_z < 3$ (breaking waves)

Figure 2. Classification of strength of block revetments

for $\xi_z < 3$ (breaking waves) and $ctg\alpha \geq 2$

and $$\xi_z = \frac{\tan\alpha}{\sqrt{H_s/L_o}} = \tan\alpha \frac{1.25\ T_z}{\sqrt{H_s}} \text{ (breaker index)}$$

where: $H_s/\Delta D$ = strenght parameter, H_s = significant wave height, α = slope angle, T_z = average wave period, D = thickness of block, $\Delta = (\rho_s - \rho_w)/\rho_w$, relative density of block and ψ = stability factor. For riprap $D = D_{n50} = (W_{50}/\rho_s)^{1/3}$, ($W_{50}$ = 50% by mass). For $\xi_z > 3$ the constant $H_s/\Delta D$- value in $\xi_z = 3$ can be kept safely (N.B. for some types of block revetment the $H_s/\Delta D$- values can increase a lot when ξ_z is large).

Note: The results of the recent large-scale tests (1987) indicated that the stability of the loosely placed blocks (cat. II), placed on the granular sublayer, can be approximated better by (Fig. 3):

$$\frac{H}{\Delta D} = \frac{\psi}{\xi_z} \quad (1 \leq \xi_z < 5 \text{ and } ctg\alpha > 2)$$

where: ψ = 3.5 - for less permeable (closed) blocks, and
ψ = 4.0 - for relatively permeable blocks (open area 5 - 20%)

The open blocks have more reserve-stability than closed blocks; the ψ = 5 value can be applied as an upper-limit (tolerable movement of blocks).

In the case of frequent double-top storms it has to be advised to reduce the $H_s/\Delta D$- values with 25% because of the long-term effect of loading.

This model works rather well for riprap on relatively impermeable core (i.e. sand) and it also gives a good approximation for some block revetments. By using this model the strength of various types of block revetments can be directly compared to the strength of riprap revetments.

Stability factor (ψ) has been determined empirically. In view of the problems on the translation of the results of small-scale models into prototype, only the large-scale tests have been taken into account (Fig. 1). The results of this have been summarized below (ψ is defined at ξ_z = 1):

I : $2 \leq \psi \leq 3$ riprap (2 layers) N $\leq$ 3000 waves; ψ = 2 denotes "no-damage" and ψ = 3 denotes max." tolerable damage
II : $3 \leq \psi < 4$ pitched stone, loose placed blocks, blocks connected by geotextile
III: $4 \leq \psi < 5$ blocks interlocked by friction, grouted blocks connected by geotextile, cabled blocks
IV : $5 \leq \psi < 6$ loose blocks directly on "good clay"
V : $6 \leq \psi$ grouted (cabled-) blocks, mechanically interlocked blocks.

Remarks:
1 The upper and lower limits within a category are generally dependent on the quality of design and construction,
2 Filter requirements of the soil have to be met by the geotextile and/or granular sublayer,
3 Blocks placed directly on geotextile and well-compacted sand: max H_s = 1.2 m,
4 "Good-clay" = according to requirements given in the Guidelines (CUR/TAW, 1984) and smooth surface (no cavities),

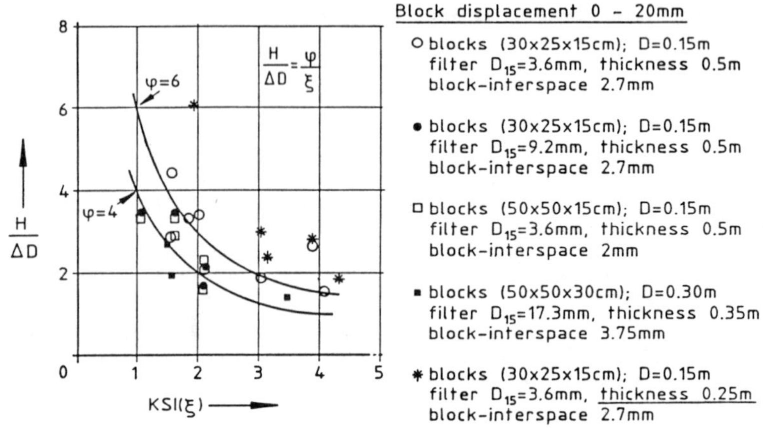

a) Stability-function for closed-block revetments

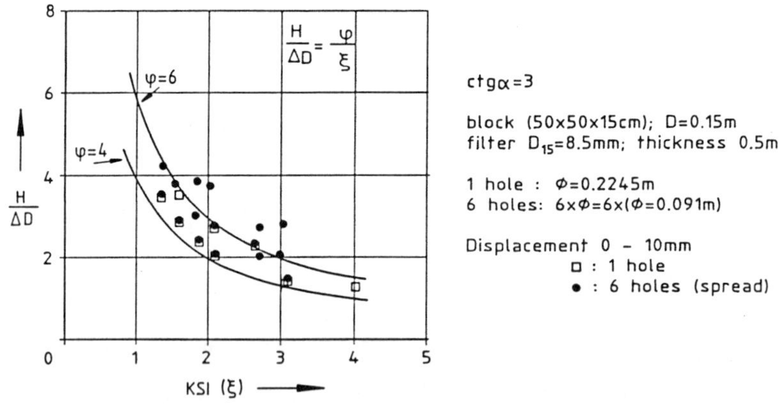

b) Stability-function for open-block revetments

Figure 3. Stability-criteria for loosely placed blocks (1987).

5 Cat. V ($\psi \geqslant 6$) must be carefully designed and examined especially regarding the stability of sublayers (large scale check of design is recommended),
6 Because of practical reasons the block thickness less than 10 cm is not recommended (for cabled blockmats 8 cm).

4. Composition of construction

There is a large number of different types of toplayer, sublayer and subsoil and, it is obvious, that there are many possible combinations that can lead to a large number of possible constructions. This does not simplify the choice of a revetment. Some possible aspects and solutions that can play a part in the choice of the construction of the revetment are mentioned below.

- Stability of top-layers strongly depend on the sort/composition of sublayers and they must therefore be regarded as a whole. As an example, from the large-test results it appears that a block revetment on a sublayer of "good clay" provides more stability than one on a permeable sublayer.
- Instability (erosion) of sublayers and/or subsoil can lead to failure of a toplayer. The stability of top-layers and sublayers must therefore be designed steadily (with an equal opportunity of failure).
- A good tuning of the permeability of the toplayer and sublayers (including geotextiles) is an essential condition for an equal design. The permeability (k) of the different parts of the construction must increase from underneath to top: k, ground $<$ k, sublayer $<$ k, toplayer.
  The strength of revetment increases with increasing the permeability of the toplayer and decreasing the permeability of the sublayer. Moreover, a thin permeable sublayer, reduces the uplift forces, but increases the internal gradients (the internal stability of the sublayers decreases).
- The granular filters are mostly expensive and difficult to realize (especially under water) within the filter-requirement limits. A substitutional solution is a geotextile (filter function) with a certain thickness of graded stone layer (with function to dump the internal hydraulic loads). A good and cheaper solution can also be realized by applying a thick layer of broadly graded waste products as minestone, slags, silex, etc. (well compacted).
- The use of blocks directly on sand-body (with geotextile in between) is restricted, at the present state of knowledge, to wave height of $H_s = 1.2$ m. The good compaction of sand is essential to avoid sliding or even liquefaction. For loads higher than $H_s = 1.2$ m a well graded layer of stone on a geotextile is recommendable (e.g. layer 0.2-0.3 m for $1.2 \text{ m} < H_s < 2.5$ m).
- For placement of blocks directly on clay subsoil/sublayer, besides the requirement of right composition and homogenity, the proper compaction and smooth surface (blocks placed as close as possible to the clay surface) are of primarily importance (CUR/TAW, 1984). In the case of "poor clay" (concerning composition and/or surface preparation) it should be recommended to protect the clay-surface by a multilayer (non-woven) geotextile and to use a lower value of ψ (i.e. $4 < \psi < 5$).
- The stability of (real) loose blocks is rather low ($\psi \approx 3$). The strenght of these blocks can be increase by introducing of the clenching forces between blocks, caused by friction and wash-in-material, or interlocking.
- The main advantage of applying block-mats and/or interlocked systems are:

- in general, higher stability of toplayer
- mechanical placement (also from water side)

and the main disadvantages (or uncertainties):

- durability/damage of binders (cables, pins, etc.)
- less flexibility and inspection problems in respect to failure sublayer
- repair problems (especially under water)
- connection of adjoining mats (especially under water)

Many of these problems will be omitted if the sublayer is designed properly.

Although the grouting of these systems (i.e. grouting/ washing in by fine broken stones) decreases the flexibility, it can be helpful for some reasons.

- Disadvantages of non-grouted systems:
  - individual blocks can move (less stability of toplayer)
  - cables/pins loaded more frequently
  - abrasion of geotextile
  - erosion sublayer/subsoil can take place more easily
- Advantages of grouted (blinded) systems:
  - higher stability of toplayer (no movement of individual blocks)
  - if (at heavy attack) the mat will be lifted out, then there is less loading of cables/pins and interlocked-elements, and less abrasion of geotextile, because such mat will move as integral unit.
- Not every system can be grouted; if the interspaces between the blocks are too large, the grouting material will be easily washed out. In this respect the basalton-system (grouted but not cabled - see Fig. 1) has more advantages than other systems, because of the tapered vertical form and lack of cabling, the blocks can still follow a limited settlement, the settlement is immediately evident, and a washing out of grouting material is limited.
- Erosion or damage often starts at joints and transitions. Therefore, an important aspect of revetment construction, which requires special attention, are the joints and the transitions; joints onto other revetments materials, and transitions onto other revetment parts or other structures. If they are inevitable the discontinuities introduced should be minimized. This holds for differences in elastic and plastic behaviour and in the permeability of the sand tightness. Proper execution is essential in order to obtain satisfactory joints and transitions.

5. Conclusions

- Stability of block revetments strongly depend on the sort and composition of toplayers and sublayers and they must be regarded and treated as a whole.
- Because of scaling problems of sublayers the small-scale models are not suitable for investigation of stability problems of revetments.
- The value of "Black-Box" approach based on large scale tests is mainly limited only to the construction tested; the extrapolation to other compositions of construction involves many uncertainties.
- The "Black-Box" model as presented in this paper can be applied only for the rough (first) approximation of necessary dimensions of revetments.
- For detail design and/or more complicated cases the mathematical model as developed in the Netherlands (Klein Breteler, 1988) or large scale tests are recommended.
- Whatever calculation method and protective system is adapted, (local) experience and sound engineering judgement play on an important part in a proper design of protective structures.

The research on dike protection and other coastal defence systems is still going on in the Netherlands. Research is now being directed towards a better probabilistic description of the design, better understanding of the failure mechanisms, application of new or alternative materials, monitoring of damage, economical aspects of design and optimal choice of constructions applied incorporating future maintenance aspects.

6. References

(1) CUR/TAW (1984). Guide to concrete dike revetments. Netherlands Centre for Research, Codes and Specification for Concrete (CUR) and Technical Advisory Committee on Water Defences (TAW), Report 119 (English version available).

(2) BURGER, A.M., et al. (1988). Design method for block revetmen, 21st Internat. Conf. on Coastal Engineering, Torremolinos, Spain.

(3) PIANC, PTC I, W.G.4 (1987). Guidelines for the design and construction of flexible revetments incorporated geotextiles for inland waterways.

(4) PILARCZYK, K.W. (1987). Sea Defences: Dutch Guidelines on Dike Protection. Rijkswaterstaat, Road and Hydraulic Engineering Division, The Netherlands, Report WB-NO-87 110, April 1987.

No. Reference	Toplayer		Sublayer	Slope ctgα	$\frac{H_s}{\Delta D}$	ξ_z	$\psi = \frac{H_s}{\Delta D}$ $\xi_z = 1$	Class
1a M1983	Riprap - no damage	L	Granular filter	2:6	full range		2:2.25	I
1b WL Pub No. 332, 1984	"Tolerable" Riprap-damage	L	thick 2D50		D = Dn		3.0	ref.
2a M2036	Vilvoordse (pitched)	L	brick rubble	3.5	2.20	1.15	2.5	
2b Sollitt	Pitched stone, D=Dn	L	crushed stone	2	2.80R	1.3÷2.6	3.0	
2c M2036	Pitched basalt	L	brick rubble	3.5	2.9÷3.5	1.10	3.5	II
3a M1795	Placed blocks	L	gravel	4+berm	3.3÷3.6	1.0	3.5	
3b M1881-IX	Placed blocks	L	gravel	3	2.8	1.7÷1.9	4	
					3.5R	1.45		
4a C.E.R.C. (US)	Gobi blocks + geo-tex.(open blocks)	L	fine gravel, 40% 2mm coarse sand	3.5	4.0R 4.55R	2.10 2.60	4.0	III
4b C.E.R.C.	Open/building blocks	L	sand 0,4 mm/ geotextile	5	4÷4.4R		4.5*	
4c M1881-XII	Placed blocks (closed)	L	sand 0,2 mm/ geotextile	3	4.16R 2.70R	2.10 3.30	5.0*	
5 M1900	Basalton (basalt)	L	crushed stone (silex)	3	4.80	1.40	5.0	
6 M2036	Placed blocks (haringman)	O	minestone + thin l. gravel	3.5	5.50	1.10	5.5	IV
7 M1910	Armorflex-blocks	L	crushed stone/	3	5.60	1.65	5.5	
8a M1795	Placed blocks	L	good clay with rough surface	4	5.60	0.95	5.5	
8b M1795	Placed blocks	L	good clay	4	6.40	1.00	6.0**	
9 C.E.R.C.	Ship-lap blocks	I	crushed stone	2	5.71R	3.40	6 a 7	
10 C.E.R.C.	Tongue-and grove "modified" (6 mm relief slot)	I	crushed stone	2	7.3R 5.30R	2.43 2.20	7 a 8	
11a Oregon (US)	Armorflex-mats (grouted)	C+G	sand 0,5 mm/ geotextile	3	5.50 7.3	2.00 1.35	7 a 8	V
11b M1910	Armorflex blocks (grouted)	G	crushed stone/ gravel	3	9.0R	2.00	8.0	
12a	Placed blocks with interspaces (grouted)	G	gravel	3	7.0R	1÷2.5	8.0	
12b	Basalton (grouted)	G	crushed/stone/ silex	3 3	5.0R 10.0R 15.0R	1.60 1.30 2.00	10.0	

Note: L = loose placed blocks, O-old; natural reinforcement/interlocking (i.e. sand, grass)
G = grouted (blinded/chocked by gravel, silex, slags, crushed stone)
C = cabled, I-interlocked blocks (artificial); R = tests with regular waves
*) Blocks/mats directly on compacted sand + geotextile only for $H_s < 1.5$ m
**) 'Good' clay: proper composition, compaction and smooth surface acc. to criteria

CHAPTER 152

RELIABILITY ANALYSIS OF RUBBLE-MOUND BREAKWATERS

Kazumasa Mizumura*
Masato Yamamoto**
Taiji Endo***
and
Naofumi Shiraishi****

To verify the effect of wave period on the motion of concrete blocks in rubble-mound breakwaters, simple physical models are employed and their motion is investigated by the numerical simulation. Finally, risk or reliability are calculated and the weight of concrete blocks for given physical condition is discussed by them.

Introduction

In Japan, many kinds of concrete blocks have been used to protect coastal structures. The weight of concrete blocks is computed by Hudson's formula for given design wave height, density of concrete blocks and breakwater slope. However, the weak points of Hudson's formula are described as follows:

1. It does not contain the effect of wave period;
2. It includes an uncertain parameter K_D;
3. Incident wave height is not constant, but probabilistic; and
4. Reliability or risk are not considered with this formula. It says only safe or not.

Especially, the item 1 has been pointed out by field engineers for a long time. To study the above points, we introduce simple physical models to express the incipient motion of concrete blocks and analyze the motion for given parameters and physical conditions which are governed by appropriate probabilistic laws. Finally, the risk is defined by the ratio of the number of concrete blocks moved to the total number of trials.

Stability Analysis of Concrete Blocks

The incipient motions of concrete blocks in the rubble-mound breakwater which frequently occur are classified into three: They are sliding, rocking and lift-up. To analyze the sliding motion of a concrete block, the sliding force f_s

----

*Professor of Civil Engrg. Dept., Kanazawa Inst. of Tech., 7-1, Ogigaoka, Nonoichimachi, Ishikawa Pref., 921, Japan.
**Assistant Manager, ***Director and ****President, Nippon Tetrapods Co., Ltd., 2-7, Higashinakanukicho, Tsuchiura, Ibaraki Pref., 300, Japan.

and the resistance force f_R are expressed by refering Fig.1(a).

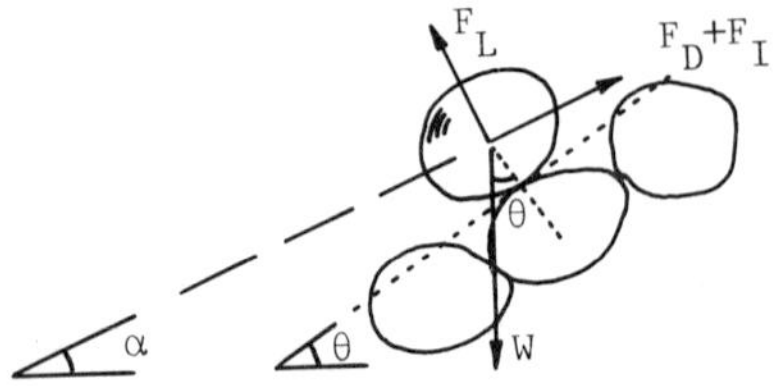

Figure 1(a). Motion of a Concrete Block

$$f_s = \varepsilon_1[(F_D+F_I)\cos(\alpha-\theta)+F_L\sin(\theta-\alpha)]-\frac{\rho_s-\rho_w}{\rho_s} W \sin\theta \quad (1)$$

$$f_R = [\frac{\rho_s-\rho_w}{\rho_s} W \cos\theta - \varepsilon_1 F_L \cos(\alpha-\theta) + \varepsilon_1(F_D+F_I)\sin(\theta-\alpha)]\mu \quad (2)$$

in which $F_D = \rho_w C_D(c_2\ell^2)u|u|/2$, (drag force); $F_I = \rho_w C_M(c_3\ell^3)du/dt$, (inertia force); $F_L = \rho_w C_L(c_2\ell^2)u^2/2$, (lift force); C_D = the drag coefficient; C_L = the lift coefficient; C_M = the virtual-mass coefficient; ρ_w = the water density; ρ_s = the density of a concrete block; θ = the repose angle of a concrete block; α = the angle of the breakwater slope makes with the horizontal; ℓ = a characteristic length of a concrete block; $c_1\ell$ = the distance from the point of support of the center of gravity of the block; $c_2\ell^2$ = the projected cross-sectional area of the block on a plane perpendicular to the direction of the velocity; $c_3\ell^3$ = the volume of the block; μ = the friction coefficient; ε_1 = the sheltering coefficient for sliding motion; and W = the weight of the concrete block. The reason why the direction of the drag and the inertia forces on concrete blocks is upward along the slope is explained as follows: Since the force due to the interlocking effect among concrete blocks is not able to be easily computed, we assume that concrete blocks do not slide down against the interlocking effect. The dropping movement of concrete blocks is observed to be caused soon after the rocking, sliding or lift-up motions of concrete blocks by the experiments. Most of the movement of concrete blocks is produced by the rocking motion. In the rocking motion (Fig.(b)), the sliding and the resistance motion M_s

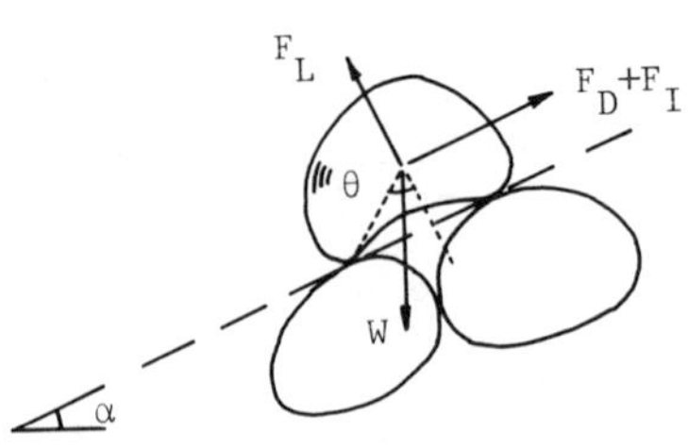

Figure 1(b). Motion of a Concrete Block

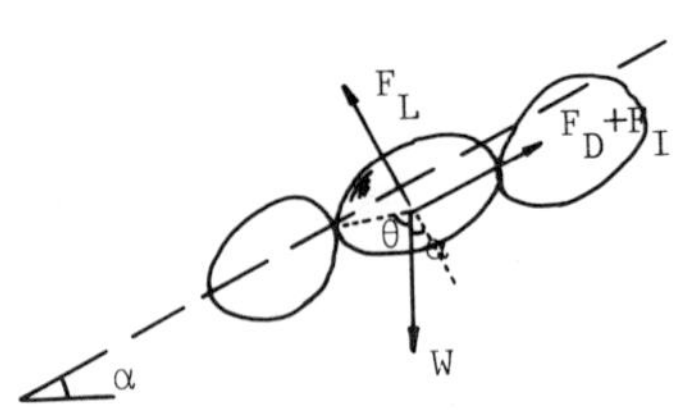

Figure 1(c). Motion of a Concrete Block

and M_R about the contact point of a concrete block for the condition of incipient motion are given (Walton et al. 1981) by,

$$M_R = \frac{\rho_s - \rho_w}{\rho_s} c_1 \ell \; W \sin\theta \cos\alpha \tag{3}$$

$$M_s = \varepsilon_2 [(F_D + F_I) c_1 \ell \cos\theta + F_L c_1 \ell \sin\theta] + \frac{\rho_s - \rho_w}{\rho_s} c_1 \ell \; W \cos\theta \sin\alpha \tag{4}$$

in which ε_2 = the sheltering coefficient for the rocking motion. In the lift-up motion, the lift force f_L and the normal component of the weight to the slope f_w are expressed as (see Fig.1(c)):

$$f_L = \varepsilon_3 [(F_D + F_I) \sin(\alpha + \theta - \frac{\pi}{2}) + F_L \cos(\alpha + \theta - \frac{\pi}{2})] \tag{5}$$

$$f_w = \frac{\rho_s - \rho_w}{\rho_s} W \cos(\frac{\pi}{2} - \theta) \tag{6}$$

in which ε_3 = the sheltering coefficient for the lift-up motion.

Movability Condition of Concrete Blocks

It is not certain that the geometrical position of concrete blocks is similar to the proposed ones in the stability analysis. We define a word "movability" when concrete blocks are in the same condition as the proposed state in the stability analysis. To connect the movability condition with the stability analysis, the following assumptions are introduced:

1. The occurrence position of a concrete block which has the movability condition is uniformly distributed from the breakwater bottom to 90% of the significant wave height above the still water level along the breakwater face;
2. The number of the concrete blocks with the movability condition per one wave is k; and
3. The movability condition of concrete blocks continues for n_i wave cycles for i=1 to k.

We consider that the process (2) and (3) are very analogous to service problems in the queueing theory. So the integers k and n_i for i=1, 2,$\cdots$, k are probabilistic variables and described by the Poisson process as follows:

$$p(k) = \frac{\exp(-\lambda) \; \lambda^k}{k!} \tag{7}$$

$$p(n_i) = \frac{\exp(-\lambda') \; \lambda'^{n_i}}{n_i!} \tag{8}$$

in which λ and λ' = parameters to govern the Poisson process.

Risk and Reliability

To calculate the risk or the reliability of the rubble-mound breakwater, let us define new variables R_s, R_r and R_ℓ as follows:

$$R_s = f_R - f_s \tag{9}$$

$$R_r = M_R - M_s \tag{10}$$

$$R_\ell = f_w - f_L \tag{11}$$

The variables on the right side in the above equations are derived in the stability analysis. The probability density functions $f(R_s)$, $f(R_r)$ and $f(R_\ell)$ are computed by the Monte Carlo method for given probabilistic variables. The failure probability or the risk of each concrete block is expressed by

$$P_s = \int_{-\infty}^{0} f_s(R_s)\,dR_s \tag{12}$$

$$P_r = \int_{-\infty}^{0} f_r(R_r)\,dR_r \tag{13}$$

$$P_\ell = \int_{-\infty}^{0} f_\ell(R_\ell)\,dR_\ell \tag{14}$$

The reliability is shown as the difference of the risk from 1. Thus, if these three motions are assumed to be independent, the occurrence probability of motion for i-th block in the breakwater slope is given by

$$P_e(i) = P_s + P_r + P_\ell - P_sP_r - P_rP_\ell - P_\ell P_s + P_sP_rP_\ell \tag{15}$$

The whole failure probability of the rubble-mound breakwater P_w is obtained by computing the failure probability of union set for the failure probability of each block $P_e(i)$ as follows:

$$P_w = 1 - \prod_{i=1}^{m}[\,1 - P_e(i)\,] \tag{16}$$

in which m = the number of concrete blocks per unit width. Since the motions treated in the stability analysis are incipient, it is reasonable to assume that the motion of each block is independent in the above derivation.

Wave Conditions

The probability density functions of wave height and period are given by

$$p_{H/\bar{H}} = \frac{\pi}{2}\left(\frac{H}{\bar{H}}\right)\exp\left[-\frac{\pi}{4}\left(\frac{H}{\bar{H}}\right)^2\right] \tag{17}$$

$$p_{T/\bar{T}} = 2.7\left(\frac{T}{\bar{T}}\right)^3\exp\left[-0.675\left(\frac{T}{\bar{T}}\right)^4\right] \tag{18}$$

in which $\bar{H}$ = the mean wave height and $\bar{T}$ = the mean wave period. The cumulative distributions of the wave height and period are obtained by integrating Eqs.(17) and (18). The significant wave height and period are expressed by using the mean wave height and period, respectively.

$$H_{1/3} = 1.60\ \bar{H} \tag{19}$$

$$T_{1/3} \simeq 1.1\ \bar{T} \tag{20}$$

In the sampling process, the waves satisfying the following condition are removed as the breaking waves before they reach the breakwater slope:

$$H/L \geq 0.142 \tag{21}$$

The condition the wave does not break on the breakwater slope is given by

$$\frac{H}{L} \leq \sqrt{\frac{2\alpha}{\pi}}\ \ \frac{\sin^2\alpha}{\pi} \tag{22}$$

For the nonbreaking waves the shallow water theory gives the flow velocity and the acceleration

$$u = \sqrt{\frac{g}{h}}\ \frac{H}{2}\ \sin\sigma t \tag{23}$$

$$\frac{du}{dt} = \frac{g\pi H}{L}\ \cos\sigma t \tag{24}$$

in which H = the wave height; h = the water depth; $\sigma = 2\pi/T$; T = the wave period; and L = the wave length. The exerted force on a concrete block due to the wave breaking p_m is calculated by Goda's formula (Iwagaki & Sawaragi, 1979). The total force P_m exerted on the concrete block is given by

$$P_m = c_2 \ell^2 p_m \tag{25}$$

In this case it is assumed that the drag force F_D has a uniform distribution between 0 to P_m. In the stability analysis, the inertia force is neglected and the lift force F_L is transformed by

$$F_L = \frac{C_L}{C_D}\ F_D \tag{26}$$

For the broken waves, the velocity u' of the water mass at any location between the SWL and the point of maximum wave runup may be approximated by (Shore Protection Manual, 1977),

$$u' = \sqrt{g h_b}\ (\ 1 - x_1/x_2\ ) \tag{27}$$

and the wave height H' above the ground surface is written by

$$H' = h_c\ (\ 1 - x_1/x_2\ ) \tag{28}$$

in which $h_c \simeq H_b/2$; h_b = the water depth at the breaking point; x_1 = distance from the still water line to the concrete block; and x_2 = distance from the still water line to the limit of wave runup. For the broken waves, the distance x_2 is suggested by

$$x_2 = 2\,H_b \cot\alpha \tag{29}$$

For the nonbreaking waves, we substitute the above equations to get the velocity of the water mass and the wave height above the breakwater surface. But the limit of wave runup is obtained by (Gunbak et al. 1979),

$$\frac{R}{H} = \frac{0.8\xi}{1 + 0.5\xi} \tag{30}$$

in which $\xi = \tan\alpha \,/\, \sqrt{H/L}$, surf similarity parameter.

Other Conditions

Several parameters included in the formulation of the stability analysis are assumed to be probabilistic variables. The shape factors of the concrete block c_1, c_2 and c_3 are supposed to be uniformly distributed between 0 and 4, 0.8 and 1.2, and 0.9 and 1.1, respectively. The value of the drag coefficient C_D, the lift coefficient C_L, and the virtual-mass coefficient C_M are also probabilistic. According to Christensen et al. (1982), in the case of a circular cylinder, these values are normal probability variables of which means and coefficients of variation are 0.60 and 0.24, 0.60 and 0.24, and 1.20 and 0.22, respectively. The friction coefficient of tetrapods μ behaves like the probability variable which is uniformly distributed from 0.5 to 0.6 (Takeda, 1981). The sheltering coefficients ε_1, ε_2 and ε_3 are also considered as the probability variables. For the sliding, the rocking and the lift-up motions, they are assumed to be uniformly distributed from 0.8 to 1, 0.4 to 0.8, and 0 to o.3, respectively. The repose angle of the concrete block is also the probability variable which is uniformly distributed from 56 to 63 degrees (takeda, 1981).

Illustrative Examples

Fig.2 compares the computed relative damage with the observed relative damage, introducing the queueing theory for different values of λ' and λ = 1. The value of λ' designates the average number of cycles for a concrete block to continue to have movability condition. From the same figure the value of λ' = 0.3 seems to be appropriate. In this study these values are used for the numerical calculations. Next, to apply our model to a designed rubble-mound breakwater, let us determine the weight of a concrete block by the following Hudson's formula:

$$W = \frac{\rho_s g H^3}{K_D\left(\frac{\rho_s}{\rho_w} - 1\right)^3 \cot\alpha} \tag{31}$$

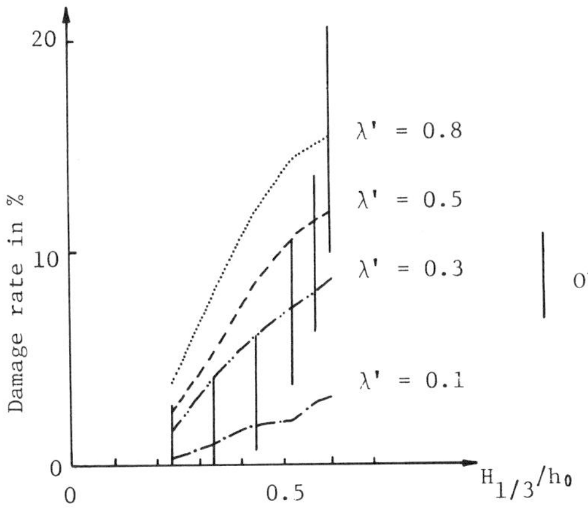

Figure 2. Comparison of Computed Relative Damage with Observed Relative Damage for Different Values of λ'

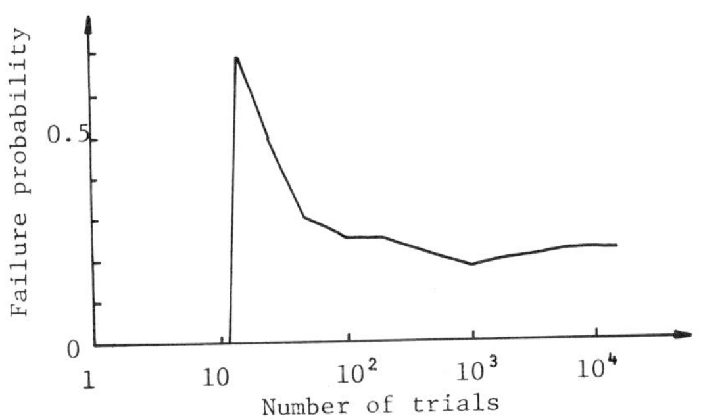

Figure 3. Effect of Number of Trials on Failure Probability

in which H = the design wave height and K_D = a parameter. The value of K_D is selected by the characteristics of a concrete block and the condition of wave breaking. For the values of ρ_s/ρ_w = 2.23, cotα = 4/3, K_D = 8.3 and H = 4m, the weight of a concrete block of 7.15 ton is computed by Eq. (31). Usually the weight of 8 ton is used by considering easiness of construction method and adding a safety factor. Our model can obtain the reliability or the risk for the arbitary value of the weight of a concrete block. The weight of a concrete block to correspond to the required value of the reliability or the risk should be selected in the field. Fig.3 represents the effect of the trial number on the failure probability. The increase of the trial number the failure probability converges to a constant value. Fig.4 indicates the effect of the significant wave height on the failure probability. As expected, the fail-

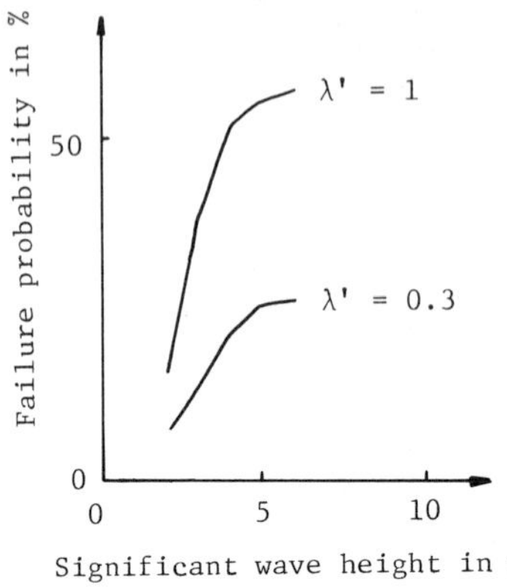

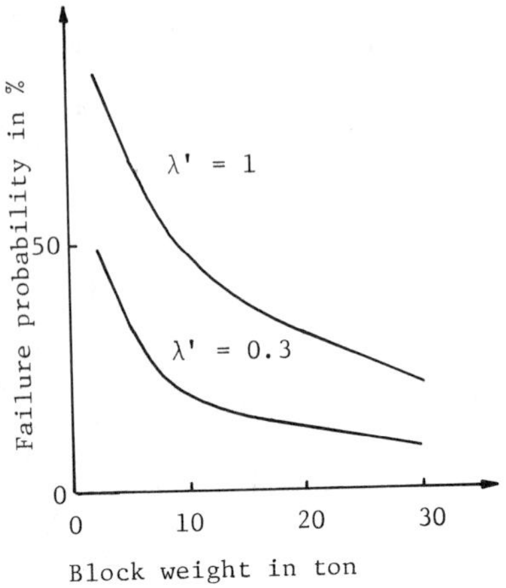

Figure 4. Effect of Significant Wave Height on Failure Probability

Figure 5. Effect of Concrete-block Weight on Failure Probability

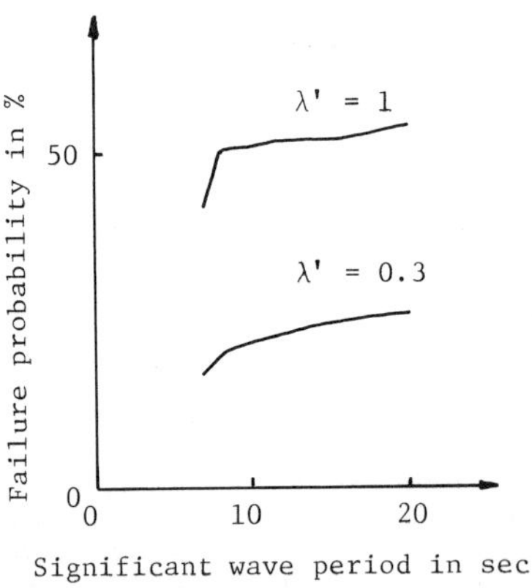

Figure 6. Effect of Significant Wave Period on Failure Probabity

ure probability increases as the significant wave height becomes large. The rate of the increase becomes small after 5m. It is explained that some incident waves break before they reach the breakwater slope. Fig.5 shows the effect of the weight of concrete blocks on the failure probability. By increasing the weight of concrete blocks, the failure probability decreases. It corresponds to the increasing of the weight of concrete blocks in the field. Fig.6 represents the effect of the significant wave period on the failure probability. The influence of the wave period, which is not included in Hudson's formula, is found. Thus, the longer wave period has a high risk. Fig.7

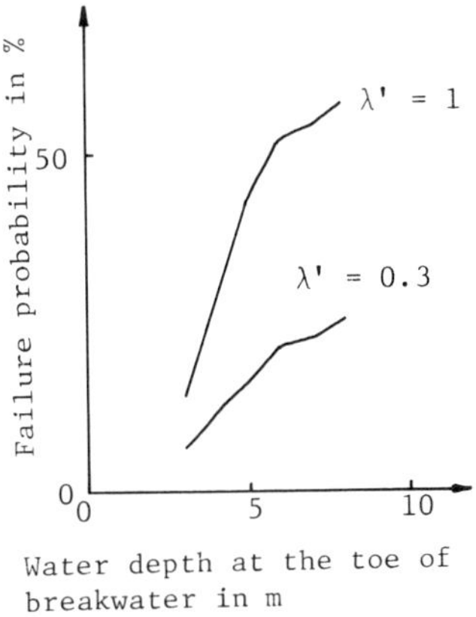

Figure 7. Effect of Water Depth at the Toe of Breakwater on Failure Probability

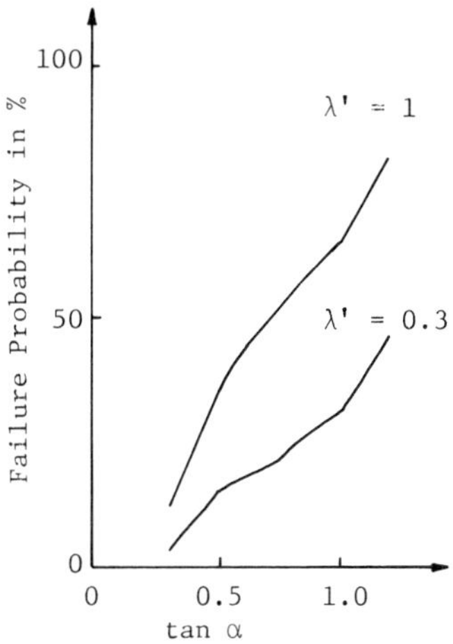

Figure 8. Effect of Breakwater Slope on Failure Probability

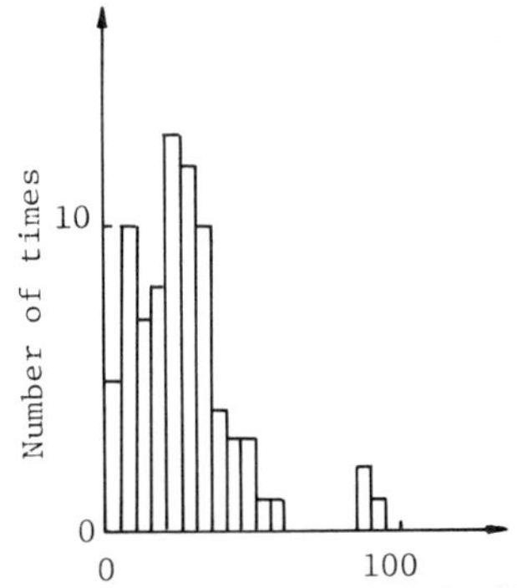

Figure 9. Failure Probability of Destroyed Rubble-mound Breakwaters

describes the effect of the water depth at the toe of the breakwater on the failure probability. Increase of the water depth gives high failure probability or risk. For the higher water level, more waves with larger wave height come on the breakwater slope. Fig.8 indicates the effect of the angle of the breakwater slope from the horizontal on the failure probability. The steep slope induces a high risk. Our model cannot consider the interlocking effect among blocks. Fig.9 shows the failure probability of destroyed rubble-mound breakwaters in Japan introducing the queueing theory when $\lambda = 1$ and $\lambda' = 0.3$. The failure probability using the weight of concrete blocks computed by Hudson's formula is between 10 % to 30 %. The destruction for the failure probability which is greater than 30 % is caused by unexpected design conditions. The destruction for the failure probability which is less than 10 % is caused by phenomena except our model.

Concluding Remarks

Classifying the motion of concrete blocks into three kinds, the failure probability or the risk of the rubble-mound breakwater are obtained. The reliability is defined by the difference of the failure probability from 1. In a result, the influence of the wave period on the failure probability becomes remarkable. That is to say, attacking of waves with longer period induces high risk to the rubble-mound breakwater as expected. For the other design variables of rubble-mound breakwaters, the computed result is reasonable and does not contradict the field observation. The relative damage computed from the failure probability appropriately predicts the observed one. Since the motion of concrete blocks is concretely investigated, we can develop a new type of concrete blocks to increase the reliability of the rubble-mound breakwaters. Finally, it is found to be rational for us to discuss the failure probability, the risk, or the reliability than K_D value included in the Hudson's formula, when one designs the rubble-mound breakwaters.

Appendix 1.--References

Christensen, P. T.(1982). Sstuctural reliability theory and its applications. Springer-Verlag.

Gunbak, A. R. and Bruun, P. M.(1979). "Wave mechanics principle on the design of rubble-mound breakwater." Proc. Port and Ocean Engrg. under Arctic Conditions, POAC 79, Nowegian Inst. of Tech., Trondheim, 1301-1318.

Iwagaki, Y. and Sawaragi, T.(1979). Coastal engineering. Kyoritsu-Shuppan Co., Ltd., Tokyo (in Japanese).

Takeda, H.(1981). "Stability analysis of concrete blocks." Summer Seminar on Hydraulic Engrg., Part B, JSCE, B.2.1-B.2.14 (in Japanese).

Tanimoto, K., Haranaka, H. and Yamazaki, K.(1985). "Damage rate of concrete blocks to irregular waves." 32nd Japanese Conf. on Coastal Engrg., JSCE, 480-484 (in Japanese).

United States Army Corps of Engineers, Coastal Engineering Research Center, Shore Protection Manual, Vol.2, United States Government Printing Office, Washington, D.C., 1977.

Walton, T. L. and Weggel, J. R.(1981). "Stability of rubble-mound breakwaters." J. WW Div., ASCE, 107(WW3), 195-201.

CHAPTER 153

Reliability Based Design of Rubble-Mound Breakwater

Masato Yamamoto*
Kazumasa Mizumura**
Taiji Endo***
and
Naofumi Shiraishi****

Abstract

The object of this present research is to study probabilistic design of armor blocks protecting composite breakwaters and to produce optimum design methodology for S-shaped breakwaters in terms of failure probability and construction cost. Failure probability in the vicinity of the still water level is greatest in the case of uniform sloped breakwaters. Therefore,S-shaped breakwaters of which the slope near the still water level is milder have a reduced risk of damage compared to uniform sloped ones. The optimum design index presents good economics and reliability in rubble-mound breakwater design.

Introduction

In recent years in Japan,as breakwaters have been constructed in deeper and deeper areas, armor block weights to protect caissons of composite breakwaters against wave action have also had to be increased. The block weights are generally determined using the Hudson formula with appropriate K_D values for each type of concrete block. However there are several problems in the usage of this formula in that,effects such as wave periods,probabilistic properties of wave heights and periods etc. can not be taken into account. Van Der Meer(1987) proposed formulae for calculation of stable armor units based on hydraulic model testing. These formulae can present the effects of wave periods. In addition he also produced probabilistic designs of rubble-mound breakwaters. When working group 12 of PIANC(1987) dealt with total reliability design for breakwaters including foundations,strength of armor units

---

*Research Engineer,Hydraulic Laboratory,Nippon Tetrapod Co.,Ltd.,2-7,Higashi-Nakanuki,Tsuchiura,Ibaraki Pref.,300,Japan
**Prof.of Civil Engrg. Dept.,Kanazawa Inst. of Tech.,7-1,Ogigaoka,Nonoichimachi,Ishikawa Pref.,921,Japan
***Director,Hydraulic Laboratory,Nippon Tetrapod Co.,Ltd.
****President,Nippon Tetrapod Co.,Ltd.,Shinjuku Dai-Ichi Seimei Bldg.,2-7,Nishishinjukuku,Tokyo,163,Japan

etc. and compared the results of levels I,II and III,the reliability theory for concrete blocks or armor stones was based on the formulae of Van Der Meer.

Mizumura et al (1988) proposed a theory for reliability design of rubble-mound breakwaters of level III. They calculated the incipient motion of an armor block using a simple physical model. According to them the stability of armor blocks depended on the wave periods.

In large wave areas the weight of blocks would have to exceed 100t in cases where the gradient of the protection is 1:4/3 or 1:1.5. In addition,in a case of a uniform sloped breakwater,the failure probability of concrete blocks would be greatest in the vicinity of the still water level obtained from the reliability analysis. As a result a complex sloped breakwater,a so called "S-shaped breakwater" should be utilized to increase stability. The failure probability of this S-shaped breakwater would be lower than that of a uniform sloped one. However, if an S-shaped breakwater were used,the cross-sectional area of armor blocks would become exceedingly large and the resulting construction cost extremely high. In this study,the authors introduce a simple index taking both the construction costs and failure probability of armor blocks of S-shaped breakwaters into consideration.

Method for Stability Analysis

In Japan,breakwaters like the one shown in Fig.1 are usually constructed.

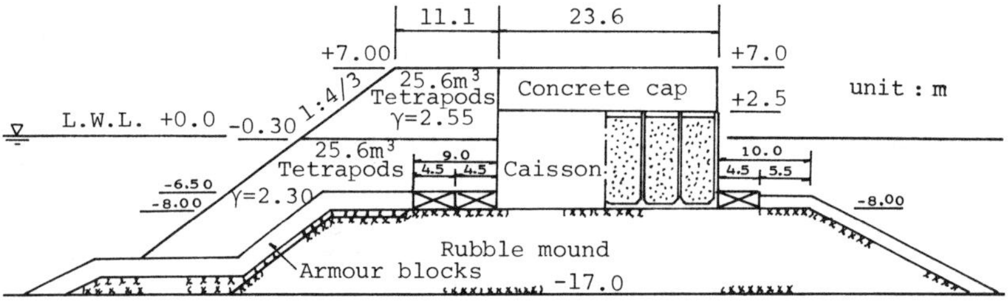

Fig.1 Typical Breakwater in Japan

In this type of breakwater,there are many possibilities for failure,as listed in Table 1.

Table 1 Risk Elements

1) Armor layer failure
   *Movement of armor unit
2) Toe scouring and collapse
3) Caisson slide and overturning
4) Damage to structural elements
   *Fatigue of materials
   *Breakage of armor units
5) Material degradation due to chemical action
6) Others

However,it is too complicated to analyse all of the risk elements theoretically but in the case of composite breakwaters protected with armor blocks,as armor unit failure or caisson slide are the principal problems, the authors have focused on armor block motion as a first step to indicate reliability of design.

The movements of armor blocks under wave action are classified into three categories sliding,rocking and lifting-up as shown in Fig.2.

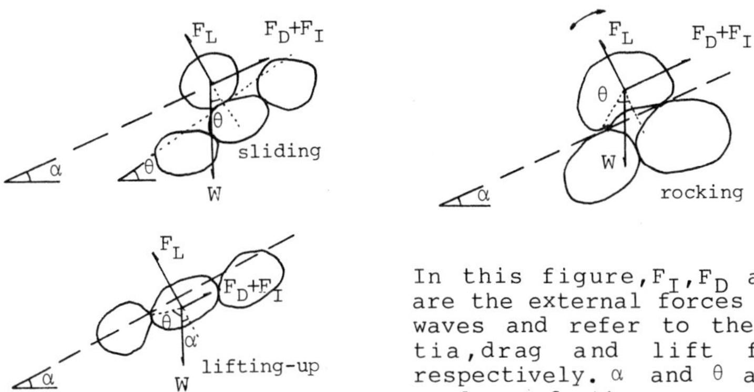

Fig.2 Motion of an Armor Unit

In this figure,F_I,F_D and F_L are the external forces due to waves and refer to the inertia,drag and lift forces respectively. α and θ are the angles of the armor block slope from the horizontal and the repose angle of blocks respectively. When a resistance force(f_R) is smaller than these external forces(f_S),an armor unit moves as expressed below.

$$\begin{aligned} &\text{Stable} \quad f_R - f_S \geqq 0 \\ &\text{Unstable} \quad f_R - f_S < 0 \end{aligned} \tag{1}$$

It is considered that the external forces acting on an armor unit are induced by two kinds of waves. One is waves breaking on the slope of the protection and the other is non-breaking waves. In the case of breaking waves,the wave force is calculated using Goda's formula (Goda 1985) while in the case of non-breaking waves,it is assumed according to the small amplitude wave theory up to the run-up wave height. Details of the theory were presented by Mizumura et al (1988).

Method for Reliability Analysis

Before mentioning risk analysis,the word "Movability" of an armor block has to be defined here. When the armor block concerned is located in the condition shown in Fig.2,the block is referred to as being in a state of movability. In other words,this armor block can not be interfered with by any block in its vicinity and resistance forces consist of only gravity and frictional forces.

The following assumptions are made to connect the movability condition with stability analysis.

(1) The movable units distribute uniformly between the bottom and 90% of the significant wave height above the still water level along the breakwater slope.

(2) The number of armor blocks falling into a movability state is K.
(3) The movability state of an armor block continues for n_i sets of a combination of the wave height and period for i being 1 to K.

In accordance with these assumptions,processes (2) and (3) above can be simulated by a queueing theory. Integers K and n_i are probabilistic variables and the process can be described by the Poisson process and expressed as eqs. (2) and (3).

$$P(k)=\frac{\exp(-\lambda)\lambda^k}{k!} \tag{2}$$

$$P(n_i)=\frac{\exp(-\lambda')\lambda'^{n_i}}{n_i!} \tag{3}$$

where λ and λ' refer to the parameters to govern the Poisson process or its mean values.

The failure probability from each motion of an armor block can be expressed by the following equation using eq. (1).

$$P=\int_0^{\infty}\int_0^{s} f(r,s)drds \tag{4}$$

where $f(r,s)=f_R-f_S$ (5)

and f(r,s) is the so-called risk function. P indicates the failure probability of any of the 3 motions as defined i.e. sliding,rocking or lifting-up.

Before the calculation of the failure probability,distribution of some coefficients,i.e.the drag coefficient,C_D,the virtual mass coefficient, C_I ,the lifting-up coefficient,C_L and the repose angle of an armor unit etc. must be decided. These coefficients and any others that are necessary must also be treated as random variables in accordance with Mizumura et al(1988). Wave heights are assumed to be according to the Rayleigh distribution and in the case of wave periods,the square also belongs to the same distribution. The Monte Carlo method with sampling a pair of wave heights and periods from their distributions is applied to obtain the distribution of the reliability function with respect to each failure due to sliding,rocking and lifting-up.

The probability corresponding to sliding,rocking and lifting-up,referred to as P_s,P_r and P_l respectively,is calculated and the failure probability,of one armor block,$P_e(i)$,can be computed by the sum of each set,P_s,P_r and P_l as below,

$$P_e(i)=P_s+P_r+P_l-P_sP_r-P_rP_l-P_lP_s+P_sP_rP_l \tag{6}$$

if P_s,P_r and P_l are independent of each other.Therefore if the number of armor units per unit width of breakwater is written with m,the total failure probability P_w can be described as equation (7).

$$P_w = 1 - \prod_{i=1}^{m} (1 - P_e(i)) \qquad (7)$$

The values of λ and λ' in eqs. (3) and (4) have to be determined. These values should be estimated using the results of the hydraulic model test performed by Tanimoto et al (1985). The trials can be made with various values of λ and λ'. Fig. 3 shows a comparison of the simulated failure probability with the result of the hydraulic model test.

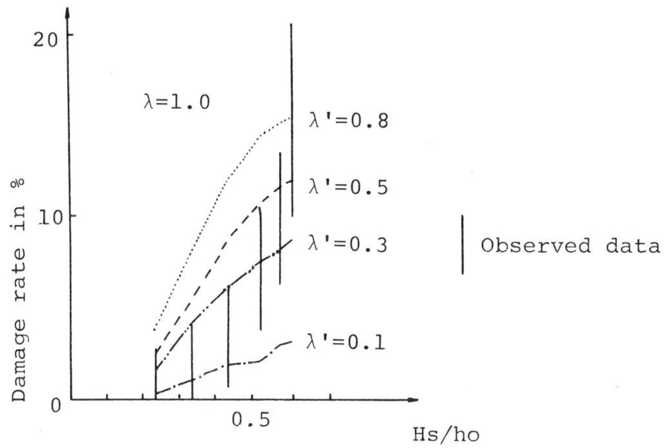

Fig.3 Comparison of Computed Relative Damage with Observed Relative Damage for Different λ' Values

From this figure,it can be seen that a λ value of 1.0 and of λ' 0.3 are appropriate and hereinafter,are used in the calculation.

Failure Probability of Uniform Sloped Breakwaters

Fig.4 shows a computed cross section of a breakwater.

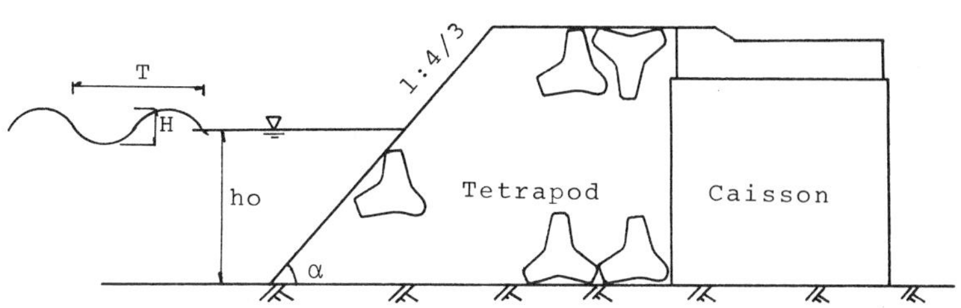

Fig.4 Sketch of Computed Cross Section of Uniform Sloped Breakwater

Tetrapods were selected as the armor blocks. The calculation conditions are summarized in Table 2.

Table 2 Calculation Conditions

Hs= 4.0 m	: $\tan\alpha$ = 0.75
Ts= 10.0 sec	: W = 7.36 t
ho= 6.0 m	
no wave overtopping	

In this Table,W indicates the weight of a tetrapod obtained using the Hudson formula.

Fig.5 shows the spatial distribution of the total failure probability along the slope of the breakwater.

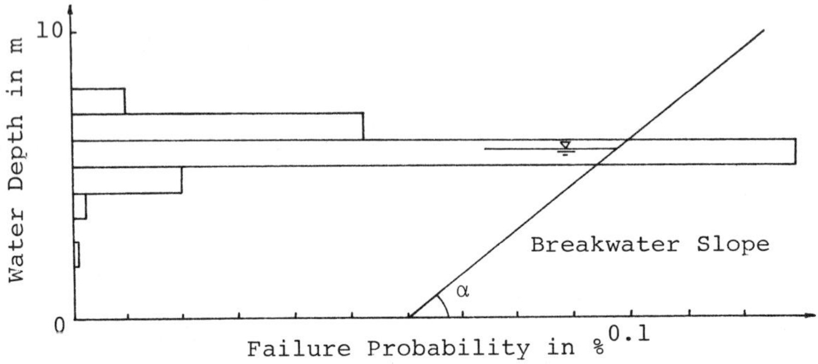

Fig.5 Spatial Distribution of Failure Probability

The risk became highest in the vicinity of the still water level as expected. This result agreed with results from hydraulic model tests and field observations conducted elsewhere except in cases of particular conditions. Therefore in any case of uniform sloped breakwaters,when armor blocks are located near the still water level,they are at more risk than in any other location.

Failure Probability of S-shaped Breakwaters

As mentioned above,the failure probability becomes a maximum near the still water level. Therefore if the failure probability in this vicinity decreased,the total risk would be reduced. An S-shaped breakwater such as the one shown in Fig.6 had to be utilized to decrease the risk of failure.

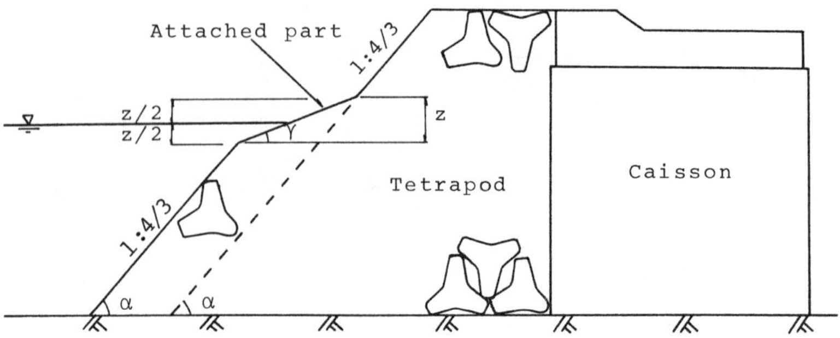

Fig.6 Sketch of S-shaped Breakwater

In this figure, γ is the angle of the attached part from the horizontal and z refers to the vertical length of the attached part.

Fig.7 shows an example of the spatial distribution of the failure probability along an S-type breakwater slope. In this calculation the values of γ and z were taken as 0.33 radians and 165cm respectively. The figure shows a failure probability significantly reduced compared with that shown in Fig.5. The beneficial effect of the S-shaped breakwater is evident.

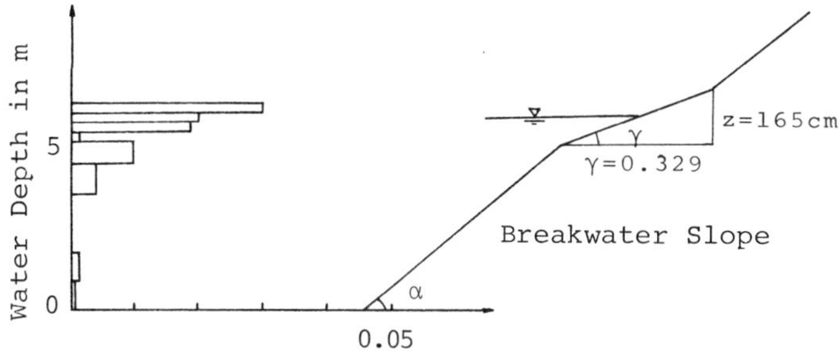

Fig.7 Spatial Distribution of Failure Probability

Fig.8 shows the failure probability of an S-shaped breakwater taking z as a parameter under the same calculation conditions as listed in Table 2.

Saville(1958) obtained run-up heights for composite sloped breakwaters,so in our calculation the run-up wave height was computed according to his method. In the range of small γ ,the failure probability of small z was larger than that of large z. This meant that in the case where the gradient of the attached slope was milder,if the slope was longer,the risk would be reduced. When γ became larger,the failure probability of a small value of z e.g. 15cm became lower and where z=165cm,it increased. Therefore in a range of large γ,it was seen that failure probabilities converge into one. Thus,the failure probability of an S-shaped breakwater varies in terms of z and γ and it becomes unclear how to decide the values of z and γ. In addition

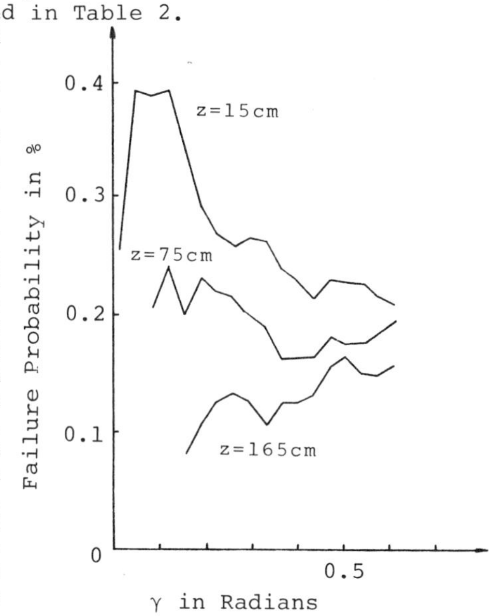

Fig.8 Failure Probability for S-type Breakwater

values of γ and z affect the area of the breakwater section and the construction cost is affected.

Optimum Design Index

An optimum design breakwater has to satisfy in terms of both reliability and economics. However the construction cost depends on the cross sectional area. From this view point,the authors employed an optimum design index,I,as expressed by the following equation under the simple assumptions that construction cost is proportional to cross sectional area.

$$I = \frac{1}{2}\left(\frac{P_f}{P_{f_o}} + \frac{A}{A_o}\right) \qquad (9)$$

in which I is the optimum design index,P_f is the failure probability of an S-type breakwater,P_{fo} indicates the same for a uniform sloped type,A shows the cross sectional area of the S-shaped breakwater and Ao is the area of a uniform sloped one.

This index is very simple and implies that the weight of armor block stays the same. The index,however,is appropriate at this stage of design. The optimum design is made by employing γ and z values making I a minimum.

A cross sectional area is shown in Fig.9. The calculation conditions are the same as mentioned previously.The figure indicates the changes in the cross sectional area against γ with fixed z value. In the case where z=165cm, the cross sectional area is much larger than that in the case where z=15cm. So,the construction cost would be extreme for the greater value of z,while the failure probability shown in Fig.8 in the case where z=165cm is lower than that of the smaller z value.

Fig.10 shows the optimum design index,I.

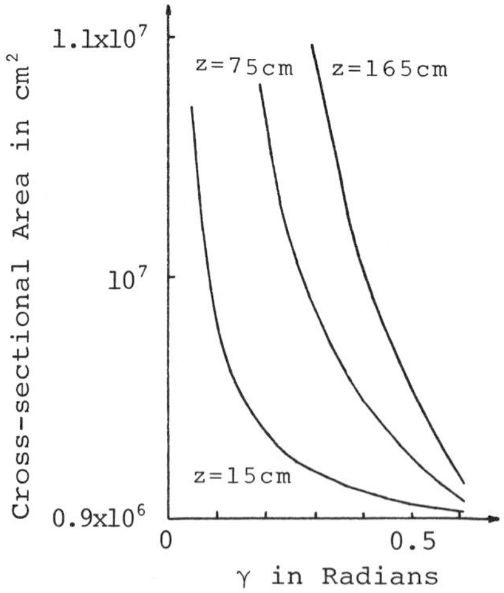

Fig.9 Cross-sectional Area for S-type Breakwater

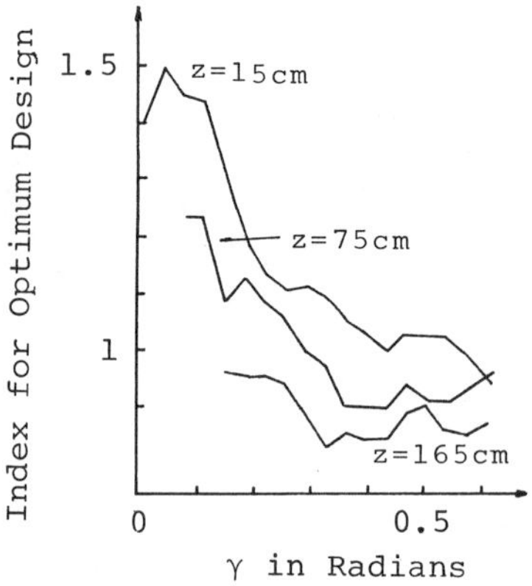

Fig.10 Index for Optimum Design

I is small for large z values. When z and γ become smaller, I becomes greater. This fact suggests that the attached part requires a gradient, i.e. the attached part is not always level. In the optimum S-shaped breakwater, the minimum value of I is 0.83 when γ is 0.33 and z is 165cm. P_f/P_{fo} and A/Ao are calculated as 0.48 and 1.18 respectively. The spatial distribution of the failure probability of an S-type breakwater has already been shown in Fig.7. Thus, from the sample calculation, although construction of an S-type breakwater requires a mere 18% cost increase, the reliability becomes twice as good as that of a uniform sloped breakwater.

Fig.11 shows relationship between the significant wave height and γ, z, the failure probability and volume of protection part when the optimum design index, I, is a minimum in the case where the wave period and block weight are fixed at 10sec and 8t respectively.

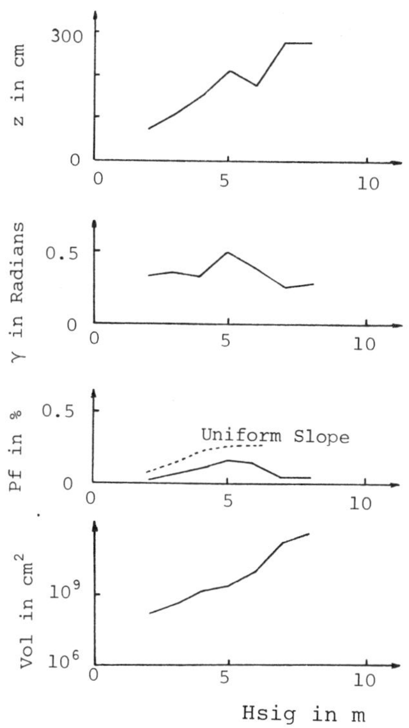

Fig.11 Relationship between H_{sig} and z, γ, P_f and Vol. when I is Minimum

In the graph of P_f, the broken line indicates the failure probability for a uniform sloped breakwater. If a uniform sloped breakwater is used, the risk of failure increases as indicated by the dotted line in terms of the incident wave height. In the case of an S-type breakwater, however, in a range of wave height higher than 5m,the failure probability would be reduced by employing γ as less than 0.5 and z greater than 200cm. Thus, this figure proves useful for the optimum design of the S-shaped breakwaters when the design wave height changes.

Fig.12 indicates changes in the failure probability,γ, z and the volume of cross section when I is a minimum in terms of the wave period under the conditions where the wave height and weight of the armor block are fixed at 4m and 8t respectively. In the condition where T_{sig} is less than 10sec,the optimum design is made by employing γ of 0.3-0.5 radians(the gradient of the attached part is 1:1.8-1:3.2) and z being 110cm-130cm. In the case where T_{sig} is 10sec-20sec,I became a minimum by keeping γ as 0.5(the gradient is 1:1.8) and z of 130cm. In this range of T_{sig},the failure probability gradually increases and is less than that of the uniform sloped breakwater. Thus even though the wave period becomes longer,it is not necessary to enlarge the protection part to ensure the same reliability for an S-Shaped breakwater.

Fig.13 shows the influence of armor unit weight to the volume,the failure probability, γ and z under the condition where I is a minimum by holding the wave height and period constant as 4m and 10sec respectively.

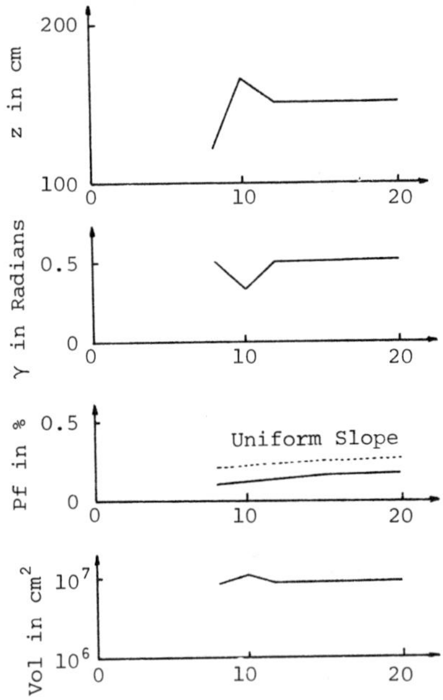

Fig.12 Relationship between T_{sig} and z,γ,P_f and Vol.when I is Minimum

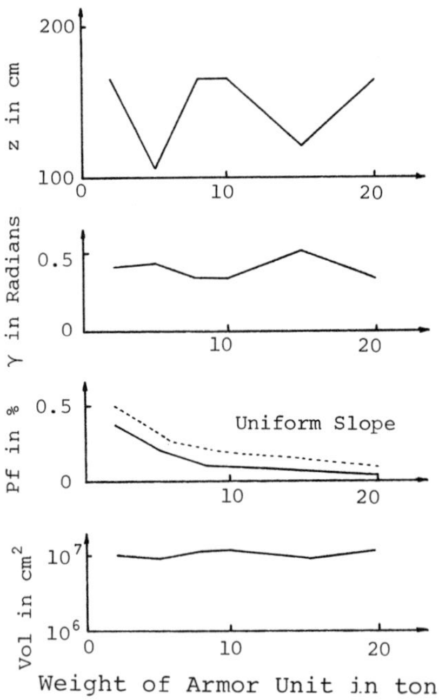

Fig.13 Relationship between Weight and z,γ,P_f and Vol.when I is Minimum

As the weight of armor unit increases,the failure probability decreases because the armor unit weight directly affects the resistance against the three kinds of block motion described previously. The volume of the protection part is almost constant when the armor unit weight becomes great. Therefore even if the weights of armor blocks increase,the construction cost does not become any higher.

As mentioned above in the optimum design,it is the wave height that most influences the construction cost and reliability compared with both the wave period and weight of armor block. Therefore if the design wave period is lengthened when designing an S-shaped breakwater,the cross sectional area need not be enlarged.

Conclusions

The following conclusions have been derived from this study.

1.The failure probability along the slope of the protection part was confirmed to be a maximum in the vicinity of the still water level in the case of a uniform sloped breakwater with armor blocks.

2.In the case of an S-type breakwater, the failure probability near the still water level became significantly lower than that of a uniform type. As a result the total risk is reduced significantly compared with that of a uniform sloped breakwater.

3.The optimum design index proposed in this study can describe both the reliability and economics of the S-shaped breakwaters. This index is most useful in producing optimum designs of breakwaters protected by armor blocks.

4.Simple calculations show that a small increase in construction expenditure results in greatly enhanced reliability.

5.The significant wave height has much more influence on construction cost than ether the wave period or block weight when the optimum design index is a minimum.

References

Goda,Y.,"Random Seas and Design of Maritime Structures", University of Tokyo press,pp.113-121,1985

Mizumura,K.,Yamamoto,M.,Endo,T.and Shiraishi,N.,"Reliability Analysis of Rubble-Mound Breakwaters",Proc. 21st International Conf. on Coastal Engrg.,Malaga,Spain,1988(to be published)

PIANC PTC II Working Group 12, "Rubble Mound Breakwaters", Meeting Document,Le Havre,France,1987

Saville,T.,JR.,"Wave Runup on Composite Slopes",Proc.,6th International Conf. on Coastal Engrg.,1958

Tanimoto,K.,Haranaka,H.and Yamazaki,K."Variation of Percent

Damage to Concrete Blocks due to Irregular Waves",Proc. 32nd Japanese Conf. on Coastal Engrg.,JSCE, pp.480-484,1985 (in Japanese)

Van Der Meer,J.W."Stability of Cubes, Tetrapods and Accropode",Proc.Breakwaters'88 Conf.,ICE,Eastbourne,UK,pp.59-68, 1988

CHAPTER 154

STABILITY AND STRUCTURAL BEHAVIOUR OF STRENGTH IMPROVED DOLOSSE

J A ZWAMBORN*, D J P SCHOLTZ** and H CLAASSENS***

ABSTRACT

Stability model tests and prototype breakage tests were done to provide data for the optimum design of dolos breakwater armouring.

Stability tests with irregular waves were done with dolosse with waist ratios of 0,33, 0,36, 0,38, 0,40 and 0,43 and breakwater slopes of 1:1,33, 1:1,5 and 1:2. All model dolosse had a mass of about 80 gr and were placed at a 'normal' density, Ø = 1,00. The results showed that regular waves with H_{reg} = 1,1 H_{mo} must be used to get similar damage. Also, a significant reduction in stability is evident for dolosse with larger waist-to-height ratios and for the shorter period waves (Figures 4 and 5).

Some 50 9 t dolosse, without reinforcing and with different types of simple rail reinforcing, were tested for breakage behaviour, using a specially developed cushioning device to closely represent breakwater conditions. The results showed that unreinforced dolosse can withstand falls up to 1,5 m, without breakage and X or double-V rail reinforced units about double this height.

The results of these tests, together with specific model tests, were used to design an effective protection for the Table Bay main breakwater. The tests indicated the use of 25 t dolosse (minimum) but, because of the shallow waterdepth, the toe dolosse of the repair slope were reinforced (double-V rail reinforcing), stabilized by 75 mm steel chains, and bitumen grout will be placed to reduce their movements.

---

*Programme Manager: Maritime and Hydraulic Structures, CSIR, P O Box 320, Stellenbosch, 7600, South Africa.

**Research Engineer:" Maritime and Hydraulic Structures, CSIR, P O Box 320, Stellenbosch, 7600, South Africa.

***Deputy Harbour Engineer, SA Transport Services, P O Box 4245, Cape Town, 8000, South Africa.

INTRODUCTION

Procedures for the design of safe and economic dolos structures have been described by Zwamborn and Scholtz (1986). Three different design approaches were described, namely:

(i) "No Movement" Design;
(ii) "No Breakage" Design; (based on 'unbreakable'units)
(iii) "Optimum" Design; using strength improved dolosse.

Low cost dolos strength improvement can be achieved by increasing the waist-to-height ratio, in other words, to make the dolosse thicker, and by using simple types of reinforcing, for example scrap rail reinforcing.

To assist with the design over 800 stability tests were done with regular and irregular waves, using model dolosse with five different waist ratios. Moreover, unreinforced 9 t dolosse and dolosse reinforced with three different types of rail reinforcing were used to check their structural behaviour under simulated 'near breakwater' loading conditions to acceptable degrees of movements for different types of dolosse.

These results, together with a series of site-specific stability tests, were used to design the major repair works for the Table Bay main breakwater at Cape Town, Republic of South Africa.

2. DOLOS STABILITY TESTS

Stability tests on dolosse with different waist-to-height ratios (r = 0,33, 0,38 and 0,43) reported by Scholtz **et al** (1982) showed a marked decrease in stability for the larger waist ratios, particularly for r > 0,38. Later tests by Burcharth **et al** (1986) indicated little influence of the waist-to-height ratio on dolos stability, particularly for the smaller damage values. To resolve this issue, a comprehensive test programme using irregular waves and 80 gr dolosse with waist-to-height ratios of 0,33, 0,36, 0,38, 0,40 and 0,43 has been carried out.

Test Facilities and Model Layout

The stability tests were carried out in a 127 m long by 3 m wide and 1,1 m deep test flume equipped with Seasim irregular wave generators, including an absorption system to reduce re-reflections. The flume was divided into three 0,75 m wide test channels leaving two narrow dummy channels on either side. Details of the test dolosse were as follows: W = 78 to 81 g, specific density 2,34 to 2,40 and r = 0,33 to 0,43.

The standard 'mean' dolos packing density was used for all the tests, that is, Ø = 1,00 (number of dolosse per unit area, $N_n = Ø_n V^{-2/3}$, where V is the block volume and n the number of 'layers', Zwamborn and Van Niekerk, 1982) and 538 dolosse were placed on each of the 0,75 x 0,75 m^2 test areas (Figure 1). The underlayers consisted of 16,5 g selected stone, that is W/5 g.

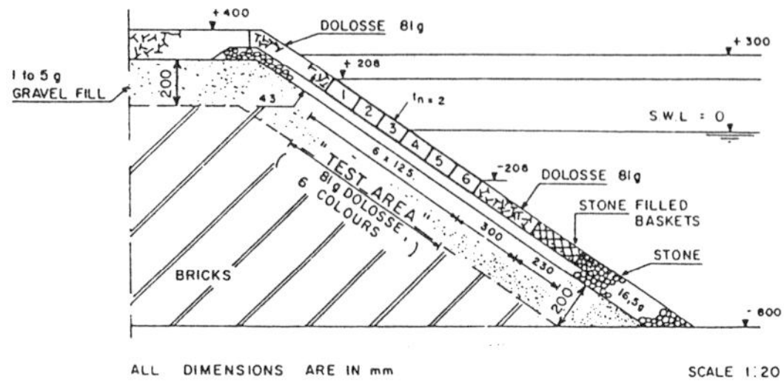

FIG. 1: MODEL CROSS-SECTION

Test Wave Conditions

Wave Spectrum	T_p (s)	Target Wave heights, H_{mo} (mm)							
Jonswap	1,25	31	50	69	90	112	129		
	1,50	30	50	69	90	109	128	155	
	1,75	30	51	72	93	113	134	152	163
	2,00	31	51	72	91	111	129	149	161
PM	1,75	31	52	72	92	113	130	146	

Wave heights were measured 5,55, 5,80 and 6,20 m away from the model slope. Re-reflections of wave energy from the wave paddles was avoided by using the Seasim wave generators' absorption system. The incident and reflected wave spectra for the above test conditions were determined by the three-point method based on the least square technique reported by Monsard and Funke (1980).

Test Procedures and Damage Criteria

The model dolosse were placed at the prescribed packing density in one operation from the bottom upwards. The tests were carried out similar to the previous tests with regular waves, that is, each wave height interval (see test wave conditions) was tested for 60 min after which the damage was determined. All test series (up to 8 wave heights) were repeated between 3 and 8 times.

Damage was determined by recording displaced units (observations during the tests and using still photography) and rocking units (observations and single exposure cine camera technique). The following damage criteria were used:

(a) displacement (movement over at least a distance h)
(b) continuous rocking or full roll-over
(c) intermittent rocking (two-thirds of the test time)
(d) occasional rocking (one-third of time or less)

The test results have been presented relative to these criteria. 'Total' damage is defined as (a) + (b) + (c).

Test Programme

The following tests were carried out using up to 8 characteristic wave heights for each test series (test numbers):

Slope	Spectrum	$T_{p(s)}$	r = 0,33	r = 0,36	r = 0,38	r = 0,40	r = 43
1:1,5	JS	1,75	W_2 to 9	W_2 to 9	A_7 to 9	W_2 to 9	A_7 to 9
	JS	2,0	P_1 to 3	P_1 to 3		P_1 to 3	
	JS	1,5	P_4 to 6	P_4 to 6	A_4 to 6	P_4 to 6	A_4 to 6
	JS	1,25	P_7 to 9	P_7 to 9	A_1 to 3	P_7 to 9	A_1 to 3
	PM	1,75	PM_1 to 3	PM_1 to 3		PM_1 to 3	
1:2	JS	1,75	A_{13} to 15		A_{13} to 15		
1:1,33	JS	1,75	A_{10} to 12		A_{10} to 12		

Test Results

The mean results of the tests for different **waist ratios** at a slope of 1:1,5 are shown in Figure 2, together with the earlier test results for regular waves. A significant reduction in stability (wave height) for dolosse with larger waist ratios is evident from this figure. When rocking units are added to the displaced units, damage was found to increase by 100 per cent for r = 0,33, 80 per cent for r = 0,36 and 50 per cent for r = 0,40.

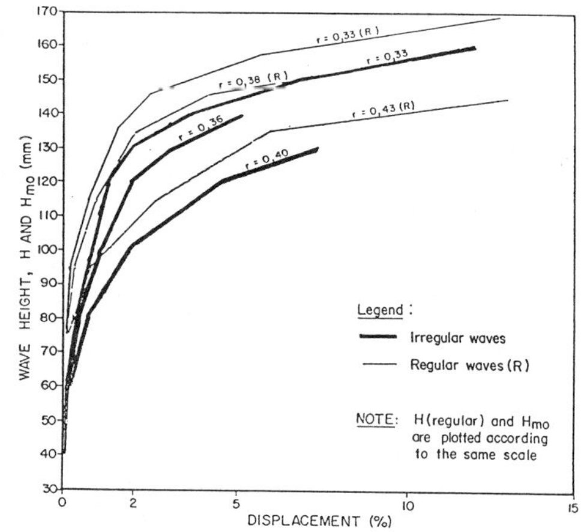

FIG. 2: EFFECT OF WAIST RATIO, REGULAR VERSUS IRREGULAR WAVES

It can also be derived from Figure 2 that **regular** waves of wave height $H_{reg} = 1,1H_{mo}$ cause the same damage as irregular waves with a characteristic wave height H_{mo}. Brorsen **et al** (1974) concluded from model tests on dolosse that similar damage is caused when $H \simeq 0,8\ H_s$ but after checking their results this was found to be incorrect; it should have been $H \simeq 1,25\ H_{mo}$ (the same error is carried through in Burcharth **et al**, 1986).

Most test were done with a **Jonswap** type of spectrum but comparative tests with a PM spectrum showed similar stability results for the various waist ratios.

A comparison of the effect of different breakwater **slopes** for r = 0,33 and 0,38 is given in Figure 3. For dolosse with r = 0,33 and 2 per cent displacement, the H_{mo} values are 116, 130 and 140 mm for slopes of 1:1,33, 1:1,5 and 1:2 respectively, thus somewhat increased stability for flatter slopes. In the case of r = 0,38, the differences in stability are seen to be much smaller.

Test results showed a significant reduction in stability for the shorter **wave period** (T_p) values, for instance, for r = 0,33 and Jonswap spectra with T_p = 2,0 and 1,25 s resulted in 2 per cent displacement K_D values of 22 and 13 respectively. Similarly, for r = 0,40, K_D values of 12 and 6 were found.

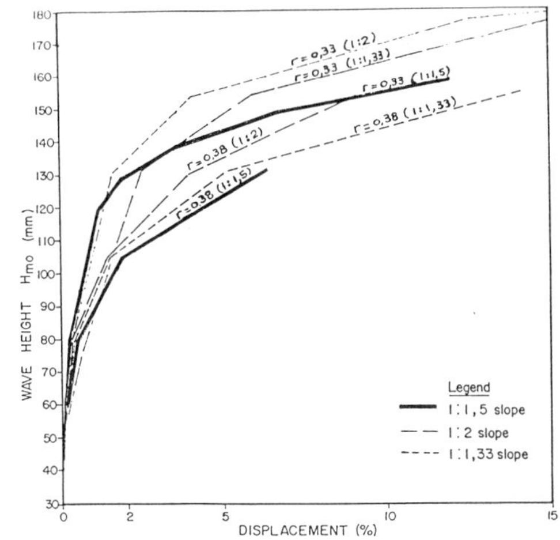

FIG. 3: EFFECT OF BREAKWATER SLOPE, IRREGULAR WAVES, T_p = 1,75 s

The overall test results are presented in Figures 4 and 5 which show the stability factors

$$K_D = \frac{\gamma_s H_{mo}^3}{W \Delta^3 \cot\alpha}$$

with H_{mo} = wave height;

$$\Delta = \frac{\gamma_s}{\gamma} - 1;$$

γ and γ_s = specific density of water and model dolosse

and α = breakwater slope angle;

versus the **Irribarren** number,

$$\xi = \frac{\tan}{(H_{mo}/L_o)^{0,5}}.$$

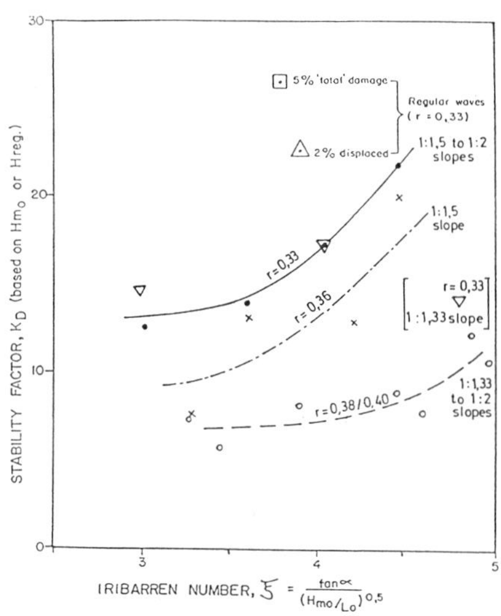

FIG. 4: STABILITY FACTORS FOR PRELIMINARY DESIGN - 2% DISPLACEMENT OR 5% 'TOTAL' DAMAGE

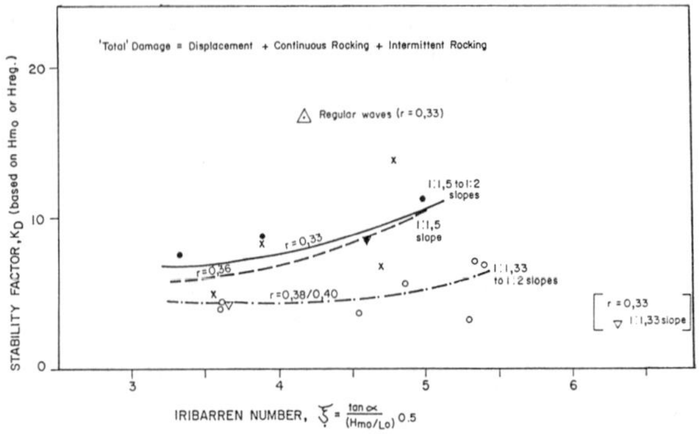

FIG. 5: STABILITY FACTORS FOR PRELIMINARY DESIGN - 2% 'TOTAL' DAMAGE

Figure 4 is based on 2 per cent displacement. This figure also represents the 5 per cent total damage quite well. Figure 5 is based on 2 per cent 'total' damage.

These figures display clearly the significant effects of the increased waist ratio's (reduced K_D values), and the wave period (low K_D values for shorter wave periods with minimum values for $2{,}5 < \xi < 3{,}5$. These figures also show the effect of irregular waves (H_{mo}) versus regular waves (H_{reg}) (Scholtz **et al**, 1983).

Figures 4 and 5 could be used for a **first design**, as a guide using Figure 4 (2% displacement or 5% total damage) for high quality strength reinforced dolos projects and Figure 5 (2% total damage) for lower quality and unreinforced dolos applications.

It should be mentioned here that the significant reduction in stability of the 'thicker' dolosse found in these tests contradicts the results reported by Burcharth **et al** (1986) which is **inter alia** thought due to different dolos packing densities and the small dolosse used by Burcharth (30 g), that is, possible scale effects.

3. STRUCTURAL TESTS ON DOLOSSE

To check on the structural strength of various types of simple reinforced dolosse, a series of breakage tests on 9 t dolosse were done at Cape Town.

Structural tests on dolosse were done elsewhere by dropping the units on a solid concrete base (drop test) or by letting a concrete ball hit the dolos (pendulum test). These test are effective in showing **relative** strengths, for instance, to show the effect of increased waist-to-height ratio's (Burcharth, 1981). However, the results of these tests cannot be used directly for the actual breakwater situation because the 'rigid'-base test is not representative for the breakwater armouring where a dolos, when moving, will drop on underlayer stone or on other dolosse that will move (yield) under the impact. A test procedure, closely representing the actual breakwater situation was therefore developed for the Cape Town tests.

Purpose of the Tests

The main purposes of the structural tests were:

(a) to develop a test procedure, closely resembling the conditions in a breakwater armour, so that **acceptable** dolos movements can be defined better;

(b) to determine the effectiveness of different types of simple reinforcement;

(c) to determine the effect of repeated impact loading; and

(d) to decide on the type of reinforcing of the 25 t dolosse for the Table Bay breakwater repair.

Test Dolosse

The test dolosse were 9 t with h = 2,9 m and r = 0,30. The concrete density was 2,4 t/m^3 and the design compressive strength was 40 MPa at 28 days. Test cubes showed strengths between 37 and 54 MPa while the equivalent cube strength based on cores, taken after 2 years (when the dolosse were tested) were found to be 23 per cent higher. The average tensile strength was found to be 1/15 of this value, that is, 4 MPa.

Unreinforced and dolosse reinforced with single rails, double-V rails and X-rails were tested (Figure 6). Scrap rails of 43 kg/m were used for the reinforcement, that is, 85, 115 and 168 kg/m^3 respectively.

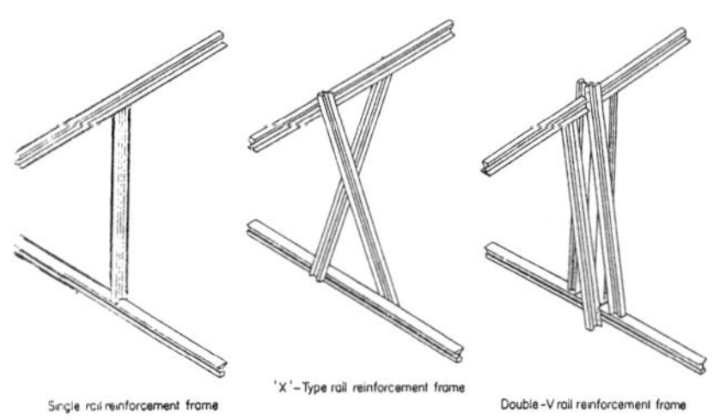

FIG. 6: TYPES OF RAIL REINFORCEMENT

Test Procedures

The basic swing test arrangement, (Zwamborn and Scholtz) was used for the tests. Various methods were tried to achieve a controlled yield and the best solution was found to be a set of three 250 mm long, 89 mm outside diameter and 4,05 mm thick pipes as an 'anvil' or 'cushioning' device (Figure 7).

The cushioning device was calibrated by comparing the deceleration of an impacting 9 t dolos in the swing test with that of a similar dolos which was dropped on 0,5 to 3 t breakwater underlayer stone and on other 9 t dolosse. The deceleration were measured with PCB Piezotronics shock accelerometers with a range of 500 g and a resolution of 0,01 g (Figure 8).

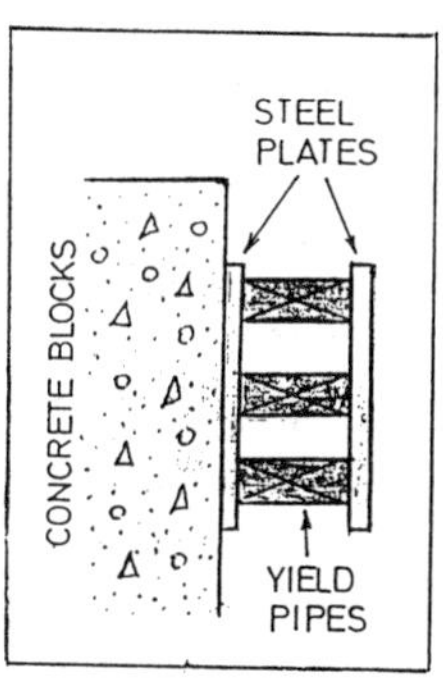

FIG. 7: CUSHIONING DEVICE

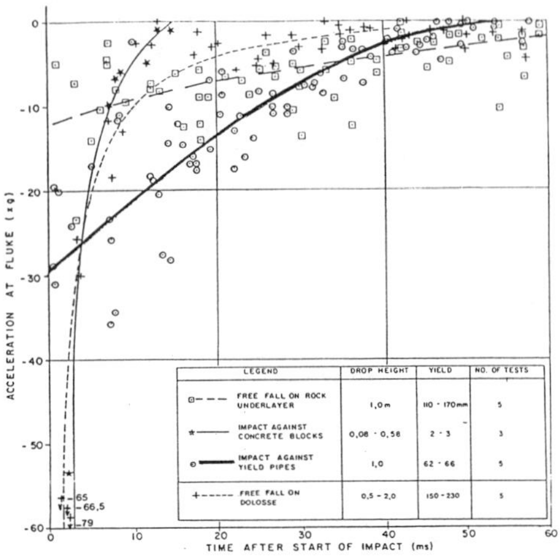

FIG. 8: CALIBRATION OF CUSHIONING DEVICE

In this way, a test procedure was developed which closely represents the conditions in a breakwater situation where a dolos is lifted by the wave action and falls back onto the underlayer or on other dolosse (see also Zwamborn and Scholtz, 1986). Normally, the first fall height was 1,1 m which was increased in steps of 0,15 m to 1,7 m with unreinforced and 1,4 m with the reinforced units whereafter the step heights were increased to 0,3 m until the dolos broke or cracks occurred of at least 0,5 mm width or a fall height of 4,1 m was reached. Repeated impact tests were done using 80 and 64 per cent of the failure fall heights repeatedly, resulting in 90 and 80 per cent of the failure stresses.

Test Programme

The test programme included free-fall tests on rock (17 tests with 0,5, 1,0, 1,5 and 2,0 m fall heights) and free-fall tests on dolosse (5 tests with similar fall heights). Preliminary swing tests included tests against solid concrete blocks (3 tests, 10 to 15 t blocks) and tests using different pipe arrangements as cushioning device (21 tests).

The actual breakage tests included (a) head on swing tests, three dolosse each, unreinforced, single rail, X-rail and double-V rail; (b) side on swing tests, two dolosse each, as above; and (c) repeated impact, three unreinforced, one single rail, one X-rail and two double-V rail. A total of 51 9 t dolosse were used for the tests.

Test Results

Whereas unreinforced dolosse in a 'rigid' drop test break for fall heights of about 0,2 m, with a realistic yield, the critical fall heights were found to be well above 2 m. The failure fall heights found in the Table Bay tests compare very well with the Gioia Tauro tests (Grimaldi and Fontana, 1984).

With regard to the **rail reinforcing**, it is concluded from the test that:

- the single rail does not increase the dolos strength significantly, although breakage is retarded;

- the X-rail and double-V rail reinforcing increase the dolos strength by about 100 per cent (double fall height) and the dolosse were damaged but did not fail;

- since the double-V reinforcing requires about 50 per cent more steel, the **X-rail reinforced dolos** is the most cost/strength effective unit.

Also, since the critical drop heights for the 9 t Table Bay and the 30 t Gioia Tauro unreinforced and double-V reinforced dolosse are seen to be of the same order, it may be concluded that the reinforcing has the same degree of effectiveness independent of dolos size, provided the steel/concrete ratio (kg/m^3) is kept constant. With regard to **repeated impact loading,** the test results are shown in Figure 9 which is based on 200 kg dolos units (Burcharth, 1984).

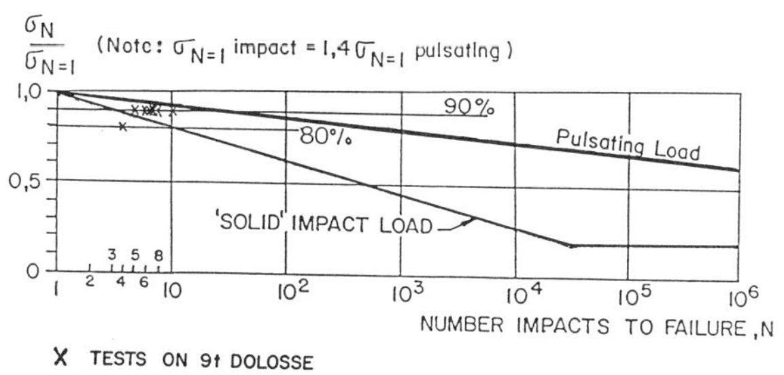

(BURCHARTH 1984)

FIG. 9: REPEATED IMPACT LOADING

Burcharth's tests are based on a 'rigid' test layout and, as expected, the Table Bay results are seen to fall between the 'solid' impact load and pulsating load case, that is, in the breakwater situation, the dolosse will be able to withstand about **twice** the number of repeat loads (for example, 6 repeat loads at 90 per cent of the ultimate stress and 25 repeat loads at 80 per cent) compared with the 'rigid' case.

Application of Results

The test results can be used in the following way for the design of dolos armouring:

- single movements of unreinforced dolosse, with adjusted waist ratio (Scholtz **et al**, 1982) up to 1,5 m can be accepted;

- occasional rocking (less than one-third of the time) up to 0,5 m can be accepted for unreinforced dolosse;
- approximately double these values can be allowed when X or double-V reinforced units are used;
- intermittent (two-thirds of the time) and continuously rocking units will break in the case of unreinforced and crack in the case of reinforced units and must thus be accepted as eventual loss (included in 'total' damage).

4. TABLE BAY BREAKWATER REPAIRS

During the past two decades the seaward slope of the Table Bay harbour's main breakwater suffered appreciable damage, particularly during winter storms when the wave climate is more aggressive. Although repairs were carried out as required on an ad hoc basis, mostly by replacing missing or damaged armour units, it was suspected that the long-term effects of this maintenance work was not satisfactory.

Over the years the situation became progressively worse as the remains of broken armour units were left behind to form part of a conglomerate of various sizes of rock, rubble, gravel and sand. It became clear that the situation should be investigated in depth to ensure that the integrity of the breakwater was maintained and to curb the escalating maintenance cost. Because of the multitude of contributory factors it was concluded that the problem could best be investigated by modeling the existing situation. The complication of the accumulation of various types of debris extending for some distance seaward of the breakwater made any alterations to the existing shallow bed slopes at the breakwater almost impossible and certainly impractical.

Model Tests

It was decided to investigate different typical sections of the breakwater separately in a 3 m wide wave flume. At the selected scale of the model of 1/53, up to 160 m long breakwater sections could be studied in the flume. The 160 m long wave flume was equipped with an irregular wave generator capable of reproducing wave conditions up to and exceeding the 1 in 100 year design wave condition, that is H_{mo} = 7,5 m.

The finally adopted design consisted of a double layer of 25 t dolosse placed on the existing breakwater slope after partial clearing of the above-water part of the existing breakwater protection (see Figure 10). This solution which included a toe row of chained dolosse, showed a total damage (displacement > 0,5 m, dolos roll over and continuous rocking) of about 3% in the shallower part of the breakwater (- 3 m MSL) but about 20% for the deeper part (- 5 m MSL). For practical reasons, 25 t dolosse were used throughout but in the deeper section, the dolosse were all reinforced and two rows of toe dolosse were chained.

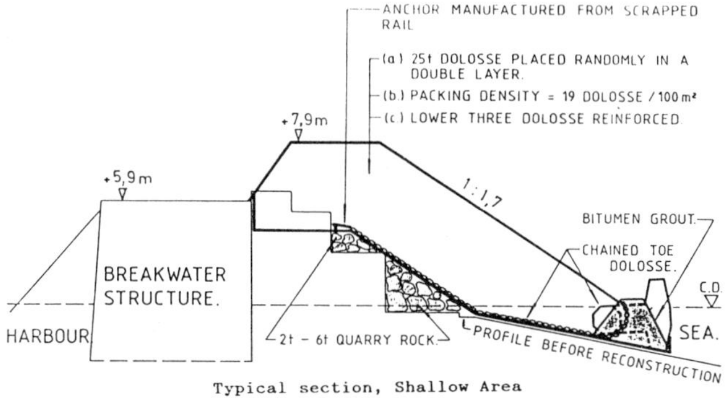

Typical section, Shallow Area

FIG. 10: TABLE BAY BREAKWATER REPAIR

Design Aspects

The dolos **waist ratio** was increased according to $r = 0{,}34\ (25/20)^{1/6} \simeq 0{,}36$ to strengthen the units.

Because of the hard nature of the sea bed it proved impractical to provide a proper toe berm for the breakwater slope and since this is a particularly vulnerable part of the breakwater, it was decided to introduce the following measures for the shallower part of the breakwater shown in Figure 10:

- rail reinforced dolosse were used for the lower three rows of the slope;
- to provide extra stability, the bottom row of reinforced dolosse were secured to the slope with chains;
- the practicality of using bitumen grout between the bottom rows would be investigated

Although the structural tests showed the X-type rail reinforcing to be most economical in terms of rail usage, the double-V type was used for the toe dolosse as the test results only became known during the construction period.

Manufacture of Dolosse

Concrete quality had to be very strictly supervised for the following reasons:

- the dolosse are expected to be subjected to higher than normal loads;
- due to practical constraints in the casting yard and to remain within the construction programme, dolosse were removed from the moulds after only 16 hours;
- to minimize the possibility of cracks which would allow the ingress of water to the rail reinforcing.

The concrete specified for the dolosse was a 40 MPa compressive strength mix with added pulverized fly ash and designed for maximum bending strength to counter the effects of early handling.

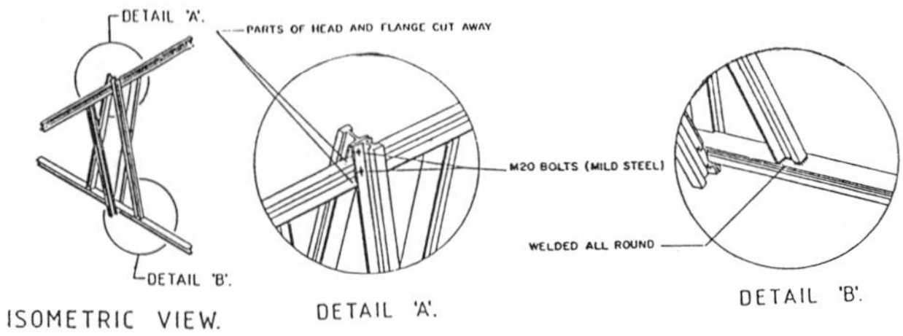

FIG. 11: RAIL REINFORCING (40 kg/m)

In most cases, 40 kg/m scrapped rails were used to manufacture the double-V type reinforcing frames (a few frames were made with 43 kg/m rails). Because of the number required (approximately 1500) the frames were built under workshop conditions using a combination of welding and bolting joints to facilitate the transport and on site assembly (see Figure 11).

Repair Work

Some of the factors that had to be taken into account in determining a placing pattern conforming to the design of the armour layer were:

(a) Only short lengths of the breakwater could be left exposed during the clearing phase of the reconstruction.

(b) As a result of the relative immobility of the rail mounted breakwater crane which operated from a movable section of track, provision had to be made for its protection against wave action during placing operations.

(c) The reach limitations of the breakwater crane had to be taken into account.

(d) The placing sequence of dolosse is important to ensure proper overlaying of the securing chains and proper interlocking between adjacent section of armour protection.

A typical placing pattern for one section, that is, one crane set-up position is depicted in Figure 12.

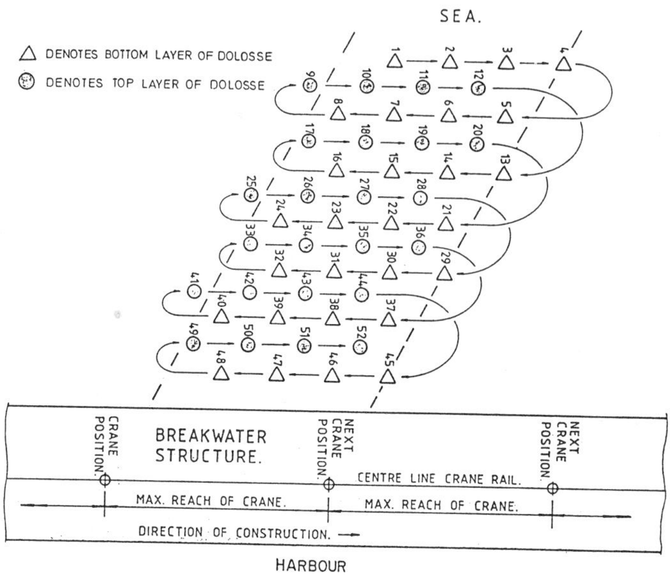

FIG. 12: DOLOS PLACING GRID

As the breakwater crane used for the reconstruction has built-in facilities for measuring angles and distance along the jib, the position of each dolos in the placing grid was translated to polar co-ordinates relative to each crane position. These polar co-ordinates were then used by the crane operator to locate the individual dolosse correctly. In practice it was found that after completing a section a visual inspection easily identified areas with obviously low packing densities, and additional units were placed where required. Generally the additional units amounted to approximately 4 per cent of the total number of units placed.

As a further precaution to reduce dolos movements in the vulnerable toe area **bitumen grout** will be placed between the bottom rows. Factors such as a suitable grout mix, placing temperature and practical considerations will be confirmed by full scale tests. During preliminary on site experimenting it was estimated that a volume placing density of around 5 m^3 per linear meter of breakwater should be sufficient to provide an acceptable degree of dolos to dolos binding.

Exact details such as underlayer profiles, dolos slope profiles and the placing position of each dolos was kept during construction for as-built record purposes. These records will be invaluable for future performance monitoring of the armour protection.

5. CONCLUSIONS

A comprehensive series of stability tests with irregular waves showed a significant reduction in dolos stability, particularly for waist-to-height ratios exceeding **0,36**. This should be taken into account when the waist ratio of larger dolosse is increased to maintain sufficient strength.

The test results showed similar dolos stability for **slopes** between 1:1,33 and 1:2 for all dolos waist ratios except for r = 0,33 and a slope of 1:1,33 for which the stability is significantly less.

Wave period was also found to have a significant influence on dolos stability, that is, lower stability was found for smaller wave periods.

The inclusion of relatively simple **rail reinforcement** increases the dolos strength (fall height) by some 100 per cent and an optimum design of a dolos armour can be achieved by selectively using increased waist ratios and/or rail reinforcing in specific danger areas. The most cost effective rail reinforcement was found to be the **X-type.**

Detailed optimization studies were done for the repair of the Table Bay breakwater. Hydraulic model tests showed that a 25 t dolos armour would not be entirely stable in the toe area of the breakwater, which was situated only 3 to 5 m below water-level. By using a waist ratio of about 0,36, double-V rail reinforcement, anchor chains and bitumen grouting, a satisfactory practical solution was found for the essential and very difficult repair work for the Table Bay breakwater.

REFERENCES

BRORSEN, M, BURCHARTH, HF and LARSEN, T (1974). Stability of dolos slopes. Proc 14th ICCE, Copenhagen.

BURCHARTH, HF (1981). Full-scale dynamic testing of dolosse to destruction. Coastal Engineering, Vol 4, No 3, Amsterdam.

BURCHARTH, HF (1984). Fatigue in breakwater concrete armour units. Report Laboriet for Hydraulic of Havnebygning, Aalborg.

BURCHARTH, HF and BREJNEGAARD-NIELSEN, T (1986). The influence of waist thickness of dolosse on the hydraulic stability of dolosse armour. Proc 20th ICCE, Taipei.

GRIMALDI, F and FONTANA, F (1984). Redesign of main breakwater of Gioia Tauro (Italy). Proc Symp on maritime Structures in the Mediterranean Sea, Athens.

MONSARD, EPD and FUNKE, ER (1980). The measurement of incident and reflected waves in random wave experiments. Proc 15th ICCE, Hawaii.

SCHOLTZ, DJP, ZWAMBORN, JA and VAN NIEKERK, M (1982). Dolos stability, effect of block density and waist thickness. Proc 18th ICCE, Cape Town.

SCHOLTZ, DJP and ZWAMBORN, JA (1983). Effect of the waist thickness on stability of dolosse. CSIR Research Report 556, Stellenbosch.

ZWAMBORN, JA and VAN NIEKERK, M (1982). Additional model tests, dolos packing density and effect of relative block density. CSIR Research Report 554, Stellenbosch.

ZWAMBORN, JA and SCHOLTZ, DJP (1986). Dolos armour design considerations. Proc 20th ICCE, Taipei, Taiwan.

CHAPTER 155

SEAWALLS and SHORELINE PROTECTION

by:
Jan van de Graaff[1)]
and
Eco W. Bijker[2)]

Abstract

All over the world shore-parallel constructions (seawalls; revetments; bulkheads) suffer from damage. It is argued that these constructions are in fact frequently built at places where they shouldn't.

1 Introduction

Coast managers are frequently faced with erosion of their sandy coasts. Valuable land is encroached by the sea. Owners of houses and hotels, bordering eroding coasts, often 'force' (sometimes by means of the politics) the coast manager 'to-do-something'. Not seldom it is then felt more important to do indeed 'something' than doing the right things.

Unfortunately, too often a seawall or a revetment is selected to overcome the direct erosion problems. However, just from the common coastal engineering practice it can be seen that seawalls or revetments do not solve the problems in the right way. Entirely destroyed seawalls can be observed along coasts all over the world. This destruction is by no means always due to the fact (that would be a legitimate reason) that the design conditions were surpassed during a single storm. No, the basic ideas behind the application of seawalls or revetments were in fact quite often wrong.

Demolished coastal constructions harm the prestige of 'coastal engineering' as a respectable profession and should be avoided as far as possible.

---

1) Senior scientific officer Delft University of Technology, the Netherlands.
2) Professor Coastal Engineering Delft University of Technology, the Netherlands.

In this paper mainly the rôles of seawalls and revetments in the game of coastline erosion are discussed. Possible alternatives come hardly up for discussion.

2 Definitions and limitations of present discussion

Seawalls and revetments are shore-parallel constructions and are meant to protect the hinterland against flooding and/or erosion. They are situated against features like scarps, embankments, cliffs, dunes and promenades. The distinction between seawalls, revetments and also bulkheads is generally not quite clear. Throughout the present paper the term 'seawalls' will be used to indicate the type of constructions under consideration. (Shore-parallel constructions.)

Sandy beaches are mainly considered in the paper. So at least under 'usual' conditions a sandy beach, for instance for recreation use, is assumed to exist in front of the seawall.

Fig. 1 shows three typical cross-sections with a shore-parallel construction. Cases like case a) and b) are considered in the present paper (sandy beaches in front of the construction under usual conditions). Cases like case c) are not considered (no beach in front of the construction).

Case a) is typical for well-developed seaside resorts. On the benefit of a prosperous development often a clear distinction between 'sea' and 'land' is desired. A seawall can provide that.

Case b) is typical for important investments which are apparently at stake. The encroaching sea obviously causes erosion of the beaches and the dunes. With the shore-parallel construction the attack of the sea is thought to be beaten off.

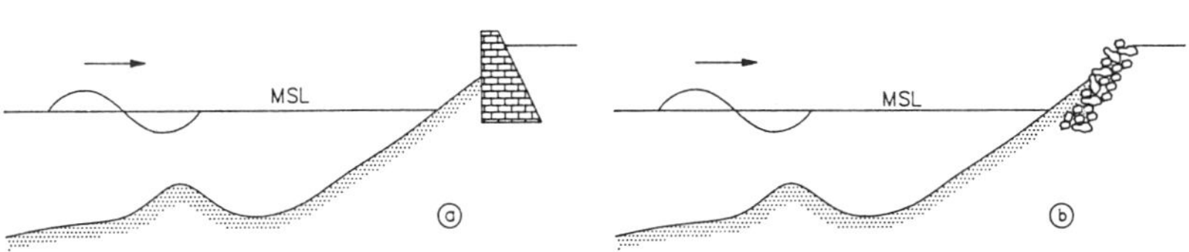

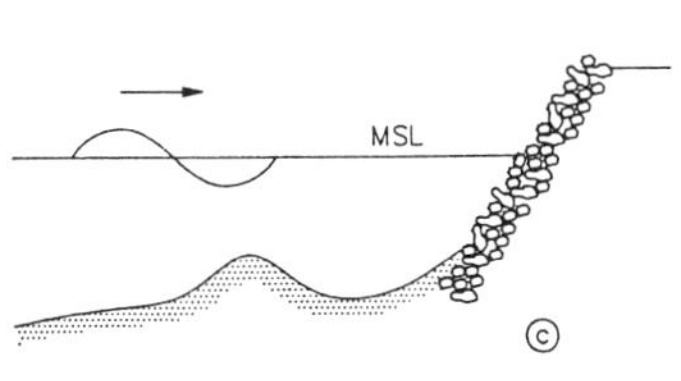

Fig. 1 Cross-sections with shore-parallel constructions

All over the world, however, damaged and even entirely demolished seawalls can be found. In many cases it is felt that in fact design errors are the main cause of the damage. Preventing of some of these errors in future is the main aim of the present paper.

3 Analysis of erosion problems

Seawalls are often built to overcome the felt 'problems' of our coasts. In principle two basic erosion problems with sandy coasts do exist, viz.:

a) erosion/recession during a storm (surge) event
b) gradual long term erosion

Fig. 2 shows in plan view a stretch of a coast at a certain moment in time (under 'usual' sea conditions). The 'problems' a) and b) can be clearly illustrated if the behaviour of cross-section A-A of Fig. 2 is considered as a function of a long time. For reasons of simplicity the behaviour of the position with respect to the reference line of the so-called dune-foot is considered to be representative for the behaviour of the entire cross-section (other characteristic profile features could also be selected). The dune-foot is the intersection point between the gentle beach slope and the steep slope of the dune front.

Fig. 3 shows three typical possibilities of the behaviour of a sandy coast.

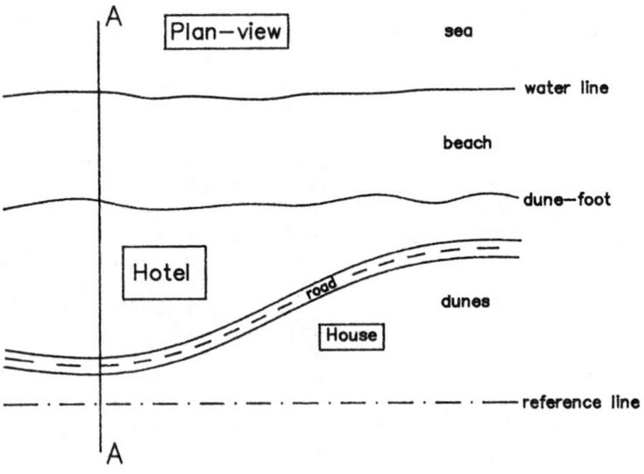

Fig. 2 Plan-view of coast

Fig. 3a represents an essentially stable coastline. However, storm (surge) events cause sudden recessions of the position of the dune-foot. Since the cross-section in consideration is, seen over a longer period, stable, a recovery of the dunes will take place in the years after the storm event. Depending on the seriousness of the storm event the magnitude of the recession can vary considerably. Fig. 3a represents in fact case a) indicated in this section.

Fig. 3b shows a gradually eroding coast (with surge events superimposed on that). The recovery after a surge is not entire. Fig. 3b represents case b).

[For the sake of completeness in Fig. 3c an accreting coast is shown (also with storm events). In the present discussions these cases are not considered since accreting coasts cause hardly 'problems' to the coast managers.]

4 Erosion/recession during a storm (surge) event

Erosion of the dunes and the upper part of the beaches can occur during a severe storm. The rate of recession during that event depends on the seriousness of the storm involved. During the storm not only the wave attack is greater than during usual conditions, but also the water level (surge level) increases to levels (far) higher than usual. Along coasts bordering oceans the increase in water level during the passage of a storm is often moderate; along coasts bordering funnel-shaped seas the increase of the water level may mount several meters.

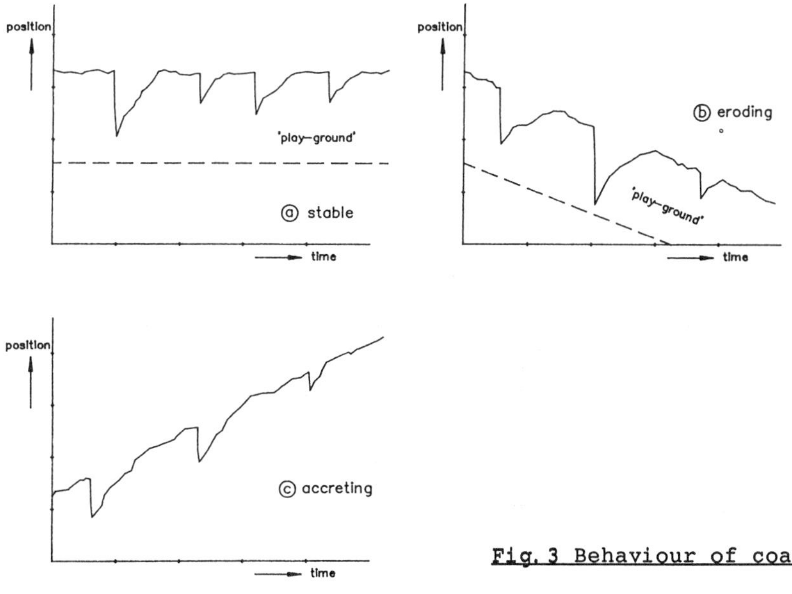

Fig. 3 Behaviour of coast

Fig. 4 shows schematically what happens during a severe surge. Material of the dunes is eroded and (mostly) settled again on the foreshore. Since the shape of the profile becomes less steep, the erosion process (the rate of erosion) slows down with time. After the surge a retreat distance RD can be observed.
Various methods are available at present to calculate the volume of dune erosion after the storm surge. [Kriebel and Dean (1985); Vellinga (1986).] These methods can be applied if the boundary conditions during the storm surge are known

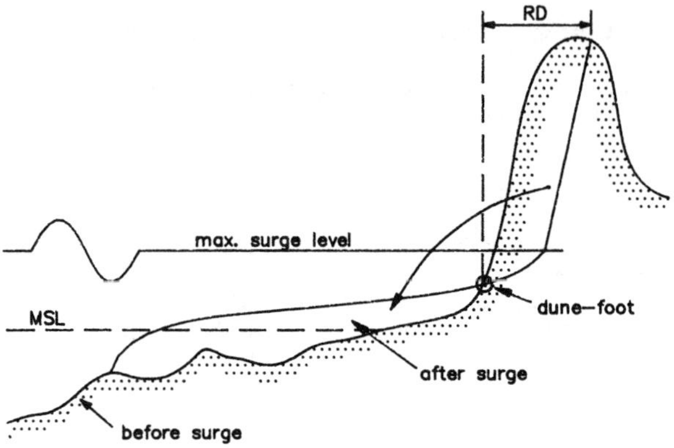

Fig. 4 Dune erosion process

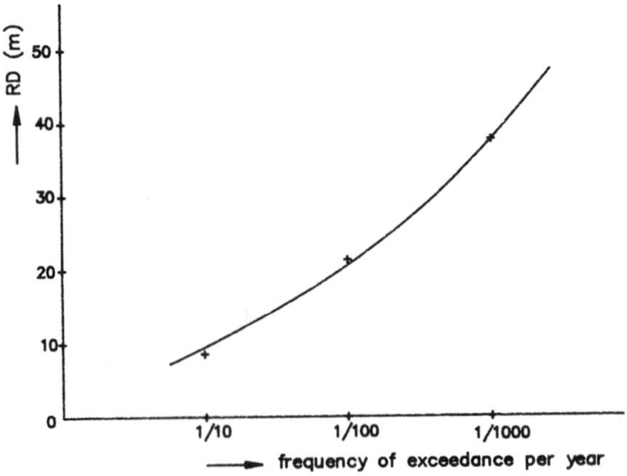

Fig. 5 RD as a function of frequency of exceedance

(wave height; wave period; maximum surge level; storm duration; particle size diameter; initial profile). Many of these boundary conditions have a highly stochastic nature. Nobody knows what the conditions will be next year. Based on long term observations and/or simulation calculations one is in fact only able to predict the probability of occurrence of a set of bad boundary conditions in a certain storm season or a certain year. With probabilistic methods [see e.g. Van de Graaff (1986)] one is eventually able to determine a figure like Fig. 5. Different retreat distances RD are expected depending on the frequency of exceedance.

In order to prevent superfluous damage the coast manager (or the 'society') has next to select an appropriate design frequency of exceedance (with a related design-RD distance). It should in fact be avoided that high investments in valuable goods like houses, hotels and roads are done in the selected design distance. From a coastal engineering point of view this seems quite logical. The sea needs a certain 'play-ground'. (In Figs. 3a and 3b arbitrary 'play-grounds' have been sketched.) Also from an economical point of view such a proceeding of a restriction policy can be simply justified. The everyday practice, however, is quite different. Partly due to a (sometimes pardonable) lack of insight in the 'tricks' of the sea, but also sometimes due to an almost blameworthy inability of the responsible authorities, too many valuable investments are done too close to the shoreline. A (apparently unexpected) rather severe storm surge may destroy consequently many buildings. After such an occasion the call for countermeasures will grow and it will be obvious that the construction of a seawall will come up as a possibility to avoid similar problems in future. It will be for sure that **in this particular case** a **well-designed** shore-parallel construction may indeed overcome the erosion problems due to a severe storm surge. A seawall can diminish the 'play-ground' of the sea. It should be stressed that a **well-designed** construction should be applied. E.g. Steetzel (1987) argues that if the erosion of the dunes is prevented due to the construction of a revetment against the dune front, severe erosion just in front of the revetment has to be expected. If the so-called 'denied' volume (= eroded volume from the dunes if the revetment was not there) is large, the depth increase in front of the construction may become quite considerable. Under Dutch design conditions, so far, erosion depths are expected of several meters below the initial beach profile. If the foundation depth of the construction is too small, stability problems for the entire construction may rise. In an essentially stable case, a recovery of the erosion pit will mostly follow soon after the storm [cf. Dean (1986)].

In conclusion it can be stated that in essential stable cases, where it is felt that the needed 'play-ground' of the sea should be restricted, a properly designed seawall might be applied.

5 Gradual long term erosion

Fig. 3b is characteristic in the case in consideration. The coast apparently recedes seen over a number of years. The yearly rate of recession depends of course on the particular situation, but is often up to some meters per year. [During real serious storm (surge) events the retreat might be up to some tens of meters per occasion (per day).]

A sneaking retreat of a built-up coastal area (with sometimes a storm surge event imposed on it), is extremely annoying to a coast manager. From time to time buildings have to be abandoned and often it can be almost precisely predicted when the next building is on its turn.

In cases like these too often the next (wrong) reasoning to build a seawall is followed:

- The coast (the built-up area) recedes.
- We have to stop this recession.
- Let us fix the dune front and the coast cannot recede further.
- Let us build a seawall to fix the dune front.

In the first years after the construction, unfortunately, all seems like a bed of roses. The erosion of the built-up area has been stopped indeed. The owners of the buildings are quite satisfied and the coast manager, happy with his great success, decides to build also a seawall along another eroding part of the coast which is under his control.

However, after the first happy years after the construction, a period with increasing problems starts. It will turn out that the erosion of the beaches in front of the seawall has continued or has even increased. The beaches in front of the seawall have lowered. The waves reach more easily and more frequently the seawall. The attack on the seawall increases. To prevent serious damage the seawall has to be reinforced or has to be 'protected'. In spite of these countermeasures the erosion of the beaches continues. After some time all the beaches have been disappeared. A situation like in Fig. 1c has been developed. There are, also under usual conditions, no beaches anymore for recreation purposes. The seawall has little by little changed in a seadike. In spite of these in fact dramatic developments, our coast manager remains often proud. The erosion by the sea has certainly been stopped. The encroachment of the sea of houses and roads has certainly ceased. That the beaches have been lost is a pitty; it is all in the game. The costs of the regular reinforcements of the seawall are paid by the tax payers, so that does harm 'nobody'. At the end, after a number of years, only a few old persons know that in past nice sandy beaches did exist in this area.

What went wrong in fact?

The basic point is that it is a gradual erosion problem (cf. Fig. 3b). To overcome this type of problems, two methods are available in principle:

1) Feed from time to time artificially the amount of material which has apparently lost along the coastline (beach nourishment).
2) Try to interfere in the sediment transport process which causes the erosion. This interference should be such that the erosion stops **in the area you like to save**.

Method 1) is almost trivial. However, this method is increasingly attractive and can be applied almost always. The dutch Manual on Artificial Beach Nourishment (1987) is recommended to consult if an actual beach nourishment scheme has to be designed.

Method 2) is more difficult to execute. First of all it should be established which sediment transports cause the erosion. If that has been established one has next to decide how to interfere actually.

Two main causes do exist for a gradual erosion of a part of a coast:

a) <u>Continuous</u> offshore sediment transport from the upper parts of the beach profile (beach and dunes) to the deeper parts of the profile.
b) <u>Gradient</u> in the longshore sediment transport.

Re a) Continuous offshore transport

Cross-shore directed sediment transports always take place in an actual cross-section of a beach profile. Depending on the wave conditions, water level and the shape of the initial profile, onshore or offshore transports take place through an arbitrary vertical cross-section. Seen over a relatively long period (from year to year for example) the shape of the beach profile often doesn't change so much (dynamic equilibrium). In a constant situation with respect to the boundary conditions (constant mean sea level and constant yearly wave climate), there is in usual cases no reason that a <u>continuous</u> offshore directed transport will be the reason for a gradual erosion of the coast. If a continuous offshore transport from the upper parts of the beach profile to the lower parts of the profile would occur, the beach profile would be flatter with time. Next it can be argued that the offshore transport rates will then slow down till an equilibrium has been reached. (The 'sea' had many centuries time to reach that!) Only if the supplied material to the deeper part of the profile is eroded again (e.g. by a gradient in the longshore transport in that region) a continuous cross-shore transport might be maintained. The real reason for the erosion of the coastline is then, however, not the offshore transport, but the mentioned gradient in the longshore transport on the deeper parts of the beach profile. The offshore sediment transport is only an intermediary mode of transport.

Only if the boundary conditions change with time, a more or less continuous offshore sediment transport might occur. This transport can be seen as 'necessary' in order to adjust the shape of the beach profile so that is fits (again) with the 'new' boundary conditions. The global sea-level rise and an abrupt change in the wave climate are examples of changes in the boundary conditions. It will be clear that if these changes in the boundary conditions occur, a simple seawall cannot solve the induced erosion problem. Under usual conditions the seawall doesn't interfere in the underlying transport processes. (Only if the seawall is directly attacked, some interference might occur.) The erosion of the upper parts of the beach will continue. The seawall will be attacked more frequently and more intensively. If one likes to interfere actually in the (offshore) sediment transports one has to think of (submerged) detached offshore breakwaters. Such a construction might eventually effect (reduce) the wave action which approaches the coast and consequently might reduce the offshore sediment transports.

Re b) Gradient in the longshore sediment transport

In many cases a gradient in the longshore sediment transport is the main reason of the erosion problems of sandy coasts. Examples are the lee-side erosion near breakwaters or jetties and the erosion of convex coastlines.

Fig. 6 shows schematically the magnitude of the longshore sediment transport (e.g. in m3/year) as a function of the position along an eroding coast. The increasing sediment transport with x is the reason of the erosion problems. The magnitude of the gradient dS/dx is a measure of the actual erosion problem. If one wants to stop the recession of the coast in stretch A-B in Fig. 6, one has to nourish volume V (see Fig. 6) along section A-B on a yearly basis or one has to 'change' in one way or another the sediment transports

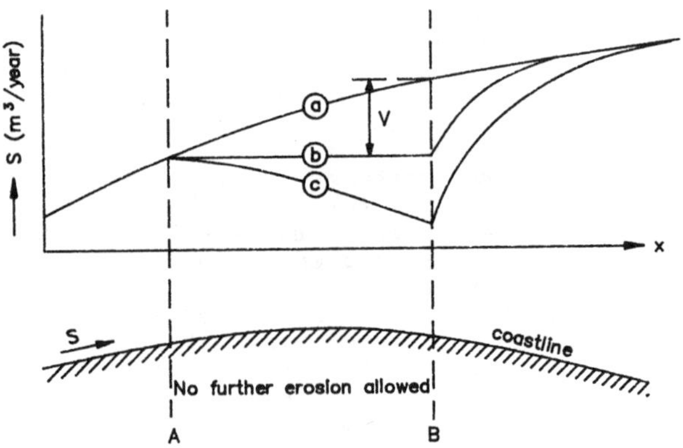

Fig. 6 Sediment transports along coast

along stretch A-B. Instead of a distribution of the sediment transports according to line a) in Fig. 6, a distribution according to line b) would in principle fulfil our requirements. The erosion in section A-B has stopped indeed (dS/dx = 0), but down stream of stretch A-B an increased erosion can be expected. (Increased dS/dx values in comparison with the initial situation.) To sketch a line like line b) is of course quite simple and is one thing, but how to acquire such a line in an actual case is another. It is for sure, however, that with a seawall by no means a situation like line b) can be reached. The basic problem was a gradient in the total longshore sediment transport and the longshore sediment transports take place (at least under usual conditions) on the foreshore and the shoreface. Since a seawall doesn't interfere in this type of sediment transport, no direct reducing effect can be expected. Only under storm conditions, when waves hit the shore-parallel construction directly, probably some reduction of the longshore sediment transports might be expected. In that case, however, it is also conceivable that, due to wave reflection, an increase of the longshore sediment transport will take place.

Since a shore-parallel construction near the dunes doesn't interfere generally in the sediment transports, the erosion continues; the shoreface, the foreshore and the beaches become deeper and at the end the attack of the sea on the seawall will intensify. Damage occurs; reinforcements of the seawall will be necessary. Nearly all seawalls built to restrict the further erosion at the lee-side of harbour breakwaters or jetties suffer from the stated problems.

As it has been indicated it is impossible to 'change' the sediment transport line a) in Fig. 6 to line b) with the help of a seawall. With the construction of groins or rows of (wooden) piles [perpendicular to the coast], or the construction of a (submerged) detached breakwater [parallel to the coast], the desired change from line a) to line b) is possible indeed. How to design these countermeasures actually, is a difficult coastal engineering problem, but from the 'physics' it can be understood at least that these constructions might help to overcome the erosion problem. (They interfere in the original sediment transports.) That a proper design of these countermeasures is indeed a difficult task can be illustrated with the help of Fig. 7. Case I line a) shows the same coastal erosion as Case II line c). The same gradients dS/dx occur, however, the magnitudes of the longshore sediment transport differ considerably. The lines b) and d) represent the ultimately desired sediment transport lines for Case I and Case II respectively. It is beyond doubt that the operation to achieve line b) from line a) calls for quite different countermeasures than to achieve line d) from line c). One has consequently to know quite precisely what the actual magnitudes of the sediment transports are, before a proper design of countermeasures can be made.

Changing the sediment transports along stretch A-B in fig. 6 from line a) to line b) is just necessary and sufficient to stop the erosion in that section. If line a) is changed in line c) accreting of the section A-B can be expected. (That is in fact an 'over-kill' operation.) Accreting might be favourable to people living along section A-B. The erosion problems at the lee-side of section A-B will, however, consequently increase.

Since so many examples of the bad behaviour of seawalls are available, one might ask oneself whether seawalls will even worsen the erosion problems of the coasts they intend to protect. From the physics no sound support can be found for this idea. Seawalls do not interfere in the sediment transports, so consequently they will not increase the erosion problems. This hold at least till the time that some beaches are still present in front of the seawall under usual conditions. When all the beaches have disappeared (a case like Fig. 1c has been reached then), a quite different situation is reached. Increased as well as decreased sediment transports are conceivable depending on the actual conditions.

6 Conclusions

- Demolished seawalls harm the prestige of 'coastal engineering' as a respectable profession.
- The application of seawalls as a means to combat erosion of coasts should be restricted to a very limited number of cases.
- Only in essential stable cases where some reduction of the so-called 'play-ground' of the sea is required, a seawall would be an appropriate solution to achieve that goal.
- Along gradually eroding coasts, seawalls shouldn't be used to prevent further erosion. A seawall doesn't 'work' in these cases since a seawall doesn't interfere in the sediment transport processes.
- Generally seawalls don't 'destroy' beaches.

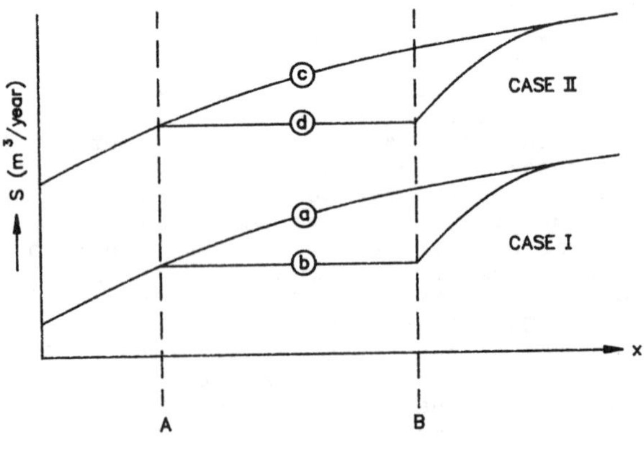

Fig. 7 Effect of magnitude of sediment transport

7 References

Dean, R. G. (1986)
Coastal armoring: effects, principles and mitigation.
Proc. 20th Intern. Conf. on Coastal Engineering, Vol. III; Chapter 135, Taipei.

Graaff, J. van de (1986)
Probabilistic design of dunes; an example from the Netherlands.
Coastal Engineering, Vol. 9, pp. 479-500; Elsevier Science Publishers; Amsterdam.

Kriebel, D. L. and R. G. Dean (1985)
Numerical simulation of time-dependent beach and dune erosion.
Coastal Engineering, Vol. 9, pp. 221-245; Elsevier Science Publishers; Amsterdam.

Steetzel, H. J. (1987)
A model for beach and dune profile changes near dune revetments.
Proc. Coastal Sediments '87; Vol. I, pp. 87-97, New Orleans.

Vellinga, P. (1986)
Beach and dune erosion during storm surges.
Ph. D. Thesis Delft University of Technology; also Delft Hydraulics Communications No. 372.

Manual on Artificial Beach Nourishment (1987)
CUR/Rijkswaterstaat/Delft Hydraulics Report 130.
ISBN 90 212 6078 6, Gouda, the Netherlands.

CHAPTER 156

QUALITY AND DURABILITY OF CONCRETE ARMOUR UNITS

L. Van Damme *
L. Taerwe **
R. Dedeyne +
J. De Rouck ++

1. INTRODUCTION

The reliability of rubble mound breakwaters not only depends on the hydraulic stability but to a large extent on the strength of the armour units as well. In the debate on the shape of armour units, which turns out quite often to be merely academic in nature, generally too little attention is paid to the question of structural strength. On the occasion of the realization of the new outer harbour at Zeebrugge (Belgium) a serious effort has been made to arrive at a better understanding in the behaviour of concrete as construction material for armour units. This contribution gives an overview of the routine tests and of some particular investigations, carried out with respect to quality and durability of concrete armour units.

2. GENERAL CHARACTERISTICS OF THE ARMOUR UNITS

2.1. Types of armour units

In the Zeebrugge harbour extension works two types of armour units have been used : the Antifer cube (±60.000 units) and the HARO® (±13.000 units) (fig.1).

During the design of the main breakwaters the Antifer cube was chosen as armour unit. Consequently it has been used on the breakwaters protecting the LNG-terminal, which were built first.

* Ministry of Public Works, Coastal Department, Administratief Centrum, Vrijhavenstraat 3, 8400 Ostend, Belgium
** Ghent State University, Laboratorium Magnel, Grote Steenweg Noord 2, 9710 Ghent, Belgium
+ SECO S.V., Aarlenstraat 53, 1040 Brussels, Belgium
++ HAECON N.V. Consulting Engineers, Deinsesteenweg 110, 9810 Ghent, Belgium.
Ghent State University.

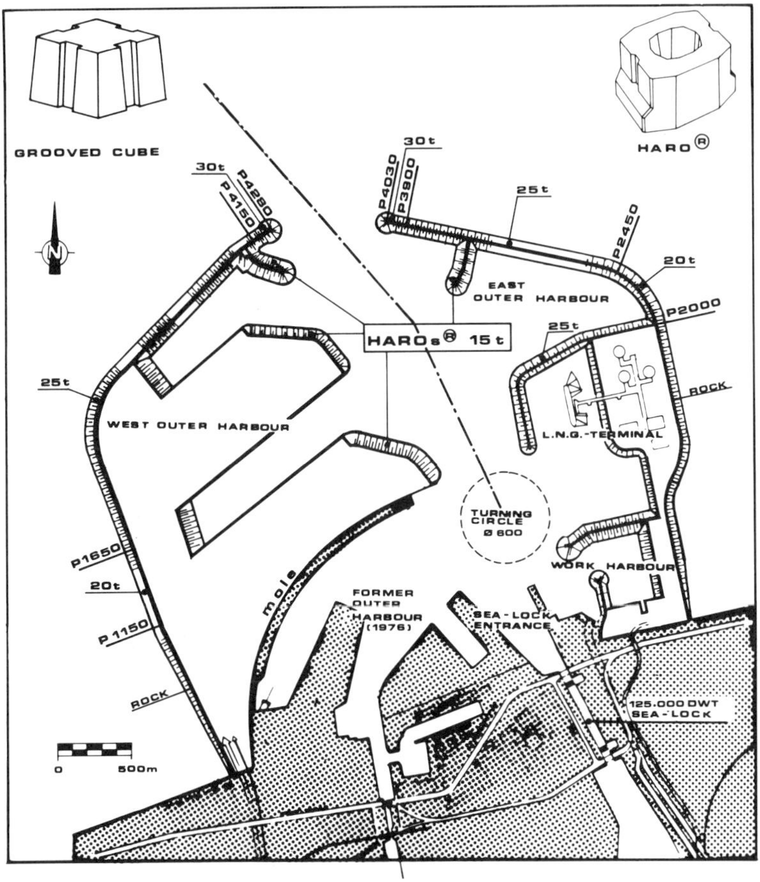

Figure 1 : Zeebrugge outer harbour, with indication of the types of armour unit

On the outer-harbuor breakwaters, a slightly adapted grooved "cube" (the height is somewhat smaller than the bottom side) has been used. During the same optimization one has looked for concrete with higher density : a mean density of 2,47 t/m³ (1434 test results) has been achieved by adding lead slag. Depending on the location the weight of the units equals 20 or 25 tonnes. The main-breakwater heads have been armoured with 30t-grooved cubes.

Fig. 1 shows the areas, where the different units have been used.

During the design of the inner-harbour breakwaters the HARO® has been developed. The HARO® is a flat (height is less than the short side in plan view) concrete block with a large central opening. Both short sides are made wider at the base, thus constituting an alternative type of block profile by weight. The corners of the block are asymmetrically tapered in plan. The general form of the block in plan is rectangular.

The criteria which have been taken into account during the development of the HARO® are :

- a high structural strength (given by its strong compact shape) ;
- hydraulic stability ($K_D > 15$) ;
- advantageous wave reflection and run-up characteristics ;
- simplicity of fabrication : no opening of the moulds and demolding in one single movement upwards ;
- simplicity in placement of the block on the breakwater slope ;
- economy : by using HAROs® the total volume of concrete needed in the armour layer decreases by more than 35% compared with grooved cubes. This has mainly been achieved by the reduced block weight ($K_D > 15$) and by the high porosity (Γ = 52 à 55%).

The inner-harbour breakwaters will be armoured with HAROs®. The specified characteristic density of the HARO®, is 2,3 t/m³.

More details on these types of armour units may be found in ref. (1) and (2).

2.2. Placement pattern

During the first construction phase (breakwaters surrounding the LNG terminal) the chosen placement pattern required very little re-positioning of the crane (fig. 2a). However hydraulic stability tests showed a rather high risk of jamming-up when tending towards the design conditions.

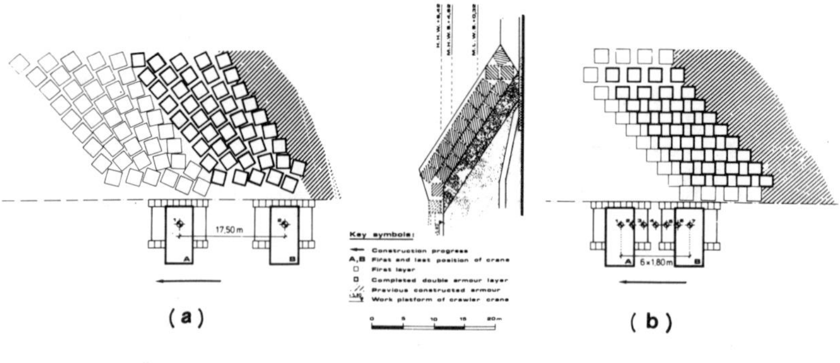

Figure 2 : Placement pattern of armour units.

A placement pattern to eliminate this phenomenon has been elaborated : Within the considered (upper or lower) layer each unit rests on the two adjacent blocks of the lower row. This placement pattern requires an almost continuous repositioning of the crane (fig. 2b). All units on the outer-harbour breakwaters and all HAROs® are placed according to this pattern.

2.3. Concrete composition

The concrete mix proportions to obtain a characteristic density of 2,3 t/m³ are as follows :

. cement LK30 300 kg/m³
. gravel (4/28) 1.300 kg/m³
. sand (0/5) 700 kg/m³

This concrete composition holds for a mean water content of aggregates of ca. 2,5 %. To obtain a characteristic density of 2,4 t/m³, as stipulated for the grooved cubes, lead slag is added at a rate of ca. 100 l/m³. The W/C ratio was lower than 0,4. A blast furnace cement (LK30), with no more than 15% Portland clinker, and low heat of hydration (200 kJ/kg at an age of 72 h) is used. The sand and gravel are marine aggregates with a sufficiently low chloride content and an acceptable shell content.

2.4. Concrete quality

A total of 2.313 compressive strength results are available over a period of 4 years. The test specimens are cubes with side length of 200 mm and cured under water until an age of 28 days. The following statistical characteristics were found :

mean	39.40	N/mm²
standard deviation	4.59	N/mm²
coefficient of variation	11.60	%
skewness	0.369	
curtosis	3.768	

On the basis of the values for skewness and curtosis, the normality hypothesis for the distribution function has to be rejected. This is due to the long term variation of the parameters which determine the strength characteristics of the concrete.

3. USE OF LEAD SLAG

3.1. Selection criteria

A higher density of concrete can be obtained by substituting heavy inert materials for part of the gravel. The selection of the most suitable aggregate requires the examination of several physical and mechanical concrete properties, such as : density, compressive and tensile strength, segregation during mixing, plasticity, porosity, shrinkage, creep, capillar absorption, permeability, workability, optimal compaction, resistance against aggressive agents.

The following aggregates have been considered : natural rock, basalt, hematite, lead slag and iron. A cost comparison showed the lead slag mix to be the most economic (table 1).

Unit weight w_c of concrete (t/m^3)	Cost
Concrete :	
w_c = 2,30	1
Concrete with basalt (w_b = 3,10 t/m^3)	
w_c = 2,41	1,02
w_c = 2,45	1,12
w_c = 2,51	1,25
w_c = 2,58	1,39
Cyclops concrete with natural rock (w_r = 2,65 t/m^3)	
w_c = 2,40	1,10
Concrete with hematite (w_h = 5 t/m^3)	
w_c = 2,40	1,16
w_c = 2,50	1,34
w_c = 2,65	1,61
Concrete with lead slag (w_l = 3,5 t/m^3)	
w_c = 2,42	1,03
w_c = 2,54	1,07
w_c = 2,65	1,11
Concrete with iron cuttings (w_i = 7,85 t/m^3)	
w_c = 2,40	1,16
w_c = 2,50	1,35
w_c = 2,65	1,63

Table 1 : Cost comparison of high-density concrete

3.2. **Tests on lead slag**

Concrete containing lead has a very bad reputation, as expansion of lead in a humid environment may give rise to splitting of concrete. However, lead slag as a residue from the lead production process has a negligible lead content.

A comprehensive analysis has been carried out in order to determine the behaviour of lead against cement and the behaviour of concrete containing lead slag against sea water.

The analysis has been conducted in the following way :

a) Mineral analysis of the slag

From slag samples, thin slices with a thickness of 30 to 40 microns have been taken. Examination of the slices showed the presence of various mineral components difficult to identify in an optical way. An analysis by electron microprobe was used to determine the composition of very small zones (1 micron). The components found are not likely to provoke serious damage for the intended utilization of the lead slag.

b) Reactivity of the slag with respect to cement

The reactivity of the slag with respect to cement has been analyzed by comparing the mechanical properties of two mortars realized with cement and two types of fine aggregates : an inert sand (quartz) and crushed slag. This part of the analysis revealed no significant difference of behaviour.

c) Simulation of sea-water attack

After three months of hardening in a humid chamber, mortar specimens were alternately immersed in and removed from sea-water. At an age of 4 1/2 months, mechanical resistance tests and microscopic examinations were carried out on specimens, subjected to sea-water attack and on reference specimens, one series being composed of normal sand and another series being composed of crushed slag.

The mechanical resistance showed higher values for both series exposed to sea-water. The corresponding results of sand and slag mortar were of the same magnitude (maximum deviation 8%). For both specimen types, no irregularities were noticed by microscopic examination, neither in the cement matrix nor at slag-cement or sand-cement interface.

d) Examination of core samples

At an age of about two years, core samples were taken from blocks on-site. The following characteristics were determined :

- density
- water absorption by immersion
- compressive strength.

The results of these tests correspond to values usually recorded for standard concrete and show no significant deviation from the reference values. The examination of ground-up slices taken from the cores showed that lead slag didn't instigate concrete disintegration.

The foregoing considerations show that the addition of lead slag, with a low lead content, to the concrete is not likely to alter the behaviour of this concrete over its design life.

4. THERMAL STRESSES

Different researchers have demonstrated the relationship between heat development due to hydration of cement and crack formation in young concrete. In order to verify the behaviour of concrete at early ages, a comprehensive investigation program was set up.

4.1. Temperature measurements in armour units

Two series of measurements on concrete blocks were carried out on the site, one in a winter period, the other during summertime.

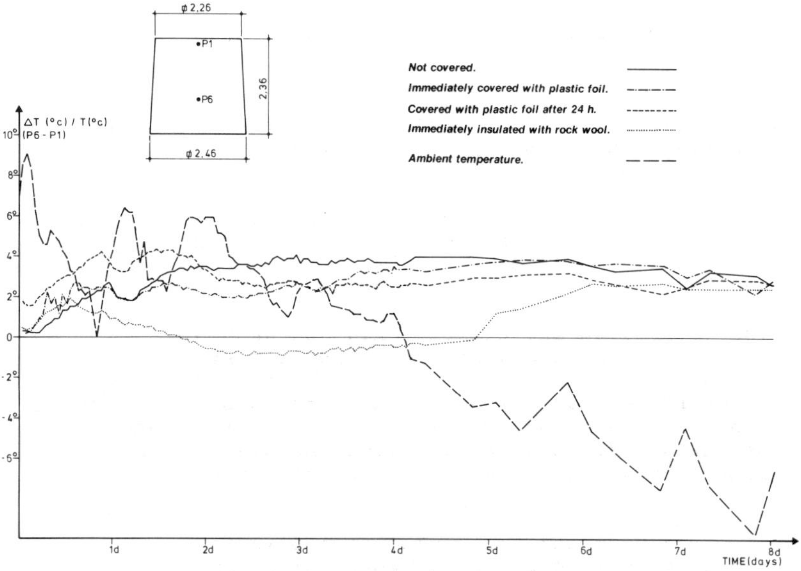

Figure 3 : Evolution of temperature difference between the core and the upper suface for 30 t cubes (winter time)

In the first test trial, four cubic blocks (30 t) were instrumented with 11 thermocouples. After removal of the mould, two blocks were covered with a plastic foil, one block was insulated with rock wool and one block was left unprotected. Fig. 3 shows the evolution of temperature difference between the core and the upper surface for these four 30t-cubes.

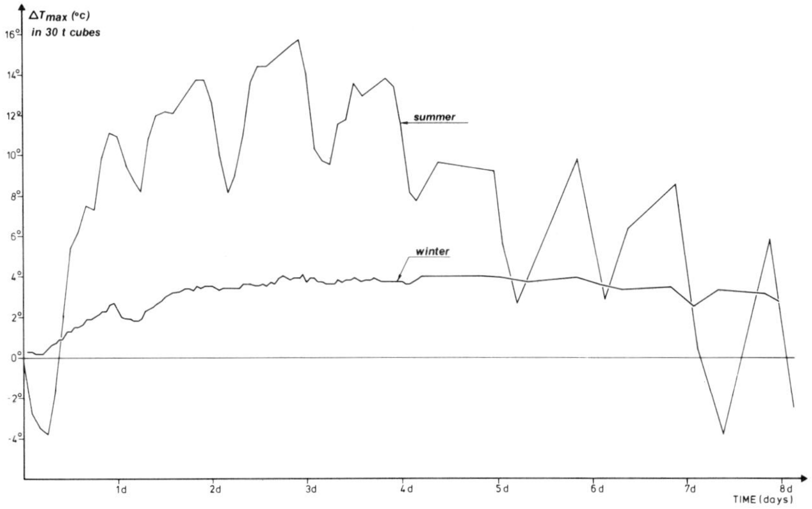

Figure 4 : Maximum measured temperature difference in the unprotected 30 t cubes

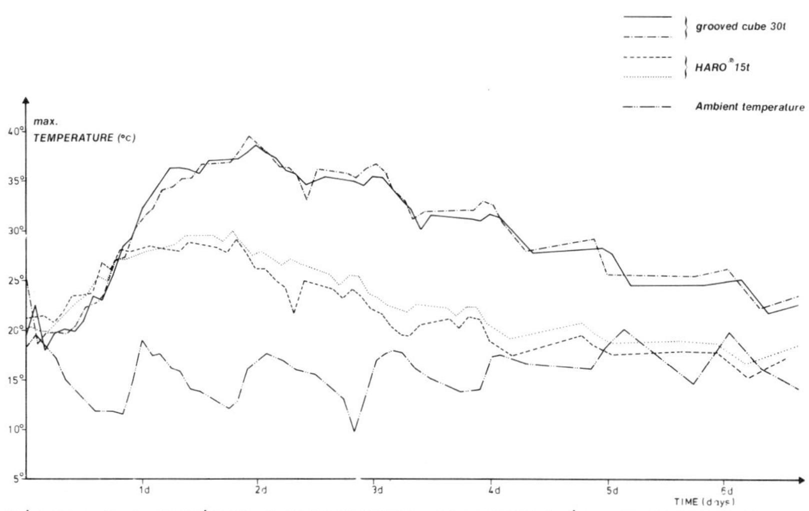

Figure 5 : Maximum temperature (measured in summer) for grooved cubes and HAROs®.

A second series of temperature measurements was carried out during a summer period on 2 cubic blocks (30t) and 2 HAROs® (15t). All blocks were left unprotected. Fig. 4 shows the maximum temperature difference between core and upper surface for the unprotected blocks i) in winter, ii) in summer. Fig. 5 also shows the maximum temperature, both in summertime, and for both grooved cube and the HARO®. Due to the central opening in the HARO® the temperature increase remains small.

4.2. Characteristics of cement and concrete

4.2.1. Properties of cement

The mechanical properties of cement type LK30 were determined on mortar prisms (40x40x160 mm). The compressive strength equals 17.4 N/mm² at 3 days, 28.8 N/mm² at 7 days and 41.8 N/mm² at 28 days.

Setting starts at 6h40 min. and ends at 9h20 min. (Vicat test).

A chemical analysis reveals the following composition :

calcium oxide	44,9 %	magnesium oxide	1,6 %
silicon oxide	30,1 %	loss on ignition	0 %
aluminium oxide	12,3 %	insoluble residue	1,5 %
ferric oxide	1,7 %	other components	0,2 %
sulphur trioxide	7,3 %		

The heat of hydration, determined by means of the conduction method, is found to be 53.7 kJ/kg at 24 h, 146.2 kJ/kg at 48 h and 198.4 kJ/kg at 72 h. It is clear that the heat of hydration of the cement used is rather low which is favourable in order to limit internal thermal stresses.

4.2.2. Mechanical properties of the hardening concrete

During the measuring trial, the compressive and the splitting tensile strength of cubes (side length 200 mm) is determined at several ages, both for on-site and laboratory cured specimens. The laboratory cured specimens are left in a humid chamber at 10°C and a relative humidity of at least 90%. Prisms for the determination of the secant modulus of elasticity are also cured in these environmental conditions.

The results for the concrete of the 30 ton cubes cast in summer are mentioned in table 2.

Age (days)	Curing	Compressive strength (N/mm²)	Splitting tensile strength (N/mm²)	Secant Modulus of Elasticity (N/mm²)
1	site	2.63	0.19	4.300
	laboratory	2.64	0.18	-
2	site	-	0.99	31.000
	laboratory	-	1.08	-
3	site	18.0	1.59	36.200
	laboratory	23.2	1.93	-
7	site	-	1.98	35.400
	laboratory	-	2.87	-
28	site	32.0	2.76	42.500
	laboratory	40.8	3.24	-

Table 2 : Mechanical properties of the concrete used for the 30-ton cubes (summertime)

The results for the HAROs® also cast in summer are of the same order.

4.3. Thermal stresses in young concrete

The measured temperature and the secant modulus of elasticity, determined on the specimens were used to calculate the thermal stresses that develop in concrete blocks by means of a finite element program.

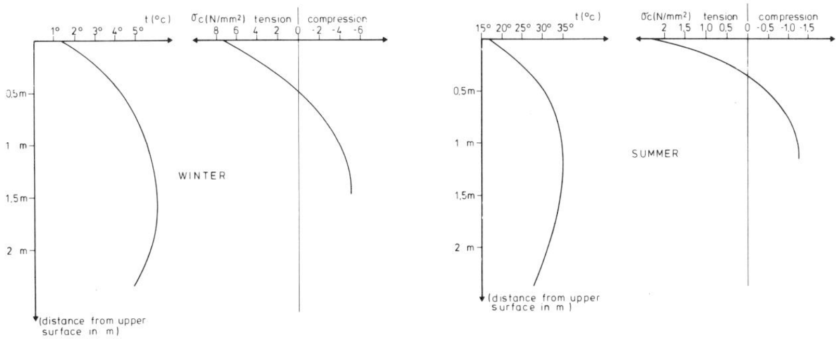

Figure 6 : Measured temperature distribution and calculated horizontal stress distribution along the vertical axis in 30 t cubes.

Fig. 6 gives the measured temperature distribution and calculated horizontal stress distribution along the vertical axis in a cubic block (30t).

The maximum tensile stress at the block's upper surface is obtained after 4 days in the winter and after 3 days in the summer.

It appears that the calculated tensile stresses due to hydration heat are comparable in magnitude with the tensile strength of the concrete at that time as shown in table 3. Consequently, cracks may arise at the block's surface.

	WINTER (4 days after casting)	SUMMER (3 days after casting)
calculated maximum tensile stress	0.75 N/mm^2	2.3 N/mm^2
tensile strength	0.4 N/mm^2	1.6 N/mm^2

Table 3 : Calculated maximum tensile stress versus available tensile strength of the young concrete

5. FROST RESISTANCE

After two severe winters, some blocks on the site showed a surface with exposed coarse aggregates due to the alternating action of freezing and thawing.

One could fear that with a repetition of that process the units could loose an important fraction of their weight or that an accelerated deterioration long before the presumed lifetime could take place and consequently cause stability problems.

In order to investigate the problem in detail, freezing and accelarated ageing tests were executed on specimens taken from the damaged blocks.

Frost resistance is related to the water-cement ratio and the air content of the concrete. Although the W/C ratio is below 0.40, the porosity at the surface, especially the upper surface, might be higher than in the core of the units. Tests on cores drilled near the surface showed water absorption ranging from 6 to more than 8 percent.

5.1. Freezing tests

A freezing test was executed on a large sample with a mass of 950 kg.

In the first stage the following environmental conditions were used :
- freezing chamber at -15°C
- laboratory conditions at 20°C
- salty water at 20°C.

A complete cycle has a duration of 24 hours. No breakage or deterioration of the blocks was observed after 3 cycles.

In the second stage, 15 cycles of two days each are performed. The first day at -15°C in the freezing chamber, the second day it is brought into salty water at 20°C. After the third cycle cracking of the block started and finally several branched cracks extending over the complete block were observed.

After the freezing test concrete quality was measured on cores drilled from the uncracked areas of the block. In comparison with reference values, a reduction of compressive and tensile strength is observed but the values are still acceptable taking into account the very severe conditions. The water absorption also decreased.

5.2. Accelerated ageing test

From several blocks, two cores are drilled. One as a reference specimen and the other one is used for the tests.

The test specimens are subjected to 12 ageing cycles over a total period of 24 hours, each consisting of consecutive periods of variations in temperature, ultraviolet and infrared radiation, artificial rain and freezing. After 12 cycles no breaking or deterioration was observed. Here too the concrete quality remains acceptable and the water absorption percentage is decreasing.

6. REMOVAL OF THE FORMWORK

Apart from thermal influences cracking may also occur while removing the formwork and as a result of drying-out of the surface.

About 16 hours after casting, demoulding takes place. To detach the mould from the young concrete, a hydraulic jack is used which pushes on a steel plate placed on top of the block. After the complete mould has slightly been lifted, it is further struck upwards in one single movement. During the whole operation, the mould is not opened.

Particularly attention is paid to avoid cracking of the young concrete, given its low resistance at the moment of striking. The steel plate is therefore as large as possible in order to distribute the load.

Special attention was paid to the curing of the concrete. A heavy duty membrane compound was used in order to keep concrete surfaces permanently wet for the whole of the curing period. Nevertheless, where cracks were noticed they were treated with a synthetic resin.

7. ALKALI SILICA REACTION

Another possible cause of crack formation is internal damage of the concrete.

As the first blocks were placed almost 10 years ago, it was found useful to make a petrographic analysis of the concrete in order to investigate whether or not internal damage occured due to the exposure to the marine environment. More particularly, alkali silica reaction (ASR) and ettringite formation are examined.

The ASR in concrete is a reaction which takes place in humid environment between aggregates (SiO_2 minerals) and the alkalis (Na, K) present in the cement in a humid atmosphere. The reaction results in a very expansive gel which causes the formation of cracks, which is accelerated when ettringite is formed in the cracks.

That aggregates that have been noticed for their sensitivity to alkalis belong to the granulometric fraction of sand and fine gravel. Some types of coarse aggregates also take an active part in the ASR e.g. flint stones, which generally have a white external layer. Also volcanic rock is known as a reacting aggregate.

The concrete, affected by ASR, is manufactured with a cement of the ordinary portland cement type.

As in many regions there is only a limited choice in aggregates and cement, this petrographic examination can provide useful information with respect to the durability of the concrete units.

8. RECOMMENDATIONS

The experience gained during the construction of the breakwaters at Zeebrugge has led to the development of the following proposals for obtaining high-quality concrete :

- A minimum cement content of 300 kg/m^3.
- Use of low heat cement.
- A low water-cement ratio (<0.4).
- Good compaction of concrete, not only in the core of the unit but also close enough to the mould.
- Careful curing (use of curing compound).
- A rigid formwork, which is easy to remove.

- Removal of the formwork at an appropriate time depending on the setting-time of the cement and on the ambient temperature.
- Dense and impermeable aggregates approved for use by prior testing.
- Limitation of the initial temperature of fresh concrete.
- Insulation of the young concrete may be envisaged in order to limit the temperature gradient.

Although the application of these proposals in practice could be limited by local technical or economic constraints, it generally does not increase production costs.

In spite of the application of these guidelines and the extra care taken, about 1% of the concrete cubes placed in Zeebrugge present significant cracking. Although these cracks are note expected to cause stability problems, they are carefully followed.

No cracks are noticed in the HARO®-units.

9. ACKNOWLEDGMENT

The investigations on the quality and the durability of the concrete were supported by the N.V. De Vreese-Simon, supplier of marine aggregates, the Research Centre of the Belgian Cement Industry and the Magnel Laboratory for Reinforced Concrete.

10. REFERENCES

(1) L.V. Van Damme, The main breakwaters of Zeebrugge, 18th International Conference on Coastal Engineering, Cape Town 1982.

(2) J. De Rouck, F. Wens, L. Van Damme, J. Lemmers, Investigations into the merits of the HARO®-breakwater armour unit.
1987 Proceedings of Coastal & Port Engineering in Developing Countries. Volume I, p. 1055-1068, Beijing, China 1987.

CHAPTER 157

Large verification tests on rock slope stability

J.W. van der Meer* and K.W. Pilarczyk**

ABSTRACT

A number of large scale tests on stability of rock slopes and gravel beaches is described and compared with small scale test results. The following topics are treated: the stability of a rock armour layer, the profile formation of a berm breakwater, the profile formation of gravel beaches, including ripple formation, and reflection and overtopping on rock slopes. The general conclusion is that scale effects could not be found.

INTRODUCTION

An extensive research program has been performed on static and dynamic stability of rubble mound revetments, breakwaters and gravel beaches. The first part was based on statically stable rubble mound breakwaters. Based on roughly 300 tests two new practical stability formulae were derived, including the wave period, storm duration, permeability of the structure and a clearly defined damage level.

The second part was concentrated on dynamic stability, i.e. the profile formation of rock slopes and gravel beaches under wave attack. About 150 tests were performed in this stage. The result was a computer program that can predict the profile for various wave conditions, including tides and storm surges.

All tests mentioned above were performed in small scale facilities with waves roughly between ten and twenty centimeters. It was stated at the beginning of the research program that the results derived in the small scale facilities should be verified on a larger scale in the Delta flume. These verification tests on scale effects are the subject of this paper.

For results on static stability one is referred to Van der Meer (1987a) and (1988a). Results on dynamic stability were presented by Van der Meer and Pilarczyk (1986) and Van der Meer (1987b). The complete research including set-up, analysis and data of tests was presented by Van der Meer (1988b).

MODEL FACILITY

The Delta flume has a length of 230 m, a width of 5 m and a depth of 7 m. The maximum significant wave height that can be generated is nearly two meters. The random wave generator was equiped with a system that measured and compensated for reflected waves from the structure. With this system standing waves and basin resonance were avoided.

* DELFT HYDRAULICS, P.O. Box 152, 8300 AD Emmeloord, The Netherlands
** RIJKSWATERSTAAT, P.O. Box 5044, 2600 GA Delft, The Netherlands

A surface profiler on a carriage was developed for the investigation. The profiler for the small scale facility was described in more detail by Van der Meer (1987a). The profiler of the Delta flume was constructed in the same way, but all length dimensions were increased by a factor 5.

RESEARCH TOPICS

Various aspects concerning rock slopes and gravel beaches were investigated in the Delta flume on a large scale. Static stability of rock slopes was investigated together in combination with run-up, run-down and reflection. A berm breakwater can be classified as initially dynamically stable and finally statically stable. A berm breakwater was tested with regard to profile formation and overtopping. Tests on dynamic stability of gravel beaches were divided into verification tests on profile formation and extrapolation tests with very small shingle.

STATIC STABILITY

The 300 tests on static stability of rock armour layers in the small scale facility resulted in two new practical stability formulae, given by:

$$H_s/\Delta D_{n50} * \sqrt{\xi_m} = 6.2\ P^{0.18}(S/\sqrt{N})^{0.2} \qquad (1)$$

for plunging waves ($\xi_m < \xi_m$ (transition)),

$$H_s/\Delta D_{n50} = 1.0\ P^{-0.13}(S/\sqrt{N})^{0.2}\ \sqrt{\cot\alpha}\ \xi_m^P \qquad (2)$$

for surging waves ($\xi_m > \xi_m$ (transition)), with:

$$\xi_m\ (\text{transition}) = \left[6.2\ P^{0.31}\ \sqrt{\tan\alpha}\right]^{1/(P+0.5)} \qquad (3)$$

where:

H_s = significant wave height
Δ = relative mass density = $\rho_a/\rho-1$
ρ_a = mass density of rock
ρ = mass density of water
D_{n50} = nominal diameter = $\left(W_{50}/\rho_a\right)^{1/3}$
W_{50} = 50% value of the mass distribution curve
ξ_m = surf similarity parameter = $\tan\alpha/\sqrt{s_m}$
α = slope angle
s_m = wave steepness = $2\pi H_s/gT_m^2$
T_m = mean wave period
P = permeability coefficient of structure:
- P = 0.1: impermeable core (lower limit)
- P = 0.4: most multi-layer breakwaters
- P = 0.5: permeable core
- P = 0.6: homogeneous structure (upper limit)

S = damage level = A/D_{n50}^2
S = 2-3: start of damage
S = 5-8: moderate damage
S = 8-15: filter layer visible (two layer system)
A = erosion area of cross-section

N = storm duration in number of waves

It was shown in the small scale tests that permeability of the structure had a large influence on stability. And especially the flow characteristics in a small scale model might be due to scale effects. Therefore tests in the Delta flume were concentrated on two different structures: a structure with a permeable core (P = 0.5) and a structure with an impermeable core (P = 0.1).

Scale effects in small scale tests on armour stability were discussed by various researchers. The effect of the Reynolds number on stability was investigated by Dai and Kamel (1969), Thomsen et al. (1972), Broderick and Ahrens (1982), Jensen and Klinting (1983), Sörensen and Jensen (1985), Shimada et al. (1986) and Burcharth and Frigaard (1987).
Although results are not througout consistent, lowest values for which no scale effects will be present are often set at $Re = \sqrt{gH_s}D_{n50}/\nu = 1.10^4 - 4.10^4$, with ν = kinematic viscosity. The range of Reynolds numbers used in the small scale tests was about $4.10^4 - 8.10^4$. Thomsen et al. (1972) found no scale effects for $Re > 2.10^5$. Shimada et al. (1986) suggest a value of $Re > 4.10^5$. The results of Thomsen et al. and Shimada et al. were both obtained in large wave flumes with monochromatic wave attack.

Eleven tests of the small scale series were repeated in the Delta flume and were scaled up according to Froude's law by a linear factor of 6.25. The stones had an average mass of W_{50} = 26.5 kg, a nominal diameter of D_{n50} = 0.214 m, a mass density of 2,700 kg/m³ and a grading of D_{85}/D_{15} = 1.38. The wave period was T_m = 4.4 s in all tests, the wave heights ranged from 0.7 - 1.2 m. The slope angle was 1 : 3. In total six test were performed on a permeable structure and five tests on an impermeable structure. The armour layer was rebuilt after each test which means that the data points in a wave height versus damage plot were independent.

The average of nine parallel profiles gives a clear picture of the structure and the damage. Figure 1 shows the profiles of the permeable structure and Figure 2 those of an impermeable structure, where the thin filter layer was placed on a concrete underlayer.

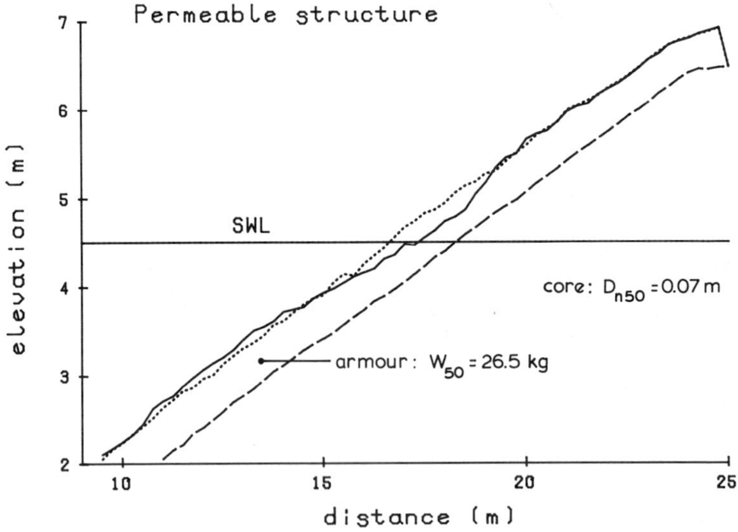

Figure 1 Cross-section with permeable structure in Delta flume

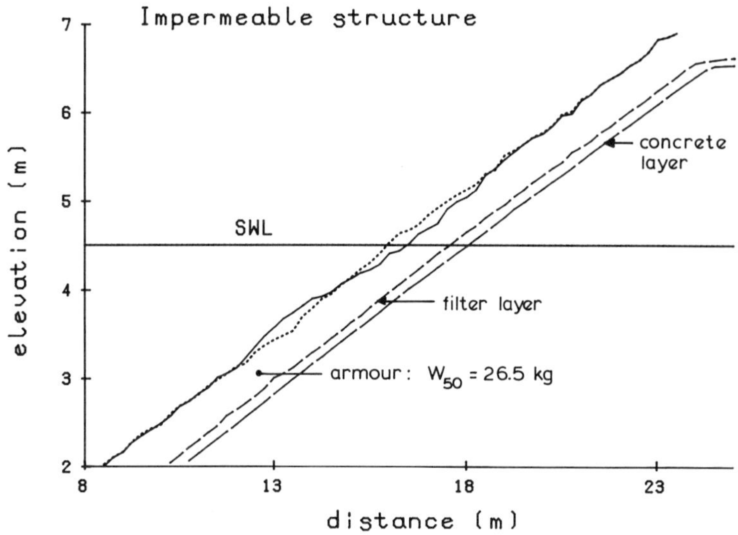

Figure 2 Cross-section with impermeable structure in Delta flume

Results of small and large scale tests can directly be compared in a dimensionless damage curve, where $H_s/\Delta D_{n50}$ is plotted versus the damage S. Figure 3 gives the results of the permeable core and Figure 4 the results of the impermeable core. Besides the different data points of the small and large scale tests, stability formula (1) was plotted in the figures (the curved line).

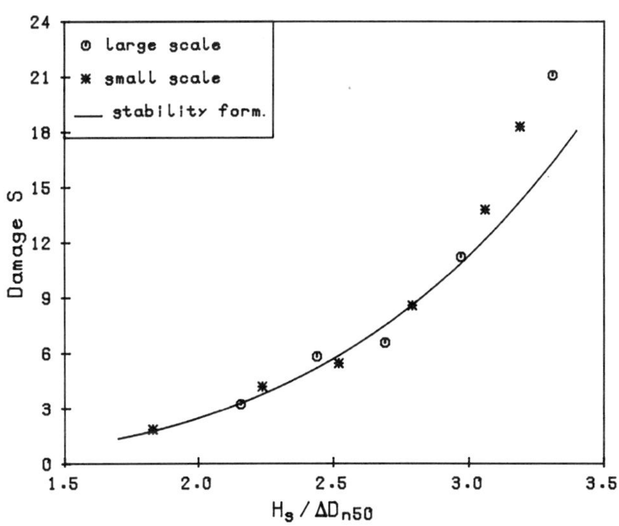

Figure 3 Results of permeable structure

From both Figures it can be concluded that the results of small and large scale tests are in close agreement. This confirms the validity of stability formulae (1) and (2). The stability curve fits well with the data, although some difference is found in Figure 3 for extreme damage levels, S > 12 (filter exposed).

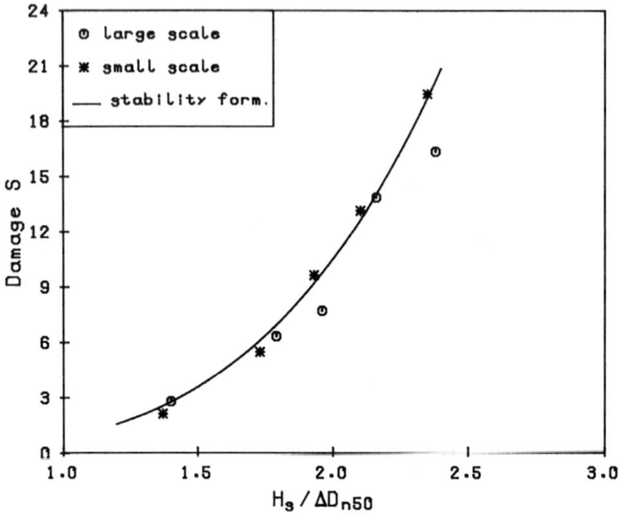

Figure 4 Results of impermeable structure

Comparison of reflection coefficients for small and large scale tests is shown in Figure 5 and coefficients are plotted versus the $H_s/\Delta D_{n50}$. The impermeable structure gives higher reflection coefficients in both cases than the permeable structure, where more energy is dissipated into the structure. The reflection for the impermeable structure is a little higher in the large scale tests. It is almost exactly the same for the permeable structure, the lower data points. In general the agreement is fair.

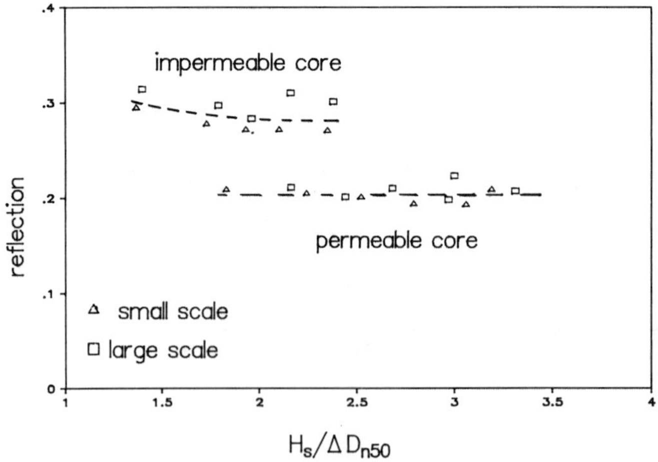

Figure 5 Reflection on rock slopes

The run-up was measured with a capacitance wire. Analysis showed, however, that the exceedance curves of the run-up gave a shift or nod about one meter above the still water level. Probably the wire was damaged at that point by a rolling stone. Therefore, run-up could not be compared with small scale tests.

The final conclusion of the large scale tests on static stability can be stated as follows: large scale model tests confirmed the validity of the small scale tests. The stability of an armour layer of rock was not influenced by the Reynolds number when Re was between 4.10^4 and 7.10^5. As these figures give the whole range of testing, the value of Re = 4.10^4 can only be regarded as an upper boundary for which scale effects on rock armour stability might start.

BERM BREAKWATER

The design concept of the berm breakwater is clearly presented by Baird and Hall (1984). A berm breakwater is a structure that behaves dynamically stable under the first storms and is statically stable further on. The berm breakwater of St. George in Alaska was extensively tested (Delft Hydraulics (1985)). The actual design was performed in a three-dimesional basin on a scale of 1 : 35. Tests on scale effects were performed before the design stage started. One test was performed in a flume on a scale of 1 : 35 and repeated in the Delta flume on a scale of 1 : 7.

The berm breakwater consisted of 2 - 10 tons rock (0.046 - 0.233 kg in the small scale wave flume and 5.8 - 29.2 kg in the Delta flume). The depth limited wave height at the structure was about H_s = 6 m (0.17 and 0.86 m, respectively). The deep water wave height was up to 11 m (0.31 m and 1.57 m). The test consisted of 8 steps with various wave height - wave period combinations, including long swell with peak periods up to 25 s.

Figure 6 gives one of the profiles of the Delta flume test, together with the corresponding profile of the small scale test, but scaled up with a factor 5. The profiles of both tests are very similar. They show the same amount of erosion and subsidence (due to lack of filter layer) at the berm and at the rear of the crest due to overtopping. Even the depth of the scour hole is the same, although shape and length of the scour hole are different.

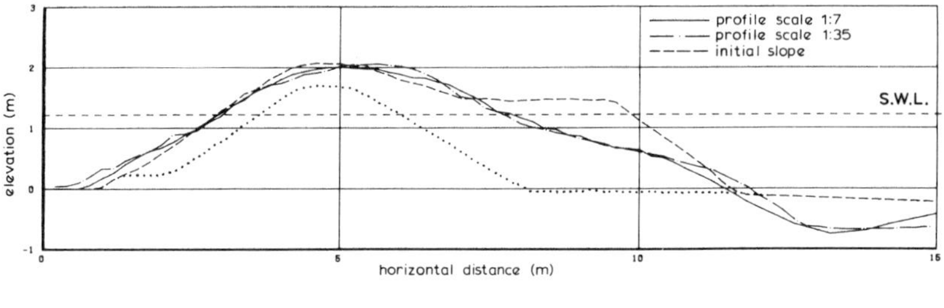

Figure 6 Profiles of berm breakwater in small and large wave flume measures on scale 1 : 7

The volume of erosion at the berm is given in Figure 7 versus the various steps of the test. The agreement between small and large scale test is very good up to step 5. In steps 6 to 8 the small scale test gives a little smaller amount of erosion.

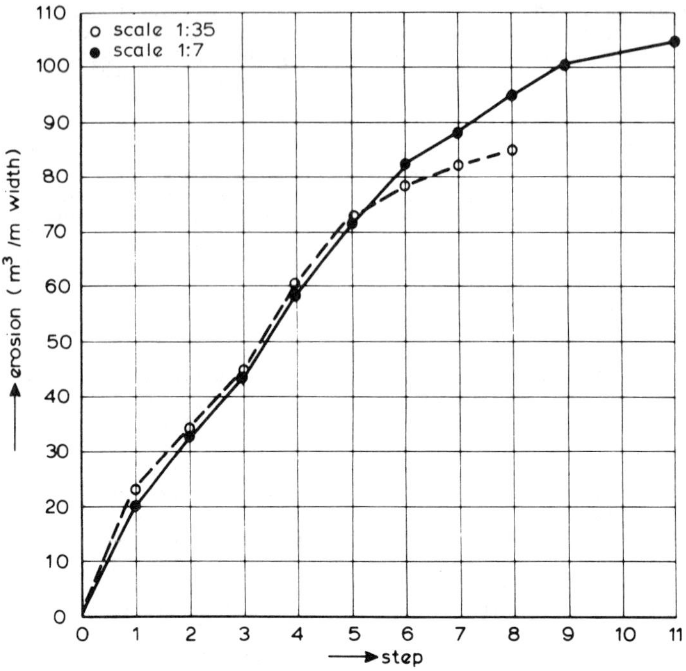

Figure 7 Erosion at berm developed during test

The comparison of overtopping in the berm breakwater tests (in prototype measures) is shown in Figure 8. The left plot shows the percentage of waves that reached the crest of the structure. The right plot gives the significant wave height behind the structure, generated by overtopping. The vertical axis gives the values of the Delta flume test and the horizontal axis the values of the small scale facility. The agreement of results is fair. Only the wave height behind the breakwater is consequently a little larger in the Delta flume.

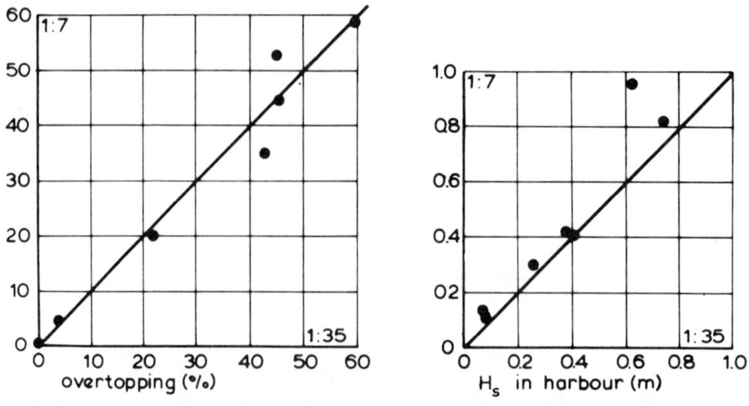

Figure 8 Results on overtopping

The complete analysis of the profiles, erosion and overtopping showed that no scale effects on stability were present in the small scale tests and it was proven that a scale of 1 : 35 for the three dimensional investigation would give reliable results.

DYNAMIC STABILITY

Dynamically stable structures are characterized by the forming of a profile under wave attack. In this case damage is not important, but the developed profile. Rock slopes can be classified as dynamically stable if the $H_s/\Delta D_{n50}$ value exceeds 3 - 4. Gravel or shingle beaches are described by $H_s/\Delta D_{n50}$ values in the order of 20 - 500.

Two topics were evaluated in the Delta flume. First the verification of some tests on gravel beaches, performed by Van Hijum and Pilarczyk (1982). Secondly the behaviour of very small shingle.

Two tests on gravel beaches were repeated on a 4.6 times larger scale. The small scale tests were presented by Van Hijum and Pilarczyk (1982 - tests 11 and 12). The diameter of the shingle was D_{n50} = 0.0187 m and the gradation D_{85}/D_{15} = 1.64. The initial slope was 1 : 5. The wave heights were respectively H_s = 0.77 and 1.00 m and the wave period was T_m = 4.3 s in both tests. This resulted in $H_s/\Delta D_{n50}$ values of respectively 26 and 33. The wave boundary conditions were not exactly the same in the small and large scale facility. Therefore it is very difficult to draw conclusions from a direct comparison of profiles. In stead of that the characteristic points of the profile, defined as dimensionless parameters by Van der Meer and Pilarczyk (1986), were compared.

Two of those dimensionless parameters are shown in Figures 9 and 10. The dimensionless crest height and step height (the point below the water level where the gentle upper slope changes into a steeper slope) are defined by respectively:

$$h_c/D_{n50}N^{0.15} \text{ and } h_s/D_{n50}N^{0.07}$$

The vertical axis of Figures 9 and 10 is the combined wave height - wave period parameter H_oT_o, defined by:

$$H_oT_o = H_s/\Delta D_{n50} * \sqrt{g/D_{n50}}\, T_m \qquad (4)$$

where:

$H_o = H_s/\Delta D_{n50}$ = dimensionless wave height parameter

$T_o = \sqrt{g/D_{n50}}\, T_m$ = dimensionless wave period parameter related to D_{n50}

Figures 9 and 10 show all test results for an initial slope of 1:5 for a range of H_oT_o = 100 - 4000. The two tests in small and large wave facility have different symbols. The vertical difference in the Figures gives the difference in wave boundary conditions, the horizontal difference is important for comparison. Figure 9 shows a good agreement of test results and established relationship (the curve). Figure 10 shows that most of the large scale tests have a little smaller value for h_s. But within the variation of the results it is acceptable. From the analysis of all profile parameters it followed that no scale effects could be found.

Finally the discussion of the tests on very small shingle. The shingle had a mean diameter of 4 mm which is almost at the transition to sand.

Such small diameters are difficult to scale for small scale tests as scale effects will definitely be present. Light weight material might be a solution, but will probably give scale effects above the water level (where the crest is formed) resulting in a too steep crest. Only almost prototype tests can give reliable information. The wave heights ranged from H_s = 0.7 to 1.7 m and the wave periods from T_m = 2.5 - 6 s. This resulted in $H_s/\Delta D_{n50}$ values of 90 - 260.

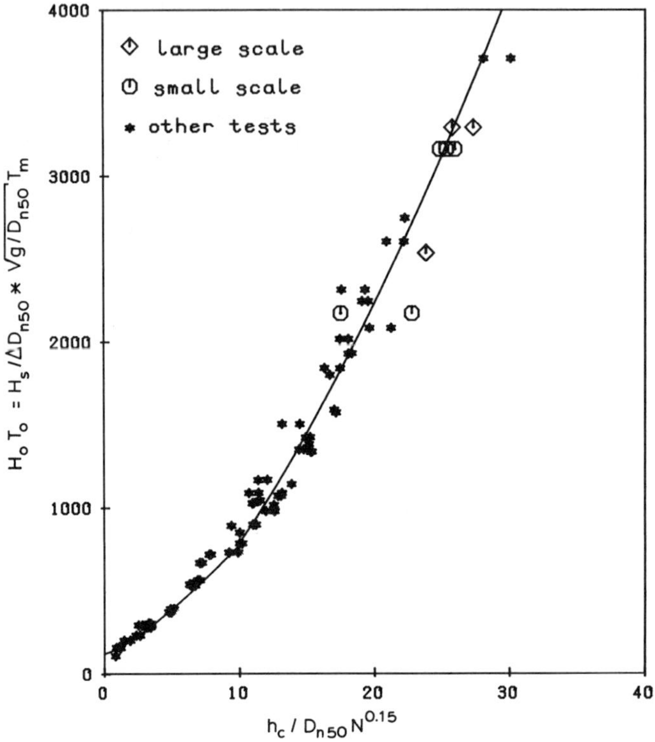

Figure 9 Evaluation of scale effects on crest height h_c

In total six tests were performed with this small shingle. The profiles were analyzed and used to develop a computer program which describes the profile formation of dynamically stable structures. Results were presented by Van der Meer (1986) and Van der Meer (1988b).

One phenomenon was never reported for shingle beach testing, although it is common for sand beaches. Namely the formation of ripples at the lower part of the slope, below the step. The authors are not aware of prototype measurements where ripple formation was reported. It will be difficult, however, to find ripples in prototype. First of all the sounding interval should be small enough to detect these ripples. Secondly, ripples are only formed for small shingle under high waves (storm conditions). They will be formed under the peak of the storm and probably with high water levels. The last part of the storm with lower water levels is able to flatten out the ripples formed before. And prototype profile measurements during the peak of the storm are rare.

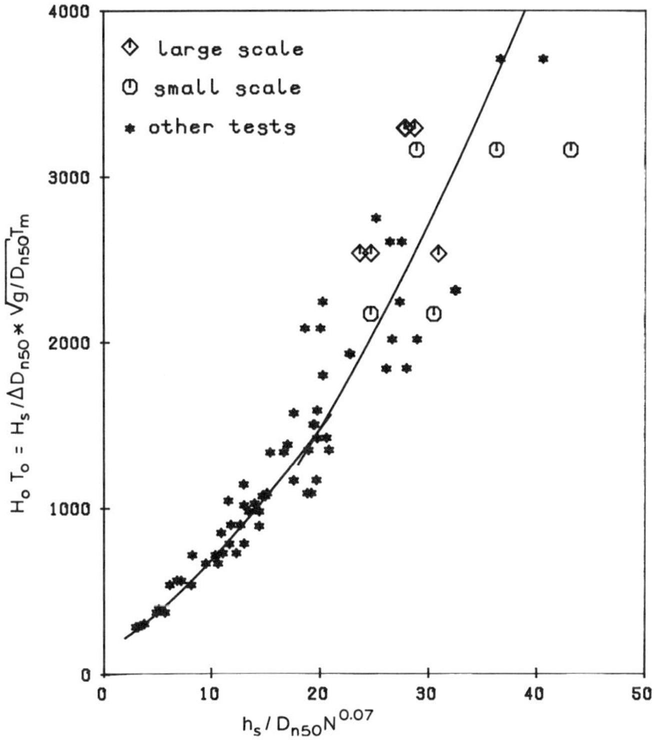

Figure 10 Evaluation of scale effects on step height h_s

The ripple phemomenon is very clearly shown in Figure 11 where 3 parallel profiles were plotted. The ripple height ranged between 10 and 40 centimeters and the ripple length ranged between 1 and 3 meters. As ripple formation has been studied for sand bottoms and slopes under wave attack and under flow conditions, it is interesting to compare the results for sand with very small shingle.

The best fit of the data was found with the results of Nielsen (1981). Nielsen defined the mobility number, ψ, which is a function of the water velocity at the bottom, u, calculated by linear wave theory:

$$\psi = u^2/\Delta gD \tag{5}$$

Ripple crest height, η, and ripple crest length, λ, were related to the wave amplitude, a = 0.5 H. Nielsen gave the following relationships between crest height and length and the mobility number ψ.

ripple height:

$$\eta/a = 0.275 - 0.22\sqrt{\psi} \tag{6}$$

ripple length:

$$\lambda/a = 2.2 - 0.345\psi^{0.34} \tag{7}$$

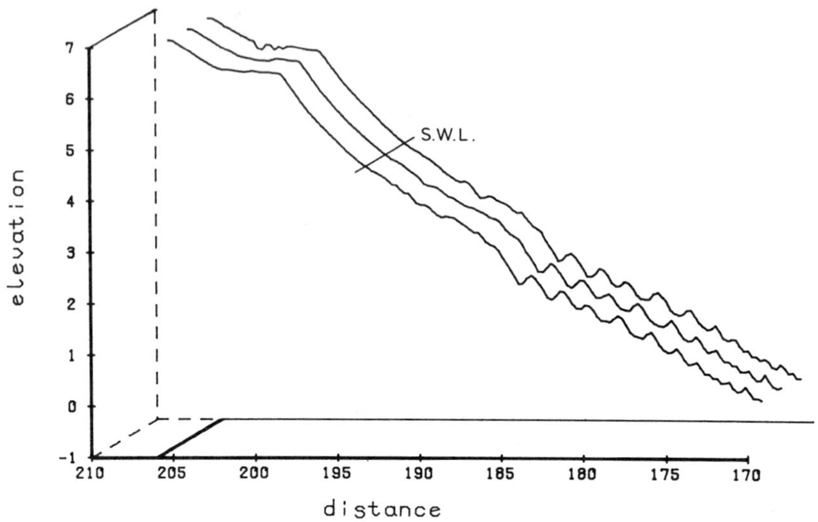

Figure 11 Ripple formation on 4 mm shingle

Equations 6 and 7 are shown in Figures 12 and 13 together with the results on shingle beaches. The general agreement is fair, although quite a lot of scatter is present. From the tests a boundary could be established for which ripple formation might start. This boundary is given by a $H_s/\Delta D_{n50}$ value in the order of 80 - 90 or a mobility number in the order or ψ = 10.

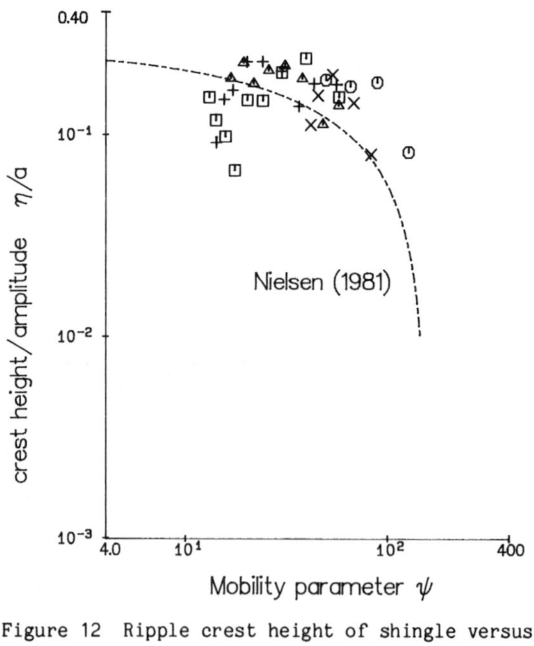

Figure 12 Ripple crest height of shingle versus mobility parameter ψ

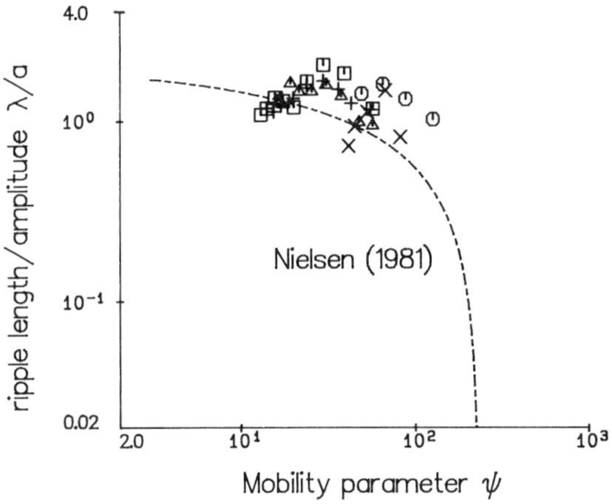

Figure 13 Ripple crest length of shingle beaches versus mobility parameter ψ

CONCLUSIONS

The conclusions derived from large scale testing of rock slopes and gravel beaches can be summarized as follows:

Small scale testing with wave heights higher than about 8 centimeters and for static stability with Reynolds numbers higher than 4.10^4, did not show scale effects for:
- rock slope stability
- berm breakwater stability
- reflection and overtopping
- profile formation.

A Reynolds number of 4.10^4 does not mean that lower values are not possible. Lower values were simply not used in the tests.

Finally, ripple formation could be described by sand ripple theory and a boundary could be established where ripple formation might start.

REFERENCES

Baird, W.F. and Hall, K.R., 1984. The design of breakwaters using quarried stones. Proc. of the 19th ICCE, Houston, USA, Ch. 173.

Broderick, L.L. and Ahrens, J.P., 1982. Riprap stability scale effects. CERC, Technical Paper No. 82-3, USA.

Burcharth, H.F. and Frigaard, P., 1987. On the stability of berm breakwater roundheads and trunk erosion in oblique waves. Seminar on unconventional rubble-mound breakwaters, Ottawa, Cananda. Proc. published by ASCE, special issue of Journal of WPC and OE.

Dai, Y.B. and Kamel, A.M., 1969. Scale effect tests for rubble-mound breakwaters. WES, Research Report H69-2, USA.

Delft Hydraulics, 1985. St. George Harbor, Alaska. Two-dimensional tests on scale effects. Three-dimensional tests on breakwater stability. Report on model investigations, M2102.

Jensen, O.J. and Klinting, P., 1983. Evaluation of scale effects in hydraulic models by analysis of laminar anf turbulent flows. Coastal Eng., 7, pp. 319-329.

Nielsen, P., 1981. Dynamics and geometry of wave-generated ripples. Journal of Geophysical Research, Vol. 86, No. C7, pp. 6467-6472.

Shimada, A., Fujimoto, T., Saito, S., Sakakiyama, T. and Hirakuchi, H., 1986. Scale effects on stability and wave reflection regarding armor units. Proc. 20th ICCE, Taipei, Taiwan, Chapter 165.

Sörensen, T. and Jensen, O.J., 1985. Reliability of hydraulic models of rubble-mound breakwaters as proven by prototype measurements. The Dock & Harbour Authority, March, pp. 155-157.

Thomson, A.L., Wohlt, P.E. and Harrison, 1972. Riprap stability on earth embankments tested in large- and small-scale wave tanks. CERC, Technical Memorandum No. 37, USA.

Van der Meer, J.W. and Pilarczyk, K.W., 1986. Dynamic stability of rock slopes and gravel beaches. Proc. 20th ICCE, Taipei, Taiwan, Ch. 125.

Van der Meer, J.W., 1987a. Stability of armour layers - Design formulae. Coastal Eng., 11, pp. 219-239.

Van der Meer, J.W., 1987b. Application of computational model on berm breakwater design. Seminar on unconventional rubble-mound breakwaters, Ottawa, Cananda. Proc. published by ASCE, special issue of Journal of WPC and OE.

Van der Meer, J.W., 1988a. Deterministic design of breakwater armor layers. Proc. ASCE, Journal of WPC and OE, Vol. 114, No. 1.

Van der Meer, J.W., 1988b. Rock slopes and gravel beaches under wave attack. Doctoral thesis, Delft University of Technology. Also published as Delft Hydraulics Communication No. 396.

Van Hijum, E. and Pilarczyk, K.W., 1982. Equilibrium profiles and longshore transport of coarse material under regular and irregular wave attack. Delft Hydraulics Publication No. 274.

CHAPTER 158

ANALYSIS OF 42-TON DOLOS MOTIONS AT CRESCENT CITY

Thomas R. Kendall, M. ASCE*

ABSTRACT

Photogrammetry has been applied to make precise measurements of post-construction displacements of 38.2-metric ton (42-ton) dolosse on the outer breakwater at Crescent City, California. Data from two storm seasons are currently available from this monitoring program which was initiated in November 1986 and which is expected to continue at least through the next three winters. Supplemented by conventional land surveys, wave measurements, aerial inspections, and side scan sonar underwater imagery, observations from the photogrammetric monitoring have led to several preliminary conclusions regarding regions of relatively large dolos movement, dolos nesting, the dependency on breakwater slope of uprush or drawdown dominated armor unit movement, and the relative importance of armor unit movement and boundary conditions in dolos breakage. This paper describes the methods used for, and results obtained from, this monitoring effort along with the application of these results to a Corps study (Howell, 1988) whose objective is to develop structural design criteria for the dolos concrete armor unit.

INTRODUCTION

The high incidence of breakage encountered in large concrete dolos armor units has led the Corps of Engineers to initiate a field data collection program aimed towards developing a structural design procedure for dolosse. The overall program being carried out includes field monitoring as well as physical modeling and numerical simulation of forces experienced by dolosse. The primary focus of this paper is on the results obtained to date from aerial photography and photogrammetric field monitoring of dolos displacements and breakage.

Field data have been collected from the outer breakwater at Crescent City, California (Figure 1), which was rehabilitated in 1986 by placing 680 steel

* Civil Engineer, U.S. Army Corps of Engineers, San Francisco District, 211 Main Street, San Francisco, CA 94105-1905

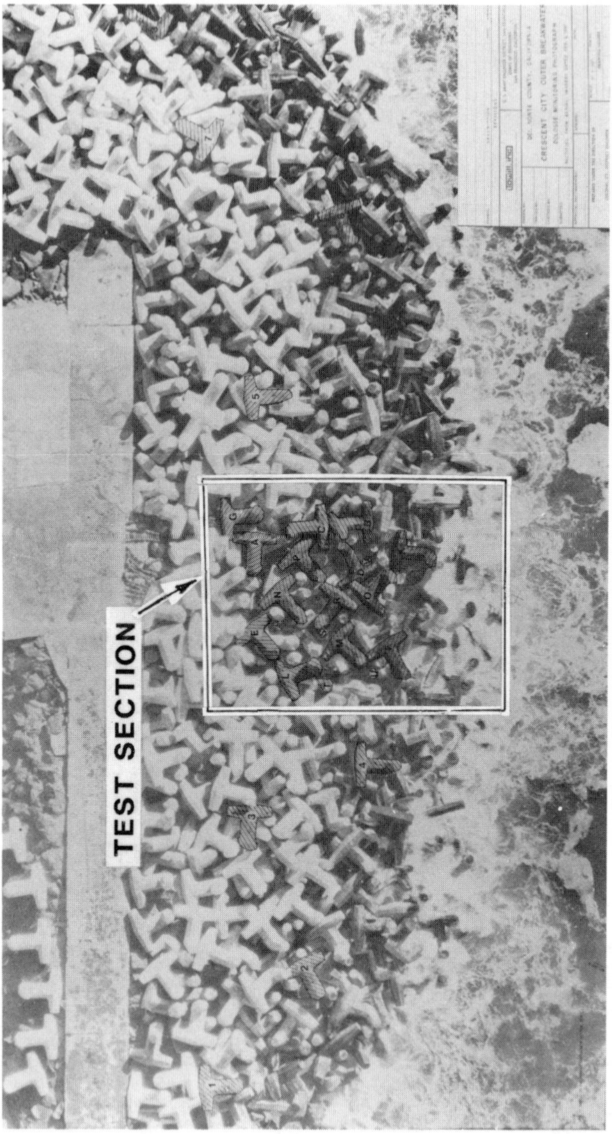

Figure 1 - Dolos section of the Crescent City Outer Breakwater showing the location of targeted dolosse and the dolos test section.

fiber-reinforced, 38.2-metric ton (42-ton) dolosse over an existing dolos section on the breakwater. The dolos section, which is approximately 100 meters in length, encompasses the seaward end of the trunk plus the bend to the breakwater's easterly dogleg extension. The section consists of a minimum of two dolos layers on a variable slope that averages between 4H:1V and 5H:1V and extends to toe depths ranging from 7.5 to 9 meters. Above the mean waterline, structure slopes are generally flatter than the average; below the waterline, the slopes are typically steeper. The bathymetry offshore of the structure is highly complex with numerous rock outcroppings.

The design wave height proposed by Hales (1985) for the rehabilitated section is a depth-limited breaking wave with a height of approximately 10.5 meters and a peak period of 15 seconds. Wave periods up to 20 seconds, however, are not uncommon and can occur with almost comparable height. The physical model study used to design the rehabilitation (Baumgartner, Carver, and Davidson; 1985) subjected the structure to depth-limited regular waves with a range of periods (up to 21 seconds) over a simulated extreme tidal cycle (maximum tide approximately 3 meters MLLW). Only during the highest tide and for the longest period waves were wave heights near 10.5 meters encountered. This condition occurred during one-twelfth of the design "hydrograph" used in the model.

As part of the 1986 rehabilitation, 17 of the new dolosse were instrumented with internal strain gages, including four with accelerometers, to measure dynamic and static structural response during placement and subsequent exposure to storm waves. These dolosse are arranged in a rectangular test section (14 in the top and three in the bottom layer) on the trunk of the breakwater. Data from the instrumented dolosse test section have been collected during the two winter storm seasons which have followed construction. Howell (1988) further discusses the features of, and data obtained from, these test section dolosse.

An important element of the field data collection program has been the assessment of dolos movement and overall structural stability, both within the test section and within the surrounding dolosse. To make this assessment, photogrammetric surveys of the breakwater's dolos section have been conducted approximately monthly over the two post-construction winter-storm seasons. While aerial photography and photogrammetry alone can only provide information about displacements, much can be inferred about dolos movements when these surveys are conducted repeatedly and supplemented by observations of dolos abrasion and breakage.

Unlike the highly unique and costly strain measurements, which were terminated after the second season, the relatively inexpensive photogrammetric monitoring, supplemented by conventional land surveys, wave measurements, aerial inspections, and side scan sonar underwater surveys, will continue to be conducted at least through the next three winters.

The data obtained from these surveys are being used to answer the following specific questions:

1. What are the differences between the laboratory projected stability of the structure (i.e. the results of physical model studies conducted during the design of the rehabilitation project) and the observed field performance?

2. What are the full scale patterns of dolos response, i.e. the nesting and consolidation processes, dominant movement types, and regions, if any, of excessive movement or breakage?

3. Is breakage necessarily associated with armor unit movement and vice versa? and

4. What quantity of dolos breakage may compromise the breakwater's integrity?

In addition to answering these questions of interest to the field engineer, the photogrammetric monitoring allows the research engineer to visualize the bigger picture when interpreting results obtained from internal strain measurements within the test section. For example, changes in the static loading experienced by an instrumented dolos can often be traced to dolos settlement or to a shift in the boundary conditions evident in the photogrammetric record. The photogrammetric and breakage inspection records also allow comparisons to be made between the performance of test section dolosse and the remainder of the visual dolos field in an effort to assess how representative the test section results are. Additionally, information on dolos orientation collected from these surveys is being used to develop finite element models of the test dolosse, and, along with extensive post-construction data on the breakwater contours and surrounding bathymetry (2-foot contouring), to construct an "upgraded" physical model of the breakwater so that more refined comparisons between full scale and hydraulic model movements and loadings can be made.

PROCEDURES

Of critical importance to any photogrammetric monitoring program is the proper positioning and accurate surveying of the ground control targets which are used as the basis of scaling, levelling and coordinates of the photogrammetric stereo models. Additionally, targets must be clearly marked and identified on the objects to be monitored, in this case dolosse. As it is not practicable to maintain targets on all visible dolosse, only select dolosse (primarily those in the test section) have been targeted. By printing rectified aerial photography of the entire dolos field onto a transparent mylar base, overlays are made each month so that movements of non-targeted dolos can also be visualized (to the nearest 15 centimeters in plan). If, in the future, a more refined and three-dimensional assessment of the movements of non-targeted dolos is deemed necessary, such an assessment can be made using the photography already taken by preparing photomaps outlining all visible dolosse using the same stereo model set up for the targeted dolosse. This photomapping approach was used to monitor dolosse movements on the jetties at Manasquan Inlet, New Jersey (Gebert and Clausner; 1984).

The twenty-six dolosse which have been marked in this monitoring project are shown in Figure 1. Eighteen of these are located in the instrumented section and eight are distributed uniformly throughout the remainder of the dolos field. Twenty two of these dolosse are located in the upper dolos layer and have been marked with three targets each which allows their movement to be described with six degrees of freedom. The remaining four, which are located in the lower dolos layer of the test section, have only one clearly visible surface each which has been targeted for monitoring.

All targets have been painted on the dolos surface with special anti-fouling paint to discourage marine growth from diminishing target visibility. Marine growth, however, has persisted on those dolosse nearest the waterline. To avoid excessive repainting of these relatively inaccessible targets, epoxy paints have recently been used. Three coats (epoxy primer, epoxy paint, and an epoxy polyurethane finish coat) have been applied to these targets for a total wet thickness of approximately 20 millimeters.

Each target is in the form of a cross, with legs approximately 8 centimeters in width by 45 centimeters in overall length. Each cross is also identified with its appropriate designation (e.g. A-1, A-2, etc.) which is visible in the photography, thereby providing positive identification as to which point is being measured.

Photography was initially taken with a 306.42-millimeter (12-inch) focal length mapping camera from an altitude of 366 meters (1200 feet) above sea level, resulting in a photo scale of 1:1200 (or 1 inch equals 100 feet). However, in order to provide more accuracy in the vertical dimension, a Wild 152.09-millimeter (6-inch) focal length precision mapping camera is now being used to take photography from an altitude of 183 meters (600 feet) above sea level, resulting in the same direct photo scale of 1:1200. While these lower altitude flights improve photogrammetric accuracy, they also make maintaining the same photo center from survey to survey quite difficult. As a result, the images produced from these lower altitude flights are not as suitable for visually tracking the gross movements of dolosse. Therefore, an additional high altitude flight is now made as part of each survey so that similar photo centers can be maintained from survey to survey and so that rectified and scaled photographic images which are produced from these each month at an enlarged scale of 1:120 on mylars are suitable for overlaying and movement detection.

A Cessna Tu-206 photo plane is the platform from which the aerial photography is taken. Black and White aerial film (Kodak Double-X Aerographic film #2405) is used. Two photographs, taken approximately 110 meters apart in a straight line pass which bisects the angle formed by the elbow of the breakwater, form the photogrammetric stereo model from which the dolos measurements are made. Additional photo passes are made, one to the left and one to the right of the designed flight line to offer additional stereomodels to choose from if shadows or overhanging dolos prevent 3-D viewing of any target.

A Qasco fully analytical (computer driven) stereoplotter is currently being used to form the stereomodels and make the individual measurements. Each measured set of tri-ordinates is recorded in a data file which is entered into a database and compared with previous measurements to detect position changes. For dolos which have not broken, results are then interpreted by converting individual target movements to dolos centroid translations along, and rotations about, each axis of the California State Plane Coordinate System. Local axes are defined on each dolos to facilitate the recording of dolos orientation.

Computer generated three-dimensional drawings are used to illustrate the dolos movements measured. Time-series graphics for each surveyed dolos are presented as "exploding diagrams" projecting from their respective dolos indicated on a base map (aerial photograph) of the dolos field (Figure 2). Multiple sheets are required to track all moving dolosse. A separate sequence is maintained for the phogrammetric surveys and the periodic land surveys which are conducted on targeted dolosse as a check on the photogrammetric results. The example sheet shown in Figure 2 has a graphic entry for the earlier two surveys (lower two rows) for all dolosse shown. However, between the second and third surveys, only dolos `L' moved beyond the theshold used for precipitating an updated orientation graphic (15 centimeters movement on any target) and, therefore, the third (upper) survey row only shows an entry for dolos `L'.

Periodic ground inspection and repainting of control survey and dolos targets is a requisite for a successful photogrammetric monitoring project, as is periodic ground surveying to compare the accuracy of the two methods.

Comparisons to date between tri-ordinates generated from simultaneous ground and photogrammetric surveys of 70 dolos targets show agreements between the two methods of approximately 3 centimeters (total three-dimensional vectorial displacement).

In addition to monitoring the movement of individual units in the dolos section, unit breakage is periodically inspected via helicopter. These aerial inspections of breakage are conducted over the entire dolos field at the time of each photogrammetric survey. Breakage is

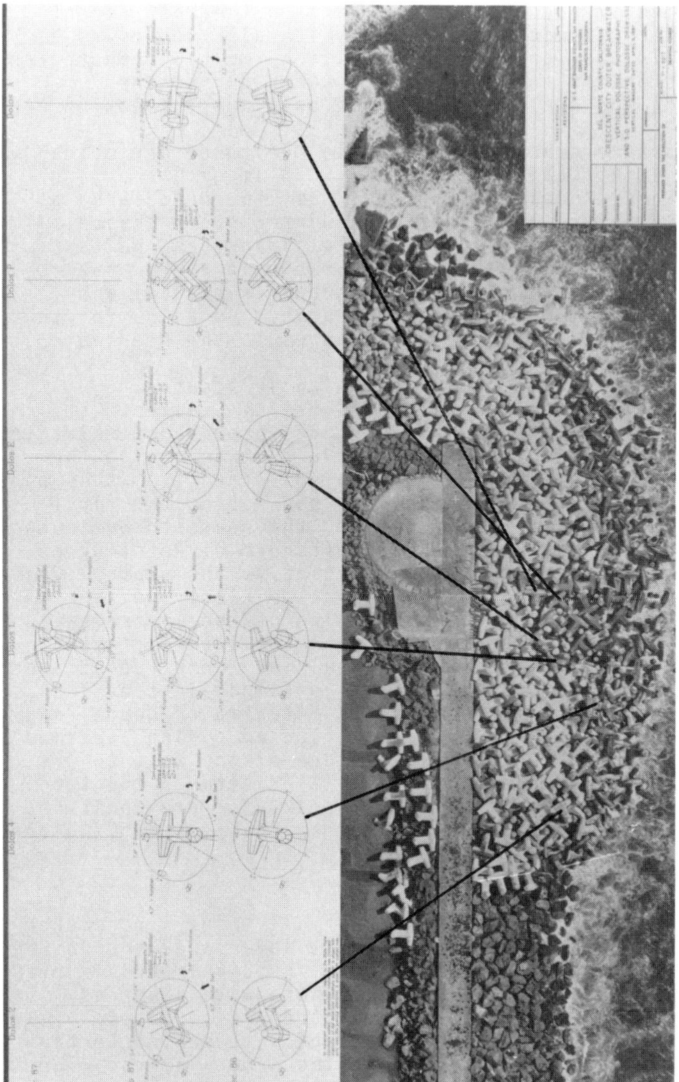

Figure 2 - "Exploding Diagrams" used to illustrate dolos displacement with time.

inventoried by recording all breaks on the most recent aerial photographic base sheet, which also serves as a breakage inspection map. In addition, multiple photographs (at least one of each break) are taken during the inspection and are indexed on the breakage inspection map.

Subsurface conditions are also monitored periodically via side scan sonar surveys. An EG&G Image Correcting Digital Sonar Model 260 is towed past the dolos area at least once before and after each winter storm season. Individual dolos breaks, structure slopes, and units displaced are not currently measurable in a repeatable, quantitative way with side scan sonar (Kucharski and Clausner; 1988), however, these surveys do allow for the detection of gross subsurface changes to the breakwater.

The Point St. George Buoy, which is located approximately ten kilometers offshore of Crescent City in approximately 60 meters of water, has provided a relatively continuous record of the local wave climate during the monitoring period. However, data collected from this buoy contained a nearly two-month long gap during the first winter season; during this period, the nearest continuously operating wave buoy was located offshore of Point Arena, which is approximately 240 kilometers to the south. While other wave gages and buoys were set up closer to the breakwater as part of the monitoring of the instrumented dolos test section, these only operated intermittently during the first season and the bulk of their second season data is yet to be reduced. Limited comparisons of the available nearshore wave data (12 meter water depth) and the Point St. George Buoy data collected during a period of typical long-period storm swells out of the west to southwest indicate strong similarities between the two records. Water levels near the breakwater are monitored using the tide station maintained at Crescent City Harbor.

OBSERVATIONS

Figure 3 relates the timing of all surveys conducted during the first two seasons to the wave power encountered. The dates of both land surveys and photogrammetric surveys are indicated on the figure. Offshore wave power as defined on the figure has been plotted (as opposed to wave energy) because of the importance of wave period to the stability of the highly porous dolos armor layers. The "reservoir effect" of the pores, as explained by Burcharth and Thompson (1983), makes dolosse more vulnerable to long-period waves.

The first winter season was relatively moderate and the second season was more severe. Based on the available wave and tidal records for the area, it is concluded that a small percentage of individual waves each season (more during the second) approached design magnitude. The design storm, however, was not experienced.

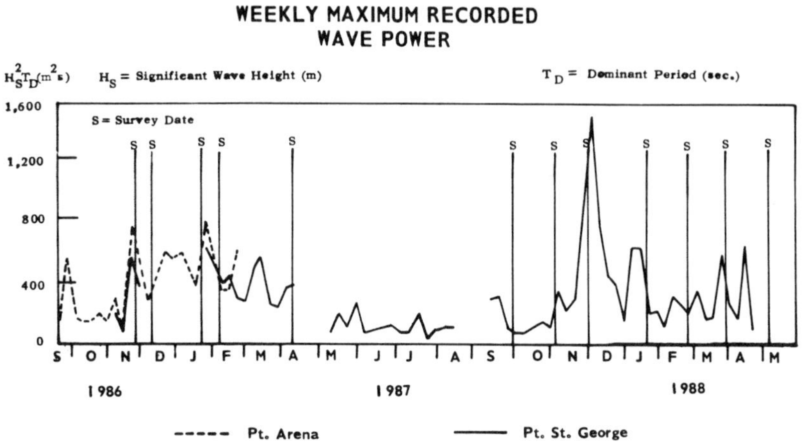

Figure 3 - Timing of dolos surveys relative to offshore wave power.

<u>Movement</u>

Dolos movement between the December 1986 and January 1987 surveys was the largest recorded to date; the subsequent storms of that first winter season and the more severe second winter season failed to produce substantial movement, even of those dolosse that had been previously mobile. This appears to indicate that the dolosse have consolidated and nested into a more stable matrix as a result of these earlier storms. The consolidating movement of the targeted dolosse produced by the series of storms that occurred in December 1986 and January 1987 is summarized in Table I.

TABLE I

CRESCENT CITY OUTER BREAKWATER
TARGETED DOLOSSE MOVEMENTS
FROM 10 DECEMBER 1986 TO 20 JANUARY 1987

	AVERAGE	MAXIMUM
TRANSLATION	27 cm	218 cm
(SETTLEMENT)*	(4 cm)	(29 cm)
ROTATION	6.1°-Z	34.5°-Z

* Settlement is a component of the total translation.

The movement of over 2 meters (almost half the unit dimension (15 feet, or 4.6 meters) of a 38.2-metric ton (42-ton) dolos) experienced during this period by dolos 'L' represents the largest recorded movement to date of all visible dolosse. Generally speaking, the upper layer dolosse experience larger movements than the more constrained lower layer dolosse; therefore, movement among the targeted dolos (85 percent of which are in the upper layer) is higher than the visible dolos field average.

Comparisons of rectified aerial photography showing the entire visible dolos field (including those dolosse visible in lower layers) indicate that most units to date have moved less than 15 centimeters, about 95 percent have moved less than 50 centimeters and about 99 percent have moved less than 120 centimeters. Of the one percent (or five of the approximately 500 visible dolosse) observed to have moved over 120 centimeters, three of these dolosse were broken dolosse, one was dolos 'L' which did not break but experienced the largest measured displacement, and one had disappeared beneath the waterline making it impossible to quantify the magnitude of its displacement. The broken dolosse which moved over 120 centimeters may have broken prior to experiencing the majority of the displacement evident between photographs.

To date, the dominant dolos movement has been upslope with slight settling plus rotation about the vertical or z-axis (yaw). Yaw was also the predominant dolos rotation observed by Gebert and Clausner (1984) at the Manasquan Inlet jetties. This type of rotation probably occurs more frequently because the upper layer dolosse are not constrained laterally as much as they are vertically; note that the settlement shown in Table I is a relatively small component of the total translation experienced.

Dolosse settlement at Crescent City was greatest during the early storms; more recent surveys show that detectable vertical motions are approximately balanced, i.e. an equal amount of elevating and settling occurs. This tends to support the notion that the dolosse have already consolidated and nested into a more stable matrix. Initial settling of the dolosse (and the associated tilting, wedging, and general constraining) corresponds to the observed test section trend of increased static loading through time for most all dolosse measured.

Jensen of the Danish Hydraulic Institute indicates that for structure slopes milder than 3.5H:1V, wave run-up dominates and the net armor unit movement is upslope; for steeper slopes, wave run-down dominates and the net armor unit movement is downslope (Sorensen and Jensen; 1986). Considering the structure slopes at Crescent City (typically flatter than 4H:1V in the zone of wave impact), the observed upslope movement of the dolosse is not surprising. Other recent full-scale monitoring of smaller dolosse on the East Coast and Great Lakes of the United States has documented downslope movements (wave run-down dominated) at structures with slopes ranging from 1.5H:1V to 3H:1V (Pope and Clark, 1983; Gebert and Clausner, 1984; Pope, 1988; Gebert, 1988). Full-scale observations of dolosse within the United States appear to support Jensen's conclusion that the critical slope which determines whether armor will move upslope with wave run-up or downslope with wave run-down is near 3.5H:1V.

At Crescent City, the greatest movement was recorded on the upper slope of the centrally located test section and in the vicinity of the waterline. Spatially averaged movements within the test section are quite comparable to those found outside the test section, however, the region of high movement within the test section is generally located further upslope. Higher movement near the waterline is not surprising and has also been observed in dolosse at Cleveland Harbor (Pope and Clark; 1983). A possible explanation for the observed high movement on the upper slope of the test section is the existance of a slight contour dip or trough in the test section region of the breakwater, as well as the fact that many of the dolosse placed in this low spot, had initial boundary conditions that did not hinder sliding. It has also been suggested, based on the historic pattern of breakwater damage, that the test section may be located in an area where wave focusing occurs.

Breakage

Since placement two years ago, six broken dolosse have been detected. Five of the dolosse broke during the nesting storm sequence; one broke during the peak storms of the second season. Seven post-placement breaks are visible (one dolos broke in two places). Five of the breaks occurred at the dolos shank-fluke interface; two occurred mid-fluke. Breakage appears to be concentrated in the relatively mobile regions of the breakwater near the waterline. Breakage, however, did not occur within the equally mobile upper test section. All visible broken dolos pieces have remained stable, and breakage, thus far, appears to not have significantly compromised the breakwater's integrity.

An examination of the data collected by the instrumented dolosse indicates that these large dolosse can be loaded to near their structural capacity while simply resting in the dolos matrix under static loading conditions, leaving little residual strength for any dynamic loads. A variety of support conditions exist for the dolosse and some are clearly more detrimental than others. In one example, a dolos was placed in a cantilevered support condition after adjacent dolosse migrated approximately one meter upslope and vertically constrained its horizontal fluke. The dolos eventually broke. There is no indication of impact loading to the cantilevered portion of the dolos nor of any displacement of the constrained horizontal fluke, which leads to the conclusion that failure was caused primarily by poor boundary conditions accompanied by some wave-generated pulsating drag loads on the cantilevered portion of the dolos. Dolos movement, however, was ultimately responsible for the failure in that it caused the critical shift in boundary conditions supporting the dolos which ultimately failed. In another example, a dolos suffered a break across its horizontal fluke after enduring continuous impacts. The horizontal fluke of this dolos had been wedged between two adjacent vertical flukes which constrained the upslope movement and yaw of the dolos. During the first season's nesting storms, approximately one meter of upslope movement was experienced by these dolosse which were in contact. Minor residual movement occurred thereafter and the wedged horizontal fluke finally broke during the peak storms of the second season.

These two examples of poor boundary conditions can be contrasted with the boundary conditions found in the dolos test section, where the placement of each dolos was coordinated by a field engineer who stood among the dolosse and radiod instructions to the crane operator. During placement, most test-section dolosse experienced a three-point landing. On the other hand, due to tong clearances, the different levels of experience of the crane operators used on the job, and the fact that a field engineer did not radio instructions to the crane operators during dolosse placement outside the test section, about 70 percent of the dolosse placed outside the test section experienced a two-point landing. These dolosse were then either wedged or clamped in by subsequently placed dolosse or were left to rock or tilt during nesting until wedged into a final support condition. This most likely resulted in boundary conditions which were more detrimental than those which resulted from the typical three-point landings experienced by dolosse in the test section.

These differences in boundary conditions may explain why dolosse have yet to break in the test section. While large movements occurred in the test section, these movements were of a much less detrimental nature and typically involved some sliding permitted by a dolos's exclusively vertical supports. Furthermore, these large movements occurred in a region relatively far from the waterline where dolosse are probably loaded less frequently.

While breakage has typically been associated with some amount of movement, either of the dolos which breaks or of adjacent units, it has not necessarily been associated with significant movement and vice versa. At the Manasquan Inlet jetties, Gebert and Clausner (1984) also observed that the largest displacements experienced were by dolosse which did not break. The dolosse at Manasquan Inlet weigh 14.6 metric tons (16 tons). For the even larger dolosse at Crescent City (which may have even less residual strength), it appears, thus far, that the extent to which movement causes a detrimental shift in boundary conditions is more important than the absolute magnitude of the movement itself.

Model and Field Performance Comparisons

To the extent that comparisons can be made between the observed performance of the breakwater physical model and the observed field performance of the structure itself, qualitative similarities between the two exist. Only very preliminary comparisons can be made for this two-year time period during which a small percentage of individual waves approached design height but the design storm, per se, was not encountered. Several differences also exist between model and prototype construction and reporting techniques.

The initial model used for designing the rehabilitation of the breakwater used fewer dolosse on a generally more uniform slope than the prototype and subjected the structure to regular waves. In addition, the model only quantified dolos translations which exceeded the dolos unit length and rotations which were greater than 180 degrees, i.e. only those dolosse which had clearly moved out of their original location were recorded. Dolos breakage could not be modelled. By contrast, field monitoring thus far has recorded dolos translations which were all less than half the dolos unit length and rotations which were all less than 40 degrees; in addition, dolos breakage has been observed in the field. Furthermore, the model study basin could be drained to assess underwater displacements with much more clarity than the side scan sonar imagery collected in the field.

Inspite of these differences, it does appear, thus far, that the use of trenching and buttressing (with stone) along the southern perimeter of the dolosse has produced a stable dolosse-stone transition in the field, similar to the model study finding. On the other hand, the northern dolosse-stone transition region, which proved to be a problem area in certain laboratory tests, has been an area where two above-water dolosse have broken and where side scan sonar records indicate some migration of the toe. Physical model testing of the breakwater is again being proposed so that various buttressing schemes, etc. may be tested for their effectiveness in stabilizing the northern dolosse-stone transition.

CONCLUSIONS

Photogrammetry has proven to be a reliable, safe, accurate, economical and repeatable method of monitoring the movement of dolos armor units on the outer breakwater at Crescent City. The photographs produced in this monitoring effort serve as a permanent record of high resolution images providing excellent visual documentation of dolos movement and breakage, even among non-targeted dolosse.

The principal observations made from the limited data available to date are summarized below:

> The greatest movement of dolosse is on the upper slope of the centrally-located dolos test section and in the vicinity of the waterline. The movement on the upper slope is thought to result from the existence of a slight contour dip or trough in this region of the breakwater and because many of the dolosse placed there had initial boundary conditions that did not inhibit sliding.
>
> Spatially averaged movement within the dolos test section is quite comparable to that found outside of the test section; however, the region of high movement within the test section is located further upslope.
>
> The dominant dolos movement is upslope with slight settling plus rotation about the

vertical axis (yaw). Upslope movement is a consequence of the breakwater's mild slope which causes wave run-up forces to dominate.

Comparison of the interim results of the Crescent City monitoring program with the results of two other full-scale photogrammetric dolos monitoring programs conducted in the U.S. suggests that the critical slope which determines whether armor units will move upslope with wave run-up or downslope with wave run-down, is near 3.5H:1V.

Storms which occurred early during the first-post construction winter season have produced the largest dolos movements to date. Reduced movement during subsequent storms indicates that the dolosse have consolidated and nested into a more stable matrix.

Breakage, while typically associated with some amount of movement, has not necessarily been associated with significant movement and vice versa. For the large dolosse at Crescent City (which can have little residual strength), the extent to which movement causes a detrimental shift in boundary conditions appears more important than the absolute magnitude of the movement itself.

The photogrammetric monitoring of dolos movement will continue at Crescent City for at least the next three years. Results from the photogrammetric monitoring conducted during the first two post-construction winters are an integral part of the Corps intensive data collection effort which will be used to develop a structural design procedure for the dolos concrete armor unit.

ACKNOWLEDGMENTS

The work described in this paper has been conducted as part of the Crescent City Prototype Dolosse Study sponsored by the U.S. Army Corps of Engineers (USACE) San Francisco District. Funding for the forth and fifth years of photogrammetric monitoring will be provided by the USACE Monitoring Completed Coastal Projects (MCCP) program. The author would like to acknowledge contributions of the following individuals to this effort: Bill Brick, Richard Lou, and Jim Wilkerson of USACE San Francisco District, the staff of Richard B. Davis Company, Ken Meme of Towill Inc., David Gilhousen of the National Data Buoy Center, Jeff Gebert of USACE Philadelphia District, Gary Howell, D.D. Davidson, and Jeff Melby of the Coastal Engineering Research Center, USACE Waterways Experiment Station, and Denise, my wife.

REFERENCES

Burcharth, H.F., and A. Thompson, 1983, "Stability of Armour Units in Oscillatory Flow," Proceedings of Coastal Structures '83, ASCE, New York, NY.

Baumgartner, R.C., R.D. Carver, and D.D. Davidson, 1985, Breakwater Rehabilitation Study, Crescent City Harbor, California, Technical Report CERC-85-8, U.S. Army Engineer Waterways Experiment Station, Vicksburg, MS.

Gebert, J.A., and J. Clausner, 1984, "Photogrammetric Monitoring of Dolos Stability, Manasquan Inlet, New Jersey," Coastal Engineering, 1984 Proceedings of the International Conference, ASCE, New York, NY.

Gebert, J.A., 1988, U.S. Army Engineer District, Philadelphia, PA, personal communication.

Hales, L.Z., 1985, Water Wave Refraction/Diffraction/Shoaling Investigation, Crescent City, California, Miscellaneous Paper CERC-85-3, U.S. Army Engineer Waterways Experiment Station, Vicksburg, MS.

Howell, G., 1988, "Measurement of Forces on Dolos Armor Units at Prototype Scale," Coastal Engineering, 1988 Proceedings of the International Conference, ASCE, New York, NY.

Kucharski, W., and J.E. Clausner, 1988, Underwater Inspection of Coastal Structures Using Commercially Available Sonars, Technical Report REMR-CO-108, U.S. Army Engineer Waterways Experiment Station, Vicksburg, MS.

Pope, J., and D.R. Clark, 1983, "Monitoring of a Dolos Armor Cover; Cleveland, OH," Proceedings of Coastal Structures '83, ASCE, New York, NY.

Pope, J., 1988, U.S. Army Engineer Waterways Experiment Station, Vicksburg, MS, personal communication.

Sorensen, T., and O.J. Jensen, 1986, "Experience Gained From Breakwater Failures," Developments in Breakwaters, Proceedings of the conference on Breakwaters '85, Thomas Telford Ltd, London.

CHAPTER 159

WAVE FORCE ON BREAKWATERS WITH CONCRETE BLOCK MOUND

by
Masataro Hattori[1], Keiji Inagaki[2],
Yuuji Noguchi[3], and Taiji Endo[4]

ABSTRACT

Analytical theory is derived for prediction of the wave force acting on a upright structure armored by concrete block mound of rectangular shape. The theoretical treatment starts from the linear long wave theory and the flow resistance in the mound is described by a linearized friction law. As a practical application of the theory, approximate method for the structure with sloping block mound is proposed. Experiments are conducted to verify the theory and approximate method.

1. INTRODUCTION

Mound-type structures consisting of rubble stones and concrete blocks have been attracted many attentions as an porous structure protecting effectively a harbor from the action of rough seas. In the past decades, analytical and experimental studies have been conducted to understand the reflection and transmission characteristics of the porous mound structure(Madsen and White, 1975; Massel and Mei, 1977). Excellent review of the previous studies up to 1972 is given by Sollitt and Cross(1975).

In addition, it has been well known among coastal engineers that mound-type structures backed by a caisson or vertical seawall reduce considerably wave forces acting on the backed-up structure as well as wave run-up and overtopping(Goda, 1985). In Japan, such the mound structures comprising randomly placed artificial concrete blocks are very common and have been constructed to reduce sever wave actions against the structures. Figure 1 shows an idealized typical cross-section of such a block mound breakwater of composite-type (or Japanese-type)(PIANC, 1988).

When the block mound is used for this purpose, it is especially important for coastal engineers to estimate the effectiveness of a given block mound on the wave force

---

1)Professor, Department of Civil Engineering, Chuo University, Bunkyo-ku, Tokyo 112. JAPAN
2)Associate Engineer, Yokkaichi City, Mie Prefecture 510.
3)Manager, R & D Section, Hydraulic Laboratory, Nippon Tetrapod Co., Ltd., Tsuchiura, Ibaraki Prefecture 300. JAPAN
4)Director, Hydraulic Laboratory, Nippon Tetrapod Co., Ltd., Tsuchiura, Ibaraki Prefecture 300. JAPAN.

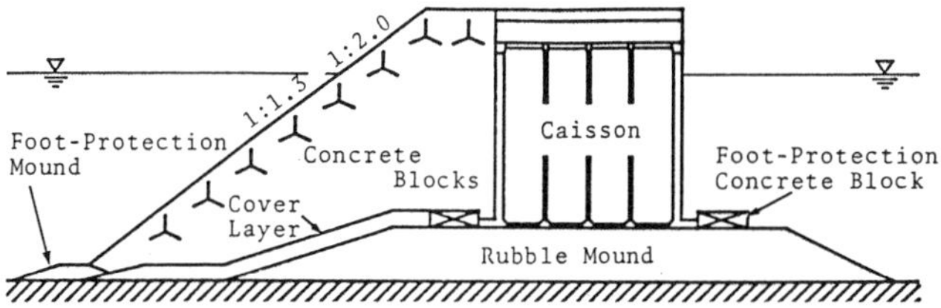

Fig. 1 Idealized typical section of a breakwater with concrete block mound.

reduction for a breakwater design. Incident wave energy is dissipated not only by wave breaking and frictional effect on the seaward slope of block mound, but also by resistance associated with oscillatory flows inside the mound. The interaction of incident waves with the block mound is a very intricate problem and relates various physical factors, such as wave properties, structural characteristics of the mound, and size and shape of concrete blocks.

Lack of adequate knowledge of these wave-structure interactions associated with the composite-type block mound breakwater has hindered from developing the analytical theory for predicting the wave force and reflection characteristics. As a result, laboratory model experiments have been recognized as one of possible and reliable approaches to the solution of such complicated problems. In Japan, more than 20 programs of laboratory experiments using regular and irregular waves have been made to deepen understanding on hydrodynamic behaviors of the composite-type structure with block mound , and to find a feasible standard design procedure for such structures. In spite of many intensive efforts, however, any empirical methods for predicting the wave force exerted on the backed-up structure have not been established yet with general confidence.

The principal aims of this study are as follows:

1. To develop an analytical theory for predicting the wave force acting on the caisson-type breakwater with rectangular block mound, and
2. To propose an approximate method for the estimate of wave force on the breakwater with sloping block mound.

An analytical solution for the present problem is derived under the following assumptions:

(1) Long waves are normally incident on the breakwater.
(2) The wave motion outside and inside the block mound can be described by linear long wave theory.
(3) The block mound structure with a rectangular cross section is homogeneous and isotropic.
(4) The flow resistance of the mound can be expressed by a linear relation in terms of the seepage velocity .

As Madsen and White discussed (1975), it is considered that wave conditions adopted in breakwater designs commonly falls in the long wave range of the spectrum. The linear long wave approximation, therefore, is used in the theoretical treatment.

An experimental program is carried out to confirm the validity of the theory and approximate method, and to determine empirical friction law.

2. THEORY

2.1 ANALYTICAL MODEL DERIVATION

The rectangular block mound with a width of l is located between x= 0 and x = l. The coordinate system is shown in Fig. 2. With the first and second assumptions, the governing equations for the wave motion outside the block mound ($x \leq 0$) are given as Eqs. (1) and (2).

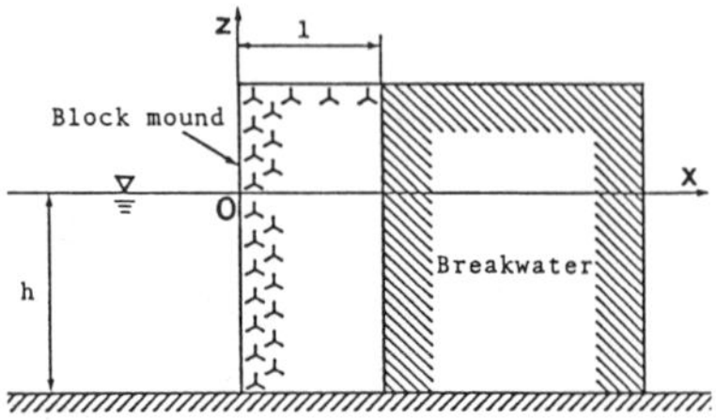

Fig. 2 Coordinate system.

$$(\partial \eta / \partial t) + h(\partial u / \partial x) = 0, \tag{1}$$

and

$$(\partial u / \partial t) + g(\partial \eta / \partial x) = 0, \tag{2}$$

in which η is the free surface elevation above the still water level, h is the constant water depth, u is the horizontal particle velocity, g is the gravitational acceleration, and t is the time.

Following to Kondo(1970) and Madsen and White(1975), the motion inside the block mound ($0 \leq x \leq l$) can be written as

$$(\partial \eta / \partial t) + (h/\epsilon)(\partial u / \partial t) = 0, \tag{3}$$

and

$$(S/\epsilon)(\partial u / \partial t) + g(\partial \eta / \partial x) + (f\omega/\epsilon)u = 0, \tag{4}$$

in which u is the seepage velocity, ϵ is the porosity of the mound, and ω is the angular frequency of the incident waves. The parameter S in the momentum equation (4) is given by Eq. (5) and represents effect of the unsteady fluid motion within the mound.

$$S = 1 + C_m(1+\epsilon). \tag{5}$$

And f is the constant friction factor of the resistance law linearized using the Lorentz' Principle of equivalent work, and is written as

$$f\omega/\epsilon = \alpha + \beta|u|. \tag{6}$$

Two terms on the right hand of Eq. (6) represent laminar and turbulent flow resistance, respectively.

Since the governing equations are linear, we assume a periodic solution for the free surface elevation and horizontal particle velocity.

$$\eta(x,t) = \mathrm{Real}[\zeta(x)e^{i\omega t}], \tag{7}$$

and

$$u(x,t) = \mathrm{Real}[U(x)e^{i\omega t}], \tag{8}$$

in which $i = \sqrt{-1}$, and the amplitudes of η and u are functions of x only.

Outside the block mound, incident waves and partially reflected waves from the breakwater form a standing wave field. The flow inside the mound consists of two waves propagating in the positive and negative x-directions.

With the two matching conditions of the free surface elevation and particle velocity at the front surface of the block mound (x=0), and with the condition of no flow through the caisson(x=l), the final expressions for the wave motions both outside and inside the block mound are obtained as follows:

For $x \leq 0$ (outside the block mound):

$$\eta(x,t) = (H_i/2)\{e^{-ik_0x} + K_R e^{ik_0x}\}e^{i\omega t}, \tag{9}$$

and

$$u(x,t) = U_i\{e^{-ik_0x} - K_R e^{ik_0x}\}e^{i\omega t} \tag{10}$$

For $0 \leq x \leq l$ (inside the mound):

$$\eta(x,t) = H_i \cosh\{ik(x-l)\}/\{\cosh(ikl) + \tau \sinh(ikl)\} \times e^{i\omega t}, \quad (11)$$

and

$$u(x,t) = -2U_i \tau \sinh\{ik(x-l)\}/\{\cosh(ikl) + \tau \sinh(ikl)\} \times e^{i\omega t}, \quad (12)$$

in which K_R is the reflection coefficient, as given by Eq.(13),

$$K_R = |H_r|/H_i = |\{1 - \tau \tanh(ikl)\}/\{1 + \tau \tanh(ikl)\}|. \quad (13)$$

The wave number k_0 and horizontal velocity amplitude of incident waves U_i are given by the long wave theory as

$$k_0 = 2\pi/L_i = \omega/\sqrt{gh}, \quad (14)$$

and

$$U_i = (H_i/2)\sqrt{g/h}. \quad (15)$$

k is the complex wave number of waves inside the mound, and is given by

$$k = k_0(S - if)^{1/2}, \quad (16)$$

and τ is the parameter for simplifying the theoretical expression.

$$\tau = \epsilon(S - if)^{-1/2}. \quad (17)$$

H_i(real) and H_r(complex) are the incident and reflected wave heights outside the block mound.

Since the wave number k is complex, the theory indicates that waves propagating to the positive x-direction through the block mound is attenuated exponentially due to the frictional effect of the mound. Thus, we presume that the wave is perfectly reflected at the vertical front surface of caisson, and the wave force is calculated by a linear wave theory.

With the linear long wave assumption, the maximum horizontal wave force per unit width of the caisson, F_{Hm}, is written by Eq. (18), a function of the maximum surface elevation at the caisson, η_{ml}.

$$F_{Hm} = (\rho g/2)[\,2h\,\eta_{ml} + \eta_{ml}^2\,]. \quad (18)$$

By putting x = l in Eq. (16), η_{ml} is written as

$$\eta_{ml} = \mathrm{Real}[H_i/\{\cosh(ikl) + \tau \sinh(ikl)\}]. \quad (19)$$

2.2 DISCUSSIONS ON THE THEORETICAL RESULTS

Practical calculations are usually made with the known mound geometry and porosity as well as the incident wave condition. Parameter S given by Eq. (5), depends not only on the shape and size of blocks, but also on the unsteady flow characteristics inside the mound. Measurements of the added mass coefficient C_m in Eq. (5) are so difficult that it is necessary to assign an appropriate value to the parameter S before discussing the theoretical results. Madsen and White

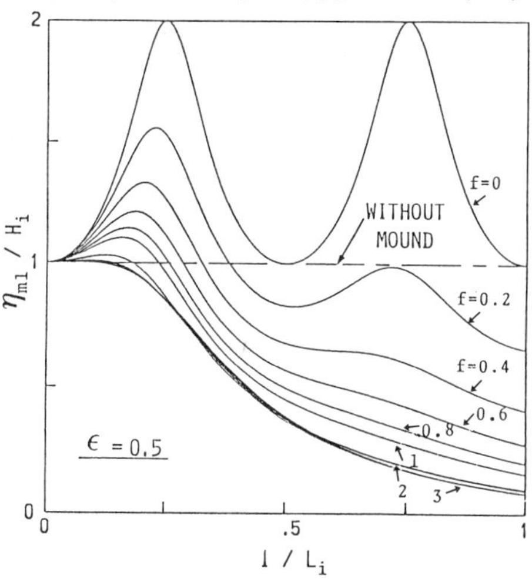

Fig. 3 Theoretical relation between η_{ml}/H_i and l/L_i.

(1975) discussed possible values of S and estimated to be $1 \leq S \leq 1.5$. They approximated it to be unity for practical circumstances. We shall, therefore, take S = 1.

The maximum wave force F_{Hm} is a function of η_{m1} only, so that dominant characteristics of the mound to the wave force reduction can be learned more clearly from the the general behavior of the solution of Eq. (19) for η_{m1}, in stead of Eq. (18) for F_{Hm}. An example of the numerical solution of Eq. (19) is presented in Fig. 3, showing the relative maximum surface elevation η_{m1}/H_i as a function of the relative mound width l/L_i for various values of the linearized friction factor f and for a mound porosity of $\epsilon = 0.5$. The broken line of $\eta_{m1}/H_i=1$ corresponds to the perfect reflection at a vertical breakwater without block mound.

When f = 0, it refers to as a fictitious mound, in which only the net flow cross section is reduced to a half of that seaside of the mound. From Eqs. (11) and (19), the maximum surface elevation for f=0 is given as

$$(\eta_{m1}/H_i)_{f=0} = [\cos^2 k_0 l + \epsilon^2 \sin^2 k_0 l]^{-1/2} \qquad (20)$$

Cyclic change of η_{m1}/H_i with respect to l/L_i indicates that standing waves with a node and antinode at the common boundary of outside and inside the block mound are formed in conditions of l/L_i=0.25 and 0.50. With the increase of the friction factor f, the cyclic change is attenuated, and η_{m1}/H_i tends to approach constant values with increase of the relative mound width, when f>2. According to Eqs. (6) and (12), increase in the friction factor may be a result either of decreasing the mound porosity or of increasing the incident wave height (Sollitt and Cross, 1972). From Fig. 3, therefore, a very interest process is found that when small waves are incident on the mound breakwater, the block mound gives rise to stronger wave force than without the mound. This results from the two following facts: (1) decrease in the net flow cross section, and (2) the resonance of the standing wave systems between outside and inside the block mound.

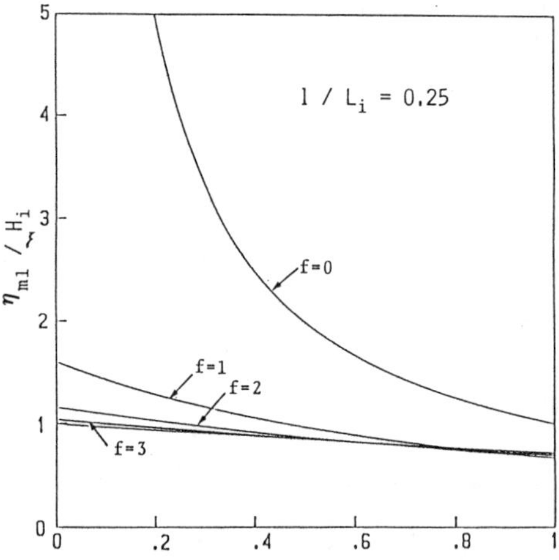

Fig. 4 Theoretical relation between η_{m1}/H_i and ϵ.

In a practical planning of coastal and harbor structures of block mound type, of a particular interest is the dependency of mound porosity on the wave force. Figure 4 is a numerical result showing the dependency of the mound porosity on the relative maximum free surface elevation η_{m1}/H_i for l/L_i=0.25. Since actual range of the mound porosity is estimated to be 40 % to 60 %, it is noticed that the amplitude of standing waves at the caisson or the wave force does not depend noticeably on the mound porosity for practical circumstances, in which the friction factor becomes usually larger than 2.

In addition to the wave force reduction, another important function of the block mound is the damping of reflected waves from the breakwater. Figure 5 is an numerical example of the relationship between the reflection coefficient K_R and relative mound width l/L_i for ϵ =0.5. As the relative mound width increases, the reflection coefficient tends to approach to a constant value for the case when the friction factor is larger than 2. Reduction in the wave reflection by the block mound can not be expected, if the relative width is smaller than 0.1.

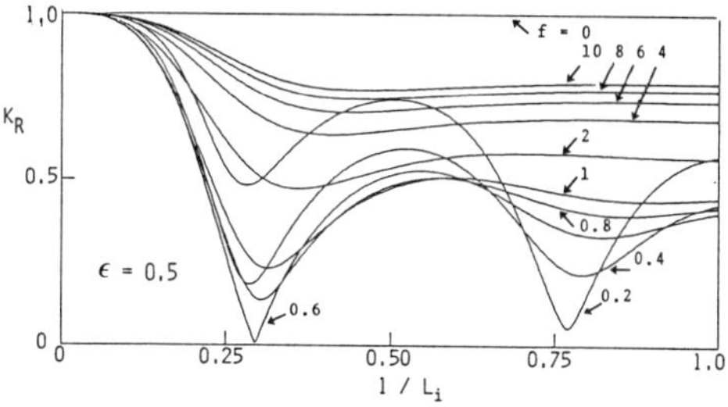

Fig. 5 Theoretical relation between K_R and l/L_i.

3. EXPERIMENTAL SETUP AND CONDITIONS

3.1 EXPERIMENTAL EQUIPMENT

Experiments were performed in a glass-walled wave flume of 1 m wide, 1 m deep and 50 m long. Figure 6 shows general arrangement of the experimental equipment. Waves were produced by a piston-type wave generator installed at one end of the wave flume. The wave flume was separated into two sections with a plywood bulkhead and equipped a steel beach of a slope of 1/50.

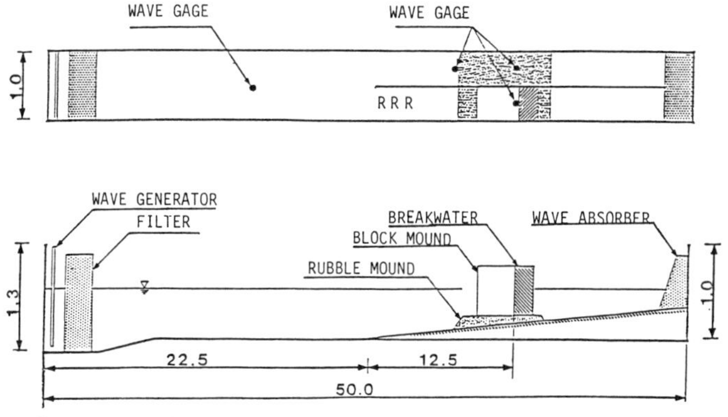

Fig. 6 General arrangement of experimental equipment. (Units:m)

In the test section with 0.50 m wide, model breakwaters comprising a steel rectangular parallelepiped as model caisson [0.466×m 0.20 m×0.40 m, weight in air: 20 kg] and rectangular or trapezoidal block mound was placed on a gravel mound of 5.0 cm high. The other section of the flume was used for measurements of incident waves as progressive waves at the breakwater location.

3.2 MODEL BLOCK MOUND

Rectangular model mounds with three different widths were comprised by model tetrapods of three different sizes, whose physical dimensions are described in Table 1. The block was contained in wire screen cribs, shaped as rectangular parallelepiped. In situ porosity was checked up by the total number of blocks used and volume of the crib.

Additional experiments were conducted to examine approximate method for predicting the wave force on the constructed with the same blocks used for the rectangular

Table 1 Dimensions of rectangular mounds(Tetrapods).

RELEVANT BLOCK SIZE*		B(cm)	5.0, 8.2, and 11.3
MOUND	WIDTH	l(cm)	22.5, 30.0, and 45.0
	HEIGHT	(cm)	40.0
IN SITU POROSITY		ϵ(%)	50.0

[*) The relevant block size is defined as $B = (abc)^{1/3}$, in which a, b, and c are long, intermediate and short sizes of a block]

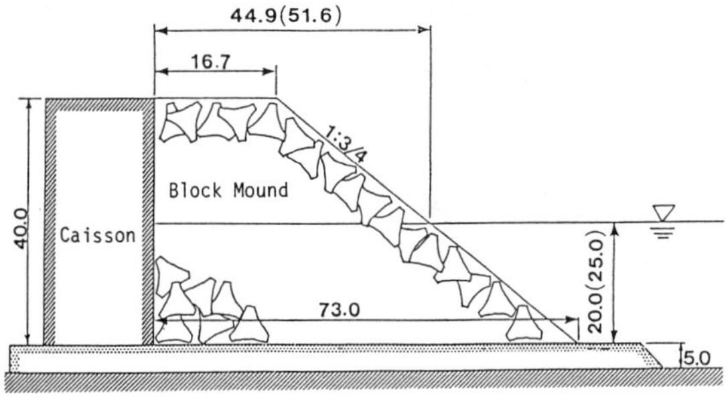

Fig. 7 Cross-section of sloping block mound(Units: cm).

mound. Figure 7 illustrates the dimensions of the trapezoidal block mound, constructed with the same blocks used for the rectangular model mound.

3.3 WAVE AND WAVE FORCE MEASUREMENT

Measurements of water free surface elevations at various locations, denoted in Fig. 6, were made using capacitance wave gages. In particular, simultaneous measurements of the free surface elevation at three offshore locations denoted by R in Fig. 6 were used for the computation of reflection coefficient of the breakwater. Resolution of incident and reflected waves was performed by means of the spectral component analysis with a simultaneous record of the free surface elevation at three adjacent locations of known relative distances (Goda, 1985).

Total horizontal wave forces acting on the model caisson were measured by using a specially designed force meter of cantilever-type, made of a steel square bar(40 mm×

70 mm × 1070 mm)(Hattori et al., 1979). The force meter was fixed on the backside of the caisson, and had a natural frequency of 13 Hz in a case when the caisson was placed in a water depth of 0.20 m. Outputs of the water surface elevation and wave force were recorded on a 7-channel analogue recorder. The data were digitized by an A-D converter at a sampling frequency of 100 Hz for extensive processing by computer.

Wave conditions are given in Table 2.

Table 2 Experimental conditions.

INCIDENT WAVE HEIGHT	H_i(cm)	1 to 17
PERIOD	T(s)	1.5 and 2.0
WATER DEPTH	h(cm)	20 and 25
BLOCK REYNOLDS NUMBER	$Re_B = BU_i/\nu$	$(1\ \text{to}\ 70)\times 10^3$

4. COMPARISONS OF THEORY AND EXPERIMENT

4.1 EMPIRICAL RELATIONSHIP FOR THE FRICTION FACTOR

The theoretical solutions indicate that the friction factor representing the hydraulic characteristics of the mound is a physically fundamental parameter for wave energy damping within the mound structures. Hence, the determination of the friction factor is a very important problem for the predictions of the wave force. To simplify the analytical treatment, the friction factor was assumed to be constant. In addition, the solutions were obtained by approximating the parameter S to be unity, because the added mass coefficient C_m of blocks could not be evaluated experimentally. It is, therefore, necessary to take into account of unsteady effect of the fluid motion in the determination of the friction factor.

In the previous studies, however, friction factor of rubble stones and concrete blocks were examined from experiments of steady and unidirectional flows in open channel or pipe (Shuto and Hashimoto, 1970; Sollitt and Cross, 1972). Based on some preliminary considerations and experiments, the friction factor was evaluated from measured maximum free surface elevation at the front of caisson breakwater, η_{ml}, with aids of Eqs. (11) and (16).

Since the friction factor is assumed to be independent of the time and space, we examine validity of the friction law of Eq. (6), in which the local seepage velocity u is replaced with the mean horizontal velocity inside the mound U_m, as given by Eq. (18).

$$U_m = (1/lT)\int_0^l\int_t^{t+T}|u|\,dx\,dt = (8/3\pi l)\int_0^l u_r\,dx, \qquad (18)$$

in which u_r is the real part of the seepage velocity given by Eq. (12). Substituting Eq. (9) into Eq. (18), U_m for a special case of f=0 is written as Eq. (19).

$$U_m = (2\epsilon U_i/k_0 l)[\cos^2 k_0 l + \epsilon^2 \sin^2 k_0 l]^{-1/2}\cdot[1-\cos k_0 l], \qquad (19)$$

Figure 8 shows relationships of $f\omega/\epsilon$ and U_m, and indicates that the linear friction law similar to Eq. (6) is valid for the mean seepage velocity. According to Eqs. (18) and (19), the mean seepage velocity U_m is proportional to the velocity amplitude of incident waves U_i and the mound

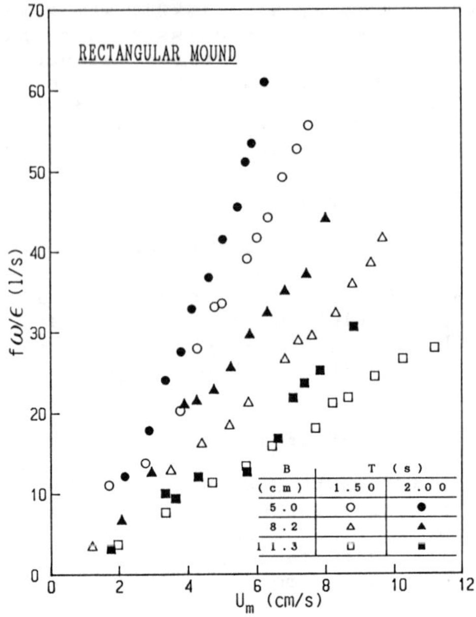

Fig. 8 Relationship between $f\omega/\epsilon$ and U_m.

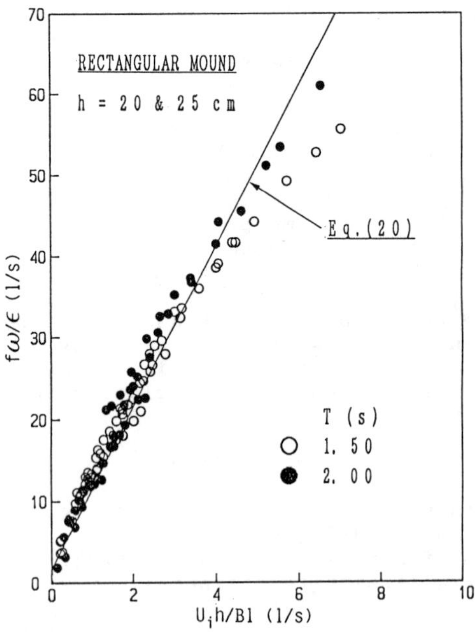

Fig. 9 Empirical friction law as a function of U_i.
[$f\omega/\epsilon$ and $U_i h/Bl$]

porosity ϵ, and is inversely proportional to the relative mound width l/L_i. Since U_m is generally a function of the friction factor, we can not explicitly calculate the friction factor from Eqs. (12) and (18) with a given incident wave condition and known mound properties.

From the standpoint of practical calculation of the friction factor, Fig. 9 is prepared for determining an empirical friction law as a function of the horizontal velocity amplitude of incident waves U_i, calculated from a linear wave theory. The best fitted line in Fig. 9 leads an empirical relationship for the friction factor of

$$f\omega/\epsilon = 2.0 + 10.4(U_i h/Bl), \qquad (20)$$

in which the factor (h/Bl) is introduced to describe the block mound characteristics.

4.2 REFLECTION COEFFICIENT

In order to examine applicability of the empirical relationship for the friction factor, experiments on the wave reflection were conducted. Figure 10 is an example of comparisons between the theoretical and experimental reflection coefficients K_R, with respect to the relative height of incident waves H_i/h for a relative mound width of l/L_i= 0.206. The friction factor is computed with the aid of Eq. (20). Although exhibiting some scatter of the data, the good agreement supports the determination of the friction factor adopted in this study, and validates the empirical equation (20).

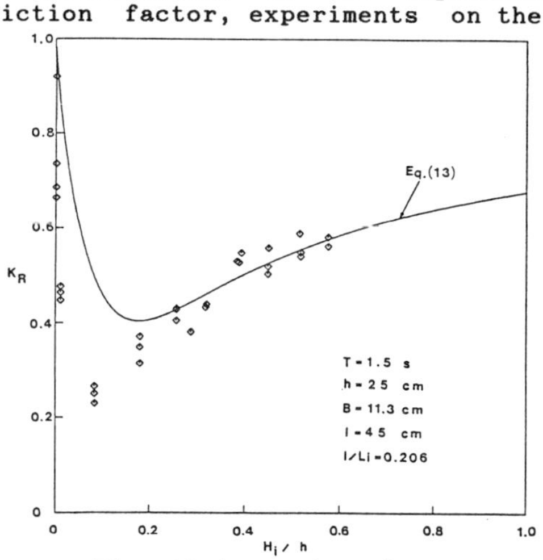

Fig. 10 Comparison between predicted and measured reflection coefficients.

4.3 MAXIMUM HORIZONTAL WAVE FORCE

As was mentioned previously, various factors influence on wave energy dissipation of the block mound. Hence, comparisons between experimental and predicted results of the maximum horizontal wave force on the upright caisson will be discussed by using the following dimensionless parameters, $\emptyset_m$ and K as given by Eqs. (21) and (22), which were derived from a dimensional analysis (Inagaki, Hattori, and Noguchi, 1986).

$$\emptyset_m = F_{mH} / \rho g H_i^2, \qquad (21)$$

and

$$K = [(1-\epsilon)/\epsilon]^{1/3}[(H_i l)/(hB)]\tanh k_0 h, \qquad (22)$$

where $\emptyset_m$ is the relative maximum wave force per unit width of breakwater, and K is the parameter combining the characteristics of an incident wave and block mound. For examples, the term of $[(1-\epsilon)/\epsilon]^{1/3}(l/B)$ on right hand of Eq. (22) represents the number of voids per unit width within

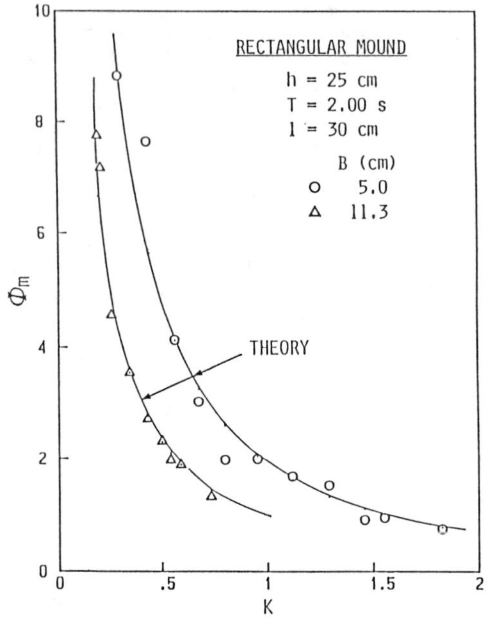

Fig. 11 Relationship between ϕ_m and K.

the block mound. And tanh k_0h is considered as a correction factor for the wave pressure distribution on the caisson.

Figure 11 is an experimental result on the relative maximum wave force ϕ_m as a function of K. For comparison the predictions by the present theory are also shown. And Figure 12 shows a comparison between measured and predicted maximum horizontal wave force for all the experiments. The agreement between the theory and experiments is fairy good and confirms the general validity of the present theory for the rectangular block mound breakwater.

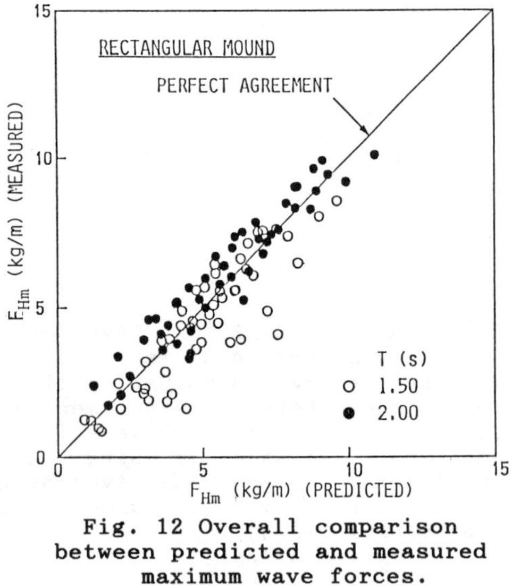

Fig. 12 Overall comparison between predicted and measured maximum wave forces.

5. APPROXIMATE METHOD FOR TRAPEZOIDAL BLOCK MOUND

5.1 PROPOSE OF APPROXIMATE METHOD

Wave motions at a block mound with steep front slope are very complicated and make the theoretical treatment more difficult than that for a rectangular block mound.

Sollitt and Cross (1972) proposed a concept of the equivalent rectangular breakwater which has the same submerged volume as that of the trapezoidal breakwater, and developed a semiempirical procedure for the prediction of reflection and transmission coefficients of permeable breakwater of trapezoidal shape. Hence, applicability of their approximate method to the present problem is examined by a comparison of all the measured maximum wave forces with the predictions based on the equivalent rectangular mound concept by Sollitt and Cross. In the predictions, energy dissipation due to the wave breaking is evaluated by the same procedure as by Sollitt and Cross. Figure 13 shows an overall comparison between predicted and measured maximum wave forces. Wc can clearly recognize a systematic trend that the predictions always underestimate the wave force and the deviation between the prediction and experiment becomes larger with increasing the maximum wave force, corresponds to the increase of incident wave height.

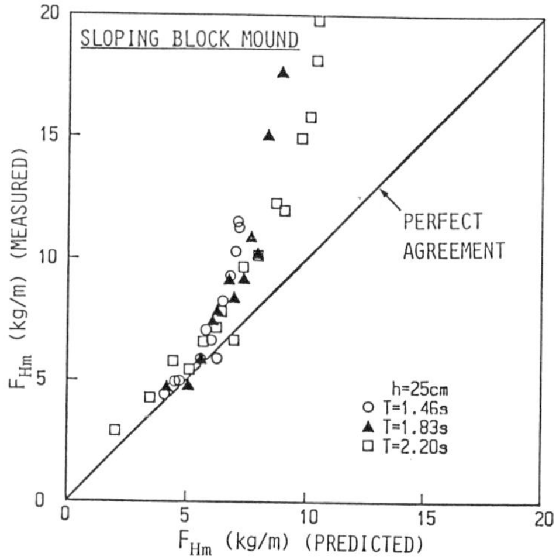

Fig. 13 Overall comparison between predicted and measured maximum wave forces.

According to experiments of trapezoidal block mounds, incident waves break and run-up on the sloping face of block mound. Water mass in the run-up wedge penetrates into the mound and flows through the upper portion of mound with narrower width than the submerged portion. As incident wave height increases, the water flow sometimes hit against the caisson and gives rise to additional wave force on the caisson. In addition, the water mass transport associated with the wave run-up on the sloping face rises the mean water level inside the mound above the still water level outside the mound. As a result, it spreads upwards the front surface area of caisson, to which wave forces are exerted.

Approximate method by Sollitt and Cross generally gives a wider mound width above the still water level than that of the actual mound. Thus, it does not reproduce reasonably well nonlinear effects of the wave motions observed both on the front slope and inside the block mound, and overestimates the energy dissipation due to flow resistance inside the mound. The experiments suggest that the following two effects of increasing the water depth around the block mound should be taken into account for the prediction of maximum wave force on a caisson backed by sloping block mound.

(1) Vertical asymmetry of the free surface elevation around the still water level associated with incident waves, and
(2) Wave set-up caused by the mass transport from the sloping face of the mound.

Consequently, we propose an approximate method that the sloping mound is replaced by an equivalent rectangular mound having the same width as that of the sloping mound at the mean water elevation including the wave set-up η_s and displacement of the mean water level from the still water level δ , as shown in Fig. 14. The mean water depth inside the block mound for determining the equivalent width of a sloping mound l_E is written as

$$h_E = h + \eta_s + \delta \quad , \qquad (23)$$

and

$$\delta = \eta_c - H_i/2, \qquad (24)$$

in which η_c is the crest elevation above the mean water level outside the mound and calculated by Cnoidal wave theory of first order(Isobe, 1985).

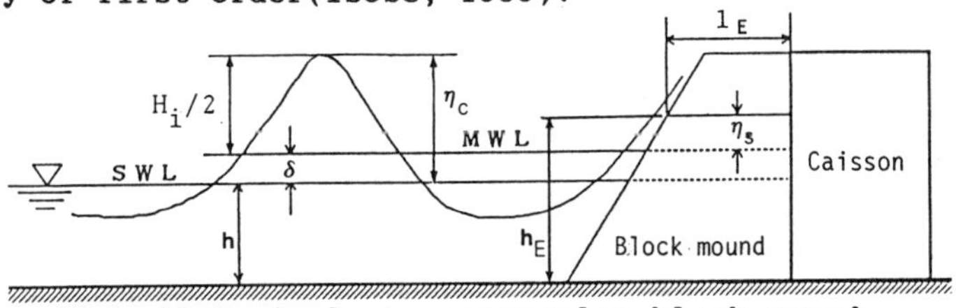

Fig. 14 Equivalent rectangular block mound.

5.2 WAVE SET-UP INSIDE BLOCK MOUND

Although many theoretical and experimental studies of the wave set-up on a natural beach have intensively been conducted, none of the published information has examined the wave set-up inside permeable structures. In order to examine the wave set-up inside the block mound, additional experiments both for the rectangular and sloping mound breakwaters were made. Deviation in the mean water level inside the mound from the still water level was evaluated from the time variation of free surface elevation measured at the front of the caisson.

Experimental results on the wave set-up inside the block mound are shown in Fig. 15, showing the relationship between the relative wave set-up η_s/H_i and the relative wave height H_i/h. Wave set-up inside the rectangular mound indicates a week dependency on incident wave height . On the contrary, as seen in Fig. 15, that inside the sloping

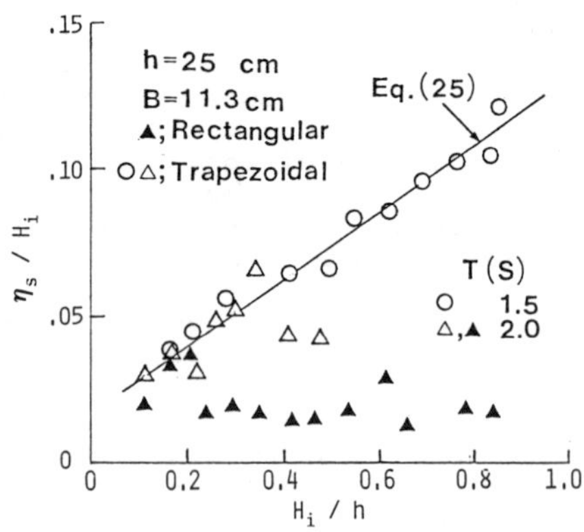

Fig. 15 Relation of η_s/H_i and H_i/h.

mound increases almost linearly with incident wave height, when $H_i/h > 0.1$. An empirical relationship between the wave set-up and incident wave height is obtained as

$$\eta_s/H_i = 0.02 + 0.11(H_i/h) \qquad (25)$$

5.3 COMPARISONS BETWEEN PREDICTED AND MEASURED WAVE FORCE

Figure 16 shows an example of comparisons between predicted and measured maximum wave force, using the relationship between the nondimensional parameters, ϕ_m and K. From the fairly well agreements as seen in Fig. 16, it is concluded that the approximate procedure proposed here is applicable to the prediction of the wave force of a breakwater with sloping mound.

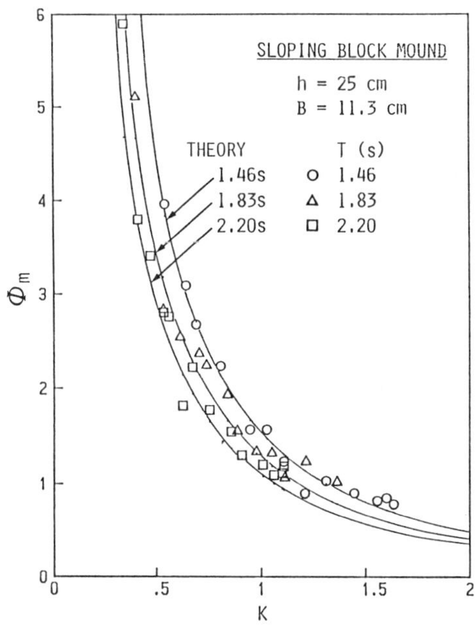

Fig. 16 Comparison between predicted and measured relative maximum wave forces ϕ_m.

To make both the analytical model and the approximate procedure generally applicable, the empirical equations for the friction law and for the wave set-up should be examined for various wave conditions and mound geometries.

6. CONCLUSIONS

In this paper, analytical theory are proposed for predicting the wave force acting on the upright caisson structure with a rectangular mound of concrete blocks of energy dissipation type. The theory is developed with approximation of the linear long wave. The wave damping inside the block mound is treated by the equivalent linearization technique after the Lorentz's Principle of equivalent work, and empirical formula is used. The validity of the analytical theory is confirmed from comparisons with laboratory experiments.

For practical applications, approximate procedure is developed for the prediction of the wave force on the caisson armored with sloping block mound. The sloping block mound problem is solved by introducing an equivalent rectangular mound which has the same mound width as that of the sloping mound at the mean water elevation taking account of two nonlinear effects, the vertical asymmetry of incident wave profile at the sloping face and wave set-up inside the mound.

Although a number of assumptions have been introduced in the approximate procedure, the good agreement between predicted and measured maximum wave forces indicates that the procedure may be valuable for practical applications.

7. REFERENCES

Goda, Y.(1985): Random Seas and Design of Maritime Structures, Univ. of Tokyo Press, 309 p.

Hattori, M. et al.(1979): Experimental study on wave force against perforated breakwaters, Proc. of 26th Japanese Conf. on Coastal Eng., pp. 396-400 (in Japanese).

Inagaki, K., M. Hattori, and Y. Noguchi(1986): Wave force reduction by armor block mound, Proc. of 33rd Japanese Conf. on Coastal Eng., JSCE, pp. 362-366 (in Japanese).

Isobe, M.(1985): Calculation and application of first-order Cnoidal wave theory, Coastal Engineering, Vol.9, No.4, pp. 309-325.

Kondo, H.(1970): An analytical approach to wave transmission through permeable structures, Coastal Engineering in Japan, JSCE, Vol. 13, pp. 31-42.

Madsen, O. S. and S. M. White(1975): Reflection and transmission characteristics of porous rubble-mound breakwaters, Tech. Report No. 207, R.M. Parsons Lab., Dept. of Civil Eng., MIT.

Massel, S.R. and C.C. Mei(1977):Transmission of random wind waves through perforated or porous breakwaters, Coastal Engineering, Vol. 1, pp. 63-78.

PIANC(1988): Analysis of rubble mound breakwater, PTC Working Group 12, pp. 1-6.

Shuto, N. and H. Hashimoto(1970): Hydraulic resistance of artificial concrete blocks, Coastal Engineering in Japan, Vol. 8, pp. 43-54.

Sollitt, C.K. and R. H. Cross(1972): Wave reflection and transmission at permeable breakwaters, Tech. Report No. 147, R.M. Parsons Lab., Dept. of Civil Eng., MIT.

CHAPTER 160

BLOCK REVETMENT DESIGN WITH PHYS. AND NUM. MODELS

A. Bezuijen[1],
J. Wouters[2],
C. Laustrup[3],

Abstract

A combination of a physical model and numerical models has been used in the design of a block revetment for the Danish North Sea coast. The wave pressure loading on the revetment during design conditions was investigated in a physical scale model. The measured wave pressures were used as a boundary condition for the numerical models. Solutions for the flow equations through the coverlayer, filter layer and subsoil were then obtained in the numerical models, taking into account the influence of turbulence. With these solutions the stability of the coverlayer and subsoil was evaluated. The paper presents a description of the various models and information about the design of the revetment.

1. Introduction

Numerical models were developed within the scope of a research programme on block revetments in order to calculate the loading on a block revetment during wave attack, Hjortnaes-Pedersen et al (1987) and Bezuijen et al(1987). This research programme was commissioned by the Dutch Department of Public works (Rijkswaterstaat) and was performed by Delft Hydraulics in cooperation with Delft Geotechnics. The numerical models were used to calculate the pressure distribution in the filter layer and subsoil below a block revetment when the pressure distribution on the revetment due to wave attack is known. Both the wave pressures and the calculated pressures underneath the revetment determine the uplift pressures on the coverlayer of the revetment. These uplift pressures can be compared with the "strength" of the coverlayer.
The wave pressures which were required as inputs for the numerical models were determined by means of physical model tests. Since only wave pressures on the slope had to be measured and the influence of wave impacts on the stability of a block revetment could be neglected, these pressures could be measured in a small scale model.

[1]. Delft Geotechnics, Delft, The Netherlands
[2]. Delft Hydraulics, Delft, The Netherlands
[3]. Kystinspektoratet, Lemvig, Denmark

The combination of a physical scale model and numerical models was used to evaluate the design of a dune toe protection which is now under construction in parts of the Danish North Sea coast.
The design of the block revetment is presented in this paper and the physical model tests described. A brief description of the numerical models is given together with the results of the calculations for this revetment. Finally the modifications in the design, based on the results of the calculations, are discussed.

2. Block revetment used as dune protection

On some parts of the Danish North Sea coast erosion has been very large, on average 3 - 4 m a year. This has caused some dunes to disappear and others to become very weak. As a result the low areas behind the dunes are open to flooding. On these stretches of coast it is necessary to stop the erosion and to re-establish the flood protection. The measures being taken against the erosion are a combination of onshore and offshore beach nourishment and low detached breakwaters parallel and close to the shore line. The flood protection is being re-established by building artificial dunes protected by a concrete block revetment. Concrete blocks are being used because:

a. They are cheap compared with other types of artificial coverlayers.

b. Denmark has no quarries close to the North Sea coast.

c. Concrete blocks look relatively atractive into the sandy coast environment.

d. Constructional procedures are relatively easy and a high quality can be achieved.

e. The concrete is very durable in a marine environment.

A sketch of the concrete block revetment is presented in Figure 1. Water level and wave conditions vary depending on the location. The conditions expected with a return period of 100 years, for the most exposed structure, are shown in Table 1.

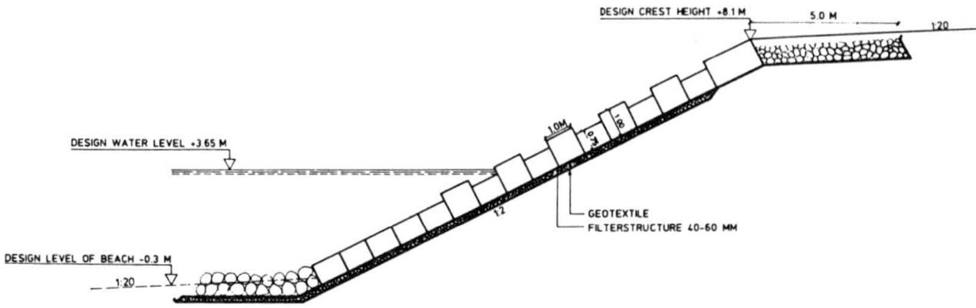

Figure 1: Original cross section for the design parameters given in Table 1

Waterlevel:		3.65 m	above datum
Waves in a depth of 19 m	H_s	= 8.2 m	
	T_p	= 12 s	
Shore profile: slope		1:20	from -0.30 to -6.00 m below datum
		1:100	below -6.00 m below datum
Revetment slope		1:2	from - 0.30 to + 8.10 m above datum

Table 1: Design conditions

The design crest height at the various locations is taken as the sum of the high water level, the wind set-up and the maximum breaking wave height above still water level. For the conditions given in Table 1 the crest height will be 8.1 m. Since waves will run-up the slope to above the level of the block revetment, the first 5 m of the crest of the structure has been protected with rubble.

The toe of the structure has been designed at such a level, that there will be no damage by scouring during the design storm.

The blocks are placed on a filter structure of 40 - 60 mm rubble between two layers of geotextile. The purpose of the upper layer is to prevent the rubble layer from being filled with sand from above. The blocks are 0.75 m high and weigh 3,000 kg. Some of the blocks are 1.0 m high in order to introduce slope roughness.

The following questions were raised when evaluating the design:

- Can block thickness be reduced?
- Is a 1:3 slope preferable to a slope of 1:2?
- What is the influence of revetment roughness on run-up produced by using blocks of different height?

3. Physical model

Small-scale model tests were carried out in a wave flume to obtain information about the wave pressure distribution on the slope. The wave flume has a depth of 0.8 m and is equipped with a system to compensate for wave reflections. The tests were carried out with irregular waves on slopes of 1:2 and 1:3. Since only the wave pressures on the slope had to be measured in a physical model test (to provide inputs for the numerical models) it was unnecessary to model the revetment itself. Only the geometry of the revetment and the foreshore were of interest in the physical model and these were modelled in concrete. The model layout is shown in Figure 2.

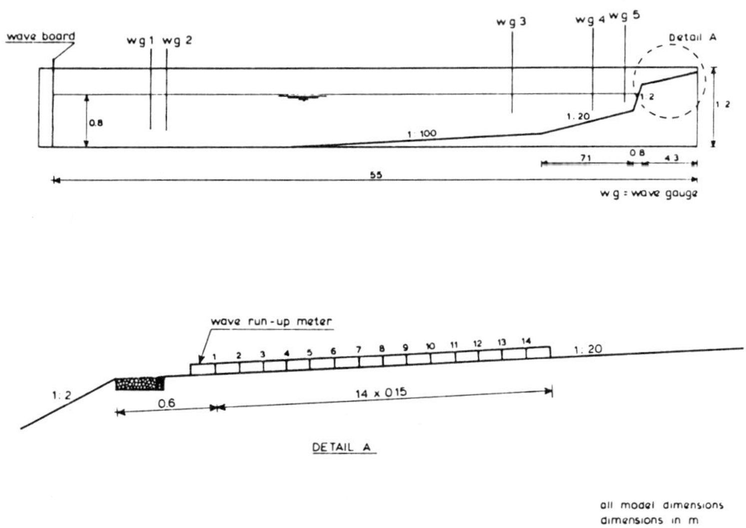

Figure 2: Model layout

Wave heights were measured on the profile with five wave gauges; run-up was measured with a wave run-up meter on the 1:20 slope, see Figure 2. The positions of the pressure gauges on the 1:2 slope are shown in Figure 3. Because wave impacts shorter than 0.2 s duration are not of importance to the stability of the revetment, the pressure signals could be filtered by a 6.25 Hz (in model 25 Hz) low pass filter. The low-pass filter is necessary to prevent interference with high frequencies. Most tests were run at a geometric scale of 1:16. The stability of the rubble on the crest was measured directly in the model tests and therefore this was modelled to the same geometric scale factor. The nearshore incident wave conditions for the model tests were calculated from the deep water conditions with the ENDEC computer program, Stive (1984).

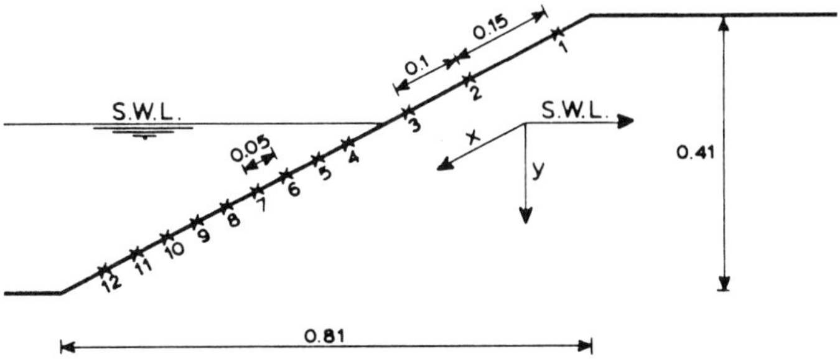

Figure 3: Location of wave pressure gauges on the 1:2 slope

The model tests indicated that the wave runup on the 1:20 slope can be considerable and during the design conditions, Table 1, and with a crest height of 8.2 m in prototype, more than 10% of the waves passed the highest indicator of the wave run-up meter. This implies a wave run-up of over 43 m up the 1:20 slope. These run-up values led to an investigation of the run-up as a function of the crest height. The results are shown in Figure 4, for the design conditions given in Table 1.

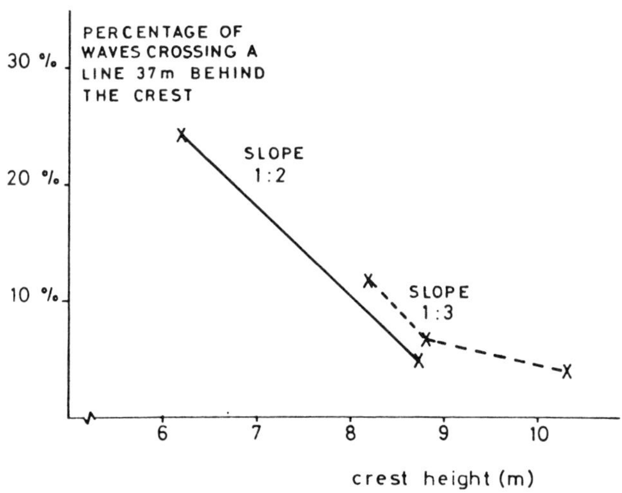

Figure 4: Percentage of waves crossing the 37 m line (see text) as a function of the crest height

On the 1:20 slope a line 37 m from the top of the revetment in prototype was chosen to indicate the extent of run-up, because information was available about the waves crossing this line also for the tests runned at different geometric scales.

More waves passed the 37 m line with the 1:3 slope than with the 1:2 slope, but the difference was small.
It appeared that the increased roughness, obtained by using blocks of different height, would led to a 10% reduction in the number of waves crossing the 37 m line during design conditions. During moderate storm conditions this reduction would be even higher. With a significant wave height of 3 m in deep water the reduction would be up to 25%.

The wave pressures measured in the physical model have been used in the numerical model. This topic is discussed in the following chapter.

4. Numerical models

In order to evaluate the stability of the revetment, it is essential to determine the uplift pressures on the blocks. These pressures are determined by the difference between the pore pressure in the filter layer under the blocks and the wave pressures on the blocks. If the mean uplift pressure on the block is larger than the pressure corresponding to the weight of the block plus the friction forces between adjacent blocks then a block can be lifted out of the revetment. The stability of the sand underneath the revetment is also important since no sliding may occur. These stability criteria cannot be investigated in a small scale model test because of the soil mechanical scale effects which occur in such a test. A large scale investigation is a possibility, but in this project it was decided to use a numerical approach to evaluate the stability criteria. The uplift pressures on the blocks and the stability of the blocks were calculated with the STEENZET/1 program, to calulate the pore pressure in the filter layer and the resulting block movement (see Section 4.1).
The pore pressure distribution in the subsoil was calculated with the STEENZET/2 finite element program which can be used to calulate the pressure distribution in both the filter layer and the subsoil, assuming no block movement (see Section 4.2). This pore pressure distribution was used in a stability calculation to evaluate the geotechnical stability against sliding. These numerical models are described briefly in the following sections.

4.1 STEENZET/1

The pressure distribution in the filter layer is determined by the flow through this layer and the flow through the joints around the blocks. The flow in the subsoil itself has no influence because of the low permeability of the sand compared to the permeability of the filter layer. Assuming a flow parallel to the slope in the filter layer, a flow perpendicular to the slope in the coverlayer and a coverlayer permeability which is concentrated in the "horizontal" joints, see Figure 5, the following formula can be derived for the potential in each joint:

$$\phi_i = \frac{1}{1 + 2\frac{kbD}{k'l^2}} \left\{ \frac{kbD}{k'l^2} \left(\phi_{i-1} + \phi_{i+1} \right) + \phi_{t,i} \right\} \qquad (1)$$

Where: ϕ_i : the piezometric head in the filter layer near joint i (m)
$\phi_{t,i}$: the piezometric head on the revetment near joint i (m)
b : the thickness of the granular sublayer (m)
D : the thickness of the blocks (m)
l : the length of the blocks (m)
k : the permeability of the filter layer (m/s)
k' : the permeability of the cover layer (m/s)

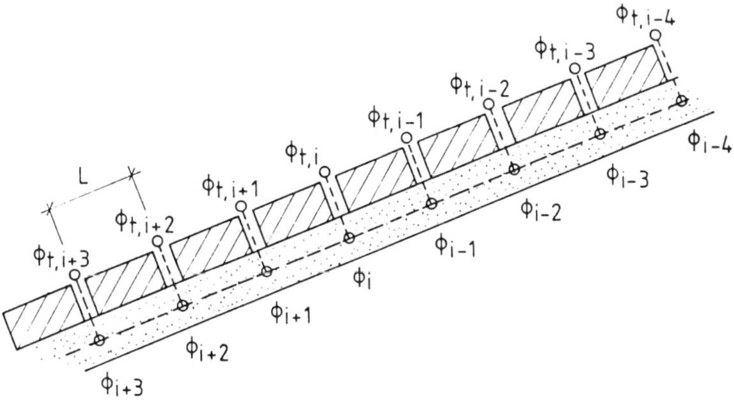

Figure 5: The STEENZET/1 finite difference scheme

At the phreatic surface the piezometric head in the filter layer is given by the position of the phreatic surface. At the lower end of the revetment the condition $\phi_{i+1}=\phi_{i-1}$ is assumed. If there is no phreatic surface, $\phi_{i-1}=\phi_{i+1}$ at the top of the revetment. The potential distribution in the filter layer can be solved numerically with Equation (1) for all joints if the different values of the piezometric head on the revetment ($\phi_{t,i}$) and the position of the phreatic surface are known. In the STEENZET/1 computer program this solution is obtained using an iterative scheme. Results of den Adel (1987) can be used to calculate the permeability of the filter layer. On the basis of permeability tests den Adel described the coefficients a and b in the Forchheimer relation ($i = av + bv^2$, with i the hydraulic gradient and v the filter velocity) as a function of the porosity and d_{15} of the granular material. The permeability of the cover-layer is determined by various flow resistances in and near the joints as described by Klein Breteler and Bezuijen (1988). Turbulence in the flow through the joints is included by adapting the permeability of the coverlayer to the calculated gradient during each iteration cycle. A linear flow condition is used in the filter layer.

The turbulent flow has therefore to be linearized. This linearisation is performed in such a way that the flow velocity calculated with the linear flow relation is the same as the flow velocity calculated using a turbulent permeability relation at a gradient equal to sin(α), with α the slope angle.

The piezometric head on the revetment ($\phi_{t,i}$) is determined from measured wave pressures by linear interpolation of the results of the pressure gauges. The wave pressures measured in the small scale model can be transferred to prototype values by using Froude's law. The measured pressures can be scaled up using the geometric scale and the time scales using the square root of that scale. This means that the 50 Hz sampling frequency in the model tests, at a scale 1:16, corresponds to a 12.5 Hz sampling frequency in prototype.

The ϕ_i values were calculated for the various time steps using different values of $\phi_{t,i}$. The position of the phreatic surface is calculated for each time step by taking the still water level as a starting position and adapting the phreatic level to the nett flow of water in the filter layer. In this way the potential distribution can be calculated for each sampling of the wave pressures over a period of several waves. The results of this program have been compared with the results of large scale model tests and show good agreement (Bezuijen et al (1987)).

If the mean calculated uplift pressure on one block exceeds the pressure corresponding to the weight of the block and the friction forces, then a block will start to move. This movement can also be calculated in the program using a simple routine. The uplift pressure, multiplied by the block area, determines the uplift force. Subtracting the weight of the block and the friction force gives the nett force F_n, which causes block movement. The acceleration of the block can then be calculated using the well known relationship:

$$F_n = M_b . a$$

Where: M_b : the mass of the block (kg)

a : the acceleration of the block (m/s²)

Double integration of the acceleration giving the block movement.

The movement calculated in this way however is too large. In reality the block movement is less, because the moving block itself causes a pressure decrease in the filter layer. A routine that includes the influence of the moving block on the pressure distribution has been developed. This routine was not used in this project because, due to lack of experimental evidence, it was not certain that the results would always be on the safe side.

4.2 Results of STEENZET/1 calculations.

Calculations were made for slope angles of 1:2 and 1:3 and various coverlayer and sub layer permeabilities. The influence of the slope on the maximum uplift pressures appeared to be small. This means that a 1:2 slope is in fact the most economical. The results of a typical calculation are shown in the Figures 6 and 7. The black area shows the measured wave pressures and the grey area the uplift pressure, but only when larger than corresponding to the weight of the blocks. The vertical height of the area represents the pressure, in metres of water, at the location were the pressure was measured. Figure 6 shows the position of the wave front at maximum wave run down; there is hardly any water on the revetment and as a consequence the wave pressures are very small.

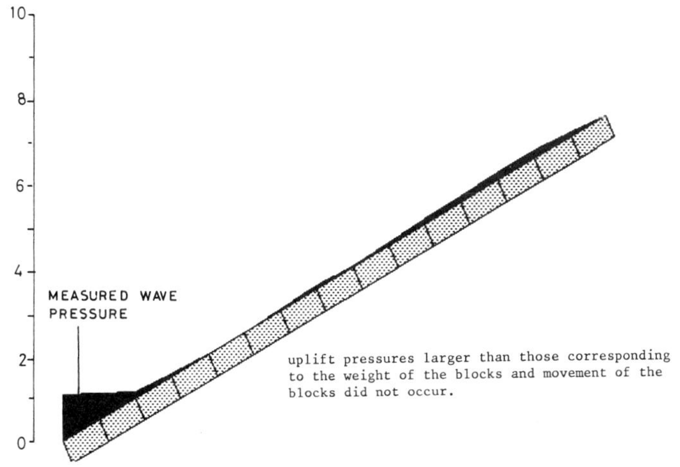

Figure 6: Wave pressure measured during wave run down. Results of STEENZET/1 calculations

Figure 7 shows the pressure distribution just after wave impact. From these figures it is clear that, for the revetment being studied the highest uplift pressures can be expected just after wave impact, when two areas of high wave pressure occur on the revetment separated of an area of low pressures. The high wave pressures are transmitted through the filter layer, leading to high uplift pressures in the area with small wave pressures.

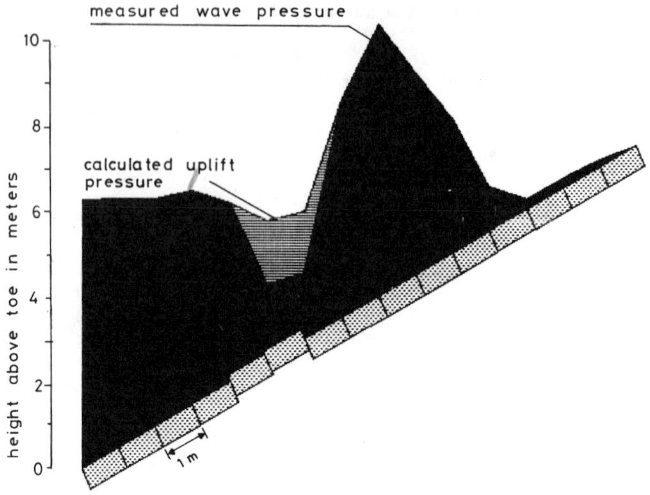

Figure 7: Wave pressure measured after wave impact. Uplift pressure when lagrger than corresponding to the weight of the bloks and calculated movement of the blocks. Results of STEENZET/1 calculations

Although during wave run down the filter layer is almost completely filled with water, the maximum wave run down does not appear to be a critical situation. The uplift pressure is never larger than the pressure that corresponds to the weight of the blocks. Due to the hydraulic gradient in the filter layer the pressure distribution in that layer is by no means hydrostatic, leading to small pore pressures and uplift pressures.

The permeabilities of coverlayer and filter layer had a distinct influence on the uplift pressures. The lowest uplift pressures were found for a minimum filter layer permeability and a maximum coverlayer permeability. The permeability of the coverlayer is determined by the permeability of the joints and is reduced due to the geotextile between the blocks and the filter layer. The design was therefore adapted and the geotextile removed. Without the geotextile some sand will be transported into the filter layer. However this sand will only reduce the filter layer permeability. Dutch experience has shown that this sand never causes trouble.

A design criterion had to be chosen for determining the block thickness. The criterion, no block shall ever move during design conditions, leads in fact to a very large block thickness and is too strict to be of practical use. The clamping forces between the blocks are completely neglected; however, due to these clamping forces, the strength of the revetment is increased significantly. Pulling tests have shown that sometimes forces more than 10 times the weight of a block are necessary to extract it from the revetment (Burger 1985). However the reliability of this extra strength decreases if calculations show that blocks immediately above and below can move at the same time. The following design criterion was therefore used: the revetment is considered to be stable if calculations with STEENZET/1 show that only one block moves and that the movement calculated is much smaller than the thickness of the block. In reality this will be a stable situation due to the clamping forces.
Apply this criterion a block thickness of 0.5 m was found, assuming a filter material with d_{15} of 20 mm and an average joint width of 5 mm.

An unexpected result was the influence of the toe permeability on the calculated uplift pressures. This was simulated in the calculations by changing the permeability of the lowest joint in the revetment; Figure 8 shows this influence. The increased permeability leads to higher piezometric head in the lower end of the filter layer and, as a consequence higher uplift pressures.

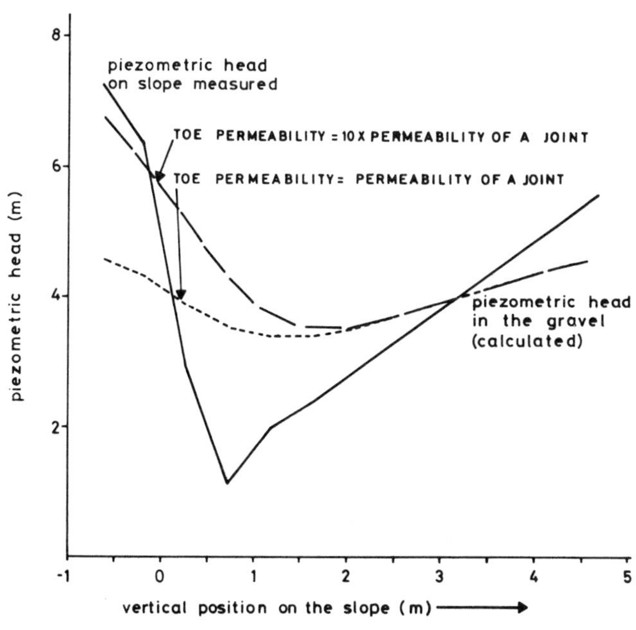

Figure 8: Influence of toe permeability on uplift pressures

4.3 STEENZET/2

The results of the STEENZET/1 calculations showed that a 1:2 slope would be more economical, because less concrete is needed. The calculation however omits the effects of sand body sliding. This kind of failure can occur locally when the grain stress in the sand is reduced by seepage.
The pore pressure distribution in the subsoil due the wave attack was therefore calculated with the STEENZET/2 program.

STEENZET/2 is a 2-dimensional finite element program specially developed to calculate the pore pressure response under revetments due to wave attack. The model is based on what is referred to as the storage equation:

$$\nabla q = C\ d\Phi/dt \tag{3}$$

Where:
- Φ : the piezometric head (m)
- q : the specific discharge (m/s)
- $C = \rho g n \beta$ (1/m)
- ρ : the mass density of the fluid (kg/m^3)
- g : acceleration due to gravity (m/s^2)
- n : porosity ( )
- β : compressibility (m^2/KN)
- ∇ : $d/dx, d/dy$

With solutions of this equation it is possible to investigate the influence of the air content of the pore water on the results. The program can handle turbulent flow on the base of the Forchheimer relation and various materials can be considered. Measured wave pressures can be used as boundary conditions in the same way as in the STEENZET/1 program. A more extended description of this program is given by Hjortnaes-Pedersen et al(1987). The pore pressure distributions calculated with STEENZET/2, were used in a stability analysis as described by Bishop. This method is well known and is described in, for instance Terzaghi and Peck (1967).

4.4 Results of STEENZET/2 Calculations

Calculations were performed for the 1:2 slope and the block thickness indicated as stable by the STEENZET/1 calculations (0.5m). The element mesh used is shown in Figure 9. The model tests showed that a large amount of overtopping can be expected during design conditions and it was therefore assumed that, during these conditions, the sand body directly behind the revetment would be completely filled with water. Outward directed hydraulic gradients perpendicular to the slope, are most dangerous for revetment stability. The wave period which led to the highest uplift pressures in the STEENZET/1 calculations was used as a (time varying) boundary condition. Very little is known about the air content of the pore water and therefore calculations were run with different values of air content. Results are shown in Figures 10 and 11. In these figures lines of equal piezometric head are shown at the moment of maximum outward directed hydraulic gradient; the wave boundary condition is also shown at this particular moment.

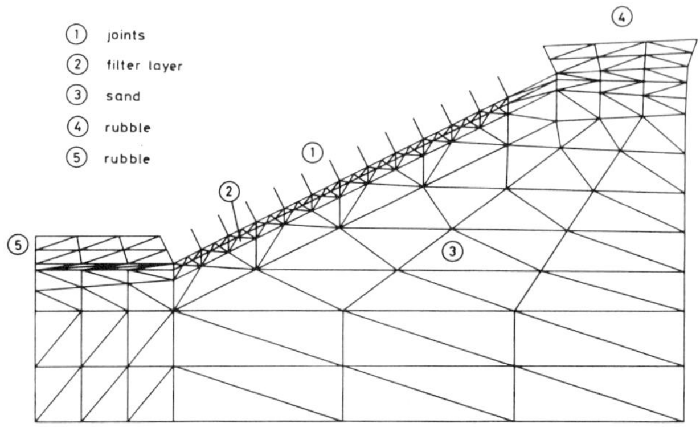

Figure 9: Finite element mesh used in STEENZET/2 calculations

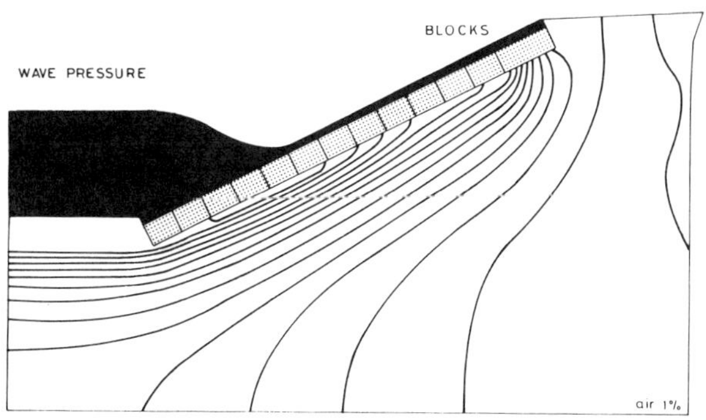

Figure 10: Measured wave pressures and calculated lines with epuipotential. Results of STEENZET/2 calculations -assumed air content in the pore water: 1%

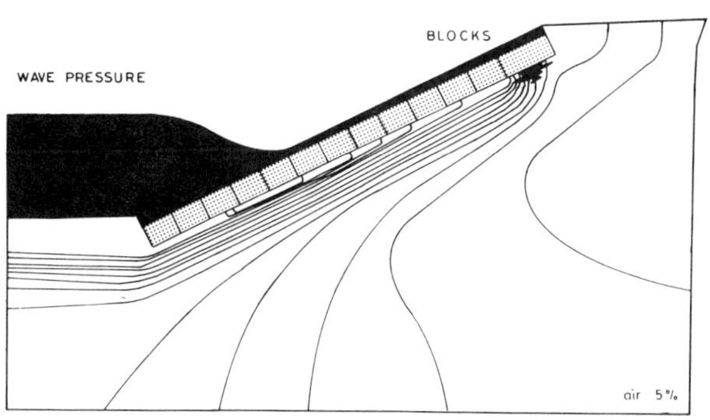

Figure 11: Measured wave pressures and calculated lines with epuipotential. Results of STEENZET/2 calculations -assumed air content in the pore water: 5%

This wave condition was measured just after a period of rapid run down. Since Equation (3) is a transient equation, the boundary condition prior to this particular moment is also of importance. The figures clearly show the influence of the air content. The higher the air content, the higher the outward directed gradient. The stability calculations showed however that even with an assumed air content of 5%, the safety of the 1:2 slope against sliding is more than 1.5 with a friction angle of 35° for the sand, 40° for the gravel and schematizing the concrete blocks to a friction material with a friction angle of 25°. From these results it was concluded that the 1:2 slope will be stable against sliding during design conditions.

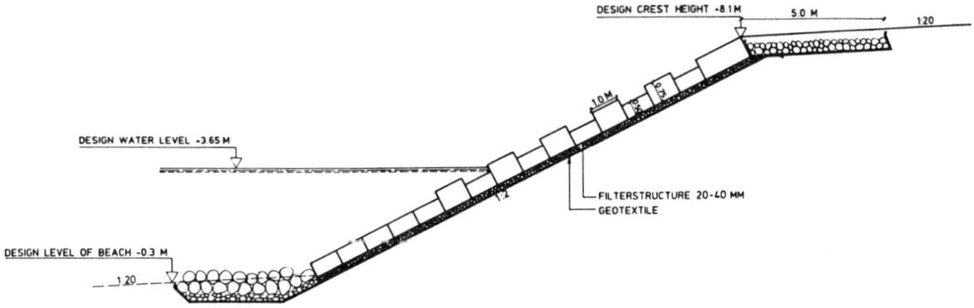

Figure 12: Adapted design based on model studies

5. Conclusions

The combination of physical and numerical models can be a valuable tool in the design of a placed block revetment. As a result of this study the following adaptations were made to the design:

- a heavier rubble is used for the crest protection.
- the geotextile between the blocks and the filter layer is removed.
- a filter layer of relatively small gravel (16-32 mm) is used.
- During the manufacturing process the surface of the blocks is made irregular so that the joints will be about 5 mm wide.
- The geotextile between the blocks and the filter layer is replaced by a filter of relatively small gravel in order to reduce the uplift pressures. This makes it possible to reduce the block thickness from 0.75 to 0.5m.

The model tests showed that increasing the roughness of the slope by varying the thickness of the blocks reduces the wave run-up, especially during moderate storm conditions. According to the stability calculations a structure with a slope 1:2 is stable during design conditions. Since the difference in calculated maximum uplift pressure and measured run-up with a 1:2 slope and 1:3 slope is only small, a 1:2 slope is preferred because less concrete has to be used. Figure 12 shows the adapted design.

A total of 11,420 m of this revetment had been built by September 1988.

References

- Terzaghi K., Peck R.B. Soil mechanics in engineering practice 2nd edition. Joh. Wiley & Sons 1967
- Stive. M.J.F., Dingemans M.W., Calibration and verification of 1-dimensional model for wave energy decay, Delft Hydraulics report. M1882 Dec. 1984.
- Burger A.M. Strength dikes along the Eastern Scheldt under concentrated wave attack (in Dutch). Delft Hydraulics Report. M2036 1985
- Hjortnaes-Pedersen, A.G.I., Bezuijen, A., Best, H. Non stationary flow under revetments using the F.E.M. Proc. 9th. Eur. conf. on Geotechnics and foundation Engineering Dublin. Sept. 1987.
- Bezuijen A., Klein Breteler, M, Bakker, K.J. Design criteria for block revetments and granular filters. Proc. 2nd Int. COPEDEC Beijing. Sept. 1987.
- den Adel A., Analysing permeability measurements with the Forchheimer relation (in Dutch). Delft Geotechnics report CO-272550/56 Nov.1987
- Klein Breteler M., Bezuijen A., The permeability of closely placed blocks on gravel. Proc. Int. SOWAS-symp. Delft Aug. 1988

CHAPTER 161

FULL-SCALE WAVE ATTACK OF UNIFORMLY SLOPING SEA DYKES

Alfred Führböter[1] and Uwe Sparboom[2]

ABSTRACT

Shock pressure phenomena due to breaking waves acting on sloping faces of sea dykes are focussed in this paper. The probability approach is pointed out and maximum shock pressure estimations are given for smooth and impermeable dyke slopes 1:4 and 1:6. An extension of the results to steeper and flatter slopes is proposed.

Results of full-scale stability tests on concrete block slope revetments are also reported in this paper. For various structural solutions with granular and geotextile filter layers stability numbers are recommended. Initial block lifting is explained physically by pressure measurements.

The full-scale experiments were carried out in the new research facility LARGE WAVE CHANNEL of the universities in Hannover and Braunschweig (Federal Republic of Germany).

1. INTRODUCTION

In the past dykes and revetments at the coastal zone of Germany were frequently destroyed by heavy wave attack during storm surge tides. Therefore, the most important objective of the new large scale facility LARGE WAVE CHANNEL is to investigate dykes and revetments with special reference to the German North Sea Coast (Führböter, 1982). The main reasons for dyke failures can be summarized as follows:

- shock pressures due to wave breaking,
- wave run-up and overtopping,
- up- and down-rush velocities,
- dyke construction (slope, cover layer, subsoil, etc) and
- local sea state characteristics.

First investigations of the fundamental research program dealt with shock pressure phenomena on dykes with uniformly sloping faces. Shock pressures due to plunging breaker types may cause very first damage of the cover layer and the subsoil (Führböter, 1966; Stephan, 1981). Comprehensive studies on this topic were firstly published by Bagnold (1939). Contributions by Skladnev and Popov (1969) dealt with impact forces and scale effects. Nearly full-scale test results on wave

---

[1] Prof. Dr.-Ing., Full Professor, Leichtweiß-Institute, Division of Hydrodynamics and Coastal Engineering, Technical University of Braunschweig, Beethovenstraße 51a, 3300 Braunschweig, Fed. Rep. of Germany

[2] Dr.-Ing., Senior Research Engineer, Large Wave Channel, Coastal Engineering Research Group SFB 205, University of Hannover, Merkurstraße 11, 3000 Hannover 21, Fed. Rep. of Germany

impacts acting on dyke slopes were reported by Stive (1984). Führböter (1986a) compared results of model and full-scale tests of wave impacts with reference to a 1:4 dyke slope.

Further full-scale investigations were focussed on the stability of concrete block slope revetments. Such permeable revetments must be designed in combination with filter layers if the dyke core exists of sand. The filter layer can be realized by a 3-dimensional granular layer, by a 2-dimensional geotextile mat (Heerten, 1984) or by a combination of both feasibilities. Stability criteria were often found by the extension of results from small-scale models to the prototype structure (Führböter et al., 1976; Whillock, 1980; Kostense and den Boer, 1984). This method is always affected by scale effects. A review on this topic was reported by Powell et al. (1985). From Dutch large-scale model tests stability numbers were published (den Boer et al., 1983; TAW/CUR-VB, 1984; Pilarczyk, 1987). Results of pressure measurements in the filter layer as well as results of numerical and analytical calculations were published by Bezuijen et al. (1986) and (1987). A theoretical treatment of the behaviour of loose blocks on slopes under wave action was given by Townson (1987). With reference to the bottom protection structure with loose blocks placed on granular sublayers at the EIDERDAMM storm surge barrier unexpected failure mechanisms were considered theoretically by Führböter (1986b).

The new research facility LARGE WAVE CHANNEL makes it possible to carry out full-scale investigations on dykes and revetments at German coastal regions. The main dimensions of the channel are: depth 7.0 m, width 5.0 m and length 324 m. Regular waves and random seas are produced mechanically by a wave generator. The maximum wave height is 2.5 m. Details about the LARGE WAVE CHANNEL were published by Grüne and Führböter (1975). Design criteria and technical works were reported by Grüne and Sparboom (1982).

2. PROTOTYPE DYKE STRUCTURES

The core profiles of the prototype dykes were constructed by sand. The compact cover layers were built by asphalt-concrete in a thickness of nearly 0.20 m. During the wave tests the core was drainaged to avoid dyke failures caused by positive water pressure below the impermeable cover layer.

With regard to modern dyke protection works the faces of the prototype dykes were constructed with uniform slopes 1:4 and 1:6 (Sparboom, 1987). The investigated asphalt dykes (Fig. 1) were supplied with measuring devices to obtain calibrated electrical signals for

- wave parameters at the toe of the dyke,
- wave impact pressures on the slope surface,
- wave run-ups on the slope and
- wave impact pressures in the subsoil.

The waves were generated regularly under absorption control. For each test nearly 200 waves were generated. Additionally, test series with narrow banded wave spectra were carried out.

A cross-section of the prototype revetment is given in Fig. 2. The area of the highest expected wave attack was subdivided into two testfields each of 10 m in length and 2.5 m in width. Out of the testfields the blocks were directly placed on a geotextile mat with a stabilization layer (needle punched fabric TERRAFIX 904 RS). The permeability of this filter was about $k_f \approx 3 \cdot 10^{-3}$ m/s and ten times greater than the estimated sand permeability ($k_s \approx 3 \cdot 10^{-4}$ m/s).

There were tested four structural variants in the testfields (Fig. 3). In all cases the blocks were placed in stretcher-bond equivalently to practice.

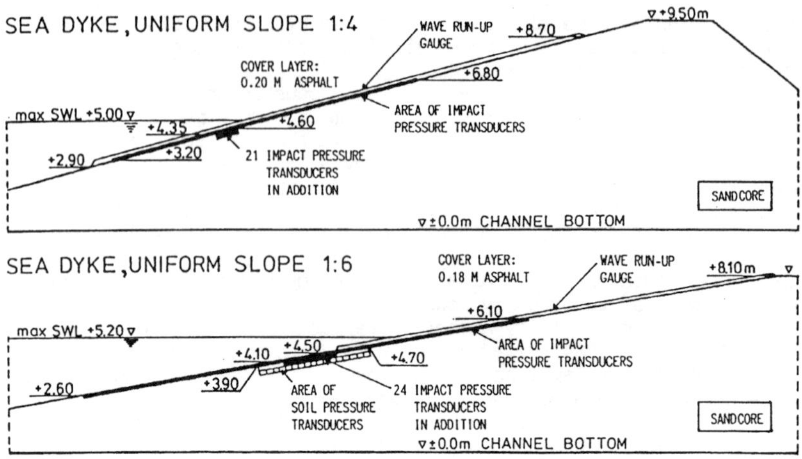

Figure 1. Investigated prototype sea dykes with impermeable top layers in the LARGE WAVE CHANNEL

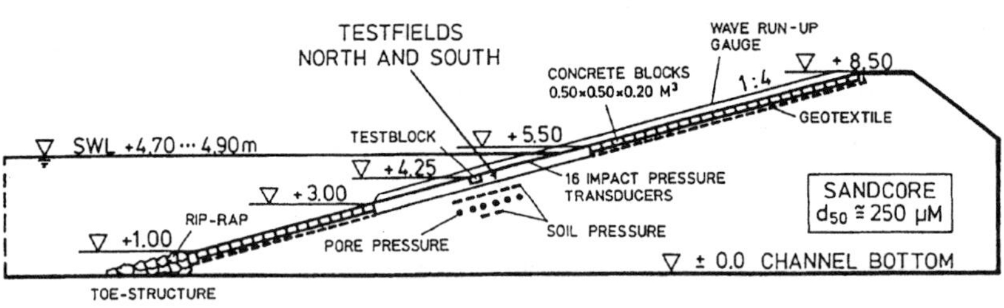

Figure 2. Prototype revetment in the LARGE WAVE CHANNEL

The measuring equipment (Fig. 2) was designed to registrate synchronous electrical signals for:

- wave parameters at the toe of the dyke revetment,
- wave impact pressures acting on the testblocks (top, bottom, each gap side),
- soil pressures beneath the top layer,
- pore pressures in the subsoil,
- block motions,
- wave run-ups on the slope and
- wave impact pressures on the slope between the testfields.

The waves were generated regularly applying an integrated absorption control system. The minimum wave number for each test was 200. After the first block failure the test was interrupted.

3. SHOCK PRESSURE INVESTIGATIONS

A typical record of a shock pressure (wave impact) due to wave breaking (plunging breaker type) on the slope can be seen in Fig. 4.

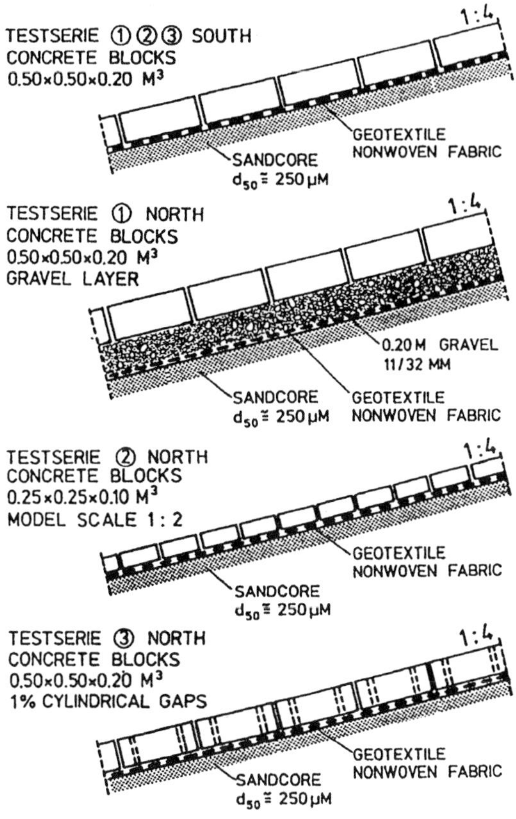

Figure 3. Cross-sections of investigated placed block revetments

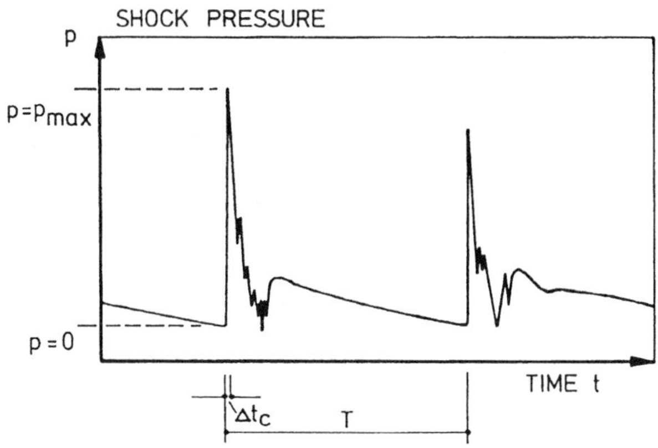

Figure 4. Typical shock pressure caused by breaking waves (plunging breaker type)

The peak pressure which is much higher than the wave or breaker height is highly effective during a relative short time. Due to the included airvolume by the plunging breaker the shock pressure occurrence can be explained by a compression phase. The time of compression Δt_c of all evaluated impacts lies between 10 and 60 milliseconds. The amplitude as well as the time history of shock pressures are strongly dependent on

- breaker process and aeration,
- angle of the sloping face,
- thickness of the backrush-water and
- wave characteristics (regular and random).

3.1 Probability theory

In the analysis of jet impacts (Führböter, 1966 and 1969) the maximum pressure p_{max} is described by

$$p_{max} = \delta \cdot \varrho \cdot v \cdot c \cdot (c/v)^{1/3} \quad (1)$$

$$\text{with } \delta = (E_a/E_w \cdot R/D)^{2/3}$$

E_a and E_w = elasticity of air and water
R = hydraulic radius of impact area
D = representative thickness of included air content
δ = dimensionless impact number
ϱ = density of water
v = impact velocity vertical to the wall
c = sound velocity in water.

In this equation the air content term expressed by D (see Bagnold, 1939) is strongly stochastic and has an important influence on the impact as well as the hydraulic radius R of the impact area and the relation E_a/E_w. The following transformation of Eq. (1) yields:

$$p_{max} = \underbrace{(E_a/E_w \cdot R/D)^{2/3}}_{\text{stochastic}} \cdot \underbrace{\varrho \cdot v \cdot c \cdot (c/v)^{1/3}}_{\text{deterministic}} \quad (2)$$

$$\log p_{max} = 2/3 \log E_a/E_w + 2/3 \log R - 2/3 \log D \;\; \text{(stochastic)} \quad (3)$$
$$+ \log (\varrho \cdot v \cdot c \cdot (c/v)^{1/3}) \;\; \text{(deterministic)}.$$

In this way a linear function is obtained for log p_{max} and log D. If the stochastic term is distributed after GAUSS, p_{max} must consequently follow a Log-Normal distribution (Führböter, 1966; see also Weggel, 1971). Comparing results on jet impacts, wave impacts from 1:10 scaled model and from prototype (scale 1:1) in the LARGE WAVE CHANNEL Führböter (1986a) concluded that the Log-Normal distribution is valid for impact pressures in general as predicted by theory. Impact pressures due to breaking waves can therefore never be described deterministically but only probabilistically by the Log-Normal distribution.

With special reference to wave impacts on slopes Führböter (1986a) extended Eq. (1) to

$$p_{max} = \varepsilon \cdot \varrho \cdot g \cdot H_B \cdot (y_B/H_B)^{1/3} \cdot ((1+2 \cdot n^2)/(1+n^2))^{1/3} \quad (4)$$

$$\text{with } \varepsilon = (c^2/2g \cdot E_a/E_w \cdot d_i/H_B \cdot 1/D)^{2/3}$$

ε = dimensionless impact number for breaking waves on slopes
H_B = breaker height
y_B = breaker crest above the bottom
d_i = length of the impact area on the slope

g = acceleration due to gravity
n = front slope 1:n .

Considering practical design methods of sea dykes the relation between wave height and impact pressure seems to be most convenient (Führböter, 1986a):

$$p_{max} = p_i = \text{const.} \cdot \varrho \cdot g \cdot H \qquad (5)$$

with i = 50, 90, 99, 99.9 % .

3.2 Shock pressure experiments

Maximum shock pressures were evaluated from each individual breaking wave. As already shown by Führböter (1986a) the maximum shock pressure values follow a Log-Normal distribution. In Fig. 5 and 6 the results for i = 50, 90, 99, 99.9 % are plotted as dimensionless relative impact pressures. Each line represents the results of a test serie with nearly 200 waves. For this wave number sufficient estimates were reached up to a level of 99.9 % probability.

In order to get informations about the spatial distribution of shock pressure occurrence the measured pressures ($p_{i\ max}$) of each transducer at the slope were determined for each test cycle. Fig. 7 represents results of two test cycles. The length of the shock pressure area is strongly dependent on the wave parameters. The mean level of the highest maximum shock pressure, evaluated relatively to the wave height is drawn in Fig. 8 versus the relative shock pressure. The mean occurrence point for the slope 1:6 can be characterized by 0.5 H below SWL. The same value was found by Stive (1984) for steeper slopes 1:3 and 1:4.

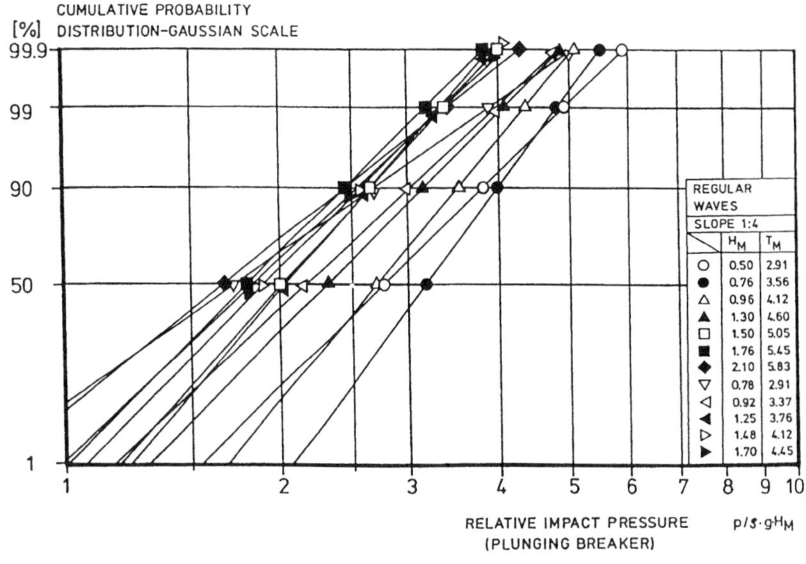

Figure 5. Shock pressure results, slope 1:4

Especially at the slope 1:6 shock pressures induced by simulated sea states were investigated. Representatively, in Fig. 9 relative shock pressures of a test with a significant wave height of 1.5 m and a peak period of 6.0 s are drawn against the well-known breaker number

(IRIBARREN-number). Maximum pressures occurred at breaker numbers from 0.6 to 1.8. The highest pressures were found at a breaker number of nearly 1.2.

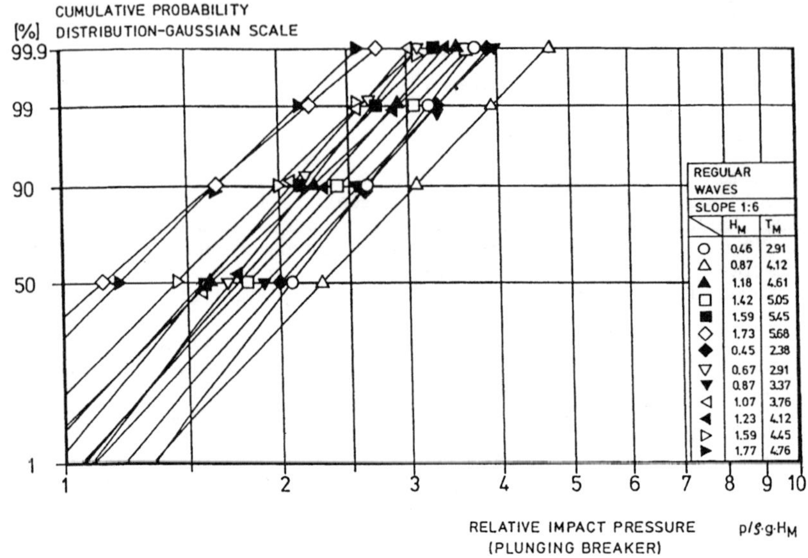

Figure 6. Shock pressure results, slope 1:6

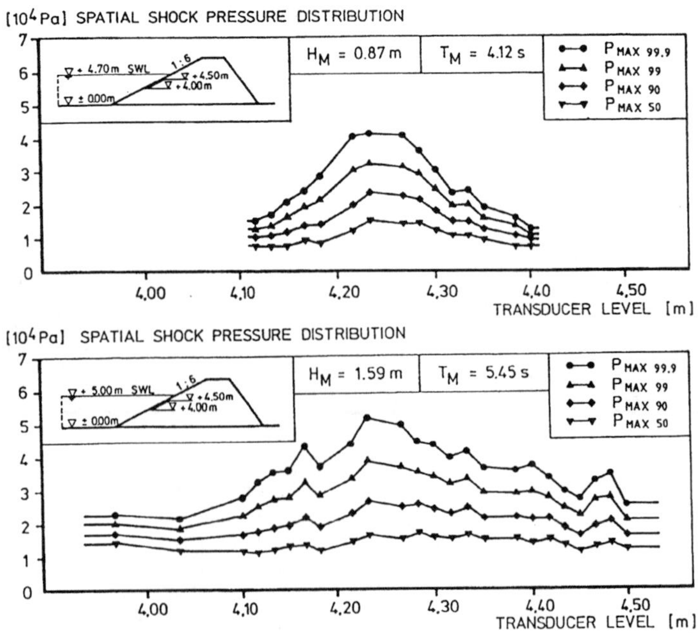

Figure 7. Spatial distributions of shock pressures on the slope 1:6

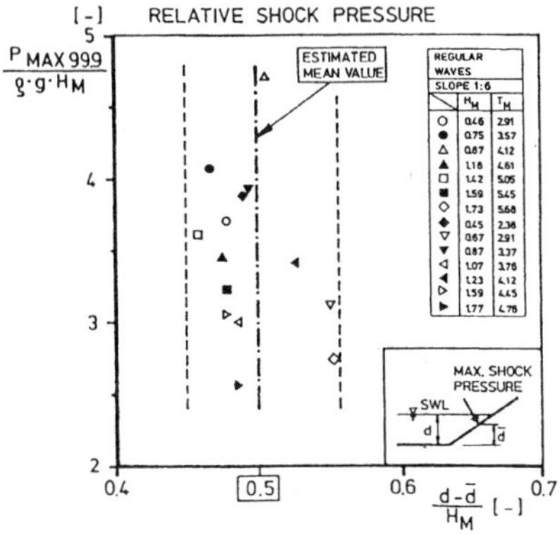

Figure 8. Area of maximum shock pressure occurrence for the slope 1:6

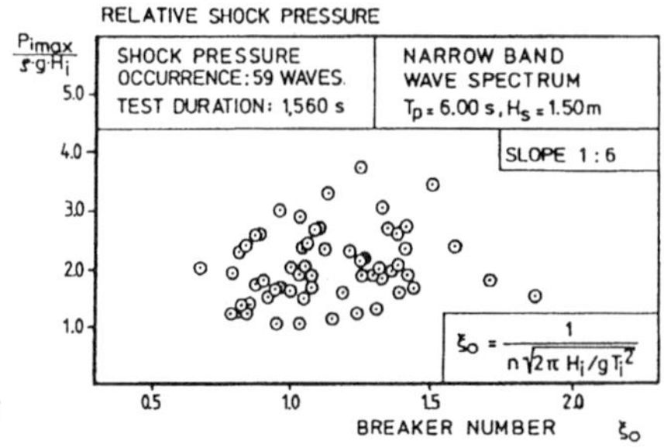

Figure 9. Random wave impact vs. breaker number, slope 1:6

3.3 Maximum shock pressure estimation

In Fig. 10 there are compared full-scale shock pressure results obtained by test cycles with regularly generated waves with those obtained by field measurements at the EIDERDAMM storm surge barrier and by test cycles with narrow banded wave spectra in the channel. Looking towards a practical approach of sea dyke design the maximum shock pressure distribution of the worst case is estimated with a tendency to the safe side (line with circled points).

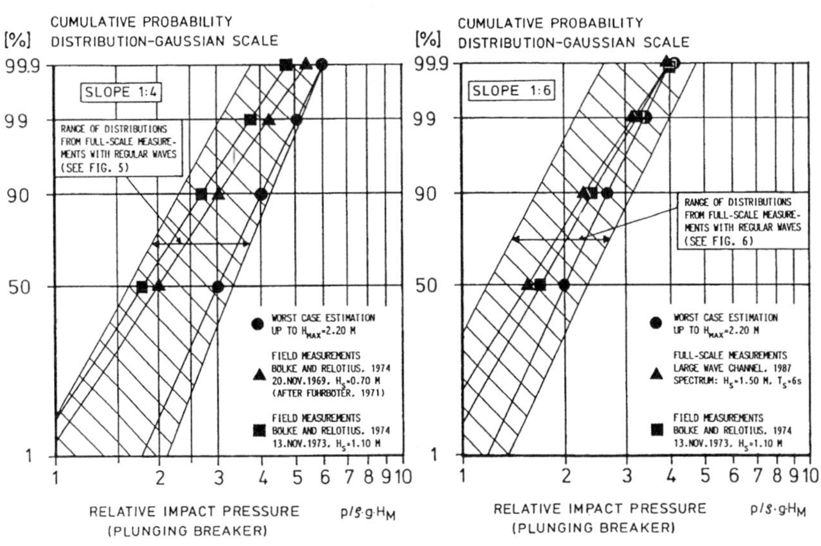

Figure 10. Full-scale shock pressure distributions

Based upon the worst case distributions (Fig. 10) for both slopes 1:4 and 1:6 it seems to be possible to establish a formula which contains the influence of the slope angle:

$$p_{max} = p_i = const \cdot 1/n \cdot \rho \cdot g \cdot H \qquad (6)$$

with i = 50, 90, 99, 99.9 % and

n = front slope 1:n.

The empirical Eq. (6) may be justified by the important damping influence of the backrush-water on the impact amplitude. The thickness of the backrush-water increases with flatter slopes and the impact pressure decreases proportionally. The application of this formula to slopes in general is proposed in Fig. 11. With respect to the real experiments with 1:4 and 1:6 slopes it should be considered that only for the range 1:3 to 1:8 a high reliability can be achieved. For the range 1:1 to 1:3 the values are extrapolated and should therefore be considered very cautiously. Based upon jet impact investigations Führböter (1966) also determined relative impact pressures for the vertical wall. Expected values for breaker heights of 1.0 m are added in Fig. 11. Future full-scale investigations on very steep slopes up to the vertical wall could be useful to proof the proposed empirical formula.

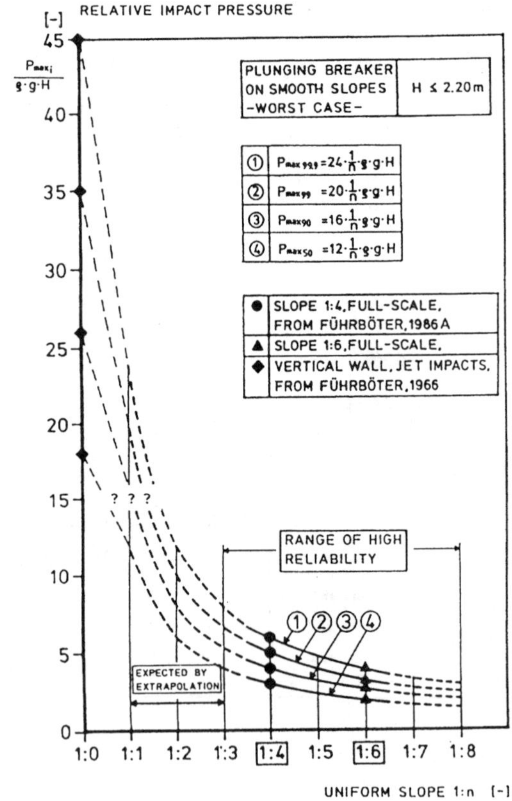

Figure 11. Generalized proposal for breaker-induced maximum shock pressures on smooth slopes

4. INVESTIGATIONS ON CONCRETE BLOCK SLOPE REVETMENTS

First experiments with permeable prototype revetments in the LARGE WAVE CHANNEL were used to explain statements of the following problems:

- Which real wave parameters (height and period) are responsible for the instability of natural revetments ?
- At which wave phase and in which location do very first failures occur ?
- How does a gravel layer (filter base) influence the stability of a revetment ?
- Which scale effects can be observed using a 1:2 model in revetment design ?
- How does an increase of the top layer permeability influence the stability of a revetment ?

4.1 Stability results

Usually, blocks or stones of seawalls are designed applying the HUDSON formula. The wave period is not considered in this approach. Recent investigations have shown clearly that the block stability depends on the breaker number ξ (Bruun, 1985). This parameter contains the influence of the wave period (or better wave steepness) and the slope angle.

Pilarczyk (1987) proposed a stability formula for concrete block revetments:

$$H_S/\Delta \cdot d_B = \varphi \cdot \cos\alpha \; / \; \xi_z^{\;1/2} \qquad (7)$$

with H_S = significant wave height
Δ = relative block density
d_B = block thickness
α = slope angle
ξ_z = breaker number
$\tan\alpha \; / \; (2\pi H_S/gT_z^2)^{1/2}$
T_z = mean wave period
φ = Stability coefficient defined for ξ_z = 1 in categories I to V.

Placed blocks which are treated in this paper should therefore be classified in categorie II ($3 < \varphi < 4$). The permeability of the top layer and of the filter layer as well as the horizontal dimensions of the single blocks are not taken into account. Special problems occur if the results of laboratory tests with regular waves are to be transferred to natural structures under real sea state conditions (wave spectra).

With regard to recent results on the block stability the parameters $H/\Delta \cdot d_B$ and ξ_o are considered here.

H = wave height (regularly generated)
Δ = relative block density
$\rho_b - \rho_w / \rho_w$
d_B = block thickness
ξ_o = breaker number in deep water
$\xi_o = 1 \; / \; (n \cdot (H \; / \; L_o)^{1/2})$
with $L_o = g \cdot T^2 \; / \; 2\pi$
n = front slope 1:n
L_o = wave length in deep water
T = wave period (regularly generated)
g = acceleration due to gravity.

Fig. 12 represents results of testseries ①, ② and ③. Safety is guaranteed up to $H/\Delta\cdot d_B = 4$. Comparing these results with those in Fig. 13 which refer to a 1:2 scaled model the same upper bound can be stated. But, failures seems to occur at higher ξ_o - values. Probably only the unit of length should be scaled up and not the unit of time simultaneously. This problem due to scaling may necessitate more systematic tests in the future.

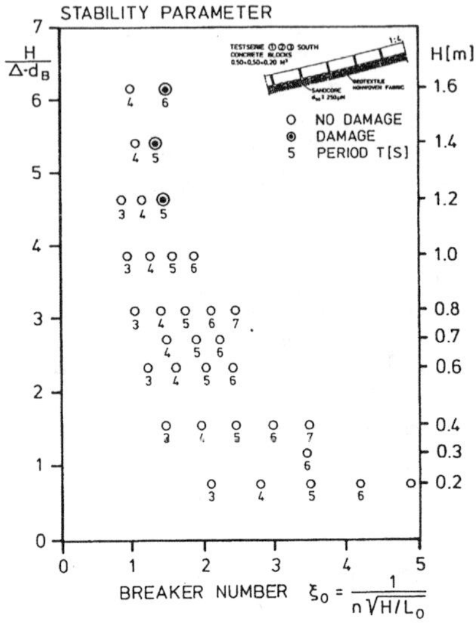

Figure 12. Stability numbers Testseries ①②③ South; Full-scale

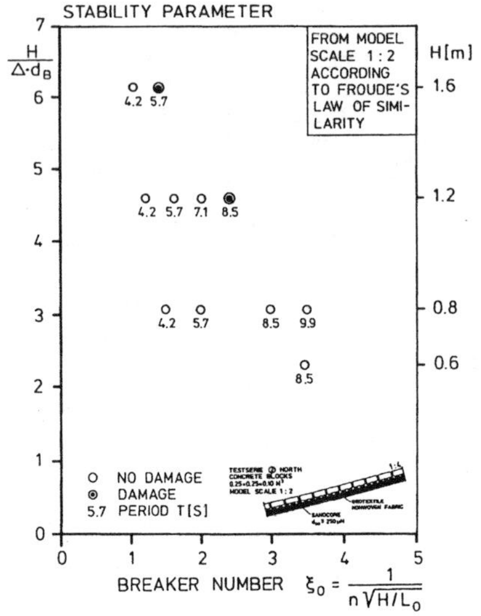

Figure 13. Stability numbers Testserie ② North; Model scale 1:2

In Fig. 14 there are plotted stability results for the structure with an added gravel layer. $H/\Delta\cdot d_B \approx 2.5$ is estimated as the maximum safe number. The influence on the stability by increasing the permeability of the top layer is seen in Fig. 15. Safe conditions can be expected for $H/\Delta\cdot d_B \approx 5.5$. In the last test vertical block motions in the range of 2 to 3 cm were observed. It was not possible to simulate the real failure because the revetment in the other testfield was totally damaged.

All observed failures occurred after about 65 to 110 waves. The location of the first block liftings can roughly be given with 0.5 H beneath SWL. Failures of the geotextile filter mats did not happen. Sandtightness was guaranteed over all test series (about 10,000 waves).

4.2 Initial block lifting

Placed block stability depends clearly on wave breaking characteristics. In the range $1 \leq \xi_o \leq 1.5$ block failure was mainly observed. This is the same range in which Führböter (1986a) found the maximum wave impacts on an impermeable 1:4 sloping asphalt dyke. Wave impacts are of short duration (milliseconds) and therefore they can not cause block liftings out of the top layer. But wave impacts probably cause a strong reduction of friction contact (or friction stress) between the single blocks. Test observations have shown that first block liftings occurred just in front of the progressive wave before breaking. In order to study pressure variations, transducers were installed in special test-

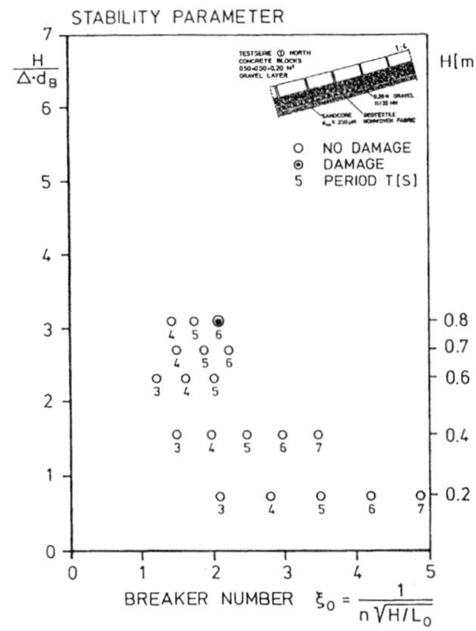

Figure 14. Stability numbers Testserie ① North; Full-scale

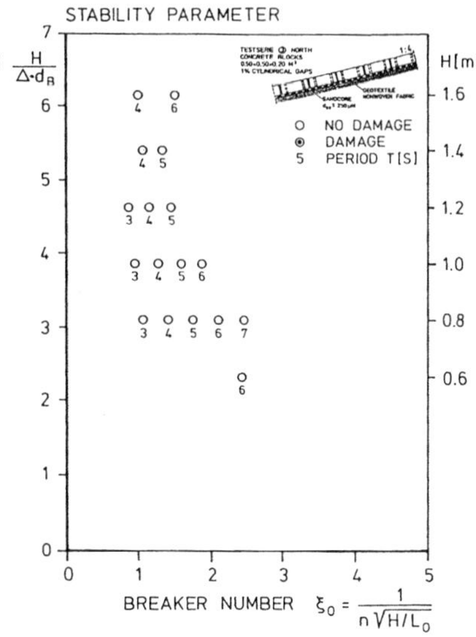

Figure 15. Stability numbers Testserie ③ North; Full-scale

blocks.

In Fig. 16 and in Fig. 17 sychronously measured pressures at top and bottom of a testblock are plotted. Additionally, the pressure difference is calculated. The positive sign indicates uplift pressures. The reaction time T_{up} is defined from the beginning of increase of Δp to the maximum of Δp before wave breaking. Dynamic impact pressures were not yet evaluated in detail. But nearly all impacts reacted in very short time and opposite to the uplift direction before wave breaking.

For about 35 waves of testserie ① (H_d = 0.70 to 0.80 m, measured at the toe of the dyke; T = 4 to 6 s) the reaction time T_{up} was evaluated (Fig. 18). The wave conditions correspond to the beginning of block liftings above the granular layer. Although a large scatter is obvious it can be seen that for both filter layers the uplift pressure amplitudes are of nearly equal amount but differ definitely in the reaction time. Increasing duration of the uplift pressure seems to be a most important phenomenon causing block failures.

In testserie ③ long reaction times were also found in the geotextile filter just before failure occurrence, but the waves were much higher. The reaction time can obviously be reduced by an increase of the top layer permeability. This was also demonstrated in testserie ③ (Fig. 15).

5. CONCLUDING REMARKS

As already stated by Führböter (1986a) shock pressure phenomena due to breaking waves acting on dyke slopes can only be described probabilistically by the LOG-normal distribution.

The thickness of the backrush-water has an important influence on the shock pressure occurrence. Prototype and field investigations (slope 1:4 and 1:6) show that a proportionality exists for shock pressures and slope angles (Eq. 6). For mostly used sloping faces of sea

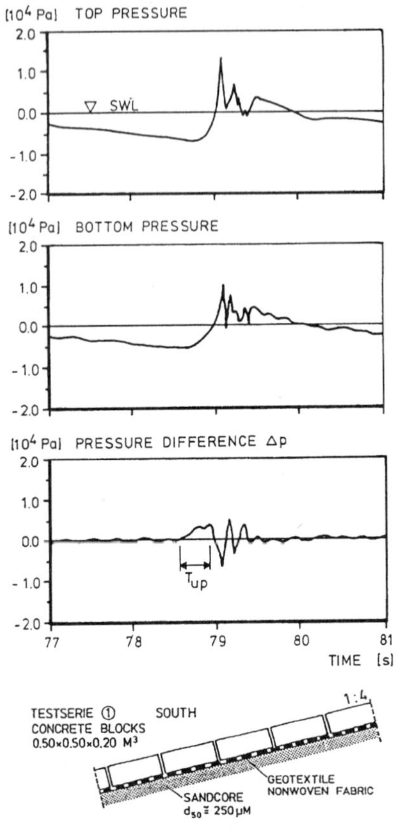

Figure 16. Block pressures Testserie ① South

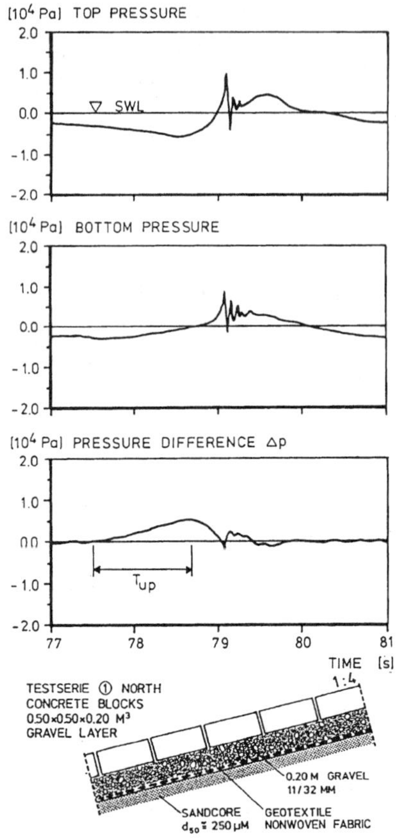

Figure 17. Block pressures Testserie ① North

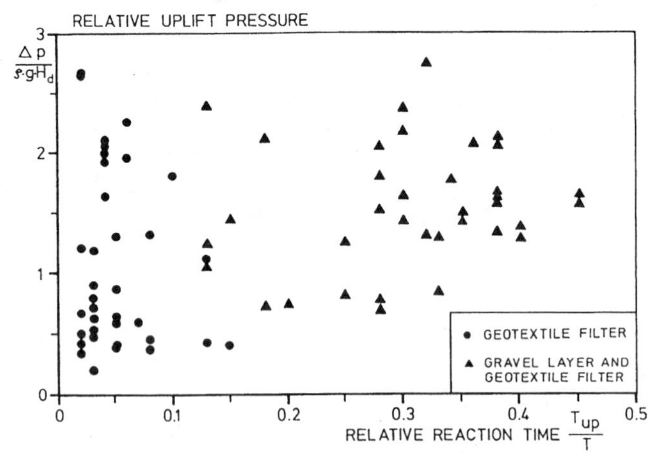

Figure 18. Relative uplift pressures versus relative reaction time (Testserie ①)

dykes in shallow water regions the following maximum shock pressures, acting nearly 0.5 H below SWL on the slope surface are to be expected:

$$\text{SLOPE 1:4} \qquad p_{max} \approx 6 \cdot \rho \cdot g \cdot H,$$
$$\text{SLOPE 1:6} \qquad p_{max} \approx 4 \cdot \rho \cdot g \cdot H.$$

Full-scale experiments on wave breaking phenomena are indispensable to avoid scale effects in small-scale modelling. Together with field investigations under real sea state conditions (Grüne, 1988) full-scale measurements guarantee highly reliable results for future design on sea dykes.

From the first analysis of the comprehensive full-scale measurements dealing with placed block revetments stability criteria are found. If the blocks are normally placed in stretcher-bond on a 2-dimensional geotextile filter, the stability parameter $H/\Delta \cdot d_B$ should not exceed 4. 1:2 scaled models can be used but variations in breaker characteristics should be considered carefully.

A 3-dimensional granular filter layer in a thickness equal to the block thickness reduces the stability considerably ($H/\Delta \cdot d_B \leq 2.5$).

Block failure occurs if the reaction time of the uplift pressure rises up to at least 1/3 of the wave period. Uplift pressure values roughly amout 2 to 3 times of the wave height measured at the toe of the dyke.

Better block stabilities can be reached by increasing the top layer permeability. For placed blocks on a geotextile filter the stability number $H/\Delta \cdot d_B \leq 5.5$ is obtained by uniformly distributed cylindrical gaps (1 % of the block surface).

Reported results can be applied to loose concrete blocks placed in stretcher-bond. Favourable stability effects due to block interlocking or connecting are not yet considered. It is proposed to carry out large-scale experiments for such block systems to find out most economical solutions.

ACKNOWLEDGEMENTS

The authors would like to acknowledge gratefully the German Research Foundation (DEUTSCHE FORSCHUNGSGEMEINSCHAFT) sponsoring the research program on safety and reliability of dykes and revetments for coastal protection (SFB 205 / A1).

REFERENCES

Bagnold,R.A.: Interim Report on Wave Pressure Research. Journal of the Institution of Civil Engineers, Vol. 12, London, 1939

Bezuijen,A.; Klein Breteler,M.; Pilarczyk,K.W.: Large-Scale Tests on a Block Revetment Placed on Sand with a Geotextile as Separation Layer. Proc. 3rd Int. Conf. on Geotextiles, Vienna, 1986

Bezuijen,A.; Klein Breteler,M.; Bakker,K.J.: Design Criteria for Placed Block Revetments and Granular Filters. Proc. 2nd Int. Conf. on Coastal and Port Engineering in Developing Countries, Beijing, 1987

Bölke,S. and Relotius,P.C.: Über die wellenerzeugten Druckschlagbelastungen von Seedeichen im Böschungsbereich zwischen 1:4 und 1:6. Mitt. Leichtweiß-Institut, Tech. Univ. Braunschweig, Heft 42, 1974 (in German)

den Boer,K.; Kenter,C.J.; Pilarczyk,K.W.: Large Scale Model Tests on Placed Block Revetments. Proc. Coastal Structures'83, Arlington, 1983

Bruun,P.: Design and Construction of Mounds for Breakwaters and Coastal Protection. Elsevier, Amsterdam, 1985

Führböter,A.: Der Druckschlag durch Brecher auf Deichböschungen. Mitt. Franzius-Institut, Tech. Univ. Hannover, Heft 28, 1966 (in German)
Führböter,A.: Laboratory Investigations of Impact Forces. Proc. Symp. Research on Wave Action, Tech. Univ. Delft, Paper No. 6, Delft, 1969
Führböter,A.: Zufallsprozesse bei der Belastung durch brechende Wellen. Jahresbericht 1970, Sonderforschungsbereich 79, Tech. Univ. Hannover, 1971 (in German)
Führböter,A.: Die Forschungseinrichtung "GROSSER WELLENKANAL" (Hannover) und ihre Aufgaben. Berichtswerk Intermaritec '82, Hamburg, 1982 (in German)
Führböter,A.: Model and Prototype Tests for Wave Impact and Run-Up on an Uniform 1:4 Slope. Coastal Engineering, Elsevier Science Publishers, 10, 1986a
Führböter,A.: Hydrodynamische Belastungen der Sohlsicherung des Eidersperrwerkes. Bauingenieur 61, Springer-Verlag, 1986b (in German)
Führböter,A.; Dette,H.H.; Grüne,J.: On the Response of Seadykes due to Wave Impacts. Proc. 15th Int. Conf. on Coastal Engineering, Honolulu, 1976
Grüne,J.: Wave Induced Shock Pressure under Real Sea State Conditions. Proc. 21th Int. Conf. on Coastal Engineering, Malaga, Spain, 1988
Grüne,J.; Führböter,A.: Large Wave Channel for "Full Scale Modelling" of Wave Dynamics in Surf Zones. Proc. Symp. on Modelling Techniques, San Francisco, 1975
Grüne,J.; Sparboom,U.: Großer Wellenkanal-Problemstellung und Lösung aus versuchstechnischer und konstruktiver Sicht. Technischer Bericht, Sonderforschungsbereich 79, Universität Hannover, 1982 (in German)
Heerten,G.: Geotextiles in Coastal Engineering-25 Years Experience. Geotextiles and Geomembranes, Vol. 1, Elsevier Applied Science, Amsterdam, 1984
Kostense,J.K.; den Boer,K.: Effect of Model Scale on the Stability on Concrete-Block Slope Revetments. Proc. Symp. on Scale Effects in Modelling Hydraulic Structures, Esslingen, 1984
Pilarczyk,K.W.: Dutch Guidelines on Dyke Protection. Report WB-NO-87110, Rijkswaterstaat, Delft, 1987
Powell,K.A.; Allsop,N.W.H.; Owen,M.W.: Design of Concrete Block Revetments Subject to Wave Action-A Literature Review. Rep. No SR 54, Hydraulics Research Limited, Wallingford, 1985
Skladnev,M.F. and Popov,J.Ya.: Studies of Wave Loads on Concrete Slope Protections of Earth Dams. Proc. Symp. Research on Wave Action, Tech. Univ. Delft, Paper No. 7, Delft 1969
Sparboom,U.: Konzeption von Deicheinbauten im GROSSEN WELLENKANAL. Technischer Bericht, Sonderforschungsbereich 205, Universität Hannover, 1987 (in German)
Stephan,H.J.: Über Schäden an Seedeichen durch Wellen- und Druckschlagbelastungen. Mitt. Leichtweiß-Institut, Tech. Univ. Braunschweig, Heft 70, 1981 (in German)
Stive,R.J.H.: Wave Impact on Uniform Steep Slopes at Approximately Prototype Scale. Proc. Symp. on Scale Effects in Modelling Hydraulic Structures, Esslingen, 1984
TAW/CUR-VB-Technische Adviescommissie voor de Waterkeringen/Stichting voor onderzoek, voorschriften en kwaliteitseisen op het gebied van beton: Leidraad cementbetonnen dijkbekledingen. Rapport 119, Betonvereniging, Zoetermeer, 1984 (in Dutch)
Townson,J.M.: A Lower Bound for the Thickness of Revetment Blocks. Proc. Institution of Civil Engineers, Part 2, Vol. 83, 1987
Weggel,R.L.: Discussion on "Shock Pressures on Coastal Structures (by Kamel)". ASCE, J. of the Waterways and Harbour Division, WW 3, 1971
Whillock,A.F.: The Stability of Revetment Blocks under Wave Attack. Rep. No. IT 195, Hydraulics Research Station, Wallingford, 1980

CHAPTER 162

FLOATING BREAKWATER RESPONSE TO WAVE ACTION

Michael Isaacson[1], M.ASCE and Ronald Byres[2]

ABSTRACT

The present paper describes a study carried out to investigate floating breakwater behavior in waves. Components of the study include a field survey of floating breakwaters in British Columbia, Canada, the development of a numerical model of breakwater behavior and the experimental testing of a particular breakwater design. The numerical model has been developed to provide breakwater motions, transmission coefficients and mooring forces. The model combines linear diffraction theory for obliquely incident waves, a mooring analysis, the inclusion of viscous damping coefficients obtained from experimental or field data, and the inclusion of drag and wave drift forces for use in the static analysis of the moorings. The experiments were carried out with normally incident regular waves of different heights and periods. Preliminary results indicate that the numerical model should prove to be a useful tool in floating breakwater design.

INTRODUCTION

Floating breakwaters have found extensive application in many areas where relatively inexpensive protection from wind- and ship-generated waves is required and where open water wave conditions are not unduly severe and water depths are relatively large. The cost of traditional bottom-founded breakwaters increases significantly with water depth, so that a floating breakwater is a relatively attractive option in deeper water.

[1]Professor, Department of Civil Engineering, University of British Columbia, Vancouver, B.C. Canada.
[2]Engineer, Sandwell Swan Wooster, 1190 Hornby Street, Vancouver, B.C. Canada.

A considerable literature on the subject has developed over the past decades, with the large variety of specific breakwater designs as well as the large number of areas of potential application having contributed to this wealth of information. Western Canada Hydraulics Laboratory (1981) carried out an extensive literature survey, covering topics ranging from analytical models of breakwater behavior to in-situ experiences with particular breakwater designs.

Numerical methods of describing breakwater response to waves have originated largely from ship hydrodynamics. Several authors have treated the two-dimensional problem of wave interaction with cylinders at or near the water surface by considering the corresponding potential flow problem and solving this to calculate the hydrodynamic coefficients necessary to determine the fluid forces on, and the motions of the cylinder. In particular, the case of oblique wave interaction with cylinders has been investigated by Bai (1975), Garrison (1969, 1984), Isaacson and Nwogu (1987) and others. Field studies involving prototype floating breakwaters are relatively uncommon. A number of authors have published field data including Nelson *et al.* (1983), Nece and Skjelbreia (1984) and Miller and Christensen (1984). Comparisons of field data with laboratory tests and numerical models have generally indicated that the response of breakwaters can be modeled quite well. However, one difficulty is that responses are lower than the inviscid theory predictions at frequencies near the resonant frequencies of the breakwater, primarily because of additional energy dissipation associated with flow separation.

The present paper describes a study carried out to investigate floating breakwater behavior in waves, with components of the study including a survey of floating breakwaters in British Columbia, Canada; the development of a numerical model of breakwater behavior which predicts transmission coefficients, breakwater motions and mooring forces for a specified breakwater/mooring system and specified incident wave conditions; and thirdly the experimental testing of a particular floating breakwater design.

FLOATING BREAKWATERS IN BRITISH COLUMBIA

British Columbia, Canada, contains relatively large areas of sheltered coastal waters, associated with protection from the open ocean provided by Vancouver Island and the presence of a large number of inlets and sounds. There is extensive use of pleasure craft in the region and a considerable number of floating breakwaters are located here. Locations of thirty more significant floating breakwaters presently in use in British Columbia are indicated in Fig. 1. In general, three categories of breakwater have been used:

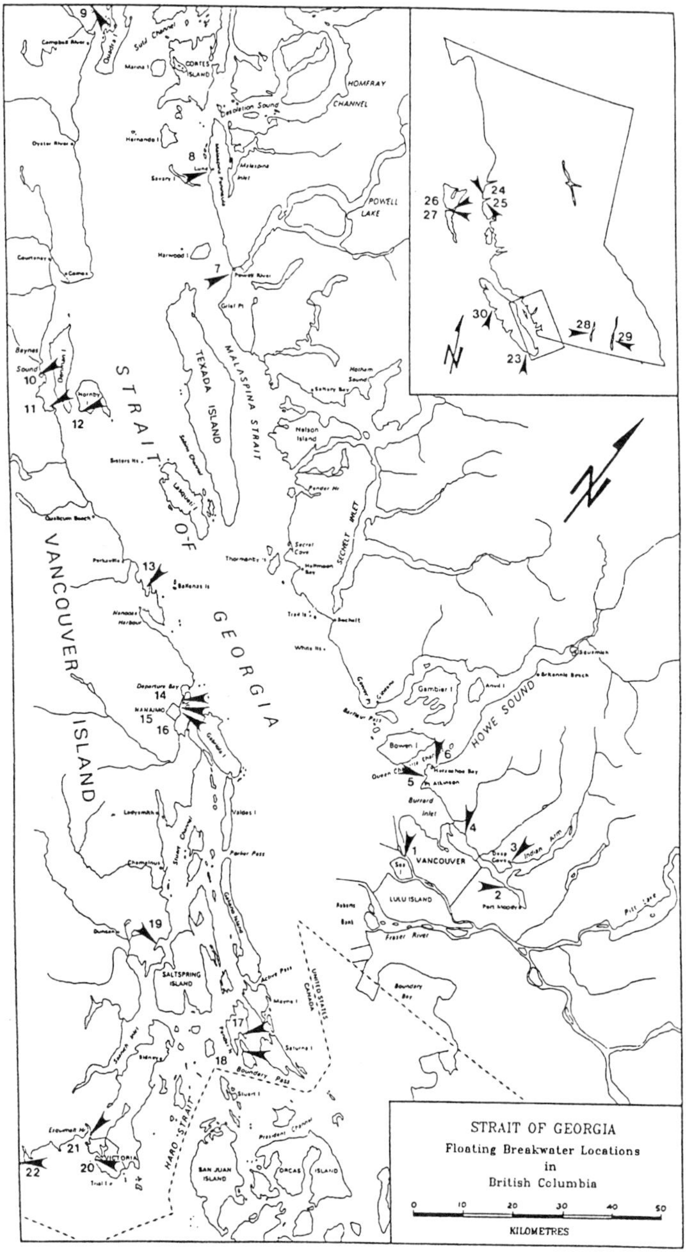

Fig. 1. Floating breakwater locations in British Columbia.

- concrete caisson breakwaters, which generally have a rectangular cross-section,
- log bundles, which generally have a circular or raft-like cross section,
- A-frame breakwaters, which generally include two pontoons and a central vertical plate.

These primarily rely on wave reflection to reduce transmitted wave heights. In addition, a scrap tire breakwater, which reduces the transmitted wave heights primarily by energy dissipation, is used at one location. Table 1 lists floating breakwaters in British Columbia corresponding to the locations shown in Fig. 1 and includes summary information on the fetch and principal wind direction for each location. Of the thirty breakwaters listed, eleven are caisson, fifteen are log and two are A-frame. Typical design wave conditions correspond to a wave period of about 4 sec and significant wave height of about 0.5 m. In addition, there is generally a large tidal range of up to about 5 m. The majority of breakwaters have generally provided quite satisfactory service. This is particularly true of the concrete caissons which tend to be more durable than the other designs and which are more amenable to secondary usage. Although they have an initially higher capital cost, the concrete breakwater systems have a lower incidence of structural damage and provide a greater degree of wave attenuation than either scrap tire breakwaters or log bundles.

As one example of a concrete caisson breakwater, a new three-module caisson breakwater was installed at Lund in May, 1987, replacing an earlier A-frame which had served at that location for 22 years. A photograph of the new breakwater during installation is shown in Fig. 2. Lund is perhaps exposed to the most severe wave climate of any caisson breakwater in the Strait of Georgia. The site is exposed primarily to waves from the southwest and northwest, with fetches of about 30 km and 12 km respectively. The layout of the breakwater is somewhat unusual in that the caisson sections are not inter-connected, but rather are staggered in the horizontal plane in an effort to avoid problems arising with inter-connections or collisions between units. The disadvantages of this arrangement include a reduction in the potential area of protection (because of the overlap of the sections), a reduction in the potential degree of protection because of wave diffraction through the gaps, and a relatively complicated mooring line arrangement.

Floating log bundles are used extensively throughout British Columbia, although in their simplest application - a mere boom of single logs - they are not particularly effective against any but the shortest period waves. An example at Reed Point Marina is shown in Fig. 3, and indicates wave reflection and diffraction around the end of the log bundle breakwater. In this case, the waves were caused by the passage of a tug and are estimated to have a height of 0. 5 m and period of 3 sec.

Fig. 2. The new caisson breakwater at Lund during installation.

Fig. 3. The log bundle at Reed Point Marina, showing wave reflection and diffraction.

The A-frame breakwater at Lund which was in use until 1987 is shown in Fig. 4. The breakwater has steel pontoons of diameter 0.76 m, a beam of 7.6 m, and a draft of 3.7 m. The timber centerboard is connected to the pontoon with a steel space-frame, and extends upwards from the still water level approximately 2 m. The identical design has been used for the A-frame breakwater at Queen Charlotte City.

Only one breakwater in British Columbia involves scrap tires and is located at Eagle Harbor. The breakwater consists of two rows of cylindrical steel pontoons between which scrap tires are strung on conveyor belting.

A unique floating breakwater at Powell River comprises of ten old concrete-hulled ships and is shown in Fig. 5. This has been in use since about 1930, having been gradually expanded to the present configuration by the use of additional ships. The ships range in length from 102 m to 128 m and are anchored with eight to ten concrete anchors, each weighing up to 14.5 tonnes.

Although the breakwaters indicated in Table 1 have generally provided satisfactory service, possible difficulties which have been reported to arise with floating breakwaters have involved their inability to provide adequate wave protection, and possible damage or failure most often associated with connections between individual units of a breakwater or with its moorings.

NUMERICAL MODEL

A numerical model of floating breakwater behavior due to wave action has been developed. Linear diffraction theory is used for a two-dimensional breakwater section to provide the breakwater motions and transmission and reflection characteristics for a regular, obliquely incident wave train. The method has been described by Isaacson and Nwogu (1987) and is based on a boundary integral equation approach deriving from Green's theorem. The breakwater is treated as an infinitely long horizontal cylinder and the fluid is assumed incompressible and inviscid and the flow irrotational so that potential flow theory is used. The velocity potential Φ of the flow is considered to be made up of components associated with the incident waves ϕ_o, the diffracted waves which would arise if the cylinder were fixed, ϕ_4, both of these components being proportional to the incident wave height H, and forced waves associated with each of three modes of motions of the cylinder, ϕ_1, ϕ_2 and ϕ_3, corresponding to sway, heave and roll respectively. The latter potentials are proportional to the amplitude of the motion ξ_j of each mode. Thus the total velocity potential is expressed as:

Fig. 4. The A-frame breakwater at Lund (removed in 1987).

Fig. 5. The floating breakwater at Powell River, comprised of ship hulls.

$$\Phi = \left[\frac{-i\omega H}{2} (\phi_o + \phi_4) + \sum_{j=1}^{3} -i\omega \xi_j \phi_j \right] \exp[i(ky\sin\beta - \omega t)] \qquad (1)$$

where $i = \sqrt{-1}$, ω is the wave angular frequency, y is the distance along the cylinder axis, β is the angle the incident wave crests make with the cylinder axis, k is the wave number, and t is time.

The equation of motion of the breakwater can be expressed in the usual way as:

$$\sum_{j=1}^{3} [-\omega^2(m_{ij} + a_{ij}) - i\omega(b_{ij} + b'_{ij}) + c_{ij}]\, \xi_j = F_i \quad \text{for } i = 1, 2, 3 \qquad (2)$$

where m_{ij} and c_{ij} are the mass and hydrostatic stiffness coefficients; and a_{ij} and b_{ij} are the added mass and damping coefficients associated with the hydrodynamic forces, and F_i are the exciting force amplitudes. b'_{ij} represents an empirical damping coefficient accounting for viscous effects which is used to modify motion predictions from those based solely on potential theory. The hydrodynamic coefficients a_{ij}, b_{ij} and F_i may be obtained in terms of the potentials ϕ_o, ..., ϕ_4. The transmission and reflection coefficients, defined as the ratio of transmitted or reflected wave heights to the incident wave height, are also of interest and are obtained by computing wave conditions on control surfaces upstream and downstream of the structure.

A suitable extension of this program provides the response to a random incident wave train, which may be obliquely incident, and which may be short-crested (see Isaacson and Nwogu, 1987). In particular, the program provides spectra of the transmitted wave train and of the component breakwater motions. These may be used to estimate the maximum transmitted wave heights and breakwater motions for a storm of specified duration.

The program may be used in conjunction with a three-dimensional mooring analysis program, which utilizes the predicted breakwater motions to provide maximum mooring line tensions.

EXPERIMENTAL STUDY

Laboratory experiments have been carried out to measure transmission coefficients and breakwater motions using two particular breakwater models subjected to normally incident, regular waves. One is a rectangular section breakwater, while the other is an A-frame breakwater. Measurements have been made of the breakwater motions and of the transmission coefficient for a range of incident wave conditions.

The experiments were carried out in the wave basin of the Hydraulics Laboratory at the University of British Columbia. This is 14 m long, 5 m wide and can accommodate water depths up to 0.45 m and wave periods ranging from about 0.5 to 1.5 sec. A variable speed electrically driven wave paddle at one end of the basin was used to provide regular long-crested waves, with the amplitude and period varied by adjusting the stroke and frequency of the paddle. A sloping beach consisting of a frame covered with artificial hair matting was located at the opposite end of the basin to reduce wave reflections. Capacitance type wave probes were used to measure the incident and transmitted wave heights. For all the tests, a vertical wall was installed in the basin to prevent waves reflecting from the model from interfering with the incident wave signal. The rectangular breakwater used was based on the concrete caisson breakwater employed in field tests reported by Nece and Skjelbreia (1984). This corresponds to a prototype beam of 4.8 m, a length of 23 m and a draft of 1.1 m, and the model was constructed to a scale of 1/15. Experiments were carried out at several different wave periods, with separate tests carried out at each wave period with waves of low and high steepnesses.

The transmission coefficient was measured by the use of wave probes, and the breakwater motions were measured by an optical system. Three reference markers were attached to the model at known positions vertically above the breakwater's center of gravity, with fixed scales located behind each target. The amplitudes of motion, but not the phases, were measured by sighting the targets and scales through a theodolite, and aligning the cross-hairs in turn on each target at its maximum horizontal or vertical excursion. Simple algebraic formulae were used to relate these motion amplitudes to those of the breakwater's center of gravity.

A preliminary comparison with numerical predictions for the case of the rectangular breakwater has been carried out and selected results are indicated in Figs. 6 and 7. In the figures, a is the half beam of the breakwater. Fig. 6 compares numerical results of the transmission coefficient with measured values, as well as with the field measurements reported by Nelson *et al.* (1983) and Nece and Skjelbreia (1984). There is considerable scatter in the measured values so that the comparison is somewhat inconclusive. The values obtained from field measurements are relatively high and relatively insensitive to wave frequency. Fig. 7 shows the heave response amplitude operator, with the numerical predictions based on various values of the viscous damping coefficient. The trend of the response to reduce with increasing wave frequency is apparent.

As expected, the numerical predictions confirm that the breakwater is most effective for higher frequency waves, corresponding approximately to $B/L > 0.3$, where B is the width of the rectangular section and L is the incident wave length. The roll and sway motions exhibit strong resonant peaks, which are quite sensitive to the value of the viscous damping

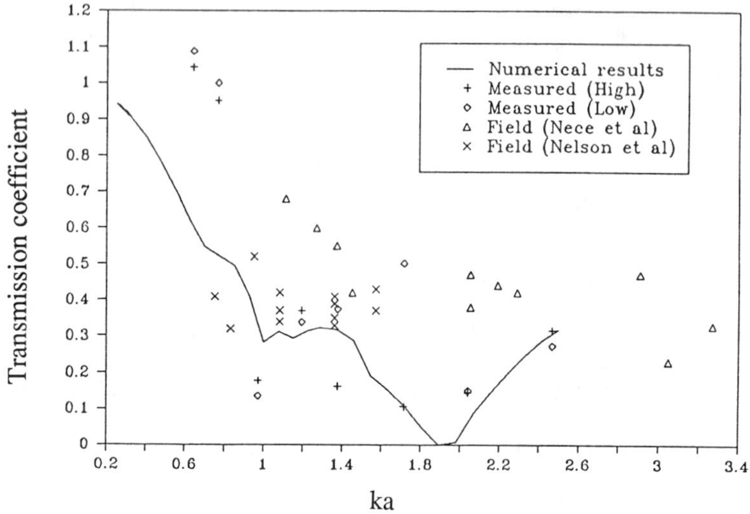

Fig. 6. Numerical, laboratory and field results of the transmission coefficient for the rectangular breakwater.

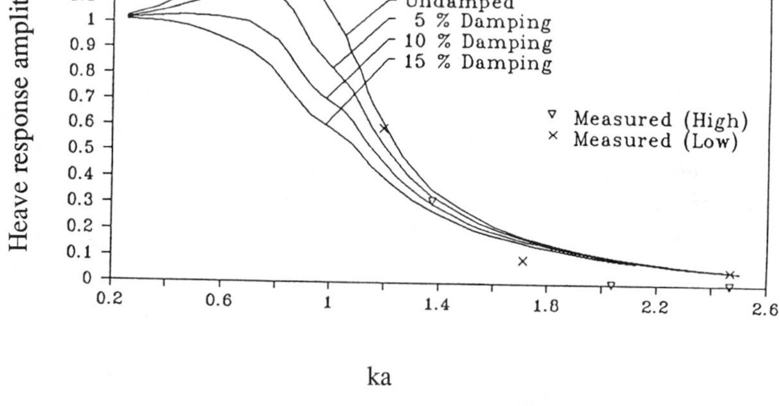

Fig. 7. Numerical and laboratory results of the heave response amplitude operator for the rectangular breakwater.

coefficient selected. With only 5% damping, these resonant peaks are all but removed, which is consistent with the observations of other studies (e.g. Miller and Christensen, 1984).

CONCLUSIONS

Floating breakwaters are a viable means of providing wave protection in areas where incident wave conditions are not too severe and water depths are relatively large. They have been widely used in British Columbia, Canada. Although experience in British Columbia has not revealed major difficulties in the use of floating breakwaters, other reported difficulties associated with their use include a possible inability to provide adequate protection from waves, and possible damage or failure most often associated with connections between individual units of a breakwater or with its moorings.

A numerical model based on linear diffraction theory and extended to include the effects of moorings, obliquely incident, random waves and additional damping has been developed. The program has been verified in part against laboratory measurements and should be quite viable for use in floating breakwater design. Laboratory tests have been carried out, but a comparison with numerical predictions is somewhat inconclusive.

APPENDIX - REFERENCES

Bai, K.J. 1975. "Diffraction of oblique waves by an infinite cylinder," *Journal of Fluid Mechanics*, **68**, pp. 513-535.

Garrison, C.J. 1969. "On the interaction of an infinite shallow-draft cylinder oscillating at the free surface with a train of oblique waves," *Journal of Fluid Mechanics*, **39**, pp. 227-255.

Garrison, C.J. 1984. "Interaction of oblique waves with an infinite cylinder," *Applied Ocean Research*, **6** (1),.pp. 4-15.

Isaacson, M. and Nwogu, O. 1987. "Directional Wave Effects on Long Structures," *Journal of Offshore Mechanics and Arctic Engineering*, **109** (2), pp. 126 - 132.

Miller, R.W., and Christensen.D.R. 1984. "Rigid body motion of a floating breakwater," *Proc. International Conf. on Coastal Engineering*, Houston, Texas, pp. 2663-2679.

Nece, R.E., and Skjelbreia. 1984. "Ship-wave attenuation tests of a prototype floating breakwater," *Proc. International Conf. on Coastal Engineering*, Houston, Texas, pp. 2515-2529.

Nelson, E.E., Christensen, and Schuldt, A.D. 1983. "Floating breakwater prototype test program." *Proc. Conf. Coastal Structures '83*, ASCE, pp. 433-446.

Western Canada Hydraulics Laboratory. 1981. "Development of a manual for the design of floating breakwaters." *Canadian Manuscript Rept. of Fisheries and Aquatic Sciences*, No. 1629, Small Craft Harbors Branch, Dept. of Fisheries and Oceans, Ottawa, 228 pp.

Table 1. Summary of Floating Breakwaters in British Columbia

No.	Location	Type	Year	Fetch (km)	Dirn.
1	Richmond	Caisson	1979	1.7	W
2	Port Moody	Log Bundle	1976	0.8	NE
3	Deep Cove	Caisson	1976	5.5	NE
4a	Burrard YC	Caisson/barge	1977	3.1	E
4b	Burrard YC (Destroyed 1983)	Scrap tire	1977	3.1	E
5	Eagle Harbour	Pontoon / tire	1977	39.1	SW
6	Horseshoe Bay	Caisson / ship		9.0	N
7	Powell River	Ship Hull	1930+	28.7	SW
8a	Lund	Caisson	1987	29.6	SW
8b	Lund (Removed 1987)	A-frame	1963	29.6	SW
9	Brown Bay	Tank Car	1983	2.8	E
10	Fanny Bay	Log Bundle		2.6	NE
11	Deep Bay	Log Bundle		2.2	W
12	Ford Cove	Log Raft		6.5	NW
13	Northwest Bay (Destroyed 1980)	Log / Styrofoam	1975	37	
14	Nanaimo			0.2	E
15	Nanaimo	Caisson	1974	5.3	SE
16	Nanaimo				
17	Pt Browning	Log		2.1	SE
18	Bedwell Hbr	Log		3.0	NW
19	Maple Bay	Caisson	1977	4.3	NE
20	Victoria Hbr	Caisson		0.5	S
21	Esquimault	Caisson		1.8	SW
22	Becher Bay	Log		1.5	SE
23	Sooke Basin	Log		1.9	SW
24	Prince Rupert	Log			
25	Prince Rupert	Log			
26	Qn Charlotte City	Log		4.4	
27	Qn Charlotte City	A-frame	1967	2.4	
28	Kelowna	Log Bundle	1978		W
29	Nakusp	Caisson	1986		
30	Tahsis	Log		2.8	SE

CHAPTER 163

SIMULATION OF WAVE FORCES ON A HORIZONTAL CYLINDER

SUSUMU TSUZUKI*
KIYOSHI HORIKAWA** F.ASCE
AKIRA WATANABE***

ABSTRACT

The characteristics of wave forces acting on a horizontal circular cylinder were investigated through numerical calculations as well as experimental findings. The laboratory data on wave forces were analyzed by the concept of wave force path and classified into two types. One is the circular type and the other one is the 8 - shaped type. In order to analyze the above phenomena, the discrete vortex method was applied with appropriate assumptions. The comparison between the numerically calculated results and laboratory data shows that the simulation model proposed in this paper seems to be favorable to predict the wave forces acting on a horizontal circular cylinder within a certain range of conditions.

1.INTRODUCTION

Prediction of wave forces acting on a circular cylinder has been investigated by number of researchers during the last thirty eight years since Morison et al. (1950) proposed their semi-empirical formula. The main interest of these researchers was to evaluate the hydrodynamic coefficients introduced in the stated formula (Koderayama et al., 1978). However, the formula cannot be applied to such a case that vorticies generated behind a circular cylinder are extremely unsymmetrical with respect to wave direction.

In the present paper, an analytical and numerical treatment was attempted to calculate the flow characteristics around a horizontal circular cylinder induced by gravity waves or oscillatory flows and the fluid forces acting on the body by using the discrete vortex

* Engineer, Tokyo Electric Co.
** Professor Emeritus, The University of Tokyo ; Professor, Saitama University, Urawa, Japan.
*** Professor, Department of Civil Engineering, The University of Tokyo, Japan.

method which was originally presented by Rosenhead (1931). The above treatment was carried out under such a restricted condition that the effects of free surface as well as of the bottom on fluid motion can be neglected. That is to say, it is assumed that the submerged cylinder is located in a certain range of water depth apart from the limited layers of both free surface and sea bed. The calculated results were compared with the experimental data reported by Sarpkaya (1975) in order to investigate the adaptability of the stated prediction method.

2. CHARACTERISTICS OF WAVE FORCES ACTING ON A HORIZONTAL CIRCULAR CYLINDER

2.1 Previous Studies

It is well known that the Morison formula is commonly applied to evaluate the wave forces acting on a cylinder, particularly on a circular cylinder. However, in the case of a submerged circular cylinder set horizontally at a certain water depth, the direction of wave force acting on the cylinder varies with time, the situation of which is different from that of a vertical cylinder. Hence the Morison formula was extended by Borgman (1958) to the vectorial form, under the assumption that the direction of drag force is the same to that of fluid velocity.

However the above assumption is not necessary to be true. Therefore Sawamoto et al. (1979) investigated the fluid force induced by oscillatory flows, and Tsuzuki et al. (1984) and Masuda et al. (1985) did by wave by introducing a phase difference. According to the result of flow visualization conducted at the University of Tokyo, it is realized that the Morison type formula can not be applied to the case in which the motion of vortices separated from the body surface is complicative.

As a theoretical approach, Sarpkaya (1968) applied the Blasius formula for unsteady flow to calculate the fluid force on a circular cylinder under uniformly oscillating flows. Detailed observations made by Sawamoto and Kikuchi (1979) in an oscillatory flume indicated that the pattern of vortex formation can be classified by using the Keulegan-Carpenter number as shown in Figure 1. The Keulegan-Carpenter number is defined by $K.C.=\hat{U}T/2R$, where $\hat{U}$ is the amplitude of oscillatory flow velocity, T the period of oscillatory flow, and R the radius of the circular cylinder. From this diagram we can realize that the unsymmetrical vortices appear under the condition of $K.C. > 8$, causing nearly regular time variation of lift force.

2.2 Experimental Findings

In order to look at the above phenomena in an oscillatory flume more clearly, let us take the wave force path defined by the diagram illustrated in Figure 2. Here two kinds of typical laboratory data were picked up and drawn in Figure 3 for demonstration. The upper diagram indicates that the wave fore path for $K.C. = 4.24$ forms a

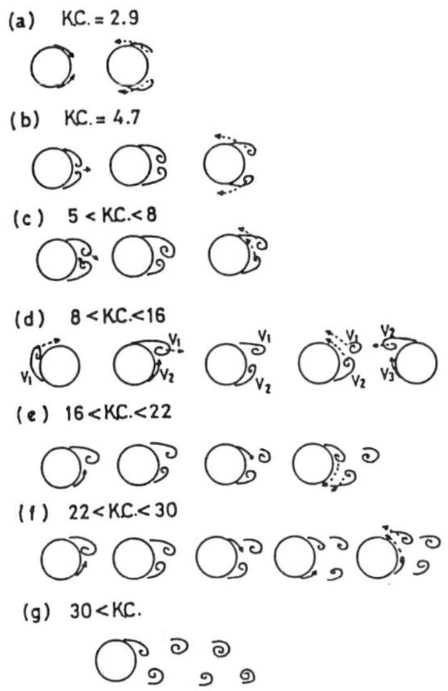

Fig.1 Vortex formation in an oscillatory flow (after Sawamoto and Kikuchi, 1979).

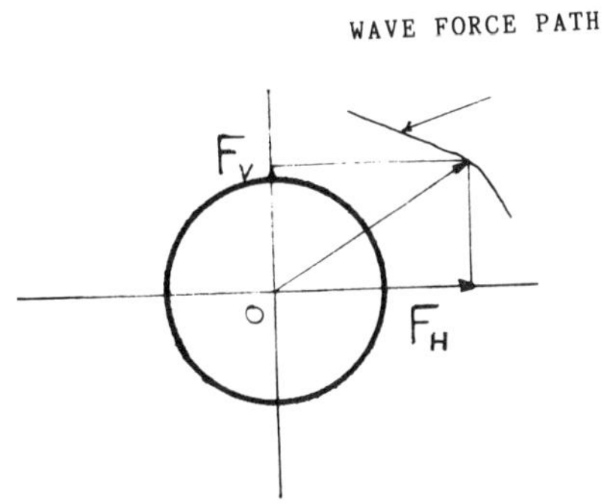

Fig.2 Definition of wave force path.

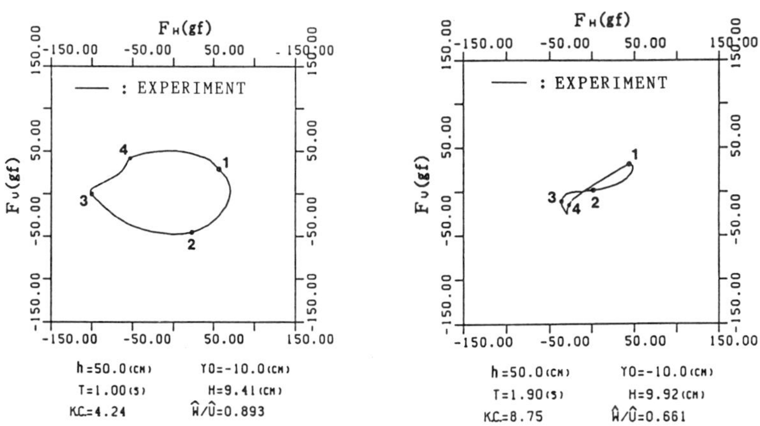

(a) circular type

(b) 8-shaped type

Fig.3 Typical examples of wave force path.

nearly circular curve, while the lower diagram indicates that the path for K.C. = 8.75 forms an 8 - shaped curve.

The situation of vortex formation in a wave field might be somewhat different from that in a uniformly oscillating flow. The laboratory data collected in a wave flume at the University of Tokyo were plotted as shown in Figure 4. The abscissa of this diagram is the K.C. number and the ordinate is the amplitude ratio between the vertical and horizontal fluid particle velocity amplitudes, $\hat{V}$ and $\hat{U}$. The open marks are for the circular type wave force path, which is called C type, and the filled marks for the wave force path with the 8 - shaped curve, which is E type. The upper part of the ordinate in this diagram corresponds to the deeper water wave condition, while the lower part corresponds to the shallower water wave condition. A demarcation curve between these two types of wave force path can not be clearly drawn. However, it can be said that E type appears under the shallower water and larger K.C. number conditions.

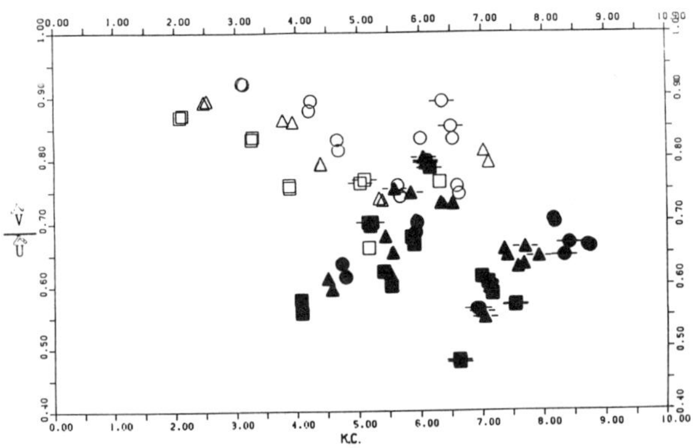

C	E	y_o (cm)	h (cm)
○	●	-10	
△	▲	-13	60.0
□	■	-16	
-○-	-●-	-10	
-△-	-▲-	-15	50.0
-□-	-■-	-20	

Fig.4

Wave force path data obtained in a wave flume.

3. SIMUMATION OF WAVE FORCES ACTING ON A HORIZONTAL CIRCULAR CYLINDER

3.1 Treatment for Simulation

(1) Discrete vortex method

The aim of this paper is to analyze the stated phenomena under appropriate conditions and assumptions. In the present treatment the discrete vortex method is applied. The following analysis is based on the potential flow theory in which the discrete vortices emitted from the horizontal circular cylinder surface are introduced. At first the Milne-Thomson circular theorem (Milne-Thomson, 1968) is used to determine the complex potential of the interested flow, however the free surface and bottom boundary make the analytical treatment extremely complex. According to Ogilvie (1963) and Chaplin (1981) it is concluded that the free surface effect on the boundary condition can be neglected when the ratio between the submerged water depth and the circular cylinder diameter is large enough, say 5. While Nath and Yamamoto (1974), and Uekita and Yamazaki (1980) investigated the effect of bottom boundary on the complex potential of the stated phenomena, and concluded that the bottom boundary effect can be neglected when the clearance between the lower portion of the cylinder and the bottom is larger than the circular cylinder diameter. Taking into account the above two results we will select an appropriate range of submergence of the circular cylinder in order to proceed the analysis in neglecting the surface and bottom boundary effects.

Now let us assume that the questioned complex potential, W, is the summation of those of potential flow surrounding a circular cylinder and of distrete vortices, W_p and W_v. That is to say,

$$W = W_p + W_v \tag{1}$$

(2) Determination of W_p

The term Wp can be expressed in the next equation by using the Milne-Thomson circular theorem,

$$W_p = W_{p0}(z) + \overline{W_{p0}(R^2/\bar{z})} \tag{2}$$

where $W_{po}(z)$ is the complex potential for the flow field where the circular cylinder does not exist, and z=x+iy. In order to express the term explicitely, we are able to apply an appropriate wave theory such as the small amplitude wave theory, the finite amplitude wave theory, and the stream function theory. In case of the small amplitude wave theory

$$W_{p0} = \frac{\omega H}{2k \sinh kh} \sin(kz + iks_0 - \omega t) \tag{3}$$

where ω is the angular frequency, k the wave number, H the wave height, h the water depth, s_o the distance of the circular cylinder center from the bed, and t the time. While in case of the stream function theory (Dean, 1965)

$$W_{p0} = -A_0 z + \sum_n A_n \sin(k_n z + i k_n s_0 - \omega_n t - \beta_n) \tag{4}$$

where A_n, β_n are coefficient and phase lag appeared in the stream function theory, k_n = nk, and δ_n = nδ.

(3) Determination of W_v

The term W_v induced by discrete vortices can be expressed in the next equation by using again the Milne-Thomson circular theorem,

$$W_v = -\frac{i}{2\pi} \sum_k \Gamma_k \log(z - z_k) + \frac{i}{2\pi} \sum_k \Gamma_k \log(z - R^2/\bar{z}_k) \tag{5}$$

where Γ_k and z_k are the circulation and the complex position of each discrete vortex.

(4) Emission of a vortex and the vortex motion

In order to clarify the emission of a vortex from the circular cylinder surface, the location and the circulation of a vortex just before separating from the circular cylinder must be evaluated. From such a view point, the separation point is neccesary to be determined by any appropriate way. As a first step, we calculated the velocity distribution within a laminar boundary layer developed along the surface of a circular cylinder. However, it is quite natural that the above treatment can not be applied to the case with large K.C. numbers and large Reynods numbers. In order to determine the separation point in the above case, we decided to adopt the fllowing two conditions. The first one is that the shear stress on the cylinder surface, τ_b, is zero at the separation point, the condition of which is commonly applied to the steady boundary layer flow. The second one is that the arcwide pressure gradient on the surface is zero at the separation point. The latter condition is adopted by the following reasons ; that is, (1) the flow separation may appear in the vicinity of the point on the surface where the pressure reaches its maximum, and (2) the stated position can be easily determined by using the potential flow velocity outside the boundary layer.

In order to check which condition is more suitable for our purpose, we applied the above two to the uniformly oscillating flow and compared the evaluted time dependent separation point with the laboratory data obtained by Grass and Kemp (1979) as shown in Figure 5. The abscissa indicates the phase of flow and the ordinate indicates the location of flow separation point on the cylinder surface. Either of these two conditions is not adequate to predict precisely the location of flow separation, however the condition for pressure gradient is better than that for shear stress. Therefore we selected the condition of pressure gradient to determine the flow separation point in

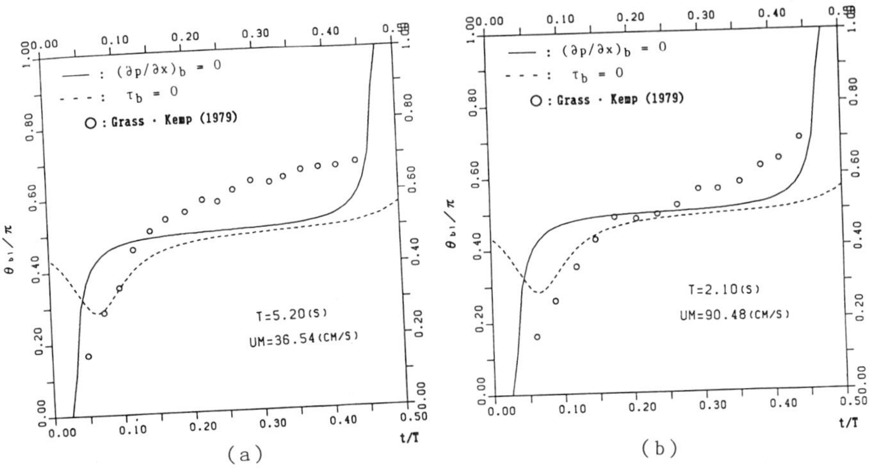

Fig.5 Comparison between predicted separation points and laboratory data

wave fields too.

The vortex flux, $\partial\Gamma/\partial t$, accumulated within the boundary layer in a unit time can be expressed by

$$\frac{\partial\Gamma}{\partial t}=\int_0^{\varepsilon}\left(\frac{\partial v}{\partial x}-\frac{\partial u}{\partial y}\right)u\,dy \tag{6}$$

where δ is the thickness of boundary layer. Considering that the term $\partial v/\partial x$ is negligible comparing with the term $\partial u/\partial y$, we applied the following equation which was proposed by Sarpkaya (1968) as an approximate expression,

$$\frac{\partial\Gamma}{\partial t}=\frac{1}{2}U_B|U_B| \tag{7}$$

where U_B is the velocity just outside the boundary layer at the separation point.

Next we assumed that the discrete vortex separated from the cylinder surface is moving with the velocity (u_k, v_k), which can be calculated by the following equation

$$u_k-iv_k=\frac{d}{dz}\left\{W+\frac{i\Gamma_k}{2\pi}\log\left(z-z_k\right)\right\} \tag{8}$$

Therefore the position of the questioned vortex at the time step, $t + \Delta t$, can be determined as

$$z_k(t+\Delta t)=z_k(t)+(u_k, v_k)\Delta t \tag{9}$$

(5) Wave forces acting on a circular cylinder

The wave force components can be obtained by the Blasius formula for an unsteady flow by using the complex velocity potential. That is

$$F_H - iF_V = \frac{i\rho}{2}\oint_C \left(\frac{dW}{dz}\right)^2 dz + i\rho\frac{\partial}{\partial t}\oint_C \overline{W} d\bar{z} \qquad (10)$$

where C is an arbitrary closed curve surrounding the questioned cylinder section. Introducing Equations (1), (2) and (5) into Equation (10), we get the next equation,

$$F_H - iF_V = 2\pi\rho R^2 \frac{\partial}{\partial t}\left(\frac{\partial W_{p0}}{\partial z}\right) + 2\pi\rho R^2 \frac{\overline{\partial W_{p0}}}{\partial z}\frac{\partial W_{p0}}{\partial z} - i\rho\sum_k \Gamma_k(u_k - iv_k) - \rho\sum_k \overline{i\frac{\partial}{\partial t}(\Gamma_k R^2/\bar{z}_k)} + i\rho\sum_k \Gamma_k \frac{\partial^2 W_{p0}}{\partial z^2}(z_k - R^2/\bar{z}_k) \qquad (11)$$

(6) Decay of vortex circulation

In the treatment of the discrete vortex method, it is assumed that the circulation of emitted vortex maintains its original value. However the above assumption can not treate the actual phenomena due to the decay of vortices by the fluid viscosity. Considering the above fact, we will introduce the decay of vortex in the following way to simulate the flow characteristecs surrounding the horizontal circular cylinder. That is to say, we assume that the discrete vortex is represented by a Rankine vortex with a core, the radius of which, r_V, can be expressed by

$$r_v = 2.24\sqrt{\nu t^*} \qquad (12)$$

where t^* is the elapsed time since the vortex emitted from the cylinder surface and ν is the kinematic viscosity of fluid. The term r_V corresponds to the core radius of Stokes vortex. In addition to the above, we also assume that the elapsed time, t^*, has its upper value T_C, which represents a kind of decay time. Hence the limited core radius, r_C, is given by

$$r_{cr} = 2.24\sqrt{\nu T_c} \qquad (13)$$

(7) Calculation procedures

The flow chart of the numerical calculation is shown in Figure 6. That is to say, emit a discrete vortex at each time step, then calculate the velocity field, wave force acting on a cylinder and the translation of the above vortex. Repeat the above calculation proceses until the time history of wave forces reaches its cyclic pattern. When two discrete vortices come in touch each other, let unite them to a single vortex in such a way to conserve the total angular momentum. When the vortex comes inside of the circular cylinder, return the vortex to the position where the vortex core comes in touch with the cylinder surface, and then let the vortex be convected by the induced flow velocity at the next time step.

3.2 Numerical Results and Verification

(1) Conditions of numerical calculations

Numerical calculations were made for the flow characterestics near the horizontally immersed circular cylinder 1) under a uniformly oscillating flow and 2) under progressive waves. Tables 1 and 2 give the various conditions of calculations for a uniformly oscillating flow and for progressive waves respectively.

Throughout the numerical calculations, the time histories of total wave force, fluid force induced by vortices, and circulation of emitted vortices were printed out in graphical forms for three to four wave cycles. In addition to these, streamlines at various time steps were illustrated to look at visually the behavior of vortices emitted from the cylinder surface.

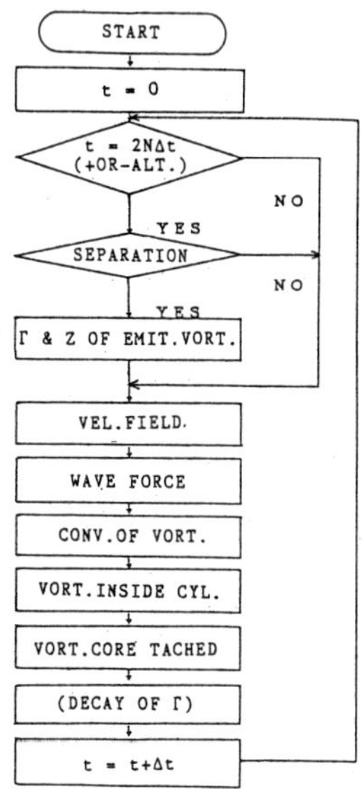

Fig.6

Flow chart

(2) Wave forces

In case of uniformly oscillating flow, it is clearly observed from the calculated results that the vortex motion becomes unsymmetrical and induces the lift force, the magnitude of which is comparable to the drag force in the direction of flow with the increase of K.C. number.

While in the case of progressive waves, the following conclusions were obtained :

1) The lift force has the magnitude comparable to the drag force and is mainly induced by the vortex motion.
2) In general the time histories of wave forces are apt to be cyclic by the introduction of vortex decay.
3) Under the deep water wave condition, the wave force has a single modal peak. But under the intermediate water wave condition, the drag force has a single peak over one wave cycle, while the lift force has bimodal peaks due to the unsymmetrical behavior of emitted vortices.
4) In Case W-1, the flow pattern and calculated wave forces were not influenced by the strength of vortex decay. However in Cases W-2 and W-3, these strongly depend on the strength of vortex decay.

(3) Wave force path

In order to investigate the effect of vortex decay, the calculated wave force pathes based on the stream function theory were compared with the laboratory data of Masuda et al. (1985) as shown in Figure 7. The solid lines are the experimental results, while the dotted lines are the calculated ones. The calculated curves in the upper

Table 1 Conditions for a uniformly oscillating flow.

(a) Calculation conditions

Case	R (cm)	1 (cm)	T (s)	U cm/s	K.C.	T/Δ t
S-1	2.5	20.0	1.0	20.5	4.1	60
S-2	2.5	20.0	1.4	22.9	6.4	80
S-3	2.5	20.0	1.9	21.8	8.3	120

R : circular cylinder diameter, 1 : cylinder length, T : oscillatory flow period, U : horizontal velocity amplitude, K.C. : Keulegan-Carpenter number, Δt : time step.

(b) Calculation runs for each case

Run	Flow pattern	Decay of vortex
1		Without
2	a) Symmetrical	With (Tc = T)
3		With (Tc = T/2)
4		With (Tc = T)
5	b) Unsymmetrical	With (Tc = T/2)

Table 2 Conditions for progressive waves.

(a) Calculation conditions

Case	R (cm)	1 (cm)	h (cm)	Yo (cm)	T (s)	H (cm)	T/Δt
W-1	2.5	20.0	50.0	10.0	1.0	9.4	60
W-2	2.5	20.0	50.0	13.0	1.4	11.0	80
W-3	2.5	20.0	50.0	10.0	1.9	9.9	120

h : water depth, Yo : cylinder center depth, H : wave height, others : same as in Table 1.

(b) Calculation runs for each case

Run	Decay of vortex
1	Without
2	With (Tc = T)
3	With (Tc = T/2)

Note : Small amplitude wave theory and stream function theory were applied for calculations.

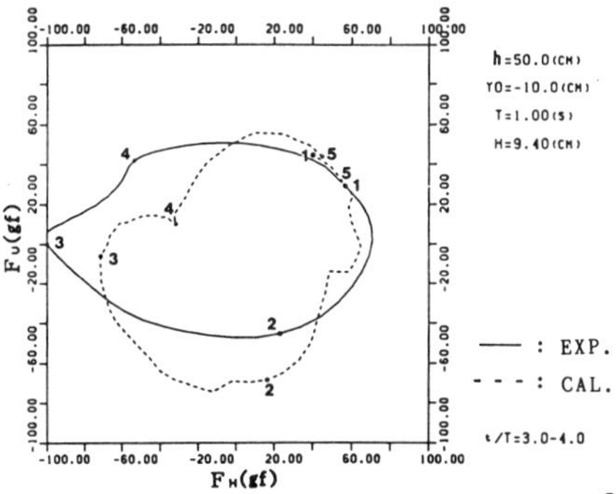

Tc = T

Fig.7 (a)

Comparison between calculated wave force path and laboratory data (C-type).

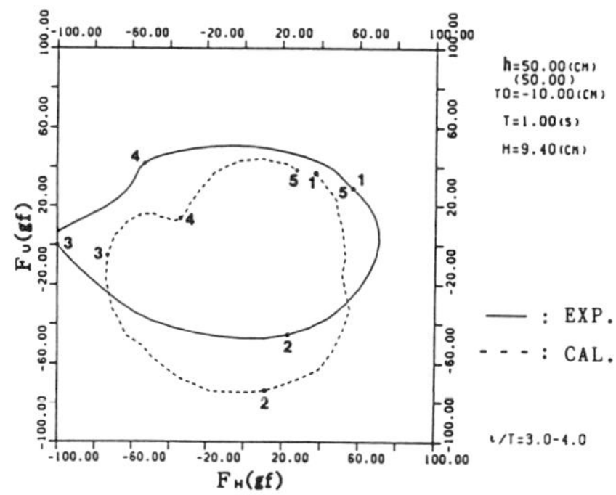

Tc = T/2

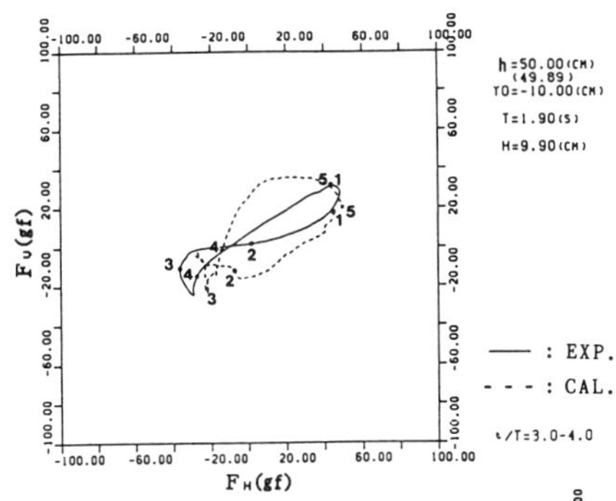

Tc = T/2

Fig.7 (b)

Comparison between calculated wave force path and laboratory data (E-type).

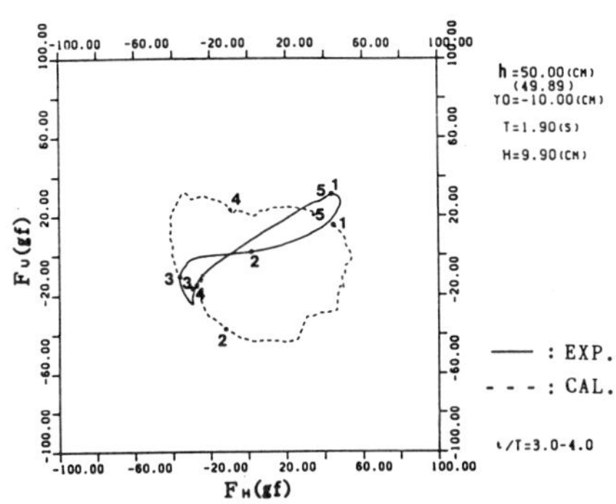

Tc = T

diagrams of Figures 7 (a) and (b) were obtained under the condition of $T_c = T$, and those in the lower diagrams were under the condition of $T_c = T/2$.

Looking at these diagrams, we can observe the following facts : 1) in the deeper water wave condition, the wave force path (C type) was simulated well under either conditions of $T_c = T$ or $T_c = T/2$, and 2) however in the shallower water wave condition, the 8-shaped type wave force path (E type) was reproduced well under the condition of $T_c = T/2$, but not under the condition of $T_c = T$. In order to make clear the reason why such a difference as stated in 2) happened, the calculated flow induced by vortices in the vicinity of a circular cylinder for Case W-3 were displayed. From these diagrams, we could observe that the unsymmetrical vortex behavior was reproduced at the time step of t/T = 3.6~3.8 for $T_c = T/2$, but not for $T_c = T$. Therefore we can conclude that appropriate decay of vortex should be introduced to simulate the interested flow characteristics.

(3) Drag coefficient, inertia coefficient, and lift coefficient

By using the calculated time histories of horizontal wave force component in a uniformly oscillating flow, the dag coefficient, C_D, and the inertia coefficient, C_M, were calculated through the following equations

$$C_D = \frac{3}{8}\int_0^{2\pi} \frac{F\sin\theta}{\rho R \hat{U}^2 \ell}\, d\theta \tag{14}$$

$$C_M = \frac{\hat{U}T}{2\pi^3 R}\int_0^{2\pi} \frac{F\cos\theta}{\rho R \hat{U}^2 \ell}\, d\theta \tag{15}$$

The obtained data for the symmetrical flow cases were plotted and compared with the laboratory data of Sarpkaya (1975) as shown in Figure 8. The abscissa in these diagrams is the K.C. number. The agreement for C_M is fairly good, however the calculated values of C_D were underestimated. The main reason why the latter discrepancy appeared is in the condition applied for determing the separation point. In this treatment the potential flow velocity surrounding the circular cylinder with the effect of vorticies was used. That is to say, at the point where the velocity is zero flow separation does not occur at any time and this fact caused rapid migration of separation point. Due to the above circumstance the development of vortex is disturbed, hence the drag coefficient results in decrease.

On the other hand two kinds of lift coefficient were calculated by using the following equations,

$$C_{L(MAX)} = \hat{F}_L / \rho \hat{U}^2 R \ell \tag{16}$$

$$C_{L(RMS)} = \sqrt{\overline{F_L^2}} / \rho \hat{U}^2 R \ell \tag{17}$$

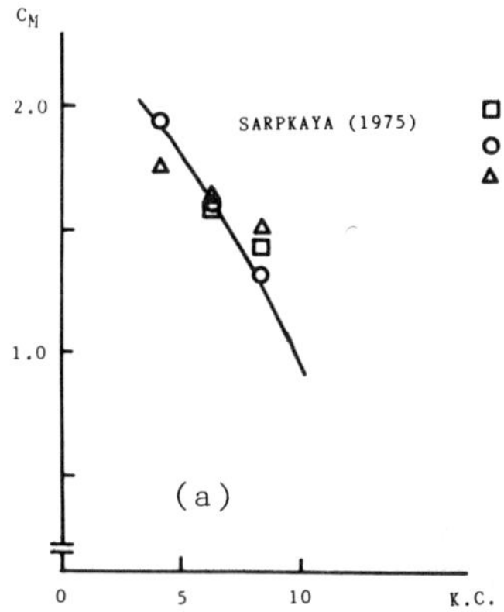

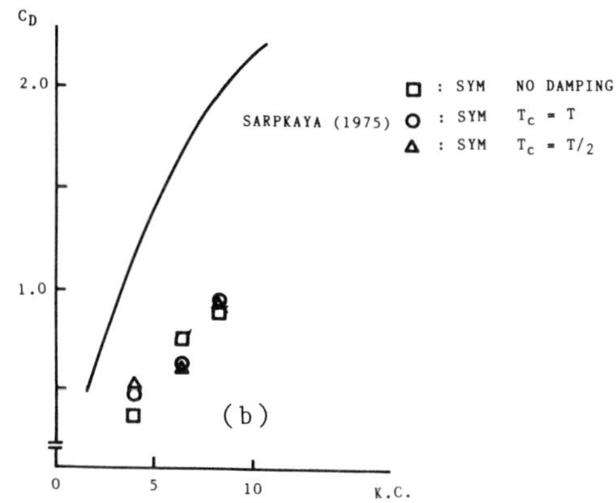

Fig.8

Comparison between calculated ones and laboratory curve for C_M and C_D.

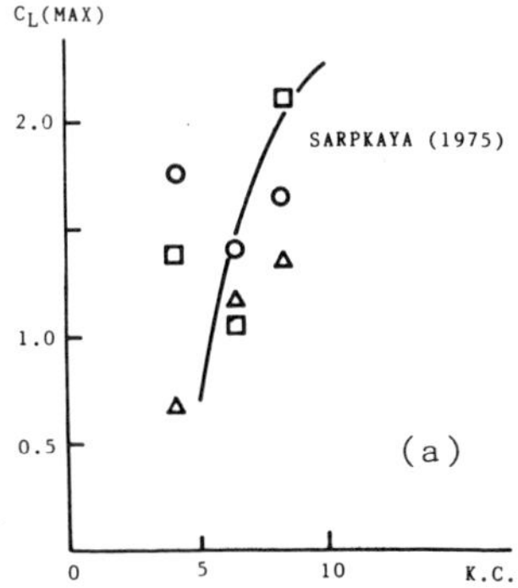

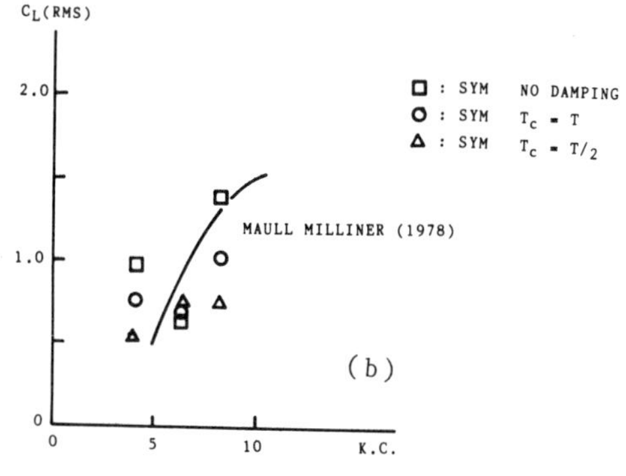

Fig.9

Comparison between calculated ones and laboratory curves for C_L.

The former and latter data for symmetrical flow cases were compared with the laboratory data of Sarpkaya (1975), and that of Maull and Milliner (1978) as shown in Figures 9 (a) and (b) respectively. The abscissa of the diagrams is again the K.C. number. The tendency of scattered data agrees with the laboratory curve in both cases.

In the same way, these three coefficients for unsymmetrical flow cases were invertigated. The unsymmetry of flow did not induce any strong influence on the coefficients of C_D and C_M, but induced a certain amount of influence on the coefficient of C_L

4. CONCLUSIONS

In the present paper was presented the numerical calculation method for predicting the wave forces acting on a horizontal circular cylinder set in a certain range of water depth. The simulation model is based on the discrete vortex method. The adaptability of the present calculation method was checked by comparing the calculated results with available laboratory data. As a conclusion of the present treatment, we found that the method presented here is applicable by introducing approprate decay of vortex circulation even to the cases where the emitted vortices induce strong unsymmetrical fluid motion.

Further study is needed to present a reasonable method to evaluate the vortex decay in the above treatment and to extend the capacity of the present simulation method for calculating the interested wave forces under more realistie or practical conditions.

REFERENCES

Borgman, L.E. : Computation of the ocean-wave forces on inclined cylinders, Trans. Amer. Geophys. Union, Vol.39, pp.885-888, 1958.

Chaplin, J.R. : On the irrotational flow around a horizontal cylinder in waves, Jour. App. Mech., Trans. ASME, Vol.46, pp.689-694, 1981.

Dean, R.G. : Stream function representation of nonlinear ocean waves, Jour. Geophys. Res., Vol.70, pp.4561-4572, 1965.

Grass, A.J. and P.H. Kemp : Flow visualization studies of oscillatory flow past smooth and rough circular cylinder, Mechanics of Wave-Induced Forces on Cylinders, Pitman, pp.406-420, 1979.

Koderayama, W. and A. Tashiro : On the wave forces acting on a submerged horizontal circular cylinder, Jour. Japanese Society of Naval Architects, No.143, pp.134-144, 1978 (in Japanese).

Masuda, S., A. Watanabe, and K. Horikawa : Wave forces acting on and velocity field of a circular cylinder set horizontally in wave, 40th Annual Conv. of JSCE, pp.II 541- 542, 1985 (in Japanese).

Maull, D.J. and M.G. Milliner : Sinusoidal flow past a

circular cylinder, Coastal Eng., Vol.2, pp.149-168, 1978.
Milne-Thomson, L.H. : Theoretical Hydrodynamics, 4th ed., McMilliam, New York, 1968.
Morison, J.R., M.P. O'Brien, J.W. Johnson, and S.A. Schaaf : The forces exerted by surface waves on piles, Petroleum Trans. AIME, Vol.189, pp.149-157, 1950.
Nath, J.H. and T. Yamanoto : Forces from fluid around objects, Proc. 14th Coastal Engrg. Conf., pp.1808-1827, 1974.
Ogilvie, T.F. : First- and second-order forces on a cylinder submerged under a free surface, J. Fluid Mech., Vol.16, pp.451-472, 1963.
Rosenhead, L. : Formation of vorticies from a surface of discontinuity, Proc. Roy. Soc., A, Vol.134, pp.170-192, 1931.
Sarpkaya, T. : An analytical study of separated flow about circular cylinders, J. Basic Engrg., Trans. of ASME, Vol.90, pp.511-518, 1968.
Sarpkaya, T. : Forces on cylinders and spheres in sinusoidally oscillating fluid, J. App. Mech., Trans. of ASME, Vol.42 No.1, pp.32-37, 1975.
Sawamoto, M., H. Oniwa, and J. Kashiwai. : A consideration on an expression of wave forces acting on a circular cylinder in an oscillatory flow, 34th Annual Conv. of JSCE, pp.II 579- 580, 1979 (in Japanese).
Sawamoto, M. and K. Kikuchi : Uplift forces acting on a circular cylinder in an oscillatory flow, Proc. 26th Japanese Coastal Engrg. Conf., JSCE, pp.429-433, 1979 (in Japanese).
Tsuzuki, S., A. Watanabe, and K. Horikawa : Wave forces acting on a horizontal circular cylinder, 39th Annual Conv. of JSCE, pp.II 303- 304, 1984 (in Japanese).
Uekita, M. and H. Yamazaki : A study on inertia forces acting on a horizontal circular cylinder, Proc. 27th Japanese Coastal Engrg. Conf., JSCE, pp.358-362, 1980 (in Japanese).

CHAPTER 164

Water Wave Scattering by Rows of Circular Cylinders

Robert A. Dalrymple[1], M. ASCE, Seung Nam Seo[2], Paul A. Martin[3]

Abstract

The scattering of waves by a finite number of rows of circular cylinders is examined. Reflection and transmission coefficients are obtained and compared to Kakuno's experimental data. Following Twersky (1962), the scattering from a single row of cylinders (or the single grating problem) is numerically solved. The wide-spacing approximation is used to find the effect of multiple gratings.

1 Introduction

The reflection and transmission of waves through rows of vertical cylinders, corresponding to the use of rows of piling as breakwaters (Wiegel, 1961), is examined. The physical situation also could represent the effects of piers and other pile supported structures on the wave environment.

Hayashi *et al.* (1966) and Mei *et al.* (1974) have examined several methods to calculate the wave field in the vicinity of a single row of cylinders, taking into consideration the loss of energy by the flow between the closely spaced cylinders. Ozsoy (1977) has empirically examined the Mei *et al.* solution and finds reasonable agreement with the theory; however there is considerable scatter in the data. Spring and Monkmeyer (1975) assumed that the flow was potential and examined the pressures and forces on an infinite row of cylinders.

Kakuno (1984,1986) has examined the same problem, and also the problem of a double row of vertical cylinders. He also considered potential flow with two assumptions: (1) the wave length is long compared with the pile spacing in each row, and (2) the wave length is short compared to the row spacing. Because of (1) he was able to give an analysis similar to Lamb (1898) for the analogous problem of acoustic waves passing through a row of cylinders. However, it is known that Lamb's method is deficient for this case: the exact solution is composed of an odd (dipole) part and an even (monopole) part, whereas Lamb's method only gives the odd part correctly. Recently Martin and Dalrymple (1988)

[1]Department of Civil Engineering, University of Delaware, Newark, Delaware 19716
[2]Korea Ocean Research and Development Institute, Ansan, Seoul 171-14, Korea
[3]Department of Mathematics, University of Manchester, Manchester, M13 9PL, U.K.

have shown how this deficiency can be rectified. Because of (2), Kakuno was able to use the "wide spacing approximation" for two rows of cylinders; see, e.g., Srokoscz and Evans (1979).

In this paper, we eliminate assumption (1) by giving an (in principle) exact numerical solution for a single row of circular cylinders. (Note that the asymptotic methods of Lamb and of Martin and Dalrymple work for cylinders with any cross-section.) We combine these results with the wide-spacing approximation and give a comparison with Kakuno's data. Finally, scattering from several rows of cylinders is calculated.

2 Scattering by a cylindrical grating

We first consider scattering of water waves by an infinite grating of circular cylinders, located along the y-axis. A plane wave of frequency ω over a constant depth h is incident upon the grating with angle θ_o (See Fig. 1).

The velocity potential, which must satisfy the Laplace equation, can be written as

$$\Phi(x,y,z,t) = Re\left\{\phi(x,y)\cosh k(h+z)e^{-i\omega t)}\right\} \tag{1}$$

where Re denotes the real part and k is the wave number, determined from the dispersion relationship, $\omega^2 = gk\tanh kh$, which relates the angular frequency ω to the wave number, the water depth h, and the acceleration of gravity, g. Factoring out the time and depth dependency, the fluid motion is governed by the Helmholtz equation

$$(\nabla^2 + k^2)\phi = 0 \tag{2}$$

On the wall of each cylinder in the grating, the no-flux condition should be fulfilled:

$$\phi_n = 0 \tag{3}$$

where subscript n denotes the outward normal derivative. The scattered potential ϕ^{sc}, defined as the total potential ϕ minus the incident potential ϕ^{in}, must satisfy a radiation condition as $|x| \to \infty$.

Twersky (1962) used the method of separation of variables for solving scattered waves by an infinite grating of equally spaced identical circular cylinders with spacing d. (See Fig. 1) The total velocity potential by an incident wave $\phi^{\mathrm{in}} = e^{i\vec{k}\cdot\vec{x}}$ can be expressed in cylindrical coordinates as,

$$\begin{aligned}\phi \;=\; & e^{ikr\cos(\theta-\theta_o)} + \sum_{n=-\infty}^{\infty} e^{in(\theta+\pi/2)}\Big\{\mathrm{A}_n \mathrm{H}_n^{(1)}(kr) \\ & +\mathrm{J}_n(kr)\sum_{s=1}^{\infty}\sum_{m=-\infty}^{\infty}\mathrm{A}_m\mathrm{H}_{m-n}^{(1)}(skd)\left[e^{-iskd\sin\theta_o} + (-1)^{m-n}e^{iskd\sin\theta_o}\right]\Big\}\end{aligned} \tag{4}$$

where J_n and $\mathrm{H}_n^{(1)}$ denote the Bessel function of first kind order n and the Hankel function of order n. The multiple scattering coefficient A_n can be obtained by imposing no-flux condition at the cylinder wall (r=a).

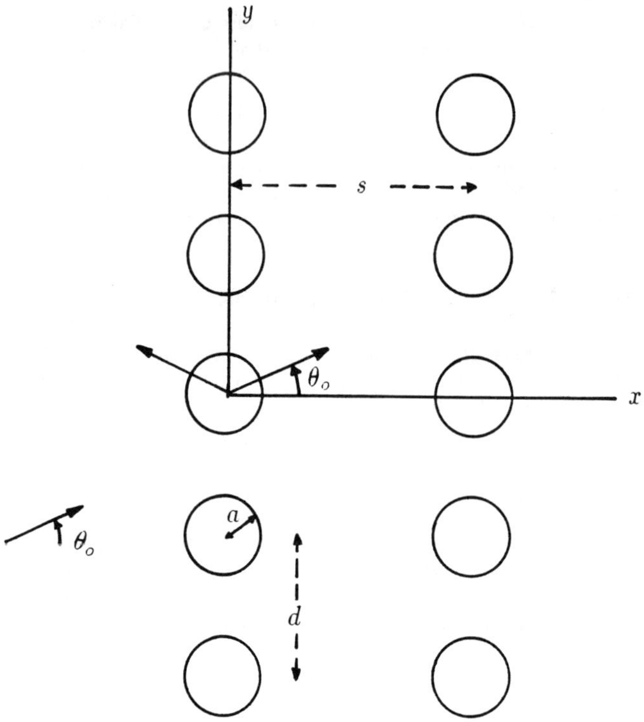

Figure 1: Schematic Diagram Showing Waves Incident on Two Rows of Cylinders

$$A_n = -\frac{J_n'(ka)}{H_n'(ka)} \left\{ e^{-in\theta_o} + \sum_{m=-\infty}^{\infty} A_m H_{n-m}^{(1)}(skd) \left[e^{-iskd\sin\theta_o} + (-1)^{m-n} e^{iskd\sin\theta_o} \right] \right\} \quad (5)$$

where prime denotes differentiation.

In order to calculate transmission coefficient K_T and reflection coefficient K_R, it is convenient to express the potential as a summation of plane wave modes in Cartesian, rather than polar coordinates. By Sommerfeld's integral representation of the Hankel functions and Poisson's summation formula, Twersky (1962) obtained the following representations: For the transmitted region ($x > 0$),

$$\phi_T = e^{i\vec{k}\cdot\vec{x}} + 2 \sum_{n=-\infty}^{\infty} A_n \sum_{\mu=-\infty}^{\infty} e^{in\theta_\mu} C_\mu \, e^{iky\sin\theta_\mu + ikx\cos\theta_\mu} \quad (6)$$

and for the reflected region ($x < 0$)

$$\phi_R = e^{i\vec{k}\cdot\vec{x}} + 2 \sum_{n=-\infty}^{\infty} A_n \sum_{\mu=-\infty}^{\infty} e^{in(\pi-\theta_\mu)} C_\mu \, e^{iky\sin\theta_\mu + ikx\cos\theta_\mu} \quad (7)$$

with

$$C_\mu = \frac{1}{kd\cos\theta_\mu} \quad \text{for } |\,Re\ \theta_\mu| \leq \frac{\pi}{2} \quad (8)$$

and the directions of the planar wave modes are given by

$$\sin\theta_\mu = \sin\theta_o + \frac{2\pi}{kd}\,\mu \quad \text{for } \mu = 0, \pm1, \pm2, \ldots \quad (9)$$

It can be seen from Eq. 7 that when $|\sin\theta_\mu| < 1$, progressive waves exist, but when $|\sin\theta_\mu| > 1$ there are evanescent modes. If $kd < \pi$, therefore, only one progressive wave exists.

Introducing a multiple scattering amplitude $G(\theta_\mu) = \sum_{n=-\infty}^{\infty} A_n e^{in\theta_\mu}$, reflection and transmission coefficients for a mode μ are

$$K_{R_\mu} = 2C_\mu G(\pi - \theta_\mu) \quad (10)$$

and

$$K_{T_\mu} = \delta_{\mu o} + 2C_\mu G(\theta_\mu) \quad (11)$$

where δ denotes the Kronecker delta. From conservation of energy flux, we have

$$\sum_{\mu=\mu^-}^{\mu=\mu^+} \left(|\,K_{R_\mu}|^2 + |\,K_{T_\mu}|^2 \right) \frac{\cos\theta_\mu}{\cos\theta_o} = 1 \quad (12)$$

where

$$\mu^- = E[-(1+\sin\theta_o)\frac{kd}{2\pi}] \quad (13)$$

$$\mu^+ = E[\ (1-\sin\theta_o)\frac{kd}{2\pi}] \quad (14)$$

and E[x] stands for the integer value of x.

Slow convergence of the series in Eq. (5), which is a family of Schlömilch series, leads to a difficulty in evaluating the multiple scattering coefficients. Twersky (1961) found a way to accelerate the convergence of the series. In the present method, Shanks transformation was used to evaluate the infinite series in Twersky's representation.

Scattering coefficients A_n were obtained by retaining up to the Mth harmonic in Eq. (5):

$$\mathrm{A}_n = -\frac{\mathrm{J}_n'(ka)}{\mathrm{H}_n'(ka)}\left\{e^{-in\theta_o} + \sum_{m=-M}^{M} \mathrm{A}_m \sum_{s=1}^{\infty} \mathrm{H}_{n-m}(skd)\left[e^{-iskd\sin\theta_o} + (-1)^{n-m} e^{iskd\sin\theta_o}\right]\right\} \tag{15}$$

for $n = -M, \ldots, M$. (Care must be taken when Bessel function J_n is calculated using a recurrence formula, because of roundoff errors; see Abramowitz and Stegun, 1965, section 9.12.)

Twersky (1962) was able to get approximate solutions when there is only one propagating mode. Present numerical solutions of $\mathrm{K_T}$ and $\mathrm{K_R}$ were compared to Twersky's results in Figs. 2 and 3. The numerical solutions satisfied conservation of energy flux up to 10 digits for all the cases tested. Twersky's solutions (Eqs. 127 and 144) are valid only if $ka < a/d < 1$. As shown in Fig. 3, Twersky's solution does not agree to the numerical solution. For a fixed radius, longer waves propagate through the grating more easily. Normal incidence waves to the grating axis are more reflective.

3 Wide–spacing Approximation

We now move to wave scattering problems by multi-row gratings, where more than one length scale is involved. We further assume that the distance between gratings is large enough, so that gratings interact with each other via the propagating waves only; hence we can neglect the evanescent modes. For a longer wave train, ($kd < \pi$), only one wave transmits through the grating with the given incident angle.

In the case of two gratings, the interference can be explained physically. When an incident wave train, proportional to $e^{i\ell x}$, where $\ell = k\sin\theta_o$, strikes the first grating at $x = 0$, part of the wave train is transmitted and part is reflected. Upon seeing the grating at $x = s$, the transmitted wave undergoes the same scattering process; part of the wave is transmitted to $x > s$ and part is back-reflected toward the first grating. This back and forth process of transmission and reflection is repeated infinitely. (See Fig. 4). Mathematically, we have

$$K_R = R_1 + T_1^2 R_2 e^{i2\ell s} + T_1^2 R_1 R_2^2 e^{i4\ell s} + T_1^2 R_1^2 R_2^3 e^{i6\ell s} + \cdots \tag{16}$$

$$= \frac{R_1 + R_2 e^{i2\ell s}(T_1^2 - R_1^2)}{1 - R_1 R_2 e^{i2\ell s}} \tag{17}$$

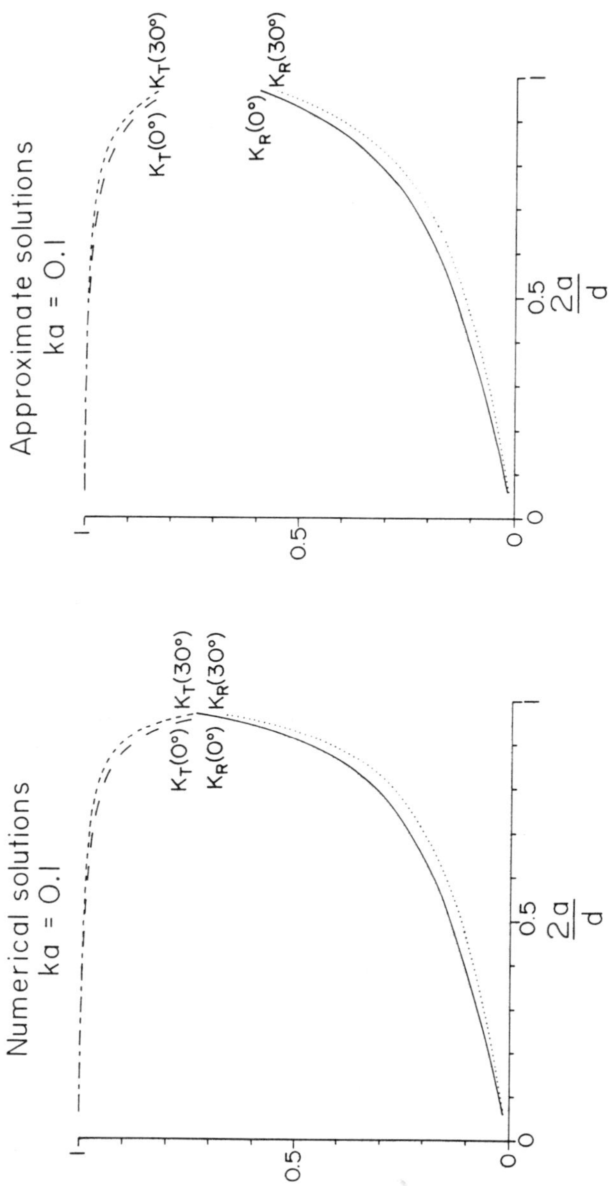

Figure 2: Comparison of Numerical and Approximate Solutions for Two Angles of Incidence for Small *ka*.

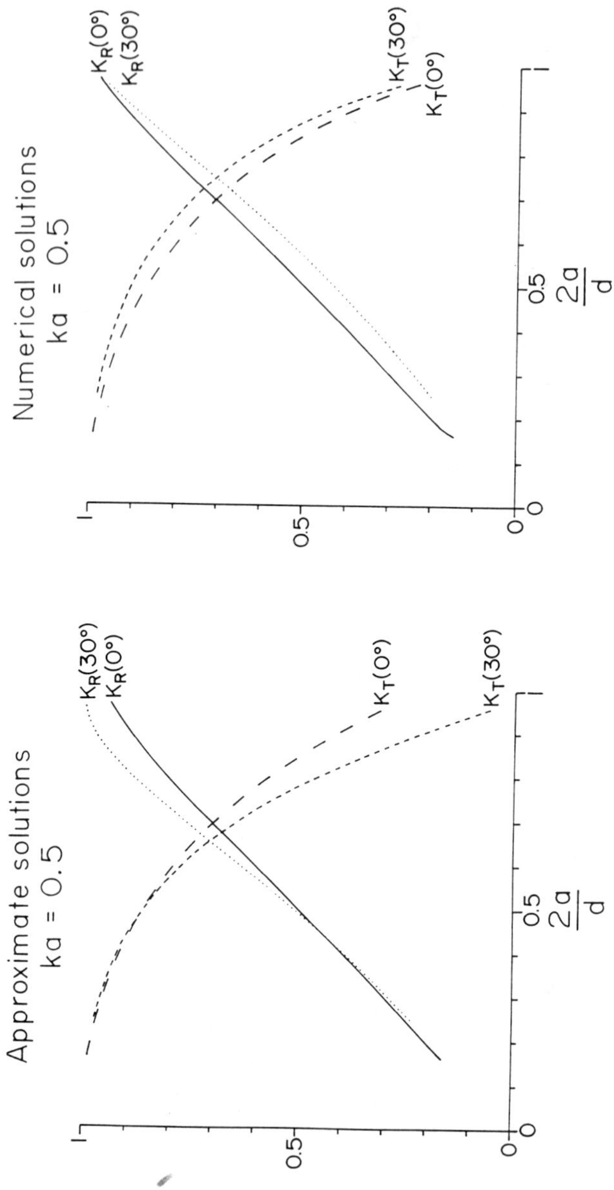

Figure 3: Comparison of Numerical and Approximate Solutions for Two Angles of Incidence for Larger ka.

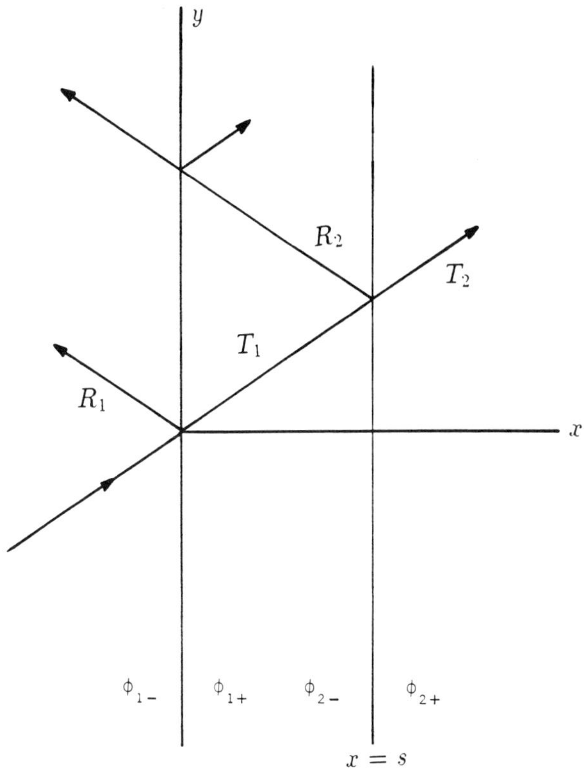

Figure 4: Reflection and Transmission by Two Gratings

$$K_T = T_1T_2 + T_1T_2R_1R_2e^{i2\ell s} + T_1T_2R_1^2R_2^2e^{i4\ell s} + \cdots \tag{18}$$

$$= \frac{T_1T_2}{1 - R_1R_2e^{i2\ell s}} \quad . \tag{19}$$

where T_i and R_i ($i = 1, 2$) are the transmission coefficient and reflection coefficient of the ith grating, respectively. Stokes (1862) used this method to find light intensity in heterogeneous media. To generalize the formulation, we solve N-row grating problems using the wide–spacing approximation. Let the solution to scattering problem for the nth grating by an incident wave $e^{il(x-x_n)}$ with $l = k\cos\theta_o$ be ϕ_{n-}. We define transmission and reflection coefficients, T_{n-} and R_{n-}, respectively, by

$$\phi_{n-} \sim \begin{cases} e^{i\ell(x-x_n)} + R_{n\_}e^{-i\ell(x-x_n)}, & \text{for } x \to -\infty \\ T_{n\_}e^{i\ell(x-x_n)}, & \text{for } x \to \infty \end{cases} \tag{20}$$

Similarly, let ϕ_{n+} solve the scattering problem for the nth grating by an incident wave from the other direction, $e^{-il(x-x_n)}$, and define T_{n+} and R_{n+} by

$$\phi_{n+} \sim \begin{cases} e^{-i\ell(x-x_n)} + R_{n+}e^{i\ell(x-x_n)}, & \text{for } x \to \infty \\ T_{n+}e^{-i\ell(x-x_n)}, & \text{for } x \to -\infty \end{cases} \tag{21}$$

Hence we can represent near field potential of the nth grating by

$$\phi = C_n^-\phi_{n-} + C_n^+\phi_{n+} \tag{22}$$

$$\sim \begin{cases} \left(C_n^- e^{-i\ell x_n}\right) e^{i\ell x} + \left(C_n^- R_{n-}e^{i\ell x_n} + C_n^+ T_{n+}e^{i\ell x_n}\right) e^{-i\ell x}, & \text{for } x \to -\infty \\ \left(C_n^- T_{n-}e^{-i\ell x_n} + C_n^+ R_{n+}e^{-i\ell x_n}\right) e^{i\ell x} + \left(C_n^+ e^{i\ell x_n}\right) e^{-i\ell x}, & \text{for } x \to \infty \end{cases} \tag{23}$$

where the $C_n^{\pm}$ denote unknown complex constants to be determined.

We now require that the near field solutions be joined smoothly in intermediate regions. For a region between the $(n-1)^{th}$ and n^{th} gratings, we have

$$\left[C_{n-1}^-(T_{n-1})_- e^{-i\ell x_{n-1}} + C_{n-1}^+(R_{n-1})_+ e^{-i\ell x_{n-1}}\right] e^{i\ell x} + \left[C_{n-1}^+ e^{i\ell x_{n-1}}\right] e^{-i\ell x} = \tag{24}$$

$$\left[C_n^- e^{-i\ell x_n}\right] e^{i\ell x} + \left[C_n^-(R_n)_- e^{i\ell x_n} + C_n^+(T_n)_+ e^{i\ell x_n}\right] e^{-i\ell x}, \quad x_{n-1} < x < x_n \tag{25}$$

Hence

$$C_{n-1}^-(T_{n-1})_- e^{i\ell\Delta_{n-1}} + C_{n-1}^+(R_{n-1})_+ e^{i\ell\Delta_{n-1}} - C_n^- = 0 \tag{26}$$

and

$$C_{n-1}^+ e^{-i\ell\Delta_{n-1}} - C_n^-(R_n)_- - C_n^+(T_n)_+ = 0 \tag{27}$$

where $\Delta_n = x_{n+1} - x_n$. This near field matching gives 2(N-1) equations for $n = 2, N$.

On the other hand, the total velocity potential due to all the gratings can be expressed by

$$\phi \sim \begin{cases} e^{i\ell x} + K_R e^{-i\ell x}, & \text{for } x \to -\infty \\ K_T e^{i\ell x}, & \text{for } x \to \infty \end{cases} \tag{28}$$

Noting that $x_1 = 0$ and $C_1^- = 1$, and comparing Eqs. (23) and (28), we have

$$R_{1-} + C_1^+ T_{1+} = K_R \tag{29}$$

Since there is no wave incident on the Nth grating from the left, C_N^+ must be zero. Comparing Eqs. (23) and (28) gives

$$C_N^-(T_N)_- e^{i\ell x_N} = K_T \tag{30}$$

Finally, we have 2N unknown coefficients: $K_R, C_1^+, C_2^-, C_2^+, \cdots, C_N^-, K_T$. Eqs. (26), (27), (29) and (30) give 2N linear equations, enough to solve 2N unknowns.

In Fig. 5, we tested the present method for two identical rows of cylinders against Kakuno's experimental data. In this case, $T_{n-} = T_{n+}$ and $R_{n-} = R_{n+}$ $(n = 1, 2)$ because depth is constant. We introduce two nondimensional parameters associated with the grating geometry; spacing parameter $\alpha = s/d$ and grating parameter $\beta = 2a/d$. In the wide-spacing approximation, we can expect $\alpha \gg 1$ and shorter spacing between gratings increases

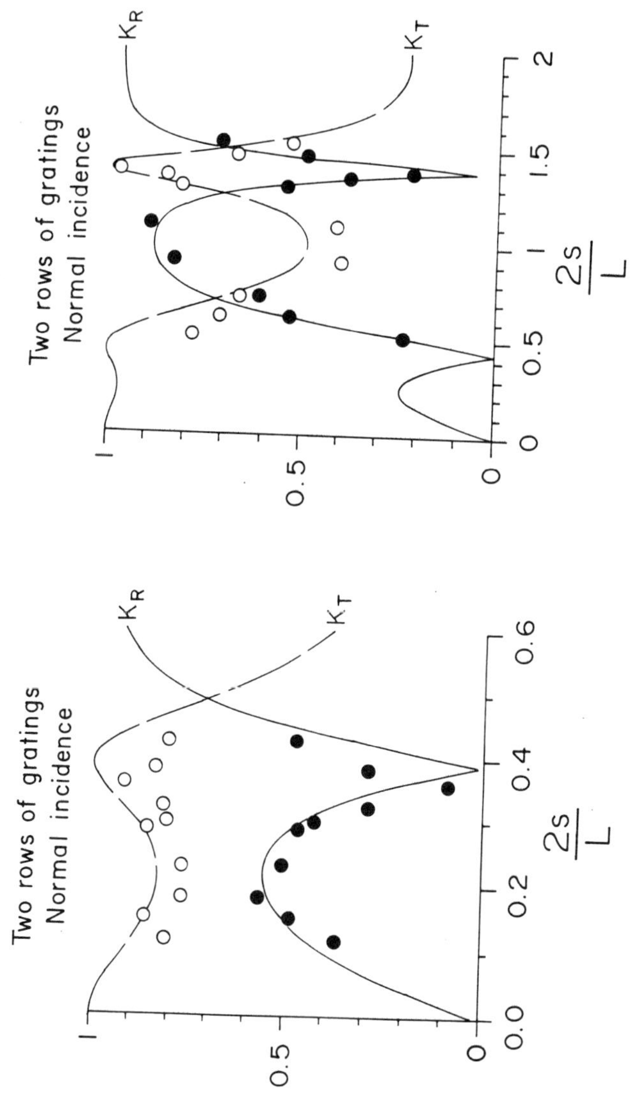

Figure 5: Comparison of Reflection and Transmission Coefficients with Kakuno's Data

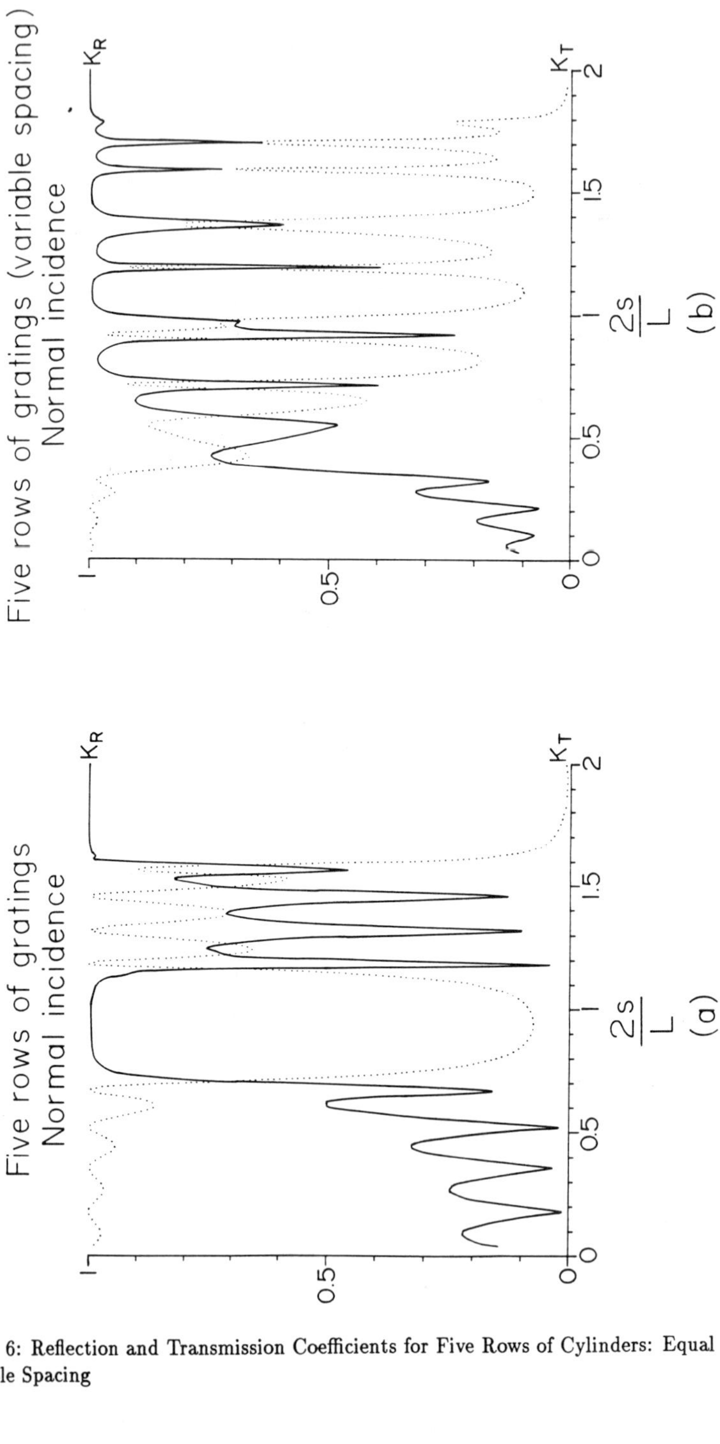

Figure 6: Reflection and Transmission Coefficients for Five Rows of Cylinders: Equal and Variable Spacing

the effect of evanescent modes. Fig. 5a shows this effect, which produces less agreeable results. For some values of α, a significant amount of waves are reflected. Under this condition, the phases of the reflected waves are the same, so that the reflected waves interfere constructively and the total amplitude is consequently the larger. This is referred to as Bragg scattering.

In Fig. 6, we presented K_R and K_T for five identical rows of cylinders. The interference of each grating is shown to be more complicated and total reflection can be observed for some values of α. In Fig. 6a, equally spaced gratings were used, while in Fig. 6b the spacing between the gratings increases from s for the first two gratings, by increments of 1.5s, to 2.5s. It is also interesting to note that some arrangements of gratings are essentially transparent to the waves.

4 Conclusion

Numerical results of wave scattering by N-row gratings are presented. Accurate numerical solutions for a single grating are constructed following Twersky (1962) procedure. For long waves ($ka << 1$), Twersky solution is identical to the numerical solution, but, as ka increases, Twersky's solution shows considerable deviation form the numerical solution. Using the wide-spacing approximation, reflection and transmission coefficients of long waves through N-row gratings are obtained.

For two identical rows of gratings, the present method shows good agreement with Kakuno's experimental data. Results for five rows of gratings are also presented.

5 Acknowledgments

Support for this research was provided by the NOAA Office of Sea Grant, Department of Commerce, under Grant NA87AA-D-SG040.

6 References

1. Abramowitz, M. and I.A. Stegun, 1965, **Handbook of Mathematical Functions**, Dover Publications, Inc., New York.

2. Hayashi, T., T. Kano and M. Shirai, 1966, "Hydraulic research on the closely spaced pile breakwater," Proc. 10th Intl. Conf. Coastal Engineering , ASCE, 873-884.

3. Kakuno, S. *et al.*, 1984, "Analysis of boundary value problems for hydraulic characteristics of vertically slotted breakwaters," Coastal Engineering in Japan (in Japanese).

4. Kakuno, S., 1986, Abstracts, 20th Int. Conf. Coastal Engineering, ASCE, Taiwan.

5. Lamb, H., 1898, "On the reflection and transmission of electric waves by a metallic grating," *Proc. London Math. Soc.*, Series 1, 29, 523-544.

6. Martin, P.A. and R.A. Dalrymple, 1988, "Scattering of long waves by cylindrical obstacles and gratings using matched asymptotic expansions," *J. Fluid Mechanics*, 188, 465-490.

7. Mei, C.C., P.L.-F. Liu, and A.T. Ippen, 1974, "Quadratic loss and scattering of long waves," *J. Waterways and Harbors Div.*, ASCE, 100, 217-239.

8. Ozsoy, E., 1977, "Dissipation and scattering by narrow openings," University of Florida, Coastal and Oceanographic Engineering Laboratory, UFL/COEL/TR-037, 137pp.

9. Spring, B.W. and Monkmeyer, P.L., 1975, "Interaction of plane waves with a row of cylinders," Proc. Civil Engineering in the Oceans III, ASCE, Newark, Delaware.

10. Srokosz, M.A. and Evans, D.V., 1979, "A theory for wave-power absorption by two independently oscillating bodies," *J. Fluid Mechanics*, 90, 337-362.

11. Stokes, G.G., 1862, "On the intensity of light reflected from or transmitted through a pile of plates," *Proc. Royal Society of London*, XI, 545-557.

12. Twersky, V., 1961, "Elementary function representations of Schlömilch series," *Arch. Rat. Mech. Anal.*, 8, 323-332

13. Twersky, V., 1962, "On the scattering of waves by the infinite grating of circular cylinders," *IRE Trans. Antennas Propag.*, 10, 737-765.

14. Wiegel, R.L., 1961, "Closely spaced piles as a breakwater," *Dock and Harbor Authority*, Sept. 1961, 150.

CHAPTER 165

WAVE TRANSMISSION THROUGH A DOUBLE-ROW PILE BREAKWATER

by

John B. Herbich, Ph.D., P.E., F.ASCE[1]
and
Barry Douglas, B.S.[2]

Abstract

Several previous investigators have conducted experiments leading to expressions for predicting the transformation of waves passing through closely-spaced pile or large cylinder breakwaters. The present study extends the earlier experiments which used a single row of piles instead of a double row of piles forming a breakwater. The experiments using the double-pile breakwater were performed in the same facility as the experiments conducted on a single-pile breakwater and employed the same method of analysis for a more meaningful comparison.

The experiments consisted of allowing waves to pass through a pile array and measuring the incident and transmitted wave heights. The variables were: depth, period, diameter, monochromatic and random waves. The experimental matrix was three water depths, four wave periods, two pile diameters, two gap dimensions between piles and four random wave spectra: Darbyshire, I.T.T.C., Pierson-Moskowitz and JONSWAP, two pile diameters and two gap dimensions between piles.

The two-row breakwater had less wave transmission than the single-row breakwater, as expected. For a gap to a pile diameter ratio, or b/D = 0.2 (where b = gap spacing, D = pile diameter), the wave transmission was reduced by 15 percent, as compared with a single-row breakwater; for a gap ratio of b/D = 0.1, the wave transmission was reduced by 5 to 10 percent.

Introduction

In recent decades there has been a greater use of vertical face breakwaters, and consequently a considerable development in technology of caisson-type breakwaters.

[1]Professor of Ocean and Civil Engineering, Texas A&M University, College Station, Texas 77843-3136, U.S.A.
[2]Graduate Student, Coastal and Oceanographic Engineering Department, University of Florida, Gainesville, FL 32611, U.S.A.

The first large cylinder-type breakwater was in all probability constructed in Hanstholm, Denmark in 1960. The cylindrical reinforced-concrete breakwater consisting of 12.5 m (41 ft) diameter units with a 0.25 m (0.8 ft) wall thickness was installed in 12 m (39 ft) of water. The breakwater at Marsa el Brega, Libya contains cylindrical shells in caissons that are 53.8 m (177 ft) long and 16 m (52 ft) wide.

A steel pipe breakwater was constructed in the Port of Osaka, Japan in 1966. The breakwater consists of 2 m (6.6 ft) diameter pipes. The spacings between adjacent steel pipes were 5 cm (0.16 ft) on the average.

A concrete pipe breakwater was constructed to protect a marina at Pass Christian, Mississippi to replace a breakwater destroyed by a hurricane. The breakwater consists of about 1.4 m (4.6 ft) diameter piles with approximately 15.2 cm (0.5 ft) spacings.

Laboratory Studies

Several laboratory investigations have been conducted to evaluate the transmission and reflection of waves from permeable-pile breakwaters. Published data include studies by Wiegel (1969), Hayashi, et al. (1966), Allsop and Kalmus (1985) and Truitt and Herbich (1987).

Hayashi, et al. derived an expression for wave transmission based on water jets discharging through the pile gaps:

$$\frac{H_t}{H_i} = 4 \left(\frac{d}{H_i}\right) E \left[-E + \sqrt{E^2 + \frac{H_i}{2d}}\right] \quad \ldots\ldots (1)$$

where H_t = height of transmitted wave,
H_i = height of incident wave,
d = water depth.

$$E = c \left(\frac{b}{D+b}\right) \Big/ \sqrt{1 - \left(\frac{b}{D+b}\right)^2} \quad \ldots\ldots (2)$$

where c = constant,
b = spacing between piles,
D = pile diameter.

Wiegel derived an expression for wave transmission based solely on geometry of the piles:

$$\frac{H_t}{H_i} = \frac{b}{D+b} \quad \ldots\ldots (3)$$

Model studies performed later showed that transmitted wave heights were larger (up to 25%) than the rate predicted by Equation 3.

Truitt and Herbich conducted experimental studies on wave transmission through a breakwater consisting of a single row of piles employing both monochromatic and random waves. Earlier experiments of other investigators had been conducted with monochromatic waves. The results of experiments with random waves indicate that the relationships developed for monochromatic waves may be applicable to random waves. Very good agreement was observed between the values of wave transmission predicted by Hayashi's equation and experimentally-obtained values with random waves, provided the coefficient employed was equal to 0.9. It was also found that two dimensionless ratios affect the transmission coefficient:

a) b/D, a ratio of breakwater spacing to pile diameter, and
b) d/H_s, a ratio of water depth to wave height.

The influence of the second ratio on the transmission coefficient is more pronounced.

Experimental Studies

The breakwater consisted of two staggered rows of piles installed in a two-dimensional wave facility, 36.6 m (120 ft) long, 0.6 m (2 ft) wide and 0.91 m (3 ft) deep (Figures 1 and 2). The wave channel is glass-walled and is equipped with a programmable SEASIM wave generator. Either monochromatic or irregular wave spectra can be generated. Wave data were obtained using resistance-type, auto-compensating wave gauges in three water depths: 61 cm (24 in.), 51 cm (20 in.), and 41 cm (16 in.). Four wave spectra were also generated: Darbyshire, I.T.T.C., Pierson-Moskowitz, and JONSWAP.

The experiments consisted of passing waves through the pile breakwater and measuring both incident and transmitted wave heights. The models were built using steel conduit to assure rigid and uniform piles. Two sizes of conduit were used: 3.0 cm diameter (1-3/16 in.) and 2.4 cm diameter (15/16 in.). Four models were employed as shown in Table I.

TABLE I. Dimensions of Pile Breakwaters

Pile Diameter (D) (cm)	(in)	b/D	Distance Between Rows (cm)	(in)
2.4	15/16	0.1	2 (D) = 4.8	1-14/16
3.0	1-3/16	0.2	2 (D) = 6.0	2-3/8
2.4	15/16	0.1	2 (D) = 4.8	1-14/16
3.0	1-3/16	0.2	2 (D) = 6.0	2-3/8

Procedure and Analysis

Wave height measurements were made with auto-compensating wave gauges. One gauge was placed upstream of the pile array and another downstream of the breakwater. For tests with lower wave reflections, the carriage-mounted upstream gauge was moved through at least one wave length to delineate the envelope of incident and reflected waves. The incident wave height was computed from the recorded wave envelope. When high reflections occurred, the

Figure 1. Wave transmission through a pile breakwater in a glass-walled wave channel.

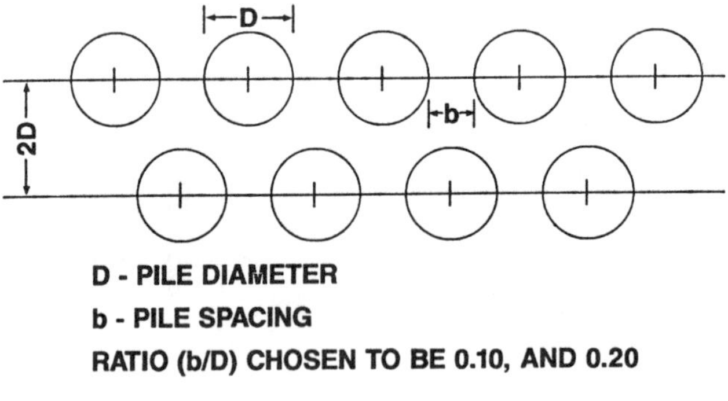

Figure 2. Definition sketch for pile geometry.

significant wave height was computed from a stationary gauge using a record of 21 waves. All of the irregular wave spectra tests were measured with a stationary gauge. Significant wave heights were computed for direct comparison with results previously obtained for a single-row breakwater.

Results

Sample plots of dimensionless wave transmission (H_t/H_i) as a function of water depth to incident wave height ratio are shown in Figures 3 and 4. The increase of wave transmission with an increase in wave period and d/H ratio (up to 25) can be seen in these figures. As anticipated, the smaller the gap ratio, the lower the wave transmission. Reducing the gap space from 20 to 10 percent resulted in transmission for a gap ratio of b/D = 0.1 being reduced to 60-80 percent of the transmission for a gap ratio of b/D = 0.2. The wave transmission was reduced even more so at lower wave periods.

Plots of dimensionless wave transmission, (H_t/H_i), as a function of d/H were also prepared for random waves (Figures 5-8). Even though scatter occurred, data for the four wave spectra tended to plot together. The curves for the two sizes of pile were similar for the same gap ratios. The reduction in gap spacing from 20 to 10 percent reduced the values of wave transmission by approximately 30 percent.

A comparison was also made between wave transmission data for one row of piles (Truitt and Herbich) and two rows of piles. Both studies were conducted in the same facilities, using the same random waves and the same pile sizes and gap ratios. Figure 9 shows a comparison for a b/D ratio of 0.2 and Figure 10 for a b/D ratio of 0.1. For these plots all data points for the same gap ratio have been plotted together regardless of pile diameter. The data for one row of piles covered a very narrow range of d/H values and the resulting curve represents linear regression of the data. For a gap ratio of b/D = 0.2, wave transmission was reduced by about 15 percent, while for a gap ratio of b/D = 0.1 it was reduced by 5 to 10 percent.

The monochromatic wave data were also analyzed using the dimensionless parameter d/gT^2 where d is the water depth, g is the gravitational acceleration and T is the wave period. Figure 11 which is a plot of wave transmission as a function of d/gT^2 indicates that the wave transmission decreases as d/gT^2 increases. Values of wave steepness H_i/L are also shown next to the individual data points. The general trend is for the wave transmission to decrease as the wave steepness increases.

These data were also compared with a similar breakwater model consisting of one row of piles (Kilpatrick, 1984). This study was performed with slightly different pile diameters: 0.63 cm (4/16 in.) and 1.9 cm (12/16 in.) but with the same gap ratios (0.1 and 0.2). It has already been shown that pile diameter has a minimal effect on transmission and that the b/D (gap ratio) is the more important variable. Several plots of wave transmission as a function of d/gT^2 for the monochromatic waves present a comparison between the single- and double-pile breakwaters (Figures 12-15).

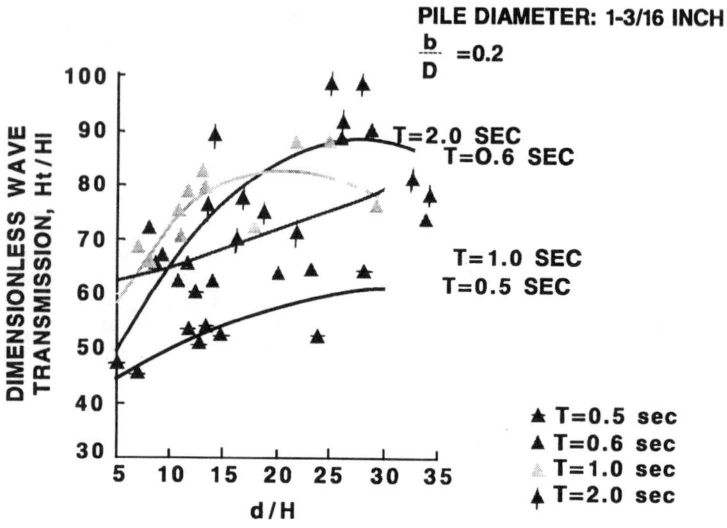

Figure 3. Dimensionless wave transmission as a function of depth to wave height ratio for four wave periods. D = 3.0 cm (1-3/16 in.), b/D = 0.2, two rows. Monochromatic waves.

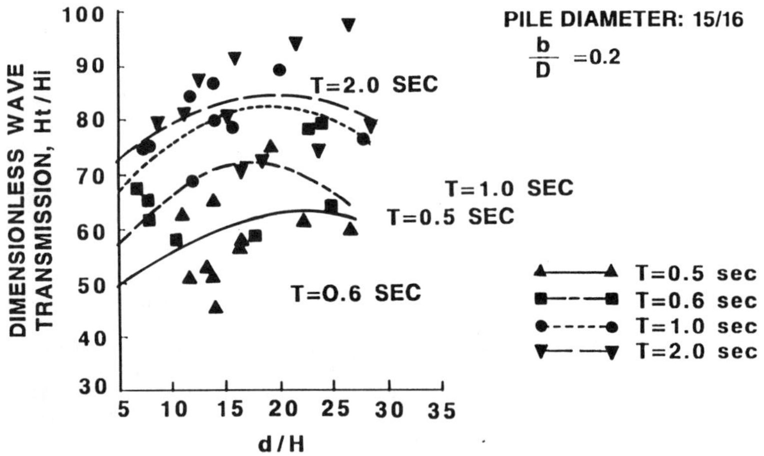

Figure 4. Wave transmission as a function of depth to wave height ratio for four wave periods. D = 2.4 cm (15/16 in.), b/D = 0.2, two rows. Monochromatic waves.

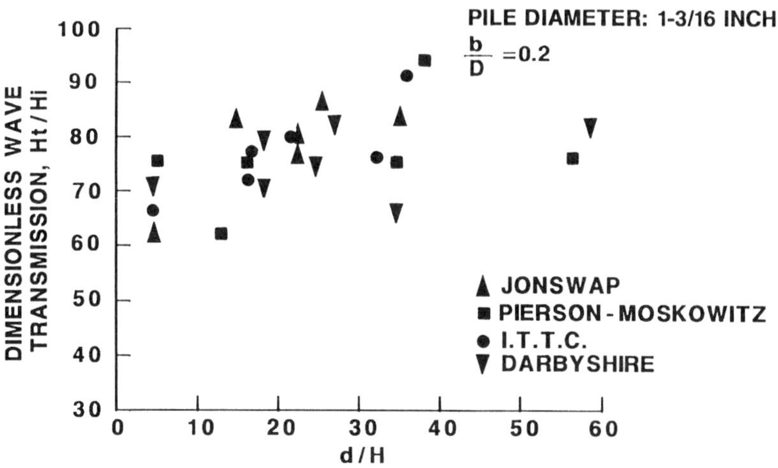

Figure 5. Wave transmission as a function of water depth to wave height ratio for random waves. D = 3.0 cm (1-3/16 in.), b/D = 0.2.

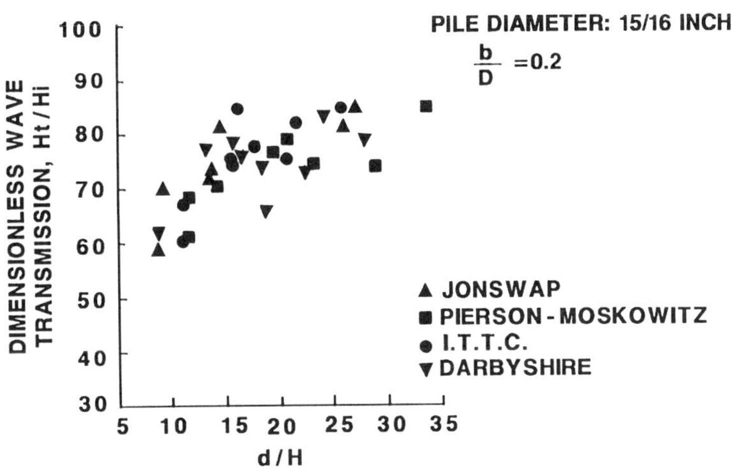

Figure 6. Wave transmission as a function of water depth to wave height ratio for random waves. D = 2.4 cm (15/16 in.), b/D = 0.2.

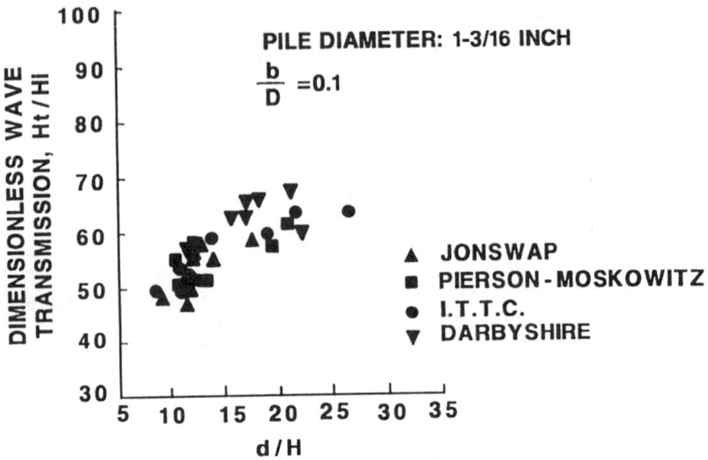

Figure 7. Wave transmission as a function of water depth to wave height ratio for random waves. D = 3.0 cm (1-3/16 in.), b/D = 0.1.

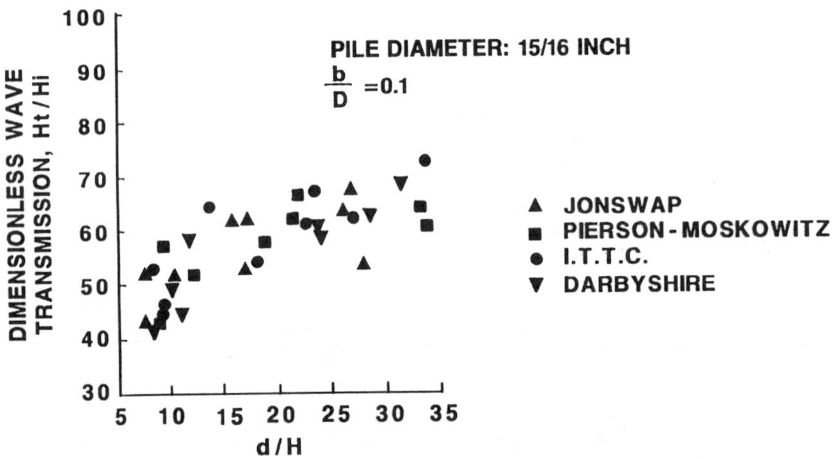

Figure 8. Wave transmission as a function of water depth to wave height ratio for random waves. D = 2.4 cm (15/16 in.), b/D = 0.1.

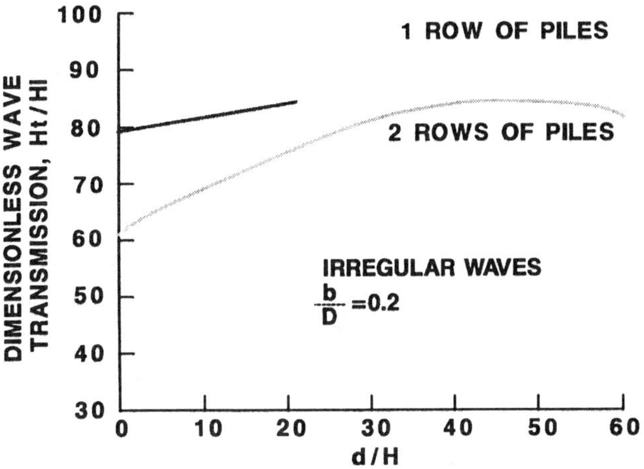

Figure 9. Wave transmission for one and two rows of piles. Random waves, b/D = 0.2.

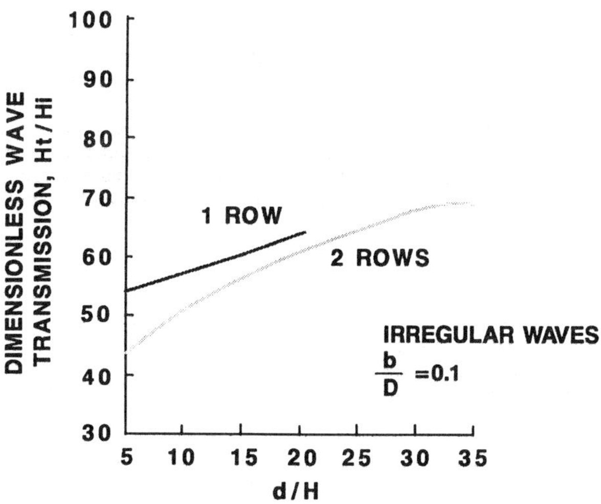

Figure 10. Wave transmission for one and two rows of piles. Random waves, b/D = 0.1.

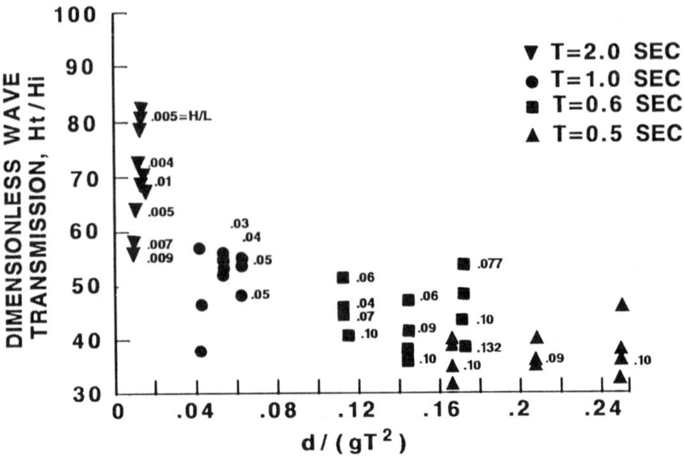

Figure 11. Wave transmission as a function of relative depth ratio for four wave periods. Monochromatic waves. D = 3.0 cm (1-3/16 in.), b/D = 0.1.

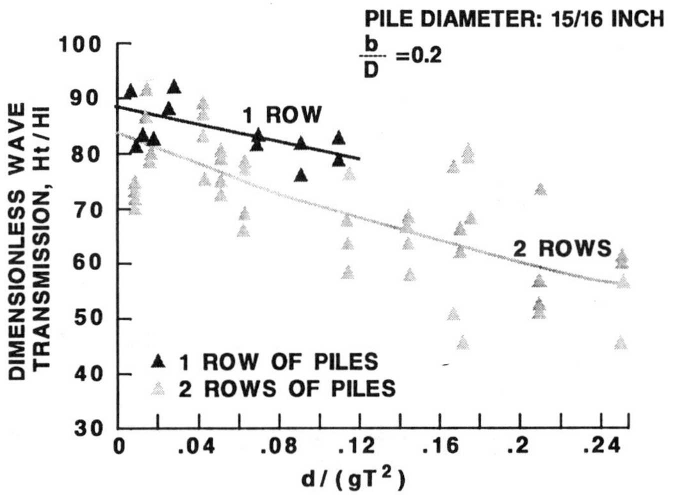

Figure 12. Wave transmission as a function of relative depth ratio for one and two rows of piles. Monochromatic waves. D = 2.4 cm (15/16 in.), b/D = 0.2.

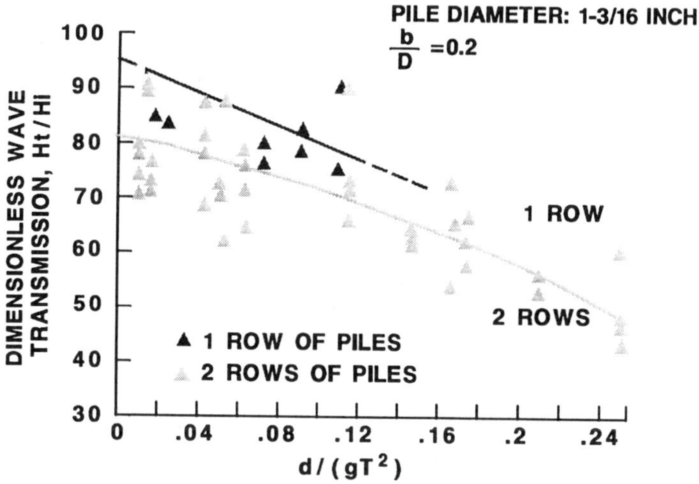

Figure 13. Wave transmission as a function of relative depth ratio for one and two rows of piles. Monochromatic waves. D = 3.0 cm (1-3/16 in.), b/D = 0.2.

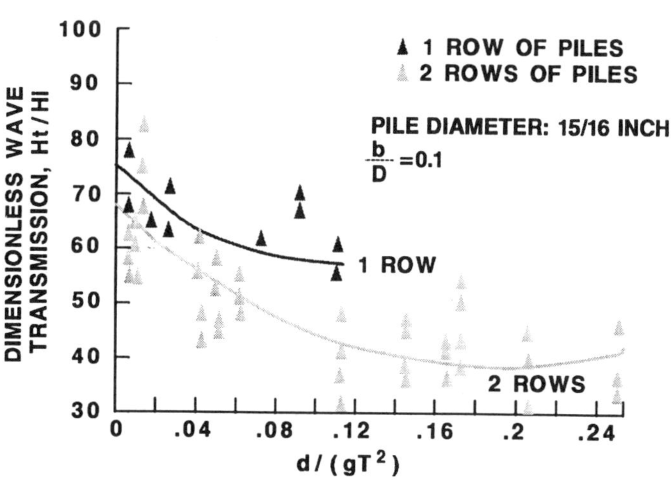

Figure 14. Wave transmission as a function of relative depth ratio for one and two rows of piles. Monochromatic waves. D = 2.4 cm (15/16 in.), b/D = 0.1.

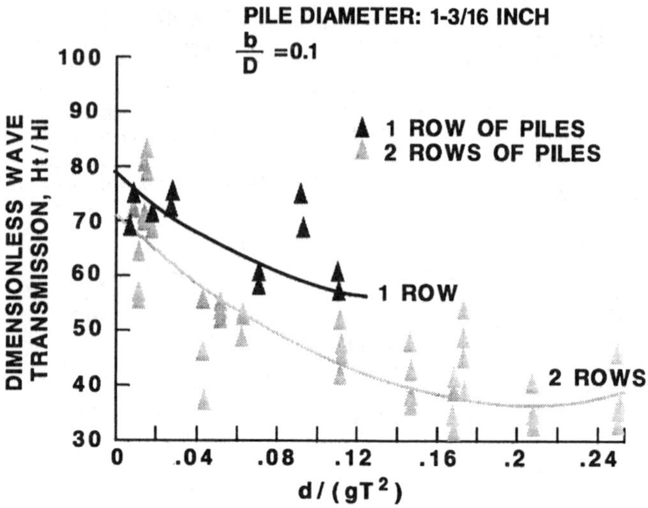

Figure 15. Wave transmission as a function of relative depth ratio for one and two rows of piles. Monochromatic waves. D = 3.0 cm (1-3/16 in.), b/D = 0.1.

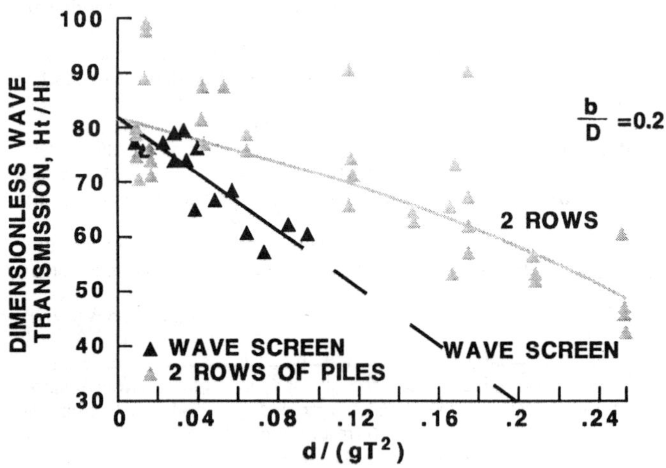

Figure 16. Wave transmission as a function of relative depth ratio. A comparison between two rows of piles and a wave screen.

The results were also compared with wave transmission data presented by Allsop and Kalmus (1985) for wooden wave screens. An arrangement of wooden wave screens was proposed to provide the wave protection for a new marina at Plymouth, United Kingdom. The wave screen breakwater consisted of vertical, rectangular slats with a gap ratio of 0.2. Experiments were performed with irregular wave spectra. The results were published for wave transmission as a function of wave frequency. These results could not be directly compared with the random wave pile experiments because the wave period was not measured for the random wave tests. However, the wave screen results were compared with the monochromatic wave pile experiments. Caution should be used when comparing these two sets of data. The wave screen experiments were conducted using irregular waves, a 8.6 meter water depth, 12 meter long slats and 3 to 12 second wave periods. The wave screen allowed less transmission than the pile breakwater. The results of both experiments agree better at smaller values of d/gT^2 and diverge as the value of d/gT^2 increases (Figure 16).

Conclusions

1. Wave transmission increases as d/H (water depth/wave height ratio) increases
2. Wave transmission for monochromatic waves increases as the wave period increases.
3. Wave transmission decreases as H/L (wave steepness) increases.
4. Wave transmission for monochromatic waves decreases as d/gT^2 (water depth/gravitational acceleration (wave period)^2) values increase.
5. For a 10 percent gap ratio, the addition of a second row of piles reduced the wave transmission by 5 to 10 percent.

References

ALLSOP, N.W.H. and KALMUS, D.C. (1985), Plymouth Marine Events Base: Performance of Wave Screens, Report No. Ex 1327, Hydraulics Research, Wallingford, U.K.

HAYASHI, T., et al. (1966), Hydraulic Research on Closely Spaced Pile Breakwaters, Proceedings of the 10th Coastal Engineering Conference, Vol. II, Chapter 50, pp. 873-884.

KILPATRICK, W.S. (1984), Wave Transmission Through a Row of Rigid, Vertical Piles, Unpublished report, Ocean Engineering Program, Texas A&M University.

TRUITT, C.L. and HERBICH, J.B. (1987), Transmission of Random Waves Through Pile Breakwaters, Ch. 169, Proc. 20th International Conference on Coastal Engineering, ASCE, Taipei, Taiwan, pp. 2303-2313.

WIEGEL, R.L. (1961), Closely Spaced Piles as a Breakwater, Dock and Harbour Authority, Vol. 42, No. 491.

CHAPTER 166

GEOHYDRAULIC INVESTIGATIONS OF RUBBLE MOUND BREAKWATERS

BÜRGER, W.; OUMERACI, H.; PARTENSCKY, H.W.

Universität Hannover
Sonderforschungsbereich 205
Appelstrasse 9A, D-3000 Hannover 1, F.R.G.

1. Introduction

Due to the increase of ship sizes in recent decades a number of harbours and terminals have been built in deeper waters. Accordingly, the structures which have to provide protection against wave action become higher, too. In most cases, these protective structures are of the rubble mound type. Under such conditions the flow induced by waves within the breakwater and the related geotechnical behaviour of the rubble mound fill become more significant for the overall stability and should be considered in the design.

In addition, it is known that the scales usually adopted in hydraulic models (1:30 to 1:60) for investigating the stability of large rubble mound breakwaters generally lead to scale effects with respect to the flow field inside the breakwater. This means that small-scale model tests are not appropriate for investigating the internal flow patterns or for evaluating the pore pressure field induced by the incident waves in the core material.

Because of the uncontrolled conditions in the prototype, and since the actual permeability of the prototype rubble mound fill cannot be predicted (segregation, settlement, variation in grading, etc.), the use of large-scale physical models seems to be the most promising method for basic investigations of this kind. Moreover, the results of such large-scale model tests may be used to validate the usual smaller scale models and to calibrate numerical models.

Therefore, it is one of the objectives of our research programme on rubble mound breakwaters, which started in 1987, to concentrate on the evaluation of the wave-induced flow and pore pressure distribution within the breakwater.

The hydraulic model investigations have been essentially conducted in the Large Wave Channel in Hannover, F.R.G., and will be terminated in December 1988. It is the main purpose of this paper to present the first results of these model tests. However, a brief literature review related to the flow within rubble mound breakwaters and its effect on the overall stability is first given in order to show the necessity of developing reliable tools for the prediction of the actual internal flow field induced by storm waves.

2. Literature review

BARENDS et al. (1983) have demonstrated the importance of the internal flow field on the geotechnical stability of rubble mound breakwaters and applied a numerical model describing this flow for a number of existing breakwaters (BARENDS, 1985). Important improvements to this model were suggested by HANNOURA and Mc.CORQUODALE (1985). However, the reliability of such models is still limited, mainly due to the difficulties in simulating the complex phenomena involved, and to the lack of proper calibration. In this respect ALLSOP & WOOD (1987) have concluded from their most interesting and comprehensive literature review that there have been significant advances in the numerical modelling of the internal flow conditions in recent years, whereas measurements are extremely sparse.

On the other hand, measurements on small-scale model are not reliable because of the scale effects resulting from the lower permeability of the model core material. LE MEHAUTE (1957/1958) was the first to demonstrate the fundamental importance of a proper similitude of the rubble core permeability when investigating wave transmission aspects. The influence of core permeability on the stability of the armour layer has been demonstrated by BRUUN (1985), TIMCO et al. (1984) and VAN DER MEER (1985).

Although a number of investigations (LE MEHAUTE, 1957; DELMONTE, 1970; WILSON, 1971; KOGAMI, 1978) have been carried out on scale effects in modelling wave-induced flow within and through rubble mound structures, there is still no reliable method for the quantification and correction of these effects. The approach using scaling factors in the FROUDE modelling of the core material, as suggested for instance by LE MEHAUTE (1957/58), JENSEN & KLINTING (1983), and OUMERACI (1984), is very limited due to the unsteadiness of the actual internal flow; i.e. it is only valid for a specific hydraulic gradient.

In addition, it is known that slip-circle stability calculations according to BISHOP's method should include the excess pore pressure build-up induced by storm waves within the rubble mound structure and underneath the breakwater toe (BARENDS, 1984; BARENDS & CALLE, 1985). The most important characteristics needed for these calculations are the shear resistance parameters of the materials composing the foundation soil and the breakwater, as well as the maximum pore pressure distribution along a potential slip surface. Prototype measurements of the pore pressures within rubble mound breakwaters (HAKIMI et al, 1984) and underneath the breakwater (VAN IMPE, 1988) have also been performed.

However, a simple reliable method for assessing maximum pore pressure distribution under extreme wave loading and which is suitable for design purposes is not yet available. Such a method should include only the most relevant input parameters, i.e. the water depth and the incident wave parameters in front of the breakwater, the geometry of the structure and the hydraulic properties of the material composing the breakwater and the subsoil.

Well-calibrated numerical and hydraulic scale models may become reliable tools for developing simple relationships between these input parameters and the resulting excess pore pressure. Since the actual permeability of the prototype core and filter material cannot be predicted, the use of large-scale models as described below is most suitable for such calibration.

3. Experimental set-up

In the present research programme much effort has been concentrated upon the development of suitable instrumentation for measuring pore and soil pressure as well as the wave motion (wave run-up) directly at the interfaces between the various layers.

After small-scale tests were performed in the wave flume of the Franzius-Institut (110 m x 2.2 m x 2.0 m) with maximum wave heights up to 0.40 m, the same model was built in the Large Wave Channel in Hannover (320 m x 7.0 m x 5.0 m) in a scale 3.7 times larger. The tests are carried out by using monochromatic waves and JONSWAP-Spectra with $\gamma = 3.3$ and a Groupiness-factor of GF = 0.77. The wave maker is equipped with a SEASIM-wave absorption control. Maximum wave heights up to 2.5 m can be generated.

The waves are recorded in deeper water as well as in front of the structure by a 3 wave gauge-system in order to determine the incident and reflected wave energy. The position of the various wave gauges used is shown in Fig. 1.

In order to determine the variation of the wave-induced stress underneath the breakwater toe, three soil pressure cells and three pore pressure cells are installed in the sand bed layer. In addition, twenty pore pressure gauges are placed inside the rubble mound structure in such a way that the pore pressure distribution and the required boundary values for the calibration of numerical models can be gained with sufficient accuracy (Fig. 2).

One of the pore pressure gauges is shown in Figs. 3 and 4.

In order to follow the incident waves propagating in the armour and under-layer and penetrating into the core material three specially designed wave run-up gauges are also installed along the slope of the tetrapod armour, of the under-layer and of the core material (Fig. 2 and Fig. 6). The grain size distribution of the core material is shown in Fig. 5.

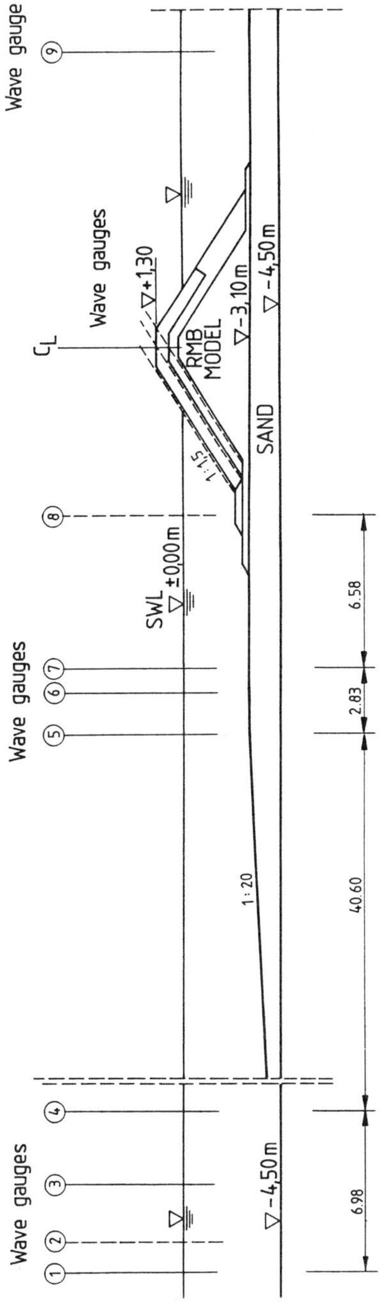

Fig. 1

Rubble mound breakwater model in the Large Wave Channel

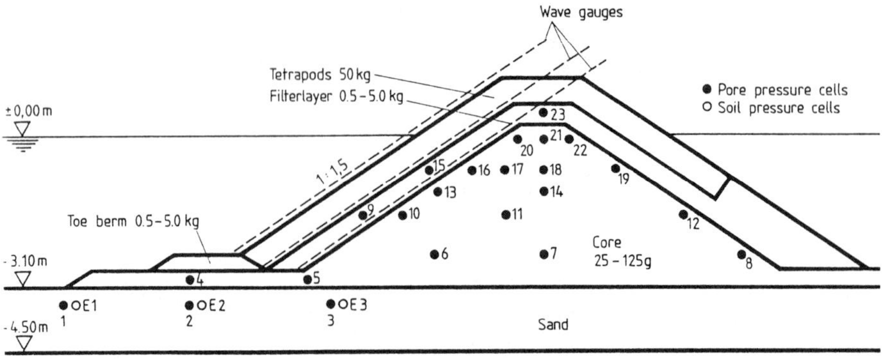

Fig. 2

Cross section of the instrumented large-scale model

4. First experimental results

Some of the first results of the performed model tests are presented in the following. A more detailed presentation and analysis of the final results will be published as soon as the test programme has been completed.

4.1 Wave reflection

Since wave reflection and energy dissipation are interrelated, and since little is known about the reflecting properties of tetrapod armours, the test results are fist analysed with respect to wave reflection. For the range of wave conditons tested ($2.5 \leq \xi < 6.0$), the following formula has been derived for the reflection coefficient K_R:

$$K_R = \frac{H_I}{H_R} = \frac{0.6\,\xi^2}{\xi^2+12} \qquad (1)$$

in which:

H_I, H_R = incident and reflected wave height, respectively
ξ = surf similarity parameter
L = wave length
α = seaward slope angle of the breakwater

The surf similarity parameter is defined as follows:

$$\xi = \tan\alpha \,/\, \sqrt{H/L} \qquad (2)$$

A reflection coefficient of $K_R = 0.20$ to 0.45 was obtained for the range of wave conditions tested.

These reflection coefficients are comparatively lower than those given by the relationship proposed by CERC (SEELIG, 1983) for natural stone armours.

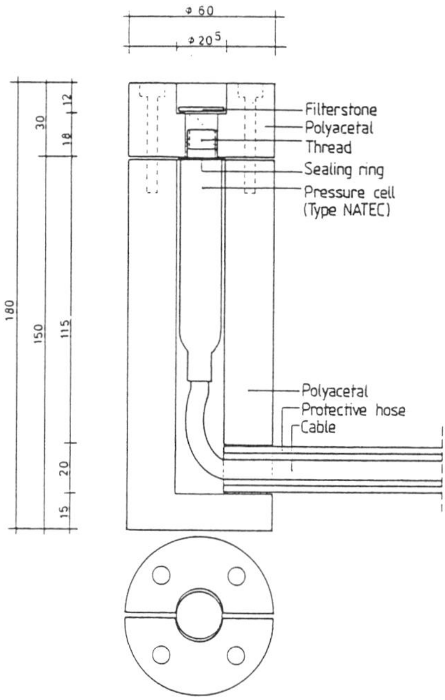

Fig. 3

Pore pressure gauge

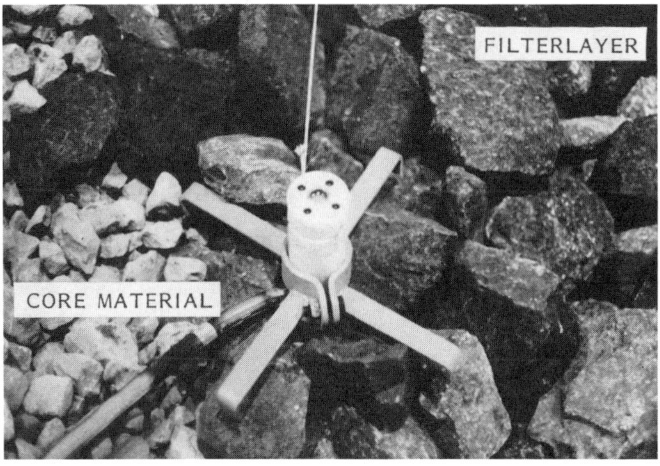

Fig. 4

Installed pore pressure gauge

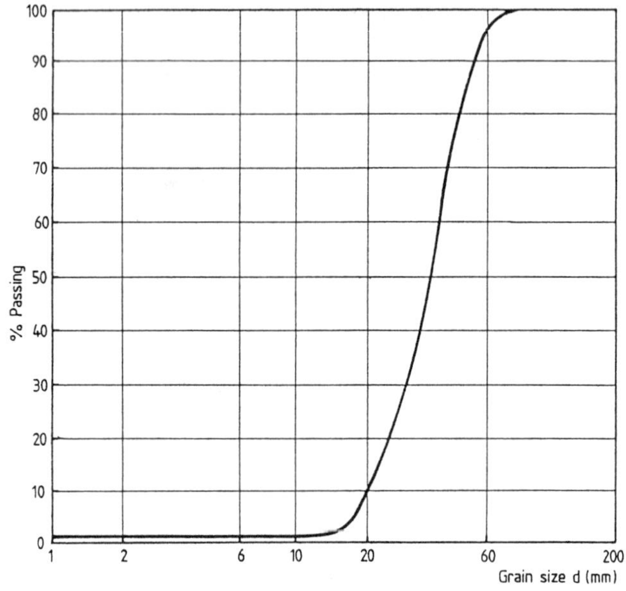

Fig. 5
Grain size distribution of the core material

Fig. 6
Pore pressure gauge and wave run-up gauge

4.2 Wave-induced pore pressure

One of the main objectives of the investigations by means of large-scale model tests is to measure and study the wave motion directly at the seaward face of the structure and the resulting wave-induced excess pore pressures inside the breakwater.

Typical examples of the waves simultaneously recorded by the 3 resistance-type wave gauges installed at the surface of the armour layer (WG 1), at the interfaces between the armour and filter layer (WG 2) and between the filter layer and the core material (WG 3), are given in Fig. 7 for incident waves with H_s = 0.73 m and T_p = 4.5 s in front of the breakwater. The corresponding pore pressure fluctuations induced in a horizontal plane and recorded at pressure cells PC 15-PC 19 (see Fig. 2 and 6) are shown in Fig. 8.

From these and further records it has been found that, in general:

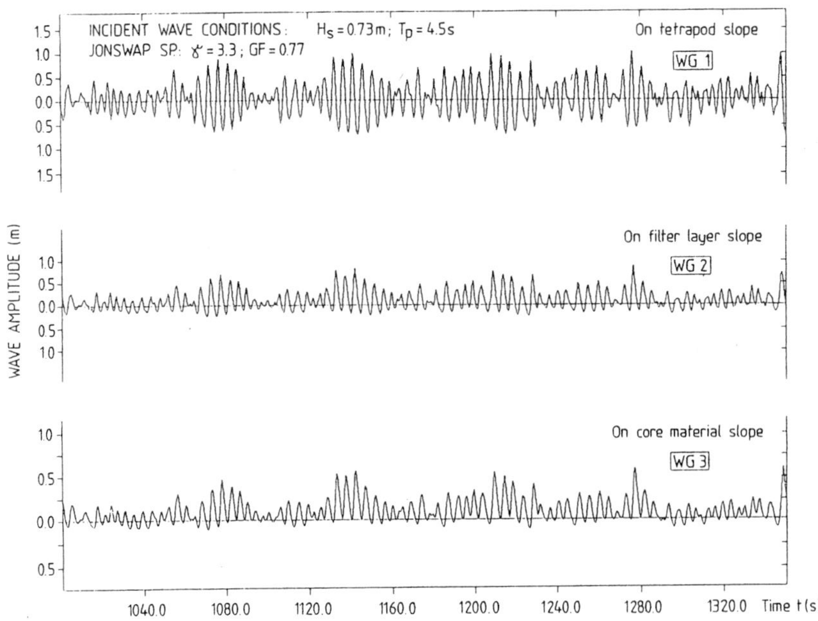

Fig. 7

Wave records at the interfaces sea-tetrapode slope, filter layer-core material (positions of WG 1, 2, and 3 see Fig. 2)

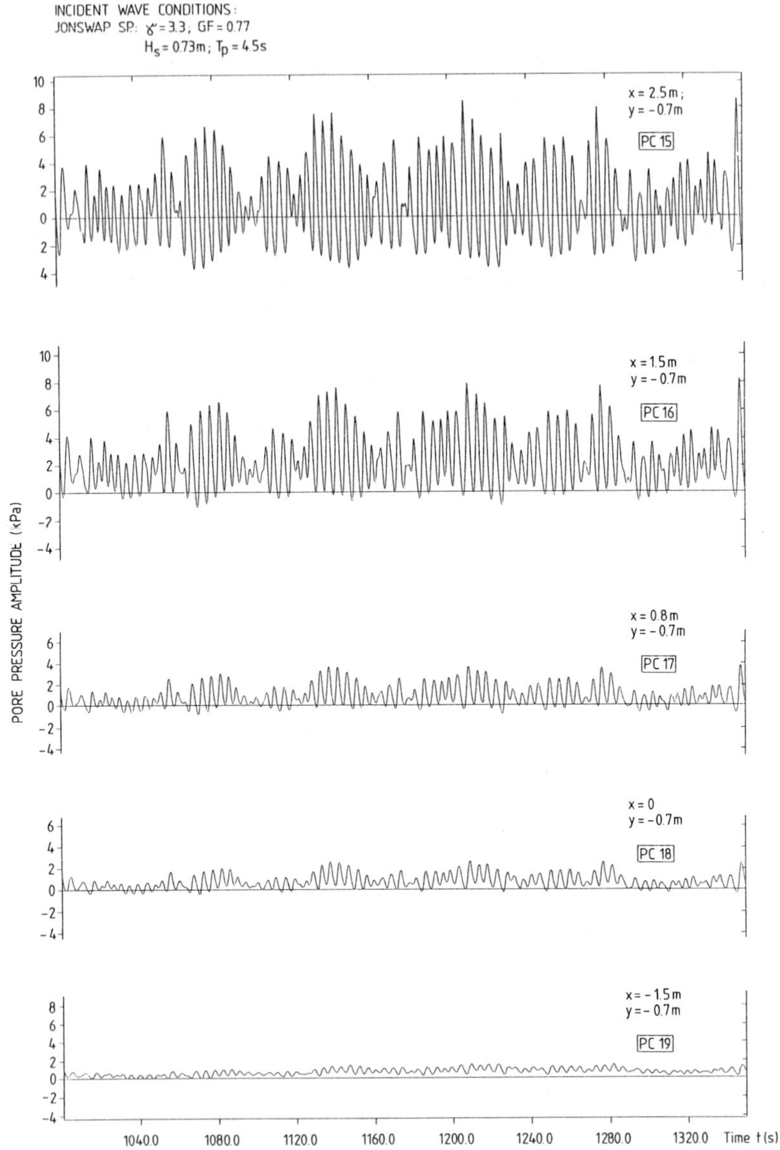

Fig. 8

Pore pressure fluctuations in a horizontal plane inside the breakwater (position of the pressure cells see Fig. 2)

- The frequency of the pore pressure fluctuations are of the same order of magnitude as the frequency of the incident storm waves;
- the incident waves induce a rise in the mean water level (setup) inside the rubble mound breakwater when compared with the sea water level outside the structure (10 - 20 % of the incident wave height);
- most of the incident wave energy is dissipated inside the armour and filter layer. The steeper the waves the more pronounced is this dissipation;
- the pore pressure amplitudes inside the breakwater decrease rapidely in the direction of wave propagation. This decay is more rapid for steeper waves.

The horizontal distribution of the maximum amplitudes of the excess pore pressure fluctuations induced by irregular waves at different elevations inside the breakwater and underneath the seaward toe is shown in Fig. 9. It can be seen that, particularly at elevations near the still water level, the maximum pore pressure amplitudes decay exponentially in the direction of wave propagation. Exponential decay of this kind has also been found theoretically by BIESEL (1950) for internal wave amplitudes. Fig. 9 and further similar results show that in deeper planes the amplitude decay becomes less pronounced.

Within the zone underneath the breakwater toe, however, the maximum pore pressure amplitudes tend to increase slightly in the direction of wave propagation. This is probably due to the more pronounced pressure build-up along the breakwater slope.

The distribution of the maximum pore pressure amplitudes at three different vertical profiles is shown in Fig. 10. From this and further results it is found that for the vertical profiles extending above still water level (SWL) the pore pressure amplitudes first increase and then slightly decrease with increasing depth. The maximum pressure amplitde is generally slightly below the still water level and depends upon the characteristics of the incident waves and the position of the vertical profile.

An interesting typical result is also shown in Fig. 11 for the distribution of the maximum pore pressure amplitudes along the slope of the core material and along another profile, parallel to the first, inside the core.

The results suggest, that:

- the pore pressure amplitudes do not decrease significantly in the direction normal to the slope;
- the position of the maximum value of the pressure amplitudes depends greatly on the depth of the profile considered.

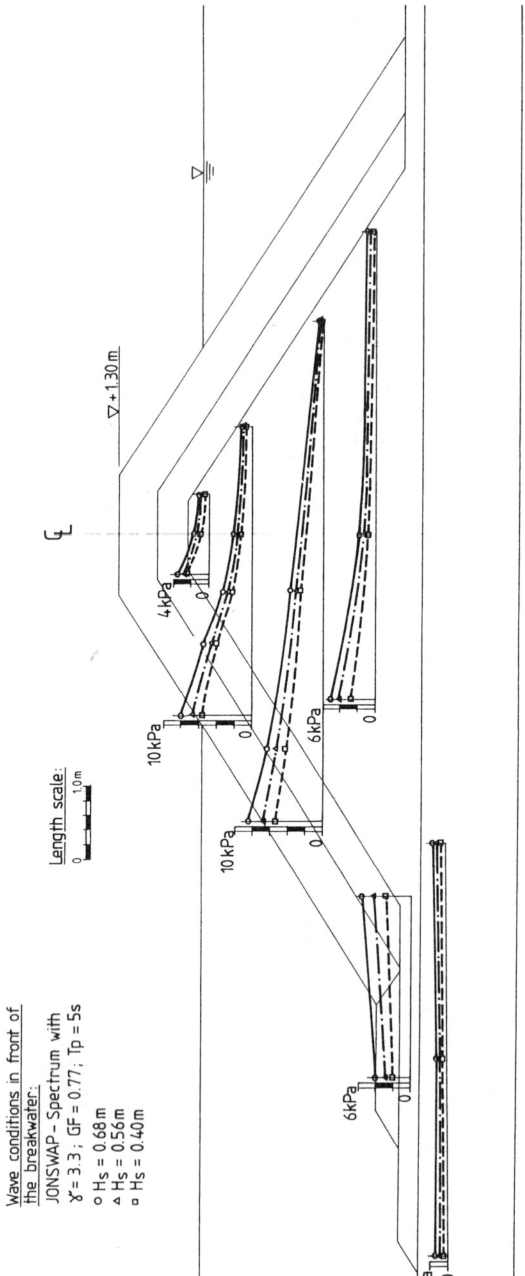

Fig. 9

Horizontal distribution of the maximum excess pore pressure amplitudes induced by irregular waves

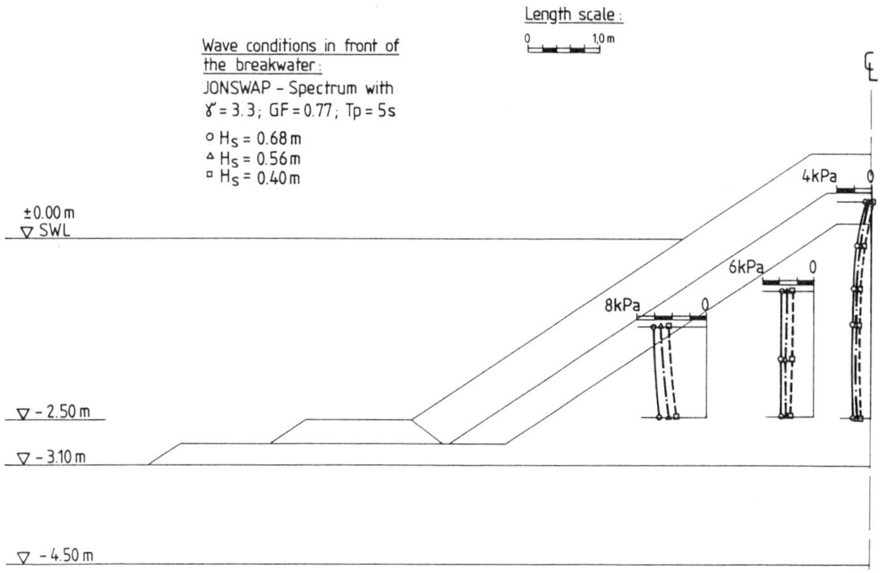

Fig. 10

Vertical distribution of the maximum excess pore pressure amplitudes induced by irregular waves

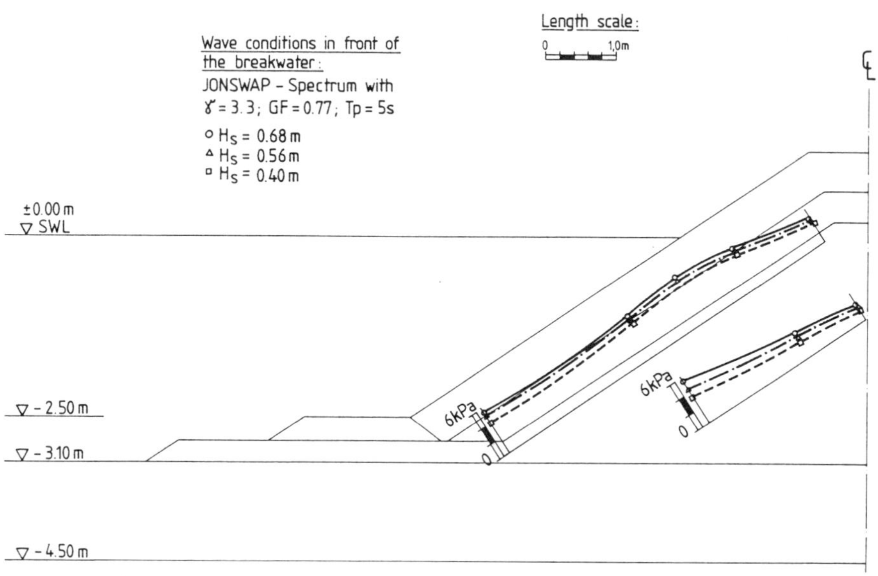

Fig. 11

Distribution of the maximum excess pore pressure amplitudes along the slope of the core material

5. Conclusion and perspectives

For the near future the problem of the wave-induced oscillatory flow inside rubble mound breakwaters will obviously not be amenable to an accurate analytical solution. This is due to a number of complex phenomena involved (breaking, air infiltration, unsteadiness and non-linear resistance of the flow, virtual mass effects, etc.) and uncertainties in the hydraulic properties of the porous media composing the breakwater.

Since a reliable method for quantifying and correcting scale effects in modelling wave-induced internal flow is still lacking and since an improvement and proper calibration of the existing numerical models by using results of large-scale model tests are needed, the analysis of the test results, which is underway, will focus particularly on these two aspects. Nevertheless, the first results and the literature review presented in this paper suggest that:

- Equation (1) for the reflection coefficient related to tetrapod armour layers would still be improved by including the results of small and large-scale model tests in the analysis;
- an empirical relationship between excess pore pressure and wave steepness (or surf similarity parameter) may be derived;
- an empirical or a semi-empirical equation for the decay of the pore pressure fluctuations within the breakwater in the direction of wave propagation should be formulated. For design application (slip-circle calculations), however, there is a need for simple formulae for approximately assessing the wave-induced pore pressures as a function of the water depth and the incident wave parameters in front of the breakwater. Unlike the approach in the hydraulic stability of armour layers, these formulae should be related to maximum wave loading;
- understanding of the interior process may be improved by including the transfer functions between wave and pore pressure fluctuations in the analysis. For the waves, use should be made of the recorded spectra at the interfaces sea-tetrapod slope and filter layer-core material.

6. Acknowledgements

This investigation is a part of a basic research programme in coastal engineering (SONDERFORSCHUNGSBEREICH 205) financed by the German Research Council (DFG). The authors are grateful to Dr. TAUTENHAIN and his staff at the Large Wave Channel, Hannover, for their help in the preparation and performance of the large-scale model investigations.

7. References

Allsop, N.W.H.; Wood, L.A.: "Hydro-geotechnical performance of rubble mound breakwaters - A literature review". Hydraulics Research Ltd. Rep. SR 98, Wallingford March 1987

Barends, F.B.T.; Kogel, H.; van der Uijttewall, F.J.; Haganaar, J.: "West-breakwater sines, dynamic-geotechnical stability of breakwaters". Proc. ASCE Conference Coastal Structures' 83, 1983

Barends, F.B.T.: "Probability of wave induced seabed instability". International Symposium on Maritime Structures in the Mediterranean Sea, Athens, 1984

Barends, F.B.T. & Calle, E.O.F.: "A method to evaluate the geotechnical stability of off-shore structures founded on a loosely packed seabed sand in a wave loading environment". Proc. of the 4th Intern. Conf. on Behaviour of Offshore Structures (BOSS' 85), Delft, 1985

Barends, F.B.T.: "Geotechnical aspects of rubble mound breakwaters". Proc. Conf. Breakwaters' 85, ICE, London 1986

Biesel, F.: "Equations de l'écoulement non lent en milieu perméable". La Houille Blanche, Mars-Avril 1950

Bruun, P. (editor): "Desgin and construction of mounds for breakwaters and coastal protection". Elsevier, Amsterdam 1985

Delmonte, R.C.: "Scale effects of wave transmission through permeable structures". ASCE Proc. of the 13th Conf. on Coastal Eng., Vancouver 1972

Hakimi, A.; Diouri, A.; Kaissoum, A.: "Contribution empirique à l'étude de la propagation de la houle à l'intérieur des digues". La Houille Blanche No 5, 1984

Hannoura, A.A.; Mc Corquodale, J.A.: "Rubble mounds: numerical modelling of wave motion". Proc. ASCE, Journal of Waterway, Port, Coastal and Ocean Engineering, Vol. 111, No. 5, Sept. 1985

Jensen, O.J.; Klinting, P.: "Evaluation of scale effects in hydraulic models by analysis of laminar and turbulent flows". Coastal Engineering, Elsevier, Vol. 7, 1983

Kogami, Y.: "Researches on stability of rubble mound breakwaters". Coastal Engineering in Japan, Vol. 21, 1978

Le Mehauté, B.: "Permeabilité des digues en enrochements aux ondes de gravité périodiques". La Houille Blanche, No.6, 1957; No. 2 & 3, 1958

Oumeraci, H.: "Scale effects in coastal hydraulic models". IAHR Symposium on Scale Effects in Modelling Hydraulic Structures, Esslingen, Sept. 1984

Seelig, W.N.: "Wave reflection from coastal structures". ASCE Proc. Conf. Coastal Structures, 1983

Timco, G.W.; Mansard, E.P.D.; Ploeg, J.: "Stability of breakwaters with variation in core permeability". ASCE Proc. 19th Conf. on Coastal Eng., Houston, 1984

Van der Meer, J.W.: "Stability of rubble mound revetments and breakwaters under random wave attack". Proc. Conf. Breakwaters' 85, ICE, London, 1985

Van Impe, W.F.: " Geotechnische Probleme beim Bau des neuen Hafens Zeebrügge". Geotechnik, 1, 1988

Wilson, K.W.; Cross, R.H.: "Scale effects in rubble mound breakwaters". ASCE Proc. of the 13th Conf. con Coastal Eng., Vancouver, 1972

CHAPTER 167

IRREGULAR WAVE LOADING ON A CONICAL STRUCTURE

Wayne W. Jamieson, Etienne P.D. Mansard and Geoffrey R. Mogridge[1]

ABSTRACT

The horizontal forces, vertical forces and overturning moments exerted by waves on a fixed model of a 45° conical structure are presented. Irregular wave loading tests were conducted for a range of conditions described by base diameter on peak period wave length D/L_p from 0.31 to 1.76, water depth on peak period wave length h/L_p from 0.11 to 0.63, and significant wave height on peak period wave length H_{m_0}/L_p up to 0.07.

Time series records, spectral densities and transfer functions for the irregular wave loading tests are used to illustrate the nonlinear nature of the measured wave loads. In most cases, similar trends in wave loading were observed for irregular and regular wave tests. For deep-water waves, the irregular and regular force measurements showed spectral peaks at the second harmonic of the wave frequency even though the waves themselves had relatively small second-order components. However, unlike the regular wave loading results, the fundamental spectral peak frequency for the irregular wave forces occurred at a frequency considerably lower than the peak frequency of the waves. Although linear diffraction theory provided a reasonable estimate of the wave forces for waves of low steepness, larger deviations were often present for higher wave steepness results. Comparison of theory and experiment for overturning moments was generally very poor for most wave conditions.

INTRODUCTION

In recent years numerous designs of offshore structures for drilling and production have been proposed for use in arctic regions. Buslov and Krahl (1983) have described sixty-one conceptual designs for arctic offshore platforms, some of which are currently being constructed, others of which are operational. Some of the designs incorporate large bottom-founded conical-shaped structures

[1] Research Officers, National Research Council of Canada, Hydraulics Laboratory, Montreal Rd., Bldg. M-32, Ottawa, Ontario, K1A 0R6, Canada.

which cause ice sheets to fail predominantly by bending rather than by buckling or crushing, in order to reduce ice loading. Of course, the relatively small area subjected to ice loading near the water surface also minimizes ice forces. The large hull of this type of gravity structure provides resistance against sliding and overturning at its base, but also provides sufficient buoyancy when deballasted so that it can be towed while floating on its hull to another site. Although none of these conical structures have yet been built, they are currently being considered for use in a range of water depths between 10 m and 80 m.

The wave loading on some of these proposed conical structures has been modelled both physically and numerically, but the results are for the most part proprietary and as a result, published data is very limited. Isaacson (1982) has published some results of numerical calculations for the case of a surface-piercing 45° conical structure utilizing the body's vertical axisymmetry to reduce considerably the computational effort required for bodies of arbitrary shape. Jamieson et al. (1985) have carried out regular wave loading measurements on a model of a surface-piercing 45° conical structure. The model was a simple axisymmetric structure which somewhat characterized the shape of proposed prototype conical structures, could easily be modelled both physically and mathematically and the results could be published without infringing on any proprietary designs. The tests indicated extremely nonlinear wave loading resulting in considerable differences between experiment and linear diffraction theory particularly for higher steepness waves. In this paper, the horizontal forces, vertical forces and overturning moments exerted by irregular waves on the same 45° conical model are presented. The irregular wave loading results are compared to those measured for regular waves and those computed by linear diffraction theory.

EXPERIMENTAL SETUP

Figure 1 shows a photograph of the conical model. It was constructed of 6 mm thick plexiglass in the shape of a 45° right circular cone, 125.5 cm in diameter at the base (see Figure 2). The model was tested in a water depth of 44.6 cm. Assuming a scale of approximately 1:100, the model corresponds to a full scale 45° conical gravity platform with a 126 m diameter at the base, resting in a water depth of 45 m. Model dimensions will be used throughout this paper. For structural considerations, the base of the model was slightly rounded rather than sharp-edged. Figures 3 and 4 show a photograph and schematic views respectively of the wave flume and the experimental setup. The model was installed 11 m from the wave generator in the centre of the 13 m wide portion of the wave flume. The outer circumference of the model base was sealed and the underside of the model was vented to atmospheric pressure so that no dynamic wave pressures were experienced underneath the model. This condition allowed the measurement of wave loading due to fluid pressures on the upper surfaces of the model, which is all that is of interest in the present study, because under full scale

Figure 1. Model of Conical Structure

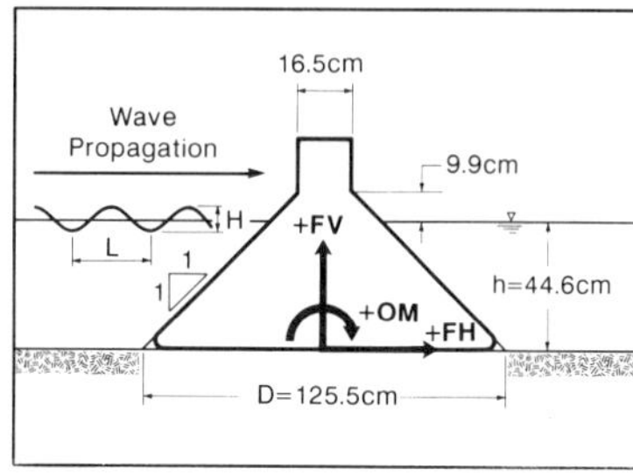

Figure 2. Model Dimensions and Sign Convention for Wave Loads

Figure 3. Photograph of Wave Flume and Experimental Setup

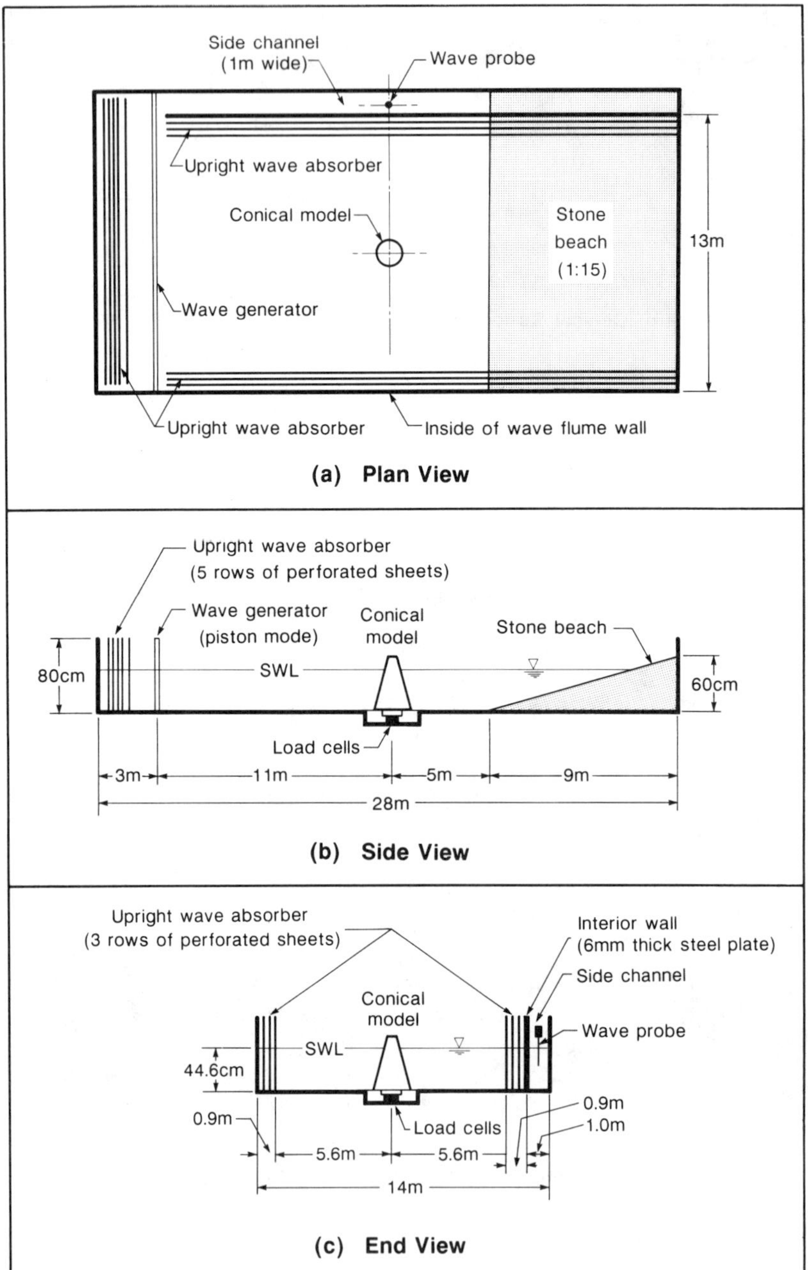

Figure 4. Schematic of Wave Flume and Experimental Setup

conditions the wave pressures underneath the structure propagate through a porous foundation rather than through an open gap as in a model study.

The base of the model was mounted to a three-component load cell dynamometer system which was in turn bolted to the bottom of a circular pit located below the concrete floor of the wave flume. The dynamometer system was designed so that three individual load cells could be bolted together in a configuration which allowed the measurement of horizontal force (FH), vertical force (FV) and overturning moment (OM). The three load cells were constructed of steel and designed by the Hydraulics Laboratory to have a very high stiffness-sensitivity product with only small cross-coupling errors. Each of the load cells used semiconductor strain gauges, which were thermally compensated and thoroughly waterproofed to minimize environmental problems caused by temperature changes and high humidity. They were initially calibrated individually in air on a calibration rig. Combined nonrepeatability, nonlinearity and hysteresis of the load cells were less than ±0.5%. During the testing program, the load cell calibrations were checked once a day by applying known loads to the model by means of a system of low-friction pulleys. Natural frequencies of vibration of the total system were measured by applying a step load to the model in both vertical and horizontal directions, and recording the voltage outputs from each load cell sampled at a rate of 500 samples/second. With 44.6 cm of water in the wave flume, the natural frequencies of vibration were 30 Hz, 13 Hz and 30 Hz as measured by the horizontal and vertical force load cells, and the torque load cell, respectively.

In this paper, the measured wave forces and overturning moments on the conical model have been related to the incident wave conditions, that is, the waves that occur at the location of the model with the model removed. To eliminate the time required to regenerate the waves with the model removed, all incident waves were measured by a capacitive wire wave probe located alongside the conical model in the 1 m wide side channel (see Figures 3 and 4). The side channel was separated from the main channel by a 6 mm thick solid steel plate wall.

Voltage signals from the test instrumentation went to a high-speed data acquisition package consisting of programmable-gain differential amplifiers, a high-speed solid-state multiplexer and an analog-to-digital converter. The output from the wave probe required no amplification; however, the two force load cell signals and the torque load cell output were amplified 200 times. A minicomputer was used for the acquisition, processing, plotting and storage of experimental data.

The waves generated by the wave machine moved along the length of the flume past the model and were effectively absorbed by the 1 in 15 sloped stone beach at the end of the flume shown in Figure 4. Wave reflection coefficients (ratio of reflected wave height to incident wave height expressed as a percentage) for this beach were less than 5% for all wave conditions tested.

Figures 3 and 4 show the placement of upright wave absorbers along the side walls of the 13 m wide main channel. Each of these side absorbers consisted of three rows of perforated vertical sheets placed over a distance of 90 cm. The porosity of the rows of sheets decreased towards the rear of each absorber. The design is based on a new type of efficient upright wave absorber developed at the Hydraulics Laboratory of the National Research Council of Canada (Jamieson and Mansard (1987)). It is particularly suitable for laboratory wave basins where (1) limited space is available for the installation of absorbers, (2) testing at variable water depths is essential without having to make adjustments to the absorber, and (3) reflected wave energy from the absorber must be minimized for a wide range of water depths, wave heights and wave periods (wave reflection coefficients of less than 5%). The placement of the rows of perforated sheets in the side absorbers is such that the waves generated from the wavemaker moved along the length of the flume through the side absorbers quite freely but the waves radiated off the conical model were effectively absorbed. The attenuation of the waves as they propagated in the main channel was negligible, so that the side channel wave probe effectively measured the incident wave heights. After a test, the waves in the flume were attenuated in less than 5 minutes compared to a settling down period of more than 30 minutes if only solid side walls were used. An upright wave absorber was also used to provide effective wave absorption over a distance of 1.5 m between the rear of the wave generator and the back wall of the wave flume.

EXPERIMENTAL METHOD

The irregular waves generated in this study were synthesized using the random phase spectrum method. It is a technique which pairs the amplitude spectrum derived from a given target spectral density with a randomly selected phase spectrum, and creates a time series by inverse Fourier transformation (Funke and Mansard (1984)). JONSWAP spectral densities with a peak enhancement factor $\gamma = 3.3$ were used as target spectra. Each time series was synthesized with a record length of 300 s, resulting in approximately 150 waves for the lowest spectral frequency and about 450 waves for the highest.

A number of sea states with different spectral peak frequencies (f_p = 0.48, 0.69, 0.80, 0.95, 1.02, 1.16, 1.33 and 1.48 Hz) were generated. For each peak frequency, the significant wave height H_{m0} of the sea state was varied in order to evaluate the influence of nonlinearities due to wave steepness. This was achieved by rescaling the time series to match the desired variance. A minimum of four significant wave heights having wave steepnesses (H_{m0}/L_p) between 0.01 and 0.07 were run for each spectral peak frequency listed above. Since only the variance of the sea state was changed, other parameters such as groupiness factor, ratio of maximum to significant wave heights etc. were kept constant.

Sampling of the data was commenced three minutes after the first waves had passed the model. For each irregular wave test, the incident wave profile in the side channel and the wave loads (FH, FV and OM) were sampled at a rate of 10 samples per second, for a total record length of 300 seconds. The water depth was maintained at 44.6 cm throughout the entire testing program. Total forces and overturning moments about the base of the model (see Figure 2) were computed and plotted along with the incident wave profile measured in the side channel.

Spectral analysis of the wave loads and the incident wave record was carried out for each test. The spectral analysis used in this study is an algorithm which performs a Fourier transform over the entire record length and computes the spectral ordinates by squaring and summing the complex Fourier coefficients. Smoothing of the spectrum was carried out by averaging the adjacent spectral estimates using a rectangular window.

Transfer functions of the irregular wave loads were calculated for each test as follows:

$$TRF(f) = [S_{WL}(f) \,/\, S_{IW}(f)]^{1/2}$$

where
$TRF(f)$ = Transfer function of the wave load (FH, FV or OM)
$S_{WL}(f)$ = Spectral density of the corresponding wave load
$S_{IW}(f)$ = Spectral density of the incident wave (H).

Regular wave tests were carried out for the purpose of comparison with the irregular wave tests. Sampling of the data was commenced after ten waves had passed the model. For each regular wave test, the incident wave profile and the wave loads were sampled at a rate of 100 samples per second for a total record length of 50 s. Regular waves were generated for a wide range of frequencies (f = 0.48, 0.60, 0.69, 0.80, 0.88, 0.95, 1.02, 1.16, 1.33 and 1.48 Hz). A minimum of six wave heights having wave steepnesses (H/L) between 0.01 and 0.10 were generated for each wave frequency listed above.

EXPERIMENTAL RESULTS AND DISCUSSION

Time Series of Wave Loading

Figure 5 shows typical time series measurements of horizontal force (FH), vertical force (FV), overturning moment (OM) and the incident wave measured in the side channel (H) for two irregular wave tests of low wave steepness (H_{m0}/L_p = 0.02). The time series results shown are representative of the tests carried out for intermediate-water depths (h/L_p = 0.21) and deep-water depths (h/L_p = 0.51). Only a 20 s portion of the 300 s record length is shown for each of the tests. The time series of the wave loading results for the intermediate-water depth (h/L_p = 0.21) results are smooth and one would expect the spectral analysis of the wave loading to give the same peak frequency as the waves. For the deep-water results (h/L_p = 0.51), a similar observation is evident for the overturning moment (OM). However, for the force records (FH and FV) both high and low frequency components are evident which are not identifiable in the wave record. Spectral

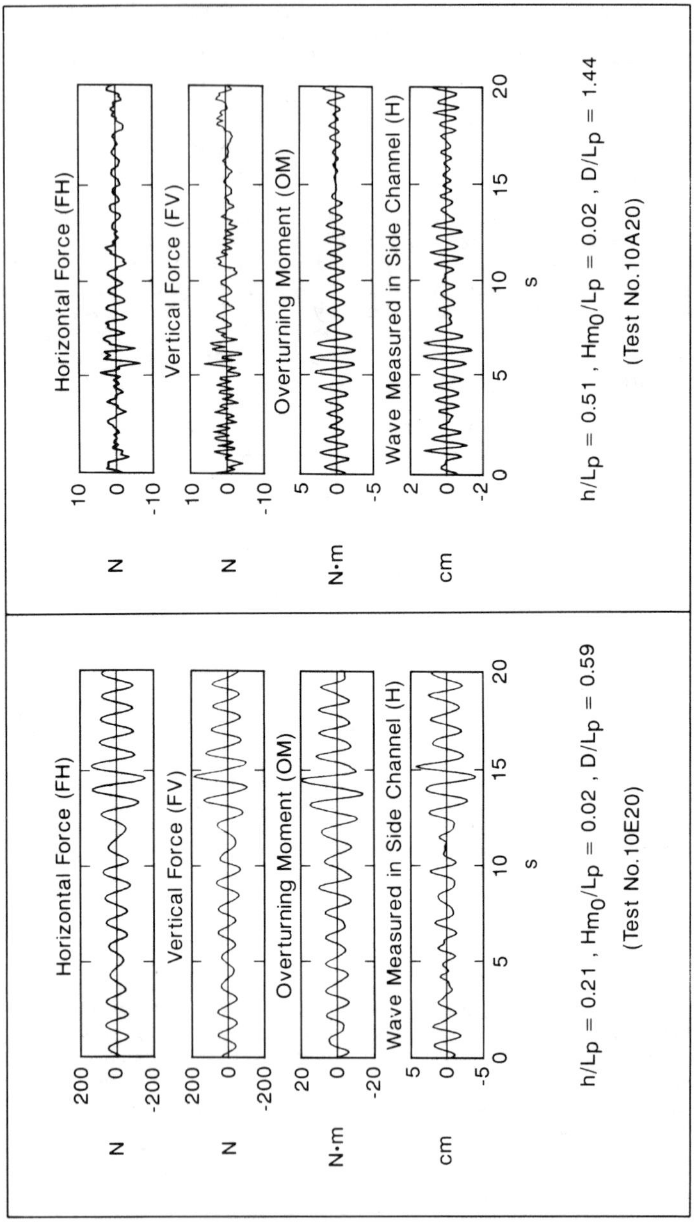

Figure 5. Typical Time Series Measurements for Irregular Wave Tests

analysis results of the data is discussed in the next section.

Spectral Densities of Wave Loading

Spectral density measurements for irregular and regular wave tests for a relative depth of 0.21 are shown in Figure 6. This illustrates the type of irregular wave loading commonly observed for intermediate-water depths, i.e. the peak frequency (0.80 Hz) of the wave forces (FH and FV) and the overturning moment (OM) is the same as that of the waves. In the time series for this test (shown in Figure 5) it can be seen that the frequency of the wave loads and the waves is about the same. The spectral density results for the regular wave test in Figure 6 also shows that for an intermediate-water depth, $h/L = 0.21$, the peak frequency of the wave loads coincides with that of the waves. The above examples are for low steepness waves for irregular and regular wave trains; the same general results occur for higher steepness waves.

Spectral density measurements for irregular and regular wave loading for deep-water waves (a relative depth of 0.51) in Figure 7 show the emergence of strong nonlinearities in the forces which are not present in the incident wave spectrum. The results for the regular wave test in Figure 7 are similar to those published previously for regular waves by Jamieson et al. (1985). For the horizontal and vertical forces, the spectral density plots show very little response at the fundamental frequency (1.34 Hz) of the incident waves; most of the response is at the second harmonic (2.68 Hz). Thus, it is not surprising that the irregular deep-water wave test ($f_p = 1.30$ Hz) results illustrated in Figure 7 indicate significant response in the spectra for the horizontal and vertical forces at frequencies approximately twice the peak frequency of the waves (2.60 Hz). However, it is not clear why the horizontal and vertical force spectral peaks for this irregular test do not occur at the same peak frequency as the waves (1.30 Hz) but at the slightly lower frequency of 1.04 Hz. It is not known why this occurs. Even though the magnitude of the wave loads for these conditions is quite small, it may still be important to understand the reason for this unusual nonlinear transfer of response. The regular wave loading results reported by Jamieson et al. (1985) for horizontal and vertical force, as well as the regular wave force results shown in Figure 7, do not show a similar shift of response because there is no interaction between frequency components. Indications of these two phenomena have already been observed in the time series plots (Figure 5). For the test results shown in Figure 7, the spectral peaks for the overturning moments occur at the same peak frequencies as the waves (1.30 Hz and 1.34 Hz for the irre-
under other wave conditions a significant amount of energy in the spectra for the overturning moment was measured at frequencies approximately twice the peak frequency of the waves.

In the previous paper by Jamieson et al. (1985) it was observed that for regular deep-water waves, the wave

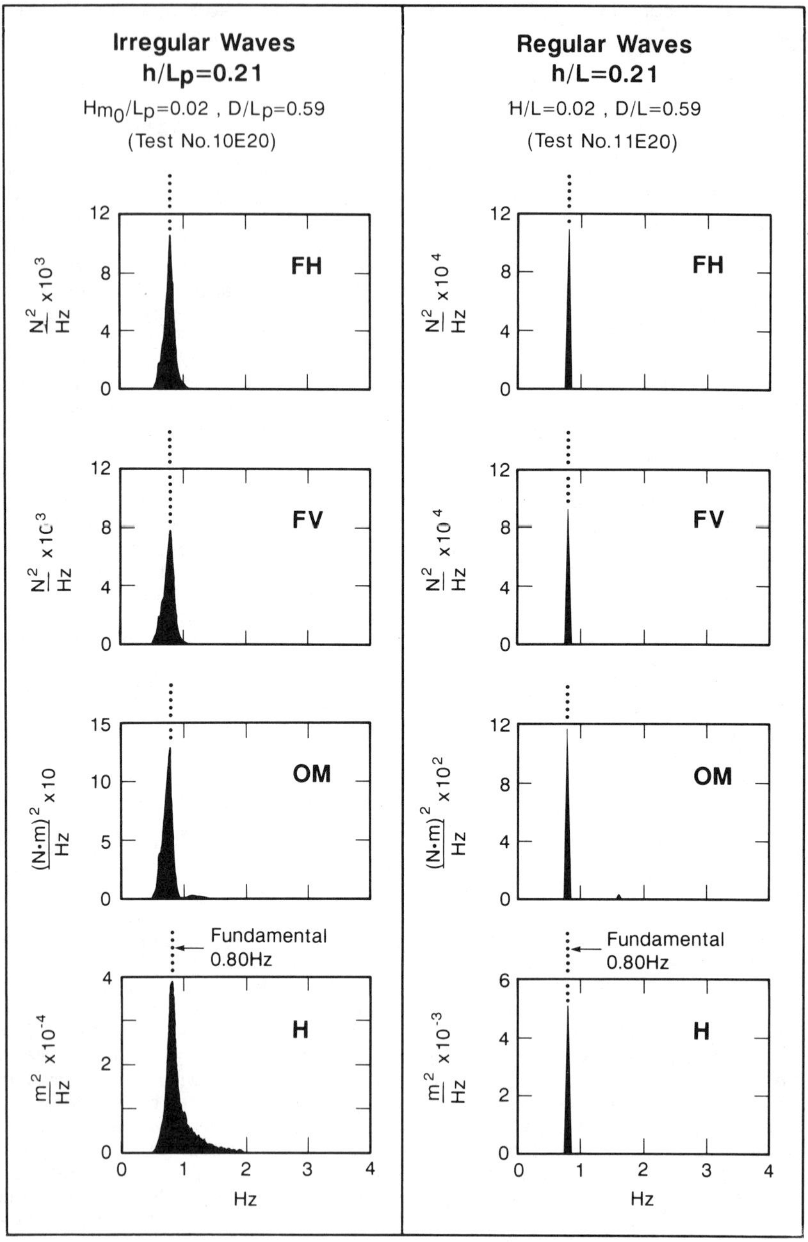

Figure 6. Typical Spectral Density Measurements for a Relative Depth of 0.21

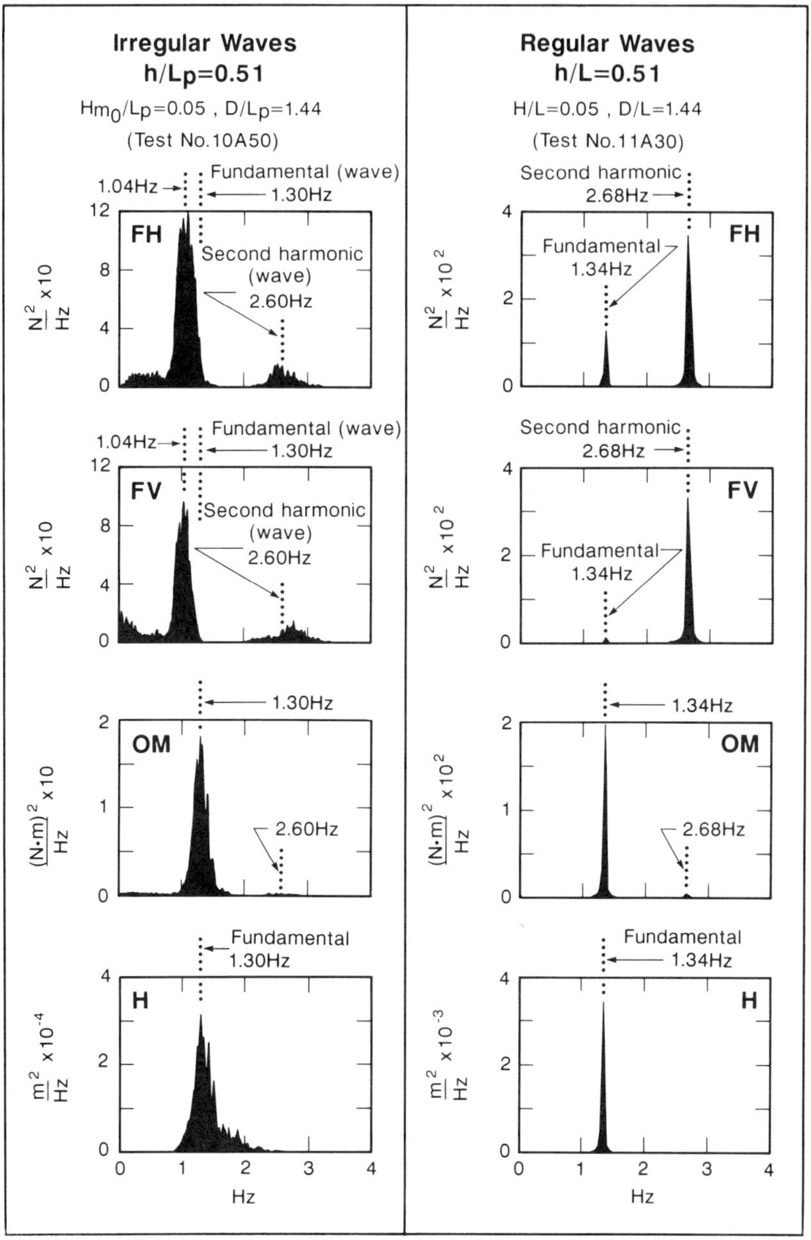

Figure 7. Typical Spectral Density Measurements for a Relative Depth of 0.51

loading was predominantly at the second-harmonic frequency of the waves (particularly for the horizontal and vertical wave forces). In that paper there was a brief discussion of previous reportings of second-order pressure fluctuations which do not attenuate with water depth, i.e. Longuet-Higgins (1950), Rundgren (1958), Hogben and Standing (1975). In the paper by Jamieson et al. (1985), it was observed that in addition to the wave frequency, the structural shape is important in determining whether second-harmonic forces and moments will appear. Thus, for a conical shape, the increasing diameter of the structure towards the bottom results in an increasing influence of the second-harmonic pressures where they are larger relative to the linear pressures. As far as the authors are aware, the shift in the irregular wave loading spectral peaks to a frequency lower than the wave excitation spectral peak frequency, has not been observed before.

Wave Load Transfer Functions

Transfer functions for horizontal force (FH), vertical force (FV) and overturning moment (OM) are plotted in Figure 8 for irregular wave tests having wave steep nesses, H_{m0}/L_p = 0.02 and 0.06. Each of these transfer functions has only been calculated and plotted over the range of frequencies where there is significant energy in the incident wave spectrum for the given irregular wave test. Thus the nonlinear transfer of response in the wave loads from the wave frequency to lower or higher frequencies is not evident in the plotted results (shown in Figure 8) because the data has been truncated. The linear transfer function used here has large errors at frequencies where the energy of the wave excitation is small or negligible.

Transfer functions for eight tests with peak wave frequencies (as outlined in the experimental method) ranging from 0.48 Hz to 1.48 Hz each with a wave steepness of H_{m0}/L_p = 0.02 have been plotted on the left side of Figure 8 for each wave load (FH, FV and OM). For these low wave steepness results, all of the transfer functions fall on the same curve for each of the wave loads.

Transfer functions for seven tests with peak wave frequencies (as outlined in the experimental method) ranging from 0.69 Hz to 1.48 Hz each with a wave steepness of H_{m0}/L_p = 0.06 have been plotted on the right side of Figure 8 for the wave loads. For these higher steepness results, the transfer functions did not all fall on the same curves but showed some scatter as indicated by the wider lines used in plotting the curves. This is particularly evident for the plotted transfer functions of horizontal force and overturning moment.

If the wave loading results were linear, then the transfer functions would be the same for all wave steepness results. However, the results for the two wave steepnesses (H_{m0}/L_p = 0.02 and 0.06) in Figure 8 show a noticeable difference between the transfer functions for the different wave steepnesses. Although not shown here, the transfer

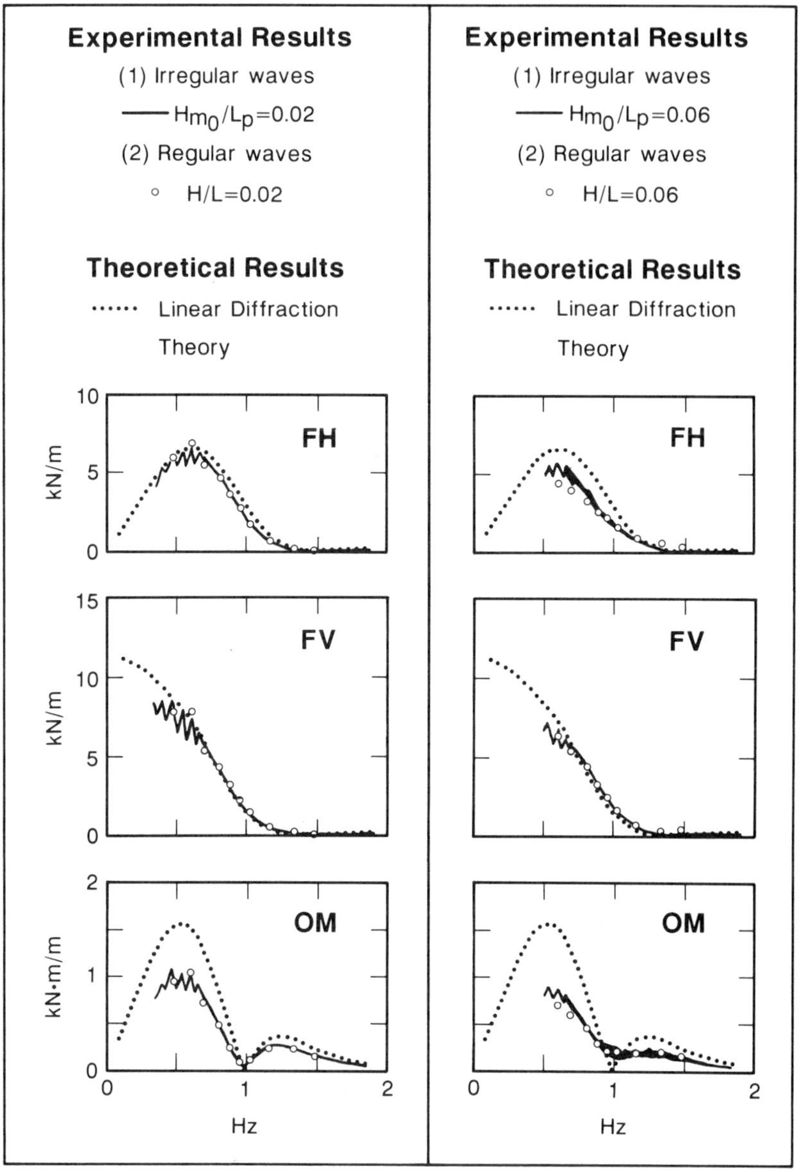

Figure 8. Comparison Between Experimental and Theoretical Wave Load Transfer Functions

functions for the other wave steepness results that were tested were also plotted and showed the same trends which are apparent in Figure 8. The variations in the transfer functions clearly indicate that the interaction between the conical structure and the incident waves is nonlinear.

Regular wave loading results are also included in Figure 8. Response amplitude operators (RAOs) were calculated for the regular wave tests previously described in the experimental method and are plotted for wave steepnesses H/L = 0.02 and 0.06. The RAOs were calculated by adding the maximum positive wave load to the maximum negative wave load (since the absolute magnitudes were generally not equal) and dividing by the incident wave height. The RAOs for the regular waves are in good agreement with the transfer functions calculated for the irregular waves for all the wave loading conditions that were tested.

The theoretical wave load transfer functions shown in Figure 8 were calculated using the linear wave diffraction program (NPLWAVE) described by Hogben and Standing (1975). The lowest wave steepness results provided the closest agreement between theory and experiment, while the largest deviations between theory and experiment were measured for the highest wave steepness results. The transfer functions for vertical forces indicate very good agreement between linear diffraction theory and experiment for a wide range of frequencies and wave steepnesses. Extremely poor agreement between theory and experiment exists for the overturning moments. Other than in the vicinity of 1 Hz, the linear diffraction theory usually overestimates the measured overturning moments by more than 50%. For low steepness waves (H_{m0}/L_p = 0.02), linear diffraction theory slightly overestimates most of the horizontal force measurements; however, for higher steepness waves (H_{m0}/L_p = 0.06) such as those shown in Figure 8, the theory generally overestimates the measured results by more than 40%. For higher frequency results, the theoretical and experimental results are very similar although in this range of frequencies the wave forces are very small in magnitude compared to the higher loads measured for the lower frequency measurements.

CONCLUSIONS

1. Horizontal forces, vertical forces and overturning moments exerted by waves on a model of a surface-piercing 45° right circular conical structure have been measured.
2. Time series records, spectral densities and transfer functions for irregular wave loading tests have illustrated the nonlinear nature of the measured wave loads.
3. Similar trends in wave loading have been observed for irregular and regular wave tests.
4. For deep-water waves, the irregular and regular force measurements have shown spectral peaks at the second harmonic of the wave frequency even though the waves themselves had negligibly small second-order components. However, unlike the regular wave loading

results, the fundamental for the irregular wave forces occurs at a frequency considerably lower than the peak frequency of the waves.

5. Linear diffraction theory provides reasonable estimates of the wave forces for waves of low steepness; however, large deviations occur for waves of higher steepness. Linear diffraction theory generally provides very poor estimates of the overturning moments. Presently there is not a theoretical nonlinear method that provides an accurate solution of the wave loading on conical structures; thus, the overall wave loading must be determined by physical model tests.

REFERENCES

Buslov, V.M. and N.W. Krahl. (1983). "Fifty-one New Concepts for Arctic Drilling and Production", Ocean Industry, Vol. 18 (1983), Parts 1, 2 and 3, Vol. 19 (1984), Parts 4, 5 and 6, 16 pp.

Funke, E.R. and E.P.D. Mansard. (1984). "The NRCC 'Random' Wave Generation Package", National Research Council of Canada, Hydraulics Laboratory Technical Report, TR-HY-002.

Hogben, N. and R.G. Standing. (1975). "Wave Loads on Large Bodies", International Symposium on the Dynamics of Marine Vehicles and Structures in Waves, edited by R.E.D. Bishop and W.G. Price, Mechanical Engineering Publications Limited, London, pp. 258-277, Authors' Replies, pp. 439-441.

Isaacson, M. de St. Q. (1982). "Fixed and Floating Axisymmetric Structures in Waves", Proc. ASCE, Vol. 18, No. WW2, May, pp. 180-199.

Jamieson, W.W., E.P.D. Mansard and G.R. Mogridge. (1985). "Wave Loading on a Conical Gravity Platform", Proc. 4th International Conference on Behaviour of Offshore Structures, Delft, The Netherlands, July, pp. 673-684.

Jamieson, W.W. and E.P.D. Mansard. (1987). "An Efficient Upright Wave Absorber", ASCE Specialty Conference on Coastal Hydrodynamics, University of Delaware, Newark, Delaware, June 29-July 1, pp. 124-139.

Longuet-Higgins, M.S. (1950). "A Theory of the Origin of Microseisms", Philosophical Transactions of the Royal Society of London, Series A, Vol. 243, No. 857, September, pp. 1-35.

Rundgren, L. (1958). "Water Wave Forces: A Theoretical and Laboratory Study", Transactions of the Royal Institute of Technology, Stockholm, Sweden, Bulletin No. 54 of the Institution of Hydraulics, 123 pp.

NRCC, Hydraulics Laboratory Technical Reports are available upon request from the Hydraulics Laboratory, Bldg. M-32, National Research Council of Canada, Montreal Road, Ottawa, Ontario, K1A 0R6, Canada.

CHAPTER 168

EXPERIMENTAL STUDY OF IMMERSED PLATES USED AS BREAKWATERS

Alain DROUIN* and Yvon OUELLET**

ABSTRACT

Under certain wave conditions, an immersed plate which is free to move horizontally produces what has already been called the oscillating water wall phenomenon. When this happens, the plate and the volume of fluid located under the plate move back and forth and then behave as a wave reflector. This type of breakwater was studied in a wave flume with monochromatic and irregular wave trains. Tests have provided a better understanding of the influence of certain parameters that define a plate and the hydraulic conditions under which the plate is used. Results show some interesting efficiencies and also transfert of the energy to harmonic frequencies.

1. INTRODUCTION

The reflective capacity of a freely moving and immersed plate has been shown in a recent study (Guevel et al., 1985). In this paper, it was demonstrated that under certain wave conditions, this kind of plate and the fluid located under it move back and forth. This movement generates in the rear part of the plate a radiation wave which interferes with the incident one; therefore, the plate behaves as a wave reflector (Figure 1). The French researchers developed a diffraction-radiation model which gave results similar to those obtained in their laboratory. However, when wave steepness was too high, the rear part of the plate became the seat of strong turbulence which seemed to generate harmonic waves, and, depending on the immersion depth of the plate, waves broke on its top. The numerical model did not account for these two phenomena.

---

* Research Assistant, Department of Civil Engineering, Laval University, Quebec, Canada, G1K 7P4.

** Professor, Department of Civil Engineering, Laval University, Quebec, Canada, G1K 7P4.

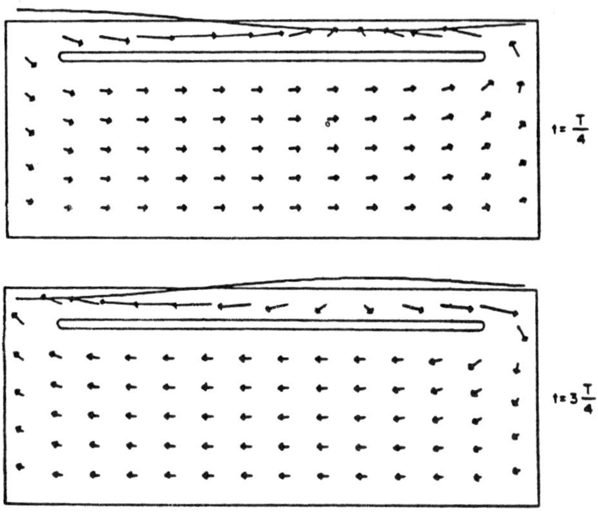

Figure 1 - Velocity distribution and free surface obtained with the oscillating water wall phenomenon (from Guevel et al., 1985)

The present work consists mainly of an experimental study on mobile and immersed plates (Drouin, 1988). The purpose was to determine if that kind of structure could be used as a breakwater under high steepness wave conditions. Plates were subjected to monochromatic and irregular waves, and some regular wave results were subsequently compared to those obtained with a fixed plate used in the same conditions. The fixed plates have been considered in many earlier studies which were well summarized in a paper written by Patarapanich in 1978.

2. EQUIPMENT AND DATA

The study was conducted in a 1.2 x 1.8 x 34.0 meter wave flume which was divided into three corridors of 0.45, 0.90 and 0.45 meters respectively. Studied structures were placed in the main corridor; therefore, the side flumes served essentially to dissipate waves reflected on breakwaters.

In order to maximize the information from the tests performed, time domain and frequency analyses were carried out. To reach the highest frequency of 4.0 Hz to which the spectrum would be evaluated, a data sampling interval of 0.125 second was chosen. In each case studied, eight records of 32 seconds were taken; therefore, the resulting increment in frequency between adjacent estimates of spectral density was 0.03125 Hz, and the stability obtained for these estimates was satisfactory.

In the time domain, the zero down crossing method was used, while the basic frequency analysis was completed with Goda and Suzuki's algorithm (Goda and Suzuki, 1976) so as to account for multiple wave reflections. For that purpose, three groups of four probes permitted the defining of incident as well as transmitted waves; the assessment of incident waves was checked with measures taken in a side flume. Finally, measurement of the horizontal displacement of each structure was also done.

3. REGULAR WAVE TESTS

A first group of tests was performed using regular wave trains. The parameters of the plate studied in tests 1 to 12 were the width, the specific mass and the mobility (Table 1), while the plate thickness was constantly maintained at 5 cm. As illustrated in Figure 2, chains placed at corners kept the mobile plates immersed, and pieces of foam were added to avoid passage of water and friction along the flume walls. A rigid support was used to fix the immobile structure studied (plate 4); therefore, the specific mass of this plate is not indicated in Table 1 because it is not significant.

Table 1 - Definition of plates studied in tests 1 to 12

PLATE IDENTIFICATION	WIDTH (cm)	SPECIFIC MASS (g/cm^3)
1	100	0.27
2	150	0.27
3	150	0.42
4	150	-

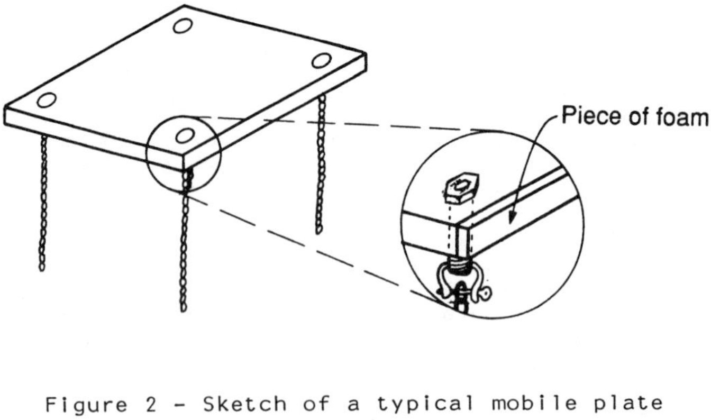

Figure 2 - Sketch of a typical mobile plate

For tests 1 to 12, plates were subjected to the fifteen wave conditions presented in Table 2, and each of these were considered with three different immersion depths of the plates (9.0, 13.5 and 18.0 cm). Water depth was kept at 75 cm; therefore, most of the waves were generated in intermediate water depth. In fact, only waves corresponding to the 0.95 second period were produced in deep water conditions.

Table 2 - Characteristics of regular waves used in tests 1 to 12

PERIOD (s)	HEIGHTS (cm)	WAVE LENGTH (cm)	WAVE STEEPNESSES
0.95	6.7, 5.3, 3.3	141	0.048, 0.038, 0.023
1.15	9.1, 7.0, 5.0	203	0.045, 0.034, 0.025
1.33	11.8, 8.8, 5.2	261	0.045, 0.034, 0.020
1.67	11.3, 8.1, 5.0	372	0.030, 0.022, 0.013
2.22	12.9, 9.9, 6.3	541	0.024, 0.018, 0.012

The effect of an increase in water depth was the object of tests 13 to 15, althought it was not possible, considering the dimensions of the flume, to raise the water level already used. In order to obtain deeper water conditions, incident waves, immersion depths and dimensions of plate 2 were reproduced at scale 1:1.4 with respect to the Froude's similitude law. The use of this scale gave an equivalent water depth of 105 cm.

Incident waves generated in tests 13 to 15 are presented in Table 3. For those tests, which were performed with plate 5, deep water conditions resulted when the two smaller periods were used. The width of the plate was 107 cm, its thickness 3.6 cm, and the plate was immersed to depths of 6.4, 9.6 and 12.9 cm.

Table 3 - Characteristics of regular waves used in tests 13 to 15

PERIOD (s)	HEIGHTS (cm)	WAVE LENGTH (cm)	WAVE STEEPNESSES
0.80	4.8, 3.8, 2.4	100	0.048, 0.038, 0.023
0.97	6.5, 5.0, 3.6	146	0.045, 0.034, 0.025
1.12	8.4, 6.3, 3.7	193	0.044, 0.033, 0.019
1.41	8.1, 5.8, 3.6	288	0.028, 0.020, 0.013
1.88	9.2, 7.1, 4.5	437	0.021, 0.016, 0.010

Although incident waves were monochromatic, transmitted ones appeared to correspond to a superposition of waves at harmonic frequencies (Figure 3). Efficiencies of tests performed in regular wave conditions, obtained using equation (1), are given in Table 4.

With the equation (1), an efficiency value near zero indicates that almost all the incident energy has been transmitted behind the structure.

$$\text{Efficiency} = \left(1 - \frac{\text{transmitted energy}}{\text{incident energy}} \right) \times 100 \qquad (1)$$

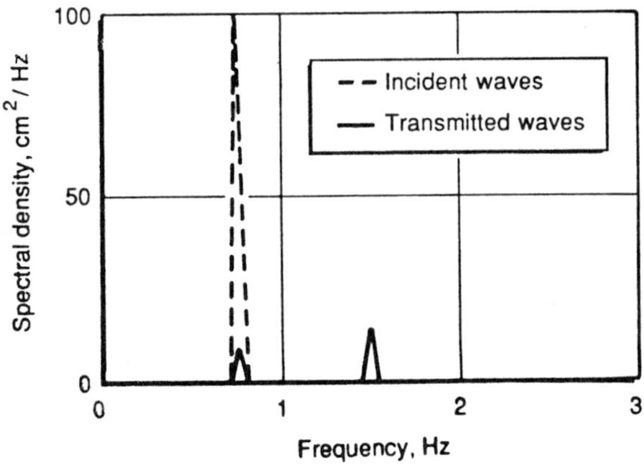

Figure 3 - Typical incident and transmitted waves spectra

Table 4 - Efficiencies obtained with regular waves

a) Immersion depth: 9 cm *

INCIDENT WAVES PERIOD (s)	HEIGHT (cm)	PLATE 1	2	3	4	5
0.95	6.7	67	86	77	77	85
	5.3	59	84	81	49	76
	3.3	23	41	36	30	53
1.15	9.1	85	77	80	79	73
	7.0	73	83	78	81	72
	5.0	44	85	76	73	68
1.33	11.8	77	87	87	87	85
	8.8	82	79	81	81	69
	5.2	68	72	72	69	54
1.67	11.3	78	96	95	91	91
	8.1	79	86	88	93	72
	5.0	55	62	61	79	55
2.22	12.9	51	75	69	79	84
	9.9	44	77	76	81	86
	6.3	76	77	80	85	83

b) Immersion depth: 13.5 cm *

INCIDENT WAVES PERIOD (s)	HEIGHT (cm)	PLATE 1	2	3	4	5
0.95	6.7	45	51	56	41	51
	5.3	25	49	37	28	53
	3.3	23	39	36	23	42
1.15	9.1	78	55	79	66	61
	7.0	48	54	67	49	57
	5.0	36	35	31	20	38
1.33	11.8	84	73	84	83	82
	8.8	69	57	62	61	73
	5.2	39	2	6	16	48
1.67	11.3	58	90	79	88	96
	8.1	44	82	82	75	74
	5.0	51	29	59	11	40
2.22	12.9	34	72	58	75	78
	9.9	64	69	54	81	80
	6.3	62	63	61	75	85

c) Immersion depth: 18.0 cm *

INCIDENT WAVES PERIOD (s)	HEIGHT (cm)	PLATE 1	2	3	4	5
0.95	6.7	10	28	38	25	30
	5.3	4	31	33	18	42
	3.3	0	35	44	14	26
1.15	9.1	56	37	40	50	43
	7.0	41	27	30	35	32
	5.0	38	31	29	32	21
1.33	11.8	80	69	91	74	77
	8.8	67	27	49	56	52
	5.2	47	27	23	41	26
1.67	11.3	61	79	74	83	74
	8.1	54	44	61	68	62
	5.0	54	23	49	16	52
2.22	12.9	32	62	38	47	74
	9.9	26	67	35	65	74
	6.3	25	61	66	83	83

* Frequencies, heights and immersion depths were reproduced at scale 1:1.4 for tests with plate 5.

Efficiencies obtained with plate 5 are illustrated in Figure 4 as a function of the kinematic energy per unit area of incident waves. Results shown in this way are representative of those obtained with other plates used.

The unusual choice of energy as a parameter to represent breakwater efficiencies can be understood by the implicit relationship between kinematic wave energy and an oscillating movement. This parameter was the only one which permitted the gathering of efficiency results in a relatively narrow range of values. Had adimensional parameters currently used, as the ratio of structure width to wave length or the wave steepness, been employed, efficiency results would have exhibited a significantly greater dispersion pattern.

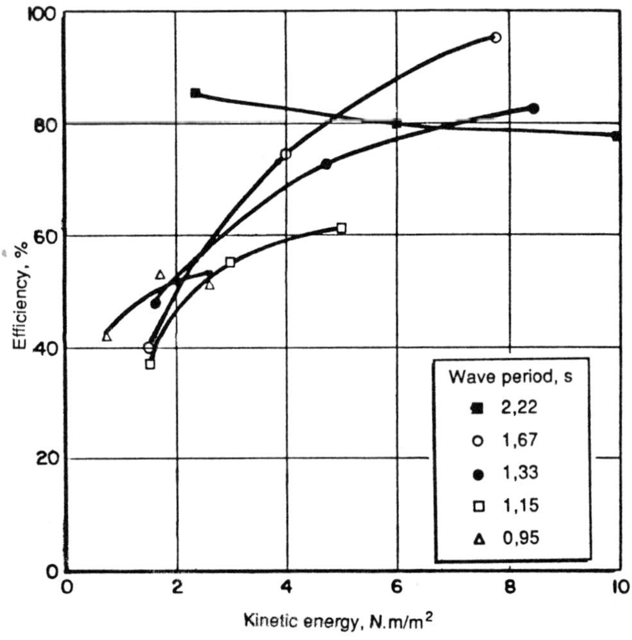

Figure 4 - Efficiencies of the plate 5 obtained with an immersion depth of 13.5 cm

4. DISCUSSION OF REGULAR WAVE RESULTS

The following analysis of results already given in Table 4, has been presented in such a way as to highlight the influence of wave and plate parameters studied.

Immersion depth

In almost all cases, efficiency decreased as the immersion depth became more important. This influence can be explained in two ways. When the immersion increased, there was less kinematic energy under the plate to initiate the water wall phenomenon as described by Guevel et al. Secondly, there was also less energy dissipated by wave breaking on the plate.

Wave height

As shown in Figure 4, with the exception of the higher period studied, the performance of plates increased as incident wave energy (proportional to height) also increased. As well as in the case of the immersion depth, results can be related to the dissipation of wave energy due to the wave breaking, and to the energy available under the plate to initiate the water wall phenomenon.

Plate width

With an immersion depth of 9 cm, plate 2 (width of 150 cm) performed better than plate 1 which was only 100 cm wide. When the plate 2 was immersed at 13.5 cm, efficiencies were equivalent to those of plate 1, except in certain cases including those with wave periods of 1.33 second. Finally, with a immersion depth of 18 cm, the second plate gave better results only with extreme wave periods studied (0.95 and 2.22 seconds); with other periods, smaller efficiencies were obtained with the larger plate.

Specific mass

With the smallest immersion depth, efficiencies calculated with plates 2 and 3 (specific masses of 0.27 and 0.42 g/cm^3) were quite similar. With other immersion depths, an increase of the mass resulted in efficiencies similar or better, except with higher waves having a period of 2.22 s.

Plate mobility

Compared to the fixed plate, the efficiency of the mobile one was better when exposed to the lower energy waves of those described in Table 2. However, the fixed structure gave better results in higher wave energy cases. With both types of plate, a part of the transmitted energy had a harmonic relationship with incident wave frequency.

Water depth

Especially when highly immersed, plate 5, which was in deeper water, gave better results than plate 2. These results and the movement of the plates help, as it will be discussed next, to understand the effect of water depth on this kind of breakwater.

As described by Guevel et al, it was observed that an upward pressure was applied under the rear part of all mobile plates during their backward movement (Figure 5). If the pressure was important enough, it seemed to limit the horizontal displacement of these plates.

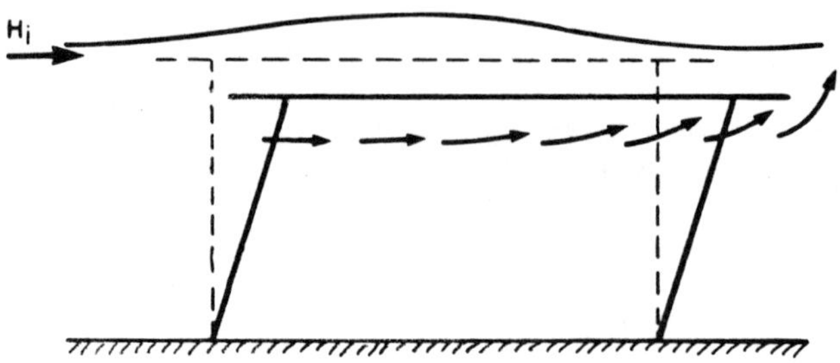

Figure 5 - Velocity distribution directly under the plate

When mobile plates 1, 2 and 3 were subjected to the longer wave periods (1.67 and 2.22 seconds), their movements were not as that described in Figure 6a, but rather like those illustrated in Figures 6b and 6c. However, when plate 5 was subjected to the same incident wave conditions, but in deeper water, its movements were as presented in Figure 6a. Especially with the two deeper immersions, the amplitude of plate 5 movements was, proportionally, bigger than those of plate 2 used in shallower water. Therefore, the horizontal displacement of the mobile plates was a function of the anchoring length.

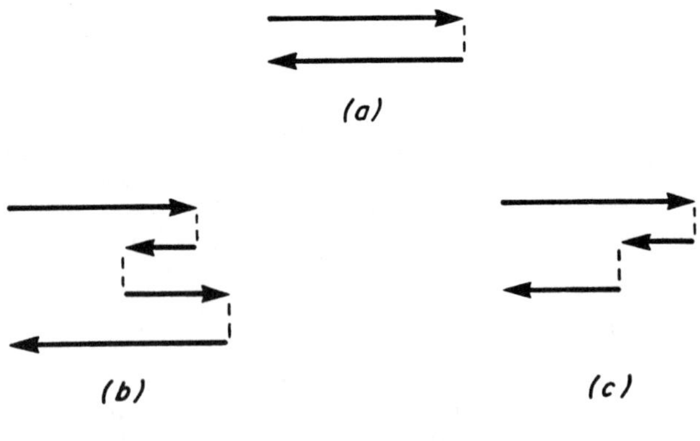

Figure 6 - Typical movements of the mobile plates

Considering geometric relations, it appears that a given horizontal displacement implies a larger vertical displacement of that structure when the anchoring length is shorter. Added to the efficiencies obtained, it seems to indicate that when a plate is limited in its horizontal oscillation, it cannot force as efficiently the horizontal displacement of the mass of water located under itself; therefore, the reflective capacity would be decreased.

It may be paradoxical to note that the fixed plate performed better than the mobile ones when subjected to the higher wave periods used. However, the mobile plates were agitated, in those cases, by strong vertical jolts which resulted, when added to the horizontal movement limitation, in lower efficiencies.

Finally, the movement of the plates described in Figures 6b and 6c seemed to create bigger strains in anchorings than that illustrated in Figure 6a; however, this assumption was not verified in tests performed, since no anchoring stresses were measured.

5. IRREGULAR WAVE TESTS

A limited amount of tests were performed using irregular incident waves. In order to permit direct comparison with regular wave tests, significant wave heights and peak periods were chosen such that they were equivalent to some of those used in monochromatic conditions. Spectra generated in the laboratory were based on the JONSWAP definition (Holmes, 1977). Efficiencies calculated with equation (1) and obtained with plates 1 and 2, already defined, are presented in Tables 5 and 6.

Considering the effect of the parameters studied, the results of irregular wave tests had a tendency to vary in the same manner as those obtained with regular wave trains. However, differences between extreme efficiencies were diminished, since lower and higher performances obtained with regular waves were more concentrated around mean values.

Table 5 - Efficiencies obtained in irregular wave conditions with plate 1

PEAK PERIOD (s)	SIGNIFICANT HEIGHT (cm)	IMMERSION DEPTH (cm)		
		9.0	13.5	18.0
1.15	9.1	65	58	50
	7.0	63	46	43
	5.0	40	34	41
1.67	11.3	74	65	51
	8.1	74	62	47
	5.0	67	57	49

Table 6 - Efficiencies obtained in irregular wave conditions with plate 2

PEAK PERIOD (s)	SIGNIFICANT HEIGHT (cm)	IMMERSION DEPTH (cm) 9.0	13.5	18.0
1.15	9.1	68	47	32
	7.0	67	42	35
	5.0	55	39	37
1.67	11.3	82	75	62
	8.1	76	66	45
	5.0	55	35	37

As well as with regular waves, transmitted energy was carried over to higher frequencies, especially to harmonics of the incident peak frequency (Figure 7).

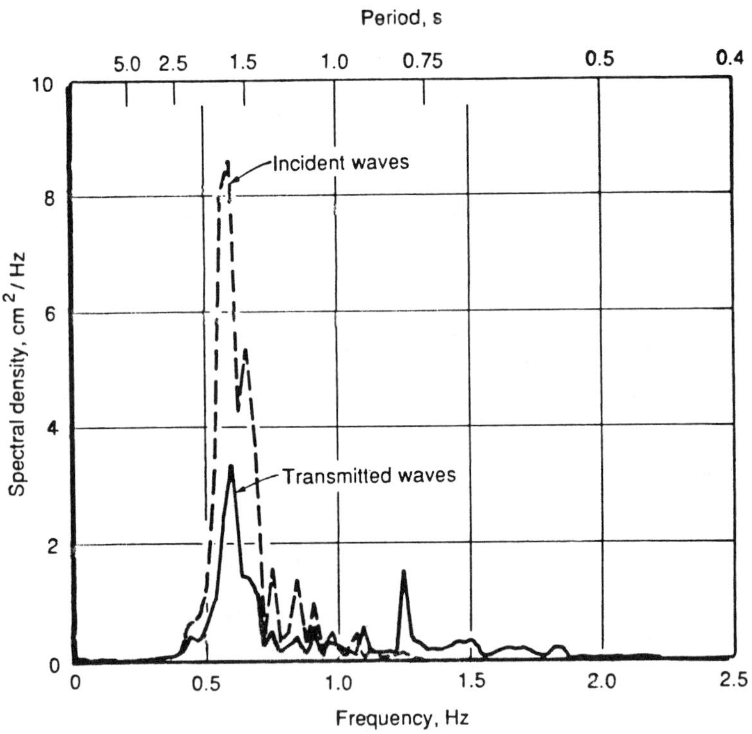

Figure 7 - Incident and transmitted waves spectra obtained with plate 1 immersed at 18.0 cm Incident waves: T_p = 1.67 s, $H_{1/3}$ = 5.0 cm

6. CONCLUSIONS

From tests performed, the plate width, the immersion depth and the distance between the plate and the bottom appeared to be the parameters having the most influence on the efficiency of a mobile plate used as a breakwater. Under severe wave conditions, horizontal displacement of the plate, and consequently its efficiency, were observed to be limited by the length of the anchoring chains (distance between the plate and the bottom) and the ascendant pressures on the rear part of the plate. In these cases, the rear part of the plate was agitated by strong vertical jolts. Under such conditions, a fixed plate of equivalent dimensions was found to be more effective.

Interesting efficiencies were obtained, even when ratios of plate width to wave length were small. These good performances occurred especially when plates were slightly immersed. With the exception of the higher wave periods studied, the efficiency of the plates as breakwaters increased as incident wave energy increased. Moreover, a variable part of this energy was carried over as a harmonic of the incident wave frequency.

More detailed studies should be done, using irregular wave conditions and variable incident wave angles, in order to better define the performance of the plates. The measurement of tensions on anchoring chains, when structures are subjected to these incident waves, will also give important information that must be known before considering the use of such mobile plates in the field as breakwaters.

REFERENCES

DROUIN, A. (1988). «Étude expérimentale de plaques mobiles utilisées comme moyen de protection contre la houle». Thèse de maîtrise, Département de génie civil, Université Laval, 142 p.

GUEVEL, P., LANDEL, E., BOUCHET, R. et MANZONE, J. M. (1985). «Le phénomène d'un mur d'eau oscillant et son application pour protéger un site côtier soumis à l'action de la houle». ATMA 1985, Principia Recherche Développement S.A. 18 p.

GODA, Y. and SUZUKI, Y. (1976). «Estimation of Incident and Reflected Waves in Random Wave Experiments». Proceedings of 15th International Conference on Coastal Engineering, ASCE, pp. 828-845.

HOLMES, P. (1977). «Wave Climate». Symposium on Design of Rubble Mound Breakwaters, Paper No 1, British Overcraft Corporation, 24 p.

PATARAPANICH, M. (1978). «Wave Reflection from a Fixed Horizontal Plate». Proceedings of the International Conference on Water Resources Engineering, Bangkok, Thailand. pp. 427-446.

CHAPTER 169

ON 3-DIMENSIONAL STABILITY OF RESHAPING BREAKWATERS

by

H.F. Burcharth
Prof. of Marine Civil Engineering

Peter Frigaard
Research Associate

University of Aalborg, Denmark

INTRODUCTION

The paper deals with the 3-dimensional stability of the type of rubble mound breakwaters where reshaping of the mound due to wave action is foreseen in the design. Such breakwaters are commonly named sacrificial types and berm types. The latter is due to the relatively large volume of armour stones placed in a seaward berm. However, as also conventional armoured breakwaters sometimes do contain a berm it is assumed that a better and more ambiguous designation would be "reshaping" mound breakwaters.

The principle of reshaping breakwaters is to use relatively fine rock material which will then be eroded to S-shape profiles if sufficient amount of material is provided, Fig. 1. This type of breakwater can be constructed and maintained without the use of expensive specialized equipment. For a detailed discussion see Baird et al., 1984.

If sufficient material is provided to prevent the top of the mound to be eroded in its full width a reshaping breakwater will always be dynamically stable in head-on waves (2-dimensional case). This principle holds for all sizes of stone materials since even a sand beach stabilizes though with a very flat profile.

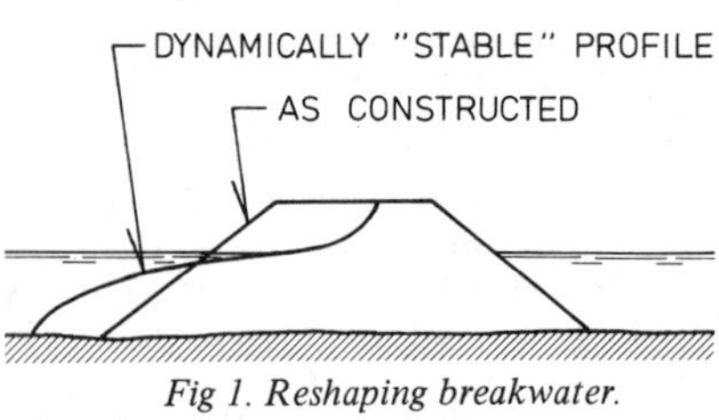

Fig 1. Reshaping breakwater.

However, in reality there are two important 3-dimensional effects which must be included in the design criteria, namely the rate of erosion (recession) of the roundhead and the rate of erosion of the trunk caused by oblique waves. The latter might cause unacceptable removal and transport of stones along the structure and thereby constitute a more restricting design criteria than head-on waves.

The paper discusses the trunk stability in oblique waves and the stability of berm roundheads and presents results from model tests.

The tests were conducted in a 3-dimensional model at The Hydraulics Laboratory, Department of Civil Engineering, University of Aalborg.

MODEL TEST SET-UP

Geometry of the model

Fig. 2 shows the cross sections of the model (before each test) and the lay-out of the model in the wave basin.

Two different lay-outs were used in order to examine the stability of breakwaters with different widths.

To avoid too many parameters a simple breakwater geometry and only one class of stones were used.

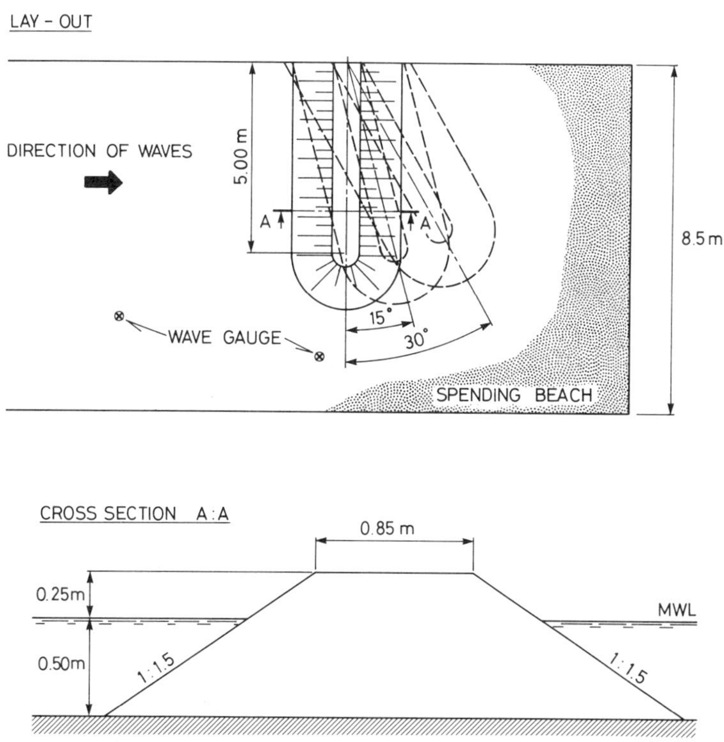

Fig 2. Lay-out and cross section of the model. (Continuied on next page).

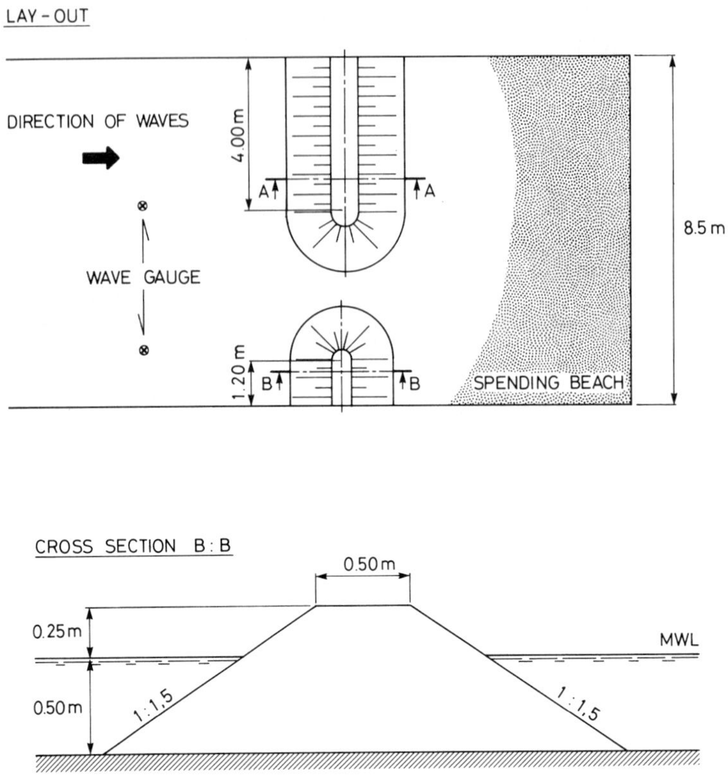

Fig 2. Lay-out and cross section of the model.

The stone material

The model consisted of one grading of crushed stones with a density $\rho_s = 2.65$ t/m^3 (metric ton) and a gradation as given in Figs. 3 and 4. No special core material was used.

It was found that the relationship between the sieve diameter (quadratic sieve) d and the stone volume V and stone weight w is

$$V = 0.7d^3 = \frac{w}{\rho_s}$$

d is regarded a characteristic diameter of the stones.

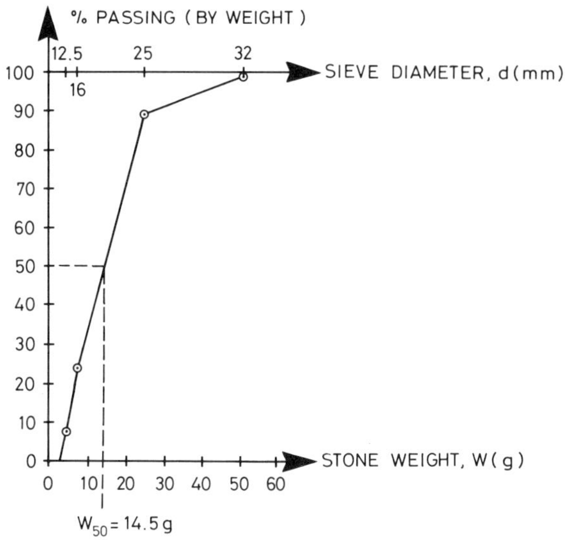

Fig 3. Gradation of model stones (Linear representation).

As seen from Fig. 3 w_{50} is found to be 14.5 g. However, it is most likely that an extra point on the gradation curve in the sieve interval 16-25 mm would have shifted the graph to the left and thereby given a w_{50} smaller than 14.5 g. This is investigated by Fig. 4.

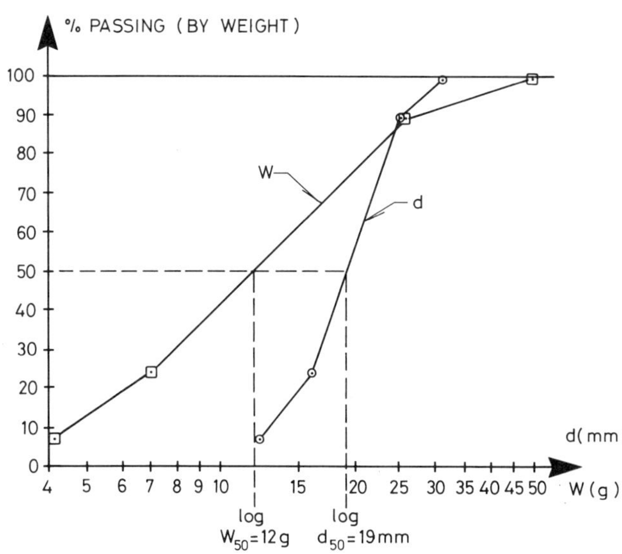

Fig 4. Gradation of model stones (Log-linear representation).

The stone weight corresponding to d_{50}^{log} = 19 mm is $w_{d_{50}} = \rho_s \cdot 0.7d^3 = 2.65 \cdot 0.7 \cdot 1.9^3 =$ 12.7g $\geqslant w_{50}^{log}$ = 12 g.

The log representation confirms that w_{50} is somewhat less than 14.5 g.

Based on the various figures a w_{50} of 12.7 g corresponding to d_{50}^{log} = 19 mm is chosen as the most correct value.

D_{n50} is a nominal diameter defined as $D_{n50} = (w_{50}/\rho_s)^{1/3}$ = 16.9 mm.

It should be noted that for the investigation of longshore transport in oblique waves samples of stones without diameters less that 16 mm were used. For these samples d_{50}^{log} = 22 mm and $w_{d_{50}}$ = 20.3 g.

Waves

All waves were irregular waves generated in accordance with a random phase JONSWAP-type spectrum (peakedness parameter $\gamma = 3.3$ and width parameters $\sigma_f = 0.10$ for $f \leqslant f_p$ and $\sigma_f - 0.50$ for $f > f_p$).

All tested sea states are stated in Tabel 1.

Spectrum no.	H_s (m)	T_p (sec.)	$H_s/\Delta \cdot D_{n50}$
A	0.10	1.50	3.5
B	0.10	2.00	3.5
C	0.15	1.80	5.4
D	0.15	2.50	5.4
E	0.20	2.50	7.1
F	0.10	1.70	3.5
G	0.13	1.80	4.6
H	0.13	2.50	4.6
I	0.15	2.00	5.4
J	0.15	2.20	5.4
K	0.15	2.50	5.4
L	0.175	2.50	6.3

Tabel 1. Tested sea states.

Two different methods of producing the wave generator control signals were used.

Control signals for the tests using spectrums A through E were calculated by use of FFT-transformation, while all other control signals were based on the method of filtering white noise.

The following three angles of wave attack were tested: $\alpha = 0°$ (head-on waves), $\alpha = 15°$, and $\alpha = 30°$.

MODEL CONSIDERATIONS

The sea states were chosen in the range from very little erosion to fast erosion of the profile under oblique wave attack.

An indication of the relative stability of the profiles can be given by the dimensionless parameter $H_s/\Delta \cdot D_{n50}$, where $\Delta = \rho_s/\rho - 1$. It is seen that $H_s/\Delta D_{n50}$ equals the stability number $N_s = (K_D \cot\alpha)^{1/3}$, where K_D is the Hudson stability coefficient and α is the slope angle. (Note that the influence of the wave period is lacking in the parameter).

According to extensive testing of rock slopes in head-on waves by DHL(Pilarczyk and Van der Meer) the values of the dimensionless parameter can be related to various types of rock slopes as follows:

$H_s/\Delta D_{n50}$	
1 – 3	Conventional breakwater layer, start of damage
2 – 5	Conventional breakwater layer, failure
3 – 7	Reshaping breakwater (Berm breakwater)
5 – 50	Rock beaches

In the present tests we have

$$\Delta = (\frac{\rho_s}{\rho} - 1) = 1.65$$

$$D_{n50} = (\frac{12.7}{2.65})^{1/3} = 1.69 \text{ cm}$$

$$\Delta \cdot D_{n50} = 2.79 \text{ cm}$$

and consequently the range of tests corresponds to

$$3.5 < H_s/\Delta \cdot D_{n50} < 7.1$$

which is the interval considered for reshaping breakwaters.

SCALE EFFECTS CONSIDERATIONS

Provided that the grading of the stones is not too wide, say $\frac{d_{85}}{d_{15}} < 3$, and provided that the amount of fine material cannot block the pores (which is usually the case if $\frac{d_{85}}{d_{15}} < 3$) it is relevant to define a Reynolds' number with the characteristic length D_{n50}

$$Re = \frac{D_{n50} \sqrt{gH_s}}{\nu}$$

ν is the kinematic viscosity = 10^{-6} m²/s at 20°C.

With H_s = 0.10 - 0.20 m we get

$$1.7 \cdot 10^4 < Re < 2.4 \cdot 10^4$$

Juul Jensen and Klinting analysed the scale effects and found that no significant viscous scale effect is to be expected if in the outer part of the structure $Re > 0.6 \cdot 10^4$. Van der Meer found no scale effects for rock slopes with characteristic stone size of 20 mm, which is approximately the stone size in the present tests. This is also the experience of the Hydraulics Laboratory at the University of Aalborg.

However, although it is believed that a viscous scale effect is present it will be either negligible or will cause the results (in terms of amount of damage) to be on the safe side.

STABILITY OF ROUNDHEAD AND TRUNK IN HEAD-ON WAVES

Test procedure

The initial cross sections profiles of the model are shown in Fig. 2.

In the rough wave conditions only the model with the large crest width was used. In tests with calmer sea conditions both models were used.

The waves were recorded throughout all the tests. Wave height, wave period and shape of spectrum were controlled. The breakwater profile was measured after 3000 waves in all tests, and in most of the tests the waveseries were repeated once or twice and then the breakwater profile was measured again.

The wave basin was equipped with two cameras and contour plots of the breakwater was made after all tests using Photogrammetry.

Moreover, the characteristics of the stone movements were found from video recordings of the movements of coloured stones.

Test results for profiles in head-on waves

In agreement with results from several other authors, i.e. Van der Meer and Pilarczyk 1986, the shape of the profiles was found to be governed by the wave heigth and the wave period. The influence of the number of waves was found to be very little for profiles exposed to more than 3000 waves. After 6000 waves the profiles seem to have reached a "state of equilibrium".

Also, it was observed that most of the reshaping took place within approximately 200 waves.

It is characteristic that a very wide range of sea states (from mild to severe) produce only slightly different trunk profiles in head-on waves.

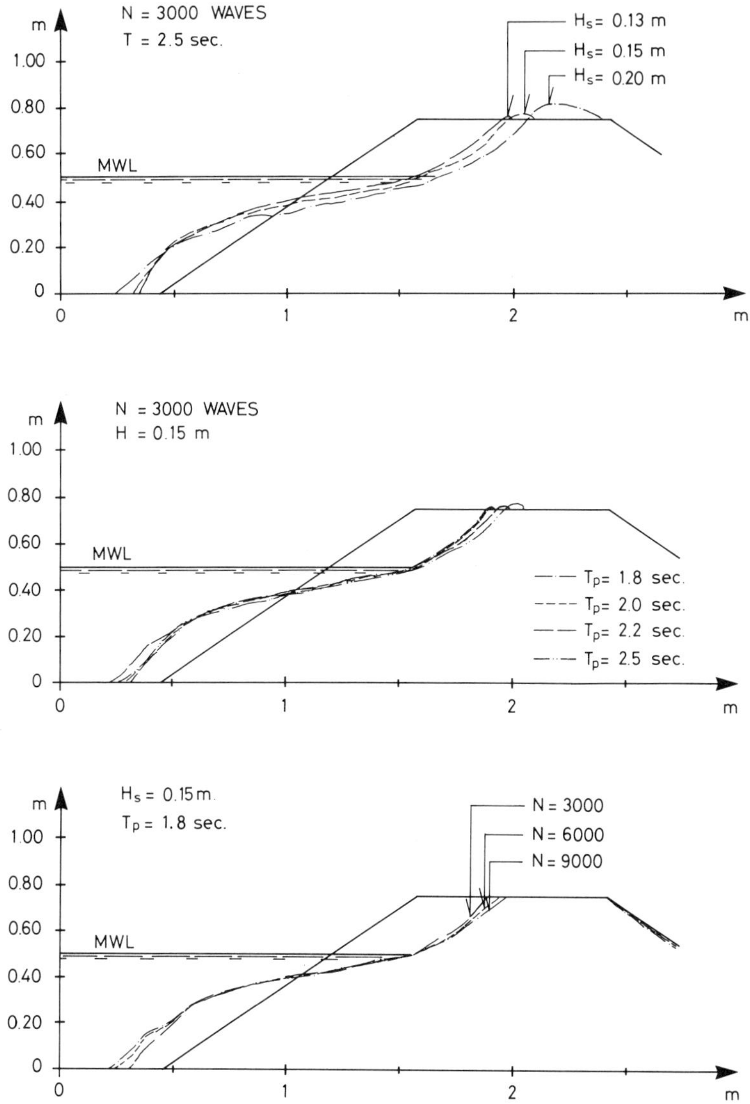

Fig 5. Examples of profiles in head-on waves
- influence of wave heigth
- influence of wave period
- influence of number of waves

The material deficit is due to settlements caused by wave compaction and material transport across the crest.

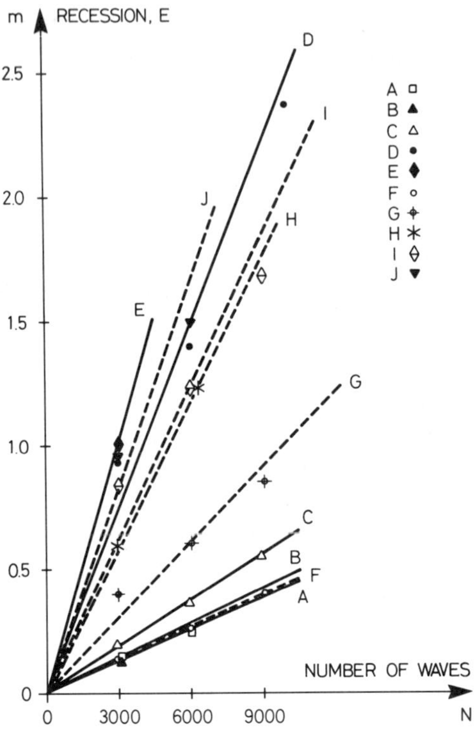

Fig 6. Recession (erosion) rate of the roundhead.

The erosion of the roundhead is expressed in terms of recession of the crest end measured a-long a longitudinal line parallel to the centerline of the breakwater, see Fig. 7.

Fig. 6 shows the recession as function of the number of waves. Note that the rate is almost constant for a certain sea state, i.e. a linear relationship between the recession of the central crest end and the time (or number of waves).

In spite of the almost linear relationship between the recession of the roundhead and the time for each sea state. Fig. 6 shows a large amount of scatter. Notice that wave spectrum G (H_s = 0.13 m and T_p = 1.80 sec.) gives larger erosion than spectrum C (H_s = 0.15 m and T_p = 1.80 sec.). It is believed that the scatter is due to different grouping in the waves caused by different methods of generating the wave control signals.

Fig. 6 shows that the roundhead erosion rate is small up to a certain sea state, characterized by both wave heigth and wave period. When this sea state is exceeded, erosion is fast and the sea state now seems to be characterized by the ability of practically every one of the waves to erode some of the stones from the roundhead and to displace (shift) them all the way across the roundhead.

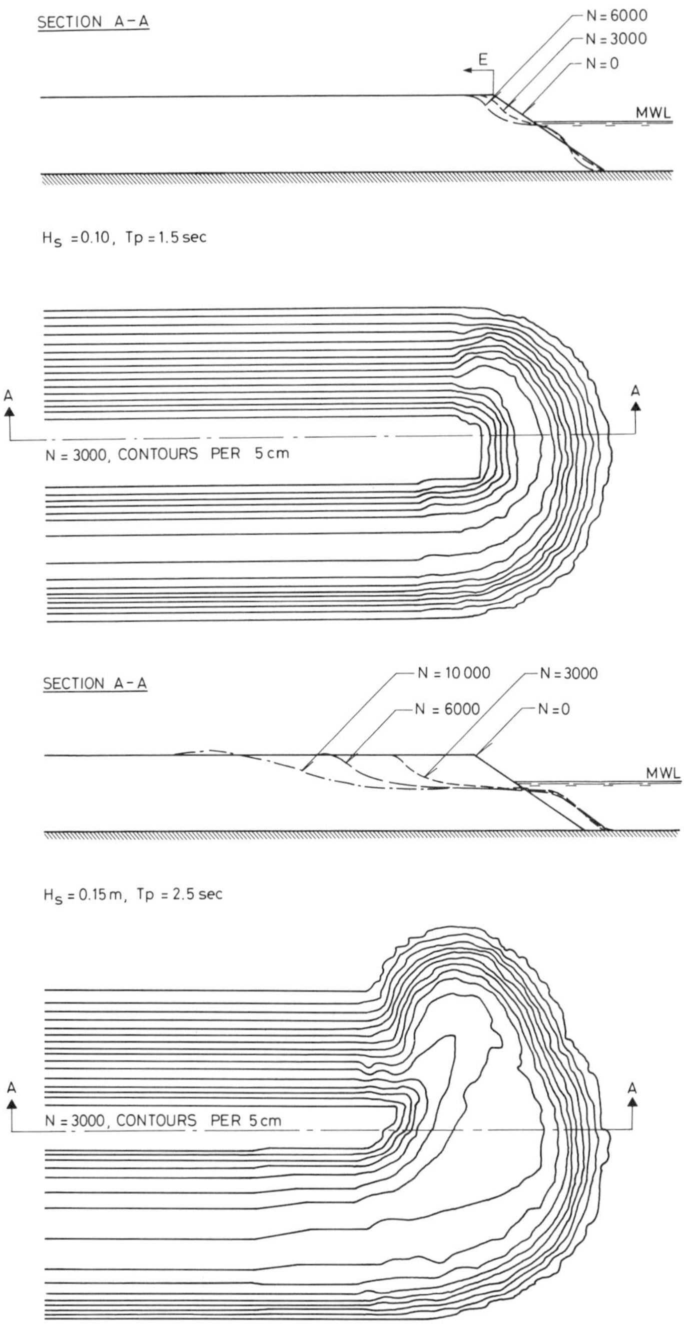

Fig 7. Examples of erosion of roundhead.

Simultaneous tests with the two lay-outs, Fig.2, indicate that erosion of the narrow breakwater will start at less severe sea states than for the wide breakwater.

Having observed in the tests that the erosion reached a level of approximately $H_s/2$ below the still water level the eroded cross sections can be evaluated.

The ratio of erosion of the breakwaters seems to have the same value as the ratio of the eroded cross sections until the sea state reaches the level where practically every one of the waves erodes some stones.

Seastate	H_s	T_p	Number of waves	Ratio cross section	Ratio recession
F	0.10	1.70	3000	1.5	1.7
F	0.10	1.70	6000	1.5	1.6
F	0.10	1.70	9000	1.5	1.5
G	0.13	1.80	3000	1.5	1.6
G	0.13	1.80	6000	1.5	1.5
H	0.13	2.50	2000	1.5	1.9
H	0.13	2.50	3000	1.5	> 2.0

Tabel 2. Ratio of recession between the narrow crested breakwater and the wide crested breakwater.

Fig. 8 shows that the maximum width of the eroded breakwater seems to be almost independent of the width of the non-eroded structure.

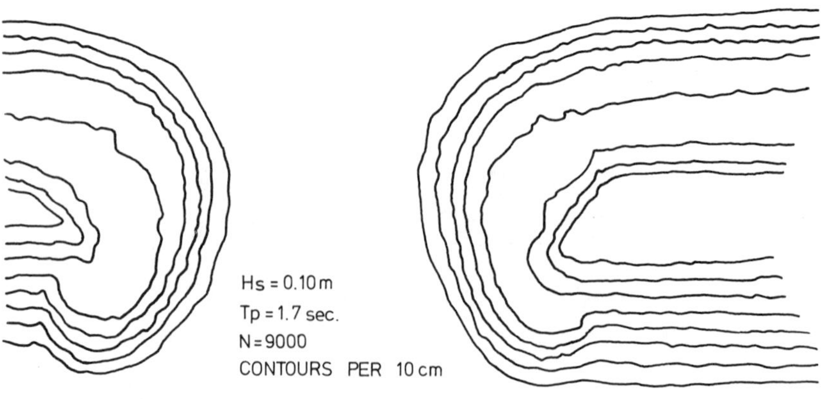

Fig 8. Example of erosion of a narrow crested breakwater and a wide crested breakwater.

The area on the breakwater where the erosion starts was found to be on the seaward side of the roundhead. The location is different from what is observed for conventional breakwaters where the erosion normally starts on the lee side of the breakwater, app. 135 degree angle with wave direction.

This difference might be explained by the larger width and larger roundhead diameter of the reshaping breakwater, which cause the stones on the leeward roundhead slope to be less exposed due to the diffraction of the waves along the relatively wide end of the roundhead. Moreover, the larger diameter provides a relatively better support of the stones in the direction of a wave attack.

Enlargement of the roundhead diameter has been used as a means of improving the stability of conventional breakwater roundheads, e.g. the Mohammedia breakwater in Morocco.

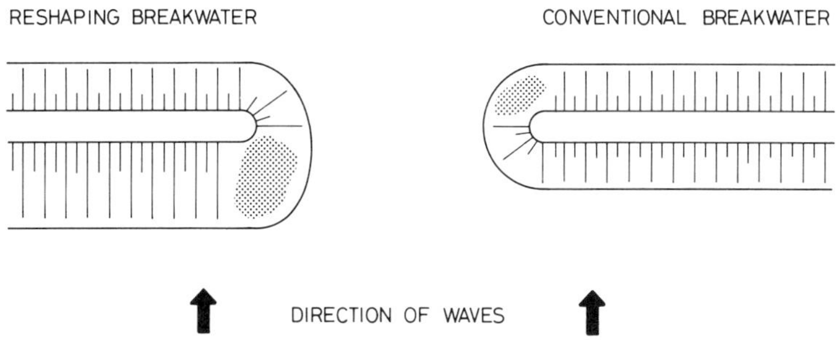

Fig 9. Area for start of erosion.

The breakwater develops into a characteristic banana shape as the erosion proceeds. Fig. 10 shows an example of the changing of the bathymetry of the roundhead.

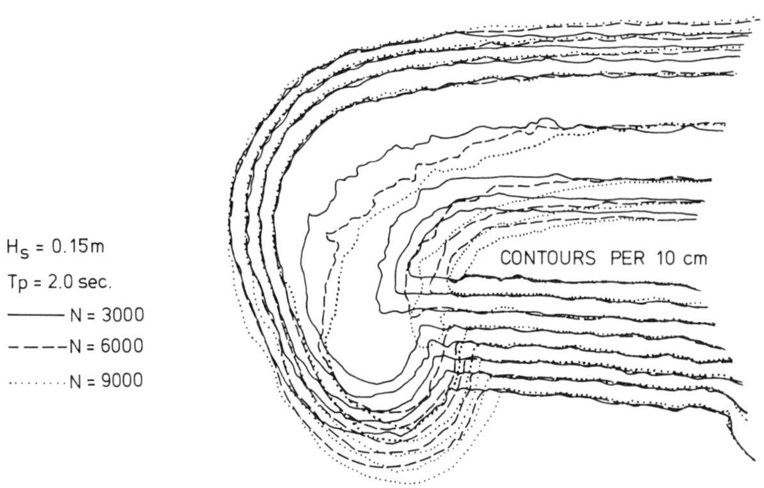

Fig 10. Time dependent development of roundhead.

STABILITY OF TRUNK AND LONGSHORE TRANSPORT IN OBLIQUE WAVES

In a trial test series it was found that for a given H_s, T_p the dynamically stable profiles in oblique waves within the tested range $\alpha \leqslant 30°$, cf. Fig. 2, were almost identical to the profiles in head-on waves.

Thus in every test with oblique waves the initial profile was chosen as the one found after 3000 head-on waves.

The longshore transport was found from video recordings of the movements of coloured stones placed in three bands over the profile. Moreover, after a specific number of waves (or time) the number, the positions and the total weight of each type of coloured stones were recorded.

The band width and the number of waves N were adjusted to the sea state in such a way that within the test period the non-coloured stones upstream the coloured bands did not pass the downstream coloured band. In this way the average transport per second (or per wave) through a cross section could be found. Moreover, by studying the distribution of the coloured stones over the profiles the maximum erosion depth (i.e. the number of stone layers within which displacements take place) could be estimated. The number of waves in each test varied between 250 waves and 900 waves.

Test results

The steady state transport of stones along the trunk was studied for two angles of wave attack, $\alpha = 15°$ and $\alpha = 30°$, c.f. Fig. 2.

Sea state	H_s (m)	T_p (sec.)	Average mass transport, Q (g/sec.)	
			$\alpha = 15°$	$\alpha = 30°$
A	0.10	1.50	0.45	0.74
B	0.10	2.00	0.68	0.92
G	0.13	1.80	3.55	2.02
H	0.13	2.50	7.18	4.29
C	0.15	1.80	11.7	16.3
I	0.15	2.00	9.13	8.0
J	0.15	2.20	9.45	13.4
K	0.15	2.50	12.2	14.2
D	0.15	2.50	7.1	17.5
L	0.175	2.50	20.2	22.9
E	0.20	2.50	32.0	35.7

Tabel 3. Steady state mass transports along the trunk.

Notice that the sea states A and B only cause very small mass transports corresponding more or less to the onset of long-structure transport. Also notice that the more severe sea states L and E result in a significant transport, which in typical prototype situations will change the geometry of the structure significantly and might endanger the stability.

From the tests it is found that the average stone weigth of the transported stones varied significantly. The range was from 14.3 g/stone (sea state H, H_s = 0.13 m, T_p = 2.50 sec., a = 15°) to 20.5 g/stone (sea state I, H_s = 0.15 m, T_p = 2.00 sec., a = 30°).

It was found that a small average stone weight corresponded to a relative large transport. Because the same stone sample was used in all the tests the observed sorting must be regarded as natural scatter of the process. The large scatter is a consequence of the relatively short duration of each test.

The results look homogeneous in the sense that the trend is that the transport increases with wave height, wave period and angle of incidence in the tested ranges. However, the large scatter in the results, makes it impossible to quantify the sediment transport accurately.

Fig 11. Example of the breakwater after longshore transport tests in oblique waves.

CONCLUSIONS

It is characteristic that a very wide range of sea states (from mild to severe) produce only slightly different trunk profiles in head-on waves, but very large differences in roundhead erosion and trunk erosion in oblique waves.

The roundhead erosion and the erosion of the trunk in oblique waves have a very strong non-linear dependency on the sea state. Below a certain sea state threshold value the erosion rates are very small, but excess of this value causes a drastic increase in the erosion. Consequently identification and consideration of this threshold value are of great importance in the design process.

Based on the presented model tests and the behaviour of some prototype breakwaters the following somewhat premature recommendations valid for permanent designs with accepted moderate damage are proposed:

	$H_s/\Delta D_{n50}$
For trunks exposed to steep oblique waves	< 4.5
For trunks exposed to long oblique waves	< 3.5
For roundheads	< 3

These values should be used as guidelines *only* if no other more qualified information is available. This is because the parameters $H_s/\Delta D_{n50}$ is insufficient as among other things it does not contain the effect of wave length and the effect of the duration of the sea. Moreover, the influence of the breakwater geometry and the relative water depth are not included.

A study of a general parametric representation of the mass transport (erosion) based on a "shields approach" using characteristic wave parameters has been tried without much success.

It is believed that in additional investigations of erosion of reshaping breakwaters it will be necessary in principle to examine and summarize the respons from every single wave instead of using characteristic parameters like H_s to characterize the sea state. This is because the character of the flow kinematics in the erosion zones is strongly dependent on the size and the steepness of the single waves.

ACKNOWLEDGEMENT

The help of Mr. Van der Meer in estimating cross section profiles during the planning of the test is gratefully acknowledged. Also many thanks to the Laboratory for Photogrammetry and Surveying, University of Aalborg, for working out the photogrammetry and the contour plots of the bathymetry of the roundhead.

REFERENCES

Van der Meer, J.W., Pilarczyk, K.W. *Dynamic stability of rock slopes and gravel beaches.* Published in proc. of the 20th Int. Conf. on Coastal Engineering, Taipei, 1986.

Juul Jensen, O., Klinting, P. *Evaluation of scale effect in hydraulic models by analysis of laminar and turbulent flows.* Coastal Eng. vol. 7 no. 4, Nov. 1983, pp 319 - 329.

Van Hijum, E., Pilarczyk, K.W. *Equilibrium profile and longshore transport of coarse material under regular and irregular attack.* Delft Hydraulics Laboratory. Publication no. 274, 1982.

Bijker, E.W. *Littoral Drift as Function of Waves and Current.* Proc. 11th Coastal Eng. Conf., London, 1968.

Baird, W.F. and Hall, K.R. *The design of Breakwaters using Quarried Stones.* Proc. of Offshore Technology Conf., Houston, Texas, 1984.

Burcharth, H.F., Frigaard, P. *Reshaping breakwaters. On the stability of roundheads and trunk erosion in oblique waves.* Presented at Seminar for Unconventional Breakwaters, Ottawa, Canada, September 1987.

CHAPTER 170

ASSESSING THE EFFECT OF ARMOURSTONE SHAPE AND WEAR

John-Paul Latham*
Alan B. Poole**

ABSTRACT

Impacts, abrasion and physico-chemical weathering which sometimes result in rapid changes of armour shape and size can have disastrous consequences for armour stability. This paper draws together the background ideas and some new techniques which have been developed to tackle the prediction of weight loss and shape change of rock armourstone during the life of a coastal structure.

The framework proposed is to measure separately the wear resisting material properties of a rock type using a tumbling mill simulation of prototype block degradation. Time on the structure is then related to mill time by a factor which accounts for initial block size and the site specific environmental conditions. A field site in Scotland is used to demonstrate the new image analysis methods for measuring rounding of prototype blocks. The shape analysis results can also be used to calibrate the equivalent wear factor for conditions at that site. Discussion of this framework is extended to dynamically stable design concepts where abrasion losses are faster.

Preliminary flume testing in collaboration with H. R. Wallingford indicates that losses in stability due to rounding could be considerable.

INTRODUCTION

It is widely acknowledged that the occurrence of sources of rock with favourable geotechnical properties of rock in areas close to sites requiring slope protection against wave attack will guarantee the widespread use of rock armoured designs into the foreseeable future. By reviewing recent progress with techniques of measurement, this paper sets out to demonstrate that for natural armourstone, both block shape and the effect of wearing processes could be important design considerations and a way forward for taking these into account is suggested.

The paper is in three parts. The first provides the background, and presents the reasons for measuring shape and wear of armourstone. The second part summarizes the laboratory techniques for simulating prototype wear and measuring weight loss and shape changes, the details of which have been published elsewhere. In the third part, photography of block outlines from a Scottish case study are used to indicate abrasion rates. These form the basis of discussion of the implications for loss of stability of rounded rocks in the light

* Research Fellow: Geomaterials: School of Engineering, Queen Mary College, London, E1 4NS. U.K.
** Senior Lecturer, Geomaterials: School of Engineering, Queen Mary College, London, E1 4NS. U.K.

of recent flume experiments which were designed to investigate the effect of shape and rounding on stability.

BACKGROUND

Armourstone Degradation

The problem of rock degradation is not restricted to coastal engineering. In highway construction, (for example) to aid the selection of appropriate rock for road aggregate, Lees & Kennedy (1975) gave an account of mechanisms of degradation. In their paper, disintegration processes were considered according to the ratio of the size of the broken product to the size of the original particle. Although informative, their discussion is only partly applicable to the special demands of rock armourstone. Fookes et al. (1988) have recently attempted an extensive review of rock weathering in engineering time but again its applicability to armourstone wear is limited.

For riprap and armourstone, Lutton et al. (1981) reported details of project experience from 38 Corps of Engineers offices in the USA. The deterioration phenomena were classified as cracking, spalling, delaminating or splitting (of foliated rock), disaggregation (of granular poorly cemented rock), disintegration (due to unavoidable occurrence of shaly or weakened rock) and dissolution. Almost half the offices reported no degradation problems in the past 10 years to 1979 and, even though it was concluded that improved testing and geotechnical supervision could further reduce problems and achieve economies, degradation problems were generally considered relatively insignificant.

In recent years, with port expansion in areas where only poor quality rock is available, there has been growing interest in degradation problems. The mechanisms of wear and degradation of rock in the marine environment have been studied in Australia, the Middle East and the UK by Poole and various co-workers. Recommendations from their field and laboratory studies were reported by Fookes & Poole (1981), Dibb et al. (1983a,b), Poole et al. (1984) and Allsop et al. (1985). They identified catastrophic fracture, abrasion and spalling as the three classes of degradation mechanisms. The presence and nature of anisotropy, pre-existing cracks, joints and discontinuities in the rock fabric of an armour block are all important considerations, particularly since they may result in numerous block failures while handling during construction. However, further discussion, is concerned primarily with wear processes.

The disruptive forces driving the decay process which have a complex, often cyclic and stochastically varying time history, are set against the cohesive forces that the rock fabric can harness.

The static loads of the block weights give compressive and tensile stress concentrations depending upon block shapes, contact areas and settlements. With tensile strengths of rock at least ten times lower than compressive strengths, tensile cracking is much the commonest catastrophic failure mode under severe dynamic loads.

Abrasion due to attrition of suspended material is a type of high frequency dynamic indentation which leads to loosening of fragments along pre-existing cracks, cleavage planes and grain-boundaries and, to chipping resulting from lateral vent tensile cracks forming sub-parallel to the surface during unloading. It is not surprising therefore that Verhoef et al. (1984) were able to show a correlation of volume loss due to sand blasting with Brazilian tensile strength. On prototype armour, the size, velocity and hardness of the impactors and their density in the water jet will influence the attrition wear rates.

The larger dynamic forces occur during construction and while rocking, rolling and sliding during storms. These can result in major breakages creating new projectiles as well as shear and tensile mode cracking on all scales down to the rock grain size. In addition to impact forces, the more gradually varying wave forces produce a pulsating form of load cycling. Both types promote the subcritical extension of tensile microcracks by the two mechanisms, stress corrosion and cyclic fatigue (see Costin and Holcomb, 1981).

The physico-chemical processes, whether salt crystallisation, hydration, differential thermal expansion or wetting and drying, can all induce cyclic pressure fluctuations which undoubtedly contribute to subcritical growth of tensile microcracks and eventual spalling of surface layers. Another cause of cracking and spalling could be the relaxation of residual tectonic stresses, an effect which can be measured using overcoring methods.

To summarize, it has been argued that wear processes at the armourstone surface, whether due to abrasion from external impacts or to disruptive pressures (e.g. from salt attack) within cracks and other surface openings, are essentially restricted by cohesive forces measured by the subcritical and critical indices of fracture toughness of the rock (which express its resistance to crack extension). Consider now the factors governing the rates of armourstone wear (i.e. weight losses) on coastal structures.

Rates of Wear

Rates of rounding were shown (Dibb et al., 1983a) to depend on the horizontal zone on the structure being considered because of the different energy associated with the various abrasion and physico-chemical forces at different positions relative to the still water level. The rates of armourstone wear as measured by the percentage of original weight remaining after a given number of years is likely to show a dependence on initial block dimensions since the rate of mass removal by abrasion is closely related to block surface area. The main factors governing wear rates on a new structure are:

(i) size of armourstone;
(ii) wave energy characteristics;
(iii) zone on structure;
(iv) size and type of water-borne attrition agents;
(v) climatic influence on physico-chemical degradation mechanisms, e.g. temperature range, rainfall, humidity;
(vi) local slope of armour layer - affects proportion of stone area exposed to attack;
(vii) armourstone displacement tolerated in design concept;
(viii) resistance of rock fabric to disaggregation under applied forces - a function of rock type and its degree of weathering.

Effects of Wear Processes

The effects of wear are to cause rounding, weight loss and some changes in void characteristics within the armourlayer surface. The number of point contacts with neighbouring blocks may even increase. The net effect will usually result in increased mobility of blocks and could possibly lead to instability. There may be a positive feedback here since increased mobility with significant rolling displacements of perhaps a few diameters per year, will further assist and accelerate wear.

Within the framework of Van der Meer's classification of rock slope structures (see Fig. 1), wear introduces two potential problems to the engineer. The first concern will be the reduction of interlock for conventional statically stable breakwaters. As the expected block mobility, crudely quantified by the dimensionless wave height parameter $H_s/\Delta D_{n50}$, is increased, a second concern becomes the weight loss. H_s is the significant wave height, Δ is the relative mass density and D_{n50} is the nominal diameter of the median armour weight. For structures with $H_s/\Delta D_{n50}$ approaching values as high as 20, seasonal profile adjustments can result in blocks moving up and down the slope depending perhaps on the timing of storms in relation to the tide cycle. The increased mobility that can result from rounding is not likely to be particularly threatening to the self-adjusting structure unless the weight losses due to size reduction and rounding become particularly marked and along-slope movements are considerable. As berm and S-shaped breakwaters and rock beach designs are increasingly being considered because of the possible economic advantages of their smaller block sizes, it is important to consider whether the increased mobility is likely to affect the performance and integrity of the structure and, if so, in what time scale. Additionally, it should be noted that the overall rate of weight loss is faster for smaller armour blocks because in relation to their weight they have comparatively high surface area.

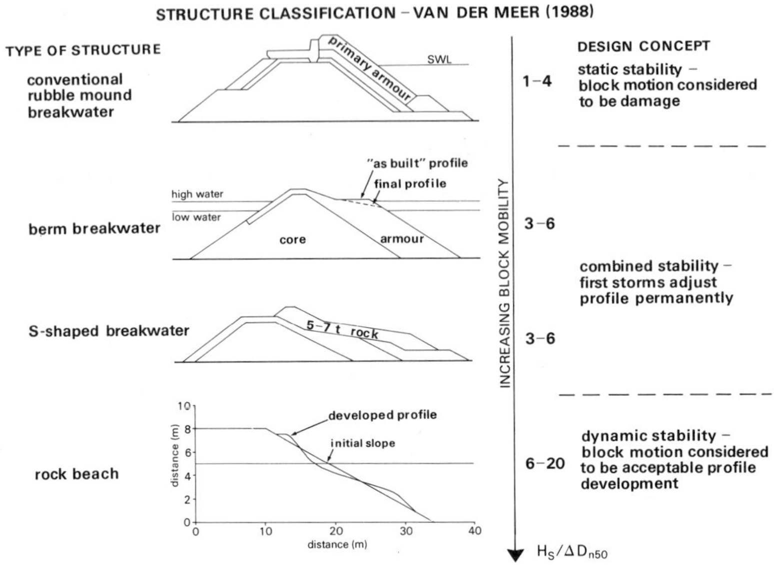

Figure 1: Types of rock armoured structures

Shape and its Controlling Factors

By shape, and in relation to armourstone, engineers will usually mean the gross shape which is often given by the axes of the enclosing cuboid. Most of the form indices such as flatness and sphericity are ratios of the X, Y and Z axes. In this paper, shape is taken to include the gross or overall shape but with details of the roundness of the corners together with the surface texture superimposed on and between the corners.

Factors at the quarry tend to govern the gross shape while those on the breakwater are a result of wear, and influence the roundness and surface texture.

The quarry may have been opened in a massive igneous intrusion yielding more equant blocks or alternatively, in a bedded sedimentary sequence giving more tabular blocks. If the rock is foliated or anisotropic, it may be impossible to obtain equant blocks with $X/Z < 2$. These are the geological controls. The structural geological setting which influences discontinuities and joint patterns as well as quarrying methods will also influence block shapes. The selection criteria (for example the exceedance value of the X/Z ratio) can also influence what is to be stockpiled and what rejected or re-blasted at the quarry.

Factors controlling block shape on the breakwater have already been discussed in relation to wear processes.

Design Implications of Block Shape

It should be stressed that it is placement technique in combination with block shape that is important in the performance of the breakwater.

For statically stable designs, interlock, macroporosity and thickness of the armour layer, which are all functions of shape are going to affect hydraulic stability. In fact, macroporosity and thickness both appear in Van der Meer's design equations through their influence on the permeability coefficient. The shape also influences hydraulic efficiency, i.e., reflection, transmission and run-up, although this may be of lesser significance.

Probably the main reason for rejecting rocks with higher X/Z ratios is a structural one. Tabular and elongate shapes, for example, have a lower resistance to fracture in bending because of the higher bending moments than for cubic blocks. However, there is some evidence emerging from the flume (see Table 1 and later discussion) that tabular blocks may have hydraulic advantages due to their higher resistance to rolling.

Purpose of Shape Quantification

To evaluate shape effects on static stability and hydraulic efficiency and thereby to improve specifications has been an overall objective of this current research. A related objective was to develop a laboratory simulation and a theory of abrasive wear in-service so that the wear during the life-time of a structure could be assessed for different rock types.

The whole question of shape effects could not be tackled with the conventional methods of shape quantification using the Krumbein/Wadell measure of roundness. The new techniques are described in a later section but first some of the evidence for the effect of rounding on static stability is reviewed.

Evidence of Shape Effects on Stability

Jensen (1984) presented two sets of results from random wave tests, one for 'quarrystone' and one for rounded 'seastones'. A lower stability was noted for rounded tests although the two sets of test conditions appear to be different, making objective comparison difficult.

Bergh (1984) using an impermeable core, composite profile and irregular waves, found that for the start of damage $H_s/\Delta D_{n50}$ was 50% lower for rounded glacial boulders than for cubic rock, implying an 8-fold increase in stone weight required for the rounded to achieve the same stability.

Van der Meer (1988) did not consider gross shape and roundness as test variables for static stability. However, significant abrasive rounding in his long test programme was suggested as a possibile explanation of certain results. (For example, for surging waves, was generally 35% lower for the older and presumably rounded test material). These explanations have been vindicated by recent research to be reported in Latham et al. (1988).

The Shore Protection Manual suggests that a different stability coefficient K_D should be used for smooth rounded rock and rough angular quarrystone. However, no clear guidance to the designer on shape effects is currently available.

It was noted that when, investigating dynamic stability, Van der Meer (1988) concluded that stone shape had no or only minor influence on the profile developed when for relatively permeable core conditions.

TECHNIQUES

Mill Abrasion Test

The mill abrasion test uses coarse aggregate pieces of crushed rock and was developed with the apparatus shown schematically in Fig. 2. For reasons set out in Latham & Poole (1987b) a number of features have been introduced which distinguish it from other tumbling mill abrasion tests used for aggregates. For example, there is a constant water level and throughput of water carrying fines and chippings away in suspension. The bulk volume fraction of the mill cylinder is always kept between 0.5 and 0.4 as the successive increments of milling are followed by making up the weight losses with fresh material. The abrasion is relatively gentle and there are no steel balls in the mill. Keeping the abrasion/attrition environment relatively constant in this way means that any non-linear response of weight or shape with time in the mill is not an artefact of non-linear test conditions. Rather, it is a reflection of the behaviour of angular particles becoming rounded and smaller while they cascade across the dimetrical free surface on each revolution. It is these weight loss and shape change relationships that are of interest in the study of wear.

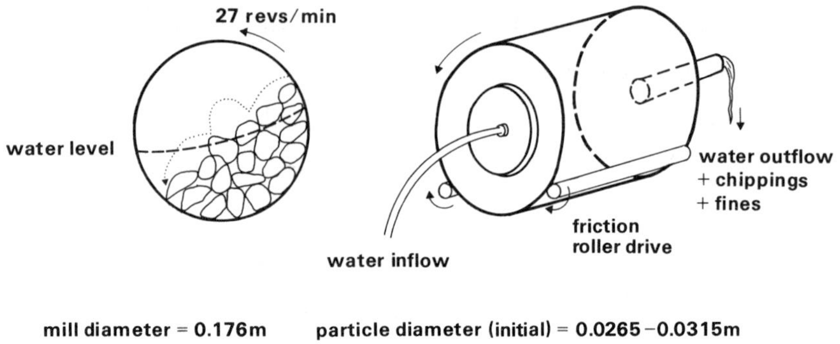

Figure 2: The tumbling mill abrasion test apparatus

The fraction of the original weight remaining when plotted against revolutions in the mill can be analysed to give two exponential coefficients, one of which is ideal for measuring and thus classifying the abrasion resistance of different rock types.

The experimental reproducibility was demonstrated using Carboniferous limestone aggregate in Latham & Poole (1988) and is reproduced in Fig. 3 together with the equation and coefficients describing the weight loss with time relationship. The coefficient k_s is the abrasion resistance index which can be used to describe the durability of the rock. The variation in test results for different rock types from a tough granite to a soft muddy limestone is indicated in Fig. 4. The values of k_s, indicated by the slopes, vary by about 2 orders of magnitude for the range of rock types likely to be encountered in coastal engineering works.

To examine the change in roundness or roughness of the blocks inside the mill, the image analysis techniques of the modern sedimentologist have proved invaluable.

Image analysis of block shape

The Fourier shape analysis methods of Ehrlich & Weinberg (1970) offer the best method for quantifying the rounding of blocks. Suppose the thin line in Fig. 5 is the outline given by the digitized coordinates (e.g. from a video camera). These coordinates can be represented by a Fourier series. The more terms or harmonics used in the series, the better the representation becomes. By the time the 10th term has been added see Fig. 5(d), the approximation, though quite good, is generally smoother than rough outlines with surface texture. To describe a rough irregular outline more precisely, the 11th to 20th harmonics are needed. In fact, the sum of the 11th to 20th harmonic amplitudes gives a very good measure of roughness, which was discussed with several other shape descriptors in Latham & Poole (1987a) and later given the name Fourier asperity roughness factor P_R where

$$P_R = (0.5 \sum_{20}^{n=11} C_n^2)^{0.5}$$

and C_n is the amplitude coefficient of the nth harmonic.

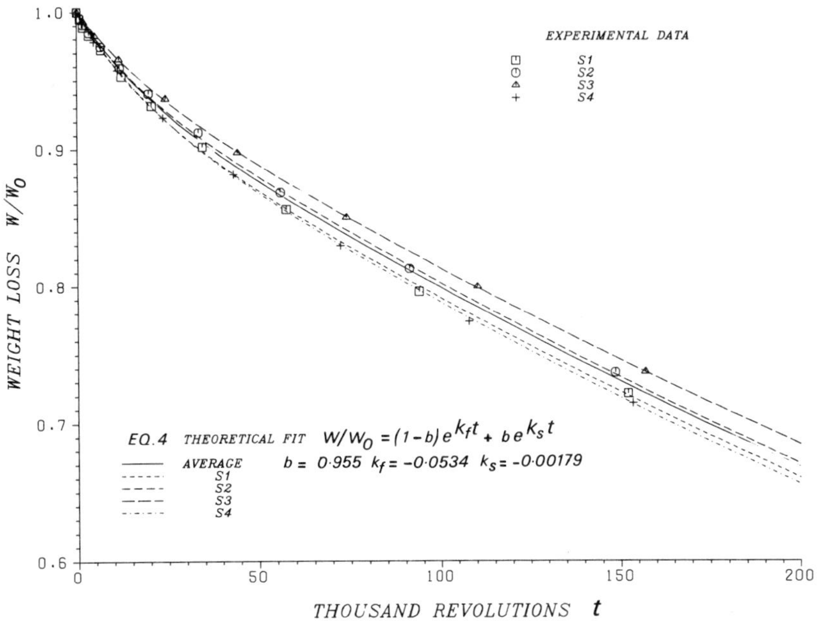

Figure 3: Reproducibility of mill abrasion test results from four limestone sample

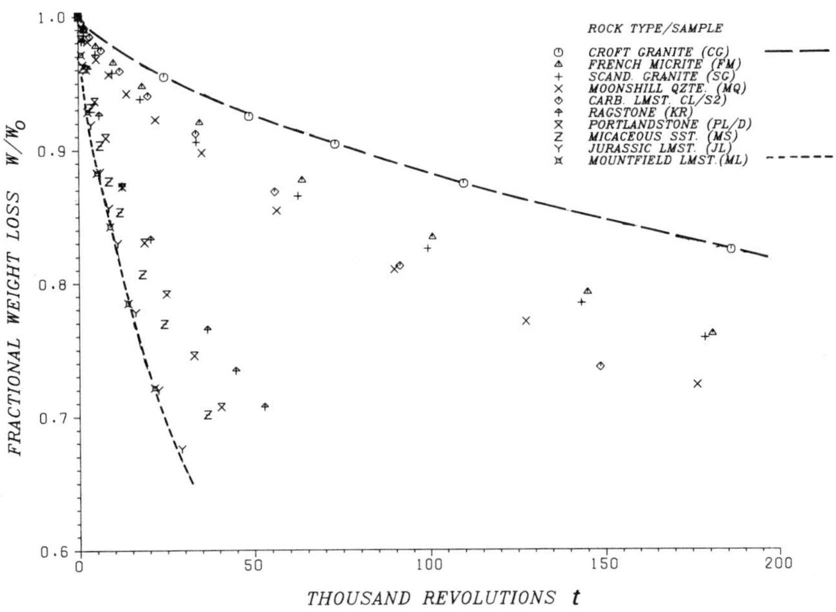

Figure 4: Variation of mill abrasion test results with rock type

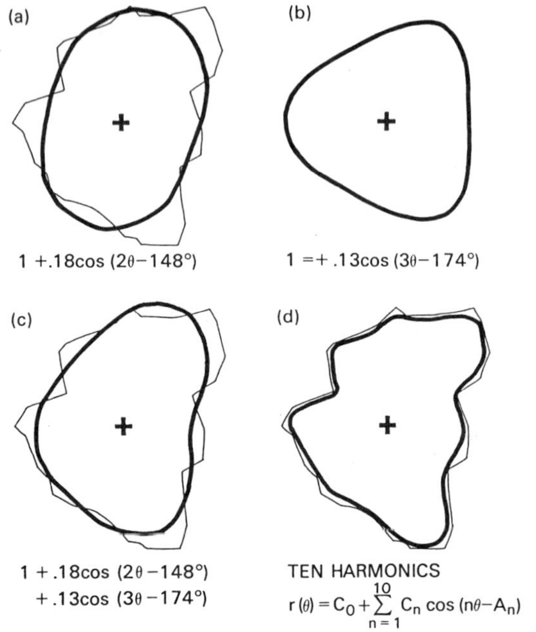

Figure 5: Contribution of Fourier harmonics to shape, after Ehrlich & Weinberg (1970)

AVERAGE SHAPE DESCRIPTOR VALUE	QMC/HR TABULAR	QMC/HR FRESH	QMC/HR EQUANT	QMC/HR S. ROUND	QMC/HR V. ROUND	DHL EQUANT
Gross shape						
X/Z	3.23	2.22	1.82	2.19	1.92	1.7
P_C	2.67	1.88	1.43	1.89	1.55	
Roughness						
P_R	.0165	.0138	.0117	.0097	.0046	
V d Meer Coeff. surging formula	1.32 (7)	– (3)	1.19 (8)	1.10 (8)	0.95 (8)	1.0

$$Hs/\Delta D_{n50} = \underset{\text{COEFF}}{1.0}\, P^{-0.13} (S/\sqrt{N})^{0.2} \sqrt{\cot\alpha}\, \xi_m^{P}$$

Table 1: Shape descriptor results for different armour shapes used in flume tests. A preliminary evaluation of a shape effect on stability is also shown.

A range of typical values of the Fourier shape descriptors is given in Table 1. The five shape classes were those tested for static stability in the flume at Hydraulics Research

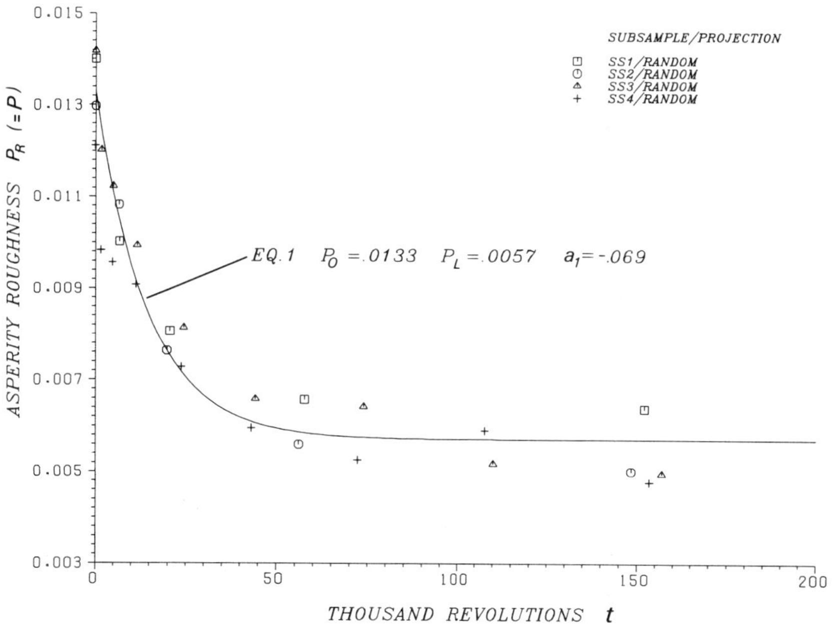

Figure 6: Roughness reduction during mill abrasion

Ltd. Each sample contained 48 blocks and the projections of the blocks were taken at random orientations to the video camera. Note P_C which indicates gross shape, is about 80% of the X/Z ratio for all 5 shapes. The asperity roughness P_R is very sensitive to the abrasive rounding. When FRESH material was put in the abrasion mill, after a 7% weight loss, P_R had fallen from 0.0138 to 0.0097 and after 23% weight loss, the VERY ROUND value was 0.0046.

Mill Abrasion Theory

To describe what is happening inside the mill, two assumptions are made. First, that a roughness parameter P will have the limiting exponential form

$$P - P_L = (P_0 - P_L)\exp(a_1 t) \qquad (1)$$

during progressive abrasion where P_0 and P_L are initial and limiting (i.e. large t) values of P, a_1 is a constant and t is measured in thousands of revolutions of the mill. Second, that the instantaneous rate of weight loss is given by how much roughness there is left for removal and by the fraction of the original weight remaining, which yields the differential equation

$$\frac{d(W/W_0)}{dt} = a_2 (P - P_L) + a_3(W/W_0) \qquad (2)$$

The solution is $$W/W_0 = \frac{a_2 (P_0 - P_L)}{(a_1 - a_3)}\exp(a_1.t) + \left\{1 - \frac{a_2(P_0 - P_L)}{(a_1 - a_3)}\right\}\exp(a_3.t) \qquad (3)$$

which is the same form as the equation given in Fig. 3 for quantifying the abrasion resistance k_s from the weight loss results. It was expressed as follows

$$W/W_0 = (1 - b)\exp(k_f.t) + b\exp(k_s.t) \qquad (4)$$

The consequence of equating these coefficients using experimental data and curve fitting to test the validity of the assumptions turns out to be favourable since a_1 is approximately equal to k_f. It appears then, that the weight loss can be directly related to shape changes when the roughness P is described by the asperity roughness P_R. For the two exponential coefficients in equation (4), k_f is the smoothing resistance index which describes the rate of the fast process of removing asperities and k_s, the abrasion resistance index, describes the rate of the slow process of overall size reduction, (b is a mixing term). Eliminating t produces an equation for W/W_o in terms of asperity roughness.

$$W/W_0 = (1 - b)\left(\frac{P - P_L}{P_0 - P_L}\right) + b\left(\frac{P - P_L}{P_0 - P_L}\right)^{k_s/k_f} \qquad (5)$$

A more detailed theoretical development was given in Latham & Poole (1988). Fig. 6 has been taken from that paper to illustrate the fit of equation (1) to the abrasion mill data for four Carboniferous limestone samples.

A new concept is now introduced so that the laboratory information and theory of the mill abrasion test can be applied beyond simple rock durability assessment.

Equivalent Wear Factor

The great number of factors that affect the rate of wear on the prototype breakwater was indicated earlier in this paper. It is suggested that all the site specific and environmental factors should be combined together and kept separate from the rock durability and that the implicit assumption be made that the progress of prototype rounding, for a given rock type, is equivalent to that which occurs inside the mill for that same rock type but that it occurs at a different rate. The EQUIVALENT WEAR FACTOR X to relate these laboratory and prototype rates was defined (Latham & Poole, 1988) as the number of years of exposure at a particular zone on a given structure that is equivalent to 1000 revolutions in the tumbling mill test described in the reference. However, in that paper, the authors neglected to point out the need also to separate the initial block weight from the combined site specific factors. The effect that initial block size will have on rate of weight loss is conveniently taken into account assuming the volume to surface area relationship. For example, if the rate of removal of two different sized blocks with the same shape is initially governed by their surface areas, then for an 8-tonne block which after 1 year loses, say, 0.20 tonnes ($W/W_o = 0.975$), a 1.0 tonne block loses 0.05 tonnes ($W/W_o = 0.950$) which is a faster rate of fractional weight loss. For a given set of site specific factors, the equivalent wear factor x will depend on the size of the prototype blocks. In Latham & Poole (1988), the suggested values of x were based on observations of rounding where the armourstones were nominally 8 tonnes. To adjust for a different initial block weight W_n in tonnes, the equivalent wear factor x becomes $(W_n/8)^{1/3}$ times its value for an 8-tonne block. Clearly a suggested value of x should refer to a particular initial block weight.

From the data obtained on statically stable breakwaters in N.E. Queensland, the Middle East and the U.K., for the most aggressive environment in the intertidal zone, a value of x = 0.5 was suggested for 8 tonne armour. Table 2 shows how the mill abrasion test data for Carboniferous limestone can be used to predict wear rates (both weight loss and rounding) on statically stable structures built with 8-tonne Carboniferous limestone armour. For example, a 12% weight loss ($W/W_o = 0.88$) would occur between 25 and 50 years in service for a relatively aggressive environment with x = 0.5 to 1. A smaller value of x implies a faster rate of wear.

Field evidence from West Bay, Dorset, reported by Clark (1988) is discussed in the report (Latham et al., 1988), where it was concluded that aggressive conditions with shingle attack corresponded to an x value of about 0.8 for 8-tonne armour.

The image analysis techniques can be applied to armour blocks from real structures. This allows for a direct measurement and comparison of the shapes and relative states of rounding of different armour blocks to be made. The results obtained can be used to help calibrate x for different site conditions as demonstrated below.

Weight loss	Asperity roughness	Revs in mill (thousand)	Equivalent Wear Factor for W_0 = 8 tonnes		
			Mild x = 5	x = 1	Aggressive x = 0.5
W/W_0	P_R	t	Years on breakwater		
0.97	0.0102	10	50	10	5
0.94	0.0083	20	100	20	10
0.88	0.0062	50	250	50	25
0.80	0.0057	100	500	100	50

Table 2: Application of the equivalent wear factor concept to Carboniferous limestone armour

APPLICATIONS

Rounding of Whin Sill Dolerite from Buckhaven, Scotland.

This revetment structure was constructed in 1975 and was reported to have been designed at a 1 in 2 slope and built with 1 to 2-tonne armourstone quarried from the Whin Sill dolerite. It is underdesigned in that storms frequently cause rocks to roll down onto the foreshore. It is one of the only U.K. examples of a structure resembling a rock beach where block mobility and shingle attrition have combined to produce dramatic rates of wear against perhaps one of the most durable rock types in the U.K. A typical angular block and a well rounded block were shown in Allsop & Latham (1987). Digitized outlines of 8 upper blocks and 8 lower blocks are shown in Fig. 7.

Although these block outlines are not statistically representative samples, for illustration the upper block values may be equated with the initial condition during constrution and the lower block values with 12 years' wear. The gross shape parameter P_C has hardly changed while the asperity roughness P_R has falled from 0.0149 to 0.0062. Mill abrasion results were not available for this rock type but it is known to have a very high fracture toughness ($K_{1c} \approx 3$MPa.m$^{0.5}$) and density ($\approx$ 3.0 tonnes/m^3).

On this basis and comparing with test results for the toughest granites, an estimate of the abrasion resistance index and other mill abrasion constants can be calculated. Applying the mill abrasion theory outlined above with the available measured and estimated data suggests that 50,000 revolutions in the mill would give the observed P_R value of 0.0062 and at a weight loss of 9%. The equivalent wear factor x for these semi-mobile blocks of about one tonne is therefore 12/50 = 0.25. However, for 8-tonne blocks, the value would be 0.5.

It might be expected that equivalent wear factors on dynamic structures would be lower (i.e. faster wear) than for statically stable structures. It is possible then that the value of 0.5 for 8-tonne blocks which was suggested in Latham & Poole (1988) for extremely aggressive environments on statically stable structures is slightly too low and is more representative of a structure which has included some mobility in the design concept (e.g. $H_s/\Delta D_{n50} \approx 6$).

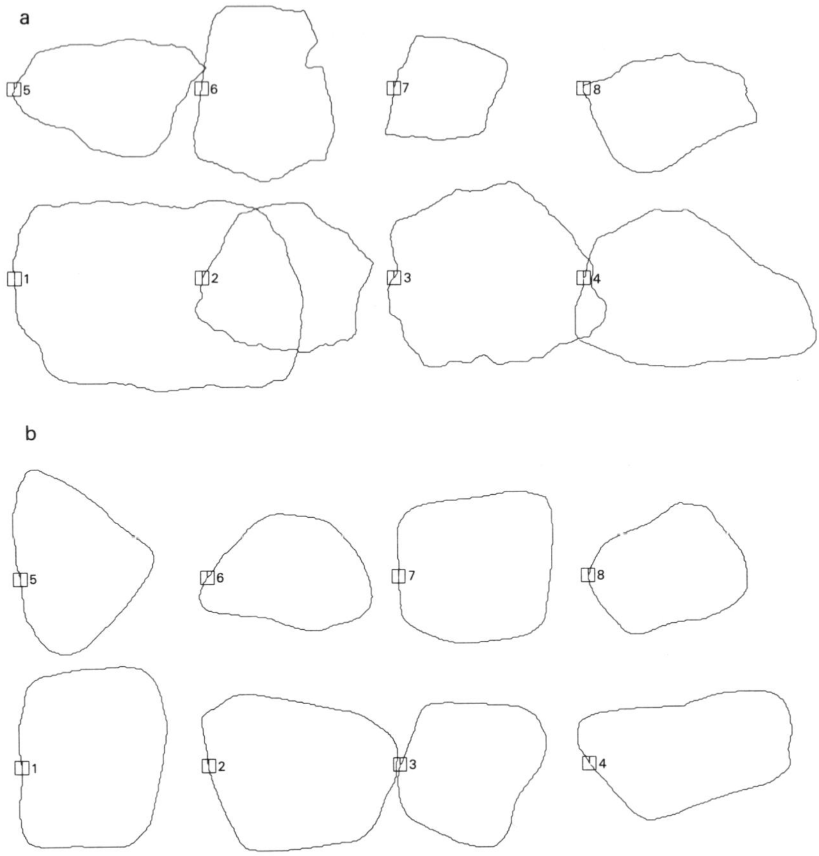

Figure 7: Digitized outlines of blocks from Buckhaven; a upper, b lower

Discussion

As a working hypothesis, it is reasonable to suggest that the material property measured by the resistance to weight loss inside the tumbling mill over, say, one day will rank different rock types correctly in terms of their weight loss performance over perhaps 10 years of episodic wear on a breakwater. The validity of the equivalent wear factor concept used in the prediction of in-service rates of wear is more tenuous and requires further validation in the future. The concept assumes that the non-linear equations (1) to (4) for mill abrasion also apply to prototype abrasion.

Two likely objections to this are (i) that wear mechanisms, particularly those of sub-critical crack growth, not properly represented in the abrasion mill may contribute to the form of the non-linear relationship on the prototype; (ii) that, while the work done per unit time is kept relatively constant in the mill, the average work done by disruptive forces on each armour block of a dynamic structure (where there is a gradual increase in block mobility due to rounding) may increase with time, thus having an accelerated effect on weight loss rates.

In response to these legitimate criticisms: (i) it was noted that the Allison and Savage (1976) prototype weight loss data gave essentially the same type of relationship observed for

abrasion inside the mill; (ii) the equivalent wear factor concept provides a framework which takes account of two of the most important reasons for relative differences in wear rates, namely material properties and initial block sizes. The framework can be refined and calibrated by future observations and tests; (iii) The relationship in equation (5) is time-independent so that, for blocks undergoing attack on all sides, it is possible to predict weight losses directly from shape changes measured on a prototype structure; (iv) for dynamic structures, the problem may be more complex although in some respects the mill simulation is more appropriate. It may be fruitful to consider Van der Meer's (1988) treatment of dynamically stable profile development in a search for the likely influence of armour block mobility on the choice of x. For structures with $H_s/\Delta D_{n50} \leq 6$, the acceleration effect of increasing mobility on wear rates may be negligible. For structures with $H_s/\Delta D_{n50} \geq 6$, a question to address is how many block diameters does the average block roll per year and does this distance depend on block shape?

Flume Test Results on Rounded and Angular Material

For static stability, the consequence of weight loss on armour stability can be calculated using Van der Meer's (1987) design equations. To predict the effect of rounding on static stability, the limited series of flume tests conducted in collaboration with Hydraulics Research Ltd, Wallingford, can provide some tentative guidance. The flume tests first reported by Bradbury et al. (1988) were intended as a preliminary series to evaluate shape effects on stability. For each different shape of rock tested, the results were fitted to Van der Meer's surging wave equation. Where Van der Meer obtained a coefficient value of 1.0, the flume tests found this coefficient, and therefore the stability, to be dependent on shape. For instance, the coefficients (see Table 1) suggest that the very round rock gives $(1.19/0.95)^5$ or 3 times more damage than the equant rock. The asperity roughness appears to account for the shape effect on stability reasonably systematically. For statically stable designs, the tests suggest that the loss of interlock due to rounding contributes more to instability than the weight loss itself accompanying the rounding. These aspects of the flume tests will be considered in detail in the report by Latham et al. (1988).

ACKNOWLEDGEMENTS

The authors wish to express their thanks to William Allsop for collecting photographs from Buckhaven and for his interest in this research. The paper has benefited from certain points raised during discussion by one questioner for which we are grateful. The research was funded by the Science and Engineering Research Council of Great Britain under grant GR/D/69921.

REFERENCES

Allison, D. M. & Savage, R. P. (1976) Tests of low-density marine limestone for use in breakwaters. U.S. Army Corps of Engineers, Coastal Engineering Research Centre, Tech. Paper No. 76-4

Allsop, N. W. H., Bradbury, A. P., Poole, A. B., Dibb, T. E. & Hughes, D. W. (1985). Rock durability in the marine environment. Hydraulic Research Ltd, Report No. SR11.

Allsop, N. W. H. & Latham, J-P. (1987). Rock armouring to unconventional breakwaters: the design implications for rock durability. Seminar on Unconventional Rubble-Mound Breakwaters, NERC, Ottawa, Canada, Paper No. 2.

Bergh, H. (1984). Riprap protection of a road embankment exposed to waves. Hydraulic Laboratory, Royal Institute of Technology, Stockholm, Sweden. Bulletin No. TRIA-VBI-123

Bradbury, A. P., Allsop, N. W. H., Latham, J-P., Mannion, M. B. & Poole A. B. (1988). Rock armour for rubble mound breakwaters, sea walls and revetments: recent progress. Report no. SR 150, Hydraulics Research

Clark, A. R. (1988). The use of Portland Stone rock armour in coastal protection and sea defence works. Q.J. Eng. Geol., London, Vol. 21, p.113-136.

Costin, L. S. & Holcolm, D. J. (1981). Time-dependent failure of rock under cyclic loading. Tectonophysics Vol. 79, p.279-298.

Dibbs T. E., Hughes, D. W. & Poole, A. B. (1983a). Controls of size and shape of natural armourstone. Q.J. Eng. Geol., London, Vol. 16, p.31-42.

Dibb, T. E., Hughes, D. W. & Poole, A. B. (1983b). The identification of critical factors affecting rock durability in marine environments. Q.J. Eng. Geol., London Vol. 16, p.149-161.

Ehrlich, R. & Weinberg, B. (1970). An exact method for characterizing grain shape. Journal of Sedimentary Petrology, Vol. 40 p.205-212.

Fookes, P. G. & Poole, A. B. (1981). Some preliminary considerations on the selection and durability of rock and concrete materials for breakwaters and coastal protection works. Q.J. Eng. Geol., London Vol. 14, p.97-128.

Fookes, P. G., Gourley, C. S. & Ohikere, C. (1988). Rock weathering in engineering time. Q.J. Eng. Geol., London Vol. 21, p.33-57.

Jensen, O. J. (1984). A monograph on rubble mound breakwaters' Danish Hydraulic Institute, Denmark.

Latham, J-P. & Poole, A. B. (1987a). The application of shape descriptor analysis to the study of aggregate wear. Q.J. Eng. Geol., London Vol. 20, p.297-310.

Latham, J-P. & Poole, A. B. (1987b). Pilot study of an aggregate abrasion test for breakwater armourstone. Q.J. Eng. Geol., Vol. 20, p.311-316

Latham, J-P. & Poole, A. B. (1988). Abrasion testing and armourstone degradation. Coastal Engineering, Vol. 12.

Latham, J-P., Mannion, M. B., Poole, A. B., Bradbury, A. P. & Allsop, N. W. H. (1988). The influence of armourstone shape and rounding on the stability of breakwater armour layers. Coastal Engineering Research Group. Report 1, Queen Mary College, London University.

Lees, G. & Kennedy, C. K. (1975). Quality, shape and degradation of aggregates. Q.J. Eng. Geol. Vol. 8, p. 193-209.

Lutton, R. J., Houston, B. J. & Warriner, J. B. (1981). 'Evaluation of quality and performance of stone as riprap or armour. U.S. Army Corps of Engineers, Tech. Report GL-81-8.

Poole, A. B., Fookes, P. G., Dibb, T. E. & Hughes, D. W. (1984). Durability of rock in breakwaters. Breakwaters - design and construction, p.31-43, Thomas Telford Ltd, London.

Van der Meer, J. W. (1987). Stability of breakwater armour layers - Design formulae. Coastal Engineering, Vol. 11, p.219-239.

Van der Meer, J. W. (1988). Rock slopes and gravel beaches under wave attack. PhD Thesis, Delft Hydraulics Communication No. 396.

Verhoef, P. N. W., Kuipers, T. J. & Verwaal, W. (1984). The use of the sand blast test to determine rock durability. Bulletin Int. Assoc. Eng. Geol., Vol. 29 PARIS, p.457-461.

CHAPTER 171

Wave Uplift Forces on Platforms with Energy Absorbers

Jiin-Jen Lee, M. ASCE [1] & C.P. Lai, A.M. ASCE [2]

ABSTRACT

Interaction of large amplitude wave with platforms or docks in a variable depth medium has been studied theoretically and numerically. The underside of the platform is equipped with energy absorbing linear spring in order to study the energy absorbing effect for reducing the uplift forces on the platform or dock.

The flow is assumed to be inviscid, irrotational without current. Laplace equation with full non-linear free surface equations are solved by isoparametric mapping technique. A Galerkin finite element model is used to model the transformed equations. The adaptive line SOR (Successive-Over-Relaxation) technique is used.

The results indicate that uplift wave forces on platforms or docks can be significantly reduced by very soft springs. Wave profile history confirms the manifested mechanism of energy absorption. The results also indicate that larger uplift forces are easier to be damped out by the prescribed energy absorbing mechanism than the smaller uplift forces.

1. INTRODUCTION

It has been found that as a finite amplitude ocean wave strikes an offshore platform which is located above the still water surface, the platform will register a positive uplift force followed by a negative force if the clearance between the platform is smaller than the wave amplitude (see El Ghamry (1963), Wang (1967), French (1969), Lai (1986), Lee & Lai (1986), Lai & Lee (1988)). Such repeated wave actions constitute a serious dynamic loading to the marine platforms.

---

[1] Professor of Civil and Environmental Engineering, University of Southern California, Los Angeles, CA 90089-0231

[2] Project Engineer, Environmental and Ocean Technology, Inc. (E.O. Tech), Suite 885, 150 N. Santa Anita Ave., Arcadia, CA 91006

One of the ways to overcome this problem would be to place the platform at a distance significantly above the sea surface so as to ensure that no such dynamic uplift forces would occur. Such a design however, might place the platform in a height that could create other negative impacts. For example, the economic concern or the dynamic vibrations. Another approach would be to develop certain energy absorption mechanism on the platform allowing certain reasonable and moderate soffit clearance. In an earlier study by Lai (1986), Lee and Lai (1986), it has been found that this hydrodynamic force is strongly dependent on relative wave amplitude. As the relative amplitude of the wave increases, the maximum uplift force increases while the maximum negative forces decreases. From the point of view of the structure safety, the increase in maximum uplift force might create significant negative impacts on the structure.

The main objective of this study is to explore ways to reduce the maximum uplift forces when a large amplitue wave stikes a platform. To accomplish this the numerical model presented by Lai (1986) and Lee & Lai (1986) has been extended to allow the treatment of significantly large incident wave amplitude and to incorporate the energy absorption mechanism at the platforms.

2. THEORETICAL DEVELOPMENT

For the mathematical model, the flow is assumed to be inviscid, incompressible, irrotational, and without ambient current. The wave-structure interaction can then be solved by making use of a potential flow theory, the velocity potential must satisfy the Lapace equation. A definition sketch is shown in figure 1. The dynamic free surface condition is that the pressure must be equal to the atmospheric pressure at the free surface. On the rigid boundaries, normal velocity is prescribed, for rigid bottom this velocity is zero, and for a body in motion the normal velocity of the fluid particle at the platform must be equal to that of the platform velocity. In the analysis, we allow physical domain to be long enough to have "no flow" conditions at both far field boundaries. This requires that the horizontal velocity be zero at lateral boundaries at infinity. Therefore, the governing equation and boundary conditions at free surface are:

$$\nabla^2 \phi = 0$$

$$h_t + \phi_x h_x - \phi_y = 0 \qquad on \quad y = h(x,t)$$

$$\phi_t + \frac{1}{2}(\phi_x^2 + \phi_y^2) + gh = 0 \qquad on \quad y = h(x,t)$$

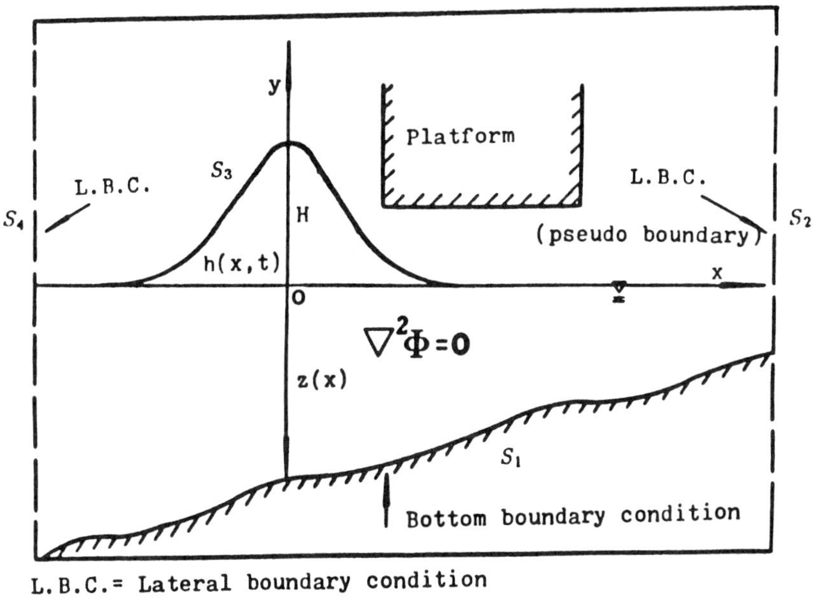

Figure 1 Definition Sketch for the Problem of a Wave Interacting with a Platform or a Dock

$$\phi_x = U = 0 \quad on \quad S_2 \quad and \quad S_4$$

$$\phi_n = 0 \quad on \quad S_1$$

$$\phi_y = V = \frac{ds}{dt} \quad (when \quad wave \quad hit \quad the \quad platform) \quad on \ S_3$$

where S_1 , S_2 , S_3 , S_4 are surface boundaries.

It is noted that the difference between the present formulation and the prior studies by the authors is to allow the underside of the platform to move at the velocity ds/dt. This motion is resisted by a distribution of linear springs assumed to exist just above the surface of the underside of the platform. A rigid platform would correspond to the case of spring constant being infinite. A softer spring therefore would yield a larger velocity of the underside of the platform.

The difficulty of the problem lies in the nonlinearity and the coupling of the boundary conditions at the free surfaces, which are time dependent. The free surface elevation is an unknown a priori and it must be a part of the solution of the problem along with the Laplace equation on the interior of the fluid region as well as the equations at the free surface. To handle a moving free surface and an irregular bottom topography, we adopt a mapping technique to transfer the fluid region and its boundaries into a regular geometry.

The physical domain is transformed into a regular computational domain by using isoparametric mapping. The transformed governing equations are more complicated than the original governing differential equations. However, the advantage is that the solution region has a regular retangular shape in the parent plane. The nonlinearity of the moving free surface is no longer a serious problem in the transformed plane. The approximate solutions of the transformed equations are sought by using the Galerkin's finite element method. For a detailed discussion of the numerical procedure, the reader is referred to Lai (1986).

The finite element model is advanced in time by integrating the dynamic and kinematic free surface equations. The Runge-Kutta method is employed with mass lumping. The mass lumping technique can eliminate the coupling between the time derivatives of the unknowns at adjacent nodes and reduce the computer storage and computational steps with only a minor loss in accuracy.

A major difficulty in numerical integration of the free surface equations is the computational instability. Approximation of spatial derivatives by numerical methods introduces truncation errors.

These errors are amplified by nonlinear terms and as they accumulate for large time, would seriously affect the accuracy of the solution. Moreover, when a wave starts to hit a platform or a dock, run-up is created by the platform or the dock, and reflected wave can be formed at the frontal edge of the platform or dock. This discontinuity in the free surface profile causes a severe oscillation at the frontal face of the platform or dock for the large time. To avoid the instability of numerical integration, an artificial viscosity term has been introduced with their value determined by numerical experiments.

An adaptive line SOR algorithm is developed and used to speed up the repeated solution of linear algebraic equations resulting from finite element discretization of the transformed Laplace equation. The idea of adaptive line SOR method is to employ adaptive parameter of relaxation in LSOR to accelerate the convergence rate. It was found that the adapted line SOR algorithm could effectively reduce the error norm by several orders of magnitude for the similar number of iteration.

3. PRESENTATION AND DISCUSSION OF RESULTS

One of the simplest energy absorbing mechanisms which could be employed for numerical experiment is to place flexible spring just above the underside of the platform where wave could hit the platform. The spring allows the underside of the platform to deform upward as the postive uplift forces occur. It would also allow it to return to the original surface as the contact between the water surface and the underside of the platform ceases to exist. A variation of different spring constant has been experimented. It is noted that the case of rigid platform corresponds to the case of infintely large spring constant K.

Figure 2 shows the numerical results of the hydrodynamic uplift force in platform with a 3 inch soffit clearance in a water depth of 15 inch. The relative wave amplitude is 0.24. The ordinate of the figure is the ratio of the total hydrodynamic uplift forces which is integrated from the computed uplift pressure over the entire platform divided by the total hydrostatic forces. The hydrostatic force is taken as the integrated weight of the water above the elevation of the platform. The computed forces based on the present numerical procedure is plotted as a function of the normalized time history. Three different spring constants are used for comparison. The rigid body would represent a spring constant of infinity. It is seen that the softer the spring the smaller the computed uplift forces. The wave amplitude used for this case is not too large, therefore the positive uplift forces differ only slightly from the hydrostatic forces. From the numerical result it is also clear that in order to meaningfully reduce the uplift force (either positive or negative) the flexible spring has to be quite

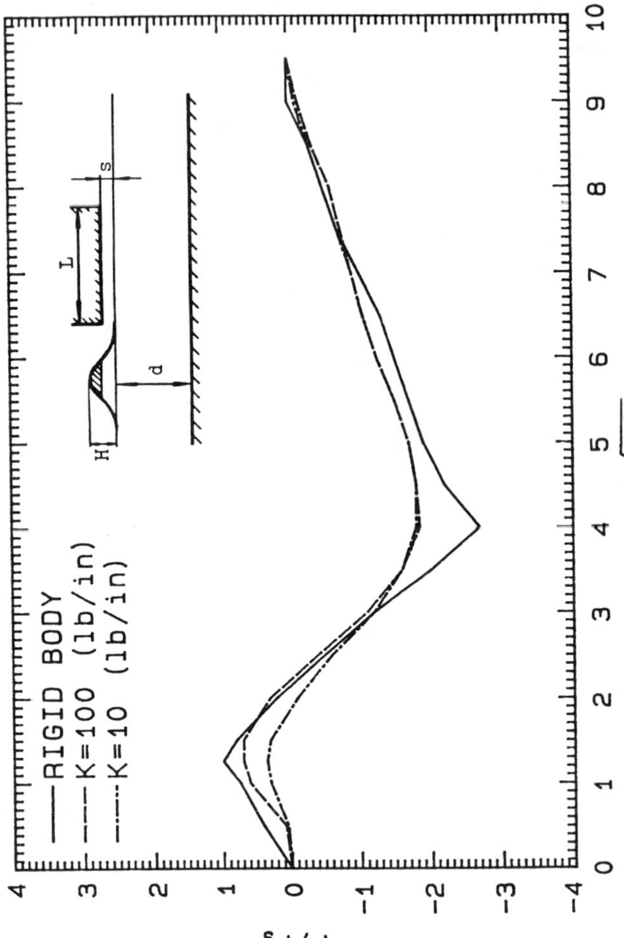

Figure 2 Normalized Total Hydrodynamic Force Per Unit Width.
(H/d = 0.24, s/d = 0.2, L/d = 4, d = 15 in.)

soft.

Figure 3 presents the force history when the incident relative wave amplitude is increased to 0.32 keeping other parameters the same as that shown in Figure 2. In this case the positive uplift force is significantly increased, this phenomena has been shown in earlier studies by Lai (1986) and Lai and Lee (1988). Based on the numerical results it is surprising to find that for moderately stiff springs the magnitude of the forces can only be modified by a small amount. In order to reduce the maximum postive uplift forces by a significant degree the spring must be quite soft. A comparison of the results for K=100 lb/in and that for K=10 lb/in show that the magnitude of the positive uplift force is significantly reduced for K=100 lb/in the magnitude of uplift force is quite close to the case of rigid platform.

The normalized uplift force per unit width in a sloping bottom is shown in Figure 4 for a rigid platform and that with two different spring constants (K=100 lb/in and K=10 lb/in). The relative incident wave amplitude is 0.32 and that the platform is placed at the back wall allowing wave reflection from the back wall. As expected, the positive uplift force is drastically large than that shown in Figure 3 signifying the effect of wave reflection from the back wall. The effect of the energy absorbing spring to dampen the large uplift forces can clearly be seen from the numerical results.

It is interesting to learn how the wave form changes as the incident wave interact with the platforms or docks. Numerical result on the wave amplitude along the direction of wave propagation is shown in Figure 5. The parameters representing the wave channel, incident waves and the dock dimensions are the same as that shown in Figure 4. In Figure 5, wave profiles along the x axis at different time steps are shown for K=100 lb/in. It is seen that the wave crest have not reached the platform after 150 time steps thus the incident wave has been propagating unchanged toward the positive x direction. After 200 time steps portion of the incident wave hit the platform (as the soffit of the platform is smaller than the wave amplitude), thus wave form is changed accordingly and the relatively soft spring allow the underside of the dock to absorb energy by compressing the spring. This process continues through 250 time steps. At 300 time steps the travelling wave is reflected back from the back wall and is travelling away from the dock region. without the confinement of the underside of the dock platform, large amplitude oscillation is seen in the wave amplitude profile presented in the result for 300 time steps. This interesting phenomena is repeated in Figure 6 at 300 time steps for two different spring constants. For K=100 lb/in the maximum clearance between the still water level and the underside of the dock is less than 0.3 as shown in Figure 6. As portion of the wave travels back and moving away from the dock site, the wave (both positive

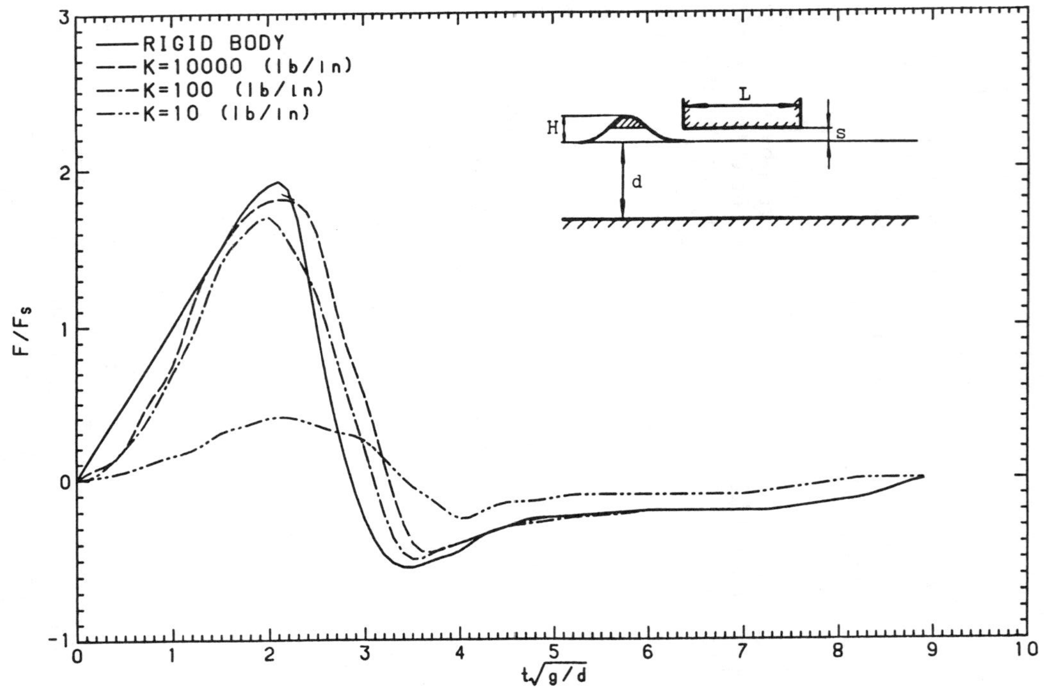

Figure 3 Normalized Total Hydrodynamic Force Per Unit Width.
(H/d = 0.32, s/d = 0.2, L/d = 4, d = 15 in.)

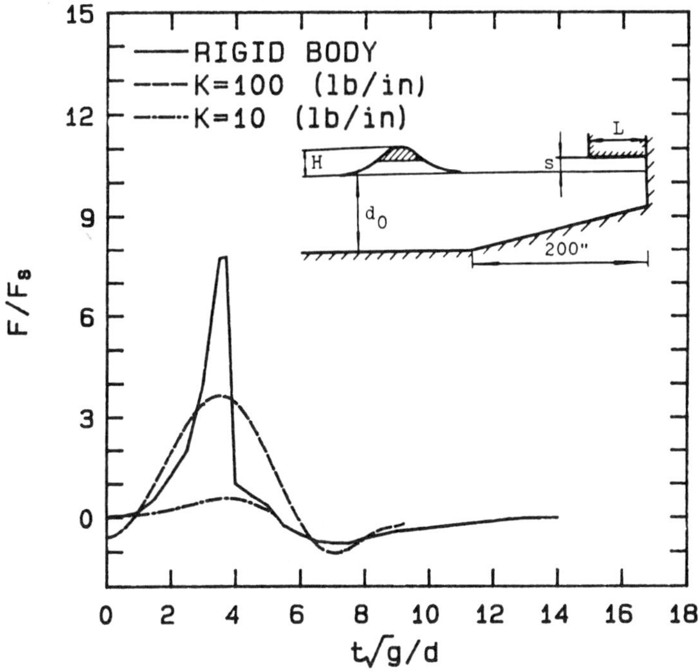

Figure 4 Normalized Uplift Force Per Unit Width (H/d_0 = 0.32, s/d_0 =0.2, L/d_0 = 4, d_0 = 15 in.)

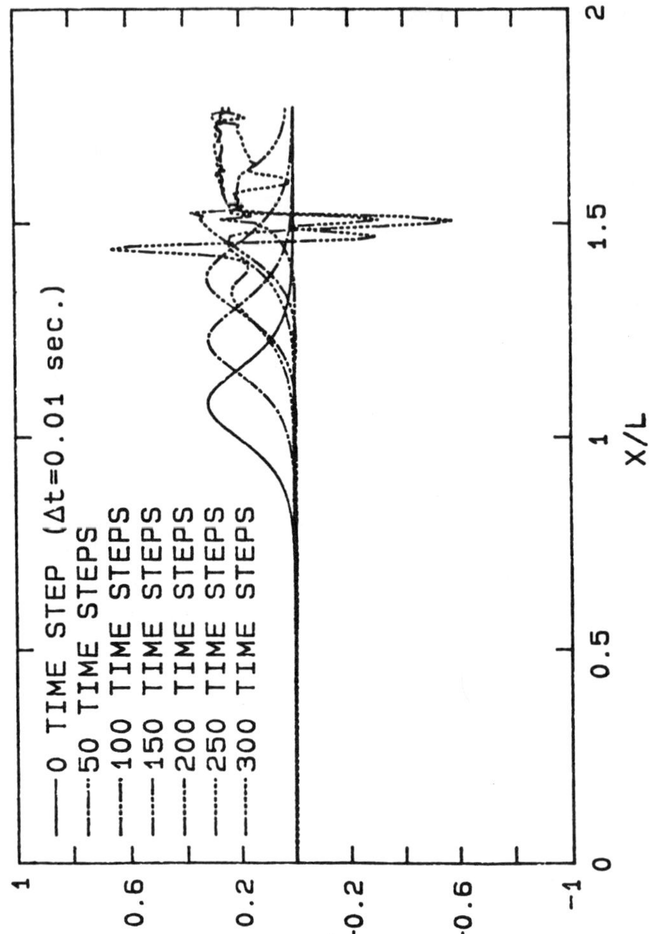

Figure 5 Wave Profiles at Different Time Step before and after Interacting with the Dock Placed at a reflecting Back Wall

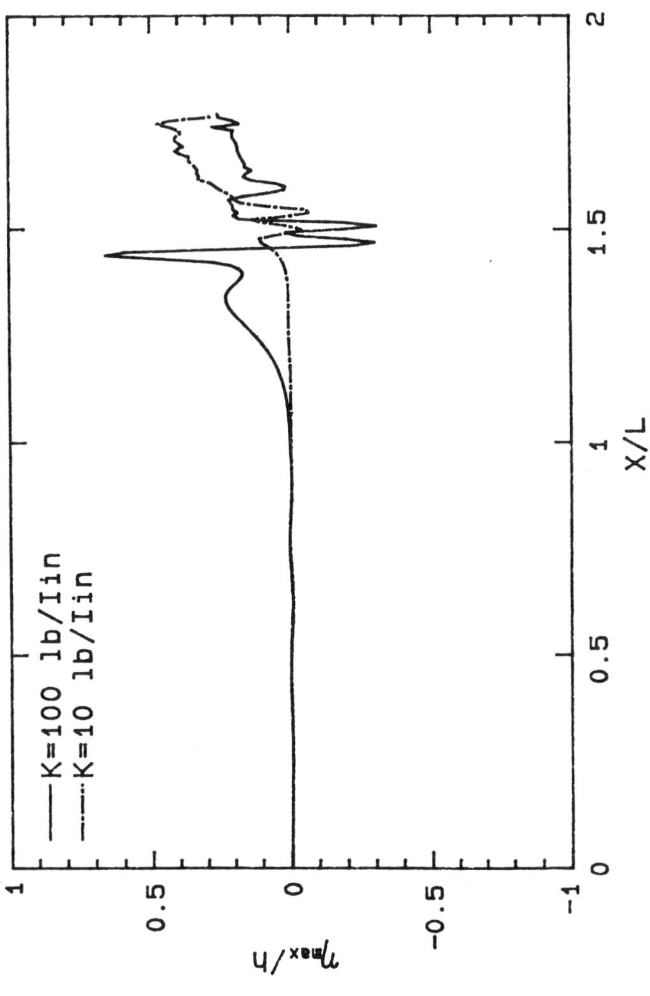

Figure 6 Wave Profile at 300 Time Steps Interacting with the Dock with Different Spring Constants Associated with the Energy Absorber

and negative) is significantly increased its amplitude. For a softer spring (K=10 lb/in) the underside of the dock is permitted to move upward thus the clearance is increased thereby absorbing a significant amount of wave energy. Therefore, wave profile for K=10 lb/in at 300 time steps is reduced. It is interesting to note the difference in wave phases for these cases for spring with K=10 lb/in, wave appears to travel at a slower speed.

4. CONCLUSION

The numerical experiments conducted in obtaining the effect of absorbing linear springs placed at the underside of the platform /or dock appears to be able to absorb certain wave energy and to reduce the positive uplift force induced by wave interacting with platforms /or docks. Significant reduction in this uplift force can only be achieved by very soft springs. For the dock problem (reflecting wall placed at the end of the dock) the energy absorbers appear to offer greater reduction in the uplift forces. The magnitude of the uplift forces experienced by the platform /or dock with spring attached at its underside can be computed using the present numerical model but can not be generalized by a simple formula due to the complexity of the problem.

5. ACKNOWLEDGMENTS

The authors appreciate the funding supports from the USC Foundation for Cross-Connection Control and Hydraulic Research (FCCCHR).

6. REFERENCES

1. El Ghamry, O. (1963), "Wave Forces on a Dock," Report No. HEL-9-1, University of California, Berkeley, California.

2. French, J.A. (1969), "Wave Uplift Pressure on Horizontal Platforms," Report NO. KH-R-19, W.M. Keck Laboratory of Hydraulics and Water Resources, California Institute Technology, Pasadena, California.

3. Lai, C.P. (1986), "Wave Interaction with Structure: Hydrodynamic Loadings on Platforms and Docks," Ph.D. thesis, University of Southern California.

4. Lai, C.P. & Lee, J.J. (1988), "Interaction of Finite Amplitude Waves with Platforms or Docks", Accepted for publication, ASCE,

Journal of Waterways Port, Coastal and Ocean Engineering Division.

5. Lee, Jiin-Jen & Lai, C.P.," Wave Uplift on Platforms or Docks in Variable Depth," Proceeding of the 20th International Conference on Coastal Engineering, November 9-14, 1986, Taipei, Taiwan, pp 2023-2034.

6. Wang, H. (1967),"Estimating Wave Pressure on a Horizontal Pier," Technical Report R546, Naval Civil Engineering Laboratory, Port Hueneme, California.

CHAPTER 172

A NEW TYPE OF BREAKWATER UTILIZING AIR COMPRESSIBILITY

Masaaki IKENO*
Naokatsu SHIMODA**
and
Koichiro IWATA***

ABSTRACT

This paper is to investigate both theoretically and experimentally the wave energy dissipation and air-pressure variation of a new type of breakwater having a pressurized air-chamber and two buoyancy tanks. The theoretical models are developed in this paper and they are shown to be in good agreement with experimental values. The new type of breakwater proposed in this paper is pointed out to attenuate the transmitted wave much better than the rectangular-shaped and concave-shaped breakwaters without the air-chamber.

1. INTRODUCTION

Development of a new type of breakwater having a high sea water exchange and wave energy dissipation is strongly required from environmental and disaster prevention viewpoints.

This paper deals with a new type of fixed breakwater which has one pressurized air-chamber and two buoyancy tanks. The great merit of the breakwater lies in the function that the resonant period can be easily adjusted only by changing the air pressure in the air-chamber without changing the geometrical shape of breakwater itself.

So far, Kiyokawa and Ohyama (1986) and Iwata et al. (1987) have discussed the wave attenuation of the breakwater with a pressurized air-chamber. There still remain, however, problems to be solved from theories and experiments.

---

* M.Eng., Research Engineer, Central Research Institute of Electrical Power Industry, Abiko, Chiba, Japan
** B.Eng., Vice-Head, General Technical Research Laboratory, Gikenkogyo Co. Ltd.,Hachiohji, Tokyo, Japan
*** D.Eng., Professor, Dept. of Civil Eng., Nagoya Univ., Nagoya, Aichi 464, Japan.

First of all, two theoretical models based on the Dividing Region and Source Distribution Function Methods are newly developed, which take account of air compressibility in the air-chamber. Secondary, laboratory experiments are performed not only to examine the validity of the proposed theoretical models, but also to reveal the characteristics of wave energy dissipation and variation of air pressure in the air-chamber.

2. THEORETICAL MODELS

The Dividing Region Method and Source Distribution Function Method are employed to evaluate changes of wave and air pressure in the air-chamber. The wave treated here is a two-dimensional and irrotational small amplitude one. The fluid is assumed to be inviscid.

2-1. Dividing Region Method

The fluid domain is divided into five regions, as shown in Fig.1. The region I includes the incident and reflected waves and evanescent mode waves, and the transmitted wave exists in the region, V. The air pressure head R in the region III changes with wave motion. Extending the analytical treatment proposed by Iwata et al. (1986), velocity potentials in the five regions are expressed by the following:

$$\phi_1 = (A_o e^{ik(x-\ell_2)} + B_o e^{-ik(x-\ell_2)}) \frac{\cosh k(h+z)}{\cosh kh} + \sum_{n=1}^{\infty} B_n e^{-k_n(x-\ell_2)} \frac{\cos k_n(h+z)}{\cos k_n h} \quad \text{---} \ (1)$$

$$\phi_2 = C_o x + D_o + \sum_{s=1}^{\infty} \left(C_s \sinh\frac{s\pi x}{(1-q_1)} + H_s \cosh\frac{s\pi x}{(1-q_1)}\right)\cos\left(\frac{s\pi(h+z)}{(1-q_2)h}\right) \quad \text{------} \ (2)$$

$$\phi_3 = (E_o e^{ik'x} + F_o e^{-ik'x})\left(\frac{\cosh k'(h+z)}{\cosh k'h}\right) + \sum_{m=1}^{\infty} (E_m e^{k'_m x} + F_m e^{-k'_m x})\left(\frac{\cos k'_m(h+z)}{\cos k'_m h}\right) + \frac{iP_o}{\rho\sigma} \quad \text{------} \ (3)$$

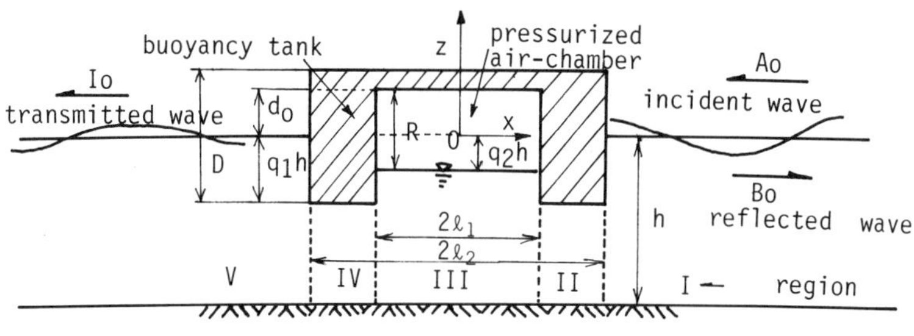

Fig.1 Shematic illustration of new type of breakawter with pressurized air-chamber and two buoyancy tanks and illustration of symboles

$$\phi_4 = G_o x + H_o + \sum_{s=1}^{\infty}\left(G_s \sinh\frac{s\pi x}{(1-q_1)h} + H_s\frac{s\pi x}{(1-q_1)h}\right)\cos\left(\frac{s\pi(h+z)}{(1-q_1)h}\right) \quad \text{---------- (4)}$$

$$\phi_5 = I_o e^{ik(x+\ell_2)}\left(\frac{\cosh k(h+z)}{\cosh kh}\right) + \sum_{n=1}^{\infty}\left(I_n e^{k_n(x+\ell_2)}\left(\frac{\cos k_n(h+z)}{\cos k_n h}\right)\right) \quad \text{---------- (5)}$$

Here, $\phi_1,\phi_2,\phi_3,\phi_4$ and ϕ_5 are, respectively, the velocity potentials in the regions, I, II, III, IV and V, Ao the complex amplitude of incident wave ($=igH_I/2\sigma$); g the gravitational acceleration, H_I the incident wave height; i imaginary unit number; σ the angular frequency ($=2\pi/T$); T the wave period, B_o the complex amplitude of reflected wave; I_o the complex amplitude of transmitted wave; $B_n, C_s, D_s, E_m, F_m, G_s, H_s$ and I_n (n=1, 2,3,--, s=m=0,1,2,--) complex unknown coefficients; h the still water depth; x the horizontal axis; z the vertical axis; k, k_n, k' and k'_m the wave numbers satisfying the relationship of $\sigma^2/g = k \tanh kh = -k_n \tan k_n h = k' \tanh k'(1-q_2)h = -k'_m \times \tan k'_m(1-q_2)h$; ρ the density of water; Po the air pressure amplitude in the air-chamber ; $2\ell_2$ the length of air-chamber; $2\ell_2$ the length of breakwater; $q_1 h$ the submerged depth of breakwater.

(b) Air pressure amplitude Po

The 5th term, $iPo/\rho\sigma$ in Eq.(3) is the pressure term added by compressibility of air in the air-chamber. In deducing Eq.(3), the air pressure, P_a in the air-chamber is expressed as

$$P_a = \overline{P}o + Poe^{i\sigma t}, \quad (\overline{P}o = \rho g q_2 h + P_{air}) \quad \text{------ (6)}$$

where, $\overline{P}o$ is the initial air pressure and P_{air} the atomospheric pressure. The linearized dynamic boundary condition on the free surface in the air-chamber is

$$\eta = -i\frac{\sigma}{g}\phi_3 - \frac{P_a}{\rho g} \quad \text{------------- (7)}$$

where, η is the free surface elevation from the still water level.

Substitution of Eqs.(3) and (6) into Eq.(7) and assumption of uniform rise and fall of water surface in the air-chamber yield

$$\eta = -q_2 h + \eta_o e^{i\sigma t},$$

$$\eta_o = \frac{1}{2l_1}\int_{-\ell_1}^{\ell_1}\left(-i\frac{\sigma}{g}(Ee^{ik'x}+Fe^{-ik'x})\left(\frac{\cosh k'(1-q_2)h}{\cosh k'h}\right)dx \quad \text{----------- (8)}$$

The volume of air, V in the air-chamber is given by

$$V = Vo - \eta_o e^{i\sigma t}, \qquad Vo = d_o + q_2 h \quad \text{-------- (9)}$$

where, Vo is the initial volume of air in the air-chamber and d_o the clearance of the upper plate from the still water level at the region, I and V (see Fig.1).

Assuming the adiabatical change of air in the air-chamber, the air pressure P_a corresponding to air volume of V is given by

$$P_a = \overline{P}o(Vo/V)^r, \quad (r = 1.4) \quad \text{-------- (10)}$$

Linearization after substitution of Eqs.(6) and (9) into Eq.(10) yields

$$Po = r\overline{P}o(\eta_o/do + q_2h)) \qquad \text{-------------} \; (11)$$

(c) Matching condition

The continuity of pressure, P and horizontal particle velocity, u is employed as the matching condition at the boundary between two regions (i.e., x = ℓ_1, ℓ_2, $-\ell_1$ and $-\ell_2$). The pressure, P_j, (j=1 - 5) and horizontal velocity of water particle, u_j, (j=1 - 5) are estimated with

$$\left.\begin{aligned} P_j(x,z;t) &= -\rho gz - i\rho\sigma\phi_j(x,z)e^{i\sigma t} \\ u_j(x,z;t) &= \frac{\partial}{\partial x}(\phi_j)e^{i\sigma t} \end{aligned}\right\} \qquad \text{-------} \; (12)$$

where, subsuffix j indicates the region, j.

$$\left.\begin{array}{lll}
\text{(i) } x = \ell_2 & P_1(\ell_2,z;t) = P_2(\ell_2,z;t) & ;\ (-h\leqq z\leqq -q_1h) \\
 & u_1(\ell_2,z;t) = 0 & ;\ (-q_1h<z\leqq 0\ ) \\
 & \qquad = u_2(\ell_2,z;t) & ;\ (-h\leqq z\leqq -q_1h) \\
\text{(ii) } x = -\ell_2 & P_5(-\ell_2,z;t)=P_4(-\ell_2,z;t) & ;\ (-h\leqq z\leqq -q_1h) \\
 & u_5(-\ell_2,z;t)=0 & ;\ (-q_1h<z\leqq 0) \\
 & \qquad =u_4(-\ell_2,z;t) & ;\ (-h\leqq z\leqq -q_1h) \\
\text{(iii) } x = \ell_1 & P_3(\ell_1,z;t) =P_2(\ell_1,z;t) & ;\ (-h\leqq z\leqq -q_1h) \\
 & u_3(\ell_1,z;t) =0 & ;\ (-q_1h<z\leqq -q_2h) \\
 & \qquad =u_4(\ell_1,z;t) & ;\ (-h\leqq z\leqq -q_1h) \\
\text{(iv) } x = -\ell_1 & P_3(-\ell_1,z;t)=P_2(-\ell_1,z;t) & ;\ (-h\leqq z\leqq -q_1h) \\
 & u_3(-\ell_1,z;t)=0 & ;\ (-q_1h<z\leqq -q_2h) \\
 & \qquad =u_4(-\ell_1,z;t) & ;\ (-h\leqq z\leqq -q_1h)
\end{array}\right\} \qquad \text{-------} \; (13)$$

Since Ao is given as $igH_I/2\sigma$ (known value), the unknowns of $B_n, C_s, D_s, E_m, F_m, G_s, H_s$ and I_s (n=s=m=0,1,2,3,---) are determined by solving Eq.(13) with help of Eqs.(1) through (12). The reflection coefficient, r_R and transmission coefficient, r_T are defined by $|Bo/Ao|$ and $|Io/Ao|$, respectively.

2-2. Sourse Distribution Function Method

The velocity potential in the fluid, $\phi(x,z)$ is expressed by the sum of the incident wave velocity potential, ϕ_I and the scattered wave one, ϕ_D and ϕ_D is given by a function of the Green function, G and the source distirbution function, f as follows (Nakamura et al.(1982));

$$\left.\begin{aligned} \phi(x,z) &= \phi_I(x,z) + \phi_D(x,z) \\ \phi_D &= \int_S f(X,Z)\cdot G(x,z:X,Z)ds(X,Z) \\ G &= \frac{(ik^2-ko^2)e^{ik|x-X|}}{k(hk^2-hko^2+ko)}\cosh k(h+Z)\cosh k(h+z) \\ &\quad -i\sum_{n=1}\frac{(k_n^2+ko^2)e^{-k_n|x-X|}}{k_n(hk_n^2+hko^2-ko^2)}\cos k_n(h+Z)\cos k_n(h+z) \end{aligned}\right\} \qquad \text{------} \; (14)$$

where, ko is the wave number in deep water; k and k_n the

wave number satisfying the relationship of σ^2/g =ko=k tanh kh =$-k_n$ tan k_nh; x the horizontal axis; z the vertical axis with origin at still water level; X and Z the source point; i the imaginary unit number; h the still water depth; s the boundary having the source points.

Once f is determined, Po, r_R and r_T are easily obtained. The detalied calculation technique is referred to reference 4). In this study, the source points were distributed at the interval of $\Delta s/L=0.02$ along the submerged breakwater surface under water and along the water surface in air-chamber, where Δs is the element length between adjacent two source points and L the wavelength at the depth of h.

3. LABORATORY EXPERIMENTS

3-1. Equipment and procedure

In the experiments, two wave tanks were used. Main parts of experiments were performed in an indoor wave tank in 0.9m height, 0.7m width and 25m length at Nagoya University. A flap-type wave generator controlled by Oil-pressure servo was installed at one end of the tank. The water is perfectly shut out from the area behind the wave board. Therefore, the input electrical signal is smoothly converted to the wave motion. At the other end of wave tank, a wave absorbing slope constructed with crush stone and stainless steel chips was placed so as to reduce the wave reflection to a minimum. A large wave tank in 1.6m height, 1m width and 50m length of General Technical Research Laboratory at Gikenkogyo Co.Ldt. was also used to examine the scale effect of model breakwater to the air pressure variation in the air-chamber.

Two model breakwaters made of acrylic plate, as shown in Fig.2, were employed to investigate the wave energy dissipation and air pressure variation in the air-chamber. The other types of breakwater such as concave-shaped and rectangular-shaped ones were also employed in order to discuss the difference in wave attenuation among the three types of breakwater. The submerged depth, q_1h and pressurized head, q_2h adopted in this study are listed in Table 1, which also includs q_1 in case of the concave-shaped and rectangular-shaped breakwaters.

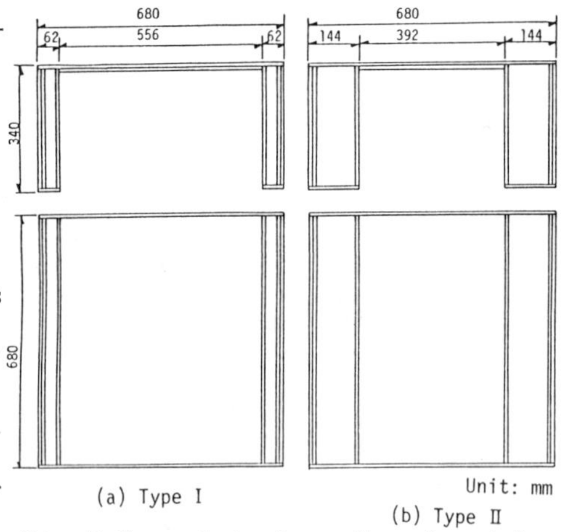

Fig.2 Geometrical scale of model breakwaters with pressurized air-chamber

Water surface profiles were measured with three capacitance-type wave gauges and they were all recorded on a magnetic tape. The transmitted wave was measured with the wave gauge, No.3 which was located at about one wave length distance from back side of

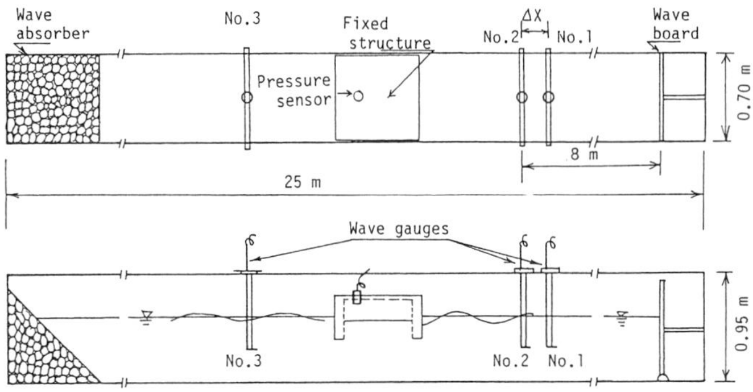

Fig.3 Shematic illustration of experimental set up

Type I pressurized-type	pressurized head q_2h (cm)	draft depth q_1h (cm)
ℓ_1/ℓ_2=0.82, 0.99	1.1	29.3
	2.1	25.3
	4.0	17.7
	5.2	13.3

Type II	pressurized head q_2h (cm)	draft depth q_1h (cm)
ℓ_1/ℓ_2=0.58	-1.7	17.7
	0.0	15.6
	1.8	13.3

Type I' concaved-shaped	pressurized head q_2h (cm)	draft depth q_1h (cm)
ℓ_1/ℓ_2=0.82	1.1	29.3
	2.1	25.3
	4.0	17.7
	5.2	13.3

Type II'	pressurized head q_2h (cm)	draft depth q_1h (cm)
ℓ_1/ℓ_2=0.58	-1.7	17.7
	0.0	15.6
	1.8	13.3

rectangular-shaped	draft depth q_1h (cm)
	29.5
	25.3
	17.7
	15.6
	13.3

Table 1 Experimental condition for three types of breakwaters (pressurized-type, concave-shaped and rectangular-shaped breakwaters)

the breakwater. The partial standing wave profiles were measured with two wave gauges, No.1 and No.2 in order to calculate the reflection coefficient, r_R. The distance, Δx between the two wave gauges was changed according to the wave period; that is, $\Delta x=(9/8)L$ for T=0.5s, $\Delta x=(7/8)L$ for T=0.6s - 0.8s and $\Delta x=(3/8)L$ for $T \geq 1.0$s, where L is the wave length at the depth of h. The air pressure in the air-chamber was measured with a pressure gauge (Kyowa PGM-0.5) attached to the top plate in the air-chamber.

All the experimental waves generated in this study were regular ones. The wave period employed was 0.5s, 0.6s, 0.65s, 0.7s, 0.75s, 0.8s, 0.9s, 1.0s, 1.2s, 1.4s, 1.6s, 2.0s, and 2.4s. The wave steepness, H/L was 0.01 for q_1=0.488 and 0.02 for q_1=0.222, 0.295 and 0.422 in order to avoid wave breaking. The still water depth was 60cm throughout the experiments.

Total number of experimental runs was about 300. The experimental set-up and conditions performed in the large wave tank of GTRL at Gikenkogyo Co.Ldt. are not described here, due to limitation of paper. As mentioned above, experiments in the large wave tank were carried out to check the model scale effect to the air pressure in the air-chamber. Based on Froude similarity law, dimensions of model breakwaters, water depth and wave conditions were determined so as to correspond to experiments in small wave tank at Nagoya university. The geometrical scale in the large wave tank was selected as 1.7 times larger than that in the small wave tank. The number of experimental runs in the large wave tank was about 50.

3-2 Data analysis

The wave reflection coefficient, r_R was calculated with two-wave-gauges method (Goda & Suzuki(1976)), using the measured wave profiles at wave gauges No.1 and No.2. The transmission coefficient, r_T was calculated using measured wave height of wave gauge, No.3. The reflection and transmission coefficients, r_R and r_T are defined by H_R/H_I and H_T/H_I, respectively, where H_R, H_T and H_I are the reflected, transmitted and incident wave heights, respectively.

3-3 Results and discussion

Using Dividing Region Method (hereafter referred to as DRM) and Source Distribution Function Method (hereafter referred to as SDFM), theoretical values of wave transmission and reflection coefficients and air pressure variation in the air-chamber were evaluated to compare with experimental values. One example is shown in Fig.4, in which experimental values are also plotted for comparison. Theoretical calculations based on the two methods are in good agreement with experimental values. Difference between theoretical values estimated with DRM and those with SDFM is recognized very small in the range of $2\ell_2/L \leq 0.6$, which is general feature in this study. Since DRM requires less computatioal time than

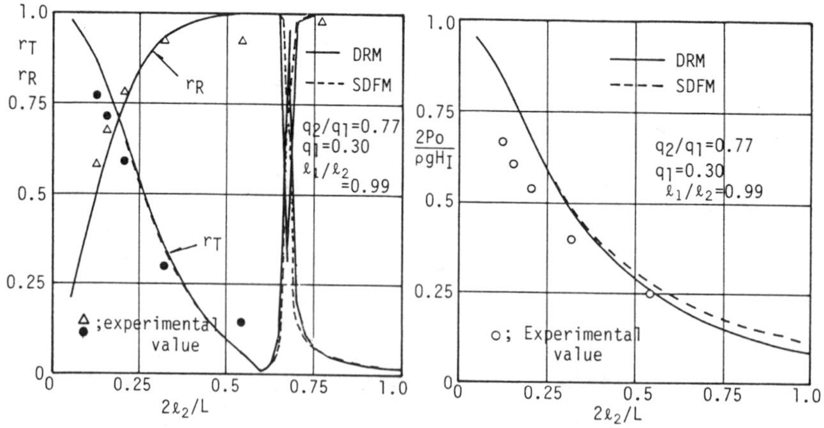

Fig.4 Comparison of theoretically estimated values between Dividing Region Method and Source Distribution Function Method.

SDFM. In addition, theoretical values of $Po/\rho gH_I$ estimated with DRM seem to be in better agreement with experimental values, as shown in Fig.4. Therefore, hereafter, theoretical values evaluated with DRM are employed to compare theory with experiments.

Fig.5 shows changes of the transmission coefficient,r_T and the reflection coefficient,r_R with $2\ell_2/L$. It is seen that the theoretical values agree well with experimental values. The transmission coeffcient,r_T increases with decreasing of $2\ell_2/L$ and q_1. r_T is expected to be less than 0.25 for $2\ell_2/L \geqq 0.3$ in case of q_1=0.422, q_2=0.035 and ℓ_1/ℓ_2= 0.82, as shown in Fig.5(c). On the other hand, the reflection coefficient,r_R decreases with a decrease of q_1 and $2\ell_2/L$. r_R is almost equal to 1 in the range of $2\ell_2/L \geqq 0.8$, which is independent of q_1, q_2, do and ℓ_2/ℓ_1, as shown in Fig.4, in this study.

The effect of initial air pressure,$\overline{P}o$ to r_R and r_T is not clearly recognized. The effect of pressurized head,q_2h to r_R and r_T is not clear too, although r_T seems to become smaller as q_2h becomes larger, as shown in Fig.4. This may be due to the reason that the experimental range of the initial air pressure,$\overline{P}o$ and the pressurized head,q_2h was too narrow to affect r_R and r_T. Concerning the effect of initial air pressure or pressurized head to r_R and r_T, we will discuss later from theoretical calculations with DRM.

Fig.6 shows two examples of variations of the dimensionless air pressure amplitude,$Po/\rho gH_I$ in the air chamber. It can be pointed out,from Fig.6,that theoretical values agree well with experimental values. $Po/\rho gH_I$ increases monotonously as $2\ell_2/L$ becomes smaller. This trend is different from those of r_R and r_T. $Po/\rho gH_I$ measured in the small wave tank is quite equal to those measured in the large wave tank. Then, the model scale effect to $Po/\rho gH$ is not recognized in the two laboratory experiments.

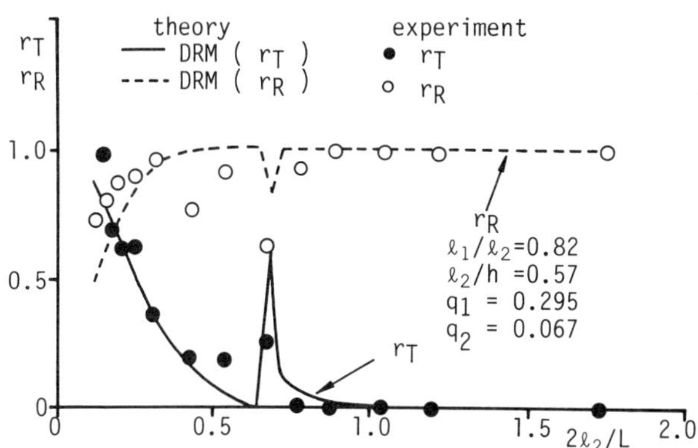

(a)

Fig.5 Comparison between theoretical values and experimental values of r_R and r_T.

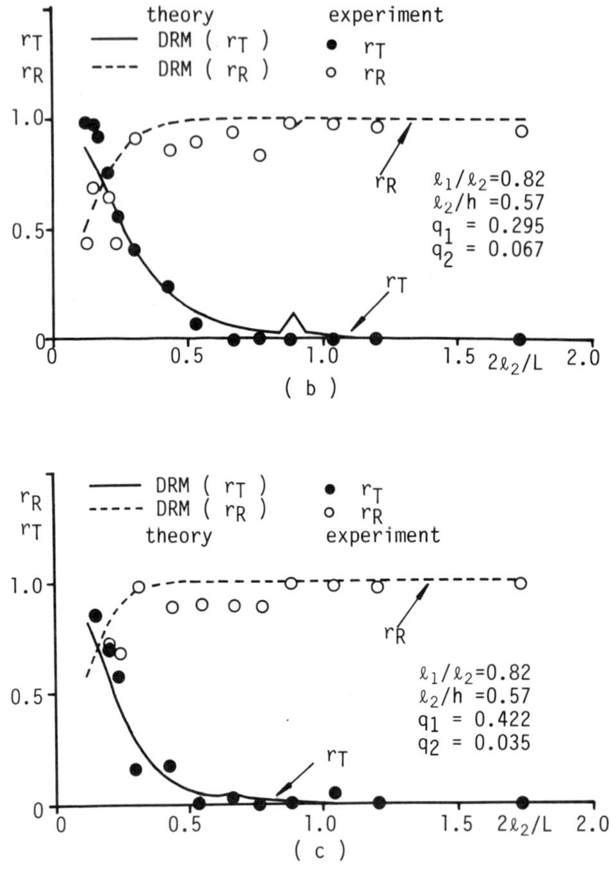

Fig.5 Comparison between theoretical values and experimental values of r_R and r_T.

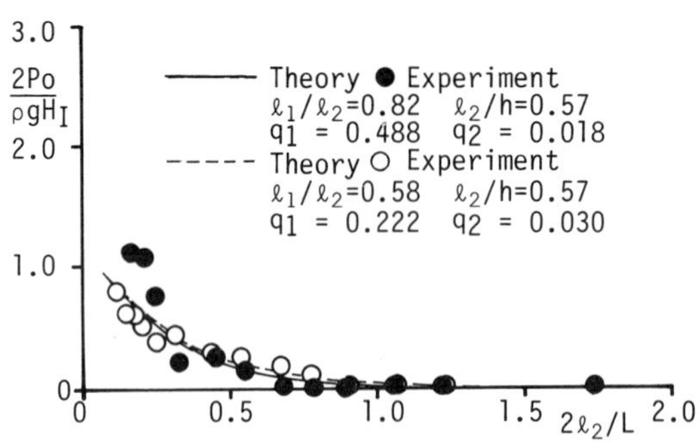

Fig.6 Comparison between theoretical values and experimental values of $Po/\rho gH_I$

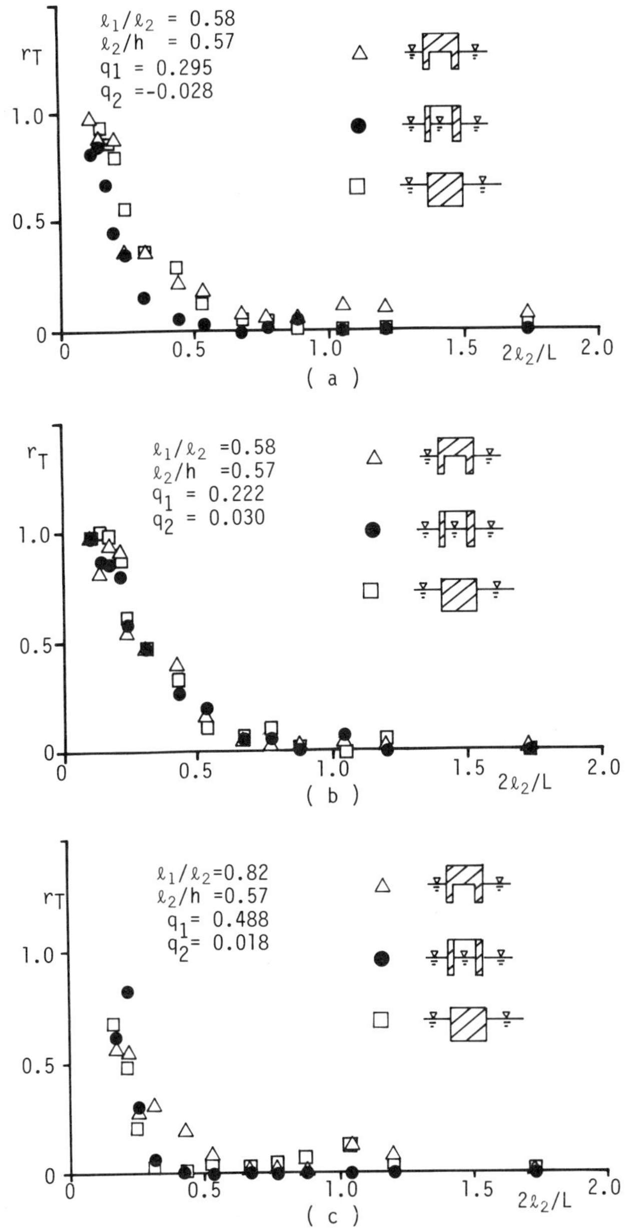

Fig.7 Comparison of transmission coefficient, r_T among pressurized-type, concave-shaped and rectangular-shaped breakwaters

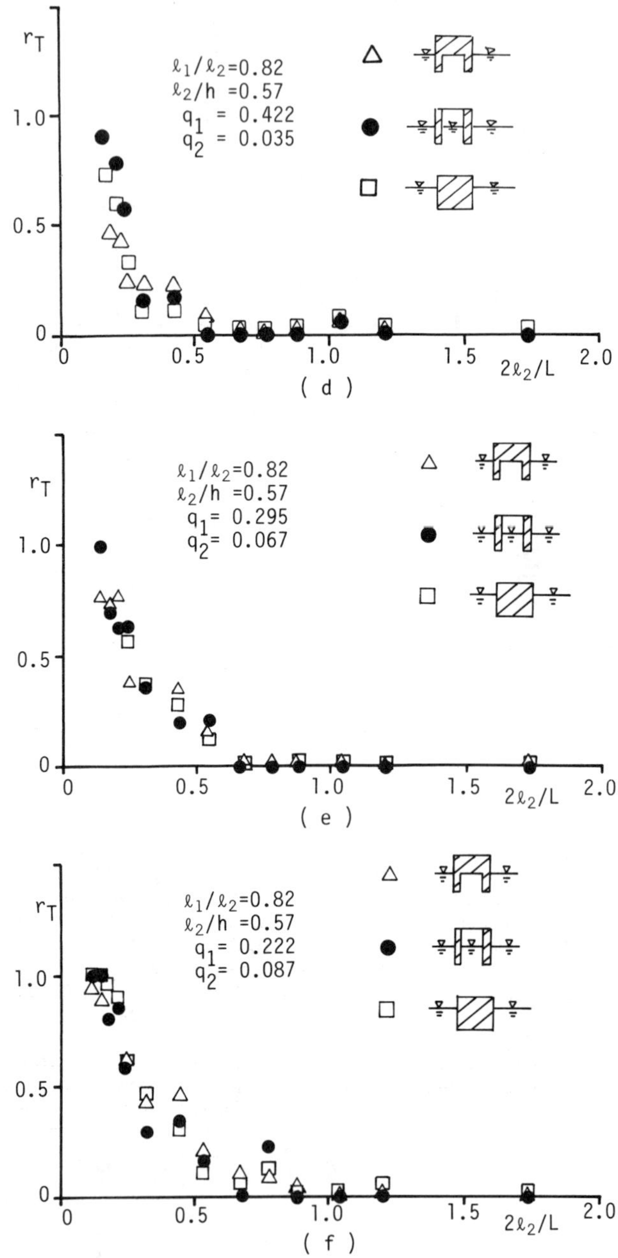

Fig.7 Comparison of transmission coefficient, r_T among pressurized-type, concave-shaped and rectangular-shaped breakwaters

Next, let us discuss the difference in the transmitted wave height due to the new type of breakwater (hereafter referred to as pressurized-type breakwater), and concave-shaped and rectangular-shaped breakwaters.

Figures 7(a)-(f) show experimental results of transmission coefficient, r_T due to the three different types of breakwaters. In each figure, the breakwater length, $2\ell_2$ and draft depth, q_1h are quite the same among the three different types of breakwaters. It is seen that the transmission coefficient, r_T increases largely with decreasing of $2\ell_2/L$ in the range of $2\ell_2/L \leqq 0.5$. As shown in Figs.(a)-(f), there are no large difference in r_T due to the pressurized-type, concave-shaped and rectangular-shaped breakwaters. However, r_T of the pressurized-type breakwater are comparatively smaller than those of the concave-shaped and rectangular-shaped breakwaters (see Figs.(a),(b), (e) and (f)). Therefore, it can be pointed out that the pressurized-type breakwater proposed in this study is better than the concave- and rectangular-shaped breakwaters in order to attenuate the transmitted wave.

From engineering viewpoint, r_T is frequently requested to be less than 0.5 for $q_1 \leqq 0.5$ and $2\ell_2/L \leqq 0.4$. As stated above, since DRM evaluates well the experimental values of r_T due to the pressurized-type breakwater, we will further discuss here, based on theoretical calculations estimated with DRM, the effect of pressurized head, q_2 to r_T and geometrical dimensions of the pressurized breakwater which attenuate r_T less than 0.5 for $q_1 \leqq 0.5$ and $2\ell_2/L \leqq 0.4$.

Figure 8 indicates that r_T becomes smaller with increasing of the dimensionless pressurized head, q_2/q_1 and r_T becomes minimal at $q_2/q_1=1$. Then, it can be said that increase of the pressurized head is effective to decay the transmission coefficient.

Figures 9(a)-(c) show some cases of optimum geometrical dimensions of the pressurized-type breakwater which can attenuate r_T less than 0.5 for $q_1 \leqq 0.5$ and $2\ell_2/L \leqq 0.5$. It is said that r_T is also affected by $(1-\ell_1/\ell_2)$ and that r_T decreases as $(1-\ell_1/\ell_2)$ becomes smaller. From Fig.9, combinations of $(1-\ell_1/\ell_2)$, q_1 and $2\ell_2/L$ which attenuate r_T less than 0.5 are found out in case of $q_1/q_2=1.0$ for three relative water depth, $h/L=0.078, 0.131$ and 0.207. Figure 9(c) shows that r_T becomes 0.3 when $q_1=0.5$, $2\ell_2/L=0.28$ and $(1-\ell_1/\ell_2)=0.01$ in case of $q_1/q_2=1.0$ and $h/L=0.207$. Thus, it is possible for this type of breakwater to attenuate r_T less than 0.3 in the case of $q_1=0.5$ and $2\ell_2/L \leqq 0.4$.

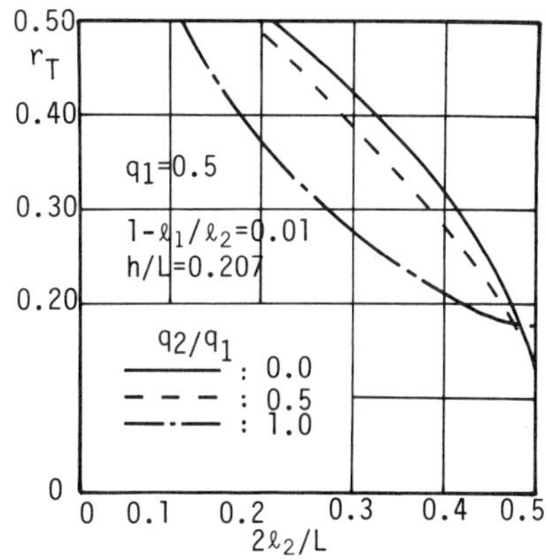

Fig.8 Effect of pressurized head, q_2 to r_T

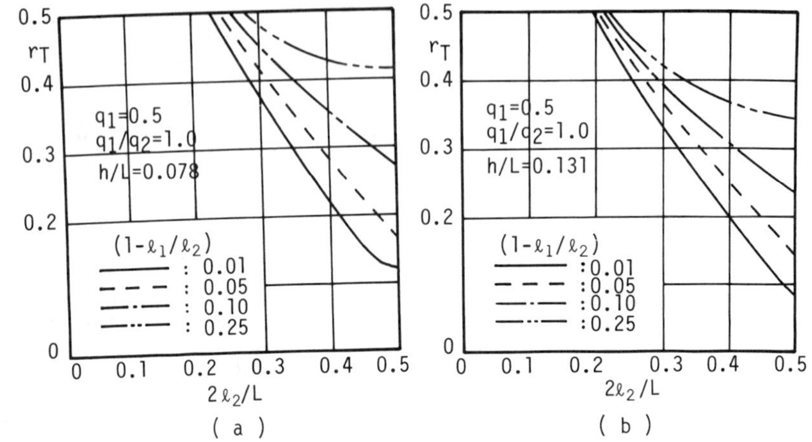

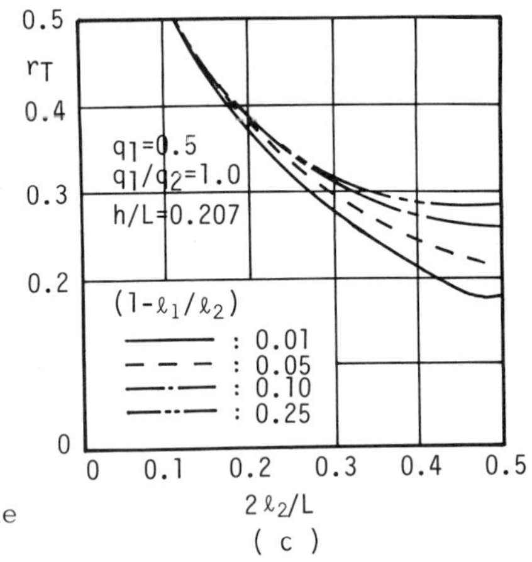

Fig.9 Combinations of $(1-\ell_1/\ell_2)$ and $2\ell_2/L$ which satisfy $r_T \leqq 0.5$ within $2\ell_2/L \leqq 0.5$ in case of $q_1=0.5$ and $q_1/q_2=1.0$ for $h/L=0.078, 0.131$ and 0.207

4. CONCLUSIONS

Wave reflection and transmission and air pressure in the air-chamber due to a new type of breakwater which has a pressurized air-chamber and two buoyancy tanks have been discussed from theory and laboratory experiments. The main results obtained in this study are summarized as follows:

(1) The theory using the Dividing Region Method is newly developed in order to evaluate the reflection and transmission coefficients and air pressure variation in the pressurized air-chamber. The proposed theory estimates well experimental values.

(2) Theoretical calculations show that the pressurized in the air-chamber plays an important role to attenuate the transmission coefficient, r_T. The transmission coefficient becomes smaller with increasing head in the air-chamber.

(3) The transmission coefficient of the pressurized-type breakwater is comparatively smaller than those of the concave-shaped and rectangular-shaped breakwaters.

(4) The effect of Po to the refelction and transmission coefficients becomes smaller as $2\ell_2/(do+q_1h)$ becomes larger.

(5) Since the wave transmission coefficient due to the

proposed new type of breakwater can be attenuated less than 0.35 for $2\ell_2/L \leq 0.3$ and $q_1 \leq 0.5$, there is high possibility that this pressurized-type breakwater will be put into practical use in the sea area where high water exchange and wave decay functions are simultaneously required.

REFERENCES

1) Goda,Y.and Y.Suzuki (1976): Estimation of incident and reflected waves in random wave experiments, Proc.15th ICCE, Hawaii,pp.828-845.
2) Iwata,K.,Oki,M.,Kitaura,S.,Okuoka,T.and M.Ikeno (1986): Motion of floating structure with a pressurized air-chamber, Proc.33rd Japanese Conf.on Coastal Eng.,pp.531-535 (in Japanese).
3) Iwata,K.,Ikeno,M.,Eguro,T.and K.Nakayama (1987): Wave deformation due to structure with a pressurized air-chamber, Abstracts of Ann. Conv., Chubu-Branch of JSCE,pp. 212-213 (in Japanese).
4) Iwata,K.,Ikeno,M.,Nakayama,K.,Ueda,T. and H.Ishii (1988): Numerical programing of motions of structure with a pressurrized air-chamber and two buoyancy tanks in floating, moored and fixed conditions, Technical Rept., Coastal and Ocean Eng. Labo, Dept. of Civil Eng., Nagoya Univ.,pp.II-1 -II-53 (in Japanese)
5) Kiyokawa,S. and T.Ohyama (1986): Wave reflection and transmission due to structure with air-pressurized curtain-wall breakwater, Proc.JSCE,No.375,pp.271-280 (in Japanese).
6) Nakamura,T.,Takahashi,K. and K.Nakayama (1982): Analytical method of wave deformation and forces around two-dimensional body using source distribution method, Proc.29th Japanese Conf. on COastal Eng.,pp.462-466 (in Japanese).

CHAPTER 173

WAVE-INDUCED SHOCK PRESSURES UNDER REAL SEA STATE CONDITIONS

by

Joachim Grüne[1]

ABSTRACT

This paper deals with a study on shock pressures, which occur on sloping seadykes and revetments due to breaking waves. Results from field measurements are presented with respect to peak pressure values as well as to characteristics of pressure-time histories.

1. INTRODUCTION

Investigations on shock pressures, especially at vertical walls, have been done since 140 years. Tests mostly have been carried out in small scale models, only a few measurements have been done in field or in laboratory full-scale (for vertical walls see e.g. Blackmore, Hewson (1984), for sloping seadykes see e.g. Stive (1984), Führböter (1986), Führböter, Sparboom (1988).

Furthermore in laboratory tests mostly regular waves were used or irregular waves were not really related to the topographic relations in front of excisting seadykes in nature.

The published data from the different investigations often show considerable differences especially with respect to maximum peak pressure values. It can be no doubt, that these differences mainly are a result of the different wave characterstic conditions. The wave characteristic conditions itself strongly influence the amount of air entrainment, therefore considerable scale effects can occur in small-scale models. Due to the more or less unsolved problems with these scale effects and due to the random characteristics of waves under real sea state conditions there still is a great need for detailed field data, to verify investigations in laboratory, even such in full scale.

2. FIELD MEASURING EQUIPMENT

The data presented in this paper, have been measured in field at two different locations at the coast of the German Bight during several storm surges (Grüne, 1982).

[1]Dipl.-Ing., senior researcher, deputy operation manager of the joint institution LARGE WAVE CHANNEL of the University Hannover and the Technical University Braunschweig, Germany

At the EIDERDAMM location two different sections were used: the original adjacent dam of the Eider-storm-surge-barrier, which has a bended slope of 1:4 in the lower and 1:6 in the upper part, and additionally a testsection with a uniform slope of 1:4 (Fig. 1). On both sections upwards from $NN + 3.00m$ (which is roughly $1.5m$ above Mean High Tide) the cover layers are made from asphalt-concrete on a sand core. On these asphalt-covers the wave-induced pressures were measured synchronously on both slopes with up to 5 pressure cells, which had a vertical distance of roughly $25cm$ from each other (on both slopes).

The section used at the east-frisian island WANGEROOGE, has also a cover layer made from asphalt-concrete on a sand core with a nearly constant slope of 1:4. Up to 25 sensors with a vertical distance from each other of only $9cm$ have been used synchronously to measure the wave induced pressures on the surface.

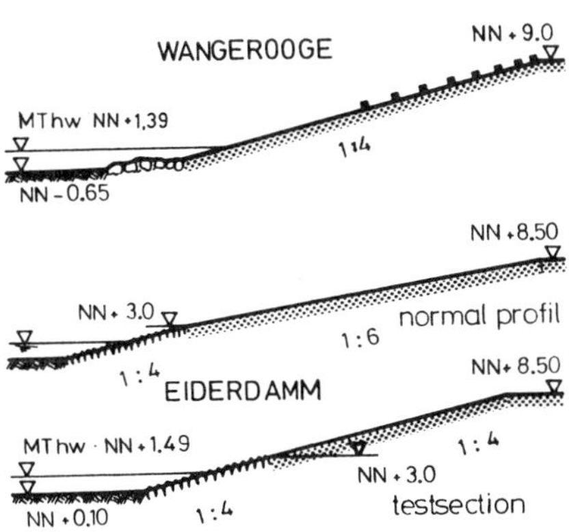

Fig. 1 Dyke profiles used for field measurements at both locations

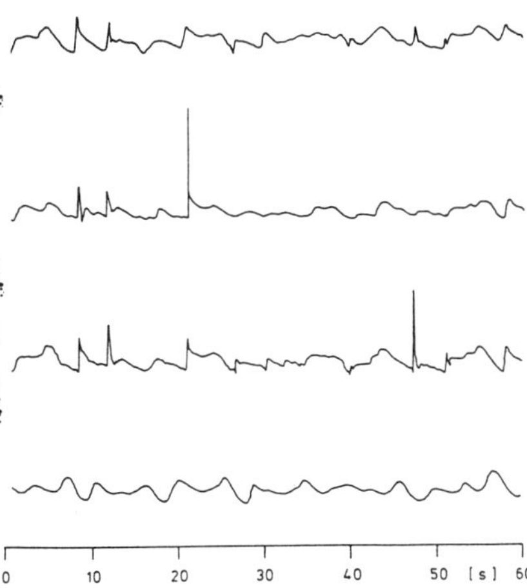

Fig. 2 Example of a synchronous record from some pressure cells on slope 1 : 4 (EIDERDAMM location)

Waves at both locations were recorded also with pressure transducers. These pressure data have been transfered to surface elevations by means of correction factors for the statistical wave parameters in time domain, found in model tests and verified by field measurements (Grüne, 1982).

3. ANALYSIS OF WAVE-INDUCED SHOCK PRESSURES

The definition of shock pressures from field records of wave-induced pressures, especially measured on slopes, is more complicated than for laboratory data with idealized boundary conditions. Due to air entrainment of breakers and the occurrence of a waterfilm from the backflow of wave run-up the shock pressures under real sea state conditions are damped more frequently (compared to those on vertical walls) and they are mixed with pressures from waves and wave run-up (Fig. 2).

In previous reports results with respect to the pressure peak values are primarily predominant, but as to be seen in Fig. 2, 3 and 4, it is often difficult to distinguish

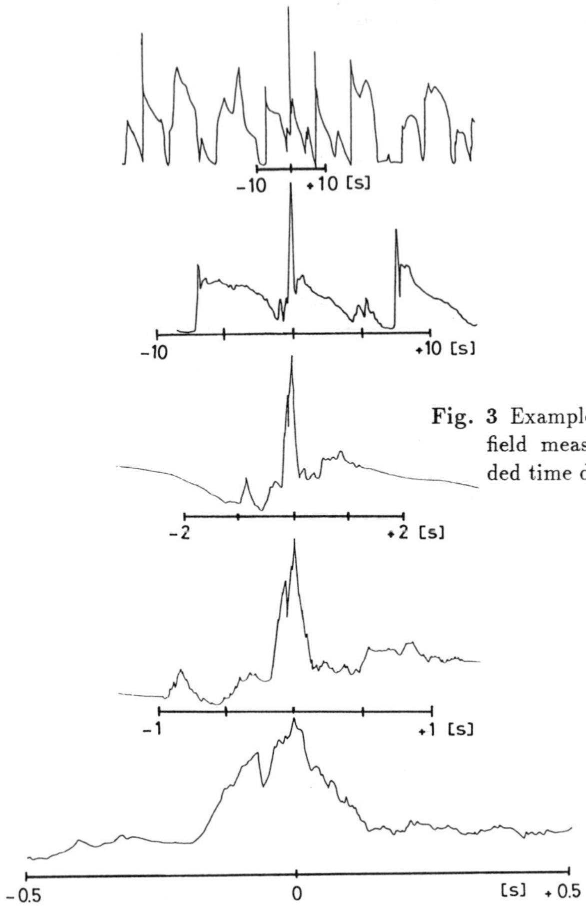

Fig. 3 Example of a record from field measurements with extended time disintegration

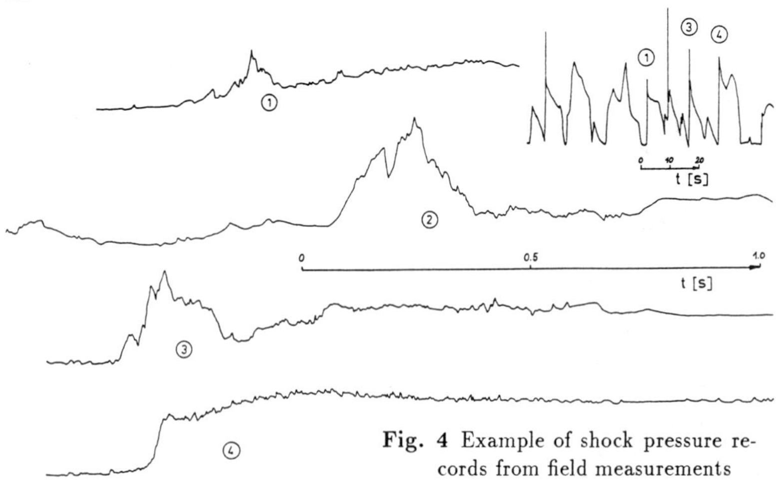

Fig. 4 Example of shock pressure records from field measurements

the real shock pressures from the steep wave fronts or the wave run-up fronts, using normal speed records. Furthermore a first review of some pressure-time histories from measured field data, which seem to fullfill classic shock pressure shapes on a normal speed record, pointed out a certain disagreement with common representations in former reports. This is illustrated in Fig. 3, where the time disintegration is extended in steps from top downwards up to 160 times. Fig. 4 demonstrates that even a series of four waves one by one produces four completely different pressure-time histories at same measuring point on the dyke surface.

Using normal speed records as shown in Fig. 2 for peak pressure analysis usually two main boundary conditions should be fullfilled to find out shock pressures:

1. the rising time up to the maximum peak pressure must be much shorter than the wave period (roughly less than 1/10 of the wave period).
2. the maximum peak value should exceed a certain multiple value of the wave height, that means a setting of a minimum value.

On the records as shown in Fig. 2, and on the top of Fig. 3 and 4 the first condition can be estimated only insufficiently. The second condition can result in neglecting real shock pressures with smaller peak values, but yet very short rising times, which might be important to the stability of seadykes. Furthermore the statistical values are affected by setting a minimum value.

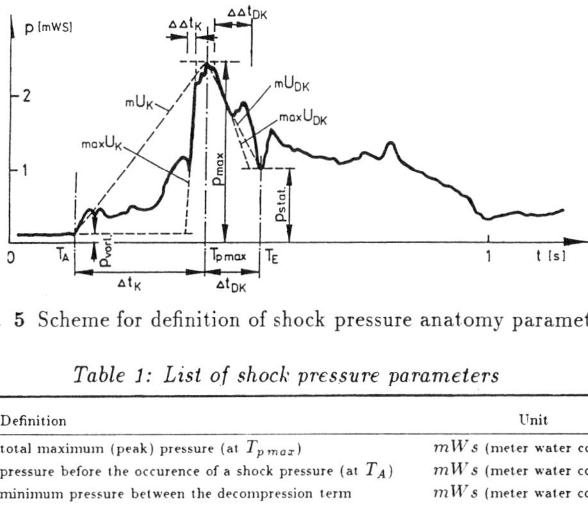

Fig. 5 Scheme for definition of shock pressure anatomy parameters

Table 1: List of shock pressure parameters

Symbol	Definition	Unit
p_{max}	total maximum (peak) pressure (at $T_{p\,max}$)	mWs (meter water column)
p_{vorl}	pressure before the occurence of a shock pressure (at T_A)	mWs (meter water column)
p_{stat}	minimum pressure between the decompression term and the quasi-static term (at T_E)	mWs (meter water column)
T_A	starting time of shock pressure	s (*second*)
$T_{p\,max}$	time at maximum pressure	s (*second*)
T_E	starting time of the quasi-static term	s (*second*)
Δt_k	total rising time of compression term, related to: $(p_{max} - p_{vorl})$	s (*second*)
$\Delta\Delta t_k$	minimum significant rising time of compression term, related to $(p_{max} - p_{vorl})$	s (*second*)
Δt_{ph}	time period to maximum pressure (at $T_{p\,max}$), related to the first $T_{p\,max}$ of all pressure cells	s (*second*)
mU_k	mean rising velocity, related to Δt_k and $(p_{max} - p_{vorl})$	mWs/s (meter water column per second)
$max\,U_k$	maximum significant rising velocity, related to $\Delta\Delta t_k$ and $(p_{max} - p_{vorl})$	mWs/s (meter water column per second)

More detailed information on shock pressures ("anatomy") can be obtained from high speed recorded pressure-time histories as shown in Fig. 3 and 4. To distinguish all these completely different pressure-time histories and to get a certain generalization, these complex shapes may be described suitable by parameters as shown in Fig. 5. The parameters in this example of a pressure-time history are seperated for compression time (index K) and decompression time (index DK). The anatomy parameters, used in this paper, are listed in Table 1. It must be remarked, that in chapter 4.1 the peak pressure values p_{max} from Table 1 are defined as p.

An example of time-pressure histories of one individual wave event, recorded at WANGEROOGE location, is shown in Fig. 6. The local distribution of the anatomy parameters, determined from these pressure-time histories as listed in Table 1, are plotted in Fig. 7 in dependence of the waveheight related distance $\Delta d/H_{1/3}$ from stillwaterlevel. It should be noted that the recorded pressure-time histories mostly have more chaotic shapes than the classic ones in Fig. 6. More aspects on analyzing high speed recorded pressure-time histories are reported by Grüne, (1988).

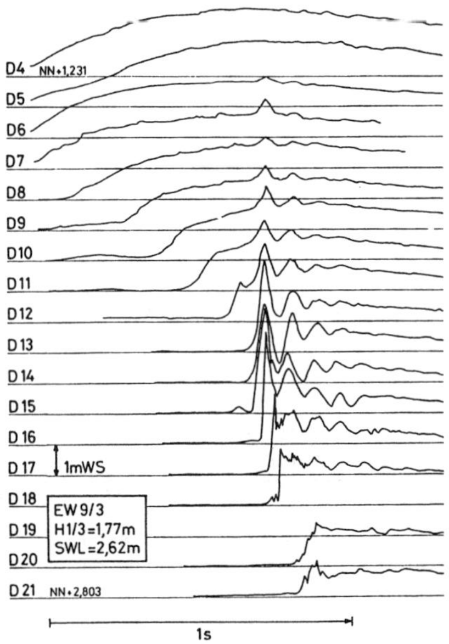

Fig. 6 Pressure-time histories for an individual breaking wave

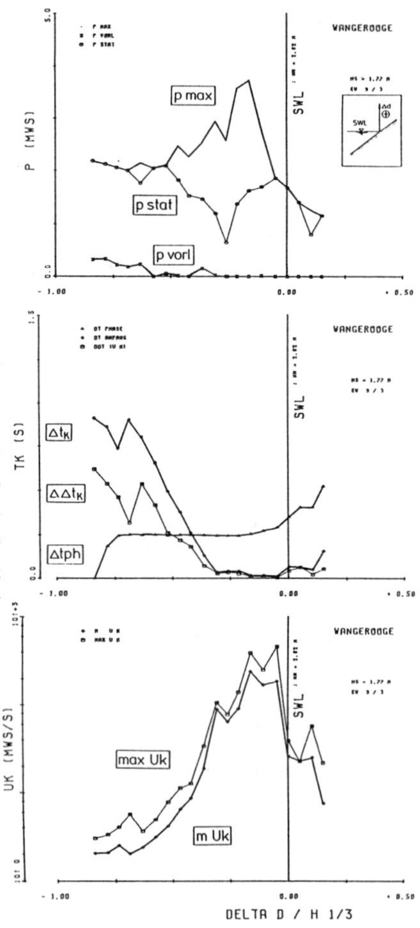

Fig. 7 Local distibution of anatomy parameters from the pressure-time histories in Fig. 6

4. RESULTS

The results, described in this paper, are analyzed from normal speed records (EIDERDAMM and WANGEROOGE location) with respect to peak pressure values as well as from high speed records (WANGEROOGE location) with respect to the anatomy of shock pressures.

4.1 Results of peak pressure values and their local distribution

Although the analysis of peak pressure values from normal speed records does not not give detailed information on shock pressure characteristics as mentioned before, this method is usefull to find the limited area of "worst case overall occurrence" of shock pressures on slopes, especially with respect to the local distribution.

An example of peak pressure data for one time interval, measured at Eiderdamm location on slopes 1 : 4, is given in Fig. 8. On the vertical axis, the statistical peak pressure values, which are determined directly from the peak pressure data, are related to significant wave height $H_{1/3}$. NP/NW gives the ratio of peak pressures to the number of individual waves (related frequency of occurence). The horizontal axis gives the local distribution on the slope as a vertical waveheight related distance from the mean stillwaterlevel $\Delta d/H_{1/3}$ of the time interval. The data of the different pressure cells are used to form an linear interpolated envelope curve.

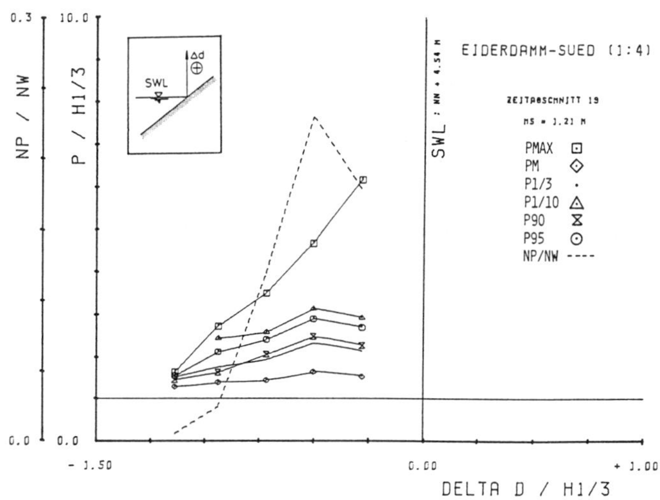

Fig. 8 Example of peak pressure data measured on slope 1 : 4 at EIDERDAMM location for one time interval

It must be remarked, that such summarized distributions of peak pressure values are influenced

first by the measuring time, which is restricted for field measurements due to changing of stillwaterlevel

second by the location of the pressure cells on the dyke referred to stillwaterlevel during storm surge events

third by the distance of pressure cells from each to another, which gives the local disintegration of the envelope curves.

Some further remarks should be made on statistical parameters, evaluated from calculated lognormal distribution. As shown in Fig. 9, the lognormal distribution of peak pressure values, measured with 23 pressure cells on the total dyke surface for one time interval, can be fitted by a least square method calculated lognormal distribution only moderate. It must be mentioned that this time interval has a length of 80 minutes with closely constant stillwaterlevel and significant wave height, which is rather rare for measurements during storm surge events.

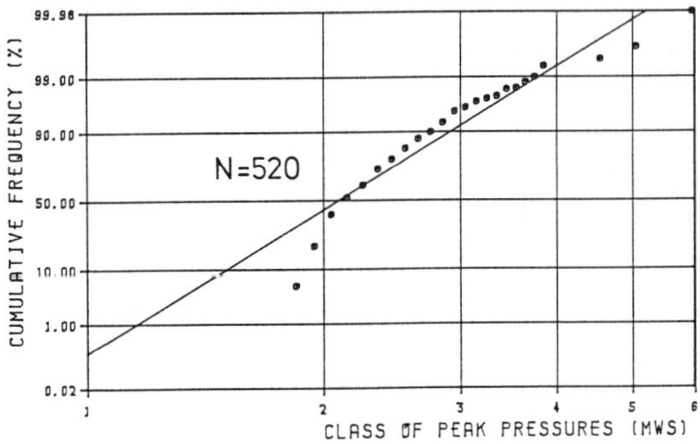

Fig. 9 Example of lognormal distribution of peak pressure values for one measured time interval

The local distributions of the statistical peak pressure parameters, evaluated from calculated lognormal distribution and related to mean peak pressure values, are plotted in the upper part of Fig. 10. It is obvious, that these parameters vary in dependance of the distance to stillwaterlevel. The local distributions of the waveheight related statistical parameters determined directly from the peak pressure values for this time interval are plotted in the lower part of Fig. 10. It is conspicious, that the measured maximum peak pressures form the statistical values extensively and that they agree quite well with the corresponding calculated values $p_{99.9}$. This comes out more clearly in Fig. 11, where for each measured time interval at Wangerooge location the calculated statistical values $p_{99.9}$ from all shock pressures are plotted versus the measured maximum ones. Due to these results and additionally due to the fact, that the time for measuring shock pressures during storm surge events in field with nearly constant stillwaterlevel and waveheight parameters is limited, the measured maximum peak pressure values have been favoured for further considerations instead of calculated statistical values.

To get the "worst case occurrence area" with a certain approach to accuracy, the envelope curves for p_{max} of the different measured time intervals are superimposed to a summarized envelope curve, which is shown in Fig. 12 as an example for one storm surge event. For the same storm surge event the summarized envelope curves of the different statistical parameters are plotted in Fig. 13. It can be seen clearly, that the occurrence area spread over a wide range on the slope, but the intensity varies considerable in relation to distance from stillwaterlevel.

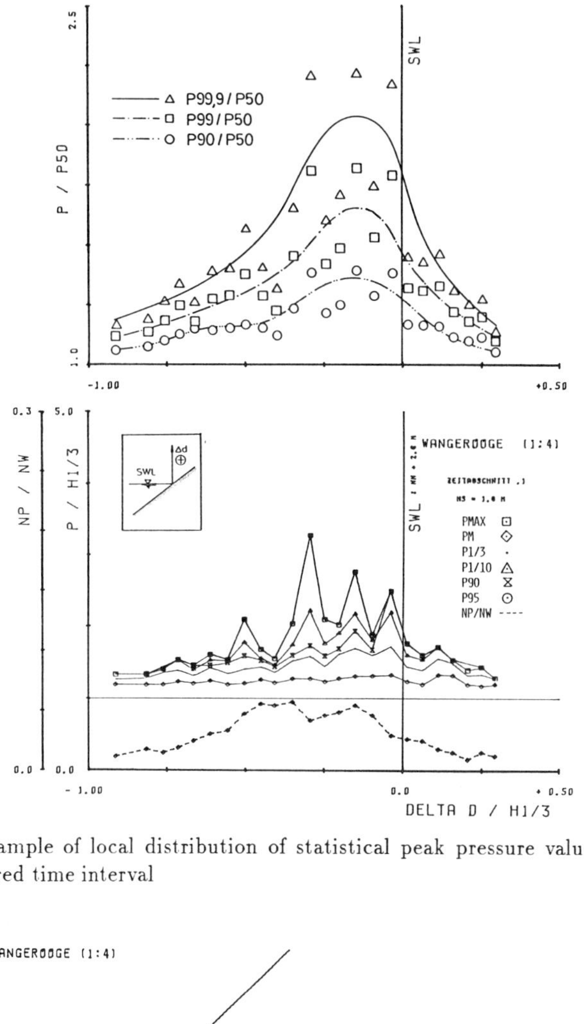

Fig. 10 Example of local distribution of statistical peak pressure values for one measured time interval

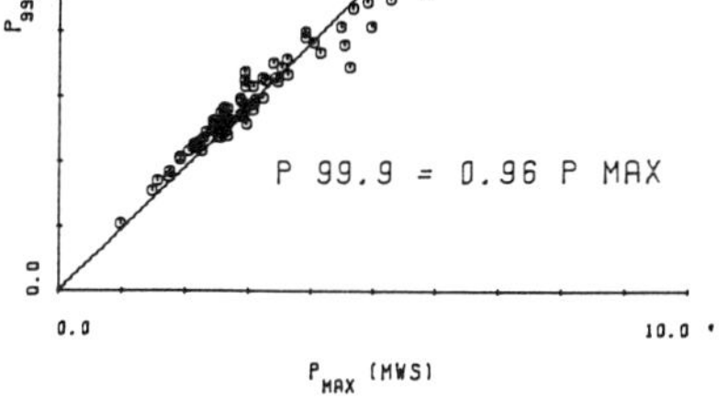

Fig. 11 Relation between statistical parameters $p_{99.9}$ and measured maximum pressure p_{max} for different time intervals, measured at WANGEROOGE location

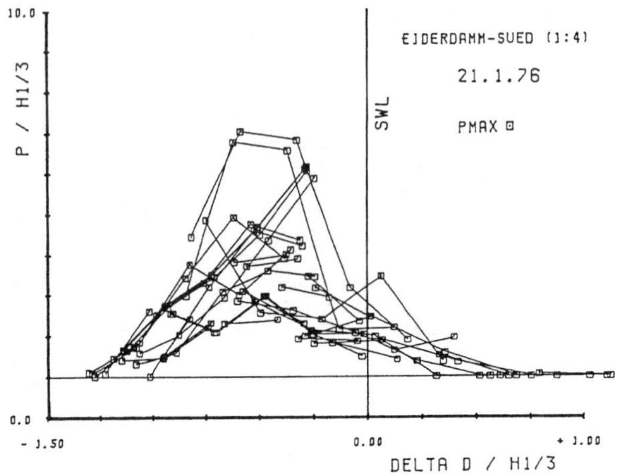

Fig. 12 Superimposed envelope curves for p_{max} of different time intervals, measured during one storm surge event at EIDERDAMM location 1 : 4

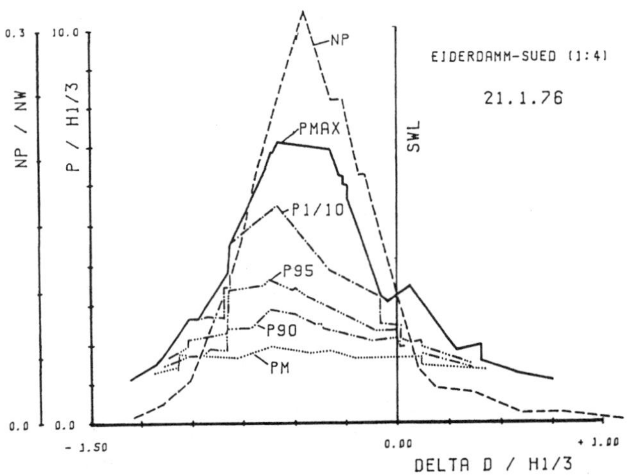

Fig. 13 Summerized envelope curves for different statistical parameters, measured during one storm surge event at EIDERDAMM location 1 : 4

The envelope curves for some parameters from all measured storm surge events at Eiderdamm location are compared in Fig. 14 for the both slopes 1:4 and 1:6. There is no significant difference of the total area of occurrence and of the mean pressure distribution between both slopes, but frequencies of occurrence and maximum peak pressures are lower on slope 1:6 due to the different wavebreaking and wave run-up characteristics.

A comparison of the peak pressure data between the two different locations with same slope 1:4 is given in Fig. 15. The total area of shock pressure occurrence is closely identical for both locations just as the mean pressure value distributions,

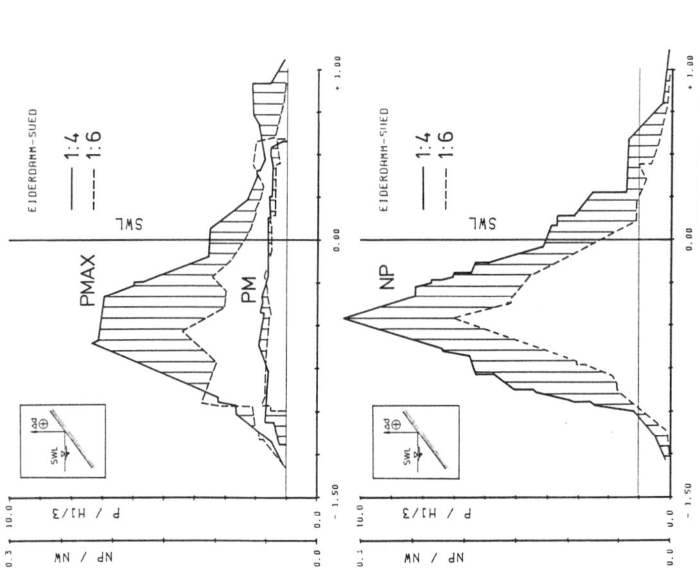

Fig. 14 Comparison of peak pressure parameters and their local distribution between 1 : 4 and 1 : 6 slope at EIDERDAMM location

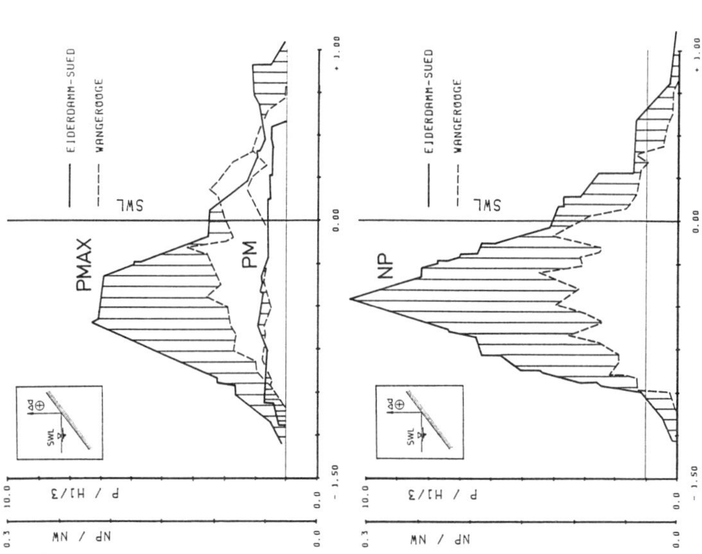

Fig. 15 Comparison of peak pressure parameters and their local distribution for slope 1 : 4 between the EIDERDAMM and WANGEROOGE location

but considerable differences occur for maximum peak pressure values and for the frequency. This is doubtless the influence of the different characteristic of local wave climate at both locations. Due to different waterdepth to waveheight relations in front of the slopes the plunger breaker type is predominant at Eiderdamm location, whereas at Wangerooge location the spilling breaker type is predominant. It should be noted, that similar effects from the wave characteristics were found also for wave run-up data, measured at both locations (Grüne, 1982).

4.2 Results of pressure-time history parameters (Anatomy parameters)

The results of parameters, evaluated from pressure-time histories as described in chapter 3, indicate very clearly, especially with respect to local distribution, that a certain gliding transition exists from the characteristics of steep wave front to those of shock pressure and then again to those of run-up front (see Fig. 7). This can be confirmed by the relations between some parameters independently from local distribution. Due to restricted place in this paper the following plots only should give an first overview of some summarized pressure-time history parameters from one storm surge event.

The relation between p_{vorl}, which represents the thickness of watersheet before shock pressure occurrence, and peak pressure values p_{max} is given in Fig. 16. The solid line gives the upper limit of peak pressures with respect to watersheet thickness and indicate the damping effect of a possible watersheet from the backflow of the preceeding wave. The mean rising time Δt_k is plotted versus the peak pressure value p_{max} in Fig. 17. The solid line represents the upper limit of the longest measured rising times, which decrease with increasing peak pressure values.

Fig. 18 gives the local distribution of the peak pressure values p_{max}. There is in fact no distinct tendency, but a broad spreading around stillwaterlevel. There is a clear tendency for the local distribution of the watersheet thickness p_{vorl} in Fig. 19; the upper limit of the thickness decrease at higher levels on the surface. All data give an impression of the numberless different shapes of pressure-time histories.

A generalized scheme for the local distribution of the pressure-time history parameters is shown in Fig. 20. This scheme is a first evaluation from a part of analyzed individual wave events. These generalized shapes were found more or less exactly for all investigated significant shock pressure events, which indicates a certain deterministic characteristic of wave-induced pressures on slopes. The shapes may be roughly classified into five different local domains (circled in Fig. 20) with different trends (listed in the table):

- the first one represents the approaching steeped wave front
- at the second domain the steeped wave front has its maximum height, this can be seen as the breaker point
- the third domain represents the area between the breaking wave front and the point, where the breaker tongue hits on the dyke surface. In this area the most chaotic pressure-time histories were found due to the enclosed air pockets with high turbulence
- the fourth one represents the area, where the breaker tongue hits on the dyke surface and where significant shock pressures with classic shapes occur
- the fifth domain represents the very steep front of wave run-up.

It must be remarked, that due to definition these parameter-shapes do not represent the conditions at same time, therefore some further investigations were made to check the phase lag of these shapes. The results from these investigations indicated, that all maximum peak pressures can occur closely within a total time lag of 50 milliseconds, referred to the maximum peak pressure $max\, p_{max}$.

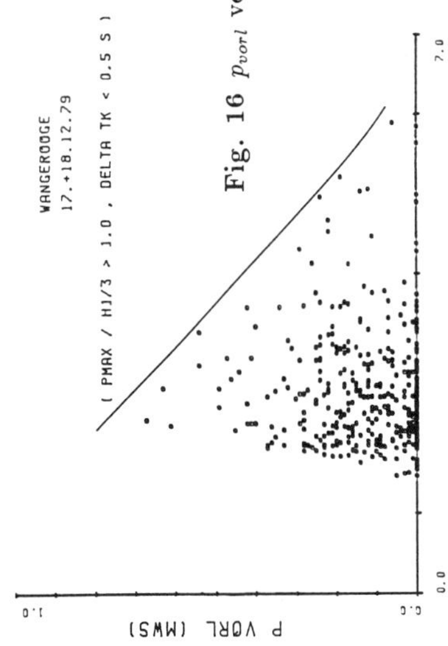

Fig. 16 p_{vorl} versus p_{max}

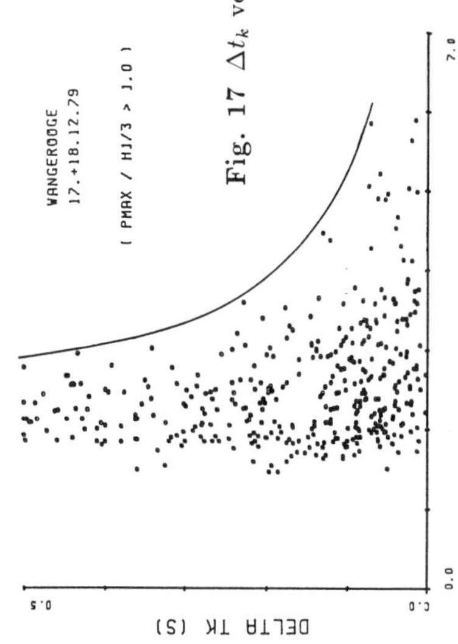

Fig. 17 Δt_k versus p_{max}

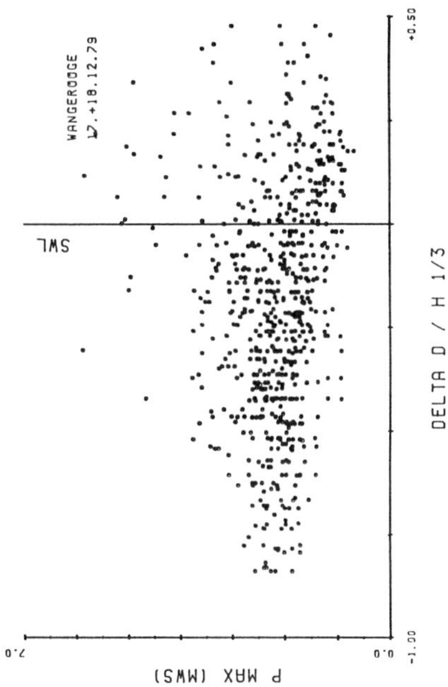

Fig. 18 Local distribution of p_{max}

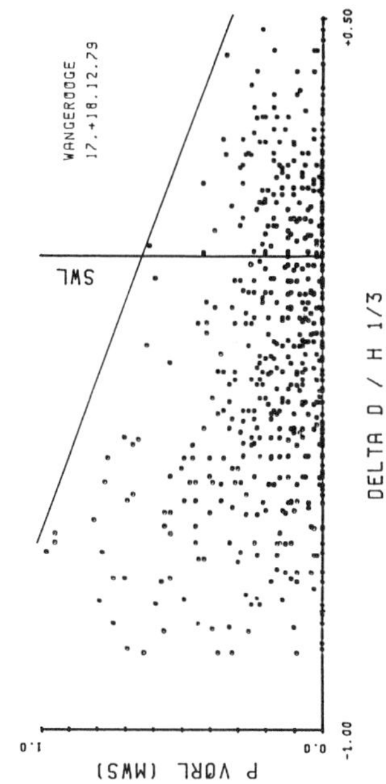

Fig. 19 Local distribution of p_{vorl}

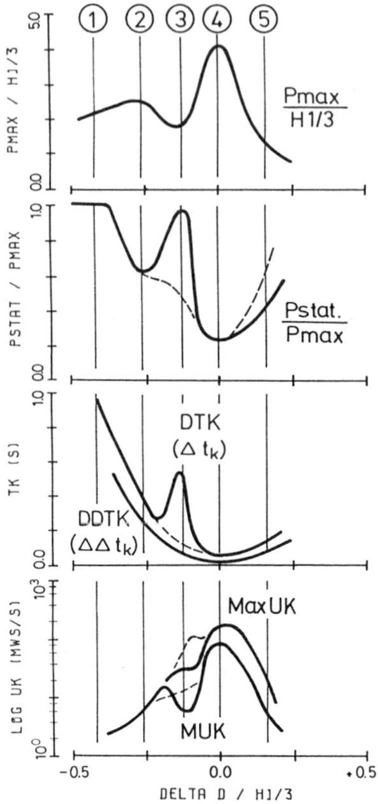

	①	②	③	④	⑤
$\frac{Pmax}{H1/3}$	< max	< max	≪max →min	max	≪max min
$\frac{Pstat.}{Pmax}$	1	< 1	→1 (< 1)	≪1 min	→1
DTK (Δt_k)	≫0	> 0	> 0 (≫0)	→0 min	> 0
DDTK ($\Delta\Delta t_k$)	≫0	> 0	> 0	→0 min	> 0
MaxUK MUK	min	> min ≪max	< max	max	< max
$\frac{MaxUK}{MUK}$	→1	> 1 (≫1)	> 1	→1	→1

Fig. 20 Generalized scheme for local distribution of anatomy parameters

This leads to the first assumption, that anatomy parameter shapes, such as shown before, can give a realistic approximation of the synchronous worst-case loadings from individual breaking waves. Such a worst-case loading model for the synchronous pressure distribution is shown in Fig. 21. The thick line is evaluated from a part of the data from WANGEROOGE location. The thin line is extrapolated to the maximum peak pressure values measured at EIDERDAMM location. The local distribution is related to the true length on the surface with zeropoint at the maximum peak pressure. It must be mentioned, that this zeropoint varies considerable with respect to distance from stillwaterlevel as shown in Fig. 19. For application this loading model may be used as an input to a numerical model, to investigate for example the bending moments of an cover layer or the penetration of shock pressures into sandcore. To get the worst case, for different applications the following parameters of the loading model in Fig. 21 should be varied systematically with certain ranges:

- the waveheight related peak pressure $max\, p_{max}$ up to 10

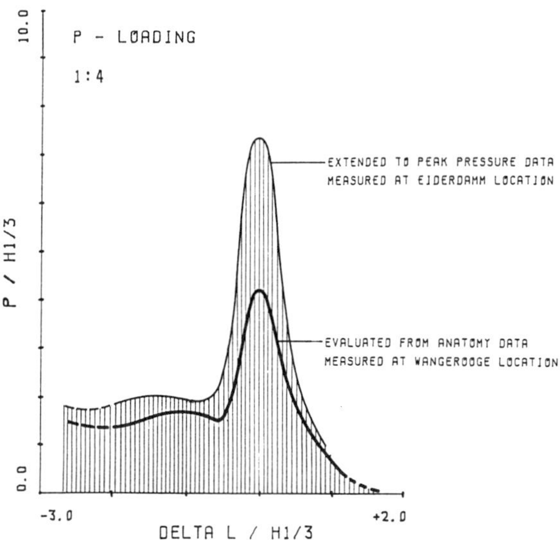

Fig. 21 Worst-case loading model (local peak pressure distribution)

- the duration time from $10ms$ up to $200ms$
- the waveheight related vertical distance from stillwaterlevel of the local acting point of $max\, p_{max}$ from -0.75 up to $+0.5$

5. CONCLUSIONS

From the results of the measurements in field the following may be stated:

- The definition of shock pressures for field data, especially measured on slopes, is more complicated than for laboratory data with idealized boundary conditions.
- Analysis of peak pressure values from normal speed records are usefull to get the local distribution of shock pressure occurrence on dyke surface.
- Shock pressures occur on slope 1:4 and 1:6 roughly within the same area on the slope, but have different magnitudes of pressure and frequency.
- The breaker type has a substantial influence on the magnitude and frequency of shock pressures, which was also found for wave run-up data. Waves with predominant spilling breaker type generate fewer and smaller shock pressures compared to waves with predominant plunging breaker type.
- The results may be described statistically, most sufficiently with a lognormal distribution, nevertheless the related ratio of the lognormal distributed pressure parameters, for example $p_{99.9}/p_{50}$, varies in dependence of the distance

from stillwaterlevel. The measured maximum peak pressure values of limited time intervals are well represented by the parameter $p_{99.9}$ from lognormal distribution.

- It has been demonstrated, that for detailed statements on the occurrence of shock pressures and the related loads on dykes an analysis of high speed pressure-time histories measured in field and in full-scale laboratory under real sea state conditions is necessary. On slopes shock pressures are damped more frequently compared to vertical walls and mixed with the pressures from waves and wave run-up, especially under real sea state conditions. Therefore the pressure-time histories mostly have very complex shapes. These complex shapes can be described suitable by parameters. If the pressure-transducers cover the area on the surface with a certain disintegration, the local distributions of these parameters for individual waves can be used for a loading model.

The research program on shock pressures will be continued with respect on combining the field data with different small-scale and full-scale laboratory data, especially those, measured in the "LARGE WAVE CHANNEL OF University Hannover and Technical University Braunschweig" (Grüne, Führböter 1975, Führböter 1986, Führböter, Sparboom 1988).

ACKNOWLEDGEMENTS

The research work partly was done by the SONDERFORSCHUNGSBEREICH 79/C4 - 205/A1 (supervision Prof. Dr.-Ing. A. Führböter) and has been supported by the GERMAN RESEARCH FOUNDTION (DFG). For their extensive and accurate help to analyze the field data the author is indepted expressly to Mr. S. Hecker and Mr. R. Kaiser.

REFERENCES

Blackmore, Hewson (1984). Experiments on full-scale wave impact pressures. Coastal Engineering, 8, pp. 331-346.

Führböter (1986). Model and prototype tests for wave impact and run-up on a uniform 1:4 slope. Coastal Engineering, 10, pp. 49-84.

Führböter, Sparboom (1988). Full-scale wave attack on uni- formly sloping sea dykes. Proc. of the 21st Intern. Conf. on Coast. Eng., Malaga.

Grüne, Führböter (1975). Large wave channel for "full-scale modeling" of wave dynamics in surf zones. Proc. of Symp. on Modeling Techniques, San Francisco, pp. 82-100.

Grüne (1982). Wave run-up caused by natural storm surge waves. Proc. of the 18th Intern. Conf. on Coast. Eng., Kapstadt, pp. 785-803.

Grüne (1988). Anatomy of shock pressures (surface and sand core) induced by real sea state breaking waves. Proc. of the Intern. Symp. on Modeling Soil-water-structure interactions (SOWAS 88), Delft, pp. 261-270.

Stive (1984). Wave impact on uniformly steep slopes at ap- proximately prototype scale. Proc. of Symp. on Scale Effects in Modeling Hydraulic Structures, pp. 7.11-1/7.11-12.

CHAPTER 174

Measurements of Forces on Dolos Armor Units at Prototype Scale

Gary L. Howell

1 Introduction

In-situ measurements of the structural bending moments and torque about the shank-fluke interface of the dolos armor unit have been made for 42-ton (36-metric tonne) dolosse at Crescent City, California jetty. The measurements include the static loads on the dolosse as well as wave induced forces. The data were obtained from internal strain gages cast into the dolos during construction along with a special data acquisition system. The measurement system was also capable of capturing impact forces caused by dolos rocking or movement. Measurements were made during the winter storm seasons from January 1987 through May 1988.

Coincident with the structural measurements, wave height and period were measured at several water depths approaching the breakwater, including a site directly in front of the dolos test section.

The Crescent City jetty is a shallow water breakwater with depth limited waves in about 10 meters of water depth. The structural measurements were made from 14 dolos units arranged in a rectangular section on the top layer of the trunk portion of the jetty. Four of these dolosse are also instrumented with an accelerometer platform to measure motion with six degrees of freedom. In addition, there are three instrumented dolosse on the bottom layer of the breakwater. These dolosse measure the static stress due to the units placed on top of them, as well as pulsating forces.

The structural and wave measurements, reported here, are supplemented with hydrostatic pore pressure measurements in the core material of the breakwater, and by aerial photogrammetric motion analysis (Kendall, 1988), land based surveys, boudary condition surveys, hydrographic surveys, and side scan sonar surveys.

2 Instrumented Dolos Test Section

Unlike the physical model of instrumented dolosse, it is not practical to relocate instrumented dolosse within the breakwater. Selection of the test section was made after a careful evaluation of different proposed stategies. The Crescent City jetty dolosse section consists of a 100 meter trunk section wrapping in to a 90 degree

*Research Engineer, Coastal Engineering Research Center, Waterways Experiment Station, U.S. Army Corps of Engineers, Vicksburg, Ms.

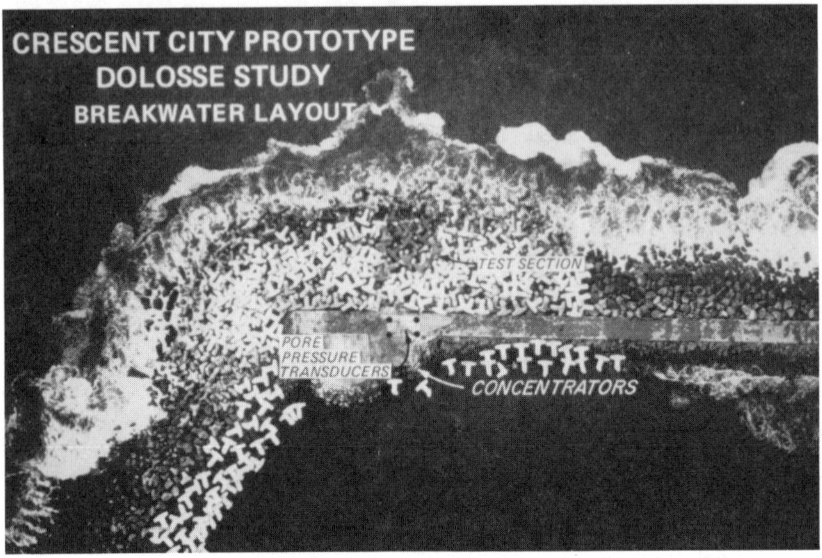

Figure 1: Crescent City Jetty Test Layout

turn to form a partial head section. Among the options considered were distributing the twenty instrumented dolosse throughout the structure, placing dolosse in three groups along the trunk and head section, purposely locating some dolosse in unstable positions, placing some dolosse on lower layers, and grouping all dolosse at a single location. Participants at the Workshop on Measurement and Analysis of Structural Response in Concrete Armor Units (Howell, 1985) discussed the trade-offs among the various options and arrived at a recommended consensus. Sixteen dolosse were placed in an approximate four by four matrix in the top layer, near the mean water line. The accelerometer equipped dolosse were in the center. The four remaining dolosse were placed in the bottom layer of dolosse under the upper instrumented units. The matrix was centered in the trunk section to permit all dolosse to receive, as much as possible, the same incident waves. The primary reason for selection of this test section was the concern that prototype measurements be useful to validate the physical model instrumentation and scaling laws. By placing the dolosse away from the head section and limiting the spatial variability of incident waves, statistical analysis of the prototype data will be comparable to the physical model. Figure 1 is an aerial photograph of the completed dolosse section and the instrumented test section. The darker colored dolosse are the instrumented units.

Figure 2: Instrumented Dolos and Chain Assembly

3 Measured Parameters

Implementation of the experiment required development of specialized instrumentation, data acquisition systems, and environmental protection (Howell, 1986). All dolos instrumentation is permanently sealed within the concrete armor unit. An internal microcomputer system for each dolos digitizes the strain at a 500 Hz sample rate, and accelerometer data at a 50 Hz sample rate. This rate is sufficient to capture all high energy modes resolvable with the strain gages located at the shank-fluke interface. Data from the dolos were retrieved by a cable connected to the underside of the unit. The cable was protected by a modified chain assembly as shown in Figure 2.

In order to gain maximum benefit from a minimum number of data channels, structural measurements were made from strain gages mounted on internal steel reinforcing bars, approximately 0.5 meters long. A rosette of these bars is located at the four faces of the section through the shank-fluke interface (Figure 3). The rosettes are 10 cm. below the surface of the concrete. Only the vertical shank-fluke interface was instrumented. Algebraic combinations of the strains allow estimates of the two bending moments, the axial thrust, and the torque about the shank-fluke interface. The data acquisition system selects any three of these estimates to be acquired simultaneously. Digital commands from the shore based computer allow

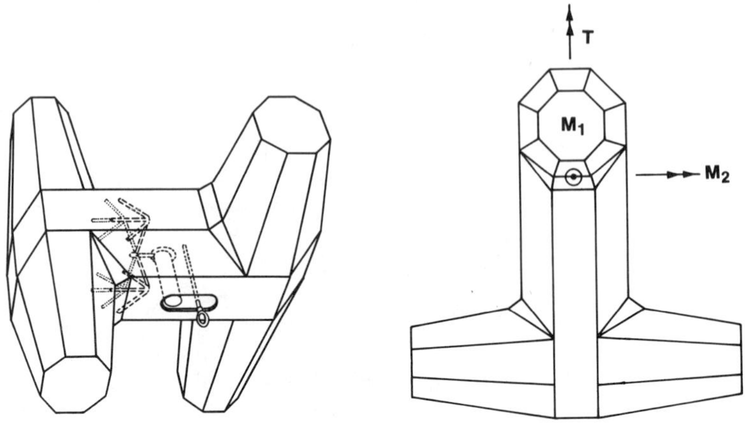

DOLOS INSTRUMENTATION **DOLOS MOMENTS AND TORQUE DEFINITION**

Figure 3: Internal Strain Gages and Moment and Torque Definitions

the selection of the signals as well as amplifier gain.

The steel bars were sized such that the strain in the bar could be assumed to approximate the strain in the surrounding concrete. The following relationships were then used to compute the moments and torques defined in Figure 3.

Labeling the rosettes clockwise, viewed from the horizontal fluke we have

A, C ... upper and lower vertical rosettes,
B, D ... right and left horizontal rosettes,
m ... the rosette bar parallel to the dolos axis,
t_1, t_2 ... the rosette bars at 45 deg angles,
ϵ ... measured strain in the bar

then,

$$\epsilon_{AC} = \epsilon_{Am} - \epsilon_{Cm}$$
$$\epsilon_{BD} = \epsilon_{Bm} - \epsilon_{Dm}$$
$$\gamma_{AC} = \epsilon_{At_1} + \epsilon_{At_2} + \epsilon_{Ct_1} + \epsilon_{Ct_2}$$

and,

$$M_v = K_{vm}\epsilon_{AC} \ \ldots \textit{vertical moment}$$
$$M_h = K_{hm}\epsilon_{BD} \ \ldots \textit{horizontal moment}$$
$$T = K_{tor}\gamma_{AC} \ \ldots \textit{torque}$$

The K factors were obtained from Finite Element Method computations for a range of typical boundary conditions.

4 Time Domain Structural Response

During the experiment period the structure was exposed to numerous storms. The most severe storms occurred from Nov. 30 to Dec. 1, 1987. The characteristics of the observed structural response resemble those hypothesized by Burcharth (1984). There is a static response which varies with the tidal period, and a pulsating response which has periods similar to the waves. Figure 4 shows a typical raw data trace from all three channels, M_v , M_h ,and T , from dolos M which is located in the center of the test section. The significant wave height in front of the structure was 3.8 meters with the spectral peak at 13 seconds.

Data from all dolosse are qualitatively similar, with the amplitude of the pulsating response proportional to the incident wave heights. Figure 5 shows an expanded trace of the M_v channel from Dolos M record along with a wave trace during the same time interval. A similar trace from Dolos H which is lower in the structure is shown in Figure 6. It would be desirable to show these traces with the waves time and celerity adjusted to coincide with the resulting moment trace, however software to permit this is not complete at this time. Therefore, the wave trace should be interpreted as only representative of the waves impacting the structure at the time of the moment trace.

The pulsating response shows a sharp peak followed by a drag-like relaxation. This response is qualitatively similar to results from physical scale model tests by Scott et al. (1986) and numerical wave force computations by McDougal et al. (1988). Dolos M and H show a double peak of the pulsating response. This feature is observed in many of the prototype dolos responses, but does not appear in the results of Scott or McDougal.

Burcharth and others predicted an impact response when dolos move and collide with each other. Although there has been dolos movement during the experiment (Kendall, 1988), no structural impact has yet been observed in the data. What has been observed are much slower changes in the mean stress such as can be seen in the M_h channel of Figure 4 and to a lesser extent in the other channels. The observations of Kendall have shown that dolos movement at Crescent City has tended to be sliding displacement upslope as would be expected from a shallow sloped (1:5) structure.

5 Static Response

Figure 7 shows mean values of T for Dolos P plotted with measured tide during the same period. Mean fluctuations of many of the dolos have been computed and show varying correlation with the tide depending on the boundary conditions or orientation of the unit in the matrix. A similar plot from Dolos H (Figure 8) demonstrates that static response is relative to the boundary condition and reference frame of the dolos. Since this record is also during a storm, the sharp bumps in the plot show the changes in the mean response mentioned previously. A somewhat surprising result is that dolosse well above the mean water level also show a static response related to tide. For example, Dolos P is located in the upper part of the test section, above the mean water level.

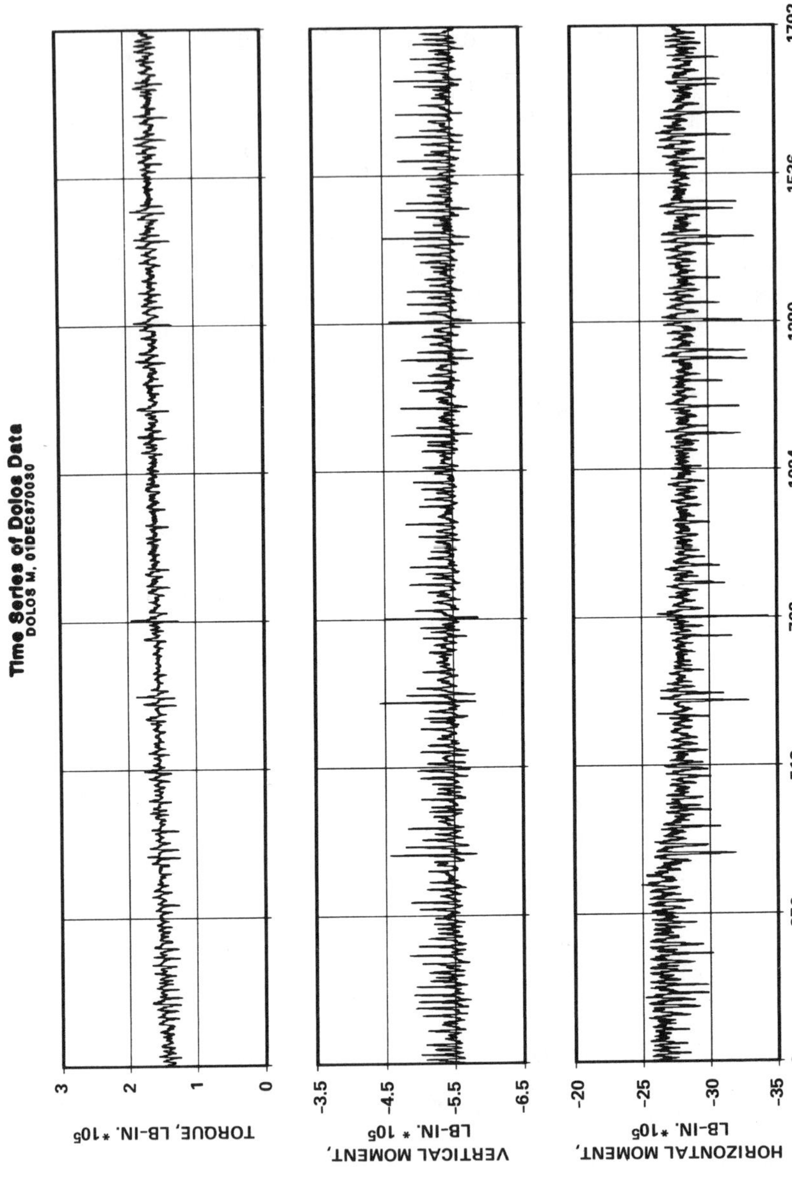

Figure 4: Raw Response Data from Dolos M

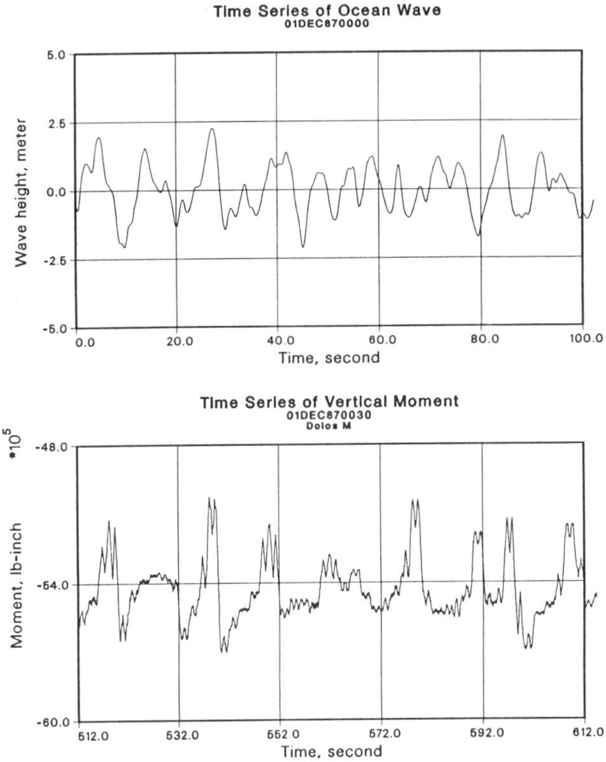

Figure 5: Dolos M Moment Response with Waves

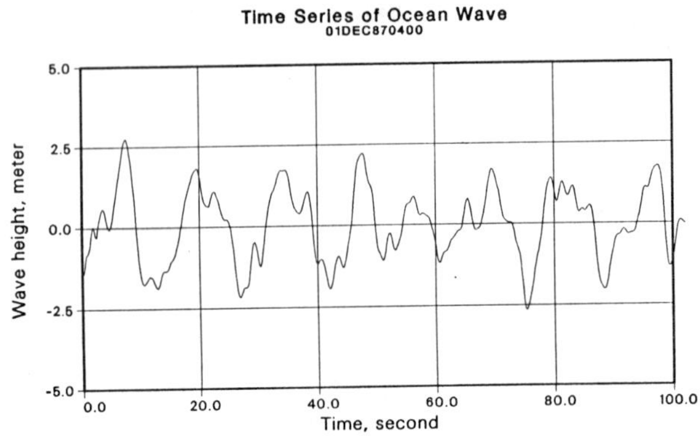

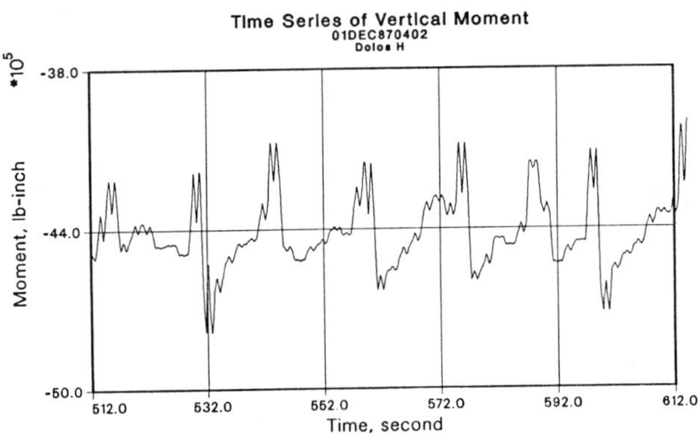

Figure 6: Dolos H Moment Response with Waves

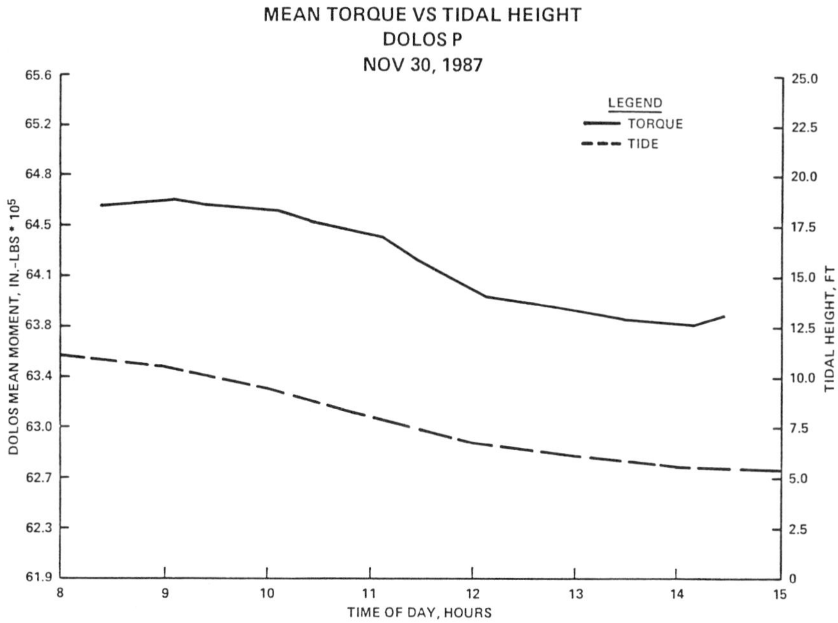

Figure 7: Dolos P Mean Response with Tide

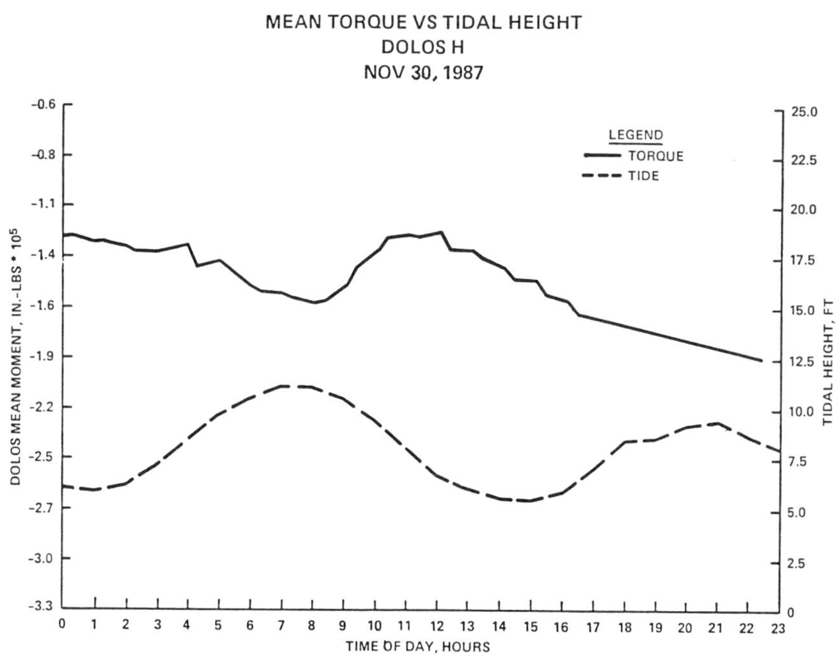

Figure 8: Dolos H Mean Response with Tide

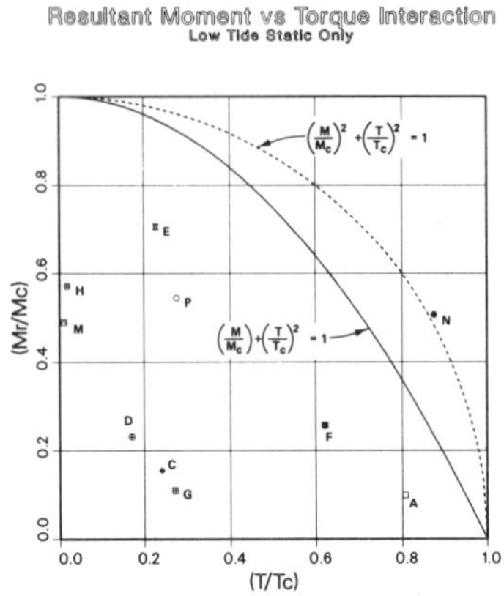

Figure 9: Static Moment-Torque Interaction Diagram

It is well known that static loads for large dolosse can be significant relative to breaking strength. However most estimates, to date, have been numerical computations based on a limited number of artificial boundary conditions. It is much more difficult to estimate static loads for individual units nested in an actual breakwater. Figure 9 shows the observed static load for 10 dolosse plotted on a moment-torque interaction diagram. The values are those observed at low tide, the case of maximum self weight load. The resultant moment and torque are normalized to the critical values, and two common design curves are plotted for the interaction breaking limit. As can be seen there is a wide distribution of loads, and all but one are within the design curves. This is consistent with the fact that none of these units are broken or cracked. The high value for dolos N is explained by its boundary condition. The vertical fluke is wedged downward into a hole with the horizontal fluke cantilevered, applying a large torque to the shank-fluke interface.

6 Pulsating Response

It is important for structural design of dolos armor units to determine the magnitude of loads due to wave attack. The details of the pulsating response are quite complex, however from a design engineer's point of view, the structural reponse is an extreme value problem. The characteristics of the stress are of little interest below breaking, but the likelihood of exceeding the breaking threshold must be predicted.

In order to develop a first look at the pulsating response, a large number of storm records from various dolosse were analyzed as follows. For a record 30 minutes long, the mean and trend were removed, and the absolute value of the maximum peak response for the record was found. For the same time period the largest wave was found. Figure 10 shows plots for Dolosse F and H which are both located on the lower portion of the test section. The horizontal axis is a wave power parameter H^2T where H is the height of the largest wave and T is its period. Although there

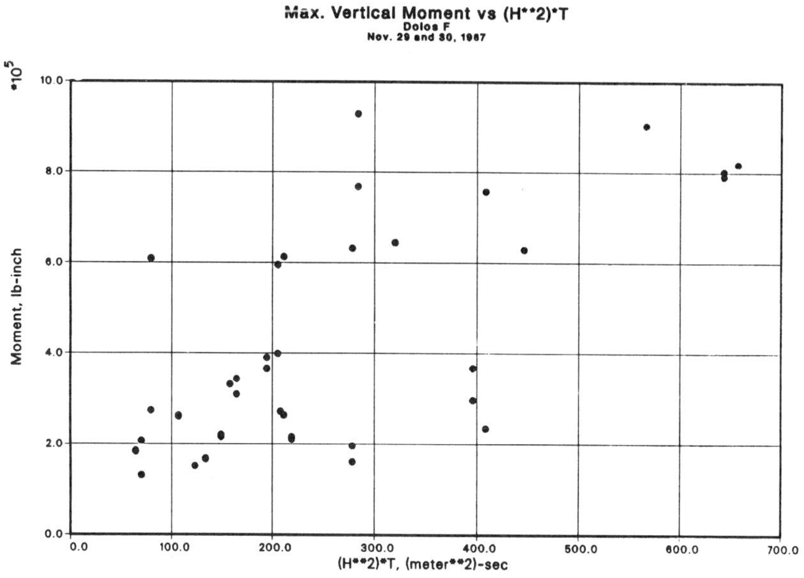

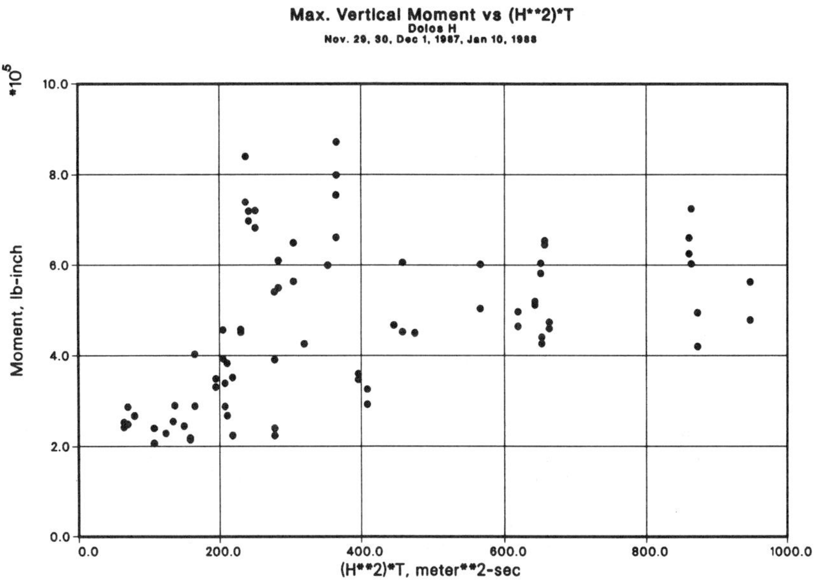

Figure 10: Maximum Pulsating Response vs. Maximum Wave Power

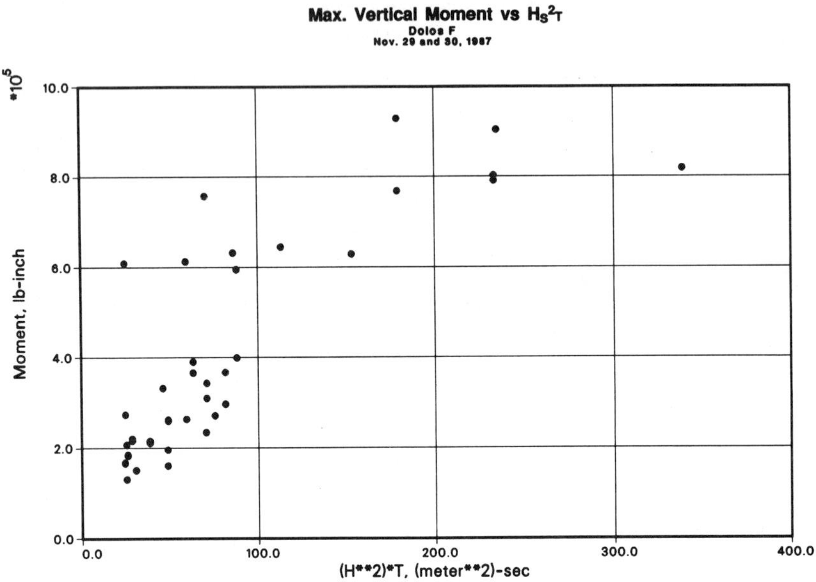

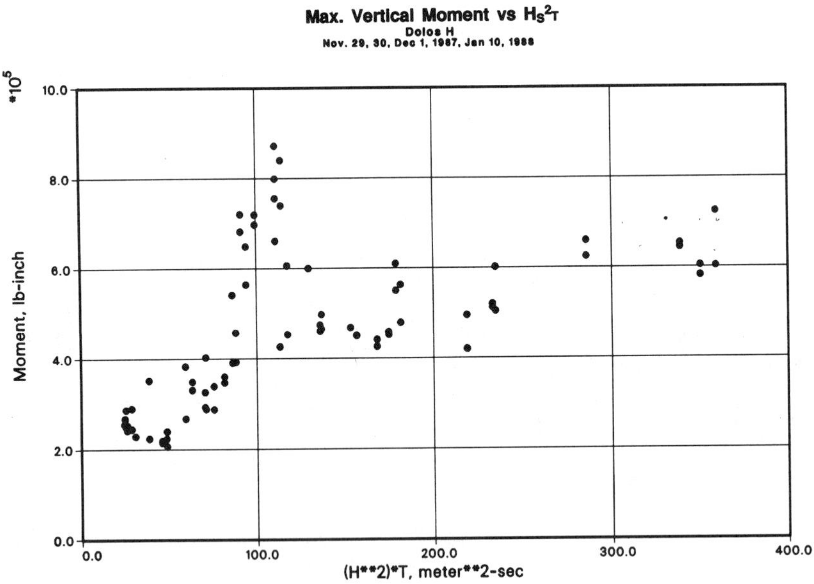

Figure 11: Maximum Pulsating Response vs. Significant Wave Power

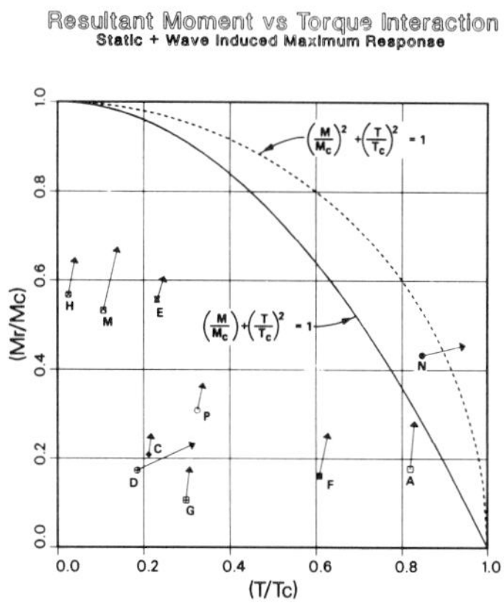

Figure 12: Combined Static and Maximum Pulsating Moment-Torque Interaction Diagram

is scatter to the data, both dolos can be plotted on the same linear scaled paper. Statistical variance can be reduced by using an averaging statistic such as $H_{1/10}$ or $H_{1/3}$ for the waves and a similar definition for the M_v. However to preserve the measured extreme value for the M_v, Figure 11 shows the same maximum M_v vs a wave power parameter computed from H_s and a T from the peak of the wave spectrum. As expected the scatter of the data is much less.

7 Preliminary results

Early in the investigation of the structural response of the dolos unit, most attention was focused on impacts as the most likely cause of dolosse breakage. It was generally assumed that wave forces would be much less important. However the magnitude of the pulsating forces observed at Crescent City are on the same order of magnitude as the static forces. It is easy to foresee an unfavorable combination of static forces and wave forces which could exceed the strength of the unit, causing failure without movement or impacts. Figure 12 shows a composite interaction diagram similar to the previous Figure 9. Here the static values are plotted at the time of the maximum pulsating response. Since this is generally at a higher water level, static forces are somewhat reduced from the low tide values shown previously. The absolute value of the maximum, resultant pulsating response is plotted as a vector extending from the base static value. The end of the arrow then represents the maximum combined static and pulsating response normalized to the breaking strength of the dolos. Dolos N is outside the conservative design range.

8 Conclusions

The work described here represents preliminary results from a unique set of measurements. Much remains to be done. However some general conclusions can focus the direction of future efforts:

- Static forces on dolosse in a real breakwater vary with mean water level and can be significant relative to the breaking strength of large dolosse.
- Wave (pulsating) forces are of the same order of magnitude as static forces.
- Wave forces compare qualitatively with scale model and numerical model results.
- Empirical relationships between wave energy and wave forces can be developed.

Acknowledgement

The work described in this paper was conducted as part of the Crescent City Prototype Dolosse Study of the Coastal Engineering Research Center, Waterways Experiment Station, US Army Corps of Engineers. Permission to publish this paper was granted by the Chief of Engineers.

Appendix - SI Unit Conversion

$1\ in. \cdot lb. = 113\ N \cdot mm = 1.13 \times 10^{-4} KN \cdot m.$

References

Burcharth, H.F. (1984) Fatigue in Breakwater Concrete Armour Units. University of Aalborg, Denmark, Oct.

Howell, G,L. (1985), ed. *Proceedings of the Workshop on Measurement and Analysis of Structural Response in Concrete Armor Units.* January 23-24; Waterways Experiment Station, Coastal Engineering Research Center, Vicksburg, Mississippi. 421 pp.

Howell, G.L. (1986) A System for the Measuremnet of the Structural Response of Dolos Armour Units in the Prototype. *The Dock & Harbour Authority,* vol. 67, number 779, May. pp.6-10.

Kendall, T.R. (1988) Analysis of 42-ton Dolos Motions at Crescent City. *Proceedings, 21st International Conference on Coastal Engineering,* 20-25 June, Torremolinos, Spain, ASCE.

McDougal, W.G. , Melby, J.A., and Tedesco, J.W. (1988) Wave Forces on Concrete Armor Units. *Journal of Waterway, Port, Coastal, and Ocean Engineering,* vol. 114, no. 5, September, pp.582-598, ASCE.

Scott, R.D. , Turcke, D.J., Baird, W.F., (1986) A Unique Instrumentation Scheme For Measuring Loads In Model Dolos Units. *Proceedings, 20st International Conference on Coastal Engineering,* Taipei, Taiwan, ASCE.

CHAPTER 175

"Double U block" and "Iblock"
——the armour blocks of two new types

S. W. TWU*, S. C. LIN**, S.G. CHANG***

Abstract

A total of eleven new types of blocks are developed for this project. After performing a series of model tests for them and having a consideration of easy casting, two types of blocks are selected and presented here in this paper. They are named "Double U block" and "I block", respectively. The two have been compared with several existing types which have been widely used in Taiwan. It is shown that the Double U block is an excellent type of block with high stability and low reflection coefficient. The I block has an advantage of easy casting, although it is not outstanding in other characteristics.

I. INTRODUCTION

In the last two or three decades, concrete armour blocks have been commonly used in Taiwan for the construction of harbors and shore-protection structures. These blocks are usually Tetrapod, Dolosse, Holtripod and Shake. However, no matter what type of blocks are used, they are destroyed very often because the waves appearing around the coast of Taiwan are so awful, either in winter monsoon seasons or typhoon periods. Furthermore, when these damage occurs, it always extend to the block-covering structures or even to the back-side areas protected by them. The back-side areas are known as wharves, land or resident areas. Taiwan is a densely-populated area, naturally, the land is very valuable. Under this circumstance it is painful to tolerate any land loss due to wave attacks. So, it is reasonable to predict that the concrete armour blocks will be used more extensively for the shore-protection structures in the future. Consequently, we are looking forward to developing one or two types of armour blocks which are superior to those used currently in Taiwan.

* Prof., Hydraulics and Ocean Engineering Dept., Cheng-Kung University, Tainan, Taiwan, R.O.C.
** Assoc. Prof, Hydraulics and Ocean Engineering Dept., Cheng-Kung University, Tainan, Taiwan, R.O.C.
*** Research assistant, Tainan Hydraulic Laboratory, Cheng-Kung University, Tainan, Taiwan, R.O.C.

II. DESIGN PROCEDURE

A good concrete armour block should share a number of characteristics, such as good stability, good interlocking effect, low reflection coefficient and easy casting. Among them the stability is the most important. It is common thing that the coastal structures are covered with concrete armour blocks. Some of them, such as off-shore breakwaters, are even piled up entirely by concrete blocks. The primary function of these blocks is to dissipate the wave energy and to scatter the wave force, so that the structure might not be destroyed by waves. If the blocks slide down or move away from its place during the wave attacks, the structure will lose its protection layer. Afterward, a damage to the structure occurs. Hence, good stability takes priority of all other characteristics of the armour blocks.

The second required characteristic is the wave-energy-dissipation ability. If the blocks are placed inside a harbour, it is expected not to produce large amount of reflection waves to the water area. Generally speaking, the lower the reflection coefficient, the better the armour block.

In addition to the two characteristics mentioned above, there is another whcih is also very important, that is the casting feasibility. Good blocks should be easy to cast as they are being formed. So that the construction work can smoothly under way. Following the above rules, both the staffs of Tainan Hydraulic Laboratory and the Department of Hydraulics and Ocean Engineering start the design work for the new types of blocks. Finally, eleven types are finished.

III. TESTS FOR STABILITY

To pick the superior ones from these eleven types of blocks, a number of model tests have to be carried out. All of these tests are performed in a wave flume made of concrete, which has a length of 75 m with 1.0 m in width,and 1.2 m in height. One side of the flume is casted with sheets of glass with 36 m in length.

Among these tests the stability test is the most important, and is conducted first. In this test, all types of blocks have the same weight, 150 gm, and are placed on the same site with a slope of 1:1.5 and with water depth of 20cm. The inclined-face model is placed on a bed slope of 1:10, which is 4 m in length and ends in connection with the horizontal flume bed where is 52.9 m away from the wave maker. The sketch for this model is shown in Figure 1.

In this test a wave period of 2 sec is adopted. The wave height is adjusted so that they break and impact just on the tested blocks. For convenience of comparison, the stability situation of these blocks are divided into five degrees as follows: (Michael, 1974; Paul, 1971)

(1) 1st degree: All the tested blocks stand still against wave attack.

(2) 2nd degree: Part of the tested blocks shake with waves, but their final mean position never change.
(3) 3rd degree: A few tested blocks are slightly moving, but the majority remains stable.
(4) 4th degree: The tested blocks move slowly and continuously until the armour layer is destroyed.
(5) 5th degre: The tested armour layer is destroyed in a short time.

For the 1st and 2nd degree, they are considered excellent in stability. In the 3rd degree a few blocks are removed at the start of the test, but later on nothing

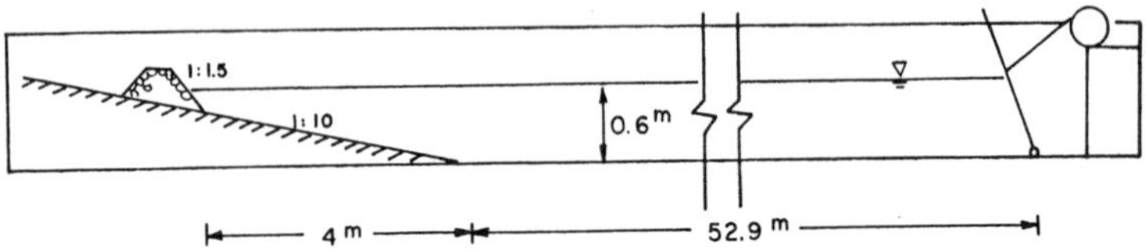

Fig 1. The sketch for the wave flume.

more happen. Therefore, the 3rd degree stability is still acceptable. If these definitions are compared with the percentage of damage. The 1st degree can be considered as zero percentage of damage. The zero percentage should fall on the transition between the 1st and 2nd degree . The 2nd degree will be equal to 0-2% damage, and the 3rd degree approximately 2-4% damage. In the 4th degree the damage percentage depends on time and therefore it will increase with increasing duration of wave attack. In the long run it will be totally destroyed. For the 5th degree total failure will occur in a short time. So, those types of blocks that reach the 4th or 5th degree in a short time are naturally not qualified. Based on the above criteria, four types of blocks are left for further tests that include tests for Kd value and reflection coefficients. After all these tests have been performed, a consultation with Chang-Ming Corporation, who financially supports this project, is held. In that meeting, we focused our attention on the stability and casting feasibility for selecting work. Consequently, two types of blocks, named "Double U block" as shown in Photo 1 and "I block" as shown in Photo 2,

Photo.1. Double U block

Photo.2. I block

are selected. To realize the superiority of the two new types, they are compared with a number of types which have been widely used in Taiwan, such as Holtripod, Dolosse and Shake. So, a number of additional tests for all these existing types have also been conducted. In case of uniform placing in two layers, Double U block and I block are compared with Holtripod and Shake in terms of the number of breaking waves needed to attack to reach the indicated stability degree. The results are shown in Figure 2. The case for uniform placing in one-layer is shown in Figure 3. For the case of pell-mell placing in two layers, the

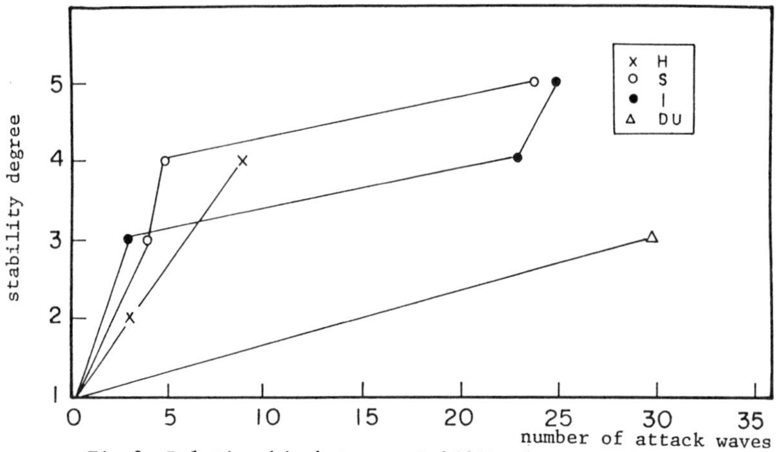

Fig 2. Relationship between stability degree and number of attack waves needed to reach that degree when blocks are uniformly placed in two layers.

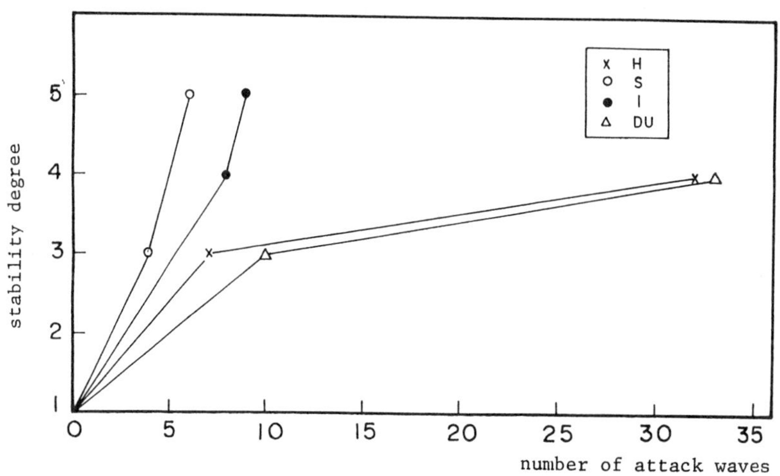

Fig 3. Relationship between stability degree and number of attack waves needed to reach that degree when blocks are uniformly placed in single layer.

new types are compared with Dolosse and the results are shown in Figure 4. It is found that Double U block is superior to all the compared ones. I block performs approximately the same as Shake and Holtripod if they are uniformly placed in two layers, and performs approximately the same as Dolosse if they are pell-mell placed in two layers. However, Figure 3 shows that I block is better than Shake but inferior to Holtripod as they are uniformly placed in single layer.

IV. TESTS FOR Kd VALUE

Kd values of armour blocks are usually calculated,by Hudson's formula

$$K_d = \frac{\gamma H^3}{W\,(S-1)^3 \cot\theta}$$

where W is the weight of individual armour block. γ and S are the specific weight and specific gravity of armour block, respectively. θ is the angle of armour-block slope with horizontal. H is the wave height.

Aside from the above-mentioned factors, Kd value for a given type of blocks is also closely related to the percentage of damage adopted for determing H in the Hudson's formula. In this study one percent of damage under the attack of nonbreaking waves is adopted. Two methods of placing are used for the test. One of them is uniform placing in two layers, and pell-mell placing in two layer is the another. In order to accurately measure the wave height the armour-block model is placed directly on the horizontal flume bed, rather than on an inclined bed, to keep the generated waves from shoaling effect. So, the water depth between the wave-maker and armour-block model is constant, i.e. 45cm. For a practical consideration, models of three armour-block slopes, i.e. 1:1.33, 1:1.5 and 1:2.0, are

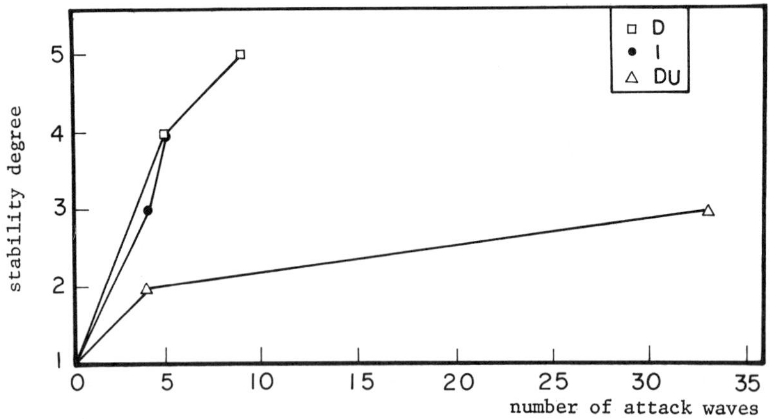

Fig 4. Relationship between stability degree and number of attack waves needed to reach that degree when blocks are pell-mell placed in two layers.

used. Wave periods of 1.5, 1.8, 2.0 sec are generated. The duration for each run is 30 mins. For a given type of blocks, the wave height in a run is adjusted by a little higher than that in the preceding run until one percent of damage to the mound of armour block meets. Dolosse has also been tested for Kd value in this study. So that they could be compared with each other under the same condition. The resulting Kd values are shown in Table 1. Five cases of them are crossed meaning that none of the waves generated in the wave flume could make any damage to them.

According to Hudson's formula, Kd is independent of wave period, but in this table it is found that Kd varies proportional to wave periods. It is indicated that Double U block has a Kd value higher than Dolosse, and I block is lower.

Table 1. Kd Vaule for the tested blocks.

method of placing	armour-block slope \ block / wave period	Double U block			I block			Dolosse		
		1.5	1.8	2.0	1.5	1.8	2.0	1.5	1.8	2.0
uniform placing in two layers	1:1⅓	20.9	29.5	30.4	18.8	23.3	26.2			
	1:1.5	18.3	24.7	31.1	21.1	24.7	28.1			
	1:2.0	/	/	/	/	/	22.9			
	average	25.8			23.6					
pell-mell placing in two layers	1:1⅓	17.8	22.2	26.2	13.9	21.1	23.8	13.4	21.4	24.1
	1:1.5	22.3	27.3	27.0	11.8	16.6	15.4	13.7	21.7	22.7
	1:2.0	14.4	19.8	20.0	13.9	14.5	14.5	13.5	21.2	17.2
	average	21.9			15.6			18.8		

The uniform placing method for Double U block shown in the table is a method named here as side-by-side method,and is shown in Photo.3. If another special uniform placing method named here as riding method, shown in Photo.4, is adopted, then the blocks would give a Kd value so high that we don't even know how high it is, because none of the waves generated in the flume can destroy them. Although the riding method could offer such a high stability, it would take much more amount of armour units to cover a given area, and produce much higher reflection coefficient, than the side-by-side method, which will be shown later.

V. TESTS FOR REFLECTION COEFFICIENT

Wave-energy-dissipation efficiency of a armour block is always indicated by the reflection coefficient and run-up value. A superior armour block should gives a lower reflection coefficient and run-up value. Two armour-block slopes, 1:1.5 and 1:2.0, are used in this test to estimate

Photo.3. side by side method for Double U blocks

Photo.4. riding method for Double U blocks

the reflection coefficient. The wave periods include 1.0, 1.5 and 2.0 sec., and five varied wave heights are generated for each wave period. The water depth in the flume, as shown in Figure 1, is 60 cm in front of the wavemaker and 20 cm at the model toe. A technique developed by Goda (1979) to resolve the incident and reflected waves from the records of composite waves is used. Two simultaneous wave records are taken at adjacent locations in front of the armour-block model, and all the amplitudes of Fourier components are analyzed. From these Fourier components the amplitudes of incident and reflected wave components are estimated. According to Goda, as long as the spacings between the model and the wave gauges are kept in a proper range, the reflection coefficient can be accurately measured by the above incident and reflected wave heights. Goda suggested

$$L_{max} = \frac{\Delta \ell}{0.05} > L > \frac{\Delta \ell}{0.45} = L_{min} \qquad \ell \geq 0.1\ L$$

Where $\Delta \ell$ is the spacing between the wave gauges, ℓ is the spacing between model and the gauge adjacent to the model, L_{max} and L_{min} are the maximum and minimum wave length of the attack waves, respectively. In this test ℓ = 60cm and $\Delta \ell$ =50cm are used, and they are in effective range as Goda suggested.

For convenience of comparison, three existing types of blocks have also been tested in addition to the new ones. Holtripod and Shake are compared with the new ones for the case of uniform placing, and Dolosse are for the case of pell-mell placing. The results are shown in Figure 5 to 10. In each case the reflection coefficients are calculated by taking average over five varied wave heights. The results are shown in Table 2. Both results of the side-by-side method and riding method for Double U block are all shown in Fig.5 and Fig.6, in which the side-by-side method is

Table 2. Reflection Coefficient

armour-block slope	wave period	uniform placing, one layer				uniform placing, two layers				pell-mell placing, two layers		
		Double U	I block	Shake	Holtripod	Double U	I block	Shake	Höltripod	Double U	I block	Dolosse
1:1.5	1.0	0.232	0.213	0.214	0.318	0.195	0.270	0.219	0.227	0.131	0.171	0.062
	1.5	0.394	0.503	0.338	0.392	0.250	0.393	0.341	0.334	0.276	0.296	0.284
	2.0	0.603	0.657	0.591	0.492	0.544	0.637	0.565	0.564	0.512	0.420	0.507
1:20	1.0	0.243	0.349	0.259	0.250	0.180	0.301	0.294	0.198	0.178	0.175	0.161
	1.5	0.368	0.374	0.330	0.360	0.239	0.289	0.305	0.273	0.164	0.255	0.237
	2.0	0.497	0.584	0.504	0.575	0.474	0.584	0.543	0.514	0.546	0.530	0.519

indicated by DU without bracket, and the riding method by DU with bracket. In the remaining figures the results of riding method for Double U block are omitted. From these datas it is found that Double U block, side-by-side method, has a reflection coefficient approximating to Shake and Holtripod as the blocks are uniformly placed in single layer. While I block has a little higher value. For uniform placing in two layers, Double U block, side-by-side method, is less than Shake and Holtripod. I block is a little higher. For pell-mell placing in two layers, Double U block and I block show approximately the same results as Dolosse. While the riding method for Double U block is adopted, it shows a reflection coefficient higher than all the compared ones.

VI. TESTS FOR RUN-UP VALUE

In this test the flume bed between the model and the wave generator is horizontal, and the water depth is 60cm. It is known that run-up value for a given type of block is primarily a function of wave steepness and armour-block slope. In addition to that, it is also related with the placing method for the blocks. With two armour-block slopes and three placing methods for the blocks, the test results are expressed as curves showing relationships between relative run-up value and wave steepness, as in Fig. 11-14. It is shown in this figures that the run-up values decrease with increasing wave steepness. And they are smaller in case of pell-mell placing than in case of uniform placing. These resulting run-up values may also be expressed by a formula, as follows

$$\frac{R}{H_0} = a - b\left(\frac{H_0}{L_0}\right)$$

where the coefficient a, b are shown in Table 3

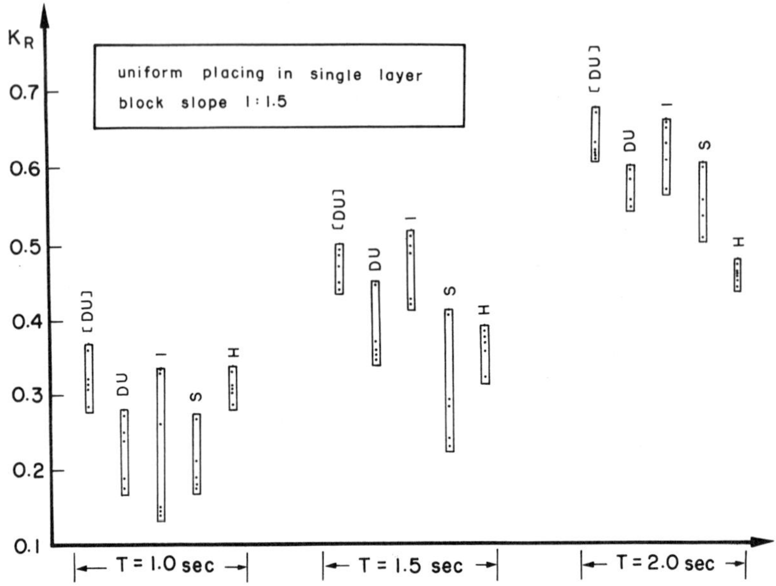

Fig.5. Experimental reflection coefficient

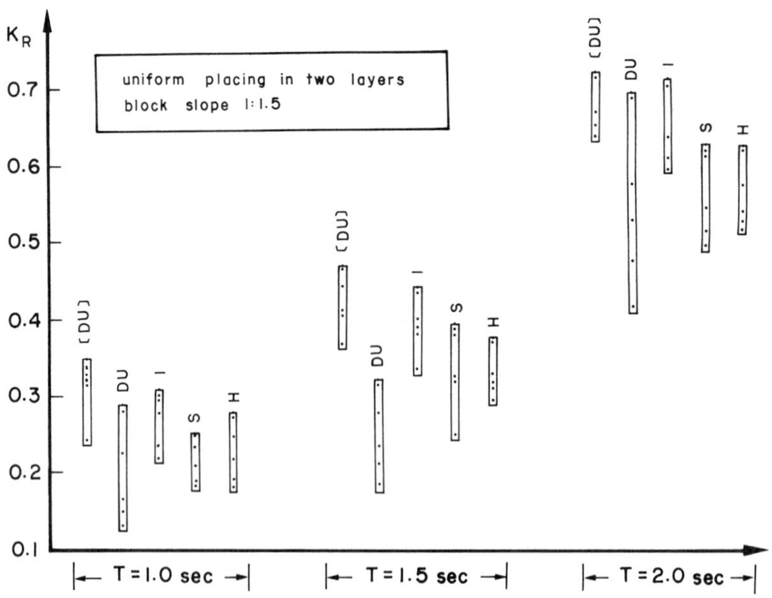

Fig.6. Experimental reflection coefficient

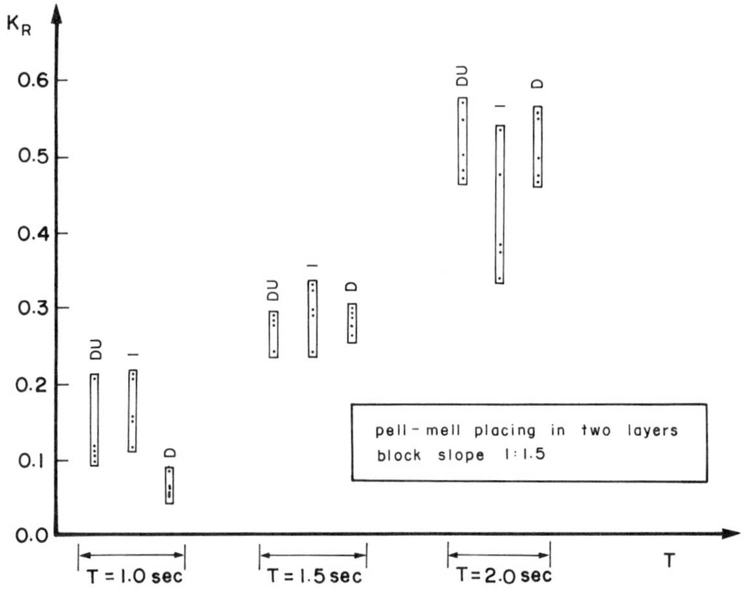

Fig.7. Experimental reflection coefficient

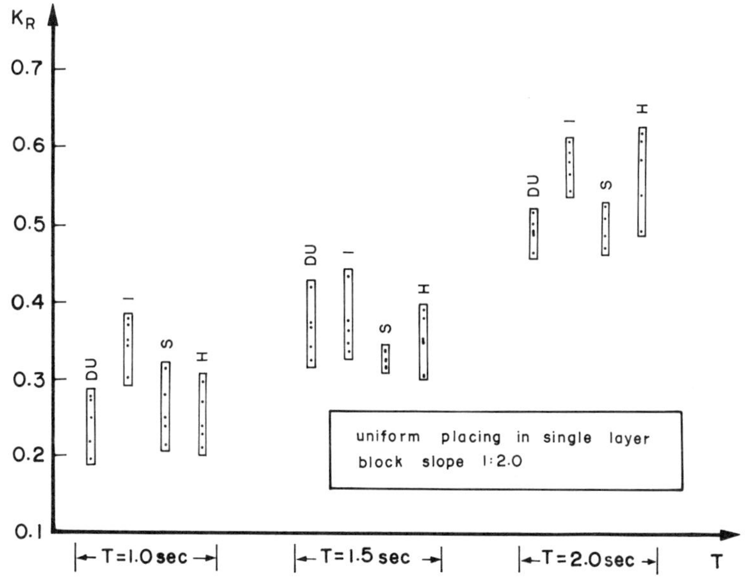

Fig.8. Experimental reflection coefficient

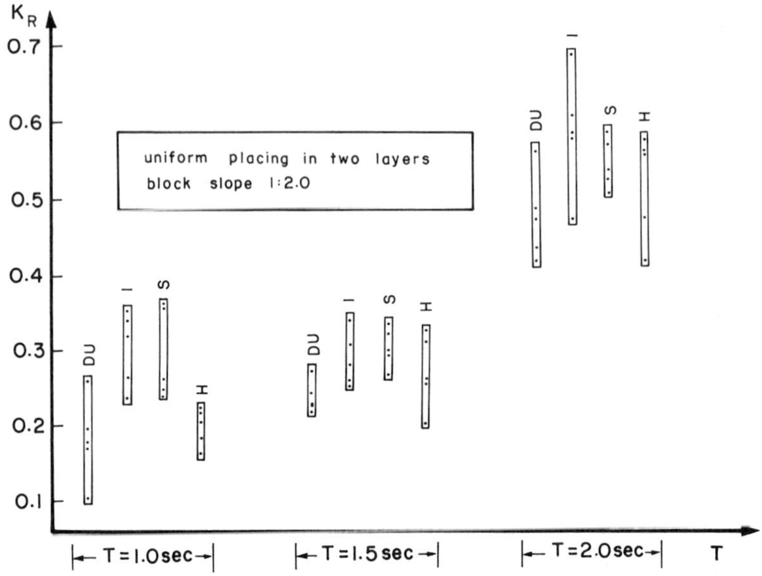

Fig.9. Experimental reflection coefficient

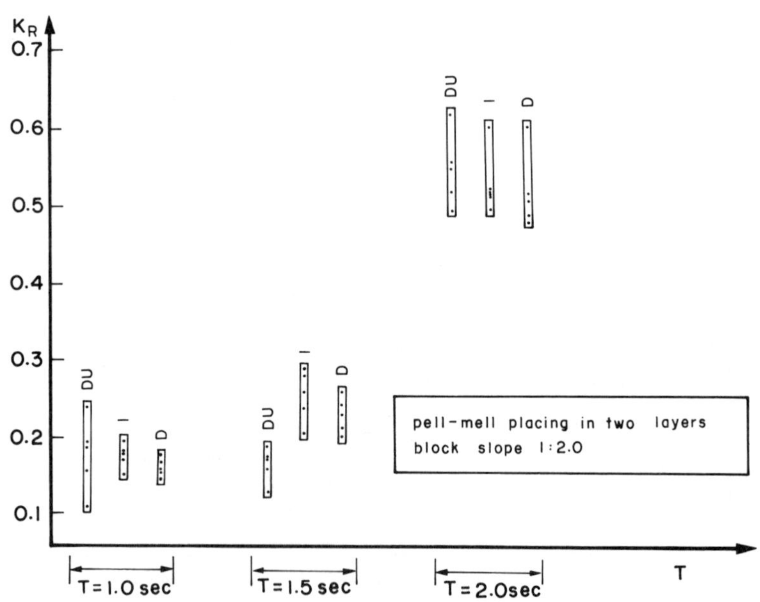

Fig.10. Experimental reflection coefficient

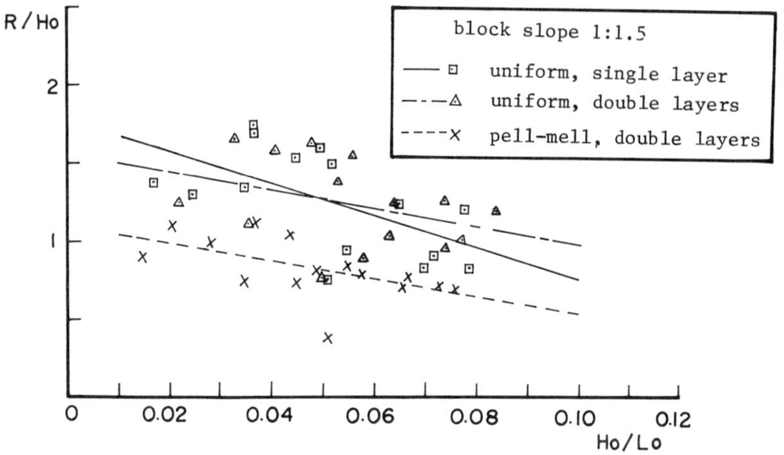

Fig 11. R/Ho∿Ho/Lo for I block

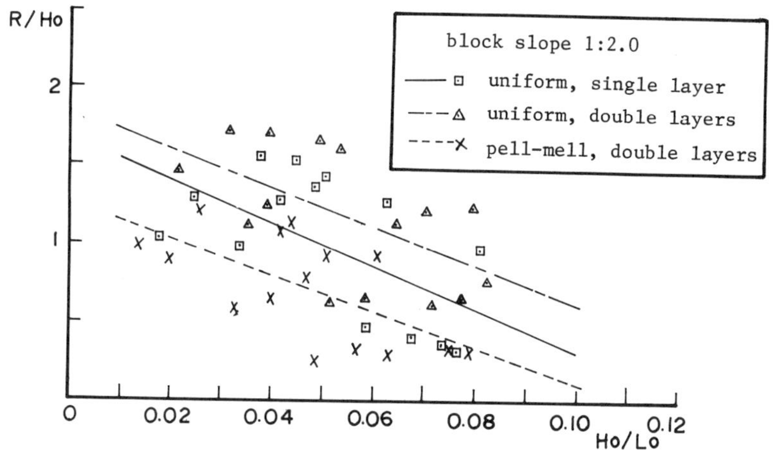

Fig 12. R/Ho∿Ho/Lo for I block

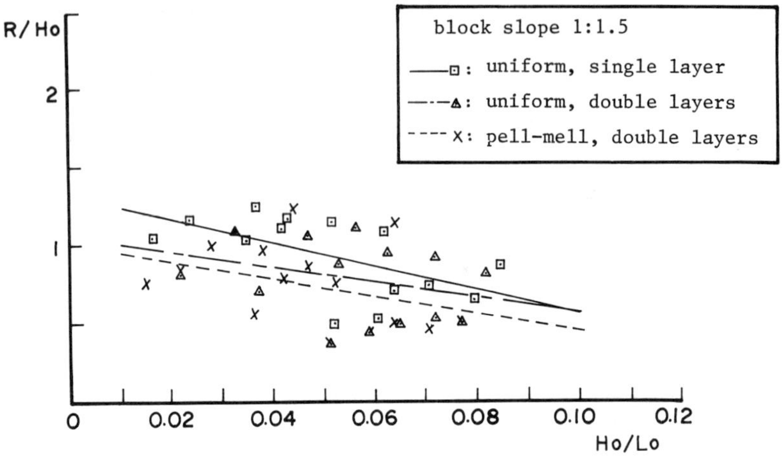

Fig 13. R/Ho∿Ho/Lo for Double U block

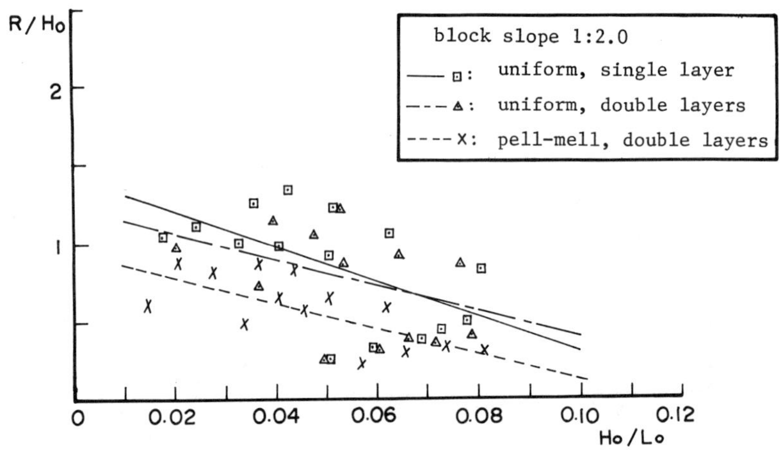

Fig 14. R/Ho∿Ho/Lo for Double U block

Table 3. a. b value for run-up

block slope	method of placing for block \ type of block / coefficient	Double U		I-block	
		a	b	a	b
1:1.5	uniform, one layer	1.319	7.52	1.775	10.19
	uniform, two layers	1.668	5.00	1.563	5.82
	pell-mell, two layers	1.017	5.69	1.100	5.72
1:20	uniform, one layer	1.408	10.90	1.677	13.52
	uniform, two layers	1.228	8.19	1.855	12.35
	pell-mell, two layers	0.947	8.15	1.259	11.41

VII. CONCLUSIONS

1. Double U block shows a stability superior to the existing blocks which have been commonly used in Taiwan. It gives a Kd value much higher than the existing types, no matter what method the blocks are placed.
2. If a riding method of placing is adopted, Double U block would give a very high Kd value, but it also produces a large reflection waves. Except that, with any other method of placing, it can always offer a low reflection coefficient in addition to the high stability.
3. Double U block is a newly developed block. The studies presented here are so far what we have done. It is easy to realize from the shape of the block that a good interlocking relation among blocks plays an important part in its high stability. Therefore, the block has to take large internal tension due to the interlocking effect. Under this circumstance, the Double U block has to be reinforced in actual field.
4. I block performs neither better nor worse than the existing blocks in stability and shows a little worse in reflection coefficient, however, it has an advantage of easy casting. So, I block is also an alternating type of block.

REFERENCE

1. Goda Yoshimi, Yasumasa Suzuki, Yasuharu Kishira, Osamu Kiruchi (1976), "Estimation of incident and reflected waves in random wave experiment." Proc. of the 15th

Coastal Engr. Conf. p.828-845.

2. Michael Brorsen, H. F. Burcharth, Torben Larsen, (1974): "Stability of Dolos Slopes", Proc. of the 14th Coastal Engr. Conf. p.1691-1701.

3. Paul, M.W., Baird, W.F. (1971): "Discussion on breakwater armour units", Proc. of 1st int. conf. on port and ocean eng. under arctice conditions. Trondheim.

CHAPTER 176

ARMOUR UNIT STRUCTURAL RESPONSE - A PARAMETRIC STUDY

C. David Anglin*, Associate Member, William F. Baird*, Associate Member, Etienne P.D. Mansard**, R. Douglas Scott*, Associate Member, David J. Turcke***

ABSTRACT

There is a general lack of knowledge regarding the nature and magnitude of loads acting on armour units used for the protection of rubblemound coastal structures. Thus, a comprehensive design procedure incorporating both the hydraulic stability and the structural integrity of the armour units does not exist.

This paper presents the results of a detailed parametric study of the structural response of armour units to wave-induced loading in a physical breakwater model. The effect of the following design parameters is investigated: breakwater slope, armour unit location, wave period and wave height.

This research has made a number of significant contributions towards the development of a comprehensive design procedure for concrete armour units. It has identified a linear relationship between the wave-induced stress in the armour units and the incident wave height. In addition, it has shown that the conditional probability of wave-induced stress given wave height can be estimated by a log-normal distribution. Finally, a preliminary design chart has been developed which incorporates both the structural integrity and the hydraulic stability of the armour units.

INTRODUCTION

Concrete armour units are often used for the protection of rubblemound coastal structures when wave conditions dictate the use of unreasonably large armour stones, or when armour stones of a sufficient size are not available. The design of these concrete armour units is primarily based on hydraulic stability, with little attention being given to the structural integrity of the individual units. The interaction between waves and rubblemound structures is a very complex process, and the wave forces acting on the armour units and the structural response of the units to these forces is poorly understood.

---

* W.F. Baird and Associates Coastal Engineers Ltd.
150-38 Antares Drive, Ottawa, Ontario, Canada K2E 7V2

** Hydraulics Laboratory, National Research Council of Canada
Ottawa, Ontario, Canada K1A 0R6

*** Department of Civil Engineering
Queen's University, Kingston, Ontario, Canada K7L 3N6

Current breakwater design procedures are based on empirical formulae, such as Hudson's equation (U.S. Army Corps of Engineers, 1977, 1984), or for larger structures, physical hydraulic modelling may be used. However, these approaches do not consider the structural integrity of the armour units themselves, but concentrate on the hydraulic stability of the armour layer. There is a general lack of knowledge regarding the nature and magnitude of loads on the armour units, and consequently a design procedure incorporating the structural integrity of the armour units does not exist.

The measurement of forces in a breakwater or stresses in an armour unit is a very complex task. However, a number of research groups have recently undertaken investigations of this nature, utilizing a variety of approaches. These have included prototype studies, physical and numerical modelling, and theoretical developments.

A number of researchers are making use of physical models to measure the structural response of armour units (for example, Scott et al, 1986a and b, Nishigari et al, 1986, Losada et al, 1988, Jensen and Juhle, 1988). In addition to these physical model investigations, a very extensive prototype study is being carried out at Crescent City, U.S.A. to measure dolos response in full scale situations (Howell, 1988), and a non-linear numerical model is being developed to determine the response of dolos in waves (Tedesco and McDougal, 1985, Tedesco et al, 1988). Other types of research, including both prototype studies (Burcharth, Feb. 1981, 1984) and physical modelling (Timco and Mansard (1982)) have also been carried out; however, these investigations have not measured the response of armour units under wave attack, but have tested units to failure. Finally, a number of investigators have utilized theoretical approaches to develop an understanding of the structural performance of armour units, and to establish design criteria (for example, Timco (1984), and Burcharth (May 1981)).

The Department of Civil Engineering at Queen's University, in co-operation with W.F. Baird and Associates Coastal Engineers Ltd., initiated a long term research and development program in 1980 to investigate the nature of forces acting on armour units in a breakwater, and ultimately to develop a comprehensive design procedure for concrete armour units. Initial investigations concentrated on a review of available instrumentation techniques for both model and prototype studies, and concluded that strain-gauging model armour units was the most viable approach (Baird, et al, 1983). Preliminary work, utilizing the concept of "strain distortion", was completed with model armour units constructed of an epoxy material and instrumented with surface-mounted strain gauges. However, the sensitivity of the instrumentation was insufficient to accurately measure the loads encountered in a typical hydraulic breakwater model (Scott, 1986). Further studies led to the development of a unique armour unit load cell, which utilizes the concept of "geometric distortion" to produce an extremely sensitive and accurate instrumentation system (Scott, 1986, Scott et al, 1986a and b).

A number of these load cells were manufactured for use in the present study, which is a parametric investigation into the structural response of armour units in a hydraulic breakwater model. The units were manufactured and calibrated at Queen's University, while the testing program was completed at the Hydraulics Laboratory of the National Research Council of Canada (NRCC). This paper describes details of the load cell, the testing program, and data reduction and analysis techniques, and finally presents a preliminary design chart which incorporates both the hydraulic stability and the structural integrity of the armour unit.

INSTRUMENTATION

The load cell, developed by Scott (1986) for his PhD degree, utilizes the concept of geometric distortion to produce an extremely sensitive and accurate instrumentation system. This force measuring device consists of a hollowed-out model armour unit (a dolos in this study) and a thin-walled aluminum tube instrumented with strain gauges which is inserted in the shank of the dolos, as shown in Figure 1. The tube is instrumented with three full strain gauge bridges, which measure two orthogonal bending moments (M_1 and M_2) and the torque (T) at the mid-shank location of the dolos.

Calibration of the load cell was conducted by rigidly holding the unit in a test rig and applying known loads to the extremities of the unit's geometry. The output from the various strain gauge bridges was plotted against the applied load to produce a calibration curve of channel output versus applied load (either moment or torque). Thus, a load cell was created such that the output from the various strain gauge circuits could be directly converted into the two moments and the torque at the mid-shank location of the dolos. The accuracy of the instrumentation system was verified by comparing various test cases against theoretical and finite element model results.

The load cell is designed to accurately measure the response of the unit to static and quasi-static loads. The response to impact loads is not correctly measured; this would require considerable effort in correctly modelling the material properties of concrete at a selected model scale. This report focuses on the quasi-static (wave-induced) loads as measured in the breakwater model.

TEST PROGRAM

The tests were completed in a 2 m wide by 60 m long wave flume at the Hydraulics Laboratory of the National Research Council of Canada (NRCC). The breakwater cross-section was constructed in a 1 m wide section of the flume. The breakwater was a conventional multi-layer design, with a primary armour layer of dolos units (2 layers, dolos weight W=482g, dolos height C=0.106m, waist ratio r=0.32, random placement, with dimensionless packing density $ø=NV^{2/3}=0.80$) placed over a filter layer of 30 to 100g crushed stone, as shown in Figure 2. The core consisted of a fine gravel.

Over 1000 tests were run to assess the effects of breakwater slope, unit location, wave period, and wave height, using both regular and irregular waves. Table 1 summarizes the range of parameters tested.

Table 1
Range of Parameters Tested

Parameter	Range Tested					
Breakwater Slope	1:1.5, 1:2.0, 1:2.5					
Load Cell Location		Number of Locations Tested Across Flume				
	elevation	+0.2m	+0.1m	0 SWL*	-0.1m	-0.2m
	top layer	2	4	4	4	2
	bottom layer		2	2	2	
Wave Period	regular waves	T = 1.25 to 3.0 s				
	irregular waves	Tp = 1.75 to 2.5 s				
Wave Height	regular waves	H = 0.05 to 0.30 m				
	irregular waves	Hs = 0.10 to 0.25 m				
Water Depth	0.80 m (non-breaking waves)					
Crest Elevation	+0.40 m above water level (no overtopping)					

*SWL = still water level

A typical test series consisted of initializing and calibrating the instrumentation, placing the load cells (two) at selected positions in the breakwater model and running a short "burst" of waves, sampling the wave and load channels for 30 s. This procedure was repeated for numerous combinations of the various breakwater parameters, as summarized above. Data were sampled at a rate of 500 Hz using a NEFF analog to digital convertor on an HP-1000 computer, and were stored on magnetic tape for later analysis.

The results presented in this paper are for a breakwater slope of 1:1.5 and regular waves only. The remainder of the data is currently being analysed.

DATA REDUCTION

Data reduction on the first series of tests (breakwater slope 1:1.5, regular waves, approximately 350 tests) has been completed using the NRCC GEDAP software system, and consisted of the following steps:

(i) demultiplex the data into individual channels,
(ii) removal of the static load component from the time series data,
(iii) filtering to remove high frequency noise and dynamic events, which as noted earlier are not correctly measured by the load cell used in this study,
(iv) calculation of combined moment, and normal, shear, and principal stresses using standard formulae (see below),
(v) identification of data peaks in the various time series with a zero-crossing analysis, and calculation of the average peak values,
(vi) calculation of mean, standard deviation, and root mean square values of the various time series,
(vii) tabulation of the various input parameters (wave conditions and breakwater geometry) and output parameters (standard deviation and average peak values of moment, torque, and principal stress),
(viii) plotting of various combinations of the input and output parameters, and regression analyses and estimation of confidence limits to identify trends.

The calculation of the various quantities is summarized below:

longitudinal stress = $\sigma_x = Mc/I$ (M = combined moment = $\sqrt{M_1^2 + M_2^2}$)
(I = moment of inertia of section)
normal stress = $\sigma_y = 0$ (c = distance to extreme fibre)

shear stress = $\tau = Tc/J$ (J = polar moment of inertia of section)

principal stress = $0.5(\sigma_x+\sigma_y)\pm\sqrt{[(\sigma_x-\sigma_y)/2]^2+\tau^2} = 0.5\sigma_x+\sqrt{(\sigma_x/2)^2+\tau^2}$

average peak value = average of the highest "n" peaks, where n is the number of waves which occurred in the 30 s sample

mean value = $\mu = (1/N)\Sigma x_i$
standard deviation value = $s = \sqrt{(1/N)\Sigma(x_i^2)-[(1/N)\Sigma x_i]^2}$
root mean square value = $rms = \sqrt{(1/N)\Sigma(x_i^2)} = \sqrt{s^2 + \mu^2}$

A typical result from one of the tests is shown in Figure 3, which shows time series plots of the load cell response, including the combined moment (calculated from the two measured moments), the measured torque, and the calculated principal stress at the mid-shank location of the dolos. The corresponding wave time series is also shown.

DATA ANALYSIS

As mentioned earlier, only results from the first series of tests were available for presentation in this paper. These results consist of approximately 350 tests conducted with regular waves and a breakwater slope of 1:1.5. In addition, wave-induced quasi-static forces only are presented here; the static load component has been removed, while dynamic events were not measured by the load cell.

A series of parametric plots has been produced to investigate the influence of various input parameters on the measured armour unit response. The response parameters of average peak stress and standard deviation stress have been plotted against wave height (H) and the surf similarity parameter ($\xi = \tan \alpha / \sqrt{H/L}$). These plots have been produced with and without distinction of the individual wave periods and armour unit locations. In addition, various statistical analyses, including regression analyses, estimates of confidence limits, and estimates of data probability distributions have been completed to identify trends in the data. A review of the various plots leads to the following general observations, as demonstrated by the referenced figures:

1) wave-induced stresses increase approximately linearly with wave height, and scatter in the data tends to increase with increasing wave height; linear regression analyses at individual locations gave correlation coefficients (p) varying from 0.34 to 0.91 for individual wave periods, and from 0.62 to 0.77 for all wave periods considered together (Figure 4 shows typical results at a single location); quadratic regression analyses gave only slightly higher correlation coefficients - the curves tend to be concave down, showing a marginally decreasing stress with wave height.

2) wave-induced stresses tend to increase with wave period to a certain point, but may increase or decrease for $T>2.25$ s depending on the load cell location (see Figures 4 and 5).

3) average peak and standard deviation wave-induced stress values show similar trends and scatter (results for standard deviation values are not presented here due to space limitations); the ratio of average peak to standard deviation stress values varies between approximately 1.8 and 3.9, but is typically around 3.2 (note that for a simple sinusoidal wave, this ratio is $\sqrt{2} = 1.414$).

4) wave-induced stresses tend to be higher in the bottom dolos layer than in the top dolos layer (see Figure 6).

5) for large waves, wave-induced stresses tend to be greatest above the still water level (SWL), while for small waves, stresses tend to be greatest at or just below the SWL (see Figure 7).

6) based on the available data, wave-induced stresses tend to be greatest for values of the surf similarity parameter between approximately 2.5 to 4 (see Figure 8); stresses tend to decrease for higher values of the surf similarity parameter, but no data is available below a value of approximately 2.2 for the 1:1.5 slope (data in this range was measured on the flatter breakwater slopes (1:2.0 and 1:2.5)); quadratic regression analyses of stress versus surf similarity gave correlation coefficients ranging from 0.37 to 0.96 for individual wave periods, and from 0.36 to 0.51 for all wave periods considered together.

7) results from all tests are presented in Figure 9, which shows the average peak wave-induced stress plotted against wave height; the conditional probability distribution of stress given wave height is skewed towards larger values, with the log-normal distribution giving a reasonable fit to the data (see Figure 10).

The trends described above tend to show up best at locations at or below the still water level (SWL), while positions above the SWL show less obvious trends and more scatter; this may be explained by the complexity of the wave-structure interaction during the wave breaking process against the structure, particularly above the swl where air entrapment and the air-water interface adds even more complexity.

These results show both consistencies and inconsistencies with previous research efforts. For example, it is widely recognized (for example, van der Meer and Pilarczyk (1987)) that armour unit stability is a minimum for values of the surf similarity parameter (ξ) in the order of 3 due to the nature of the wave breaking process. Thus, larger wave forces and wave-induced stresses in the armour units are to be expected under these conditions. This is confirmed by the results of this study, in which stresses in the armour units were largest for values of ξ between approximately 2.5 and 4, as shown in Figure 8. It is also generally accepted that forces on the armour layer reach a maximum just below the still water level, as shown by the concentration of damage at this location in prototype structures. In (apparent) contrast, the results of this study suggest that the largest wave-induced forces occur above the water level under severe wave conditions, as shown in Figure 7; however, it is important to note that these results do not include the static load component, which may be a significant factor in the total load acting on an armour unit. Clearly, the magnitude and distribution of static loads in the armour layer must be defined and incorporated in any design procedure. A detailed investigation of static loads in a breakwater armour layer is currently underway.

APPLICATIONS/DESIGN

The primary objective of this study was to complete a parametric investigation of wave-induced loading on breakwater armour units in order to establish statistically significant relationships between the stresses in the armour units and various design parameters. However, this study is also a major component in an ongoing long-term research effort with the overall objective of developing a design procedure for concrete armour units which incorporates both hydraulic stability and structural integrity. Towards this end, two applications for the presentation of the data have been developed, a moment-torque interaction diagram, and a preliminary design chart.

The moment-torque interaction diagram, shown in Figure 11, is based on an interaction equation relating failure in the dolos shank under the combined effects of bending and torsion, and was developed by Scott (1986). Based on the results of finite element modelling, moments measured at the mid-shank location of the dolos load cell are increased by a factor of 2.5 to account for the stress concentration at the fluke-shank interface. The resulting plot shows the location of the measured moments (factored) and torques relative to the theoretical failure line, and clearly shows the structural performance of the unit under the test conditions. This diagram is useful for the presentation of data from a specific breakwater study, for example, the assessment of an existing structure, or the design of a new structure. The interaction equation can be readily modified to account for unit design changes, such as size, geometry, reinforcement, and concrete strength.

The preliminary design chart, derived from the results of this study, is shown in Figure 12. This presentation consists of a plot of the maximum principal stress at the fluke-shank interface versus the dimensionless ratio H/C (wave height/dolos height) for various prototype dolos weights (W), thus allowing the designer to select the required concrete strength for a given application (H/C and W).

Data for all tests, encompassing all wave conditions and armour unit locations tested, was used to estimate the relationship between the maximum principal stress at the mid-shank location and the incident wave height. A linear regression analysis was completed, and the standard deviation of stress at a given wave height was calculated, as shown in Figure 9, in order to derive a reasonable upper limit for the model data. For the purpose of this study, the upper limit was selected as a line one standard deviation above the linear regression line.

The model stress data, measured at the mid-shank location, was increased by a factor of 2.5 to represent stresses at the mid-shank location. This factor is based on the results of a series of finite element model analyses (Scott, 1986), in which the factor varied from approximately 1.7 to 2.7 depending on the loading and boundary conditions assumed. The factored model stresses were scaled to prototype based on the geometric scale of the model.

The hydraulic stability limit shown on the design chart was derived from Hudson's formula, and is defined by:

$$H/C = 0.537(Sr-1)(Kd\cot\alpha)^{1/3}$$

For example, with Sr=2.4, Kd=25.0 (77 SPM, trunk, non-breaking waves), and cot α=1.5, this results in a hydraulic stability limit of H/C=2.52. The Kd value used here is for presentation purposes only and does not reflect the state of the art concerning rubblemound breakwater design.

As mentioned, this design chart is preliminary in nature and is subject to a number limitations, thus emphasizing the considerable research effort required to develop a design procedure incorporating both hydraulic stability and structural integrity. The following comments illustrate the major limitations:

1) The stresses are wave-induced only; static stresses are not included. Scott (1986) has shown that up to 50% of the internal strength of the dolos unit can be used up in resisting the static forces in the armour layer, depending on the concrete strength, unit size and unit location. Dynamic forces may also be significant due to unit motions and inter-unit impacts.

2) Concrete fatigue effects have not been included. Burcharth (1984) has shown that the "endurance limit" (ultimate stress range for n cycles/ultimate stress range for 1 cycle) is approximately 0.6 under quasi-static (pulsating) loads and 0.2 under impact loads.

3) The model data is based on regular waves only; as mentioned earlier, tests have been completed with irregular waves, but analysis of this data has not yet been completed

CONCLUSIONS

This paper has described a parametric study of wave-induced loading on breakwater armour units. The effects of unit location, wave period, and wave height have been investigated and presented. Additional tests with different breakwater slopes and irregular waves have been completed, but the analysis of this data has not yet been completed. Applications of the data include an interaction diagram to demonstrate the structural performance of an armour unit under given conditions, and a preliminary design chart incorporating both hydraulic stability and structural integrity.

SUMMARY OF RESULTS

The following points summarize the results of this study :

- linear relationship between wave-induced stress and wave height
- wave-induced stress tends to increase with increasing wave period
- wave-induced stresses are greater in the bottom armour layer than the top armour layer

- wave-induced stresses are greatest above the still water level for large wave heights
- wave-induced stresses are greatest for values of the surf similarity parameter between 2.5 and 4
- the ratio of average peak stress to standard deviation stress values varies from approximately 1.8 to 3.9, but is typically around 3.2
- the conditional probability distribution of wave-induced stress given wave height can be described by a log-normal distribution

ONGOING AND FUTURE RESEARCH

The reduction and analysis of data from the remaining tests involving regular waves with breakwater slopes of 1:2.0 and 1:2.5 and irregular waves with a slope of 1:2.0 is currently underway. This will also involve a further examination of data distributions and confidence and tolerance limits. In addition, a series of static load tests is underway to provide additional data to the existing quasi-static results.

In the immediate future two issues will be addressed: (i) the relationship between the response of armour units subjected to regular and irregular waves, using the results of the tests currently being analyzed; and (ii) the relationship between mid-shank and fluke-shank stresses in the dolos unit, using both numerical and physical modelling.

In the more distant future it is planned to carry out a study on scale effects in measuring armour unit forces and to develop instrumentation to measure forces in more "current" armour units.

CONTRIBUTION OF THIS STUDY

Significant contributions of this study with respect to the development of a comprehensive design procedure incorporating both hydraulic stability and structural integrity are:

(i) the identification of a linear relationship between wave-induced stress and wave height

(ii) the identification of a log-normal conditional probability distribution of wave-induced stress given wave height

REFERENCES

Baird, W.F., Hall, K.R., Turcke, D.J., and Chadwick, D.C., "Development of a Procedure for the Design of Concrete Armour Units for Rubble Mound Breakwaters", Report prepared for Public Works Canada, 1983.

Burcharth, H.F., "Full Scale Dynamic Testing of Dolosse to Destruction", Coastal Engineering, Vol. 4, No. 3, pp. 229-251, Feb. 1981.

Burcharth, H.F., "A Design Method for Impact-Loaded Slender Armour Units", Proceedings of the ASCE Convention, ASCE, New York, May 1981.

Burcharth, H.F., "Fatigue in Breakwater Concrete Armour Units", Proceedings of the 19th International Conference on Coastal Engineering, pp. 2592-2607, Houston, Texas, Sept. 1984.

Losada, M.A., Medina, R. and Alejo, M., "Wave Forces on Armour Blocks", 21st International Conference on Coastal Engineering, Spain, June 1988.

Howell, G., "Measurements of Forces on Dolos Armour Units at Prototype Scale", 21st International Conference on Coastal Engineering, Spain, June 1988.

Jensen, O.J., and Juhle, J., "Results of Model Tests on 2-D Breakwater Structure", Breakwaters 1988, Eastbourne U.K., May 4-6, 1988.

Nishigari, W., Endo, I., and Shimada, A., "On Stress in Tetrapods under Wave Action", 20th International Conference on Coastal Engineering, pp. 2119-2132, Taipei, Taiwan, Nov. 1986.

Scott, R.D., Baird. W.F., Readshaw, J.S., and Turcke, D.J., "A Procedure for the Analysis and Design of Concrete Armour Units", Proceedings of the 20th International Conference on Coastal Engineering, pp. 1739-1753, Tapei, Taiwan, Nov. 1986a.

Scott, R.D., Turcke, D.J., and Baird, W.F., "A Unique Instrumentation Scheme for Measuring Loads In Model Dolos Units", Proceedings of the 20th International Conference on Coastal Engineering, pp. 2212-2223, Taipei, Taiwan, Nov. 1986b.

Scott, R.D., The Analysis of Concrete Armour Units in a Breakwater, PhD Thesis, Queen's University, December 1986.

Tedesco, J.W. and McDougal, W.G., "Non-linear F.E.M. Analysis of Concrete Armour Units", Presented at the Workshop on Measurement and Analysis of Structural Response in Concrete Armour Units, Vicksburg, Miss., Jan. 23-24, 1985.

Tedesco, J.W., McGill, P.B., and McDougal, W.G., "Structural Response of Dolos in Waves", 21st International Conference on Coastal Engineering, Spain, June 1988.

Timco, G.W., "On the Stability Criterion for Fracture in Dolos Armoured Breakwaters", Coastal Engineering, Vol. 8, No. 2, pp. 161-170,1984.

Timco, G.W., and Mansard, E.P.D., "Improvements In Modelling Rubblemound Breakwaters", Proceedings of the 18th International Conference on Coastal Engineering, pp. 2047-2061, Cape Town, S.A., Nov. 14-19, 1982.

U.S. Army Corps of Engineers, Coastal Engineering Research Center, Shore Protection Manual, 3rd Edition, 1977.

U.S. Army Corps of Engineers, Coastal Engineering Research Center, Shore Protection Manual, 4th Edition, 1984.

van der Meer, J.W., and Pilarczyk, K.W., "Stability of Breakwater Armour Layers - Deterministic and Probabilistic Design", Delft Hydraulics Communication No. 378, February 1987.

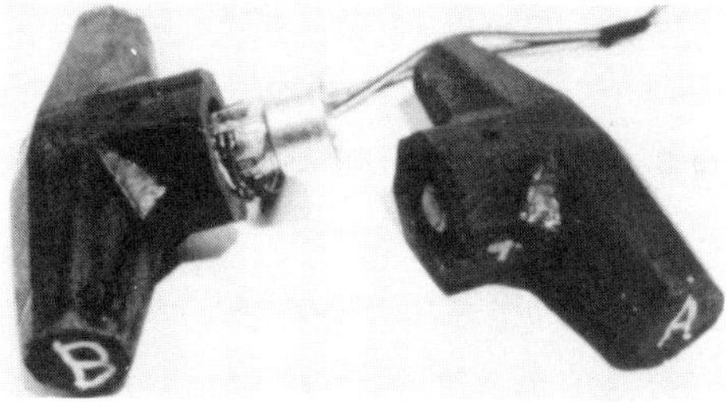

Figure 1. THE LOAD CELL

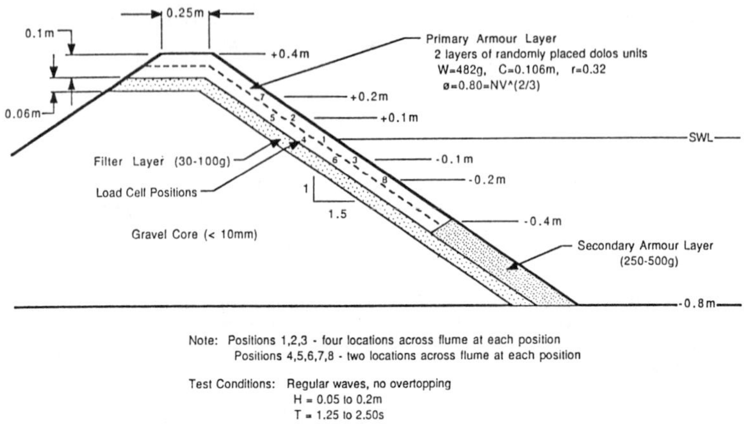

Figure 2. MODEL BREAKWATER CROSS-SECTION

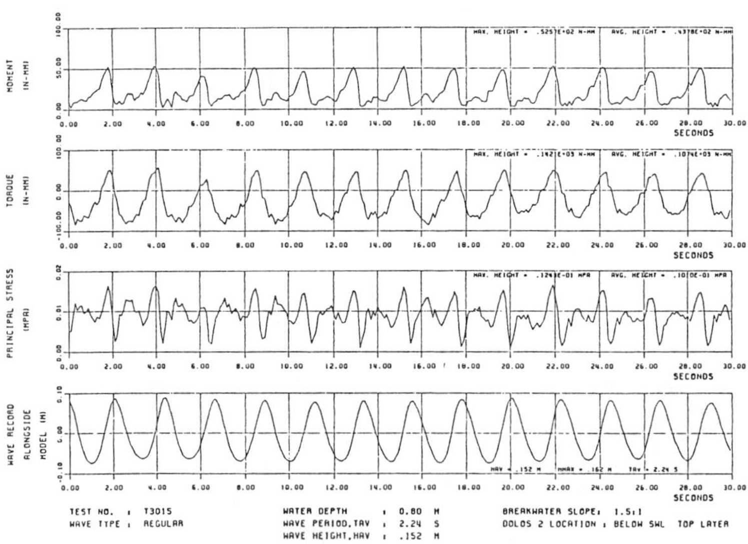

Figure 3. TYPICAL TEST RESULT SHOWING TIME SERIES RESPONSE OF THE LOAD CELL

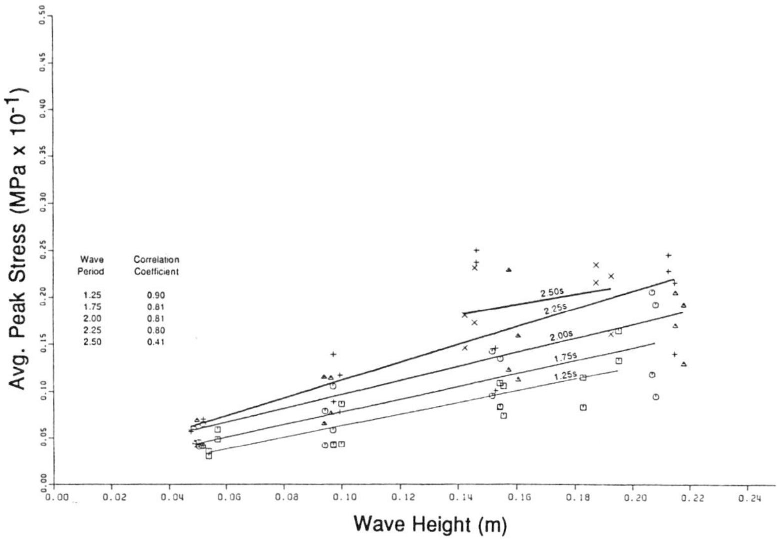

Figure 4. AVERAGE PEAK STRESS VS WAVE HEIGHT
Location 3 (0.1m below SWL, top layer)
Data Sorted by Wave Period

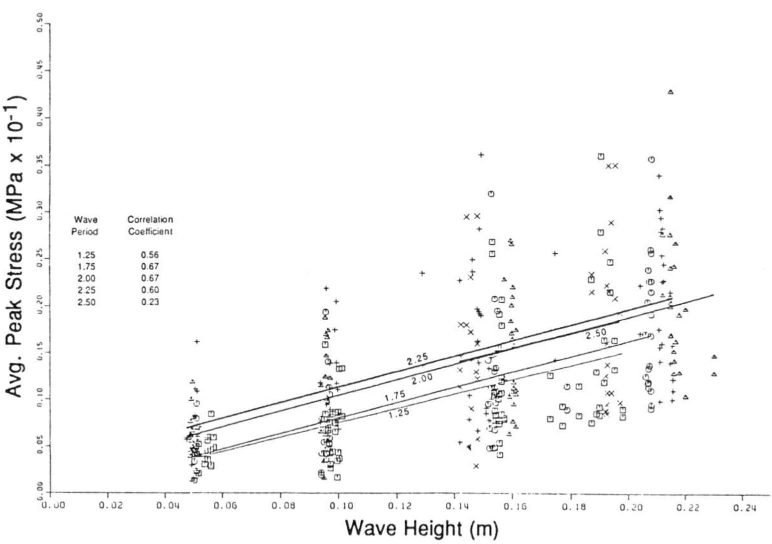

Figure 5. AVERAGE PEAK STRESS VS WAVE HEIGHT
All Locations and Wave Periods
Data Sorted by Wave Period

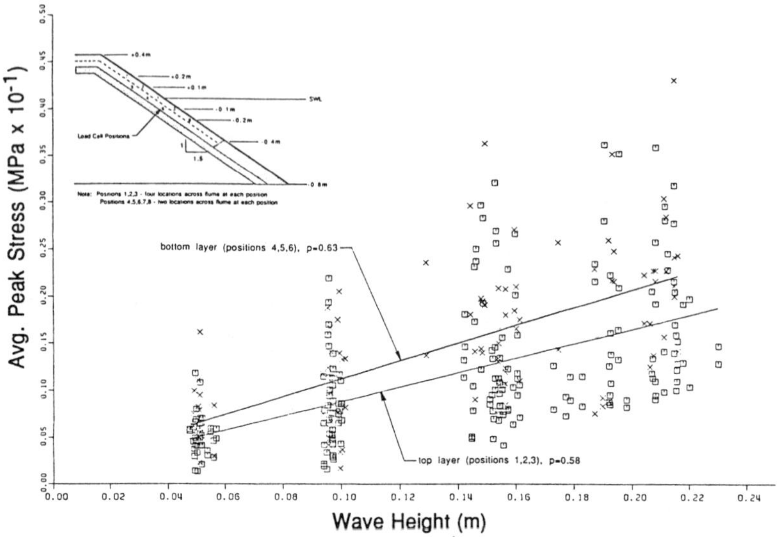

Figure 6. AVERAGE PEAK STRESS VS WAVE HEIGHT
All Locations and Wave Periods
Data Sorted by Layer

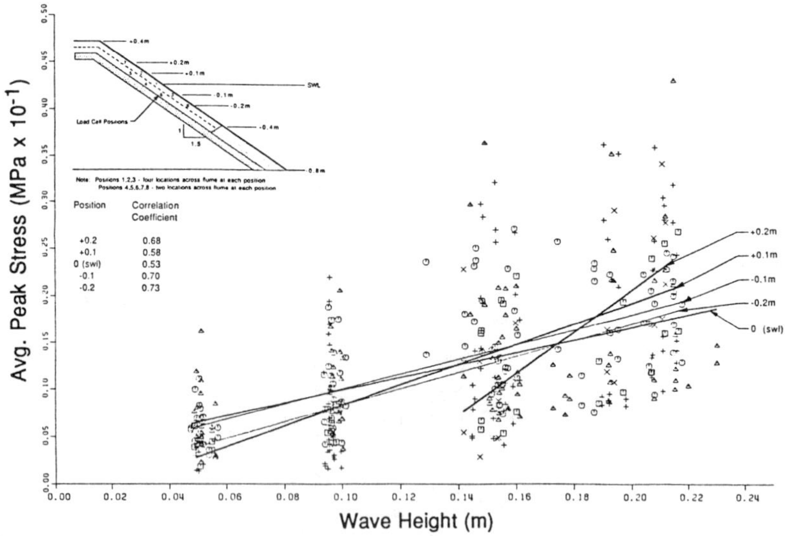

Figure 7. AVERAGE PEAK STRESS VS WAVE HEIGHT
All Locations and Wave Periods
Data Sorted by Position Relative to SWL

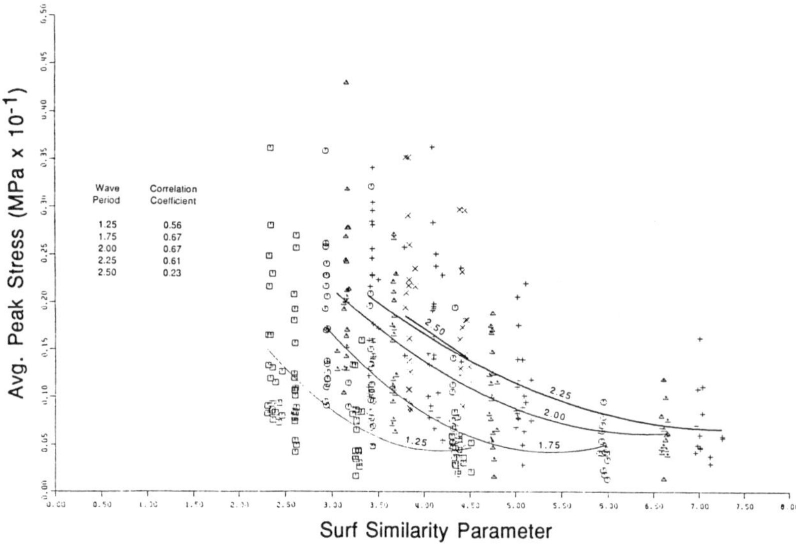

Figure 8. AVERAGE PEAK STRESS VS. SURF SIMILARITY PARAMETER
All Locations and Wave Periods
Data Sorted by Wave Period

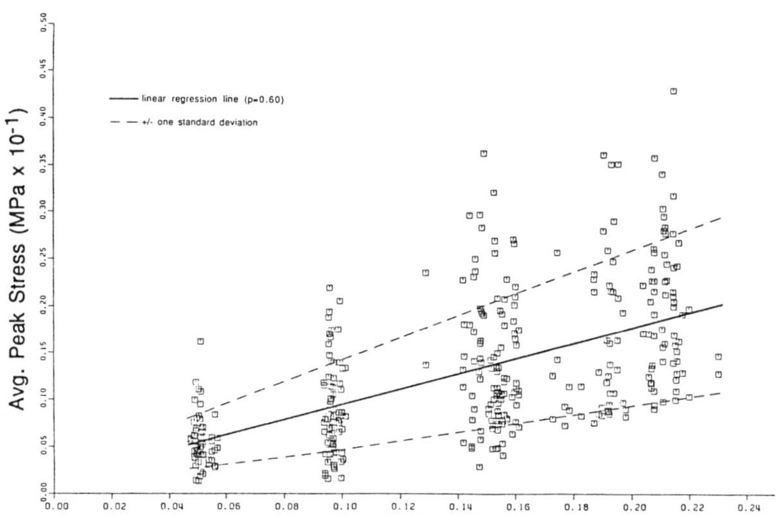

Figure 9. AVERAGE PEAK STRESS VS WAVE HEIGHT
All Locations and Wave Periods

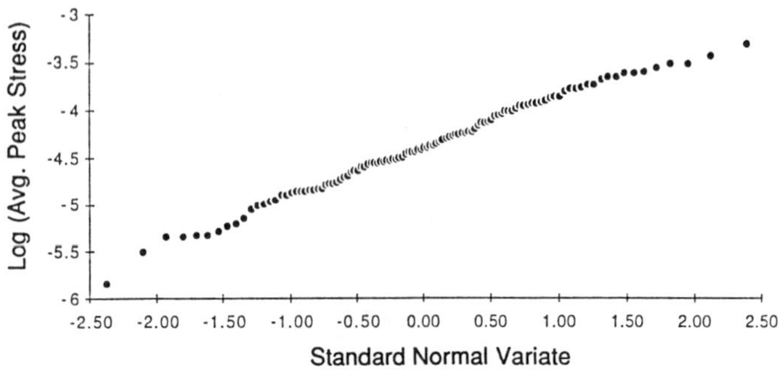

Figure 10. CONDITIONAL PROBABILITY DISTRIBUTION OF AVG. PEAK STRESS
H = 0.15m, All Locations and Wave Periods
Log-Normal Probability Paper

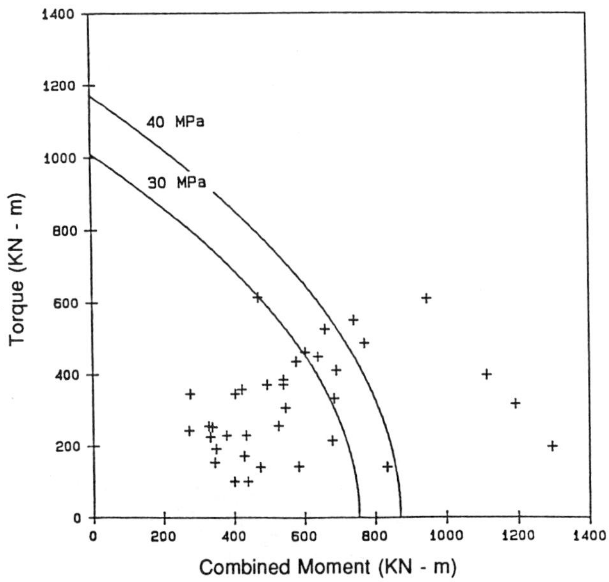

Figure 11. MOMENT TORQUE INTERACTION DIAGRAM
Avg. Peak Forces at Fluke-Shank Interface

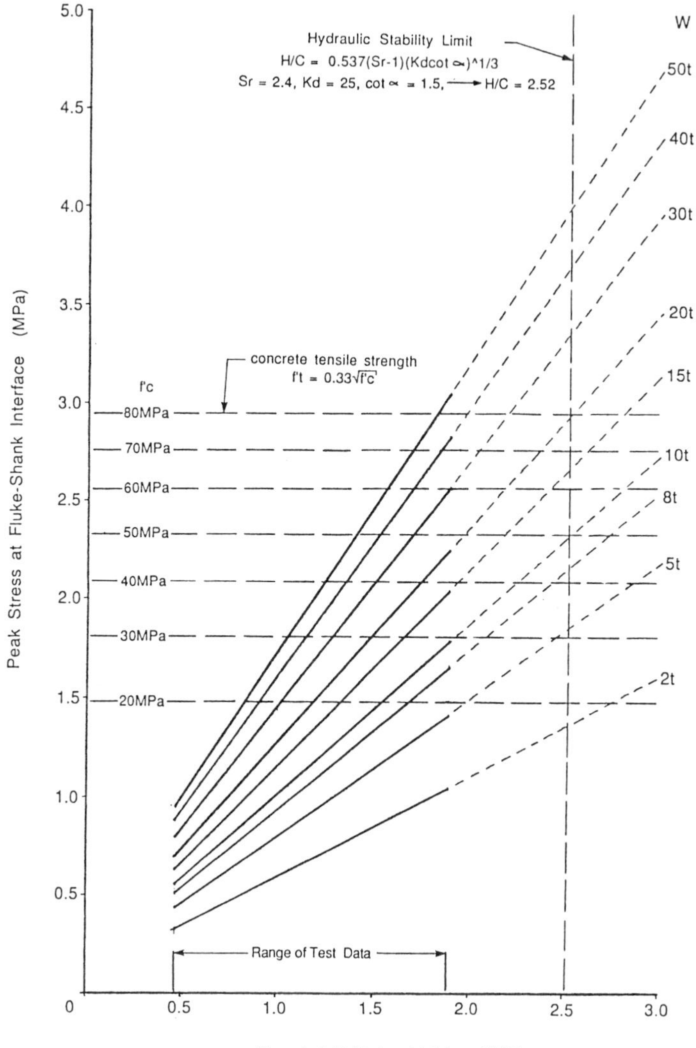

Notes:
1) stress values represent one standard deviation above the model data regression line.
2) stresses factored to mid-shank location by increasing by a factor of 2.5.
3) stresses scaled to prototype by model geometric (length) scale.
4) stresses are wave-induced only - no static stresses are included.
5) concrete fatigue effects have not been included.

Figure 12. PRELIMINARY ARMOUR UNIT DESIGN CHART
Dolos Unit, Breakwater Slope 1:1.5

CHAPTER 177

ARTICULATED CONCRETE MAT SLOPE PROTECTION

Craig B. Leidersdorf, M. ASCE[1]
Peter E. Gadd, M. ASCE[1]
William G. McDougal, M. ASCE[2]

ABSTRACT

This paper provides guidance for the design, fabrication, and installation of articulated concrete mat slope protection. Although articulated mat armor has been utilized for many years in low-energy wave environments, it has been extended recently to accommodate intermediate wave energies and severe ice loads. The development of the concept is discussed, after which hydraulic design considerations, material design considerations, and fabrication and installation techniques are presented. Prototype performance is reviewed. It is concluded that articulated concrete mat armor is capable of providing effective slope protection in intermediate-energy wave environments, and that additional research is required relating to hydraulic stability and failure modes under wave and ice loading.

INTRODUCTION

Articulated concrete mats, consisting of precast blocks connected by flexible linkages, have been used for many years as an alternative to rubble armor in low-energy wave environments. During the past decade, however, the technology has been extended to accommodate intermediate wave energies and severe ice loads in response to the slope protection requirements of artificial islands constructed in the Alaskan Beaufort Sea.

The mat armor system which has evolved, an example of which is depicted in Plate 1, incorporates the findings of conceptual design studies, large-scale model tests, and prototype test sections. Based upon its performance to date, the concept appears promising not only for arctic applications, but also for use as slope and toe protection in temperate environments.

---

1 Principal, Coastal Frontiers Corporation, 2193 Roosevelt Ave., Altadena, CA 91001-3510, U.S.A.
2 Associate Professor, Department of Civil Engineering, Oregon State University, Corvallis, OR 97331, U.S.A.

Plate 1. Concrete Mat Panel

This paper provides guidance for the design, fabrication, and installation of articulated concrete mat armor based on experience gained over the past eight years. The sections which follow present information on the development of the mat concept, a discussion of hydraulic and material design considerations, a discussion of mat fabrication and installation, a review of prototype performance, and recommendations for future applications and additional research.

CONCEPT DEVELOPMENT

The primary advantages of the articulated mat concept, when compared with discrete-unit armor systems such as quarrystone or individually-placed blocks, include:

- enhanced stability, by virtue of the interconnection between adjacent blocks;
- the ability to accommodate changes in the subgrade without permitting catastrophic fill losses; and
- suitability for rapid installation in pre-assembled, modular panels (Ref. Plate 1).

Additional benefits with more limited applicability include resistance to displacement by ice, and the potential for modular removal at the end of the project life.

A potential disadvantage of the mat concept is the critical function performed by the linkage system. If the linkages are weakened by abrasion, corrosion, or fatigue, the ability of the mat to withstand wave and/or ice forces is significantly reduced. Other drawbacks include the relatively high runup elevations which occur on the smooth surface of the mat, the need for mild sea conditions during placement, and the lack of well-established design procedures which are based on prior experience. It is noteworthy, however, that considerable research has been devoted to concrete block revetments in recent years (Bakker and Meijers, 1988; Bezuijen, et al., 1987; Bezuijen, et al., 1988; Galvin, 1988; Klein Breteler, et al., 1988; McDougal and Atkinson, 1987; Pilarczyk, 1988; Toyoshima, 1988). Although the majority of the research addresses placed block revetments, many of the findings are also relevant to the design of articulated mat armor.

The concept of articulated concrete mat armor dates back to the early 1900's (Hawkes, 1907). Prior to the 1980's, however, its use was confined primarily to low-energy environments such as rivers, lakes, and estuaries, in which the design wave height was on the order of 1 m. A number of proprietary and non-proprietary designs were developed to address this application, typically utilizing square or rectangular blocks ranging from 0.3 to 0.6 m in maximum horizontal dimension and 0.15 to 0.3 m in height. In keeping with the relatively modest design loads anticipated, the blocks were not reinforced. Linkages consisted of wire rope, synthetic rope, steel rods, or attachment to a geotextile.

The advent of offshore oil operations in the Alaskan Beaufort Sea engendered slope protection requirements which were significantly more demanding than those of previous articulated mat applications. To support oil exploration and production activities, gravel islands were constructed in water depths to 15 m. Key design considerations included significant wave heights of 3 to 6 m, impact loads from moving ice features, and an extremely short open-water season (typically 75 days) in which to install the slope protection system. Although large sand bags were used to protect many of the initial islands (Gadd, 1988), the need for a more resilient armor system was apparent from the outset (Leidersdorf, et al., 1982).

In view of the lack of quarrystone and high cost of importing non-native materials to northern Alaska, articulated concrete mats emerged as a promising candidate for island slope protection. The linkage concept was attractive from the dual standpoint of reducing the weight of the armor layer and improving resistance to ice forces.

Three major design modifications were implemented to adapt the articulated mat concept to its intended use in the arctic: the provision of high-strength, corrosion-resistant linkages, the introduction of reinforcing steel into the blocks, and the use of larger block sizes. A representative example of the resulting design is presented in Figure 1. Developed for use on Northstar Island, an oil exploration island constructed in 1985 (Hayley, et al., 1987), the mat is composed of 1.2 m square, precast concrete blocks which are reinforced with two layers of wire mesh. Linkage is provided by a combination of forged steel rods cast into each block, and interconnecting chains and shackles. Block thickness was adjusted in accordance with the design wave conditions; in this particular application, 230 mm was specified.

The selection of a 1.2 m square as the plan view outline of the block represents a compromise between the conflicting requirements of functional performance, fabrication, handling, and cost. Factors which favor a large block size include a reduced number of linkages per unit area, and an increased resistance to displacement by environmental forces. Factors which favor a reduced block size include ease of fabrication and handling, the ability to articulate into scour depressions rather than spanning across them, and a higher permeability to alleviate the hydraulic pressure differentials which may develop across the thickness of the mat and cause uplift.

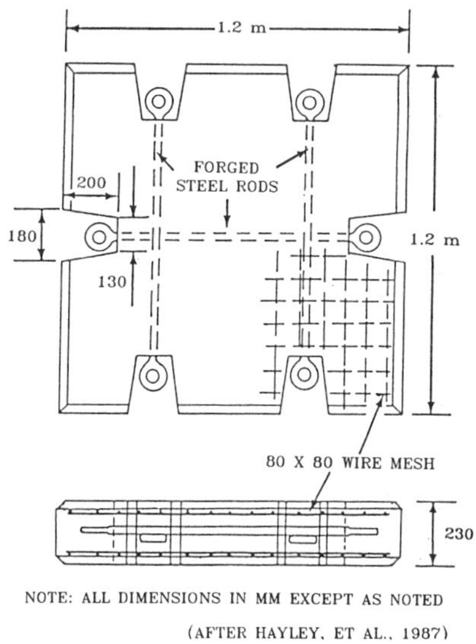

Figure 1. Concrete Block Design

Experience to date has indicated that the 1.2 m square block size approaches the upper bound for effective articulation. An additional benefit of this configuration is the fact that two blocks placed side-by-side are compatible with a standard 2.4-m (8-ft) wide truck bed. To improve the permeability of the mat system, as well as to protect the linkage components, recesses are provided in the perimeter of each block at the linkage points (Ref. Figure 1).

HYDRAULIC DESIGN

Stability

Although the current understanding of articulated mat stability is far from complete, significant insights have been gained from large-scale model testing. Based upon observations made at scales ranging from 1:6 to 1:3, the predominant hydraulic loads exerted on mat armor are as follows (Ref. Figure 2):

(1) The time-dependent wave pressure, which tends to induce uplift by creating a pressure differential between the top of the mat and the subgrade;

(2) The hydrostatic pressure head resulting from a setup of the water level in the structure core, which tends to induce uplift;

(3) The shock pressure from the breaking wave, which can produce uplift forces on the blocks adjacent to the plunge point;

(4) The shear forces from wave runup and especially wave rundown, which tend to displace the blocks in a downslope direction.

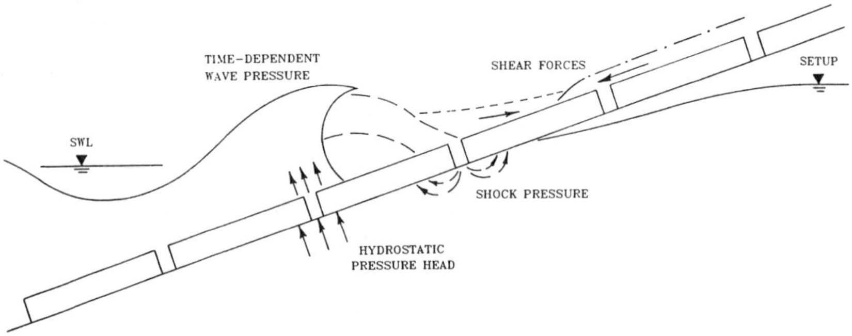

Figure 2. Hydraulic Loads on Articulated Mat Armor

The threshold wave height which initiates block uplift varies not only with the armor weight and slope steepness, but also with the wave period and the permeability of the mat and underlying filtration medium. The provision of adequate permeability is of particular importance in optimizing the performance of a selected mat thickness. In the absence of specific design guidelines, it is recommended that the permeability of each successive layer should increase from the structure core to the mat armor. For the artificial islands constructed to date in the Beaufort Sea, a minimum of 15% open area has been provided in the mat surface to accommodate the anticipated outflow from the permeable gravel core material.

A key feature of articulated mat armor is its ability to withstand wave conditions in excess of the threshold for block uplift. Although deformation of the subgrade will occur, catastrophic failure can be avoided provided that the linkage system and underlying filtration medium remain intact (Potter and Sun, 1983). Both model test and prototype experience have indicated that episodes of block uplift lead to the formation of an S-shaped slope profile, with flattening at the waterline and commensurate steepening above the limit of wave breaking and below the limit of wave rundown.

To resist the shear force exerted by wave rundown, as well as the tendency for blocks to migrate downslope during uplift episodes, the mat must be anchored at the top of the slope. Prototype experience has indicated that deadman anchors buried in the structure core material and linked to the upslope edge of the mat are an economical and effective means of fulfilling this requirement.

At present, there are no analytical models which accurately predict uplift thresholds and anchor loads for articulated mat armor. Large-scale model testing is therefore strongly recommended for the design of all major projects.

Wave Runup

The elevation of wave runup on concrete mat armor varies with the incident wave conditions, the structure slope, and the permeability of the mat and subgrade. In general, however, relatively high runup elevations are observed on the smooth surface of the mat.

Representative large-scale model test data reported by Gadd, et al. (1988), are reproduced in Figures 3 and 4. The data were acquired for breaking waves on a 3(H):1(V) gravel slope covered with a permeable geotextile and an articulated mat containing approximately 15% open area. Relative runup, R/H_o', is displayed as a function of wave steepness, H_o'/gT^2, for two values of relative water depth: $d_s/H_o' = 2$ in Figure 3, and $d_s/H_o' > 3$ in Figure 4. The notation is defined as follows:

- R = runup elevation relative to the prevailing still water level;
- H_o' = unrefracted deepwater wave height;
- g = gravitational acceleration;
- T = wave period; and
- d_s = water depth at structure toe.

Runup predictions for concrete blocks derived from the U.S. Army Corps of Engineers' Shore Protection Manual ("SPM"; 1984) are included in each figure. Although the predictions evidence good agreement with the model test

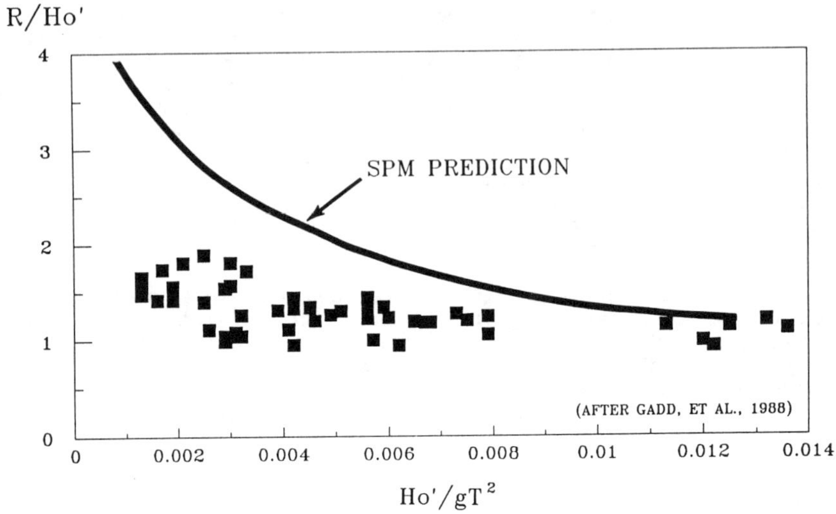

Figure 3. Wave Runup on 3(H):1(V) Slope, $d_s/H_o'=2$

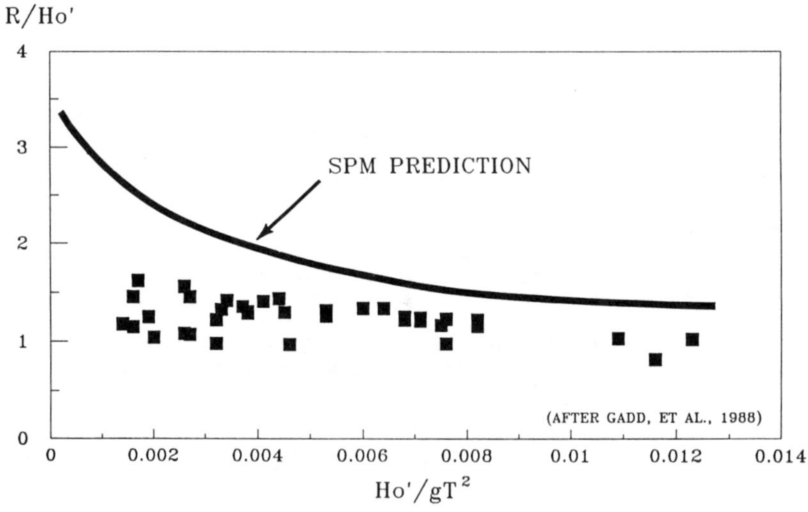

Figure 4. Wave Runup on 3(H):1(V) Slope, $d_s/H_o'\geq 3$

data at high values of wave steepness, they significantly overpredict the measured runup elevations at the lower steepness values. In consequence, the use of the SPM method to predict wave runup may be expected to yield conservative values in most instances. As in the case of hydraulic stability, large-scale model testing is recommended to evaluate the runup characteristics of all major projects.

Large-scale testing has also provided insight into measures which can reduce the runup elevations on articulated concrete mat armor. Increased roughness, as might be created by using blocks of different thicknesses, was found to have minimal effect in the region of wave breaking. If placed higher on the slope, however, in the zone of swash or sheet runup, the increased roughness significantly reduced the maximum runup elevation. Compound slope profiles which incorporated a bench near the waterline were also found to be effective. As reported in more detail by Leidersdorf, et al., (1984), a bench width of 12 m typically decreased the runup elevation by 30 to 40% relative to a straight 3(H):1(V) slope.

MATERIAL DESIGN

Concrete Blocks

Because the concrete blocks in an articulated concrete mat must withstand impacts with both the subgrade and each other during uplift episodes, they must be more durable than those used for unlinked revetments. Additional material design considerations may include abrasion from littoral sediments, corrosion and fatigue of exposed steel, and, in the case of cold climates, ice impacts, ice abrasion, and freeze-thaw cycling.

As discussed previously, steel reinforcement was incorporated into the block design when the mat technology was adapted to the more severe wave and ice conditions of the arctic. Prototype experience to date suggests that a steel content of approximately 0.5% of the block cross-section is sufficient to maintain structural integrity except in the case of impact loads exerted by very large ice features. The reinforcement should extend throughout the plan view outline of the block, including the arms between the linkage recesses (Ref. Figure 1), to insure that the block will remain intact even if the concrete is fractured. Hot-dip galvanization of all steel is recommended to reduce its susceptibility to corrosion.

The concrete should be of high strength to resist cracking. In the case of the block design presented in Figure 1, a 28-day compressive strength of 27 MPa was specified (Hayley, et al., 1987). Low permeability, which may be obtained through the use of a low water-cement ratio, a high cement factor, and the use of pozzolan or silica fume, is desirable to reduce chloride penetration.

Aggregate should be abrasion-resistant, especially when repeated exposure to littoral sediments and/or ice is anticipated. Additional design considerations for arctic service are presented by Leidersdorf (1988).

Linkage System

A properly designed linkage system will not only allow the mat to function under wave conditions which exceed the uplift threshold for individual blocks, but also facilitate rapid installation through the placement of large, pre-assembled panels. Both the placement loads and the in-place loads should be considered in determining the requisite strength of the linkage system. Additional design considerations include flexibility (to allow articulation), corrosion, abrasion, and fatigue over the design life of the armor system, and the ability to allow replacement of damaged blocks.

Kevlar rope, wire rope, and chain-and-shackle linkage systems have been tested at prototype scale with varying degrees of success. Although attractive from the standpoint of imperviousness to corrosion, Kevlar rope proved to be too elastic to permit accurate placement of large panels. Wire rope can be used for small panel sizes, but becomes difficult to handle at the large sizes required for heavy lifts. Additional disadvantages of wire rope include its susceptibility to corrosion (due to the large surface area of the component strands), and the difficulty associated with replacing individual blocks after the mat has been installed.

Chains and shackles, in conjunction with forged steel rods cast into the blocks, have emerged as the preferred linkage alternative (Ref. Plate 2). Connections between adjacent blocks are readily made, damaged blocks can be removed and replaced as needed, and corrosion is less problematical than with wire rope due to the reduced surface area. If the corrosion rate in the project area can be estimated, the linkage components can be oversized accordingly to provide for sufficient strength at the end of the design life. Hot-dip galvanization of all linkage hardware is also recommended.

Filtration Medium

The performance of an articulated mat armor system depends not only upon the characteristics of the mat itself, but also upon those of the underlying filtration medium. In addition to providing adequate permeability to relieve pore pressures in the structure core, the filter must be capable of retaining the core material during episodes of mat uplift.

The absence of suitable granular fill material has necessitated the use of geotextiles for filtration on all articulated mat armor installed in the Beaufort Sea to date. The advantages of fabric include the ability to

Plate 2. Chain-and-Shackle Linkage System

achieve a specified permeability, ease of installation, and low cost. The primary disadvantages relate to long-term performance: a susceptibility to abrasion by the concrete blocks, and progressive deterioration from exposure to the ultraviolet (UV) component of sunlight. Measures taken to ameliorate these effects have included the specification of UV-stabilized fabrics, the application of an abrasion-resistant resin coating, and the use of an overstrength fabric in anticipation of deterioration over the project life (Munday and Bricker, 1987). Although performance has been satisfactory to date, longer periods of service will be required to evaluate their long-term effectiveness.

In the absence of conclusive evidence regarding the long-term durability of geotextiles, their use is recommended for projects with a limited design life, or at which frequent mat uplift is not anticipated. When severe wave conditions and/or exposure to intense sunlight are expected, particularly over an extended design life, the use of graded stone or manufactured concrete rubble is recommended. Although rubble has not been used as a filtration medium for articulated mat armor at prototype scale, large-scale model tests have yielded satisfactory results (Leidersdorf, et al., 1984). To prevent washout, the filter layer immediately beneath the mat should consist of rubble which is larger than the largest openings in the mat surface.

FABRICATION AND INSTALLATION

Both wet and dry casting techniques have been used to fabricate the individual concrete blocks which comprise the mat. In the case of dry casting, a block molding machine is utilized to inject concrete with a low water-cement ratio into the forms. Although set-up of the molding machine and adjustment of the concrete mix can be time-consuming, the semi-automated block production process which results appears warranted for large projects. Fabrication rates of up to 20 blocks per machine-hour have been reported (Munday and Bricker, 1987).

When the blocks have cured sufficiently to be handled, they can be assembled into large panels in preparation for placement by crane. To minimize the number of underwater linkages required between panels, it is recommended that each panel extend from the lower terminus of the armor to the waterline. The width of the panel can be adjusted in accordance with the lifting capacity of the crane.

To insure that proper alignment is maintained during placement, the centerline position of each panel is monitored continuously using a theodolite. Any adjustments which are required are relayed to the crane operator. Once the panel has been lowered into position, a diver verifies that the desired side-by-side relation with the preceding panel has been achieved before releasing the mat from the crane. The diver then shackles the new panel to its predecessor, and the cycle is repeated. Based upon prototype experience to date, the underwater shackling is relatively straightforward, and is typically completed before the next panel has been lifted by the crane.

Continuous coverage of non-planar surfaces, such as island corners, can be achieved through the use of trapezoidally-shaped blocks (Ref. Plate 3). Fabrication and placement procedures are analogous to those for square blocks, although survey control of the placement operation assumes greater importance in maintaining proper alignment.

The maximum permissible wave height for mat installation depends not only upon the tolerance of the divers, but also upon the extent to which unevenness can be accepted in the finished grade. Because the mat represents a relatively thin veneer on the structure, any wave-cut scarps which develop prior to mat placement will be reflected in the mat surface.

Placement rates for articulated mat armor vary with crane capacity, wave conditions, water depth, and structure geometry. Reported rates have ranged from 44 m^2/h for an exposed location with a maximum placement depth of 6 m (Hayley, et al., 1987), to 110 m^2/h for a more sheltered location with a maximum placement depth of 3 m (Munday and Bricker, 1987).

PERFORMANCE ASSESSMENT

Articulated concrete mat armor was first installed in the Alaskan Beaufort Sea in 1980 (Leidersdorf, et al.,

Plate 3. Island Corner Armored with Trapezoidal Blocks

Plate 4. Articulated Concrete Mat on the Endicott Project

1982). Subsequent projects have utilized more than 57,000 m^2 of mat armor in coastal and offshore locations. The primary application has been as slope protection for gravel islands, although mats have also been used as toe protection for seawall structures.

The most noteworthy projects to date which employ mat armor are the Endicott Project, an oil production facility consisting of two islands and a causeway in a maximum depth of 3.7 m, and Northstar Island, an oil exploration island located in a depth of 13.7 m. On the Endicott islands, approximately 40,000 m^2 of mat were installed from just above the waterline to the sea floor (Ref. Plate 4). To accommodate the anticipated design life of 25 years, a woven polyester filter fabric with an extremely high tensile strength of 2.1 kN/cm was specified (Munday and Bricker, 1987). Installation of the slope protection system was completed in 1986, one year ahead of schedule. Since that time, the mat has remained in sound condition.

Northstar Island, designed for a minimum service life of three years, is distinguished by its exposure to more extreme environmental conditions than the Endicott Project. A band of concrete mat was installed around the circumference of the hexagonal island from +1.5 to -6.1 m, the region most subject to wave and ice impacts (Hayley, et al., 1987). In September, 1986, one year after the island was completed, it was exposed to a severe storm with estimated wave heights of 3 to 4.5 m. Although the mat experienced uplift, no significant adverse effects were noted. Damage was inflicted, however, by a late-1987 storm which drove large, deep-keeled multi-year ice features against the island. The damage consisted of block fractures and displacements in areas which received direct impacts from the ice.

CONCLUSIONS AND RECOMMENDATIONS

Based upon the experience gained to date, articulated concrete mat armor is capable of providing effective slope protection in intermediate-energy wave environments. Promising applications include not only the arctic projects for which it was developed, but also selected non-arctic uses such as:

1. Long-term slope protection against waves and ice in low-corrosion environments (such as the Great Lakes);

2. Rapidly-deployed emergency protection for coastal areas threatened by storm erosion; and

3. "Last Line of Defense" protection, in which the mat and an underlying geotextile are buried behind the beach face and exposed only during major storm events.

In the case of the last suggested application, the relatively thin nature of mat armor will facilitate its concealment in the backshore. If relocation becomes necessary to accommodate a moving shoreline (occasioned, for example, by long-term erosion or beach renourishment), the mat can be moved in modular panels.

Despite the promise demonstrated by the mat technology, additional research is required before a standardized design methodology can be elucidated. Areas which warrant further investigation include failure modes under wave and ice loading, quantification of hydraulic stability in terms of the primary design parameters, hydraulic stability on non-planar surfaces such as island corners, and geotextile performance over an extended design life or when frequent uplift is experienced. Until a more comprehensive understanding of mat performance is achieved, the final design process for all major projects should include large-scale hydraulic model testing, and, when feasible, observation of existing prototype structures.

REFERENCES

Bakker, K.J., and P. Meijers, 1988, "Stability of Sliding of Flexible Revetments", Proceedings of Modelling Soil-Water-Structure Interactions, Delft, The Netherlands, pp. 163-172.

Bezuijen, A., M. Klein Breteler, and K.J. Bakker, 1987, "Design Criteria for Placed Block Revetments and Granular Filters", Proceedings of 2nd International COPEDEC, Beijing, China.

Bezuijen, A., J. Woulters, and C. Laustrup, 1988, "Block Revetment Design with Physical and Numerical Models", Abstracts Book, 21st International Conference on Coastal Engineering, Torremolinos, Spain, pp. 121-122.

Gadd, P.E., 1988, "Sand Bag Slope Protection: Design, Construction, and Performance", in Chen, A.C.T., and C. B. Leidersdorf, Arctic Coastal Processes and Slope Protection Design, American Society of Civil Engineers, pp. 145-165.

Gadd, P.E., V. Manikian, and J.L. Machemehl, 1988, "Wave Run-Up Prediction vs. Model Test Results: A Comparison", Proceedings of the 21st International Conference on Coastal Engineering, American Society of Civil Engineers.

Galvin, C., 1988, "Stability of Tabular Concrete Revetments", Abstracts Book, 21st International Conference on Coastal Engineering, Torremolinos, Spain, pp. 163-164.

Hayley, D.W., P.E. Gadd, and D.A. Horn, 1987, "Advances in Design and Construction of Slope Protection for Gravel Islands, Beaufort Sea, Alaska", Proceedings of the 9th International Conference on Port and Ocean Engineering under Arctic Conditions (POAC'87), Fairbanks, Alaska.

Hawkes, F.W., 1907, "Riprap", Patent No. 869,566, United States Patent Office, Washington, D.C.

Klein Breteler, M., A.M. Burger, L. Banach, and A. Bezuijen, 1988, "Analytical Design Method for Block Revetments", Abstracts Book, 21st International Conference on Coastal Engineering, Torremolinos, Spain, pp. 159-160.

Leidersdorf, C.B., 1988, "Articulated Mat Slope Protection for Arctic Applications", in Chen, A.C.T., and C. B. Leidersdorf, Arctic Coastal Processes and Slope Protection Design, American Society of Civil Engineers, pp. 166-189.

Leidersdorf, C.B., R.E. Potter, B.C. Gerwick, Jr., and Y.Y. Hsu, 1982, "Modular Slope Protection for the Arctic Environment", Proceedings of the 14th Annual Offshore Technology Conference, Houston, Texas, pp. 689-704.

Leidersdorf, C.B., R.E. Potter, and C.J. Sonu, 1984, "Study of Slope Protection Works for the Arctic Ocean", Proceedings of the 31st Annual Coastal Engineering Conference, Japan Society of Civil Engineers, pp. 552-556.

McDougal, W.G., and F.S. Atkinson, 1987, "Concrete Armor Mats: Large Scale Wave Tank Tests", Proceedings of Coastal Zone' 87, American Society of Civil Engineers, pp. 1617-1629.

Munday, J.P., and W.F. Bricker, 1987, "Endicott Slope Protection Design and Construction", Proceedings of the 6th International Conference on Offshore Mechanics and Arctic Engineering (OMAE '87), Houston, Texas, pp.47-54.

Pilarczyk, K.W., 1988, "Design Aspects of Block Revetments", Abstracts Book, 21st International Conference on Coastal Engineering, Torremolinos, Spain, pp. 195-196.

Potter, R.E. and J. Sun, , 1983, "Arctic Slope Protection Development", Proceedings of the 7th International Conference on Port and Ocean Engineering under Arctic Conditions (POAC '83), Helsinki, Finland.

Toyoshima, O., 1988, "Gentle Slope Seawalls Covered with Armor Units", Abstracts Book, 21st International Conference on Coastal Engineering, Torremolinos, Spain, pp. 159-160.

U.S. Army Corps of Engineers, Department of the Army, 1984, Shore Protection Manual, Vols. I and II, U.S. Government Printing Office, Washington, D.C.

CHAPTER 178

DESIGN AND CONSTRUCTION OF BERM TYPE BREAKWATERS

Rubble Mound Structures Committee
Waterways, Port, Coastal & Ocean Division,
American Society of Civil Engineers

Baird, William F., W.F. Baird & Associates, Ottawa, Canada
Davidson, D.D., Coastal Engineering Research Center, Vicksburg, MS, USA
Edge, Billy, Cubit Engineering Ltd., Charleston, S.C., USA
Magoon, Orville T., Committee Chairman, P.O. Box 279, Middleton, CA, USA
Rauw, Charles I., Winzler & Kelly, San Francisco, CA, USA
Treadwell, Donald, Geomatrix Consultants, San Francisco, CA, USA

The standard design for a rubble-mound breakwater as reported in recent text books and manuals has consisted of a core of rubble quarry-run that is protected from wave action by two layers of relatively large quarried stone or concrete units. Filter layers of intermediate size stone are recommended beneath the armour layer to prevent loss of the core material through the armour. The pioneering work of Hudson provides coefficients for a formula that allowed widespread use of this design concept throughout the world.

A breakwater of this form tends to require the smallest volume of stone; however, it may not represent the least cost structure for a specific location because of the cost and availability of local materials.

Recently, design engineers and hydraulic laboratories have given considerable attention to alternative forms of rubble mound structures. The objective of this work has been to minimize cost while maintaining the same or improved level of stability when the breakwater is subjected to extreme wave conditions.

Principally, emphasis has been given to the utilization of locally available quarried stone and to maximizing the use of the full yield of a local quarry. This has required the design of breakwaters using smaller armour stone and a wider range of sizes than was used in the more conventional breakwater. These designs have also given consideration to the realities of construction and the limitation of construction equipment on the assumption that cost savings will be achieved with relatively simple construction methods.

Clearly the use of smaller armour stones requires a change in other properties or characteristics of the armour layer, compared to a conventional design, in order to achieve the same stability. The principal characteristics of a berm breakwater are the high permeability of the armour-layer and the significantly larger volume of armour. Berm breakwaters generally consist of a relatively large and permeable mass of armour stone (of smaller size than required for a conventional design).

Breakwaters range from coarse sand barrier beaches to a more conventional breakwater protected by one layer of large armour stones. For the breakwaters using the smaller material, movement of the stone is an important consideration. For breakwaters with larger stones where significant movement of stones does not occur the permeability of the mass of stones is an important consideration. In all structures the durability and strength of the armour stones is important.

The recent work towards "non-conventional" structures is based on extensive tests in hydraulic laboratories. Basically, a trial and error approach in undertaken and the success or failure of a structure is based on the observations of the model.

In September 1987, a two day workshop was held in Ottawa, Canada to discuss "non-conventional" breakwaters (1). Participants from Australia, Canada, Denmark, Netherlands, United Kingdom and the United States discussed design concepts, construction, materials and the performance of as-built structures. The workshop was sponsored by the American Society of Civil Engineers, Canadian Society for Civil Engineering and the National Research Council of Canada.

For the most part, the "non-conventional" breakwaters discussed consisted to a larger mass of smaller stone replacing the two layers of armour stone in a conventional breakwater. These breakwaters were referred to as mass armoured breakwaters or berm breakwaters.

Considerable discussion took place on the process by which a wave interacts with the mass of stone and processes by which stability is achieved.

It was concluded that very successful non-conventional breakwaters have been built in recent years. These structures have resulted in considerable cost savings to the owner while providing the required protection and stability when exposed to extreme waves.

However, these designs have all been supported by considerable engineering and model tests in a hydraulics laboratory. While numerical methods are being developed to assist with this process and to compare one design with another, it is unlikely that the site specific design work that can be undertaken with a physical model will be replaced by numerical methods in the near future.

Based on the results of the Ottawa workshop, it is concluded that properly designed and constructed Berm Type Breakwaters are an important and cost effective means of protecting exposed coastal harbors and locations from severe wave attack at specific locations.

The Rubble Mound Structures Committee acknowledges the contribution of Mr. Dave Willis and the Ottawa Hydraulic Research Laboratory of the National Research Council of Canada for assistance in providing the berm breakwater laboratory demonstration and venue for the workshop.

REFERENCES

(1) Berm Breakwaters: Unconventional Rubble-Mound Breakwaters. Proceedings of a workshop at The Hydraulics Laboratory, National Research Council of Canada, Ottawa, Canada, September 15-16, 1987. Published by American Society of Civil Engineers.

CHAPTER 179

SIMPLIFIED MODELS FOR MEASURING ARMOUR UNIT FORCES

R.D. Scott*, D.J. Turcke**, C.D. Anglin*, W.F. Baird***

ABSTRACT

Damage of concrete armour units on a number of breakwaters around the world has brought into question the design methodology for these structures.

The interaction of waves with a rubblemound breakwater is a complex process typified by the random nature of the wave loadings and of the placement and orientation of the armour units.

The objective of this paper is to present the development of a more simplistic physical model for establishing a realistic upperbound for the forces occurring on concrete armour units, to more readily understand the response of the unit to various wave parameters, and to explore the effect of carrying out tests at different scales.

The results of these investigations demonstrate the applicability of employing simplified models to assist in understanding the response of armour in a breakwater and the consistency of the instrumentation at different scales.

INTRODUCTION

Damage of concrete armour units on a number of breakwaters throughout the world has brought into question the design methods of these structures. In certain cases, such as at the spectacular failure of the breakwater at Sines, Portugal, it is believed that damage of the structure may have been initiated by the structural failure of the individual concrete armour units. Generally, these breakwaters have been designed on the basis of physical model studies that assessed the hydraulic stability of the proposed breakwater but did not reproduce the mechanical strength of the concrete armour units. Recently, considerable research effort has been directed towards developing an understanding of the structural strength of armour units and of the environmental forces that act on these units.

* Senior Engineers, W.F. Baird and Associates, Coastal Engineers
38 Antares Drive, Suite 150, Ottawa, Ontario, Canada, K2E 7V2

** President, W.F. Baird and Associates, Coastal Engineers
38 Antares Drive, Suite 150, Ottawa, Ontario, Canada, K2E 7V2

*** Professor of Civil Engineering, Queen's University,
Kingston, Ontario, Canada, K7L 3N6

The interaction of waves with a rubblemound breakwater is a complex process typified by the random nature of the wave loadings and of the placement and orientation of the armour units. Recent work by the authors [Scott et al (1986), Anglin et al (1988)] has focused on the measurement of the response of concrete armour units to external forces using physical modelling. A considerable number of tests have been carried out using specially instrumented model dolos units placed in a model breakwater and subjected to simulated wave attack. This paper presents the results of some initial research work undertaken to:

1. Develop a more simplistic physical model that will establish a realistic upperbound for the forces occurring on concrete armour units.

2. Test the instrumented armour units under more simplified conditions so that the response of the armour unit to changes in various wave parameters may be more readily understood.

3. Provide further confidence in the use of such instrumentation by carrying out tests at various scales.

RECENT RESEARCH

Over the past few years a number of investigations have been initiated into the development of improved breakwater design procedures which incorporate both hydraulic stability and structural integrity.

In order to develop a more appropriate design procedure for concrete armour units, it is important to quantify the nature of the forces acting on the units in a breakwater. Although this represents a complex task at the present, prototype studies, physical and numerical modelling are all playing an important role in trying to solve this difficult problem. A number of researchers have made use of physical models to measure the force response of armour units. [Scott (1986), Scott et al (1986, 1987a, 1987b) Anglin (1988), Baird et al (1986), Endo (1985), Losada et al (1988), Jensen (1988)].

Attempts have been made to create design charts which can serve as preliminary tool for a design engineer. One such chart has been developed for a dolos unit in Baird et al (1986). In this work moment-torque interaction diagrams are developed which are graphical representations of the measured moments and torques scaled to prototype relative to a theoretically derived interaction equation describing failure of the concrete in the dolos shank. This type of representation of data is potentially useful for both the analysis and design of a breakwater armoured with concrete units.

More recently, a preliminary effort to provide a design chart which incorporates both the hydraulic stability and structural integrity was presented by Anglin (1988). This design chart is a plot of peak stress versus the ratio of wave height to dolos height for various dolos weight. It allows the designer to select an appropriate dolos size given the wave height and concrete strength.

In addition to the physical model studies, a very extensive prototype study is being carried out at Crescent City by the U.S. Coastal Engineering Research Council, to measure dolos response in an existing breakwater [Howell (1988)].

Finally, a non-linear numerical model is being developed to determine the response of dolos to waves [Tedesco and McDougal (1985)].

All of the above studies are extremely complex, time consuming and in most cases, very costly.

THE ARMOUR UNIT LOAD CELL

After extensive research [Scott, Turcke and Baird (1986), Scott (1986)] a unique armour unit "load cell" was developed for measuring the response of concrete armour units to forces in a breakwater. Utilizing the concept of geometric distortion, a thin walled aluminum tube instrumented with strain gauges was inserted at the mid-shank location of a dolos unit, as shown in Figure 1. The tube was extremely thin in order to maximize the strains so that they were measurable under even the smallest loads. By calibrating the output voltages of the strain gauge circuits against known loading conditions, an armour unit load cell was created which measured the internal forces at the mid-shank location of the dolos unit.

Subsequent to this earlier work, a number of significant improvements have been made to the unit. The major changes have been to the wiring system so as to minimize the effects of the lead wires on the accuracy of the load cell. The newer load cell has been designed to be of more modular construction so it can be rapidly dissassembled. In addition, the unit was made more durable to withstand repeated testing in a hydraulic flume, particularly at the attachment points where the inner instrumentation tube connects with the armour unit itself.

Physical modelling of the behavior of concrete armour units in a breakwater subjected to wave attack is exceptionally difficult. The very nature of the environment presents difficult problems in the development of instrumentation to measure armour unit forces.

There are scaling problems in that both Froude and Cauchy scaling criteria cannot both be satisfied simultaneously by such a model using typical materials. The response of the model units to static and quasi-static forces may be readily determined but dynamic forces, particularly impact forces, are not easily measured or scaled to prototype.

Coupled with these difficulties is the random nature of the armour orientation and placement in the breakwater and the wave loading which, consequently, requires a large number of costly tests to be carried out in order that a statistical analysis of the data can be performed. This suggests the need for the development of a more simplified testing program.

SIMPLIFIED MODEL TESTS

Given the significant number of difficulties in determining forces in concrete armour units within a hydraulic model, an investigation was carried out to develop a simplified model for measuring these forces.

As a result, a model test that may be easily carried out has been developed which provides a realistic upper limit to the forces measured in a breakwater. The force measuring device selected was the dolos load cell as developed in Scott (1986), thoroughly described by Scott, Turcke and Baird (1986), and shown in Figure 2. The test set-up is presented in Figure 3. The instrumented dolos unit of height "d" is rigidly fixed on a plywood ramp. Both the position of the unit on the ramp represented by the parameter "h", and the ramp slope can be varied.

The tests reported here included three sets of regular waves with heights of 15, 18 and 24 cm and all having a period of 1.7 sec. The ramp was set at a slope of 1.5:1. Figure 5 illustrates a typical moment - time history plot which demonstrates good repeatability from one wave to another. Figures 5 and 6 show plots of the average peak moment verses the position of the unit on the slope for the 15 cm and 24 cm waves respectively. It is interesting to note that the results of these tests clearly demonstrated that the maximum moment occurs just below the still water level and as expected larger moments are experienced by the unit when subjected to the higher wave heights, independent of its position on the slope.

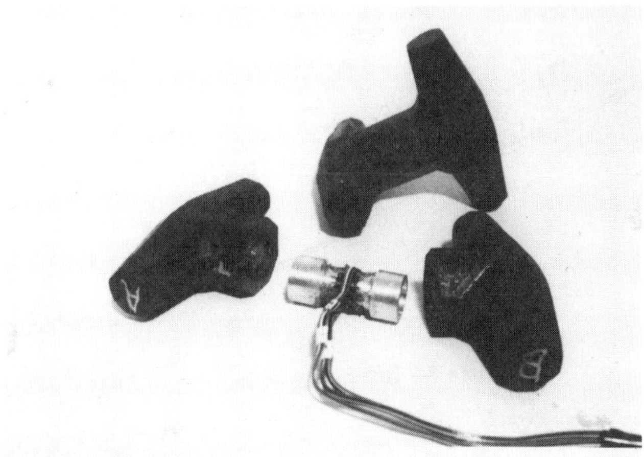

Figure 1 - Original "Load Cell"

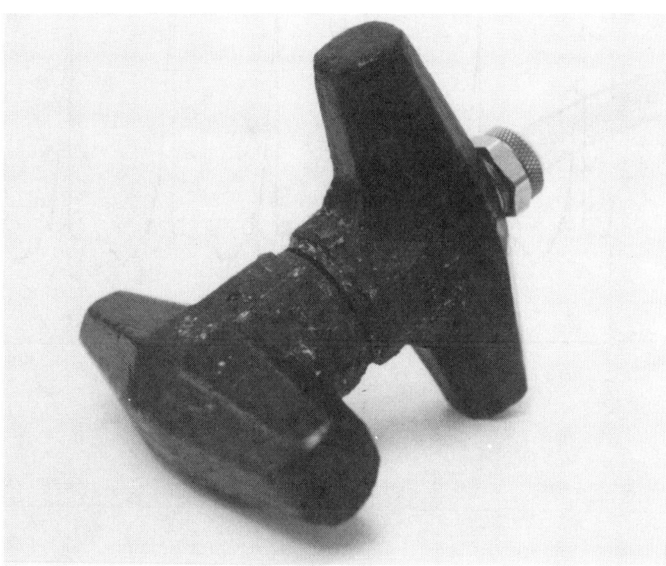

Figure 2 - Current "Load Cell"

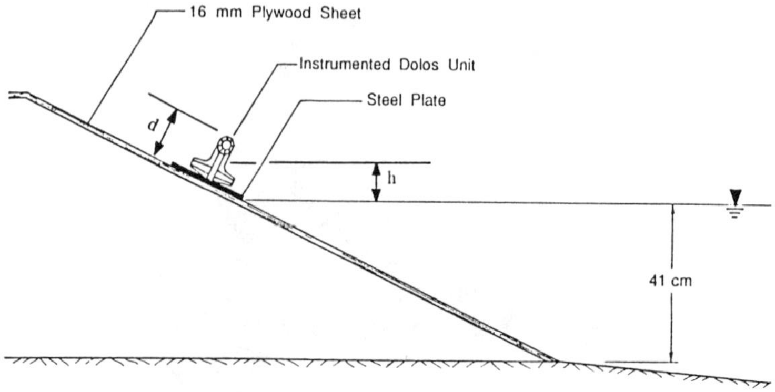

Figure 3 - Test Setup

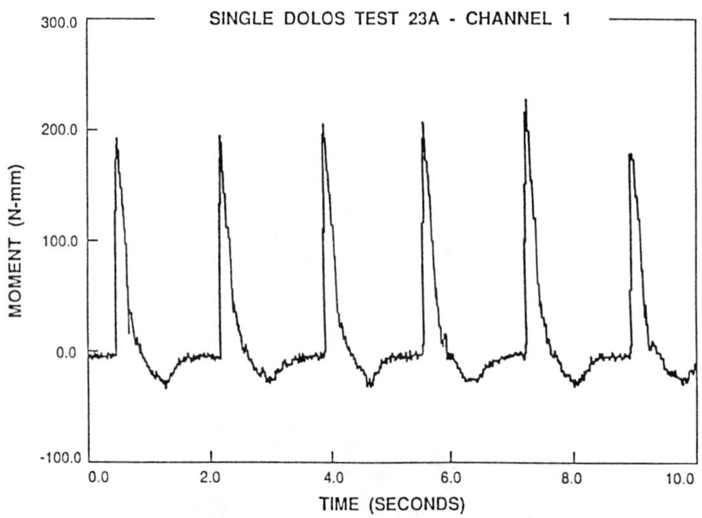

Figure 4 - Moment – Time Response History Plot

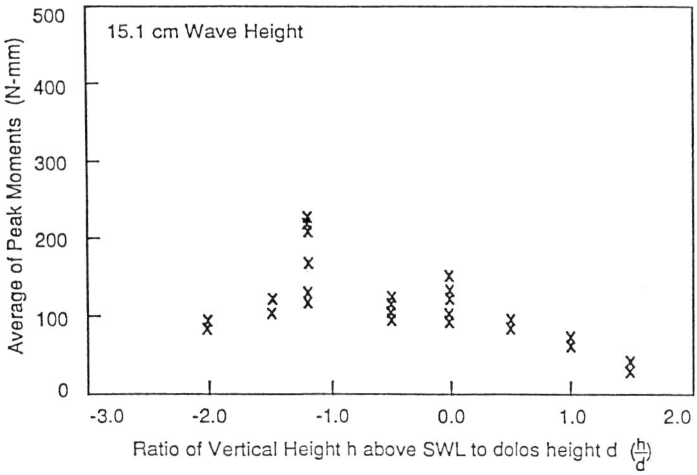

Figure 5 - Average Peak Mount versus h/d for 15.1 cm Wave Height

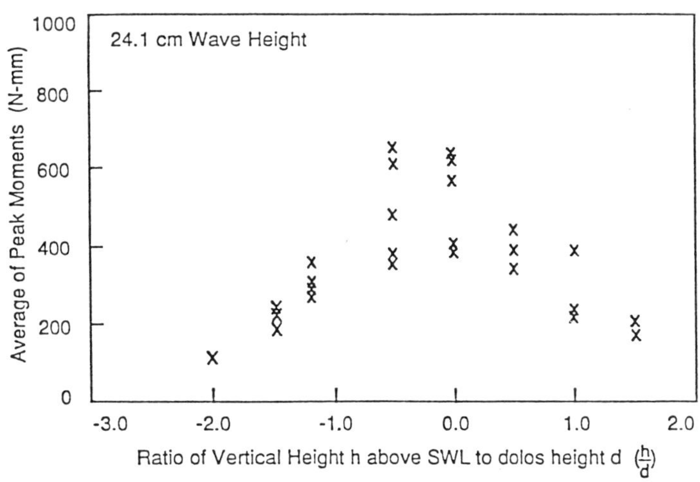

Figure 6 - Average Peak Moment versus h/d for 24.1 cm Wave Height

From these results the maximum stress considering all locations for a given wave height and period can be plotted against the ratio of wave height to dolos height, as shown in Figure 7. Even though a non-linear curve provided globally the best fit to the data, within the range of the measured data (H/d between 1.5 and 2.4), the results are strongly linear.

In order to see how well this result provides a realistic upperbound relationship, it can be compared to the research by Anglin (1988) in which an extensive parametric study of wave-induced loading on breakwater units was carried out. Over 1000 tests were performed and involved variable slopes, unit locations and wave periods and wave heights. Figure 8 shows the relationship developed here and put in a form so that it could be superimposed on the results of Anglin (1988). In this plot the average peak stress has been plotted against wave height. The data points represent the stresses in an armour unit at 22 different locations on a breakwater having a slope of 1.5:1 and subjected to waves with a period of 1.25 to 3.0 seconds and heights of 5 to 30 cm.

As presented in this figure, the simplified model has provided, with one simple test, a realistic upperbound to the set of 300 complex tests carried out by (Anglin 1988). This uncomplicated test can provide an indication of the maximum stress level in an armour unit and a global check on further more complex studies being carried out.

LARGER SCALE VERIFICATION TESTS

The purpose of this set of tests was to examine the repeatability and reproducibility of the load measurements with the instrumented armour units at various scales. Ideally, it would be useful to have data on prototype armour units tested under simplistic load and boundary conditions so that direct comparisons may be made with the physical model. Lacking this prototype data, some initial exploratory work was carried out utilizing model dolos unit load cells of three different sizes.

Three load cells were manufactured with total lengths of 106,203 and 305 mm. Each of the armour units was instrumented and calibrated so that the bending moment and shear in one plane of bending could be readily measured at the mid-shank location of the units.

A simple test was developed in which the instrumented unit was rigidly held to a steel plate then immersed below the still water level in a wave flume, as illustrated in Figure 9. The flume had an overall length of 26m, a large wave paddle capable of producing waves with a height in excess of 60 cm (depending on water depth and wave period) at one end and a series of wave absorbers at the opposite end of the flume. A series of five capacitance-type wave probes were used to measure the water surface elevation and a full wave reflection analysis was carried out.

Each of the mounted units was subjected to a series of regular waves during which the load cell output voltage was sampled at 500 H_Z over a period of 30 seconds. The wave heights, wave periods and water depth were scaled exactly according to the overall lengths of the armour units. Table 1 provides a summary of the test conditions :

TABLE 1 – TEST CONDITIONS

DOLOS SIZE (cm):	10.6	20.3	30.5
DEPTH OF WATER (cm):	52	100	150
WAVE HEIGHTS (cm):	10-20	20-40	30-60
WAVE PERIODS (s):	1.44	2.04	2.50

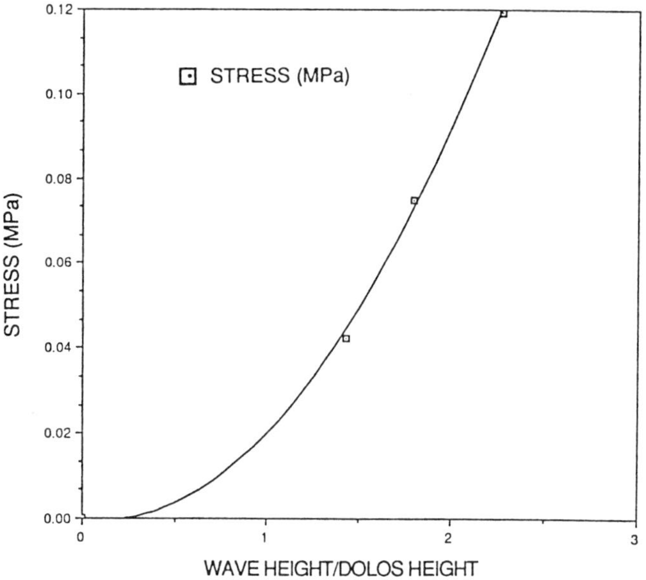

Figure 7 - Stress as a Function of H/d

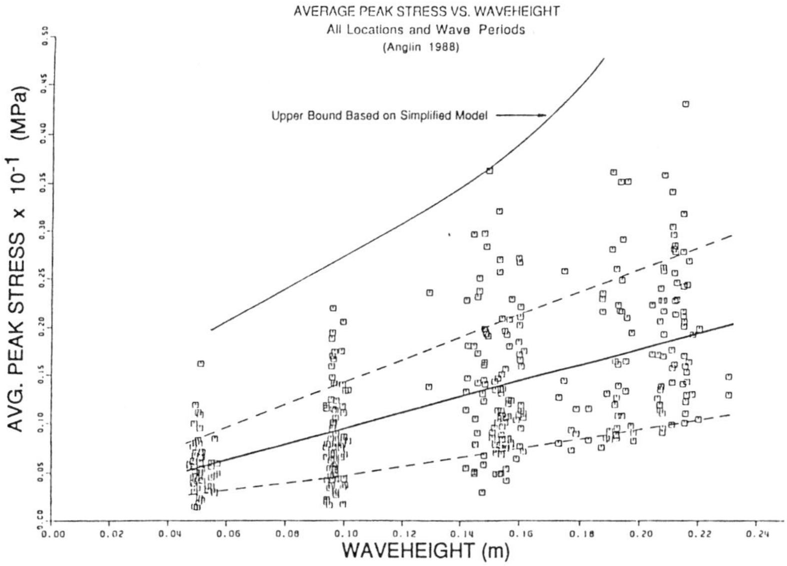

Figure 8 - UpperBound to Peak Stress Versus Wave Height Data

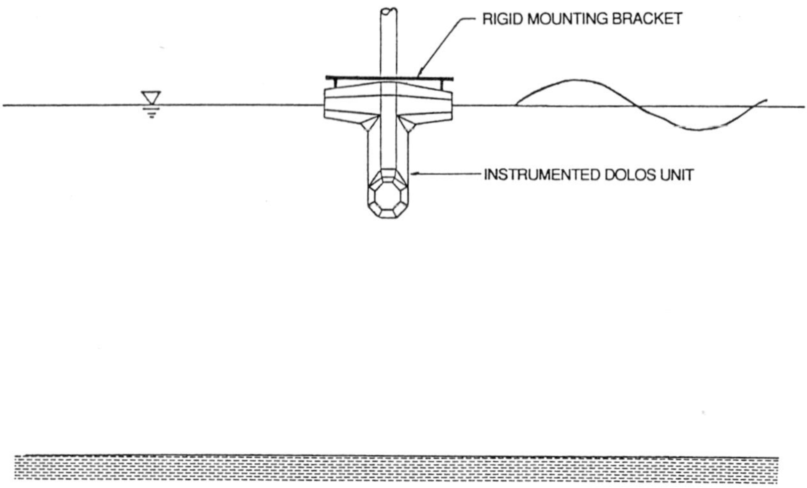

Figure 9 - Experimental Setup for Large Scale Verification Tests

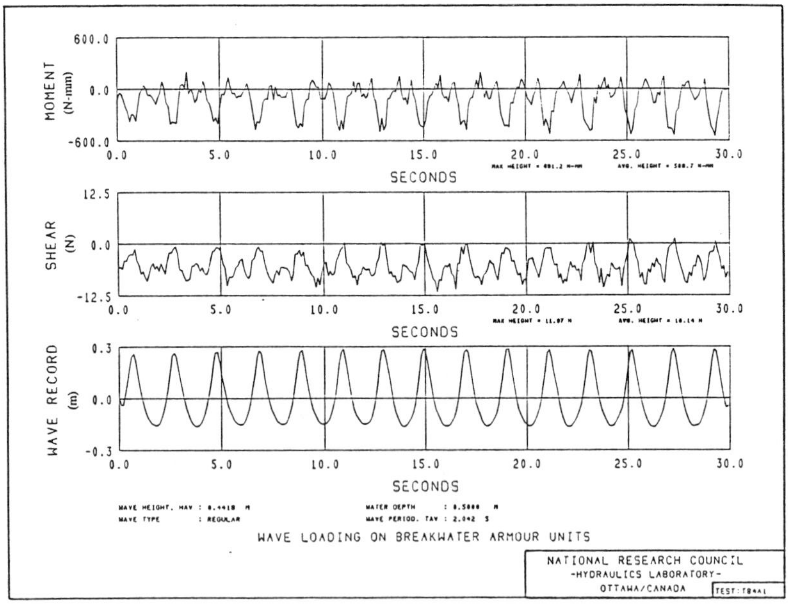

Figure 10 - Typical Response

Each test at a specific wave period and height was repeated three times. In one test series, the load cells were positioned so that the lower fluke was located at the still water level while in a second set of tests the amour units were immersed so that the still water level was at the upper fluke.

As three different sizes of dolos units were utilized in the test program, this provided three different scaling relationships for comparison of the measured responses of the load cells. That is, length scales of 1.5 (20.3 cm unit compared to 30.5 cm unit), 1.92 (10.6 cm and 20.3 cm units) and 2.88 (10.6 cm and 30.5 cm units) were used. Theoretically, the relationship between the measured bending moment of the units will vary with the fourth power of the length scale, or:

$$(M_L / M_S) = (d_L / d_S)^4 = \lambda^4$$

where, M_L is the bending moment in the larger unit having length d_L; M_S, the amount in the smaller unit having a length of d_S; and λ is the geometric length scale. The above scaling relationship indicates that, ideally, a log-log plot of moment ratio to geometric length scale will lie on a line of slope 4.

A typical response of a plot is presented in Figure 10 in which time histories of the measured bending moment, the shear and the water surface elevation are shown. The regularity of the load cell output as compared with the wave period may be noted in the figure.

In Figure 11, the average of the peak moments for each of the tests conducted on the 20.3 cm dolos unit with the upper fluke located at the still water level have been plotted against the wave height divided by the dolos length (H/d). A 3rd order polynomial was fit through the data points to examine the change in response with wave height.

In the final stage of the analysis, the peak moments measured for each of three tests repeats at a given wave height were averaged together and the ratios of bending moment and dolos length were compared. The log of the moment ratios was plotted versus the log of the geometric scales as given in Figure 12. A best fit line through the data points had a slope of 3.9 which compared remarkably well to the theoretical value of 4.0.

Further tests were carried out with waves of longer period but these tests gave inconsistent results. Subsequent analysis showed that considerable reflection within the flume was occurring with the longer waves leading to erroneous load cell measurements.

CONCLUSIONS AND RECOMMENDATIONS

This paper has presented the results of two sets of tests carried out using specially instrumented model armour units. These tests examined two issues that have arisen in the measurement of armour unit forces through physical modelling.

1. The development of a more simplified model of armour unit forces to compare with actual model breakwater tests.

Subjecting a single instrumented model armour unit positioned on a smooth slope to simulated prototype wave attack has been shown to provide a reasonable upper bound for the maximum quasi-static stress levels that may occur in a concrete armour unit. Additional research work is required to further refine this simplified test. Such refinement may include introducing porosity into the smooth slope to provide a more realistic simulation of fluid velocities in a rubblemound breakwater.

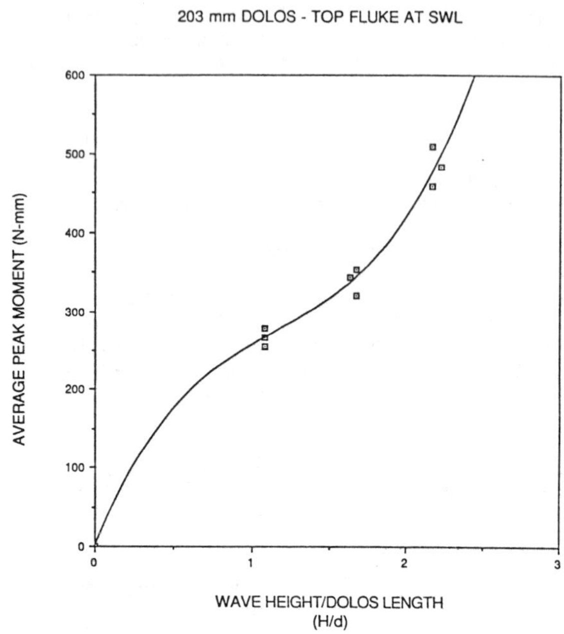

Figure 11 - Results for 203 mm Dolos Unit

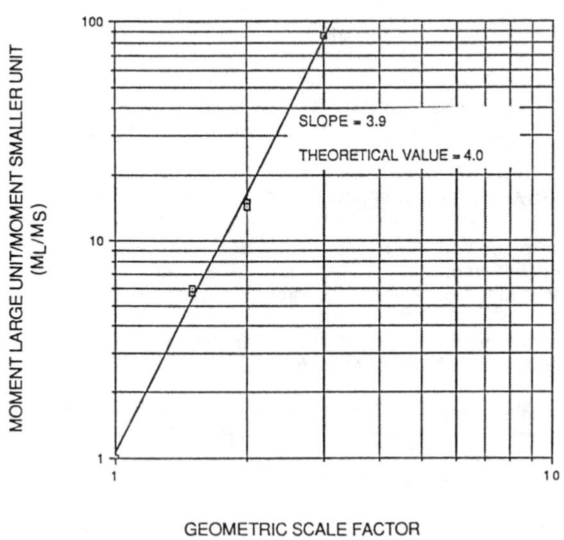

Figure 12 - Log-Log Plot of Moment Ratio to Scale Factor

2. The reliability of the instrumentation measurements if different model scales are employed.

The verification tests provided further confidence in the use of the load cell instrumentation for measuring static and quasi-static waves in physical models of concrete armour units. Three different sizes of dolos load cells were tested independently and all yielded similar results when appropriately scaled.

REFERENCES

Anglin, C.D., Turcke, D.J., Scott, R.D., and Baird, W.F., "A Parametric Study of the Structural Response of Armour Units in a Breakwater", 21st Int'l. Conference on Coastal Engineering, Spain, June 1988.

Baird, W.F., Scott, R.D., and Turcke, D.J., "An Analysis and Design of Armour Units", 20th International Conference on Coastal Engineering, Taiwan, Nov. 9-14, 1986.

Endo, T., "Outline of our Studies on the Problems of Structural Strength of Armour Units", Presented at the Workshop on Measurement and Analysis of Structural Response in Concrete Armour Units, Vicksburg, Miss., Jan. 23-24, 1985.

Howell, G., "Measurements of Forces on Dolos Armour Units at Prototype Scale. 21st Int'l. Conference on Coastal Engineering, Spain, June 1988.

Jensen, O.J., and Juhle J., "Results of Model Tests on 2-D Breakwater Structure", Breakwaters 1988, Eastbourne, UK., May 4-6, 1988.

Losada, M.A., Medina, R. and Alejo, M., "Wave Forces on Armour Blocks". 21st Int'l. Conference on Coastal Engineering, Spain, June 1988.

Scott, R.D., "The Analysis of Concrete Armour Units in a Breakwater", Ph.D. Thesis, Queen's University, Kingston, Ontario, Canada, 1986.

Scott, R.D., Turcke, D.J. and Baird, W.F., "A Unique Instrumentation Scheme for Measuring Loads in Model Dolos Units", 20th International Conference on Coastal Engineering, Taiwan, Nov. 9-14, 1986.

Scott, D., Turcke, D.J., et al "The Design of Armour Units for Breakwaters", The Dock and Harbour Authority Journal, Vol. LXVII, No. 786, pp: 203-208, January 1987a.

Scott, D., Turcke, D.J. and Baird, W.F., "Armour Protection for Rubble Mound Breakwaters", The Fifth Symposium on Coastal and Ocean Management, May 26-29, 1987b.

Tedesco, J.W. and McDougal, W.G., "Non-linear F.E.M. Analysis of Concrete Armour Units", Presented at the Workshop on Measurement and Analysis of Structural Response in Concrete Armour Units, Vicksburg, Miss., Jan. 23-24, 1985.

Tedesco, J.W., McGill P.B., and McDougal, W.G., "Structural Response of Dolos in Waves, 21st Int'l. Conference on Coastal Engineering, Spain, June 1988.

CHAPTER 180

Wave Force on a Curtain-Wall-Type Breakwater Due to Obliquely Incident Waves

by

Tsunehiro Sekimoto*, Kosuke Kondo**, Tetsunori Ohshimo*

ABSTRACT

Assessment of the characteristics of the wave force on the curtain-wall-type breakwater, due to obliquely incident waves, was conducted through numerical calculation, laboratory and field experiments. In the numerical calculations, the Boundary Integral Equation Method (BIEM) for obliquely incident waves was used. From the results of calculation and laboratory experiment, design graphs for estimating design wave force were established and the validity of the diagrams was verified through comparison with the results of laboratory and field experiments.

1. INTRODUCTION

The curtain-wall-type breakwater is a structure which has a vertical or inclined rigid thin barrier extending from above the water surface to some distance below. An example of the curtain-wall-type breakwater is shown in Fig. 1. The curtain-wall is supported by a jacket which is attached to the front side steel pile.

A curtain-wall-type breakwater is often selected as an inlet structure in the harbors of thermal or nuclear power plants in Japan, because they need to take in a great amount of water. The breakwater is also used as a simple wave eliminator in order to provide safe anchorage for vessels in large harbors or fishing ports. In these cases, the incident wave direction to the breakwater is usually restricted since breakwaters are constructed within the harbor. It is very important to know the characteristics of the wave force on the breakwater in relation to the direction of incident waves. However, little is known about the characteristics of the wave force.

In this study, two types of laboratory experiments and two series of field experiments were conducted and the numerical calculation based on Boundary Integral Equation Method was performed. Based on these experiments and calculation, the characteristics of wave force on the curtain-wall-type breakwater due to obliquely incident waves were discussed, and design graphs for estimating design wave force for a wide range of incident wave directions were established.

* Engineer, Design and Engineering Department, Penta-Ocean Construction Co., Ltd., 2-2-8 Koraku Bunkyo-ku Tokyo 112 Japan.

** Principal Engineer, Design and Engineering Department, Penta-Ocean Construction Co., Ltd.

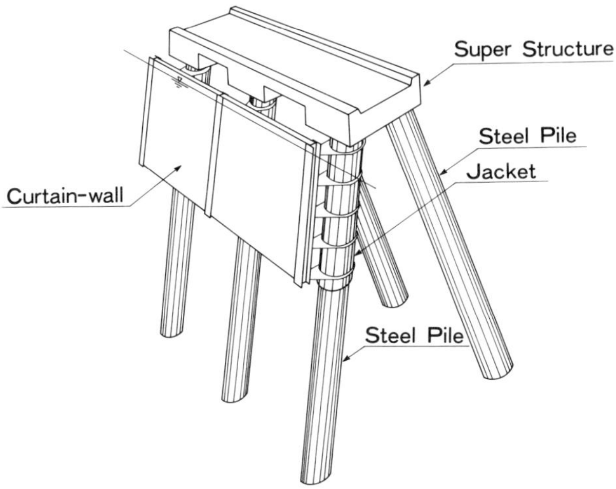

Fig. 1 A curtain-wall-type breakwater

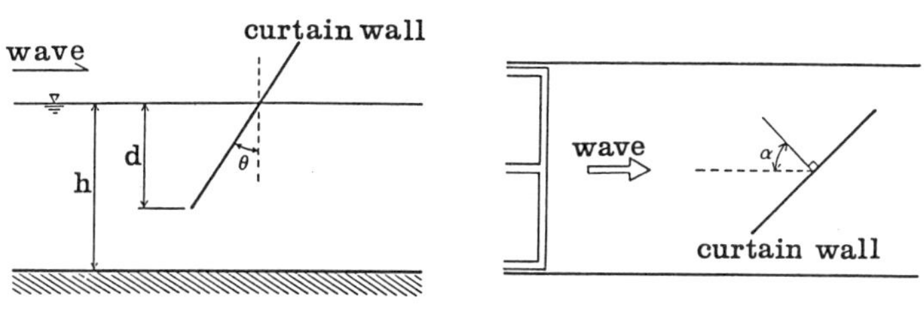

Cross-section Plan

Fig. 2 Definition sketch of a curtain-wall-type breakwater

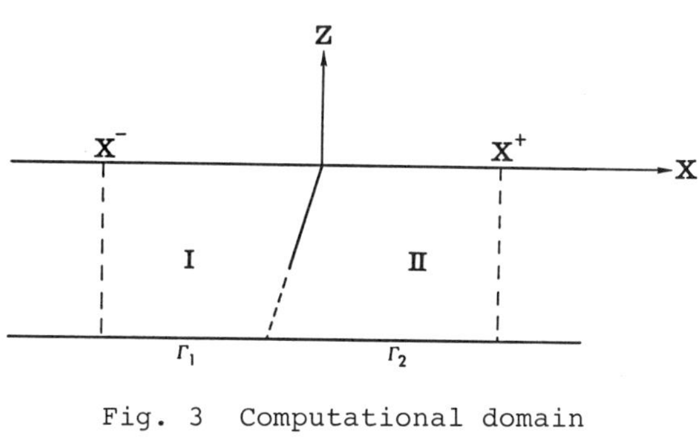

Fig. 3 Computational domain

2. NUMERICAL CALCULATIONS

Numerical calculations, which are based on the Boundary Integral Equation Method (BIEM) proposed by Liu and Abbaspour (1982), are extended to the cases of obliquely incident waves on the assumption that the length of the breakwater is infinitely long and its shape does not change along the breakwater.

A definition sketch of the curtain-wall is shown in Fig. 2. In these figures, α means incident angle, θ is curtain-wall inclined angle, d is the submerged curtain-wall depth and h means the water depth.

The sketch of the computational domain is shown in Fig. 3. Two auxiliary boundaries X^+ and X^- are introduced to utilize the BIEM technique. The procedure proposed by Liu and Abbaspour is characterized by two main points. One is the introduction of a velocity singularity at the tip of the breakwater, to aid accurate calculation. The second is the division of the calculation region into two subregions, Region I and Region II as shown Fig. 3, for simplification in the treatment of the thin barrier. The procedure of Liu and Abbaspour is as follows.

The boundary integral equation of the Laplace's Equation is

$$\frac{Cj}{2\pi}f(xj,zj) = \int_{\Gamma j}\left[f^*\frac{\partial f}{\partial n} - f\frac{\partial f^*}{\partial n}\right]ds; \quad j=1,2 \tag{1}$$

Γj is the Boundary in each region. Cj is the interior angle at the base point (xj, zj), $\partial/\partial n$ is the normal differential taken positive outwards, r is the distance between a base point to another point on the boundary. When the incident wave direction is normal to the Breakwater, f is the spatial variation of the velocity potential and f is defined by

$$\Phi(x,y;t) = f(x,z)\exp[-i\sigma t] \tag{2}$$

Where Φ is the velocity potential, σ is the angular frequency, x, z is the horizontal and vertical coordinates respectively and t is the time. The Green's function of the Laplace's Equation f* is

$$f^* = \frac{1}{2\pi}\ln\frac{1}{r} \tag{3}$$

By discretizing and integrating equation (1) numerically and introducing the boundary conditions, the solutions can be obtained.

The boundary conditions are as follows

The condition on the surface;

$$\frac{\partial f}{\partial n} = \frac{\sigma^2}{g}f \tag{4}$$

The condition along the bottom and surface of the structure;

$$\frac{\partial f}{\partial n} = 0 \tag{5}$$

The condition along the auxiliary boundaries;

$$f = f^{\pm} \tag{6}$$

$$\frac{\partial f}{\partial n} = \frac{\partial f^{\pm}}{\partial n} \tag{7}$$

f^+ and f^- are presented by

$$f^+ = A_0^+ \frac{\cosh k(z+h)}{\cosh kh} \exp[ikx^+] + \sum_{n=1}^{\infty} A_n^+ \frac{\cos K_n(z+h)}{\cos K_n h} \exp[-K_n x^+] \tag{8}$$

$$f^- = A_0^- \frac{\cosh k(z+h)}{\cosh kh} \exp[-ikx^-] + \sum_{n=1}^{\infty} A_n^- \frac{\cos K_n(z+h)}{\cos K_n h} \exp[K_n x^-] + f_I \tag{9}$$

Where K_n is the nth imaginary root of the dispersion relationship, f_I is the velocity potential of the incident wave, $A_0^{\pm}$ are the amplitude of the transmission and the reflection wave and $A_n^{\pm}$ are the amplitude of the scattering waves.

Using a previous assumption, that the length of the breakwater is infinitely long and its shape does not change along the breakwater, the three dimensional Laplace's equation can be reduced to the two dimensional Helmholtz's equation. The boundary integral equation of Helmholtz's equation is also represented by equation (1). Assuming the y axis corresponds to the direction along the breakwater, f is defined by

$$\Phi(x,y,z;t) = f(x,z)\exp[i(ky\sin\alpha-\sigma t)] \tag{10}$$

The Green's function of Helmholtz's equation is described by the modified Bessel's function.

$$f^* = \frac{1}{2\pi}K_0(kr\sin\alpha) \tag{11}$$

Introducing the boundary conditions previously used and using the equations (12) and (13) instead of equations (8) and (9), the solutions can be obtained in the case of obliquely incident waves.

$$f^{+} = A_0^{+} \frac{\cosh k(z+h)}{\cosh kh} \exp[ikx^{+}\cos\alpha] \tag{12}$$

$$f^{-} = A_0^{-} \frac{\cosh k(z+h)}{\cosh kh} \exp[-ikx^{-}\cos\alpha] + f_I \tag{13}$$

The wave force and the water surface elevations were calculated from the potential f.

3. LABORATORY AND FIELD EXPERIMENTS

One of the experiments is a fundamental study with the curtain-wall in an uniform water depth. A wave-tank which has 34m length, 5m width and 1.2m depth was used. The incident and reflected waves, the transmission waves, the water surface elevation just in front of and just behind the curtain-wall and the wave pressure acting on the breakwater are measured using capacitance type wave gages and small size pressure gages. The models of the breakwater were made from a model unit which consists of steel frame and plywood curtain-wall. In order to reduce the effect of reflected waves from the side-wall and wave-paddle, the partition wall and the wave absorber were arranged for various conditions of incident wave angle. The experimental conditions are shown in Table 1. In the regular wave experiments, four kinds of wave period were used with three kinds of wave height. In the irregular wave experiments, three kinds of significant wave period were used with significant wave height of 8 cm only. Incident wave angle varied in six different directions from 0 to 90 degrees. Three kinds of inclined curtain-wall angle and two types of relative submerged depth d/h were tested. The water depth was set at 60cm in all cases. We carried out more than 200 runs using both regular and irregular waves.

Table-1 The experimental condition

	Regular wave	Irregular wave
Wave Period T, $T_{1/3}$ (sec)	1.0, 1.5, 2.0, 2.5	1.0, 1.5, 2.0
Wave Height H, $H_{1/3}$ (cm)	5.0, 8.0, 10.0	8.0
Incident wave angle α (deg)	0, 30, 45, 60, 75, 90	
Inclined angle θ (deg)	0, 20, 45	
Relative submerged depth d/h	0.4, 0.6	

The field experiments were carried out from January through March of 1985, and January through February of 1986, at the port of Kashiwazaki-Kariwa nuclear power plant, Niigata, Japan. Kashiwazaki-Kariwa is located on the Sea of Japan in central Niigata Prefecture and is shown in fig. 4.

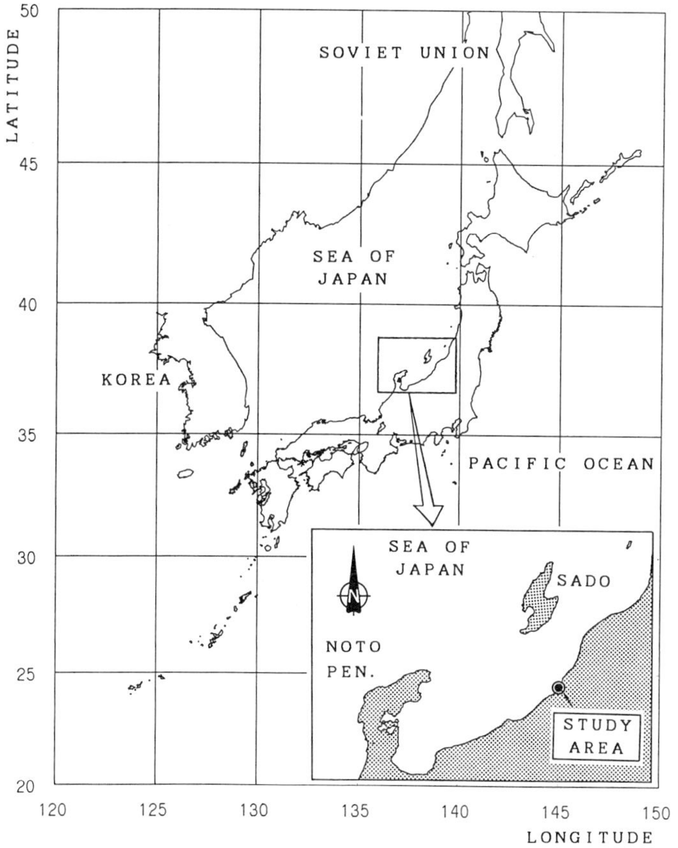

Fig. 4 The location of the investigation site

Fig. 5 is the shape of the port. The north breakwater and the north jetty are on the right hand side and the south breakwater and south jetty are on the left hand side of this figure. The curtain-wall-type breakwater is located in the middle of the south breakwater. The inclined angle and relative submerged depth of this curtain-wall are 15 degrees and 0.6 respectively. The water depth below the curtain-wall is 13 m. The water surface elevations just in front of and just behind the breakwater and the wave pressure on the breakwater are measured in the center section of this curtain-wall-type breakwater. The measurement of the water surface elevations and the wave pressures are performed using two capacitance type wave gages and eight pressure gages respectively.

Another experiment is an 1/80 scale model test of the port of Kashiwazaki-Kariwa nuclear power plant, in a 40 m length and 30 m width wave-tank. Both the water surface elevation and wave force were measured. In this experiment, unidirectional irregular waves, which have the Bretschneider-Mitsuyasu type power spectra, and JONSWAP Spectra, that approximate the field observed power spectra, were used. The water surface elevations were measured using capacitance type wave gages and the wave force measurement were performed by measuring the strain in the steel arm which supports the measuring part, when cut off from the other part of the curtain-wall.

The result of the calculations and the laboratory and field experiments were compared.

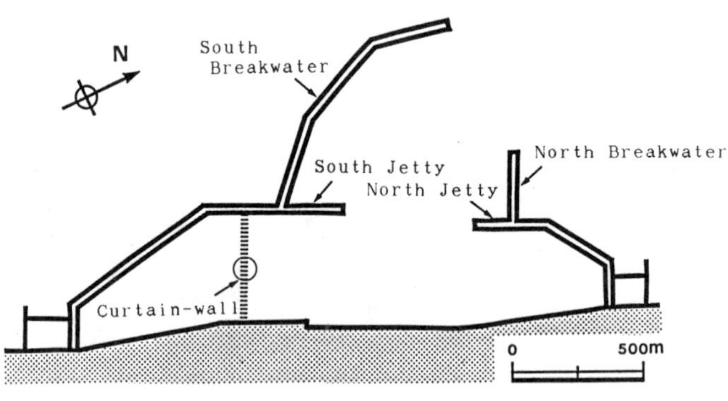

Fig. 5 Kashiwazaki-Kariwa port

4. COMPARISON BETWEEN RESULTS OF EXPERIMENT AND CALCULATION

Fig. 6 is the comparison of numerical predictions by BIEM with experimental results using regular waves for wave force coefficient as a function of the incident wave angle, with the condition that h/L = 0.201, θ = 20 degrees and d/h = 0.6. The wave force coefficient is defined as the ratio between the wave force per unit curtain-wall length and the incident wave height. The experimental results agree with calculation results with the exception of the large value of α. One of the reasons for this anomaly is that the incident wave energy is transformed into a wave running along the curtain-wall.

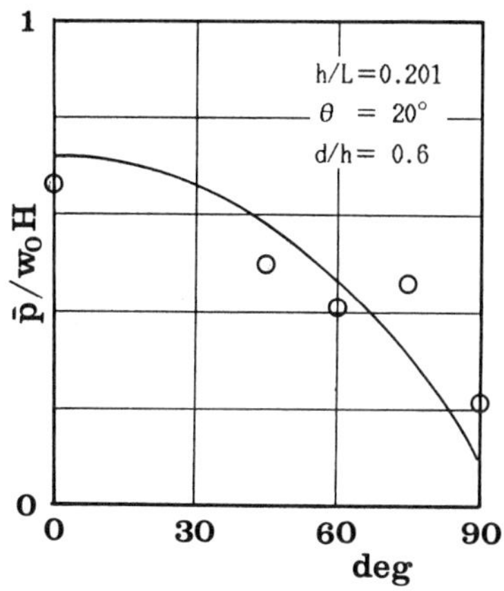

Fig. 6 The comparison between numerical and experimental results for wave force coefficient (regular wave)

The comparison results of irregular waves are shown in Fig. 7. The spectrum analysis is used for analysis of irregular waves. The wave force coefficient is defined as the transfer function of two times surface elevation of the incident wave to the wave force. The relative water depth is calculated from the wave length of each component wave. In the figures, a circle is the result using a wave with 1.5 sec significant wave period and a triangle means a wave of 2.0 sec. These are the results of plotting the wave force coefficient against the relative water depth in the case where θ = 20 degrees and d/h = 0.6.

We obtain relatively good agreement between both, except for the lower range of h/L in the case where α = 30 degrees. In this experiment, for the small incident wave direction, the distance between the edge of the model and the measuring point was short. Thus the effect of the scattering wave from the edge appeared in the result for the case where α = 30 degrees.

For the case of inclined angle θ = 45 degrees, agreement between results was not good because of the effect of the wave runup.

Fig. 8 are the figures showing the ratio of the wave force to the difference of the water level, on both sides of the curtain-wall, against h/L when α = 45 degrees. The results from BIEM fit very well with the experiment results except for the case where θ = 45 degrees. This is due to the effect of the wave runup. It is considered that the effective difference of the water levels is smaller than the measured difference. The ratio for different values of α lies almost on the same line. This fact was also confirmed in model tests and field experiments. Fig. 9 is the example of the result of an 1/80 model test and Fig. 10 is the result of field experiments. Good agreement between the calculation results and the experimental results is found in these figures. This implies that the wave force can be estimated accurately by BIEM if the difference of the water levels on both sides is given. And this fact also means the relations between the wave force and the difference of the water levels on both sides of the curtain-wall are hardly affected by the incident wave angle.

The difference of the water surface elevation on both sides can be separated into the amplitudes of the water surface elevation on both sides and the phase difference of both. Thus investigation followed on the phase difference and the ratio of the amplitude of the water surface elevation just behind the curtain-wall to the one just in front of it.

Fig. 11 shows the results of calculations and experiments on the phase difference between water levels as functions of h/L. It is found that the phase difference decreases as the incident angle of the waves increases. This is due to the increase in the apparent wave length. Agreement of the results of both calculation and experiment is good.

Fig. 12 shows the ratio of the water surface elevation just in front of to just behind the curtain-wall. When α = 30 degrees, the experimental results agree well with the calculation results. However, as the incident wave angle increases, experiment results do not increase although the calculation result increases. This may be due to the fact that a part of the incident wave transformed into the wave running along the curtain-wall, although this is difficult to estimate quantitatively. However, it is found that the amplitude of the water surface elevation just in front of the curtain-wall decreases as the incident wave angle increases, and that the water surface elevation just in front of the curtain-wall and the ratio of the amplitudes of the both sides of the curtain-wall cancel the difference between calculation and experiment.

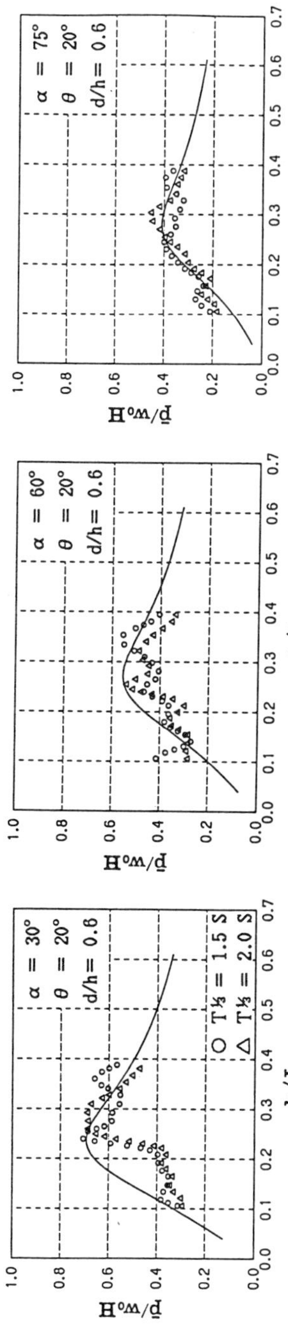

Fig. 7 The relation between wave force and incident wave height

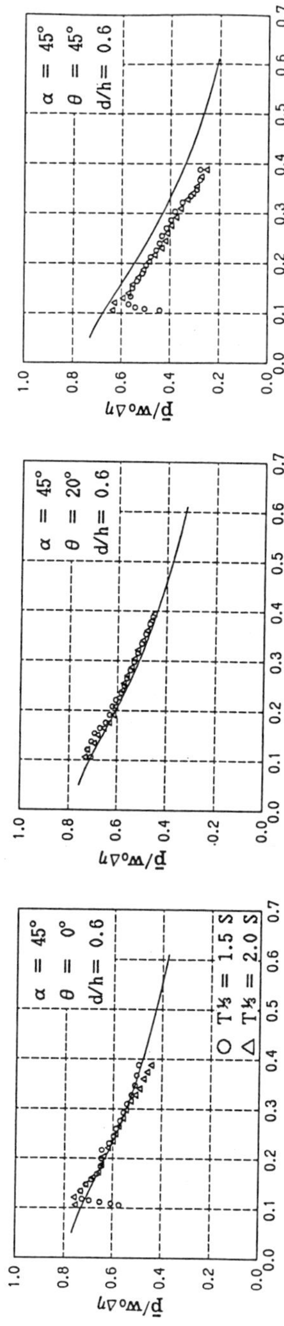

Fig. 8 The relation between wave force and the difference in water surface elevations

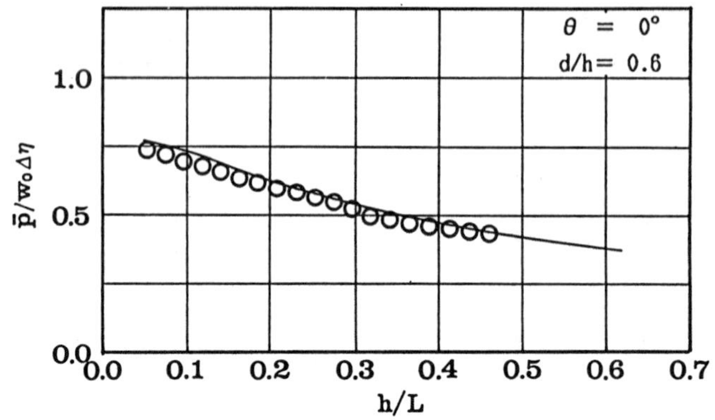

Fig. 9 The relation between wave force and the difference in water surface elevations (1/80 model test)

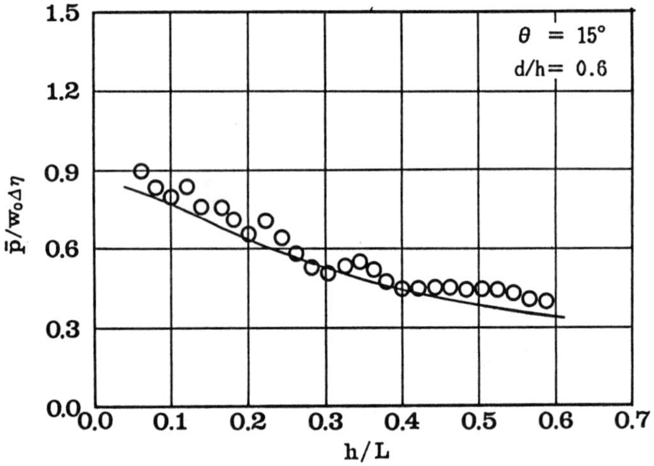

Fig. 10 The relation between wave force and the difference in water surface elevations (field experiment)

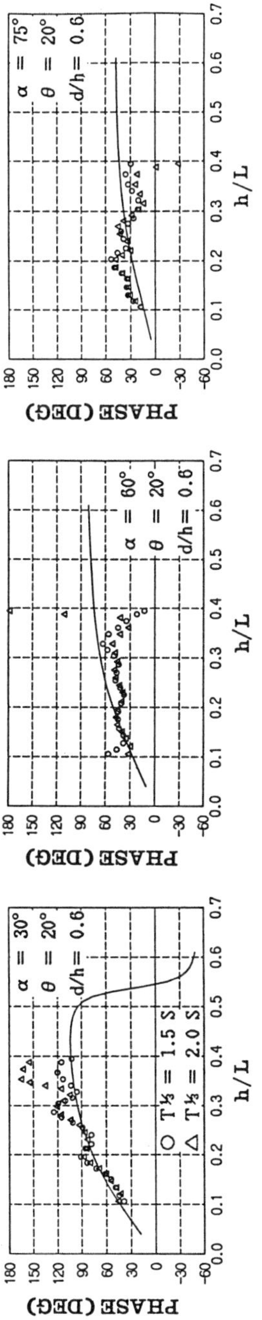

Fig. 11 The phase difference between the water surface elevations

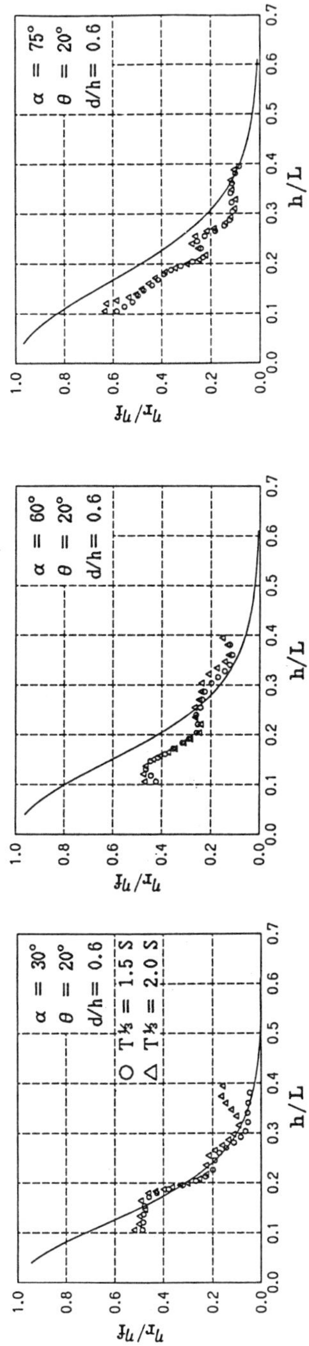

Fig. 12 The ratio between the amplitudes of water surface elevations

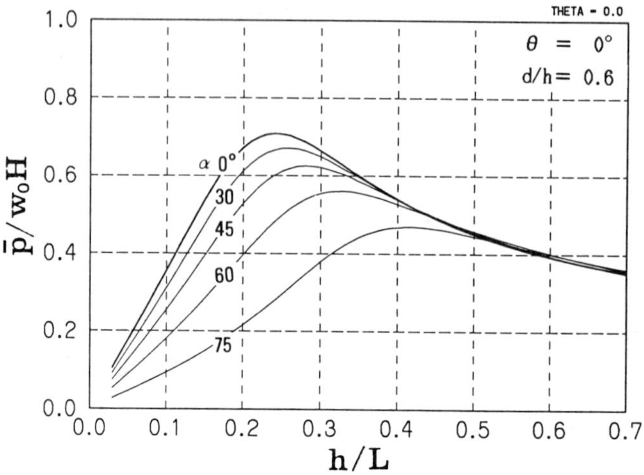

(a) θ = 0 degree

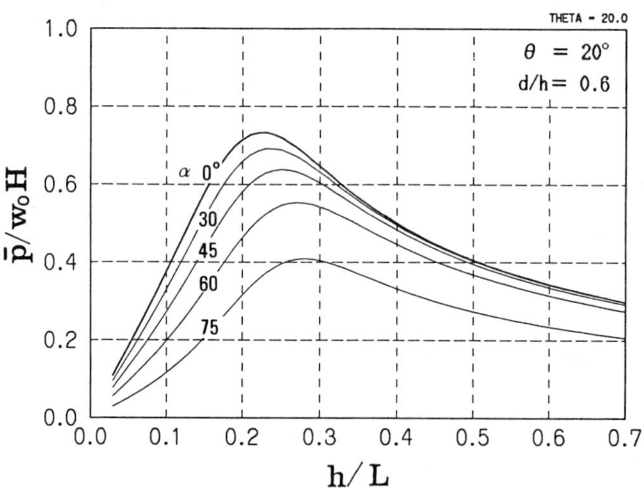

(b) θ = 20 degrees

Fig. 13 The diagram for estimation of design wave force

5. DESIGN WAVE FORCE

The wave force coefficient for irregular waves can be regarded as a transfer function. In the case of using the representative wave height as the design wave height, the moving weighted average of calculation results using BIEM, where 'weight' is a wave power spectra, is required. However, in consideration of the uncertainty about the mechanism of wave force and the accuracy of the calculation, BIEM results alone can be employed for estimation of design wave force. The diagrams for estimation of design wave force are shown in Fig. 13.

The comparison of numerical predictions with the experimental results of an 1/80 model test on wave force is shown in Fig. 14. In numerical predictions, Green's function method was used for calculation of directional spectra in a harbor [Kondo et al (1987)]. This is an extension of the procedure proposed by Gaillard (1984) in order to estimate the wave height and the wave direction at the measured point of the wave force. We obtained relatively good agreement between the calculation and the experimental results, even although the incident wave angle varied from 45 to 70 degrees.

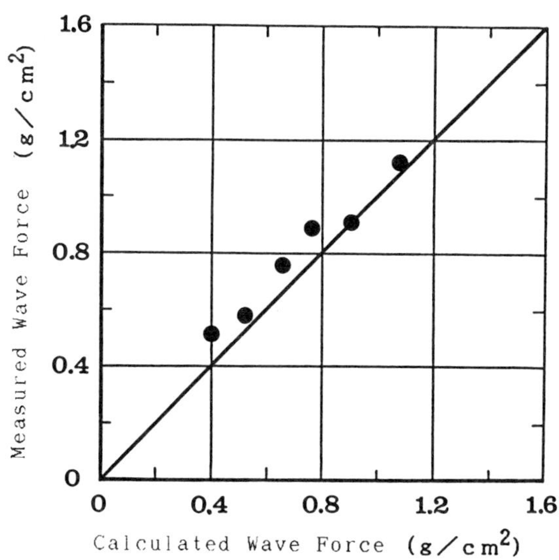

Fig. 14 The comparison of numerical predictions with the experimental results of an 1/80 model test on wave force

6. CONCLUSIONS

The characteristics of the wave force on the curtain-wall-type breakwater may be determined from the results of calculations based on BIEM and experiments. As a result, diagrams for estimating design wave force accurately for a wide range of incident wave direction were established. The applicability of this method are confirmed from the results of the 1/80 model test and field investigations.

ACKNOWLEDGMENTS

We would like to thank Mr. S. Imai of Tokyo Electric Power Co., Inc., and Mr. M. Nakamura of the Tokyo Electric Power Service Co., Ltd., for their support and cooperation.

REFERENCES

[1] GAILLARD, P., Combined refraction-diffraction calculations with directional spectra, Proc. 19th Coastal Engineering., ASCE , pp.1040-1056, 1984.

[2] KONDO, K., SHIMIZU, T. and YAMADA, K., Green function method for calculation of directional wave spectrum in a harbor, Proc. 34th Japanese Conf. on Coastal Eng., pp.106-110, 1987 (in Japanese).

[3] LIU, P.L. and ABBASPOUR, M., A wave scattering by a rigid thin barrier, J. Wtrway., Port, Coast and Ocean Engrg., ASCE, 108, PP.479-491, 1982.

CHAPTER 181

Structural Response of Dolos in Waves

Joseph W. Tedesco[1], A.M. ASCE, Preston B. McGill[2],
Jeffrey A. Melby[3], A.M. ASCE, and William G. McDougal[4],
A.M. ASCE.

ABSTRACT

Several numerical models for estimating wave forces on dolosse are presented. These models are incorporated into a finite element method (FEM) analysis of a 42-ton unit. The results of the analysis permit identification of unit structural failure with respect to wave conditions.

INTRODUCTION

One means of protecting breakwaters and shorelines from wave-induced forces has been through the deployment of dolos concrete armor units. A typical dolos armor unit is shown in Figure 1. These randomly placed units form an interlocking yet porous armor layer providing a high degree of hydrodynamic stability. However, in the past, dolos armor units in severe wave environments, such as the breakwaters in Crescent City, California and Port Sines, Portugal, have experienced structural failures (Magoon, 1974, and Magoon and Baird, 1977). The unsatisfactory behavior of these units has prompted investigations into the wave loading on dolos armor units and an assessment of the subsequent dynamic structural response of the dolos unit.

This paper summarizes recent research conducted on the structural response of dolos armor units subject to wave forces. Several mathematical models (Tedesco and McDougal, 1985, and McDougal et al, 1988) developed for estimating wave forces on a single armor unit are presented. These wave force models were incorporated into a finite element method (FEM analysis of the dolos unit). Two types of analyses were conducted: 1) direct application of a critical wave to the armor unit restrained against rigid body motion, and 2) application of impact forces generated when wave-induced rigid body motion occurs causing dolos impacts.

SIMPLE WAVE-SLAMMING MODEL

In this model, the force on the dolos is estimated from the slamming of the wave with the structure. Only the horizontal component of

[1]Assoc. Prof., Dept. of Civil Engr., Auburn University, AL 36849.
[2]Res. Asst., Dept. of Civil Engr., Auburn University, AL 36849
[3]Res. Hydr. Engr., Dept. of Army, Waterways Experiment Station, Corps of Engineers, P.O. Box 631, Vicksburg, MS 39180.
[4]Assoc. Prof., Ocean Engr. Program, Dept. of Civil Engr., Oregon State University, Corvallis, OR 97331.

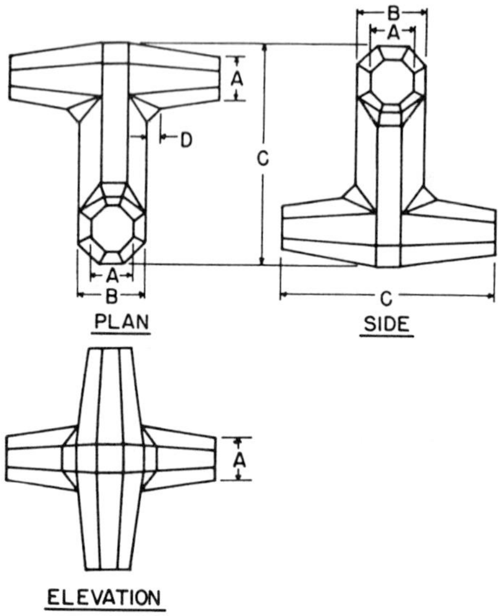

Fig. 1. Dolos Armor Unit.

this impact force is examined, which for the assumed geometry, is larger than the vertical component. Slamming is an important consideration in the design of ocean structures in the splash zone. As a result, a small body of information exists on wave-impact forces on horizontal cylinders. Kaplan (1979) and Kaplan and Silbert (1976) have developed models for both the horizontal and vertical forces due to impact. The models include the effects of buoyancy, pressure gradients, momentum flux (including added mass), and drag. Results are in reasonable agreement with measurements for the horizontal force but not for the vertical.

Experimental results suggest that the impact force, F, may be expressed in a simpler form

$$F = \frac{1}{2}C_s \rho d l U^2, \tag{1}$$

in which ρ is the fluid density; d is the diameter of the cylinder; l is the length of the cylinder; U is fluid velocity; and C_s is a slamming coefficient (Sarpkaya, 1978). The slamming coefficient is a function of the immersion depth of the cylinder and, therefore, a function of time. A time-dependent slamming coefficient has been empirically developed (Sarpkaya, 1978):

$$C_s = \frac{5.15}{1 + 19\frac{Ut}{d}} + \frac{0.55\ Ut}{d}. \tag{2}$$

The maximum value for C_s at $t = 0$ is 5.15 and this value is appropriate if a static structural analysis is performed (Sarpkaya and Isaacson, 1981). For dynamic analyses, a value of 3.2 is suggested. Therefore, eq. (2) is accordingly scaled for use in the present dynamic analysis.

The foregoing formulation is, of course, only valid when some portion of the cylinder is immersed. However, when the cylinder is totally immersed, the formulation is inappropriate. Experimental results of Sarpkaya (1978) indicate that this formulation is only valid up to the point where the top of the cylinder is just below the free surface. Therefore, at this depth (when the slamming coefficient is at minimum) the forces are assumed to no longer be impact dominated and a drag formulation is adopted.

$$F = \frac{1}{2}C_D \rho S d l U^2, \tag{3}$$

in which S is the fraction of the cylinder which is immersed and C_D is a drag coefficient. It is noted in eq. (2) that the minimum value for the slamming coefficient is 0.5. This is of the same order as the drag coefficient for a smooth cylinder. Therefore, for purposes of calculation, this minimum value of C_S will be used for C_D. Several representative time histories of force are shown in Fig. 2. For a given wave and cylinder size, the duration of the force is a function of the position of the cylinder relative to the stillwater level. The peak impact force is not a function of elevation because the horizontal velocity is assumed to be constant. The tailing off of the force depends upon the ratio of the cylinder diameter to the wave height.

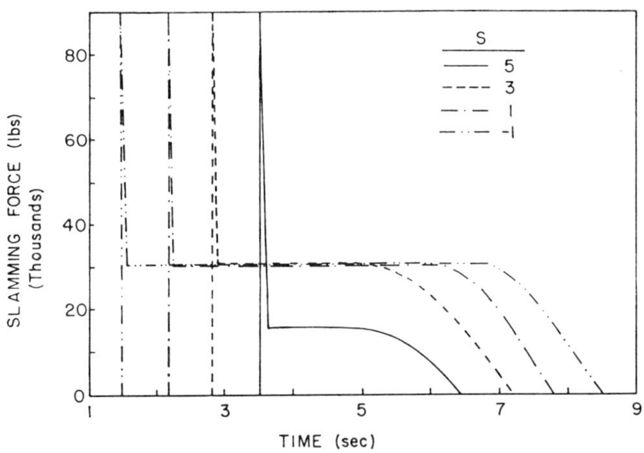

Fig. 2. Time history of slamming force. (S is height above SWL in feet)

An FEM (ADINA, 1984) analysis was conducted on three different sizes of dolosse subject to the simple wave slamming model. An isometric view of the model is depicted in Fig. 3. The analyses were performed for concrete compressive strengths, f'_c, of 5000 psi and 4000 psi for each model. The dolosse were assumed to be unreinforced, and the tensile strength of the concrete was assumed to be equal to the split cylinder strength, f_{ct}, of the concrete (where $f_{ct} = 6.7\sqrt{f'_c}$).

The failure of the dolosse depends on the maximum slamming load. This load, in the present analysis, is only a function of the wave conditions for a specific dolos unit. This load is calculated as discussed above and regions of structural stability are identified as a function of the wave conditions. These results are shown in Fig. 4 for a 42-ton dolos in which T is the wave period, T_ζ is the resonant period, H is the wave height, and H_H is the maximum stable wave height for the unit. Stable and failure regions are identified for 4000 psi and 5000 psi concrete. It is noted that structural failure would occur for the 4000 psi concrete at wave conditions for which the unit is hydrodynamically

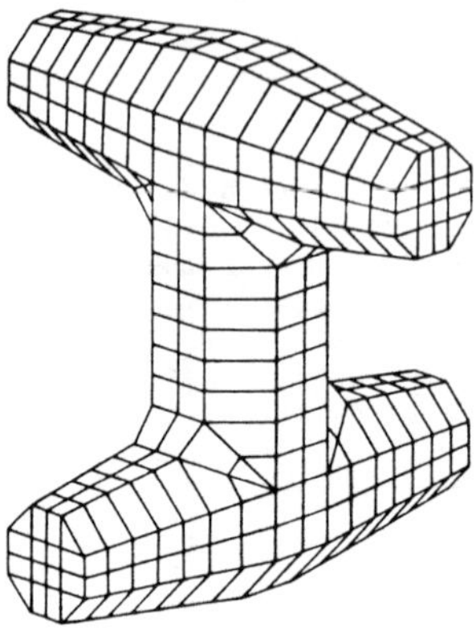

Fig. 3. Isometric view of FEM Model

stable. The higher strength concrete unit is very nearly at failure at these wave conditions.

MULTI-COMPONENT WAVE FORCE MODEL

The precise specification of the wave loads in the breaker zone is extremely difficult. However, there are many structures built in this zone and several approximate analysis methods have been developed. Kaplan (1979) and Kaplan and Silbert (1976) developed Morison type equations (Morison et al, 1950) for a horizontal cylinder in the "splash zone" where the structure is not in continuous contact with the wave. This model includes the slamming of the fluid on the structure as the structure penetrates the free surface. A primary assumption is that the wave length is large with respect to the structure diameter. Kaplan and Silbert (1976) showed that the vertical force per unit length on a horizontal cylinder for a normally incident wave can be expressed as the sum of the buoyant, inertial, kinetic and drag forces as follows:

$$F_{VH} = \rho g A_i + (m_3 + A_i)\ddot{\eta} + K_{HV}\frac{\partial m_3}{\partial z}\dot{\eta}^2$$

$$+ \frac{\rho}{2}\dot{\eta}|\dot{\eta}|d(\frac{z}{r})C_{DV}(\frac{z}{r}), \tag{4}$$

where the first subscript on F denotes the force direction and the second indicates the cylinder orientation. The force component notation is shown in Fig. 5. Also, K_{VH} is an empirical force coefficient, ρ is the mass density of the fluid, g is the acceleration of gravity, A_i is the immersed area, $\ddot{\eta}$ and $\dot{\eta}$ are the vertical wave velocity and acceleration, respectively, d(z/r) and $C_{DV}(z/r)$ are the depth dependent diameter and drag coefficient, respectively, and m_3 is the vertical added mass.

The horizontal force on a horizontal cylinder can similarly be stated as the sum of inertial, kinetic and drag forces:

$$F_{HH} = (m_2 + \rho A_i)\dot{u} + K_{HH}\frac{\partial m_2}{\partial z}\dot{\eta}u$$

$$+ \frac{\rho}{2}u|u|h(\frac{z}{r})C_{DH}(\frac{z}{r}), \tag{5}$$

where K_{HH} is an empirical force coefficient, u and $\dot{u}$ are the horizontal water particle velocity and acceleration, respectively, and m_2 is the horizontal added mass. For the case of a vertical cylinder, the previous equation for horizontal force can be directly applied to yield

$$F_{HV} = (M_4 + \rho A_i)\dot{u} + K_{HV}\frac{\partial m_4}{\partial z}\dot{\eta}u + \frac{\rho}{2}u|u|DC_{DH} \tag{6}$$

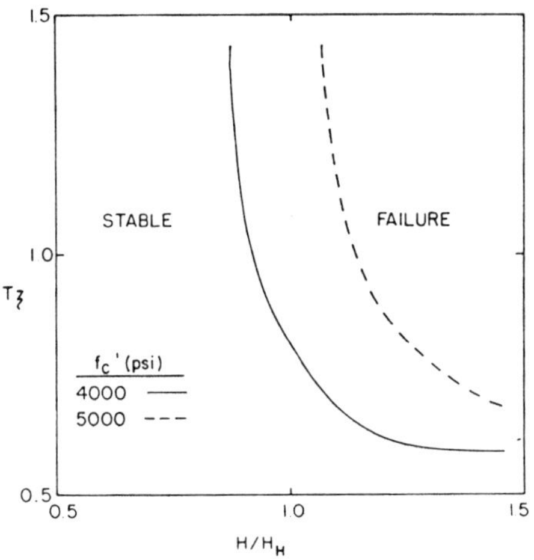

Fig. 4. Failure envelope for 42-ton dolos unit.

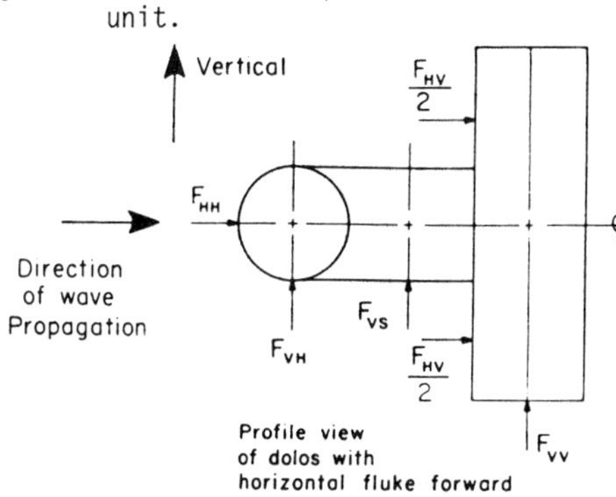

Fig. 5. Dolos orientation and wave force component notation.

in which K_{HV} is an empirical force coefficient. The vertical force on a vertical cylinder can be expressed as the sum of form drag and buoyancy acting on the lower end of the cylinder to yield

$$F_{VV} = \rho g \pi r^2 + \frac{\rho}{2} v|v| \pi r^2 C_D, \qquad (7)$$

where v is the vertical water particle velocity at the cylinder base. Finally, the vertical force on a horizontal cylinder parallel to the direction of wave propagation is assumed to be buoyancy dominated and thus given by

$$F_{VS} = \rho g A_i. \qquad (8)$$

These force equations are similar in form to the Morison equation. However, potential theory is employed to determine the kinetic term and the coefficients, m_2, m_3, m_4 and C_D, which are in terms of the geometric properties as well as flow and cylinder roughness characteristics.

In this study, linear wave theory was used to determine the values of wave variables for the design wave given in Table 1. The wave height corresponds to the maximum stable wave height for a 42-ton dolos on a 1V:2.5H slope as determined from Hudson's formula. The wave period was selected by choosing an Iribarren number of 2.5. The Iribarren number is the ratio of

Table 1. Wave and dolos specifications

Wave		Dolos	
Height	33.4 ft	Elevation	12.0 ft
Period	12.77 sec	Fluke dia.	3.98 ft
Depth	42.8 ft	Fluke length	15.29 ft
		Shank dia.	4.89 ft
		Shank length	7.33 ft

the structure slope to the square root of the wave steepness. A value of 2.5 corresponds to a critical condition in which the wave breaks when rundown is at its lowest point (Bruun and Gunbak, 1977). The depth corresponds to the depth limited breaking for this wave height.

The dolos configuration is as shown in Fig. 5 with the horizontal fluke forward. The methodology developed for estimating wave forces is also applicable to the case with the vertical fluke forward. In both cases the shank is fixed in the horizontal plane and in line with the direction of wave propagation.

The multi-component wave force model was incorporated into an FEM analysis of a 42-ton dolos. In the analysis, the dolos unit was positioned among other units so that it is restrained only at the bottom of the vertical fluke which is perpendicular to the wave crest. The horizontal fluke is forward (seaward) and parallel to the wave crest.

For the specified wave loading and structure restraint conditions, the critical section in the dolos unit was located in the bottom portion of the vertical fluke, at the fluke-shank juncture, just below the fillet. The maximum tensile stress at the critical section was calculated to be 760 psi. This value exceeds the ultimate tensile strength of 743 psi (the modulus of rupture), thereby causing a structural failure.

IMPACT ANALYSIS

Damage to dolosse is primarily due to the impact forces resulting from the collision of units due to wave-induced rolling and rocking motions (Burchart, 1984). Knowledge of the rigid body motions is required for an appropriate design. Once the wave loads have been specified and the dolos orientation is known, the equations of motion can be solved numerically. The equations of motion for a dolos constrained as described in the previous section reduce to a single rotational degree of freedom. The equation of motion is (McDougal et al, 1988)

$$I\ddot{\theta}(t) + C\dot{\theta}(t) + K\theta(t) = M(t) \tag{9}$$

where θ, $\dot{\theta}$, and $\ddot{\theta}$ are the angular displacement, velocity, and acceleration, respectively, I is the mass moment of inertia, C is damping, K is stiffness, and M is the applied moment. In the analysis, C = 0 and the applied moment is given by the product of the time varying moment arms and the respective Kaplan forces. This equation was integrated using a fourth order Runge-Kutta algorithm.

In the present study, the case investigated is depicted in Fig. 6. The horizontal fluke is seaward and parallel to the wave crest. Under wave action, the dolos is permitted to pivot about the vertical fluke which is perpendicular to the wave crest. Damage to the unit occurs when the dolos falls back to its initial position and impacts another unit.

The FEM representation of the scenario is depicted in Fig. 7. The unit is assumed to impact a rigid contact surface at a contact velocity

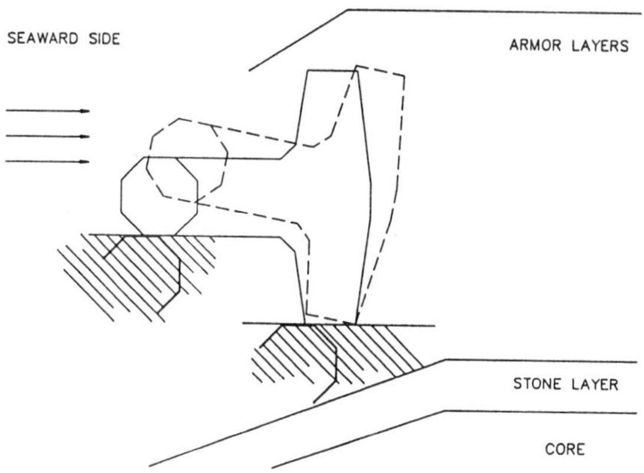

Fig. 6. Dolos orientation for impact analysis.

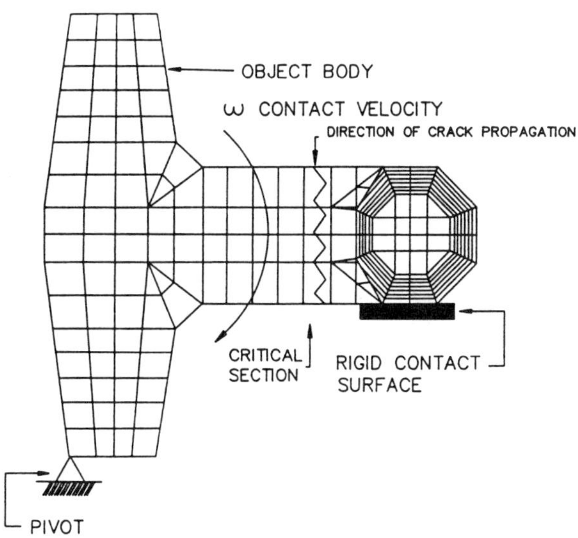

Fig. 7. FEM representation of impact analysis.

ω. The critical section in the unit is located on a vertical plane through the shank (Fig. 7). The dynamic stress history at the critical section is presented in Fig. 8.

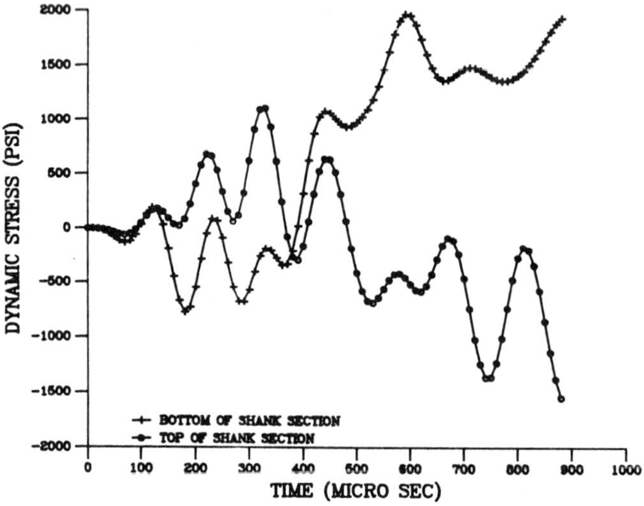

Fig. 8. Dynamic stress history for impact analysis.

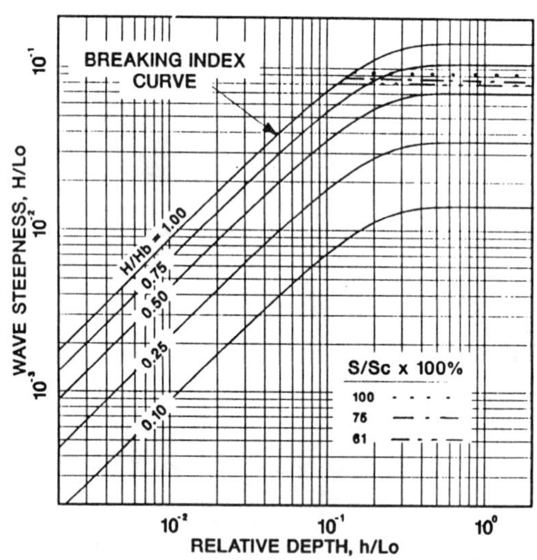

Fig. 9. Structural stability curves for a 42-ton dolos.
(10 sec. wave period)

It is quite useful to be able to determine the wave conditions which induce a structural failure in the armor unit. Structural failure of a 42-ton dolos unit is depicted in Fig. 9 for a wide range of wave conditions at a constant period. Wave conditions producing 75% and 60% failure stresses are also presented in Fig. 9.

REFERENCES

ADINA, (1984), "A Finite Element Computer Program for Automatic Dynamic Incremental Nonlinear Analysis," Report 84-1, ADINA R&D, Watertown, MA.

Bruun, P. and Gunbak, A., (1977), "Stability of Sloping Structures in Relation to Risk Criteria Design," Coastal Engineering, 287-322.

Burchart, H.F., (1984), "Fatigue in Breakwater Concrete Armor Units," Aalborg University, Denmark.

Kaplan, P., (1979), "Impact Forces on Horizontal Members," Proceedings of Civil Engineering in Oceans IV, 716-731.

Kaplan, P. and Silbert, M. N., (1976), "Impact Forces on Platform Horizontal Members in the Splash Zone," Proceedings of the Offshore Technology Conference, 169-176.

Magoon, O. T., (1974), "Damages to Coastal Structures," Proceedings of the 14th Coastal Engineering Conference.

Magoon, O. T. and Baird, W. F., (1977), "Breakage of Breakwater Armor Units," Proceedings of the Symposium on the Design of Rubble-Mound Breakwaters, Isle of Wright.

McDougal, W. G., Melby, J. A., and Tedesco, J. W., (1988), "Wave Forces on Concrete Armor Units," Journal of the Waterway, Port, Coastal and Ocean Division, 114, 5:582-598.

Morrison, J. R., O'Brien, M. P., Johnson, J. W., and Scaaf, S. A., (1950), "The Force Exerted by Surface Waves on Piles," Petroleum Trans., 189, TP 2846.

Sarpkaya, T., (1978), "Wave Impact Loads on Cylinders," Proceedings of the Offshore Technology Conference, 169-176.

Sarpkaya, T. and Isaacson, M., (1981), Mechanics of Wave Forces on Offshore Structures, Van Nostrand Rheinhold, New York.

Tedesco, J. W. and McDougal, W. G., (1985), "Nonlinear Dynamic Analysis of Concrete Armor Units," Computers and Structures, 21:189-201.

ACKNOWLEDGEMENT

The work described in this paper was conducted as part of the Crescent City Prototype Dolosse Study of the Coastal Engineering Research Center, Waterways Experiment Station, U. S. Army Corps of

Engineers. Permission to publish this paper was granted by the Chief of Engineers.

CHAPTER 182

LONG TERM EXPERIENCE WITH SEAWALLS ON AN EXPOSED COAST

by Orville T. Magoon[1], Joan L. Pope[2],

Robert L. Sloan[3], and Donald D. Treadwell[4]

Appropriately designed, constructed, and maintained rubble mound seawalls are an efficient and cost-effective means of protecting erodible sections of exposed coastline. This conclusion is supported by more than 25 years of satisfactory seawall performance along the Pacific Ocean coast of Santa Cruz County, California, U.S.A. Important factors in satisfactory seawall performance include a clear understanding of the oceanographic and geologic design parameters, a vigorous inspection program during construction, and continued observation and maintenance of the structures.

INTRODUCTION

The coastal city of Santa Cruz, in Santa Cruz County, California, U.S.A., is located about 75 miles south of San Francisco, along the northern shore of Monterey Bay (Figure 1). For more than a century, the beaches and shores in this area have served as a major recreational resource (Figure 2) for residents of the San Francisco Bay Area and the California Central Valley.

---

1 President, American Shore and Beach Preservation Association, P.O. Box 279, Middletown, California 95461

2 Chief, Coastal Structures and Evaluation Branch, Coastal Engineering Research Center, U.S. Army Corps of Engineers, Waterways Experiment Station, Vicksburg, Mississippi 39180

3 Program Manager, Planning Division, U.S. Army Corps of Engineers, South Pacific Division, 630 Sansome Street, San Francisco, California 94111

4 President, Treadwell & Associates, Inc., 447 Battery Street, Suite 300, San Francisco, California 94111

The authors are Members of the American Society of Civil Engineers.

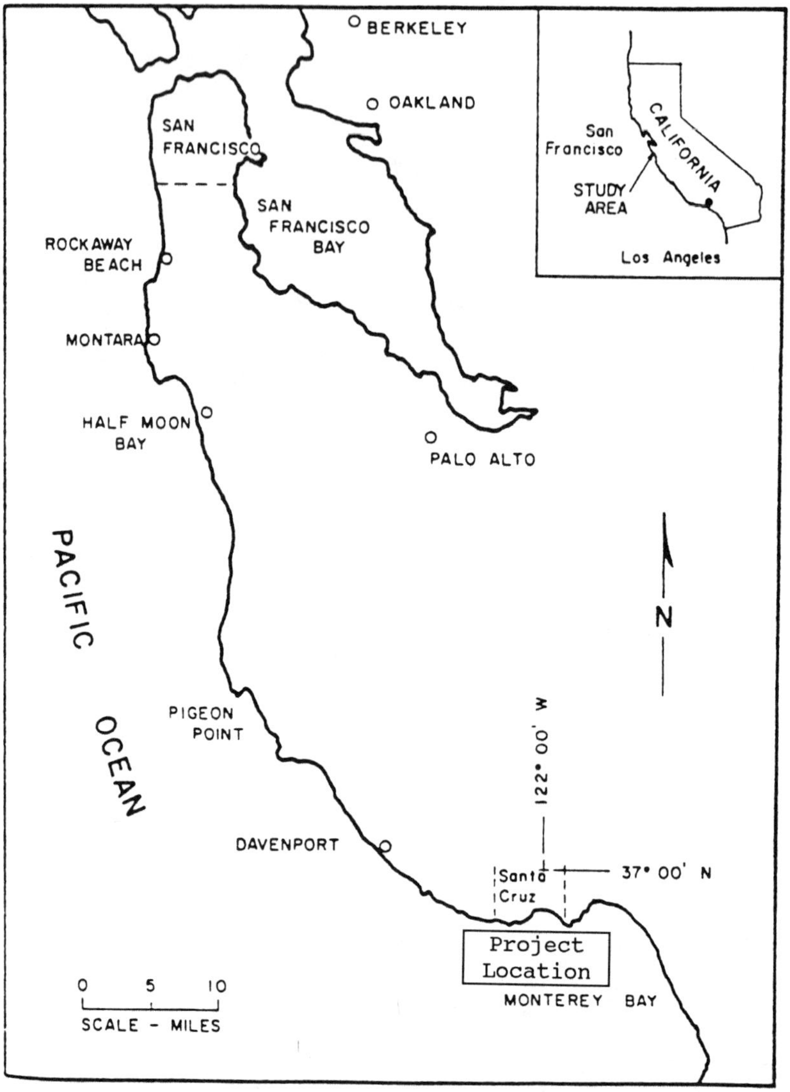

Figure 1 - Project Location

Figure 2 - Using the Beach at Capitola

The California coast in the Santa Cruz area is subject to very high waves, with maximum deep water waves often exceeding 50 feet in height. The water depths immediately seaward of the toe of the bluffs are sufficiently shallow to cause large waves to break well seaward of any protective structures under most conditions. The resulting waves at the shoreline are affected by both the depth of water as well as the complex bathymetry and topography immediately offshore.

The erosion of the beaches and cliffs on the northern side of Monterey Bay between the west limits of the City of Santa Cruz and New Brighton Beach State Park had been progressive for many years (Figure 3). The recession of the cliffs had cut through public streets, destroyed both public and private land, and threatened to destroy residences and other buildings. Individual efforts by local interests to combat the erosion problem were historically insufficient in scope or failed in their intended purpose.

GEOLOGIC SETTING

Coastal cliffs are a common landform along large sections of the geologically young California coast (Figure 4). The cliffs are the result of active erosion, being subject to periodic retreat during stormy periods. Many of the cliffs consist of relatively soft sandstones, siltstones, and shales, which are highly susceptible to erosion from wave action and surface runoff. The problem of coastal cliff erosion and retreat is well understood qualitatively and described in the literature. However, because the episodes of retreat are relatively infrequent and the overall rate is deceptively slow, property owners tend to develop a false sense of permanency and security.

The actual process and rate of cliff erosion and slope retreat depend to a large extent on the type of material involved. The weathering and erosion of the predominantly sandy rocks in California lead to the accumulation of talus and the formation of sandy beaches. The talus and the beach protect the toe of the slope from wave erosion between major storms. In areas underlain by stronger, more cohesive rocks, there is often not enough sandy material to form the beaches and the waves tend to lap directly at the toes of the cliffs. In both cases, however, the rates of cliff retreat tend to be relatively slow in the periods between large storms.

Failures, when they occur, tend to be the result of large episodic events such as major storms. For slopes protected by sandy beaches, the storm waves and tides must be sufficiently large to remove the protective sand before the waves can attach the slopes themselves. Once this occurs, the rate of cliff erosion can be very rapid. Cliffs that are not protected by beaches will obviously be exposed to significant wave action more often. Even

Figure 3 - Beaches and Cliffs near West Cliff Drive

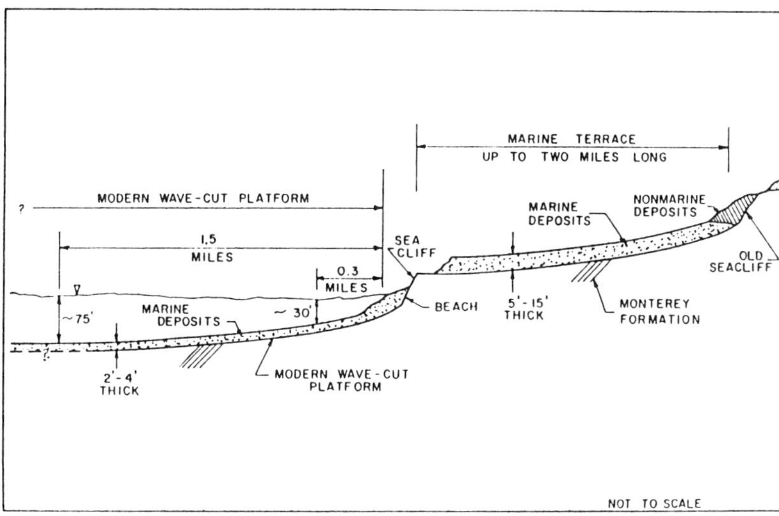

Figure 4 - Typical Profile Normal to Coastline

then, major failures will be relatively infrequent and more likely to occur during large storms. Overall, these infrequent failures result in a slow but steady average retreat of coastal cliffs, often at a rate of less than one foot per year. More important, a lack of erosion over a period of a few years is not evidence that the retreat of coastal cliffs has been halted.

Santa Cruz County lies along the northern coast of Monterey Bay in central California. Uplifted marine terraces flank most of the northern bay and also the open coast farther to the north. The seacliff varies in height from about 20 to 90 feet and is generally composed of Santa Cruz Mudstone and the Purisima Formation (siltstone and sandstone). These sedimentary rocks usually lie almost horizontal or dip gently seaward and are often capped by 6 to 20 feet of unconsolidated marine and non-marine terrace deposits.

The Santa Cruz Mudstone is predominantly a diatomaceous siliceous mudstone. It is thin-to-thick-bedded and individual beds vary from several centimeters to a meter in thickness. Joints and fractures give most outcrops a blocky appearance. The large number of rockfalls and block landslides which occur in the mudstone indicate its susceptibility to failure on steep slopes such as canyon walls and seacliffs. The Purisima Formation consists of thick bedded, poorly to moderately indurated siltstones and sandstones with occasional interbeds or lenses composed almost entirely of mollusk shells. This formation is jointed, faulted, and warped. The influence of individual stratigraphic units within the Santa Cruz Mudstone and in the Purisima Formation on local geomorphology and erosion rates is significant.

Coastal erosion or seacliff retreat is caused by both marine and terrestrial processes. Surf action is usually the dominant agent, producing both wave impact and abrasion. The rate of seacliff retreat is dependent upon the following natural factors: (1) available wave energy and exposure (including the presence or absence of a protective beach at the base of the cliff); (2) the lithology of seacliffs and their resistance to erosion; (3) geologic structure including joints, faults, and folding; and (4) the height and slope of the seacliff. Runoff and human activities are factors that can also add significantly to the rate of cliff retreat.

Within Monterey Bay, the seacliffs are generally protected from direct wave attack. The predominant waves from the west-northwest are refracted almost 90 degrees before striking the coast, and wide sandy protective beaches begin to appear. The coastal cliffs throughout most of the city of Santa Cruz are composed of erodible sediments of the Purisima Formation. Rapid erosion has cut back the cliffs, changing the trend of the coastline and creating the embayment known as northern Monterey Bay (Figure 5). Although the bay configuration protects this area from

Figure 5 - Monterey Bay (looking East)

direct wave attack, erosion rates in these sandstone and siltstone beds (Purisima Formation) are still greater than the rates in the Santa Cruz Mudstone along the open coast to the north. The erosion rate of the Purisima Formation is influenced by the varying hardness of different lithologic units within the formation, the orientation of well developed joint sets, and the presence of faults.

Cliff height sometimes exerts an indirect control on erosion rates. The quantity of material produced by a given amount of coastal retreat is a direct function of cliff height. For example, cliffs have been undercut and have subsequently failed, and large blocks have broken out along joint sets and fallen to the beach below. To the extent that this material remains in place at the foot of the cliff, it serves as temporary riprap to buffer the cliffs from direct wave attack. However, the large sandstone and siltstone blocks produced by breakdown of the Purisima Formation last only a few years in the surf zone.

PROJECT HISTORY

Studies of the beach and cliff erosion problem in Santa Cruz County were conducted in the mid to late 1950's by the U.S. Army Corps of Engineers in cooperation with the State of California, Santa Cruz County, and the City of Santa Cruz. These studies concluded that the most practicable plan for protection of the West Cliff Drive and Opal Cliffs-Capitola reaches would be rubble mound seawalls with top elevations ranging from 14.0 to 17.5 feet above Mean Lower Low Water (MLLW). It was further concluded that the irregular alignment and rocky nature of the shoreline would make placement of beach fill economically infeasible. The numerous projecting points or minor headlands that function as groins do not impound sand in sufficient quantity to protect the cliffs, indicating that groins, either with or without fill, would not correct the problem.

In order to reconstruct and restore West Cliff Drive to serve the large public demand at this important resort, the City of Santa Cruz, the City of Capitola, the County of Santa Cruz, the State of California, and the U.S. Army Corps of Engineers cooperated in planning, designing, and constructing more than 4,000 feet of rubble-mound seawalls in about twenty sections. In their studies, they found that the cliffs in the area had been progressively eroded for many years.

Further south, the Opal Cliffs-Capitola reach, also about 2.5 miles long, is characterized by an irregular shoreline backed by cliffs ranging from 35 to 75 feet in height. Except for narrow beaches found during the summer in shallow embayments and at the mouth of Soquel Creek, the greater length of this section is devoid of beach material and the cliffs are exposed to wave attack all year. During the winter months, wave action strips the beaches of sand leaving the underlying bedrock exposed.

The beach erosion control project for these two reaches that evolved from the Federal studies was authorized by the River and Harbor Act of 1958. This act provided, in part, for Federal participation by contribution of funds toward the cost of construction of: (1) rubble mound seawall units with an aggregate length of 4,700 feet, along West Cliff Drive; and (2) rubble mound seawalls about 870 feet in length at Cliff Drive in the Opal Cliffs area of the City of Capitola, about 10 miles to the southeast.

The plan of protection for West Cliff Drive provided for rubble-mound, or riprap seawalls and/or revetments at critical reaches, totaling about 4,150 feet in length (Figures 6 and 7). These seawalls have a maximum width of 10 feet at an elevation of 17.5 feet above mean lower low water, and a seaward slope of 1 on 1-1/2 after anticipated settlement. The individual face or cap stones have a minimum weight of 4.5 tons so as to resist an anticipated wave 9.5 feet high. In accordance with then current design practices, the core of the seawalls consisted of quarry run stone from 10 to 1,000 pounds with 50 percent greater than 500 pounds. A stone filter blanket was placed prior to placement of larger stone when the thickness of the sand layer at the seaward toe of the seawalls exceeded one foot. The design was predicated on the premise that the (Federal) project could only protect the bluffs from that component of the erosion caused by direct wave action, and that erosion caused by other causes (i.e., surface runoff, wave splash erosion due to geologic causes, etc.) could not be a part

Figure 6 - Seawall during Construction (April 1962)

Figure 7 - Seawall Construction (May 1961)

of the Federal project. Erosion due to causes other than waves, currents and tides would thus be a non-Federal responsibility and were not included in these designs.

Because of funding considerations, the City of Santa Cruz chose to construct the shore protection improvements in stages over a three-year period. Stage one, accomplished in 1961, consisted of the construction of thirteen small units or sections totaling 1,950 feet in length. Stage two was completed in 1962 and consisted of seven units of a total length of 1,050 feet. Stage three was completed in 1964 and consisted of three sections aggregating 1,150 feet in length, a total of 4,150 feet.

RECENT OBSERVATIONS

An inspection of these rubble mound seawalls and revetments in early 1988 indicated that over ninety percent of them are in a very serviceable or nearly as-built condition (Figures 8 and 9). Damage or potential damage to the remaining five to ten percent appears to be caused by the following: (1) erosion and/or collapse of the bluff material landward of the seawalls with subsequent loss of backing or support which allowed rubble to be displaced; (2) erosion of the bluffs at the terminal ends of the seawall, allowing lateral movement, loss of core material, and subsequent damage; (3) erosion of material under the toe or foundation and subsequent settlement of wall (for relatively small displacements of the toe, remedial action of placing relatively minor quantities of additional armor stone appeared satisfactory); and (4) inadequate design wave height, related to difficulties in determining wave

Figure 8 - Seawall near Santa Cruz (May 1988)

Figure 9 - Seawall near Capitola (May 1988)

heights at the structure due to highly irregular bathymetry and topography immediately offshore.

The large majority of the Santa Cruz West Cliff seawalls were built as single armor layer revetments. Single armor layer structures have very little tolerance to any shifting or movement of the armor cover. The cover is not a "flexible" layer in the same manner as a multi-layer armor cover. Minor movement of the armor layer as the structure settles is not fatal in a two-layer armor where the shifting units can fall into the structure to heal gaps and still provide cover to the underlayer. With single layer construction any shifting or settling of the armor (even if there is no initial unit loss) may expose the underlayer to erosion.

Three distinctive modes of failure or structure unraveling were noted in those West Cliff sections where damage was observed. Even though the foundation material is bedrock, it is susceptible to erosion.

Flanking of the individual revetment sections can occur if the adjacent bluff continues to retreat; in these cases, the armor may slide toward the unsupported flank, exposing the core; minor losses of core material cause additional shifting of the armor layer and the revetment continues to unravel until enough of the crest armor collapses down to form a multiple layer flank which effectively seals off the core from wave action.

The toe of the seawall may not in all cases have been keyed down to the bedrock. Some of the revetments are built at the back of coves where a local pocket beach covers the bedrock with a thin veneer of sand; erosion of the pocket beach or possibly even some abrasion and scour of the bedrock may undermine the toe, causing stone to shift. In a multiple layer revetment, damage caused by the down-slope migration of the toe stone is usually minimized as an upper layer unit collapses into the lower layer. This may cause an apparent steepening of the toe but shifted units will often seal the cover without exposing underlayer further up slope. The multiple layer cover heals itself as it collapses. Not so with a single layer cover, where the displaced toe exposes the core and the downward migration of the cover continues without healing. Units can only migrate down slope by sliding on the core until there is enough loss of core to allow collapsing. In these cases, the revetment is reduced to a pile of armor which has collapsed into the void left as the core was eroded.

Erosion of the back bluff was observed in several cases. Runup and overtopping of the revetment or surface water runoff may have caused further retreat of the bluff behind the revetment and loss of support to the structure. Backward sliding crest stones again exposed the underlayer or core resulting in an unraveling of the structure from the top down. Sections exhibiting this type of damage usually experienced only minor loss of the structure's integrity unless the retreat of the supporting bluff was significant.

Although several design limitations with single layer armor construction could be identified at this site, the satisfactory performance exhibited by most of the sections illustrates the acceptability of single layer construction in certain geologic settings. Single layer construction can be used in areas of a stable foundation where bedrock or erosion-resistant soil limits toe scour or retreat of the back support. However, if this shore protection system were to be designed and built today, advances in design standards and the information gathered from years of observation would lead to some differences in approach. These would include a more comprehensive investigation of foundation conditions, erosion rates, and the effects of extreme water levels and storm conditions. A formal operation and maintenance plan would also be prepared and implemented. And, double-layer construction with a filter fabric backing would probably be selected as the final design.

ACKNOWLEDGEMENTS

The authors gratefully acknowledge the cooperation of the City of Santa Cruz, the County of Santa Cruz, the State of California, and the U.S. Army Corps of Engineers in providing records of the original design and construction. The assistance of Jon A. Rosso of Treadwell & Associates during the final preparation of the paper is also appreciated.

SELECTED BIBLIOGRAPHY

Bradley, W. C. and G. B. Griggs, 1976, Form, genesis, and deformation of central California wave cut platforms: Bulletin Geological Society of America, v. 87, p. 433-449.

Dormurat, G., 1978, "Selected Coastal Storm Damage in California, Winter of 1977, 1978," Shore and Beach 46 (No. 3, July):15-20.

Fitzgerald, D. M., D. Sullivan and A. D. Magee, 1981, "Effects of rip rap on onshore-offshore sediment transport," Shore and Beach, v. 49, No. 4, p 19.

Griggs, G. B. and R. E. Johnson, 1976, Effects of the Santa Cruz Harbor on coastal processes of northern Monterey Bay; Environmental Geology, v. 1, p. 299-312.

Griggs, G. B., and R. E. Johnson, 1979, "Coastal Erosion, Santa Cruz County," California Geology 32 (No. 4, April):67-76.

Griggs, G. B. and R. E. Johnson, 1983. "Impact of 1983 storms on the coastline of northern Monterey Bay." California Geology v. 36, No. 8:163-174.

Kuhn, G. G., and F. P. Shepard, 1981, "Should Southern California Build Defenses Against Violent Storms Resulting in Lowland Flooding as Discovered in Records of Past Century," Shore and Beach 49 (No. 4, October):3-11.

Moffat and Nichol, Engineers, 1983. Construction materials for coastal structures. U. S. Army Corps of Engineers Coastal Engineering Research Center Special Rept. No. 10.

National Marine Consultants, 1970, Wave statistics for seven deep water stations along the California coast: Report prepared for the U.S. Army Corps of Engineers, Los Angeles and San Francisco Districts, 20 p.

Smith, A. W. and D. M. Chapman, 1982. The behavior of prototype boulder walls. Proc. ASCE 18th Coastal Engineering Conf. (Capetown) V.III:1914-1928.

U. S. Army Corps of Engineers, 1956. Beach erosion control report on the cooperative study of the Santa Cruz Area, Serial No. 25, San Francisco, CA Jan. 20, 1956.

U. S. Army Corps of Engineers, 1977, Coastal Engineering Research Center, Shore Protection Manual V.1 and 2, Third Edition.

U. S. Army Corps of Engineers, 1983, Coastal Storm Damages: 1983, U. S. Army Corps of Engineers, Los Angeles District.

U. S. Army Corps of Engineers and Dames and Moore, 1971, National shoreline study — California regional inventory, South Pacific Division, San Francisco, 103 p.

CHAPTER 183

DYNAMIC STUDIES ON CAISSON-TYPE BREAKWATERS

Gao Ming* Dai Guanying* Yang Jihua**

Abstract

In this paper, studies on dynamic characteristics and responses under waves of caisson-type breakwaters by field and model testings as well as dynamic analyses are described. It is found that the gravity-type structure-foundation system can be simplified as a rigid body on a linear or non-linear elastic foundation for evaluating its dynamic characteristics and responses. The dynamic displacements of the caisson are found related to the wave heights, the water depth as well as the foundation soil properties.

Introduction

Along with the increase of the ship tonnage, the construction of breakwaters, either for the development of a new port or for the extension of the existing port requires that the structures are to be built in deeper coastal areas, and these breakwaters are being subjected to severer wave actions. A problem is raised as to the dynamic behaviours of the breakwaters under the tough environmental conditions. Based on such needs, we performed a series of studies on the dynamic characteristics and responses under waves through field testing, model testing and numerical analyses for the caisson-type breakwater, either rectangular or cylindrical. This paper presents these investigations, mainly the test results and the comparison between the test results and the analytical results with different methods of analysis. (Gao et al 1988, Dai et al 1985,1987)

----

* Senior engineer, Nanjing Hydraulic Research Institute, Nanjing, China
** Engineer, Nanjing Hydraulic Research Institute, Nanjing, China

Field Test of the Rectangular-Caisson

The field testing site was chosen at a newly built port located in the southern part of Shandong province of China. The breakwater is one for composite use, i.e. with its inner side for ship berthing. Figure 1 shows the plane view and the profile of the breakwater. The rectangular caisson has a weight of 13,000t with a deck and a 2.1m high exterior parapet wall.

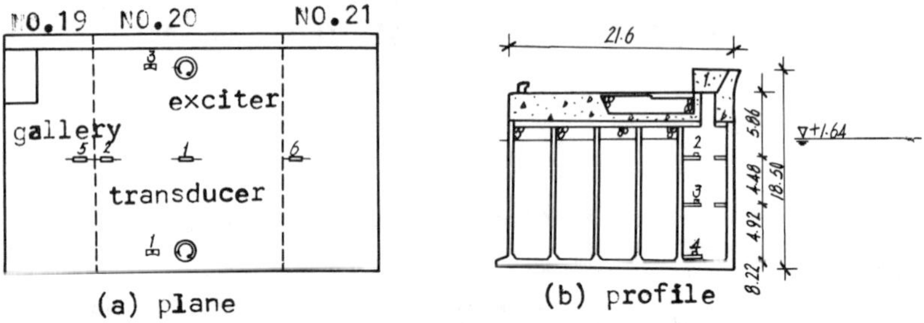

Fig. 1 Test Caisson

Two mechanical-type exciters of 6t capacity, with frequency range of 0-30Hz and frequency resolution of 0.01Hz, were adopted to excite the caisson. The exciters were placed on the top of the No.20 caisson, and the displacement transducers were fixed on the top surface of the three caissons and on the vertical gallery of the No. 19 caisson. The test results are listed in Table 1, and the mode shapes and amplitude-frequency curve are shown in Figure 2.

Table 1 Test results

Meas. pt. No.	Caisson No.	First order freq.(Hz)	phase	damp.	2nd order freq.(Hz)	phase
1	20	3.8	Ref.	.0579	8.01	Ref.pt.
2	20	3.8	same	.0592	8.48	same
3	19	3.8	same	.0329	8.48	same
4	21	3.8	same	.0619	8.48	same

From the tests, it shows that:
1. Under the exciting focre of 10-12KN, the three caissons have the same 1st and 2nd order natural frequencies and nearly the same values of damping ratio. It illustrates that due to the surface deck connection, the breakwater caissons vibrate simultaneously with good integrity.
2. From the vibration modes measured, it illustrates that the vibration is coupled, with lateral vibration as the primary and rocking as secondary for the fundamental mode

and rocking as primary and lateral vibration secondary for the 2nd order mode.

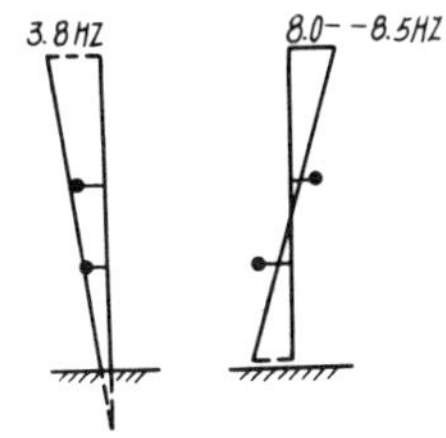

Fig. 2(b)

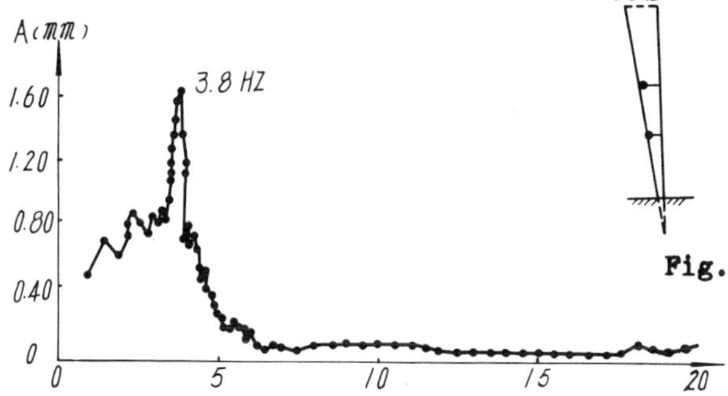

Fig.2(a) Resonance curve of pier caisson

It means that the vibration of caisson-type structure can be regarded as the vibration of a rigid body on elastic foundation, linear or nonlinear, for evaluating its dynamic characteristics and dynamic responses.

Using corresponding equations of vibration, it is feasible and easy to determine the compression and lateral coefficients of the subgrade by measuring the fundamental frequencies of the structure undergoing vertical or horizontal-rocking vibration and then to determine the bearing capacity of the subgrade.

Model Tests of the Rectangular-Caisson

The model tests were performed on a similar regular caisson-type composite breakwater at a model scale of 1 to 87. The model was not simulated exactly to the above existing structure but was for a new caisson-type breakwater in design stage. Plexiglass and weighted rubber were used as modelling materials for caisson structure and subgrade foundation. A portion of foundation, with a depth nearly equal to the height of the caisson and a width equal nearly to 3 times the width of the caisson was included in the model.

The models were tested for various restraint conditions. Table 2 shows the influence of the different foundation conditions of restraint and sizes on the natural frequency of caisson. The results given are in prototype scale.

Table 2

Foundation size M^3	Restraint condition of the foundation	Nat. freq. (Hz)	
		1st	2nd
71.3*62.3*26.1	Restricted on four side	1.18	1.85
71.3*62.3*26.1	Non-restraint	1.06	1.93
71.3*44.0*26.1	Non-restraint	0.93	2.02

Shown in Fig. 3 is the resonance curve and mode shapes of the rectangular caisson from model tests.

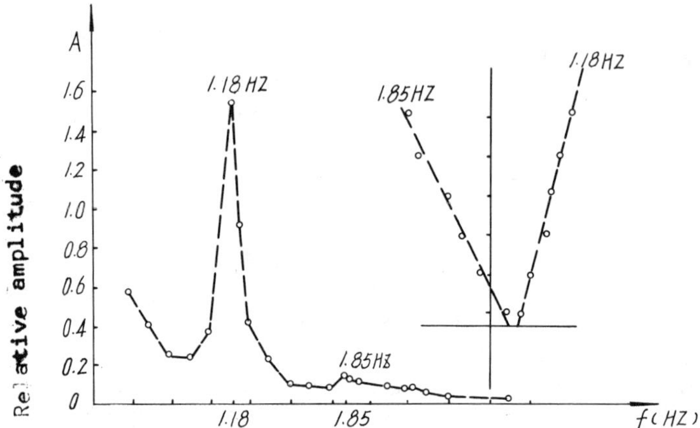

Fig. 3 Resonance curve and mode shapes

Tests for Different Water Levels

The model was tested for different water levels to investigate the added mass coefficient. Fig.4 shows the response-frequency curves with various water levels. The natural frequency of the system lowers with the rise of water level. At the checking high water level the fundamental natural frequency lowered down about 9%. From this it can be estimated that the added mass coefficient of the water for the tested

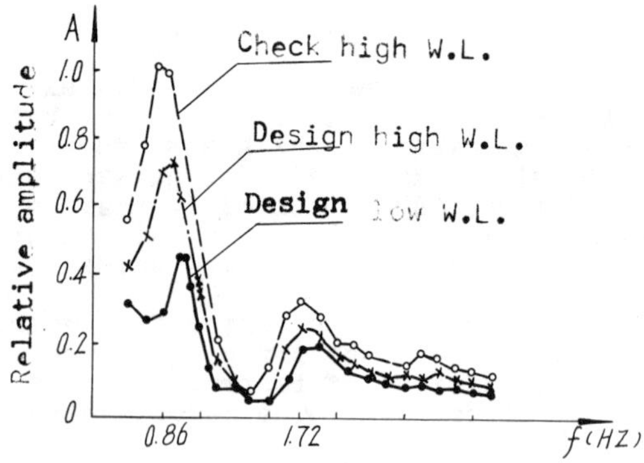

Fig. 4 Resonance curve for various water levels

caisson is around 0.6, about the same as that recommended by Dr. Hallam (Hallam 1977).

Table 3

Water level	Nat. Freq. (Hz)		Damping ratio	
	1st	2nd	1st	2nd
Without water	1.06	1.93	0.024-0.038	0.024-0.065
Design low w.l.	1.01	1.82	0.08	0.084
Design high w.l.	0.94	1.72	0.134	0.091
Check high w.l.	0.86	1.72	0.163	0.094

Dynamic Responses under Waves

The model with foundation was placed in a wave basin to test the dynamic responses under regular waves of various wave heights and wave periods with three different water levels. From these test results, it can be seen that the dynamic displacements increase with the increase of wave height as well as to water depth, the dynamic displacement is also increasing with the increase of wave period in general. The uplift forces measured show an evident linear relationship with wave heights.

Field Dynamic Experimental Studies on Cylindrical Caisson-type Breakwater

Field dynamic measurements were also carried out on cylindrical caisson of an island breakwater in the same port mentioned above. The caissons were 14m in diameter, 13.5m in height and connected together with a block at the top. Each caisson weighs 5000t. The same exciters as aforementioned were used, and the method of ambient vibration was also adopted for excitation. Fig. 5 shows the frequency-amplitude curve and the power spectra curve. The main results are listed in Table 4.

Table 4

Caisson No.	Ambient test freq. (Hz) with cap	no cap	Reson. test freq. (Hz) with cap	damping ratio
15	1.95		1.90	0.047-0.071
16	1.95			
40		2.18	2.20	0.032-0.075
41		2.18		

In order to determine the vertical dynamic characteristics, the caisson was also tested by the method of applying shock load vertically and the fundamental natural frequency for vertical vibration measured is 4.3-5 Hz..

Model Tests for Cylindrical Caisson-Type Breakwater

A model made of plexiglass for a cylindrical caisson of 14m in diameter and 26.5m in height was tested for a practical pier design at a scale of 1 to 30, also with a portion of foundation included. The foundation is simulated to different compression stiffness. The model is tested

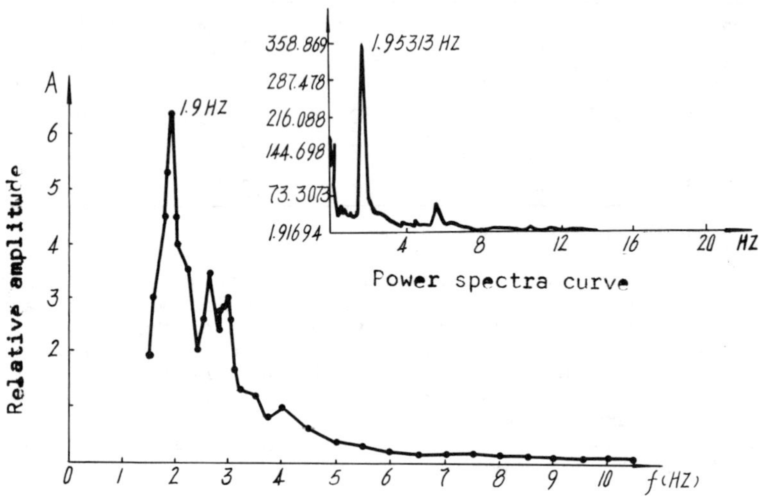

Fig. 5 Frequency-amplitude curve

in air by an actuator as well as by the method of hammering. The tested frequency-amplitude curve and the response spectrum, transfer function and coherence coefficient are shown in Figs. 6 and 7.

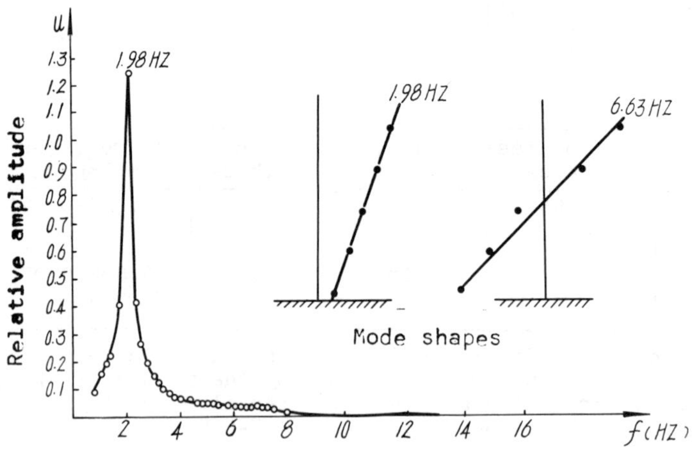

Fig.6 Frequency-amplitude curve

It agrees well with the results obtained from the field tests.

The model was then placed in a wave basin to test its respones under regular and irregular waves in different water depths. The irregular wave spectra used are the Pierson-Moskowitz spectrum and the local spectrum evaluated by Nanjing Hydraulic Research Institute. The input wave height spectrum as well as response spectrum are shown in Figs. 8(a), (b). Fig. 8(c) is the displacement response spectrum. The displacement of the caisson top versus wave height are plotted in Fig. 9 by solid line for various water depths and wave periods. The Tested results show that the displacement of the caisson is in nearly linear relation to the wave height and are much small than th those for rectangular-caisson under the same wave condition. In Fig. 9 it also shows in dotted lines the calculated values of the responses without taking into account the dynamic effect. It can therefore illustrate that the dynamic effect can be neg-

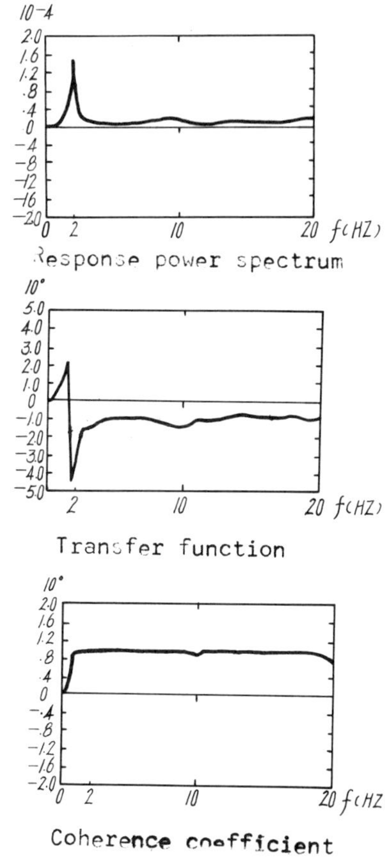

Fig. 7

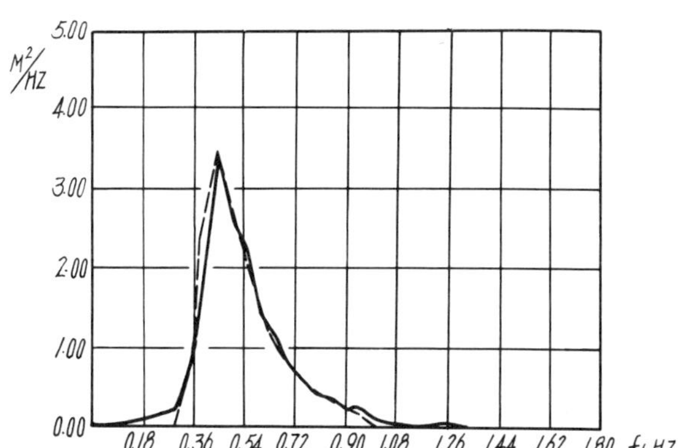

Fig. 8(a) Local wave spectrum

lected under ordinary wave condition.

Dynamic Analysis

The dynamic analysis for the caisson-type breakwater has been carried out. Two methods are adopted in the analysis: (1) The method of concentrated parameters: The system can be simplified as a 3 degree-of freedom rigid body

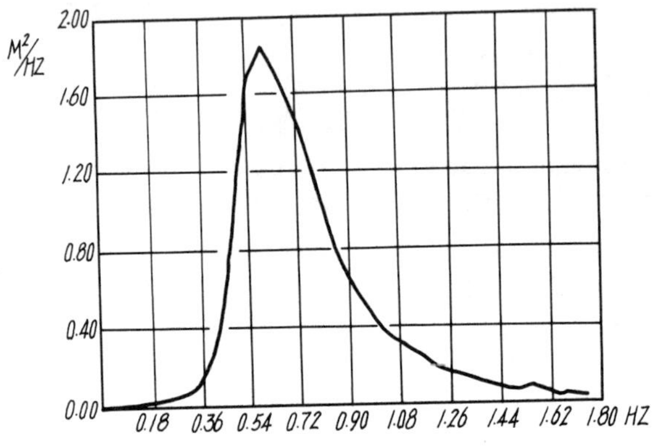

Fig.8(b) P-M wave spectrum

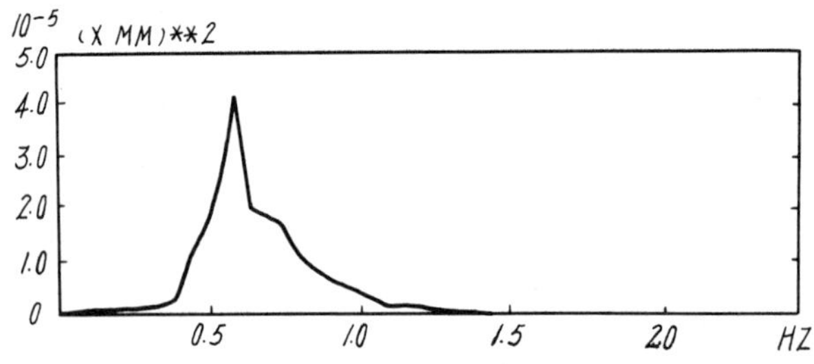

Fig.8(c) Displacement response spectrum

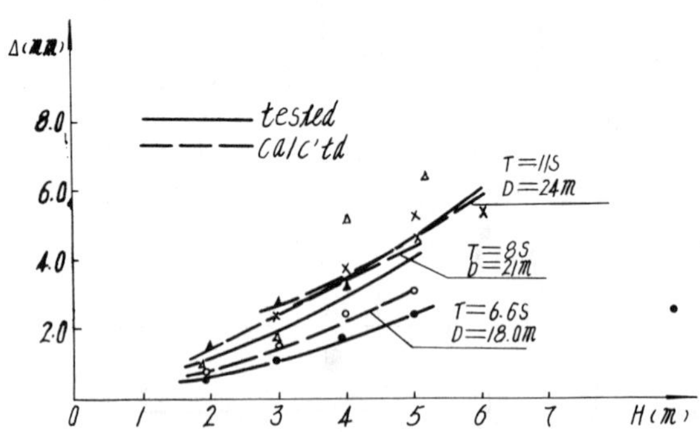

Fig. 9 Displacement versus wave height

on elastic foundation, treated as a series of vertical and lateral springs, with lateral, vertical and rocking spring constants denoted by C_x, C_z and C_ϕ respectively, as shown in Figure 10; (2) The finite element method: For the rectangular caisson, together with the foundation included, the system can be treated as a 2-dimensional problem. The calculated results for natural frequency are given in Table 5, and the test results are also listed for comparison. It can be seen from Table 5, that good agreements are reached by applying these methods.

In the finite element the dynamic responses of the rectangular caisson under different wave heights are evaluated, using the wave pressure formula recommended by Miche-Biesel and the results are shown in Table 6. It shows that, while the wave height ranges between 7 to 8m, the test results are also listed for comparison, and it can be found that the a agreement is satisfactory.

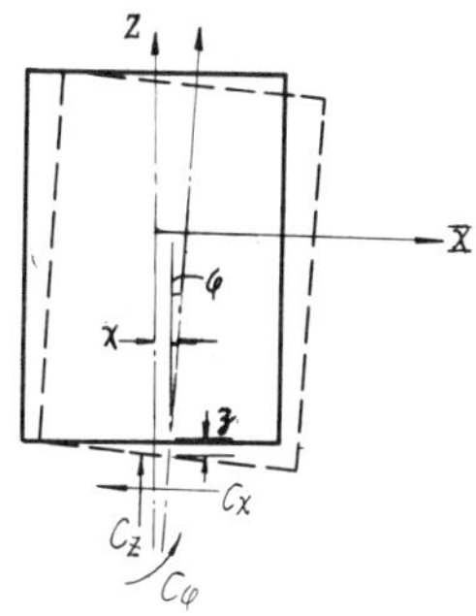

Fig. 10

Table 5

Caisson Type	Cz value of foundation (KN/M³)	1st Nat. Freq.(Hz) Test	CPM	FEM	2nd Nat. Freq.(Hz) Test	CPM	FEM
Rect.	5,150	0.98-	0.95	1.07	1.83-	2.81	2.31
	68,000	1.06			1.93		
	102,000						
Cylind.	76,820	1.92	1.91		6.66	6.85	
	71,280	1.83	1.81			6.50	
	46,530	1.44	1.46			5.25	

C_z - the vertical compressive coefficient of the subgrade.

Table 6 D/H 0.766

Wave height (m) \ Wave period (s)	8.0		10.0		12.0		14.0	
	test	cal.	test	cal.	test	cal.	test	cal.
6.86	3.05	2.96	3.48	3.65	5.22	4.18	6.00	4.70
7.76	3.57	3.39	3.74	4.34	5.48	4.96	6.79	5.66

D - water depth, Hc - height of caisson.

Concluding Remarks

Through the field testing, the model testing and the dynamic analysis using different methods, a quite thorough investigation of the problem of caisson-type breakwater has been made on its dynamic characteristics and dynamic responses to waves, and the following can be concluded:

1. The caisson-type breakwater, either rectangular or cylindrical, can be regarded as a system of rigid body on linear or nonlinear elastic foundation for evaluating its dynamic characteristics and dynamic responses.

2. The foundation stiffness has a great effect on the dynamic characteristics and responses, and any weak soft layer underlain should better be dredged out.

3. Under strong waves, the top displacement responses of the rectangular caisson will be much larger then those of the cylindrical caisson.

4. In general, the dynamic magnification factor under waves is very close to 1, and therefore, in the calculation of the stability of the caisson, it can be treated as a static problem.

References

1. Dai, G. Y. et al, 'The Dynamic Characteristics and Dynamic Responses of Gravity-type Caisson Port Structures', Jr. Nanjing Hydraulic Res. Inst., 1985, No.4 (in China)
2. Dai, G. Y. et al, 'Full Scale Dynamic Testing of Caisson Type Gravity Harbor Structures and the Determination of Load-bearing Capacity of the Foundation', Proc. of Symposium on Structural Assessment Based on Full and Large Scale Testing, Apr. 1987, London, U. K.
3. Gao, M., et al, 'The Dynamic Investigations and Aseismic Studies on Gravity-type Port Structures', 'Waterway Transportation Engineering', 1988 No. 2 (in China)
4. Hallam, M. G, et al, 'Dynamics of Marine Structures', CIRIA Underwater Engrg. Group, 1977

CHAPTER 184

Wave Forces on Armor Blocks

by

M.A. Losada*, R. Medina* and M. Alejo**

ABSTRACT

Experimental measurements of hydrodynamic forces on a cubic block near the bottom under solitary waves were carried out. Horizontal and vertical forces were recorded and instantaneous and averaged values of hydrodynamic coefficients C_D, C_M and C_L for different boundary conditions, gap between block and bottom, e, and two or three-dimensional flow, were obtained. Horizontal and vertical forces were found to depend strongly on e/D, where D is the block side . Instantaneous values of hydrodynamic coefficients vary considerably during the wave passage and differ appreciably from the averaged coefficients.

* Universidad de Cantabria. Spain.

** Universidad Politécnica de Las Palmas. Spain.

INTRODUCTION

In the last decades great efforts have been done to understand the causes of failure of sloped breakwaters with special attention to the "lift-out" of armor units from the slope by the uprush-downrush flow. Most of the research work has been oriented to evaluate the flow conditions on the slope. Minor contributions have been made in the analysis of pressure distribution around the block or resultant forces on the block. Recently Kobayashi et al., 1985, studied the stability conditions against sliding and lifting of rip-rap, expressing drag, inertia and lift forces in function of drag, C_D, inertia, C_M and lift, C_L, coefficients. This analysis similar to others presented in the past is a force equilibrium, based on Morison's equation.

The lift-out of armors (particularly in the case of cubes) generally starts with a rotation around a point or axle as experimentally may be checked. The moment equilibrium equation for an armor unit neglecting reactions by friction or interlocking and considering the flow forces applied in the gravity center of the block, can be expressed as follows:

$$W_a = \frac{1}{2} \rho_w C_D D^2 u^2 [(1 + 2 \frac{C_M}{C_D} D \frac{du/dt}{u^2} + \frac{C_L}{C_D}) \qquad (1)$$

where W_a is the armor weight and u and du/dt the flow velocity and flow acceleration respectively.

Thus, if the flow characteristics (u and du/dt) and the hydrodynamic coefficients, C_D, C_M and C_L are known, the behavior of an isolated block under waves may be predicted. Kobayashi et al., 1985, used a similar equation in their rip-rap stability prediction, and adopted constant values of the hydrodynamic coefficients for all the wave cycle.

To study the evolution of the forces and the values of the hydrodynamic coefficients under a wave cycle a set of experiments in a wave flume were run. In the following, a general description of the experimental work and of the test results is given. It is clearly shown that boundary conditions strongly affect the evolution of forces on the block and that the values of the hydrodynamic coefficients vary considerably during the wave cycle.

EXPERIMENTAL WORK

The hydrodynamic behavior of isolated cubic units under wave action was analyzed. Smooth cubic blocks with side length, D = 5 cm, were subjected to the action of

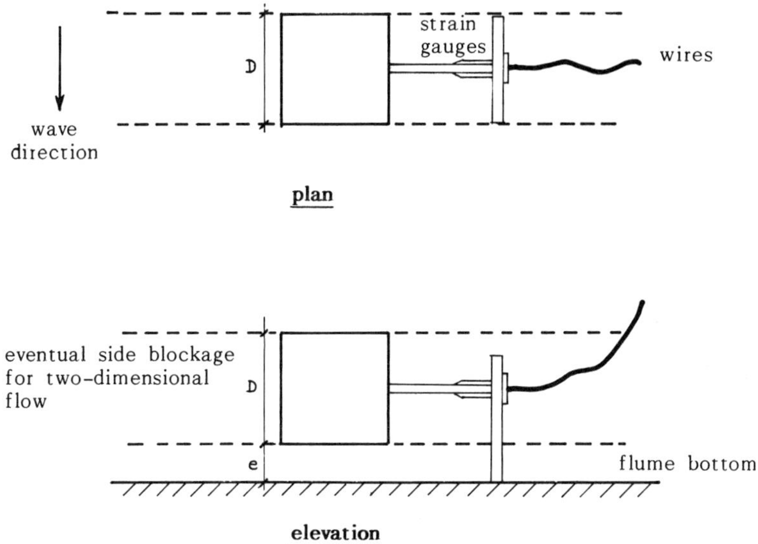

Fig. 1.- FORCE MEASURING DEVICE

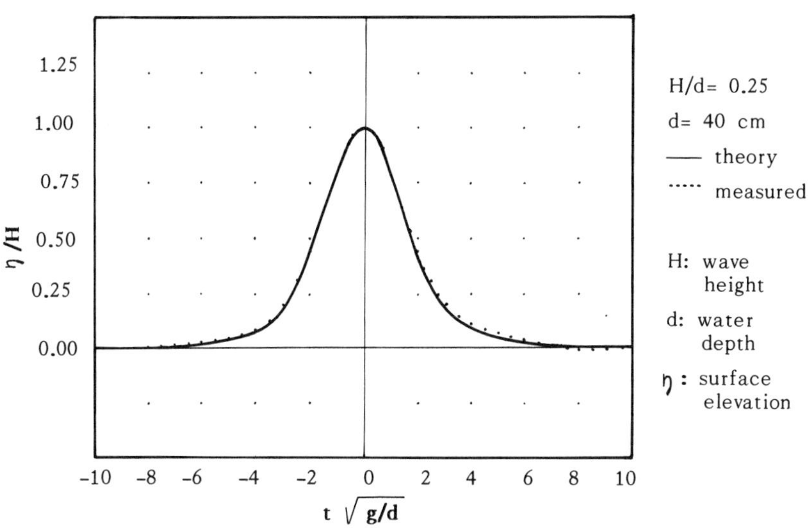

Fig. 2.- SOLITARY WAVE

solitary waves for the following conditions

e/D = approx. 0.0, 0.02, 0.1, 0.2 and 0.4

where e = gap between cube and bottom (see fig. 1). Due to the force measuring technique the case of block resting on the flume bottom was not sealed. Some pressure transmission under the block might had occur, "therefore" the indication approx. 0.0.

The type of waves considered were solitary waves (fig. 2). The distance of the test section to the paddle was approximately 45 m and the wave height varied in the range 2 < H(cm) < 20, thus $10^3 < Re < 10^5$, where $Re = uD\rho/\mu$ is a Reynolds number, μ = dynamic viscosity. The flow velocity was not recorded during the tests.

The cubes were placed horizontally and parallel to the main axes of the flume (fig. 1). Different test conditions were used, varying e, and establishing two or three-dimensional flow by incorporating the cubic block as a section of a square prism or not.

In all cases, cube size and gap were small when compared with water depth, in order to neglect the influence of the free water surface, and to allow the consideration of horizontal flow velocity.

Measurements of horizontal and vertical forces were made with pairs of strain gauges (see fig. 1). Maximum deviation expected in vertical and in horizontal force measurements is about seven per cent.

THREE-DIMENSIONAL FLOW UNDER SOLITARY WAVES

Figures 3, 4 and 5 show horizontal and vertical measured forces under a solitary wave for three cases: $e/D \approx 0$, 0.1 and 0.4 respectively. The most relevant characteristics are:

(1) The horizontal forces attain their maxima under the wave crest and take negative values (opposite to that of wave motion) associated to flow deceleration.

(2) Vertical forces are negative first (down to the bottom) and then positive for $e/D \neq 0$.

(3) For $e/D \approx 0$ vertical forces are always positive, showing two or more peaks. The first one is commonly much higher.

(4) For $e/D \geq 0.4$ the influence of the gap (indicated by the vertical force variation and magnitude) may be neglected.

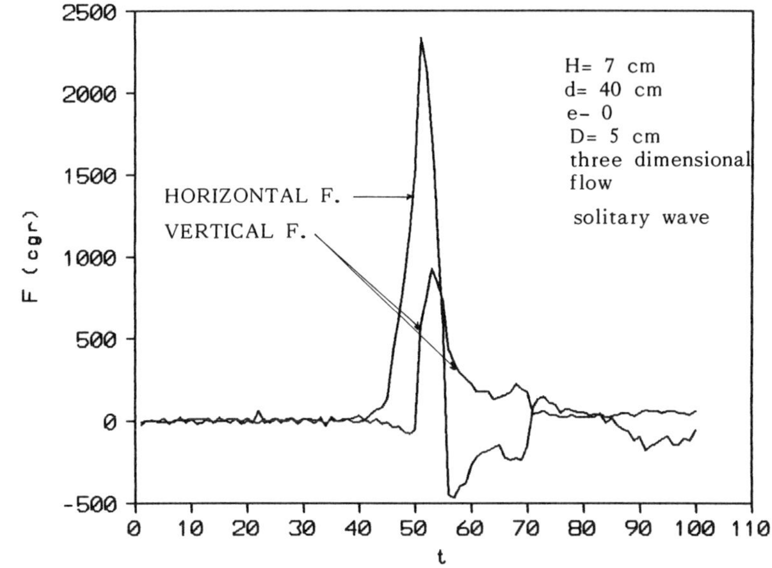

Fig. 3.- HORIZONTAL AND VERTICAL FORCES e/D- 0 . Free cube.

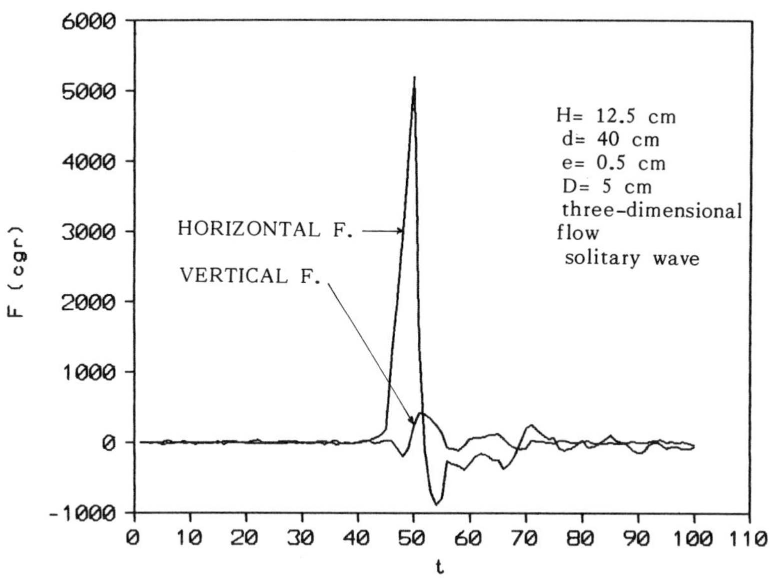

Fig. 4.- HORIZONTAL AND VERTICAL FORCES e/D = 0.1 . Free cube.

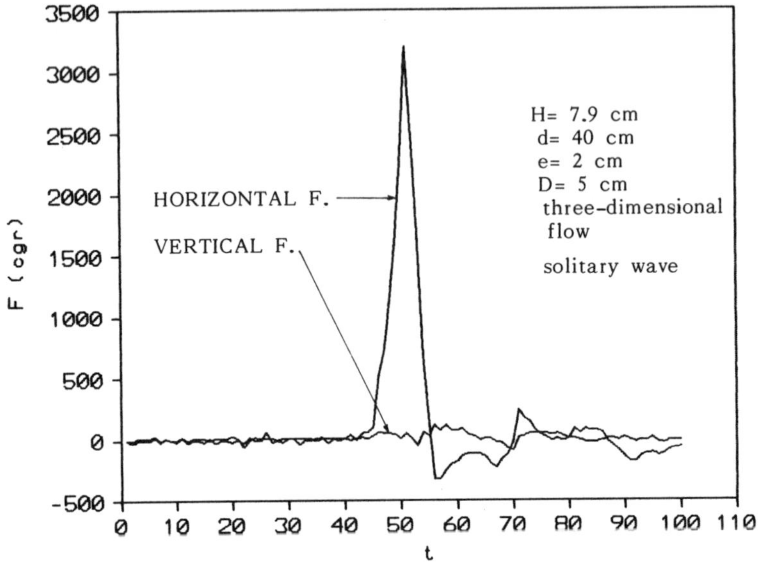

Fig. 5.- HORIZONTAL AND VERTICAL FORCES e/D= 0.4 . Free cube.

TWO-DIMENSIONAL FLOW UNDER SOLITARY WAVES

Figures 6 and 7 show horizontal and vertical measured forces under a solitary wave for e/D = 0.1 and 0.4 respectively. The most relevant characteristics are:

(1) Maximum vertical and horizontal forces are almost in phase.

(2) Vertical forces are always larger than for three dimensional flow and present two or more peaks.

INSTANTANEOUS AND AVERAGED VALUES OF C_D, C_M AND C_L

Following Sarpkaya and Isaacson, 1981, from the measured horizontal forces the instantaneous values of C_D and C_M were obtained, solving from a set of Morison-type equations at two consecutive instants assuming that C_D and C_M remain constant in the time interval. Figure 8 shows C_D and C_M with the wave passage for e/D = 0.1. Several observations can be made, which agree with those given by Sarpkaya and Isaacson, 1981, for cylinders

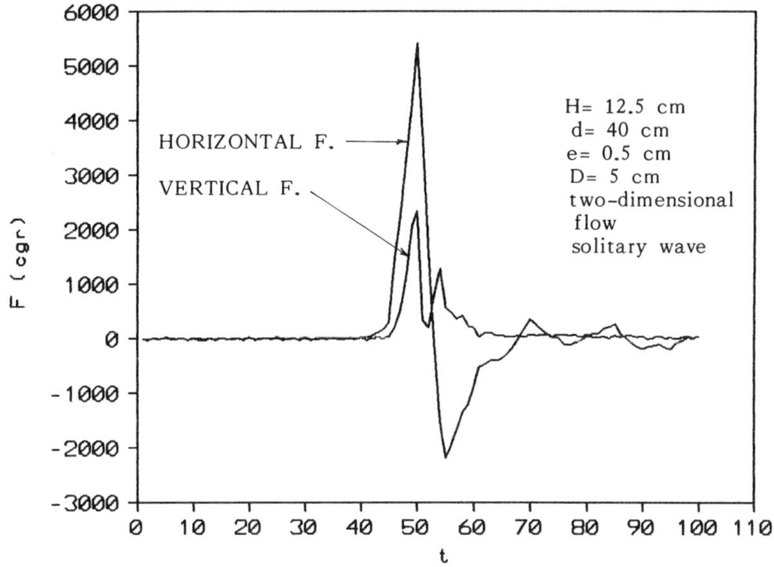

Fig. 6.- FORCES FOR TWO-DIMENSIONAL FLOW e/D= 0.1 . Cube as a part of a prism.

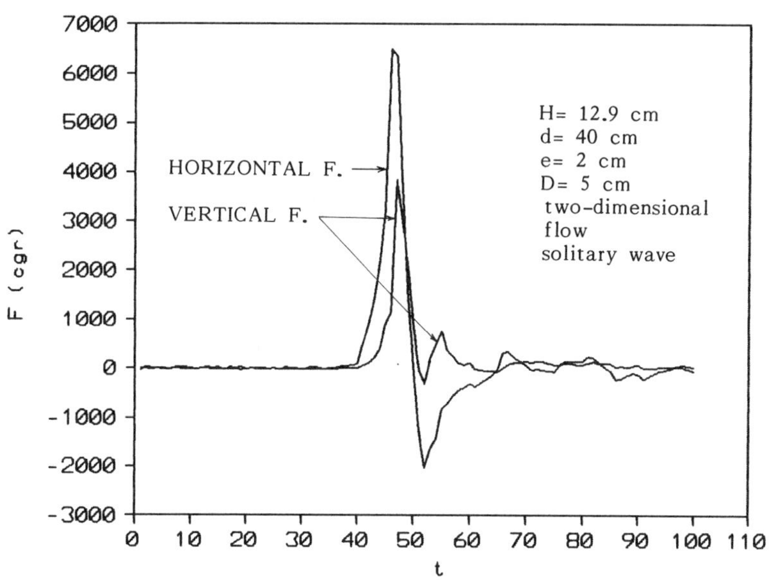

Fig. 7.- FORCES FOR TWO-DIMENSIONAL FLOW e/D= 0.4 . Cube as a part of a prism.

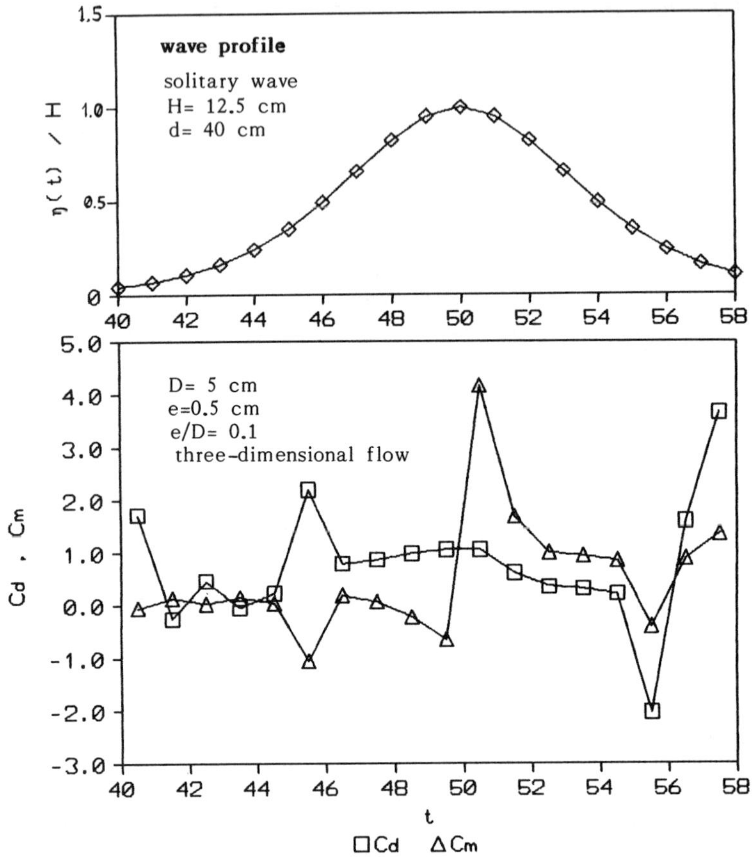

Fig. 8.- INSTANTANEOUS COEFFICIENTS Cd AND Cm

under oscillatory flows.

(1) The averaged C_M is less than unity and the added mass coefficient, C_a, is then negative, as $C_a = C_M - 1$.

(2) Neither C_M, C_D nor C_L are symmetrical with respect to the wave crest, showing a difference between accelerated and decelerated flow.

(3) C_M, C_D and C_L exhibit relative large variations during the wave passage.

From test results average hydrodynamic coefficients C_D, C_M and C_L were obtained. The average values of C_D, C_M and C_L were obtained through the use of the method of least squares by minimizing the square error between the measured and calculated forces. For a solitary wave 12.5 cm high, e/D = 0.5 and three-dimensional flow, the following values were calculated:

$$C_D = 0.65 \qquad C_M = 0.87 \qquad C_L = 4.7 \cdot 10^{-2}$$

When comparing these values with the instantaneous ones shown in fig. 8 it can be concluded that the former differs considerably from the latter.

CONCLUSIONS

In the present study an experimental investigation on wave forces exerted on an isolated armor unit was conducted. The experimental results lead to the following conclusions.

(1) The behavior of the instantaneous horizontal and vertical forces under solitary waves depend on e/D ratio, the relative distance from the block to the bottom. For $e/D \neq 0$ both the horizontal and vertical forces present a positive and a negative peak. For $e/D \approx 0$ the vertical force is always positive.

(2) The maximum positive forces for $e/D \neq 0$ and three-dimensional flow are out of phase. For $e/D \approx 0$ both forces are in phase, as well as for two-dimensional flows.

(3) The instantaneous hydrodynamic coefficients C_D, C_M and C_L vary considerably during the wave passage and differ appreciably from the averaged coefficients.

ACKNOWLEDGEMENTS

This research has been partially supported by the Comisión Asesora Científica y Técnica, project nº PA85-0176 and partially by the Dirección General de Puertos y Costas, M.O.P.U.

REFERENCES

Kobayashi, N. and B.K. Jacobs. Rip-rap stability under wave action. The Journal of Waterways, Port, Coastal and Ocean Engrg., ASCE, Vol. 111, No. 3, pp. 552-566.

Sarpkaya, T. and Isaacson, M., 1981. Mechanics of wave forces on offshore structures. Van Nostrand Reinhold Co., pp. 651.

CHAPTER 185

Hydrodynamic Characteristics of Wave-power-extracting Caisson Breakwater

Shigeo Takahashi*

ABSTRACT

The wave power extracting caisson breakwater can convert the wave energy into usable energy. The function and the stability of the caisson as a breakwater are investigated experimentally. It is found that wave reflection and wave transmission of the breakwater are relatively small and that the stability of the breakwater caisson against storm waves is high. The design method of the caisson is proposed considering the impact air pressure in the air chamber.

1. INTRODUCTION

Perforated wall caissons are increasingly adopted as the structures of seawalls and breakwaters since Jarlan (1961) invented the perforated wall caisson. This is because the perforated wall caissons have low wave reflection and overtopping characteristics, and is highly stable due to their wave absorbing capability. The wave power extracting caisson breakwater such as the one illustrated in Fig.1 is a kind of the perforated wall caisson because it can absorb the wave energy. However the wave power extracting caisson can actually convert the wave energy into usable energy. The converted energy can be used for many purpose in the sea and in the land near the breakwater.

Fig.1 Wave Power Exracting Caisson Breakwater

This breakwater is consisted of the wave power extracting caisson on a rubble foundation. The caisson has an air chamber which is attached to the ordinary caisson. The waves enter into the air chamber from the opening under the front curtain wall, and cause the vertical oscillation of the water surface in the air chamber. Then the air in the upper half of the air chamber is

* Chief of Maritime Structures Lab., Port and Harbor Res. Inst., 3-1-1, Nagase, Yokosuka, Japan 239

compressed and is expanded. The air motion generates the high-velocity flow through the nozzle which activates the air turbine and generates electricity. Thus the caisson functions as a wave power converter. A combination of the wave power converter and the vertical breakwater is attractive from the viewpoint of economical feasibility of wave power extraction, because the construction cost can be jointly borned by the accounts for power generation and harbour protection.

The studies on the wave power conversion have been conducted and the design of the caisson as a wave power converter has already been established(Ojima 1984, Takahashi 1986, 1987). In the present paper, the function and the stability of the caisson as a breakwater are investigated.

2. THEORY OF WAVE POWER CONVERSION

The conversion efficiency from wave power to air power in the air chamber is the most important factor not only as a wave power converter but also as a breakwater, because the higher conversion efficiency yields the lower wave reflection and transmission coefficient, and probably yields the higher stability.

The theory of wave power conversion by the wave power extracting caisson was originally given by Ojima et al.(1984) and was extended by Takahashi et al.(1987). The theory was named as **the thermodynamics and wave kinematics method.** The theory comprises two thermodynamics equations in the air chamber and one wave-kinematics equation at the air chamber opening under the curtain wall as follows:

$$\frac{dp^*}{dt} = \gamma \frac{p^*+p_0}{D_0-\eta}\left\{\frac{d\eta}{dt} - c_d \varepsilon w_p\right\} \tag{1}$$

$$\frac{dT^*}{dt} = \frac{T^*+T_0}{D_0-\eta}\left\{-\frac{d\eta}{dt} + \frac{D_0-\eta}{p^*+p_0}\frac{dp^*}{dt} + c_d \varepsilon w_p\right\} \tag{2}$$

$$2\eta_I = \frac{1}{\rho_w g} f(kh,\ kd_c) p^* + \frac{kB}{\tan kB}\eta + \frac{kB}{\sigma}\frac{d\eta}{dt}\frac{1}{\cos\theta_M} \tag{3}$$

The main unknown variables of the equations are

- p^* : oscillating fraction of the air pressure inside the air chamber
- η : oscillating fraction of the water surface elevation
- T^* : oscillating fraction of the air temperature inside the air chamber

The air flow velocity w_p in the air nozzle on the air chamber is defined by

$$w_p = \pm \varphi \sqrt{\frac{2}{\rho_a}} \sqrt{|p^*|} \tag{4}$$

The term $f(kh, kd_c)$ is given by

$$f\,(kh,\ kd_c) = \frac{2\cosh kh \sinh k\,(h-d_c)}{\cosh k\,(h-d_c)\,\sinh k(h-d_c) + k\,(h-d_c)} \qquad (5)$$

The above equations can be solved by step-by-step numerical calculations when the following dimensions and characteristics of the incident waves and the air chamber are given:

- η_I : incident wave profile
- σ : wave angular frequency for the significant wave period
- k : wave number for the significant wave period
- θ_M : instantaneous wave angle
- h : water depth
- B : width of the air chamber
- D_o : height of the air chamber above the still water level
- d_c : submerged depth of the front curtain wall
- A_w : nozzle opening area
- ε : nozzle opening ratio (= $A_w/l_B B$)
- l_B : length of air chamber in the breakwater direction
- c_d : contraction coefficient of the nozzle (= 1.0 in the calculations)
- φ : nozzle velocity coefficient

The constants in the equations are

- γ : specific heat ratio (= 1.4)
- ρ_w : density of water (= 1030 kg/m^3 in salt water)
- g : gravity acceleration (9.8 m/s^2)
- ρ_a : density of air (= 1.225 kg/m^3)
- T_a : temperature in open air (15 C = 288 K)
- p_a : pressure in open air (1 atm = 1.013x10^{5}N/m^2)

The wave power W_I is converted into air power. The instantaneous air power W_a, the average air power $\overline{W}_a$ and the conversion efficiency $\overline{EFF}_a$ are expressed by

$$W_a = p^* w_p\ A_w \qquad (6)$$

$$\overline{W}_a = \frac{1}{T_L}\int_0^{T_L} W_a\,dt \qquad (7)$$

$$\overline{EFF_a} = \overline{W}_a/\overline{W}_I\ \ell_B \qquad (8)$$

where T_L is the time duration of the wave action.

The theory mentioned above describes only the conversion from wave power to air power by the air chamber with an air nozzle. When a turbine and a generator are installed instead of the nozzle an equation of motion of the turbine and generator should be given and the characteristics and dimensions of the nozzle should be replaced by those of the turbine. However only the conversion from wave power to air power is discussed in this paper.

3. EXPERIMENTS

A series of model experiments were conducted in a large wave flume. Four caisson breakwaters were tested in the experiments, namely,

a) Wave power extracting caisson with a sloped front face as in Fig.2(a), (named as the S-type caisson and denoted by WPEC-S),
b) Wave power extracting caisson with a vertical front face as in Fig.2(b), (the V-type caisson , WPEC-V),
c) Ordinary caisson with a sloped front face, (the sloped front wall caisson, SFWC),
d) Ordinary caisson covered with wave dissipating blocks as in Fig.2(c), (the caisson with wave dissipating blocks, SFWC-B).

The S-type caisson was the main caisson in the experiments and the V-type caisson was employed to compare the function and the stability by the difference of the air chamber shape. The V-type caisson was used in the previous studies on the wave power conversion. The S-type caisson was designed to improve the characteristics of the V-type caisson. The two other caissons were also examined to compare the hydrodynamic characteristics with those of the wave power extracting caissons.

The S-type caisson has a width B_t of 135 cm, a crown height h_c of 55 cm, a caisson depth h' of 80 cm and an air chamber width B of 35 cm. The front wall is inclined 45 degrees from the point 2.5 cm above the still water level. The armour units for the rubble foundation have a weight of 1.23 kgf each. The foot protection concrete block has a height of 7.5 cm and a width of 22.5 cm. The submerged depth d_c of the curtain wall of the caisson is 20 cm in the figure. The submerged depth was changed to 15 cm and 25 cm by changing the curtain wall plate. The caisson has an air nozzle with a diameter and length of 4.70 cm and 5.7 cm. The opening ratio ε of the nozzle to the horizontal area in the air chamber is 0.0075.

The air chamber width of S-type caisson is set at 13 % of the wavelength for T = 1.34 s. The air chamber width and the opening ratio are set to have a relatively high conversion efficiency with a small air chamber width. These values are dependent on the wave and other conditions and should be selected considering the whole wave conversion system. The submerged depth of the curtain wall is determined to be about 0.25 H_{max}. The upper deck of the turbine room is design to be higher than 0.5 H_{max}. The crown height of the caisson is designed to be 1.0 $H_{1/3}$. The values of H_{max} and $H_{1/3}$ used in the design of model caisson are 75 cm and 55 cm. The submerged depth of the curtain wall is set so as not to receive severe impact air pressure. However, if the submerged depth is too large, the wave power conversion efficiency becomes small. The upper deck height of the turbine room is determined so as not to suffer from water intrusion from the nozzle. The crown height is set to be able to reduce the wave transmission as in conventional breakwaters. These values are determined by a preliminary study and should be examined further. Also, these values should be determined considering the entire wave power conversion system.

Six wave gauges (W1 -W6) were used in the experiments to measure the incident and reflected waves (W1 and W2), the water surface elevation in front of the air chamber (W3), the passing wave in the side flume (W4), the water surface elevation in the air chamber (W5), and the transmitted waves (W6). Six pressure transducers (P1 - P6) were installed in the model caisson to measure the wave and air pressure. The transducers have a diameter of 10 mm and were fixed in the caisson walls. Only one pressure transducer (P1) was fixed in the front wall and five transducers (P2 -P6) were set in the air chamber. Among them transducers P2 and P3 usually measured the pure air pressure, and the others measured the air pressure plus the wave pressure. Ultrasonic displacement meters (D1 - D3) were also installed to measure the displacement of the caisson. These meters were newly developed to measure the displacement up to 1 m with a resolution of 0.1 mm and 100 Hz.

Irregular waves were used in the experiments. The target wave spectrum was the Bretschneider-Mitsuyasu spectrum(Goda 1985) and the target wave periods $T^{*}_{1/3}$ were 3.35, 2.24, 1.79, 1.57, and 1.34 s. The water depth h was 97.5 cm. Five wave groups A, B, C , D and E were prepared in the experiments for each significant wave period. Each wave group contained more than 450 waves. The relative water depth ranged from 0.10 to 0.356 and the ratio of the air chamber width to the wave length ranged from 0.036 to 0.128.

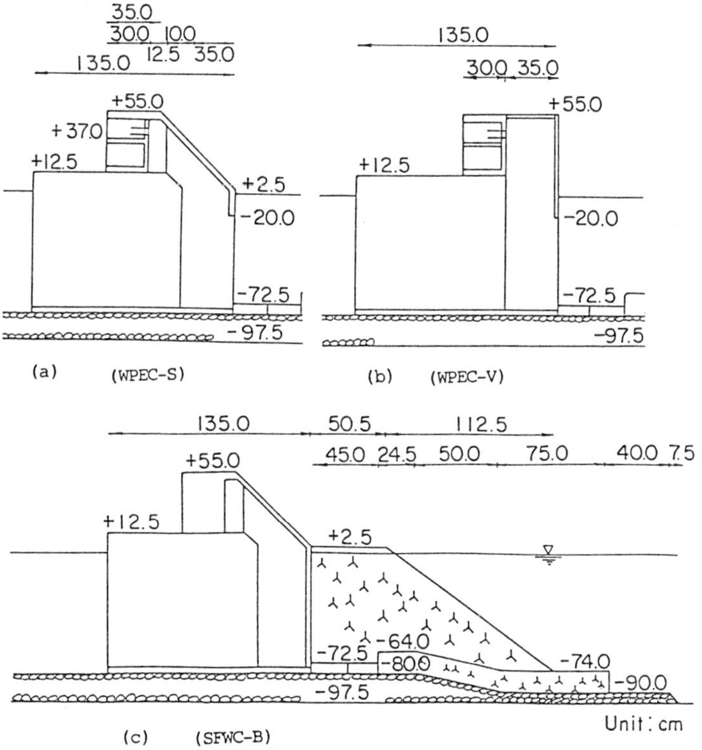

Fig. 2 Cross Sections of Model Caissons

4 FUNCTIONS OF CAISSONS

Conversion Efficiency of the Wave Power Extracting Caissons

The wave power is converted into air power in the air chamber with the nozzle opening ratio set at 0.0075. Figure 3 shows the conversion efficiency of the V-type wave power extracting caisson. The efficiency is from 0.55 to 0.7 where the significant wave height is less than 15 cm. The efficiency is relatively high in this region. The efficiency is lower than 0.5 where the significant wave height is larger than 25 cm, because the wavelength is relatively large and the air leaks from the opening under the curtain wall. The lines in the figure indicate the calculated results. The calculated results agree with the experimental results well.

Figure 4 shows the conversion efficiency for the S-type wave power extracting caisson. The conversion efficiency is changed by the submerged depth of the curtain wall. The efficiency increases with the decrease of the submerged depth clearly where the significant wave height is less than 20 cm. The efficiency of the S-type caisson of d_c = 20 cm is around 70 % where the significant wave height is less than 15 cm. The efficiency for the S-type caisson of d_c = 20 cm in the figure is larger than that for the V-type caisson in Fig 3.

Reflection Coefficient

Figure 5 shows the experimental results of the reflection coefficient of the S-type caisson together with those of the sloped front wall caisson and the caisson with wave dissipating blocks. The reflection coefficient of the S-type caisson is about 0.45 where the wave height is relatively small. The reflection coefficient is about 0.55 even when the wave height is large. The reflection coefficient of the S-type caisson is much smaller than the sloped front wall caisson, but larger than that of the caisson with wave dissipating blocks.

Wave Transmission and Overtopping

The wave transmission coefficients of the four caissons are compared in Fig. 6. For example, the wave transmission coefficient of the S-type caisson is 0.15 when the significant wave height is about 45 cm. The transmission coefficient of the V-type caisson is 0.13 for the same wave height. The transmission coefficient of the S-type caisson is a little larger than that of the V-type caisson. However, the transmission coefficient of the S-type caisson is lower than that of the sloped front wall caisson and is lower than that of the caisson with wave dissipating blocks. When the wave length and height are large, the waves climb up on the wave dissipating blocks. Therefore, the transmission coefficient of the caisson with the wave dissipating blocks is not less than that of the S-type caisson. The wave transmission in the experiments includes the wave transmission through the gap between the model caisson and the flume walls in addition to the wave transmission through the rubble foundation.

It is said that a sloped front wall caisson needs to have a crown height of 1.0 $H_{1/3}$ to reduce the wave transmission to a level as low as that by an ordinary vertical wall caisson with a crown height of 0.6 $H_{1/3}$. The wave transmission of the S-type caisson with the crown height of 1.0$H_{1/3}$ is probably lower than that of the ordinary vertical caisson with the crown height of 0.6$H_{1/3}$. The wave transmission characteristics of the four caissons can be observed also by the wave overtopping rate although the experimental results are not shown here.

<u>Water Surface Elevation</u>

The maximum and the 1/3 highest values of the water surface elevation η^+ in the air chamber of the S-type caisson are plotted in Fig. 7. The maximum and the 1/3 highest values of $\eta^+/H_{1/3}$ increase with the increase of the significant wave period and decrease with the increase of the significant wave height. The maximum value ranges η^+ from 0.6 to 1.0 $H_{1/3}$ and is above 37 cm when the significant wave

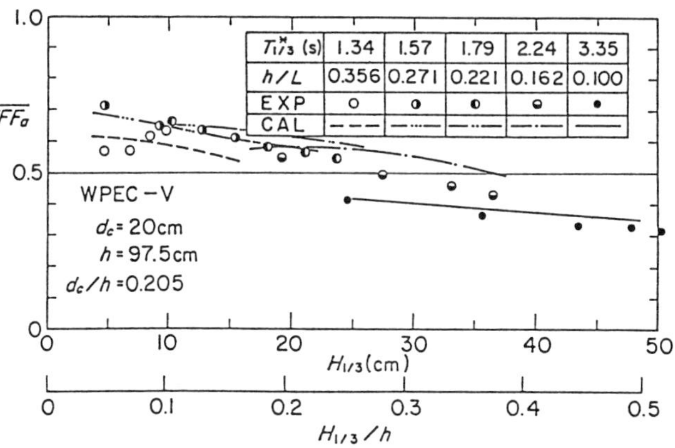

Fig.3 Conversion Efficiency of V-type Caisson

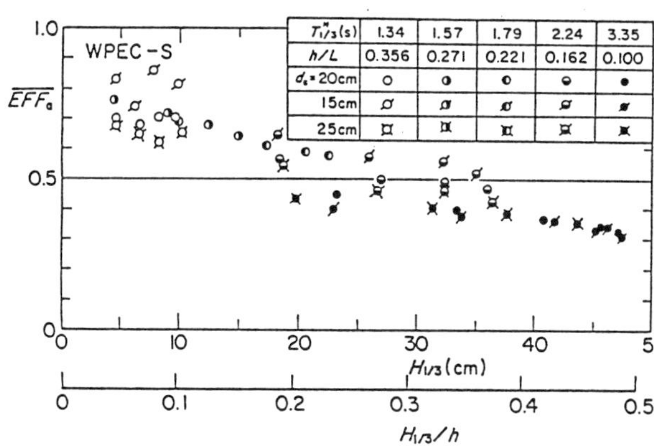

Fig. 4 Conversion Efficiency of S-type Caisson

period is 3.35 s and the significant wave height is larger than 40 cm. The upper deck of the machine room of the caisson is 37 cm above the still water level, which is designed to be $0.5H_{max}$. When the water level goes higher than the upper deck level, the water might intrude into the turbine opening. However, only the waves with a wave height near the maximum wave height can go up above the level and the velocity of the water surface elevation is very small when the water level reaches the upper deck level. It should be noted that the water surface elevation in the air chamber increases with the increase of the nozzle opening ratio.

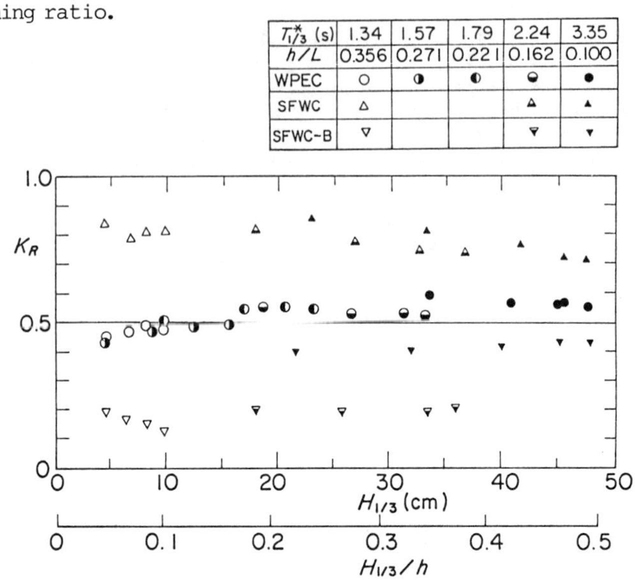

Fig. 5 Wave Reflection Coefficients

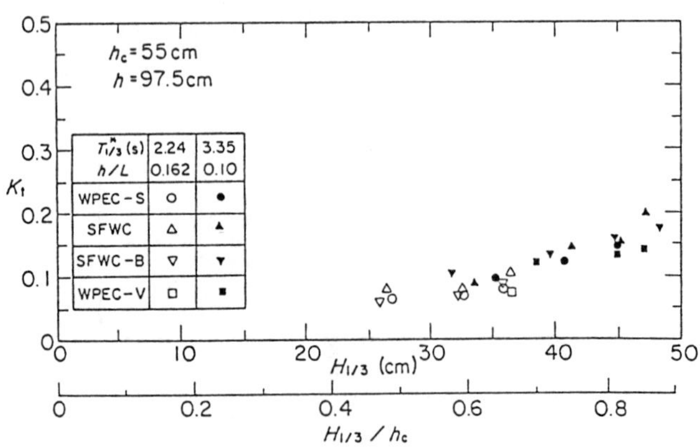

Fig. 6 Wave Transmisson Coefficients

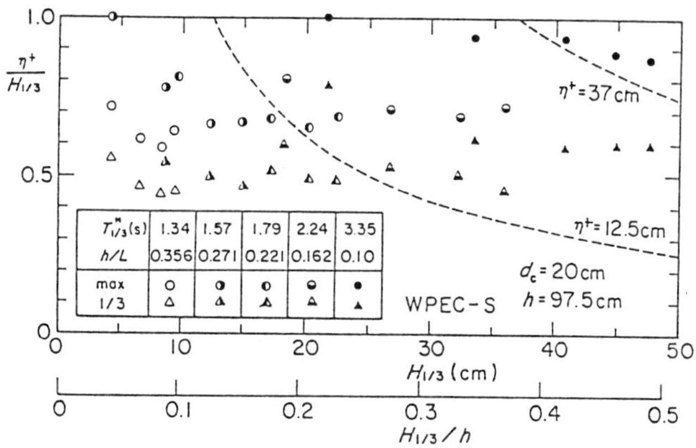

Fig. 7 Water Surface Elevation in the Air Chamber

5. WAVE PRESSURE AND WAVE-INDUCED AIR PRESSURE

Sample Analogue Data

Figure 8 shows sample analogue data when irregular waves of $T_{1/3}$= 3.35 s and $H_{1/3}$= 47.8 cm act on the S-type wave power extracting caisson. The water surface elevations η_F in front of the caisson, η_{pass} in the side channel and η in the air chamber are shown in the figure. The wave pressure p_1 on the sloped wall, the air pressure p_2 in the air chamber and the horizontal displacement R_S of the caisson are also shown in the figure. The air pressure becomes almost zero when the water surface elevation in the air chamber is below the curtain wall depth. Impact air pressure appears when the air in the air chamber is compressed by the uprising water surface. The impact air pressure occurs only when the water surface elevation in front of the chamber goes down significantly below the curtain wall depth. Impulsive pressure also appears on the sloped front wall. This pressure is generated by the direct impact of the water surface of waves and is characterized by its sharp rise and very short duration. Takahashi and Tanimoto (1983) named the air compression pressure Bagnold pressure and the direct wave impact pressure Wagner pressure. Bagnold (1939) and Wagner(1932) gave the theories on these pressures respectively.

Positive and Negative Peak Values of the Air Pressure

Figure 9 shows the positive peak values of the air pressure in the air chamber of the S-type caisson. The value of the air pressure p^+_{23} is the average of the values obtained by the two pressure sensors P2 and P3. The 1/3 highest peak values and the maximum peak values which are divided by $w_oH_{1/3}$ are plotted in the figure. The maximum and the 1/3 highest values of $p^+_{23}/w_oH_{1/3}$ increase with the increase of the significant wave height as a whole. However, the maximum value decreases where the significant wave height is larger than 40 cm. This is because the wave height is limited by the wave breaking. The maximum value fluctuates very significantly, and therefore additional

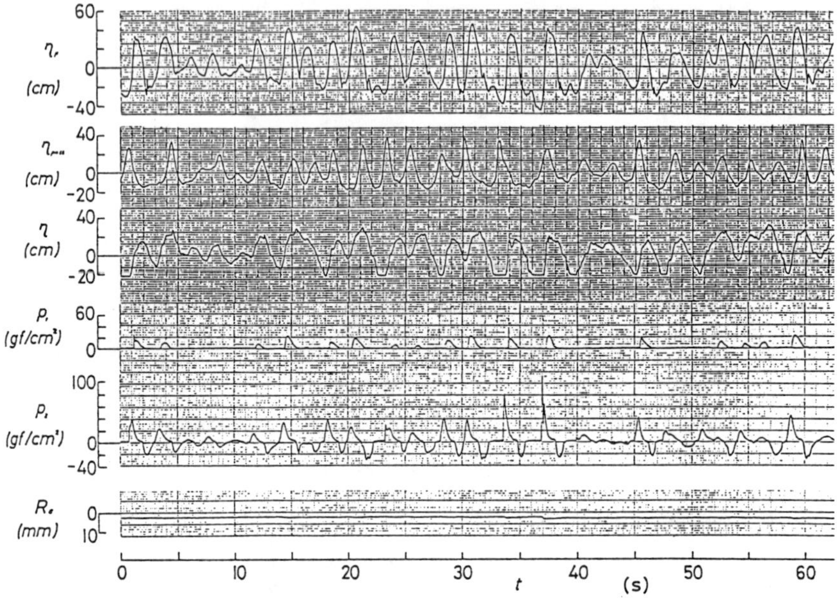

Fig. 8 Sample Analogue Data ($T^{*}_{1/3}$=3.35s, $H_{1/3}$=47.8cm, WPEC-S)

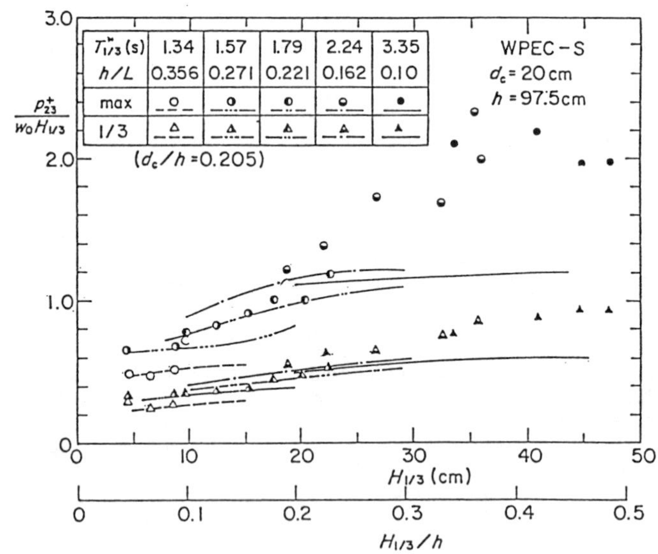

Fig. 9 Positive Air Pressure

measurements were conducted. Even though the incident wave train in the experiment is the same, the maximum value varies greatly. For example, the average of the maximum values for eight repeated measurements is 92.0 gf/cm^2 and their standard deviation is 10.4 gf/cm^2, while the average value of the significant wave height is 47.3 cm and the standard deviation is 0.6 cm. The average of the maximum values is 1.35 w_oH_{max} and equals 1.94 $w_oH_{1/3}$. The calculated value of the air pressure is also shown by the lines in the figure. The calculated value is much smaller when the impact air pressure occurs because the calculation cannot include the impact air compression.

Figure 10 shows the negative peak p^-_{23} of the air pressure. The maximum and the 1/3 highest absolute value of $p^-_{23}/w_oH_{1/3}$ increase with the increase of the significant wave height. However the maximum value decrease and the 1/3 highest value becomes constant when the significant wave height is larger than approximately 20 cm. This is because the water surface in front of the air chamber goes below the curtain wall and then the air pressure becomes the atmospheric pressure. The maximum absolute value of p_{23}^- is 0.50 w_oH_{max} when $T^*_{1/3}$= 3.35 s and $H_{1/3}$= 47.4 cm.

The measurement of the air pressure was also conducted for the V-type caisson. The air pressure of the V-type caisson shows almost the same tendency as that of the S-type caisson. The submerged depth of the curtain wall was changed in the experiments. It was found that the air pressure increases with the decrease of the submerged depth.

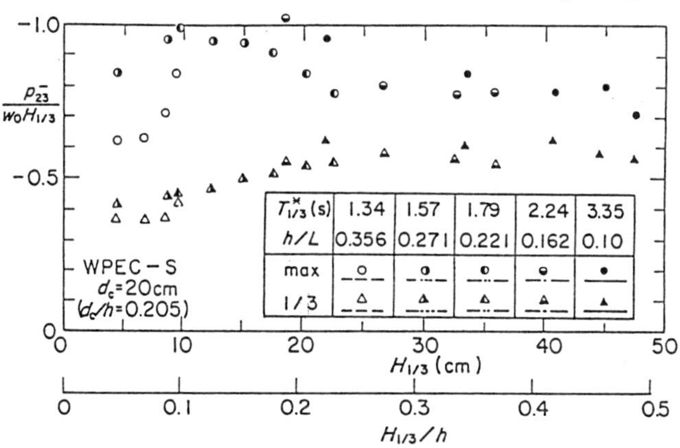

Fig. 10 Negative Air Pressure

Theory of the Impact Air Pressure

The uplift force on the ceiling slab of a wave dissipating caisson is generated by the compression of an air layer by the uprising water surface. This phenomenon is very similar to the impact air compression in the air chamber. The uplift force was investigated by Takahashi and Tanimoto(1985) employing the theory introduced by Bagnold(1939) and extended by Mitsuyasu(1966). The theory comprises the equation of motion of the effective water mass which compresses the air and the equation of air pressure. The theory can be applied for the impact air compression in the air chamber.

The equation of the air pressure derived by Bagnold assumes an adiabatic air compression. However, the thermodynamic equations used in 2 can also describe the impact air compression even without the assumption of the adiabatic change. That is, Eqs.(1) and (2) express the thermodynamics in the air chamber. Then the water surface elevation can be expressed by the following equation:

$$\rho_w k_e \frac{d^2\eta}{d t^2} = -p^* \tag{9}$$

where, k_e is the thickness of the effective water mass, which is given by

$$k_e = 0.25 \pi B \tag{10}$$

The initial conditions of the equations are given by

$$\left.\begin{array}{l} \dfrac{d\eta}{dt} = u_o \\ \eta = -d_c \end{array}\right\} \text{ at } t = 0 \tag{11}$$

where u_o is the uprising velocity at t = 0 and $\eta = -d_c$. The initial thickness of the air layer is $d_c + D_o$. The equations mentioned above are almost the same as those in 2 but Eq.(9) instead of Eq.(3) is used. The equations can be solved by step-by-step numerical calculations.

Sample Calculations and Similitude Law

Sample calculations were made for the impact air pressure in the model air chamber. The value of the effective thickness of the water k_e is 27.5 cm, and the initial thickness D_o+d_c of the air layer is 65 cm. For example the peak value of the air pressure p_c becomes 95.8 gf/cm^2 when the uprising water surface speed is 194.4 cm/s and the nozzle opening ratio is 0.0075. The time t_p from the start of the compression to the peak is 0.0457 s in this case.

If the model scale is 1/20, the peak air pressure p_c is 787 gf/cm^2 when the uprising velocity u_o is 8.96 m/s and ε = 0.0075. This means that the the peak air pressure becomes only 7.49 times that for the 1/20 model. The peak impact air pressure is 1/2.43 that predicted by the model experiments. That is, the correction factor λ of the experimental results is 2.43 in this case. This model scale effect comes from the compressibility of the air as described by Takahashi and Tanimoto(1985).

Prediction of the Impact Air Pressure

The impact air pressure for the prototype and model air chamber can be predicted by the calculation mentioned above. The effective thickness of the water mass k_e and the initial thickness of the air are evaluated by the dimensions of the air chamber. Estimation of the uprising velocity of the water surface is very difficult. However the maximum value of the uprising velocity in the experiment can be roughly estimated by the following procedure:

a) The maximum absolute value of negative peak of water surface elevation $|\eta^-_{pass}|$ is given by the wave height $H_{1/3}$ for incident wave train as follows:.

$$\left|\left(\frac{\eta^-_{pass}}{H_{1/3}}\right)\right|_{max} = -1.2\frac{H_{1/3}}{h}+1 \quad : H_{1/3}<0.417h \tag{12}$$
$$= 0.5 \quad : H_{1/3}\geq 0.417h$$

b) The maximum absolute value of the water surface elevation in front of the air chamber $|\eta^-_F|$ is obtained from $|\eta^-_{pass}|$ as follows:

$$|(\eta^-_F)|_{max} = 1.4|(\eta^-_{pass})|_{max} \quad : |(\eta^-_{pass})|_{max}<d_c/1.4$$
$$= 3.5|(\eta^-_{pass})|_{max}-30 \quad : d_c/1.4\leqq|(\eta^-_{pass})|_{max}<d_c \tag{13}$$
$$= 2.0|(\eta^-_{pass})|_{max} \quad : |(\eta^-_{pass})|_{max}\geqq d_c$$

c) The maximum value of the uprizing velocity u_o is given by $|\eta^-_F|$ as follows:

$$u_0=\sqrt{g|\eta^-_F|}\{1-(d_c/|\eta^-_F|)^2\}^{1/2} \tag{14}$$

The calculated values by the above procedure agree with the experimental value as a whole. However, Eqs(11),(12) and (13) are only derived from the experimental results. Further study should be made to confirm the above procedure.

6. DESIGN OF CAISSON

Design of the Walls in the Air Chamber

The walls in the air chamber including the curtain walls should be designed against the wave and wave-generated air pressure. The following three phases should be considered in the design:

a) Peak of the positive air pressure
The impulsive air pressure can be calculated by the procedure described in 5. It should be noted again that the prototype value becomes much smaller than that predicted from the model value. If the impact air pressure is small, the slowly-varying air pressure is important for the design. The value 1.0 $w_o H_{max}$ in the figure is recommended for prototype air chambers of ordinary size and with an ordinary nozzle opening ratio. It should be noted that the impact pressure may exceed 1.0 H_{max} when the air chamber is small and the wave height is small.

b) Peak of negative air pressure
The value of 0.5 $w_o H_{max}$ is recommended for ordinary air chambers. This is a conservative value. The negative air pressure is not impulsive.

c) Peak of impulsive wave pressure
When the impulsive wave pressure acts on the front walls the air pressure is still positive. However the air pressure becomes relatively small for the prototype air chamber. Therefore, the air

pressure can be assumed to be zero when the impulsive air pressure acts. The value of 1.0 w_oH_{max} is recommended for the impulsive wave pressure on the front wall.

Stability Design of the Caisson

Goda proposed a wave pressure formula for the upright section of vertical breakwaters, which is now a standard formula for the caissons of composite breakwaters. The formula can be applied even for the wave power extracting caissons with a little modification. That is, the coefficient α_2 in the Goda formula is set to be zero for the wave power extracting caissons because the caissons absorb the wave energy. (For details, see Goda 1985 pp.107 - 144.)

When the front face of the caisson is inclined as the S-type caisson, the wave pressure on this part contributes to the forces in the horizontal direction and in the vertical direction. This vertical force acts downward and increases the stability of the caisson. The vertical force P_{sv} and the horizontal force P_{sh} are usually defined as follows:

$$P_{sh} = P_s \sin^2\theta_s$$
$$P_{sv} = 0.5\, P_s \sin 2\,\theta_s \qquad (15)$$

where θ_s is the angle of the inclination of the sloped wall and P_s is the wave force above the beginning of the sloped wall.

Results of the Sliding Tests of the Caissons

Sliding Tests of the caissons were conducted to examine the pressure formula and to compare the stability of the caissons. The weight of the caisson was changed and the sliding distance of the caissons was measured. It can be concluded that the application of Goda's pressure formula with the modification is valid for the caissons examined in the experiments. The weight needed to stabilize the caisson changes greatly depending on the shape of the caisson. The S-type wave power extracting caisson requires less weight than the V-type wave power extracting caisson and than the sloped front wall caisson.

7. CONCLUSIONS

The functions and stability of the wave power extracting caisson breakwaters are investigated experimentally. Two wave power extracting caissons are tested in the experiments. One is the wave power extracting caisson with the sloped front wall (the S-type caisson) and the other is the caisson with the vertical front wall (the V-type caisson). However the S-type caisson is mainly discussed. Other caissons such as a sloped front wall caisson and a sloped front wall caisson with wave dissipating blocks are also tested in the experiments for comparison. Major conclusions are as follows:

1) The wave power extracting caisson can convert wave power into air power with relatively high efficiency, and the wave power extracting

caisson has low wave reflection and low wave transsmision characteristics.

2) Large impact air pressure occurs in the model air chamber of the wave power extracting caisson. The calculation method of the impact air pressure is presented. The impact air pressure in the prototype air chamber predicted by the calculation is much smaller than that evaluated with the Froude similitude law from the experimental result.

3) The pressure distribution for the design of the walls of the air chamber is proposed. The pressure distribution for the stability design of the S-type wave power extracting caisson is proposed, which is confirmed by the sliding tests of the caisson.

4) The stability of the S-type caisson against storm waves is high compared with other caissons.

The First Port Construction Bureau of the Ministry of Transport has decided to conduct field experiments of the wave power extracting caisson breakwater in Sakata Port from fiscal 1987 to 1991. A prototype wave power extracting caisson breakwater will be installed with a turbine and a generator.

The author wishes to express his deep appreciation to Dr. Y. Goda, Mr. T. Asada and Dr. K. Tanimoto for their precious comments on the present study.

REFERENCE

Bagnold, M.R.A.(1939): Interim report on wave pressure research, Jour. of Institute of Civil Engineers, Vol.12, pp.201 - 226.

Goda, Y.(1985): Random Seas and Design of Maritime Structures, Univ. of Tokyo Press.,322p.

Jarlan, G.E.(1961): A perforated vertical wall breakwater, Dock and Harbour Authority, pp.394 - 398.

Mitsuyasu, H.(1966): Shock pressure of breaking wave, Proc. of 10th Conference on Coastal Engineering, pp.268 -283.

Ojima, R., Suzumura,S., and Goda, Y.(1984): Theory and experiments on extractable wave power by an oscillating water-column type breakwater caisson. Coastal Engg. in Japan, Vol.27, pp.315- 326.

Takahashi, S.,Tanimoto, K., and Suzumura, S.(1983): Generation mechanism of impulsive pressure by breaking waves on a vertical wall, Rept. of Port and Harbour Res. Inst., Vol. 22, No.4. pp.3 - 31, (in Japanese).

Takahashi, S., and Tanimoto,K(1985): Uplift forces due to compression of enclosed air layer and their similitude law, Coastal Engineering in Japan, Vol.28, pp.191 -206.

Takahashi, S., Suzumura,S., and Myose, K.(1986): Turbine power by wave power extracting system with vertical breakwaters, Proc. of 5th OMAE Conference, ASME, pp.553-559.

Takahashi, S.,Myose,K., Yoshimoto,Y., and Tanaka, S.(1987): Variation of wave power extraction due to incident angle and directional wave spreading. Rept. of PHRI, Vol.26, No.3. pp.3-39.

Wagner, H.(1932): Über Stoss-und Gleitvorgänge and der Oberfläche von Flussigkeiten, Zeishrift für Angewandte Mathematik und Mechnik, Band 12, Heft 4, pp.193 -215.

CHAPTER 186

DYNAMIC FORCES DUE TO WAVES BREAKING AT VERTICAL COASTAL STRUCTURES

by

Hans-Werner PARTENSCKY

Franzius-Institute, University of Hannover, F.R.G.

1. Introduction

In the past 20 years, considerable effort has been devoted to replacing the widely used approaches of HIROI, MINIKIN, NAGAI, PLAKIDA and others /1,2,3,4/, for the design of vertical breakwaters under the impact of breaking waves, with improved and more exact calculation methods. However, almost all new theoretical and empirical approaches lacked the support of prototype measurements or test results from model measurements at a larger scale. The difference between the proposed design criteria and classical approaches is sometimes so great that engineers do not have a reliable method for the design of a vertical or composite breakwater. Figure 1 shows the resulting wave forces per unit width due to different theories as a function of the design wave height H.

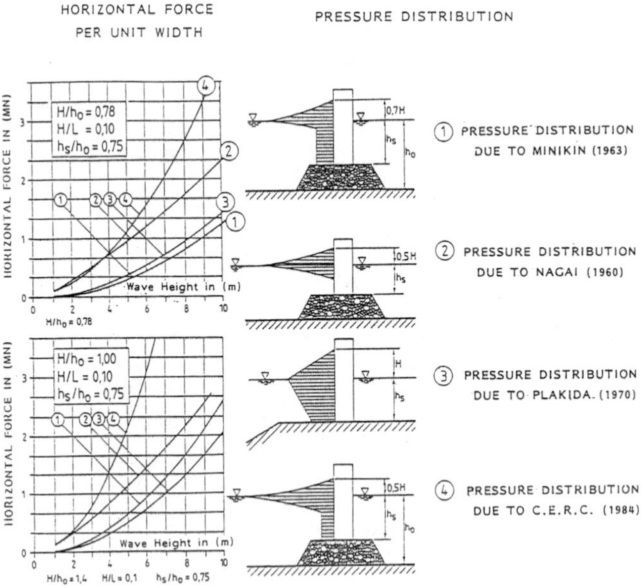

Fig. 1

Horizontal wave forces per unit width due to breaking waves according to different theories /5/

The results show clearly how much the different theories deviate from one another. Therefore, there seems to be an urgent need to re-examine the existing approaches and to support them by means of prototype or large-scale model measurements. The photos of Figures 2 and 3 show some classical examples of waves breaking against coastal structures.

Fig. 2
Impact of waves on the south west breakwater of Colombo/Sri Lanka

Fig. 3
Wave breaking against Catania breakwater on Sicily/Italy /4/

Although there have been no spectacular failures of vertical breakwaters in recent years, the existing and widely used design criteria fail to include some newly defined and recognized factors which may promote or cause damage to structures of this kind.

There factors include:

- the simplified assumption of the dynamic pressure distribution on the structure caused by breaking waves,

- the imprecise determination of the design wave height,
- the scale effects of past model tests with respect to the air entrainment of breaking waves, and
- incorrect dimensioning criteria for waves approaches the structure at an angle (Mach reflection).

There are basically three possible major failure modes for vertical and composite breakwaters (Fig. 4). These are sliding and overturning of the upright section as well as failure of the foundation.

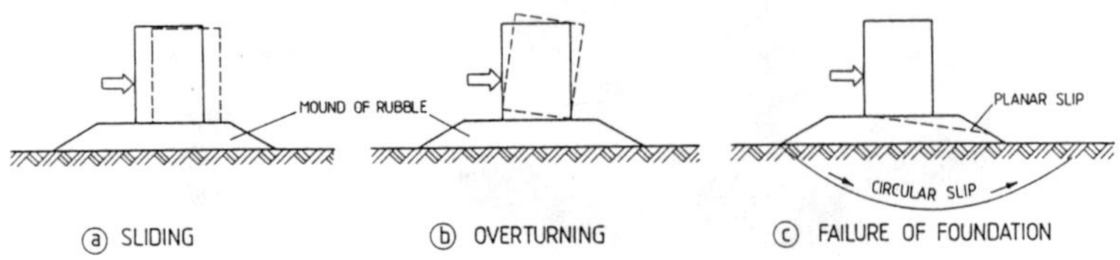

Fig. 4
Possible modes of failure of vertical and composite breakwaters

In all three cases, the impact of the breaking wave at the vertical face of the structure is one of the main causes for the damage. It is therefore extremely important to determine the instantaneous pressure distribution at the vertical wall due to the breaking wave.

2. Modes of wave impact

The resulting pressure distribution at the structure due to a breaking wave depends to a high degree on the mode in which the incoming wave collides with the wall. In this respect, two different cases must be distinguished.

In the first case, the wave begins to break in front of the wall and the tongue of the plunging breaker hits the vertical face enclosing a certain amount of air (Fig. 5).

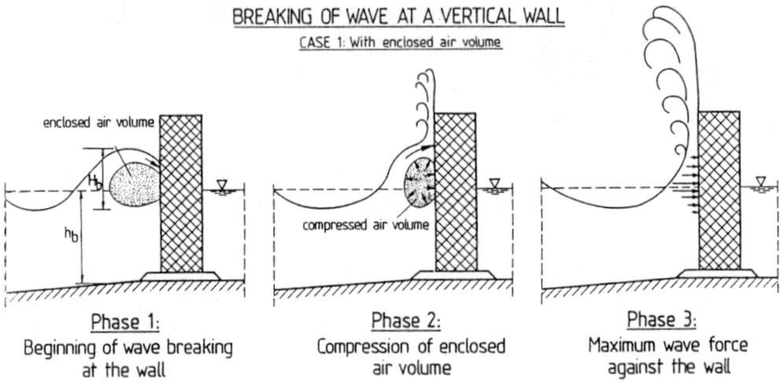

Fig. 5
Wave breaking against a vertical structure with an enclosed air volume

This air volume is quickly compressed during the impact process before the uprushing water carries it upward and away. Figure 6 shows the impact of breaking waves at the sea wall of Westerland/Sylt (F.R.G.) during the stormflood of January 4, 1976. In this case, a considerable amount of air is enclosed in the uprushing water mass which dampens the instantaneous wave pressure exerted at the wall.

Fig. 6

Impact of waves at the Westerland sea wall, Island of Sylt, F.R.G., during the stormflood of January 1976 /5/

In the second case, the wave collides with the structure with an almost vertical front face and only little air is enclosed between the wall and the wave front (Fig. 7).

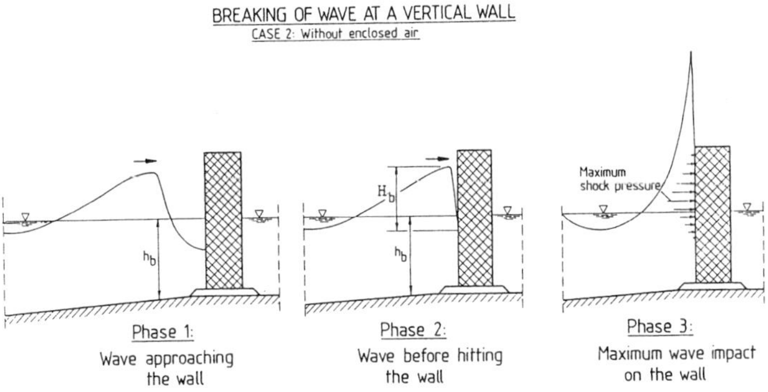

Fig. 7

Wave breaking against a vertical wall without enclosed air

In this case, the forward momentum of the impinging wave is fully transformed into a pressure force which acts on the vertical face of the structure. Measurements showed that the highest pressure values would be exerted at the wall in this case. Figure 8 shows an example of a breaking wave with little air entrainment.

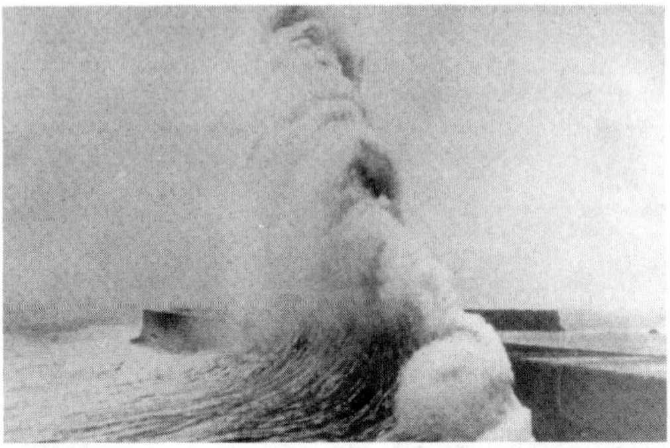

Fig. 8

Wave breaking against Catania breakwater in Sicily/Italy

Preliminary results of measurements obtained in the wave flume of the Franzius-Institut and in the Large Wave Channel showed that the force resulting from the instantaneous pressure distribution at the wall can rise to more than 10 times the hydrostatic pressure force corresponding to the breaker height. This result is in good agreement with observations made by GODA et al. /7,8/.

The peak pressure at the structure is exerted at some distance above the mean water level depending upon the breaker type. It is of very short duration (0.01 s to 0.03 s), and can easily reach values of several hundred kN/m², depending upon the height of the breaking wave and the amount of air entrapped.

According to OUMERACI /9/, the impact pressure depends essentially on the air content and can be determined by the following equation:

$$p_{dyn} = K_L \cdot \rho \cdot g \cdot H_b \quad (1)$$

in which: p_{dyn} = maximum dynamic pressure due to the wave impact (kN/m²)
H_b = breaking height of the wave (m)
ρ = density of the water (kN·s²/m⁴)
g = acceleration of gravity (m/s²)

K_L is a kind of air content coefficient which is given by:

$$K_L = 5.4 \left(\frac{1}{\kappa_L} - 1\right) \quad (2)$$

with κ_L = relative air content (Fig. 9).

Using equation (1), the dynamic pressure exerted at a vertical wall by a breaking wave of 1.5 m of height with an almost vertical front face (little air content, κ_L = 0.1 to 0.2) would amount to p_{dyn} = 300 to 700 kN/m², depending upon the volume of air enclosed. The order of magnitude of these values is in good agreement with the results of measurements obtained from tests in the Large Wave Channel in Hannover (see Table 1).

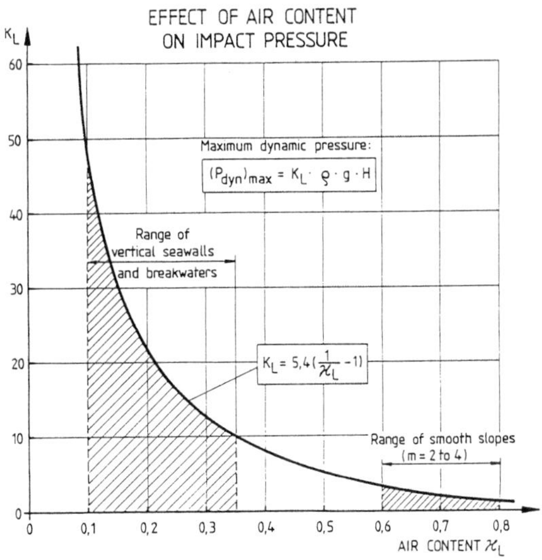

Fig. 9

Effect of air content on impact pressure due to breaking waves

3. Theoretical approach

In order to determine the instantaneous pressure distribution at a vertical wall due to a breaking wave (vertical face assumed), the momentum exchange between the water mass in motion and the rigid assumed structure must be determined. The resulting force per unit width is then given by (Fig. 10):

$$F = m \cdot du/dt = \int_{-h_s}^{h_c} \int_{0}^{t_1} p(z,t) \cdot dt \cdot dz \qquad (3)$$

where m = participating water mass,
u = $f(z)$ = velocity distribution under the breaking wave and
t_1 = duration of impact (0.013 s $\leqq t_1 \leqq 0.02$ s).

The difficulty is that none of the existing wave theories describes with a sufficient accuracy the velocity field under a breaking wave /10/. A simplified appraoch must therefore be used. For $z \leqq 0$, linear wave theory is applied for $u_1(z)$, whereas for $z > 0$ an approximation is used for $u_2(z)$ with $u_c \cong 0.5\ c$ at the wave crest (Fig. 10). For waves just before breaking, this is in accordance with measurements carried out by LE MEHAUTE and al. /11/, by OCHI and TSAI /12/ and WATANABE, HARA and HORIKAWA /16,17/.

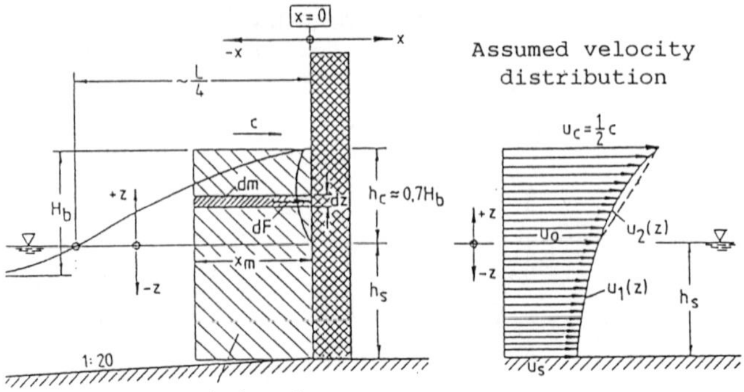

Fig. 10

Definition sketch

It is difficult to determine the exact water mass participating in the momentum exchange during the short duration of the wave impact. At the mean water level, the thickness of the water mass in motion can be approximated to be $x_m \approx 1/2 \cdot L/4 \approx H_B$. Although deeper lying water layers contribute less to the momentum exchange, for reasons of simplification the value of x_m was used over the entire water depth in the following calculations (Fig. 10).

The momentum exchange during the wave impact is of a very short duration with $t_1 \leq 0.02$ s (Fig. 11).

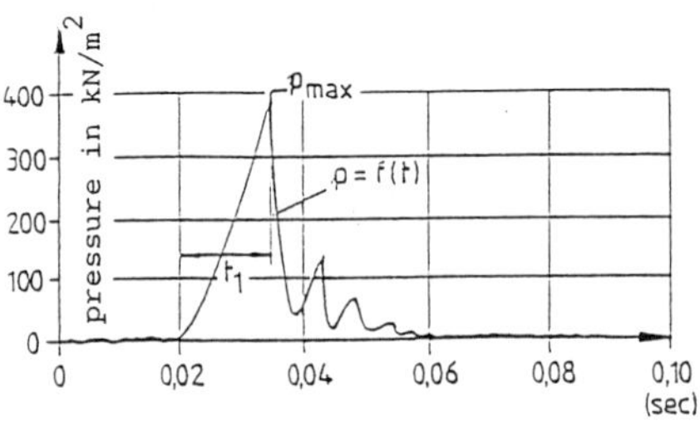

Fig. 11

Increase in pressure during the wave impact at a given location at the wall

Assuming a linear increase in pressure for $0 \leq t \leq t_1$ at a given location z on the wall, equation (4) holds:

$$p(z,t) = \frac{p(z)_{max}}{t_1} \cdot t \tag{4}$$

For the maximum pressure at a location z on the wall follows:

$$p(z)_{max} = \frac{2\rho x_m}{t_1} \cdot u(z) \approx \frac{2\rho H_b}{t_1} \cdot u(z) = f(z) \quad . \tag{5}$$

Using linear wave theory, the horizontal velocity components under the wave crest are obtained at the mean water level (z = 0):

$$u_o = \frac{\pi \cdot H}{L} \cdot c \tag{6}$$

and at the bottom ($-z = h_s$):

$$u_s \cong \frac{\pi \cdot H}{L} \cdot \frac{c}{\cosh kh_s} \quad . \tag{7}$$

At the wave crest ($z = h_c$), it is assumed (Fig. 10):

$$u_c = \frac{1}{2} c = \frac{1}{2} \sqrt{gh_s} \quad . \tag{8}$$

By using the velocity distribution as shown in Figure 10, the instantaneous pressure distribution at the wall is obtained from equation (5). The peak pressure at $z = h_c$ can be determined by the follwing equation:

$$p_c = \frac{\rho \cdot H_b}{t_1} (gh_s)^{1/2} \quad . \tag{9}$$

The pressure at z = 0 is obtained by:

$$p_o = \frac{2\pi\rho \cdot H_b^2}{t_1 \cdot L} (gh_s)^{1/2} \tag{10}$$

and at the bottom ($-z = h_s$) by:

$$p_s = \frac{2\pi\rho \cdot H_b^2}{t_1 \cdot L \cdot \cosh kh_s} (gh_s)^{1/2} \quad . \tag{11}$$

Figure 12 shows the resulting peak-pressure distribution at the wall due to the wave impact as well as the simplifying linear approxi-

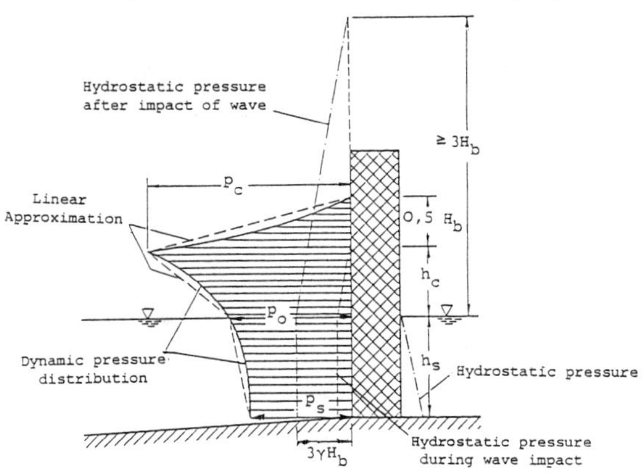

Fig. 12
Resulting peak-pressure distribution

mations. Measurements showed that the uprushing breaker tongue at the wall could reach a height of $3H_b$ or more after the wave impact (Fig. 12).

4. Research program of the Franzius-Institute

The research program of the Franzius-Institute concentrated on the impact of waves breaking against a vertical, solid structure. Portions of the systematic tests were performed in a smaller wave flume with significant wave heights of up to 0.40 m. The most important investigations, however, have been carried out in the Large Wave Channel of the Universities of Hannover and Braunschweig having an overall length of 320 m and a width of 5.0 m. This wave flume is equipped with an hydraulically driven wave maker by means of which monochromatic and random waves can be produced with wave heights up to 2.50 m.

In both flumes, the instantaneous pressure distribution at a vertical wall due to breaking waves was measured by means of a number of high sensitive pressure cells having a natural frequency of 28 kHz with an operating pressure range of up to 5 bar.

In the Large Wave Channel, the vertical wall was composed of seven stop logs, each with a height of 1.0 m and about 25 pressure gauges installed over the face of the wall (Fig. 13).

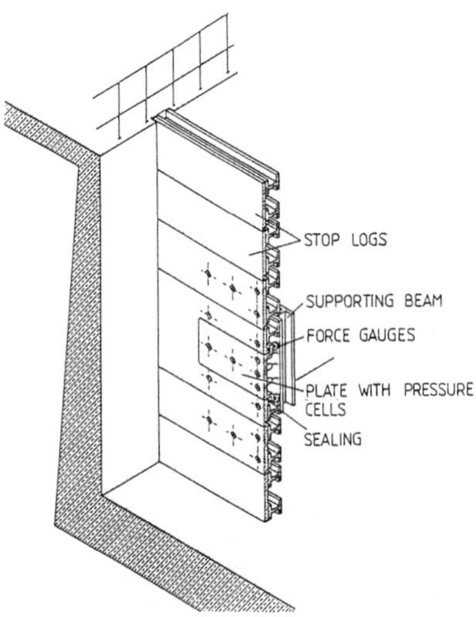

Fig. 13

Installation of vertical wall in the Large Wave Channel

Fig. 14

Wave breaking against vertical wall in the Large Wave Channel

The measurements showed that the maximum dynamic pressure at the wall is exerted at a distance of $h_c \approx 0{,}7\ H_b$. Above the mean water level, its value depending upon the breaker type and the amount of air enclosed. Figure 15 shows some time series of impact pressures obtained at different measuring points in the Large Wave Channel for a breaking wave height of 1.50 m.

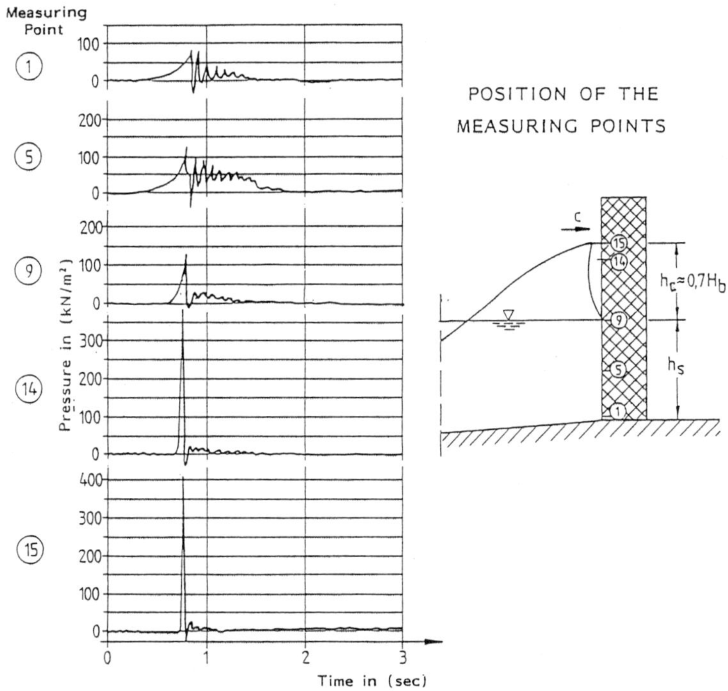

Fig. 15

Time series of impact pressures and location of measuring points

The evaluation of the measurements carried out in the Large Wave Channel is not yet finished. The first results, however, show clearly that the criteria used so far for the design of vertical breakwaters under the impact of breaking waves underestimate considerably the peak pressure and resulting wave forces.

Peak pressures at $z = h_c$ for a breaking wave height of $H_b = 1.50$ m measured in the Large Wave Channel exceeded, for example, the value resulting from the MINIKIN approach /13/ by a factor of 4 (Table 1). Based on the maximum pressure values obtained at different measuring points at the wall, the resulting force per unit width exceeded the corresponding CERC-value by a factor of 7.0 and the overturning moment about the toe of the wall was approximately

12 times greater than that obtained from the current design criteria (Table 1).

Peak pressures measured in the Large Wave Channel for given breaker heights were in good abgreement with the distribution of maximum pressures as shown in Fig. 12 /5/.

In Table 1, a comparison of test results with existing theories and the new approach is made for a breaking wave height of H_b = 1.50 m. Based on the peak pressure values, the calculated wave force per unit width as well as the overturning moment about the toe of the wall were in good agreement with the theoretical values obtained by the new approach (Table 1).

Theoretical approach	Maximum dynamic pressure (kN/m^2)	Force per unit width (kN/m)	Overturning moment (kNm/m)
HIROI (1903)	23,0	124,6	191,2
NAGAI (1968)	20,0	45,7	62,6
PLAKIDA (1970)	23,0	80,3	117,6
C.E.R.C. (1984) (Based on MINIKIN, 1950)	97,0	63,1	86,7
GODA (1985)	15,3	100,0	134,0
FRANZIUS-INSTITUTE (1987) (Measurements at Large Wave channel)	395,0	496,1*	1016,8*
PARTENSCKY (1987) Based on maximum pressure values with t_1 = 0,0167 s:	361,0	541,0	1020,0
Based on maximum pressure values with t_1 = 0,02 s:	301,0	455,0	855,0
Recommended design values with t_m = 0,04 s:	75,2	130,0	230,0

* Calculation based on maximum pressure values

Table 1

Comparison of test results with existing theories and new approach (Example for breaking wave height of H_b = 1.50 m with T = 7.8 s)

5. Recommendations for the design of vertical coastal structures

As can be seen from the time series of impact pressures in Figures 11 and 15, the duration of the peak pressure at the different locations is extremely short (t_1 < 0.02 s). Due to its inertia, the coastal structure will therefore hardly show reactions during the short duration of the momentum exchange.

Prototype measurements at CAISSON-like breakwaters in Japan showed that the natural period of oscillation of these types of coastal structures lies between 0.1 and 0.3 s /14/, which is one order of magnitude higher than the duration of the peak pressure during the wave impact.

In addition, from the recorded pressure time series it can also be seen that the maximum pressures at the different locations along the wall do not occur simultaneously, the phase shifts, however, are small.

The dimensioning of a coastal structure on the basis of the peak pressure distribution of Figure 12 therefore appears to be too conservative. It seems more realistic to replace the resulting peak pressure distribution by an averaged pressure distribution, in which the mean dynamic pressure $p_m = f(z)$ over the total impact duration $t_m = t_1 + t_2$ is applied to each point of the structure (Fig. 16) /15/.

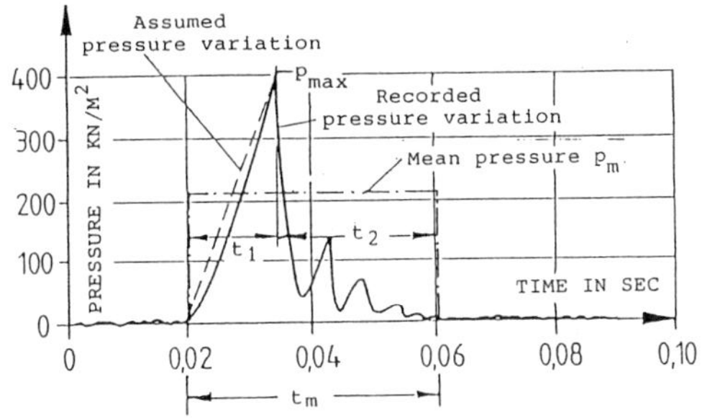

Fig. 16

Mean pressure p_m during the momentum exchange due to the wave impact

With the simplifying assumption of a linear increase and decrease of the pressure during the impact duration t_m, it follows for the mean pressure p_m that:

$$p_m(z) \approx \frac{1}{2}\, p_{max}(z) \qquad (12)$$

By using this relationship, the following reduced pressure values result from equations (9), (10), and (11):

$$\text{At } z = h_c: \quad (p_c)_m = \frac{\rho \cdot H_b}{2\, t_m} (gh_s)^{1/2} \qquad (13)$$

$$\text{At } z = 0: \quad (p_o)_m = \frac{\pi \cdot \rho \cdot H_b^2}{t_m \cdot L} (gh_s)^{1/2} \qquad (14)$$

$$\text{At } -z = h_s: \quad (p_s)_m = \frac{\pi \cdot \rho \cdot H_b^2}{t_m \cdot L \cdot \cosh kh_s} (gh_s)^{1/2} \qquad (15)$$

The time series of the impact pressures recorded at the Large Wave Channel show that the impact duration is $t_m \approx 0.04$ s at and above the mean water level, whereas below the mean water level t_m increases with depth (Fig. 15).

Figure 17 shows the recommended distribution of average impact pressures which might serve as a new approach for the future design of vertical coastal structures subject to the impact of breaking waves.

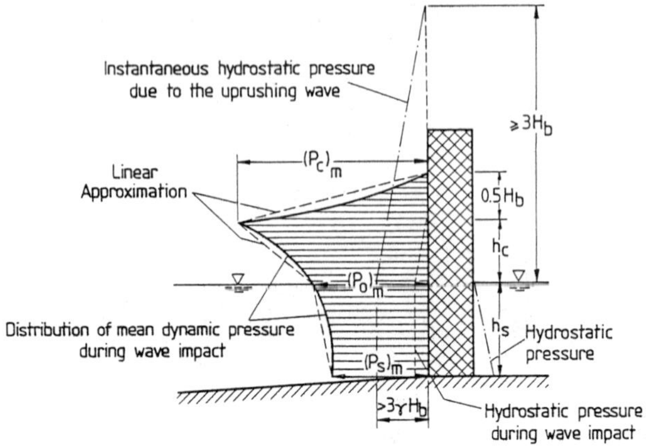

Fig. 17

Recommended pressure distribution for the design of coastal structures subject to the impact of breaking waves

In Figure 18 the dynamic pressure distributions at a vertical structure for a breaking wave height of H_b = 1.50 m are represented as they result from the classical MINIKIN-approach recommended in /13/ and from the new approach.

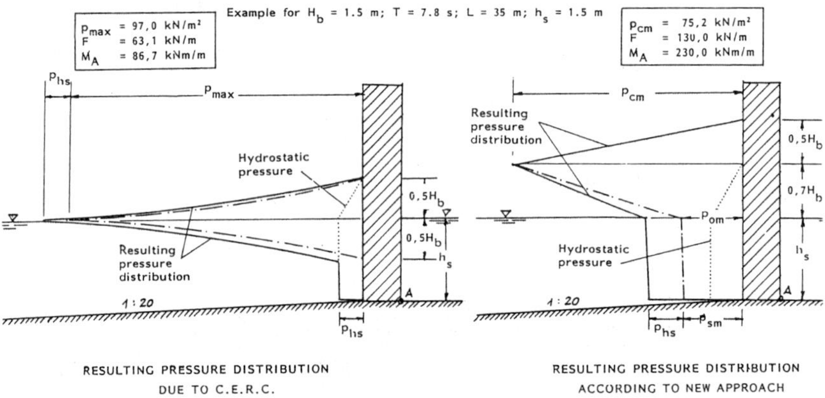

Fig. 18

Comparison of dynamic pressure distributions according to MINIKIN (C.E.R.C.) and the new approach

Although the recommended averaged peak pressure p_{cm} at the wave crest is somewhat smaller than that resulting from MINIKIN's theory, the effective wave force per unit width is twice as high and the overturning moment about the footpoint A of the structure is almost three times greater than that obtained from the classical theory (see also Table 1).

6. Conclusions

The results of the investigations show clearly that the design criteria used so far for the dimensioning of vertical coastal structures such as sea walls, vertical and composite breakwaters underestimate considerably the resulting wave forces.

The proposed pressure distribution of Figure 17 could therefore serve as a more realistic approach for the design of coastal structures under the impact of breaking waves.

To what extent the effect of strucural elasticity as well as the type of foundation must be considered in the design and overall stability of the structure has not yet been investigated. It will, however, be the subject of further studies.

In addition, a numerical analysis of the impact process is planned in order to determine more precisely the time- and space-dependent phases of the momentum exchange at the coastal structure.

7. Acknowledgement

The research work decribed in this paper has been supported since 1983 by the German National Research Council (Deutsche Forschungsgemeinschaft) in Bonn, F.R.G. This support is highly appreciated by the Special Research Group (SFB 205) of the University of Hannover as well as by the author.

8. References

/1/ Partenscky, H.W.: Breakwater design problems. Proc. of European Workshop on Coastal Zones, Techn. Univ. of Athens, 1985

/2/ Nagai, S.: Shock pressures exerted by breaking waves on breakwaters. Trans. of ASCE, Vol. 126, No. 3261, 1961

/3/ Plakida, M.E.: Pressure of waves against vertical walls. Proc. of 12th Coastal Eng. Conf. Vol. 3, 1970

/4/ Minikin, R.R.: Wind, waves and maritime structures. Charles Griffin, London, 1963

/5/ Partenscky, H.W.: Neue Bemessungskriterien für senkrechte und geschüttete Wellenbrecher, Jahrbuch der HTG, Band 42, 1987

/6/ Partenscky, H.W.: New investigations on vertical and rubble mound breakwaters. Proceedings of the second Chinese-German-Symposium, Vol. II, University of Hannover, 1987

/7/ Goda, Y.: Random seas and design of maritime structures. University of Tokyo Press, 1985

/8/ Technical standards for port and harbour facilities in Japan, 1980

/9/ Oumeraci, H.: Belastung von Betonplatten-Deckwerken und ihre Berücksichtigung bei der Bemessung. Dissertation, T.U. Dresden, 1981

/10/ Longuet-Higgins, M.S.: The unsolved problem of breaking waves. Keynote address, Proc. of 17th Conf. con Coastal Eng., Sydney, 1980

/11/ Le Méhauté, B., Divoky, D. and Lin, A.: Shallow waterwaves, a comparison of theories and experiments. Proc. of 11th Conf. on Coastal Eng., Vol. I, 1968

/12/ Ochi, M.K. and Tsai, Chen-Han: Prediction of impact pressure induced by breaking waves on vertical cylinders in random seas. Appl. Ocean Research, Vol. 6, No. 3, 1984

/13/ US-Army, Corps of Engineers: Shore protection manual, Vol. 2, Coastal Eng. Research Center, Vicksburg/Miss., 1984

/14/ Muraki, Y.: Field investigation on the oscillation of breakwater caused by wave action. Coastal Eng. in Japan, Vol. 9, 1966

/15/ Partenscky, H.W.: Impact of waves breaking at vertical structures. Proceedings of the 21nd Conference on Coastal Engineering, Torremolinos/Spain, 1988

/16/ Watanabe, A., Hara, T. and Horikawa, K.: Study on breaking conditions of compound wave trains. Coastal Engineering in Japan, Vol. 27, 1983

/17/ Horikawa, K. (Editor): Nearshore dynamics and coastal processes. University of Tokyo Press, Tokyo, 1988

San Ciprian—Lugo

PART IV

COASTAL, ESTUARINE

Estarit—Gerona, costa Brava

CHAPTER 187

MIXING OF THERMAL DISCHARGES IN COASTAL WATERS

By N.M. Ismail[1] A.M. ASCE, R.L. Wiegel[2] F.ASCE.
P.J. Ryan[3], M. ASCE and S.W. Tu[4], M.ASCE

ABSTRACT: Mixing of thermal effluents, being discharged from thermal power plants on coastlines and which head into surface waves was investigated by analyzing extensive field and laboratory data on plume and ocean ambient conditions. Emphasis was given on the effect of waves and surf zone currents on the modifications of plume surface area and vertical temperature profile in the near-field area.

The results of this investigation showed that large opposing waves increase the plume surface area, in the vicinity of the outfall, for all cases of tide level and wave direction. Moreover, waves focused cold bottom currents on the discharge outlet and consequently the temperature of the released warm water was decreased at the surface and near the bottom. Wave-induced cross flows decreased the plume cumulative surface area which corresponded to fractional excess temperature ranging between 0.8 and 0.5 normalized values. This decrease was shown to be contingent that there is no interaction between the far-field and near-field plume waters. Gradient of wave momentum flux across surf zone was found to be necessary parameter to characterize the incident wave field.

INTRODUCTION

The growing interest in recent times in the pollution of natural water bodies through the discharge of thermal effluents has advanced considerably the state-of-the-art knowledge on the mixing of surface buoyant jets in the aquatic environment. Thermal

---

1. Engineering Specialist, Enppi, Cairo, Egypt; formerly, Bechtel Inc., San Francisco. CA.
2. Professor of Civil Engineering, University of California, Berkeley, CA.
3. Chief Engineer, Hydraulics/Hydrology Group, Bechtel Civil, Inc., San Francisco, CA.
4. Research Engineer, Department of Engineering Research, Pacific Gas and Electric Company, San Ramon, CA.

pollution studies usually involve a field survey in conjunction with either a physical and/or a mathematical model. When warmed water is discharged into the surf zone in the open sea, field surveys reveal that the thermal plume often shows a complicated horizontal shape. This phenomenon may be caused by the influence of waves in shoaling water and by nearshore currents. Such effects were reported by Grider (2) in 1973 to be important, based on the field survey of surface temperatures in the thermal plume discharged into the nearshore zone from the Encina Power Plant, Carlsbad, California. After examining the order of magnitudes of offshore currents and wind velocities, Grider concluded that heavy surf conditions play a major role in determining the size and the shape of the plume. Although a correlation between significant wave height and size, of the plume was shown to exist, some contradictory data were reported. Grider concluded that additional study of surface effects on horizontal surface plumes and detailed analysis of wave effects are warranted.

In 1974 (Wiegel et al., 1976) a 1:75 undistorted scale model of the Diablo Canyon Power Plant was constructed and used to predict the behavior of a 4000 cfs heated discharge into a shallow, rocky cove west of San Luis Obispo, California. In 1984-6 extensive field data were obtained for a wide range of plant operating and receiving water conditions. The data included horizontal and vertical temperature profiles and surface temperature isotherms for over 50 field tests. Key tests were selected, and the data used to calibrate and verify the physical model. The results demonstrated that a physical model is a reliable tool for prediction of prototype behavior as long as key parameters are simulated correctly in the model. Three parameters which are not normally emphasized in jet mixing models were found to be important, including details of the bottom topography, bottom roughness, and incident wave characteristics. The plant layout is shown in Fig. 1-a and an expanded view of the discharge cove is illustrated in Fig. 1-b. The lattest figure shows the main features of the bottom bathymetry of Diablo discharge cove.

The field and model data showed that deep water-significant wave height and its associated period were sufficient parameters to characterize the incident wave field. An increase of height of waves, coming at 255° with north, from 4.5 to 6 ft destratified the south edge of the plume, and higher waves shifted the plume to the north. Further (Ismail et. al., 1987), it was found that the asymetric distribution of the horizontal temperature profiles of the plume observed for all field tests conducted under large waves (> 6 ft), of 255° angle of approach, was due to a reverse eddy current. This eddy is generated in the discharge cove by wave breaking along the shallow plateau near the cove west entrance. These wave effects were similar at both low and high tides. Further it was observed that when the wave approach was changed in the model from 255° to 278°, there was a corresponding trend

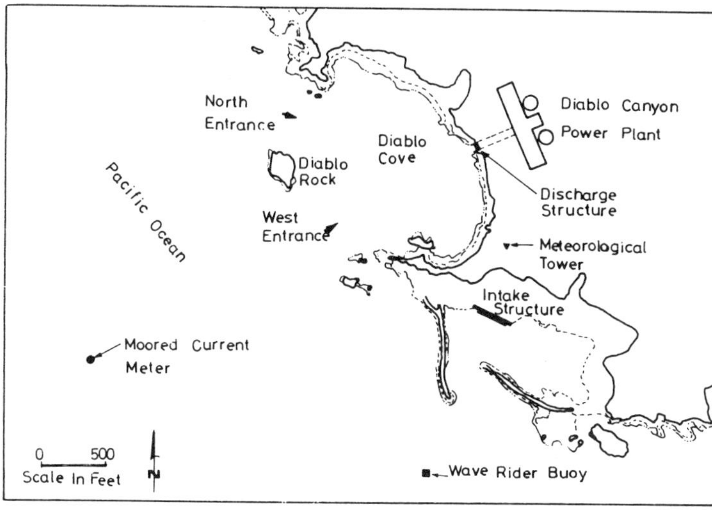

FIG 1-a DIABLO CANYON POWER PLANT AND RECEIVING WATER

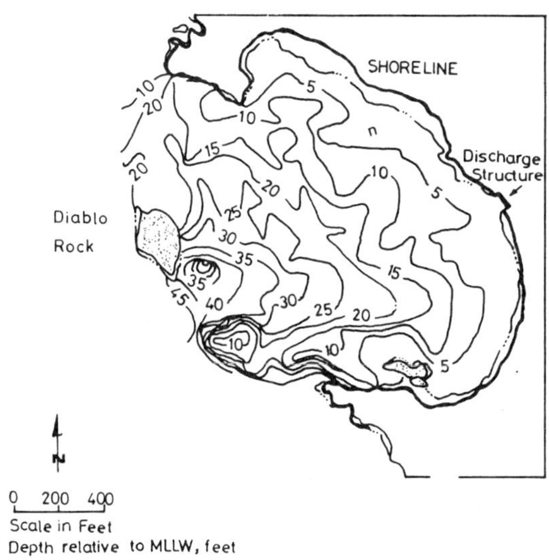

FIG. 1-b BOTTOM BATHYMETRY OF THE DISCHARGE COVE

of recirculation of warm water within Diablo Cove. The extent of this recirculation depended on the incident wave height and tide level. This recirculation had flattened the horizontal temperature profiles in the cove and increased the plume surface area within isotherms close to the discharge point. However, there was no attempt made to crystallize the effects of wave parameters and its induced-nearshore circulation on the plume surface area and the accompanied vertical temperature profiles.

Objective - The main purpose of the present work was to determine effects of wave parameters such as height, direction on the isotherm surface area in the near-field zone of thermal plumes discharged in coastal waters. In addition, the corresponding modifications of the plume vertical temperature profile will be determined. Emphasis was given to the effects induced by the interaction of the incident wave field with the nearshore bathymetry/topography. The results will be derived from examining the field data available on the thermal discharges from three power plants. Two of these plants; namely Fukushima in Japan and Encina in Calif. are located on straight long beaches. The third plant which is Diablo Canyon in Calif., is located on a partially closed embayment. The derived results from the field data will be supplemented with experimental prediction reduced from a systematic set of experiments which was conducted on the physical model of Diablo Canyon. This set of experiments will focus on effects of wave direction and modifications of vertical temperature profiles.

MIXING OF JETS/PLUMES IN WAVE-CURRENT FLOW

Most predictive models of thermal diffusion in the coastal zone consider an offshore current and/or tidal current, but do not consider the influence of surface waves and/or nearshore currents due to breaking waves. Among the few predictive models in the literature is the mathematical model developed by Tanaka and Wada (17) in 1982. Their evaluation of the numerical model was based on a comparison between the model predictions and data collected on a physical model. The authors came to the conclusion that in order to improve the capabilities of their model, it is necessary to acquire more field data on both the nearshore currents and diffusion patterns of large discharges of warm water under strong wave conditions. In addition, virtually all past hydraulic model studies of jet mixing problems have been performed in the absence of surface gravity waves.

The studies of Ismail and Wiegel (Ref. 4-7, 1980-83) which were performed for the case of non-buoyant jets were of considerable assistance in understanding the effect of waves during the analysis of the field and model data for Diablo Canyon thermal discharges. This is particularly valid in the zone close to the discharge area (near-field zone) where advection and free turbulence created by the shearing action, of the discharge causes jet diffusion.

Ismail has found that the interaction of waves with jet driven flows modified the undisturbed coastal circulation of each and the combined bottom mass transport is always focused on the jet outlet (6, 1982). In addition, it was shown experimentally and theoretically that momentum jets expand at a greater rate than in the absence of waves (7, 1983). Also, velocity measurements at the jet-longitudinal-axis have shown that mean velocities of the current near the bottom are decreased with a tendency of the current shear near the water surface to increase as it is illustrated in Fig. 2. The effects of waves on jet spreading and bottom mass transport are shown in the sample photographs of Fig.3

FIELD DATA OF DIABLO CANYON

A total of 56 field tests were performed from late 1984 to mid 1986, including tests for a wide range of power levels (300-2280 Mwe), pump flows [500 - 4000 cfs (14.2 - 114 cms)] and ambient conditions. The ambient conditions included extreme low to extreme high tide, strong and weak currents both upcoast and downcoast, and low to high waves. Thirteen field tests, from the total of 56, were used to verify the model. The vertification tests incorporated one unit, two unit and heat treatment operation, two and four pump flows (including co-flowing warm and cold jets), and a wide range of ambient conditions.

During each field test,temperature profiles were taken at 4 depths in the top 10 ft (3m), along 5 transect lines in Diablo Cove, ranging from approximately 400 ft (122m) from the discharge structure (Line 4) to just outside the west entrance [1400 ft (429m)] from the discharge (Line 14). Surface temperatures were measured outside Diablo Cove to approximately the $1^{o}F$ ($0.5^{o}C$) contour, typically a distance of 2-4 miles (3.2-6.5 km). Vertical temperature profiles were taken at 9-10 locations in the cove and 3-6 locations offshore. currents were measured both with drogues and moored current meters. Waves were measured with a wave rider buoy. The duration of each field test was 2-3 hours.

PHYSICAL MODEL OF DIABLO CANYON

The model is an undistorted densimetric Froude model with a scale of 1:75, and covers a prototype area 6400 ft (1960 m) along the coast by 4800 ft (1468 m) normal to the coast. The model basin is 85 ft by 64 ft by 2.5 ft deep (26 m x 19.6 m x 0.76 m). Model bathymetry was based on prototype hydrographic studies plus aerial photography. Coastal currents parallel to the coast were generated using a pump/manifold system. Dynamic tides were simulated using a computer controlled system. Monochromatic waves were generated using a piston-type wave maker. Temperature measurements were taken using a 45 sensor array, spaced at 50 ft (15.2 m) (prototype) centers in the horizontal, and 4 ft (1.2m) in the

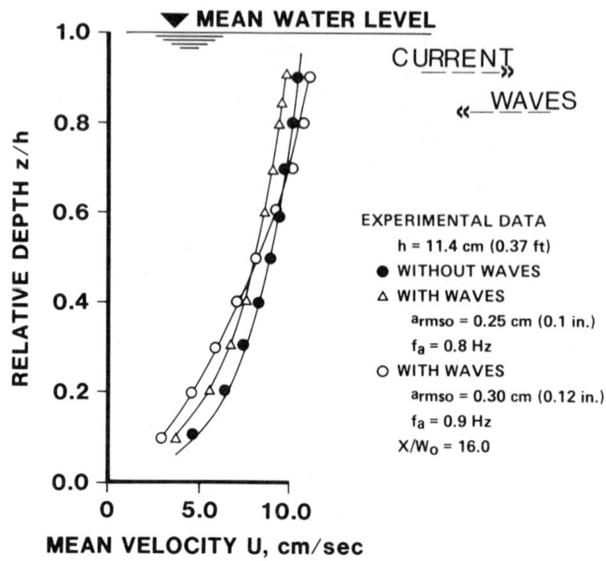

FIG.2 VELOCITY MODIFICAT. OF CURRENTS
Ismail (1980)

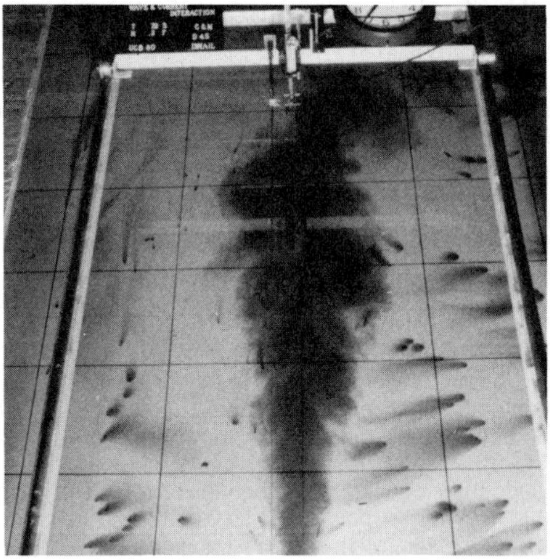

FIG. 3A FLOW VISUALIZ. OF SURFACE JET

After ISMAIL (1981)

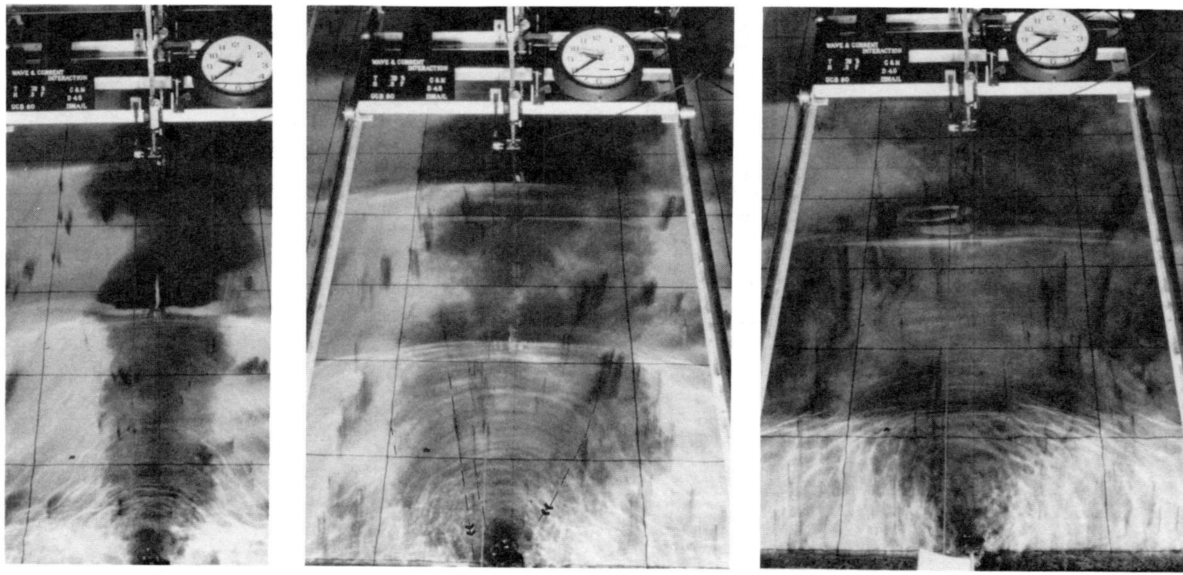

t=3 sec   t = 15 sec   t=45 sec

FIG. 3.b Momentum Jet In Presence Of Opposing Waves

wave period 0.8 sec; wave height $a_0 \approx 0.9$ cm

vertical. The entire array could be moved in 3 directions, allowing the 3 dimensional plume field to be measured.

Experimental procedures- Detailed model operating procedures for the model were developed and strictly adhered to. Water level, currents and waves were established to simulate field conditions. Discharge flow conditions were established and temperature of the discharge was set so that the density difference in the model was the same as during the field test. Temperature probes and wave gages were calibrated prior to each test. The system was allowed to come to quasi-steady prior to taking any temperature measurements.

FIELD AND LABORATORY RESULTS

One the simplest methods of parameterizing plume spreading and decay is in terms of surface areas within isotherms. For Diablo Canyon surface temperature data in the field were used to develop isotherm contours. The far-field isotherm contours refer primarily to the area outside Diablo Cove. The near-field isotherms correspond to the area within Diablo Cove. The cumulative surface area for the complete field versus normalized temperature is shown on Fig. 4 for six field tests which reflect the effects of wave height and tide level. On the same figure the available field data for Encina and Fukushima power plants are plotted. To establish a comparative basis to examine wave effects on the thermal discharges from the three power plants, wave parameters in the surf zone were evaluated from deep water wave conditions. Figure 5 displays the shoreline configuration for the power plant sites. On the same figure, values of wave breaker height and surf zone width and flow pattern of surf zone currents are illustrated.

The results shown in Fig. 4 indicate that higher waves in general increase the plume total surface area in the near-shore zone. However, for Diablo Canyon and at low tide, the increase of surface area is limited close to its discharge structure . Thereafter , the initial region, entrainment is inhibited due to the restricted bottom bathymetry of the shallow plateau in the south cove. And further, the plume is shifted to the north by the induced offshore cool water ingress and circulation in the cove. This decrease of plume surface area in the presence of higher waves and at low tide was confirmed in the laboratory as it is illustrated in Fig. 6.

For extreme high tide, it is noticed on Fig. 4 that , under data for high waves, there is a slight trend of decline of the total surface area in the medium range of excess temperature. Examining this trend in the laboratory indicated that the decrease of the total surface area is mainly contributed by the action of the cool water return current. The extent of this decrease of the surface area depends greatly on the angle of wave incidence as it is demonstrated in Fig. 7. The corresponding modifications of the

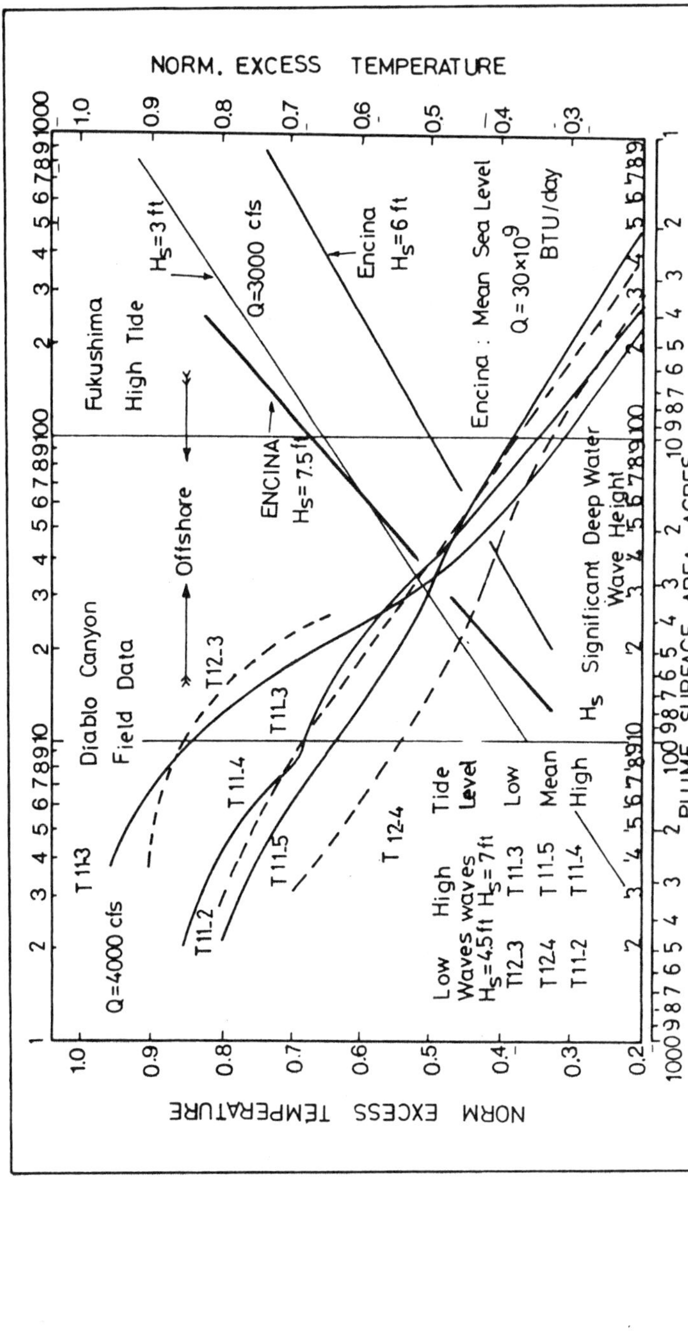

FIG. 4 COMPARISON OF DIABLO CANYON SUREACE ISOTHERM AREA TO FIELD DATA FROM OTHER POWER PLANTS ON A STRAIGHT SHORELINE

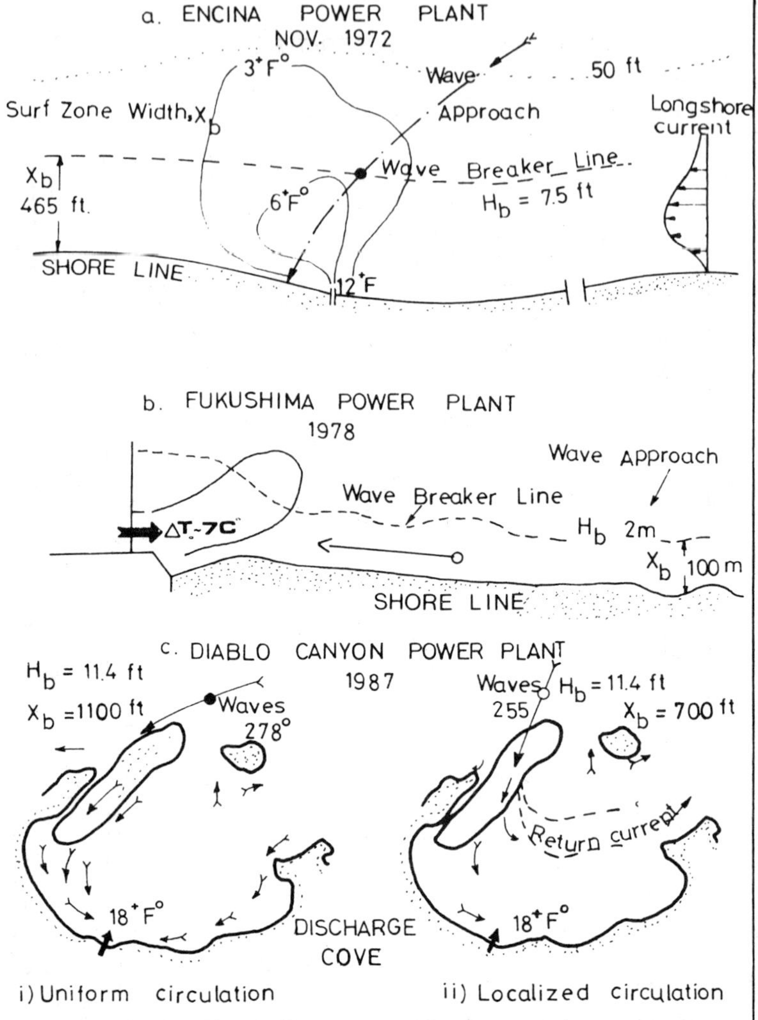

FIG.5 PATTERNS OF WAVE-INDUCED CIRCULATION VS. SHORELINE CONFIGURATIONS

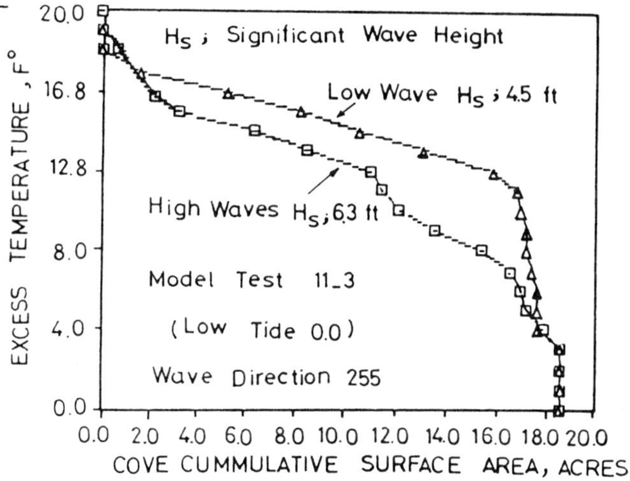

FIG. 6 PLUME SURFACE AREA VS. WAVE HEIGHT

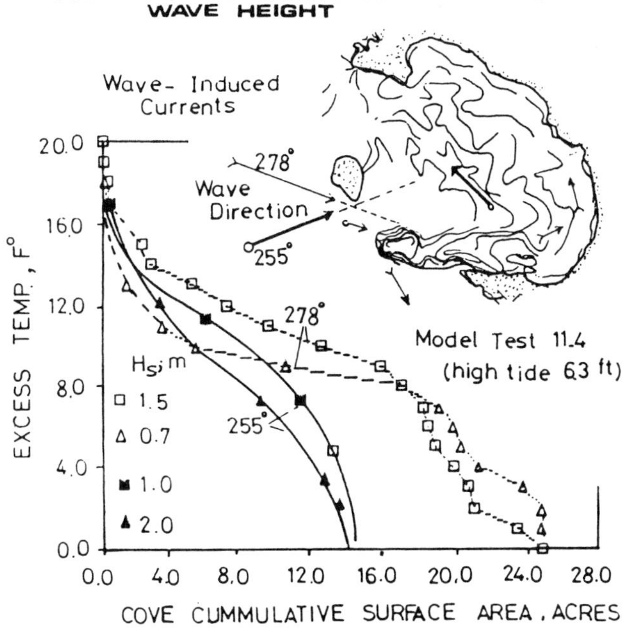

FIG. 7 WAVE PARAMETERS VS. PLUME SURFACE AREA

vertical temperature profile as a result of high waves are shown in Fig. 8 for cases of low and high tides.

The surface isotherm data for Diablo Canyon, obtained within Diablo Cove, and defined in terms of fractional excess temperature, are somewhat of less extent than surface areas for thermal discharges into open water. For thermal discharges on straight long beaches, it is possible to describe the surface area within the isotherms as it increases exponentially with decreasing relative excess as shown on Fig. 4. The slope of such a functional correlation was found to depend greatly on the gradients of wave momentum fluxes across the surf zone. Therefore, deepwater conditions of water waves are not generally sufficient parameters to characterize the incident wave field.Instead, wave characteristics in the surf zone and its associated currents should be evaluated from statistical wave data at deep water locations

CONCLUSIONS

Extensive field data on horizontal and vertical temperature distributions of thermal plumes, discharged horizontally from thermal electric power plants into coastal waters, were analyzed with a view to obtain basis of qualitative prediction of wave and currents effects on near-field mixing. Information on winds, swells/seas, currents, tide, shoreline configuration and sea bottom bathymetry were avaialble for the analyzed data. Two of these plants Fukuskima in Japan and Encina in southern California are located on straight long beaches. The third plant is Diablo Canyon and is located on a partially closed embayment. In addition, a series of sensitivity laboratory tests was conducted on the 1:75 physical model of Diablo Canyon to provide complementary data on the effect of waves and its induced currents on plume surface area and vertical temperature profiles. The results of the study lead to the following conclusions:

(i) Cumulative surface area for the overall isotherm contours, of excess temperature ranges in the near field of thermal plumes, increases in the presence of opposing waves of finite height. This increase might be inhibited by restrictions due to water depth and side boundaries. Further, this trend of surface area increase might not apply uniformly over all ranges of excess temperature in the near field. This is due to the different mechanisms by which surface waves affect plume mixing in the surf zone and the plume densimetric Froude number.

(ii) Within the upper range of normalized excess temperature, ≥ 0.8, opposing waves increase the plume surface area due to the excess longitudinal momentum flux of waves. Further it is also seen that the wave-induced cold bottom currents are entrained into the plume upper layer, close to the outfall, and thereby destratifies the vertical temperature profile.

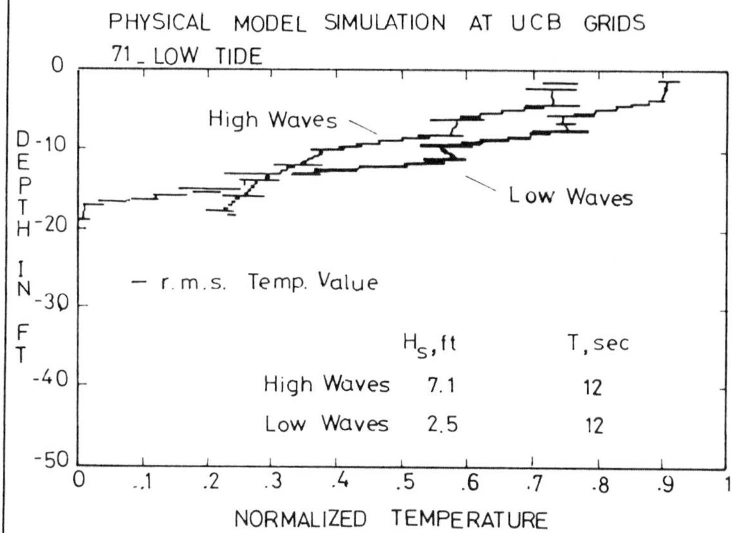

FIG 8-a MODIFICATION OF VERTICAL TEMP. PROFILES

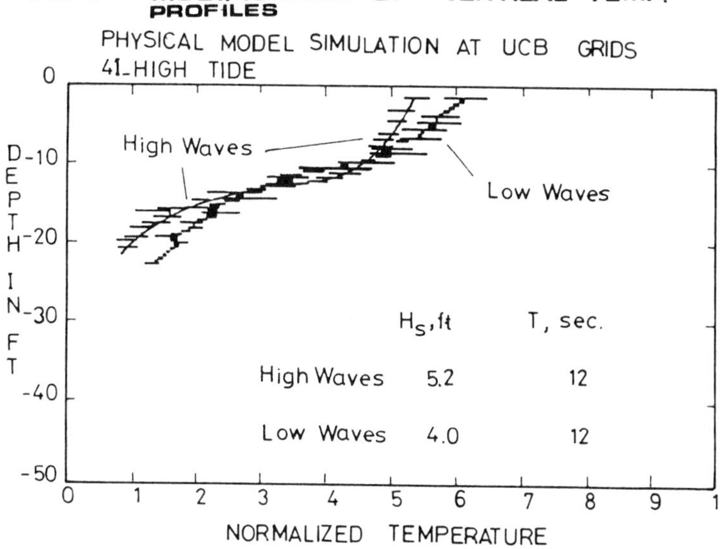

FIG8_b MODIFICATION OF VERTICAL TEMP. PROFILES

Longshore or return rip currents, generated on straight long beaches and within shoreline embayments reduces plume surface area for the medium range of (0.8-0.5) normalized excess temperature in the near field zone.

(iii) Interaction of incident oblique-wave field with a partially enclosed embayment induces a basin secondary circulation with an offshore wave-breaking current. This flow pattern forces a part of far-field plume warm water to return to the discharge area and be rentrained in the plume near-field. Thus waves under certain conditions may interact with the nearshore topography leading to a much higher effluent concentrations and larger surface areas for the high and medium ranges of excess temperature.

(iv) Gradient of wave momentum flux across the surf zone is an important parameter in shaping the slope of functional relationship between plume surface area versus fractional excess temperature in coastal waters.

APPENDIX 1. REFERENCES

1. Adams, E.E., "Discussion of Reynolds Number Effects on Buoyant Surface Jets", ASCE, Journal of Hydraulic Engineering, Vol. III, No. 8, August 1985.
2. Grider, G.W., "Some Effects of Heavy Surf on a Horizontally Discharged Thermal Plume, Proc.Offshore Technology Conference, Paper 1721, Houston, TX, May 1973.
3. Horikawa, K., Lin, M-C and Sasaki, T.O, " Mixing of Heated Water Discharged in the Surf Zone", Proc. 16 th Int. Conf. Coastal Engineering, Ch. 155, Hamburg, Germany, Aug. 27-Sept. 3, 1978.
4. Ismail-Awadalla, N., "Wave-Current Interaction", Technical Report HEL 27-7, Hydraulic Engineering Laboratory, University of California, Berkeley, CA, October 1980.
5. Ismail, N.M., "Effects of Opposing Waves on the Mixing of a Horizontal Surface Momentum Jet", Technical Report HEL 27-9, Hydraulic Engineering Laboratory, University of California, Berkeley, CA, June 1981.
6. Ismail, N., "Effect of Wave-Current Interaction on Littoral Drift", Shore & Beach, Vol. 50, Jan. 1982.
7. Ismail, N.M., and Wiegel, R.L., "Opposing Wave Effect on Momentum Jets Spreading Rate", Journal of Waterways, Port, Coastal and Ocean Engineering, Vol. 109, No. 4, Nov. 1983.
8. Ismail, N.M., Ryan, P.J. and Tu, S.W. "Role of Wave Transformation on Mixing of Buoyant Jets", ASCE Speciality Conf. on Nearshore Hydrodynamics, Delaware, June 29-July 1, 1987.
9. Kuhlman, J.M., "Reynolds Number Effects on Buoyant Surface Jets". Journal of Hydraulic Engineering, ASCE, Vol. 110, No. 6, June 1984.
10. Leighton, J.P., Tu, S.W., Petroccitto, A.A., Eastman, L.K., "Characterization of Receiving Water Temperatures During Power Accension Testing of Unit 1 Diablo Canyon Power Plant", Report

No. 420-85, Pacific Gas and Electric Company, Department of Engineering Research, March 1986.
11. Phillips, O.M., "The Dynamics of the Upper Ocean", Cambridge University Press, Cambridge, England, 1977.
12. Porter, B.A., "Model Verification Study of the Thermal Plume Discharge Geometry and Temperature within Diabio Cove", Report No. UCB/HEL 27-14, U.C. Berkeley College of Engineering, June 1985.
13. Ryan, P.J., Tu, S.W., Leighton, J.P., and Wiegel, R.L., "Hydraulic Model Tests for Unit 1, Diablo Canyon Power Plant, Report No. 420-86. 557, Department of Engineering Research, Pacific Gas and Electric Co., November 1986.
14. Ryan, P.J., Ismail, N.M., Lou, R. Tu, S.W. and Wiegel, R.L., "Hydraulic Model Verification Tests for Units 1 and 2 Diablo Canyon Power Plant", Report No. 420-DC-87. 15, Department of Engineering Research, Pacific Gas and Electric Co., March 1987.
15. Ryan, P.J., Kloepper R., Ismail, N., Tu, S.W., and Wiegel, R.L., "Hydraulic Model Tests for Heat Treatment Conditions at Diablo Canyon Power Plant", Report No. 420-DC-87.16, Department of Engineering Research, Pacific Gas and Electric Co., April 1987.
16. Safaie, B., "Mixing of Buoyant Surface Jet Over a Sloping Bottom", Journal of Waterways, Harbors, and Coastal Engineering Division, Proc. ASCE, Vol. 105, No. WW4, 1979.
17. Tanaka, H. and Wada, A., "Prediction of the Diffusion of Discharged Warm Water Influenced by Nearshore Currents", Coastal Engineering in Japan, Vol. 25, 1982.
18. Tu, S.W., Leighton, J.P., White, C.O. Hsu, C.C. "Surface Buoyant Jet Characteristics of the Thermal Discharge Plume at Diablo Canyon Power Plant", Report No. 420-86, 475, Department of Engineering Research, Pacific Gas and Electric Company, November 1986.
19. Wiegel, R.L., Harms, V.W., Safaie, B., Cumming, J.D., Della, R.P., C.B., Leidersdorf, and C. Young, "Report on Model Study of Cooling Water System of Pacific and Electric Company Nuclear Power Plant Located at Diablo Canyon, California", Technical Report HEL 27-2, Hydraulic Engineering Laboratory, University of Califirnia, Berkeley, CA, January 1976.

CHAPTER 188

STOCHASTIC ANALYSIS OF OFFSHORE CURRENTS

Michel K. Ochi* and Richard I. McMillen*

ABSTRACT

This paper presents the results of stochastic analysis of offshore currents. The purpose of this study is to clarify the stochastic properties of offshore currents, and to develop a method to predict the magnitude of the extreme currents and their direction from measured data. Statistical analysis as well as spectral analysis are carried out on current velocities measured over a period of several months off Newfoundland, Canada. The effect of water depth on the statistical characteristics of current velocity is also clarified. The extreme current velocity expected to occur in 4- months and its direction are estimated by applying order statistics and the results are compared with the measured data.

INTRODUCTION

Current velocity measured in offshore areas, in general, results from currents primarily related with astronomical tide and those induced by local wind. Current velocity associated with ocean circulation and storm surge, etc., may also be significant depending on geographical location. All of these components can exist simultaneously, and hence the measured data demonstrate that the magnitude of the current velocity fluctuates in a random fashion with frequencies covering a wide range from 0.0008 to 0.08 cycles per hour (period range from 12 to 1200 hours).

Furthermore, the direction of the current velocity changes with time; it rotates clockwise in the northern hemisphere due to the rotation of the earth and the coriolis force. The angular frequency of rotation, however, is random. It varies from 0.31 to 0.70 radians per hour. Thus, the current velocity is considered to be a random process with varying frequency and direction.

Several studies have been carried out on the fluctuating properites of offshore currents as listed, among others, in References [1] through [8]. The results demonstrated in these references suggest that currents may be considered as a stochastic process consisting of various frequency components; (a) a tidal component with a frequency of 1 cycle per day (cpd) or 2 cpd depending on the location, (b) a wind-induced component with frequencies ranging from 0.10 to 0.20 cpd, and (c) seasonal variations of extremely low-frequency.

---

* Coastal & Oceanographic Engineering Department
University of Florida, Gainesville, Florida, U.S.A.

The purpose of the present study is to clarify the stochastic properties of offshore currents, and to develop a method to predict the magnitude of the extreme currents and their direction from measured data. For this, a statistical analysis is carried out on current velocities measured over a period of several months off Newfoundland, Canada. Statistical analysis as well as spectral analysis are carried out and the relationship between the area under the spectral density function and the variance evaluated from the time history is clarified. An analysis is also made of the data to examine the effect that water depth has on the statistical characteristics of current velocity. Finally, the extreme current velocity expected to occur in 4-months, and its direction, are estimated by applying order statistics and the results are compared with the measurred data.

OFFSHORE CURRENT DATA AND ANALYSIS PROCEDURE

Data

Statistical analyses are carries out on approximately four months of continuous current data. The current meter mooring was located in a water depth of 84 meters. The mooring had three current meters sampling at 10 minute intervals at depths of 26, 47 and 75 meters. The current meter data are hourly mean velocities obtained by removing the high-frequency noise with a digital filter and sub- sampling the resulting smoothed signal to hourly values.

A dominant hourly current direction was obtained for each hourly current velocity from the directional data which was obtained from compasses moored with the current meters at the three depths. Here, the current direction is defined as that to which the water is flowing.

Analysis Procedure

The time history of the near-surface current velocity is first decomposed into an East-West and North-South direction. An example of the time history of hourly current velocities as well as the decomposed East-West and North-South time histories are shown in Figure 1. This example represents the data obtained for two weeks from October 13 to October 27, 1981.

A statistical analysis is then performed on the East-West and North-South time histories in which the probability distributions applicable for deviations from the mean and the maximum (positive and negative variations) are obtained. After the statistical analysis is made for the East-West and North-South time histories, a Fast-Fourier-Transform (FFT) technique is used to obtain the energy spectrum for each time history.

In general, it is somewhat difficult to differentiate the measured current velocities into various components such as the wind- generated, tidal, and ocean current components, etc. Hence, it may be appropriate to decompose the measured currents into a high and low-frequency component. Here, the high-frequency component is considered to be the currents associated with tides, while the low-frequency components are those attributed to all other environmental conditions which may be called the residuals. The FFT is used to obtain the high-frequency (frequency $>$ 0.035 cph) and the low-frequenct (frequency $<$ 0.035 cph) current velocity time histories in the East-West and North-South directions. These are called the tidal and residual currents, respectively, in the present study.

As an example, Figure 2 shows the time histories of the high-frequency (tidal) and the low-frequency (residual) current fluctuations for the East-West component which was shown in Figure 1.

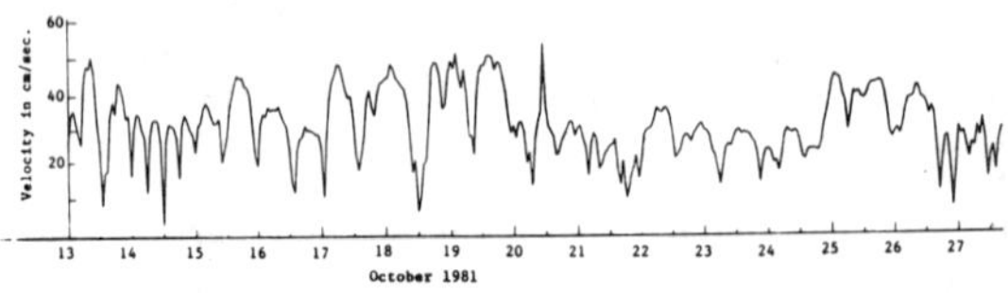

Current velocity

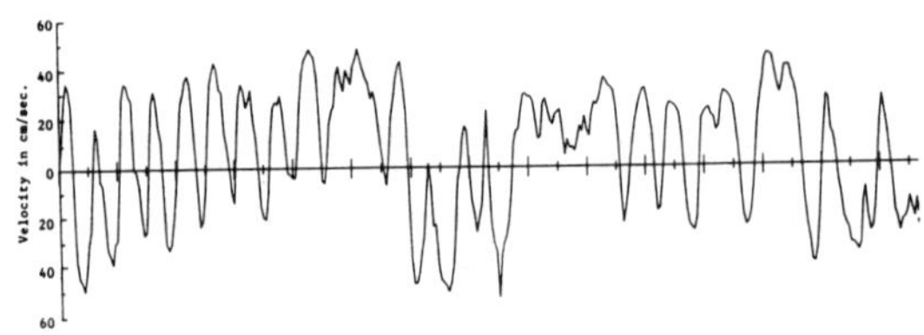

East-west component of current velocity

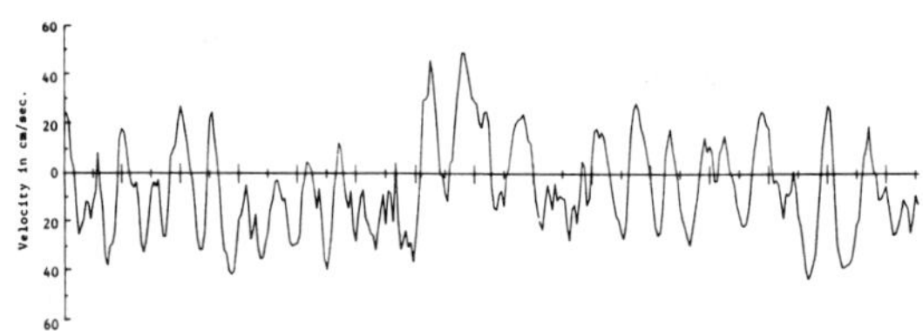

North-south component of current velocity

Figure 1: Time histories of current velocity, and East-West, and North-South components of velocity.

RESULTS OF STATISTICAL ANALYSIS OF CURRENTS

Figure 3 shows an example of a polar diagram indicating the magnitude and direction of current velocity as a function of time. The data pertains to a 20-hour observation made on October 10th. The current velocities rotate clockwise in the Northern hemisphere through 360° in aproximately 20 hours (a circular frequency of 0.31 radians per hour in this example. However, further analysis of the data indicates the circular frequency of the near- surface current is not constant, but instead varies from 0.31 to 0.70 radians per hour (9.0 to 20 hours). The circular frequencies of the current at 47 meter and 75 meter water depths are also examined. It is noted the preferred directions of motion water throughout the water column appears to be consistent with Ekman veering arguments.

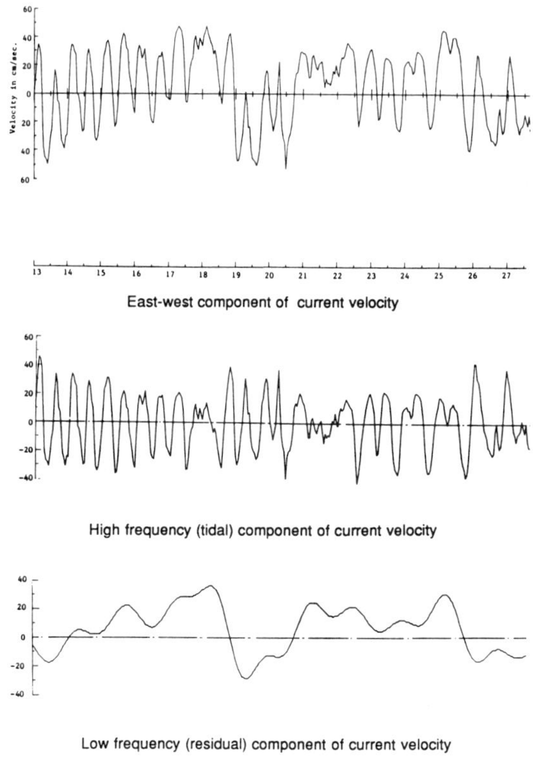

Figure 2: Time histories of East-West component of current velocity, and high frequency (tidal) and low-frequency (residual) components.

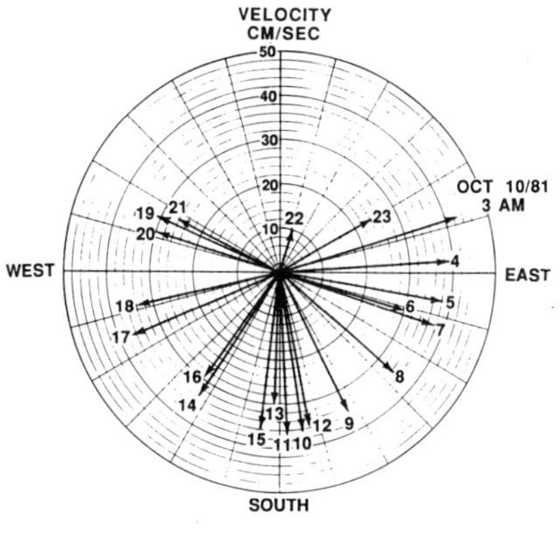

Figure 3:
Current velocity magnitudes rotating in a clock-wise direction.

Distribution of Fluctuating Current Velocities

Figure 4 shows the histograms of the deviations from the mean constructed from the time history for the East-West components. The deviations include both tidal and residual velocity components.

Included also in the figures are the normal (Gaussian) and non-Gaussian distributions for comparison with the histograms. The non-Gaussian probability density function is given by the following form of a Gram-Charlier expansion (Edgeworth [9] and Longuet-Higgins [10]):

$$f(x) = \frac{1}{\sigma\sqrt{2\pi}}\exp\{-\frac{x^2}{2\sigma^2}\}[1 + \frac{\lambda_3}{3!}H_3\left(\frac{x}{\sigma}\right) + \frac{\lambda_3^2}{72}H_6\left(\frac{x}{\sigma}\right) + \frac{\lambda_4}{4!}H_4\left(\frac{x}{\sigma}\right) + \ldots], \quad (1)$$

where σ^2 = variance of current velocity, λ_3 = skewness of the distribution, λ_4 = kurtosis − 3, $H_n\left(\frac{x}{\sigma}\right)$ = Hermite polynomial of degree n.

As can be seen in the figure, the distribution of the deviations from the mean value appears to follow a non-Gaussian probability distribution with parameter λ_4 which is associated with the kurtosis of the distribution. The same trend can also be observed for deviations from the mean in the North-South current component.

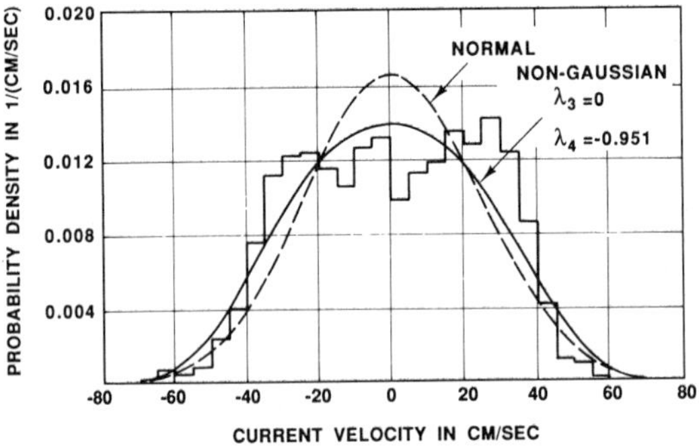

Figure 4:
Histogram of deviations from the mean of East-West current velocity.

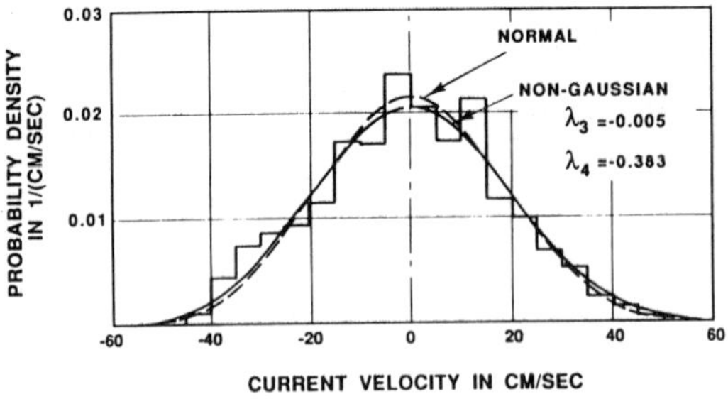

Figure 5:
Histogram of deviations from the mean of tidal current velocity in East-West direction.

The correlation coefficient between two components (the East-West and North-South velocity components) is extremely small, on the order of 0.005. This implies that a statistical analysis of current velocities can be done independently in two directions.

Figures 5 and 6 show the histograms of deviations from the mean value for the tidal and residual components, respectively, in the East-West direction. Included also in these figures are the normal and non-Gaussian distributions for comparison with the histograms. As can be seen in these figures both the tidal and residual currents may be assumed to be Gaussian random processes. However, it will be shown later that the tidal component can be considered a narrow-band random process, while the residual component appears to be a non-narrow-band random process. This same trend can also be observed for currents in the North-South direction.

The results of computation show that the correlation coefficient between the tidal and residual currents is extremely small for both East-West and North-South directions. Therefore, it may safely be assumed that the tidal current and the residual current velocity components are statistically independent.

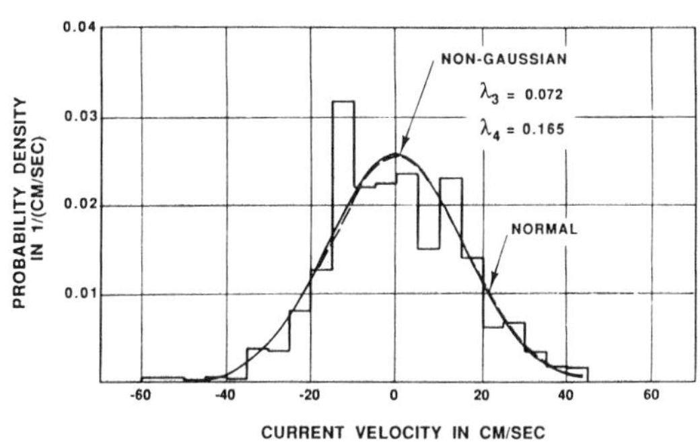

Figure 6:
Histogram of deviations from the mean of residual current velocity in East-West direction.

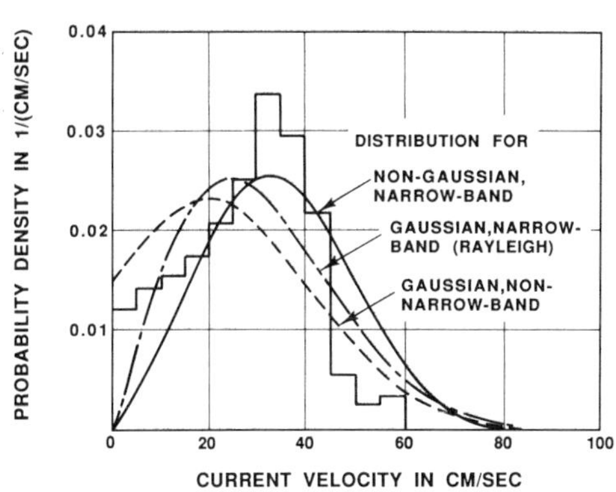

Figure 7:
Histogram of maxima of current velocity in the East direction.

Distribution of Maxima

Figure 7 shows the histogram of the maxima in the East direction constructed from current velocities including both tidal and residual components. Included also in the figure are the probability distributions applicable for the maxima. These include the probability density functions of the maxima applicable for a (a) narrow-band Gaussian random process, (b) non-narrow-band Gaussian random process, and (c) narrow-band non-Gaussian random process given below (Ochi & Wang [11]):

$$f(\xi) = \frac{1}{L}\left[\frac{\xi}{\sigma^2}\exp\left\{-\frac{\xi^2}{2\sigma^2}\right\}\left\{1+\frac{\lambda_4}{4!}H_4\left(\frac{\xi}{\sigma}\right)\right\}\right] \tag{2}$$

where σ^2 = variance of current deviation from the mean
ξ = maxima, $0 < \sigma < \infty$, L = normalization factor.

As can be seen in the figure, the histogram of the maxima, as expected, is not represented by either of the two distributions associated with a Gaussian (normal) process. The representation of the histogram by the probability distribution of the maxima for a non-Gaussian random process is marginally acceptable. A possible explanation of this discrepancy between the histograms and the distribution for the maxima of a non-Gaussian random process is as follows:

The distribution applicable for the maxima of a non-Gaussian random process (Equation 2) is developed based on the assumption that the process is narrow-banded. However, the time history of offshore currents including both tidal and residual components encompasses an extremely wide range of frequencies (from 0.0008 to 0.08 cycles per hour). This indicates that the maxima of the offshore currents (including both tidal and residual components) may obey the distribution applicable for a non-narrow-band, non-Gaussian random process.

Spectral Analysis

Spectral analysis is carried out on measured current velocities to examine (a) frequency components which carry significant energy with the current flow, and (b) the relationship between spectral energy (area under the spectral density function) and the variance of the current fluctuations which can be evaluated from the time history of the current velocity. Figures 8 and 9 show the energy spectra for the tidal and residual components in the East-West direction, respectively. A Fast-Fourier-Transform technique is used to obtain the tidal and residual time histories in both the East-West and North-South directions. A cutoff frequency of 0.035 cycles/hour (28.4 hour period) is used to separate the time histories into their respective tidal and residual components.

The area under the spectral density function, denoted by m_o, and the bandwidth parameter, denoted by ε, are evaluated from the spectra and tabulated in Table 1. Included also in the table are the variances, σ^2, obtained from the time history of current velocity. It may be of interest to compare the variance, σ^2, with the moment, m_o. These should agree, in general for a weakly stationary, ergodic random process with zero mean. The offshore current velocity measured over a period of several months, however, cannot be considered as a stationary random process. Nevertheless, as can be seen in the table, the agreement between σ^2 and m_o is reasonable in both the East-West and North-South directions. The difference between them is approximately 8 percent in the East-West direction, and 4 percent in the North-South direction. The agreement between σ^2 and m_o for the decomposed tidal and residual components is excellent. This is because the time histories of the tidal and residual components are obtained through FFT filtering techniques which retain the statistical properties of random phenomena including phase information as well.

The results obtained above substantiate the fact that the spectral density function of the current velocity in any direction is interrelated with the variance of current fluctuations. Hence, the probabilistic prediction of the magnitude of current velocity may be made through spectral analysis.

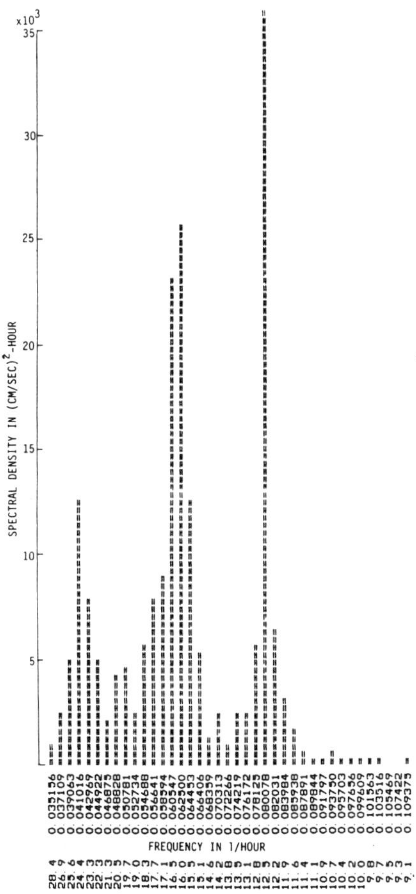

Figure 8:
Energy spectrum of tidal (High frequency) current velocity in East-West direction.

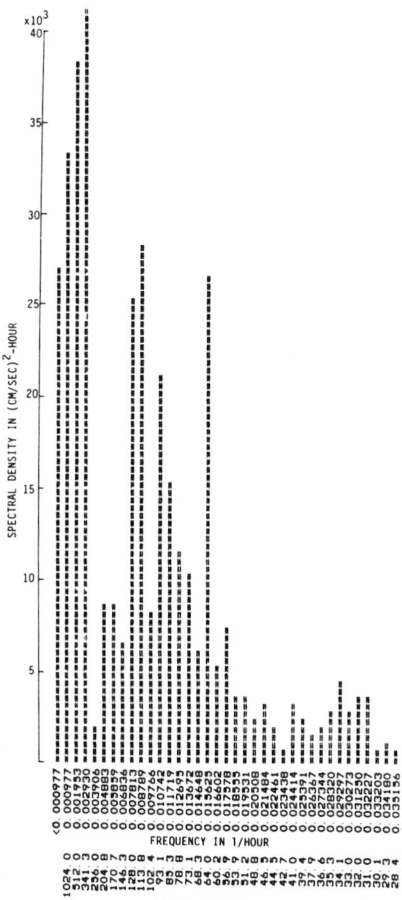

Figure 9:
Energy spectrum of residual (low frequency) current velocity in East-West direction.

Table 1: Variance evaluated from time history of current velocity, σ^2, area under the current energy spectrum, m_o, and spectrum band-width parameter, ε.

Current velocity components	Variance σ^2	Area under the spectrum m_o	Spectrum band-width parameter ε
East-West direction			
Total	627.5	583.2	0.735
Residual	248.6	248.7	0.845
Tidal	334.2	334.3	0.494
North-South direction			
Total	529.0	508.1	0.735
Residual	214.2	214.2	0.828
Tidal	293.6	293.7	0.505

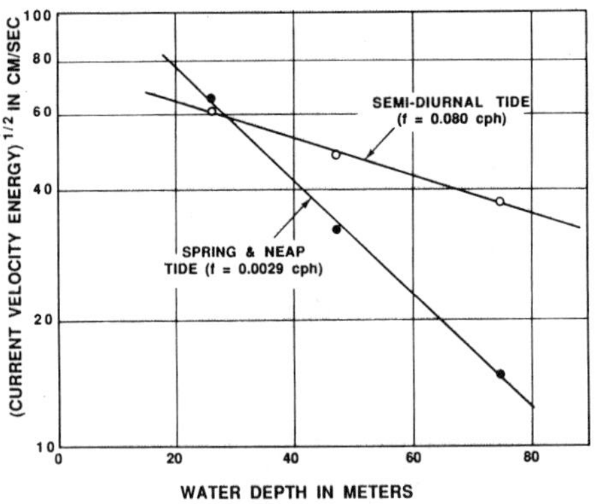

Figure 10: Square-root of current energy density for frequencies 0.080 cph (12.5 hour period) and 0.0029 cph (14.2 day period).

Effect of Water Depth

The same statistical analysis carried out on the near-surface current data was also made on data obtained at two deeper depths; 47 and 75 meters. These are called the mid-depth and near-bottom locations in the following analysis.

The results of the analysis indicate that almost all statistical properties of the fluctuating current velocity obtained from analysis of data at the deeper depths appear to be nearly the same as those obtained from the current velocities at the near surface. However, the severity of current fluctuations is drastically reduced with increase in depth, and the non-Gaussian random process characteristics observed for the near-surface current velocity are substantially less pronounced at deeper locations.

In order to examine the reduction of current velocity with increase in water depth, Figure 10 was prepared. The figure shows the square root of the current energy density (the vector sum of the East-West and North-South components) for the frequencies 0.080 cph (period 12.5 hours) and 0.0029 cph (period 341 hours) plotted on semi-log paper as a function of water depth. Here, the frequency 0.080 cph corresponds to that for the semi-diurnal tide, and the tidal current (high frequency) spectrum shows an extremely large energy density at this frequency. The frequency 0.0029 cph corresponds to that for the spring and neap tide, and the residual current (low frequency) spectrum shows considerable energy density at this frequency.

As can be seen in the figure, the square-root of current energy, which is proportional to the current velocity, reduces exponentially with increase in water depth. The reduction rate of the current velocity for the spring and neap tide (low-frequency) is much larger than that for the semi-diurnal tide (high-frequency component). This result indicates that the tidal and residual current energies both decrease substantially with increase in water depth. However, the rate of decrease for the residual currents is much larger than that for the tidal currents.

Estimation of Extreme Value

It was shown that the current velocity fluctuations can be considered as a non-Gaussian random process with the parameter λ_4, and that the East-West component and the North-South component are statistically independent. This being the case, it can be proved that the current velocity for an arbitrary direction is also approximately a non-Gaussian random process as derived in the Appendix. That is, the probability density function of current velocity fluctuations (deviation from the mean) for an arbitrary direction θ can be given by Eq. (1) with $\lambda_3 = 0$ and with the following variance, σ^2, and parameter, λ_4, which are now a function of angle θ:

$$\begin{aligned} \sigma^2(\theta) &= \sigma_e^2 \sin^2\theta + \sigma_n^2 \cos^2\theta \\ \lambda_4(\theta) &= \frac{\lambda_{4e}\sigma_e^4 \sin^4\theta + \lambda_{4n}\sigma_n^4 \cos^4\theta}{\left(\sigma_e^2 \sin^2\theta + \sigma_n^2 \cos^2\theta\right)^2} \end{aligned} \qquad (3)$$

where θ = angle measured from North to East, σ_e^2, σ_n^2 = variance of the East-West and North-South direction, respectively, $\lambda_{4e}, \lambda_{4n}$ = parameter λ_4-value for the East-West direction and North-South direction, respectively.

This results in the probability density function of the maxima having the same form as shown in Eq. (2) but with the variance, $\sigma^2(\theta)$, and parameter, $\lambda_4(\theta)$, as given in Eq. (3). That is,

$$f(\xi, \theta) = \frac{1}{L}\left[\frac{\xi(\theta)}{\sigma^2(\theta)} \exp\left\{-\frac{1}{2}\left(\frac{\xi(\theta)}{\sigma(\theta)}\right)^2\right\}\left\{1 + \frac{\lambda_4(\theta)}{4!} H_4\left(\frac{\xi(\theta)}{\sigma(\theta)}\right)\right\}\right] \qquad (4)$$

where $L = 1 + \lambda_4(\theta)/3!$

Then, the extreme current velocity which is most likely to occur in n-peaks for a given angle θ, denoted by $\bar{\xi}(\theta)$, is given as the value which satisfies the following equation:

$$\int_{o}^{\bar{\xi}(\theta)} f(\xi,\theta)d\theta = F(\bar{\xi}(\theta)) = 1 - \frac{1}{n} \tag{5}$$

Numerical computations were carried out for various angles between 0 and 90 degrees, with variance $\sigma_e^2 = 583.2(\mathrm{cm/sec})^2$ and $\sigma_n^2 = 508.1(\mathrm{cm/sec})^2$, and $n = 175$ in a 4-month observation period.

Figure 11 shows the computed extreme current velocities for various directions plotted in polar coordinates. As can be seen in the figure, there is no substantial difference in the magnitude of the extreme current velocities in various directions for data used in the present analysis. The magnitudes of the extreme current velocities in the shaded domain in the figure are nearly equal (70.3 - 69.6 cm/sec). The magnitude of the extreme current velocity observed in the same time period is 74.0 cm/sec in the South-West direction.

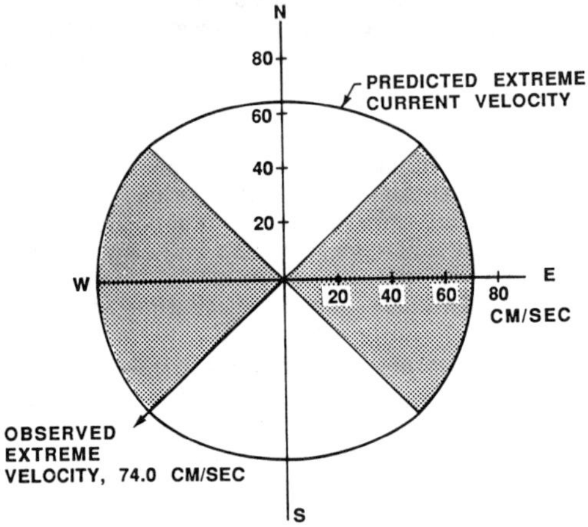

Figure 11:
Computed extreme current velocities expected in a four-month observation for various directions.

CONCLUSIONS

From the results of the stochastic analysis of offshore current data the following conclusions can be drawn:

(1) Deviations from the mean value of the current velocity for the East-West and North-South components (including both tidal and residual velocity components) obey a non-Gaussian probability distribution with parameter λ_4 which is associated with the kurtosis of the distribution.

(2) The correlation coefficient evaluated for the East-West and the North-South current velocity time histories is extremely small; hence they are considered to be statistically independent.

(3) When the East-West and North-South velocity components are decomposed into high-frequency (tidal) and low-frequency (residual) components, the tidal and residual components of the current velocity are found to be statistically independent. Both the tidal and residual currents may be assumed to be Gaussian random processes. In particular, the residual component appears to be a non-narrow-band Gaussian random process.

(4) Histograms of maxima (peak values) of fluctuating currents in all four directions (East, West, North, and South) indicate that the maxima (including both tidal and residual components) can be marginally represented by the distribution of the maxima for a narrow-band non-Gaussian random process. It appears that the maxima obey the distribution applicable for a non-narrow-band non-Gaussian random process.

(5) The area under the spectral density function agrees well with the variance obtained from the time histories for both the East- West and North-South velocity components.

(6) The statistical properties of current velocity obtained from analysis of current data at deeper depths are nearly the same as those obtained from near-surface current data. However, the non- Gaussian random process characteristics stated in Item (1) are substantially less pronounced at deeper depths.

(7) The square-root of the current energy, which is proportional to the current velocity decreases exponentially with an increase in water depth. The rate of decrease is much larger for the residual current velocity than that for the tidal current velocity.

(8) An approximate method for estimating the magnitude of extreme current velocity is developed by applying order statistics to the probability distribution of a non-Gaussian random process. The results of computations made for a 4-month observation period show satisfactory agreement with observed data.

ACKNOWLEDGEMENTS

This research was sponsered by the Office of Naval Research, Ocean Technology Program, through contracts N00014-86-k-0537 to the University of Florida. The authors would like to express their appreciation to Dr. E.A. Silva for his valuable technical discussions received during the course of this project. The authors are grateful to Dr. Max Sheppard for his kind help in obtaining the current data used in this study. The authors also wish to express their appreciation to Ms. Laura Dickinson for typing the manuscript.

APPENDIX: PROBABILITY DISTRIBUTION OF CURRENT VELOCITY FOR AN ARBITRARY DIRECTION

The standardized non-Gaussian probability density function with the parameter λ_4 can be written as

$$f(z) = \frac{1}{\sqrt{2\pi}}\exp\{-\frac{z^2}{2}\}\left[1 + \frac{\lambda_4}{4!}H_4(z)\right] \qquad \text{(A.1)}$$

where $H_4(z) = z^4 - 6z^2 + 3$.

mean, variance σ^2, and parameter λ_4 becomes

$$\Phi_z(t) = \exp\{-\frac{\sigma^2 t^2}{2}\}(1 + \frac{\lambda_4}{4!}\sigma^4 t^4) \tag{A.3}$$

Next, let us consider two statistically independent non-Gaussian random variables x_e and x_n, representing the East-West current velocity and the North-South current velocity, respectively. These are,

$$\begin{aligned} x_e &: \text{non-Gaussian}(0, \sigma_e^2; \lambda_{4e}) \\ x_n &: \text{non-Gaussian}(0, \sigma_n^2; \lambda_{4n}) \end{aligned} \tag{A.4}$$

We may then write the current velocity for an arbitrary direction θ as

$$x(\theta) = x_e \sin\theta + x_n \cos\theta \tag{A.5}$$

where θ = angle measured from North to East.

Since x_e and x_n are statistically independent, the characteristic function of $x(\theta)$ can be obtained by applying Eq.(A.3) as

$$\begin{aligned} \Phi_{x(\theta)}(t) &= \Phi_{xe}(t\sin\theta) \cdot \Phi_{xn}(t\cos\theta) \\ &\sim \exp\left\{-(\sigma_e^2 \sin^2\theta + \sigma_n^2 \cos^2\theta)\frac{t^2}{2}\right\} \\ &\qquad \times \left[1 + \frac{1}{4!}(\lambda_{4e}\sigma_e^4 \sin^4\theta + \lambda_{4n}\sigma_n^4 \cos^4\theta)t^4\right] \\ &= \exp\left\{-(\sigma_e^2 \sin^2\theta + \sigma_n^2 \cos^2\theta)\frac{t^2}{2}\right\} \\ &\qquad \times [1 + \frac{1}{4!}\frac{\lambda_{4e}\sigma_e^4 \sin^4\theta + \lambda_{4n}\sigma_n^4 \cos^4\theta}{(\sigma_e^2 \sin^2\theta + \sigma_n^2 \cos^2\theta)^2} \\ &\qquad \times (\sigma_e^2 \sin^2\theta + \sigma_n^2 \cos^2\theta)^2 t^4] \end{aligned} \tag{A.6}$$

From a comparison of Eqs.(A.3) and (A.6), it can be proved that the random variable $x(\theta)$ has approximately a non-Gaussian distribution with the following variance, $\sigma^2(\theta)$, and parameter, $\lambda_4(\theta)$:

$$\begin{aligned} \sigma^2(\theta) &= \sigma_e^2 \sin^2\theta + \sigma_n^2 \cos^2\theta \\ \lambda_4(\theta) &= \frac{\lambda_{4e}\sigma_e^4 \sin^4\theta + \lambda_{4n}\sigma_n^4 \cos^4\theta}{(\sigma_e^2 \sin^2\theta + \sigma_n^2 \cos^2\theta)^2} \end{aligned} \tag{A.7}$$

REFERENCES

1. Beardlsey, R.C., and B. Butman, "Circulation on the New England Continental Shelf: Response to Strong Winter Storms," Geophysical Research Letter, Vol. 1, 1974, pp. 181-184.

2. Smith, R.L., "A Description of Current, Wind, and Sea Level Variations During Coastal Upwelling Off the Oregon Coast, July-August 1972," Journal of Geophysical Research, Vol. 79, 1974, pp. 435-443.

3. Lee, T.N. and D.A. Mayer, "Low-Frequency Current Variability and Spin-Off Eddies Along the Shelf Off Southeast Florida," Journal Marine Res., Vol. 35, 1977, pp. 193-218.

4. Chuang, W-S., D-P. Wang and W.C. Boicourt, "Low-Frequenct Current Variablity on the Southern Mid-Atlantic Bight," Journal of Physical Oceanography, Vol. 9, 1979, pp. 1144-1154.

5. Hayes, S.P., "Variablity of Current and Bottom Pressure Across the Continental Shelf in the Northeast Gulf of Alaska," Journal of Physical Oceanography, Vol. 9. 1979, pp.88-103.

6. Mayer, D.A., D.V. Hansen and D. Ortman, "Long-Term Current and Temperature Observations on the Middle Atlantic Shelf," Journal of Geophysical Research, Vol. 84, 1979, pp.1776-1792.

7. Noble, M. and B. Butman, "On the Longshelf Structure and Dynamics of Subtidal Currents on the Eastern United States Continental Shelf," Journal of Physical Oceanography, Vol. 13, 1983, pp. 2125-2147.

8. Butman, B. and R.C. Beardsley, "Long-Term Observations on the Southern Flank of Georges Bank. Part 1: A Description of the Seasonal Cycle of Currents, Temperature, Stratification, and Wind Stress," Journal of Physical Oceanography, Vol. 17, 1987, pp. 367-384.

9. Edgeworth, F.Y., "On the Representation of Statistical Frequency by a Series," Journal, Royal Statistical Society Series A, Vol 70, 1907, pp. 102-106.

10. Longuet-Higgens, M.S., "The Effect of Non-Linearities on Statistical Distribution in the Theory of Sea Waves," Journal of Fluid Mechanics, Vol. 17. Part 3, 1963, pp.459-480.

11. Ochi, Michel K. and W-C. Wang, "Prediction of Wave Height of Wind-Generated Seas in Infinite Water Depth," Tech. Rep. UFL/COEL/TR-052, University of Florida, 1984.

CHAPTER 189

Entrainment due to Mean Flow in Two-Layered Fluid

Masaru Ura*
and
Nobuhiro Matsunaga**

Abstract

The entrainment phenomena have been investigated across an interface between two-layered stratified flow induced by wind shear stress. The velocities of mean flow, turbulence and entrainment have been measured under three different conditions of water surface by using a wind-wave tank. When the entrainment velocity u_e is expressed on the basis of the turbulent quantities at the interface, the turbulent entrainment coefficient E ($\equiv u_e/u$) is given by $E = A\cdot(\varepsilon gl/u^2)^{-3/2}$ ($A = 0.7$). Here εg, u and l are the effective buoyancy, the turbulence intensity and the integral lengthscale of turbulence at the interface, respectively. This result coincides with the relationship of entrainment due to oscillating grid turbulence, in which the mean flow does not exist. When, for the practical purpose, the estimation of u_e is made by using the mean velocity U_m and the depth h of mixed layer, E_m ($\equiv u_e/U_m$) $= A_m\cdot(\varepsilon gh/U_m^2)^{-3/2}$ is derived from the transformation of $E = A\cdot(\varepsilon gl/u^2)^{-3/2}$. There holds $A_m = A\cdot T_f$ between A_m and T_f, T_f being a turbulence factor given by $(u/U_m)^4\cdot(l/h)^{-3/2}$. It has been found that this relationship is also valid in various types of two-layered stratified flows as well as the wind-induced two-layered flows.

Introduction

The density stratification is often observed in lakes or estuaries. When the wind blows over the surface of such a water body, the mixing phenomena occur frequently owing to turbulent flow induced by wind shear stress. Many laboratory studies on two-layered stratified flows have been made in order to understand this mixing phenomena. The mixing is characterized by entrainment process. Let us focus attention on the two-layered flows in which the shear stress τ is exerted on the water surface. The previous experimental works may be classified into four categories

* Professor, Department of Civil Engineering, Kyushu Institute of Technology, Tobata-ku Kitakyushu 804, Japan.
** Research Asocciate, Department of Civil Engineering, Kyushu University, Higashi-ku Fukuoka city 812, Japan.

as shown in Figure 1. They are as follows:

(i) the case when flow near the interface has the same direction as that of action of τ and wind waves are present (Tsuruya et al., 1984),

(ii) the case when shear-induced flow has the same flow pattern as the case (i) without the presence of wind waves (Kantha et al., 1975; Deardorff and Willis, 1982),

(iii) the case when flow near the interface has the opposite direction to the surface shear stress and the wind waves are present (Wu, 1973; Kit et al., 1980; Ura, 1984; Kranenburg, 1985), and

(iv) the case when flow pattern is the same as the case (iii) without the presence of wind waves (Kit et al., 1980).

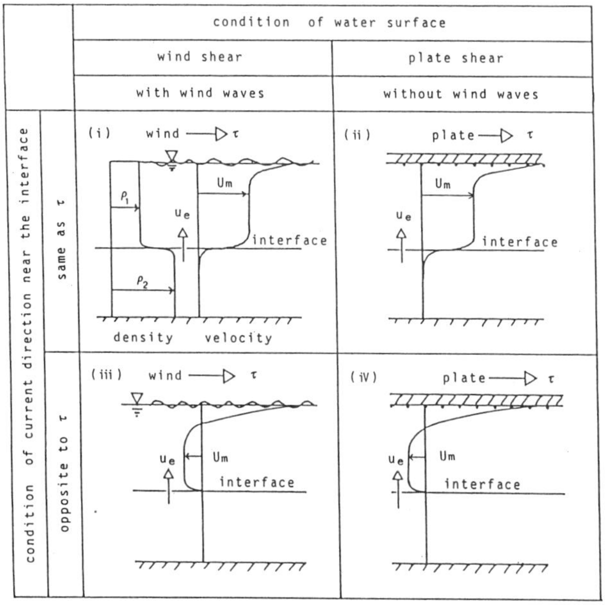

Figure 1. Classification of previous experimental works.

The researchers introduced in cases (i) to (iv) discussed the dependence of the entrainment coefficient E_* ($\equiv u_e/u_*$) on the Richardson number Ri_* ($\equiv \varepsilon gh/u_*^2$). Here, u_e, u_*, εg and h are the entrainment velocity, the friction velocity acting on the water surface, the effective buoyancy and the water depth of mixed layer, respectively. Their results are plotted together in Figure 2. We can support the relationship of $E_* = A_* Ri_*^{-3/2}$, but the proportional factor A_* varies from 700 to 0.13 corresponding to cases (i) to (iv). This wide variation is closely related to whether mean flow direction near the interface is same or opposite to the surface shear stress and whether wind waves are present or not. Using u_* as a characteristic velocity may not be proper in order to explain consistently these data. This reason is that four cases appear under a given magnitude of u_* at the least as far as referred here and that u_* includes no information about whether the reverse flow forms or not. Therefore, it seems appropriate to use the mean velocity U_m of uniform flow induced in the mixed layer rather than u_*.

On the other hand, it has been proposed by Ura et al. (1987b) that the entrainment process depends on the characteristic quantities of turbulence such as turbulence intensity u and integral lengthscale l of turbulence at the

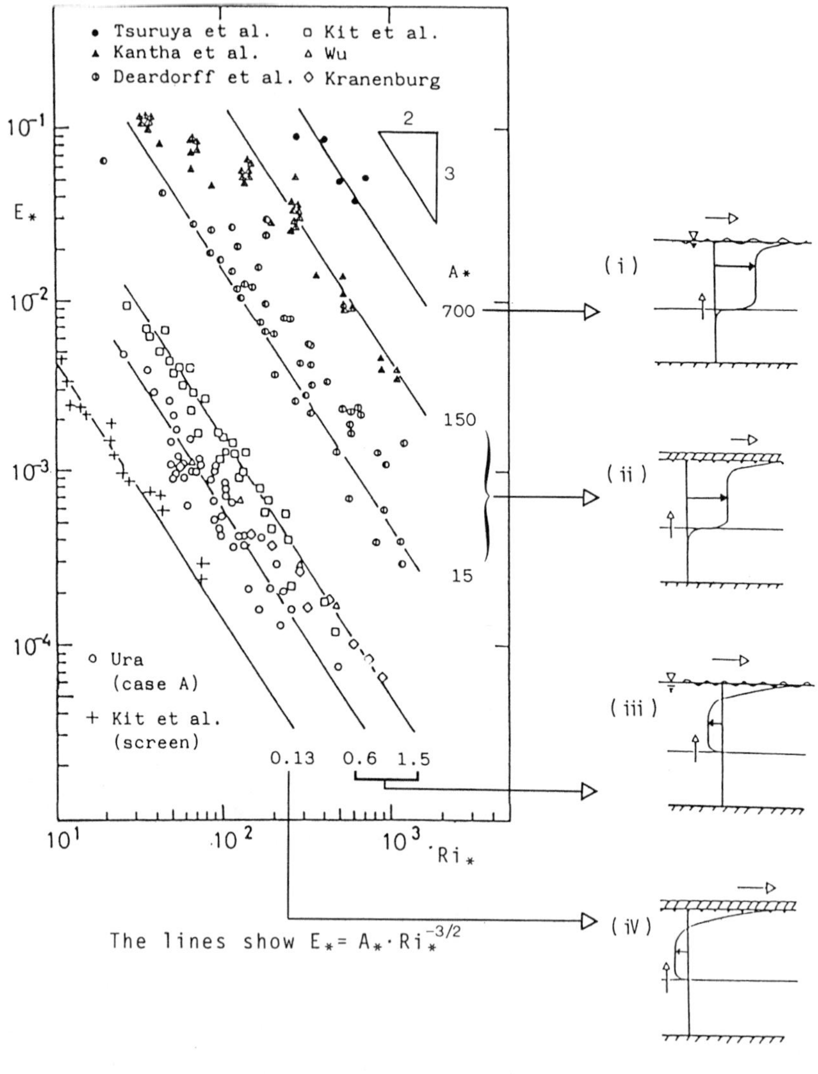

Figure 2. Ri_* - dependence of E_* .

interface. In fact, the turbulent entrainment coefficient E ($\equiv u_e/u$) is given by

$$E = A\ (\varepsilon gl/u^2)^{-3/2} \qquad (1)$$

for various types of two-layered stratified flows. If equation (1) is rewritten by using h and U_m, the overall entrainment coefficient E_m ($\equiv u_e/U_m$) becomes

$$E_m = (u_e/u)\cdot(u/U_m)$$
$$= A\ \left[(\varepsilon gl/u^2)\cdot(u/U_m)^2\cdot(h/l)\right]^{-3/2}\cdot(u/U_m)$$
$$= A\ (u/U_m)^4\cdot(l/h)^{-3/2}\cdot(\varepsilon gh/U_m^2)^{-3/2}$$
$$= A_m\cdot Ri_m^{-3/2} \quad , \qquad (2)$$

where

$$A_m = A\cdot(u/U_m)^4\cdot(l/h)^{-3/2} \quad \text{and} \quad Ri_m = \varepsilon gh/U_m^2 \ .$$

It is seen from equation (2) that the proportional factor A_m is determined by the characteristic quantities of turbulence at the interface. Here, let us call $(u/U_m)^4\cdot(l/h)^{-3/2}$ the 'turbulence factor', i.e.,

$$T_f = (u/U_m)^4\cdot(l/h)^{-3/2} \ . \qquad (3)$$

Equation (2) may also be applicable to general two-layered flows whose surface the shear stress does not act on.

The purpose of this paper is to explain consistently the entrainment coefficient of various types of two-layered flows on the basis of equation (2).

Experimental Methods

The wind-wave tank used in this experiment is schematically shown in Figure 3. It was 5.0m long, 0.2m wide and 0.59m high. The wind tunnel was 0.2m high with an approach section 1.0m long. A fetch of the wind on the water surface was 4.0m. The tank was filled with a two-layered fluid. The ordinate z is taken vertically downward from the mean water level.

A 4-pole conductivity probe has been used for the measurement of the fluid density. Density profiles were

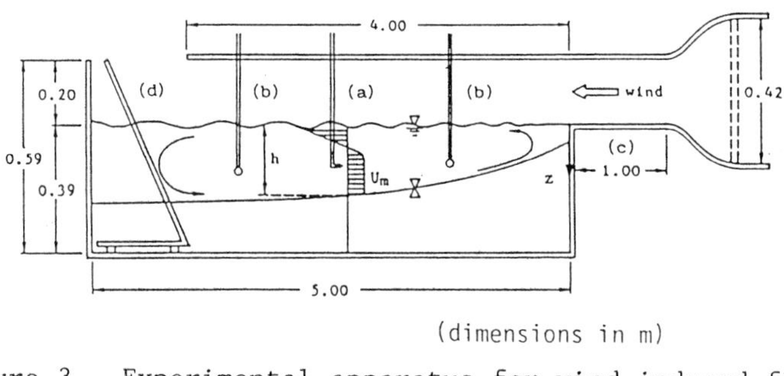

(dimensions in m)

Figure 3. Experimental apparatus for wind-induced flow. (a) Hot-film anemometer; (b) conductivity probes; (c) approach section; (d) wave absorber.

measured at the positions 1.42, 2.42 and 3.40m from the upstream end of the tank. The position of an interface is defined as the location where the mean density of fluid is equal to $(\rho_1+\rho_2)/2$, where ρ_1 and ρ_2 are the fluid densities of upper and lower layers, respectively. Entrainment velocity u_e has been estimated from time variations of the depth h from the water surface to the interface level (i.e., $u_e = \partial h/\partial t$).

The wind shear stress, which induces a drift current and compensating return flow as sketched in Figure 3, acts on the free surface by blowing the wind into the tank. The vertical velocity distributions of the wind-induced flow were obtained at the middle of the tank in the longitudinal direction by using a single-component hot-film anemometer. The friction velocity u_* was evaluated from both the log-law profile of mean velocity under the water surface and the relation

$$\rho_1 u_*^2 = \gamma \rho_a u_{*a}^2 \quad ,$$

where ρ_a is the density of air, u_{*a} is the air friction velocity and γ is a coefficient, which take a value between 1.0 to 0.9 (Ura, 1984). The intensity u and lengthscale l were calculated from the horizontal velocity fluctuation. The lengthscale was based on either the auto-correlation technique or the empirical formula proposed by Townsend (1976), i.e.,

$$l = 0.8\, u^3/\varepsilon_d \ .$$

Here, ε_d is the rate of energy dissipation calculated from the -5/3-power law of wavenumber spectrum.

Three kinds of experiments were conducted by changing the condition of water surface as shown in Figure 4.

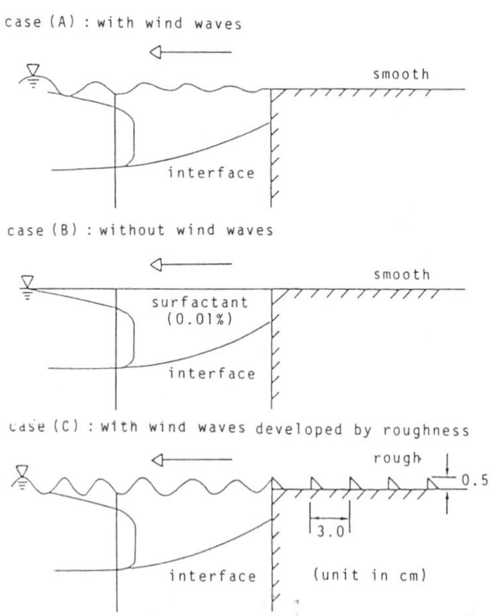

Figure 4. Experimental conditions of water surface.

Case (A): the two-layered flow with wind waves, which developed gradually along the fetch because of the smooth approach section (i.e.,case(iii)).

Case (B): the formation of wind waves was suppressed by adding surfactant(sodium dodecyle sulfate, concentration $1.1x10^{-2}$ %) in the water (similar to case(iv)).

Case (C): the wind waves formed at the beginning of the fetch because of the existence of roughness elements on the approach section (i.e., case (iii)).

Flow Properties

Figures 5 (a) to (c) show vertical distributions of mean velocity, turbulence intensity and integral lengthscale, respectively for cases (A), (B) and (C). In these figures z/h = 0 indicates water surface and z/h = 1.0 the

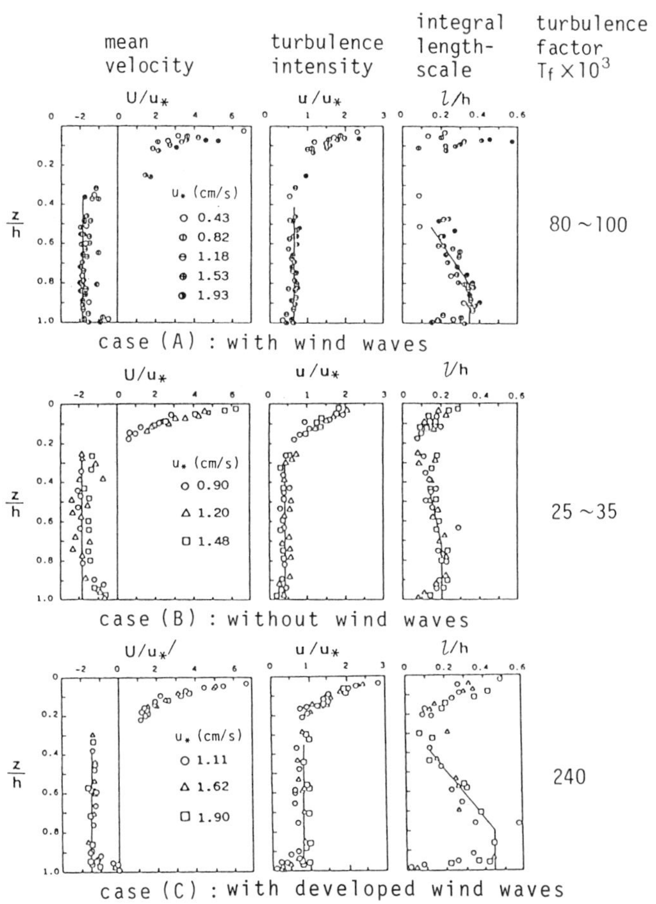

Figure 5. Vertical distributions of mean velocity, turbulence intensity and integral lengthscale.

interface. The values of these physical quantities at z/h = 0.9 are used as the representative values, because on the interface they are influenced significantly by the buoyancy effect. It is seen from the mean velocity profiles that a strong drift current occurs near the water surface by the wind, and that return flow with nearly uniform velocity forms in the range 1.0 > z/h > 0.25. Denoting this uniform velocity by U_m, U_m/u_* equals about 1.85 in cases (A) and (B) and 1.50 in case (C). The relative turbulence intensities u/u_* have almost constant values in the return flow region and the intensities are approximately given by

$u = 0.65u_* = 0.35U_m$ for case (A),
$u = 0.39u_* = 0.21U_m$ for case (B), and
$u = 0.82u_* = 0.52U_m$ for case (C).

The integral lengthscales near the interface are given by

$l \fallingdotseq 0.35h$ for case (A),
$l \fallingdotseq 0.20h$ for case (B), and
$l \fallingdotseq 0.45h$ for case (C).

From the results of these measurements, the values of the turbulence factor T_f are found to take 0.08 ~ 0.10, 0.025 ~ 0.035 and 0.24 in cases (A), (B) and (C), respectively. These results reveal that the condition of the water surface influences the value of T_f considerably.

Entrainment Coefficient

Figures 6 (a) and (b) are typical examples of the vertical density profiles, measured as the time proceeds, at the fetches F of 3.40 and 2.42m. The time-variations of h based on Figure 6 are shown in Figure 7. It is seen that the entrainment velocity $\partial h/\partial t$ is constant during the time of the measurement.

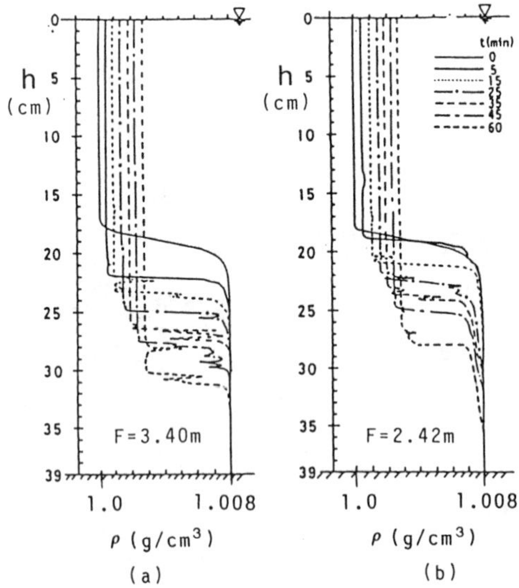

Figure 6. Time variations of vertical density profiles.

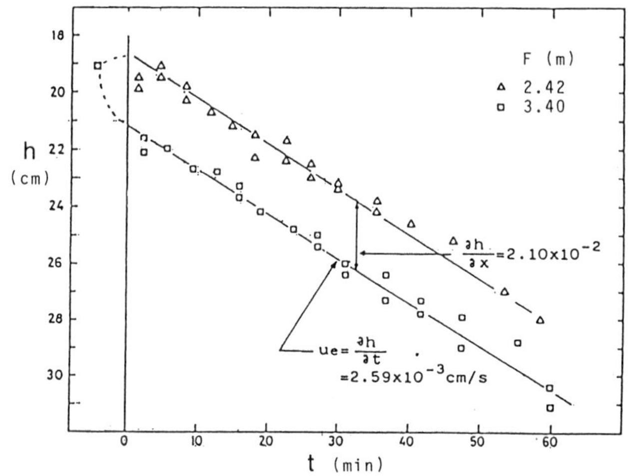

Figure 7. Time variations of the depth of the mixed layer.

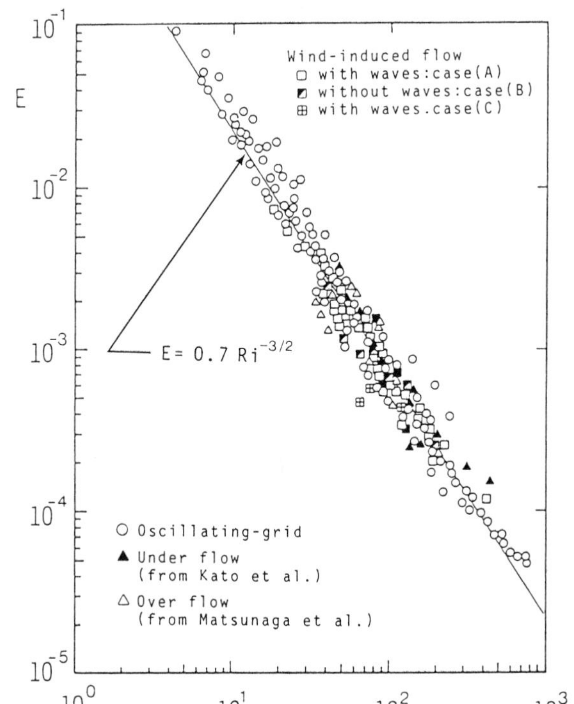

Figure 8. Ri - dependence of E.

In Figure 8, the turbulent entrainment coefficients $E = u_e/u$ are plotted against the local Richardson number $Ri = \varepsilon gl/u^2$ for the three cases shown in Figure 4. The experimental results obtained from the oscillating grid turbulence (Ura et al., 1987a) and the over flow (Matsunaga et al., 1984) are also plotted in this figure. We confirm the unified expression of E based on Ri (Ura et al., 1987b), i.e.,

$$E = 0.7 \cdot Ri^{-3/2} \tag{4}$$

for various types of two-layered flows.

Figure 9 shows the relationships between bulk parameters E_m and Ri_m, which are obtained by many other researchers and authors. The values of T_f are also represented, being calculated on the basis of the data by the authors and Kato

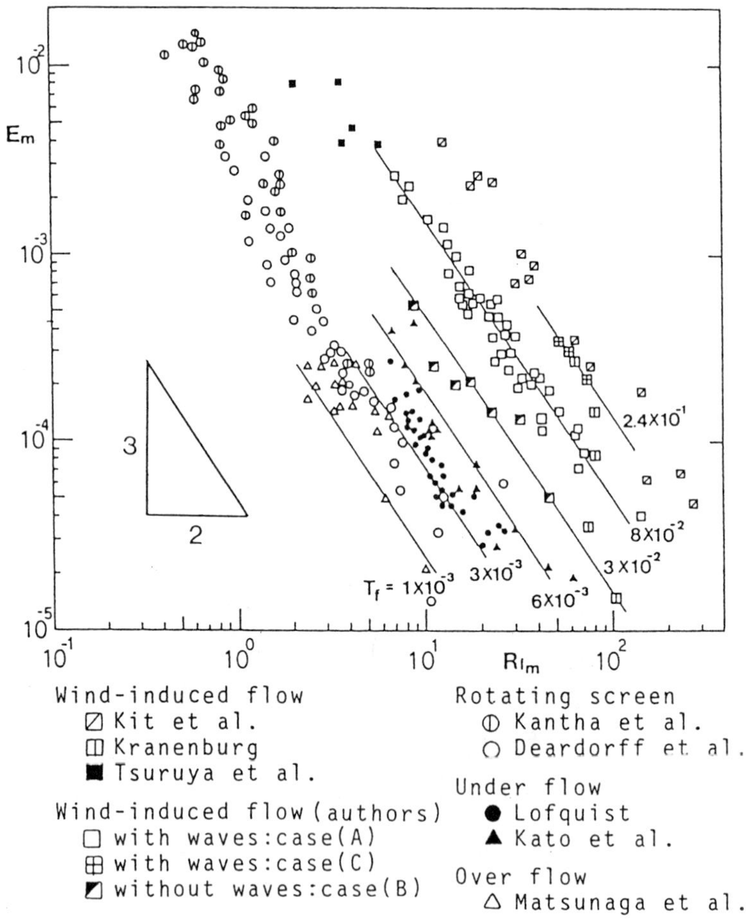

Figure 9. Relationships between E_m and Ri_m obtained by considering turbulence factor T_f.

et al. (1981). The values of T_f by the other researchers are unknown since they did not report the turbulence quantities. From this figure, however, we find that E_m is proportional to $Ri_m^{-3/2}$ and the proportional factor A_m consistently depends on T_f.

In order to confirm the validity of equation (2), it is necessary to show that A_m is proportional to T_f. Figure 10 shows the relationship between A_m and T_f. A_m is clearly proportional to T_f and the proportional constant takes 0.7. Therefore, we conclude that the entrainment relationship

$$E_m = A_m \cdot Ri_m^{-3/2} \quad ,$$

$$A_m = 0.7 \cdot T_f$$

holds for the various types of two-layered stratified flows.

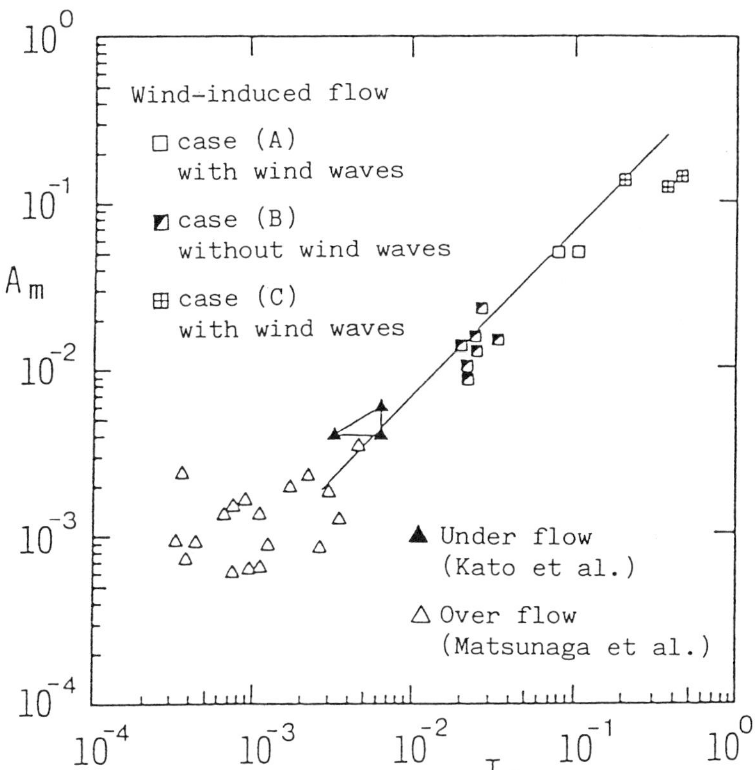

Figure 10. Relationships between A_m and T_f.

Acknowledgment

The authors thank Professors T. Tsubaki and T. Komatsu for stimulating discussions and Mr T. Awaya for technical help.

References

Deardorff, J. W. and Willis, G. E. (1982). "Dependence of mixed-layer entrainment on shear stress and velocity jump." J. Fluid Mech., 115, 123-149.
Kantha, L. H., Phillips, O. M. and Azad, R. S. (1975). "On the turbulent entrainment at a stable density interface." J. Fluid Mech., 79, 753-768.
Kato, H. and Ikeda, T. (1981). "An experiment of lower density current(1)." The 28th Japanese Conf. Coastal Eng., 520-524,(in Japanese).
Kit, E., Berent, E. and Vajda, M. (1980). "Vertical mixing induced by wind and rotating screen in a channel." J. Hydr. Res., 18(1), 35-58.
Kranenburg, C. (1985). "Mixed-layer deepening in lakes after wind setup." J. Hydr. Div., ASCE, 111(HY9), 1279-1297.
Lofquist, K. (1960). "Flow and stress near an interface between stratified liquids." Phys. Fluids, 3, 158-175.
Matsunaga, N., Ura, M., Namikawa, T. and Tsubaki, T.(1984). "Flow properties near an interface in upper-layer flow." Proc. 28th Japanese Conf. Hydr., 345-351(in Japanese).
Thompson, S. M. and Turner, J. S. (1975). "Mixing across an interface due to turbulence generated by an oscillating grid." J. Fluid Mech., 67, 349-368.
Townsend, A., A. (1976). The structure of turbulence shear flow (2nd ed.), Cambridge University Press. Cambridge, London.
Tsuruya, H., Nakano, S. and Ichinohe, H. (1984). "Vertical mixing and turbulent characteristics of wind-driven two-layered flows." Proc. 31st Japanese Conf. Coastal Eng. 655-659, (in Japanese).
Ura, M. (1984). "Interfacial waves and entrainment in a wind-induced stratified flow." Proc. 4th Cong. APD- IAHR, 403-417.
Ura, M., Komatsu, T. and Matsunaga, N. (1987a). "Entrainment due to oscillating grid turbulence in two-layered fluid." Turbulence Measurements and Flow Modeling, Chen, C. J., Chen, L. D. and Holly, F. M. Ed. Hemisphere Publishing Corporation, New York, N.Y., 109-118.
Ura, M., Tsubaki, T., Matsunaga, N. and Namikawa, T. (1987b)."A turbulence-based expression for the entrainment coefficient for two-layered stratified flows." J. Hydroscience and Hydr. Eng., 5(1),22-37.
Wu, J. (1973). "Wind-induced turbulent entrainment across a stable density interface." J. Fluid Mech., 61, 275-287.

CHAPTER 190

BOUNDARY CONDITIONS IN FINITE ELEMENT MODELING OF STRATIFIED COASTAL CIRCULATION

by George C. Christodoulou[1], M.ASCE
and George D. Economou[2]

ABSTRACT

The effect of boundary conditions on numerical computations of stratified flow in coastal waters is examined. Clamped, free radiation and sponge layer conditions are implemented in a two-layer finite element model and the results of simple tests in a two-layer stratified basin are presented.

INTRODUCTION

Numerical models of coastal circulation require the specification of boundary conditions at the sea boundaries, where the computational domain is artificially terminated. In 2-D depth-averaged models conditions are normally imposed on the vertical displacement of the free surface at the grid nodes located on the boundary. Most commonly, the surface displacements are prescribed in a desired manner, according to either a known tidal input or to an elevation time series available from field data. This type of b.c. is referred to as clamped, since the position of the surface at the boundary is absolutely defined a priori for the entire computation. Other types of b.c. often employed include the Sommerfeld free radiation conditions in various forms, which allow disturbances created within the domain to travel freely outwards, or sponging functions, which serve as filters absorbing outgoing waves. The importance of boundary conditions has been recognized in recent years and the development of improved formulations as well as their implementation in discretized form has been the subject of several investigations (Orlanski, 1976; Enquist and Majda, 1977; Miller and Thorpe, 1981; Roed and Cooper, 1986). Evidence from recent research (Blumberg and Kantha, 1985; Roed and Cooper, 1986) indicates

1 Assoc. Prof., Dept. Civil Engrg., Natl. Tech. Univ. of Athens, Greece.
2 Grad. Stud., Dept. Civil Engrg., Natl. Tech. Univ. of Athens, Greece.

that the type of boundary conditions used may have an appreciable influence on the computed circulation patterns, especially for transient problems.

During the summer, many coastal water bodies become stratified, primarily due to substantial temperature differences over the water column. Stratification may also develop in estuaries receiving large freshwater inflows, due to salinity differences. The simplest way to study such water bodies is by a two-layer idealization. However, in a two-layer system the problem of boundary conditions becomes more complicated, since conditions have to be specified not only for the free surface, but also for the interface, where data are usually lacking. The effect of the treatment of b.c. on the resulting circulation of layered systems has received little attention (Lorrenzzetti et al, 1985), although it may be at least as important as in the 2-D case.

In an effort to assess the importance of boundary conditions in hydrodynamic modeling of stratified coastal waters, the present paper examines the influence of some common types of b.c. on the wind-driven circulation of a two-layer stratified basin. In the following, the implementation of the various boundary conditions and the results obtained for some simple test problems are presented and discussed.

GOVERNING EQUATIONS AND NUMERICAL MODELING

For the purposes of this study an available two-layer finite element circulation model, code named CAFE-2 (Wang and Connor, 1975; Pagenkopf et al, 1976) was employed as a tool, after some suitable modifications for handling the various types of b.c. considered. It is felt, however, that the results obtained would be equally applicable, at least qualitatively, to other numerical models as well. The model used solves the following set of governing equations for the two-layer system, where the subscript 1 denotes the bottom layer and the subscript 2 the top layer (see Fig. 1):

$$\frac{\partial H_1}{\partial t} + \frac{\partial q_{1x}}{\partial x} + \frac{\partial q_{1y}}{\partial y} = q_{01} \qquad (1)$$

$$\frac{\partial q_{1x}}{\partial t} + \frac{\partial}{\partial x}(\bar{u}_1 q_{1x}) + \frac{\partial}{\partial x}(\bar{u}_1 q_{1y}) - f q_{1y} = -\frac{\partial F_{1p}}{\partial x} + \frac{\partial F_{1xx}}{\partial x} + \frac{\partial F_{1xy}}{\partial y} + \frac{1}{\rho_1}(\tau_x{}^i - \tau_x{}^b) + \frac{1}{\rho_1}\left(p^i \frac{\partial \zeta_1}{\partial x} - p^b \frac{\partial \zeta_o}{\partial x}\right) \qquad (2)$$

$$\frac{\partial q_{1y}}{\partial t} + \frac{\partial}{\partial x}(\bar{u}_1 q_{1y}) + \frac{\partial}{\partial y}(\bar{v}_1 q_{1y}) + f q_{1x} = -\frac{\partial F_{1p}}{\partial y} + \frac{\partial F_{1xy}}{\partial x} + \frac{\partial F_{1yy}}{\partial y} + \frac{1}{\rho_1}(\tau_y{}^i - \tau_y{}^b) + \frac{1}{\rho_1}\left(p^i \frac{\partial \zeta_1}{\partial y} - p^b \frac{\partial \zeta_o}{\partial y}\right) \qquad (3)$$

$$\frac{\partial H_2}{\partial t} + \frac{\partial q_{2x}}{\partial x} + \frac{\partial q_{2y}}{\partial y} = q_{02} \tag{4}$$

$$\frac{\partial q_{2x}}{\partial t} + \frac{\partial}{\partial x}(\bar{u}_2 q_{2x}) + \frac{\partial}{\partial y}(\bar{v}_2 q_{2x}) - f q_{2y} = -\frac{\partial F_{2p}}{\partial x} + \frac{\partial F_{2xx}}{\partial x} + \frac{\partial F_{2xy}}{\partial y} + \frac{1}{\rho_2}(\tau_x{}^s - \tau_x{}^i) + \frac{1}{\rho_2}(p^s \frac{\partial \zeta_2}{\partial x} - p^i \frac{\partial \zeta_1}{\partial x}) \tag{5}$$

$$\frac{\partial q_{2y}}{\partial t} + \frac{\partial}{\partial y}(\bar{u}_2 q_{2y}) + \frac{\partial}{\partial y}(\bar{v}_2 q_{2y}) + f q_{2x} = -\frac{\partial F_{2p}}{\partial y} + \frac{\partial F_{2xy}}{\partial x} + \frac{\partial F_{2yy}}{\partial y} + \frac{1}{\rho_2}(\tau_y{}^s - \tau_y{}^i) + \frac{1}{\rho_2}(p^s \frac{\partial \zeta_2}{\partial y} - p^i \frac{\partial \zeta_1}{\partial y}) \tag{6}$$

where $\zeta_o, \zeta_1, \zeta_2$ are the elevations of the bottom, interface and free surface from an arbitrary datum, respectively.

H_1 = $\zeta_1 - \zeta_o$ = bottom layer thickness

H_2 = $\zeta_2 - \zeta_1$ = top layer thickness

ρ_1, ρ_2 are the layer densities.

$\bar{u}_i, \bar{v}_i$ are the x,y components of the mean layer velocity (i=1,2)

q_{xi}, q_{yi} are the x,y fluxes ($q_x = \bar{u}H$, $q_y = \bar{v}H$)

q_{oi} is any external input, eg. rainfall

f = $2\omega \sin\varphi$, is the Coriolis parameter

p^s, p^i, p^b denote the pressure at the free surface, interface and bottom, respectively.

τ^s, τ^i, τ^b denote the shear stress at the free surface, interface and bottom, respectively, which are modeled by quadratic laws.

F_{pi} is the total pressure force over the layer thickness

$F_{xxi}, F_{xyi}, F_{yyi}$ are the integrated internal stresses due to molecular viscosity, turbulent velocity fluctuations and spatial velocity nonuniformity over the vertical, which are modeled by the eddy viscosity approximation.

By employing linear triangular elements for the spatial discretization, the above partial differential equations are reduced to a system of ordinary differential equations, which is further discretized in time as follows (i=1,2):

$$\underset{\sim}{M} \frac{\underset{\sim}{H_i}^{n+\frac{1}{2}} - \underset{\sim}{H_i}^{n-\frac{1}{2}}}{\Delta t} = \underset{\sim}{P_i} \, (H^{n-\frac{1}{2}}, Q^n) \tag{7}$$

$$\underset{\sim}{M}' \frac{\underset{\sim}{Q}_i^{n+1} - \underset{\sim}{Q}_i^{n}}{\Delta t} = \underset{\sim}{P}_i'(H^{n+\frac{1}{2}}, Q^n) \tag{8}$$

where $\underset{\sim}{H}_i$ are the vectors of layer thickness at the nodes

$\underset{\sim}{Q}_i(\underset{\sim}{q}_{xi}, \underset{\sim}{q}_{yi})$ are the vectors of fluxes at the nodes

$\underset{\sim}{M}, \underset{\sim}{M}'$ are matrices related to the geometry of the grid

and $\underset{\sim}{P}, \underset{\sim}{P}'$ include all other terms, i.e. pressure gradient, friction, eddy viscosity, Coriolis, convective terms etc.

The time-stepping scheme employed solves for the dependent variables at alternate time levels, i.e. for the layer thicknesses $\underset{\sim}{H}_1, \underset{\sim}{H}_2$ at times $n+1/2$, $n+3/2$ and for the fluxes $\underset{\sim}{Q}_1, \underset{\sim}{Q}_2$ at times n, $n+1$, $n+2$ etc.

IMPLEMENTATION OF BOUNDARY CONDITIONS

For the numerical tests an idealized basin was considered, shown in Fig. 2. It has a rectangular shape, with dimensions about 40x60 Km, and an island near the center. The water depth was assumed constant, at 100 m, and the initial position of the interface was set at a depth of 35 m. The density difference between the layers was assumed to be 2‰. As shown in the figure, a regular grid with 86 nodes and 129 triangular elements was established. Three types of boundary conditions were considered for the wind-driven circulation, as follows:

(a) Clamped boundaries: Zero displacement was specified for both the free surface and the interface at all nodes on the sea boundary, so that the layer thicknesses were fixed:

$$H_1 = \bar{H}_1 = 65\text{m} \tag{9a}$$

$$H_2 = \bar{H}_2 = 35 \text{ m} \tag{9b}$$

(b) Free radiation: Extending the well-known Sommerfeld radiation condition (Roed and Cooper, 1986) and considering as celerity that of long gravity waves, the following equations are applicable for the displacements of the free surface and the interface at the boundary nodes:

$$\frac{\partial \zeta_2}{\partial t} + \sqrt{g(H_1+H_2)}\ \frac{\partial \zeta_2}{\partial \eta} = 0 \tag{10a}$$

$$\frac{\partial \zeta_1}{\partial t} + \sqrt{g' \frac{H_1 H_2}{(H_1+H_2)}}\ \frac{\partial \zeta_1}{\partial \eta} = 0 \tag{10b}$$

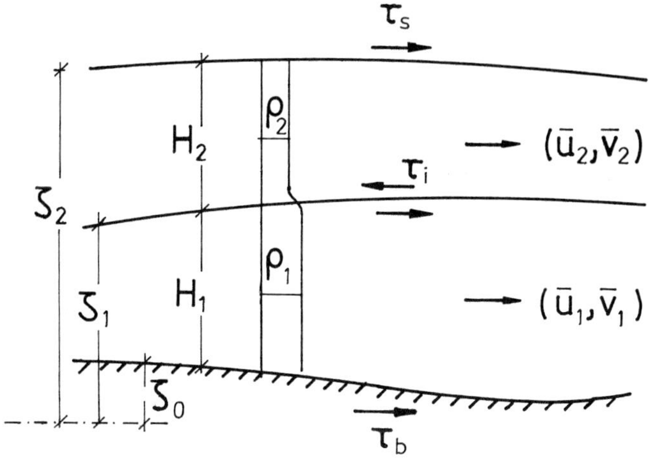

Figure 1. Notation sketch

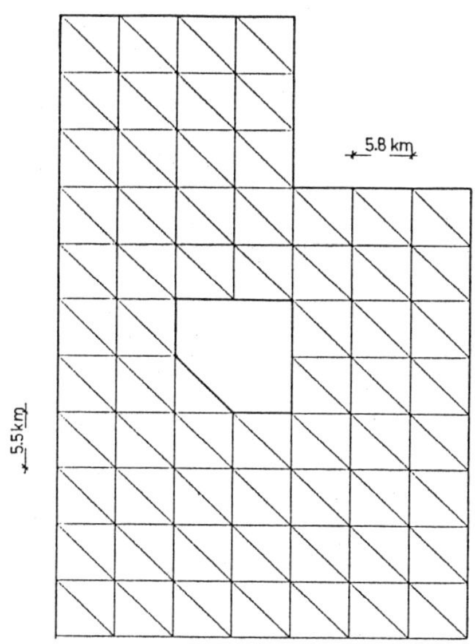

Figure 2. Idealized basin for the numerical tests

where $g' = g(\rho_1-\rho_2)/\rho_1$

and η is the direction of the outward normal to the boundary.

To obtain the boundary values at each time step, the above equations were discretized by employing forward differences in time and backward in space, as follows.

$$\frac{\zeta_{2,b}^{n+1} - \zeta_{2,b}^{n}}{\Delta t} + \sqrt{g(H_{1,b}^{n} + H_{2,b}^{n})}\ \frac{\zeta_{2,b}^{n} - \zeta_{2,a}^{n}}{\Delta\eta} = 0 \qquad (11a)$$

$$\frac{\zeta_{1,b}^{n+1} - \zeta_{1,b}^{n}}{\Delta t} + \sqrt{g'\frac{H_{1,b}^{n}\, H_{2,b}^{n}}{(H_{1,b}^{n} + H_{2,b}^{n})}}\ \frac{\zeta_{1,b}^{n} - \zeta_{1,a}^{n}}{\Delta\eta} = 0 \qquad (11b)$$

where the subscript b denotes the boundary node, the subscript a refers to the adjacent node to the interior of the domain, and $\Delta\eta$ is the distance between these two nodes.

(c) Sponge layer: The displacements of the boundary nodes were set to zero, as in case (a), but the elevation fields were sponged within a layer adjacent to the boundary, following the scheme proposed by (Lorrenzetti et al, 1985). To accommodate the sponge layer of a chosen thickness $L=2\Delta s$ the grid was locally refined in this case, so that four element were formed in this zone. The integration scheme (eq.7) was modified as follows:

$$\underset{\sim}{M}\ \frac{\underset{\sim}{H}_i^{n+\frac{1}{2}} - \underset{\sim}{H}_i^{n-\frac{1}{2}}}{\Delta t} + \underset{\sim}{K}\ \frac{\underset{\sim}{H}_i^{n-\frac{1}{2}} - \underset{\sim}{\bar{H}}_i^{n-\frac{1}{2}}}{\Delta t} = \underset{\sim}{P}_i \qquad (12)$$

where $\underset{\sim}{\bar{H}}$ includes the specified depths at the boundary

and $\underset{\sim}{K}$ is a damping function depending on the ordinate η normal to the boundary:

$$K(\eta) = A\left(\frac{\eta-\eta_o}{L-\eta_o}\right)^x \qquad \text{for} \qquad \eta_o < \eta \leqslant L \qquad (13a)$$

$$K(\eta) = 0 \qquad \text{for} \qquad \eta \leqslant \eta_o \qquad (13b)$$

The behaviour of the sponge layer depends to some extent on the values of the parameters A, x, in eq. (13a). The results shown below correspond to values of $A=1.0$ and $x=0.5$.

RESULTS AND DISCUSSION

The b.c. described above were used for studying the circulation in the rectangular basin of Fig. 2 under uniform wind forcing. For these tests, the non-linear convective terms and the Coriolis terms were omitted in the computations. Preliminary experiments indicated that considerable oscillations of the interface persist for several days after a sudden application of the wind field. Therefore to reduce the impact of sudden forcing, the wind was introduced gradually, using a linear increase from zero to 20 knots over a period of 90,000 sec; the wind speed was kept constant thereafter.

Figures 3, 4, 5 show the circulation patterns calculated at 360,000 sec for a 20-knot wind normal to the open boundary, directed inwards, for each of the three types of b.c. described previously. It is seen that in all cases an inflow is established in the top layer and outflow in the bottom layer, as expected for this particular wind direction. However, there are significant differences between the three cases. For the clamped boundaries, an anticlockwise gyre around the island forms in the top layer, with stronger velocities on the right side of the basin, and a reverse gyre in the bottom layer. Somewhat similar results are observed for the sponge layer conditions, but with comparable intensity of flow on both sides of the basin. However, the free radiation conditions generate a markedly different flow field, with nearly uniform inflow over the entire top layer and outflow at the bottom layer. Substantial differences are also noted in the displacement of the interface, which under the free radiation conditions descends much lower throughout the domain compared to the other two tests.

Fig. 6 presents the evolution of the differential volume of each layer relative to the initial position of the free surface and the interface. It is evident from the figure that under clamped or sponge boundary conditions, the resulting flow fields attain macroscopically nearly the same steady state after about 2 days of wind forcing, while no steady state is approached after 4 days for the free radiation conditions. In fact, there seems to be a gradual filling of the top layer and a corresponding emptying of the bottom layer. This result is qualitatively similar to that of Blumberg and Kantha (1985), who also observed in their 2-D model a continuous emptying of the domain when using free radiation conditions.

Figures 7, 8, 9 show the circulation pattern for a 20-knot wind parallel to the open boundary. Under clamped b.c., more intense currents are observed in this case. The flow generally tends to follow the wind direction on both sides of the island and in both layers, and outflow occurs on the opposite end of the sea boundary. Rather similar results are obtained under the sponge conditions. But again, the free radiation conditions yield a quite different flow pattern, where there is inflow along the entire opening in the top layer and outflow in the bottom layer. The behaviour of the differential volumes in this case is shown in Fig. 10.

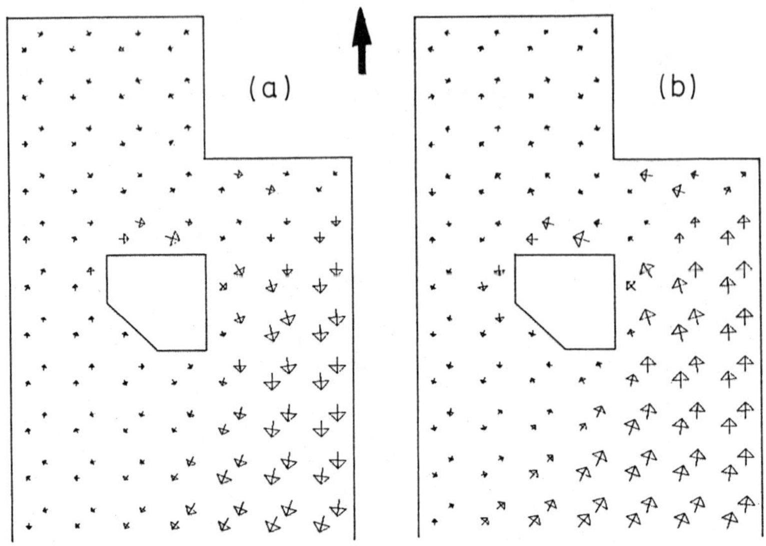

Figure 3. Flow field for wind normal to the open boundary, with clamped conditions: (a) Bottom, (b) Top layer

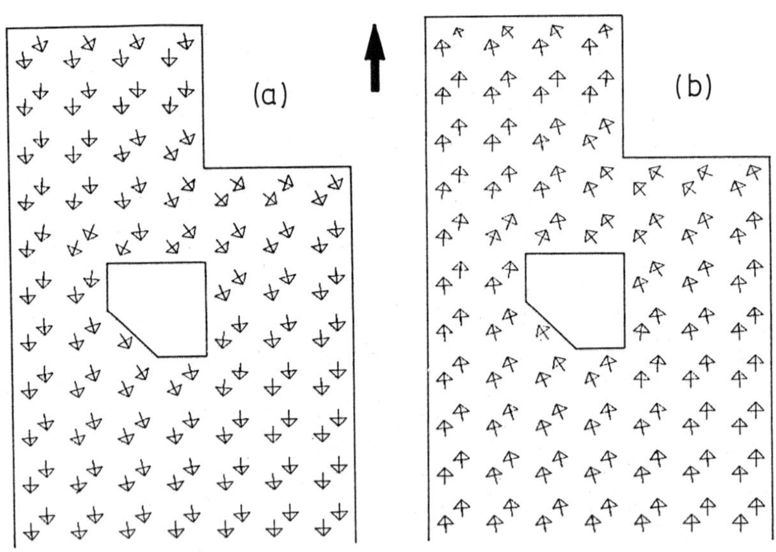

Figure 4. Flow field for wind normal to the open boundary, with free radiation conditions: (a) Bottom, (b) Top layer

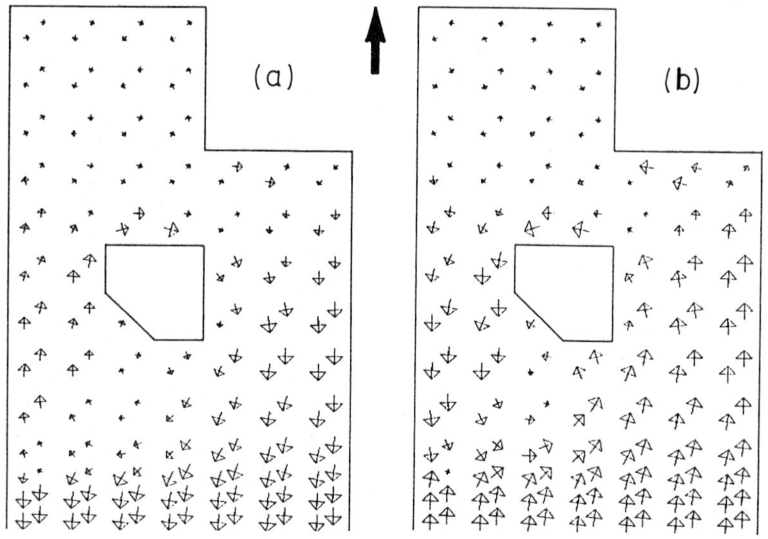

Figure 5. Flow field for wind normal to the open boundary, with sponge layer conditions: (a) Bottom, (b) Top layer

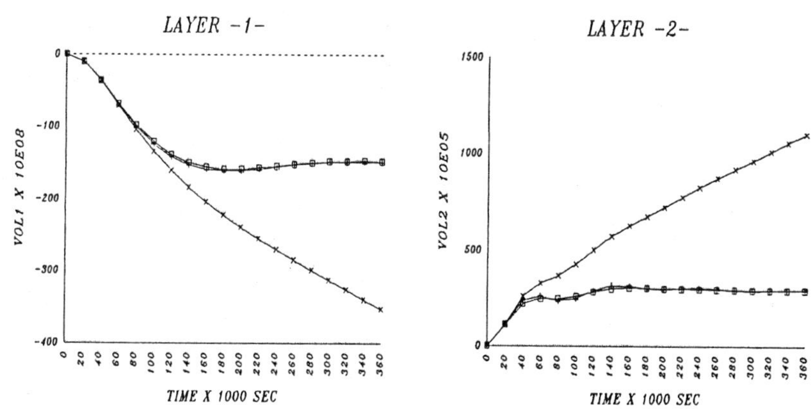

Figure 6. Time change of layer volumes for wind normal to the open boundary; + clamped conditions, x free radiation, □ sponge layer.

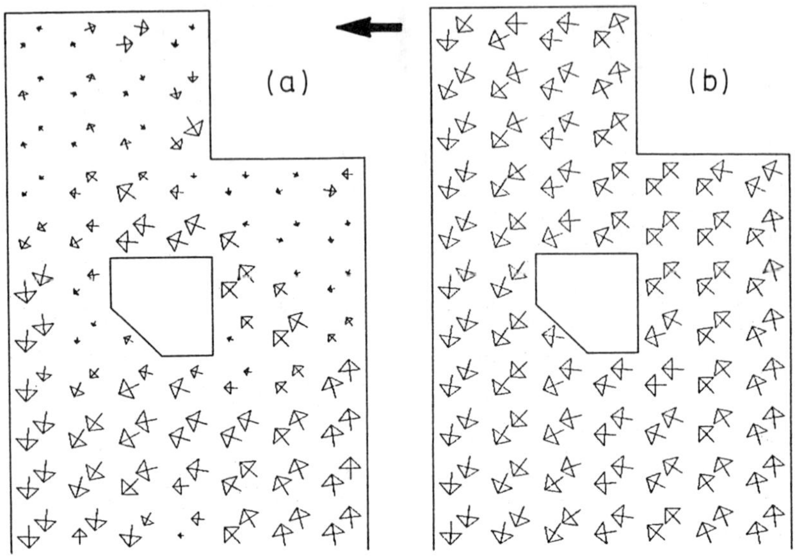

Figure 7. Flow field for wind parallel to the open boundary, with clamped conditions: (a) Bottom, (b) Top layer

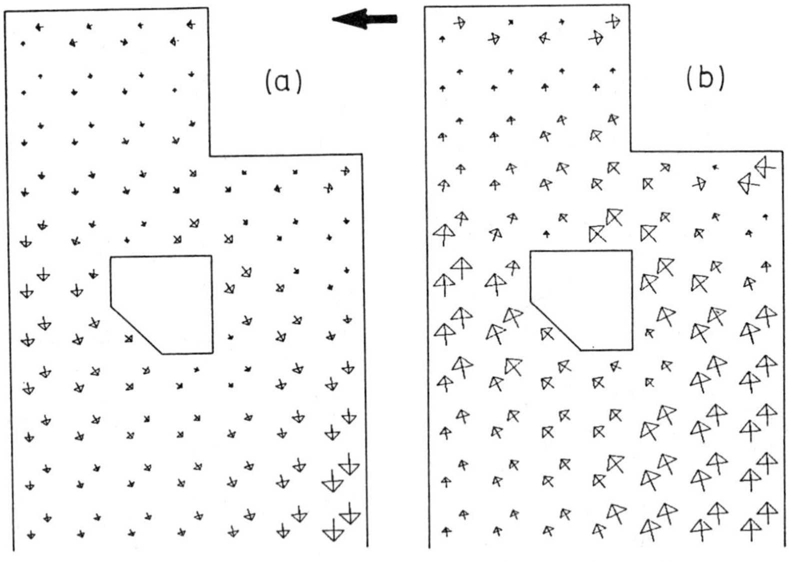

Figure 8. Flow field for wind parallel to the open boundary, with free radiation conditions: (a) Bottom, (b) Top layer

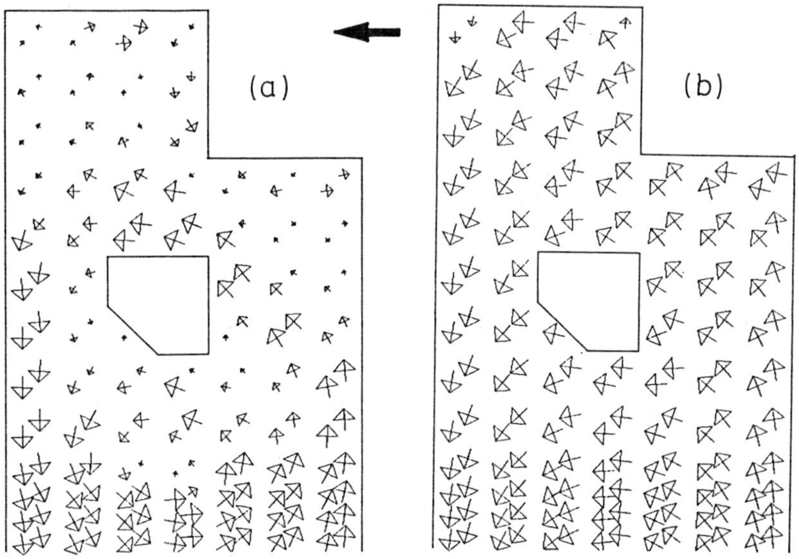

Figure 9. Flow field for wind parallel to the open boundary, with sponge layer conditions. (a) Bottom, (b) Top layer

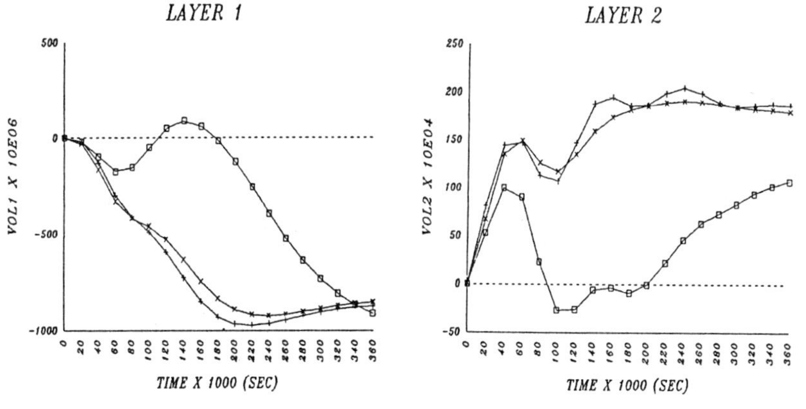

Figure 10. Time change of layer volumes for wind parallel to the open boundary: + clamped conditions, □ free radiation, x sponge layer

Substantial oscillations are observed for both layers, which under clamped conditions persist for almost 4 days, while stabilization occurs much earlier for the sponge conditions. Still, the results of these two types of b.c. are close to each other, both during transient and near steady state. On the contrary, the free radiation conditions yield much different global features and, after about 2 days of oscillatory transients, seem to establish again a progressive emptying of the lower layer and filling of the top layer.

It is evident from the above that the way of forcing has also some effect on the response of the two-layer system. Most boundary conditions are ideally designed to handle disturbances travelling normal to the opening. Thus, the enhanced oscillations observed in the second test are conceivably due to the wind direction being parallel to the boundary, indicating the difficulty of the system to accommodate the oblique incidence of waves.

The degree to which the above observations depend on the morphological features of the basin is currently being studied by additional tests made on a natural water body, namely the Saronic Gulf in Greece, which has dimensions of the same order of magnitude as those of the idealized basin but a variable bathymetry and complex geometry. Preliminary results obtained with the Coriolis terms included show also major differences for the wind-driven circulation under the three types of b.c., indicating qualitatively the same behaviour as above.

Therefore, it may be concluded that the clamped and sponge conditions give approximately the same results concerning the wind-driven circulation of a two-layer system. The former is simplest to implement, but the latter has the advantage of faster damping of the initial transient oscillations, so that the computation time to steady state is reduced. The free radiation conditions could be more appropriate for transient problems since they allow interior disturbances to exit, however they should not be used when a steady state solution is sought. Still, the computation of steady state might often be of academic rather than of practical interest. Taking into account the slow time response of a large stratified basin, it is practically impossible for a meteorological forcing to prevail for the duration necessary to create a steady flow field. In simulating events of finite duration, a combination of free radiation and clamped conditions of the form proposed in 2-D problems by Blumberg and Kantha (1985) may prove successful for stratified problems as well. In any case, the observed substantial influence of the b.c. on the computed flow patterns points out the need for further research into the subject.

CONCLUSIONS

Three different types of boundary conditions that may be specified along the sea boundaries of numerical models for stratified coastal circulation were examined, namely clamped, free radiation and sponge layer conditions. These boundary conditions were implemented in an idealized two-layer basin and numerical experiments for wind forcing were carried out. It was found that the b.c. generally have a very pronounced effect on the flow field, not only close to the boundary, but essentially over the entire domain. Results obtained for clamped and sponge boundaries are to some extent similar, but free radiation conditions give very different patterns. Besides, in the latter case steady state is not reached for a long time, even after gradual introduction of the wind forcing.

ACKNOWLEDGMENT

Partial financial support by the Ministry of Environment, Planning and Public Works is gratefully acknowledged.

REFERENCES

Blumberg, A.F., and Kantha, L.H. (1985), "Open boundary conditions for circulation models", J. Hydr. Engrg., ASCE, 111, (2), 237-255.

Enquist, B. and Majda, A., (1977), "Absorbing boundary conditions for the numerical simulation of waves", Math. Comp., 31, 629-651.

Lorrenzzetti, J.A., Wang, J.D., Lee, T.N., and Pietrafesa, L.J. (1985), "On the modeling of the summer circulation in the South Atlantic Bight by a two-layer finite element model", RSMAS Report No. 85002, Univ. Miami.

Miller, M.J. and Thorpe, A.J. (1981), "Radiation conditions for the lateral boundaries of limited-area numerical models", Quart. J.R. Met. Soc., 107, 615-628.

Orlanski, I. (1976), "A simple boundary condition for unbounded hyperbolic flows", J. Comp. Phys., 21, 251-269.

Pagenkopf, J.R., Christodoulou, G.C., Pearce, B.R. and Connor, J.J. (1976), "A user's manual for CAFE-2, a two-layer finite element circulation model", R.M. Parsons Lab. Report No.220, Mass. Inst. of Tech., Camgridge, Mass.

Roed, L.P. and Cooper, C.K. (1986), "Open boundary conditions in numerical ocean models", Adv. Phys. Ocean. Modeling, J.J. O' Brien (ed), Reidel.

Wang, J.D. and Connor, J.J. (1975), "Mathematical modeling of near coastal circulation", R.M. Parsons Lab. Report No. 200, Mass. Inst. of Tech., Cambrdige, Mass.

CHAPTER 191

A MODELING SYSTEM FOR COASTAL OIL SPILL RISK ANALYSIS

Shiao–Kung Liu and Jan J. Leendertse

INTRODUCTION

At present, the Prodhoe Bay oil field in Alaska contributes a substantial amount of the domestic oil production of the United States. Oil is also expected to be present on the continental shelf of Alaska, and it is estimated that approximatedly 28 percent of the total U. S. reserve is located beneath the shallow ice covered seas of the Alaskan continental shelf. To expolre and to exploit these oil rich resources, engineers are confronted with hostile oceanographic conditions such as high tides, waves, strong currents and sea ice. The same area is also rich in fishery resources. Being one of the most productive fishing grounds in the nothern Pacific, the potential ecological impact due to an oil spill is of a major concern. This paper describes the methologies used for the development of a modeling system for the oil risk analysis. The system is designed with generality in mind so it can be used for other coastal areas.

The development of three dimensional models used in the modeling system described here have been published in the earlier International Coastal Engineering Conferences (Liu and Leendertse, 1982, 1984, 1986) and a report published recently by RAND (Liu and Leendertse, 1987). In the oil–spill risk analysis, these three dimensional hydrodynamic models are coupled to a two–dimensional stochastic weather model and an oil weathering model.

Before decisions are made concerning which specific offshore areas to lease for exploration or exploitation, the reponsible government agency must balance orderly resources development against the protection of human, marine, and coastal environment, to ensure that the public receives a fair return for these resources. In studies made for this purpose, the impact of hypothetical oil spill are considered. Not only are oil spill pathways required for the impact analysis, in certain instances knowledge about the extent of oil spill is required as well as about the oil concentration that would occur in the water column. This paper presents an efficient way for making the risk analysis while maintaining the essential dynamics of the processes involved.

The RAND Corp., Santa Monica, CA. 90406, USA

THE MODELING APPROACH FOR LONG DURATION WIND DRIVEN CURRENTS

As the three-dimensional models made of the different offshore areas of Alaska simulate the movements of water, and as a model is available to simulate wind sequences offshore, it would be logical to use these models in the computation of oil spill movements. To use these models effectively, we have developed a method to compute wind driven currents. This method retains the dynamic detail of the three-dimensional model and yet is approximately two orders of magnitude more efficient than the simulations with the three-dimensional model; it is called the "wind-driven response function method." In essence, the method extends the basic idea of the "drift ratio" between the wind speed and current speed except that the ratio changes in time and over space and is derived from the three-dimensional model.

The traditional, simple fixed drift-ratio method has many difficulties when applied in the Alaskan coastal waters. It is applicable only for cases of steady wind with constant speed blows over water with finite depth and with no boundaries. However, the concept of the "drift ratio" is a good one—but we need to include more dynamics in it.

In examining the fundamental dynamics of wind-driven currents, even under the assumption of steady (in time), constant (over space) wind and an infinitely long straight coastline, wind-driven currents over water of finite depth do vary both in direction and speed at the surface and at different levels to satisfy the law of conservation of mass. Using information on the distance from shore, wind direction, and local depth, Ekman (1905) worked out the variabilities of drift currents by using highly simplified terms in the equations of motion. On the other hand, to include more terms would require the solution of the complete three-dimensional model.

In our study, time-varying response functions under various wind conditions were developed using wind stress associated with the marine wind speed. Reverse procedures (convolution) were then used during the oil trajectory simulation; therefore, they are not linear with respect to local wind speed. Since the response functions for all layers are derived from the three-dimensional model, time-varying effects (such as a moving storm, deepening of a mixed layer, and inertia components) are included in the oil spill trajectory computation. The method is very efficient, however, the oil spill trajectory model was programmed so that the drift ratio and deflection angle from field observations under various conditions over an entire area can, as an option, still be used for the computation of oil movement.

Wind driven currents over stratified waters vary with the degree of vertical stability associated with the stratification. To illustrate this point we use a simple case where the time series of water movement at two nearby locations in Norton Sound is plotted (Fig. 1), and wind from the east is applied for a duration of 12 hours (close to the inertial period) over the water. The response of surface water at two nearby locations is not the same to satisfy the continuity principle of water within a bay. Response functions over the water column of the entire modeled area are calculated by the three-dimensional model.

To generate the complete set of response functions, five computer simulation runs are needed. One computer run is without wind but with tide. The other four computer runs are with tides and with wind from each of four directions. The four response functions set are derived from the difference between them and the one with tide as the only forcing function. The level of tidal currents at different areas produces variable wind responses under the same wind, so the tide has to be included when deriving wind response functions, otherwise they will be overestimated. This is why in a coastal area with strong tidal currents the drift ratio would be lower than in the open ocean because of the quadratic nature of the bottom friction.

When response functions are saved in discrete time intervals (30 minutes was used) the drift velocity at a certain time is computed by numerical convolution.

$$U_{ij}(n\,\Delta t) = \Delta t \sum_{k=0}^{n} (W^2) h_{ij}(n\,\Delta t - k\,\Delta t) \qquad (1)$$

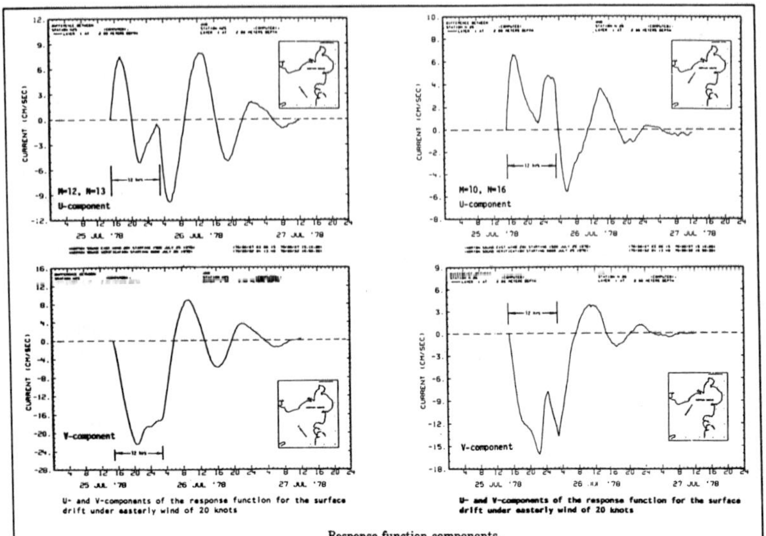

Response function components.

Fig 1

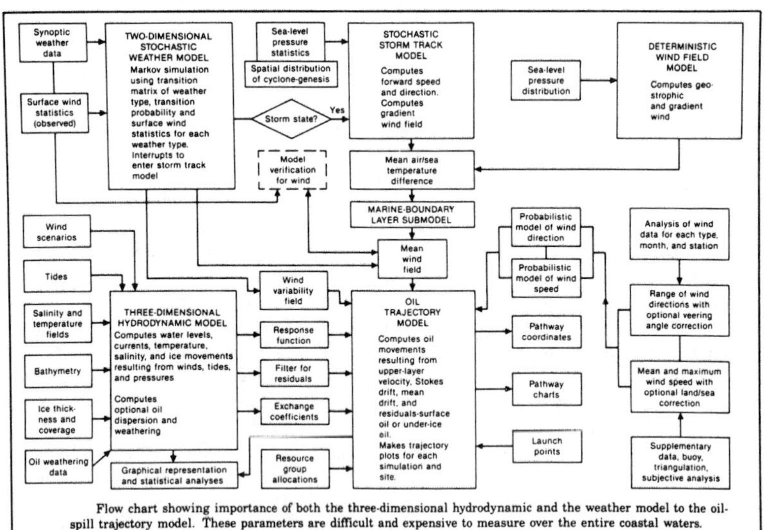

Flow chart showing importance of both the three-dimensional hydrodynamic and the weather model to the oil-spill trajectory model. These parameters are difficult and expensive to measure over the entire coastal waters. Observed wind roses, if available, can still be used to drive the trajectory model in its simplest mode.

Fig 2

where W = wind speed from a certain direction,
U_{ijk} = velocity at a particular point (i, j, k), and
h_{ijk} = time domain response function between squared wind speed and velocity at point (k, j, k).

With this formula the velocity at point i, j, k can be determined if the wind speed from a specific direction is known, as well as the response function.

The same principle applies for complex wind scenarios, then the vectorial decomposition is involved.

COMPUTING OIL BEHAVIOR UNDER ICE

In the absence of a current, oil released in a water column will rise and be trapped underneath the ice. Under porous young ice during formation, oil will initially undergo a certain degree of vertical migration through the vertical brine channels. Most oil is initially in the form of droplets until a lateral slick is formed. The oil sheet tends to spread with an obtuse contact angle. For typical Alaska Prudhoe Bay crude oil, the average observed values of interfacial surface tension, density, and the contact angle are 31 dynes/cm, 0.911 g/cc, and 20°, respectively (Kovacs et al., 1980). The static thickness of the same oil is approximately 1.2 cm. Since the dynamic pressure exerted by a moving current on an oil slick of finite length tends to balance between the front and the back faces, the equilibrium thickness should be the same for both unaccelerated and static slicks.

The bottom roughness of ice not only determines the amount of oil that may be trapped beneath it, it also influences the speed of oil movement under moving currents. Using a radar echo sounding system, Cox et al. (1980) made extensive measurements of ice bottom morphology and found the standard variation of ice thickness to be 3.1 cm over a mean thickness of 1.53 m in an undeformed shore-fast ice zone near Prudhoe Bay. The thickness of ice was also found to be inversely proportional to the thickness of snow cover over it. The snow acts as an insulator that reduces heat exchange from the sea water through the ice to the atmosphere and thus retards the growth rate. Consequently, a substantial quantity of oil can be retained underneath the pack ice. Under weak currents, trapped oil will travel with the pack ice. The movement of oil under this condition would be identical to the computed movement of ice described above but the shear stress coefficients between water and ice are reduced.

Under strong "relative currents" (between water and ice), oil will travel at a speed different from the ice and currents. To compute the movement of oil under these conditions, in the three-dimensional model we adapted a method developed by Cox et al. (1980) with parameters evaluated from laboratory tests. This method involves the evaluation of a critical relative velocity between ice and water. Using ρ_w to represent the density of sea water in the surface layer of our computation, the critical velocity for the incipient motion with large roughness is approximately:

$$u_{\text{critical}} = 1.5\left[\left[\frac{\rho_o + \rho_w}{\rho_o \rho_w}\right]\left[\sigma_{o/w}\, g\,(\rho_w - \rho_o)\right]^{1/2}\right]^{1/2} \qquad (\ 2)$$

in which ρ_o, $\sigma_{o/w}$ are, respectively, the density of oil and the surface tension at the oil/water interface. With the aforementioned typical values observed in the Beaufort Sea, this critical value is approximately 21 cm/sec.

Equation (2) is developed considering the formation of Kelvin-Helmholtz type instability, which exerts a limit on the thickness of an oil slick near the head region. A multiplier of 1.5 on the right-hand side of Eq. (2) was used considering the actual velocity that would cause droplet tearing. Ultimate slick failure occurs at about twice the critical velocity.

According to Cox et al. (1980), the critical velocity at which oil begins to move relative to the water when the relative velocity between ice and water is:

$$u_{oil} = u_{water}\left[1 - \left[\frac{\kappa}{A\,F_\delta^2 + B}\right]^{1/3}\right] \qquad (\ .3)$$

where κ is the amplification factor and F_δ is a slick densimetric Froude number defined by:

$$F_\delta = \frac{u_{water}}{\sqrt{[(\rho_w - \rho_o)/\rho_w]\,g\,\delta}} \qquad (\ 4)$$

in which δ represents the equilibrium slick thickness.

Equation (3) is derived from the momentum balance between form drag, oil/water interfacial shear stress, and the retarding of oil/ice frictional force. Constants A and B in the equation contain the effects of frontal shear and plane shear, as well as the normal force from oil's buoyance against the ice. For the field conditions of the model areas, the values of these coefficients are 1.75 and 0.115. The amplification factor κ equals unity for a hydrodynamic smooth region and is greater than zero for rough surfaces. For the field conditions in the model areas, the factor was given a value of 1.105. To determine the equilibrium oil slick thickness from the density of the oil, we used an empirical relation. The empirical difference is:

$$\delta = 1.67 - 8.5\,(\rho_w - \rho_o)(\text{cm}) \qquad (\ 5)$$

In the three-dimensional model the local density of water is evaluated by the equation of state of sea water. The density of oil can be computed by a table look-up procedure.

In our computation, the local density of sea water associated with the ice formation/salt rejection and advection was evaluated and updated. The results of oil movement beneath the ice under various wind conditions, in the form of response functions computed from the three-dimensional simulation, were recorded on magnetic tape as subsequent inputs to the oil spill trajectory computation.

We found that oil will generally move with ice except under two conditions that cause it to travel at a different speed. The first condition is beneath the shore-fast ice in an area where tidal currents are strong. The second condition is when pack ice is located very close to a passing storm center, when drifting ice abruptly changes direction. Under this condition a high relative velocity between the water and ice can be reached..

Because of the pronounced nonlinear vertical shear coupling, and at high latitude, the direction of an oil movement appears to be extremely variable. Therefore, the vertical shear coupling should be included in the computation even though spilled oil beneath the ice may not seem to be in constant motion with appreciable magnitude.

MODELING OIL SPILL TRAJECTORIES

Oil spill trajectory computations involve two parts—the first part calculates the movement of oil mainly by advective transport, and the second part calculates the movement of dispersive mechanisms, including weathering, diffusion, and dissolution processes. In this section we will describe the modeling of advective transport only.

Oil transported by advective mechanisms contains several major components. The method used to compute each component is as follows:

Oil Transport by Mean Wind Drift. During this computational step, oil movement resulting from wind stress at the surface layer and at different levels in the water column is calculated by the response function technique. The response function represents local advective transient response to a given wind stress. If a three-dimensional model is used to develop these response functions, the effects of transient inertia, bottom, shoreline, and vertical stratification are all included. The computed movement using this response technique gives only the movement near the middle of the surface layer (typically 5 meters) schematized in the 3D model. For the surface movement the results are extrapolated for speed and direction near the surface using an analytical solution of the Ekman type assuming constant density within that surface (mixed) layer.

Stokes' Transport. When wind blows over the water surface it generates Stokes' transport in addition to the mean wind-driven current. This transport is caused by the non-linear residual orbital motion associated with the local wind waves field. The magnitude of this transport is a function of the intensity and age of the wave field. The direction is nearly identical to the wave-propagating direction. In the oil spill trajectory model, a special subroutine is used to compute the direction and speed of this transport. According to measurements in the field and in the laboratory, Stokes' transport is approximately 1.6 percent of the local wind speed if the wave field is not limited by wind duration and fetch length. The wind used to compute the Stokes' transport is obtained from the wind field model described in (Liu and Leendertse, 1987)

Tidal and Baroclinic Residual Component over the Alaskan Outer Continental Shelf Area. Because of the complex tidal regime and density field, tidal residual and baroclinic circulation components are quite essential. We have discussed their dynamics in great detail in (Liu and Leendertse, 1987)

To simulate a number of trajectories with the trajectory model, many data are needed from other models that we have previously described. Figure 2 gives an overview of the data flow between these models. As illustrated in the diagram, when computing the oil movement, the oil/trajectory model plays the role of data synthesizer. As physical parameters involved in calculating oil movements are difficult and expensive to measure over the entire Alaskan waters, the model is programmed with flexibility in mind, so that any field data, if available, can be used to drive the model in its simplest mode. On the other hand, the trajectory model would link results from the other models. To perform this task, it contains the basic physical parameters of the entire lease area as well as the grid network of the entire model and submodels within the system.

During the study period, spill trajectory analyses were made on a lease-area basis. For each lease sale, approximately 30 to 40 launch points were selected by the Minerals Management Service according to the potential petroleum resource. The movements of oil were then tracked for a period of time, typically a month during the summer period, to as long as six months during winter.

From each launch location 40 to 60 trajectories are computed under different weather scenarios. For each trajectory, half-hourly positions are computed and landfall locations are recorded where possible. As described above, the wind-driven component of the oil movements is computed using the wind-driven response function technique through the convolution procedure. To maintain accuracy, each response function has half-hourly weighting elements for each wind direction, each computational grid, each layer, and each season. One magnetic tape is required to store all response functions from each of four wind directions. For the computation of oil spill trajectories, this information is transferred to disk storage.

Results from a typical simulation are presented in Fig. 3. In the figure, the computational grid of the three-dimensional model of the Beaufort Sea is superimposed over the oil trajectory model, which also covers the eastern portion of the Chukchi Sea. The response functions and net-current field over that area are averages obtained from the two models.

On top of the graph, computed 12-hour wind vectors sampled at Point Barrow are plotted. The mean winds and half-hourly varying winds from the simulation are also presented in the form of wind roses for speed and direction, also at Point Barrow. The wind direction rose represents the frequency of occurrence of wind direction toward which wind is blowing. A wind speed rose represents the average marine wind speed associated with each of the 16 wind directions mentioned above. The plotting scale of the highest speed in the rose is 12 knots as indicated under the rose.

Each dot in the oil trajectory represents daily displacement originating from the launch point, which is marked by a number. When examining the trajectories one would notice the following interesting aspects:

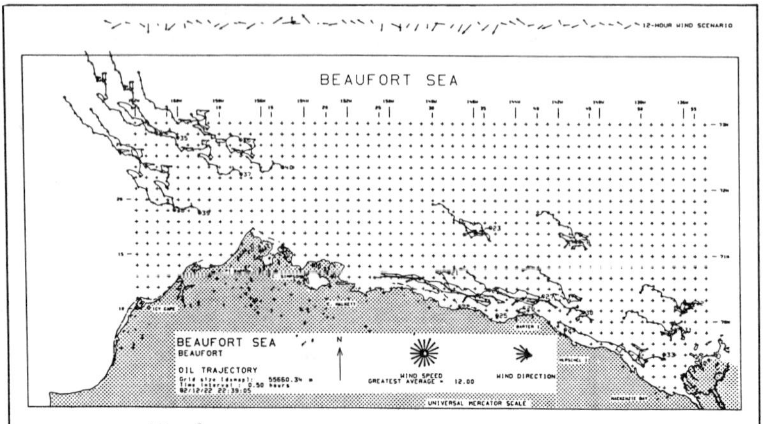

Fig 3 Daily movements of oil originated from launch points 21 through 40.

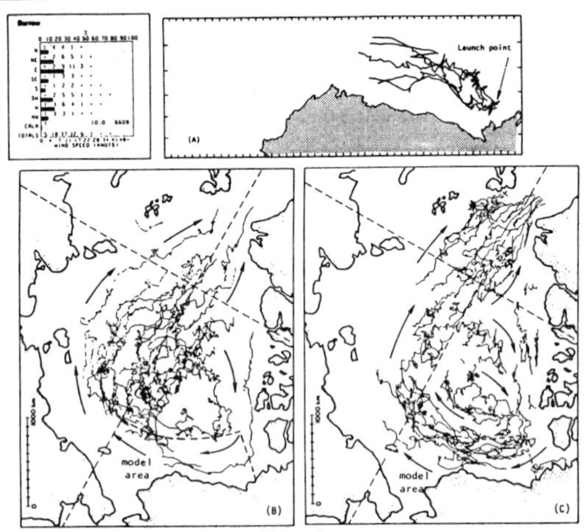

Fig 4
Three groups of ice drift data. (A) Coast Guard drogues for 3 months during the summer of 1979; (B) composite trajectories compiled using data from 1893-1972 by Colony and Thorndike (1984); and (C) trajectories of automatic data buoys (1979-1982), also from Colony and Thorndike (1984).

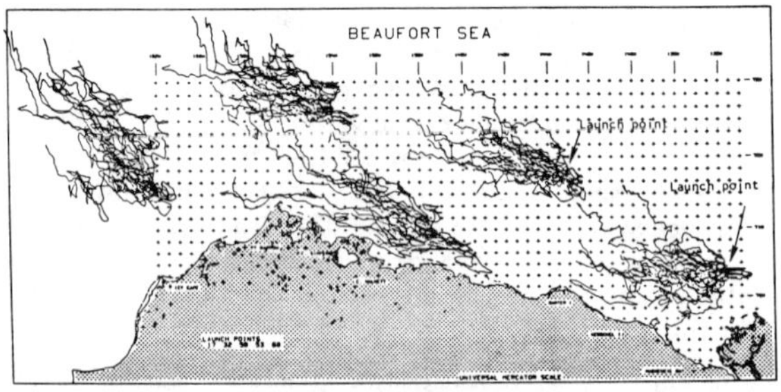

Fig 5 computed by the RAND oil trajectory model and the two-dimensional stochastic weather simulation model.

1. The predominant wind during a summer period is from the east-southeast.
2. Oil spilled closer to the shore travels faster, in a downwind position.
3. Oil spilled offshore moves in a more random direction and has a larger deflection angle. This can be attributed to the greater water depth and the existence of ice floes.
4. Oil spilled further offshore travels in a direction approximately the same as the Arctic Gyre (Colony and Thorndike, 1984, U.S. Coast Guard buoy data, Fig. 4). The simulation in Fig. 4 was made in December 1982.

The trajectories shown in Fig. 3 represent oil movements under a given 30-day weather scenario. In Fig. 5 comparison between satellite-tracked buoys (Murphy et al., 1981) and trajectories computed using the coupled trajectory-weather model is shown. During the summer period, the average observed movement of ice is approximately 140 nautical miles per month. The same is found in the computed monthly average displacement. The observed and the computed trajectory patterns in the Mackenzie Bay are quite irregular. This may be due, in part, to the cyclonic local eddy described above.

Without observed wind fields and the variability of winds, tracing the deterministic motion of a particular ice floe is not as desirable as comparing a group of observed trajectories to a group of computed trajectories using a weather model. The same weather model will be used for the statistical trajectory analyses below.

In the trajectories it can be seen that the impact of a moving storm can sometimes be seen as a loop in a trajectory. The size and shape of the loop vary because of their location relative to the moving storm.

The computed trajectories for the winter season have the similar direction of predominant movement. Figure 6 shows the general direction of movement launched from three selected points. The residence time within the modeled area is approximately two to three months. If all launch points for a given season are considered, one can assess the oil spill risk by counting the number of contact occurrences within each square area whose length is 10 nautical miles in the north-south direction (Fig. 7). In Fig. 7 the size of a circle represents the spatial distribution of landfall frequencies. If oil is trapped in a near-shore lagoon, a continuous contact is assumed for the remaining period. In preparing the map, analysis is made using two-hour counting method. Plotting scale for the circle is 21211, two-hour exposure periods equals one latitudinal grid spacing for the radius of the circle.

If the near-shore entrapments are excluded, a similar diagram (Fig. 8) gives the spatial distribution for the marine resource contact frequencies. In this case, each latitudinal grid spacing equals 1872 two-hour contact period for the radius of a circle. From graphs like Fig. 7 and 8, one would be able to obtain a general assessment of contact risk associated with the oil spill. However, sometimes it is more desirable to estimate the concentration of oil, if a contact is made.

DETERMINING THE OIL CONCENTRATION FIELD

When released in water, fresh crude oil will undergo major changes in its composition while being transported and dispersed. The spreading of oil at the surface is mainly due to mechanisms associated with viscosity, surface tension, and inertia. As time progresses the major process responsible for the spreading of spilled oil are advection and turbulent dispersion. While oil is being advected and dispersed, its concentration decreases as a result of evaporation, photochemical degradation, and biodegradation. These processes are called weathering.

In modeling an oil concentration field, advection, dispersion, and weathering are considered as well as the transport of oil. The rates of evaporation and the bio- and photochemical degradation were evaluated under field conditions by other investigators (Payne et al., 1983). The oil decay rates for the simulation were estimated by these investigators on the basis of turbulence levels determined by means of the three-dimensional models for the different areas.

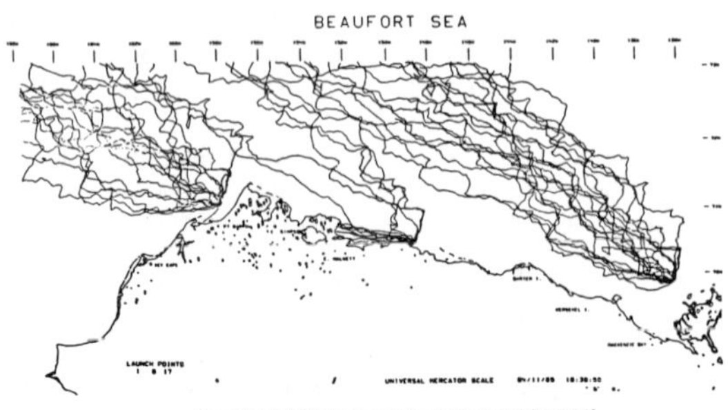

Composite oil spill trajectory map showing the general direction of movements launched from stations 1, 8, and 17 during winter. The staggered launch scheme represents an equally likely chance of spill during the long winter period. The residence time within the modeled area is approximately two to three months.

Fig 6

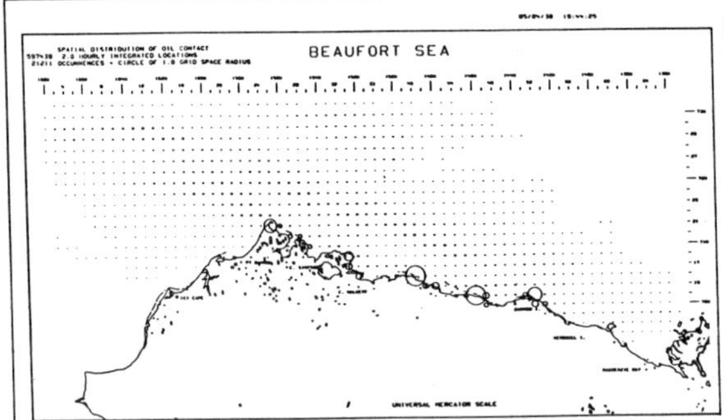

Spatial distribution of oil contact frequencies. If oil is trapped in a near-shore lagoon, a continuous contact is assumed for the remaining period. Analysis is made using two-hour counting method. Plotting scale for the circle is 21211, two-hour exposure periods equal one latitudinal grid spacing for the radius of the circle.

Fig 7

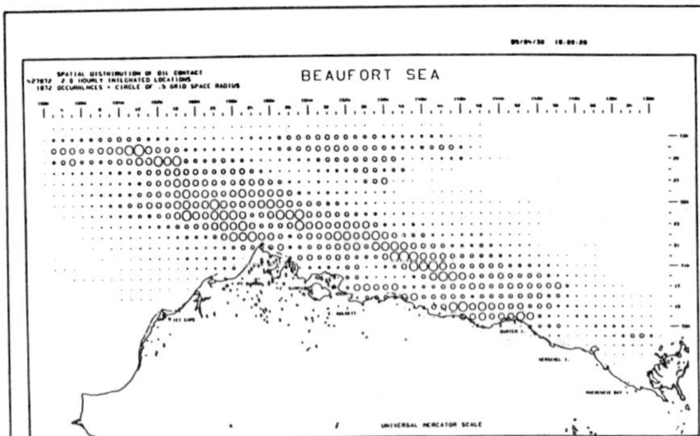

Spatial distribution of oil contact frequencies excluding near-shore entrapments. Latitudinal grid spacing equals 1872 two-hour contact periods for the radius of a circle.

Fig 8

To illustrate the diffusion process induced by the turbulent oscillating flow, it is more convenient to demonstrate the magnitude of diffusion over the vertical plane in the absence of surface energy input from the wind. In other words, in this illustration the turbulence is generated primarily at the bottom by tide. To show the turbulent diffusion processes one hundred particles are released in each vertical layer near the Pribilof Island and half-hourly displacements are plotted there for a period of 24 hours (Fig. 9). The elevations for the eight layers are 2.5, 7.5, 12.5, 17.5, 25, 35, 85, and 240 meters, respectively.

In each layer the movements of particles are caused by advection and diffusion processes. For instance, the hourly displacement of the particle group in the first layer starts from the upper position, gradually moving with the tidal motion. As the group moves the distance between the particles increases because of the turbulent diffusion. In a stricter sense, the separation of particles is the combined result of diffusion and the nonuniformity of the current field. The combined process is called the turbulent dispersion. In shear flow, such as the one illustrated here, the major mechanism responsible for dilution of a soluble is dispersion. This is evident from the amount of dispersion experienced by the particle groups in the lower layers where the velocity gradient is much sharper than in the upper five layers. These five layers are located above the sharp pycnocline, which partially isolates the upward momentum transport.

Also of interest are the distances between the first and the last particle group within each layer. They represent the net displacement over a period of two days. The changes in net transport over the vertical are quite common in the coastal area, to satisfy the law of conservation of mass.

The example presented here illustrates the dispersion mechanism associated primarily with bottom stress and nonuniform velocity distribution. Dispersion effects can also be induced by shore line irregularity through the variability of the velocity field. For each of the large modeled areas, submodels are used to compute near-shore oil movements (Fig. 10). The turbulent diffusion coefficients averaged over ten tidal cycles, as computed by the three-dimensional model for each large area and for each layer, are stored on magnetic tapes. These diffusion coefficients became very useful for diffusion analysis in a limited near-shore area. Figure 11 represents the results of oil dispersion analysis in which crude oil is released instantaneously from five locations near the Bering Strait. Displacement of the one-part-per-billion concentration envelopes are plotted every five days. The influence of the shoreline and the variability of local advective/diffusive mechanisms (Fig. 11) are quite evident, as seen by the changing speed and direction of the oil movement.

Under a scenario of continuous release, the distribution of surface oil concentration is presented in Fig. 11. When oil moves through the Bering Strait, the strong local current tends to elongate the oil. Also notice the cumulation effects when the oil reaches the coastal area, where the on-shore current components drop and the along-shore currents strengthen. This near-shore effect tends to redirect the oil while slowing it down. The speed of an oil transport can be seen from the top diagram of Fig. 12, where daily displacements of the advancing plume are plotted. The daily traveling speed of oil is governed by the evolutionary weather state as well as the local circulation pattern.

To illustrate the effects of weather and local baroclinic circulation, a group of six trajectories are launched from five hypothetical spill locations in the Chukchi Sea/Barrow Arch lease area (Fig. 13). The net displacement for the northern trajectories during the eight-month period ranges between 3–5° latitude, which represents a daily movement of 1.4 to 2.31 km (Fig. 14). Oil launched near Point Hope travels substantially slower than its northern counterpart, which moves predominantly within the Arctic Gyre. The simulated direction and speed of ice movements within the Chukchi Sea agree with the observed values reported by Gordienko (1958), and Hibler (1979), who made computations using ice models including nonlinear plastic flow effects by means of viscous-plastic constitutive law.

Perhaps it is more illustrative to show and analyze a group of trajectories launched near an embayment so that the shore effects can be seen. The launch point is located between St. Lawrence Island and the Gulf of Anadyr, USSR (see the insert map of Fig. 15). Twenty-one

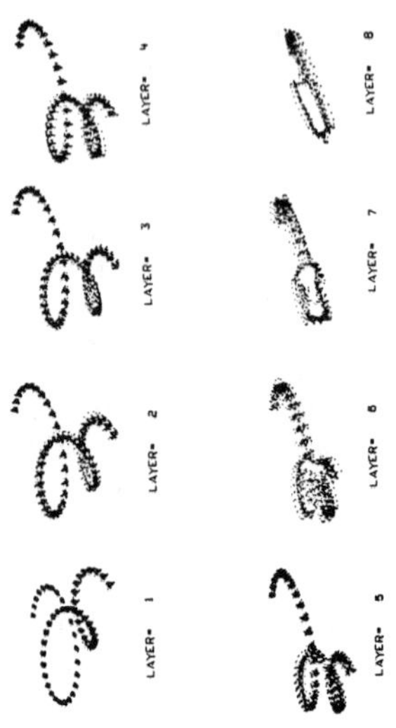

Fig. 9—Pathways of particle groups released in different layers of the model. Partially insulated by the pycnocline, 24-hour trajectories of particle groups released in the upper five layers experience less turbulent diffusion than the lower layers when the system is forced only with tidal energies.

Fig 9

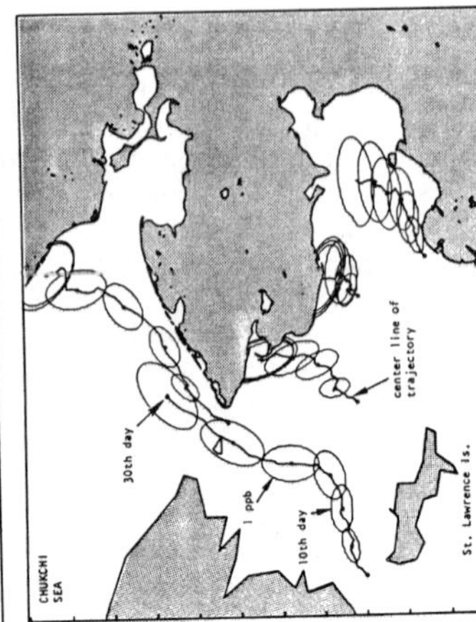

Fig 11

Fig. 11—Displacements of the one-part-per billion concentration envelope every five days for an instantaneous release of 700,000 barrels of crude oil from five hypothetical launch points during the summer. The variability of local advective and diffusive mechanisms is illustrated by the changing speed and direction of the oil's movements.

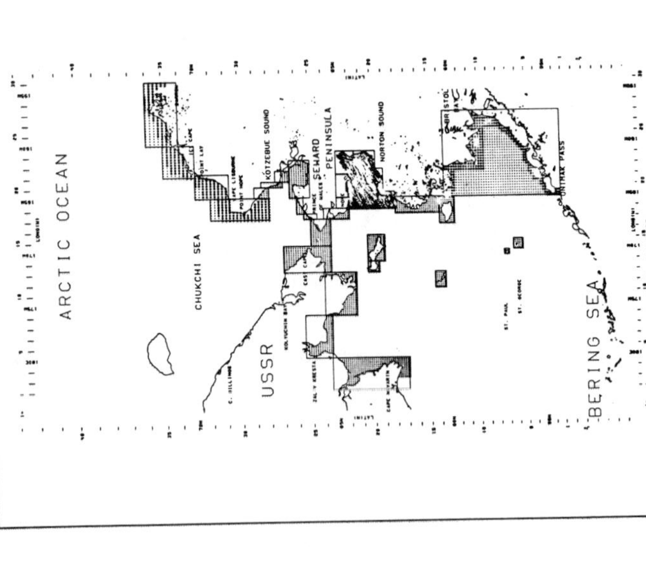

Fig. 10—Submodels are used to improve the near-shore resolution of the trajectory computation. The computational grids of the submodel are nested within the larger model's computational grid (Liu and Leendertse, 1984).

Fig 10

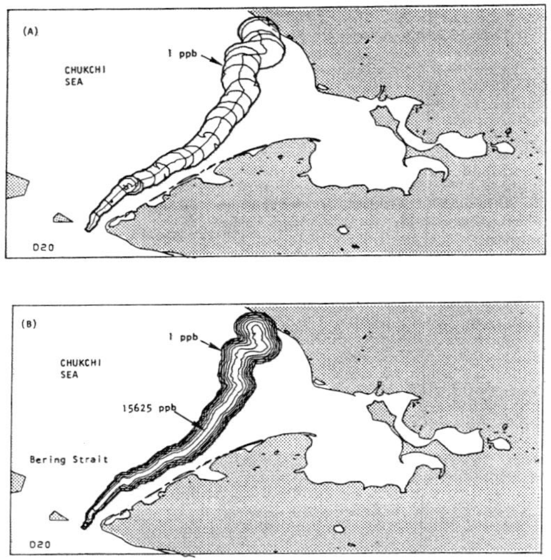

Fig. 12—Oil spill trajectory launched in the middle of the Bering Strait under a 30-day stochastic weather scenario during the summer. (A) illustrates the progressive daily displacements of the 1 part per billion concentration envelope for the continuous discharge of 2000 barrels of crude oil. The traveling speed of oil is governed by the evolutionary weather state as well as the local circulation pattern. (B) illustrates the concentration contour of the oil. Notice the cumulation effects when the oil reaches the coastal area where the on-shore current components drop and the along-shore currents strengthen. This near-shore effect tends to redirect the oil while slowing it down somewhat.

Fig 12

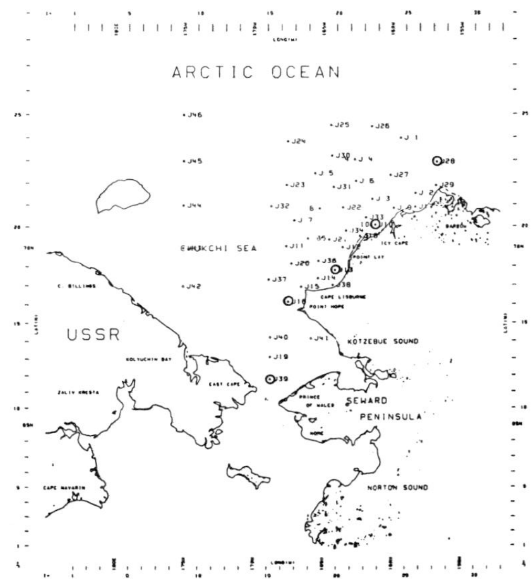

Fig. 13—Location of 46 launch points specified for the oil spill trajectory computation. Barrow Arch lease area (No. 109).

Fig 13

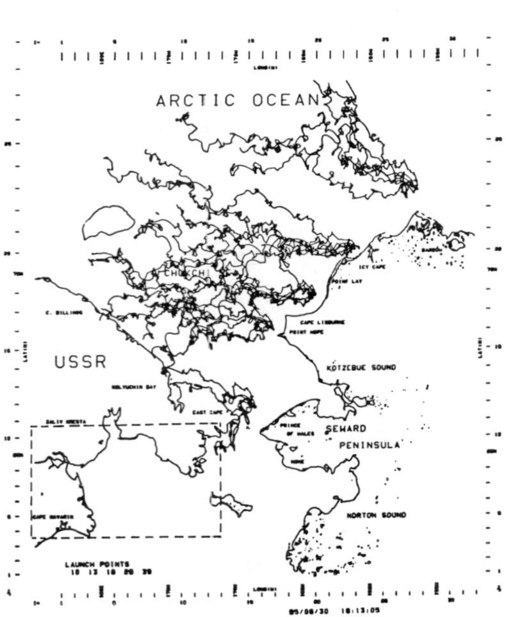

Fig. 14—Five sets of randomly selected trajectories from launch points 10, 13, 16, 28, and 39; under the staggered-launch mode, each spill has about equally likely chance of occurrence during the winter. Spill launched near the Bering Strait can move either south or north depending on the weather sequence.

Fig 14

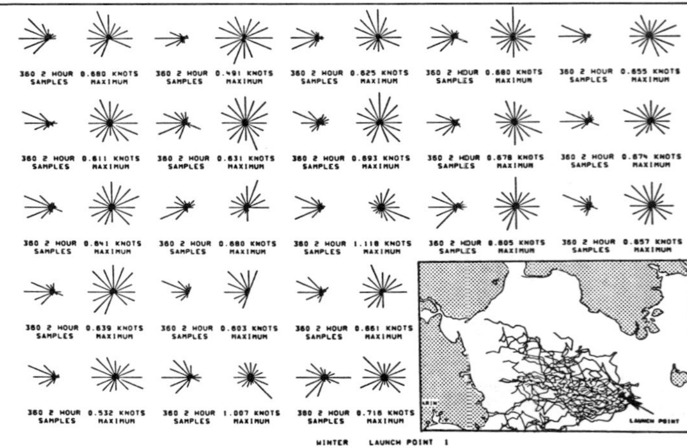

Fig. 15—Current roses analyzed from twenty-one 30-day oil spill trajectories for winter, launched near a point between St. Lawrence Island and the Gulf of Anadyr, USSR (see insert map). Local residual and the relative distance from the shore-fast ice make the oil containing ice move differently from the free-drifting pack ice. Consequently, there are more random ice motions under the local wind stress than the oil movements in the water column resulting from inertia and momentum filtering effects.

Fig 15

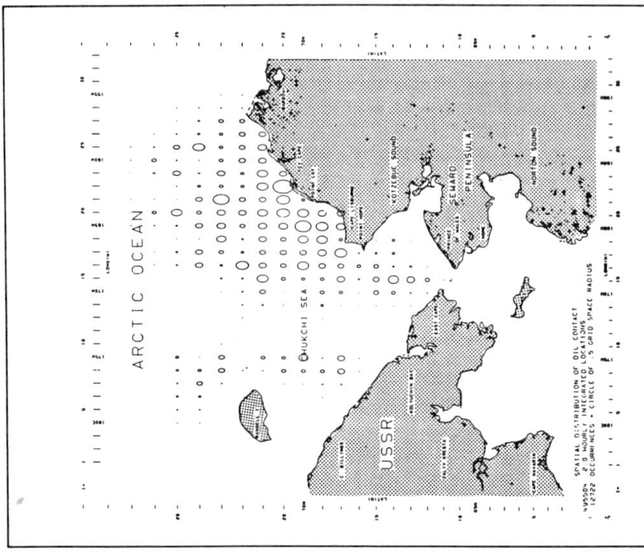

Fig. 17—Spatial distribution of oil contact frequencies during summer. Latitudinal grid spacing equals 12722 two-hour contact periods for the long axis of the ellipse.

Fig 17

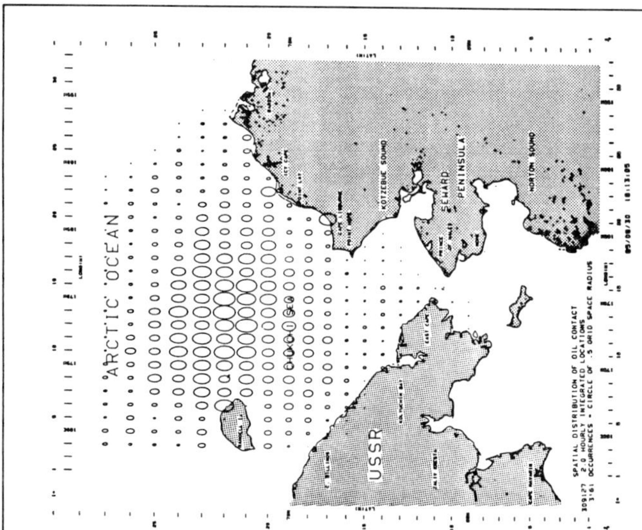

Fig. 16—Spatial distribution of oil contact frequencies during winter. Latitudinal grid spacing equals 3161 two-hour contact periods for the long axis of the ellipse.

Fig 16

groups of 30-day oil spill trajectories are sampled every two hours. These sampled data are then analyzed for their direction and speed. For a 30-day duration there are 360 two-hour samples for current directions and speeds. Since the directions and speeds of currents at every two hours are located over a different area, local residual circulation and the relative distance from the shore-fast ice make the movement of oil contained in ice different from that of free-drifting pack ice. Consequently, there are more random ice motions under the local wind stress than the oil movements in the water column resulting from inertia and momentum filtering effects.

Figure 16 indicates that most oil spill trajectories move in a predominantly northwesterly direction. For the summer period (Fig. 16), however, most contacts are closer to the eastern shore. It should be noted that the plotting scale of Fig. 17 is four times that of the one shown in Fig. 16. The western Chukchi Sea receives much less impact during the summer season than in the winter period. On the average, oil travels a shorter distance and moves more randomly under the summer winds. During summer, winds have higher variabilities and, as a consequence, the inertial current components have a substantial contribution toward the overall current direction. The area of greatest impact is located near Icy Cape.

REFERENCES

Colony, R., and A. S. Thorndike, (1984): A estimate of the main field of the Arctic ice motion, J. Geophy. Res. Vol. 89, No. 6., pp. 10623–10629.

Cox, J. C., L. A. Shultz, R. P. Johnson and R. A. Sheley, (1980): The transport and behavior of oil spilled in and under sea ice, Arctic Fac. Rep. No. 460C.

Ekman, V. W. (1905): On the influences of the earth's rotation on ocean currents, Ark. Math. Astr. Fysik. Kong, Su. Vetenskap. Akad., Stockholm, Vol. 2, No. 11, p. 52.

Gorgienko, P., (1958): Arctic ice drift, Proc. Conf. on Arctic sea ice, Published by the National Academy of Sciences and the National Research Council, Washington, D. C., pp. 210–220.

Hibler, W. D., (1979): A dynamic thermodynamic sea ice model, J. Phys. Oceano. Vol. 9, No. 7., pp. 815–846.

Kovacs, A., R. M. Morey, and D. Cundy, (1980): Oil pooling under sea ice, Environmental Assessment of the Alaskan Continental Shelf, Ann. Rep. Vol. 5, pp. 333–339.

Liu, S. K., and J. J. Leendertse, (1982): A three–dimensional model of Bering and Chukchi Sea,COASTAL ENGINEERING Vol. 18, ASCE. N. Y.

Liu, S. K., and J. J. Leendertse, (1984): A three–dimensional model of Beaufort Sea, COASTAL ENGINEERING Vol. 19, ASCE. N. Y.

Liu, S. K., and J. J. Leendertse, (1986): A three dimensional model of the Gulf of Alaska, COASTAL ENGINEERING Vol. 20, ASCE, N. Y.

Liu, S. K., and J. J. Leendertse, (1987): Modeling the Alaskan Continental Shelf Waters, RAND R–3567–NOAA/RC, 136 pages.

Murphy, D. L., P. A. Tebeau, and L. M. Lissuer, (1981): Long–term movements of satellite–tracked buoys in the Beaufort Sea, Office of Research and Development, U. S. Department of Transportation, CG–D–48–82.

CHAPTER 192

Bore front modeling in terms of Burgers equation and its numerical calculation method

Takao Yamashita [1]

Yoshito Tsuchiya [2]
and
David R. Basco [3]

Abstract

Bore front hydraulics are investigated in terms of Burgers equation to clarify the dynamics of a moving discontinuity in water flow. Burgers equation has been derived from the one dimensional open channel equation with the horizontal turbulent diffusion term. The derived equation system consists of Burgers with respect to dynamic characteristic and hyperbolic equation with respect to water surface elevation, which satisfys Jeffery-Vedernikov condition $F_r = 2$ through discontinuity. It has been verified from the experiments that the condition $F_r = 2$ is a good approximation of bore front dynamics. The numercal calulation method of Burgers equation employing Cole-Hopf transformation and QUICKEST algorithm was also proposed and confirmed its efficiency.

1. INTRODUCTION

Bores, which can often be observed in rivers, estuaries and coastal zones, i.e. hydraulic jumps, flood waves, breaking waves, tidal bores etc., are well known for their complicated flow structures consisting of typical time scales of order of mean flow, waves and turbulence. From the coastal engineering view point, developing our knowledge of bore dynamics may lead to a better understanding of breaker dynamics and flood wave hydraulics. From the view point of numerical calculation for bore front propagation on beaches, dry lands or rivers, more theoretical studies for its propagation speed and hydraulics are needed. Example numerical calculations include flooding simulation of tsunamis, storm surges and run-up of coastal waves. Sometimes an artificial (physically meaningless?) moving boundary condition is forced to apply to numerical models of this type. The so-called threshold depth is defined in the numerical cells in order to establish an artificial moving boundary.

---

1) Instructor, Disaster Prevention Research Institute, Kyoto Universuty, Kyoto 611, Japan
2) Professor, Disaster Prevention Research Institute, Kyoto Universuty, Kyoto 611, Japan
3) Professor of Civil Engineering and Director of Coastal Engineering Institute, Old Dominion University, Norfolk, VA 23529-0242, U.S.A.

The basic concept in treating this problem is made by employing the hydraulic jump analogy, where a weak solution of the physical conservation laws is allowed. There are, in general, three conservation laws, mass, momentum and energy, from which we get three sets of jump conditions. For the derivation of shallow water theory, the incompressible fluid approximation is assumed and only two conservation laws are used through the discontinuity.

On the other hand, the so-called Burgers equation is well known as an approximation for dissipative fluid systems, and has been used in analyzing shock-like fluid motions. In this equation, nonlinearity of propagation and diffusive effects are balanced. The main aim of this paper is to determine a method to apply this simple equation to the problem of modeling bore front propagting on the river and dryland.

It is also well known that the advantage in using Burgers equation is that it can be easily solved by transforming it into the linear heat equation by applying the Cole-Hopf transformation. In this paper, we attempt to describe bore front motion by application of Burgers equation, and we propose a solution method which is wiggle-free with no numerical viscosity. Consequently, the simple formulation of bore motion by Burgers equation, and its transformed linear version, advection/diffusion equation, can be solved accurately by the QUICKEST algorithm.

We carried out experiments of bore propagation as a dam-break problem. Our purpose was to examine the bore front profile while focusing on the criterion of undular bore creation in the relationship to bore amplitude and water depth, including extremely small depth (dry land condition). Furthermore, the derivation of Burgers equation, with respect to the propagation speed, is accomplished by using the equations of mass and momentum consevation, in which the Gardner-Morikawa (G-M) transformation is employed Finally, we proposed an effective numerical calculation method for the Burgers equation by employing both the Cole-Hopf transformation and the QUICKEST numerical algorithm.

2. BORE FRONT PROPAGATION

2.1 Flood Wave Theories

Assuming a wide and rectangular channel, where the hydrostatic pressure law is valid, the 1-D open channel flow equations are,

$$h_t+(hu)_x = 0 \tag{1}$$

$$u_t+uu_x+g'h_x = g'S-C_f\frac{u^2}{h} \tag{2}$$

where subscripts indicate partial derivatives, $u(x,t)$ and $h(x,t)$ are depth-averaged current, and water depth, respectivly, and $g'=g\cos\theta$, $S=\tan\theta$, θ is bottom slope angle(see Figure 1).

First of all, some consideration on flood waves with

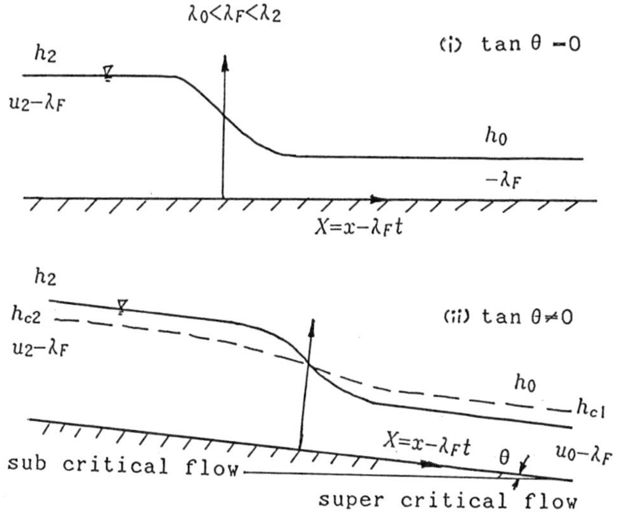

Figure 1. Coordinates and definition of variables

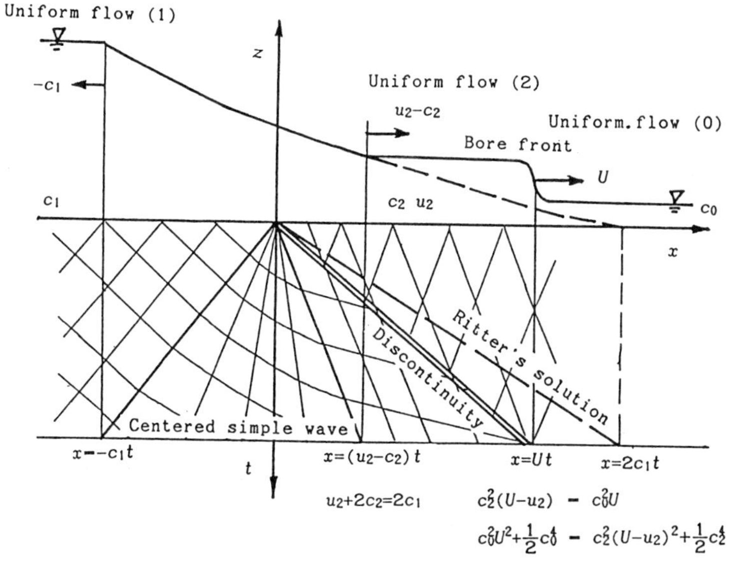

Figure 2. Characteristics of dam break problem.

smooth surface profile are made. If flow motion varies gradually, the kinematic wave approximation is derived, which is called Kleitz-Seddon formula.

$$h_{0t}+\frac{3}{2}u_0 h_{0x} = 0, \qquad u_0 = \sqrt{\frac{g'Sh}{C_f}} \tag{3}$$

Whitham(1973) showed the monoclinal flood wave by assuming boundary conditions of kinematic approximation at both up and down stream ends, in the reference frame moving at the speed of λ_F as,

$$h = h(X), \qquad u = v(X), \qquad X = x-\lambda_F t \tag{4}$$

Eqs.(1) and (2) can be written by

$$\frac{dh}{dX} = -\frac{(Q-\lambda_F h)^2 C_f - g'h^3 S}{g'h^3-Q^2} \tag{5}$$

where $Q = h_1(\lambda_F-u_1) = h_2(\lambda_F-u_2)$ is total discharge, subscripts 1 and 2 indicate the uniform state quantities in the up and down stream directions, respectively. The behavior of the solution of Eq.(5) depends on the sign of the denominator $g'h^3-Q^2$. The numerator must have roots of $h = h_1, h_2$ to be zero at both boundaries, which are the two roots of the cubic. The third root is

$$h_3 = \frac{C_f Q^2}{Sg'h_1 h_2} = \frac{h_1 h_2}{(h_1^{1/2}+h_2^{1/2})^2} \tag{6}$$

We are seeking the solution in the range of $h_1 < h < h_2$. Moreover, since $h_3 < h_1$ and $\lambda_F > u$, the sign of dh/dX depends on the denominator. At the stage of $h_2 \rightarrow h_1$, we get the front propagation speed as,

$$\lambda_F = \frac{3}{2}u_1 \tag{7}$$

This means the smallest amplitude flood wave is the kinematic wave, and a smooth surface solution (monoclinal flood wave) exists when the denominator is positive, which is satisfied in the condition of

$$\frac{3u_1}{2} < \lambda_F < u_1+\sqrt{g'h_1} \tag{8}$$

The smooth disturbance propagates faster than the kinematic wave but slower than the dynamic wave in the flow ahead.

When the disturbance's speed exceeds the downstream dynamic wave's

$$\lambda_1 = u_1+\sqrt{g'h_1} < \lambda_F < \lambda_2 = u_2+\sqrt{g'h_2} \tag{9}$$

no smooth surface solution exists. Flow changes from supercritical to subcritical in the reference frame (see Figure

1), which means the initiation of a breaking front (bore front).
Now we consider the higher order contibution of the advection term by introducing a small parameter ε perturbed from the kimematic solution of u_0, h_0, and assuming constant S and C_f,

$$u = u_0+\varepsilon u^{(1)}+\varepsilon^2 u^{(2)}+\ldots..$$

$$h = h_0+\varepsilon h^{(1)}+\varepsilon^2 h^{(2)}+\ldots.. \quad (10)$$

In the order of $O(\varepsilon)$, we get

$$h_t^{(1)}+u_0 h_x^{(1)}+h_0 u_x^{(1)} = 0 \quad (11)$$

$$u_t^{(1)}+u_0 u_x^{(1)}+g' h_x^{(1)}+g' S\left(\frac{2u^{(1)}}{u_0}-\frac{h^{(1)}}{h_0}\right) = 0 \quad (12)$$

From Eqs.(11) and (12) eliminating $u^{(1)}$, we get the single equation for $h^{(1)}$ in the form

$$\left(\frac{\partial}{\partial t}+c_+\frac{\partial}{\partial x}\right)\left(\frac{\partial}{\partial t}+c_-\frac{\partial}{\partial x}\right)h^{(1)}+\frac{2g'S}{u_0}\left(\frac{\partial}{\partial t}+\frac{3}{2}u_0\frac{\partial}{\partial x}\right)h^{(1)} = 0 \quad (13)$$

where $c_\pm=u_0\pm\sqrt{g' h_0}$. In Eq.(13) the first term indicates dynamic characteristics and the second term the kinematic one.

If the approximation of $\partial/\partial t \sim 3u_0/2\,\partial/\partial x$ is applied in Eq.(13), we get the linear Burgers equation as,

$$h_t^{(1)}+\frac{3}{2}u_0 h_x^{(1)} = \frac{u_0}{2g'S}\left(g' h_0-\frac{u_0^2}{4}\right)h_{xx}^{(1)} \quad (14)$$

The right hand side of Eq.(14) is the diffusion term. If its sign is a positive, solution of this equation is a diffusive type, and the negative sign leads to an explosive solution which means there is no solution balancing advection and diffusion. This criterion $F_r \gtreqless 2$ is called Jeffery-Vedernikov condition, where F_r is Froude number defined as,

$$F_r = \frac{u_0}{\sqrt{g' h_0}} = 2 \quad (15)$$

In other words, this condition depicts the criterion of the balancing of nonlinearity and dissipation as well as the initiation of the bore front.

After the discontinuity appears, the weak solution is assumed to the conservation laws, such as the hydraulic jump analysis. However, we sometimes need information of flow quantities inside the discontinuity. For this we need a more complicated constitution equation system including turbulence dynamics.

On the other hand, the simplest equation which describes flow field inside the discontinuity is Burgers equation. As long as the depth-averaged quantities are allowed, it might be easy-to-use the field equation. From the consideration of higher order effects in the flood wave, it is supposed that Burgers equation should describe the local balance of nonlinearity and dissipstion effects through the discontinuity.

2.2 Dam break problem

To make sure the bore front propagation speed we considered here the dam break problem shown in Figure 2. Assuming the invisid fluid flow with hydrostatic pressure distribution on the horizontal bottom, the open channel flow equations (1) and (2) can be rewritten by,

$$\left\{\frac{\partial}{\partial t}+(u\pm c)\frac{\partial}{\partial x}\right\}(u\pm 2c) = 0 \tag{16}$$

where $c = \sqrt{gh}$, u is the depth-averaged flow velocity. If $c_0 = 0$ (dry land), the Ritter's solution is obtained, which consists of two regions, zone of silence and centered simple wave. In the case of $c_0 \neq 0$, the only possible configulation consists of three uniform flow regions and a centered simple wave, as shown in Figure 2. The discontinuity must occure in the region between uniform flow regions (0) and (2) shown in the figure. For determination of the unknown propagation speed of the discontinuity $\lambda_F = U$, c_2 and u_2 in the uniform flow region (2), we need additional conservation law. Other than mass conservation we could use the momentum or energy conservation law but both. Our choice for this problem is that of momentum.

One more relation can be obtained on the characteristic of $x = (u_2-c_2)t$ where the Riemann invariant $u+2c = 2c_1$ in the region of centered simple wave is given by the known quantity c_1. Consequently we get three equations, as

$$u_2+2c_2 = 2c_1 \tag{17}$$

$$c_2^2(U-u_2) = c_0^2U \tag{18}$$

$$c_0^2U^2+\frac{1}{2}c_0^4 = c_2^2(U-u_2)^2+\frac{1}{2}c_2^4 \tag{19}$$

Combining Eqs.(18) and (19), third order polynomial equation with respect to U is obtained.

$$U^3-2u_2U^2+\{(u_2)^2-c_0^2\}U+\frac{1}{2}u_2c_0^2 = 0 \tag{20}$$

From Eqs.(17) and (18), the third order polynomial equation with respect to u_2 is derived.

$$u_2^3-(4c_1+U)u_2^2+4c_1(c_1+U)u_2-4U(c_1^2-c_0^2) = 0 \tag{21}$$

To solve the above set of two equations, we consider Eq.(20), the second order polynomial equation with respect to u_2 and two its solutions $u_{2(\pm)}$ are

$$u_{2(\pm)} = \frac{1}{4U}\{4U^2 - c_0^2 \pm c_0\sqrt{c_0^2+8U^2}\} \tag{22}$$

At the limit of $c_1 \rightarrow c_0$, u_2 should be zero. This condition is satisfied by $u_{2(-)}$. Substituting $u_{2(-)}$ into Eq.(22) and sloving the resultant equation numerically in the range of $c_0 < U < 2c_1$, we get the configulation of solutions of U, u_2, c_2 as shown in Figure 3, where three regions of breaking front with undular component, strong bore front and Burgers front are zoned in terms of $\beta = c_0/c_1$. In addition to these other information such as $F_r = U/c_2$, $F_{r2} = u_2/c_2$ and U/c_0 are shown.

2.3 Experiment of bore front propagation

Experiments were conducted as a dam break problem by using a flume and gate system with dimensions of 0.7m in width and height (Figure 4). Surface profiles were measured by 5 wave gauges of capacitance type. Flow visualization techniques were employed for measuring the mean velocity fields near the bore front. The initial water depth in front of the gate was varied in order to examine the effects of downstream boundary condition to the bore propagation. The experimental conditions are presented in Table 1, where h_0, h_1 and h_2 are initial water depths of up/down-stream sides of the gate and water depth in the uniform region formed just after bore front, *i.e.* uniform flow (2) in Figure 2.

Table 1 Experimental conditions

Run No.	h_1 (cm)	h_0 (cm)	c_0/c_1	h_2 (cm) crest-mean	c_2/c_1 crest-mean
1	19.8	1.0	0.225	5.14	0.557
2	20.9	3.8	0.426	7.71-8.0	0.608-0.62
3	22.6	9.8	0.659	6.00-9.0	0.515-0.63
4	25.6	12.8	0.707	6.94-10.5	0.521-0.64
5	19.3	2.2	0.338	6.51	0.581
6	24.8	0.8	0.180	6.64	0.517
7	14.7	0.8	0.233	6.43	0.661
8	8.0	0.9	0.335	3.94	0.702

Figure 5 shows two typical examples of measured bore profiles above the initial still water level (Run No.3 and 6). It can be seen that undular bores are observed in the wave profiles of the case whose initial still water is deep, however there is no undular bore in the case of an extremery shallow initial depth. This fact can be explained by using the results shown in Figure 3. From the consideration of characteristics in the uniform region (2), the undular bore criterion is $u_2 = c_2$, which corresponds to the intersection of the curves c_2/c_1 and u_2/c_1. Circles in Figure 3 indicate the experimental results of U/c_1 and c_2/c_1. Numbers indicate Run No. in Table 1. Solid circles represent the cases of undular bores and white ones are strong bore fronts without undular components. Figure 6 shows Froude numbers F_{r2} calculated by measured currents and bore heights. It can be confirmed that Froude numbers exist in the range, $1 < F_{r2} < 2$.

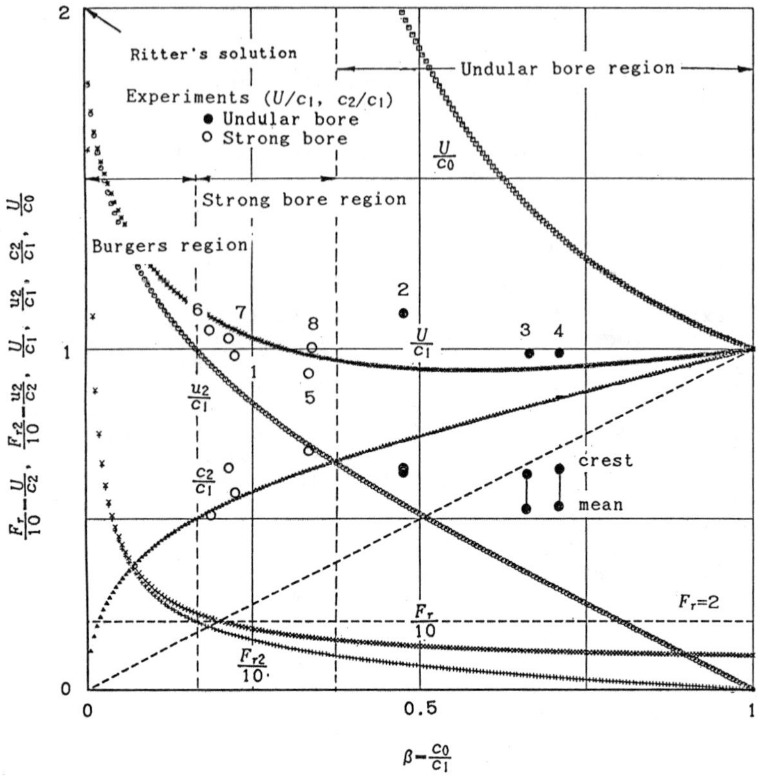

Figure 3. Solutions of dam break problem.

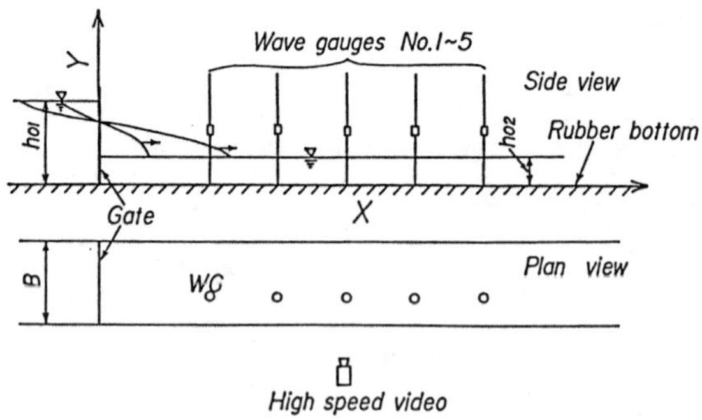

Figure 4. Set-up of experiment.

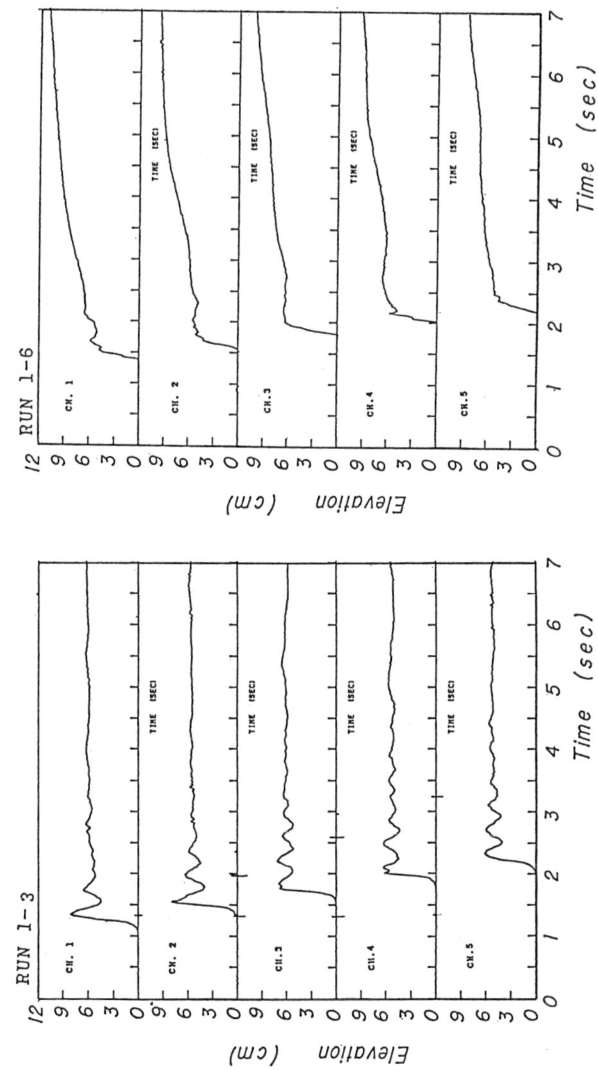

Figure 5. Examples of bore front surface profiles. (above the still water level)

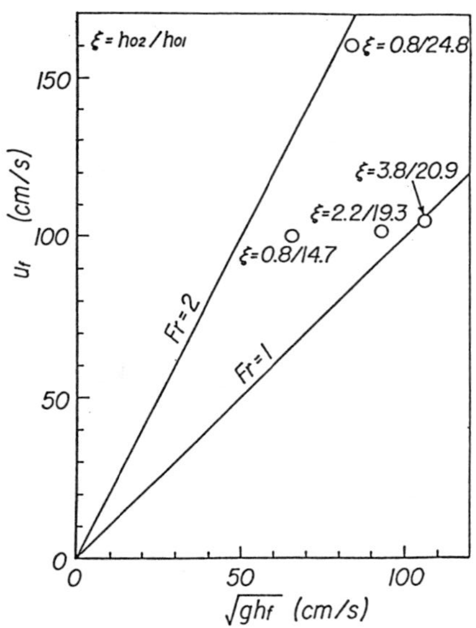

Figure 6. Experimental results of bore front Froude number.

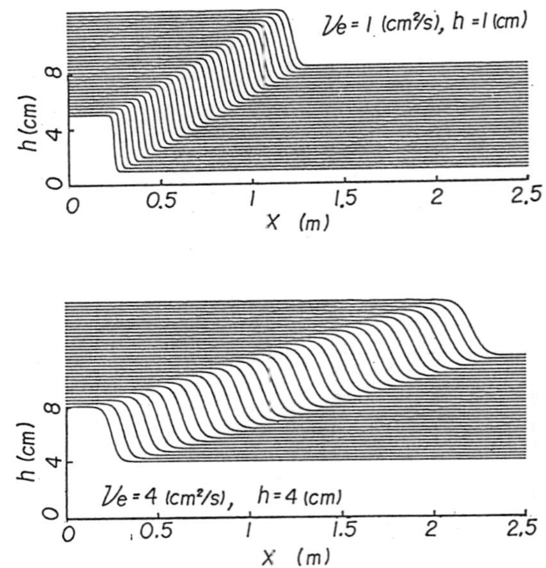

Figure 7. Test runs of numerical calculation of Burgers eq.

In this figure it should be noted that bore propagating speed U/c_1 keeps almost constant up to the point of $F_{r2} = 2$, which is corresponding to Jeffery-Vedernikov condition. Futhermore experimental results of U/c_1 have good agreement with the curve calculated by nonlinear long wave theory, especially in the region of strong bore. Unfortunately there is no data in the region of $F_{r2} > 2$, however, it might be that the nonlinearity of the propagation and turbulent mixing effects balance each other to keep the relation of $F_{r2} = 2$ in this region. This is the basic conception of Burgers approximation which will be elaborated on in next section. It is also mentioned in the figure that the dispersion effect can not be neglected in the region of $F_{r2} < 1$, in which experimental reults in c_2/c_1 are smaller than the theoretical one. This result indicates that undular bore components sustain their energy in the form of wave motion while some of them are transported in the backward direction.

3. BORE FRONT MODELING BY BURGERS EQUATION

3.1 Derivation of Burgers Equation

To introduce the balance effect of nonlinearity and dissipation, the horizontal momentum diffusive effect, $\nu_e u_{xx}$, is considered in the momentum equation (2).

$$u_t+uu_x+g'h_x = g'S-C_f\frac{u^2}{h}+\nu_e u_{xx}, \tag{23}$$

where ν_e is the horizontal turbulent eddy viscosity.

First of all, for the 0-th order(basic) flow condition, uniform flow, still water or dry land must be defined. For the open channel flow balancing between gravitational force and bottom friction the kinematic wave condition is the lowest order of the flood wave formulation called Kleitz-Seddon law. In addition to this it is possible to apply Burgers equation to the analysis of a discontinuity propagating on the 0-th order condition of still water or horizontal dry land. To achieve this aim the higher order effects of nonlinearity of propagation and turbulent diffusivity should be considered in the momemtum equation, which should be described by Burgers equation. Employing the formulation of nonlinear long wave theory in terms of Riemann invariant, two variables of the propagation speed of disturbance defined by $c = \sqrt{gh}$ and the characteristic speed defined by $\lambda = u+c$ are choosen. The basic equations are rewritten using these variables as,

$$c_t+(\lambda-\frac{3}{2}c)c_x+\frac{1}{2}c\lambda_x = 0 \tag{24}$$

$$\lambda_t+(\lambda-\frac{1}{2}c)_x+\frac{3}{2}cc_x=g\tan\theta\frac{C_f u_0^2}{h_0}+\nu_e\lambda_{xx}+\nu_e c_{xx} \tag{25}$$

Based on the 0-th order solutions (kinematic waves or still water) 1-st order (Burgers discontinuity) and higher order components are introduced by

$$\lambda = \lambda_0+\varepsilon\lambda^{(1)}+\varepsilon^2\lambda^{(2)}+\cdots$$

$$c = c_0+\varepsilon c^{(1)}+\varepsilon^2 c^{(2)}+\cdots \quad (26)$$

where ε is defined by the parameter of magnitude of discontinuity, *i.e.* amplitude of discontinuity

$$\varepsilon = \frac{(\lambda_{-\infty}-\lambda_{\infty})}{\lambda^{(1)}_{-\infty}}. \quad (27)$$

From the similitude of the Gardner-Morikawa transformation balancing equation, the nonlinearity of propagation and diffusivity are achieved by the following transformation (Taniuchi & Nishihara (1977)).

$$\xi = \varepsilon(x-\lambda_0 t), \qquad \tau = \varepsilon^2 t \quad (28)$$

Applying the G-M transformation to the fundamental equations (24),(25) and substituting Eq(26), we get Burgers equation

$$\lambda^{(1)}_{\tau}+\lambda^{(1)}\lambda^{(1)}_{\xi} = \frac{2\nu_e}{3}\lambda^{(1)}_{\xi\xi} \quad (29)$$

and hyperbolic equation of $s^{(1)}$

$$c^{(1)}_{\tau}+\lambda^{(1)}c^{(1)}_{\xi} = 0, \qquad \lambda^{(1)} = \frac{3}{2}u^{(1)} \quad (30)$$

These equations (29) and (30) are the Burgers equation system derived here for the problem of bore front propagation. Notice that discontinuity h or c propagates by the speed of $3u/2$ and the dynamic characteristic speed λ is described by Burgers equation (29). This also means that λ satisfys the Jeffery-Vedernikov condition which describes the intersection of the dynamic and kinematic characteristics through the discontinuity. As the discontinuity becomes greater, horizontal turbulent diffusivity increases in order to decrease the propagation speed, which is the balancing of effects of nonlinearity of propagation and horizontal diffusivity of momentum. The bottom friction does not contribute to the propagation of Burgers components.

For the practical purpose, we can calculate the front propagation speed $\lambda = \lambda_0+\varepsilon\lambda^{(1)}$ and surface elevation h and current u by using the resultant set of equations when the up-stream boundary condition $\lambda_{-\infty}$ and 0-th order flow (basic flow) condition are given. If a bore propagates on the horizontal dry land and uniform flow exists in the up-stream region, $\xi = -\infty$, applying the basic flow condition, $\lambda_0 = 0$, the bore front speed can be given by $\lambda_F = \lambda_{-\infty}/2$.

The bore front formulation by using Burgers equation mentioned here might have following advantages.
(1) Continuous flow structures through the discontinuity can be calculated by the depth-averaged quantities.
(2) Bore front propagation speed and profile (front length) are also evaluated.
(3) Burgers equation can be transformed into the linear

advection/diffusion equation by the nonlinear transfomation, which makes the numerical calculation easy; this will be refered in the following section.

4. NUMERICAL CALCULATION METHOD OF BURGERS EQUATION

Burgers equation has an interesting property, in that it can be transformed into a linear equation (heat equation) by the Cole-Hopf transformation

$$h = -2\nu\varphi_x/\varphi, \quad \nu = \frac{2\nu_e}{3} \tag{31}$$

Applying this transformation to the derived Burgers equation (35) in the fixed coordinates, the advection/diffusion linear PDE is obtained

$$\varphi_t + \lambda_0\varphi_x = \mu\varphi_{xx}, \qquad \mu = \frac{2\nu_e}{3} \tag{32}$$

Leonard(1979) has developed a third order accurate numerical scheme for this equation called the QUICKEST algorithm. Basco(1984) has shown that using the quardratic upstream interpolation operator for the advection term of Eq.(5) along with a forward time operator and centered second derivative operator, is identically equal to a forward-time and centered advection operator(FTCS) plus removal of truncation error by employing a special technique. The resulting difference equation is

$$\varphi_j^{n+1} = \varphi_j^n + \{\rho(1-C_r)-\frac{C_r}{6}(C_r^2-3C_r+2)\}\varphi_{j+1}^n - \{\rho(2-3C_r)-\frac{C_r}{2}(C_r^2 -2C_r-1)\}\varphi_j^n + \{\rho(1-3C_r)-\frac{C_r}{2}(C_r^2-C_r-2)\}\varphi_{j-1}^n + \{\rho C_r+\frac{C_r}{6}(C_r^2 -1)\}\varphi_{j-2}^n \tag{33}$$

where $C_r = c_0\Delta t/\Delta x$, $\rho = \mu\Delta t/\Delta x^2$ and $c_0 = \sqrt{gh_0}$.
The abovementioned method of solving Burgers equation is wiggle-free, accurate and low cost.

Example testing of this method was performed by using propagation test of a bore whose profile was given by the kink solution of

$$h = h_0 + a\tanh\{-\frac{a}{2\nu}(x-x_0)\} \tag{34}$$

where a is half amplitude x_0 is initial position of bore, and the corresponding transformed profile is given by

$$\varphi = C\exp\{-\frac{1}{2\nu}\int h dx\} \tag{35}$$

Using this initial condition and the boundary conditions of constant water depth in the far field, example solutions of Burges equation were calculated as shown in

Figure 7.

5. CONCLUSIONS

Bore front hydraulics were investigated in terms of Burgers equation and experiments were carried out to make clear the dynamics of moving discontinuity of water flow. The maim results obtained are:
(1) Burgers equation has been derived from the one dimensional open channel flow equation with horizontal momentum mixing term. The derived equation system consists of Burgers with respect to dynamic characteristic and hyperbolic equation with respect to water surface, which satisfys the Jeffery-Vedernikov relation $F_{r2} = 2$ through discontinuity.
(2) It is made clear from experiments that in the region of $1 < F_{r2} < 2$ nonlinear long wave theory has a good applicability for bore front propagation, however, in the region of $F_{r2} > 2$ turbulent diffusivity of momentum has to be considered in terms of Burgers equation. Moreover, in the region of $F_{r2} < 1$ the dispersion effect is not negligible in the observed undular bores.
(3) An accurate numerical calculation method to solve Burgers equation has also been proposed by employing the Cole-Hopf transformation and QUICKEST algorithm.

Acknowledgements

The authors would like to express their appreciation to Mr. M.Imashio, Civil Engineer of Kawasaki Steel Co.Ltd., for his help in the experiments. Part of this investigation were funded by a Grant-in-Aid for Scientific Research (No.59020023) from the Ministry of Education, Science & Culture of Japan.

References

Basco,D.R.:An implicit wiggle-free and accurate upstream finite-difference algorithm for the one-dimensional advection-diffusion equation, USAE, Waterway Exp. Sta., M.P. EL-84, 49p.,1984.

Leonard,B.P.:A stable and accurate convection modelling procedure based on quardratic upstream interpolation, Computer Meth.in App. Mech. Eng., vol.19, pp.59-98, 1979.

Taniuchi, T and K. Nishihara: Nonlinear Wave Theory, Iwanami-shoten, p. , 1977.

Whitham G.B.: Linear and Nonlinear Waves, John Wiley & Sons, p.636, 1974.

CHAPTER 193

Nearshore Circulations on a Wavy Coast

Ming-Chung Lin* and Sheng-Yeh Hwang**

1. Introduction

Nearshore circulations, produced by wave-induced radiation stress gradicents, form different circulation patterns under different wave characteristics and topographical conditions. Although numerous studies of nearshore circulations, such as Bowen (1969), Miller and Barcilon (1978), Dalrymple & Lozano (1978) and so on, have been appeared in the literature, it seems that little attention is paid on the case of a non-straight shoreline. O'Rorske & Leblond (1972) have investigated the wave-induced longshore currents in a semicircular bay, while Uda (1982) has used the coordinate transformation to treat nearshore circulations on a circular-arc shaped coast. Lin and Lee (1982), introduced a small perturbation quantity of wave set-up and set-down, induced by the non-straight shoreline, into the mean total water depth to obtain a governing partial differential equation by which they investigated the nearshore circulations on a cuspate coast. In Lin and Liou (1986), a more general equation was deduced on the orthogonal curvlinear coordinate system to unify the diversities among the related theories, and moreover, to investigate the rearshore circulations on the arc-shaped coast.

2. Mathematical formulation

With the wave field there will be associated a mean transport velocity which we will lump together with the secondary mean currents that we wish to describe under the velocity vector U, assumed horizontal and depth purely independent. The mean momentum density of a water column is

$$M = \rho d U \tag{1}$$

where ρ is the water density and d the total water depth. For a steady wave field the mean mass conservation equation reduces to

$$\nabla \cdot M = 0 \tag{2}$$

and then the linearized mean momentum equation is expressed as (O'Rourke and Leblond, 1972)

---

* Professor, Dept. of Naval Architecture, National Taiwan Univ., Taipei, Taiwan, Rep. of China.
** Graduate Institute of harbor and River Eng., National Taiwan College of marine Science and Technology, Keelung, Taiwan, Rep. of China.

$$\frac{1}{\rho d}\frac{\partial S_{ij}}{\partial X_i} = -g\frac{\partial \zeta}{\partial X_i} - \frac{fU_i}{\rho d} \tag{3}$$

where f is the friction coefficient, a function of total water depth and wave amplitude; S_{ij} is the radiation stress of the waves, g is the acceleration of gravity. In this equation, the lateral friction is neglected. Define

$$T_i = \frac{1}{\rho d}\frac{\partial S_{ij}}{\partial X_j}, \quad F_i = -\frac{fU_i}{\rho d} \tag{4}$$

then, in surf zone, equation (3) can be rewritten as

$$\nabla \times T = \nabla \times F \tag{5}$$

In linear model (O'Rourke and Leblond, 1972), in the outside surf zone:

$$\nabla \times F = 0 \tag{6}$$

and in the surf zone

$$\nabla \times T = \nabla \times F \tag{7}$$

The coordinate system used and the domain interested are duplicated as Fig. 1. in which ϕ is the angle of wave incidence, A and B, denote the region inside and outside the surf zone respectively.
the shoreline is assumed as

$$Y = \alpha \cos kx, \quad k = 2\pi/L \tag{8}$$

where L is spatial wave length of shoreline, and α shoreline wave amplitude. Applying conformal mapping (Lyne, 1971)

$$z = x+iy, \quad \zeta = \mu_1 + i\mu_2 = z - i\alpha e^{ikz} \tag{9}$$

it yields, up to $O(\alpha k)$

$$\begin{aligned} \mu_1 &= x + \alpha e^{-ky} \sin kx \\ \mu_2 &= y - \alpha e^{-ky} \cos kx \end{aligned} \tag{10}$$

The continuity equation (2) then becomes

$$\frac{\partial}{\partial \mu_1}(J^{-1/2}\, du) + \frac{\partial}{\partial \mu_2}(J^{-1/2}\, dv) = 0 \tag{11}$$

where J is Jacobian:

$$J = \left|\frac{d\zeta}{dz}\right|^2 = 1 + 2\alpha k e^{-ky}\cos kx + (\alpha k)^2 e^{-2ky} \tag{12}$$

By equation (11) the velocity components u and v are

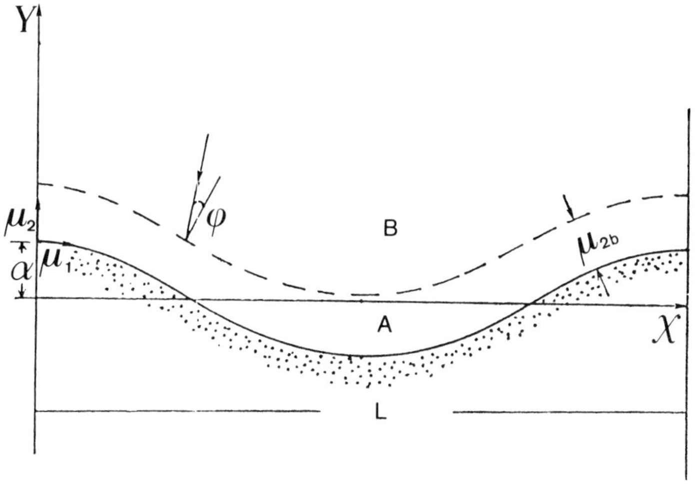

Fig. 1 Schematic diagram of coordinate system.

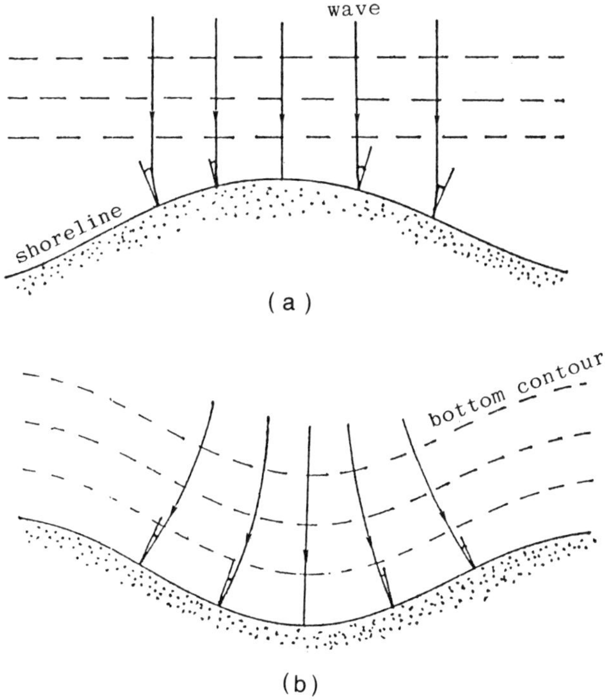

Fig. 2 Cases designed for calculations.

represented in terms of the transport stream function x as

$$u = \frac{J^{1/2}}{d} \frac{\partial X}{\partial \mu_2}, \quad v = \frac{J^{1/2}}{d} \frac{\partial X}{\partial \mu_1} \tag{13}$$

In surf zone, we express radiation stress components in (μ_1, μ_2) system to be ($T\mu_1$, $T\mu_2$) then

$$\nabla \times T = J[\frac{\partial}{\partial \mu_1} (J^{-1/2}.T\mu_2) - \frac{\partial}{\partial \mu_2} (J^{-1/2}\, \mu_1)] \tag{14}$$

where $T\mu_1$, $T\mu_2$ are defined as

$$T\mu_1 = \frac{J}{\rho d} [\frac{\partial}{\partial \mu_1}(J^{-1/2}.S\mu_1\mu_1) + \frac{\partial}{\partial \mu_2}(J^{-1/2}\, S\mu_1\mu_1)]$$

$$T\mu_2 = \frac{J}{\rho d} [\frac{\partial}{\partial \mu_1}(J^{-1/2}.S\mu_2\mu_1) + \frac{\partial}{\partial \mu_2}(J^{-1/2}\, S\mu_2\mu_2)] \tag{15}$$

Substituting equation (15) into equation (5), and expanding into Taylor's series, up to O(αk), we obtain

$$J\, [\frac{\partial}{\partial \mu_1}(\frac{f}{\rho d^2} \frac{\partial x}{\partial \mu_1}) + \frac{\partial}{\partial \mu_2}(\frac{f}{\rho d^2} \frac{\partial x}{\partial \mu_2})] =$$

$$J\, \{\frac{\partial}{\partial \mu_1} [\frac{J}{\rho d} \frac{\partial}{\partial \mu_1}(J^{-1/2}.S\mu_1\mu_2) + \frac{J}{\rho d} \frac{\partial}{\partial \mu_2}$$

$$(J^{-1/2}.S\mu_2\mu_2)] - \frac{\partial}{\partial \mu_2} [\frac{J}{\rho d} \frac{\partial}{\partial \mu_1} (J^{-1/2}.S\mu_1\mu_1)$$

$$+ \frac{J}{\rho d} \frac{\partial}{\partial \mu_2} (J^{-1/2}.\, S\mu_2\mu_2)]\} \tag{16}$$

where J_1 can be expanded by Taylor's series, up to O(α k)

$$J^{-1/2} = 1-\alpha k e^{k\mu_2} \cos k\mu_1 \tag{17}$$

In the region outside the surf zone, equation (6) can be similarly transformed to be

$$\nabla \times F = J\, [\frac{\partial}{\partial \mu_1} (\frac{f}{\rho d^2} \frac{\partial x}{\partial \mu_1}) + \frac{\partial}{\partial \mu_2} (\frac{f}{\rho d^2} \frac{\partial x}{\partial \mu_2})]$$

$$=0 \tag{18}$$

we further assume that in the seaward zone the mean currents do not reach to the bottom, but to about the water depth at the breaker line, as assumed by Liu and Lennon (1978). Thus the total water depth $d = \zeta + h_b$, where h_b is the water depth at the breaker line, will be taken as a constant. Assume $d = d_b$, we get

$$\frac{\partial^2 X}{\partial \mu_1^2} + \frac{\partial^2 X}{\partial \mu_2^2} = 0 \tag{19}$$

Let

$$\mu_1 = \frac{2\pi}{k}\xi , \quad \mu_2 = \frac{2\pi}{k}\eta \tag{20}$$

We can get the nondimensional equation:

$$\frac{\partial^2 X}{\partial \xi^2} + \frac{\partial^2 X}{\partial \eta^2} = 0 \tag{21}$$

3. General solution

In the seaward zone, the stream function satisfies $\nabla^2 X = 0$. We will label the stream function in each one of the two regions with a corresponding letter subscript. Now we consider some boundary conditions. The normal velocity at the shore line must vanish, that is

$$X_A(\xi, 0) = 0 \tag{22}$$

Further, the normal velocity must be matched at the boundaries between the two regions. By matching the stream function across those boundaries, this requirement is achieved:

$$\begin{aligned} X_A(\xi, 1) &= X_B(\xi, \delta) \\ X_{A,B}(0, \eta) &= 0 \\ X_{A,B}(1, \eta) &= 0 \\ X_B(\xi, \infty) &= X_\infty \end{aligned} \tag{23}$$

Set

$$X_A = X_1 + \delta X_2 \tag{24}$$

for δ is a small perturbation magnitude. In surf zone, we shall assume that the total mean depth is a linear function of the distance from shore

$$d = m(\xi).\eta \tag{25}$$

where $m(\xi) = m_o (1 + p \cos 2\pi\xi)$, m_o, p are small quantity. Set a simple, but reasonable form for ϕ_b, we take

$$\phi_b = -\phi_o \sin 2\pi\xi \tag{26}$$

Substituting equation (25) and (26) into equation (16) and make it nondimensional with

$$X_{A,B} = \frac{\pi g^{1/2} \gamma m_b^{5/2}}{C} \cdot X^*_{A,B} \tag{27}$$

we finally get the solutions for the region inside the surf zone:

$$X_A^* = \frac{\phi_o}{6}\left(\text{Sin } 2\pi\xi + \frac{5}{4}\text{p Sin } 4\pi\xi\right)\cdot\eta^3$$

$$+\frac{2\pi}{7}\delta\text{p}\left(\text{Sin } 2\pi\xi + \frac{3}{4}\text{p Sin } 4\pi\xi\right)\cdot\eta^{\frac{7}{2}}$$

$$+\frac{7\pi\phi_o^2}{24}\delta\cdot\text{Sin } 4\pi\xi\cdot\eta^4 + (\alpha k)\{\frac{\phi_o}{4}(\text{Sin } 4\pi\xi$$

$$+\frac{5}{4}\text{p Sin } 2\pi\xi + \frac{5}{4}\text{p Sin } 6\pi\xi)\cdot[\frac{\eta^3}{3} - \frac{k\mu_{2b}}{4}\cdot\eta^4$$

$$+\frac{(k\mu_{2b})^2}{10}\cdot\eta^5] - \frac{5\pi}{4}\delta(\text{Sin } 2\pi\xi + \frac{5}{4}\text{p Sin } 2\pi\xi)$$

$$\cdot[\frac{2}{7}\eta^{\frac{7}{2}} - \frac{2k\mu_{2b}}{9}\cdot\eta^{\frac{9}{2}} + \frac{(k\mu_{2b})^2}{11}\cdot\eta^{\frac{11}{2}}]$$

$$+\frac{\pi\delta\text{p}}{2}(\text{Sin } 4\pi\xi + \frac{3\text{p}}{4}\text{Sin } 2\pi\xi + \frac{3}{4}\text{p Sin } 6\pi\xi)$$

$$\cdot[\frac{2}{7}\eta^{\frac{7}{2}} - \frac{2k\mu_{2b}}{9}\cdot\eta^{\frac{9}{2}} + \frac{(k\mu_{2b})^2}{11}\cdot\eta^{\frac{11}{2}}] + \pi\phi_o^2\delta$$

$$\cdot(\text{Sin } 2\pi\xi + \text{Sin } 6\pi\xi)\cdot[\frac{\eta^4}{6} - \frac{2k\mu_{2b}}{25}\cdot\eta^5 + \frac{(k\mu_{2b})^2}{42}\cdot\eta^6]$$

$$+\frac{\pi\phi_o^2\delta}{4}(\text{Sin } 2\pi\xi + \text{Sin } 6\pi\xi)\cdot[\frac{\eta^4}{4} - \frac{k\mu_{2b}}{5}\cdot\eta^5$$

$$+\frac{(k\mu_{2b})^2}{12}\cdot\eta^6] + \frac{\pi\phi_o\delta}{4}\cdot(\text{Sin } 4\pi\xi + \frac{5}{4}\text{p Sin } 2\pi\xi$$

$$+\frac{5}{4}\text{p Sin } 6\pi\xi)\cdot[\frac{\eta^4}{4} - \frac{k\mu_{2b}}{5}\cdot\eta^5 + \frac{(k\mu_{2b})^2}{12}\cdot\eta^6]\} \tag{28}$$

Equation (21) for the region outside the surf zone is the Laplace equation which can be solved easily. The final nondimensional form of the solution is

$$X_B{}^* = e^{-2\pi(\eta-\delta)} \cdot \{ \frac{\phi_o}{6} [\mathrm{Sin}\ 2\pi\xi + \frac{5}{4} p\ \mathrm{Sin}\ 4\pi\xi$$

$$\cdot e^{-2\pi(\eta-\delta)}] + \frac{2\pi}{7} \delta p [\mathrm{Sin}\ 2\pi\xi + \frac{3}{4} p\ \mathrm{Sin}\ 4\pi\xi$$

$$\cdot e^{-2\pi(\eta-\delta)}] + \frac{7\pi}{24} \delta \phi_o{}^2 \cdot \mathrm{Sin}\ 4\pi\xi \cdot e^{-2\pi(\eta-\delta)}$$

$$+ (\alpha k) \{ \frac{\phi_o}{4} [\mathrm{Sin}\ 4\pi\xi \cdot e^{-2\pi(\eta-\delta)}$$

$$+ \frac{5}{4} p\ \mathrm{Sin}\ 6\pi\xi \cdot e^{-4\pi(\eta-\delta)} + \frac{5}{4} p\ \mathrm{Sin}\ 2\pi\xi]$$

$$\cdot [\frac{1}{3} - \frac{k\mu_{2b}}{4} + \frac{(k\mu_{2b})^2}{10}] - \frac{5\pi\delta}{4} \cdot [\mathrm{Sin}\ 2\pi\xi$$

$$+ \frac{5}{4} p\ \mathrm{Sin}\ 4\pi\xi \cdot e^{-2\pi(\eta-\delta)}] \cdot [\frac{2}{7} - \frac{2}{9} k\mu_{2b} + \frac{(k\mu_{2b})^2}{11}]$$

$$+ \frac{\pi}{2} \delta p [\mathrm{Sin}\ 4\pi\xi \cdot e^{-2\pi(\eta-\delta)} + \frac{3}{4} p\ \mathrm{Sin}\ 2\pi\xi$$

$$+ \frac{3}{4} p\ \mathrm{Sin}\ 6\pi\xi \cdot e^{-4\pi(\eta-\delta)}] \cdot [\frac{2}{7} - \frac{2k\mu_{2b}}{9}$$

$$+ \frac{(k\mu_{2b})^2}{11}] + \pi\phi_o{}^2\delta [\mathrm{Sin}\ 2\pi\xi + \mathrm{Sin}\ 6\pi\xi \cdot e^{-4\pi(\eta-\delta)}]$$

$$\cdot [\frac{1}{6} - \frac{2}{25} k\mu_{2b} + \frac{(k\mu_{2b})^2}{42}] + \frac{\pi\phi_o{}^2\delta}{4} [\mathrm{Sin}\ 2\pi\xi$$

$$+ \mathrm{Sin}\ 6\pi\xi \cdot e^{-4\pi(\eta-\delta)}] \cdot [\frac{1}{4} - \frac{k\mu_{2b}}{5} + \frac{(k\mu_{2b})^2}{12}]$$

$$+ \frac{\pi\phi_o\delta}{4} \cdot [\mathrm{Sin}\ 4\pi\xi \cdot e^{-2\pi(\eta-\delta)} + \frac{5}{4} p\ \mathrm{Sin}\ 6\pi\xi$$

$$\cdot e^{-4\pi(\eta-\delta)} + \frac{5}{4} p\ \mathrm{Sin}\ 2\pi\xi] \cdot [\frac{1}{4} - \frac{k\mu_{2b}}{5}$$

$$+ \frac{(k\mu_{2b})^2}{12}] \} \} \qquad (29)$$

The solutions X_A and X_B (A indicates inside surf zone; B outside) finally obtained include two parts, one concerned with (αk) and the other independent on (αk), where (αk) denotes the steepness of wavy shoreline.

4. Numerical results

If we neglect $O(\alpha k)$, then

$$X^*_A = \eta^3 \{ \frac{\phi_o}{6} [\sin 2\pi\xi + \frac{5}{4} p \sin 4\pi\xi] + \frac{2\pi}{7} \delta p \sin 2\pi\xi \cdot \eta^{1/2} + \frac{7\pi}{24} \delta\phi_o^2 \sin 4\pi\xi \cdot \eta \} \qquad (30)$$

$$X^*_B = e^{-2\pi(\eta-\delta)} \{ \frac{\phi_o}{6} [\sin 2\pi\xi + \frac{5}{4} p \sin 4\pi\xi \cdot e^{-2\pi(\eta-\delta)}] + \frac{2\pi}{7} \delta p \sin 2\pi\xi + \frac{7\pi}{24} \delta\phi_o^2 \sin 4\pi\xi \cdot e^{-2\pi(\eta-\delta)} \} \qquad (31)$$

Set

$$\Psi^*_{A,B} = X^*_{A,B} \times 10^2 \qquad (32)$$

we take the term up to $O(\alpha k)$, in order to find the difference from that up to $O(\alpha k)^o$ done by Uda (1982). Two cases of topographical conditions, as shown in Fig. 2, are designed. In case A, we take $\phi_o = 0.35$, $p=1/3$; in case B, $\phi_o = -0.35$ $p=1/3$ are used. Then take the wave length of wavy shoreline L=1000m, 500m and 300m in turns, the amplitude α=50m and 80m, and μ_{2b}=100m. Some typical results are shown in Fig. 3-10. Table 1 shows the corresponding between those Figures and cases considered.

Table 1. Corresponding between Figures and Varions Cases.

α =50m				
	up to $O(\alpha k)$		up to $O(\alpha k)^o$	
case / L=	A	B	A	B
1000m ($\alpha k/2$ =0.05)	Fig. 3	Fig. 6	Fig. 7	
500m ($\alpha k/2$ =0.1)	Fig. 4			
300m ($\alpha k/2$ =0.16)	Fig. 5		Fig. 8	
α =80m				
	up to $O(\alpha k)$		up to $O(\alpha k)^o$	
case / L=	A	B	A	B
1000m ($\alpha k/2$ =0.08)	Fig. 9	Fig. 10		

From Fig. 3, Fig. 4 and Fig. 5, we can find that the shorter the length of one wavy shoreline is, the smaller the magnitude of circulations are going to be; in similarity, from Fig. 3 and Fig. 9, only reducing the amplitude α, the magnitude of the scope of circulations are smaller. Consequently, we can indicate that the steepness of a giant cuspate shoreline has influence to nearshore circulations. By comparing Fig. 3 with Fig. 7. It appears that term O(αk) has not much influence, while from Fig. 5 and Fig. 8, the term $O(\alpha k)$ has more influence. In addition it is supposed the positive values of $X^*_{A,B}$ denote counterclockwise of circulations, while minus values of $X^*_{A,B}$ mean that the circulations is in clockwise. Comparing to the definition of ϕ_o, we get the results that in case (A), the direction of circulations are counterclockwise, while in case (B), it is in the clockwise, since we just take a half of one wave length of shoreline into account, with symmetry we can get the whole associated nearshore currents.

5. Concluding remarks

Although the situations considered here may be too simple, two important conclusions are made concerning with the effects of the steepness of a wavy shoreline on the nearshore circulations:

(1) The smaller the steepness of a periodic shoreline, the larger the scope of the nearshore circulation.

(2) It is common that there appear two even three circulations, when the steepness becomes larger to some

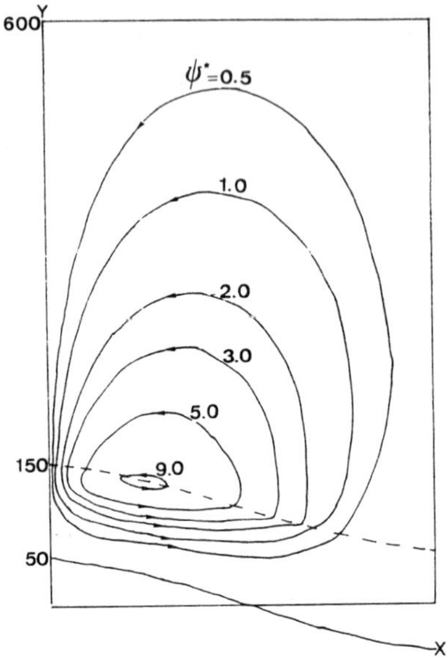

Fig. 3 Circulation pattern, Case(A), αk/2π=0.05, L=1000M, α=50M.

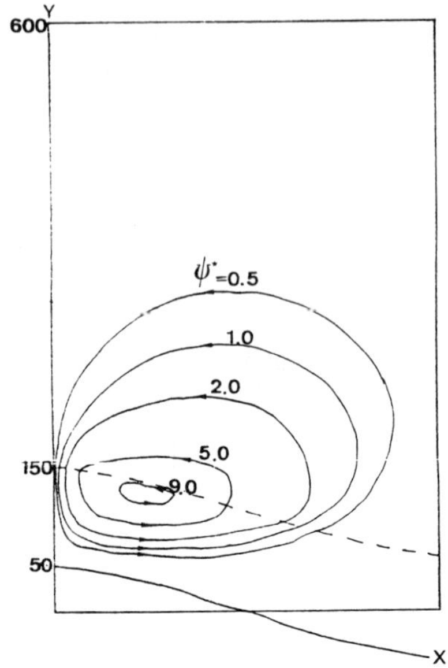

Fig. 4 Circulation pattern, Case (B), αk/2π=0.1, L=500M, α=50M.

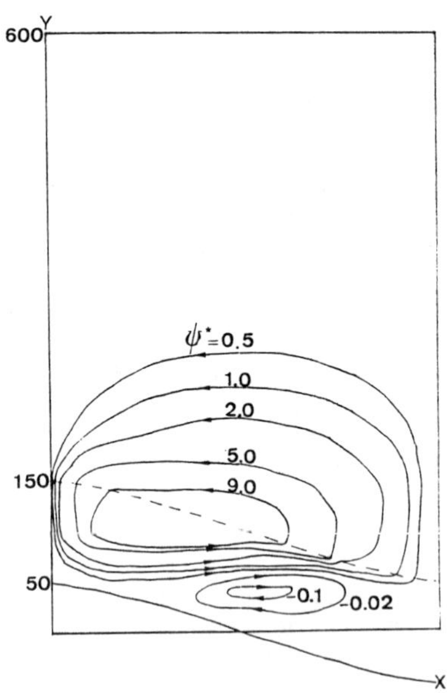

Fig. 5 Circulation pattern, Case(A), αk/2π=0.16, L=300M, α=50M.

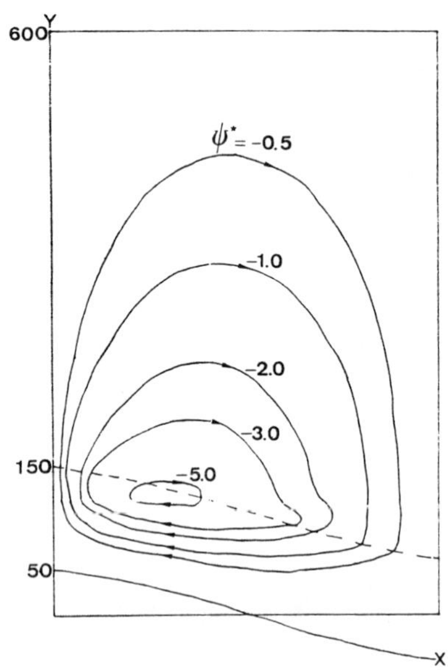

Fig. 6 Circulation pattern, Case(B), αk/2π=0.05, L=1000M, α=50M.

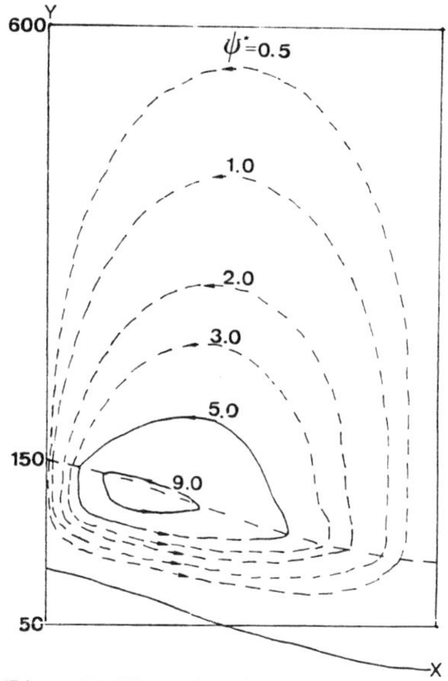

Fig. 7 Circulation pattern, Case(A), O(αk) term neglected, αk/2π=0.05, L=1000M, α=50M.

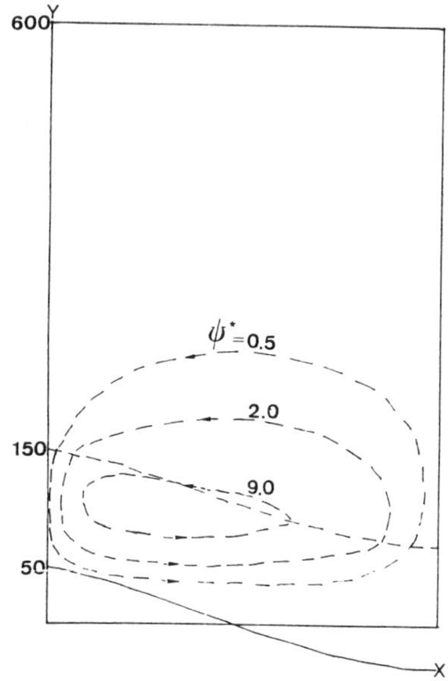

Fig. 8 Circulation pattern, Case(A), O(αk) term neglected, αk/2π=0.16, L=300M, α=50M.

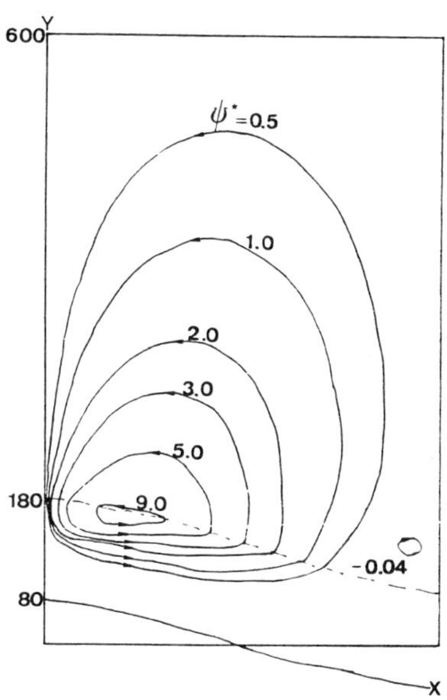

Fig. 9 Circulation pattern Case(A), αk/2π=0.08, L=1000M, α=80M.

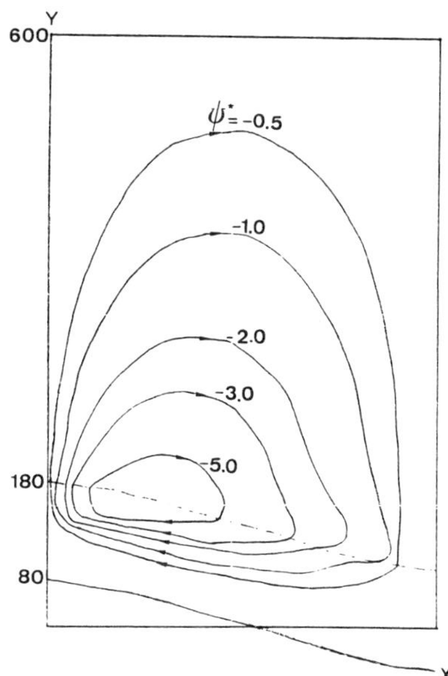

Fig. 10 Circulation pattern, Case(B), αk/2π=0.08, L=1000M, α=80M.

amount at extent.

6. Reference

1. Bowen, A.J. (1969) "Rip currents 1. theoretical investigations", J. Geophys. Res., Vol. 74, No. 23, pp.5467-5478.
2. Dalrymple, R.A. and C.J. Lozano (1978), "Wave-current interaction models for rip currents", J. Geophys. Res. Vol.83, pp.6063-6071.
3. Lin, M.C. and Y.L. Lee (1982) "Nearshore circulations on a cuspidal coast" Proc. 6th Conf. on Ocean Eng. in Taiwan, pp.37-1-37-21. (in Chinese).
4. Lin, M.C. and J.Y. Liou (1986) "Analytical study of wave-induced longshore currents", Journal of the Chinese Institute of Engineers, Vol. 9, No.5, pp.437-449.
5. Liu, P.L.F. and G.P. Lennon (1978): "Finite element modeling of nearshore currents", J. waterways, Port, Coastal and ocean Div., ASCE, vol.103, pp.175-189.
6. Lyne, W.H. (1971): "Unsteady viscous flow over a wavy wall", J. Fluid Mech., Vol.50. pp.33-48.
7. Miller, C and A. Barcilon (1978): "Hydrodynamic instability in the surf zone as a mechanism for the generation of horizontal gyre", J. Geophys. Res., Vol. 83, No.C8, pp.4107-4116.
8. O'Rourke, J.C., and P.H. LeBlond (1972) "Longshore currents in a semicircular bay", J. Geophys. Res., Vol.77, No.3, pp.444-452.
9. Uda. T. (1982):"Nearshore circulation around a circular spit and the cuspate coast", Proc. JSCE, No.326, pp.63-72 (in Japanese).

CHAPTER 194

STUDY ON THE VARIATIONS OF AN EUTROPHIC ECOSYSTEM FROM THE SPANISH MEDITERRANEAN LITTORAL: DATA TO HAVE IN MIN IN THE NUTRIENT DUMPING TO THE SEA.

by

E. Soler T. (1); **J.G. del Río** (1) **and J.J. Díez** (2).

ABSTRACT:

Cullera Bay is a neritic ecosystem placed on the Spanish Mediterranean Littoral largely influencied by the Jucar River, that brings about lower salinities than sourrounding waters, and broad variations of its values. An extensive research, with 9 samplings throughout the year, was carried out, measuring both physical and chemical parameters, and the planktonic communities. The trophic status of the ecosystem, the spatial and temporal variations of the nutrients and the planktonic communities were studied, evaluating the influence of the river loads and the littoral dynamics. Some essential basis to allow a suitable emplacement of waste waters disposals along the Valencian littoral are set up in order to minimize the gradual eutrophication of this coast.

INTRODUCTION:

According to the classic literature (until fifties), the incidence of human wastes on the trophic status of the littoral communityes was almost negli-

----

(1) Biologist. Cát. Ing. Sanitaria, Dpto. Ing. Hidráulica y Medio Ambiente, E.T.S.I.C.C. y P., Univ. Politécnica, Valencia, SPAIN.
(2) Civil Engineer. Cát. Oceanografía, E.T.S.I.C.C. y P., Univ. Politécnica, Madrid. SPAIN.

geable. It was thus likely because of the lack os a systematic study of the waters from the continental shelf and the peculiar conditions of the ecosystems where these surveys were carried out. Nevertheless, after current studies, coastal ecosystems have appeared as much more complex and diverse than offshore ones.

Numerous processes of eutrophication caused by domestic waste waters have been monitored, and both politicals and scientists have found a lot of objections to standarize their study, monitoring and management, because of their broad variations according to local parameters -coastal shape, littoral dynamics, bulk and kind of loads, etc.

Thus, it is very important to keep in mind the particular characteristics of these areas where waste waters are dumped to the sea, and to conduct specific mionitoring programmes to minimize their ecological consecuences. In other wise, the uncontrolled spills may cause unpredictible and damaging consecuences to the ecosystem. Regrettable examples are well known: the recent "red tides" generated by blooms of toxic phytoplankton species (Baltic Sea, Adriatic Sea, Galician Esuaries, etc.), the gradual degradation of benthic communities, the sanitary problems of beaches, etc.

In an extensive study carried out in Cullera Bay (eastern Spain), the combined effects on the neritic ecosystems caused both by the unpurified disposals of waste waters and the particular hydrodynamics conditions into the bay were evaluated throughout a year.

SAMPLING AREA:

Cullera Bay is placed on the Spanish Mediterranean Littoral (0 13'-0 15', W; 39 8'- 39 12', N), on the middle of the so called Valencian Oval (Figs.: 1A and 1B). A particular environmental problem is raised in this bay because of:

- The Jucar River's presence, which outflows at the southern beaches of the bay.

- The development of the tourism in this area, reaching more than 250.000 inhabitants in summer. The domestic wate waters are directely dumped into the bay through an insufficient marine outfall.

- The importance of the agriculture, involvong a great amount of waste waters with high levels of nutrients, detritus, pesticides, etc.

The particular hydrodynamics within the bay, largely conditioned by some docks built in the mouth of the river and along the northern beaches. This allows the formation of a semi-enclosed area at the norht, where the most serious environmental problems are noticed.

18 sampling stations were spread over the bay: 9 along the beaches and 9 within the bay (Fig.: 1C). 9 samplings were done throughout a year, from April-1983 to March-1984.

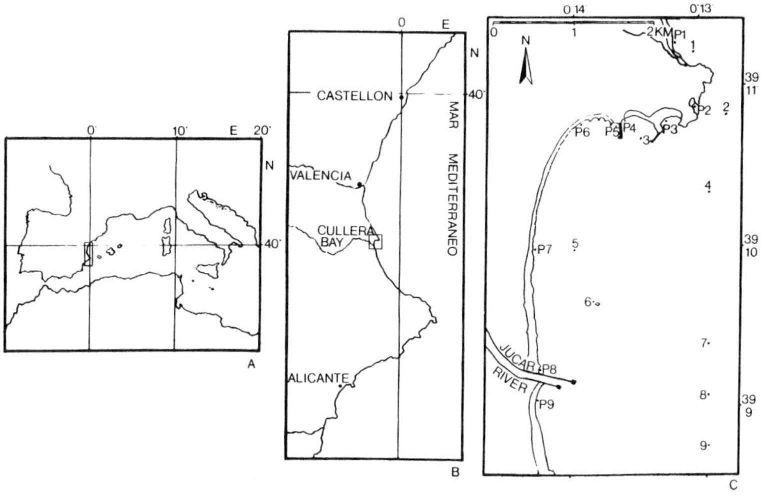

Fig.: 1.- Geographical location of Cullera Bay and sampling stations. A: Western Mediterranean Basin; B: Valencian Littoral (Valencian Oval); C: Cullera Bay with sampling stations (P=Beaches).

MATERIAL AND METHODS:

The parameters measured in ecah sample were:

.- Quantitative and qualitative evaluation of phytoplankton communities.

.- Sea temperature.

.- Salinity.

.- Winds and waves.

.- Nutrients: NO2, NO3, NH4, PSR and SiO4.

Samplings were done with a Van Dorn bottles of 2x4 l, at 0 m depth at beach stations, and at 0 and 5 m depth at bay stations. Phytoplankton samples were fixed with lugol and studied at the laboratory according to Uthermöhl techniques. Salinity was also determined at the laboratory with a salinometer by induction GRUNDY ENVIRONMENTAL SYSTEMS, Inc., 6230 N (The International Oceanographic Tables Vol. I, UNESCO). Because of the broad temporal ans spatial variations of surface salinities, the average values of the salinity at the stations placed at 5 m depth have beenused in the present study, because they show more accurately the mean salinity values.

Nutrients were determined following basicaly the Strickland and Parsons' (1965) and Grashoff's (1975) methods, with some modifications by Solorzano (1969), Mullin and Riley (1955) and Murphy and Riley (1962).

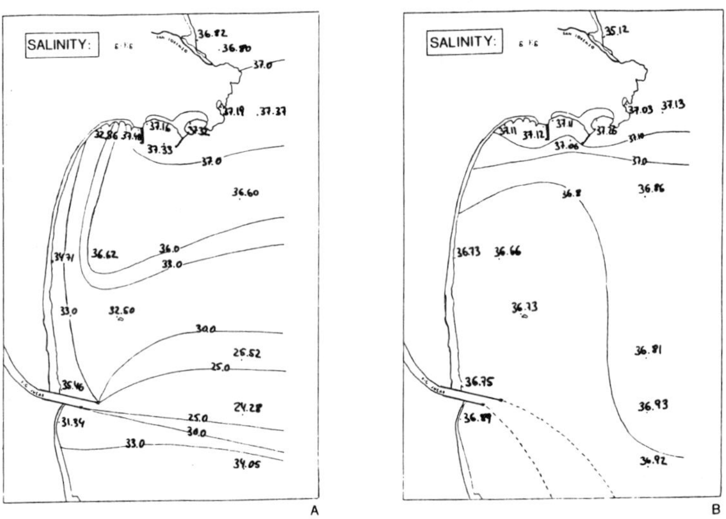

Fig.: 2.- Variations in surface salinities caused by changes in wind direction. A: North winds; B: South winds (G. del Río, 1987).

RESULTS:

The influence of the river gives rise to lower salinities than sourrounding waters, with remarkable spatial and temporal variations. It is possible to differenciate between the variations caused by momentary changes of water direction and those originated by changes of average values in the water body. In the former, the local winds are the determinant factor, being less important the waves and the river flow.

These variations involve esentialy to the upper layer, showing a bigger influence of freshwaters when the south wind prevails, and of offshore waters when it is the north or east wind the prevailing one (Figs.: 2A and 2B).

		S g/Kg	P.S.R. µatg/l	NH_4^+ µatg/l	NO_2^- µatg/l	NO_3^- µatg/l	SiO_4^{-3} µatg/l	N.I.D. µatg/l
10-5-83	$\bar{X}$	37,90	0,36		0,07	11,3	0,7	
	C.V.	0,2	41,7		42,9	61,9	57,1	
4-7-83	$\bar{X}$	37,25	0,10	0,12	0,06	7,7	3,2	7,81
	C.V.	0,2	20,0	50,0	66,7	24,7	40,6	24,7
19-7-83	$\bar{X}$	37,43	0,03	1,26	0,08	1,9	2,1	3,18
	C.V.	0,2	66,7	70,6	12,5	52,6	23,8	38,4
25-8-83	$\bar{X}$	37,43	0,33	1,02	0,05	4,7	1,0	5,77
	C.V.	0,2	21,2	73,5	40,0	34,0	20,0	37,3
10-10-83	$\bar{X}$	37,61	0,30	15,97	0,34	7,8	5,6	25,75
	C.V.	0,6	16,7	96,3	20,6	73,3	41,1	72,3
29-11-83	$\bar{X}$	37,54	0,38	2,13	0,53	6,2	4,7	8,31
	C.V.	0,1	10,5	16,4	7,6	32,3	10,6	25,5
17-1-84	$\bar{X}$	38,03	0,23	1,67	0,12	3,4	0,7	5,07
	C.V.	0,1	26,1	20,4	16,7	88,2	28,6	63,7
31-1-84	$\bar{X}$							
	C.V.							
21-2-84	$\bar{X}$	38,09	0,14	1 42	0,37	14,9	3,2	16,32
	C.V.	0,3	35,7	20,8	59,5	59,7	75,0	56,5
28-3-84	$\bar{X}$	37,88	0,18	1,90	0,45	15,5		17,47
	C.V.	0,8	44,4	57,9	64,4	34,2		34,0

Table: 1.- Mean valalues (X) and coefficient of variation (C.V.) of salinity and nutrients at 5 m depth (P.S.R. = Soluble Reactive Phosphorus; N.I.D. = Dissolved Inorganic Nitrogen) (G. del Río, 1987).

Several factors are involved in the mean salinity of the bay, being the most importants:

- The river discharge.
- The evaporation.
- The littoral dynamics.

Atending to the first and second ones, and assuming a maximum evaporation in summer and a rainfall pattern with summer and winter as dry seasons and rains in spring and, mainly, in autumm, the highest salinity records should be expected in summer and the lowest ones in autumm. Nevertheless, our data do not agree with that pattern (Table: 1), showing that it is the littoral dynamics which is the essential factor to explain this kind of variation.

		FITOPLANCTON nº cél./1.x10^6	CIANOFICEAS nº cél./1.x10^6	COCOLITOFORIDOS nº cél./1.x10^6	DIATOMEAS nª cél./1.x10^6	DINOFLAGELADOS nº cél./1.x10^6	EUGLENALES nº cél./1.x10^5	FLAGELADOS nº cél./1.x10^6
10-5-83	$\bar{X}$	0,80	--	--	0,58	0,06	--	0,15
	C.V.	57,5	--	--	69,0	70,9	--	100,0
4-7-83	$\bar{X}$	0,61	0,16	--	0,22	0,05	0,02	0,16
	C.V.	54,1	106,3	--	18,2	9,3	150,0	243,8
19-7-83	$\bar{X}$	0,79	--	--	0,59	0,20	--	0,15
	C.V.	63,3	--	--	62,7	60,0	--	253,3
25-8-83	$\bar{X}$	2,55	0,03	--	2.38	0,06		0,09
	C.V.	30,5	133,3	--	32,8	50,0	--	77,8
10-10-83	$\bar{X}$	3,76	5,44	--	2,74	0,28	--	0,25
	C.V.	51,1	51,5	--	66,8	46,4	--	84,0
29-11-83	$\bar{X}$							
	C.V.							
17-1-84	$\bar{X}$	3,55	0,12	--	2,15	0,08	--	1,17
	C.V.	36,3	58,3	--	61,9	50,0	--	29,1
31-1-84	$\bar{X}$							
	C.V.							
21-2-84	$\bar{X}$	1,97	--	--	0,72	0,28	--	0,89
	C.V.	39,6	--	--	41,7	28,6	--	39,3
28-3-84	$\bar{X}$				0,95			
	C.V.				41,1			

Table: 2.- Mean values (X) and coefficient of variation (C.V.) of phytoplankton community at 5 m depth (G. del Río, 1987).

The salinity of the bay results from the mixing of waters with different salt concentration. Firstly, the inshore waters from the bay, seccondly the continental freshwaters from the river, and finally the more salted offshore waters. The ratio of each one of them is largely depending on the littoral dynamics.

Two levels in water mixing have to be distinguished between the effects caused by the waves. On the first hand, a low intensity level, which do not generate a quantitatively significant mixing in the water column within the bay. On the other hand, a level of higher intensity, when the autumm and winter storms produce the mixing not only of the water coumn, but also the dilution of the bay water in the large bulk of open sea waters removed by these storms; thus, an increase in the salinity is observed, reaching its maximum values in this period. The clear deviation from the average value of the sample taken on 29th November was on account of the intense storm and flood suffered some days before of sampling.

The waters taking part in the mixing have distinct salinities at the same time as very distinct trophic status. Those from the Jucar river are very eutrophiced, those from the bay have an intermediate eutrophication levels, and those coming from open sea are clearly less eutrophic and cleaner. According to the ratio of each one of them in the mixing a neat accumulation or a loss of nutrients will be produced, respectively.

Thus, it is possible to suppose that an accumulation of nutrients is taken place during summer and a loss in winter (Table: 1). This can be observed by comparing the samples on 10th October (before the storms) and on 17th January (at the heigh of winter); not only a higher phytoplankton standing stock (Table: 2), but also higher mean values of the nutrient concentrations (G. del Río, 1987). This winter loss is favoured by the bulk of offshore waters displaced by these storms to the bay, which release and redilute the benthic nutrients from the sediments.

Several dense phytoplankton blooms have been observed in Cullera throughout the year. This is a characteristic feature of eutrophic coastal ecosystems, where the shallowness of the waters allows the release of benthic nutrients in the water column with a low intensity mixing of waters, because no strong termocline is formed. Only in calm waters like in the sampling of July, the flow of nutrients from sediment to water column ceases, and the ecological succession beguins in

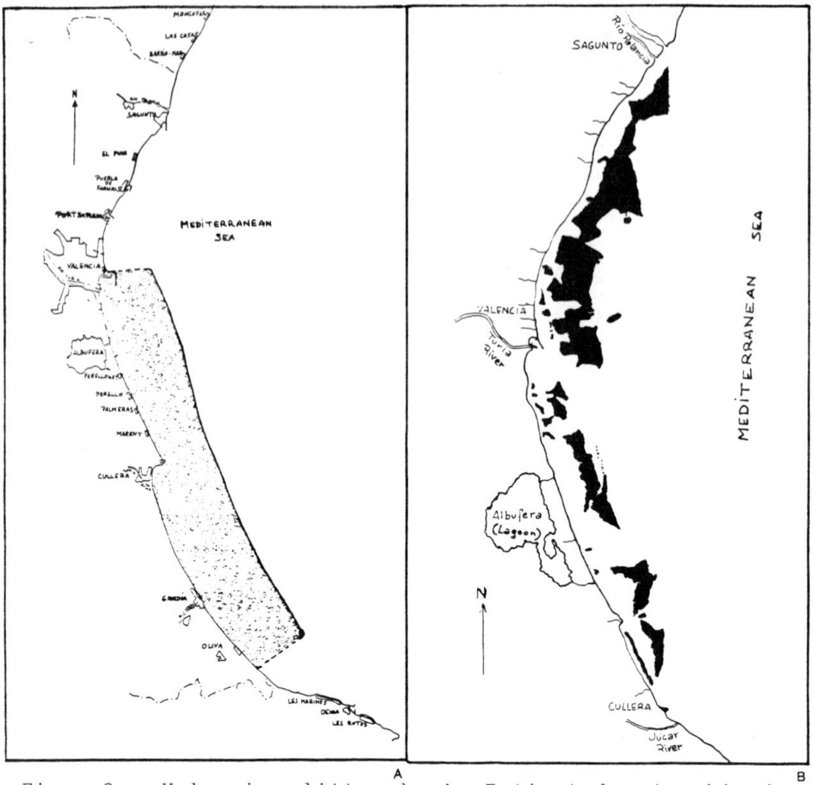

Fig.: 3.- Valencian littoral. A: Estimated eutrophiced area (dotted zone), where the Posidonia oceanica has almost disappeared and the plankton based ecosystem is the dominant one; B: Posidonia oceanica sea grases over forty years ago spread over the Valencian coast (black zone).

plankton populations (G. del Río, 1987). Furthermore, the differences between the phytoplankton communities of the northern beaches are sharper in this period. similar fenomenons have been noticed in Kiel Bight (Smetacek, 1985), Rumanian coasts (Minhea and Cuingioglu), Narragansett Bay (Smayda, 1983), the estuaries of St. Lawrence (Sinclair, 1978) and Puget Sound (Winter, 1975), Anka Bay (Iverson et al., 1979) among others.

This kind of neritic ecosystems with highly

eutrophiced waters is not restricted to this place of the Valencian Oval. In a survey carried out that winter (1983-1984) between Valencia Harbour and Cullera Bay, very similar qualitatively and quantitatively phytoplankton communities appeared. But the area with a high eutrophic level is clearly delimited by the dramatic regression of the *Posidonia oceanica* sea grass. It comprises the coast between Valencia Harbour and the southern limit of the province (Fig.: 3A), where a very intense and fast industrialization and an increase of the resident and non-resident population have happened, and where vast sea grasses of this marine phanerogam spread over the coastal mobile bottoms of this section of littral have disappeared in a few years (Fig.: 3B). Because of that, and an intense overfishing by trawling, an oligotrophic benthic based ecosystem has been transformed into an eutrophic and plankton based one, with limited diversity and very low economic profitability, and a drastic decrease in the number of benthic and demersal species.

This new and unbalanced situation maakes possible the unpredictable and uncontrolable proliferation of opportunistic species, like the massive blooms of the benthic Rodophyta *Ceramium fastigiatum* during several years along the Valencian Oval, accumulating large amounts of this red algae on the beaches, mainly in summer, owing to local hydrodynamic features (Cullera Bay, Puebla de Farnals, etc.), and causing sanitary and aesthetic problems in these tourist beaches. But in the same way as a non-toxic algae has proliferated, other toxic phytoplankton species can do it too, causing environmental problems as the "red tides".

The high richness of waters and, indirectly, the increase of water turbidity (phytoplankton blooms, suspended particles, detritus, etc.) seems to be the main reason for the regression of the sea grasses and the proliferation of the red algae in this littoral.

CONCLUSIONS:

Marine outfalls obviously can not be considered the definitive solution to the problem of the disposal iof waste waters into the sea but, in any case, they must be contemplated as a suitable choice to mitigate the uncontrolable and impredictable impact caused by the loads in the same shoreline. Nevertheless, not only sanitary aspects must be taken into account in the project and management of sea dumplings, but also the eutrophic conditions of the ecosystem and the problems in regard to them. It seems hence suitable to consider as conclusions of the present work four suggestions in order to keep the Valencian coasts as clean and as healthy as possible.

1.- To preserve the Posidonia oceanica sea grasses of the northern part of the littoral, placing the disposal points as far away as possible from them.

2.- To asses the real spreading of the affected area by the eutrophication, and to evaluate the loss of benthic and demersal communities in regard to the volume and pollutant charge of the disposasls, specially on P. oceanica.

3.- To favour the correct dilution of the nutrients in the open sea along the coast, removing them from the shoreline.

4.- To keep in mind that some coastal constructions (harbours, docks, etc.) are both sand traps and nutrient-organic matter too, because of the special hydrodynamic conditions generated around and inside of them, favouring hence mass blooms of opportunistic species, some of them of damaging effects for the ecosystem, like the "red tides".

REFERENCES:

G. DEL RIO, J., 1987. Problemas de eutrofización litoral. El caso de la Bahía de Cullera. Ph. Thesis (unpubl.). Universidad de Valencia. 514 pp. and two attached vols.

GRASSHOF, K., 1976. Methods of seawater analysis. Verlag Chemie. Weinheim. New York. 317 pp.

IVERSON, R.L.; T.E. WHITLEDGE and J.J. GOERING, 1979. Chlorophyll and nitrate fine structure in the southeastern Bering Sea shelf break front. Nature, London. 281: 664-666.

MINHEA, J.B. and E. CUINGIOGLU, 1982. Particularités des phénomènes de "floraison". Cercetari. Mar. 15: 27-57.

MULLIN, J.B. and J.P. RILEY, 1955. The spectrophotometric determination of silicate-silicon in natural waters with special reference to seawater. Annal. Chem. Acta. 12: 162-170.

MURPHY, J. and J.P. RILEY, 1962. A modified single solution method for the determination of phosphate in natural waters. Annal. Chem. Acta. 27: 31-26.

SINCLAIR, M., 1978. Summer phytoplankton variability in the lower st. Lawrence Estuary. J. Fish. Res. Bd. Can. 35: 1171-1185.

SMAYDA, T.J., 1983. The phytoplankton of estuaries. Estuaries. (ed. B.H. Ketchum). 65-102.

SMETACEK, V., 1985b. The annual cycle of Kiel Bight plankton: a long-term analysis. Estuaries. 8: 145-157.

SOLORZANO, L., 1969. Determination of ammonia in natural waters by the phenol hypoclorite method. Limnol. Oceanogr. 14: 799-803.

STRICKLAND, J.D.H. and T.R. PARSONS, 1965. A practical handbook of seawater analysis. Bull. Fish. Res. Can. 167: 1-310.

WINTER, D.F.; K. BANSE and C.G. ANDERSON, 1975. The dynamics of phytoplankton blooms in Puget Sound, a fjord in the northwestern United States. Mar. Biol. 29: 139-176.

CHAPTER 195

FLUSHING BEHAVIOUR OF A COASTAL MARINA

R. A. Schwartz * and J. Imberger **

Abstract

The flushing characteristics of a newly constructed marina were determined from a comprehensive field study which included a dye dilution experiment together with measurements of barotropic and baroclinic forcing parameters. The results to be presented show that the major exchange process in this particular marina is a baroclinic circulation, and that the exchange rate is controlled by the harbour entrance geometry.

1.0 Introduction

The continued growth of population centres along the Western Australian coast implies that the existing recreational and commercial boating facilities will not satisfy future demand. To increase the number of sheltered boat moorings, state and local authorities are constructing new marinas and enlarging the existing facilities. State regulations require that such projects be preceded by an Environmental Review and Management Planning (E.R.M.P.) report, wherein the potential impact on water quality must be assessed. Since the quality of the water in a coastal harbour largely depends on the natural ability of the harbour to flush out trapped pollutants, any water quality forecast requires an understanding of how the water is exchanged with the adjoining sea.

The circulation of water between a marine harbour and the connecting sea is generally the result of natural parameters such as the local tidal range, wave climate, meteorological conditions, and water density differences. Depending on the geographical location, one or more of these variables can dominate the exchange properties of a harbour. Pollutants such as pesticides, heavy metals, and organic compounds enter a marina basin via surface run-off, leaching from boats' anti-fouling paint, and spillages from service facilities located within the marina boundaries. If pollutant concentrations are allowed to increase above critical levels, then the result is sub-standard water quality as characterized by a reduction in dissolved oxygen and the presence of algal blooms. Detrimental water quality within a marine harbour can be avoided by reducing the potential sources of pollutants,

* Graduate student, Department of Civil Engineering, University of Western Australia, Nedlands, W.A., 6009, Australia.
** Professor of Civil Engineering, and Director of the Centre for Water Research, University of Western Australia, Nedlands, W.A., 6009, Australia.
Environmental Dynamics reference ED - 88 - 259

and by maintaining an optimal flushing rate. The water exchange rate is affected by factors such as the planform geometry, entrance dimensions and number of entrances (Nece, 1984). So, once the geometrical effects and locally dominant flushing parameters are understood, a marina can be designed to have an optimal water exchange rate.

To understand the flushing characteristics of a marine harbour situated on the Western Australian coastline, the Centre for Water Research (CWR) in conjuction with the State Department of Marine and Harbours conducted a series of field experiments in a newly constructed marina. The specific aims of this project were to quantify the flushing rate as a function of time and space, identify the dominant flushing mechanisms, and measure fine-scale dissipation rates of turbulent kinetic energy. The results will be used to develop a strategy for water quality monitoring in all marinas under the control of the Department of Marine and Harbours and aid in designing future marinas.

The marina chosen for the field experiments is known as Hillarys Boat Harbour. It is located on the southwest Australian coast approximately 26 km north of Fremantle. The tidally averaged depth of Hillarys Boat Harbour is approximately 5.5 m. With a water surface area of about 2.46 x 10^5 m^2, its tidally averaged volume can be estimated as 1.3 x 10^6 m^3. The harbour was designed to contain 1,000 boat moorings. It has a single asymmetric entrance which is 90 m wide by 5 m deep, and is oriented to the north as shown in Figure 1.

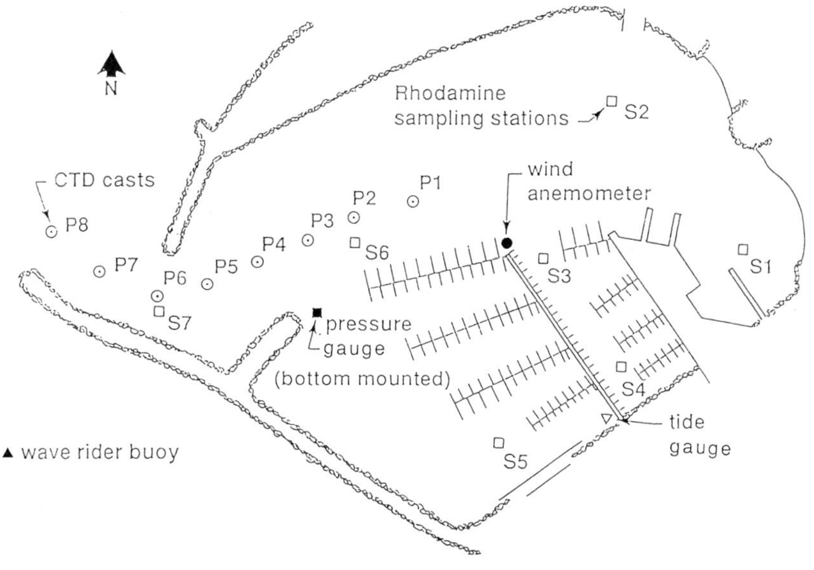

Figure 1 Hillarys Boat Harbour planform together and sampling sites.

The flushing time of Hillarys Boat Harbour was determined by uniformly seeding the harbour with a fluorescent dye and subsequently monitoring its dilution as a function of time and space. Simultaneous to the dye dilution experiment, vertical profiles of salinity and temperature were measured with Conductivity Temperature Depth (CTD) probes, and a microstructure probe was deployed to estimate the dissipation of turbulent kinetic energy. The *Djinnang III* was the largest of the two CWR research vessels used for collecting data. Its onboard instruments included a CTD probe, a microstructure probe, and a mini-ranger system for establishing the coordinates of the sampling stations. The other research vessel, known as the *Trail Craft,* also had a CTD probe onboard and was outfitted with a fluorometer which measured the concentration of the fluorescent dye. These vessels and their respective instruments were used for three weeks to monitor the response of the water body to natural forcing conditions.

Wind speeds, tidal amplitudes, and wave statistics were measured with remotely operated devices positioned as shown in Figure 1. Wind directions were obtained from the Bureau of Meteorology's weather station at Ocean Reef, located approximately 5 km north of the harbour. A CWR mini-logger was installed to measure wind speeds. The mini-logger was attached to RIMCO® cup anemometer and recorded a sample every ten minutes. Other remote recording instruments were installed and monitored by the Department of Marine and Harbours. These included a float/counter-weight type tide gauge mounted within the marina, a bottom-mounted Weslog® logger for recording wave actions inside the marina, and a Datawell® series 6000 wave-rider buoy located outside the marina.

2.0 Experimental Programme

Early on 29 April 1987 (Julian day 119), 60 litres of a 20% Rhodamine WT aqueous solution were diluted with seawater and injected into the harbour from a small dinghy. The dye mixture was pumped through a diffuser end-pipe which was continuously raised and lowered while the dinghy slowly criss-crossed the interior of the harbour. Immediately following the injection of the dye, a monitoring cycle was initiated at seven interior stations (S1 - S7 shown in Figure 1) to determine the dilution rate of the Rhodamine WT tracer as a function of time and space. At each station, tracer concentrations were measured just below the surface, at mid-depth and 1 m from the bottom. Water was selectively withdrawn through a special disc inlet (Imberger and Berman, 1986) with an effective vertical resolution of 0.30 m, and pumped up to a debubbler mounted one metre above a Turner Designs® model 10 series fluorometer. A gravity outflow from the debubbler supplied the fluorometer with a bubble-free stream of water, which was analysed and recorded on a Multiterm® micro-computer before being discharged overboard. The natural fluorescence of the marina water was measured prior to injecting the dye, and the fluorometer was tuned to zero out this 'blank' concentration. Each station was initially sampled several times a day. However as the experiment progressed, it was only necessary to sample once a day or once every other day.

In conjunction with the dye dilution experiment, free-falling and rising probes were deployed from the *Djinnang III* and the smaller *Trail Craft* to obtain density, salinity and temperature profiles inside and outside of the harbour. The free-falling CTD probe (Fozdar, 1983) consisted of a pressure transducer for determining depth, a SEA-BIRD SBE-3 oceanographic thermometer for temperature measurements and a SEA-BIRD SBE-4 conductivity cell for determining salinities. Water parcels were sampled at 50 Hz as the probe descended to the bottom with a fall-rate of 1 ms^{-1}. This is equivalent to a spatial resolution of 2 cm. The vertically rising microstructure profiler (RMP) (Carter and Imberger, 1986) was used to measure fine-scale temperature gradients from which rates of turbulent kinetic energy dissipation were estimated. As the RMP rises at 0.1 ms^{-1} from a pre-set depth, its pressure transducer, two fast-response thermistors and two specially designed conductivity cells sample at 100 Hz. This results in a spatial resolution of 1 mm, which represents a 200% increase in resolution over the CTD probe.

3.0 Data Analysis

Profiles representing the measured dye concentrations at every sampling station for each circuit showed that a non-uniform dye distribution existed throughout the harbour for 24 hours after the dye was injected. This initial depth and spatial variation is primarily a reflection of the logistics involved in manually seeding such a large volume of water to obtain a uniform distribution. A well-mixed condition was eventually achieved on Julian day 120, one day after the dye was introduced into the system.

Dilution Rate

The rate at which the tracer concentration decreased with time is shown in Figure 2. The plotted data were best described by an exponential curve fit of the form

$$C = 169 \times 10^{(-0.0917\,t)} \qquad (1)$$

where C is the concentration value for a particular Julian day value t. Using an initial Rhodamine WT concentration of 1.35×10^{-9} on Julian day 120, an e-folding time was calculated as the time required for the initial concentration to drop to C / e. Solving equation (1) for t yielded a corresponding Julian day value of 125. Thus, the measured dilution rate of Rhodamine WT had an e-folding time of 5 days!

Rhodamine WT is quite stable with respect to photochemical decay, sediment absorption, and pH; however, salinity and temperature can significantly affect the fluorescence of Rhodamine WT. Smart and Laidlaw (1977) reported contradictory results from two independent tests which produced radically different decay rates for the fluorescence of Rhodamine WT with long-term exposure (~10 days) in saline environments. To evaluate the effect of salinity on the fluorescence of Rhodamine WT in the harbour, a laboratory experiment was conducted under controlled conditions in which samples with dye concentrations of 1×10^{-9} were monitored for two weeks. The results showed a 3% reduction in the initial dye fluorescence for a typical seawater chlorinity of 19 $^o/_{oo}$ (salinity 35 $^o/_{oo}$). The effect of temperature on fluorescence intensity was determined from an empirical equation of the form

$$F = F_o\, e^{-0.027\,t} \qquad (2)$$

where F is the fluorescence at a standard temperature, F_o is the fluorescence at the sample temperature, and t represents the difference between the standard and sample temperatures (^{o}C). The error in fluorescence intensity which can be attributed to temperature variations was estimated to be less than 5%. Because the combined error in fluorescence intensity due to salinity and temperature was less than 8%, it was concluded that the observed dye loss in the harbour was caused by flushing actions and not chemical degradation.

Tidal Motions

The tides of the southwestern Australian coast are unusual because they are generally diurnal with amplitudes so small that astronomical and meteorological forces produce changes in the sea level of the same order of magnitude. Similar characteristics were exhibited in the sea level fluctuations measured during the three week field study, namely, an overriding diurnal pattern with amplitudes less than one metre.

The overall contribution to flushing from the oscillating sea level was determined by using the tidal prism method. This method assumes that the exchange flow is only a function of the tidal range, and any mass removed on the ebb tide does not return on the following flood tide. These assumptions were proven to be valid for Hillarys Boat Harbour by aerial photographs taken on Julian day 119. The photographs showed that wind-induced circulations outside the harbour quickly dispersed any dye which was flushed out with the ebbing tide. Given that the sea level prism is about 0.5 m, an e-folding time for the tidal flushing mechanism was calculated to be 11 tidal cycles. This corresponds to 11 days since the tidal cycle was primarily diurnal. A comparison between the analytical e-folding time of 11 days and the measured e-folding time of 5 days revealed a discrepancy which was too large to be attributed to experimental error. Therefore, the tide was not the dominant mechanism for driving an exchange flow between the harbour and the sea.

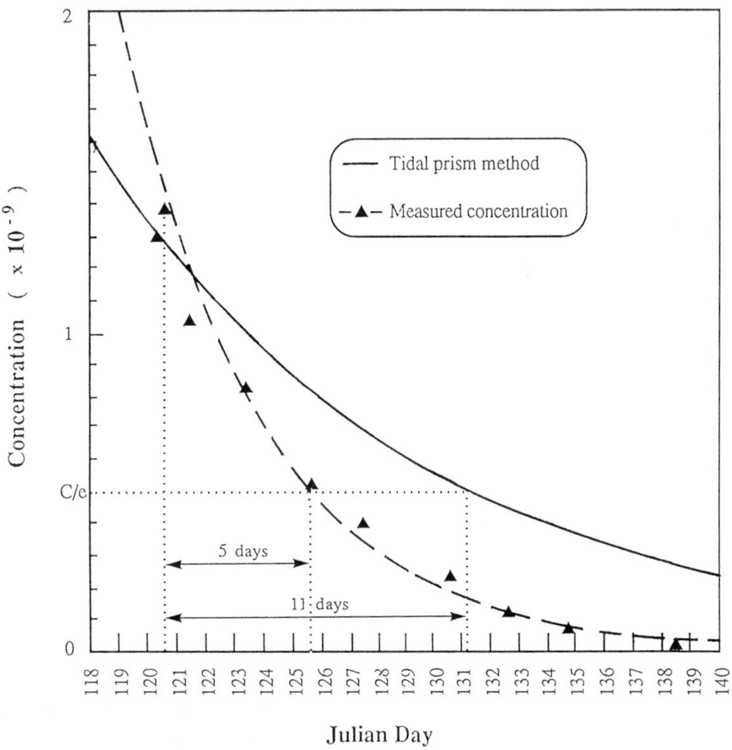

Figure 2 Comparison of the tidal prism method with the measured dilution rate.

Stratification

Hillarys Boat Harbour is situated on the coastal border of an inland freshwater mound which rises 70 m above sea level. Submarine groundwater discharges along the coast in this area were first reported by Johannes (1980). Freshwater patches were detected by the CTD probe near the bottom and along the shoreline inside the marina. The quantity of freshwater discharging into the marina was estimated to be about 2×10^3 m^3 day^{-1}.

Knowledge about the density structure of a water body is useful for quantifying the internal transport processes associated with advection, mixing and diffusion. The CTD and the vertically rising microstructure profiler (RMP) were deployed at various points in the vicinity of the harbour to measure conductivity and temperature characteristics. From these measurements, practical salinities and densities were calculated using the practical salinity scale and equations of state given in UNESCO (1981). The raw data were processed to account for the different response characteristics of the probes' sensors and their vertical separation. In this way, all the data relating to the same parcel of fluid were matched with each other. Failure to properly match the conductivity, temperature and depth signals results in unrealistic spiking in the salinity and density calculations. After the CTD and RMP data were adjusted for vertical separation, a recursive time series technique developed by Fozdar *et al.* (1985) was used to match the sensor response times.

Data collected from a series of equally spaced CTD drops were used to produce contour plots of salinity, temperature and density profiles. An interesting example of an isopycnal plot is shown in Figure 3. The density structure shown in this figure is indicative of a baroclinic circulation caused by a gravitational adjustment. Only a few of the isopycnal plots revealed such a pattern, while the remainder showed density structures characteristic of well-mixed to longitudinally stratified conditions. Thus, no single steady-state stratification condition existed within the harbour during this study. In Table 1, the various isopycnal plots were classified according to the degree of stratification exhibited along with measured wind speeds and tidal conditions. The noteworthy feature in this table is that an internal frontal structure resembling a standing wave was observed to occur under light wind conditions (less than 3 ms^{-1}) and predominantly on the ebb tide. This observation is supported by Simpson and Nunes (1981), who reported similar frontogenesis behavior occurring on the ebb tide in several different estuaries. Dyer (1973) also described frontal formations in a partially mixed estuary as a consequence of a distinct two-layer flow system.

Wind Effects

Air currents moving across an exposed body of water exert a drag force on the water surface in the form of a shear stress. This mechanism generates surface waves which act to transfer momentum and mechanical energy from the wind to the water surface, thereby contributing to the circulation dynamics of the water body. Effects such as topographic gyres, differential deepening of the mixed-layer and subsequent gravitational adjustments are all directly related to the input energy flux from the wind.

Imberger (1985) defined the surface mixed-layer in a stratified water body as that part of the water column which is directly influenced by the momentum and turbulence introduced by a surface wind stress. Production of turbulent kinetic energy (TKE) at the base and surface of the mixed-layer causes the mixed-layer depth to fluctuate. TKE produced at the base of the mixed-layer is known as "shear production" and is primarily caused by seiching motions. At the surface, TKE production is commonly referred to as "stirring". As the wind-induced TKE is dissipated down through the water column by viscous forces, heavier water from below the mixed-layer is entrained. The result is that the mixed-layer deepens, the potential energy of the water column increases, and a longitudinal density gradient is established. A subsequent reduction in TKE allows the density structure to "relax", thereby creating a gravitational current which further redistributes the mass within the water body.

Table 1 Classification of density contour plots.

Julian day	Time	Stratification	Wind Speed (ms^{-1})	Tide
119A	11:20 - 11:27	V.S.	< 3	Ebb
119B	13:12 - 13:27	V.S.	< 3	Ebb
119C	16:24 - 16:45	F	< 3	Ebb
119D	18:03 - 18:12	F	< 3	Ebb
119E	18:53 - 19:07	F	< 3	Ebb
120A	08:12 - 08:56	M	8	Flood
120B	12:44 - 12:59	M	10	Ebb
120C	14:28 - 14:59	M	12	Ebb
121A	08:38 - 09:19	L.S.	9	Flood
121B	14:40 - 15:09	L.S.	7	Ebb
123A	10:04 - 10:19	M - L.S.	4	Flood
123B	13:08 - 13:25	V.S.	4	Flood
123C	15:03 - 15:17	V.S.	5	Ebb
125A	11:01 - 11:14	V.S.	5	Flood
125B	17:29 - 17:43	F - V.S.	< 3	Ebb
127A	09:19 - 09:44	L.S.	6	Flood
127B	14:07 - 14:30	V.S.	5	Flood
130	11:44 - 12:00	F - L.S.	4 - 7	Ebb
132	11:19 - 11:38	F	3	Ebb
134	16:44 - 17:00	F	3	Ebb
135	00:16 - 00:38	F	< 3	Flood

V.S. = Vertically Stratified M = Well-mixed condition
L.S. = Longitudinally Stratified F = Frontal structure evident

The efficiency of the energy transfer process depends on the wind strength, the stability of the atmospheric boundary layer over the water surface, and the basin geometry relative to the wind direction. Of these influences, the wind speed is the dominant factor that determines the actual stress exerted on the water surface, and the response of the wind mixed-layer is a function of this kinematic shear stress.

Wave Actions

A Draper-Tucker wave analysis was performed to extract significant wave heights and periods from the data collected by the two Marine and Harbours wave recording instruments. The surface buoy, which was mounted outside the harbour in 8 m of water, measured an average significant wave height of 0.71 m and a mean period of 7.29 s. The bottom-mounted pressure transducer inside the harbour recorded an average significant wave height of 0.11 m and a 39.52 s mean period. The difference between the larger waves outside the harbour and the smaller waves inside is a measure of the protection provided by the breakwaters. The contribution to flushing by surface wave actions outside Hillarys Boat Harbour was considered to be negligible; however, the surface wave actions within the harbour were important for mixing.

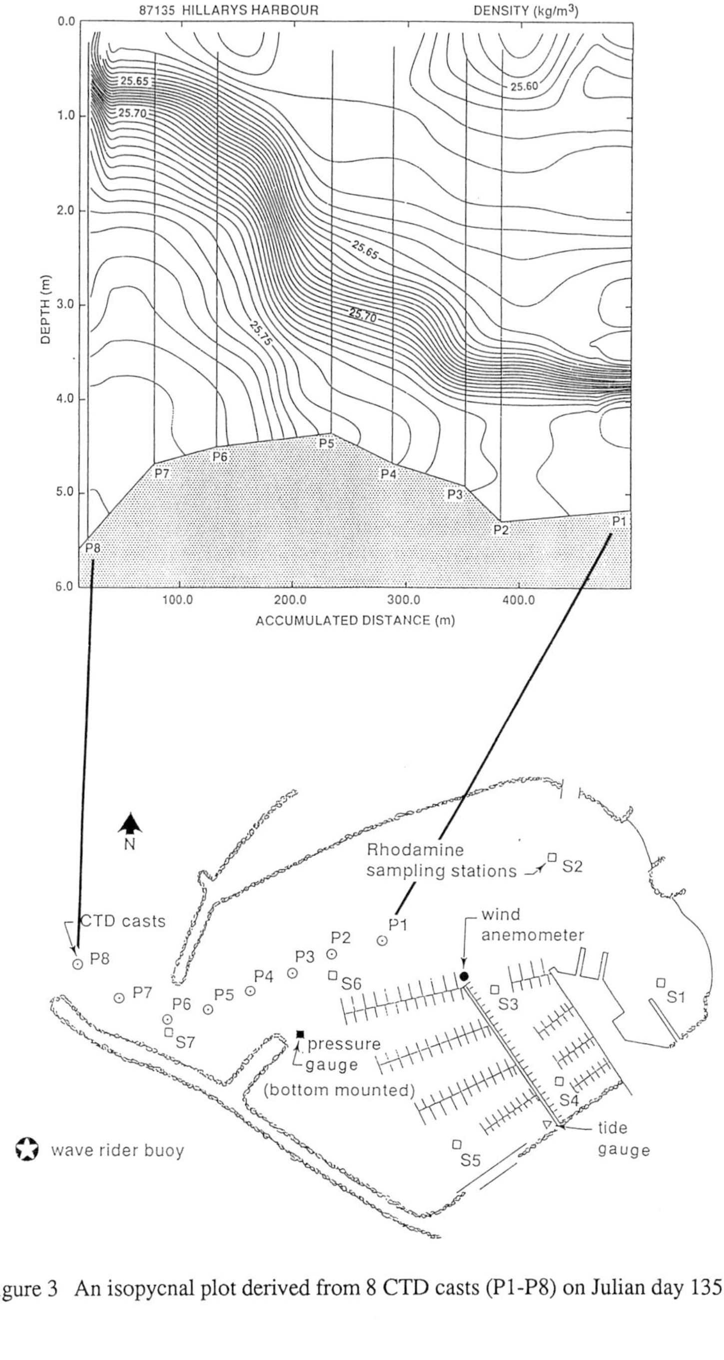

Figure 3 An isopycnal plot derived from 8 CTD casts (P1-P8) on Julian day 135.

Turbulent Mixing

It was necessary to assess the efficiency of turbulent mixing processes because the effectiveness of tidal flushing is dependent on the rate at which the tidal inflow is mixed with the harbour water. The transport of mass by turbulent motions is a function of the kinematic viscosity and the supply rate of kinetic energy. Energy enters a water body at large scales of motion which correspond to the basin dimensions, and it is transferred to smaller scales of motion by random velocity fluctuations and turbulent diffusion. In his universal equilibrium theory, Kolmogorov (1941a,b) formulated the following expression for the smallest length scale which is characteristic of turbulent motion,

$$\eta = (\varepsilon^{-1} \nu^3)^{1/4} \qquad (3)$$

where ε is the turbulent kinetic energy dissipation rate and ν is the kinematic viscosity. Typically, η is on the order of a few millimeters. At this level, viscosity becomes important and kinetic energy is dissipated into heat by internal friction forces. From knowing the rate at which kinetic energy is dissipated, the transport rate associated with the turbulent motions can be determined.

An estimate of the turbulent kinetic energy dissipation rate can be determined either directly from measurements of the velocity-shear spectrum, or indirectly from fine-scale temperature gradient measurements (Dillon and Caldwell 1980, and Oakey 1982). The current study made use of the latter approach, which is based on the premise that the spectrum of small-scale temperature variations is related to the energy spectrum by the same wavenumber. Batchelor (1959) used Kolmogorov's (1941) hypothesis to predict the existence of a length scale beyond which no further refinement of the temperature distribution can occur. The smallest length scale for temperature fluctuations is derived by balancing heat advection with thermal diffusion and is known as the Batchelor length scale. The reciprocal of the Batchelor length scale is the Batchelor wavenumber κ_B written as

$$\kappa_B = (\varepsilon \nu^{-1} D^{-2})^{1/4} \qquad (4)$$

where D is the molecular diffusivity of heat. Gibson and Schwarz (1963) used Batchelor's results to derive the one-dimensional Batchelor wavenumber spectrum whose upper limit represents the onset of viscous or diffusive effects. Thus, equation (4) can be solved to give an estimate of the turbulent kinetic energy dissipation rate after the maximum Batchelor wavenumber has been identified.

To identify the maximum Batchelor wavenumber, a Wigner-Ville distribution was applied to the microstructure temperature gradient signal from the vertically rising microstructure profiler (RMP) instrument in accordance with the method described in Imberger and Boashash (1986). The Wigner-Ville transform method is an analysis technique whereby the energy content of a non-stationary signal, such as the microstructure temperature gradient, is distributed in a time-frequency domain. This technique yields the instantaneous and maximum frequencies of the energy spectrum. The maximum temperature fluctuation frequency is then converted by a dispersion relationship into a wavenumber representing the maximum Batchelor wavenumber.

The RMP probe was only used on Julian day 119 at two different locations within the harbour. The probe was deployed three times at each location to increase statistical confidence in the collected data trends. The respective dissipation rates calculated from the temperature measurements are listed in Table 2 along with the corresponding times of deployment and wind speeds.

Table 2 Dissipation rate estimates from RMP temperature measurements.

Julian Day	Time	U_3 (ms^{-1})	ε ($m^2 s^{-3}$)
$119A_1$	10:31	2.3	3.1×10^{-7}
$119A_2$	10:41	2.3	4.4×10^{-7}
$119A_3$	10:48	2.6	4.2×10^{-7}
$119B_1$	12:08	3.1	3.8×10^{-7}
$119B_2$	12:16	2.3	3.0×10^{-7}
$119B_3$	12:30	3.0	3.7×10^{-7}

Scaling arguments were used to estimate a time scale for the turbulent diffusion process. The time scale associated with horizontal transport due to turbulent motions can be written as

$$T_x \sim (L_x)^2 / K_x \qquad (5)$$

where K_x is a horizontal diffusion coefficient which represents the non-advective mixing processes; therefore, its real value is highly dependent on the magnitude of these non-linear processes. Fischer *et al.* (1979) depicted a unified diagram of patch diffusion (after Okubo, 1974) from which it is possible to estimate a diffusion coefficient given the size of a tracer cloud. Given that the width of the harbour is of order 1×10^4 cm, K_x was found to have a value of 1×10^2 $cm^2 s^{-1}$. Evaluating (5) with $K_x = 1 \times 10^{-2}$ $m^2 s^{-1}$ and $L_x = 100$ m gives a value for a horizontal transport time scale of order 10 days, which was representative of the turbulent diffusion processes associated with wind speeds of about 2.5 ms^{-1}. However, it was shown that the harbour remained stratified when wind speeds were less than 3 ms^{-1}, and a baroclinic exchange existed under these conditions. Therefore, it was desirable to know the horizontal transport time scale for the diffusion mechanism under more turbulent conditions than those measured on Julian day 119.

A relationship between the dissipation rate of mechanical energy in the surface mixed-layer and the energy flux from the atmosphere was reported by Oakey and Elliott (1982). They recorded turbulence levels under a variety of wind-wave conditions at a site near Emerald Basin on the Scotian shelf. The correlation between their dissipation and atmospheric forcing data showed a linear relationship between ε_I and $(U_{10})^3$. Here, ε_I is an average value of the dissipation rate integrated over depth, and U_{10} is an average wind speed measured at 10 m above the water surface for the hour prior to the dissipation measurements.

For the data presented in Table 2, ε_I was equal to about 6×10^{-6} $m^3 s^{-3}$ and U_{10} cubed was equal to approximately 30 $m^3 s^{-3}$. There was good agreement between these data and those presented by Oakey and Elliott (1982). It was therefore reasonable to assume that the dissipation rate in Hillarys Boat Harbour goes as the 10 m wind speed cubed. Given that ε goes as $(U_{10})^3$, then doubling the wind speed halves the horizontal transport time scale associated with the diffusive mechanism. Thus, a wind speed equal to 6 ms^{-1} could introduce enough TKE for eddy-diffusion processes to disperse a tracer cloud the size of the harbour in a time of order 5 days. Since the time for dispersion by turbulent motions was greater than the diurnal period of the tide, this indicated that the contribution to flushing by barotropic forcing was limited by mixing and not the size of the tidal prism. Therefore, the 11 day flushing time derived from the tidal prism method was an optimal estimate, which means the actual barotropic flushing time was probably several days longer.

4.0 Discussion of Results

The interpretation of the processed data was simplified by viewing Hillarys Boat Harbour as a small estuary. A partially stratified estuary was defined by Imberger (1976) as "one which possesses a quite definite longitudinal salinity gradient from the mouth to the head of the estuary, but only a very weak vertical or transverse salinity structure". Such a stratification condition requires a fresh water source near the head of the estuary and a vertical mixing mechanism which will overcome the vertical stratification caused by the fresh water input.

It was shown that wind generated turbulence in Hillarys Boat Harbour was sufficient to break down the vertical structure caused by the groundwater seepage, thereby establishing a longitudinal density gradient. Such a density gradient causes a pressure gradient force which is directed from the mouth towards the head of the estuary. A second pressure gradient force results from a surface slope and is generally directed towards the mouth of the estuary. Consequently, a net circulation is established in which a barotropic force due to a surface slope drives a seaward flow in the upper layer, and a baroclinic force caused by a longitudinal density gradient drives a landward flow in the lower layer.

Two-Layer Flow

The concept of critical flow in a two-layer system was first studied by Stommel and Farmer (1952, 1953). They showed that the discharge at the mouth of an estuary can be limited by an interfacial Froude number F_i equal to unity. The interfacial Froude number was defined as

$$F_i = v_1 / (g' y_1)^{1/2} \qquad (6)$$

where the subscript 1 refers to the upper layer. Critical flow occurs in the upper layer maintaining a critical depth so that the compensating inflow in the lower layer is restricted. Wood (1968) discovered that there are actually two points of control, one at the narrowest section and another at a neighbouring upstream location. The latter control point is called the virtual point of control, and its position is determined by the barotropic flow. The results by Wood (1968) were extended by Armi (1986), who parameterized the flows in terms of the internal Froude numbers in each layer and presented solutions to the non-dimensional continuity and energy equations in a Froude number plane. This approach parallels the specific energy method in that the conditions for critical flow in a two-layer system can be determined from a Froude number diagram just as unstratified critical flow conditions are determined from a specific energy diagram.

For a two-layer flow through a contraction, the non-dimensional continuity equation can

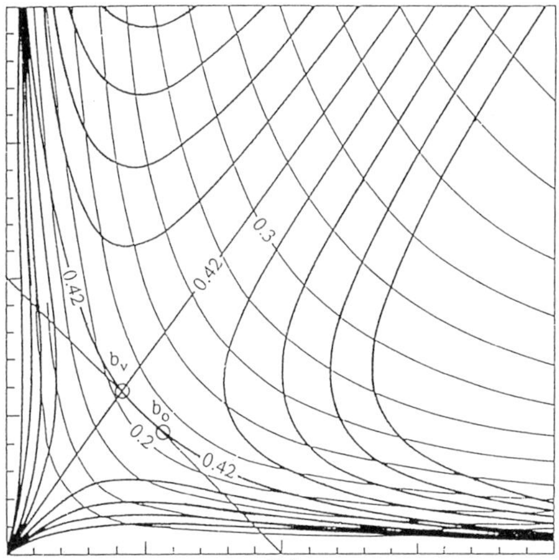

Figure 4 Critical flow conditions on Julian day 135 with $Q_r = 1.4$.

be written as

$$Q_r^{2/3} F_1^{-2/3} + F_2^{-2/3} = (Q_2' / b')^{-2/3} \qquad (7)$$

where Q_2' is the non-dimensional lower layer discharge, b' is the width and $Q_r = Q_1 / Q_2$. The solutions to equation **XX** define the locus of possible Froude number pairs for any given Q_r and Q_2'/b'. For two-layer flow through a contraction, the non-dimensional energy equation in Froude number space is given by Armi (1986) as

$$\Delta H' = \frac{F_2^{-2/3}(1 + 0.5 F_2^2) - 0.5 Q_r^{2/3} F_1^{-2/3} F_1^2}{Q_r^{2/3} F_1^{-2/3} + F_2^{-2/3}} \qquad (8)$$

where $\Delta H'$ is the dimensionless energy difference between the two layers. For small internal Froude numbers, $\Delta H'$ is the same as y_2' measured at the narrowest section. Armi and Farmer (1986) described all the flow properties for maximal two-layer exchange using nine dimensionless equations, nine unknowns and a non-dimensional barotropic flow.

The aforementioned method was applied to Hillarys Boat Harbour using data from Julian day 135 measurements. The barotropic flow for this case was calculated to be 2.85 m^3s^{-1}. Dividing this value by $(g')^{1/2} b_o (y_1 + y_2)_o^{3/2}$, where $g' = 0.0014$ ms^{-2}, $b_o = 90$ m, and $(y_1 + y_2)_o = 5$ m, yielded a non-dimensional barotropic flow component $U_o' = 0.09$. This was used to solve for $y_2'_o \approx 0.42$, $b_v' \approx 1.05$, $Q_1' \approx 0.30$, $Q_2' \approx 0.21$, $v_1' \approx 0.56$, and $v_2' \approx 0.44$. The subscripts $_o$ and $_v$ refer to the narrowest section and location of the virtual control, respectively. The results are schematically shown in Froude number space in Figure 4. The line connecting (1,0) with (0,1) represents the locus of composite Froude numbers ($G^2 = F_1^2 + F_2^2 = 1$) that separate subcritical flow ($G^2 < 1$) from supercritical flow ($G^2 > 1$). On Julian day 135, a composite Froude number $G^2 = 0.64 + 0.40 = 1.04$ identified the existence of internal critical flow conditions with a normal control at $b_o = 90$ m and a virtual control at 94.5 m.

Flushing

The tidal prism method was used to calculate an optimal tidal flushing time on the order of 11 days. By superimposing the tidal prism decay rate onto the measured decay rate, it became apparent that at times the Rhodamine WT concentration was dropping at a rate similar to the tidal flushing rate; while at other times, the tracer concentration decayed much faster. This can be seen in Figure 2. An explanation for the variable flushing rate is that a baroclinic flushing mechanism was dominant during calm weather conditions, and the barotropic tide coupled with diffusion processes were responsible for flushing during active mixing periods. A similar case study was reported by Bienfang (1980) in which continual groundwater infiltration produced harbour flushing rates six to ten times those calculated for tidal flushing alone.

The flushing rate due to an internal flow driven by a density variation within Hillarys Boat Harbour was determined using $v_1'_o$ & $v_2'_o$ from Julian day 135. It was shown that $v_1' \approx 0.56$, and $v_2' \approx 0.44$. Dividing $v_1'_o$ & $v_2'_o$ by $[g'(y_1+y_2)]^{1/2}$ gives their dimensional form as an outflowing upper layer velocity $v_{1o} \approx 0.045$ ms^{-1}, and an inflowing lower layer velocity $v_{2o} \approx 0.035$ ms^{-1}. These values appeared to be consistent with drogue measurements conducted on Julian day 119. Defining the smaller of the two velocities as the baroclinic exchange rate, the time required for an uninterrupted baroclinic exchange flow to renew the 1.3 x 10^6 m^3 harbour volume was calculated to be on the order of 2 days! However, it was shown that wind stirring was capable of destroying the vertical stratification, thereby inhibiting the flushing effectiveness of an established baroclinic circulation. Since the total flushing time and the barotropic component were known, it was possible to estimate the percentage of time for which a baroclinic circulation was active.

Substituting a measured total flushing time $T_f = 5$ days, a barotropic component of 11 days, and a baroclinic flushing time of 2 days into the expression

$$1 / T_f = 1 / T_{Tide+Diffusion} + X / T_{Baroclinic} \qquad (9)$$

indicates that the baroclinic flushing mechanism was active during approximately 20% of the experimental period. It should be remembered that the barotropic flushing time used to solve equation (9) was an optimal estimate. Therefore, the baroclinic flushing process was probably active for a bit more than 20% of the time during the experiment.

5.0 Conclusions

The primary conclusion of this investigation is that within Hillarys Boat Harbour a baroclinic circulation caused by a fresh groundwater influx enhances the barotropic flushing. An overall flushing time of 5 days was measured during the experimental programme, and an optimal tidal flushing time of 11 days was calculated using the tidal prism method. The discrepancy was too large to be explained by experimental error, which meant that another process was accelerating the ongoing tidal exchange flow.

Fine-scale salinity and temperature measurements revealed a range of stratification conditions within the harbour due submarine groundwater discharges. When a vertical stratification condition existed, an internal two-layer flow was established. This flow was shown to be critical near the harbour entrance, which acted as a control on the depth-discharge relationship. An uninterrupted flushing rate for the observed baroclinic exchange was calculated to be on the order of 2 days. However, a variable wind field was shown to be capable of destroying the gravitational circulation. As a result, the gravitational circulation was estimated to have been active during at least 20% of the experimental period, which was sufficient to cause a two-fold improvement over the tidal flushing time.

Acknowledgment

The Hillarys Boat Harbour flushing study was conducted by the Centre for Water Research, University of Western Australia on behalf of the Department of Marine and Harbours of Western Australia. The writers are grateful to the members of these two organizations for their financial support and cooperative assistance. The authors would also like to thank Dr. L. Armi for suggesting the baroclinic flushing as a possibly important mechanism.

References

Armi, L. (1986) "The Hydraulics of Two Flowing Layers with Different Densities". J. Fluid Mech. **163**:27-58.

Armi, L. and Farmer, D. M. (1986) "Maximal Two-Layer Exchange Through a Contraction with Barotropic Net Flow. J. Fluid Mech. **164**:27-51.

Batchelor, G.K. (1959) "Small Scale Variation of Convected Quantities Like Temperature in Turbulent Fluid". J. Fluid Mech. **5**:113-133.

Bienfang, P. (1980) "Water Quality Characteristics of Honokohau Harbor: A Subtropical Embayment Affected by Groundwater Intrusion". Pacific Science, **34**:279-291.

Carter, G. D. and Imberger, J. (1986) "Vertically Rising Microstructure Profiler". J. Atmospheric and Oceanic Tech. **3**:462-471.

Dillon, T. M. and Caldwell, D. R. (1980) "The Batchelor Spectrum and Dissipation in the Upper Ocean". J. Geophys. Res. **85**:1910-1916.

Dyer, K. R. (1973) *Estuaries: A Physical Introduction.* John Wiley & Sons, pp. 133.

Fischer H. B., List E. J., Koh R. C., Imberger J. and Brooks N. H. (1979) *Mixing in Inland and Coastal Waters.* Academic Press, New York.

Fozdar, F. M. (1983) "Mixed Layer Probe Technical Report". Env. Dyn. Ref. ED-83-039, Univ. of Westn. Aust., 68 p.

Fozdar F. M., Parker G. J., and Imberger J. (1985) "Matching Temperature and Conductivity Response Characteristics". Amer. Meteor. Soc. pp. 1557-1569.

Gibson, C.H. and Schwarz W.H. (1963) "The Universal Equilibrium Spectra of Turbulent Velocity and Scalar Fields". J. Fluid Mech. **16**:365-384.

Imberger, J. (1976) "Dynamics of a Longitudinally Stratified Estuary". Proc. 15th Coastal Eng. Conf. pp. 3108-3123.

Imberger, J. (1985) "The Diurnal Mixed Layer". Limnological Oceanogr. **30**:737-488.

Imberger, J. and Berman T. (1986) "An Optimized Pumping System for Vertical Profiling in Aquatic Environments". Env. Dyn. Ref. EFM-1, Univ. of Westn. Aust.

Imberger, J. and Boashash, B. (1986) "Application of the Wigner-Ville Distribution to Temperature Gradient Microstructure: A New Technique to Study Small-Scale Variations". J. Phys. Oceanography. **16**:1997-2012.

Johannes, R. E. (1980) "The Ecological Significance of the Submarine Discharge of Groundwater". Marine Ecology Progress Series. **3**:365-373.

Kolmogorov, A. N. (1941a) "The Local Structure of Turbulence in an Incompressible Viscous Fluid for very Large Reynolds Numbers". C.R. *Acad. Sci.,* URSS, **30**: 301.

Kolmogorov, A. N. (1941b) "Dissipation of Energy in Locally Isotropic Turbulence". C.R. *Acad. Sci.,* URSS, **32**: 16.

Nece, N. E. (1984) "Planform Effects on Tidal Flushing of Marinas". J. Waterway, Port, Coastal and Ocean Eng. Div., A.S.C.E., pp.251-269.

Oakey, N. S. (1982) "Determination of the Rate of Dissipation Turbulent Energy from Simultaneous Temperature and Velocity Shear Microstructure Measurements". J. Phys. Oceanography, **12**:256-271.

Oakey, N. S. and Elliott, J. A. (1982) "Dissipation Within The Surface Mixed Layer". J. Phys. Oceanography, **12**:171-185.

Okubo, A. (1974) "Some Speculations on Oceanic Diffusion Diagrams". Rapp. P. Reun. Cons. Int. Explor. Mer. **167**:77-85.

Simpson, J. H. and Nunes, R. A. (1981) "The Tidal Intrusion Front: An Estuarine Convergence Zone". Estuarine and Coastal Shelf Science. **13**:257-266.

Smart, P.L. and Laidlaw, I.M.S. (1977) "An Evaluation of Some Fluorescent Dyes for Water Tracing". Water Resources Research. **13**:15-33.

Stommel, H and Farmer, H. G. (1952) "Abrupt Change in Width in Two-Layer Open Channel Flow". J. Mar. Res. **11**(2):205-214.

Stommel, H and Farmer, H. G. (1953) "Control of Salinity in an Estuary by a Transition". J. Mar. Res. **12**:13-20.

UNESCO (1981) "The Practical Salinity Scale 1978, and the International Equation of State 1980". Tenth rep. of the joint panel on oceanographic tables and standards. UNESCO Tech paper in Mar. Sci. **36**:13-21.

Wood, I. R. (1968) "Selective Withdrawal from a Stably Stratified Fluid". J. Fluid Mech. **32**:209-223.

CHAPTER 196

Steep Slope Seawater Supply Pipeline

Dr. Lloyd F. Lewis[1] Dr. Joseph Van Ryzin[2] Dr. Luis Vega[3]

Abstract

The State of Hawaii's Ocean Science and Technology (HOST) Park, the U.S. Department of Energy (DOE), and the Pacific International Center for High Technology Research (PICHTR) sponsored the construction and installation of an expanded seawater supply system at the Natural Energy Laboratory of Hawaii (NELH). This effort included the installation of a 1.0m diameter high density polyethylene pipe capable of delivering 840 l/s of cold seawater, representing the longest (2,060m) large diameter pipe traversing the steepest slope ever spanned. Acceptance testing of the system was completed in June 1988 and the design service life is 10 years.

Introduction

Since 1975, the State of Hawaii, DOE, and others have sponsored projects using relatively pristine seawater found adjacent to NELH at Keahole Point, Hawaii (Towill, 1976). Figure 1 illustrates the location of the NELH site. With the successful installation in 1981 of a bottom-mounted 0.3m diameter polyethylene pipe (PICHTR, 1988), experiments have had a constant supply of cold (7-8°C), nutrient-rich, pollution- and pathogen-free seawater as a resource. The State of Hawaii and others supported research in both warm and cold water mariculture (Daniel, 1985) and DOE has focused its support on both closed- and open-cycle ocean thermal energy conversion (OTEC) research (Penney, 1987). By 1985, with the formation of the High Technology Development Corporation and HOST Park adjacent to NELH, the State of Hawaii developed plans to increase the existing 130 l/s cold seawater supply and to install a 0.7m diameter pipeline to pump up to 430 l/s of cold seawater to several large-scale mariculture demonstration projects (HTDC, 1985). At the same time, DOE began designing a similar size line to supply up to 410 l/s of cold seawater (as well as a 0.7m line for 600 l/s of warm water) to a planned DOE-PICHTR net-power producing open-cycle OTEC experiment at NELH (SERI, 1985; Lewis, 1987). As plans and funding availability were evaluated, it became clear that a cost-effective solution would be to install a single line of 1.0m diameter to supply the required 840 l/s of cold seawater along with a 600 l/s warm seawater line.

---

1. U.S. Department of Energy, Washington, D.C. 20585
2. Makai Ocean Engineering, Kailua, HI 96734
3. Pacific Int'l. Ctr. for High Tech. Research, Honolulu, HI 96814

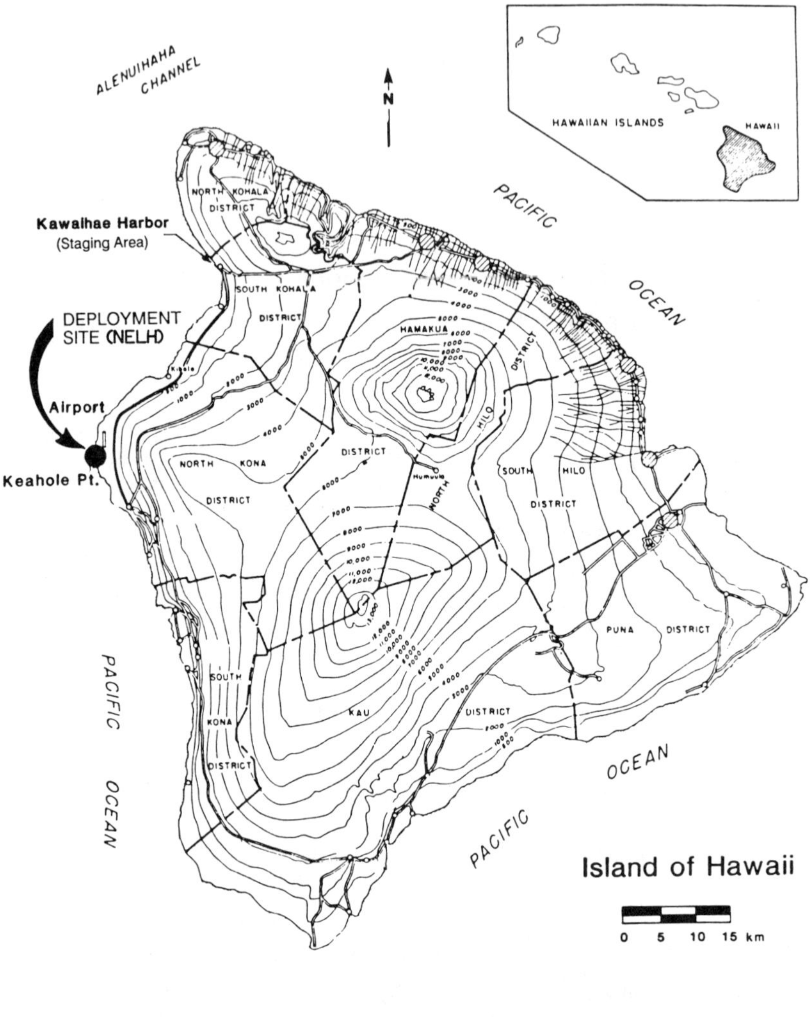

Figure 1. Seawater Supply System Deployment Site

Keahole Point Seawater Pipelines

The 0.3m diameter bottom-mounted cold ocean water pipe has been operating since 1981. In addition to the 1.0m seawater system described here, three other bottom-mounted polyethylene pipelines were installed at Keahole Point. Hawaiian Abalone Farms installed two 0.4m lines and the State of Hawaii installed a 0.5m line (PICHTR, 1988), all in 1987. These lines were small enough to be installed by a direct pull from shore. Pump installation and final inspection for all three lines were recently completed.

Site Conditions

Site-specific oceanographic, bathymetric, and environmental studies have been conducted at and near Keahole Point since 1975 in anticipation of OTEC and mariculture research (Bathen, 1975). Although Keahole Point is located on the leeward side of the Big Island of Hawaii relative to the predominant trade winds, during the November-March time frame there is a strong possibility of Kona storms (i.e., from the unprotected south and west) at the site (Rocheleau, 1979). Waves from those storms break suddenly on the very narrow island shelf and are known to move large boulders in the shoreline area. In addition, tsunami runup has been recorded at 3.7m (Cox, 1982) causing significant site inundation. Open ocean waves for the 100-year event have been estimated to exceed 14.3m (Bretschneider, 1978). Ocean currents off Keahole Point have been monitored by several researchers since 1975 (Bathen, 1975; Noda, 1979; Frye, 1981; Lobel, 1985; Noda, 1986). In the presence of a large-scale eddy often formed west of the Big Island, local currents can reach 2.6 m/s at the surface, 1.1 m/s at 150m depth, and 0.4 m/s at 300m depth. The bathymetry and sediment cover off Keahole Point have been studied in detail both with surface supported (Campbell, 1979) and submersible supported (Noda, 1979; Makai, 1983; Fornari, 1984) geophysical and geotechnical equipment. Although the distance offshore and therefore the pipe length to the cold water resource is less than 1.5 km, the slopes are very steep, occasionally exceeding 50°. In places, large outcroppings of irregular lava occur. Fig. 2 schematically illustrates the bottom slope and thin sediment condition. Environmental studies have pointed out that although there are no resident threatened or endangered species near Keahole Point, endangered marine mammals and sea turtles are known to migrate nearby in the winter months (HTDC, 1985; MSG, 1986; NELH, 1987).

Pipe Design

Considering the site-specific and environmental conditions described above, the design chosen for the 1.0m pipeline was the partially bottom-weighted and partially inverted buoyant catenary configuration shown in Figs. 2 and 3 (Horn, 1980; State of Hawaii, 1986; Vuillemot, 1988). Table 1 lists the principal design requirements of the pipeline for the DOE-PICHTR OTEC experiments (Lewis, 1987). In addition to the 410 l/s of cold seawater supplied to the OTEC experiments, another 430 l/s are simultaneously pumped to HOST Park for mariculture projects (Table 2). Additional design requirements included having only relatively inert surfaces contacting seawater and a design service life of 10 years.

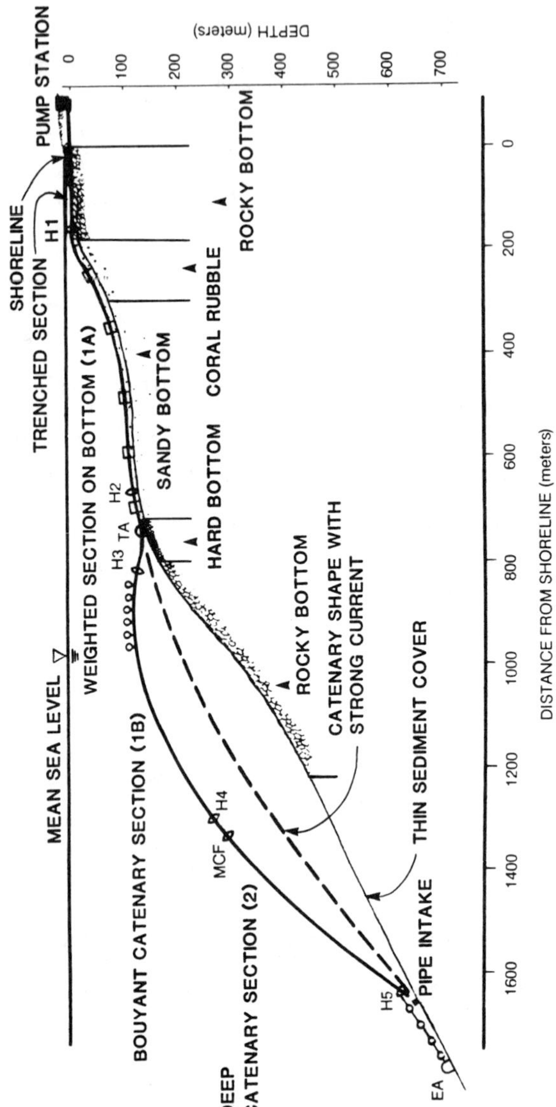

LEGEND:

- O - Holdfast (H1 - H5); Mid-catenery Flange (MCF)
- □ - Anchor Blocks (sect 1A, 51 @ 18.3 m on centers)
- ⚬ - Added Bouyancy (H3 to just past high point)
- ▢ - Transition Anchor (TA); End Anchor (EA)

DEPTH TABLE (meters):

H1	- 18	MCF	- 290 (estimate)
High point	- 116	H5	- 668
H3	- 126	Pipe Intake	- 670
H4	- 260	EA	- 707

Figure 2. Profile View of Seawater Supply System

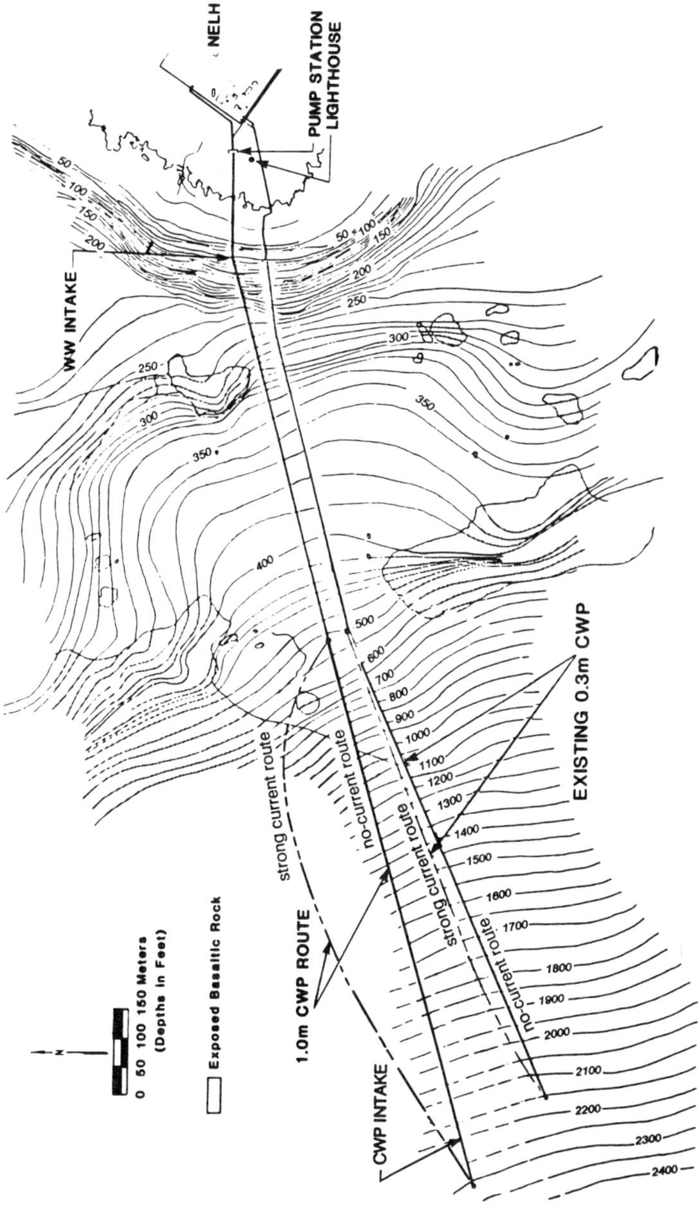

Figure 3. Plan View of Seawater Supply System

High density polyethylene was chosen as the pipe material for several reasons. Its natural buoyancy in seawater offers the option of reversibility in the deployment scenario. Its inertness satisfies the user requirements. It is sufficiently flexible yet strong in tension and bending in the chosen diameter and wall thickness to facilitate towing, controlled submerging, and long-term flexure in the deployed configuration. Fabrication is facilitated by thermal butt welding (Fig. 4) resulting in welds as strong as the extruded pipe. Finally, it has demonstrated success in prior bottom-laid configurations on steep and rugged slopes (PICHTR, 1988).

The buoyant, inverted catenary prevents the pipe from contacting the rough and steep ocean floor that it spans. As illustrated in Figures 2 and 3, however, the catenary does allow the pipe to sway in the locally strong currents (as much as 150m horizontally and 75m vertically). An 18,000kg gravity anchor below the cold water intake fixes the seaward end of the catenary as shown in Fig. 5. At a depth of about 150m, the pipe transitions from the floating catenary to a bottom-weighted configuration through bridles and a sliding sleeve (Fig. 6). Another gravity anchor (36,000kg) restrains the shoreward end of the catenary through these bridles. Shoreward of this transition section, the slope is less steep and generally sediment-covered with only minor outcrops along the route. The weighted section of pipe is restrained from downslope movement by chain bridles linked to rock-bolted anchor plates installed at the shore end of the pipe. At a depth of 15m (just above the warm water intake), the cold water pipe enters a trenched and backfilled section (Fig. 7) along with the warm water pipe and two spare conduits, all of which continue to the onshore pumphouse invert at approximately 5.5m below sea level. Submersible pumps draw warm and cold water from the pumphouse sump and force these supplies through a distribution network to the NELH and HOST Park sites.

Makai Ocean Engineering (MOE) designed the ocean portion of the pipe system. R. M. Towill Corporation designed the land portion of the system and, assisted by MOE, also served as the resident engineer for the State of Hawaii during construction. Construction of the seawater system was done by Kiewit Pacific Company under contract to the State of Hawaii.

Site Preparations

The principal site preparation for installing the seawater system involved trenching for the nearshore, surf zone, and onshore pipe route. Other predeployment activities included installing the nearshore rock anchors and clearing the chain bridle paths to the pipeway. Blasting and excavation were precluded during the winter marine mammal migration period. Pre- and post-construction surveys were conducted to assess the impact on, and recovery time of, local nearshore aquatic biota. In addition, these diver surveys were coordinated with the National Marine Fisheries Service and Hawaii Department of Land and Natural Resources. A shore-based navigation and positioning triangulation system was installed along with visual range markers. Numerous radio frequencies were selected for voice communications among the deployment vessels and shore-based personnel.

Figure 4. Polyethylene Pipe Thermal Butt Welding

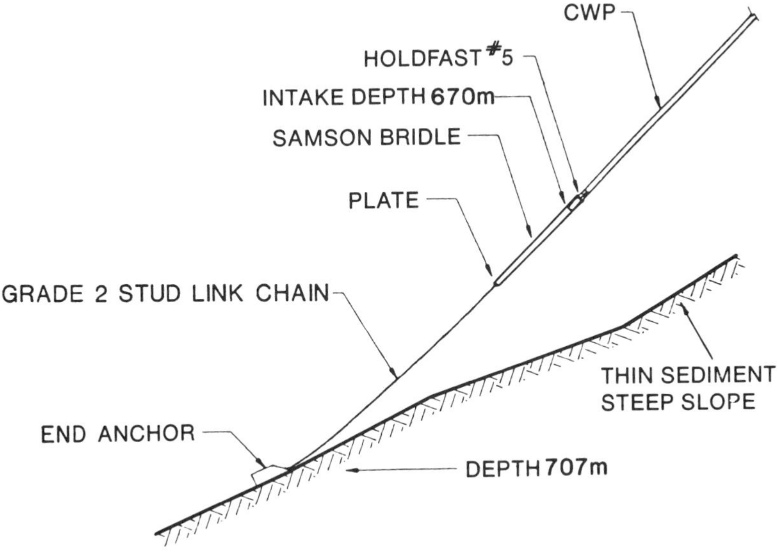

Figure 5. Pipe Intake Assembly

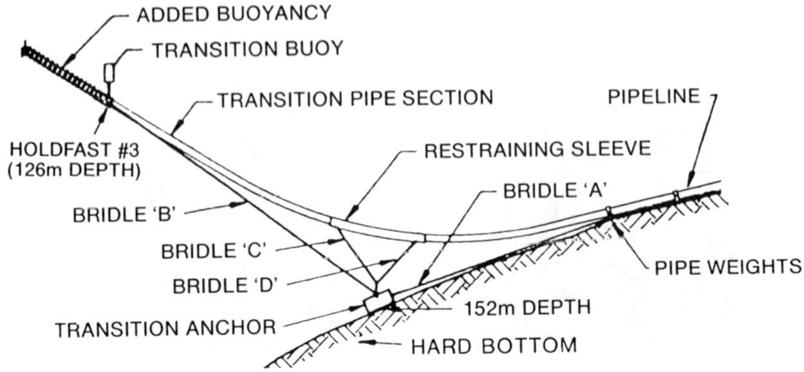

Figure 6. Pipe Transition Area

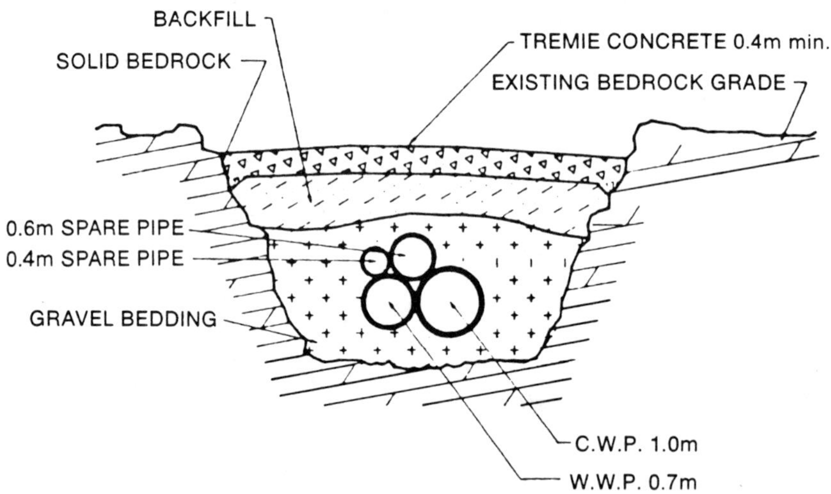

Figure 7. Section of Offshore Trench, Typical

Weather windows were selected with regard to each of the planned deployment operations. The deployment guideline (State of Hawaii, 1986) called for towing in sea states less than 4, pipe alignment in sea states less than 3 with surface currents less than 0.8 m/s, and pipe lowering in sea states less than 3 with currents less than 0.5 m/s. Weather conditions were monitored daily from the National Weather Service and by a Kiewit-employed meteorologist. An in situ telemetering current meter was monitored continually during critical phases of the deployment. None of the condition limits were exceeded during deployment.

Fabrication, Assembly, and Deployment

The high density polyethylene pipe was shipped by the manufacturer, DuPont of Canada, in 12m sections and off-loaded at the Kawaihae Harbor staging area. DuPont and Kiewit personnel then formed sections 1A, 1B, and 2 (Fig. 2) using a thermal welding machine (Fig. 4). Holdfasts, floats, weights, flanges, certain bridles, and other attachments (Fig. 2) were installed and the pipe sections floated into the harbor (Fig. 8) These sections were anchored to prevent movement due to winds and were also restrained to prevent twisting during holdfast installation. Sections 1A and 1B were pressure tested by running a pig through the pipe and pressurizing to 125% of the maximum expected deployment pressures (e.g. 620kPa) for 24 hours. Lights were attached at 60m intervals for night tows.

Deployment of the ocean portion of the seawater system off Keahole Point was initiated in late July, 1987. Sections 1A and 1B were mechanically flanged together (Fig. 9), pressurized to 140kPa, and towed as a unit to Keahole Point. The tow tug transferred holdfast 1 to the crane barge for controlled lowering and connection by divers to the nearshore anchor plate bridles. After aligning section 1 along its deployment corridor (Fig. 10), hoses were connected to the shoreward end, and the 36,000kg transition anchor and associated bridles were connected near holdfast 3 (Fig. 6). In order to cool the pipe and thus reduce bending strains, it was necessary to spray water on the floating pipe during this period. The pipe was pressurized to 350kPa to prevent pipe collapse during deployment. Then the pig was forced through the pipe by water pumped from the shore end, causing the pipe and concrete anchor blocks to sink; however, pressure equilibrium was maintained in case the pipe had to be raised during this period. The offshore tug, pulling up to 220kN tension, maintained the pipe alignment and acceptable bend radii. As the final anchor blocks were submerging, the crane barge lowered the transition anchor into place. After about 600,000 liters were pumped, lowering was stopped and holdfast 4 was kept on the surface.

The University of Hawaii's manned submersible PISCES V was launched to inspect the pipe from about anchor block 30 through the transition area to confirm placement of weights, condition of bridles, clearance from the existing 0.3m line, avoidance of rock outcrops, etc. After the inspection, pipe section 2 was towed from Kawaihae and connected to section 1 at the mid-catenary flange. The crane barge moved to the pipe intake, and the catenary end anchor and bridles were attached. Pumps were started again and the pig

Figure 8. Pipe Haul, Kawaihae Harbor

Figure 9. Connecting Pipe Sections 1A and 1B, Kawaihae Harbor

Figure 10. Aligning Pipe Section 1, Keahole Point

displaced air from the pipe while the barge lowered the end anchor. Since the distance between the two catenary anchors was critical in determining catenary shape, the end anchor was acoustically positioned into its touchdown location. A few days after this six-day pipe installation was completed, PISCES V made another dive to confirm the as-deployed condition of the pipe over its full length. The entire deployment was documented by helicopter-, surface-, and submersible-supported video taping and still photography.

Acceptance Testing

In June 1988, after completion of the surf zone and onshore pipes and pumphouse components, a full set of acceptance tests was successfully completed, including operation of the eight cold and four warm water pumps (Table 2). The stainless steel submersible pumps are rated for service to 20m depth. They have alarms and shutoffs for overheating, excessive motor current, vibration, and low water levels in the pumphouse sump. The pumps are rotated during low flow use periods to maintain uniform wear.

Additional Effort

The HOST Park portion of the system is presently operating with discharges into large capacity, deep injection wells. It is not anticipated that the NELH portion of the system will be fully used until 1990-1991 when the DOE-PICHTR large-scale, net-power producing open-cycle OTEC experiment is operational. Presently, a shallow trench is being designed and, when constructed, will serve to collect and disperse the OTEC discharges.

The pipe will be instrumented at holdfast 3 with recording depth gauges to monitor depth changes as a function of ocean currents, which will also be monitored on a separate nearby mooring. It is expected that this data will assist in verifying predictive pipe motion models and in designing future catenary pipe systems.

References

Bathen, K.H. (1975). "A Further Evaluation of Oceanographic Conditions Found off Keahole Point, Hawaii." Department of Ocean Engineering, University of Hawaii, Honolulu, HI.

Bretschneider, C.O. (1978). "An Evaluation of Extreme Wave Climates at Keahole Point, Hawaii." Proceedings, 16th International Conference on Coastal Engineering, Hamburg, Germany.

Campbell, J.F. and Erlandson, D.L. (1979). "OTEC-1 Anchor Site Survey." HIG-79-C. Hawaii Institute of Geophysics, University of Hawaii, Honolulu, HI.

Cox, D.C. (1982). "Tsunami Hazard at Keahole Point, Hawaii." Special Report. 0028, Environmental Center, University of Hawaii, Honolulu, HI.

Daniel, T.H. (1985). "Aquaculture Using Cold Water." Proceedings, Marine Technology Society OCEANS '85, San Diego, CA.

Fornari, D.J. (1982). "Deep Sea Vehicle TURTLE Dives off Keahole Point, Hawaii." ETEC N285-058ZKX. Energy Technology Engineering Center, Canoga Park, CA.

Frye, D., Leavitt, K. and E.K. Noda (1981). "Current Data from the Keahole Point, Hawaii OTEC Benchmark Sites, June 1980-April 1981." DOE/NBM-2015777. National Technical Information Service, Springfield, VA.

Gibbs & Cox (1985). "Seacoast Test Facility Upgrade Project Conceptual Design Study." ETEC STFU-XR-0002. Energy Technology Engineering Center, Canoga Park, CA.

High Technology Development Corporation (1985). "Final Environmental Impact Statement and Development Plan for HOST Park and Expansion of NELH." High Technology Development Corp., Honolulu, HI.

Horn, H.M. and Van Ryzin, J.C. (1980). "Deep Ocean Polyethylene Pipe Installations." Proceedings, International Pipeline Technology Exhibition and Conference, (Interpipe 80) Vol. I, Houston, TX.

Lewis, L.F., Trimble, L. and J.L. Bowers (1987). "Open-Cycle OTEC Seawater Experiments in Hawaii." Proceedings, Marine Technology Society OCEANS '87, Halifax, Nova Scotia, Canada.

Lobel, P.S. and Robinson, A.R. (1985). "Summary of Wind and Ocean Current Patterns Along the Kona Coast, Hawaii." Marine Sciences Group, University of California, Berkeley, CA.

Makai (1983). "Offshore Engineering Survey at Keahole Point, Hawaii, Using the Submersibles MAKALI'I and TURTLE." Makai Ocean Engineering, for the Department of Business and Economic Development, State of Hawaii, Honolulu, HI.

MSG (1986). "Environmental Report, Construction of an Expanded Facility for the Seacoast Test Facility, Stage 2 of NELH." Marine Sciences Group, University of California, Berkeley, CA.

NELH (1987). "Final Supplemental Environmental Impact Statement, Alternative Methods of Seawater Return Flow Disposal." Natural Energy Laboratory of Hawaii, Honolulu, HI.

Noda, E.K. and Brower, R.E. (1979a). "Bottom Current Survey, Keahole Point, Hawaii." Look Lab-79-47. University of Hawaii, Honolulu, HI.

Noda, E.K. (1979b). "Manned Submersible Reconnaissance Survey, OTEC Seacoast Test Facility Proposed Pipeline Route, Keahole Point, Hawaii." UNIHI-SEAGRANT-TP-79-02, University of Hawaii, Honolulu, HI.

Noda, E.K. (1986). "Oceanographic Criteria for the Design and Deployment of the Cold Water Pipe System." EKN-1071-R-1 E.K. Noda & Associates, Honolulu, HI.

Penney, T.R. and Bharathan, D. (1987). "Power from the Sea." Scientific American, Vol. 256, No. 1, pp. 86-92.

PICHTR (1988). "Bottom Mounted OTEC Seawater System Workshop." Pacific International Center for High Technology Research, Honolulu, HI. (in press).

Rocheleau, R.E. (1979). "Evaluation of Extreme Wind and Wave Climate at Keahole Point, Hawaii." Sea Grant Working Paper 42, University of Hawaii, Honolulu, HI.

State of Hawaii (1986). "HOST Park, First Increment and Seacoast Test Facility Upgrade, Keahole, North Kona, Hawaii." Job. 01-26-8018 Plans and Specifications. Department of Accounting and General Services, State of Hawaii, Honolulu, HI.

Towill (1976). "Environmental Impact Statement, NELH, Keahole Point, Hawaii." R.M. Towill Corp. for the Research Corporation, University of Hawaii, Honolulu. HI.

Vuillemot, F.L., Van Ryzin, J.C. and A.M. Resnick (1988). "The HOST-Seacoast Test Facility (OTEC) Project in Hawaii, Planning, Design and Construction." Presented at Marine Technology Society Pacific Congress on Marine Science & Technology (PACON'88), Honolulu, HI.

TABLE 1

OPEN-CYCLE OTEC SEAWATER
SUPPLY SYSTEM CHARACTERISTICS

Characteristic	Cold Water	Warm Water
Primary Flow (l/s)	410	600
Secondary Flow (l/s)*	19	19
Head at Test Pad (m)	6.1	6.1
Intake Depth (m)	670	20
Particulates Screened (cm)	1	1

*continuous flow

TABLE 2

SEAWATER PUMP CHARACTERISTICS

User	Type	Number*	Power (kW)	Flow (l/s)	Head (m)
HOST	Cold	5	66	110	37
DOE/NELH	Cold	3	57	205	19
DOE/NELH	Warm	4	57	200	19

*each use has one redundant pump

CHAPTER 197

Coastal and Estuarine Hydrodynamic Modeling in Curvilinear Grids

by

Y. Peter Sheng[1], M. ASCE

T.S. Wu[2]

P.F. Wang[2]

ABSTRACT

A time-dependent three-dimensional numerical model of coastal and estuarine circulation, which allows the use of generalized curvilinear grids to resolve the complex shoreline geometry and bathymetry, has been developed. Model formulation is briefly presented, followed by example model applications to such water bodies as Lake Okeechobee, James River and Chesapeake Bay.

INTRODUCTION

Numerical models are routinely used for scientific investigations and management studies of circulation and transport processes in coastal and estuarine waters. However, a number of complexities are associated with the numerical modeling of coastal and estuarine circulation and transport. First of all, the geometry and bathymetry of water bodies are often quite complex (see, for example, Fig. 1). Secondly, circulation and transport in coastal and estuarine waters are generally turbulent and time-dependent, and often three-dimensional. Last but not least, circulation and transport are generally affected by wind, tide, wave, and often density stratification. A thorough discussion of these various aspects and a review of numerous existing hydrodynamic models were presented in Sheng (1986a). This paper focuses on the aspect of complex geometry and bathymetry, and highlights the recent development and application of a curvilinear linear-grid numerical model of coastal and estuarine circulation and transport.

[1]Professor, Coastal and Oceanographic Engineering Department, University of Florida, Gainesville, Florida 32611

[2]Postdoctoral Associate, Coastal and Oceanographic Engineering Department, University of Florida, Gainesville, Florida 32611

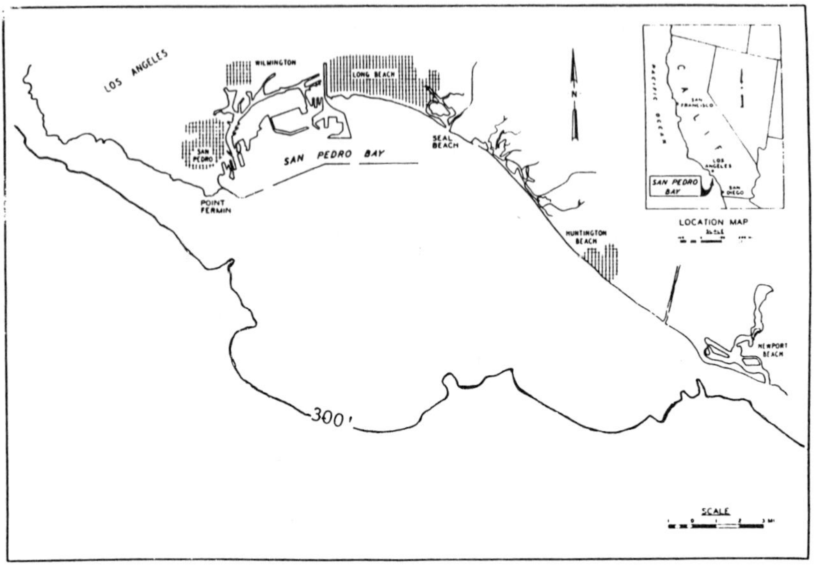

Figure 1. Los Angeles and Long Beach Harbor.

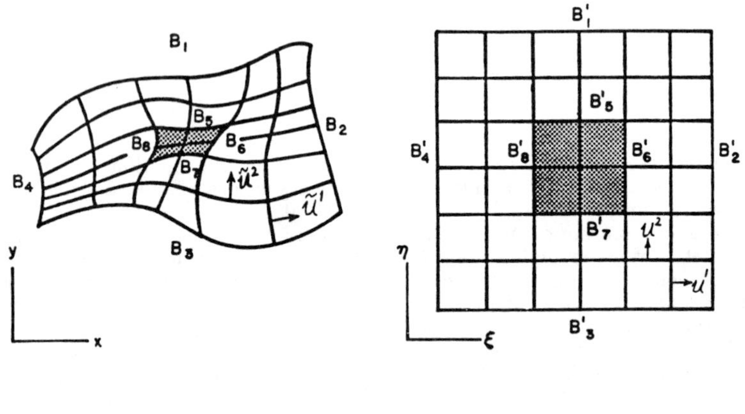

Figure 2(a). Generalized curvilinear (boundary-fitted) grid used in the horizontal directions of the 3-D model.

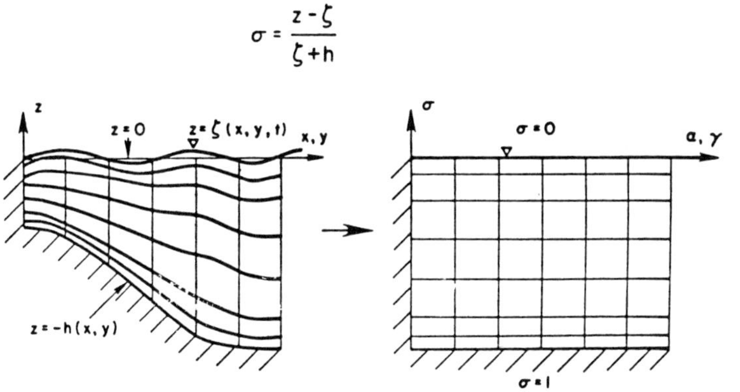

Figure 2(b). Vertical stretching of the coordinates in the 3-D model.

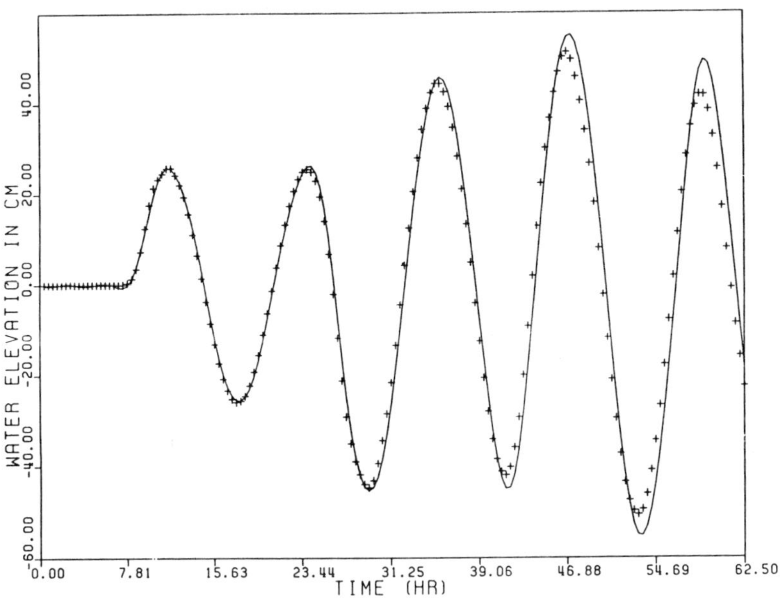

Figure 3. Transient response of water level near the closed end of an initially quiescent basin of 300 km x 29 km x 10 m, subject to a tidal forcing with 1 m amplitude and 12.5 hr period. (—— model results, +++ analytical results)

Traditional finite-difference models use uniform or non-uniform rectangular grid to approximate the complex geometry of water bodies. Consequently, very fine grid is often required to resolve a shoreline or an internal feature such as a structure or navigation channel. As a larger number of rectangular grid points are employed, the "stair-step" model shoreline can better represent the actual shoreline, at the expense of much increased computational cost. To alleviate some of these problems, simple curvilinear grids such as conformal grid (Warnstrath, 1977) and orthogonal curvilinear grid (Bennett and Schwab, 1981) have been used for modeling water bodies where geometries are not so complex. Despite the relative simplicity of these curvilinear grid models, it is extremely difficult (if not impossible) to generate conformal or orthogonal curvilinear grids for most coastal and estuarine water bodies, such as the Los Angeles and Long Beach Harbors shown in Figure 1. Not only is the grid generation process extremely difficult, but the final grid may also be unnecessarily fine in some areas while very coarse in others. In addition, the final grid may not be very smooth. Thus, for coastal and estuarine applications, a hydrodynamic model that can resolve arbitrary shoreline and bathymetry must be developed. Numerical model must be able to accept not only conformal grids or orthogonal grids, but also generalized curvilinear grids which are often non-orthogonal. Such a numerical model will allow the modeler added flexibility in the design of numerical grids with optimal smoothness, resolution, and orthogonality.

This paper presents a generalized curvilinear-grid model for coastal and estuarine circulation and transport. The model resolves the contravariant components of horizontal velocity vectors, thus significantly simplifies the resulting equations of motion and boundary conditions in the transformed grid. In the limiting cases of conformal grids or orthogonal grids, the model equations become substantially simplified.

MODEL FORMULATION

Governing Equations in Curvilinear Grid

The basic circulation and transport model is developed in terms of the contravariant velocity components (u^1 and u^2) in the horizontal directions (x^1 and x^2) and the vertical velocity (ω) in the vertical direction of a σ-stretched (Sheng and Butler, 1982) curvilinear grid (Fig. 2). The basic equations of motion in dimensionless and tensor-invariant forms can be written as (Sheng, 1986b):

$$\varsigma_t + \frac{\beta}{\sqrt{g_o}} \frac{\partial}{\partial x^k} \left(\sqrt{g_o} H u^k \right) + H \frac{\partial \omega}{\partial \sigma} = 0 \tag{1}$$

$$\begin{aligned} \frac{1}{H} \frac{\partial H u^k}{\partial t} = & - \varsigma !^k - g_{nj} \varepsilon^{kj} u^n \\ & - \frac{Ro}{H} \left[(H u^\ell u^k)_{,\ell} + \frac{\partial H u^k \omega}{\partial \sigma} \right] \end{aligned} \tag{2}$$

$$
\begin{aligned}
&+ \frac{E_v}{H^2}\frac{\partial}{\partial\sigma}\left(A_v\frac{\partial u^k}{\partial\sigma}\right) + E_H A_H u^k_{,m}!^m \\
&- \frac{Ro}{Fr_d^2}\left[H\int_\sigma^0 \rho!^k d\sigma + H!^k\left(\int_\sigma^0 \rho d\sigma + \sigma\rho\right)\right]
\end{aligned}
$$

where $k = 1$ and 2, $(x^1, x^2) = (\xi, \eta)$ represent the coordinates in the computational domain which is always consisted of uniformly spaced rectangular grid, $\partial/\partial x_k$ is the partial derivative, $g_{\ell n}$ is the metric tensor while g_o is the determinant of the metric tensor, ς is the free surface displacement, $H = h + \varsigma$ is the total depth, u^k is the contravariant velocity component, $\sigma = (z - \varsigma)/(\varsigma + h)$, $(\,)_{,\ell}$ represents the covariant spatial derivative, $!^k$ represents the contravariant spatial derivative, and ε^{kj} is the permutation tensor, $Ro = u_r/fXr$ is the Rossby number, $E_v = A_{vo}/fH^2$ is the vetical Ekman number, $E_H = A_{Ho}/fL^2$ is the lateral Ekman number, $Fr_d = u_r(gD\Delta\rho_o/\rho_o)^{0.5}$ is the densimetric Froude number, $\beta = (Ro/Fr)^2 = gD/f^2L^2$, f is the Coriolis parameter, and (u_r, D, L) represent the reference (velocity, depth, horizontal length).

The above Eq. (2) can be expanded into 2 equations for u^1 and u^2. For example, denoting (u^1, u^2) by (u, v), the following equation represents the momentum equation in the ξ direction (Sheng, 1986b):

$$
\begin{aligned}
\frac{1}{H}\frac{\partial Hu}{\partial t} = &- \left(g^{11}\frac{\partial\varsigma}{\partial\xi} + g^{12}\frac{\partial\varsigma}{\partial\eta}\right) + \frac{g_{12}}{\sqrt{g_o}}u + \frac{g_{22}}{\sqrt{g_o}}v \qquad (3)\\
&- \frac{Ro}{H}\left[\frac{\partial}{\partial\xi}(Huu) + \frac{\partial}{\partial\eta}(Huv) + (2D^1_{11} + D^2_{12})Huu\right. \\
&+ \left.(3D^1_{12} + D^2_{22})Huv + D^1_{22}Hvv + \frac{\partial Hu\omega}{\partial\sigma}\right] \\
&+ \frac{E_v}{H^2}\frac{\partial}{\partial\sigma}\left(A_v\frac{\partial u}{\partial\sigma}\right) \\
&- \frac{Ro}{Fr_d^2}\left[H\int_\sigma^0\left(g^{11}\frac{\partial\rho}{\partial\xi} + g^{12}\frac{\partial\rho}{\partial\eta}\right)d\sigma\right. \\
&+ \left.\left(g^{11}\frac{\partial H}{\partial\xi} + g^{12}\frac{\partial H}{\partial\eta}\right)\left(\int_\sigma^0 \rho d\sigma + \sigma\rho\right)\right] \\
&+ E_H A_H \cdot (\text{Horizontal Diffusion of } u)
\end{aligned}
$$

where D^i_{jk} represents the Christoffel symbol of the second kind:

$$
\begin{aligned}
D^i_{jk} &= g^{in}D_{njk} \qquad (4)\\
&= \frac{1}{2}g^{in}(g_{nj:k} + g_{nk:j} - g_{jk:n})
\end{aligned}
$$

The above equation appears to be more complicated than the momentum equation in a Cartesian grid with (x, y, σ) or (x, y, z) as coordinates. Most of the terms are doubled, while nonlinear and lateral diffusion terms are more than doubled. However, if an orthogonal grid is used, the following equation will result from Eq. (3):

$$\begin{aligned}\frac{1}{H}\frac{\partial Hu}{\partial t} = & - g^{11}\frac{\partial \varsigma}{\partial \xi} + \frac{g_{22}}{\sqrt{g_o}}v \\ & - \frac{Ro}{H}\left[\frac{\partial}{\partial \xi}(Huu) + \frac{\partial}{\partial \eta}(Huv) + \frac{\partial Hu\omega}{\partial \sigma}\right] \\ & + \frac{E_v}{H^2}\frac{\partial}{\partial \sigma}\left(A_v\frac{\partial u}{\partial \sigma}\right) \\ & - \frac{Ro}{Fr_d^2}\left[H\int_\sigma^0 g^{11}\frac{\partial \rho}{\partial \xi}d\sigma + g^{11}\frac{\partial H}{\partial \xi}\left(\int_\sigma^0 \rho d\sigma + \sigma\rho\right)\right] \\ & + E_H A_H\left(g^{11}\frac{\partial^2 u}{\partial \xi^2} + g^{22}\frac{\partial^2 u}{\partial \eta^2}\right)\end{aligned} \quad (5)$$

As indicated before, such simplified equation cannot be used unless the shoreline geometry is so simple such an orthogonal grid can be generated. Even then, the numerically computed metric coefficients, which represent the geometric effects in the transformed equations of motion, do not necessarily guarantee the satisfication of conservation laws in the discrete system.

The transport equations for temperature, salinity, and sediment concentration are simpler and can be written as:

$$\begin{aligned}\frac{1}{H}\frac{\partial H\phi}{\partial t} = & \frac{E_v}{Pr_v}\frac{\partial}{H^2\partial \sigma}\left(K_v\frac{\partial \phi}{\partial \sigma}\right) \\ & - \frac{Ro}{H\sqrt{g_o}}\left(\frac{\partial\sqrt{g_o}Hu\phi}{\partial \xi} + \frac{\partial\sqrt{g_o}Hu\phi}{\partial \eta} + \frac{\partial\sqrt{g_o}H\omega\phi}{\partial \sigma}\right) \\ & + \frac{E_H K_H}{Pr_H}\left(g^{11}\frac{\partial^2 \phi}{\partial \xi^2} + 2g^{12}\frac{\partial^2 \phi}{\partial \xi \partial \eta} + g^{22}\frac{\partial^2 \phi}{\partial \eta^2}\right)\end{aligned} \quad (6)$$

where ϕ can be temperature (T), salinity (S), or suspended sediment concentration (C), (Pr_v, Pr_H) represent the vertical and lateral turbulent Prandtl numbers, and (K_v, K_H) are the dimensionless vertical and horizontal turbulent diffusivities.

Other Model Features

The above system of equations are solved in conjunction with the proper boundary conditions at the free surface, the bottom, and the lateral boundaries. At the bottom, the velocity profile asymptotically approaches that of the law-of-the-wall, with a drag coefficient which depends on the bottom roughness.

Turbulence parameterization used in the present model is similar to the one described in Sheng (1987). While the horizontal turbulent eddy coefficients are prescribed as uniform values which decrease with the grid spacing, the vertical eddy coefficients are determined from simplified versions of a second-order closure model of turbulent transport (Sheng and Chiu, 1986; Sheng and Villaret, 1988).

Numerical Algorithm

To allow efficient solution of the 3-D model, the mode-splitting technique described in Sheng et. al. (1978) and Sheng and Butler (1982) is also used here. Vertically-integrated equations obtained by integrating Eqs. (1) and (2) represent the so-called "external mode", while the internal mode consists of equations obtained by subtracting the vertically-averaged equations from the u and v equations. An ADI scheme, which is somewhat more complicated than that for the Cartesian equations, is used for the external mode of the present model. Either two-time-level or three-time-level scheme can be used for the time integration of the 3-D model.

The numerical model described above constitutes the main body of a flow code, CH3D. In addition to CH3D, a grid generation code based on the elliptic grid generation technique (Thompson and Johnson, 1985) and a preprocessor code (Sheng, 1986) which computes the metric coefficients must also be used when solving practical problems.

At the present time, we do not adjust the grid during a particular model simulation. Adaptive grid technique, which allows the adjustment of grid resolution according to the evolution of dynamic flow, is being developed (Sheng and Luo, 1988).

Model Applications

Numerous model simulations have been performed. Model applications to a number of idealized basins were presented in Sheng (1987). Additional model sensitivity tests have been performed. For example, Figure 3 shows the transient response of an initially quiescent rectangular basin (300 Km x 29 Km x 10 m) to a suddenly imposed tidal forcing with 1 m amplitude and 12.5 hr period. Since no diffusion or bottom friction was used in the model, the model results faithfully simulated the analytically computed transient response which consists of higher harmonics in addition to the forcing harmonic.

Model application to the James River estuary (Fig. 4) produced mean tides which agreed well with data, both in terms of amplitude (Fig. 5) and phase shift (Fig. 6). Model application to the Chesapeake Bay (Fig. 7) produced results which compared well with mean tides. Significant spatial variation is found in the model computed velocity field, which is being compared with measured data now. The same basic model is being applied to the upper Chesapeake Bay (Johnson, 1988).

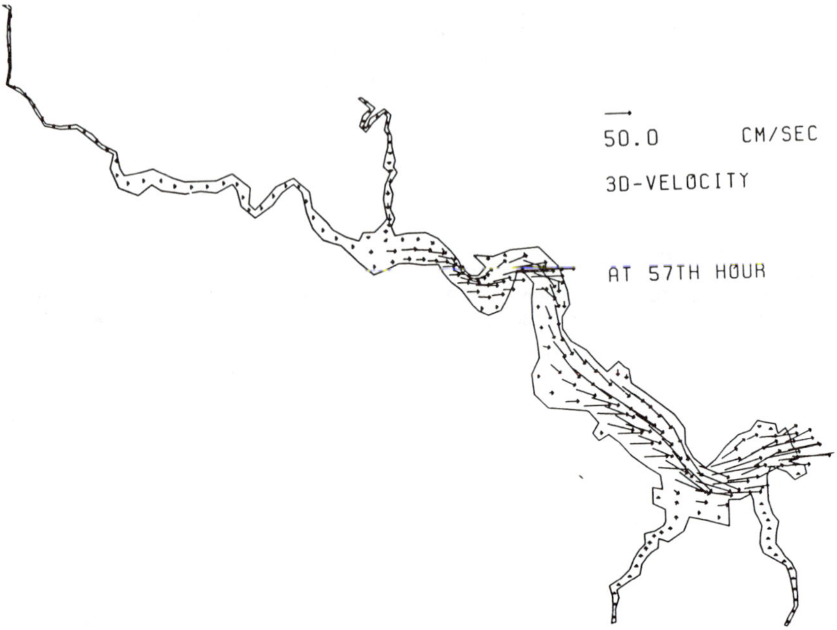

Figure 4. Near-surface current in the James River driven by a M_2 tide of 38 cm amplitude at the open boundary.

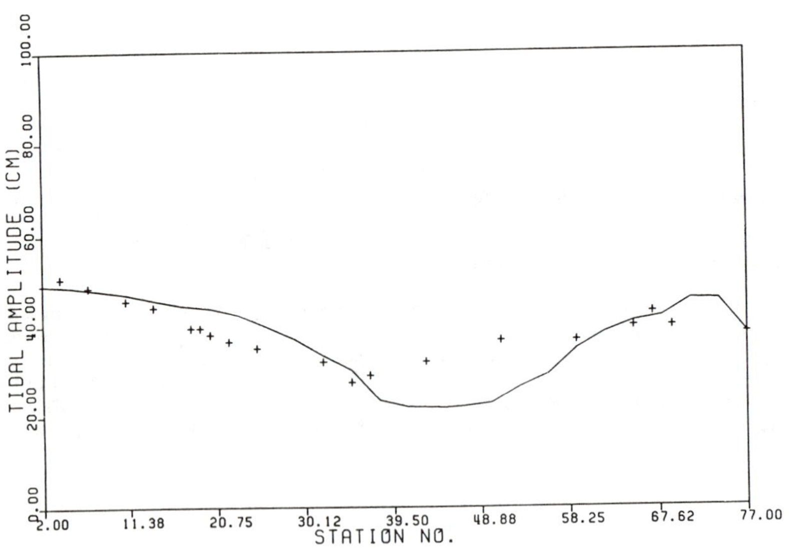

Figure 5. Mean M_2 tidal amplitude in the James River (—— model results, +++ data). Amplitude at open boundary = 38 cm.

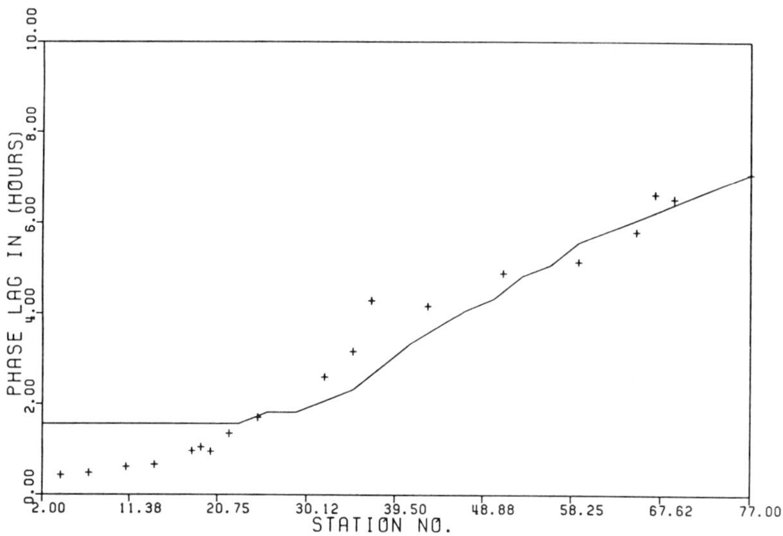

Figure 6. Mean M_2 tidal phase lag in the James River (—— model results, +++ data).

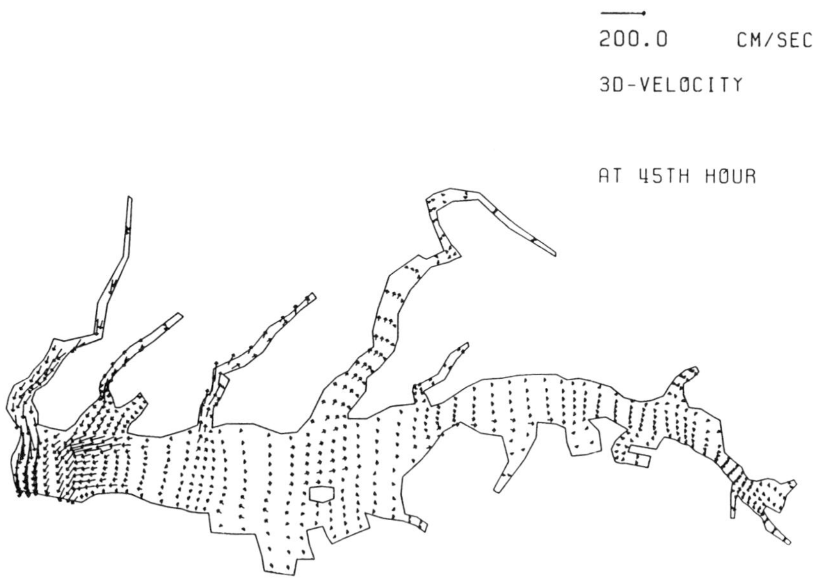

Figure 7. Near-surface current in the Chesapeake Bay driven by a M_2 tide of 60 cm amplitude.

Model application to Lake Okeechobee showed that, despite the shallowness of the lake (average depth of 3m), significant return flow is produced near the bottom due to a 5 ms^{-1} wind from the east (Figs. 8 and 9). The vertically-averaged velocity field clearly showed the existence of 2 topographically induced circulation gyres.

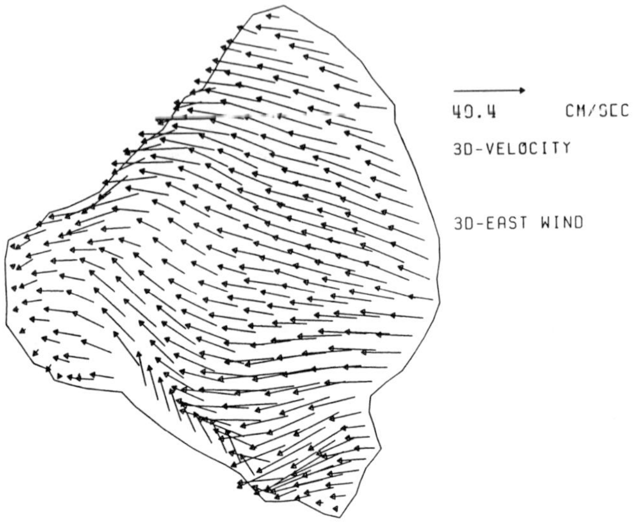

Figure 8. Near-surface wind-driven currents in Lake Okeechobee due to a wind of 5 m/sec from the east.

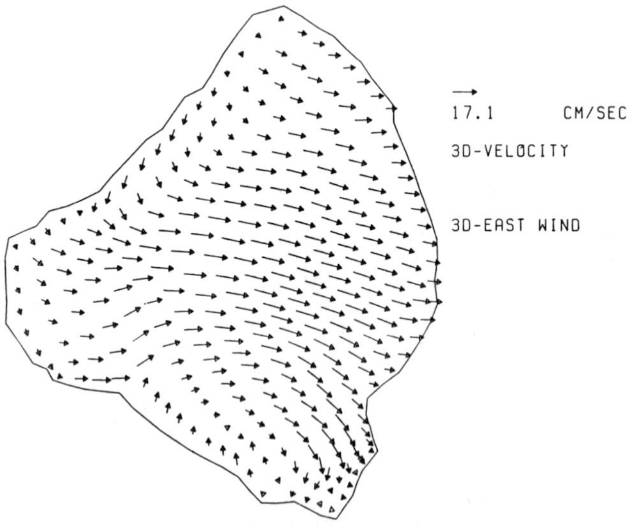

Figure 9. Near-bottom wind-driven currents in Lake Okeechobee due to a wind of 5 m/sec from the east.

ACKNOWLEDGEMENT

Supports from U.S. Army Engineer Waterways Experiment Station, Virginia Institute of Marine Science, and South Florida Water Management District are acknowledged.

REFERENCES

1. Bennett, J.R. and D.J. Schwab, 1981: "Calculation of the Rotational Normal Modes of Oceans and Lakes with General Orthogonal Coordinates", Journal of Comp. Phys., vol. 44, pp. 359-376.

2. Johnson, B.H., 1988: "An Approach for Modeling the Upper Chesapeake Bay", Proc. of the 1988 National Conference on Hydraulic Engineering, ASCE, pp. 588-93.

3. Sheng, Y.P., 1986a: "Finite-Difference Models for Hydrodynamics of Lakes and Shallow Seas", in Physics-Based Model of Lakes, Reservoirs, & Impoundments (W.G.Gray, ed.), ASCE, pp. 146-228.

4. Sheng, Y.P., 1986b: "A Three-Dimensional Mathematical Model of Coastal, Estuarine and Lake Currents Using Boundary-Fitted Grid", ARAP Technical Report No. 585, Aeronautical Research Associates of Princeton/California Research and Technology/Titan Systems, Princeton, N.J.

5. Sheng, Y.P., 1987: "On Modeling Three-Dimensional Estuarine and Marine Hydrodynamics", in Three-Dimensional Models of Marine and Estuarine Dynamics (J.C.J. Nihoul and B.M. Jamart, eds.), Elsevier Oceanography Series, Elsevier, pp. 35=54.

6. Sheng, Y.P., W. Lick, R.T. Gedney, and F.B. Molls, 1978: "Numerical Computation of Three-Dimensional Circulation in Lake Erie: A Comparison of a Free-Surface Model and a Rigid-Lid Model", J. Phys. Oceanog., Vol. 1, pp. 713-27.

7. Sheng, Y.P. and H.L. Butler, 1982: "Modeling Coastal Currents and Sediment Transport", Proc. of the 18th ICCE, ASCE, pp. 1127-48.

8. Sheng, Y.P. and S.S. Chiu, 1981: "Tropical Cyclone Generated Currents", Proc. of the 20th ICCE, ASCE, pp. 737-51.

9. Sheng, Y.P. and C. Villaret, 1988: "Modeling the Effect of Suspended Sediment Stratification on Bottom Exchange Processes", Submitted to the Journal of Geophysical Research.

10. Thompson, J.F. and B.H. Johnson, 1985: "Development of an Adaptive Boundary-Fitted Coordinate Code for Use in Coastal and Estuarine Areas", Misc. Paper HL-85-5, U.S. Army Engineer Waterways Experiment Station, Vicksburg, MS.

11. Warnstrath, J.J., 1977: "Nearshore Numerical Storm Surge and Tidal Simulation Model", T.R. H-77-17, U.S. Army Engineer Waterways Experiment Station, Vicksburg, MS.

CHAPTER 198

HYDRAULIC ASPECTS OF WETLAND DESIGN

James R. Walker[1], M. ASCE
Andrea Bertolotti[2], M. ASCE
Reinhard E. Flick[3]
C. Robert Feldmeth[4]

ABSTRACT

Tidal wetland preservation, restoration and creation have become requisites for coastal development projects in the United States. A basic approach to design of tidal wetlands is presented, stressing cooperation between regulatory agencies, biologists, engineers, and developers. Basic principles of wetland functions are explained and presented as criteria for engineering design. A description of wetlands is given to identify biological features relevant to design. Also, some key features of tides are summarized as they affect wetland design. A numerical model was used to demonstrate how tidal wetlands may be designed to conform with criteria developed by the agencies and biologists. This approach has been used on wetland designs in California, but the approach may be applicable to other areas of the world.

1.0 INTRODUCTION

Wetlands must be designed based on sound engineering principles to meet given biological requirements. Commercial and residential development along the coastal areas of the United States has increased pressure on many wildlife habitats. State and Federal wildlife agencies have been charged with preservation, enhancement and replacement of wetlands. Now, new developments in coastal areas be must done in conjunction with restoration or creation of wetlands.

The agencies must define the habitat type desired and biologists must define the conditions which will create the desired habitats. The objective is to create habitat that will support desired animal and plant life. After a site is

---

[1] Vice President, Moffatt & Nichol, Engineers, Long Beach, CA 90807

[2] Coast Engineer, Moffatt & Nichol, Engineers, Long Beach, CA 90807

[3] Oceanographer, Cal. Dept. of Boating and Waterways, Scripps Institution of Oceanography, La Jolla, CA 92013

[4] Professor of Biology, Joint Science Center, The Claremont Colleges, Claremont, CA 91711

selected, hydraulic studies are conducted in order to determine how best to achieve the desired biological setting. The wetland habitat depends in part on the tide elevations relative to the ground elevation. The ocean tide, which drives the water exchange in the wetland, has peculiar properties that influence the behavior of the water levels in the wetland. Tide control structures regulate the flow of water in and out of a wetland. Hydraulic models can simulate the flow regime in the wetland and through tide control structures. The effect of tide control structures to change wetland tide elevations relative to ground elevations is also an important design consideration.

2.0 BIOLOGICAL CHARACTERISTICS OF WETLANDS

Southern California coastal wetlands occur around bays and in estuaries and are generally restricted to narrow river valleys that were cut during periods of lower sea level in the Pleistocene era. Many of these wetlands are exposed to daily tidal action. The rise and ebb of water in tidal marshes produces a zonation of vegetation types determined by both physical and biotic factors. Other wetlands persist where the tidal range is muted by culverts or restricted by tide gates. Wetlands exist because the soils remain saturated throughout most of the year. These wetland areas are characterized by an assemblage of plants and animals, an ecological community, adapted to the particular conditions of this transition area between salt water and land.

At a particular ground elevation in a wetland, there is an associated duration of inundation by tides and salinity in the soil. Wetland plants grow in areas where they are best suited to resist the inundation by tides and soil salinity. Mahall and Park (1976) examined the transition between cordgrass (low wetland) and pickleweed (mid wetland) in wetlands in northern San Francisco Bay and found tidal inundation to be the most important factor determining their distribution.

A typical Southern California coastal wetland is composed of a low wetland community characterized by dense stands of cordgrass (Spartina foliosa). Cordgrass normally occurs between mean lower high water (MLHW) and mean higher high water (MHHW) (Zedler, 1982), as shown schematically in Figure 1. At lower elevations, mud flats may support eelgrass (Zostera marina) and often mats of various algal species such as sea lettuce (Ulva).

At elevations above the low wetland, a mid and high wetland vegetative association is present. Common pickleweed (Salicornia virginica) is by far the most important plant species in terms of biomass. It normally ranges from MHHW to about 0.6 meters (2 feet) above extreme high water (EHW). Common pickleweed is abundant down to the low wetland and can often occur with cordgrass in soils that are wet by each high tide.

Also present in the mid wetland association are a number of other plants such as saltwort (Batis maritima), annual pickleweed (Salicornia bigeovii) and Jaumea carnosa. High wetland associations include common pickleweed and sea blite (Suaeda californica). These assemblages extend up to an elevation of about 0.6 meters (2 feet) above MHHW.

The transition from mid wetland to high wetland is less distinct. Soil moisture decreases in the high wetland and soil salinity increases. Shoregrass, saltgrass, and glasswort appear to be more tolerant of the drier soils of this higher wetland zone. A summary of wetland vegetation with respect to the elevation of the tide ranges is given in Table 1.

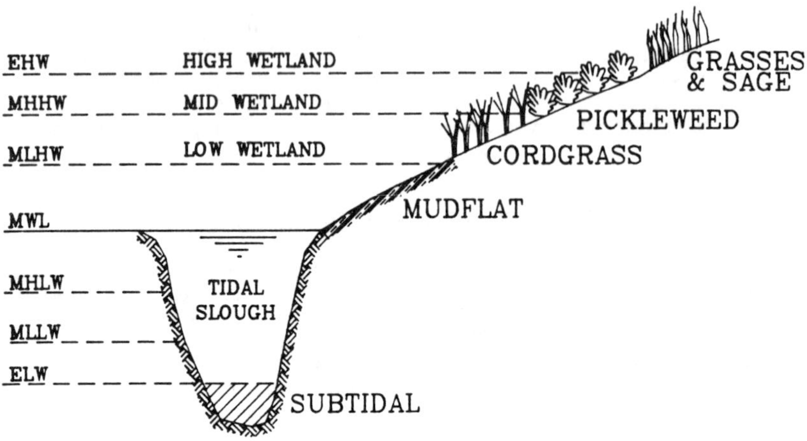

Figure 1 Typical Southern California Wetland

TABLE 1

Division of Wetland Vegetation by Elevations of Tide Ranges

TIDE RANGE	HABITAT NAME	TYPE OF HABITAT
Above EHW	High Wetland	Grasses and Coastal Sage Scrubs
MHHW to EHW	Mid Wetland	Pickleweed, Salt Wort, Annual Pickleweed and Sea Blite
MLHW to MHHW	Low Wetland	Cordgrass
MSL to MLHW	Mudflat	Eelgrass, Algae and Sea Lettuce
Below MSL	Subtidal	Marine habitat

In the mid and high wetland areas, marsh plants provide cover and foraging material for many invertebrates. Birds, such as the Clapper Rail, also utilize this cover in order to forage on burrowing crustaceans. Small snails occur on the substrate surface. Crabs and shrimp burrow in the substrate within the vegetational cover.

At high tide, mudflats are a foraging area for fishes and large wading birds, such as herons and egrets. At low tide, the exposed mudflat becomes a foraging area for shore birds. Fishes forage in the mudflat zone and up into the low wetland. These fishes form an important source of food for fish-eating birds, such as herons, egrets, and terns.

Horn (1976) showed that the fish species diversity for a bay or an estuary is proportional to its size. The relationship of bay size to species composition is shown in Figure 2. Fish diversity is an index of habitat quality; the more fish species present, the better the system is as an environment for bay-wetlands plants and animals. Other factors such as water depth, habitat diversity, pollution load are also important. Bay systems of various sizes, where the tidal regime is muted by culverts or other restrictions, can be evaluated in terms of their fish assemblages (Feldmeth, 1980 & 1986).

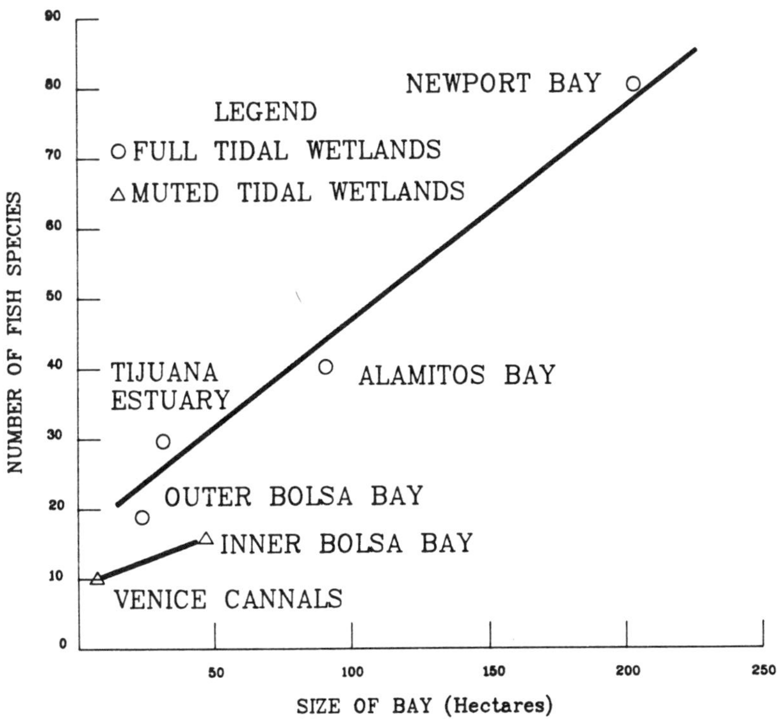

Figure 2 Number of Fish Species versus Area of Bay

3.0 TIDE CHARACTERISTICS

General tide characteristics in Southern California are presented to outline features relevant to the hydraulic design of wetlands. Tides are a well studied phenomenon and are only summarized here to illustrate their influence on design of wetlands.

Tides along the California coast are mixed semidiurnal and show a marked diurnal inequality and extreme tide ranges approach 3 meters (10 feet). A common characteristic of California tides is that the lower-low water almost always follows the higher-high water after a time interval of about 7 hours. The semidiurnal peak breaks the rising tide such that there are about 17 hours between lower-low water and higher-high water. In Figure 3, the average rate of water elevation change is plotted versus the elevation change between high and low tide for the 1983 tide record at San Diego, California. As shown in the figure, the greatest water level changes and highest tide ranges occur during ebb period. The peak flow velocities in a bay entrance will be greater during ebb flows than flood flows. This may have an effect on sediment transport and on altering the water elevations in a wetland because of the non-linear nature of tidal hydraulics.

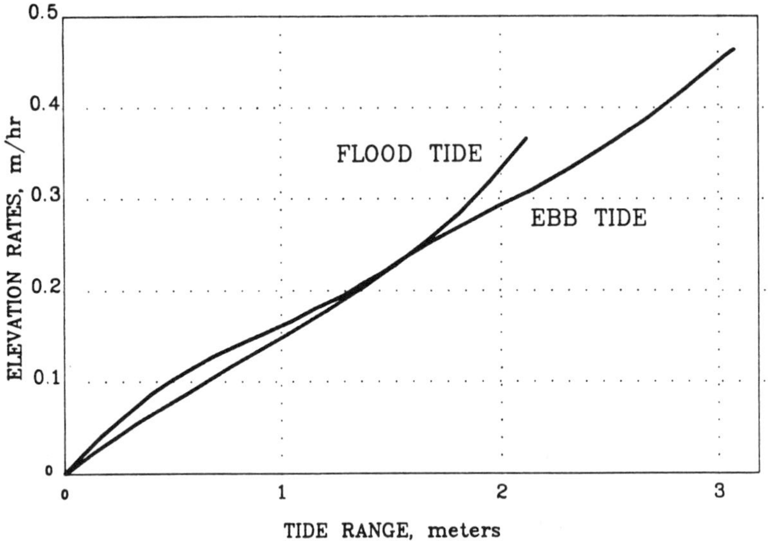

Figure 3 Water Elevation Change Rates versus Tide Range for Southern California

Variations in the tide also have a yearly cycle. Higher-high tides tend to occur in the morning during winter and in the evening during summer. Changes in the times of day of high and low tide during the year lead to significant changes in wetlands exposed to sunlight. Figure 4 shows monthly averages of the percentage of daylight hours (taken as the time from sunrise to sunset) the tide remains below MHHW, MSL and MLLW. Maximum intertidal exposure to sunlight occurs from February through April, with tide below MSL about 65% of daylight hours. There is a marked decrease in exposure to a minimum of about 25% during September.

Zetler and Flick (1985a, b) describe the cyclical variations of predicted maximum monthly high tides for the period 1983-2000. Besides the summer-winter cycle, they identify a 4.4 year variation and an 18.6 year cycle that characterize the west coast tides. The 4.4 year period arises from a modulation of the lunar perigee cycle while the 18.6 year variation is associated with the lunar node cycle. These periods increase the tide range about 0.3 and 0.15 meters (1 and 0.5 foot) respectively. During peak tides within the 4.4 year cycle, the winter and summer maximum ranges can exceed the spring-autumn ranges by about one meter (3 to 4 feet). This effect also tends to obscure the variation in seasonal mean sea level which tends to be low in April and higher by about 0.15 meter (0.5 foot) in September (Flick and Cayan, 1984)

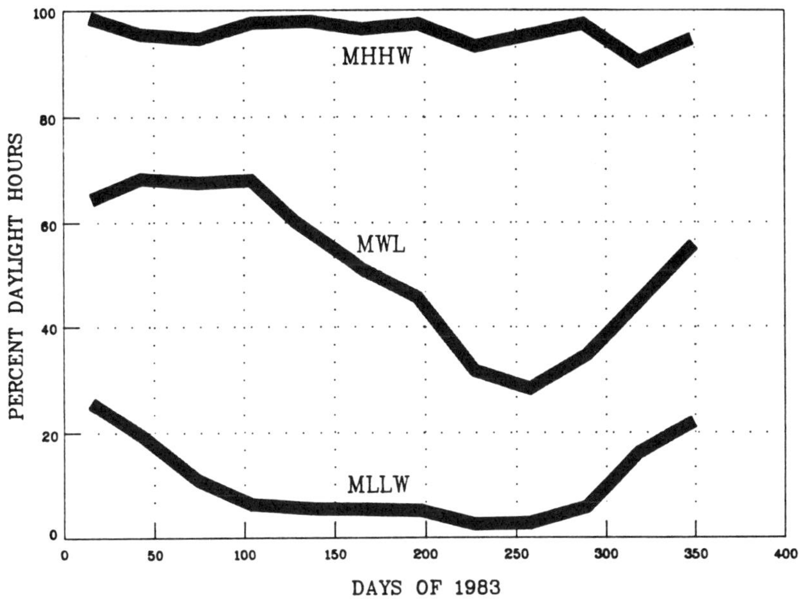

Figure 4 Percent of Daylight Hours that the Tide Remains below MHHW, MSL and MLLW.

4.0 DESIGN ANALYSIS AND METHODS

4.1 Principles

Perhaps the most important design consideration is the relationship of the tide range with respect to the ground elevation inside the wetland. Changing either the tide regime or the ground elevations can significantly effect the nature of the wetland. Scour and siltation of wetland channels is also an important design consideration that includes an evaluation of flow regimes and of soil types. In wetlands that have reduced tidal flushing or are near a source of pollutants, water circulation and quality need to be considered. Finally, in newly created wetlands, colonization or plantation of flora and fauna need to be evaluated in view of the particular restoration goals.

In California, wetland restorations are founded on goals specified by wildlife agencies such as the Department of Fish and Game, National Marine Fisheries and the Fish and Wildlife Service. Other agencies that can participate are the California Coastal Commission, the Audobon Society and the Army Corps of Engineers. While no specific agenda exists for developing design goals and criteria, some guidelines can be followed such as those proposed by Smith (1983) and Williams and Harvey (1983).

As described previously, the relationship between tide range and ground elevations provides a zonation of vegetation that characterizes a wetland. Wetlands can have tide regimes and sea levels that are different from those in the open ocean. This is an essential consideration in wetland design. Friction and flow restrictions induce energy loses in tidal flow that give rise to reduced amplitudes and phase lags in the tides of inland water bodies. Some harbors or bays experience resonant conditions in which the amplitude of the tides is larger than in the ocean. In muted tidal wetlands, however, the energy dissipations at the tide control structure tend to suppress resonant oscillations and, in general, larger reductions in amplitude are accompanied by greater phase lags. Under some conditions, the mean water level in the wetland will be different than in the ocean.

The ground elevations in a wetland should be designed considering stable soil slopes and the possibility for scour and siltation of the channels. Myrick and Leopold (1963) applied the concept of equilibrium profiles developed for inland rivers to describe the hydraulic geometry of tidal channels. These equations relate the water surface width, mean depth and mean velocity to the discharge. Pestrong (1965) used the same equations to characterize the drainage patterns of tidal marshes in South San Francisco Bay, California. In the design of muted wetlands, channels that are initially sized close to the equilibrium profile can be expected to have less changes over time than channels that are of other sizes. Adding new wetland areas to existing wetland can induce scour in existing channels.

Wetlands with a small tidal prism in relation to the total water volume, or wetlands that are near a source of pollutants, can promote unfavorable water quality and circulation for the animal and plant communities. The accumulation of nutrients such as nitrogen and phosphorous used in fertilizers can promote undesirable algal blooms.

In newly created wetlands, the flora can be established by either natural colonization or by planting. The desired fauna, instead, can only be introduced by natural colonization. The type of hydraulic connection between the new wetland and the ocean should consider the passage of plant seeds and marine animals, such as fish and crustaceans. This is important were natural plant colonization is desired and were the wetlands are to function as spawning areas for fish.

4.2 Hydraulic Modeling

A numerical model based on work by Fisher (1977) was developed to analyze tide elevations and flow velocities in wetlands subject to an ocean tide. The hydraulic system is divided into a series of basins and channels. The water level in the basins and the flow velocity in the channels are related through the conservation of mass and conservation of momentum equations. These equations are solved for one-dimensional flow by the method of finite differences. The model performs a time domain simulation and calculates tide elevations and flow velocities at specified time step intervals.

The model, called Hydrodynamic Circulation Model (HCM), employs an implicit routine, where the water elevations and velocities are related to each other and to their present values through a system of simultaneous non-linear equations. The equations are solved iteratively using a Newton-Raphson algorithm. One-dimensional differential equations are used to express the conservation of mass and momentum in an open channel. The equation are as follows:

Mass: $b(dy/dt) + A(du/dx) = 0$
Momentum: $du/dt + g(dy/dx) + gku|u| = 0$

where:
t = time
x = horizontal distance
y = water surface elevation
u = velocity in x direction
g = acceleration due to gravity
b = surface width
A = cross-sectional area
k = friction coefficient

The convective acceleration term, u(du/dx) which is part of the conservation of momentum equation, was omitted in the model. Including this term significantly increases the computation load and does not have a significant effect on the results.

The friction coefficient is derived from Manning's open channel equation and has been modified to include an entrance and exit energy loss term as follows:

Friction Coefficient: $k = Kx + Ke + (n/1.486R^{2/3})^2$

where:
Kx = 1.0 (common exit loss factor)
Ke = entrance loss factor
n = friction coefficient
R = hydraulic radius

The hydraulic features of an area are included in the model through a number of user-selected parameters. These include the length, width and depth of channels, the friction associated with the type of channel bed, the slope of the channel banks, the storage curves of basins, the tide properties of the ocean or driving body of water, and the type and geometry of the tide control structure.

The friction of the channel beds and concentrated head losses of restrictions to flow can be derived from published values. However, calibration of these parameters with field measurements from a hydraulic area similar to that being modeled provides more reliable values. The HCM model has been used for design of Anaheim Bay and Bolsa Chica wetlands in California (Moffatt & Nichol, Engineers, 1987a and 1987b)

4.3 Tide Control Structures

Tide control structures can be used in wetland restoration projects to control the ocean tide and produce specific tide properties within the wetland. Tide control structures can vary from a simple pipe culvert through a dike to more complex systems that have flap gates and overflow weirs that require manual or automatic operation. Usually, the complexity of the tide control system is directly related to the amount and type of change induced to the ocean tides.

A basic muted wetland comprises a dike, which separates the wetland from the source of ocean water and protects surrounding lowland areas from flooding, and one or more culverts, which allow for tidal flow between the ocean and the wetland. In this system, the flow area of the culverts and the invert elevations can be selected to produce the desired response. Check valves, typically in the form of flap gates, can be used in tide control structures to change the mean water elevation in the wetland. Flap gates can be installed on one or several culverts comprising the tide control structure depending on the amount of sea level control desired.

4.4 Maintenance of Tide Control Structures

Tide control structures are subject to gradual deterioration by being exposed to environmental conditions. The maintenance of tide control structures is important in order to keep the water elevations in the wetland within the desired range. Mussels, barnacles and other marine organisms will grow at the entrance as well as inside the culvert. This increases friction to flow and reduces the flow area. For culverts with flap gates, marine organisms can grow on the flap and culvert entrance, allowing water to leak when the flap gate should be closed. Flap gates have tendencies to restrict low flow out and to remain open with leaks for inward flow. Pinch check valves may be considered as an alternative with more positive control.

4.5 Determination of Tidal Elevations

Tidal elevations are measured by the National Oceanic and Atmospheric Administration (NOAA). Data are averaged over an 18.6-year tidal epoch to establish reference elevations such as MHHW, MLLW and MSL at the gauge station. Tide elevations in wetlands are a function of the connection with the ocean and can be different than tide elevations in the ocean.

A technique was developed to predict reference tide elevations in wetlands with tide ranges different than those in the ocean. The tide data for the Los Angeles station are used as an example. The distribution of tide elevations for a 18.6-year epoch was determined from data published by NOAA. The data was plotted as percent time of exposure versus tide elevation as shown in Figure 5; this format can be useful to wetland biologists. For the purpose of numerical modeling, a hypothetical 14-day mixed tide was developed so that it would reproduce the distribution of a 18.6-year tidal epoch. The mix between high and low tides as well as diurnal and semi-diurnal features was varied to achieve the desired distribution. This 14-day tide was used in the model analysis to determine the tidal distribution inside a wetland. In this manner, non-linearities with storage curves and hydraulic losses would be included in the prediction.

From the distribution of a 18.6-year period of tide elevations given by NOAA, it can be determined that ground at an elevation of MHHW is exposed 95 percent of the time, at MSL it is exposed 50 percent and at MLLW 5 percent. In a tidal wetland, these percentages give rise to the habitat zonations previously described.

The distribution of tide elevations inside a wetland can be defined using numerical models such as the HCM. The MHHW elevation inside a wetland is then identified as the elevation at which the ground is exposed

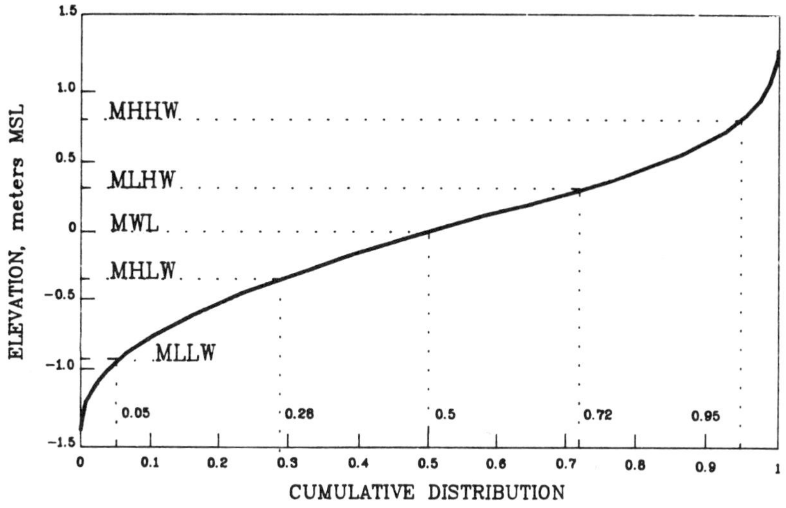

Figure 5 Percent Occurrence of Water Elevation At or Below a Given Level

95 percent of the time. The definitions of MWL and MLLW follow a similar form, where ground at MWL is exposed 50 percent of the time and at MLLW is exposed 5 percent of the time. In this method, shown in Figure 6, the tidal elevations in a wetland are related to the type of hydraulic connection between the wetland and the ocean.

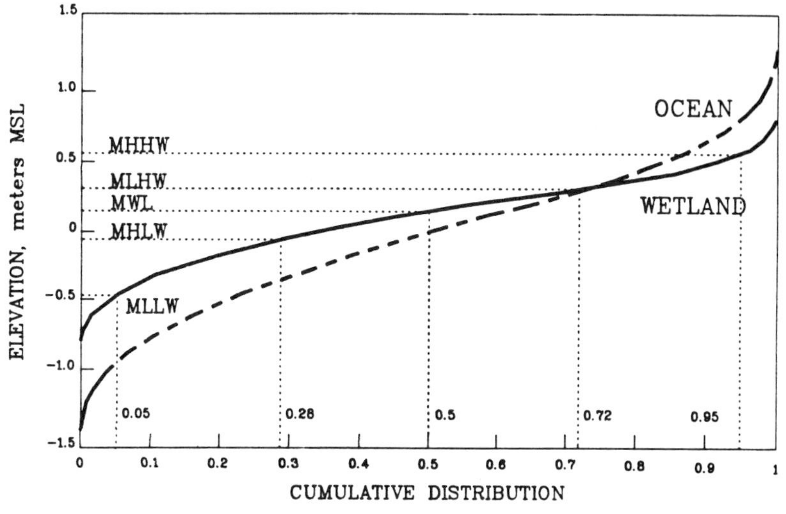

Figure 6 Method for Determining Tide Reference Elevations in a Wetland

4.6 Effects of Tide Control Structures

The numerical model HCM was applied to demonstrate the effects of changing control structure on wetland tides. The parameters of control structures investigated were culvert area, invert elevation, culvert roughness, entrance losses and use of flap gates.

A hypothetical wetland, connected to the ocean by a channel and tide control structure, was used in the simulations. The ground elevations and storage curve of the wetland were similar to those of an existing muted tidal wetland located at Bolsa Chica, 20 km south of Los Angeles, California.

4.6.1 Culvert Size

The flow area of a culvert connecting the ocean to the wetland was varied to evaluate the effects on wetland tides. The culvert height was kept constant, with an invert elevation low enough so that the culvert would always flow with a full section even at low tide. The width of the culvert was then changed to increase the flow area. The culvert used was 2 meters high, 10 meters long, an invert at -5.5 meters, with an entrance loss factor of 0.5 and a Manning's "n" of 0.05 to represent marine growth inside the culvert. The results, shown in Figure 7, indicate a strong effect of flow area on tide range in the wetlands. It can be noted that the mean water level inside the wetland is similar to that in the ocean and is not affected by the flow area of a culvert that is always submerged.

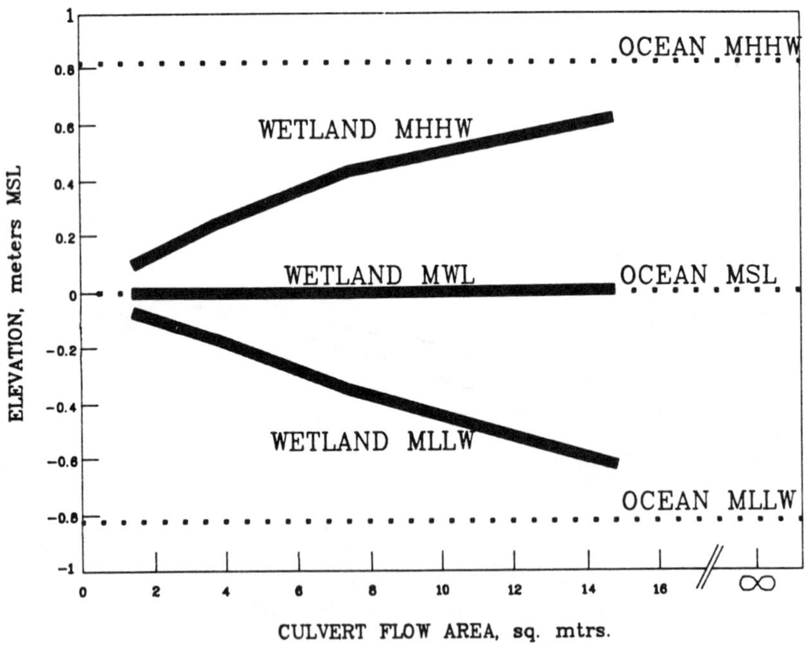

Figure 7 Effect of Culvert Flow Area on Wetland Tide Elevations

4.6.2 Invert Elevation of Culverts

The invert elevation of a culvert is the distance from the bottom of the flow area to the elevation datum. The invert elevation determines the minimum elevation that a wetland can drain and also whether a given culvert flows with a free surface. The effect of varying the invert elevation of wetland tide is shown in Figure 8. When the invert of the culvert is fairly deep, the culvert has a full flow section throughout the tide range. However, as the invert elevation is raised, the culvert will flow partially full with a free surface at low tide. The creation of a free surface affects first the elevation of the MLLW in the wetland by raising it higher than with a full section. The elevations of MWL and MHHW do not respond to changes in invert elevation as soon as MLLW. Further raising the invert elevation affects the MWL elevation in the wetland by raising it above MSL in the ocean. At higher invert elevations, the MHHW elevation in the wetland is reduced.

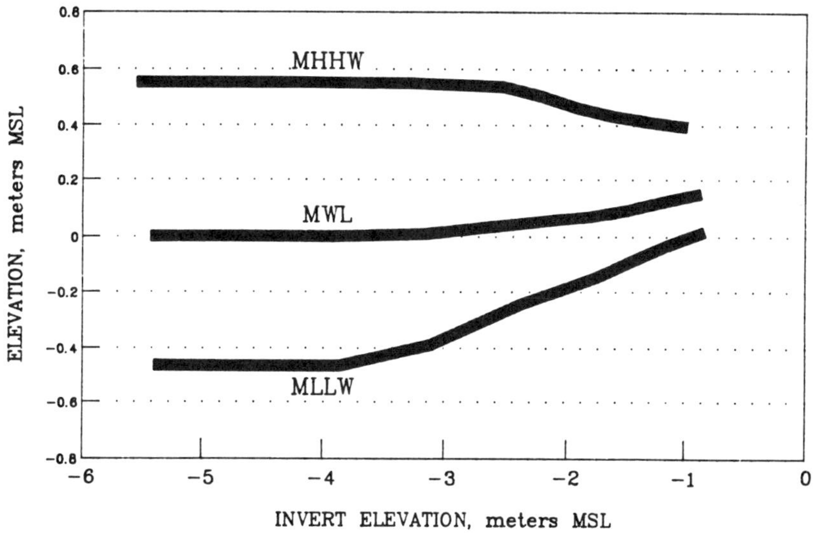

Figure 8 Effect of Culvert Invert Elevation on Wetland Tide Elevations

4.6.3 Culvert Roughness

The friction factor for a culvert depends on the type of material used in construction as well as on marine growth inside of the culvert. Values of friction factors for new culverts can be found in many references (such as Chow, 1959). However, increase in friction due to organic growth inside the culvert is more difficult to determine. An analysis of the effect of the friction factor on the wetland tide is shown in Figure 9. For a new concrete culvert, for example, Manning's "n" can vary from 0.011 to 0.013 (Chow, 1959). Experience with analysis of culverts colonized by marine growth indicated a friction factor of 0.04 to 0.06 (Moffatt & Nichol, Engineers, 1987b). The growth also reduces the culvert flow area which further restricts the flow.

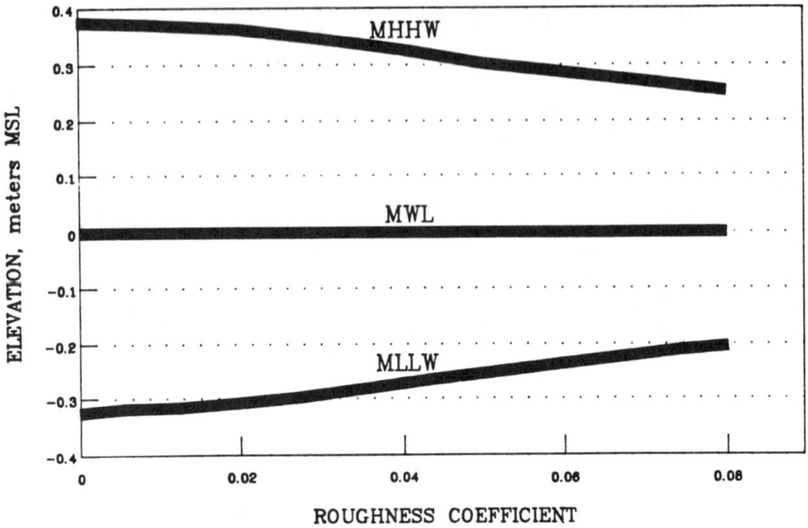

Figure 9 Effect of Culvert Roughness on Wetland Tide Elevations

5.0 FINDINGS

- In wetland creation or restoration, the goals and criteria must be established by the agencies in charge. Then biological criteria and sound engineering principles can be used in the creation or restoration of wetlands.

- Wetland creation or restoration goals should be defined by habitat type rather than by ocean tide ranges.

- Wetland habitats can be defined by the relative tide and ground elevations as well as soil conditions.

- Tide elevations inside a wetland can be defined from a water elevation distribution in the wetland. Ground at MHHW will be exposed 95 percent of the time, at MSL it will be exposed 50 percent of the time and at MLLW 5 percent of the time.

- Hydraulic numerical models are an essential tool for the analysis of wetland projects. The models provide an accurate and cost-effective method for evaluation of tide ranges in wetlands with varying ground elevations and type of hydraulic connections and controls.

- Tide characteristics should be evaluated to assess the impact of tide control structures on wetlands. In particular, during field survey and tide measurements, the various cycles of the tides should be considered to include seasonal and longer-term influences.

- o The size and invert elevation of culverts as well as the topography and channel dimensions are very important parameters in the design of muted tidal wetlands.

- o The long term effect of a wetland design should be investigated. Factors that will need to be considered include sea level rise, land subsidence or upheaval, sedimentation, and biofouling of tide control structures.

ACKNOWLEDGEMENTS

The design approach and model development presented in this paper came from designs of wetlands and field studies of existing wetlands. It is a summary of a larger wetland design study. The designs include the Anaheim Bay Mitigation Project for the Pier J expansion at the Port of Long Beach and the Bolsa Chica wetland development project for Signal Landmark. The authors extend their gratitude to Dr. Andrew Gram of Moffatt & Nichol, Engineers for his participation and development of the numerical model.

REFERENCES

Feldmeth, C.R. "Survey of inner and outer Bolsa Bays: recolonization and recruitment study", Unpublished report submitted to Signal Bolsa Corp., 1980, 30pp.

Feldmeth, C.R., "Biological resources of the Venice Canals", Report to Phillips Brandt Reddick for the City of Los Angeles Planning Department. 1986, 18pp.

Fisher, H. B., "A Numerical Model of the Suisun Marsh, California", Proceedings from the 17th Congr. Int. Assoc. Hydraul. Res. held at Baden-Baden, Germany 1977.

Flick, R.E. and D.R. Cayan, "Extreme Sea Levels on the California Coast", Proc. 19th Coastal Eng. Conf., Amer. Soc. Civil Eng., 1984, p. 886-898.

Horn, M.H., "Numbers of species and faunal assemblage of marine fishes in California Bays and Estuaries", Bulletin of S. Calif. Acad. Sci., 1976, 75:159-170.

Mahall, B.E. and R.B. Park, "The ecotone between Spartina foliosa Trin. and Salicornia virginica L. in salt marshes of northern San Francisco Bay. I. Biomass and production. II. Soil water and salinity. II. Soil aeration and tidal immersion", J. Ecol., 1986, 64:793-820.

Moffatt & Nichol, Engineers, "Anaheim Bay Mitigation Project" prepared for the Port of Long Beach, March, 1987.

Moffatt & Nichol, Engineers, "Bolsa Chica Project: Hydraulic Design Approach for Muted-Tide Wetlands", prepared for Signal Landmark, Inc., December, 1987.

Smith, David D., "A Practical Evaluative Procedure for the Design of Coastal Wetland Restoration Projects", Presented at the Coastal Zone 83 conference in San Diego, California, June, 1983, Vol II, pp. 1477-1487.

Williams, Philip B. and H. Thomas Harvey, "California Coastal Salt Marsh Restoration Design", presented at the Coastal Zone 83 conference in San Diego, California, June, 1983, Vol II, pp. 1444-1456.

Zedler, J.B., "The ecology of Southern California coastal salt marshes: A community profile", U.S. Fish and Wildlife Service, Biological Services Program. Washington D.C. FWS/OBS-81/54, 1982, 110pp.

Zetler, B.D. and R.E. Flick, "Predicted Extreme High Tides for California, 1983-2000", Jour. Waterway Port, Coastal and Ocean Eng., Amer. Soc. Civil Eng., 1985a, v. 111, n. 4, p. 758-765.

Zetler, B.D. and R.E. Flick, "Predicted Extreme High Tides for Mixed Tidal Regimes", Jour. Phys. Oceanog., 1985b, v. 15, n. 3, p. 357-359.

CHAPTER 199

A COMPARISION BETWEEN GERMAN AND NORTH AMERICAN TIDAL INLETS

REINHARD DIECKMANN *
MANUELA OSTERTHUN **
HANS WERNER PARTENSCKY ***

ABSTRACT

Initial results are presented relating to an investigation of geomorphological parameters from 26 of a total of 28 tidal inlets and 11 coastal structures similar to inlets along the German Bight. The following parameters were investigated:

- tidal prism - inlet area relationship,
- cross-sectional form and depth,
- location of the channel within the cross-section,
- ebb tidal deltas and
- the structures of the back barrier regions.

The most important parameter governing the shaping process of these coastal structures is considered to be the tidal volume. The results of the investigations were compared with data from American publications. Despite the differences between the regions studied, in overall terms, surprizingly good agreement was obtained between the parameters investigated. Notable differences exist only in relation to the morphological structure of the back barrier regions (tidal flats, salt marshes, open water lagoons). In respect of the latter differences, a short account is given of the biological and climatological influencing factors.

1. INTRODUCTION

Barrier islands and inlets are one of the most commonly occurring coastal formations worldwide. In particular, the American barrier island - inlet systems are widely dealt

* Dr.-Ing., Wasser- und Schiffahrtsamt Bremerhaven, Am Alten Vorhafen 1, D-2850 Bremerhaven, West Germany

** Dipl.-Ing., Research Engineer,

*** Prof. Dr.-Ing. Dr.phys., Director,

both at

Franzius-Institut für Wasserbau und Küsteningenieurwesen der Universität Hannover, Nienburger Str. 4, D-3000 Hannover 1, West Germany

with in literature. In contrast, the corresponding inlets along the coast of the German Bight have not as yet been geomorphologically investigated in their entirity. An exception to this are three inlets on the southern North Sea coast (WALTHER, 1972; LUCK, 1976; FITZGERALD et.al., 1984).

Initial results will be presented concerning a comprehensive investigation of 26 from a total of 28 tidal inlets and 11 inlet-type structures along the coast of the German Bight between Den Helder in the Netherlands and Skallingen in Denmark.

Basic differences between the coast of the German Bight and the southeast American Atlantic coast are highlighted by NUMMEDAL and FISCHER (1978). In this context, it should be noted that the German Bight lies on the same latitude as the southern part of Hudson Bay (James Bay) in Canada.

2. TIDAL INLET SYSTEMS ALONG THE GERMAN BIGHT

2.1. COASTAL OVERVIEW

Over a length of 450 km, the inner part of the German Bight is bordered by a large tidal flat area of 7,500 km^2 (see Fig. 1) with a barrier island chain in front of it.

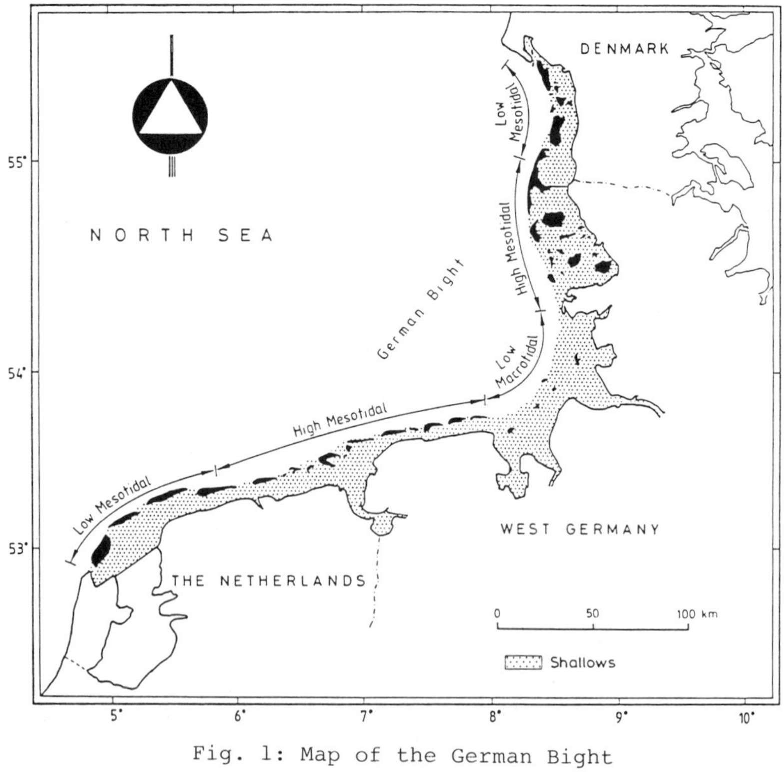

Fig. 1: Map of the German Bight

If one considers the investigations by HAYES (1979) concerning the relationships between coastal morphology, tidal range and wave climate, it becomes evident that the coast of the German Bight is a typical barrier island - inlet region.

According to the tidal and wave conditions, the majority of the German Bight region belongs to the "tidally dominated mixed energy coasts". The barrier islands on the south coast frequently have a well-defined drumstick shape with distinct ebb tidal deltas owing to the littoral drift. Flood tidal deltas, on the other hand, are not present as usual under meso- and macrotidal conditions. Along the south coast, a downdrift offset configuration is apparent, whilst in the case of the east coast, a negligible or small updrift offset is detectable.

The inlets on the south coast are strongly influenced by natural forces. Owing to the action of westerly winds, waves and currents parallel to the coastline, the islands and hence the inlets undergo continuous displacement from the west to the east. Since about 1890, the westerly heads of the islands have been fortified to avoid further drift. The east ends of the islands, however, remain exposed to natural forces and are thus in a continuous state of change. The currents in the inlets are so large as to prevent sedimentation and there are no jetties available which might reduce or alter the direction of the littoral drift.

On the east coast, the tidal inlets are less clearly defined and little is known concerning the barrier island - inlet systems on the east coast. Preliminary depth chart evaluations indicate however that these inlets are relatively stable. Consequently, the need to fortify or protect the heads of the islands has never arisen.

2.2. DATA COLLECTION

A total of 37 inlets and inlet-type coastal structures along the German Bight were investigated. For each inlets investigated, a data set containing the following information was established:

- tidal heights at MHW, MLW and the half-tide water level (HTWL),
- cross-sectional area of the inlet at MHW, MLW and HTWL,
- width of the inlet at MHW, MLW and HTWL,
- drainage area size at MHW, MLW and HTWL,
- maximum and mean depths of the inlet,
- tidal volume.

Since each inlet of the German Bight has a precisely defined drainage area, the mean tidal volumes (tidal prisms) could be obtained with high accuracy from the drainage basin hypsometric curves. The evaluation technique used for this purpose is described in detail in DIECKMANN & PARTENSCKY, 1985.

3. US AMERICAN TIDAL INLETS

In relation to US American tidal inlets, a vast amount of material has been published. A comprehensive review is given in the GITI Reports published by the US Army, Corps of Engineers and Waterways Experiment Station as well as in "Stability of Tidal Inlets" by PER BRUUN (1978). Further details will be given there.

4. COMPARISION OF RESULTS

4.1. TIDAL PRISM - INLET CROSS-SECTIONAL AREA RELATIONSHIP

Relationships between the minimum throat cross-sectional area below mean tide level and the tidal prism of an inlet were first reported by O'BRIEN (e.g. 1969). More recent evaluations were presented by JARRETT (1976). On the basis of such relationships, estimates may be made of the long-term changes in the inlet cross-section up to the point at which a mean state of equilibrium is attained.

In Fig. 2, data relating to the inlets of the German Bight are presented in the form of diagrams after JARRETT. All data points lie within the 95% confidence limits and very close to the regression curve given by JARRETT.

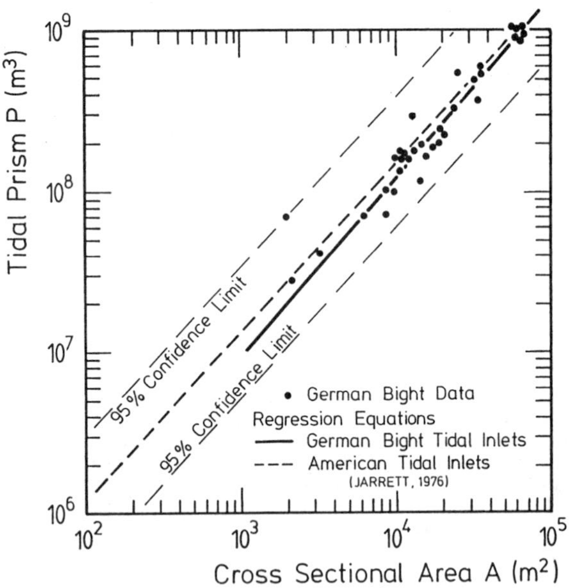

Fig. 2: Tidal prism vs. cross-sectional area for American inlets and inlets along the German Bight.

The corresponding regression equations in metric units are as follows:

German Bight: $A_c = 3.720 \cdot 10^{-4} \cdot p^{0.915}$ (1)

Atlantic coast: $A_c = 3.039 \cdot 10^{-5} \cdot p^{1.050}$ (2)

Gulf coast: $A_c = 9.311 \cdot 10^{-4} \cdot p^{0.840}$ (3)

Pacific coast: $A_c = 2.833 \cdot 10^{-4} \cdot p^{0.910}$ (4)

where:

A_c = minimum cross-sectional area below MSL = HTWL (m^2)

p = tidal prism (m^3)

As a result, the best agreement is achieved between the inlets of the German Bight and the Pacific coast as may be seen from the equations (1) to (4). The good agreement between the above relationships indicates that the minimum cross-sectional area of an inlet is primarily determined by the inflowing and outflowing tidal volume. For the German Bight no particular influence of the different coastal and tidal range regions could be detected. This result is surprizing in so far as significant differences exist between the US American inlets and tidal flats and the corresponding regions in the German Bight.

4.2. GEOMETRIC PARAMETERS

4.2.1 SHAPE AND DEPTH OF THE INLET CROSS-SECTIONS

An inlet is characterized in the terms of its geometry by the shape of its cross-section and the mean or maximum depth. The cross-sectional shapes of the inlets in the German Bight may be subdivided into 3 groups (see Fig. 3):

- the wide synclinal profile will tend to form in the case of small tidal volumes in combination with relatively wide inlets,
- the triangular profile is characteristic of narrow inlets exposed to large tidal volumes and
- the composite profile may be taken to represent the normal case.

		shape of tidal inlet		

occurence	-	5	12	20
width/depth ratio	-	›300	50-150	90-250
mean depth	m	3-8	›9	›7
maximum depth	m	5-16	12-48	10-37

Fig. 3: Different types of tidal inlet shape.

An account of American inlet cross-sections by VINCENT & CORSON (1980) indicates that the triangular shaped cross-

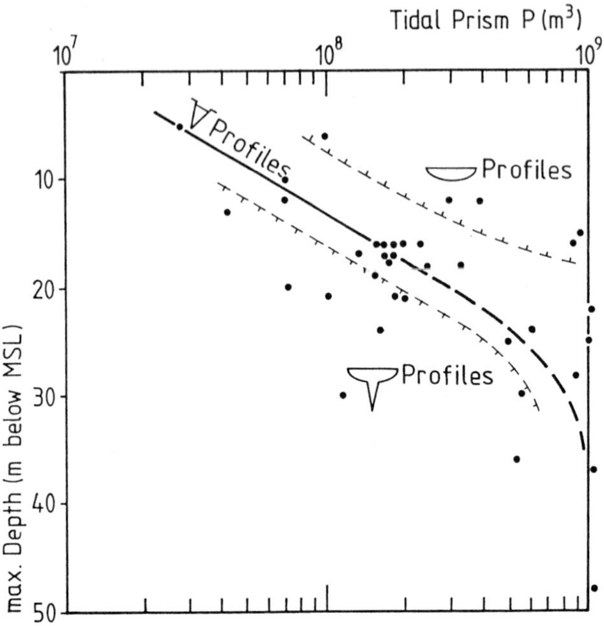

Fig. 4: Relationship between the tidal volume and the maximum depth of inlets in the German Bight.

section is clearly predominant whereas the synclinal shape seldom occurs.

The maximum water depth is also dependent upon the tidal volume. The corresponding relationship for the inlets of the German Bight is illustrated in Fig. 4.

Furthermore, a relationship also exists between the mean and maximum depth in a inlet cross-section. The data for the German Bight inlets have been added to another diagram of VINCENT & CORSON, as shown in Fig. 5. As may be seen in the Figure, the scatter in the latter data is much larger than that of the American data. On the whole, the German Bight inlets are characterized by larger maximum depths.

An interesting aspect concerns the ratio of the tidal volume to the inlet cross-sectional area which lies between $1.0 \cdot 10^4$ and $1.8 \cdot 10^4$, i.e. with a mean value of $1.4 \cdot 10^4$. From a total of 37 inlets investigated, only 5 showed a deviation from the latter ratio. Deviations above this value clearly characterize wide inlets with several channels, whilst deviations below are indicative of narrow inlets with a composite cross-section and a narrow, deep channel (> 20 m). Further investigations may quite possibly lead to the characterization of inlet cross-sections by simple numerical values.

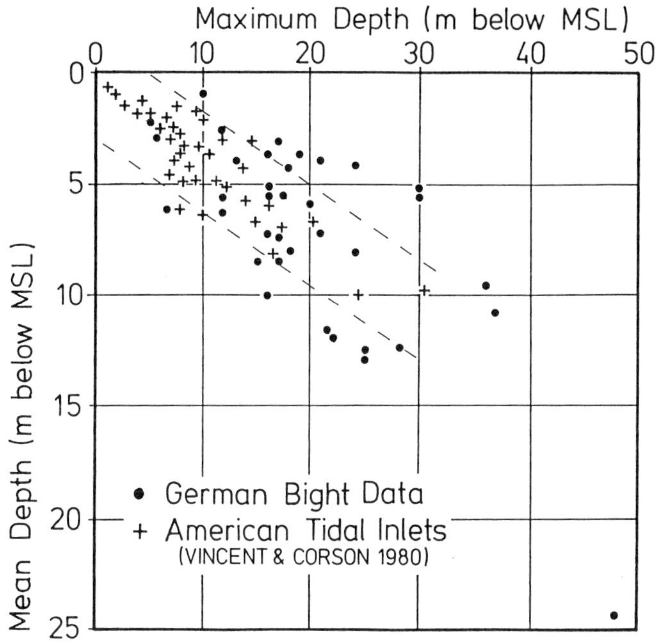

Fig. 5: Relationship between the mean and maximum depth in an inlet cross-section.

4.2.2. LOCATION OF THE CHANNEL WITHIN THE CROSS-SECTION

The geometry of the cross-sectional area and the location of the deep channel within the inlet cross-section in the region of the German Bight are different from the correspon-ding parameters arising from American investigations (FITZ-GERALD & FITZGERALD, 1977). In the case of the German Bight, these factors are less dependent upon meandering of the channel bed, the form of the coastline outside the inlet mouth and the dominant longshore transport direc-tion, but are mainly governed by the shape of the drainage area and the location of the latter with respect to the inlet. These conditions are particularly noticeable along the south coast of the German Bight in the vicinity of the East Frisian islands. A distinction must be made between the following cases:

a) The drainage area is located symmetrically behind the inlet and the deep channel lies in the middle of the inlet (symmetrical cross-section).

b) The drainage area lies for the most part behind an island (barrier island), e.g. in the downdrift direction relative to the inlet, whilst the deep channel of the inlet is orientated in the updrift direction. This situation is due to the fact that the tidal water volume associated with the mophologically active ebb phase must flow behind the barrier island before the tidal inlet is reached (non-symmetrical updrift cross-section).

c) In the case of receding coastlines, two main channel arms often meet together within the inlet. One arm usually runs parallel to the barrier island whilst the other runs directly towards the inlet. If the tidal volume of the channel orientated at right angles to the inlet is sufficiently large, the channel running parallel to the island will be forced in a downdrift direction against the barrier island (non-symmetrical downdrift cross-section). The development of inlets with double channel systems is also possible.

On the east coast of the German Bight, the inlet drainage areas are orientated at right angle to the coastline. For this reason and also because of the lower degree of sediment transport, the channels extend a long way seawards and are located centrally within the inlet cross-section. In the water inlets, side channels frequently occur.

4.2.3. EBB TIDAL DELTAS

General forms of ebb tidal deltas, which develop due to differences in flow velocities and longshore sediment transport rates, have been compiled by OERTEL (1975) for the mesotidal coast of Georgia (Fig. 6).

Similar structures are also apparent in the German Bight. Of particular significance are the curved sand bars of the tidal inlets along the East Frisian chain of islands which have been dealt with in detail by LUCK (1976).

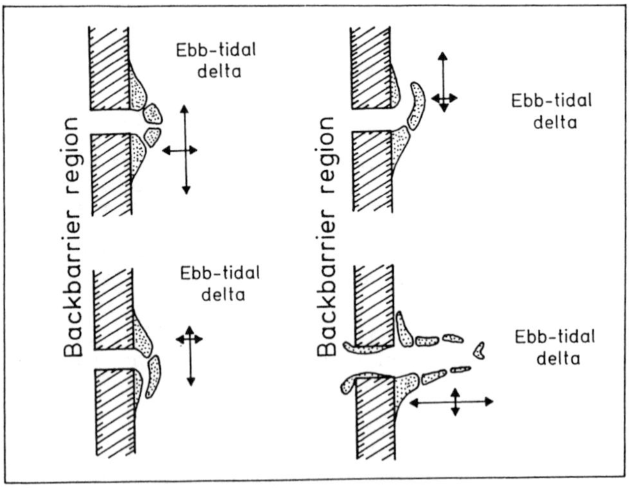

Fig. 6: Ebb tidal deltas of tidal inlets (with arrows indicating the relative magnitudes of onshore, longshore and offshore currents).

4.3. BACK BARRIER BATHYMETRY

The differences between the back barrier bathymetry of the American south coast and the German Bight were first pointed out by NUMMEDAL & FISCHER (1978). The American intertidal salt marshes are mainly comprised of areas overgrown with spartina; only about 20% of the total surface area consists of open water (tidal creeks). In contrast to this, the geomorphological structure of corresponding areas in the German Bight is totally different. Below the MHW line, no vegetation exists. The main factor which governs the morphological structure of the tidal flats is the tidal range (DIECKMANN & PARTENSCKY, 1985). Hypsographic curves published in the literature for American coastal sections in the mesotidal range are presented in Fig. 7 together with curves for corresponding regions in the German Bight. As may be seen in Fig. 7, the differences are clearly apparent.

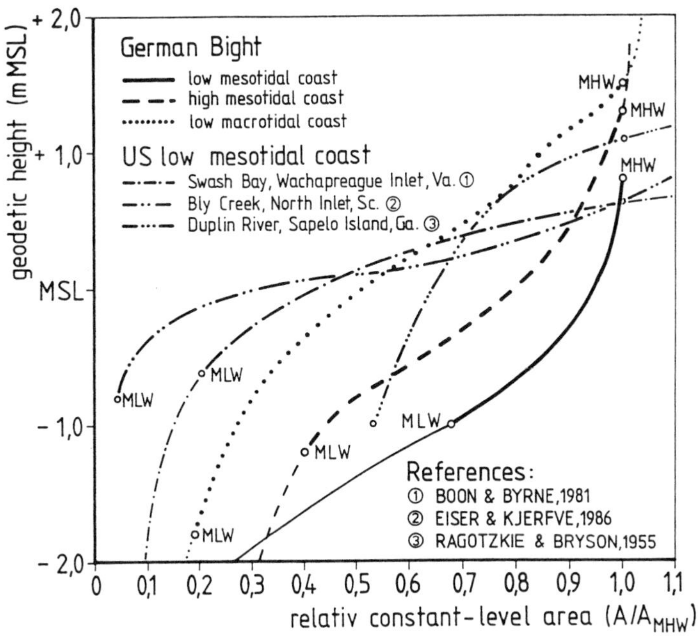

Fig. 7: Hypsographic curves for several American coastal sections (mesotidal range) and the basic types of different tidal range regions in the German Bight.

It should be noted that the given American hypsographic curves represent relatively small coastal sections (0.6 to 27.4 km^2) whereas those for the German Bight are derived from larger regions (95 to 260 km^2). Differences in the back barrier bathymetry do not govern the size and shape of the tidal inlets, which are themselves determined by the tidal volume, but rather the form of the flow velocity profiles within the inlets.

4.4. OTHER PARAMETERS

In addition to the parameters mentioned above, there are no doubt other parameters worthy of comparison such as sediments, flows and biological effects relating to different climatic conditions.

For example, NUMMEDAL & FISCHER make mention of the fact that in inlets adjoining open water lagoons in the back barrier region, flood flow dominance is evident whereas for tidal flats and salt marshes in the back barrier region, ebb flow dominance is present. Similar conditions also appear to apply in the case of the German Bight region. For 3 inlets on the East Frisian coast with tidal flats in the back barrier region, a definite ebb flow dominance could be ascertained (PFENNIG, 1978).

5. CONCLUSIONS

From a comparision of American inlets with those in the German Bight, it was found that in the case of all parameters which describe hydrodynamic processes, good agreement exists. In contrast, significant differences were apparent in relation to the morphological structure of the tidal flat salt marshes as well as climatic influences (vegetation). In the context of coastal engineering science, however, these aspects are of no particular importance.

6. REFERENCES

BOON, J.D.III; BYRNE, R.J.: On Basis Hypsometry and Morphodynamic Response of Coastal Inlet Systems. Marine Geology, vol. 40, p. 27 - 48, 1981.

BRUUN, P.: Stability of Tidal Inlets. Theory and Engineering. Developments in Geotechnical Engineering no. 23, Elsevier Scientific Publishing Company, Amsterdam, 1978.

DIECKMANN, R.; PARTENSCKY, H.W.: Quantitative Geomorphology of Tidal Flat Areas. In: SIGBJARNARSON, G. (ed.): Proc. Iceland Coastal and River Symposium, Reykjavik/Iceland, p. 53 - 62, 1985.

EISER, W.C.; KJERFVE, B.: Marsh Topography and Hypsometric Characteristics of a South Carolina Salt Marsh Basin. Estuarine, Coastal and Shelf Science, vol. 23, p. 595 - 605, 1986.

FITZGERALD, D.M.; FITZGERALD, S.A.: Factors Influencing Tidal Throat Geometry. Proc. Coastal Sediments '77, publ. by ASCE, p. 563 - 581, 1977.

FITZGERALD, D.M.; PENLAND, S.; NUMMEDAL, D.: Control of Barrier Island Shape by Inlet Sediment bypassing: East Frisian Island, West Germany. Marine Geology, vol. 60, p. 355 - 376, 1984.

HAYES, M.O.: Barrier Island Morphology as a Function of Tidal and Wave Regime. In: LEATHERMAN, S.P. (ed.): Barrier Islands. Academic Press, New York, 1976.

JARRETT, J.T.: Tidal Prism - Inlet Area Relationship. General Investigations on Tidal Inlets (GITI), report no. 3, US Army Engineer Waterways Experiment Station, Vicksburg/Miss., 1976.

LUCK, G.: Inlet Changes of the Eastfrisian Islands. Proc. 15th Int. Conf. on Coastal Eng., Honolulu/Hawaii, p. 1938 - 1957, 1976.

NUMMEDAL, D.; FISHER, I.A.: Process Response Models for Depositional Shorelines: The German and the Georgia Bights. Proc. 16th Int. Conf. on Coastal Eng., Hamburg/West Germany, p. 1215 - 1231, 1978.

O'BRIEN, M.P.: Equilibrium Flow Areas of Inlets on Sandy Coasts. Jour. of the Waterways and Harbors Div., ASCE, no. WW1, p. 43 - 52, 1969.

OERTEL, G.F.: Ebb Tidal Deltas of Georgia Estuaries. In: CRONIN, L.E. (ed.): Estuarine Research, vol. II: Geology and Engineering. Academic Press, New York, p. 267 - 276, 1975.

PFENNIG, A.: Bewertungsmodelle zur Untersuchung transportwirksamer Reststromsysteme. Sonderübung, Franzius-Institut, University of Hannover/West Germany, unpubl., 1978.

RAGOTZKIE, R.A.; BRYSON, R.A.: Hydrography of the Duplin River, Sapelo Island/Georgia. Bull. of Marine Science of the Gulf and Caribbean, vol. 5, p. 297 - 314, 1955.

VINCENT, C.L.; CORSON, W.D.: The Geometry of selected US tidal inlets. GITI Report no. 20, US Army Engineer Waterways Experiment Station, Vicksburg/Miss., 1980.

WALTHER, F.: Zusammenhänge zwischen der Größe der ostfriesischen Seegaten mit ihren Wattgebieten sowie den Gezeiten und Strömungen. Forschungsstelle Norderney der niedersächsischen Wasserwirtschaftsverwaltung, vol. 23, 1972.

CHAPTER 200

FLOW DEPENDENCY OF BOTTOMSTRESSES IN TIDAL MODELS

Jan J. Leendertse [1]
A. Langerak [2]

In modeling studies the bottomstress parameter may not be taken as a constant for making predictions. This is particularly true when large changes are being made in the water bodies. Engineering judgement is than necessary as studies to a relation between the bottomstress parameter and tidal flow velocity are not conclusive.

Introduction to the project

In 1976 the Netherlands Rijkswaterstaat embarked on constructing the final stage of the Deltaplan with the Eastern Scheldt-project. Three structures had to be built: a 3 kilometer long stormsurge barrier in the mouth of the estuary - to be closed during storms - and two so-called compartimentation dams (fig. 1). The dams, which created a lake in part of the estuary, had to provide (a) a tide-free shipping lane from the city of Antwerp to the river Rhine, and b) storage of fresh water flowing into the estuary. The latter was necessary to maintain high salinity in the estuary without large fluctuations for the ecological system as the stormsurge barrier reduced the tidal currents and thus the mixing.

The Deltaplan as a whole (fig. 2) enhanced the protection from the stormsurges in the Rhine-Meuse delta. The plan comprised closure of all estuary branches except for the New Waterway, which is the shipping lane to Rotterdam, and the Western Scheldt, which is the shipping lane to Antwerp. In 1976, the Deltaplan was completed except

---

[1] The RAND Corporation, Santa Monica, California, USA

[2] Tidal Waters Division, Rijkswaterstaat, Zierikzee, The Netherlands

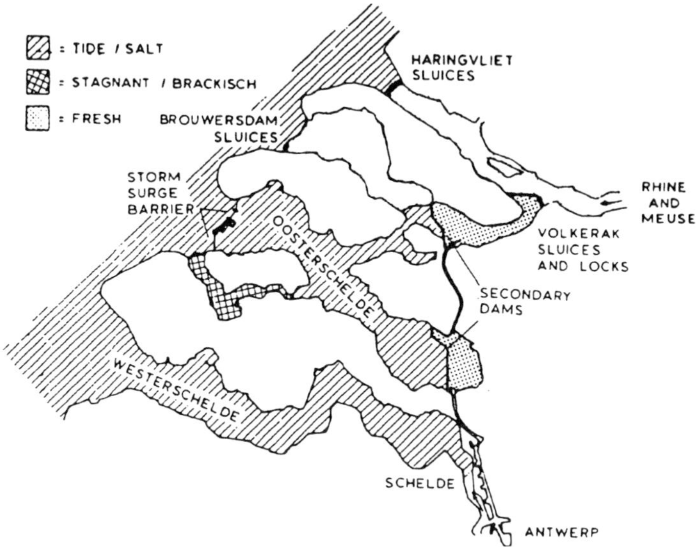

Fig. 1 Overview of the Eastern Scheldt project with location of the Stormsurge Barrier and Compartimentation Dams.

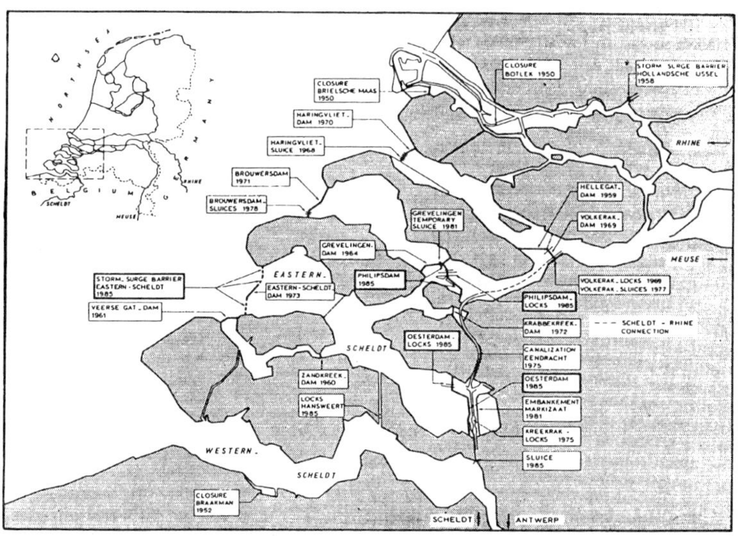

Fig. 2 View of the Deltaplan, including the Eastern Scheldt project.

for the most difficult part, namely the construction of the stormsurge barrier and the compartimentation dams in the Eastern Scheldt.

Function of the models

Large hydraulic engineering projects like the construction of the stormsurge barrier and the compartimentation dams require reliable short term predictions of the weather and of the currents and waterlevels. The predictions are needed to schedule and adjust the progress of the construction. Predictions over longer periods are also required to evaluate different options in the design and the construction methods.

The contract between the government and the construction-consortium for the project contained a clause which required the Rijkswaterstaat to provide the consortium with accurate tide and surge predictions. If the predictions deviated from observed values by a certain amount, extra payments were due for expected damages. As a result, the required accuracy of the models was high. The waterlevels had to be predicted within 5% of the observed values on an average, and currents had to be predicted within 10% of the observed values on an average. The prediction period ranged from 24 hrs.- for predictions of the combined tide and surge - to 4 yrs., the latter for predictions of tide only.

For the short term predictions, a one-dimensional model was used. This model was based on the numerical solution of the St. Venant-equations (fig. 3). For the long term predictions of tidal heights and currents, the one-dimensional model was also used, as the computation time on a HP1000-A was only one minute for every 24 hrs. real time. However, for the prediction of current distributions and intensities to be used in the design and construction planning, the one-dimensional model had insufficient resolution and a family of depth averaged two-dimensional models (fig. 4) was used with a spatial solution ranging from 45 m. for the detail models to 400 m. for the overview model. Some of the high resolution simulations required 20 hours on a main frame computer.

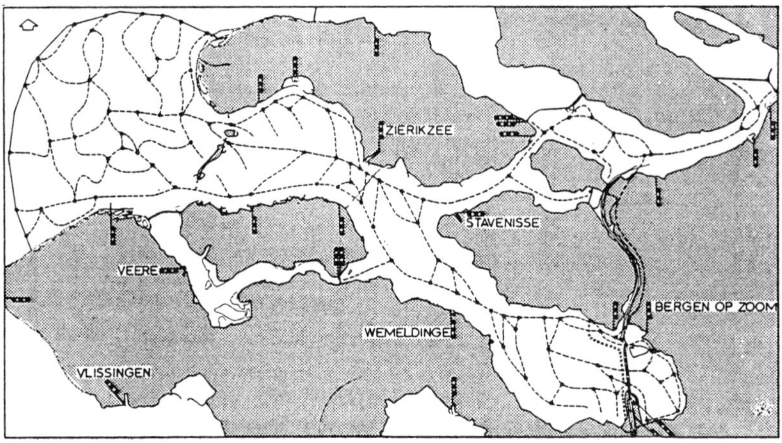

Fig. 3 One-dimensional tidal model used to make long and short term predictions.

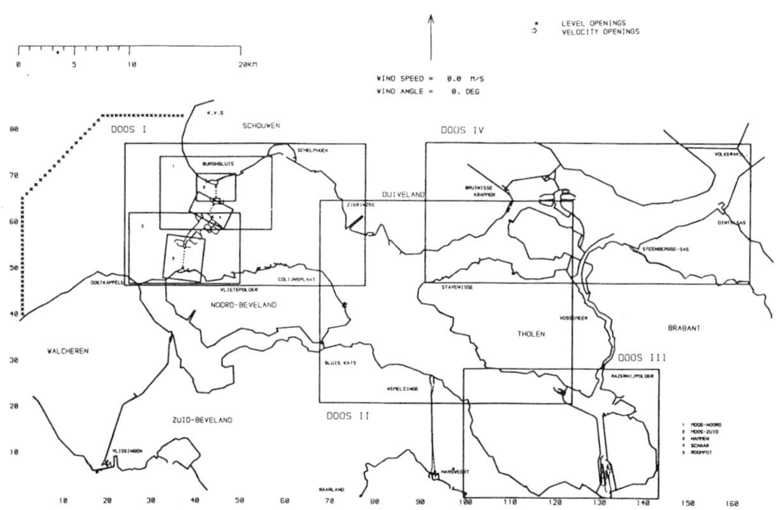

Fig. 4 Two-dimensional tidal models used in the evaluation of the design and construction methods.

In the beginning of the planning and execution of the project, hydraulic scale models were used together with numerical models. But after comparing hydraulic scale model results and numerical model results with observed data it was shown that numerical model results were more accurate (Leendertse 1984). As a numerical model was also more flexible and cost-effective, during the final construction phases of the project the design, the preparation and the actual construction depended completely on the use of numerical models.

Importance of the bottomstress representation

To fulfill the requirements of accuracy, all aspects of the models like the boundary conditions, bottom-schematisation and friction factors were carefully reviewed. A large number of experiments were made in a two-dimensional model of the estuary (Leendertse, 1988) to investigate the effects of the advection term approximation in the hydrodynamic equations and to investigate the comparative importance of the timestep size, the gridsize, the depth accuracy, the roughness estimates, and the horizontal momentum exchange. Similar experiments were made for the one-dimensional model.

From these experiments, it was concluded that the depth approximation, the boundary conditions and the bottomstress coëfficients were the most significant factors to obtain accurate predictions. For the depth approximation about a man-year's effort was expended to present the geometry of the estuary with high accuracy in the models. For the depth approximation, extensive use was made of computer programs which accessed the huge bathymetric data files. To verify the accuracy of the approximation, graphs were prepared with the depth in every cross-section of the model in comparison with the data of the bathymetric files. For the boundary conditions, a new technique was developed for estimating boundary tide levels from existing waterlevel recording stations in the vicinity of the boundary. A high accuracy of the estimates of these boundary conditions was obtained, but it appeared more expedient to measure the waterlevels at the model seaward boundary by installing off-shore recording stations and transmitting the data to the computer which made all the predictions. For verification purposes of the one-dimensional model, measurements of levels and currents were made at many locations in the estuary and these were also available on-line for verification.

The bottomstress parameters were found by adjusting the model for existing conditions. After making verification simulations, a project review indicated that all required standards were met for the existing conditions. However, as the flow conditions in the estuary would change due to the construction of the barrier and the dams, it was to be expected that the bottomstress parameters would change also. Thus for the long term predictions, considerable uncertainty was introduced particularly as the tidal amplification in the estuary is very sensitive to the bottomstress.

Bedforms and tidal systems

If significant changes in a hydraulic system of an estuary with a sandbottom are made, then such changes have an impact on the bedform. However, little is known about bedform changes in estuaries. This is in contrast to rivers where estimates are available on the impact of discharge changes on the bedform and the frictional losses. For estuaries where the changes in flow are much more rapid such estimates are not yet possible as the research directed toward establishing relations between flow and bottomstress is not conclusive. However, a relationship between flow and bottom roughness for quasi stationary systems can, under certain conditions, be used to obtain an insight in the behaviour of the bedform and bottomstress in an estuary.

The key parameter in this case is the bed response time to hydraulic changes. The bed-response time will be short if the bottom consists of loose packed fine sands and the currents are strong. The bed is then easily erodable and new bedforms appear in a matter of hours. In such a case, the relationships for quasi stationary flow can be expected to be applicable. However, if the bottom consists of densely packed sands or consists of coarse sands or gravel, the development of new bedforms can be a matter of months. In the first case, with loose packed fine sands and relatively strong currents, the tidal currents will have a direct one to one relationship with the bedform and the full cycle of bedforms depicted in Fig. 5 might be seen during a springtide/neaptide cycle or even during one tidal cycle. In the second case, the tidal currents have an impact on bedforms, but the extent of the changes can not yet be predicted.

The Eastern Scheldt estuary resembles more or less to the first case as it consists for a large part of the fine (D50 is 300 mu or less) loose packed sands and the currents are strong. Thus, we may expect a "rapid" response to changes in the hydraulic system.

Effect of the stormsurge barrier on bottomstress parameters

An extensive analysis of field data and a very extensive modeling effort permitted an accurate prediction of the modification of the tidal amplification in the Eastern Scheldt due to the construction of the stormsurge barrier.

Part of the modification was due to the obstruction of the flow by the barrier, in part it was due to modification of the bottomstress. In engineering investigations, it is common to assume that the bottomstress parameters are invariant, but before the barrier was constructed, it was noted from comparisons of one- and two-dimensional model results with observed tidal heights and currents that the bottomstress parameters depended on the flow conditions during previous tidal cycles. For example, if the model was adjusted for average tide conditions then the computation of a springtide would show too much tidal amplification and computations for neaptide would underestimate the tidal heights (Leendertse 1988).

To analyze the phenomena, the amplification of the tide in the estuary was determined from observations at two stations. One station was located at the seaward extremity of the estuary, the other far inland. The tidal amplification was determined by cross-spectral analysis from records of about 100 hours during springtide, during neaptide, and during average tide. The tidal amplification of the semi-diurnal tide appeared strongly dependent on the tidal heights.

For example, during neaptide amplification is 1.31, but during springtide the amplification is only 1.17. The phase lag is also influenced by the tidal amplification.

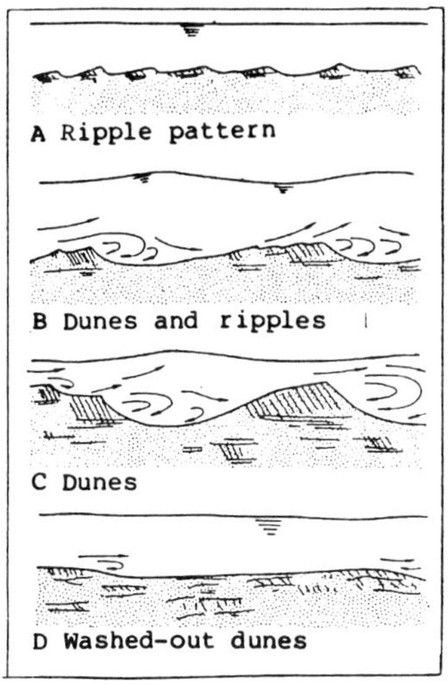

Fig. 5 Various bedforms at increasing velocity.

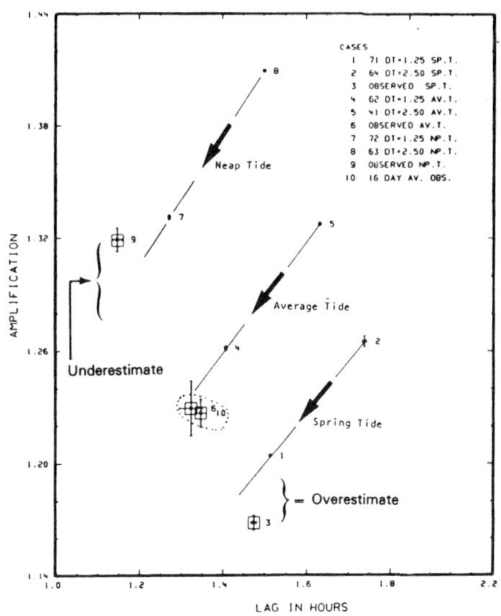

Fig. 6 Computed and observed amplification and phase lag of the semi-durinal tide over the estuary.

The amplification and phase-shift of the tide in a model may depend on one hand on the resolution in space and time of such a model and on the other hand on the bottomstress parameters such as the Chézy- or Manning's-value found during calibration of the model. In the numerical experiments with a two-dimensional model the three mentioned tide conditions were simulated and the tidal amplifications were determined by cross-spectral analysis. Two sets of experiments were performed, one with a time-step of 1.25 min. and the other with a time-step of 2.5 min.

As in numerical models, the time-step influences the propagation speed and amplitude of the tide, the two sets of experiments deter mined the influence of the resolution in time upon the results. The results of the analysis are summarized in fig. 6. The arrows in the graph show the effect of the reduction of the time-step. Increased resolution brings the amplification of tide in the model within the confidence band of the observations of the average tide, but this is not the case for the simulations of the springtide and the neaptide.

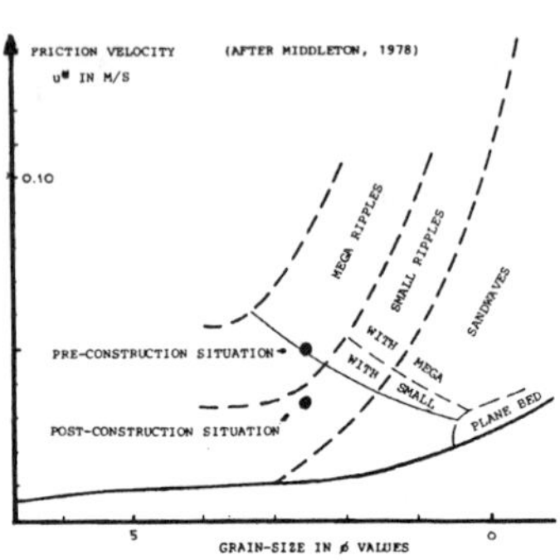

Fig. 7 Bedforms during the pre-construction and the post-construction period.

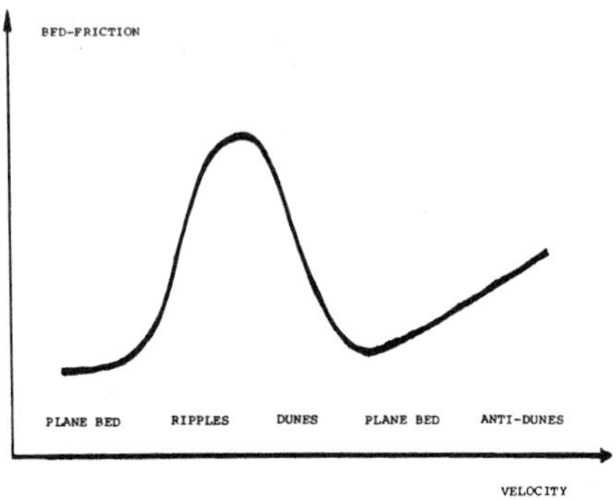

Fig. 8 Bedforms and their impact on bed friction.

From the experiments with the two-dimensional model and similar experiments with the one-dimensional model, it was concluded that the magnitude of the tidal flow reduction due to the barrier had to be taken into account in determining the bottomstress parameter. The impact of the structure as a whole on the hydraulic system was con siderable as a decrease of the tidal currents at least to 70% of the original value was predicted; thus not only needs the bottomstress parameter to be adjusted for neap- and springtide, but also for the long term predictions.

From literature for stationary systems, it was evident that the bed forms for fine sands like in the Eastern Scheldt (D50 is 200 mu) would vary with the flow speed according to the following ranges:

- flat bed 0,00 to 0,40 m/s
- bed with small ripples 0,40 to 0,70 m/s
- bed with mega-ripples or dunes 0,70 to 3,00 m/s

The friction is deemed to vary accordingly and a literature study indicated that the bedform with the highest bottomfriction is centered around the transition from mega-ripples to small ripples.

If the Eastern Scheldt would be a stationary system or a system with a very short response time, the bedforms before the construction should be a transition from mega-ripples to small ripples (fig. 7). Side looking sonar surveys showed indeed small ripples on top of mega-ripples as the dominant bedform in the non-stationary Eastern Scheldt.

As this bedform has the highest bedfriction (fig. 8), lower frictions could be expected when the construction proceeded. The tidal velocities during construction would drop at least to 70% of their original value, leading to a predominant bedform with small ripples and thus a smaller bedfriction.

Two different approaches were used to predict the modification of the bottomstress parameter. In the first approach, the Chézy-coëfficient in the numerical model was adjusted for neaptide and for springtide and a relation was established between the Chézy-coëfficient and the tidal range. The next step was to extrapolate the available data on the neaptide/springtide cycle to the future situation after the construction of the stormsurge barrier. We assumed a linear relation between the tides and the change in bedfriction and so an extrapolation was made for the future situation with lower tidal amplitudes, lower velocities and thus a lower bedfriction. A 12% higher Chézy-value (12% lower bedfriction factor) based on this analysis was expected.

For the second approach to predict the modification of the bottomstress parameter, we used the results of experiments in a sand flume of the Delft Hydraulics Laboratory. The Netherlands Rijkswaterstaat had commissioned the Laboratory to study the interaction between sand and flowing water. The studies resulted in a relation between velocity and bedfriction for stationary systems at different grain sizes. The results (Van Rijn 1982) are depicted on a prototype scale in fig. 9.

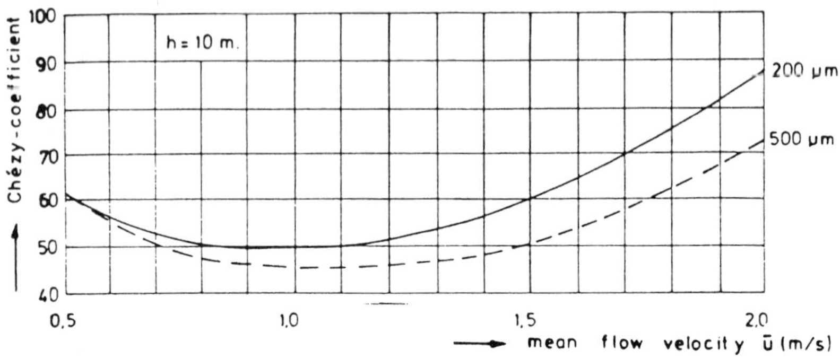

Fig. 9 Relation between the velocity and the Chézy-value in a flume at constant velocity for two grainsizes.

As fine and loose packed sands are present in the Eastern Scheldt, we assumed the bed-response time of this estuary to be short and certainly shorter than the construction time of the barrier. The results of the flume investigation indicated that the velocities which were present before the construction would generate mega-dunes with small ripples, and these were indeed present. As from numerical model studies, the tidal changes after construction were as a first estimate known, we could derive from fig. 9 that the Chézy-values would increase by 10 to 12%.

As both approaches indicated about the same increase, we estimated that a 10 to 12% higher Chézy-value was to be expected after the construction of the stormsurge barrier. This modification was then used for the post-construction predictions. After the barrier was constructed, a comparison was made again between observed tidal heights and currents and those obtained from the one- and two-dimensional models.

The agreement was excellent and this confirmed that the Chézy-value had indeed increased by 12% because of the construction of the bar rier. We found it somewhat remarkable to achieve such a good esti mate, because predicting changes in morphology is still more an art than a science.

Effect of dam construction on bottomstress parameters

The two dams near the inland extremities of the estuary were con structed to form two lakes (Fig. 1). These two lakes were connected by a ship channel. The dams were not constructed at about the same time, but the tidal openings in the dams were closed in sequence.

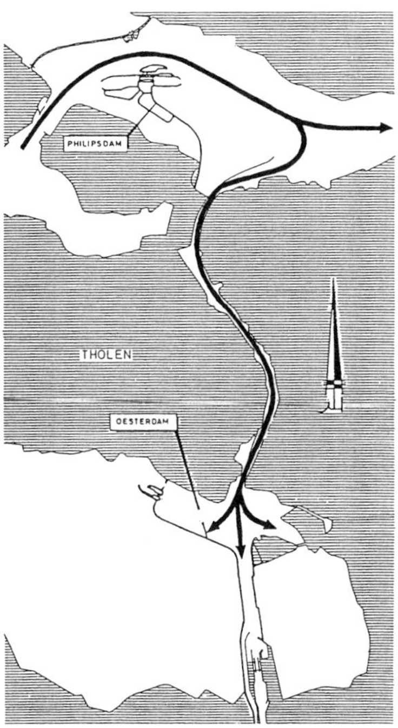

Fig.10 Flooding of the (southern) lake via the adjoining ship-channel.

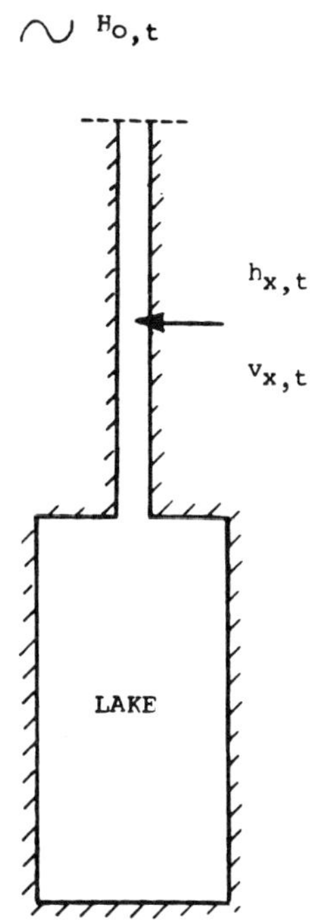

Fig.11 Schematic representation of the southern lake and the adjoining channel.

First the opening in the southern dam was closed and about a half year later the remaining opening in the northern dam was closed. Before the dam construction and also before the openings in the dams were closed, the tidal amplitudes and phases were about the same at the northern and southern end of the channel. As the surface area of the channel is fairly small, the tide generated only weak currents in the channel. During flood, water flowed southward in the northern part of the channel and northward in the southern part. Near the middle currents were very weak and variable (slack). This rather unusual situation precluded estimation of the bottomstress parameter with adjusted numerical models.

It was to be expected that after closure of the southern dam, the tidal flow in the channel would change very significantly. The area behind the southern dam would then receive its ebb and flood waters through the channel (Fig. 10). To predict the currents in the channel and the tidal heights in the area behind the southern dam for the condition with one dam closed, a series of model experiments were made. In the simulations the same bottomstress parameter as in other tidal channels with similar bottom material and currents was used, but shortly after the closure of the southern dam, it appeared that the measured currents were about 20% larger than predicted. Apparently, the bedfriction in the channel was reduced significantly. These high currents introduced erosion at several locations and required the installation of more revetments and bottom protection devices then originally expected. The high currents impeded shipping also.

After the closure, the tidal system consisted of a basin connected by a narrow channel to the tidal waters at the northern end of the channel (Fig. 11). In the channel, the velocities would reach their maximum value about 10 minutes after high water or low water slack and the maximum velocity remained constant until 10 minutes before the next change of tide.

Further analysis indicated that the tidal flows (maximum velocities) and the bedform were in a weak unstable equilibrium. If the maximum velocities increase temporarily because of a passing storm, then the bottom roughness could be reduced slightly by a feed back system as depicted in fig. 12 and in following tidal cycles higher velocities could occur. If the relation between bedform (flow resistance) and maximum velocities was taken into account in the analysis, it appeared that the velocities would increase gradually untill the velocities would exceed 1,65 m/s.

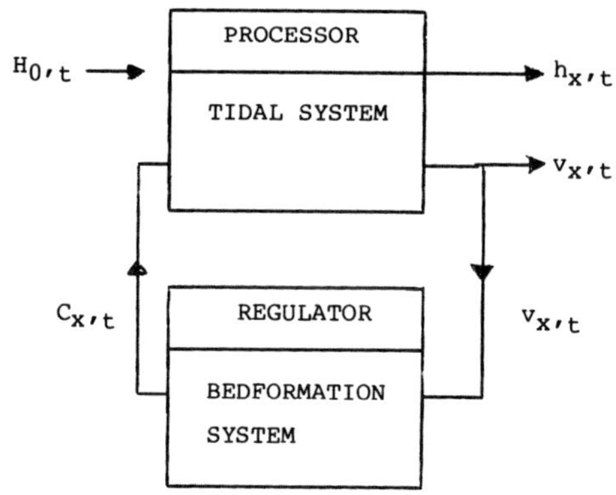

Fig. 12 The ship channel depicted as a control system

If then an increase would occur by a storm, the bottom roughness might be reduced significantly and the velocities would become very high. However, a stable equilibrium would finally be attained with maximum velocities of 2.0 m/s. From measurements in the channel just before the closure of the northern dam, it appeared that the maximum velocities were 1.60 m/s. This just below the critical transition.

Fortunately, the very high velocities of 2,0 m/s were not observed. It is not clear if the gradual increase had not reached the critical transition of about 1,65 m/s or that the relations used in the analysis where not valid. For example the analysis used a relation between velocity and bottomstress obtained from data which show considerable deviations from the mean. Also the sediment of the channel

contained some clay, thus the bottom material was not exactly the same as used in the analysis. Furthermore, the installation of revetments and bottom protection, which all have rough surfaces,might have increased the flow resistance locally.
A significant change in an estuary due to a temporarily unstable flow/bedform equilibrium can not often be observed. However, it is expected that such temporarily unstable bedform conditions are occurring quite often in estuaries. The instabilities are, however, mostly not obvious as the changes in the tidal system occur gradu ally. The "unstable" system than adapts itself by increased sedimen-tation or erosion till a new equilibrium is reached. The occurrance of such events may by the way be one reason for the deviations be tween observed and computed water levels and currents of a well adjusted model.

Conclusions

In modeling studies, the bottom stress parameter may not be taken as a constant for making predictions. This is particularly true when large changes are being made in the water bodies that are modeled.

References

Leendertse, J.J., "Verification of a model of the Eastern Scheldt"; The Randcorporation; R-3108-NETH; april 1984.

Leendertse, J.J., "A summary of experiments with a model of the Eastern Scheldt"; The Randcorporation;R-3611-NETH; juni 1988.

Van Rhijn, L.C., "The Prediction of Bedforms, Alluvial Roughness and Sediment Transport"; The Delft Hydraulics Laboratory; S 487; oktober 1982.

CHAPTER 201

NUMERICAL MODELLING OF SUSPENDED SEDIMENT TRANSPORT IN THE LOIRE ESTUARY

TEISSON Ch.*, FRITSCH D.*

ABSTRACT

In order to evaluate silt displacements and deposits in the outer part of the Loire Estuary over one year, a two-dimensional depth-integrated model of suspended sediment transport has been carried out. It takes into account transport and dispersion of suspended sediment by tidal currents, and processes of erosion, deposition and bed consolidation. Erosion laws and consolidation laws are specific of the Loire Estuary and come from laboratory experiments.

Representative tidal conditions and river Loire flowrates have been identified : results of the model for these typical cases will be tentatively extrapolated to a one year evolution.

1. INTRODUCTION

As part of environmental studies involved in a future nuclear power plant near the mouth of the Loire Estuary, the Laboratoire National d'Hydraulique has been commissionned to look into hydraulic and sediment problems and to focus on silt displacement in the outer part of this estuary, located on the west coast of France (fig. 1).

The main feature of this estuary is a turbidity maximum which lies most of the time in the inner part ; but every year, during flood discharge, about one million tons of suspended sediments are partially expulsed seawards (fig. 1).

It is therefore of great interest to know if a silt particle, coming from the inner part of the estuary and possibly polluted with radionucleide, will be either quickly ejected seawards or trapped off the mouth of the estuary in one of the sensitive bays devoted to tourism or oyster farming.

* Research Engineers, Laboratoire National d'Hydraulique. Electricité de France, 6 quai Watier - 78400 CHATOU, FRANCE

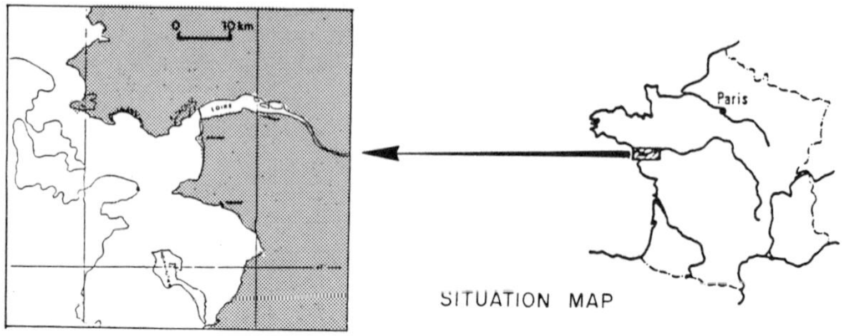

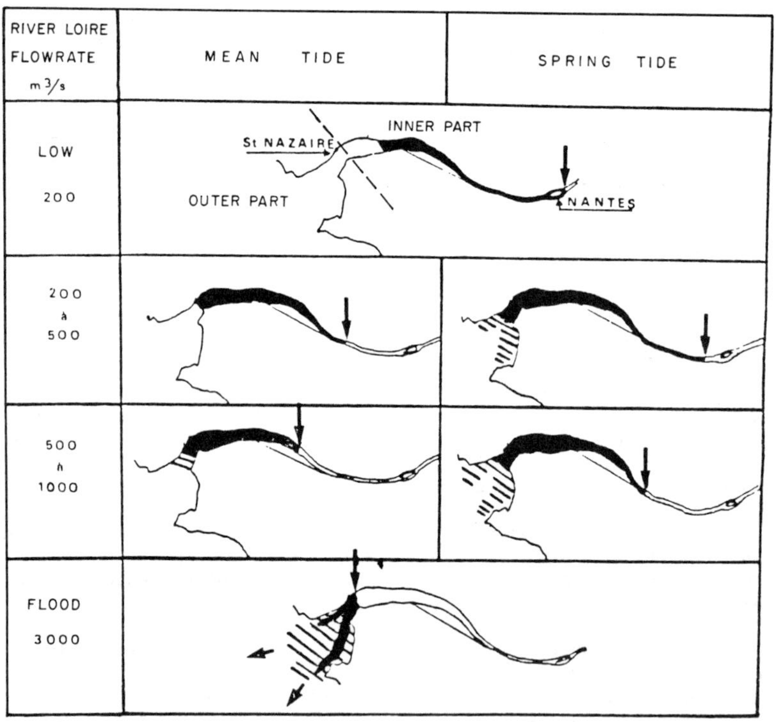

Figure 1 _ OSCILLATIONS OF THE TURBIDITY MAXIMUM IN THE LOIRE ESTUARY, ACCORDING TO RIVER FLOWRATES AND TIDAL RANGE (after GALLENNE 1974) _

After a presentation of the sediment model and processes of mud involved, hydrodynamic results, turbid plume extent, bed evolution are shown and lead to a better understanding of sediment dynamic.

2. SEDIMENT MODEL

2.1 The depth-averaged equation and its hypothesis

Flow structure and sediment circulations in coastal waters and estuaries are complex phenomena and need often a three- dimensional model in order to simulate really accurately their motions and patterns. In fact, in case of morphological computations as maintenance of navigation facilities, a prediction over a period of several months or years is asked and a three-dimensional model is not well adapted due to prohibitive computational work.

Fortunately, it is not always necessary to model the complete structure of the flow. Provided the transport quantity is well mixed throughout the water depth, it can be sufficient to consider depth-averaged equations : this approximation is particularly appropriate for a wash-load which consists of fine cohesive particles with very low settling velocities, and when sediment concentration remains moderate. Such depth-integrated models have been extensively developped in the past years (ARIATHURAI and al, 1977 ; HAYTER, 1983 ; THOMAS and MC ANALLY, 1985).

In this way the use of a monodimensional vertical model of suspended sediment transport developped at LNH has enabled to outline first criteria concerning the vertical homogeneity allowing depth integration : advection time scale (linked to the fall velocity) has to be at least twice the diffusion time scale, and furthermore must be much larger than the slack duration (TEISSON, 1984).

The bidimensional depth-averaged equation for transport of suspended sediment is then the following :

$$\frac{\partial C}{\partial t} + u \frac{\partial C}{\partial x} + v \frac{\partial C}{\partial y} = \frac{1}{h} \frac{\partial}{\partial x}\left[hK_x \frac{\partial C}{\partial x} \right] + \frac{1}{h} \frac{\partial}{\partial y}\left[hK_y \frac{\partial C}{\partial y} \right] + \frac{S}{h} \qquad (1)$$

in which :

C = depth-integrated concentration of the suspended sediment

u, v = depth-integrated components of velocity

Kx, Ky = dispersion coefficients

h = water depth

S = source - sink term accounting for erosion or deposition

The main hypothesis of the numerical modelling of suspended sediment transport is the decoupling between hydrodynamic and sedimentology. In the equation for transport of suspended sediment, hydrodynamic conditions - i.e. velocity and water depth fields (U, V, h) - are known, being previously computed by a numerical model solving the equations of tide propagation. It is assumed that sediment transport does not affect the hydrodynamic behaviour of the flow ; a new computation of velocity components occurs only if bottom evolution is relevant, due to erosion or deposition.

2.2 Mud processes

2.2.1) Settling velocity

For very low concentration, the settling velocity W_c is assumed constant. For higher concentration the settling velocity can depend on the concentration, in the range C = 1 - 10 g/l, as for example :

$$W_c = K\,C^{4/3} \tag{2}$$

where K is a constant about 10^{-5} (S.I. units), depending on sediment characteristics (from measurements in the Loire Estuary made in 1969).

2.2.2) Deposition rate

The deposition rate can be given by the KRONE's formulation (1962).

$$Q_d = P_d\,W_c\,C \tag{3}$$

in which P_d is the probability of particles sticking to the bed and not being reentrained by the flow, given by :

$$P_d = 1 - \left[\frac{u_*}{u_{*d}}\right]^2 \qquad \text{for } u_* < u_{*d} \tag{4}$$

where u_* is the shear velocity and u_{*d} the critical shear velocity under which deposition occurs.

2.2.3) Erosion rate

Erosion rate is represented by an experimental formulation obtained for the Gironde Estuary (CORMAULT, 1971) :

$$Q_e = M\left[\left(\frac{u_*}{u_{*e}}\right)^2 - 1\right] \qquad \text{for } u_* > u_{*e} \tag{5}$$

where u_{*e} is the critical shear velocity above which erosion occurs and M is determined experimentally :

$$M = 0.55\left(\frac{C_s}{1000}\right)^3 \qquad \begin{array}{l} C_s \text{ in g/l} \\ M \text{ in kg/m}^2\text{/s} \end{array} \tag{6}$$

where C_s is the concentration of the outcropping deposit.

Other formula of erosion rate depending of soil consolidation (MEHTA et al, 1982) may also be tested.

2.2.4) Bed profile

Consolidation of the bed is simulated at each node in the numerical model by a pile of ten layers of increasing concentration, each layer of various thickness being characterized by a maximum residence duration (after which mud of the layer goes into the underlocated more concentrated layer) and a critical shear stress for erosion related to the layer density.

The duration stay accounts for consolidation and is based upon experiences of MIGNIOT (1980) (fig. 2).

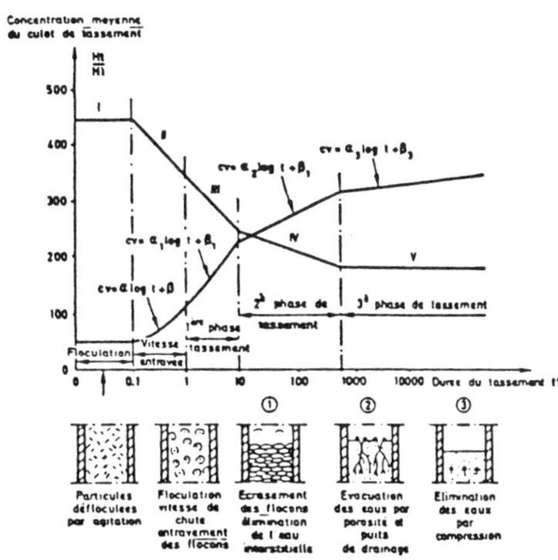

Fig. 2 - Graphical representation of bedding-down phases

For instance, for the Loire Estuary, the mean concentration of settled sediment in g/l is given by :

$$C(t) = 136.2 \log_{10}(t + 5.42) \quad 0<t<24h \quad t \text{ in hours}$$
$$C(t) = 200 + 70 \log_{10}(t) \quad t>1 \text{ day} \quad t \text{ in days} \qquad (7)$$

Furthermore MIGNIOT (1981) has related the critical shear for erosion to the concentrations of deposit. For the Loire Estuary :

for partially consolidated deposits

$$(C_s<240 g/l) : u_{*e} = 3.2\ 10^{-5}\ (C_s)^{1.175}$$

for consolidated deposits

$$(C_s>240\ g/l) : u_{*e} = 5.06\ 10^{-8}\ (C_s)^{2.35} \qquad (8)$$

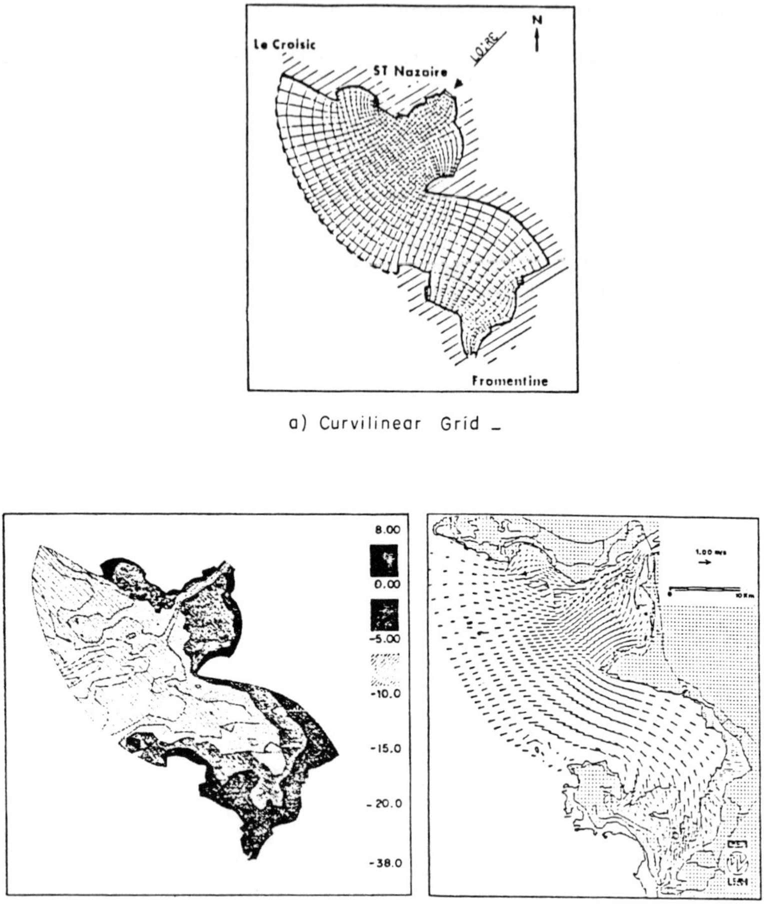

a) Curvilinear Grid _

b) Bathymetry _

c) Current Field (Ebb)

Figure 3 _ MODEL OF THE RIVER LOIRE ESTUARY. SPRING TIDE . LOIRE FLOWRATE : 1000 m^3/s .

With these data, and for a type of mud corresponding to the one of the Loire Estuary, the bed has been modelled as shown in table 1 :

n° Layers	Concentration (g/l)	Duration stay (h)	u_{*e} (cm/s)
1	100	2	0.72
2	118.6	2	0.88
3	132.7	2	1.00
4	144.1	4	1.10
5	161.8	6	1.26
6	181.3	8	1.44
7	200	16	1.62
8	215.5	32	1.77
9	233.4	54	1.94
10	250.4	infinite	2.19

Table 1 : Layer characteristics for bed modelling

When deposition occurs, the deposit always goes into the first layer, the thickness of which increases consequently. After 2 hours, if no erosion occurs, the content of the first layer goes into the underlocated more consolidated layer and is added to the existing content (the same operation is repeated for other layers).

When bed shear exceeds critical shear velocity of the top layer, then erosion occurs ; each successive layer is similarly tested for potential erosion.

Finally at each time step, the source - sink term is computed :

$$S = Q_d + Q_e$$

from the formulations (3) and (5) above.

2.3 Numerical solution

Given the hydrodynamic data, the numerical model of suspended sediment transport computes horizontal advection and dispersion by the current by means of a time splitting method.

The first intermediary step solves the advection terms on the left hand of equation (1), with a characteristics method.

The second and third intermediary steps solve dispersion in the x and y direction by means of an implicit scheme which takes into account the source term S (TEISSON, 1985).

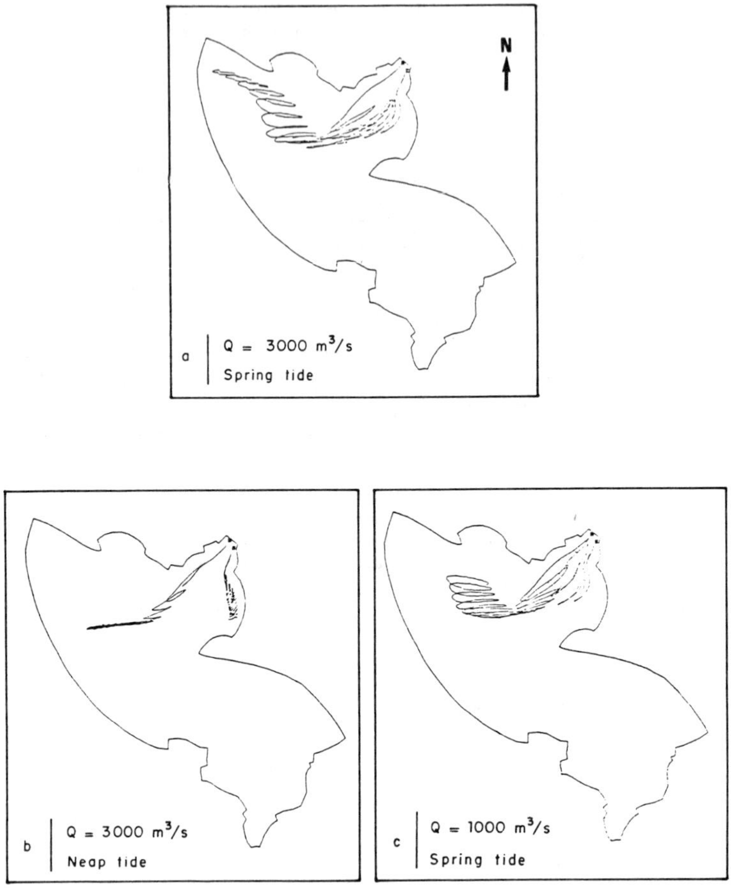

Figure 4 _ INFLUENCE OF THE TIDAL RANGE (a - b) AND OF THE LOIRE FLOWRATE (a-c) ON LAGRANGIAN DRIFTS _

Every 2 hours the residence times of deposits are compared to the duration stay of the corresponding layers : when this duration stay is reached, the contents of the layer go into the underlocated more consolidated layer and the total thickness of deposit is reduced consequently.

Both hydrodynamic and sediment depth-integrated models use rectangular or curvilinear orthogonal grid, finite difference scheme, time - splitting method.

3. APPLICATION TO THE LOIRE ESTUARY

Running of the hydrodynamic and sediment models have been performed for various tidal conditions and discharge flows of the Loire river. Assuming that the most interesting phenomena for suspended sediment occur for strong discharge flows (see fig. 1), two flow rates have been chosen - 1000 m^3/s and 3000 m^3/s - associated to three different tidal range - spring tide, neap tide and mean tide - in order to obtain six representative environmental conditions.

The curvilinear grid comprises 1146 nodes ; the size of the mesh varies from less than 300 meters in refined areas (navigation fairway) to 2500 m for the largest (fig. 3). Bathymetry has been very accurately represented, taking into account navigation fairway, intertidal flat, isolated islands. Time step is 240 s, both for hydrodynamic and sediment model.

3.1 Results of the hydrodynamic model

The hydrodynamic modelling is performed using the two-dimensional, depth-integrated flow model CYTHERE ES1 (BENQUE and al., 1982). This model solves the classical St-Venant equation for tide propagation and incorporates intertidal flats.

The corresponding fields of current and water depth (u, v, h) have been calculated for the six environmental conditions, and calibrated against natural measurements (TEISSON and al., 87). The results are good and do account with the complex problem of intertidal flats : these flats take up 10% of the area at low water.

As an example figure 3 presents current field during the ebb of a spring tide for a Loire flowrate of 1000 m^3/s.

Lagrangian calculations have also been made and keeping tracks of floats during 10 tides from hydrodynamic result displays an offshore drift, magnified in flood, which saves the bays (fig. 4).

3.2 Input data for the sediment model

In addition to velocity and water depth fields at each time step, the model requires values for sediment parameters derived from consolidation, deposition and erosion laws already presented in Section 2.2, as also boundaries and initial conditions.

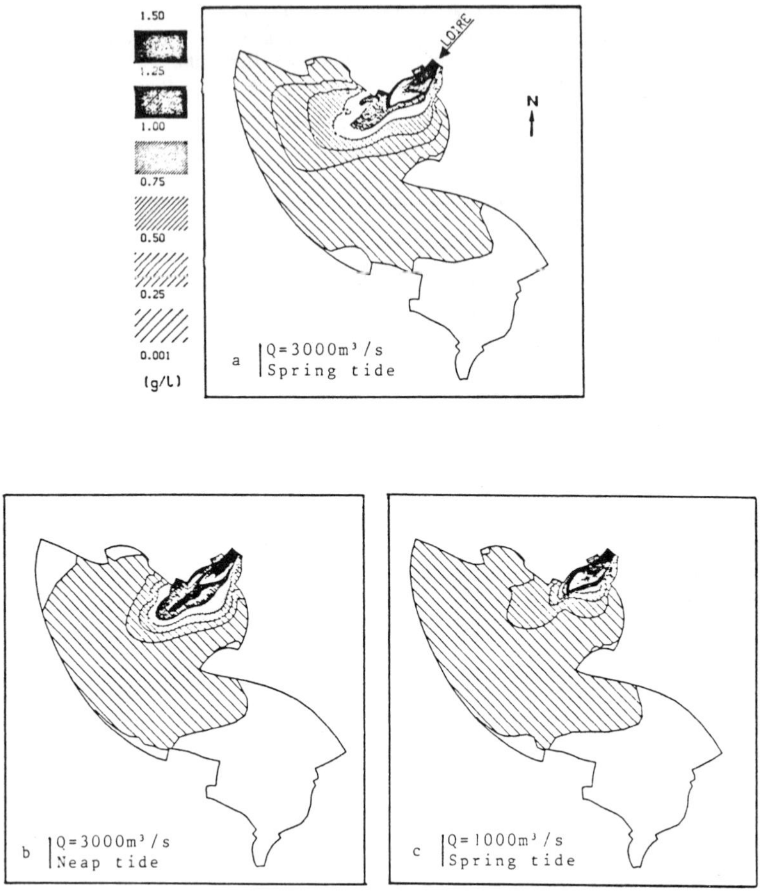

Figure 5 - INFLUENCE OF THE TIDAL RANGE (a-b) AND OF THE LOIRE FLOWRATE (a-c) ON THE PLUME EXTENT (Low Water)

At the upstream boundary, extensive fields surveys have been made in four stations in the narrowest section, and provide suspended sediment concentrations throughout the tide period for various tidal range and river flowrate : these concentrations can reach up to several grams per liter. At the seaward boundary, suspended sediment concentration varies from 0 to 5 mg/l from scarce measurements.

Initial conditions are the following : there is no sediment in suspension or at the bottom, as the aim is to investigate silt displacement coming from the inner part of the estuary.

3.3 Results of the sediment model

Numerous model runs, constituting a sensitivity analysis, have been conducted in order to evaluate the influence of various parameters as horizontal dispersion coefficient, critical shear stress for deposition, dependance of fall velocity upon suspended sediment concentration. For each typical case, representative of one tidal range and one river flowrate, computations were performed over a period of ten identical tides in order to load the model and reach an equilibrium state.

The fig. 5 presents the influence of the tidal amplitude (compare a - b) and of the Loire flowrate (a - c) on the suspended concentration : the flowrate appears to be the main parameter for plume extent. The plume extent is canalized by the navigation fairway where the highest concentrations of suspended sediment are found : this is a feature common to the prototype.

Although suspended sediment concentration is governed by the river flowrate, sediment deposition appears to be controlled by tidal range : thicknessess of deposit are 8 times greater in neap conditions than in spring conditions (compare a-c on fig. 6). Bed evolutions throughout the tide can be followed : deposit that occurs in the fairway during slack at high water is swept out by strong currents during ebb conditions (a-b, fig. 6). Beside the navigation fairway, the first results show good qualitative results ; the difficulty is the lack of reliable and evenly distributed in situ measurements to accurately assess the results for deposit.

3.4 Modelling of a typical year evolution

As it would be computer time consuming to represent the whole year (705 tides), modelling of a typical year evolution is attempted by use of the six typical environmental conditions of the hydrodynamic model. A fortnightly tidal cycle is schematized by a succession of the three basic tidal ranges, for a constant Loire flowrate (either 1000 or 3000 m^3/s). The typical year evolution is approached by the combination and extrapolation of various fortnightly tidal cycles results, with or without mud injection at the upstream boundary depending on season (winter-summer).

This stage of the experimentation is going on at the present time : results after fortnightly tidal cycles are available but it remains to combine and extrapolate these results to one year.

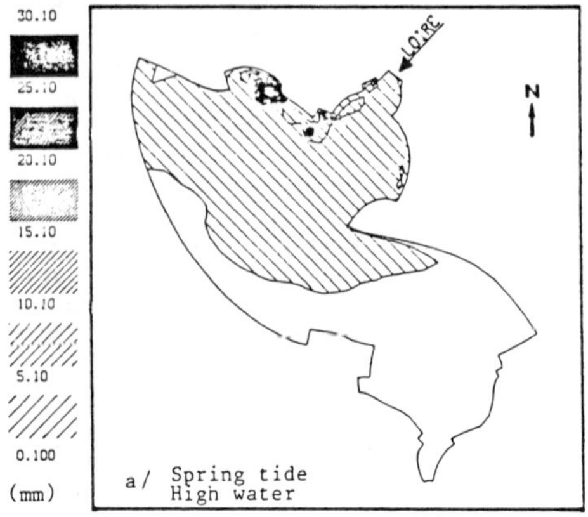

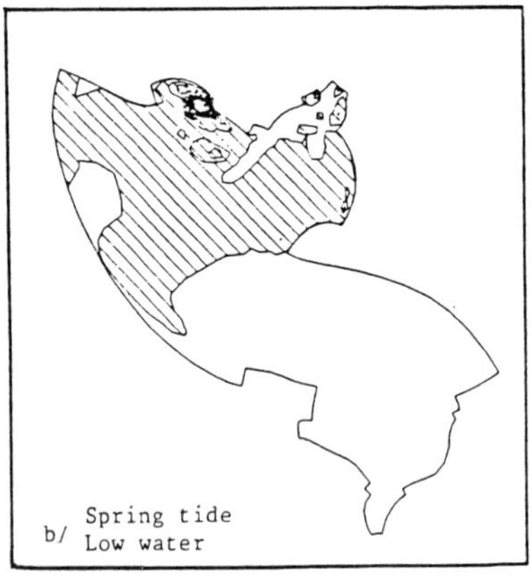

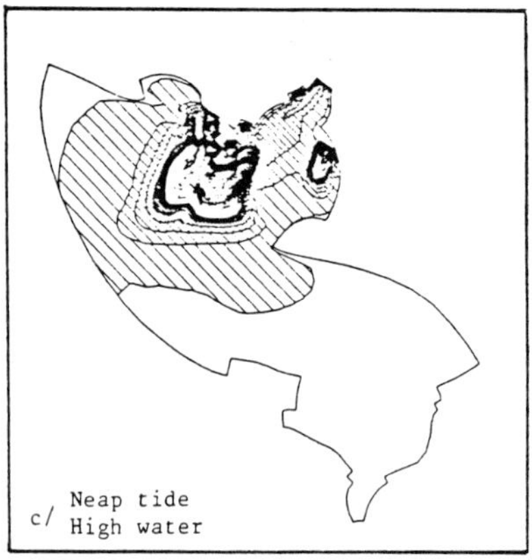

Figure 6 - TOTAL THICKNESS OF DEPOSIT
$Q = 3\ 000\ m^3/s$

CONCLUSION

The objective of the study was to evaluate silt displacement and deposit in the outer part of Loire estuary in an attempt to represent typical year's evolution.

The hydrodynamic model CYTHERE ES1 has enabled to compute very satisfactorily six typical current and water depth fields related to different tides and flowrates of the Loire River.

For each of these six typical cases, the 2D numerical model of suspended sediment transport has led to a better understanding of the processes which govern sediment dynamic. Suspended sediment concentration is under the influence of river Loire flowrate, while deposit is controlled by tidal range. Importance of spring-neap tidal cycle has been pointed out, and this fortnightly cycle has been chosen as an intermediary step before the extrapolation to one year's evolution which is currently under progress.

REFERENCES

ARIATHURAI R., MAC ARTHUR R.C, KRONE R.B. (1977 : Mathematical model of estuarial sediment transport. T.R. D77.12 - U.S. Army Engineer - Waterways Exp. Station P.O. Box 63 Vicksburg, Miss 39180.

BENQUE J.P., CUNGE J.A., FEUILLET J., HAUGUEL A., HOLLY F.M. Jr. (1982) : "New Method for Tidal Current Computation", Journal of the Waterway, Port, Coastal and Ocean Div., ASCE, vol 108, No. WW3, Aug. 1982.

CORMAULT P. (1971) : "Détermination expérimentale du débit solide d'érosion des sédiments fins cohésifs", 14e Congrès de l'A.I.R.H., Paris.

GALLENNE B. (1974) : "Les accumulations turbides de l'estuaire de la Loire. Etude de la crème de vase. Thèse de 3ème cycle. Université de Nantes.

HAYTER E.J. (1983) : Prediction of cohesive sediment movement in esturial waters PHD Thesis UFL/COEL-83/010 Coastal and Oceanographic Engineering Department. University of Florida, Gainesville, Florida 32611.

KRONE R.B. (1962) : "Flume studies of the Transport Sediment in Estuarial Shoaling Process", Tech. Resp., Hydr. Eng. Lab., University of California at Berkeley.

MEHTA A.J., PARCHURE T.M., DIXIT J.G., ARIATHURAI R. (1982) : Resuspension potential of deposited cohesive sediment beds Estuarine comparison, U.S. Kennedy, Ed, Acad. Press. New York pp. 591-609.

MIGNIOT Cl. (1980) : "Dynamique sédimentaire estuairienne - Matériaux cohésifs et non cohésifs", Oceanis, vol. 6, Fasc. 4, pp. 359-432.

MIGNIOT Cl. (1981) : "Erosion et sédimentation en mer et en rivière" in "La Pratique des Sols et des Fondations", Edition Moniteur 1981, 103 p.

TEISSON Ch. (1984) : "Transport en suspension en mer ouverte. Modélisation bidimensionnelle verticale", LNH HE/42/84.30.

TEISSON Ch. (1985) : "Transport en suspension en mer ouverte - Modélisation bidimensionnelle horizontale", LNH HE/42/85.57.

TEISSON Ch., GRAFF M., FRITSCH D. (1987) : "Etudes hydrosédimentaires de l'estuaire externe de la Loire liées au projet de Centrale du Carnet - 1ère partie : Modèle hydrodynamique", LNH HE/42/87.15.

THOMAS, W.A., and W.H. Mc ANALLY (1985) : A generalized numerical modeling system for open channel flow and sedimentation, TABS-2, "IR HL-85-1, U.S.A. WES, Vicksburg, Miss.

Puerto de Barcelona

PART V

CASE STUDIES

San Ciprian—Lugo

CHAPTER 202

ENGINEERING STUDY FOR A NEW SEAWATER INTAKE SYSTEM

Billy L. Edge* and Roger S. Schlueter**

ABSTRACT

Environmental assessment, engineering studies and designs were completed for a new 26.5 m^3/s seawater intake system in the Persian Gulf. The original intake facility consisted of a curved, 60m breakwater with one end attached to the shoreline, a settling basin immediately adjacent to the shoreline and dredged to a maximum depth of approximately 5m, and a pumphouse structure located on shore such that the seaward wall formed one side of the settling basin. The facility located on an island in the Gulf, which served multiple seawater uses, had experienced both structural and operational problems, the latter consisting principally of excessive ingestion of sediment and seaweed. These factors plus the requirement for additional demands for seawater beyond plant capacity caused the owner to initiate a study of alternative intake systems, produce a design for the most effective solution and construct the new intake system.

INTRODUCTION

The intake system supplies cooling water primarily to a large LNG processing facility. Cooling water enters from the Gulf on the north side of the island through a small 5m deep sedimentation basin into the front bay of a large pump station with seven pumps. The cooling water circulates through the condensers and then back into the Gulf on the east side of the island. The site location is shown on Figure 1. Through field studies and numerical analysis it was shown that the heated discharge did reach the intake at some stages of the tide causing poor cooling performance.

A second problem has occurred seasonally for several years. In particular, at certain times of the year Colpomenia cf. sinuosa (a macro algae) appear in great quantity and clog the cooling water intake. In front of the pump bays, a rotating screen operates behind a bar screen. On several occasions the quantities of macro algae were so great as to totally clog the system and to cause the plant to shut down for several days. The economic consequences of this are obvious.

The third problem which required solution was the identification of the source and elimination of the sediments being ingested into the plant. These sediments were not only filling in the deposition basin but were also going through the pumps causing excessive wear on the cooling system.

Lastly, there was a desire to increase the capacity of the intake plant by 50% to include increased cooling requirements as well as fire water and desalinization needs.

---

*Cubit Engineering, 4431 Embarcadero Drive, West Palm Beach, FL

**AMI Systems Inc., Santa Barbara, CA

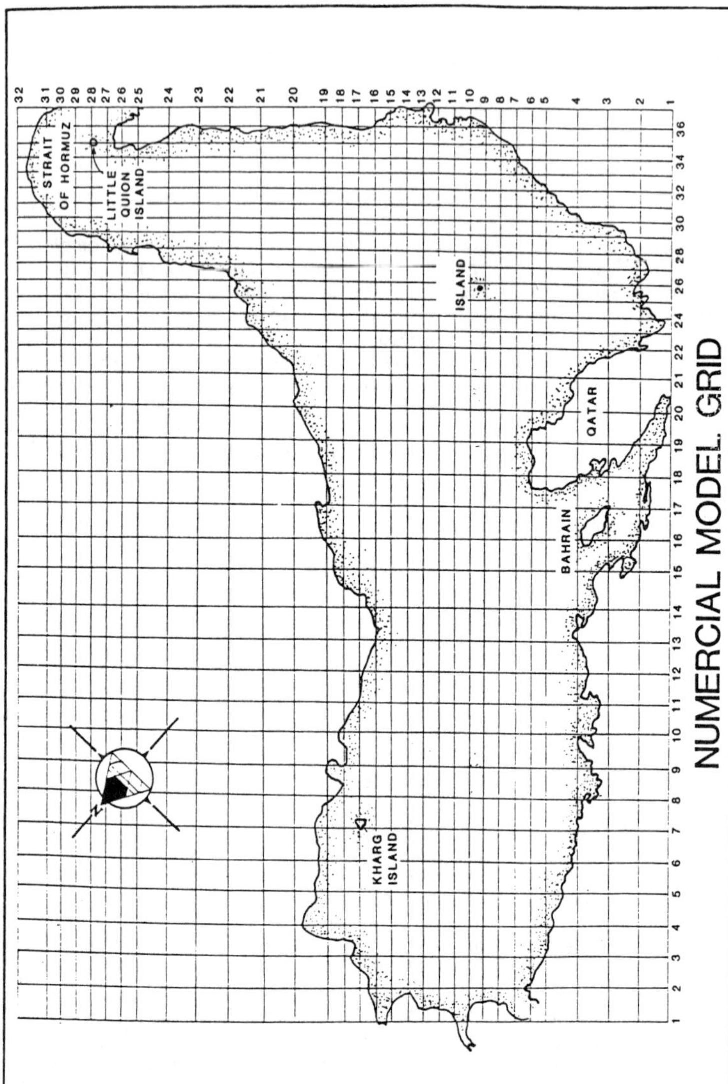

Figure 1. Location of project site in Persian Gulf.

A program was developed to analyze the problem, to identify appropriate solutions and to carry those solutions through design to construction. The study and analysis included:

- o Field data collection and analysis covering site physical oceanography, meteorology, coastal geomorphology, biofouling, and thermal dispersion;
- o Engineering studies including thermal recirculation, bio-fouling control, and development of design criteria;
- o Formulation and evaluation of alternative conceptual layouts for a new intake facility with particular attention directed to construction and operational integration with the existing plant;
- o Recommendations for any additional field studies and hydraulic modeling prior to initiation of the final design process;
- o Preparation of final design and tender documents and completion of construction.

FIELD AND ENGINEERING STUDIES

Winds

Winds are of primary concern in the design of virtually any coastal or offshore structure because of the direct influence they have upon the wave climate, current circulation, and storm surge phenomena. The dominant extreme wind regime in the Persian Gulf is the winter "shamal", characterized by strong northwesterly winds which generally persist for a period of several days. A statistical analysis of wind velocity maxima was made on the basis of continuous speed and direction measurements during the period 1958 to 1967 (IMCOS Marine, 1969). The analysis indicates that about 50 percent of the time winds are from the northwest. Return probabilities of storm wind occurrence at the site were estimated from this data and are shown below:

Table 1. Extreme wind statistical analysis

Return Period (years)	6-hr Mean Velocity (m/sec)
100	23
25	21
10	19

In support of these figures it should be noted that ship observations in the southern Persian Gulf tabulated by the U.S. Naval Weather Service Command (USNWS, 1971) indicate that during the period 1963 to 1971 no winds occurred that were in excess of 24 m/sec from any direction.

A recent study performed for ARAMCO (Ocean Weather, Inc., 1983) reviewed the storm meteorology of the Persian Gulf since 1963. Historical wind data were compiled, annotated, and reduced to 15 extreme events over the study period. This data base was used to develop design wave and water level criteria.

Waves

A statistical definition of storm-generated wave heights and periods was required to determine (1) the degree of attenuation necessary to protect shoreline pump bays and (2) the structural requirements necessary to ensure the survival of intake structures and breakwaters during adverse weather conditions. Design wave parameters have been previously prepared by A.H. Glenn Associates (1982).

Long-term records of wave data directly relevant to the site are virtually non-existent. Consequently, a description of the design wave condition were made through hindcast techniques. In this analytical technique, the more extensive wind data over the Gulf region is used to reconstruct historical storms of record. Numerical and empirical models are then employed to calculate or "hindcast" the corresponding wave characteristics at the point of interest.

Table 2. Predicted wave heights

Return Period (years)	Water Depth (m ISLW)	Sig. Wave Ht. (m)	Sig. Wave Per. (sec)	Max. Wave Ht. (m)	Max. Wave Per. (sec)
10	4.7	3.4	7.3	---	---
	13.5	4.2	7.4	7.9	8.2
	25.0	4.4	7.4	8.2	8.2
100	6.0	4.4	8.6	4.6	9.6
	13.5	5.4	8.7	10.0	9.7
	25.0	5.7	8.7	10.7	9.7

Based upon 15 extreme wind events compiled in a recent study for ARAMCO (Oceanweather, Inc., 1983), a storm considered to be the largest event over the last 25 years was analyzed for wave incidence at the intake site. This shamal occurred over the period January 19 to 21, 1964 and is estimated to represent a 25- yr event. Hindcast analysis of this storm using the discrete spectral model of Resio (Resio et al., 1981), which has been calibrated for this region of the Persian Gulf, gave a peak significant wave height of 4.5 m and a spectral peak period of 10 seconds.

Wave refraction analysis was performed to evaluate the transformation of deeper water waves to the near shore region on the northern side of the island. Wave behavior for waves with periods of 6, 8, 10 and 12 seconds approaching the site from the northwest, north and northeast sectors was analyzed. These conditions were chosen as representative of those generated by the shamals.

Water Levels

Anticipated extreme water levels over the design life of the intake system determine requirements for pump machinery protection. Furthermore, they govern the maximum wave height which can propagate to the intake. Conversely, the minimum water level to be expected at the site is an important consideration in terms of insuring enough water depth to maintain the design cooling water flow rate. Astronomical tidal information relative to the site is summarized below:

Table 3. Astronomical tide summary
(elevations in meters relative to Chart Datum)

Level	Elevation (m)	Reference
Highest astronomical tide	1.7	A.H. Glenn, 1982
Mean Higher High Water	1.2	U.S. NOS, 1980
Mean Tide Level	.8	A.H. Glenn, 1982
Mean Lower Low Water	.5	U.S. NOS, 1980
Lowest Astronomical tide	-.1	A.H. Glenn, 1982

Storm surge and seiche resulting from sustained winds blowing over the Gulf can create an additional rise or fall of the water level. The northwesterly shamal is capable of generating storm surges via wind stress effects over the axis of the Gulf. IMCOS Marine (1969) predicted a still water level of +3.05 m and it is assumed that this value represents a 100-yr event.

To confirm this value, one and two-dimensional numerical storm surge models were run over the entire Persian Gulf using hindcast wind fields obtained from Oceanweather, Inc. Surge was assumed to be coincident with a Mean Higher High Water tide and results indicated that a maximum still water level on the order of +2.76m and possibly lower may be a more likely expectation over the project design life.

Southerly winds which persist for some duration are capable of lowering the still water level at the southern end of the Persian Gulf. A simplified one-dimensional drawdown calculation was made using bathystrophic storm tide theory. Preliminary and uncalibrated results indicate that drawdown on the order of 0.75 m may be expected over the project life of the intake structures.

Currents

Determination of predominant flow characteristics provides important operational design information related to net flow rates of fouling organisms or other material toward the intake as well as thermal plume recirculation. No historical, site specific data were available on currents in the vicinity of the site. Consequently, a brief field measurement study to record in situ current speed and direction was undertaken.

Thermal Recirculation

The hydrodynamic and thermal characteristics of the LNG intake and discharge structures are reasonably complex due to the poorly defined advection and mixing conditions and complex bathymetry. The discharge volumes and velocities of the proposed new discharge channel are large as is the temperature difference between the discharge and ambient waters. The discharge is into shallow waters which are subject to complete current reversal as the tide sweeps around the island.

The field study results of the vertical profiles of temperature and the synoptic temperature plots indicate that the discharge plume remains attached to the surface and shows relatively little vertical dispersion. Field studies also indicate that during conditions of tidal cycle which carries the discharge plume towards the intake, the observed temperature rise was primarily confined to the upper water layer.

Because of the complexities stated above it was decided to divide the analyses into two phases, the first analyzing the near-field plume and the second the far-field effects. The near-field is defined as that region in which the discharge plume characteristics are dominant. The far-field is defined as the region in which the larger scale oceanic forces dominate.

Modeling of the far-field, where tidal and wind forces are dominant, poses different problems from the near-field. Due to the location of the site, a conventional hydrodynamic modeling effort is complicated by the difficulties to determine appropriate boundary conditions. With the surface plume stratification, use of a three-dimensional or stratified model would be desirable. However, given the lack of information regarding vertical mixing and the far greater amount of boundary information required by a three-dimensional model, such an approach is not justifiable. Therefore, it was decided to use a depth averaged model and apply various sets of assumptions which would bound the extreme case conditions. Boundary conditions were determined by running a two-dimensional model for the entire Gulf as schematized in Figure 1.

Sediments

Required seawater quality in terms of undissolved solids for the new intake system were defined as 1.5 ppm for calm conditions and 5 ppm for storm conditions. In addition, it was stated that water for the new firewater system should not contain solids which will block a 3.5-mm diameter spray nozzle. These requirements essentially define the limitations on sediment ingestion by the new seawater intake system.

Significant sedimentation within the existing pump forebay area was reported but no quantitative information adequately delineated the problem. Therefore, it was necessary to conduct an investigation to determine the origin and magnitude of this sedimentation problem.

While the rate of intake of sediments rises during storms, it is reported to be a minor problem. The sediments appearing just on the outside of the breakwater are not in sufficient quantity to create an ongoing problem. In addition, because of their coarseness, they are not readily mobilized and do not appear to be moving toward the mouth of the intake at a rate to create a problem. Moreover, these sediments do not appear to have been transported into this area from a remote distance.

Field studies determined that the breakwater was responsible for the majority of sediment build-up. Improperly graded filter stone was placed beneath the 15-ton tetrapod armor units and significant leaching of the core material occurred. Furthermore, poor stone materials, including breccia from the island, were used and a breakdown toward fine material began via water induced deterioration and wave generated abrasive stresses.

Geotechnical Factors

The site is one of a number of diapiric structures which occur in the Persian Gulf and which are part of a major province of salt diapirism centered in southern Iran. During the early Pleistocene epoch, tectonic pressures within the crustal rocks at depth beneath the Gulf initiated a redistribution of a major evaporite horizon within the Cambrian Hormuz series. The low specific gravity and viscosity of this evaporite as compared with the surrounding rock types resulted in the upward movement of the evaporite mass through the overlying strata forming a salt plug.

The northern part of the island is the gypsum cap of the salt plug. Severe disturbance of surrounding rocks occurred during the upward movement of the evaporites. Large masses of rock have been detached and reoriented and a wide variety of rock fragments were mixed into a gypsum matrix. The subsequent dissolution of the salt plug by seawater has resulted in a chaotic aggregation of evaporite, sedimentary and igneous rocks around numerous domes of gypsum and gypsum breccia, and today, no orderly geological sequence can be established within the rocks. However, it is estimated at this time that the near shore region sub-bottom most likely consists of a layer of coral rock underlying breccia or sandstone.

From an engineering standpoint, offshore excavation will be difficult and require powerful dredging equipment as some of the coral and breccia can be moderately strong (unconfined compressive strengths for coral and breccia reaching values of up to 5 MN/m^2 and 17 MN/m^2, respectively). Breakwater protection can most likely be placed directly on the sea bottom without any risks for a base type failure. The rock will need to be imported to the island.

EVALUATION OF ALTERNATIVE SYSTEMS

Several onshore and offshore alternative cooling water intake system concepts were evaluated during the course of this study. Each concept was formulated to satisfy the following basic design requirements:

1. Provide the necessary cooling water volume
2. Eliminate sediment ingestion
3. Eliminate entrainment of biological foulants
4. Not interfere with the existing intake during construction
5. Afford practical construction methodology
6. Minimize thermal recirculation of discharge water.

In developing alternative system configurations, the primary concern was system reliability since the present facilities provide unacceptable levels of downtime associated with fouling and unacceptable rates of wear and high maintenance due to sediment ingestion.

Because of the large expense associated with shutdowns of the seawater system, a more reliable system having a higher initial construction cost may provide a more economical long-term solution when total costs are considered. Consequently, hydraulic performance and antifouling characteristics were heavily weighed evaluating the alternatives. Secondarily, construction logistics were considered in terms of the ease of building the various schemes at this site in the Persian Gulf.

Table 4. Cooling water intake alternatives

Onshore Concepts

1. Extension of West Breakwater to create a calming zone and construction of new intake pump chamber near existing structure.
2. Porous dike.

Offshore Concepts

3. Offshore pipeline and intake; pipeline laid on bottom surface.
4. Offshore pipeline and intake; pipeline buried below bottom surface.
5. Offshore tunnel and intake.
6. Offshore pipeline trestle and suction riser.

Onshore Concepts

An onshore intake withdraws seawater from the entire nearshore water column because of the relatively shallow entry depth compared to the large pumping demand. Any onshore system will be exposed to the same loading stresses which the existing pump station has experienced. Therefore, facility improvements are required so that sand entrainment, biological fouling, and pollution contamination can be handled with no loss of plant efficiency.

Two onshore concepts were reviewed. The first concept is shown in Figure 2 and is essentially an extended modification of the existing intake layout with the breakwater running parallel to the shoreline. The stilling basin would be designed to provide sufficient wave protection and a sufficiently low current that sediments would be induced to settle out in the deepened basin. This would allow a longer accumulation time between maintenance dredging.

Entrained Colpomenia and other biota would still require mechanical screen removal. However, it is possible that some improvement over the present design could be obtained through careful specification of filter systems and screen wash management.

The second onshore alternative incorporates the use of a porous dike--a rubblemound structure constructed of highly permeable material--to completely enclose the onshore pump station. In this scheme, seawater filters through the structure's voids en route to the pump station and undesired foulants and sediments are virtually precluded from entry.

The principal drawback to porous dikes are their uncertain history regarding long-term clogging. The structure must be designed to anticipate a degradation of porosity. Calculations were made to determine the length of

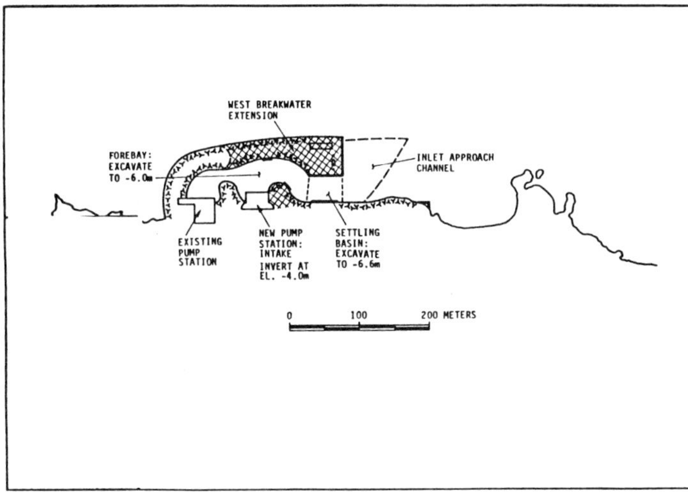

Figure 2. Onshore intake with parallel breakwater.

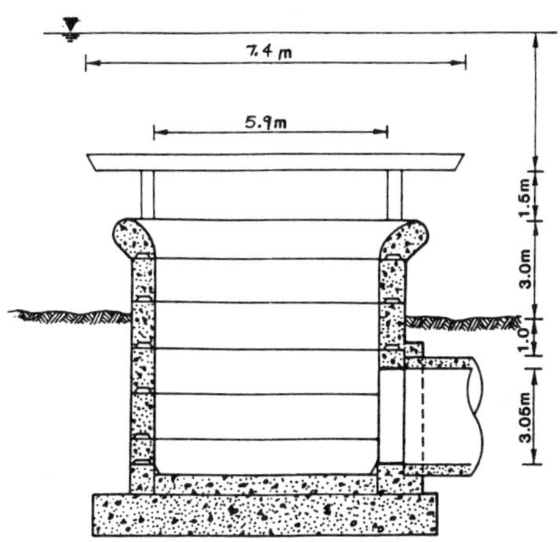

Figure 3. Offshore submerged intake riser.

dike needed to provide the design flow rate. Assuming a cross section constructed with fine gravel, a dike length of at least 1,200 m would be required. Furthermore, the ability of the system to prevent infiltration of floating pollutants is unknown. For these reasons, the concept was rejected.

Offshore Concepts

An offshore intake is often employed to withdraw large quantities of water when onshore intakes are not acceptable. Furthermore, an offshore intake can be used to access ambient water with cooler temperature characteristics. Consequently, this concept has important, attractive features that allow it to meet the design specification for the cooling water system.

Offshore intakes have been successfully used in a variety of marine cooling water applications. By submerging the point of water withdrawal sufficiently below still water level, operation can continue uninterrupted even during severe storm wave activity.

Based on available information, the offshore intake concept was recommended for design. Because of its proven operating capabilities, it offers the highest potential for alleviating the problems which are presently being experienced at the existing onshore intake.

The cooling water intake for an offshore system consists of three components:

1. The pump station and forebay;
2. The seawater intake; and
3. The connecting corridor between the pump forebay and seawater intake.

The following sections discuss desirable design features of each component in formulating a complete offshore system.

Pump Station and Forebay

The pump station should be designed so that the area is sufficiently protected from destructive waves, affords the opportunity to clarify the intake water prior to pump suction, and prevents fouling from nearshore waters. All three objectives can be met by completely enclosing the pump station within a small stilling basin. Little difference is expected between the design for the pump station with an offshore intake and that for an onshore intake.

Of primary importance, incorporation of a stilling basin provides a quiescent area for settling of any sand ingested at the offshore intake. This settling basin limits the entrainment of particles to sizes and concentrations required by design requirements. Construction of the settling basin can be achieved by extending the west breakwater to the east or building an entirely new breakwater for the new pump station. This will also preclude oil pollution, wave action, sediments, and seaweed from directly entering the basin. In short, the enclosed forebay greatly enhances pump station efficiency and reduces maintenance activity.

Seawater Intake

The primary function of the intake is to eliminate ingestion of large quantities of foulant which would overload pump station screens or harm internal cooling water components. The intake should be located in areas which exhibit reduced concentrations of Colpomenia populations and minimal sediments.

Offshore seawater intakes are simple structures which rest on the sea bottom and are elevated sufficiently above the seabed to minimize entrainment of sediment or bottom dwelling organisms. Figure 3. illustrates the chosen concept. The intake consists of prefabricated annular sections, normally of reinforced concrete, which are pieced together onsite. The assembled structure is then floated to the site and sunk into position. The horizontal velocity cap atop the structure induces lateral inflow of seawater thereby minimizing vortex generation and entrainment of bottom or surface floating material. The entrance slot is sized so that the velocity through the intake is no greater than .3 m/sec to minimize fish entrainment.

Intake Corridor

The most standard scheme is an offshore buried pipe. In this design, shown schematically on Figure 5, a trench is excavated to accommodate a large diameter conduit which is then covered for wave protection purposes. The diameter of the pipe must be sufficient to maintain velocities and hydraulic head losses within reasonable limits. A single conduit with a 4.27 m internal diameter (ID) would be required to satisfactorily convey the required flow from the offshore intake structure to the nearshore forebay.

Burial of a large diameter pipe involves extensive excavation of the marine bottom. The trench must then be lined with a gravel bedding to provide uniform foundation support for the heavy pipe segments. Select backfill is then placed to the crown elevation of the pipe and the trench is capped with suitable armor material.

An unburied pipe section would be fully exposed to the hydrodynamic loadings of storm generated waves and currents. Estimated wave forces would be of such magnitude that they could not be resisted by conventional pipeline anchors. An exposed conduit is therefore not recommended for further consideration.

Because of the hard bottom material found at the site and a sensitivity to the use of explosives to aid in excavating such strata, a tunneled conduit was considered. A tunnel can be considerably less disruptive to the marine environment and operation of the present intake. The total length requirement of 200m for the tunnel proved too costly for implementation of this alternative.

The third alternative, an over-water pipe system, is one in which water is drawn from an offshore location and pumped/siphoned via several smaller conduits to the depressed forebay. The siphon would be elevated over water by a conventional trestle-type structure as depicted on Figure 4. This concept can make use of smaller, lightweight, and hydraulically smooth Fiberglas reinforced pipe. Based on the flow requirements of the site, three 2.44-m ID pipes would satisfactorily deliver th^ cooling water from the intake to the

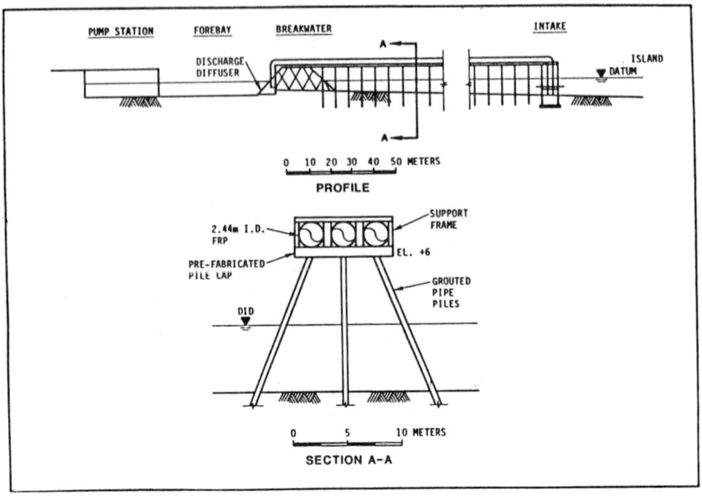

Figure 4. Pump/siphon water intake on trestle.

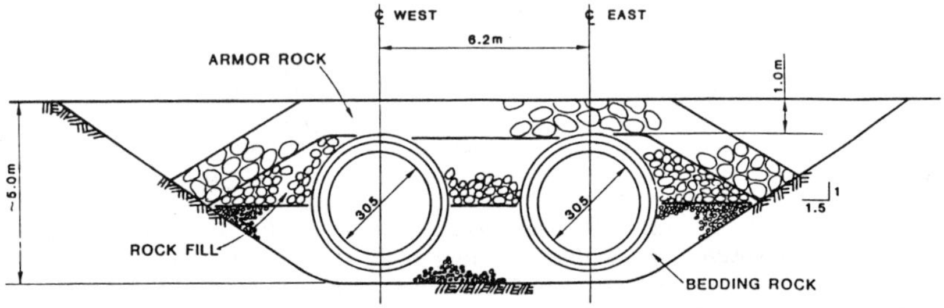

Figure 5. Buried concrete pipe design with backfill.

forebay. Additional advantages include ease of construction and an exposed pipe which facilitates operational maintenance. The elevated trestle allows waves to pass beneath it such that only the piling requires wave force design.

ENGINEERING DESIGN

Based on the above criteria, an intake system comprised of an onshore pumping facility and an offshore intake structure best met all relevant criteria and provided a proven, workable system. Key characteristics of the recommended system are given below.

Intake Design

1) Two partially buried, cylindrical offshore intake structures 5.88m in diameter are located approximately 175m offshore at the 8m contour. Each riser structure including velocity cap is approximately 9m high with the intake gap protruding 3.5m above the seabed.

2) Each riser is connected to a shoreline stilling basin via a buried concrete pipe 3.05m in diameter as shown in Figure 5.

3) The stilling basin is formed by a horseshoe shaped breakwater designed around the new pump station. The entrance to the 30m long stilling basin is a 20m long concrete diffusor structure which transitions the flow from the concrete pipe(s) into quiescent flow in the stilling basin. The enclosed stilling basin provides the desired water quality and protects all shoreline structures from surge and waves.

4) The intake stilling basin is dredged to -5.0m and will require only infrequent dredging for maintenance purposes.

5) Construction of the system can be achieved through positive isolation of nearly all of the new facilities from the present operating facility. Changeover from the old facility to the new is required to be achieved in a smooth transition period without loss of cooling water service to the island.

Hydraulic Model and Design Criteria

As part of the preliminary design, an extensive series of two and three dimensional model tests were conducted at the Delft Hydraulics Laboratory. Horizontal and vertical loadings on the intake risers were measured and the required pipeline armor stone size was determined. The wave tests were also used to confirm the design horseshoe breakwater. Separate model tests were conducted for the stilling basin diffusor and pump bay.

The allowable damage for the breakwater was 3% for Tr = 50 years. The allowable damage for the armor cover and the pipeline was 1% for Tr = 100 years. Damage was defined as movement of an armor unit more than 1.5 units away. One of the interesting results of the model tests was that the breakwater responded best to the placement of a berm at the toe as shown in Figure 6. This allowed the tetrapod layer to be reduced from 15 ton to 10.6 ton units and eliminate the observed Mach stem effect at wave breaking.

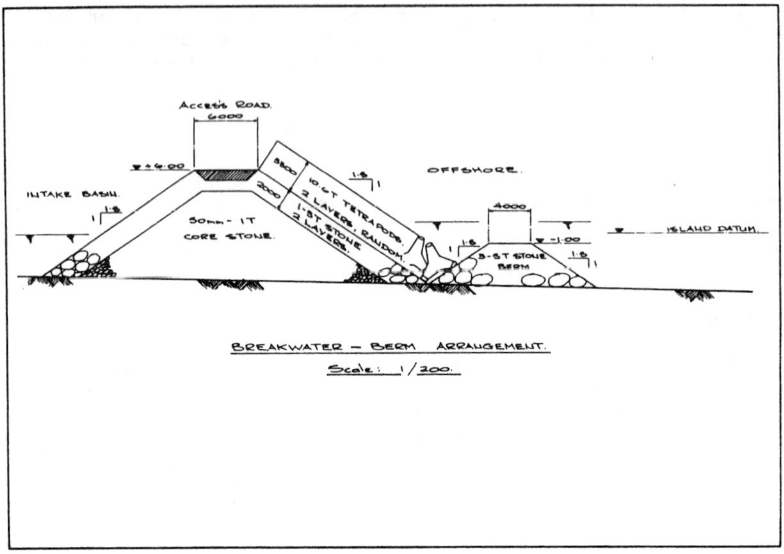

Figure 6. Breakwater cross-section showing berm and tetrapods.

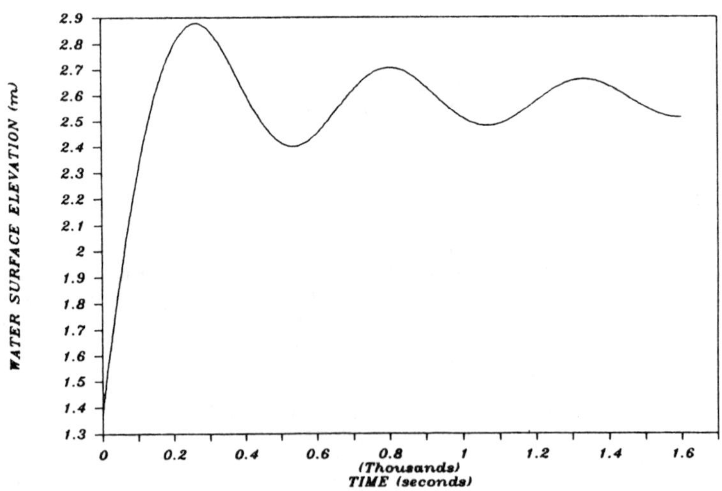

Figure 7. Surge analysis for pump shut-down conditions.

In addition to the above there were several additional specific criteria which were used in the design and the model studies. In the two and three dimensional wave tank tests a Jonswap spectrum was used having the following characteristics:

γ = 3.3
λ_a = 0.07
λ_b = 0.09

Hydraulic performance of different components of the system were tested at different design water levels. For example, the breakwater was tested at a HWL of +2.67 representing a 50 year event whereas the pipeline armor was tested at a HWL of +2.76 representing a 100 year event. The significant wave height for the breakwater is Hsd = 5.23m for a return period of Tr = 50 years.

During the hydraulic performance tests three discharges were applied:

- half normal flow: 13.25 m^3/s through 1 pipe
- normal flow: 26.5 m^3/s through both pipes
- maintenance flow: 25.4 m^3/s through 1 pipe

Both numerical and physical tests were performed to ensure that the settling basin would not be subject to surging problems due to external wave conditions.

SUMMARY

No difficult or insurmountable final design, construction or operation issues were revealed in the selected alternate for this replacement of the cooling water intake. Nevertheless, there were a series of critical design issues which required satisfaction in the final design.

Construction of this facility is currently proceeding to completion. The offshore parts of the new intake system is expected to be completed in 1988.

CHAPTER 203

ACCRETION - EROSION IN THE BEACHES OF THE CANARY ISLANDS (SPAIN).

JESUS MARTINEZ MARTINEZ

Member of the Spanish Geological Society

Facultad de Ciencias del Mar. Apartado 550. Las Palmas. España.

ABSTRACT

In the beaches under study on the island of Grand Canary, we are identifying and measuring the intertidal annual cyclic processes of accretion and erosion, starting from the topographical movements of the strand.

These processes:

a) We are analyzing and interpreting these in the framework of the physical and geographical interdependences, and as responses to the more internal evolutions of the morphodynamics, in the sedimentary environments in question,

b) and permitting the definition and calculation of some indices of sedimentary gains and losses, of interest to coastal planning and management.

INTRODUCTION AND OBJECTIVES

In the opinion of Charlier (1987), the mathematical calculation and managment of the sedimentary volumes, through observation of the topographical movements is, actually, the optimum method for the identification, measurement, and interpretation of the accretion and erosion processes (gains and losses) in sandy beaches.

If the topographical movements refer to the intertidal zones, as it happens in this work, we are obtaining answers to the more internal evolutions of the beach environment morphodynamics, in a greement with the outline of Wright and Short (1983).

The sedimentary dynamics of the beaches develop, to a very important proportion, between the shoreline and the breaker line. With a range of the astronomical tide of up to 2.82 m. (as is the case in the Canaries), the intertidal zones have sufficient amplitude to support truly valid investigations, referring to the accretion - erosion processes, in the beaches under study.

The accretion - erosion processes represent effects resulting from some determined causes. Knowing and understanding the "effect - cause" duality, we can predict "causes - effects" through the numerical model.

The modelling of these processes develops various conceptual stages. We will order these from botton to top, depending on the best level of abstraction and generalization:

a) In the first phase, the works in the field aimed at identifying a physical model, in which predominates the selection of basic qualitative characteristics.

b) The translation of qualitative data to quantitative data constitutes a complex process, which passes through intermediate partial modelling steps. The representation of the selected descriptive parameters, of adequate scale, culminates in the establishment of a quantitative stage.

c) And in the quantitative phase, we apply, essentially, the mathematical method. This establishes an abstract problem which serves, preferentially, for the analysis of general questions.

The actual work, in the framework described, is situated between the qualitative phase and an intermediate semi-quantitative phase.

On the other hand, the estimations and indices of the sedimentary gains and losses in the beaches are useful, in the present form, in coastal planning and management. Nevertheless, the making of decisions needs to be based on an analysis of the whole of the physical processes and handicaps which appear in these situations.

GEOGRAPHICAL SETTING

The sandy beaches under study (figure 1) are located on the island of Grand Canary (Spain). The descriptions of the same are presented in condensed form in table 1.

ACTUAL TECHNIQUES

The study is based on the following techniques:

1. Elevation of the topographical profiles of the monthly studies, mostly in 1986. We are using the method of geometric levelling (by altitudes), as developed and described by Martínes et al (1987 a, 1987 b, and 1987 c).

2. Calculation of the relative volume of the sands, by the trapezoidal method, in agreement with Puig Adam (1979). The volumes correspond to an intertidal strip, previously determined, and from a convenient base level. The contrast of the series of relative volumes, in relation to the lowest value, measures the sand gains or losses.

3. Oceanographical analysis of the annual gains and losses of the beach sand. For this we have recourse to:

- the surface meteorological predictions
- and visual observations of the surf from ships "en route", in the Canary environment. The data is taken from the "Oceanographic Atlas of the North Atlantic Ocean", in its sucessive revisions.

With observations of the surf we analyze, for the greatest part of the Grandcanary coast:

- the maximum possibility of presentation/altitude, indefinite depth, of the significant "swell" and "sea" waves in a certain direction (pattern of the directional surf),

Beach	Location	Morphology	Approximate dimensions		Genetic classification after Suárez Bores (1980)
			Length in m.	Maximun width in m.	
Sardina del N.	Costa N. (Gáldar)	in cove	85	45	$GG^{o}_{o,o}$
Las Cante_ ras	Costa N. (Las Pal_ mas)	subcoves in a large cove	3100	90	$GDdGGDdG^{1}_{o,o}$
El Hombre	Costa Oriental. (Telde)	in cove	224	91	$GG^{o}_{o,o}$
El Burrero	Costa Oriental. (Agüimes)	in cove	205	30	$GG^{o}_{o,o}$
El Inglés	Costa Meridional (S.Bartolomé)	rectilinear	2700	100	$Gg^{o}_{o,o}$ (2)
La Bajeta	Costa Meridional (S.Bartolomé)	convex towards the ocean	470	100	$g^{1}_{o,2}$
Maspalomas	Costa Meridional (S.Bartolomé)	rectilinear	2103	50	$gD^{2}_{o,2}$

Table 1

Descriptions of the beaches of Grand Canary (Spain) under study.

- as well as the annual percentages.

Once the annual processes of accretion and erosion have been measured and interpreted, we can define two operative indices:

- estimation of the sedimentary support capacity,
- and the sedimentary variability.

INDEX OF SEDIMENTARY SUPPORT CAPACITY (S.S.C.I.)

This index defines the difference between the actual and potencial volumes, derived from topographical profile levelling in situ and theory, respectively, for unity of the observed intertidal surface, and in a certain period of the annual cycle. This is expressed with the formula:

$$S.S.C.I. = \frac{Cr - Cp}{Su}$$

in which:

Cr = actual volume
Cp = potencial volume
Su = observed surface

The theoretical topography is calculated based upon the slope, obtaining:

a) In agreement with:

- the median diameter and densities of the beach sand,
- and the surf characteristics.

Formulas of the type of Sunamura (1984) are considered.

b) Or by employing empirical curves of the equilibrium between beach-face slope and the median diameter of sand.

Among others, Bascom (1959), Komar (1976), and Martínez (1986) describe curves of this type.

Using this method the estimation error is larger.

The potential volume for the operative total of a sandy beach, are deduced, in immediate form, with the equation:

$$Cp = L\,A\,(H - \frac{A}{2}\,\text{tag}\ a)$$

in which:

Cp = potential volume
L = operative length of the beach
A = amplitude of the observed strand
H = average vertical length from the head of the profiles till reference basis
a = beach - face slope.

The index in question qualifies the beaches according to:

- equilibrium,
- deficiencies
- or in surplus,

as is shown in table 2.

INDEX OF SEDIMENTARY VARIABILITY (S.V.I.)

This index is defined as the estimate of the average altitude of erosion to accretion, in an annual cycle, that overtakes the surface of an intertidal delimeted strip, in a sector of or in all of the beach. To sum up, it evaluates the maximum "effective" sedimentary deposition during the annual cycle under study.

In the calculation of this index:

- for a determined transverse profile, with a convenient length and width,
- or for the whole of the intertidal delimited strip of the beach;

we divide the volume of the sand into increments, during the period of significant accretion, for the basic profile section area, or for the operative strip.

The index of sedimentary variability qualifies the beaches as:

- changing attenuatedly (with low indices),
- changing moderately (with intermediate indexes)
- or changing strongly (with high indices),

as shown in table 3.

This index allows us to represent the longitudinal topographic movements of the beaches quantitatively, in scale block diagrams.

A third index that may be fundamental in temporal series analysis of the strand volumes, with data from successive campaigns during a significant number of years under study. The beaches are classified as stable, hyper-stable, and unstable, after the work of Suárez Bores (1980).

All of these indices need to be considered:

- in the design of projects, for the optimization and recuperation of sandy beaches,with maritime works and/or artificial sustenance,
- and in the continueing analysis of these beaches, when we have already had experience with the same, for verifying the conduct of the sedimentary process, and to adapt, in each case, the necessary corrective measures.

RESULTS

Table 4 shows the sand gains, in cubic meters, of the beaches under stu

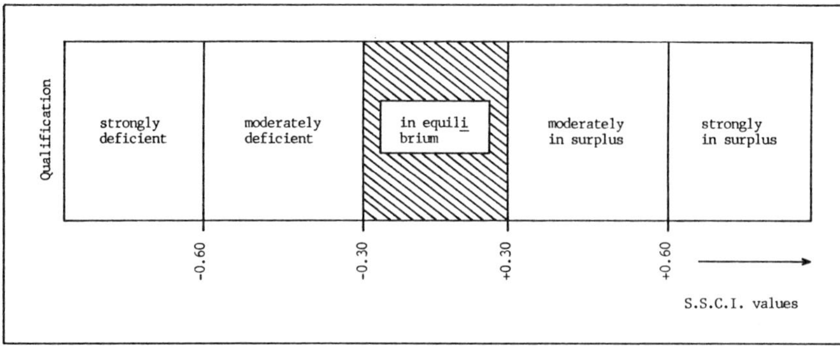

Table 2

Tentative scale for qualifying the sandy beaches according to the index of sedimentary support capacity.

S.V.I. values	Qualification
of 0.00 to 0.75	changing attenuatedly
of 0.75 to 1.50	changing moderately
> 1.50	changing strongly

Table 3

Tentative scale for qualifying the sandy beaches, according to the index of sedimentary variability.

Date \ beach	Sardina del N	Las Canteras	El Hombre	El Burrero	El Inglés	La Bajeta	Maspalomas
	Sand gains, in m^3, in relation to the significant erosion of the annual cycle under study						
	D:66.8x9.0	D:2169.0x10.0	D:200.6x10.0	D:97.1x12.0	D:2111.0x15.0	D:471.5x15.0	D:2102.5x15.0
1985/12	303.14	05745.23	3285.30	689.49	32642.96	10680.09	04192.74
1986/1	272.50	03816.09	3585.86	334.70	35704.79	12365.50	06562.86
1986/2	038.38	04095.64	2893.17	313.86	40266.28	12024.72	00000.00
1986/3	100.58	00000.00	0271.30	300.93	38369.33	13671.17	01729.70
1986/4	000.00	03511.69	0000.00	166.31	35401.32	12476.15	01829.60
1986/5	070.63	05861.03	2779.44	292.80	29423.33	11574.44	04696.84
1986/6	266.69	09953.13	3434.31	251.82	17331.43	10148.07	14806.72
1986/7	494.01	09075.08	2686.86	226.63	16042.21	06671.15	16841.51
1986/8	489.18	10687.70	2959.49	132.51	14070.81	04225.79	28591.80
1986/9	327.72	09794.63	3192.79	091.71	12653.42	05070.47	30345.58
1986/10	444.91	04845.50	3096.40	000.00	08796.16	04499.33	37778.00
1986/11	343.37	---	3347.16	023.23	01338.24	02208.58	28671.13
1986/12	445.95	---	3051.68	---	00000.00	00000.00	10215.95

D = dimensions, in meters, of the intertidal delimited strip

Table 4

Sedimentary balances, during an annual cycle, in various sandy beaches on the island of Grand Canary (Spain).

dy, in relation to the most extreme erosion of the same. This corresponds to an annual cycle (1986).

In the Canaries are presented, habitually, certain surface meteorological situations. These form three groups, depending on the prevailing wind direction:

1. Atlantic anticyclones (trade winds), in combination with the occasional influence of Saharic thermal depressions, during the spring and summer. These involve winds from the N-NE.
2. Deep Atlantic storms with a nucleus over the Gulf of Vizcaya, and at times Atlantic anticyclones greatly displaced to the south, during the autumn and the end of winter-beginning of spring. These involve winds from the W-NW.
3. North-african and South-european anticyclones, Saharic thermal depressions, and Atlantic storms displaced to the south, from autumn to the beginning of spring. These involve winds from the E-S.

In figure 2 are recompiled the directional components and the frequency of these winds, depending on the aforementioned situations, that provoke the dominant and prevailing surfs on the Canary Island coasts.

In the selected beaches. The values of the sedimentary indices are recopied in table 5.

DISCUSSIONS

1. The beaches of Grand Canary are classified based on:

- their geographical location
- and the seasonal incidence of the surf.

We propose the fallowing groups:

Group I:

- beaches in the north cornice, under the influence of diffracted trade wind surf (from the N-NE), and exposed to storms from the W-NW,
- beaches of the south, protected from the trade wind surf.

Group II:

- beaches in the north cornice, directly exposed to the trade wind surf,
- beaches on the east coast, exposed to the trade winds.

Group III:

- beaches on the east coast, protected from the trade winds and open to the SE.

Group IV:

- beaches on the west coast.

We would like to present these transitional cases.

Beach	Period under study	S.S.C.I. (accretion)	Qualification	S.V.I. (annual)	Qualification
Sardina del N. (Galdar)	1985-86	-0.13	in equilibrium	0.82	changing moderately
Las Canteras S. (las Palmas)	1985-86	+0.15	in equilibrium	0.84	changing moderately
Playa Chica (Las Canteras)	1985-86	+0.24	in equilibrium	0.62	changing attenuatedly
Las Canteras N. (Las Palmas)	1985-86	+0.46	moderately in surplus	0.58	changing attenuatedly
El Hombre (Telde)	1985-86	+0.01	in equilibrium	1.79	changing strongly
El Burrero (Ingenio)	1985-86	-0.01	in equilibrium	0.59	changing attenuatedly
El Inglés (S. Bartolomé)	1985-86	+0.89	strongly in surplus	1.27	changing moderately
Punta de La Bajeta (S.Bartolomé)	1985-86	+0.74	strongly in surplus	1.93	changing strongly
Maspalomas (S.Bartolomé)	1985-86	+0.20	in equilibrium	1.20	changing moderately

Table 5

Estimations of the sedimentary support capacity and of the annual inter tidal sedimentary variability, in various beaches under study in the Grand canary environment.

2. The significant accretions and erosion (table 4) of the beaches under study, inside of a yearly cycle, have a strong dependence on the proposed beach groups.

In the Group I sandy beaches the erosion takes place preferably during the winter and spring months, and the accretion during the summer and the first part of autumn.

In the Group II beaches the erosion culminates in summer, and the maximum accumulations occur at the end of autumn-beginning of winter.

In the Group III beaches the erosion develops, above all, during autumn, while the accretion takes place from the end of autumn until the begin ning of spring.

The Group IV beaches are under actual study at this time.

Figures 3 and 4 show the performance described.

3. At this time we establish a correlation between:

- The most frequent of the different surface winds (figure 2), co rresponding to the period of taking the beach data.
- The greatest probability of presentation and the highest maxi mims, to indefinite deph, of the surf in relation with these winds (with data from visual observations of ships en route).
- And the significant erosion processes in the beaches open to the surf in question (table 4).

In the correlations:

a) During the dominant time of the trade winds, in the summer, we see the greatest probability of the presentation of the "swell" surf from the N-NE.

In this period, and in the north, east and suthern coasts, predominate,- with various exceptions, wave heights between 1.83 and 3.66 m. Also are presented, although in low percentages, waves with heights greater than 3.66 m.

b) On the north coast, the maximum probability of the presentation of the "sea" surf, from the N-NE, coincides with the dominance of the tra de winds. In this case the dominant waves are of little height, less than 0.91 m. They do not surpass a height of 2.44 m.

c) On the east and southern coasts, the maximum probabilities of the presentation of the "swell" and "sea" surfs from the E, SE and S take pla ce during the autumn and winter and coincide with the most frequent winds determined by:

- South-european or North-african anticyclones,
- Saharic storms,
- or Atlantic storms displaced to the south.

The "sea" surf from the NE does not depend on the trade winds, but on the first two meteorological conditions mentioned above.

In this period waves of low height are predominate; smaller than or equal to 1.83 m. With the exception of "sea" surfs from the NE, they do not ap proach 3.66 m.

d) When the winds come from the W-NW, from autumn to spring, we observe the maximum probabilities of the presentation of the "swell" and "sea" surf from the NW.

During this period prevailing waves of heights between 1.83 and 3.66 m., and a few others that are higher, come to the nothern coast, which is directly affected.

4. In agreement with the described previous aspects, the physical model of the accretion - erosion processes, designed for the sandy beaches of Grand Canary, is formulated as fallows:

a) When the trade wind surf predominates and takes its main energy (in the summer season).

- This produces the most extreme erosion in Group II beaches.
- And develops the most important accretions in Group I beaches.

b) During the decline of the trade winds:

- The Group II beaches attain the process of accretion (autumn-winter).
- Surfs from the E,SE, and S make their appearance. The dominant and most energetic of these, in autumn, are connected with Group III beach erosion.
- And the Atlantic storms come from the W - NW (from autumn to spring). These same do not impede the most important accretions in the Group III beaches (winter) and determine, in those situations of high surf energy, the significant erosion in the Group I beaches (winter-spring).

5. The index of sedimentary support capacity qualifies for all the beaches under study between in equilibrium and strongly in surplus (table 5).

Forthwith, aside from anomalous very energetic oceanographic situations, and if we do not wish to enlarge the surface in use, the beaches do not have the capacity to accept artificial sustenance for the optimization of these sedimentary environments.

6. The annual sedimentary variations of the beaches are adjusted by a continuity function:

- the "net" processes of erosion are determined by the seasonal impact energies of the surf,
- and the "effective" sedimentary contributions (predominately gains over losses) occur when these impacts subside.

With respect to the dominant trade wind surf, the exposed beaches under study (El Hombre and Punta de La Bajeta) have higher indices of sedimentary variability than the protected ones (Sardina del Norte, Las Canteras, El Burrero, El Inglés and Maspalomas), as we see in table 5. The first group takes the title of "attenuating" and "moderately changing" while the second group takes the title "strongly changing".

From this we deduce that, on the island of Grand Canary, the surf from the N-NE is the most erosive in relation to the others.

On a single beach (Las Canteras for example), the sector most exposed to the indicated surf shows the largest sedimentary variation (Las Canteras South).

In a single beach or a sector of the same, with an important annual sedimentary variation, the construction of an adequate breakwater, nevertheless determines a reduction in the contributions, would produce a considerable reduction in the erosion processes. The result of the above conveys a net sand gain, with consequent growth of the dry-intertidal area.

CONCLUSIONS

The Grand Canary beaches are classified into four groups. For this we take into account:

- their geographic location,
- and the seasonal incidence of the surf.

The annual accretion and erosion processes are different for each beach group.

In the analysis of the sedimentary balances, we satisfactorily establish correlations between the selected significant variables.

The beaches under study are situated, for the one part, between in equilibrium and strongly in surplus, and for the other part, between attenuating and strongly changing, depending on the scales proposed for the sedimentary indices here defined.

ACKNOWLEDGMENT

The author wishes to express his appreciation to Dr. José M. Pacheco Castelao, the Dean of the Facultad de Ciencias del Mar (Universidad Politécnica de Canarias), for his critical reviewing and contribution of suggestions in the development of this work.

REFERENCES

Bascom,W.N 1959. The relationship between sand size and beach-face slope. Am. geophys. union trans., 32 (6), 866-874.

Charlier,R.H. 1987. General discussion and round table. European Intensive Course on Land-use Problems, Planing and Mamagement in the Coastal zone. Bilbao (Spain), october, 8-17.

Enriquez,F.y Berenguer,J.M. 1986. Evaluación metodológica del impacto ambiental en las obras de defensa de costas. Publicación del CEDEX.Madrid.

Komar,P.D. 1976. Beach processes and sedimentation. Ed. Prentice-Hall.Englewood Cliffs, New Jersey.

Martínez,J. 1986. Estabilidad-inestabilidad en los depósitos de arenas de las playas canarias: Relaciones entre pendientes topográficas y granulométricas. Boletín del Instituto Español de Oceanografía, 3(2), 87-96.

Martínez, J.; Sastre, J.; Alemán, G.; Castro, J.; Martín, A. y Robayna, D. 1987 a. Los movimientos de las superficies topográficas en las playas de arena: método de investigación e interpretación. Revista de Obras Públicas, julio-agosto, 469-483.

Martínez, J. y Cárdenes, M. 1987 b. Los cambios topográficos y sedimentológicos en las playas arenosas de El Inglés y Maspalomas (Gran Canaria-España). Actas de la VII Reunión sobre el Cuaternario, AEQUA, 223-226.

Martínez, J.; Navarro, T.; Roldán, A. y Rosario, M. Del. 1987 c. Cuantificación e interpretación de los procesos de acreción-erosión, en la playa arenosa de El Hombre (Gran Canaria-España). Actas de la VII Reunión sobre el Cuaternario, AEQUA, 227-230.

Puig Adam, P. 1979. Cálculo Integral. Editorial Gómez Puig. Madrid.

Suárez Bores, P. 1980. Formas Costeras. Publicación de la Escuela Técnica Superior de Ingenieros de Caminos, Canales y Puertos de Madrid.

Sunamura, T. 1984. Quantitative predictions of beach-faces slopes. Geological Society of America Bulletin, 95, 242-245.

US Naval. 1970. Oceanographic Atlas of the North Atlantic Ocean. S. Naval Oceanographic Office. Washington.

Wright, L. y Short, A. 1983. Morphodynamics of beaches and surf zones in Australia. In: P.D. Komar (Editor), C.R.C. Handbook of Coastal Processes and Erosion, C.R.C. Press, Boca Raton, Fla., pags 35-64.

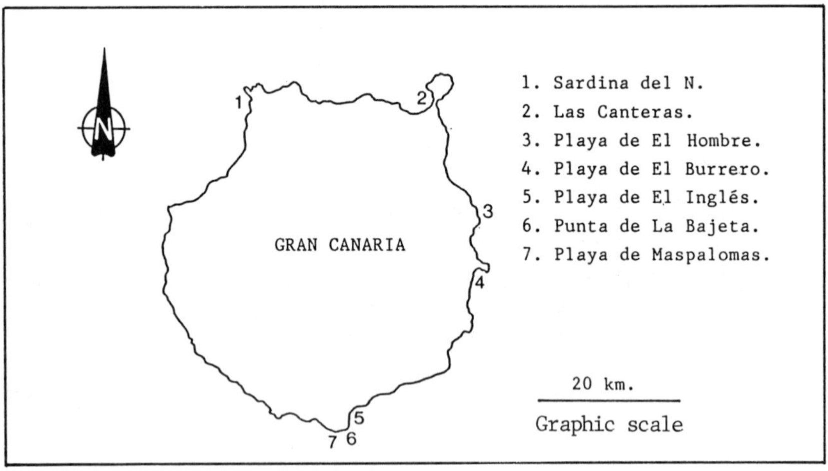

Figure 1

Geographical location of the Grandcanary beaches under study.

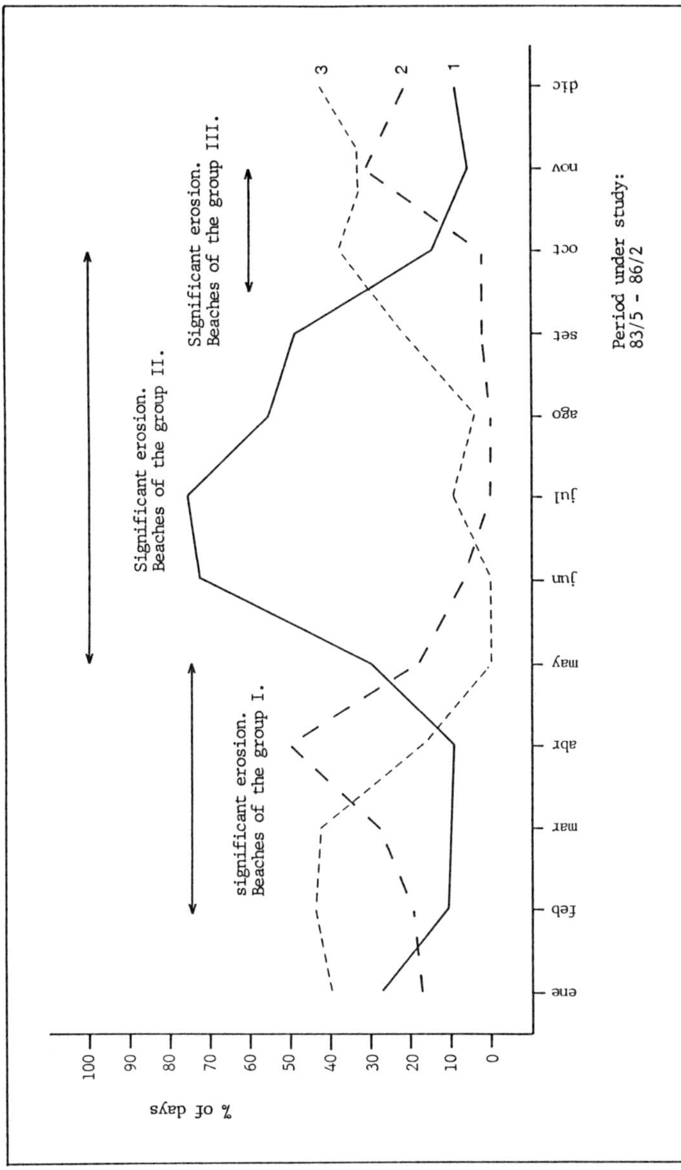

Figure 2

Frequency of the winds in the environment of the island of Grand Canary (Spain). 1: winds from the N - NE,, 2: winds from the W - NW,, 3: winds from the E - S.

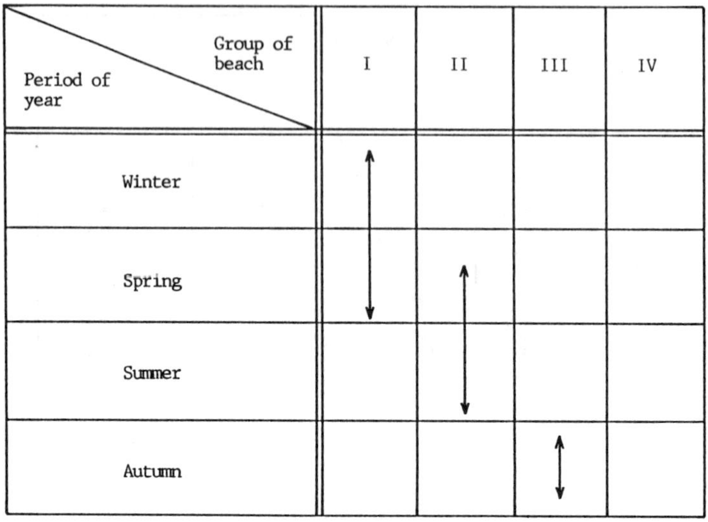

Figure 3

Localization of the extreme processes of annnual erosion in the sandy beaches of Grand Canary.

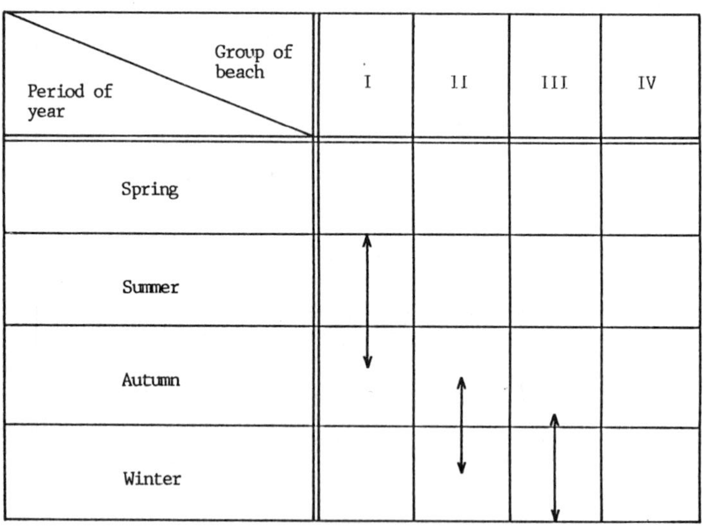

Figure 4

Localization of the most important annual accretion processes in the sandy beaches of Grand Canary.

CHAPTER 204

SUBMARINE SIPHONS FOR ATHENS SEWERAGE SYSTEM

Y. Eisenberg, M.ASCE
Th. C. Gofas, M.ASCE
R. A. Fasano, M.ASCE
F. S. Hindes

ABSTRACT

The submarine siphons with an overall length of almost 1300 meters and an ultimate capacity of 27 cubic meters (m^3) per second (about 600 million gallons per day) will be a major element of the new wastewater conveyance and treatment system presently under construction in Athen, Greece. This will help alleviate the present condition where an average of more than 6 m^3 per second (130 million gallons per day) of untreated domestic and industrial wastewater are discharged into the sea near Athens. Construction of the submarine siphon pipes started in late 1984 and was completed in early 1987.

Description of the data collection for and the design, manufacture and construction of these submarine siphons are presented in this paper.

INTRODUCTION

In 1983, the Greek Government undertook a $100 million program dealing with the collection, transportation, treatment, and disposal of sewage of the greater Athens area serving a population of over three million people. The program involves the collection of sewage from the mainland and its transportation to the small island of Psyttalia through two 1300 meter long submarine siphon pipelines with an inside diameter of 2.8 meters. A treatment plant on Psyttalia will process the sewage prior to its ultimate disposal through a 1600 meter long diffuser outfall into the 60 meter deep water of the Gulf of Saronikos.

The submarine siphon pipelines, whose design and construction are described in this paper, have a hydraulic design capacity of 27 cubic meters per second (about 600 million gallons per day). They traverse a shipping channel carrying heavy traffic to Elefsis Bay near Piraeus Harbor (see figure 1 - Vicinity Map and figure 2 - Project Site) through a troublesome layer of sludge created by previous sewage dumping and are placed in a protective trench at depths reaching almost 50 meters below sea level (see figure 3).

Y.Eisenberg, Parsons Brinckerhoff, San Francisco, California, U.S.A.
Th.C.Gofas, Th. Gofas and Partners Ltd., Athens, Greece
R.A.Fasano, AWD Technlogies, Wayne, New Jersey, U.S.A.
(formerly with Woodward-Clyde Consultants, Wayne, NJ, U.S.A.)
F.S.Hindes, Lafayette, California U.S.A.

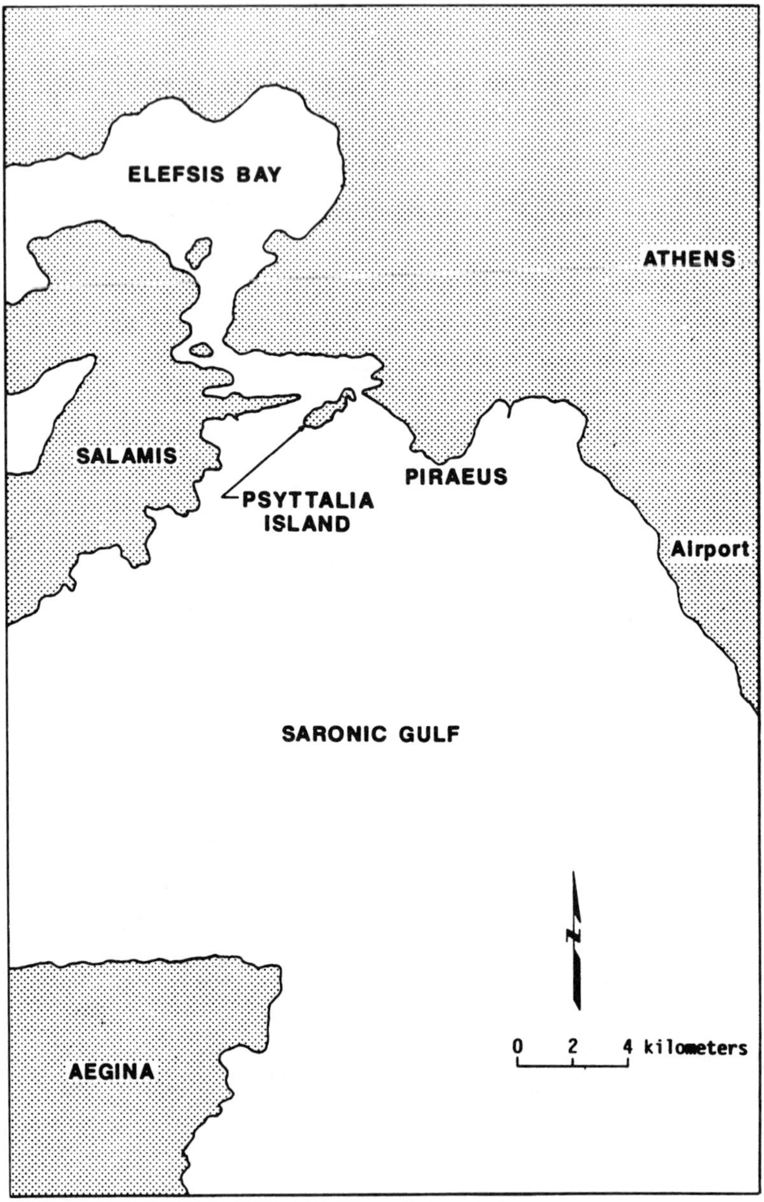

VICINITY MAP Figure 1

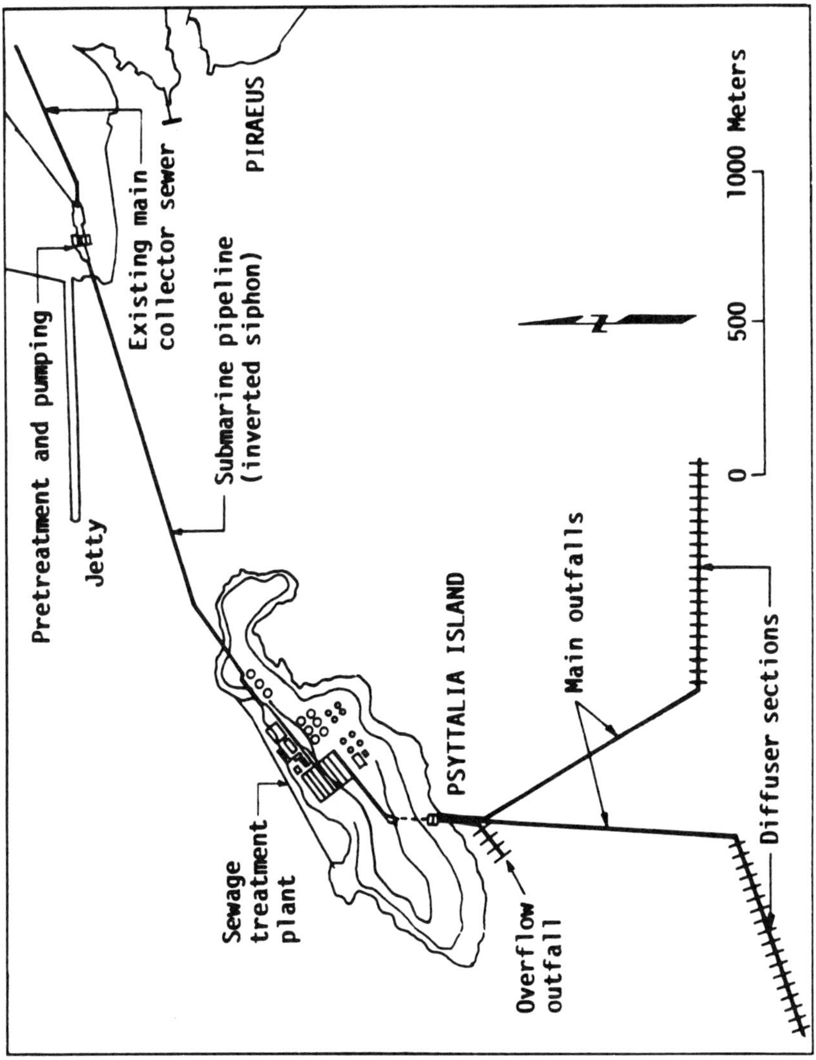

SITE PLAN

Figure 2

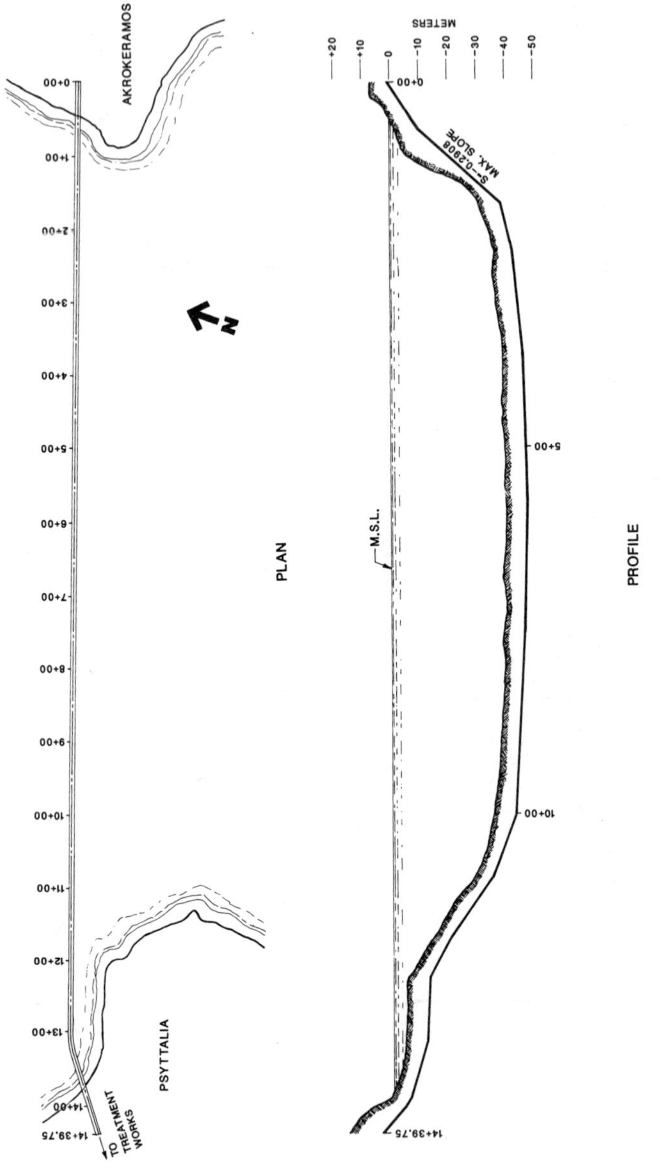
AKROKERAMOS
PSYTTALIA
TO TREATMENT WORKS
PLAN
PROFILE
M.S.L.
MAX. SLOPE
S=0.2908
METERS
+20
+10
0
-10
-20
-30
-40
-50
0+00
1+00
2+00
3+00
4+00
5+00
6+00
7+00
8+00
9+00
10+00
11+00
12+00
13+00
14+00
14+39.75
N

Figure 3

DATA COLLECTION

The geotechnical conditions along the route of the pipeline were investigated by a combination of borings, probes, and marine geophysics.

A total of 14 borings were made at 100 meter intervals to depths of approximately 15 meters below the sea bottom. The borings were supplemented by a total of 30 probes driven to "firm" conditions below the sea bottom. The purpose of the probes was to provide additional information on the depth, extent, and consistency of the sewage sludge discharged into the sea from the main interceptor sewer, and the thickness of loose natural sediments lying on the sea bottom.

Marine geophysical methods included bathymetric surveys and an attempt at sub-bottom profiling. The bathymetric surveys, combined with depth measurements from the borings and probes, were sufficient to map the surface of the sea bottom in an approximately 500 meter wide corridor along the route of the pipeline. Sub-bottom profiling was not successful over most of the alignment due to the presence of methane gas bubbles in the sewage sludge. The gas bubbles tended to cause absorption of the acoustic energy of the signal, rather than having it reflected back to the sensor by successive harder layers below. The results of the geotechnical investigation indicated that the sewage sludge covered the sea bottom over approximately two-thirds of the pipeline route from the discharge of the interceptor sewer on the east side of the channel to near the end of a jetty on the north side of the channel. The sludge varied in consistency from very loose or "liquid" near the surface to medium stiff at its deepest point. The gain in consistency with depth is attributed to consolidation of the lower, older portions of the sludge under the weight of more recently deposited sludge near the surface. The thickness of sludge varied from 0 near the edges of the deposit to approximately 6 meters near the center of the deposit.

The natural sea bottom consists of loose to medium dense marine sand with a high silt and clay content. The sand exhibited some natural cementation which varied in location and depth. The thickness of the sand varied from 0 near the shorelines to greater than 15 meters near the center of the channel.

Beneath the marine sand is rock which is classified as Neogene limestone. Based on inspection of cones and observations of outcrops of rock on the mainland and on Psyttalia Island, the rock is characterised by the presence of solution cells and cavities. Some of the smaller cells are filled with calcite. As observed on land, the dimensions of cavities vary from 1 to 2 cm up to a few meters. The formation of these cavities is attributed to the existence of joints and faults in the limestone and to solution activity. It was the presence of these cavities which lead to the decision to transport the sewerage from the mainland to Psyttalia Island via siphon pipes laid in a trench in the sea bottom, rather than via a sewer tunnel in rock beneath the sea bottom.

DESIGN

The submarine siphon pipelines are situated in a seismically active zone. More than 40 seismic events with intensities exceeding 5.0 on the Richter scale have been recorded in the last 80 years. This ground shaking as well as seismically induced liquefaction were among the major design concerns. The siphon pipelines location under a major shipping channel with all the attendant risks such as anchor dragging required special remedial protective measures.

The siphon pipelines were designed to withstand all forces to which they may be subjected during manufacture, transportation, installation, testing, backfilling and operation.

The pipe finally selected has 2.80 meter inside diameter, was made of reinforced concrete with an upset bell and spigot. The pipe has a wall thickness of 350 millimeters (mm) and a spigot 370 mm long. All pipe spigots contain two grooves for 32 mm diameter O-ring gaskets (see figure 4). The two gaskets allow the pipe joint to be tested under water for integrity. The long spigots will minimize the likelihood of a pipe pull-out during a seismic event. All pipes were coated with coal tar epoxy, both internally and externally. In order to protect the siphons from the twin hazards of dragging ship anchors and seismically induced liquefaction, they were placed in a 7 meter deep protective trench and covered with rock (see figure 5). Specially designed double bell sections were designed for the lowest point of the siphons and pipe laying proceeded from the double bells in both directions (see figure 6 and figure 7).

One unique feature of the siphon design was the requirement that they be able to withstand an evacuation test, allowing for a visual inspection of the pipes after the completion of construction.

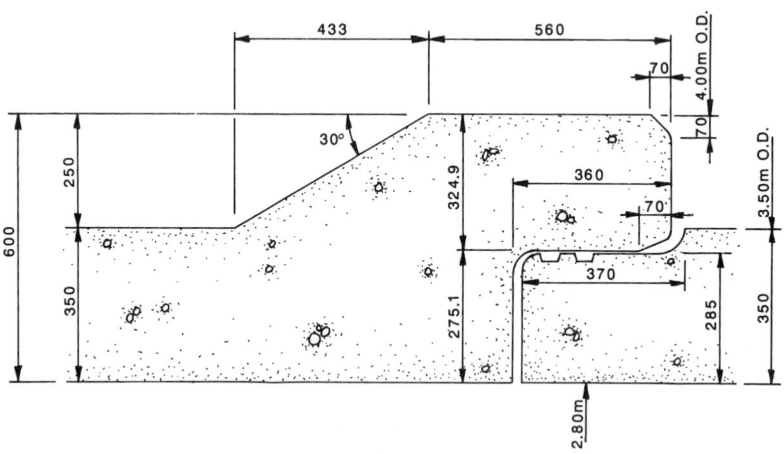

BELL AND SPIGOT DETAIL

Figure 4

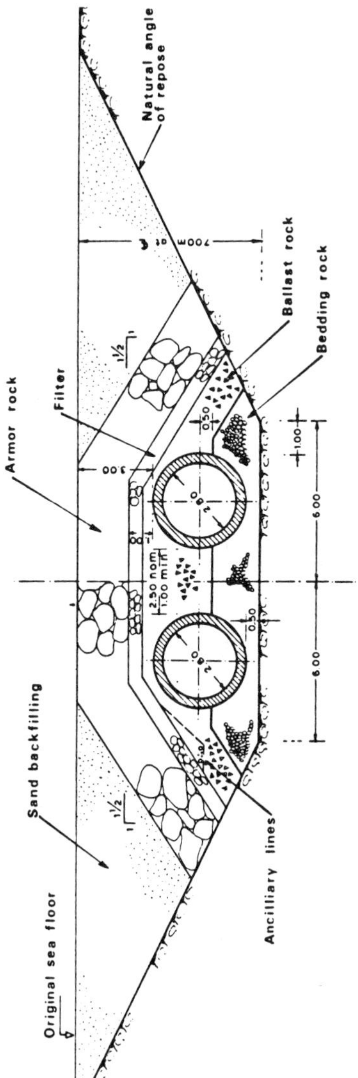

Bedding rock : 9 to 60 m.m

Ballast rock : 90 to 240 m.m

Filter : 150 to 250 kg

Armor rock : 1.5 to 2.5 tonnes

BACKFILL DIAGRAM

Figure 5

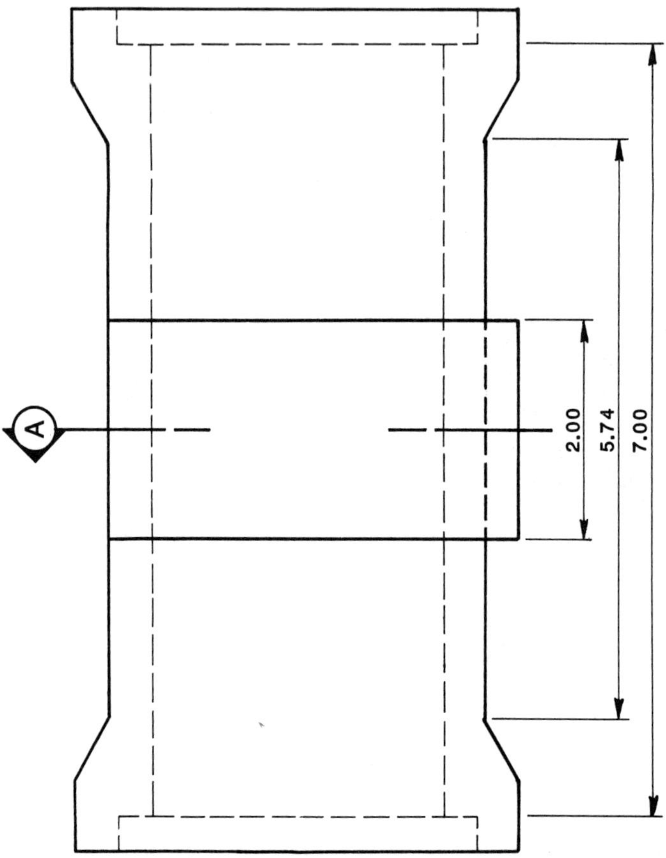

DOUBLE BELL PIPE AT STATION 5+43

Figure 6

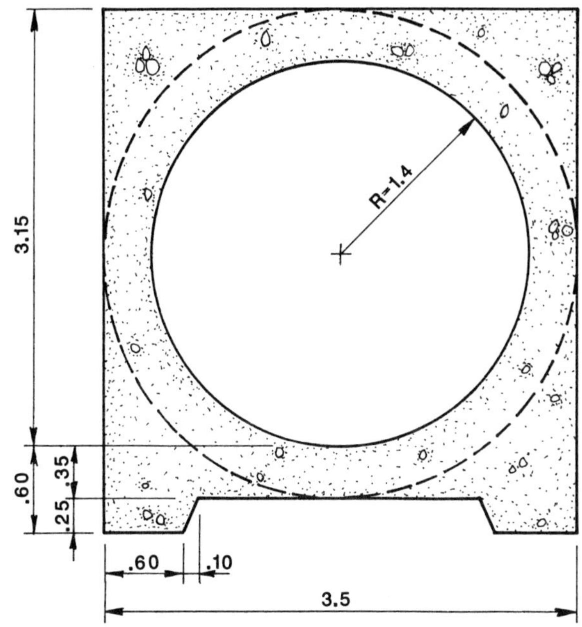

Ⓐ SECTION OF DOUBLE BELL PIPE

Figure 7

PIPE MANUFACTURE (See Figures 8, 9, 10 & 11)

The 8-meter lengths of pipe were cast at the jobsite. PIPE, Inc. of Tacoma, Washington, were retained as consultants to the Contractor and were vital in minimizing the problems that did occur. Casting large Reinforced Concrete Pipe is a much more sophisticated operation with unique skills than would appear on the surface to the uninitiated.

The pipe was cast-in-place vertically with the raised bell down. Three sets of forms were fabricated locally with some of the problems which should be expected. The tolerances on the bell and spigot machined rings are critical in order to have interchangeability between pipe cast in different forms with the very close dimensional clearances in the joint design.

The double reinforcing cages were fabricated at the jobsite using PIPE's imported machine, which they had used on San Francisco's Southwest Ocean Outfall Project.

The spigot is the "tenderest" part of the pipe, and any spalling or cracking of the front edge usually results in a test failure at installation, which is a costly expense to the Contractor. Unfortunately, due to the nature of the spigot being cast up, this concrete has the least consolidation, most segregation, air entrapment and laitence formation requiring extraordinary care in concrete placement, vibration and finishing. Unfortunately, this lesson is not appreciated until major problems are encountered with the laying, at which time a considerable amount of defective pipe has already been cast.

Another problem arose when the too-small crane used to strip the too-tight fitting and untrue interior form would rock back and forth as the form would break loose and then stick, resulting in a non-vertical pull on the form, which in turn had a prying effect on the spigot end, invariably resulting in spalling the spigot. The result was that every pipe spigot had to be patched. The O-ring grooves required extensive finishing work ending up with every spigot epoxy-coated to provide the necessary crack-free, smooth, uniform surface and uniform groove depth which varied due to lack of attention in maintaining the forms.

Every pipe was internally hydrostatically tested, revealing very few cracks and giving a good check on the sealing configuration of the spigot. Additionally, after months of unexplained joint test failures during field installation, each pipe was joined with what was going to be its mate, and the joint was hydrostatically tested in the yard before being sent out for installation.

The epoxy painting of the interior and exterior was an unusual requirement and made for a very high-quality finish, which should more than pay for itself in increased longevity and flow characteristics. A side benefit was that the light sandblast prior to application of the paint revealed many air-entrapped pockets under the grout surface, particularly on the interior of the pipe, including high porosity conditions at the rebar chairs. These areas were epoxy patched before painting.

The concrete was batched on the job into ready mix trucks and then craned into place. As the local aggregates are not high stength, the Contractor elected to use generous proportions of cement, resulting in exceedingly high compressive strengths. The Greek Ministry of Public Works maintained a complete and competent inspection force at the jobsite.

PIPE PLANT Figure 8

PIPE SPIGOT DETAILS Figure 9

DOUBLE BELL SECTION Figure 10

COAL-TAR EPOXY FINISH ON PIPE Figure 11

CONSTRUCTION

The construction of the submarine siphon described below was divided into the following components:

1. Excavation
2. Pipe Laying and Bedding
3. Rock Backfill
4. Surveying
5. Utility Line Installation
6. Final System Leakage Tests

1. **Excavation** (See Figure 12)

The trench was dredged in its entirety using three 2400 Lima barge-mounted cranes clamming into small dump scows (i.e., 250 m^3) and disposing of the material a short distance away.

The first order of business was to remove the sewer sludge which overlay a third of the alignment on the mainland side and varied up to 7 meters in thickness. This material was the deposited result of the many, many years of discharge of Athen's raw sewage at the shoreline adjacent to the project and was only diverted 100 meters away for the convenience of construction. The material consisted of very low shear strength "black mud" and posed no problem in excavation other than its odorous nature. However, due to the long delay (1-½ years) between the initial excavation and pipe placement in the excavated trench, the sludge crept back into the trench, which required extensive clean up before pipe installation and again before cover rock placement.

The 7-meter deep trench in the seabed was founded generally in hard clayey sand, posing considerable problems for the Contractor. One-half of the material lies below a 40-meter depth and almost three-quarters below 30 meters.

The dredging was conducted in two shifts, six days a week, and the average production was in the range of 250-300 m^3 per dredge per shift.

Rock was encountered on both ends of the alignment, and after unsuccessful attempts at drilling and shooting, the rock was removed using chiseling and scraping with the buckets.

Upon experiencing the hardness of the seabed material, the Contractor elected to help offset the adversities by excavating the trench to the minimum width (12 meter) with the side slopes standing competently at near vertical. this, of course, minimized the quantity of material to be excavated and the quantity of manufactured rock backfill to be produced and placed. The minimum width combined with the vertical side slopes did cause some difficulty in trying to maintain adequate clearances for the pipe laying, as one can imagine the difficulties in trying to clam a 12 meter-wide groove 40 plus meters below you in the blind!

In total, the excavtion of the 7 meter deep trench entailed the removal of 125,000 m^3 of sludge overburden in addition to 200,000 m^3 of other material.

PIPE TRENCH EXCAVATION Figure 12

2. Pipe Laying and Bedding (See Figures 13, 14, 15, 16 & 17)

The pipe was lowered into the trench, mated with the previously laid joint and held in alignment while being bedded in the conventional manner using a "Horse" (see Figure 14). This device is like a four-legged table under which the pipe is strapped, and, once set on the trench bottom, is independent of the barge-mounted 4600 Ringer lowering crane above. The four legs telescope independently such that grade and slope can be controlled. The pipe is fastened to a carriage within the "table", allowing for longitudinal (2 meters) translation for joining and independent transverse movement (±1 meter) of either end of the pipe for alignment. The hydraulic controls for these movements are all on the surface and directed via communication link from a diver on the bottom. The station positioning of the "Horse" during lowering and setdown was controlled by an inhaul winch on the beach. The cable was marked at 8-meter intervals, which allowed the "Horse" to be positioned at just the right place so as not to set on the previously laid pipe nor too far away to be "out-of-stroke" for joining. The inhaul winch also held the "Horse" from sliding down the 25% slopes on the shore ends of the alignment.

As there was no visibility on the bottom due to the existing sewage discharge and/or silt from the bedding operation, all pipe laying and jointing was done "in-the-dark" by feel.

The dual pipelines were laid concurrently with the joints of one line, staggered a half-length (4 meters) from the adjacent line in order to give the maximum clearance at the raised bells. The narrowness of the trench gave surprisingly few problems to the laying operation, and, in fact, forced the Contractor down the "straight and narrow" such that he didn't get into any steering/gap tolerance problems.

The bedding was accomplished by bringing an anchored gravel barge with a long tremie pipe over the top of the "Horse" once the pipe was joined. A 1-½ m^3 rubber-tired loader would dump gravel into an overboard hopper which fed the tremie pipe. The tremie was "spotted" over the centerline of the pipe and gravel was fed down bedding the leading end of the previously placed pipe and about 3/4 of the one being laid and held in position by the "Horse". Transport of the gravel under the bottom of the pipe was aided by divers using a hand-held water jet and later by a jetting system plumbed into the "Horse" itself. The water was provided by a small diesel-powered jet pump on board the derrick barge.

Adequate placement and consolidation of the bedding under the barrel of the pipe is always one of the more controversial aspects of pipe laying, and this job was no exception. If the bedding is not adequately placed and consolidated under the pipe, voids can exist which may cause differential settlement when the pipe is covered with rock, resulting in the joints articulating.

3. **Rock Backfill**

The remainder of the ballast rock, that above the springline and covering the pipe, was placed similarly to the bedding portion by tremie pipe fed by a front end loader on the rock barge. The section was controlled by leadline sounding and later proofed with fathometer readings. Low spots were later supplemented with rock placed with a clamshell derrick.

The ½-meter cover of filter rock was placed using the "bombing" method -- discharged from a split-hull dump scow. In the deep water this method worked fairly satisfactorily, as the material has a chance to disperse on the way down. However, the material, if discharged too rapidly, will have a tendency to "blow away" from the center of impact leaving highs on the edges and lows in the middle.

The armor rock was very laboriously placed using the 2400 Lima clamshell rigs -- up to four rigs operating at a time, which in the end ran the rock quarry dry.

In total, the backfill operation entailed the placement of 30,000 m^3 of ballast, 10,000 m^3 of filter material and 40,000 m^3 of armor rock.

PIPE CARRIER Figure 13

PIPE LAYING "HORSE" Figure 14

PIPE LAYING BARGE Figure 15

BACKFILL BARGE Figure 16

ON-SHORE TERMINUS OF SIPHON PIPES

Figure 17

4. Surveying

Alignment for all the operations was controlled by normal vertical-fan shore mounted lasers. As almost all of the marine work was on a tangent, this system worked satisfactorily.

Grade was controlled strictly by manual tide gages. The tide gages (floats) were housed in stilling wells which negated the influence of the short period waves (chop). The grade on the pipe laying was similarly monitored by attaching a floating tide gage, also in a stilling well, by cable to the center of the frame on the leading end of the pipe. Given the low tidal currents in the area, this tensioned leg gage also served for monitoring the transverse position of the pipe.

Due to the steep-sidedness of the trench excavation, meaningful fathometer readings were unobtainable so lead line soundings were used for the dredging cross-sectioning.

5. **Utility Line Installation**

A multi-utility line bundle consisting of one 12" water line, one 6" gas line, two fiber optic cables and one electrical power line was placed in the trench. The water line and gas line were cement-coated at the jobsite to provide proper ballast for the pull. The bundle was prefabricated into 100-meter lengths with a continuous timber plant bottom. The bundle was pulled across the 1400 meters by a winch placed on the opposite shore. The electric cable required one mid-pull splice, which required a two-day delay in the pull.

6. **Final System Leakage Tests**

Both siphons successfully passed the system-wide leakage tests which were completed in May 1987. No pipe evacuation tests were carried out as of the writing of this paper (June 1988). They are presently scheduled to take place during the summer of 1988.

ACKNOWLEDGEMENTS

The Owner of the facility is "The Special Service for the Sewerage and Sewage Treatment Authority of Greater Athens" (E.Y.D.E.) of the Greek Ministry of Environment, Regional Planning and Public Works. Consultants to E.Y.D.E. is a Joint-Venture of Th. Gofas and Partners, Athens, Greece; Haiste International, Leeds, U.K. and Woodward-Clyde Consultants, San Francisco, California, U.S.A. Parsons Brinckerhoff, San Francisco, California, U.S.A. is under subcontract to Woodward-Clyde Consultants.

The Contractor was a Joint-Venture of Athena Hellenic Engineering, Industrial and Touristic Co. S.A., Athens, Greece, and Raymond International, Houston, Texas, U.S.A.

REFERENCES

Theofanis C. Gofas, Yiannis Tselentis, Hugh A. Allan, Ralph A. Fasano, Yair Eisenberg and Donald D. Treadwall. Athens Siphon and Outfall - Coastal Zone '87, ASCE, Seattle, Washington, U.S.A., May 1987.

CHAPTER 205

Quantification of Longshore Sand Transport and Sand Bypassing at South Lake Worth Inlet Palm Beach County, Florida

John S. Yeend, P.E.[1]
Darryl J. Hatheway, E.I.[2]

1.0 INTRODUCTION

1.1 Abstract

The intent of this paper is to express the quantities of sand bypassed by a sand transfer facility and transported in the littoral drift at the South Lake Worth Inlet, Palm Beach County, Florida. Discussion on littoral transport volumes will include estimates of the annual average drift from 1948 to 1987, and discussions of sand bypassing quantities will include improvements in the estimation of pumping rates.

1.2 History

The South Lake Worth Inlet District was established by the Florida Legislature in 1915 (Ref. 1), for the purpose of creating an Inlet flushing canal from the Atlantic Ocean to help alleviate serious pollution in freshwater (slightly brackish) Lake Worth, resulting from rapid development of the area after the turn of the century.

The United States Army Corps of Engineers (COE) issued a permit in April of 1924, authorizing the District to construct and maintain the Inlet, including stone jetties on each side of the Inlet and allowing for the construction of a fixed bridge across the Inlet (Ref. 2). The Inlet was completed in July 1927.

Substantial improvements resulted from a 1964 model study by the University of Florida (Ref. 3). These included a new Inlet configuration with curved north and south jetties extended from their original lengths and a widening of the Inlet by removing a bridge abutment. Also, a training wall was constructed west of the bridge to improve the flow and self-maintenance of the Inlet channel and to redirect the suspended sediment flow into a sand trap area. The Sand Transfer Facility was relocated approximately 100 feet seaward of the MHW line on the north jetty (Fig.1).

----

1) Manager, Coastal Engineering Group, GEE & JENSON Engineers-Architects-Planners, Inc.,West Palm Beach, FL

2) Coastal Engineer, GEE & JENSON Engineers-Architects-Planners, Inc., West Palm Beach, FL

Fig. 1 Aerial Photo of South Lake Worth Inlet (photo circa 1984)

The STF's function is to transfer the sand as it reaches the north side of the north jetty to the downdrift beaches. Sand not transferred, generally flows around the Inlet in the ebb tidal shoal, acting as a conveyor belt, and is deposited on the downdrift beach to the south (Fig. 2).

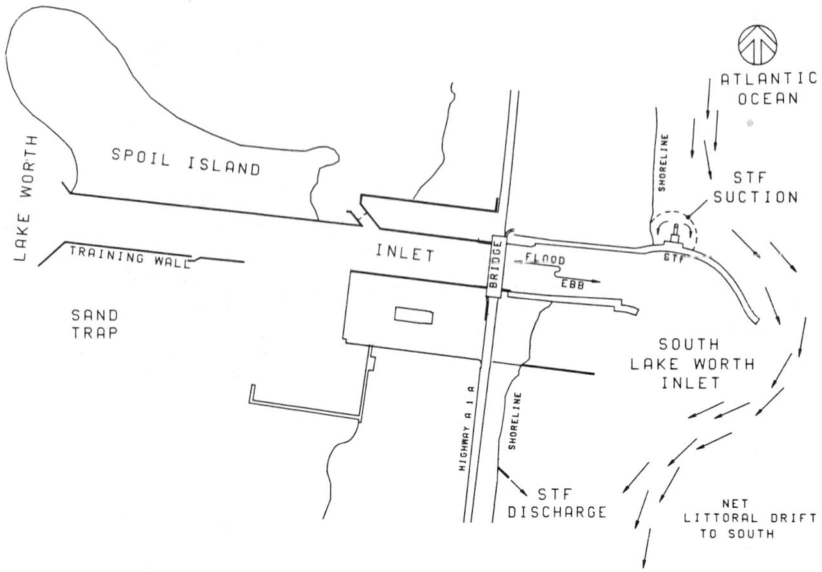

Fig. 2 Present Configuration of South Lake Worth Inlet, Palm Beach County, Florida

2.0 QUANTIFICATION OF LITTORAL TRANSPORT

2.1 Littoral Environment Observations

In October 1984, monitoring of the littoral environment was initiated utilizing the (COE) Littoral Environment Observation (LEO) Data Collection Program (Ref. 4). The LEO beach monitoring site is located about 1,000 feet north of the South Lake Worth Inlet. The LEO Program was developed by the Coastal Engineering Research Center (CERC) branch of COE. The purpose of the LEO Program is to record visual observations of breaker conditions (height, angle of approach, and tides), local winds, longshore currents, rip currents and beach geometry on a daily basis for at least one year or more (Ref. 5). The reliability and credibility of the LEO observation results and analysis depend on the number of observations taken and the availability of trained observers.

The application of the LEO data at the South Lake Worth Inlet is for the purpose of predicting longshore sand transport rates utilizing observed waves breaking at a recorded angle to the shoreline. The quantities of sand transported along the shore are a result of waves creating a longshore current and the breaking wave turbulence that suspends the sediment. Empirical equations in the Shore Protection Manual (Ref. 6) utilize LEO data in the empirical relationship to relate the longshore transport to wave conditions. Using a second method in 1980, Todd L. Walton Jr. (Ref. 5) applied LEO current observations to compute the longshore transport. The annual summary of

LEO data by CERC uses these two separate, although not completely independent, methods for determination of sediment transport rates.

The data summary in Table 1 presents two years of LEO longshore transport estimates. Review of the annual estimated littoral transport rates clearly show the net movement of sand is to south, but a true determination of the annual rate will need more years of data collection and observations. The LEO estimates for 1987 have been deleted due to data collection errors resulting in inaccurate transport rates.

TABLE 1

A. CERC LEO estimated annual net southerly sediment transport volumes (Jan-Dec.)

	Method 1	Method 2
1985	379,731 cy	110,987 cy
1987	385,168 cy	126,996 cy

B. South Lake Worth Inlet average annual estimated sand bypassing volumes (dredge years 1968-87)

Based on 100 CY per hr of pump operation	50,140 CY
Based on 125 CY per hr of pump operation	62,675 CY
Based on 160 CY per hr of pump operation	80,224 CY
Based on 200 CY per hr of pump operation	100,280 CY

C. South Lake Worth Inlet average annual measured sand bypassing volumes (1985-88) 74,409 CY

2.2 Annual Average Drift

There have been many estimates of the annual average drift for the purpose of determining the net littoral drift to the south (Table 2). In 1948, the COE did a study which concluded that during the fourteen (14) year period immediately following completion of the Inlet and the jetties, material was impounded at a rate averag-ing 150,000 to 225,000 cubic yards (CY) per year (Ref. 7). In a 1953 study by the COE, an estimate of 200,000 CY per year was used as the "southward drift at the South Lake Worth Inlet to the 27 ft. depth" (Ref. 8). In 1964, the University of Florida Coastal Engineering Department. by con-tract for the South Lake Worth Inlet District, used an es-timate of 200,000 CY as the net drift to the south (Ref.3).

In 1966, Per Bruun's book entitled "Inlets and Littoral Drift" (Ref. 9), credited the COE, Jacksonville District for its determination of the predominant drift at the South Lake Worth Inlet as being 150,000 to 200,000 CY per year. In a later report dated July 21, 1965, Per Bruun, assumed that the normal rate of drift at the Inlet is 225,000 CY per year (Ref. 10).

TABLE 2
Estimates of Annual Average Drift

Year	Agency/Individual	Rate (Net to South)
1948	U.S. Army Corps of Engineers (COE)	150,000 to 225,000 cy/yr. Drift to south at area of South Lake Worth Inlet
1953	COE	200,000 cy/yr.
1964	University of Florida	200,000 cy/yr.
1965	Per Bruun	225,000 cy/yr.
1966	Per Brunn	150,000 to 200,000 cy/yr.
1985	COE/WES	270,000 cy/yr. based on wave data collected in Atlantic Wave Information Study (1956-1975)
1987	Olsen	270,000 cy/yr.

The COE Waterway and Experiment Station (WES) in Vicksburg, Mississippi utilized twenty (20) years of wave hindcast data (1956 to 1975) to determine the potential average annual net southerly littoral transport rate of 270,000 CY in the area of South Lake Worth Inlet, which was later reported in the 1985 General Design Memorandum for Beach Erosion Control Projects in Palm Beach County (Ref. 11). The wave data utilized in the determination was supplied by the 1976 Atlantic Ocean Wave Information Study (Ref. 12). In 1987, Olsen & Associates prepared a report for Palm Beach County entitled "Guidelines for the Review and Consideration of Coastal Permits" and utilized the same 1985 WES estimate (Ref. 13).

2.3 Sediment Transport Rates

Most of the above annual littoral transport quantities are based on estimates rather than the observed data interpretation that the LEO Program attempts to correlate. The LEO Program allows for the variation in weather conditions (i.e., calm periods, when less sand will be moving, as opposed to the very active northeastern and hurricane periods, with the larger sand movement). It would be erroneous to expect any value of estimated net littoral drift to be constant year after year.

As stated in CETA 81-5 (Ref. 4) "the usefulness of the LEO program is best expressed in the statistical descriptions of the environment, often in areas where no other data exists, and in the inexpensive estimation of longshore transport rates."

2.4 Discussion

The estimates of net littoral transport have a common flaw in that there is an assumption of no interruption in the littoral transport system. The nearest interruption to the South Lake Worth Inlet littoral drift system is located fourteen (14) miles to the north at the Lake Worth Inlet. The Lake Worth Inlet, at its present location, was cut in the Barrier Island in 1917 and currently has a project depth of 35 ft. and channel width of 400 ft.

In a reference to maintenance dredging by Robert Dean (Ref. 14), it was suggested that the placement of sand in water deeper than 12 ft. does not allow for the sand to re-enter the longshore system by natural forces. This would mean that any sand naturally bypassed by the Lake Worth Inlet around the north jetty would have entered the wide channel where depths are greater than 12 ft., or deposited in the channel to be carried offshore with the ebb tide currents, or into the channel during the flood tide currents and deposited in the Intracoastal areas.

Therefore, since 1958 when the sand transfer plant was constructed on the north jetty of the Lake Worth Inlet, the only constant littoral drift across the Lake Worth Inlet was the approximately 70,000 CY per year pumped by the sand transfer plant.

All of the above information presented would make estimates of the net littoral drift at the South Lake Worth Inlet questionable. It would appear that the only way the South Lake Worth Inlet has been able to maintain the transferring, both artificially and naturally, of large quantities of sand is by tapping the supply between the two inlets. Therefore, there is a range of uncertainty in all of the above estimates.

3.0 QUANTIFICATION OF SAND BYPASSING

3.1 History

Prior to 1937, the South Lake Worth Inlet configuration with stone jetties extending perpendicular from the center line of Ocean Boulevard, approximately 350 ft. in length, allowed large quantities of sand to drift into Lake Worth through the Inlet (Ref. 15). In order to mitigate these sand losses, a sand transfer plant was constructed on the north jetty through a cooperative effort between the Inlet District and private property owners. This sand transfer plant is believed to have been the first of its kind in the United States.

Because of the fuel shortage during World War II, the plant did not operate between the years 1942 and 1945. Even though resumption of the plant operation following the war significantly restored the beaches to the south of the Inlet, it was found that a considerable amount of sand was still being trapped inside the Inlet in Lake Worth. Accordingly, the sand transfer plant was enlarged in 1948. This pump remained in operation until major renovations were made in 1967 in response to the physical model study and recommendations by the University of Florida, Coastal Engineering Lab (Ref.3) and previously discussed.

The primary purposes for these modifications were to prevent the loss of sand to the interior shoals of the Inlet, improve the sand transfer capabilities of the Inlet, and to make the Inlet channel self-flushing.

3.2 STF Quantities

The sand transfer facility today is a fixed hydraulic suction dredge with a rotating boom and 12 in. diameter suction intake line and 10 inch diameter discharge line. The dredge pump is driven by a 400 hp. diesel Caterpillar engine rated to pump 4,000 gpm with 20 percent solids in suspension. The discharge line extends west from the sand transfer facility to the road, then south across the bridge to the discharge location approximately 700 feet south of the south jetty.

Records of the quantities of sand bypassed and dredged at the South Lake Worth Inlet date back to the 1960's. For this paper, the review of quantities of sand bypassed will be for periods following the 1967 renovations and newly constructed inlet configuration. The estimated quantities of sand bypassed by hydraulic dredge have been based on rates varying from 100 to 200 CY per hour of pump operation from dredge years 1968 to 1988 (Ref. 16). Current sand bypassing estimates of 125 CY per hour of pump operation were determined by comparing estimated quantities with the measured quantities from a nuclear density measuring unit and Doppler flow meter combination that comprises the production monitor unit at the South Lake Worth Inlet.

The production monitor is a liquid slurry density system developed by Kay-Ray Inc. Industrial Process Control Equipment. The unit was installed in late 1985 to work with the Taylor Disk Pressure Recording System installed in the pumphouse of the sand transfer facility. The measuring unit was placed inside the engine room on the 10 in. discharge line and linked to the recording and equipment panels located in the room above. The control panel and lever room are located in the upper level where a digitized readout for production monitor measured quantities can be recorded by the dredge operator. The system is calibrated to read out in cubic yards of dredged material discharged through the system.

3.3 Discussion

The production monitor data indicated that during peak pumping periods, typically September to March, a larger volume of sand was being bypassed than previously estimated. It was in October of 1986 that the conversion of sand bypassing estimated cubic yards per hour of operation was increased from 100 to 125. Data in Table 1 for measured quantities reflect adjusted totals to account for periods when the production monitor was not operating due to repairs.

Prior to the introduction of the production monitor, the method of determining the dredge volumes bypassed by the sand transfer facility were to assume a constant quantity per unit of time. Changes in the estimated cubic yard per hour totals for plant operation have varied from 100 to 200 cubic yards from 1968 to 1988.

The graph (Fig. 3) presents the data collected from the production monitor since 1985. The vertical axis of sand transfer operating hours per week and the horizontal axis of measured cubic yards per week from the production monitor system illustrates the variation of weekly available material for bypassing.

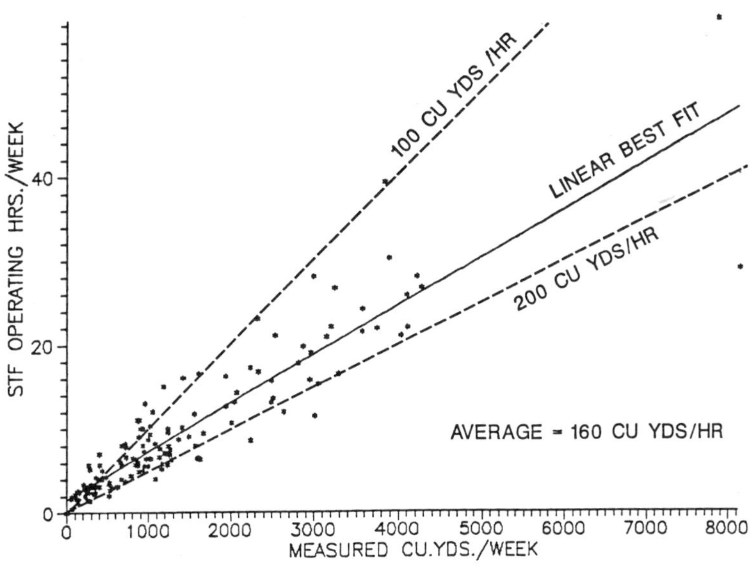

Fig. 3 Weekly Production Monitor Quantities (1984-1988)

During less active weather climates and operating conditions typical of summer and southeast winds and waves, the average measured bypassing totals are near an estimated 75 cubic yards per hour of plant operation. During active transport periods, with maximum sand availability, the plant has shown the capability to pump at a higher capacity for longer periods of time. Totals of measured bypassing in excess of 250 cubic yards per hour of plant operation have been recorded. The data in Table 3 lists the three peak periods of operation and the bypassing quantities measured during that dredge week from 1984-1988.

TABLE 3

Peak Measured Bypassing Volumes
(1984-1988)

Weekly Hours	Total Volume	Hourly Rate
39.2	3,832 CY	98.75 CY/Hr.Operation
58.6	7,816 CY	133.39 CY/Hr.Operation
28.9	8,113 CY	280.73 CY/Hr.Operation

Dashed lines on Fig. 3 represent the estimated rates of 100 and 200 cubic yards per hour to show the variation of measured bypassing rates around the upper and lower previously estimated rates of the sand transfer facility.

With a computerized linear best fit line, interpretation of the line was accomplished by averaging the data for operating periods of 10, 20 and 40 hours per week. The resulting average rate (based on these measured dredge rates) would be approximately 160 cubic yards of material per hour of plant operation.

The quantities of dredged material will vary depending upon the weather climate in the area of the South Lake Worth Inlet. As reported in an earlier paper, "Inlet Dredging Production as a Function of Sea State" (Ref. 17), a correlation was made between wave height and direction and the estimated quantities of material that could be dredged per hour of plant operation. It was determined that the problem was much more complex than simply averaging volumetric quantities for varying weather conditions.

In this paper, we also present the average annual estimates of sand bypassing volumes from 1968 to 1988 based on four (4) different estimates for bypassing: 100, 125, 160 and 200 CY per hour of pump operation (Table 1).

The graph (Fig. 4) shows that annual variations in the operating time for the sand transfer facility are apparent. The period of 1982-83 had 1,140 hours of operation and the period of 1971-72 had 231.5 hours of operation.

The graph shows a decrease in annual operating hours since 1982. This reflects a change in operation procedures and better efficiency in pumping as a result of the production monitor data (measured data) being available.

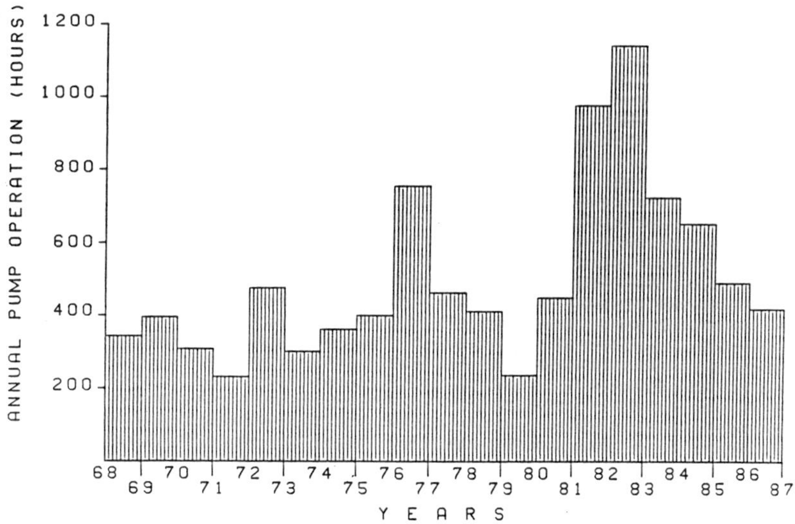

Fig. 4 Sand Transfer Facility Annual Operating Hours (1968-1987)

4.0 SUMMARY

4.1 Littoral Transport

Littoral transport is generally defined as the movement of sediments in the near shore zone by action of waves and currents. The Longshore Sand Transport component of the littoral transport is the volume of sand moving parallel to the shore. The review of the littoral drift and estimated sediment transport rates and associated volumes as presented in this paper were based on estimates, wave hindcast data, and LEO's two Shore Protection Manual methods.

The variations in the estimates of wave angle, surf zone width, and weighting of calm wave height observations in the littoral transport rate determination produce a wide range of estimated annual volume totals. As stated earlier, there is a range of uncertainty in all of the above estimates. The past estimates of littoral drift presented here documents the longshore transport rates variability.

4.2 Sand Bypassing

The sand bypassing operations at the sand transfer facility of the South Lake Worth Inlet show a great variability in the total annual volumes of material pumped from the north beach to the south beach. The combined presentation of the bypassing data and estimated LEO transport rates illustrates the variation in the dredge volumes due to the seasonal fluctuations of the wave and wind climate that create the transport events.

REFERENCES AND BIBLIOGRAPHY

1. Chapter 7080, Laws of Florida, 1915

2. War Department Permit to Construct Inlet, Executed 30 April 1924

3. Coastal Engineering Laboratory, University of Florida, Coastal Engineering Study of South Lake Worth Inlet, Gainesville, Florida, May 1964

4. Schneider, C., "The Littoral Environment Observation (LEO) Data Collection Program," CETA 81-5, U.S. Army, Corps of Engineers, Coastal Engineering Research Center, Ft. Belvoir, Virginia, March 1981

5. Walton, TL, Jr., "Computation of Longshore Energy Flux Using LEO Current Observations," CETA 80-3, U.S. Army, Corps of Engineers, Coastal Engineering Research Center, Ft. Belvoir, Virginia, March 1980

6. U.S. Army, Corps of Engineers, "Shore Protection Manual," Coastal Engineering Research Center, 1977.

7. House Document No. 772 - 80th Congress, 2nd Session, December 1948

8. Watts, G.M., A Study of Sand Movement at South Lake Worth Inlet, Florida - Beach Erosion Board, Corps of Engineers TM-42, October 1953

9. Per Bruun, Tidal Inlets and Littoral Drift, Stability of Coastal Inlets, Vol. 2, 1966

10. Bruun, P., The Effect of South Lake Worth Inlet and Its Bypassing Plant on Adjacent Shores, Present and New Installations for the TIIF, University of Florida, July 1965

11. U.S. Army, Corps of Engineers, Jacksonville District, "Beach Erosion Control Projects for Palm Beach County, General Design Memorandum with Environmental Impact Statement," April 1987

12. U.S. Army, Waterways Experiment Station, "Atlantic Coast Hindcast, Shallow-Water, Significant Wave Information," 1976

13. Olsen Assoc., Inc., Guidelines for the Review and Consideration of Coastal Permits, for Palm Beach Beaches and Shores Council, October 1987

14. Dean & O'Brien, Florida's East Coast Inlets Shoreline Effects and Recommended Action, December 1987, for Florida Department of Natural Resources, Division of Beaches and Shores.

15. Caldwell, J.M., Bypassing Sand at South Lake Worth Inlet, Florida, Proceedings of First Conference on Coastal Engineering, October 1950

16. South Lake Worth Inlet Dredging Records, 1960 to Present

17. Yeend, J.S. and Hatheway, D.J., "Inlet Dredging Production as a Function of Sea State," Proceedings of the Fifth Symposium on Coastal and Ocean Management, Coastal Zone '87, American Society of Civil Engineers, 1987

CHAPTER 206

The beach of Vinaroz. Its origin, evolution and future

JOSE SERRA PERIS*
JOSE JAVIER DIEZ GONZALEZ**

INTRODUCTION

The great physiografical unity named "Ovalo Valenciano" is outstanding in Western Mediterraneum. The Oval is limited by River Ebro, in the North, and Cape San Antonio in the South. On its Northern, margined by River Cenia, in the north, and Tombolo of Peñiscola in the South, is the subunity called Peñiscola and there stand the beach of Vinaroz, extending itself from the harbour of Vinaroz to river Cervol mouth.

COASTAL EVOLUTION

The beach lies on a steep coast, the cliff being higher all around River Cenia mounth and disappearing on the border line between Benicarló and Peñiscola districts the steep rock is composed of conglomerates made from multi-sized pebbles, cemented together in a matrix of finer materials, even clays. Sea action has given shape to a littoral where small bays, having pebbly and sandy beaches, alternate with relict debris cones (Figure, 1)

Cavanilles, giving us an extremely clear description, writes: Una descripción muy clara la encontramos en Cavanilles, quien dice: "En estas costas las olas por lo comun baten contra un terreno duro pocas veces de piedra, y muchas de un hormigón endurecido, compuesto de chinas, cantos y marga arcillosa roxa con algunas arenas. A fuerza de los choques de las olas y de renovarse la humedad, se ablanda la base que sostiene el cortezón y se descarna: presenta al principio hacia el mar excavaciones y cuevas; cayendo después al agua por su propio peso masas considerables. Mientras permanecen allí caídas, sirven de parapeto a las furias de las olas, y defienden por algún tiempo la porción con la cual estuvieron unidas; pero

*University Titular Professor. Escuela Técnica Superior de Caminos Canales y Puertos de Valencia. Camino de Vera s/n, 46071 VALENCIA. Spain.

**Professor of University. Escuela Técnica Superior de Caminos Canales y Puertos de Madrid. Ciudad Universitaria, 28040 MADRID. Spain.

cediendo en fin quedan expuestas a igual suerte las que en otros siglos existieron sin riesgos, y de este modo va haciendo el mar lentas conquistas en la costa" (Cavanilles, 1975)

The beach of Vinaroz, born backing on the harbour breakwater, developed itself beneath a low cliff which constitutes now the foundation of the maritime promenade. It receives materials mainly from nearby sources: River Cervol, Rambla Barbiguera, River Cenia and erosive materials coming from cliffs located on River Cenia shores. River Ebro, being a further source, supplies fines to be deposited on the submerged beach; they are easily removed and conveyed southward to settle on more receptive beaches. The beach of Vinaroz has been defined as a typical gravel one because of the local character of its sources of materials and the grain size characteristics of the stuff.

SEDIMENT TRANSPORT

The medium yearly direction of the littoral dynamic is North-South, seasonally being South-North sometimes. Carriage ability moves from 170.000 m3 to 510.000 m3 per year (Serra, J., 1986), Figure, 2).

The beach of Vinaroz has enveloped in accordance with its genesis; the harbour breakwater acted partially as a barrier to conveyance and so allowed a backed up beach to be born, no longer than one kilometer but well dimensioned in with. An aggregates explotation was set beside River Cervol which cut off stuff supply towards the littoral. Even when the explotation came to an end, River Cervol is no longer a source of materials because of its few functionalism; adding to this scarce mobility of products coming from cliff's erosion and the decrease of Rivers Cenia and especially Ebro's apports, the result is a recessive shore.

Even when it only brought thin stuff River Ebro allowed the existence of a submerged beach trending to a profile of equilibrium with heterogeneous materials sand and gravels which formed a somerhour stable beach. As in the earlier paragraph has been told, the conjuction of all the afore mentioned aspects brings forth as a result a recessive beach whose characteristics are a weak accumulation by the harbour dyke (breakwater) and strong erosion in the rest of the berm sand and in the whole of the submerged beach (see figures 3 and 4).

CONCLUSIONS

Two actions trending to regenerate the beach have been undertaken in the last years: building a T-groin in the Northern end of the beach and artificial feeding.

A T-groin was built in 1985, giving birth to a sheltered beach leeward, but this came to accentuate erosion in the central stretch of the beach as materials gravitate towards sheltered zones in the break-water and in the harbour causing a natural selection of sediments, the coarses swelling the backed up beach and the media and fine ones going to the sheltered beach. The results of these conditions had they been maintained, would have been a sandy beach sheltered by the break water; the birth of a pebbly beach backed up to the harbour; and strong erosion in the center of the beach which would endanger the stability of the maritime promenade.

Erosion in the central stretch of the beach has been palliated since 1986 by artificial feeding but this caused two serious objetions: in the first place, since their origin was pounding, material used nailed into bathers' feet, the water getting a peculiar colour because of high contents of clays and for the same reason the beach "swallowed" bathers beside the shore. Secondly, the beach retained very few materials, due to high contents in thin ones, the rest escaping to the sheltered zone, besides off-shore loses of media and coarses stuff caused be steep slopes proper from pebbly beaches.

Several solutions can be adopted in order to regenerate the beach of Vinaroz. Among them, two are to be mentioned:

Exclusive artificial feeding; the united effect of the T-groin and artificial feeding would be satisfactory, provided that a series of permises come together. Apports should be grain-size distribution appropiate, with periodical reloading in amounts to be predetermined. Feeding with natural, not pounded, sand should be requires as we cannot hypothesice that because of surge action might get typical round shape proper from marine sediments after one year stay in the shore.

Isolated breakwaters; artificial apports can be retained by mean of isolated breakwaters, since they act as reducers of surging energy, diminishing its ability to convey and avoiding deposits gravitation towards the harbour or the T-groin, and so helping to the maintenance of apported materials. It would be not necessary that these breakwaters we propose obey to classical conceptions such as emergent or semi-emergent dykes; as a solution could by adopted that of the artificial reef, according to expres wishes of the area fishermen whom by mean of the afore mentioned system intend to protect and strengthen their fishing reservoir. This solution requires artificial feeding too, in the same qualitative conditions said in the former paragraph.

REFERENCES

-Cavanilles, A.J., 1795-1797. "Observaciones sobre la historia natural, geografía, agricultura, poblaciones y frutos del Reyno de Valencia". Imprenta Real. Madrid.
-Copeiro, E., 1980. "Sobre la progresiva ruina de nuestras

costas arenosas". Rev. de Obras Públicas. Madrid, pp 307-319.
-Díez, J.J., Arenillas, M., Serra, J., 1982. "Shore protection in Almazora coast" (Castellón, Spain). 4th Inter. Congress I.A.E.G. New Delhi.
-Serra, J., Esteban. V., Díez, J.J., 1984. "The performance of different shore protection systems in the East-Spanish Mediterranean coasts". Inter. Sym. on maritime struc. in the Mediterranean Sea. Athenas.
-Serra, J., 1986 "Procesos litorales de las costas de Castellón". Tesis Doctoral Universidad Politécnica de Valencia.
-Serra, J., Medina, J.R., 1987. "Arrecifes artificiales en el litoral español. Experiencia valenciana." Rev. de Obras Públicas. En prensa.

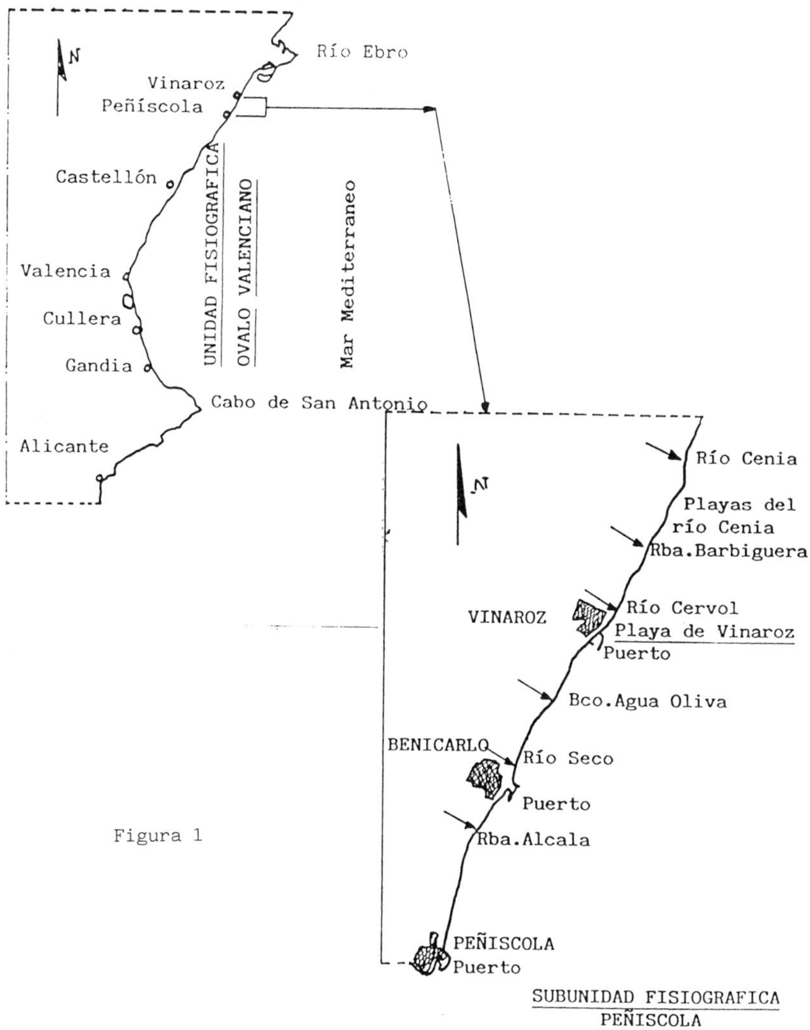

Figura 1

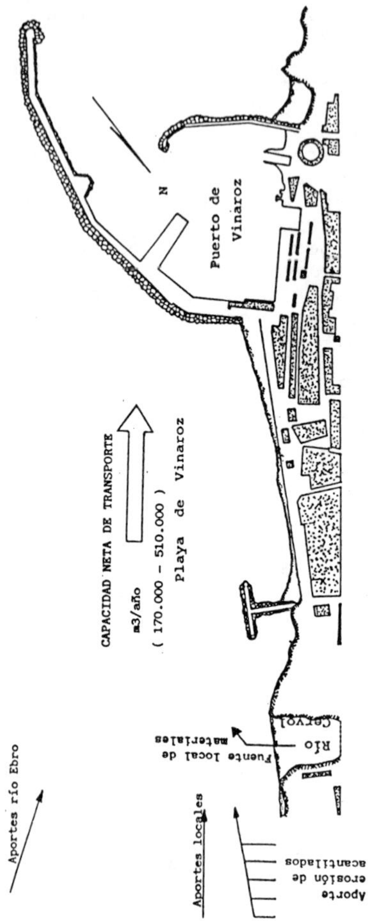
N
Puerto de Vinaroz
CAPACIDAD NETA DE TRANSPORTE
m3/año
(170.000 - 510.000)
Playa de Vinaroz
Río Cervol
Fuente local de materiales
Aportes río Ebro
Aportes locales
Aporte erosión de acantilados

Figura. 2

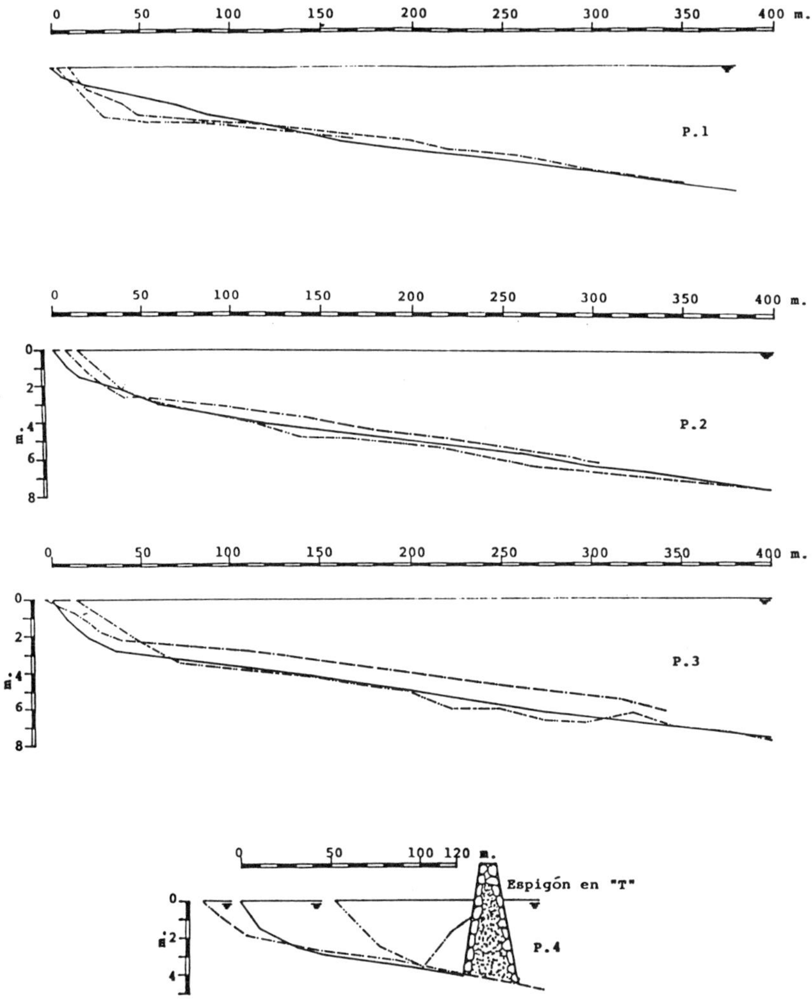
0
50
100
150
200
250
300
350
400 m.
P.1
P.2
P.3
m.
0
2
4
6
8
120 m.
Espigón en "T"
P.4

Figura 3

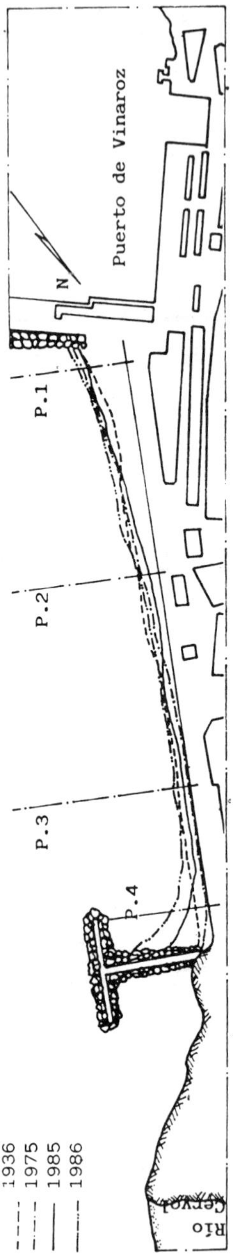

Figura 4

CHAPTER 207

STABILIZATION OF COAST BY CONSTRUCTION OF HEADLANDS ON THE KASHIMANADA COAST, JAPAN

Uda, T.*, M. Sumiya**, and H. Sakuramoto***

ABSTRACT

Eleven artificial headlands were constructed on the Ohno•Kashima Beach facing the Pacific Ocean to prevent further extension of beach erosion and to form the dynamically stable beach. This paper summarizes the results of the follow-up study of the new works applied on the actual coast. Field investigations such as bottom sounding around headlands and aerial photographing were conducted. It was found that the periodic changes of the shoreline configuration correpsonding to seasonal change of the wave direction exist on this coast and these variations of the shoreline position can be reduced by the effect of these headlands.

I. INTRODUCTION

In recent years beach erosion has been considerably severe along Japan's coasts, so that various types of coastal protective measures have been carried out; seawalls, wave-dissipating breakwaters, groins or detached breakwaters have been constructed. Most of these facilities are constructed continuously or with small intervals along the shoreline. As a result of these works natural sandy beaches decrease year by year. Natural beach posesses not only prominent function of disaster prevention but also great advantages in view of landscape. Therefore, it is desirable for preventive works to be carried out while maintaining natural sandy beaches as much as possible.

Along the Kashimanada Coast in Ibaraki Prefecture, beach erosion has been severe in recent years (Uda et al., 1986 a,b). For a measure against beach erosion, construction of the artificial headland was planned by the Ibaraki Prefectural Government since 1983 on the

---

* Dr. Eng., Head, Coastal Eng. Div., Public Works Res. Inst., Ministry of Construction, Ibaraki 305, Japan.

** River Div., Public Works Dep., Ibaraki Pref. Office, Ibaraki 310, Japan.

*** INA Civil Eng. Consultants Co., Ltd., Tokyo 162, Japan.

Kashimanada Coast facing the Pacific Ocean, and a group of 11 headlands have already been constructed up to the present. Artificial headlands have been applied in Japan for the first time, and at present there remain some problems to be solved in its applications. This paper summarizes the results of the follow-up study of the new works applied on the actual coast. Present headlands have not been completed up to its final design, but further field investigations are to be conducted to attain the reasonable form of the headland system through actual field tests.

Regarding the headland system for the beach stabilization method some previous studies have been already conducted by Silvester (1976) or Tsuchiya (1984). These methods can be applied to the beach where waves obliquely approach to the shoreline with large angle. On the coasts in Ibaraki Prefecture, however, the direction of wave approach varies seasonably and periodically, so that their methods are not applicable. One of the authors has proposed a shoreline control method --artificial headland-- for the field with periodic changes of the incident wave direction (Uda et al., 1987). Since both methods have similar meaning with respect to a point that a long extensive beach is changed stable by separating into several segments by these works, the measures carried out in this study will be called by the same name-headland.

II. OVERALL FEATURES OF BEACH CHANGES ALONG THE KASHIMANADA COAST

The study site (the Ohno•Kashima Beach) is located on the north side of Kashima Port in the middle of the Kashimanada Coast (Fig. 1). On this beach the direction of longshore sediment transport is changing according to the seasonal changes of incident wave directions, but

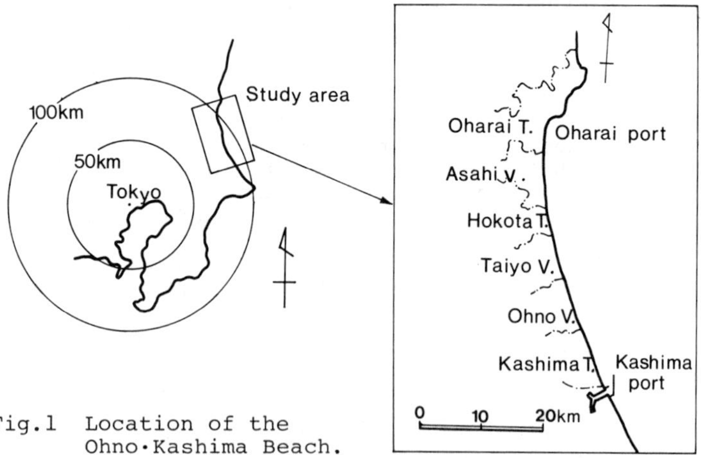

Fig.1 Location of the Ohno•Kashima Beach.

northward component of the wave energy flux dominates, so that severe beach erosion is observed on the north side of the Kashima Port, obstructing northward longshore sand transport (Uda et al., 1986).

The longshore distribution of the shoreline change, median diameter of beach materials and foreshore slope are shown in Fig. 2, illustrating the shoreline change until 1985 and 1987 with reference to the one measured in October, 1984. The abscissa is the longshore distance from the origin located at Oharai Port (refer to Fig. 1). In addition, median diameter and foreshore slope were measured both in 1985 and 1987. It is clearly understood that the shoreline has retreated considerably in recent years on the Ohno·Kashima Beach 25km to 39km south of the origin, and also beach materials and foreshore slope have become

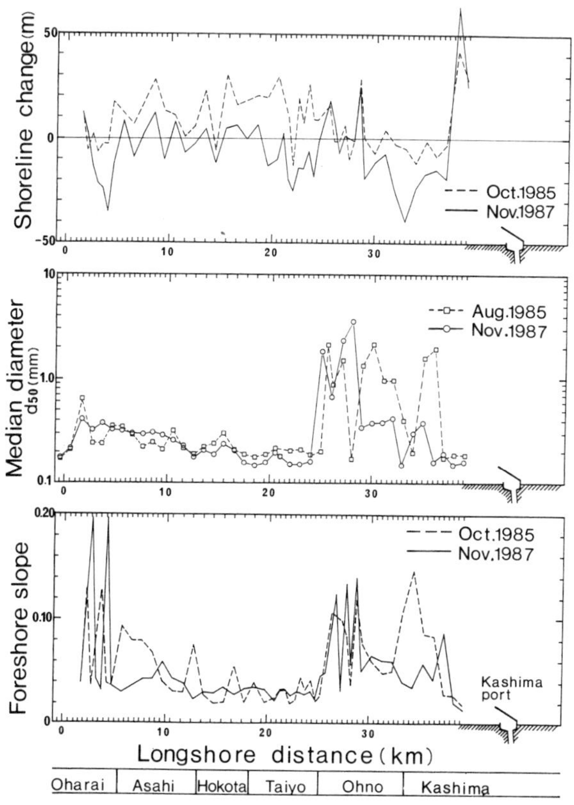

Fig.2 Longshore distribution of shoreline change, median diameter of the foreshore materials and foreshore slope along the Kashimanada Coast.

coarser and steeper at the same locations, respectively. The same characteristics as mentioned above are found on the south beach adjacent to Oharai Port. In order to prevent these beach erosions and to form stable beaches on this coast the construction of headlands was proposed.

III. CONSTRUCTION OF HEADLANDS

Headlands were constructed at the interval of about 1km based on the aerial photographs taken in August, 1984, assuming that the shoreline retreat should be less than about 30m, because the original shoreline width at the planning time had already been narrow enough and the shoreline retreat over 30m was not permissible for the stability of various coastal facilities. The construction sites of headlands were decided, based upon the simulation of the shoreline evolution by using one-line theory with the present shoreline data. The arrangement of each headland is shown in Fig. 3. Eleven headlands have already been constructed up to the present and the two are to be set in future. The construction of No. 1 and No. 4 headlands out of 11 headlands began in December, 1985 and were completed in March, 1986. The construction period of the remains is presented in Table 1. In addition, Fig. 4 represents details of headland plan; the crown width is 6m, the crown height is +2.2m above the M.S.L., its length is 100m and it was built by natural stones over 1 ton. The length of headland is about 80m at present.

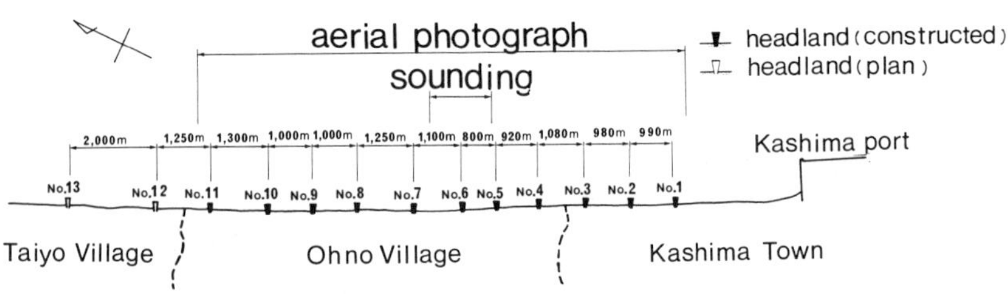

Fig. 3 Location of construction site of headlands on the Ohno·Kashima Beach.

Table 1 Dates of measurement of the bottom sounding and aerial photographs.

Kind \ Date	1986				1987												1988
	Sep.	Oct.	Nov.	Dec.	Jan.	Feb.	Mar.	Apr.	May	Jun.	Jul.	Aug.	Sep.	Oct.	Nov.	Dec.	Jan.
Sounding	O			O	O	O		O			O		O		O		O
Aerial Photo		O		O	O	O		O			O		O		O		O
Construction Period of Headlands	← No.3, No.5~7 → (Sep.–Feb.)				← No.8 ~10 → (Jan.–Apr.)											←No.2, No.11	

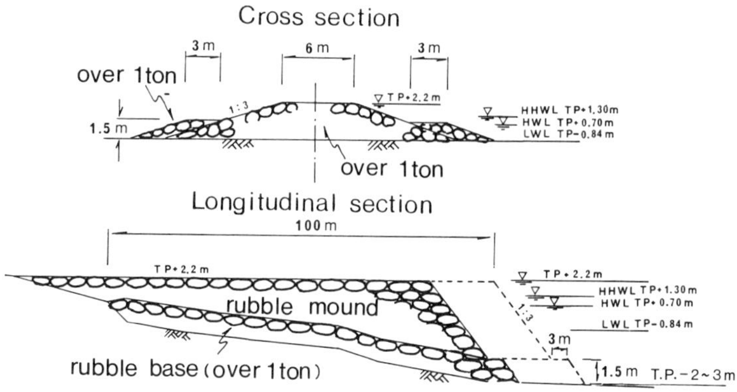

Fig. 4 Configuration of headland.

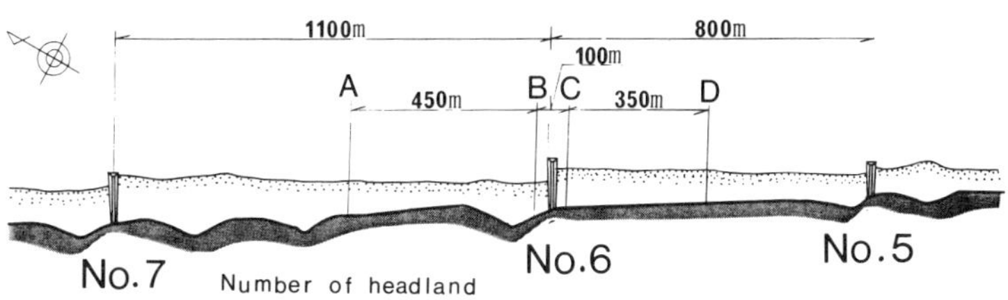

Fig. 5 Arrangement of survey lines of bottom sounding.

IV. METHOD OF FIELD INVESTIGATIONS

Field investigations consist of bottom soundings and aerial photographing, whose schedules are summarized in Table 1. Bottom soundings were carried out 9 times in total. Both observations were tried to be conducted at the same day, but there was a little time lag in execution due to the weather conditions.

The soundings were carried out along 20 survey lines of 200m in length, covering the sites of No. 5 to No. 7 headlands. The investigation of the profile changes were conducted in detail only with the representative 4 survey lines such as the north lines (A, B) and the south ones (C, D) of No. 6 headland as shown in Fig. 5, because beach changes caused by the construction of headlands can be well examined by the comparison of beach profiles near the structure with those far from that. Comparison with the beach profiles along the survey lines B and C near the structure enables to evaluate the influences of the headland for the surrounding beach topography.

Figure 6 represents the wave data during the field survey in terms of significant wave height ($H_{1/3}$) and wave

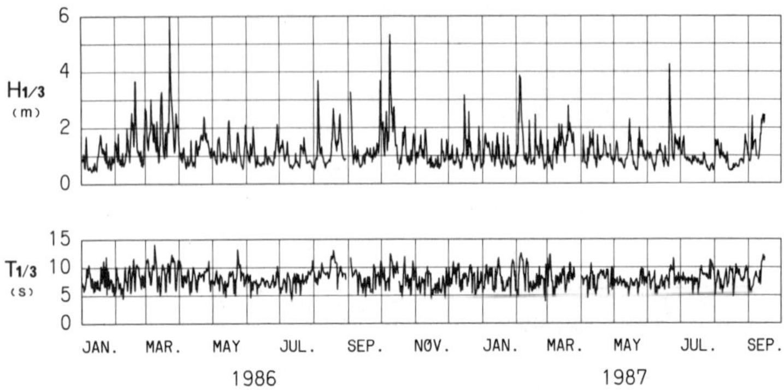

Fig. 6 Temporal changes of significant wave height and wave period off Oharai Port.

period ($T_{1/3}$) observed at a location of 21m deep off Oharai Port in January, 1986 through September, 1987. Wave condition of this coast is usually rough in February, March and October. On the contrary, from April to August calm sea condition generally prevails. There exists a weak correlation between significant wave height and wave period as shown in Fig. 6, and wave period tends to be longer with the increase of wave height.

At Oharai Port wave directions have been observed. The joint distribution of significant wave height and wave direction is summarized in Table 2, in which the data are summed up separately in 1986 and 1987. On both tables four predominant direction with high wave energy level are apparent, and in addition these wave directions are roughly divided into two groups of NE, ENE and SE, SSE. Assuming the same wave incidence at a location off the Ohno·Kashima Beach, the longshore sediment transport on this site

Table 2 Joint distribution of significant wave height and wave direction.

Jan. 1986 ~ Dec. 1986 (%)

$H_{1/3}$(m) \ Direction	N	NNE	NE	ENE	E	ESE	SE	SSE	S	Total
0.00 ~0.49										
0.50 ~0.99		0.6	9.0	11.3	4.8	0.3	1.0			27.0
1.00 ~1.49		0.6	7.7	15.1	1.9	2.3	6.8	3.9		38.3
1.50 ~1.99		1.0	4.2	7.1	2.9		5.8	1.3		22.2
2.00 ~2.49		0.3	0.6	1.6	1.3	0.6	2.6	0.3		7.4
2.50 ~2.99				0.6	1.3	0.6	0.3			2.9
3.00 ~3.49				0.3	0.3		1.0			1.6
3.50 ~3.99							0.3			0.3
4.00 ~				0.3						0.3
Total		2.6	21.5	36.3	12.5	3.9	17.7	5.5		100.0

Jan. 1987 ~ Sep. 1987 (%)

$H_{1/3}$(m) \ Direction	N	NNE	NE	ENE	E	ESE	SE	SSE	S	Total
0.00 ~0.49				0.6			0.6	0.6		1.7
0.50 ~0.99		1.1	8.0	10.9	2.3		1.1	5.1	0.6	29.1
1.00 ~1.49			14.3	9.7	1.7		1.7	10.9	0.6	38.9
1.50 ~1.99			5.7	5.7				7.4		18.9
2.00 ~2.49			1.7	1.1				4.0		6.9
2.50 ~2.99			1.1					0.6		1.7
3.00 ~3.49								1.1		1.1
3.50 ~3.99			0.6							0.6
4.00 ~										
Total		1.1	31.4	28.0	4.0		3.4	30.9	1.1	100.0

dominates southward by wave incidence from NE, and northward from SE or SSE, respectively because the direction normal to the mean shoreline is ENE. As is mentioned, this coast is characterized by seasonal changes of wave direction and to this reason typical changes of the beach configurations can be seen around coastal structures.

V. RESULTS OF INVESTIGATION

5.1 Shoreline Evolution

Shoreline changes around headlands between September 8, 1986 through January 7, 1988 are expressed in Fig. 7. In this figure comparisons are made with subsequent shoreline configurations. The locations of No. 1 through No. 7 headlands are also shown at the bottom of the figure. From September 8 to October 28, the shoreline retreated on almost all area except adjacent part north of No. 1 and No. 4 headlands which had been already constructed at that time. This recession of the shoreline is thought to be caused mainly by the cross-shore sediment transport by high waves which attacked the beaches with the significant wave height over 5m on October 9 as shown in Fig. 6. On the other hand, the local advance on the north part of No. 1 and No. 4 headlands may be caused by the accretion due to the southward longshore transport by waves approaching from counterclockwise direction

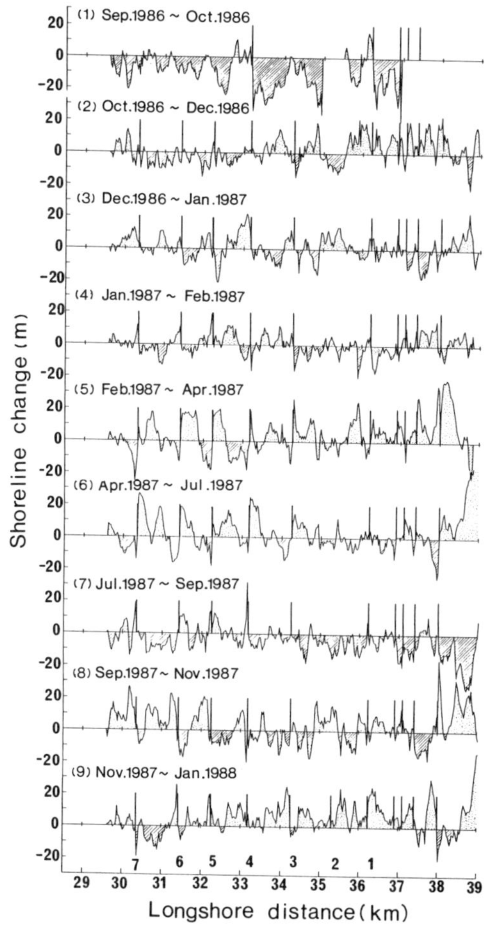

Fig. 7 Shoreline change between September, 1986 and January, 1988 measured from the aerial photographs.

to the normal of the shoreline. Between October 28 and December 24, any typical feature of shoreline evolution did not appear, only accompanying with some local erosion or accretion. From December 24 to January 17, as seen especially around No. 3, No. 4 and No. 7 headlands, the shoreline retreated on the south side in contrast to the advance on the north side of these headlands. These phenomena of the shoreline change were induced by the southward longshore sediment transport resulting from the wave incidence from the counterclockwise direction to the normal of the shoreline, corresponding to the wave approach from NE to ENE on the study coast. On the contrary, the shoreline changes occurred between February 13 and April 13, and between April 13 and July 23 are entirely reversed to the above, i.e., the shorelines advanced greatly on the south (right) side but retreated on the north side of each structure. This is due to the wave approach from the predominant direction of SE to SSE through spring to summer. Then, remarkable changes of the shoreline did not appear between July 23 and September 12, but from September 12 to November 6 the shoreline advanced again on the north side and retreated on the south side of each structure. These characteristics of shoreline changes are the same as the ones between December 24, 1986 and January 17, 1987. Consequently, it is found that the shoreline between headlands are dynamically stable with periodic fluctuations of the amplitude of about 40m.

The above-mentioned periodic fluctuations of shoreline configuration between headlands are recognized on the north coast in the same Ibaraki Prefecture (Uda et al., 1986 a,b). Taking into consideration of these facts, it is considered that this shoreline evolution represents a general features along the Ibaraki Coast facing the Pacific Ocean.

A reversal mode was observed on south and north sounding lines near a certain headland. This change is due to the periodic change of incident wave direction. In the following these phenomena will be examined in detail on the basis of temporal change of shoreline position on both survey lines of 25m away from No. 6 headland (Fig. 8). In the figure solid and open circles represent the shoreline positions on the south and north lines, respectively, including the shoreline position before the construction plotted as a broken line. There is little difference in the shoreline change before the construction on both lines, whereas the shoreline position advanced on the north side and retreated on the south side between January and February, 1987 after the construction of the headland. Then, the opposite mode of the shoreline change was observed until August resulting in large advance of the shoreline on the south side. Moreover, between August and December appeared again the recession of the shoreline position on the south side. The reasons of above mentioned cyclic shoreline changes could be explained as follows; from autumn to winter the southward longshore sand

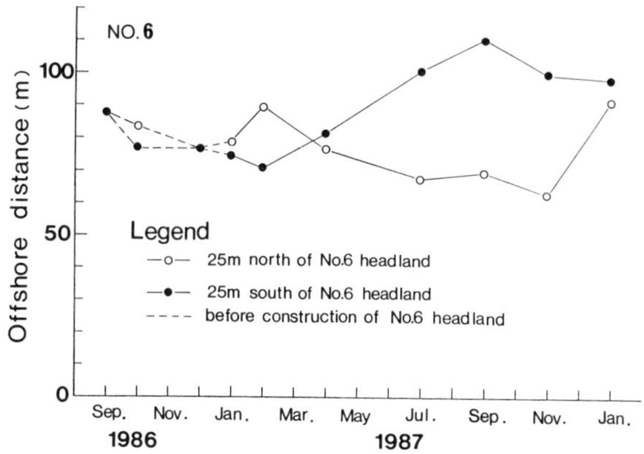

Fig. 8 Temporal change of shoreline position in the vicinity of No.6 headland.

transport prevails by the incident waves approaching from NE or ENE to this coast and then the obstruction of longshore sediment transport by the structures produces the progression of the shoreline on the north side. On the contrary, from spring to summer northward longshore sediment transport is caused by the predominant waves from SE or SSE, which is clockwise to the normal of the coastline. Then shoreline position progrades on the south side of the structure.

5.2 Comparison of Aerial Photographs Around Headlands

Shoreline configurations will be investigated in detail, selecting No. 7 headland as a typical example. Aerial photographs were taken five times between January 17 and September 12, 1987 around No. 7 headland.

The shoreline on January 17, 1987 is shown in Photo 1. Taking into account of the headland length of about 80m, it is seen that the sea is very calm and the surf zone width is narrow. This calm wave condition coincides with the characteristics observed in Fig. 6. Comparing the shoreline configuration on both sides of the headland, the shoreline on the right (south) side retreats a little more than that on the left (north) side. It shows a resultant beach change caused by the weakly-oblique wave incidence from the counterclockwise direction to the normal of the shoreline. On February 13 wave conditions turned rough and the whole picture were covered by the surf waves (Photo 2). This photograph is thought to represent a typical rough wave condition in winter, because wave height in February is high as shown in Fig. 6. In addition, the shoreline south of the headland retreats greatly compared with that north of headland. This fact indicates the same feature

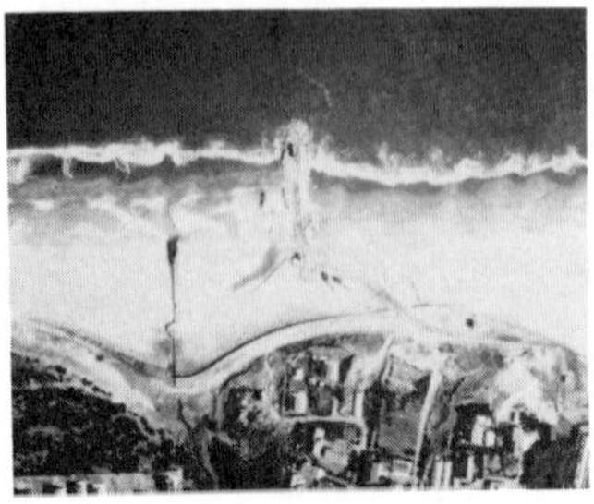

Photo 1 Aerial photo of No.7 headland (January 17, 1987)

Photo 2 Aerial photo of No. 7 headland (February 13, 1987)

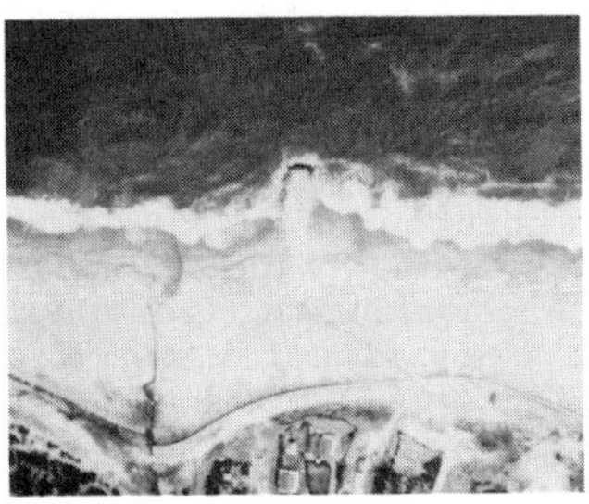

Photo 3 Aerial photo of No.7 headland (April 13, 1987)

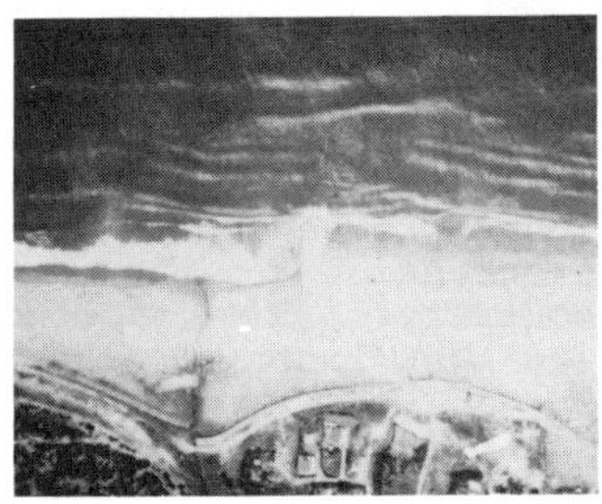

Photo 4 Aerial photo of No.7 headland (July 23, 1987)

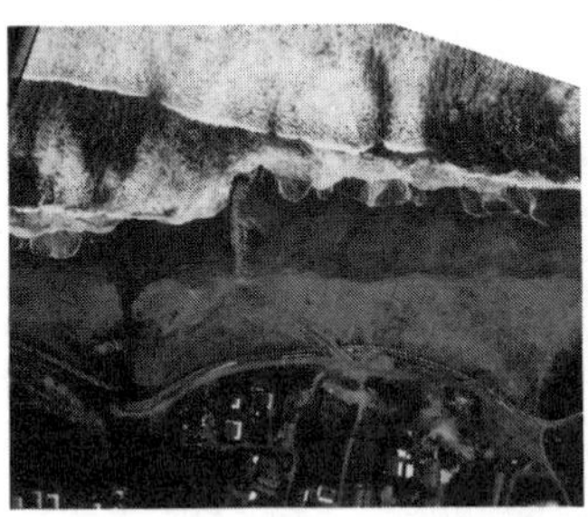

Photo 5 Aerial photo of No.7 headland (September 12, 1987)

observed in Photo 1. In other words, incident wave approaches from counterclockwise direction to the normal of the shoreline.

Two month later (Photo 3), the shoreline configuration on both sides became almost symmetry to the headland, and there disappeared the distinct shoreline retreat on the south side in Photo 2. It shows that the direction of incident wave turned clockwise with respect to the normal to the shoreline during these two months. In addition, low waves are supposed to be approaching to the coast judging from the occurrence of narrow surf zone. Actually low wave conditions appeared in April as shown in Fig. 6, so that Photo 3 is thought to show the typical state of the shoreline form in this season. On the other hand the shoreline on July 23, 1987 (Photo 4) represents remarkable changes; great advance on the south side and recession on the north side. The characteristics of the shoreline changes on this date are totally opposite to those on February 13. Furthermore, the wave conditions are also calm and the swell waves are clearly noticeable to approach in the offshore zone of the headland from a clockwise direction. Finally it is found that morphological changes between April and July are caused by the clockwise wave incidence.

Photo 5 represents the beach circumstances on September 12. Entire region of the photograph is covered by the surf waves, thereby high waves were clearly incoming to the beach at that time. Also in the offshore of the headland the waves approach clockwise at an angle of about 12° to the normal of the shoreline and even the tip of the headland locates within the surf zone. Since the northward longshore sediment transport dominates under the wave climate, the shoreline advanced on the south side, whereas the shoreline retreated on the north side of the structure. Consequently, it is concluded that the periodic variations of the incident wave direction throughout a year determine the direction of the longshore transport on this coast. This means that the seasonal changes of wave directions are the most important factor to consider beach deformation on this coast. Having seen in Photos 2 and 5, the tip of the headland becomes contained occasionally within the surf zone under high wave conditions. It is, therefore, possible for the longshore sediment to turn around the tip of the structures to the adjacent beaches. This means that the headland of present dimension can reduce the rate of sediment, but can not prevent from the decrease of sand volume between headlands. Taking the present situation into account, it is necessary to elongate the headlands in future.

5.3 Beach Profile Changes

Here will be described time series of the beach profile changes based on the data surveyed almost at the same time as the shoreline changes as shown in Fig. 7. Figure 9 expresses the beach profiles along the section A, where the influence of the existence of the headlands is minimum because it is located in the middle of No. 6 and No. 7 headlands. Since this section is located far from the structure, the characteristics of topographic changes due to cross-shore sediment transport can be investigated in detail.

First, from September 8 to December 24, 1986 the foreshore was eroded to form a bar-trough topography off the coast. According to Fig. 6, very high waves beyond $H_{1/3}$ = 5m attacked the coast on October 9, so that the bar-trough topography might be formed due to the offshore sand movement. Thereafter the beach had not changed so greatly untill April 14, but changed remarkably between April 14 and July 8; offshore bar receded as well as sand accumulated on the foreshore. This topographic change has a reverse mode against the pattern occurred between September 8 and December 24. Furthermore the critical depth to which the erosion or accretion occurs near the foreshore agrees well with about 3m in both cases. During the period between April and July mainly calm sea conditions continued with the significant wave height less than about 2m except on June 19 with the height of $H_{1/3}$ = 4.3m. Considering these calm wave conditions, it may be concluded that the profile change illustrated in the figure was caused by the onshore sand transport. After that, the whole beach profiles tend to retreat slightly until November 6.

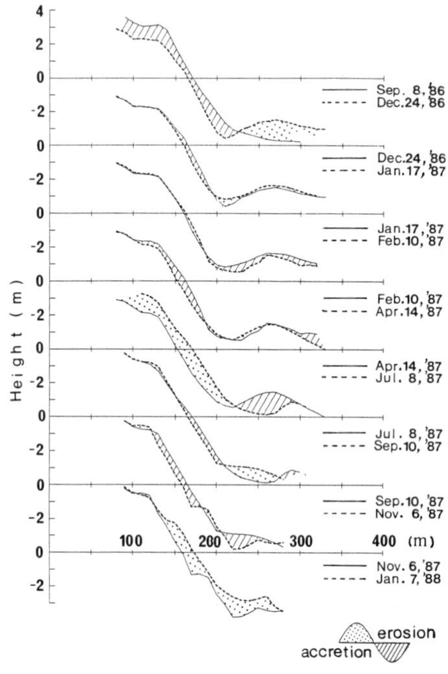

Fig. 9 Temporal profile change along section A.

Comparing the profile on September 8, 1986 with the one on September 10, 1987, both shoreline positions and beach profiles are almost the same each other and therefore it is found that the eroded beach retrieved into the previous state after one year. It should be noted that on this

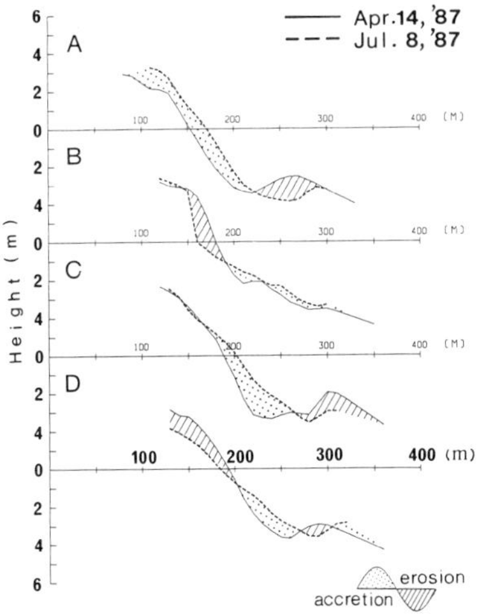

Fig. 10 Spatial and temporal profile changes along section A through D measured between April and July in 1987.

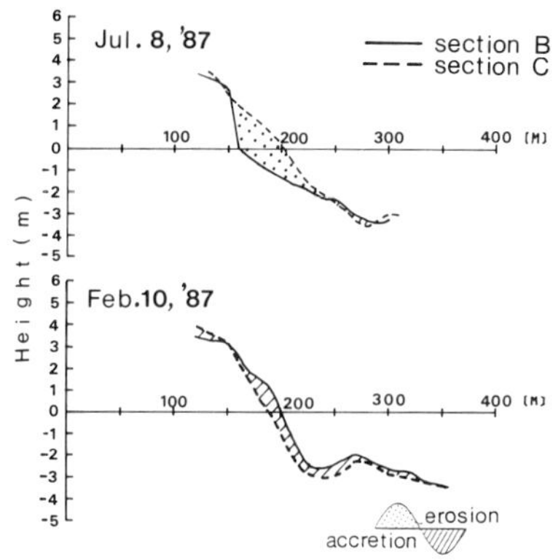

Fig. 11 Comparison of beach profiles on both sides of No.6 headland between February and July in 1987.

coast the following two types of sand movement can be observed; the variation of the longshore sediment transport caused by the periodic changes of wave direction and the change of cross-shore sediment transport due to wave climate change.

As shown in Fig.7, the shoreline on the north side of the structure retreated between April 13 and July 23, 1987. The typical profile changes in this period are illustrated in Fig. 10. On the section C just south of the headland of No. 6, sand accumulated near the shoreline to form the gentle slope of the foreshore, and in addition the accreted area extends to the depth of about 3m. On the contrary, on the section B located in the symmetric position with respect to the headland, the area shallower than the water depth of 1m was eroded to form a steep slope of the foreshore, whereas beach slope of the offshore zone becomes mild. These accretion-erosion pattern of topographic changes having opposite features on both sides of the headlands indicate that the beach changes are caused mainly by the longshore transport. On the other hand, on the section D, 375m south from No. 6 headland, the foreshore was eroded to form a gentle slope similar to the section C.

In order to examine the extent of the obstruction of longshore sand transport by the headlands, it is preferable to compare the spatial profiles on both sides of the structure rather than to examine the temporal changes of profiles as shown in Fig. 10. Because the influences of the structure to surrounding topography can be well understood by superimposing each others, since the two sections next to the headland should have the same profile before the construction of the headlands.

Figure 11 represents the superimposed profiles the sections B and C measured on February 10 and July 8, 1987. The former date is in the period when the southward longshore sediment prevails as shown in Fig. 7. The shoreline position on the section B north of the headland advances more than that on the section C, and the sand obviously accumulates in the region shallower than the depth of 3m deep, but at further offshore zone both profiles nearly coincide with each other. Taking the tip depth of about 2m into account, the above accretion shows that the effect of the headland extends to the region somewhere offshore from the tip of headland.

The similar comparison will be made with regard to the profiles on July 8, when the northward longshore transport was predominant. The shoreline position on the section C south of the structure is located further offshore than that on the section B, resulting in the quite opposite to the case of February 10. The foreshore slope on the section B is steeper than that on the section C, and also the region shallower than 2m deep is eroded, but almost coincides with the section C in further offshore zone.

Consequently, it is concluded that the headlands on this coast are effective for controlling sediment movement in a zone shallower than 2m because the tip of the headland is about 2m, although sediment may pass through around the tip of the headland. Nevertheless, it can be said that this headland is useful at present stage in reducing the longshore sediment transport rate.

VI. CONCLUSIONS

(1) Eleven headlands were constructed in order to prevent further extention of beach erosion and to form the dynamically stable beach on the Ohno•Kashima Beach. The variation of the shoreline position can be reduced by the effect of the construction of these headlands in the field where seasonal change of the wave direction dominates.

(2) The periodic changes of the shoreline configuration corresponding to seasonal change of the wave direction have been observed along the other coast facing Pacific Ocean in Ibaraki Prefecture such as the Joban Coast or the Ajigaura Beach. From these facts, the characteristics mentioned above may be considered as a typical example of the beach changes along the Eastern Coast of Japan.

(3) The depth at the tip of the headland is about 2m. Comparing beach profiles on both sides of headland, there exist differences in the region shallower than about 3m. Therefore the region where longshore sediment transport is remarkably obstructed by the

headlands is thought to be the region up to a location of about 3m deep.

ACKNOWLEDGEMENT

The authors wish to express their gratitude to Ibaraki Prefectual Sanpin Port Office with offering the wave data.

REFERENCES

Silvester, R.(1976): Headland defense of coasts, Proc. 15th Coastal Eng. Conf., ASCE, pp.1394-1406.

Tsuchiya, T.(1986): Beach erosion control, Coast, Vol.24, pp.100-109. (in Japanese)

Uda, T., M. Sumiya and Y. Kobayashi (1986 a): Beach change along the Ibaraki Coast, Japan, Trans. Japan. Geomorphological Union, Vol.7, No.3, pp.141-163. (in Japanese)

Uda, T., M. Sumiya and Y. Kobayashi (1986 b): Analysis of beach erosion around large-scale coastal structures, Proc. 20th Coastal Eng. Conf. ASCE, pp.2329-2343.

Uda, T. and K. Yamamoto (1987): On the beach stabilization method of sandy beach, Civil Eng. Jour., Vol.29, No.9, pp.21-26. (in Japanese)

CHAPTER 208

Results from the Atlantic City, NJ Beach Nourishment Monitoring Program

R. M. Sorensen* F. ASCE
S. L. Douglass** M. ASCE
J. R. Weggel*** F. ASCE

Abstract

A beach nourishment project was carried out at Atlantic City, NJ during the summer of 1986. One million cubic yards of sand was dredged from adjacent Absecon Inlet and placed over a 9000 foot long section of the beach. The authors monitored the fill for a period of eighteen months, including two winters, after placement of the fill.

This paper briefly summarizes the fill project and monitoring program. Key results of the monitoring program are presented and compared with results of the monitoring program for previous nourishment projects at the site in 1963 and 1970. Differences in the behavior of the 1986 fill vis-a-vis the 1963 and 1970 fills are discussed and possible reasons for these differences are given.

Introduction

It is still difficult to fully predict the behavior of a planned beach nourishment project. Season to season variations in wave climate, the impact of existing structures and proposed structure modifications, and the frequency and mode of fill placement can have a significant impact on the behavior of beach fill. Monitoring of completed beach nourishment projects will hopefully increase our understanding of fill behavior and continue to improve our ability to design such projects. The 1986 beach nourishment project at Atlantic City, NJ was monitored and results of the monitoring program are compared with the behavior of earlier fill projects at Atlantic City.

Project Site

Atlantic City is situated on the northern third of Absecon Island, a barrier island located 40 miles north of the southern tip of the Jersey coast. It is separated from Brigantine by the jettied Absecon Inlet (Figure 1). The shoreline has a general northeast-

----

* Professor, Department of Civil Engineering, Lehigh University, Bethlehem, PA 18015

** Project Engineer, Pare Engineering Corporation, Pawtucket, RI 02860

*** Professor, Department of Civil Engineering, Drexel University, Philadelphia, PA 19104

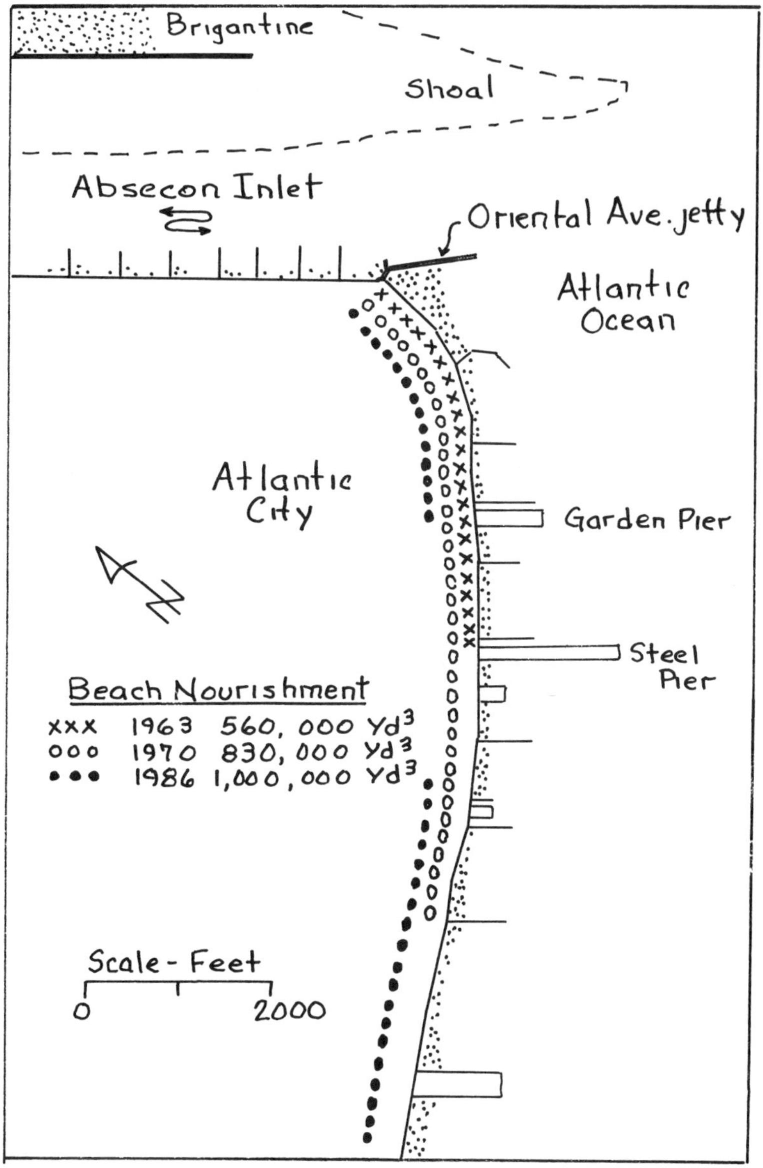

Figure 1 - Project site

southwest orientation.

The beach at Atlantic City is backed by wooden bulkheading and a boardwalk. There are five commercial piers which act as permeable groins as well as nine wood and stone groins varying from 100 ft to 800 ft (but typically 500-600 ft) in length. Beach berm widths vary from zero up to 500 feet (at Oriental Avenue jetty) out from the boardwalk. However, along several important beach segments, the berm width is typically fifty feet or less.

The U.S. Army Corps of Engineers (1974) estimates that there is a net southwestward transport of 150,000 cu.yd./yr. at Atlantic City (from a gross transport of 250,000 cu.yd./yr. northestward and 400,000 cu.yd./yr., southwestward). Visual wave observations indicate dominant southwestward transport throughout the year except for some months during the summer. The dominant southwestward transport drives the Absecon Inlet Channel up against the Atlantic City side of the inlet. And it causes a large shoal to form on the Brigantine side of the inlet and extending over a thousand yards seaward of the head of the Brigantine jetty (Figure 1).

Although there is a general net southwesterly longshore transport in the study area, there is a local reversal of transport on the Atlantic City shore near the inlet. Observations of beach offsets at the groins taken from historic air photos and from onsite observations at different times of the year (Sorensen and Weggel, 1985) showed a diverging nodel zone that is consistently located between Garden Pier and Steel Pier (Figure 1). The nodal zone was generally located around Garden Pier during the spring to fall months, but it shifted to the vicinity of Steel Pier during the winter to early spring months.

Beach Nourishment Projects

Major beach nourishment projects were carried out at Atlantic City in 1948, 1963 (February-May), 1970 (June-July) and 1986 (April-June). The fill volumes and locations for the 1963, 1970 and 1986 projects are shown on Figure 1. The gap in the 1986 fill placement was due to ongoing construction work in the vicinity of Steel Pier.

The borrow area for all of the fill projects was the shoal area on the north side of Absecon Inlet. Sand was hydraulically dredged from various points along the channel side of the shoal and carried to the fill areas by pipeline. The four fill projects were essentially bypassing operations.

1963, 1970 Nourishment Projects

As a direct outcome of the damaging "five high" storm in March 1962, the U.S. Army Corps of Engineers established a beach monitoring program at Atlantic City which lasted from October 1962 to May 1973. Beach profiles, to wading depth, were measured at seven locations (Figure 2) on approximately a monthly basis (more frequently in winter, less frequently in summer). Note that profiles 1-3 are in the 1963 fill area; profiles 1-4 are in the 1970 fill area. The profile data are presented by McCann (1981) and discussed by Everts et al. (1974) in relation to the 1963 and 1970 fill projects. Owing to the truncation of profiles at wading depths, changes discussed by Everts et al. (1974) refer only to the portion of the beach profile above MSL.

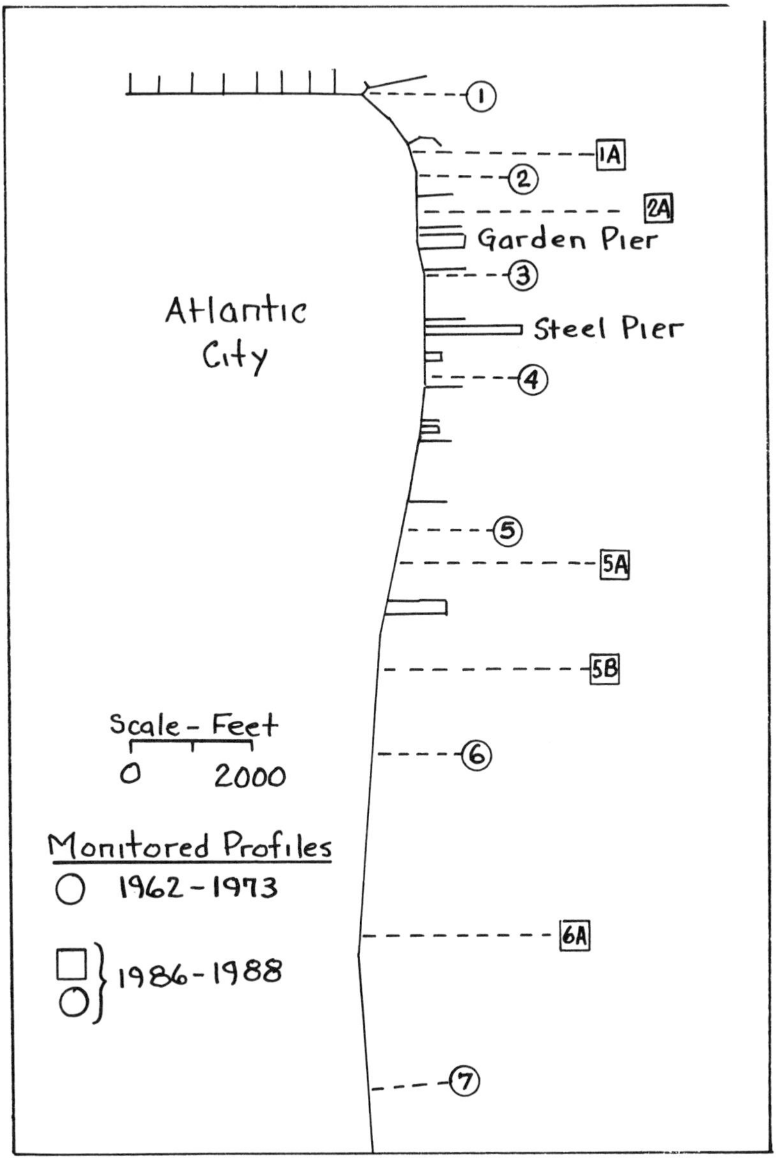

Figure 2 - Monitoring program profile locations

At profile 1, which is adjacent to the inlet, there was a rapid and constant loss of sand until the profile returned to its pre-fill volume. Ninety percent of the fill volume was lost in the first six months in 1963 and within eight months in 1970 (i.e. by December 1963 and February 1971 respectively). At profile 2, located about 1500 feet south of the inlet, there was an initial rapid loss similar to profile 1; but, when the volume loss reached seventy percent of the fill volume, the rate of loss dropped sharply to less than half the initial rate. Still, essentially all of the fill placed above MSL at profile 2 was removed during the first year of both fill projects. On profile 3, the loss rate was uniformly lower and large natural recoveries were observed during following summers. Four years passed before ninety percent of the 1963 fill volume above MSL was lost at profile 3. The profile lines south of the 1963 fill (lines 4-7) showed initial net erosion but within a period of one to three years after the fill, they showed net accumulation.

The trend in beach sediment losses at the successive profiles (1 to 7) indicated that sand moved alongshore to the southwest from the inlet (as well as having a net movement offshore). This was consistent with wave observations (1962-1967) which showed that the mean direction of wave approach during all months (except April when there was a reversal) was from the northeast.

Based on their evaluation of the profile data for the two fill projects, Everts, et al. (1974) offered the following interpretations and recommendations:

1) Predominant offshore loss of sand appears to occur between September and March, so fill should be placed in the spring for optimum residence time for the fill. Also, the larger the volume of fill placed at a given location, the faster it is lost; smaller volumes of fill placed more frequently would increase total fill residence time.

2) Fill should be placed in a localized feeder beach adjacent to the inlet (vicinity of profile 1) - because fill from that area moves along the shore to the southwest to nourish downdrift beaches and because that area requires the shortest pumping distance for fill material taken from the inlet.

3) Placement of fill in a localized feeder beach adjacent to the inlet should be done after April, when longshore current reversals are at a minimum and all of the fill material that moves alongshore can be expected to nourish beaches to the southwest.

Shore Structure Modifications, 1983-1984

During 1983 and 1984 four of the shore structures at Atlantic City were modified. These modifications, which are summarized in Figure 3, appear to have had an impact on the behavior of the 1986 fill project vis-a-vis the 1963 and 1970 fill projects.

The Oriental Avenue jetty, which had an original seaward crest elevation of +7 feet MLW and which had deteriorated some since its construction during 1946-1961 was rehabilitated with a raised crest elevation of +11 feet MLW. The first groin to the south was extended 200 feet (to yield a C rather than just a dog-leg alignment) and

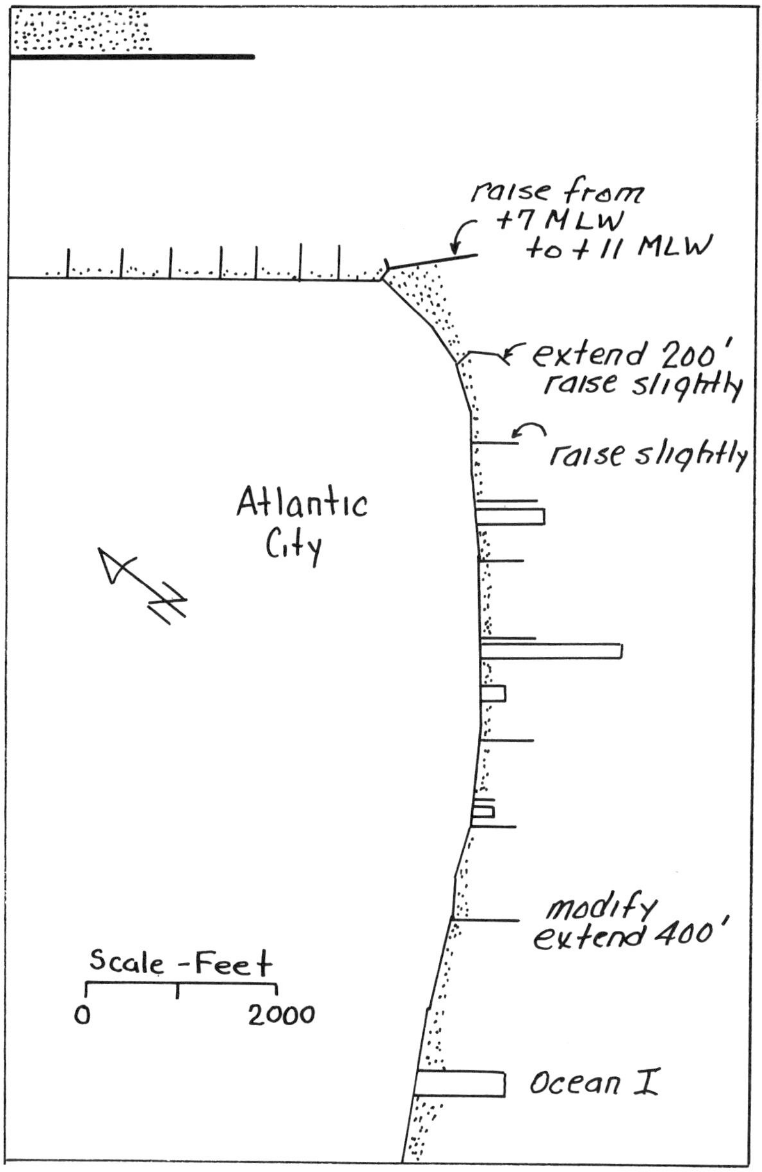

Figure 3 - Structure modifications: 1983 - 1984

raised slightly; and the second groin was raised slightly. A groin further to the southwest, which had been a timber groin, was reconstructed 400 feet longer with a stone seaward section. Also, in the early eighties, the Million Dollar Pier was removed and replaced with a new structure - Ocean I. Ocean I has a high density of relatively small support piles and, consequently, acts as a permeable groin.

Monitoring Program, 1986 Nourishment Project

The 1986 beach nourishment project was funded by the State of New Jersey and conducted through the New Jersey Department of Environmental Protection. NJDEP contracted with the authors to monitor the project for a year and a half period following completion of the fill. The monitoring program ran from August 1986 to March 1988, and included the collection of sand samples, visual wave observations, and beach profile data.

Sand samples were taken from the dredge discharge line and from the beach face during and after the nourishment operation. LEO visual wave observations (Schneider, 1981) were taken at a site in the fill area and at a site southwest of the fill area. Beach profiles (to wading depth) were taken at the seven profile locations used by the Corps of Engineers during 1962 and 1973, plus an additional five profile lines added by the authors. The locations of these twelve profile lines are shown in Figure 3. Profiles were taken on approximately a monthly basis from August 1986 to March 1988.

An NJDEP nearshore hydrographic survey was conducted in December 1985, approximately four months prior to the start of the nourishment project. Data from this survey do not cover all of the twelve profile lines, but data were taken sufficiently close to profile lines 1 through 6 to define the profile depths at these locations. Since the NJDEP survey was conducted four months prior to the start of the nourishment project, an allowance for this four month interval must be made when looking at the profile data.

1986 Monitoring Program Results/Discussion

The beach profile data were analyzed to determine sequential volumetric changes between profiles. Key results are summarized in the Table where volumetric changes are given in cubic yards per yard of beach and in percent change from the volume tabulated in column two.

The second column in the Table gives the profile volume changes between the December 1985 NJDEP survey and the end of the fill in August 1986, when the first monitoring program profiles were obtained. This column thus gives the fill volume plus changes in the beach profile during the previous four months. Fill was placed at profiles 1, 1A, 2, 2A, 5 and 5A; and some of the fill may have been placed in the vicinity of profile 3. Thus, the accumulation of between 40 and 65 cu.yd./yd. at profiles 4, 5B and 6 gives an indication of the natural change from December 1985 to August 1986. The accumulations at profiles 1, 1A, 2, 2A, 5 and 5A should be reduced by about that amount to give a truer indication of the fill placed at each profile line. This yields values of between 400 and 500 cu.yd./yd.

The third column gives the volume changes during the first year after fill placement (and the percentage of the second column). Note that both gains and losses occurred at the unfilled profiles. The

Table - Profile Unit Volume Changes - Yd^3/Yd and (%)

Profile No.	Fill* (12/85 - 8/86)	Change Yr 1 (8/86 - 8/87)	Change Yr 2 (8/87 - 3/88)	Σ (%)
1	465.0	-39.0 (-8.4)	-105.0 (-22.6)	(-31.0)
1A	460.0	-200.0 (-43.5)	-20.0 (-4.3)	(-47.8)
2	505.0	-200.0 (-39.6)	-5.0 (-1.0)	(-40.6)
2A	437.2	-81.5 (-18.6)	-51.3 (-11.7)	(-30.3)
3	162.9	-21.4 (-13.1)	-55.6 (-34.1)	(-47.2)
4	43.2	+54.4 (+125.9)	-18.8 (-43.5)	(+82.4)
5	561.2	-72.4 (-12.9)	-30.0 (-5.3)	(-18.2)
5A	567.3	-95.4 (-16.8)	-40.9 (-7.2)	(24.0)
5B	65.7	-15.0 (-22.8)	-7.5 (-11.4)	(-34.2)
6	50.0	-84.3 (-168.6)	-51.3 (-102.6)	(-271.2)
6A	-	+24.5 -	-21.8 -	
7	-	+62.2	-72.2 -	

* Includes any natural profile changes between 12/85 and end of fill project in 8/86.

fourth column gives the volume change for the remainder of the monitoring program - essentially these values represent changes that occurred during the second winter season after fill placement. The final column gives the percent change during the monitoring program compared to the approximate fill volume (in the second column).

The major difference in the behavior of the 1986 fill and the 1963 and 1970 fills is the greatly reduced rate of loss of material above MSL, north of the nodal point. Profiles 1 through 2A lost a total of 30 to 40 percent of their volume after two winters compared to a loss of essentially all of their volume during the first winter during 1963 and 1970. Four possible reasons for the improved behavior of the fill in this area are: improved quality of sand fill, less fill placed at a given location (less berm extension and beach face steepening), milder wave climate, and the structure modifications. Each is discussed below.

Pipeline outfall sand samples showed very little difference in median particle diameter from the contemporary native sand on the beach face. Earlier studies (McMaster, 1954, Ramsey and Galvin, 1977 and Sorensen and Weggel, 1985) showed that there is very little difference in the range of sand sizes at the inlet borrow area and on the beach face.

Making allowances for the nonfill changes included in the tabulated values in column 2 of the Table and for differences in the profile volume analyses techniques for the 1963/1970 fills and the 1986 fill, it appears that the volume of fill placed at profile lines 1 and 2 was approximately the same during all three nourishment projects. Comparison of the volumes of fill and the lengths of beach filled for the three projects (see Figure 1) confirms that the volumes of fill per unit length of beach were about the same.

It appears that differences in the wave climate partially account for the observed differences in beach fill behavior. No uniform set of wave data is available for the year following the 1963 and 1970 fills and the two years following the 1986 fill, from which a definite comparison can be made. Hindcast wave data (Jensen, 1983) are available for Atlantic City for the period 1956-1975. From these data, the winters of 1963-1964 and 1970-1971 can be compared to the average 1956-1975 data to see how relatively mild or severe they were. This can then be compared with the experience of the authors and others active along the New Jersey shore during 1986-1988.

Comparing mean monthly significant wave heights for September through April in 1963-1964, 1970-1971, and 1956-1975 (Figure 4) indicates that the winters following the 1963 and 1970 beach fills were essentially average for the twenty year period. Also shown on Figure 4 are the average significant wave heights for March and April of 1962 when the major beach scour that instigated the 1963 fill occurred. These monthly averages are significantly higher than any of the monthly averages in 1963-1964 or 1970-1971. The 1986-1987 winter had two big storms that were coincident with unusually high astronomical tides. Overall, however, this winter was average or slightly milder than average. The winter of 1987-1988 had essentially no major storms until April which was just after the last beach monitoring profiles were taken. Comparison of the first and second year changes in the Table confirms the comparitively milder conditions in 1987-1988 versus 1986-1987.

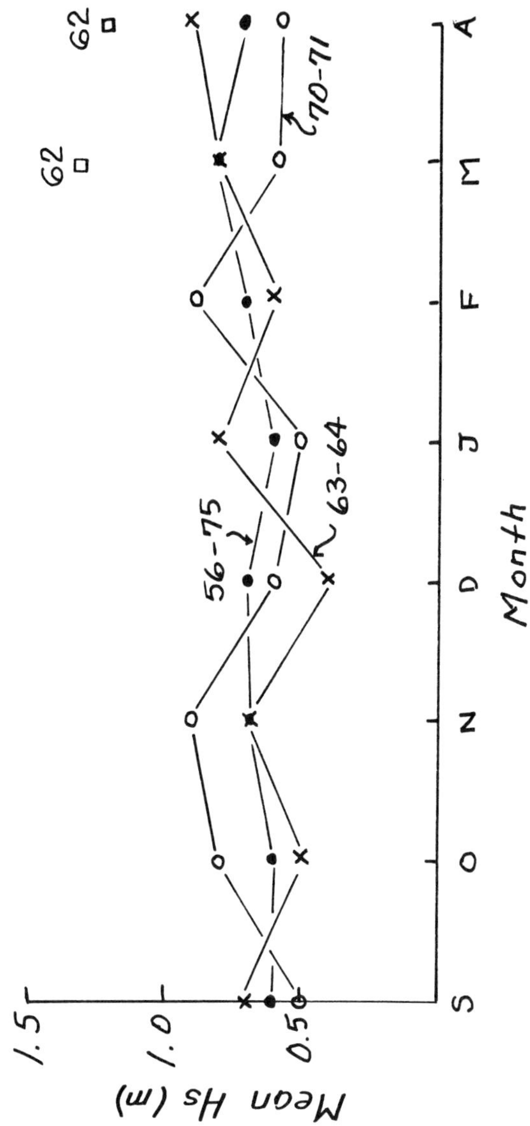

Figure 4 - Atlantic City wave hindcast data

Thus, it appears that the winters following the 1986 fill were somewhat milder than those following the 1963 and 1970 fills. However, there is another cause for the improved behavior of the 1986 fill versus the fills in 1963 and 1970, in the vicinity of profiles 1 and 2. That is the structure improvements made in 1983-1984.

Although Everts et al. (1974) suggest that beach fill placed between the Oriental Avenue jetty and Garden Pier (see Figure 1) moves to the southwest (and offshore), the observed nodal zone position suggests otherwise. Much of the fill place in this region in 1963 and 1970 probably moved toward the northeast, over and around the jetty, and into the inlet. The 1983-1984 structure modifications (Figure 3) limited this movement and subsequent loss. Everts et al. (1974) also suggest that after April dominant transport each month is to the southwest. However, LEO observations made in 1987 showed northeastward littoral transport to dominate in June.

The fill placed at profiles 5 and 5A suffered an immediate and consistent but small rate of loss. The rate of loss was about half that which occurred at profiles 1 through 2A. In 1963, no fill was placed at profile 5 which showed little net change after the first year and then showed a net accumulation (over the condition at the time of fill) for the next several years - likely because of the transport of sand from upcoast. Compared to the post-1986 fill behavior at profiles 1 through 2A and to the post-1963 behavior at profile 5, the post-1986 fill behavior at profiles 5 and 5A was excellent. Again, the probable causes for this behavior were the milder wave climate and the structure modifications.

Summary and Conclusions

Overall, the 1986 beach nourishment project at Atlantic City has been successful to date. Comparison of beach fill behavior at the six fill profile lines during the first and second winters after the fill (see Table) shows the importance of the milder wave climate during the second winter. Part of the reduced rate of loss is due to the typical exponential loss rate that occurs for beach fills; however, most of the difference is likely due to the milder wave climate.

Comparison of the behavior of the 1963 fill and the 1986 fill (given the differences in wave climate) demonstrates the improvement in fill behavior owing the improvement of some of the shore structures, which reduced the loss of sand alongshore to the inlet and to the southwest. The jetty, groins and piers do significantly improve the behavior of any beach fill placed at Atlantic City.

As Everts et al. (1974) recommend, future fills should be placed as often as possible rather than in larger volumes at longer intervals. Beach nourishment at Atlantic City should be thought of as a sand bypassing operation. The shore stabilization structures should not be allowed to deteriorate, and the fill should be placed over the entire beach length northeast and southwest of the nodal zone.

References

Everts, C. H., A. E. DeWall and M. T. Czerniak (1974), "Behavior of Beach Fill at Atlantic City, New Jersey" Proceedings, 14th International Conference on Coastal Engineering, Copenhagen, Denmark, June 1974.

Jensen, R. E. (1983), "Atlantic Coast Hindcast, Shallow-Water Significant Wave Information", U.S. Army Waterways Experiment Station, WIS Report 9, January 1983.

McCann, D. P. (1981), "Beach Changes at Atlantic City, New Jersey (1962-73)" Misc. Report 81-3, U.S. Army Coastal Engineering Research Center, Ft. Belvoir, VA, March 1981.

McMaster, R. L. (1954), "Petrography and Genesis of the New Jersey Beach Sands", Bulletin 63, Geologic Series, New Jersey Department of Conservation and Economic Development, Trenton, 1954.

Ramsey, M. D. and C. J. Galvin (1977), "Size Analysis of Sand Samples from Southern New Jersey Beaches", Miscellaneous Report No. 77-3, U.S. Army Coastal Engineering Research Center, Ft. Belvoir, VA, March, 1977.

Schneider, C. (1981), "The Littoral Environment Observation (LEO) Data Collection Program", CETA 81-S, U.S. Army Coastal Engineering Research Center, Ft. Belvoir, VA, March 1981.

Sorensen, R. M. and J. R. Weggel (1985), "Evaluation of Beach, Behavior and Coastal Structure Effects at Atlantic City, NJ" Fritz Engineering Lab Report, Lehigh University, April 1985.

U.S. Army Corps of Engineers (1974), "New Jersey Coastal Inlets and Beaches, Barnegat Inlet to Longport", Interim Report, Philadelphia District, September 1974.

CHAPTER 209

BERM BREAKWATER FAILURE AT ST. PAUL HARBOR, ALASKA

Javier Weckmann,[1] Associate Member, ASCE
George M. Watts,[2] Fellow Member, ASCE

1.0 INTRODUCTION

St. Paul Island, Alaska, is located at 50°10'N latitude and 170°15'W longitude in the south central Bering Sea. It is the most northward and largest island of the Pribilof Island group. The area of the island is about 70 square miles (180 square kilometers), with the city and harbor of St. Paul located at a cove (Village Cove) on the southern coastline. The Pribilof Islands are of volcanic origin and are generally hilly with much of the coastline consisting of precipitous rocky cliffs. Moderate to strong winds are characteristic throughout the year, causing the island to be treeless. It is predominantly covered with grasses, sedges, and wild flowers.

The Pribilofs are a natural haven for a variety of flora and fauna. More than a quarter of a million seabirds nest each year along the coastal cliffs. About two-thirds of the world's population of northern fur seals migrate annually to the Pribilofs for mating purposes. The Pribilof Island area of the Bering Sea is also one of the most abundant and richest seafood grounds in the world. Due to a recent moratorium, the harvest of fur seals in the Pribilofs has been discontinued. In order to maintain existing cultural and economic resources, the City of St. Paul has elected to construct a harbor facility at Village Cove to provide services to commercial fishing vessels operating in the central Bering Sea. The maximum natural water depth in the Village Cove area is 26 feet (7.9m) relative to mean lower low tide datum (MLLW=0.0). Mean higher high tide level is 3.2 feet (1m) above MLLW, with extreme high tide during storm periods being estimated at between 5.0 and 6.0 feet (1.5 to 1.8m) above MLLW. Waves approaching from the southwest sector have the most effect on St. Paul Harbor. During the winter months, breaking waves with heights of 25 feet (7.6m) and 13-16s periods can be expected at Village Cove several times each year.

---

(1) Senior Engineer, Tetra Tech, Inc., 630 N. Rosemead Blvd., Pasadena, CA, 91107, U.S.A.
(2) Director, Tetra Tech, Inc., 1911 Fort Myer Dr., Arlington, VA, 22209, U.S.A.

2.0 1984 BREAKWATER CONSTRUCTION

2.1 ORIGINAL DESIGN

The St. Paul Harbor was to be constructed in Village Cove using a shore-connected breakwater commencing at the base of Village Hill, and extending approximately 2,000 feet (600m) in a north-northwest direction. The original design of the rubblemound breakwater followed a conventional 3 layer system. This consisted of a quarry stone core protected by 2 layers of 8-17 ton stone (Class II). The seaward slope of this structure would have been 1 vertical to 2.5 horizontal and have a crest elevation of +30 feet (9m) MLLW. The head section was to be constructed using a heavier armor layer of 17-24 ton stone (Class I) with a slope of 1:3. A typical cross-section is illustrated in Figure 1.

Due to a low and insufficient production of Class I and II armor stone at the selected island quarry sites, the contractor and engineer agreed to redesign the breakwater to better suit the quarry stone production. The modified design was then based on the berm breakwater concept, hereby referenced to as the "A-B" design. In theory, this approach to breakwater design would have maximized the use of all quarry stone by minimizing stone by-product, and therefore resulting in a more economical unit stone production cost. The breakwater was then completed to a length of approximately 870 feet (260m) by early October of 1984 using the "A-B" design.

2.2 MODIFIED BREAKWATER DESIGN ("A-B" DESIGN)

The "A-B" design called for an outer layer of "A" stone which ranged from 0.75 tons to 8 tons with a median stone size of 1.5 tons. A 60-foot (18.2m) wide berm was placed on the seaward side. The core material ("B" stone) also composed an outer berm section, with a gradation similar to the original Class V core stone. The crest height of the "A-B" breakwater was +28 feet (8.5m) MLLW (see Figure 1). This design was based on 2-D tests conducted at the Danish Hydraulic Institute (DHI). The tests used wind, wave, and storm parameters developed in a Pribilof Island wave study prepared by DHI in June 1982.

The breakwater was completed with a temporary head section consisting of "A" and "B" stone only. No armor stone of the Class I or II type was placed on the head as an armor layer. It was assumed at that time that construction would continue the following year.

2.3 BREAKWATER PERFORMANCE WITH "A-B" DESIGN

Substantial damage to the breakwater resulted from storms occurring on 13 November and 7 December 1984. A hindcast analysis showed that the first storm produced a deepwater significant

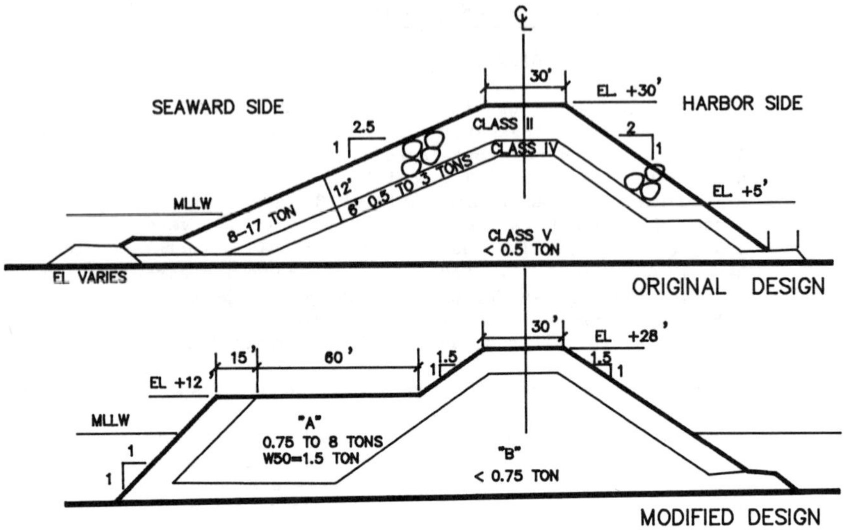

FIGURE 1 BREAKWATER CROSS SECTIONS

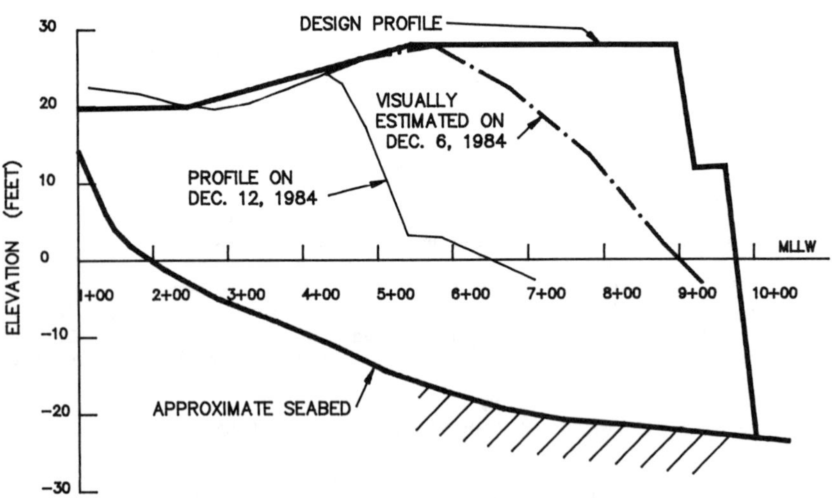

FIGURE 2 COMPARISON OF CENTERLINE PROFILES

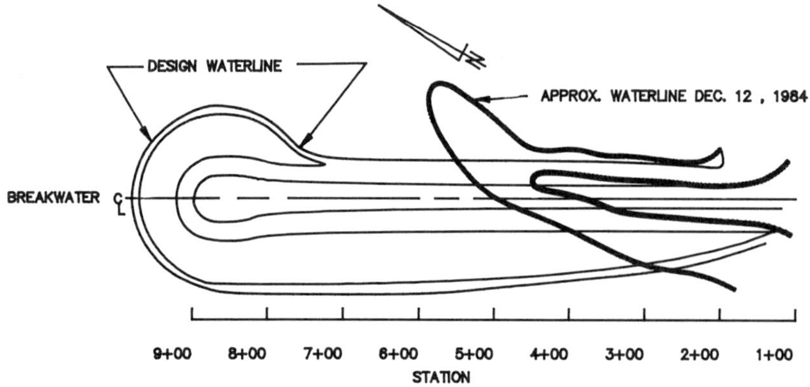

FIGURE 3 BREAKWATER PLANVIEW

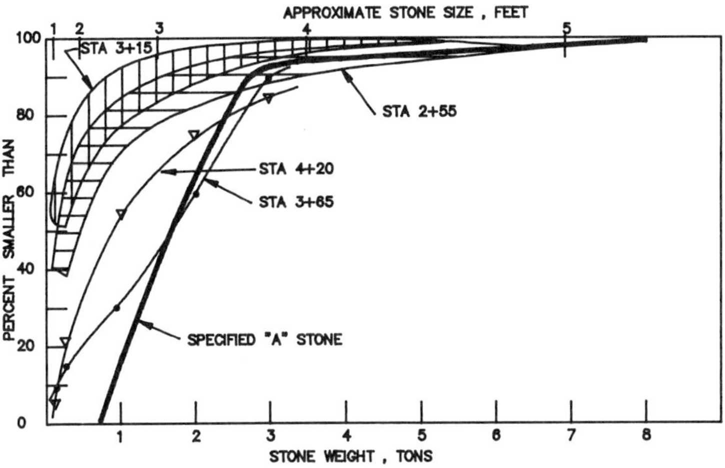

FIGURE 4 ESTIMATED GRADATION OF "A" STONE

wave height, H_s, of 30 feet (9.1m) with peak periods of 16 seconds, and the second storm had an H_s of 22 feet (6.7m) with 13 second peak periods. Storms of these magnitudes can be expected to occur several timess a year in the Bering Sea. Tide levels were estimated at +3.5 to +4.0 feet (1.1 to 1.2m) MLLW for both storms. The 13 November storm resulted in considerable redistribution of the "A" stone along the entire structure length. Approximately 300 feet (91m) of the original crest and 100 feet (30m) at the waterline were lost. Data gathered after the 7 December storm indicated that an additional 200 feet (60m) of crest and 250 feet (75m) of waterline were lost, for a total damage of 500 feet (150m) of crest and 350 feet (100m) at the waterline. A comparison of centerline profiles before and after the storms is shown in Figure 2. It was evident that material at the head section was transported into the harbor area and formed a low, wide underwater mound east of the initial centerline, producing a reef type structures. Plan views of the before and after structure are shown in Figure 3.

Independent gradation estimates were made at various locations along the breakwater above the waterline. The results of this work showed that the in-place "A" stone outer layer was generally finer than the specified gradation for "A" stone (see Figure 4). Some signs of slumping on the harbor side south of STA 4+50 slope were also visible.

2.4 DISCUSSION OF DAMAGE

During the field investigation, slope measurements indicated substantial adjustment of the outer "A" stone layer, similar to that experienced in the DHI 2-D tests. However, north of STA 4+50 and with respect to the construction of the head section, the excessive damage was due to the fact that the "A" stone size used was inadequate for the design wave conditions of 20-25 foot (6-7.6m) breaking waves with periods ranging from 13-16 seconds. At the head, the direction of wave attack will be at an angle which will cause displaced stones to travel laterally and into the harbor and entrance channel. The head then receded until the water depth limited the wave height to less than 10 feet (3m), where a 1.5-2 ton stone on a 1:5 slope can be stable.

Breaching of the breakwater trunk was also witnessed by City of St. Paul officials during the 13 November 1984 storm. The reduced porosity of an extremely well-graded material, such as "A" stone in combination with "B" stone, probably increased the run-up potential of the design wave conditions. This, in turn, resulted in excessive overtopping, and finally breaching of the "A-B" breakwater.

3.0 BERM BREAKWTER DESIGN CONSIDERATIONS

The investigation on the construction of this berm breakwater, as well as the events and circumstances that culminated in its completion, have raised various items of concern when designing and constructing these types of structures. Several of these items are presented below.

1. The procedures for production and inspection of armor stone for any rubblemound type structure are extremely important. It may be more difficult to determine if an armor stone class with a wide gradation meets design specifications, as compared to conventional armor stone with a narrow gradation. The performance of a berm breakwater, as well as the potential degree of damage, would also have to be assessed if the median stone size in the berm armor stone is less than that specified. Does a small reduction in median stone size or skewed gradation result in a disproportionate degree of damage under design conditions?

2. The determination of potential long shore movement of material along a berm type structure should be addressed in order to evaluate its long-term maintenance needs, and consequently estimate the annualized maintenance costs. A design with a significant savings in capital costs may not be the most economical if the maintenance costs are excessive.

3. Due to the inherit capability for the seaward slope of berm type structures to adjust in direct relation to the impinging sea state, some guidelines need to be established for the definition and assessment of potential damage.

These are a few of the items that need to be addressed by the engineer and planner when evaluating berm type breakwaters.

4.0 BREAKWATER - 1985 TO PRESENT

Following the "A-B" breakwater damage in 1984, the Alaska Department of Transportation hired Tetra Tech, Inc. to develop a breakwater damage assessment. The City then retained Tetra Tech, Inc. to provide assistance in re-designing the breakwater, designing a 200-foot (60m) length dock, preparing plans and specifications, and advertising and awarding construction contracts. A 25-foot (7.6m) breaking wave was selected for the re-design, which required armor stone of 14-ton on the breakwater trunk section and 18-ton on the head section. Seaside trunk slope was 1:2.5 and 1:3 for the head. This redesign also incorporated the criteria of near-zero percent annualized maintenance (minimum 50 year design life). Since the 200-foot (60m) dock, and future extension thereof to 1000 feet (300m) in length, would be positioned along the harbor side of the breakwater, the breakwater crest elevation was designed for no wave over-topping and established at +37 feet (10.1m) MLLW. The 200-foot (60m) long by 40 feet (12m) wide dock was a pre-stressed, pre-cast concrete caisson design. The caisson was constructed in Tacoma, Washington, towed to St. Paul Island, Alaska and placed on a specially constructed foundation. During the period May 1985 to January 1986, the breakwater was constructed. The concrete caisson/dock was installed, and a 200-foot (60m) wide, 300-foot (90m) long channel dredged in the summer of 1986.

The breakwater with armored head has been subjected to the design wave a number of times during the 1985, 1986 and 1987 winter seasons. Visual inspections indicate no armor stone displacement along the trunk or head of the breakwater. The dock system has been utilized extensively for vessel off-loading of cargos which previously had to be lightered. St. Paul Harbor is a Federally Authorized Project and since December 1986, the U.S. Army Corps of Engineers (Alaska District) and the City of St. Paul have been carrying out further studies of the existing design for the St. Paul Harbor Project. This has included 2 and 3 dimensional hydraulic model studies, and complete re-evaluation of economic benefits for the project. At this time, findings of the hydraulic model tests indicate the seaward slope of the breakwater can be steepened to 1:2 using 18-ton armor stone. Presently it is planned to construct additional navigation features in 1989. This includes the extension of the breakwater length to Station 18+00, adding 700 feet (210m) of dock, constructing a 1000-foot (300m) second detached breakwater and final excavation of a mooring basin.

CHAPTER 210

COASTAL GENESIS

J.H. de Vroeg *, E.S.P. Smit *, W.T. Bakker **

ABSTRACT

In order to investigate the effect of enchanced sea level rising (greenhouse effect), of sand mining in the North sea and the large-scale effects of beach nourishment, the Rijkswaterstaat (Governmental Board for Ways, Waterways and Harbours) started the project "Coastal Genesis". This paper deals about research, carried out at the Delft University of Technology in the scope of this project.
The behaviour of the Holland coast at three timescales, i.e. 15 years, 1000 years and 6000 years is considered. Use is made of line models. New techniques are developed for finding the coastal constants of these models (the use of Kalman filtering) and for simulating the effect of sea level rise on coastal accretion and erosion. An extensive set of data on coastal topography, the JARKUS dataset is analyzed.
It is concluded that the effect of cross-shore transport becomes more and more dominant compared to the longshore transport when the time scale increases. A physical conception is displayed, which explains more or less the accretion of the Dutch coast between 4000 Before Christ and the Roman times, and the erosion later on.

I. INTRODUCTION

The coastal management of Rijkswaterstaat requires a strategy of overall coastal defence, starting from the general idea that the Netherlands should be kept in shape, however not against unlimited expense. According to a statement of the Dutch queen "the coast is safe" since the stormsurge barrier in the Eastern Scheldt came in operation. What should be done to keep it safe and how should aspects of increasing importance of nature preservation and conservation of areas needed for water purification and recreation be encountered? How to cope with enchanced sea level rise?
Good predictions are necessary. Understanding of how the present was generated from the past might be of help, and for this goal Dutch geologists, geographers and civil engineers of Rijkswaterstaat, Delft Hydraulics Laboratory and a number of universities are collaborating in the project "Coastal Genesis". For a further description may be referred to Vellinga and Zitman (1988).
This paper deals with research at the Delft University of Technology carried out in the scope of this project. The behaviour of the

*) Researchers,Faculty of Civil Engineering,Delft University of Technology,The Netherlands.
**) Principal scientific officer,Faculty of Civil Engineering, Delft University of Technology,and coastal specialist Rijks-waterstaat (Public Works Department),the Hague,The Netherlands.

uninterrupted central part of the Dutch coast in 6000 years is considered. Although "the" truth never will be found, it will be seen, that the effect of onshore-offshore transport dominates more and more the effect of longshore transport when larger time scales are considered. In this paper the general conception is captured by the use of line models. Constants occurring in these models are found from hindcasting coastal behaviour. Another paper (Bakker et al.,1988) also gives the relations between these constants and the wave characteristics. Ch.II gives a description of the Dutch coast in the last 15 years and ch.III describes a newly developed way of finding the coastal constants. In ch.IV a one-line model concerning the behaviour in the last 1000 years is given, leading to constants, comparable with the ones for the last 15 years. In ch.V the qualitative effects of sea level rise are modelled (only considering cross-shore transport) and finally ch.VI contains some conclusive remarks.

II. 15 YEARS

Coastal behaviour during the last two decennia can be investigated rather accurately due to the Jarkus dataset, which contains yearly measurements of coastal profiles along the entire dutch coast, up to a depth of 5 a 6 m below MSL. The distance between the ranges is 200 m. A more detailed description is given by Bakker and de Vroeg (1987,1988[a,b]).
This dataset can be used to study the bottom topography and changes of the coastal profile. Also, the measurements provide a useful source for verification and calibration of coastal models.

First the bottom topography is considered. Fig 9[a] shows the situation. Fig 1[a] shows a three-dimensional picture of the coastal area between Den Helder and IJmuiden (see also fig 9[a]), as measured in the year 1984. The scales are distorted: the distance between Den Helder and IJmuiden is 55 km, while the width of the considered coastal zone is 800 m. The landward boundary is chosen near the dune foot. In the north, at Den Helder, the influence of the Texel inlet is visible. At Petten a large seawall prevents the coast from moving landward. Due to a structural regression of this part of the coast the wall is now situated seaward compared to the rest of this region. At IJmuiden there is a physical boundary, formed by large harbour moles.
The picture clearly shows the coast is not uniformly shaped. Not only differences in the slope of the profile are visible, but also shape and number of breaker bars vary along the coast. It appears that a gentle beach slope results in a great number of bars. The same can be concluded from fig 1[b], which shows the coastal area south of IJmuiden over a distance of 59 km. (At the harbour of Scheveningen no measurements were available).
As a result of the breaker bars the coast is constantly transforming, which can be derived from fig 2[a,b,c]. These figures show "3-D-time diagrams" in which the changing of one coastal profile during 15 years is displayed by plotting the profile in successive years next to each other, thus resulting in three-dimensional figures, however, with the time instead of the distance along the coast as third dimension. These diagrams enable the investigation of the migration of breaker bars. It appears that the bars are moving in a seaward direction, however, with strongly varying velocities. In the north, between Den Helder and Callantsoog, the bars are very stable (fig 2[a]). Between Callantsoog and IJmuiden a seaward velocity of 15 a 20 m/year is found (fig 2[b]). South of IJmuiden the bars are moving very fast: 50 m/year (fig 2[c]). No correlation is found between the structural movement of the coast and the migration of the bars.

The changes of the profile due to the breaker bars can be expressed as a fluctuating cross-shore transport within the surfzone. In fig 3 this transport is compared to other sources of transverse transport. Without any doubt some interaction between the tidal zone and the inshore exists, however according to present estimates it is of no great importance where the behaviour of the beach on a time scale of decennia is concerned. On this time scale gradients in the wave induced longshore transport give a possible explanation of the development of the curved Dutch coastline (although not the only possibility). Fig 4 shows an estimate of the magnitude of this transport, based upon Bakker (1971) and Bakker and de Ras (1971). Some simple models based on this longshore transport exist, and will be discussed in this paper.

Though many of the observed characteristics are not yet fully understood, knowledge derived from this "phenomenological approach" can be made of use for statistical purposes. The 3-D-time diagrams in fig 2 show it is difficult to determine any structural movement of the profile. This structural movement is small compared to the fast movement of large amounts of sand due to the breaker bars. This is a problem when trying to use the Jarkus dataset for the calibration of models. One can treat the motions of the bars as a periodical feature or as a random noise, each treatment giving its specific problems (Bakker and de Vroeg (1987,1988[b]). Kalman filtering appears to be a tool for finding the vital characteristics.

III. THE LINEAR KALMAN FILTER

The Kalman filter combines the advantages of the deterministic and the statistical approach, and has proven to be very suitable for recursively updating parameter estimates or time-varying model coefficients, based on the latest data information. It is an algorithm which uses measurements, a physical model of the system, and knowledge of the inaccuracy (noise) of both the measurements and the model, in order to achieve an optimal estimate of system behaviour (Kalman,1969). The method is based on two linear equations:

System Equation : $\underline{x}(t) = F(t)\ \underline{x}(t-1) + \underline{w}(t)$ (1)

Measurement Equation : $\underline{z}(t) = H(t)\ \underline{x}(t) + \underline{v}(t)$ (2)

where $\underline{x}(t)$ is the system state at time t.
- $\underline{z}(t)$ is a measured state, not necessarilly the same as the system state, however, it has to be linearly related to the system state (in the linear filter).
- F(t) is a transformation matrix, transforming the system state at time t-1 to the state at time t.
- H(t) is a transformation matrix, describing the linear relation between the system state and the measured state.
- $\underline{w}(t)$ is the white noise, affecting the state $\underline{x}(t)$.
- $\underline{v}(t)$ is the white noise, affecting the observations $\underline{z}(t)$.

F(t) and/or H(t) contain the physics of the system. Both $\underline{w}(t)$ and $\underline{v}(t)$ refer to additional noise, generated during one time step.
Using the sytem equation it is possible to estimate the system state $\underline{x}(t-)$ one time step ahead (- denotes an estimate before the measurements are taken into account). The term $\underline{w}(t)$ is added because the transformation through F(t) is not perfect: an error will occur which is not known at every time. Only the statistical characteristics of this error are known or can be estimated. Assuming a Normal distribution of this noise with mean 0 and variance Q(t), this can be denoted as $\underline{w}(t) \approx N(0,Q(t))$. This variance Q(t) is used in the filter

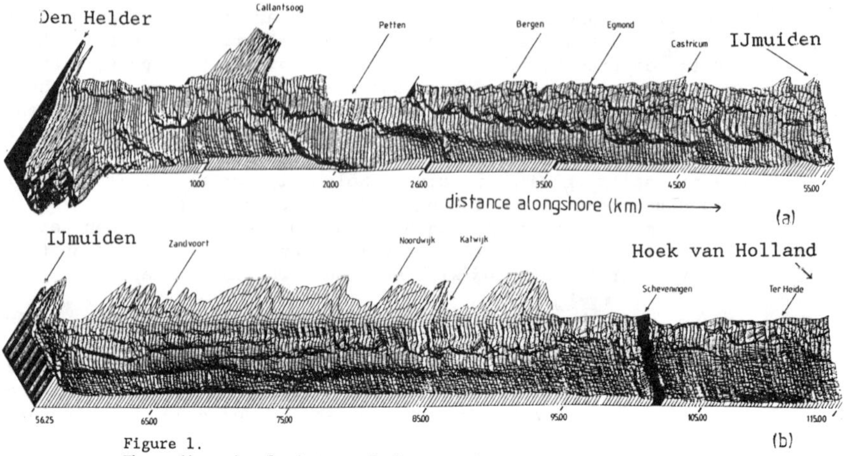

Figure 1.
Three dimensional picture of the coast between
(a) Den Helder and IJmuiden (b) IJmuiden and Hoek van Holland
(see fig 9[a] for situation Holland coast)

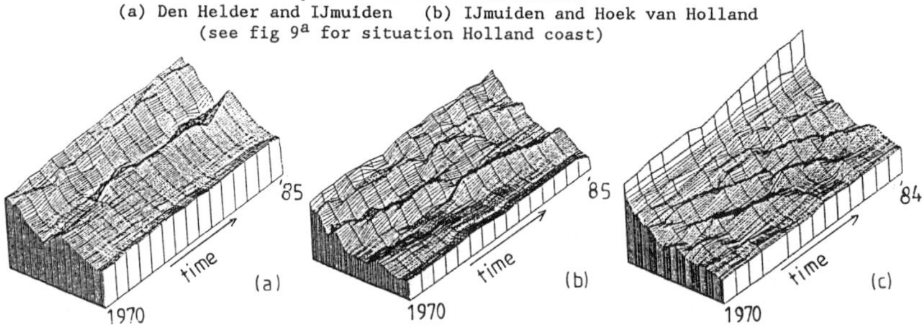

Figure 2.
3-D-time diagram of profile between (a) Den Helder and Petten
(b) Petten and IJmuiden (c) IJmuiden and Scheveningen

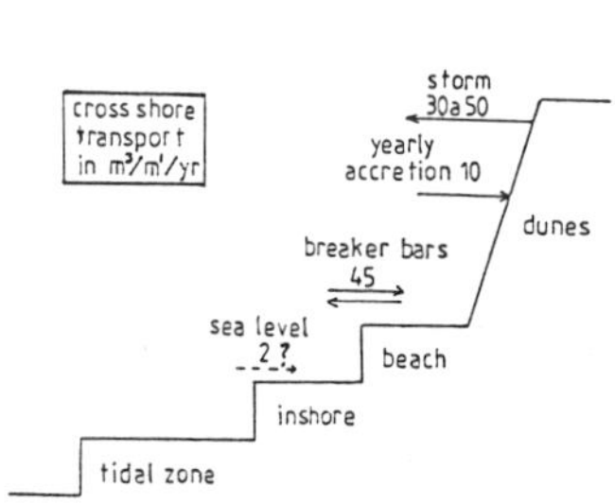

Figure 3.
Estimated cross-shore transport along the central coast of Holland (storm erosion is balanced by on-shore windtransport during quiet conditions)

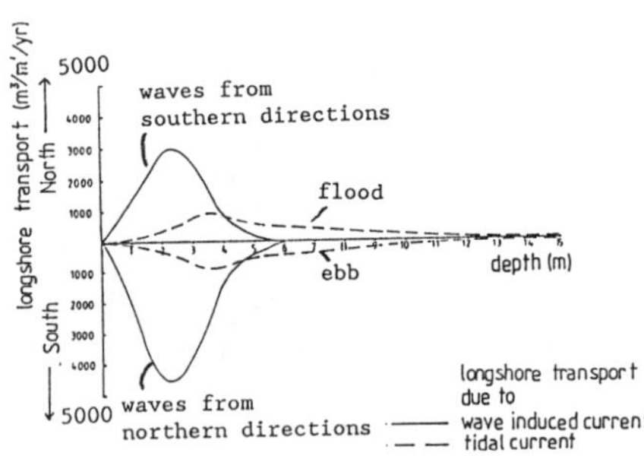

Figure 4.
Longshore transport, along the coast of Holland as a function of depth.

algorithm. Like all other noise terms it is used in the form of a covariance matrix, so it is possible to take into account not only variances but also noise covariances.
Applying the measurement equation (disregarding the noise) results in an expectation of the measured state at the considered time. This equation is also affected by errors, which are characterized by $\underline{v}(t) \approx N(0,R(t))$. The variance R(t) will appear in the filter algorithm.
When the measurements $\underline{z}(t)$ actually become available, these are taken into account to get an improved estimate $\underline{x}(t+)$, using the equation:

$$\underline{x}(t+) = \underline{x}(t-) + K(t) \{ \underline{z}(t) - H(t)\, \underline{x}(t-) \} \qquad (3)$$

where $\underline{x}(t+)$ is the optimal estimate.
$\underline{x}(t-)$ is the estimate of the state at time t before the measurements are taken into account.
$\underline{z}(t) - H\,\underline{x}(t-)$ is the difference between the measurements and their expected value.

K(t) is the "gain matrix", the essence of the Kalman filter. It has two functions: $\underline{1}$: it re-transforms the correction on $\underline{z}$ to a correction on $\underline{x}$, and $\underline{2}$: it gives a weight to the importance of the correction, varying from zero (perfect theory not to be disturbed by sorded measurements) to one (What You See is What You Get). The expressions for K(t), depending on H(t), Q(t) and R(t) are given in the Appendix, which may be written more or less symbolically as:

$$K(t) = \frac{1}{H(t)\left\{ 1 + \dfrac{\text{noise in measurements}}{\text{noise in latest estimate}} \right\}} \qquad (4)$$

The method has been applied to parts of the Dutch coast, using as physical model the line-theory and estimating the model coefficients. The physical conception involved is the existing of a (set of) coastal constant(s) $\underline{x}$, stationary in time. This means that F(t) in (1) should be the unit (identity) matrix. Improved estimates of $\underline{x}$ are made by checking a calculated coastline (or for the two-line theory: lines of beach and inshore) against measured values $\underline{z}(t)$. This application is best demonstrated by the following example, based on the single-line model which is described in the next chapter, see equation (8). The successive steps of the procedure are shown in fig 5.
In order to find the coastal constant s in (8) the system equation (1) becomes:

$$s(t) = s(t-1)$$

$$\text{with inaccuracy characterized by} \quad w(t) \approx N(0,Q) \qquad (5)$$

Due to the noise w(t) this "constant" is given the opportunity to change a bit each time step. This noise is thought to have a Normal distribution (Gaussian) with mean 0 and variance Q. In this particular case Q is assumed to remain stationary and uniform along the coast.
Using the Jarkus dataset "measured coastlines" can be determined in each year, giving in each range the mean of the positions of the various contourlines between NAP -5m and NAP +15m (fig 8)(NAP≈MSL). The measurement equation (2) provides the link between s(t) and these measured coastlines, using an explicit difference scheme for the Pelnard-Considère equation (8). This can be rewritten as

$$y_i(t) = y_i(t-1) + B \{ y_{i-1}(t-1) - 2y_i(t-1) + y_{i+1}(t-1) \}\, s \qquad (6)$$

where y denotes the seaward coordinate of a coastline (fig 8).
i denotes the distance along the coast, t the time.
$B = \frac{1}{h}\frac{\Delta t}{\Delta x^2}$; Δx = distance between two adjacent ranges.

Thus for the one-line theory the measurement equation (2) becomes:

$$\begin{bmatrix} y_2 \\ y_3 \\ \cdot \\ \cdot \\ y_{n-1} \end{bmatrix}_t = \begin{bmatrix} B\,(\,y_1 - 2y_2 + y_3\,) & y_2 \\ B\,(\,y_2 - 2y_3 + y_4\,) & y_3 \\ \cdot & \cdot \\ \cdot & \cdot \\ B\,(\,y_{n-2} - 2y_{n-1} + y_n\,) & y_{n-1} \end{bmatrix}_{t-1} \begin{bmatrix} s \\ 1 \end{bmatrix}_t \tag{7}$$

with inaccuracy characterized by $\underline{v}(t) \approx N(0,R)$

H(t) (the matrix $[\ldots]_{t-1}$ in (7), see equation (2)) is derived from the measured coastlines at time t-1.
$\underline{v}(t)$ is the Gaussian noise with mean 0 and variance R, in this particular case containing two sources of error: measurement errors and inaccuracy of the model. It appears that it is difficult to determine this variance on physical grounds. So this noise is estimated after study of the measurements and assumed to remain constant with time. This variance can be of magnitude of several tens of m^2 to several hundreds of m^2.

If the matrices Q and R and an initial estimate of s are known the filter procedure can be started, resulting in a "time history" for the observed model coefficient s. The result shown in fig 6 (s = 1 a 1.2 * 10^6 m^3/year) is based on measurements for the coast between IJmuiden and Scheveningen. Results for the two-line model and the use of those coastal constants for making hindcasts and predictions will be exposed by Bakker et al. in another paper (1988).

IV. 1000 YEARS

Fig 7 shows the accretion and erosion of the coast of Holland, as follows from ancient registrations of the site of the dune foot in the last 100 years according to Edelman (1961). Edelman considers the coast as principally accreting, however with "sinks" at the Northern and Southern boundary. A possible explanation for the coastal behaviour, as indicated already by Edelman and Eggink (1963) is based upon the conception of Pelnard-Considère (1954). This involves that a concave coast accretes and a convex coast erodes. The ruling equation is:

$$\frac{\partial y}{\partial t} = \frac{s}{h}\frac{\partial^2 y}{\partial x^2} \tag{8}$$

where x,y are the coordinates of the coastline (x in longshore direction, y in seaward direction), t is the time, h is the thickness of the layer of the coast, which is supposed to participate in the process of accretion and erosion (fig 8) and s is the rate of change of coastal transport (m^3/year/radian) as function of the coastal direction. Equation (8) is based on the continuity equation and a transport equation, which read respectively:

$$h\frac{\partial y}{\partial t} + \frac{\partial S}{\partial x} = 0 \tag{9}$$

$$S = S_0 - s\frac{\partial y}{\partial x} \tag{10}$$

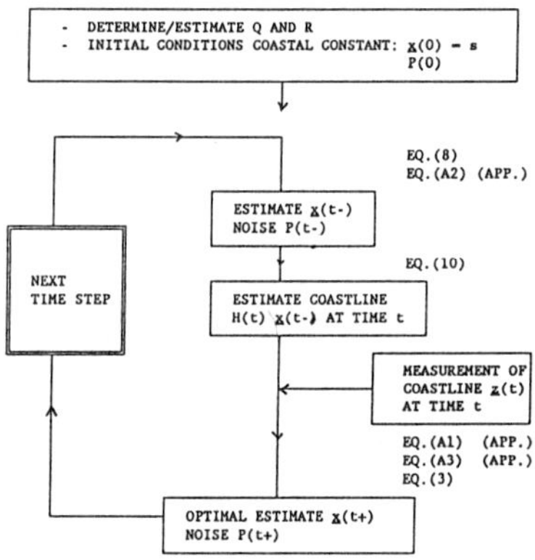

Figure 5.
Procedure Kalmanfiltering.

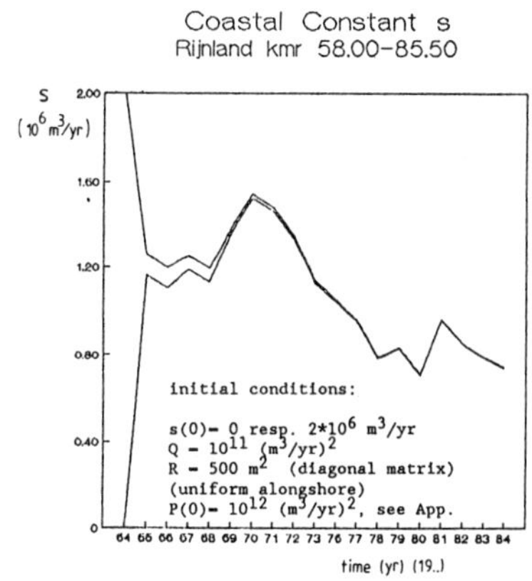

Figure 6.
Value of coastal constant s, found by Kalman filtering

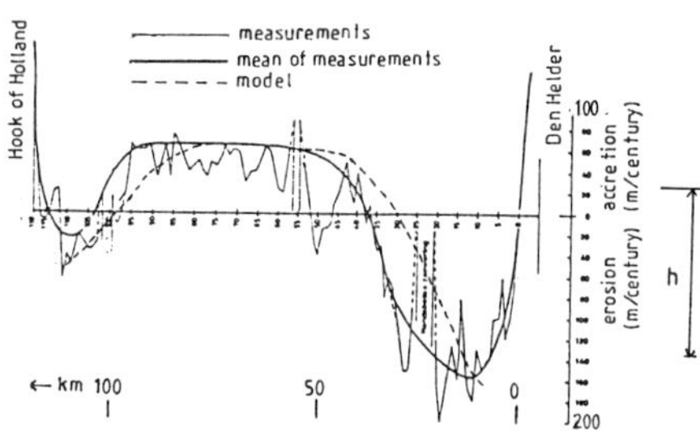

Figure 7.
Accretion and erosion of the coast of Holland during the last 100 years, after Edelman (1961)

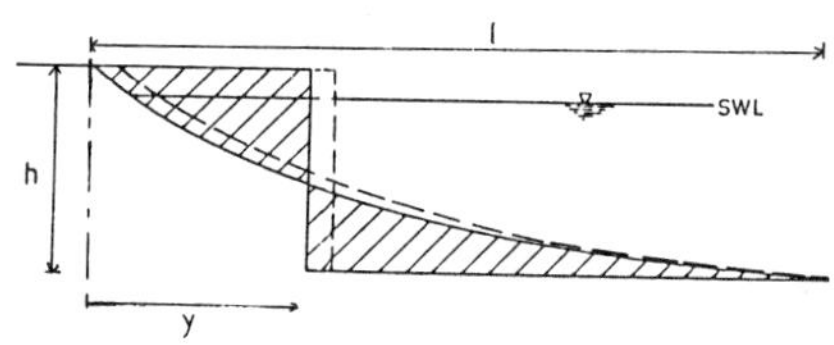

Figure 8.
Line schematization of a coastal profile

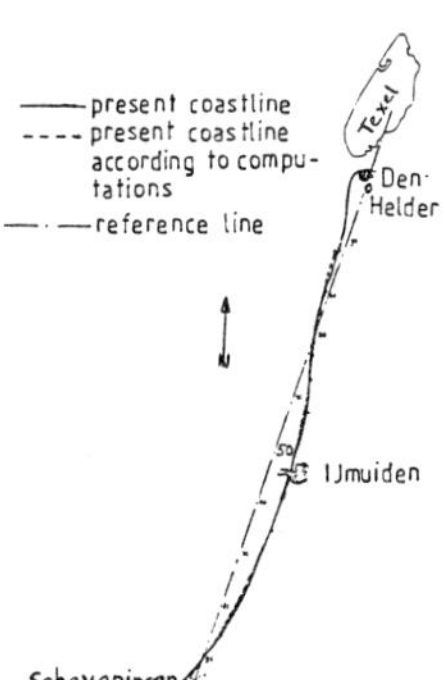

Figure 9a.
Model-computed coastline, compared with real coastline

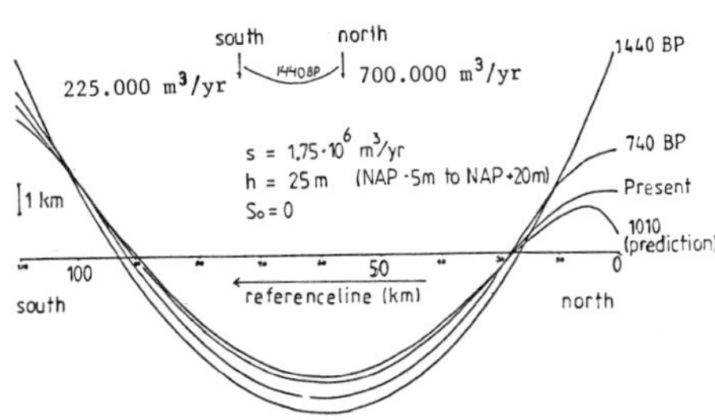

Figure 9b.
Result line-model and used constants on a 1000 years time scale

where S is the littoral transport and S_0 is the littoral transport along a coast, parallel to the x-axis.
This physical concept allows for a simplified one-line model concerning the behaviour of the Holland coast in the last 1000 years (between km 10 and km 100) starting from the following initial and boundary conditions (fig 9^b):
a. the initial coast is assumed to have a parabolic shape;
b. at the boundaries the coast is assumed to erode according to $\sqrt{t}$ following the Pelnard-Considère solution for a "negative delta" (1954).
This implies, that the coastal development is schematized according to a parabolic shape, stretching itself, combined with two "negative delta's" on both sides.
Implicit assumptions are a constant subtraction of material at the boundaries, starting more or less abruptly. Physically, this can be justified at the Northern side because of the developing Texel inlet, leading to the Wadden shallows and, at the Southern side, sand subtraction by the emerging outlet of the river Rhine. Development of the Texel inlet was triggered because of the burning and dewatering of existing peat layers (Edelman,1974) which caused the strong development of this inlet, starting about a millenium ago. Since then, the Wadden shallows tend to keep themselves at mean sea level (despite sea level rise). The reason for this is a subtle mechanism of sand, being brought to this shoals during flood tide with rather high flood velocities and coming to rest on this shoals during the ebb with rather low velocities. Already in 1949 Van Bendegom reported about this socalled "sand hunger" of the Wadden shallows.
Fig 9^a shows the present coastline according to the line model and in reality. Fig 9^b gives constants used in the applied model. A value of $s/h = 71 * 10^3$ m^2/year is found, leading with a depth h of 25 m (20 m dune height + 5 m water depth) to $s = 1.75 * 10^6$ m^3/year. The dashed line in fig 7 gives the present coastal erosion/accretion according to the model, compared with the Edelman data. Details are given by Bakker and Smit (1987).
However, the following objection against the model can be made. It appears (Ligtendag en Borger,1987) that the middle part of the dutch coast showed erosion from 1600 A.D. to 1800 A.D. instead of accretion, as found in the last century. Probable cause is wind erosion which is not taken into account in the Pelnard-Considère theory. Only since the 19th century the dune front is protected by planting marram grasses etc.

V. 6000 YEARS

Where for "short" terms longshore transport (only) can provide some explanation of coastal behaviour, for a 6000 year period sea level rise and cross-shore transport become vitally important.
Fig 10 shows the sea level rise in the course of time since 10000 Before Present (BP) according to Jelgersma (1979). Fig 11, from Jelgersma et al. (1985), shows the submerging of the North Sea in the last milennia. Fig 12 shows the development of the Dutch coast according to Jelgersma et al. (1985). Apart from a regression during the period of fast sea level rise a transgression took place between ca. 5500 BP and 3000 BP (van Straaten (1965), Zagwijn (1974), Jelgersma et al. (1985), Roep (1984)). Afterwards, erosion started again.

Is it possible to make a conceptual model of this process of sea level rise? A good model should be able to reproduce as well the coastal transgression from 5500 to 3000 BP as the erosion afterwards.
In the present preliminary stage, it has only been attempted to make a

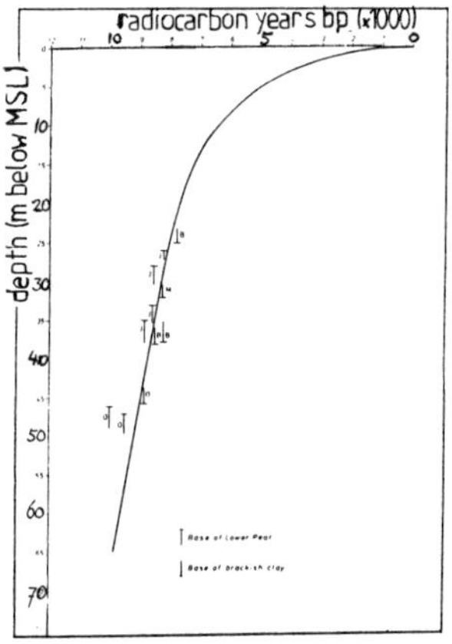

Figure 10.
Sea level rise during
the last milennia,
after Jelgersma (1979)

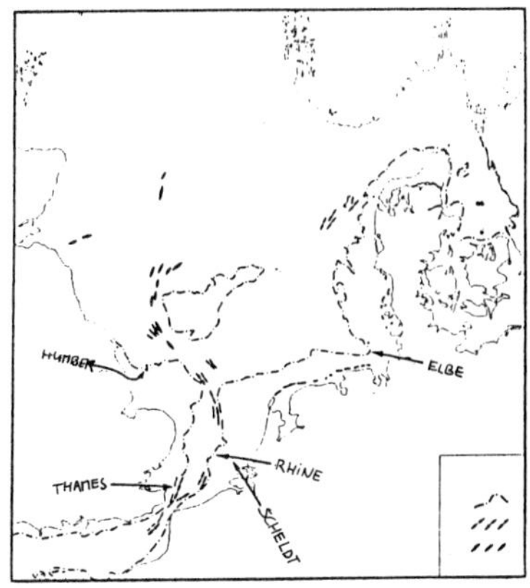

Figure 11.
The submerging of the North Sea
in the last milennia,
after Jelgersma et al.(1985).
–·–·-Coastline 8000 B.P.
sealevel 30m below present
level.
/// Old sediment ridges

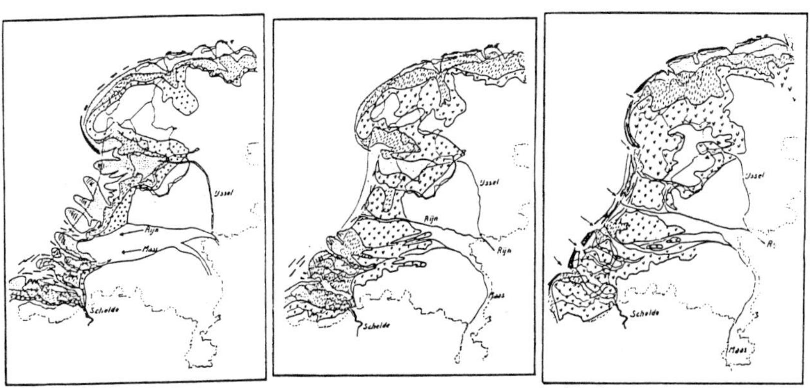

Figure 12.
Development of the Dutch coast according to Jelgersma et al.(1985)

Time (years Before Present) :	(a) 5500	(b) 4600	(c) 2000
Sealevel (m Below Present Level) :	7	3.5	1

model which gives qualitative agreement, without paying too much attention to quantitative accuracy. Before explaining this model, some literature concerning this subject will be reviewed.

A very simple model for the response of a coast on sea level rise is provided by Bruun ((1983) and earlier papers) (fig 13). He starts from the assumption, that the sea level rise is so slow, that the upper part of the coast - up to a certain socalled "closure depth" - will always be able to adjust itself and will keep a cross-section according to an equilibrium profile. From simple geometry and continuity it shows, that the coastal erosion will be equal to the cotan of the mean slope (l/h in fig 13) times the mean sea level rise a. Here h denotes the socalled "closure depth", i.e. the depth up to which eroded material settles itself.

According to the authors it will always cost time to attain the equilibrium profile. They agree with Wright et al. (1984), that short-term processes take place with a small closure depth, where a larger part of the profile will be involved in long-term processes; the time necessary to reshape the profile to an equilibrium profile will be longer. Fig 14, after Wright et al. elucidates this idea.

The time-dependent adapting of a coastal profile to its equilibrium profile is simulated mathematically by Bakker (1968) and Swart (1974). These authors schematized the coastal profile essentially to two areas, a beach and an inshore, each represented by a line in plan view (fig 15). Beach and inshore are separated by a horizontal flat between. The region seaward of the inshore is assumed to be horizontal; here no cross-shore transport takes place. According to their schematization and according to experiments of Swart, if longshore transport plays no role, the line of beach and inshore will tend to reach an equilibrium distance W in a negative-exponential way.

The model starts from the assumption, that the cross-shore transport S_y depends on the overall coastal slope:

$$S_y = s_y \{y_1 - (y_2-W)\} \tag{11}$$

where S_y is positive in seaward direction, y_1 and y_2 denote the position of the line of beach and inshore (fig 15, i.e. the mean of the positions of the equidistant contourlines of the beach and the inshore respectively) and W denotes the equilibrium distance.

Basic physical thought behind the model is, that a slope determines the cross-shore transport; that two counteracting mechanisms determine this slope, where one mechanism increases the offshore transport when the slope increases (gravity mostly) and the other determines the onshore transport and increases, when the slope becomes flatter (for instance the asymmetry of the orbital motion). Making a Taylor series of the transport in the vicinity of the equilibrium slope gives a dynamical equation of the kind as given by (11). Implicitely it is assumed that the cross-shore transport is not affected by local changes in slope of beach and inshore separately, as long as the mean position of the beach, c.q. the inshore remains the same. Some motivation for this assumption can be found from fig 2: like a huge rake the breaker bars correct differences in large-scale slope every three to four years and the local slope is a very short-term feature, effected by the breaker bars and has no essential importance.

This philosophy also more or less determines the lower edge of the inshore, which can be considered as a kind of "closure depth" as well. The theory only holds when the profile of beach and inshore respectively can adapt itself much faster to changes than at the time scale of the coastal process to be schematized by the two-line theory. Looking to fig 3 it might be expected that cross-shore transport below the level, where the breaker bars decay (say 5 m below MSL) will be much smaller than above this level. Therefore, this provides a logical closure depth. The characteristic time scale T_0 for exponential decay according to the two-line theory equals:

$$T_0 = (h_1 \; h_2)/(s_y \; h) \qquad (12)$$

where h_1 and h_2 are the thickness of the layer of beach and inshore respectively and h equals $h_1 + h_2$.
In the case of sea level rise, cross-shore transport will take place between the inshore and the layer below, however at a larger time scale (slower) than between beach and inshore. Furthermore, it might be expected that the slope between the inshore and this lower layer to a great extent determines the cross-shore transport between those layers. In other words: the local slope will be more of importance than the overall slope and therefore a two-line approach will not be sufficient. The ideas of Wright (variable closure depth) and of Bakker and Swart (negative-exponential tending to an equilibrium profile) can be combined in the following way. To simulate sea level rise, a 6-line program has been made, starting from 6 layers with seaward distance with respect to a reference point of y_1, y_2,... y_6, each layer having a height of 3 m; the cross-shore transport S_{y1}, S_{y2}... S_{y5} between the first and the second, the second and the third.. etc. layer is related to y_1-y_2, y_2-y_3,... in the following way:

$$S_{yn} = s_{yn} \{(y_n - L_n) - (y_{n+1} - L_{n+1})\} \qquad (13)$$

where n = 1 to 5 and L_n denote the position of y_n in an equilibrium position. The characteristic time T_n is defined accordingly:

$$T_n = (h_n \; h_{n+1})/s_{yn} \qquad (14)$$

Fig 16 shows the schematization and the assumed time scales T_n. The values of T_4 and T_5 are pure guesses: for some justification of the values of T_1 to T_3 is referred to Bakker and Delver (1986).
It is assumed, that the present profile is an equilibrium profile (fig 17[a]) and that 6000 BP the rising of the sea level was so fast, that no time was available to bring the profile in an equilibrium shape; the profile of 6000 years ago is assumed to be the present profile, in which the water level is 6 m lowered (fig 17[c]); the water level at 6000 BP was 6 m lower than the present level (fig 17[b]). This "sudden" exposure of the present coastal area to the waves is caused also for a large part by the degradation and erosion of former coastal barriers in the North Sea, which became submerged. With respect to this conception is referred to Niederoda et al. (1985) for the behaviour of coastal barriers in general and to van Straaten (1985) for the behaviour of the Holland coast. Fig 18[a] (in which the third dimension denotes the time) shows how, according to the model, the coast would have been accreted, if the sea level had not risen since 6000 BP; in fig 18[b] this sea level rise has been taken into account. In fig 18[a] the coast accretes, because the profile of fig 17[c] is much flatter than the equilibrium profile; in fig 18[b] erosion starts again because the upper part of the profile has formed nearly an equilibrium profile and erodes according to the Bruun conception (see above). For details concerning the way of computation is referred to Smit (1987).
Thus it shows, that the model gives qualitative agreement with the features, found in nature.

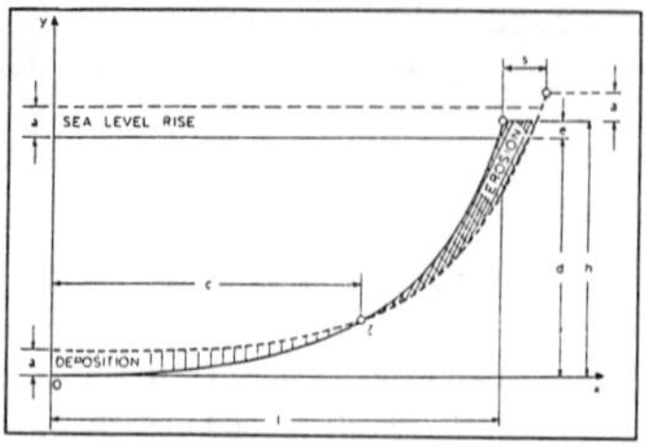

Figure 13.
The Bruun model, translation of coastal profile due to sea level rise.

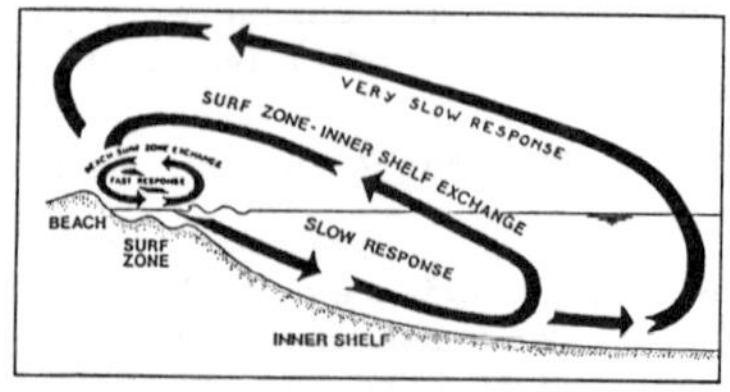

Figure 14.
Interaction of different parts of the profile, after Wright et al. (1984)

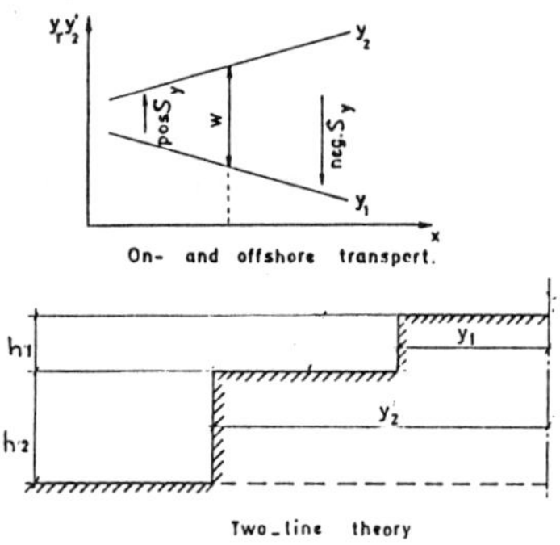

Figure 15.
Schematization in beach and inshore according to Bakker and Swart.

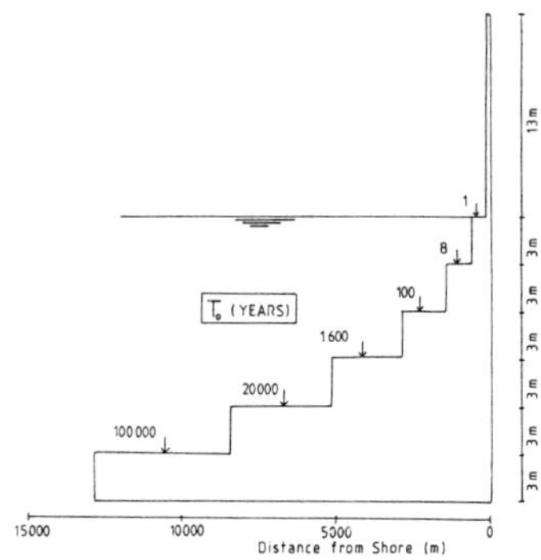

Figure 16.
Schematization n-line model and time scales.

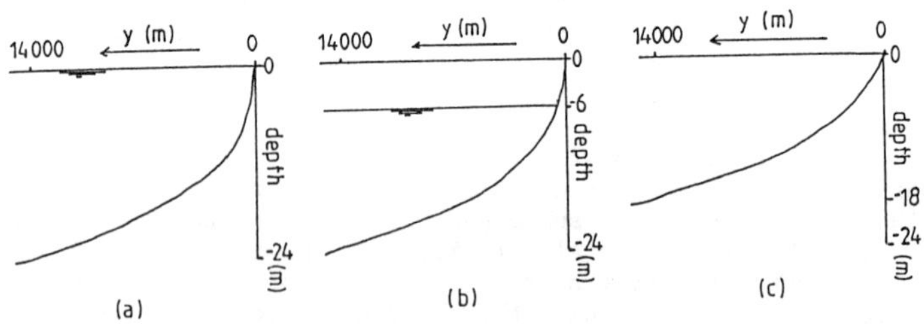

Figure 17.
Intial profile of the coast.
(a) Present profile
(c) Assumed profile of 6000 BP

VI. CONCLUSION

If coastal behaviour is considered at increasing time scales, beginning at a time scale of decennia and going to milennia, the effect of sea level rise and cross-shore transport becomes more and more pronounced. Line models may provide (a rough) insight at every time scale.

For the coast of Holland, coastal constants derived on one hand from data concerning the last 15 years (with the aid of Kalman filtering) and on the other hand from data with regard to the last 100 to 1000 years are not contradictory. For larger time scales, a schematization of the coast in one or two layers (one- or two-line model) will not be sufficient. Small amounts of cross-shore transport from deeper water to the inshore, as well as the local slope of these deeper regions become important at this time scale. In this paper, for a case like this the use of for instance a six-line model is proposed. This simulates as well a closure depth dependent on the time-scale of the feature, corresponding to the measurements of Wright (1984) as a negative-exponential trend to an equilibrium profile according to ideas and experiments of Bakker (1968) and Swart (1974). The method is more sophisticated than the Bruun rule (1983). Accretion of the Holland coast between 4000 years and the Roman times and erosion later on can be simulated this way.

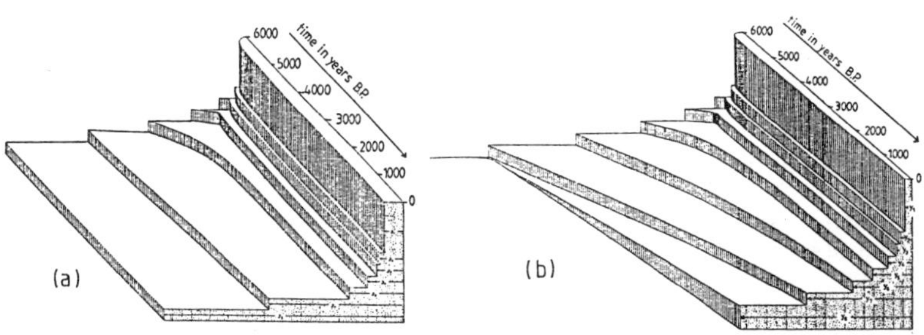

Figure 18.
Development of the profile of 6000 BP,
(a) if no sea level rise is taken into account.
(b) taking sea level rise into account.

APPENDIX

The equation for the Kalman gain matrix K(t) is

$$K(t) = P(t-)\, H^T(t)\, \{ H(t)\, P(t-)\, H^T(t) + R(t) \}^{-1} \quad \text{(A1)}$$

Where

$$P(t-) = F(t)\, P((t-1)+)\, F^T(t) + Q(t) \quad \text{(A2)}$$

is the inaccuracy (variance) belonging to $\underline{x}(t-)$ (The righthand side of (A2) denotes old noise + new noise, so P(t-) involves the total noise as a result of all preceeding time steps)
and

$$P(t+) = \{ I - K(t)\, H(t) \}\, P(t-) \quad \text{(A3)}$$

is the inaccuracy belonging to $\underline{x}(t+)$.
I denotes the unit (identity) matrix.
H^T and F^T denote the transposed of H and F respectively.

By thinking of the matrices as scalars (notation small letter instead of capital) equation (A1) can be rewritten as:

$$k(t) = \frac{1}{h\{1 + \frac{1}{h^2(t)} \frac{r(t)}{p(t-)}\}} = \frac{1}{h\{1 + \frac{\text{noise in measurements}}{\text{noise in latest estimate}}\}} \quad \text{(A4)}$$

So the Kalman gain matrix K(t) contains the inverse of the matrix H(t), thus transforming $\underline{z}(t)$ to $\underline{x}(t)$, and a matrix of weight-factors (derived from system- and measurement noise, through P(t-) and R(t) respectively), determining the weight given to the physics relative to the measurements.
Kalman proved that this algorithm results in an optimal estimate $\underline{x}(t+)$ in least squares sense on time t (Kalman, 1969).
Present estimates are computed in a recursive way, meaning that all previous measurements are taken into account in a sequential way (filtering by a "moving window"). Unlike batch processing of measurements, in this recursive filter there is no need to store past measurements (Gelb,1974). The weight given to the previous measurements depends on the noise characteristics of the model and the measurements.

REFERENCES

Bakker, W.T. (1968), The dynamics of a coast with a groyne system, Proc. 11th Conference on Coastal Engineering, London, ch. 31,pp 492-517
Bakker, W.T. (1971), Computation of the wave-induced littoral drift according to the Svasek-method, adapted to parallel depth contours, Rijkswaterstaat, Dept. for coastal res., Rep. WWK71-2
Bakker, W.T., de Ras, M.A.M. (1971), Computation of sand transport by waves and tide, using an adapted Bijker method (in Dutch), R.W.S., Dept. for coastal res., Rep. WWK71-4
Bakker, W.T., Delver, G. (1986), Coastal changes caused by a shallow sanddam in front of the Delfland coast (in Dutch), Delft Univ. of Techn., Hydraulic Engineering Group
Bakker, W.T., Smit, E.S.P. (1987), Possible truth in coastal behaviour (in Dutch), R.W.S. nota GWAO-87.004

Bakker, W.T., de Vroeg, J.H. (1987), Coastal modeling and coastal measurements in the Netherlands, Proc. seminar Sogreah: Prototype Measurements to validate Numerical Models of Coastal processes, Grenoble, Oct 14-15

Bakker, W.T., de Vroeg, J.H. (1988[a]), Looking at the Holland coast (in Dutch), de Ingenieur (mei)

Bakker, W.T., de Vroeg, J.H. (1988[b]), Is the coast safe ? (in Dutch), R.W.S. Tidal Waters Dept., Report GWA088.017

Bakker, W.T., v.d. Kerk, C., de Vroeg, J.H. (1988), Determination of coastal constants in mathematical line models, 2nd European workshop on coastal zones as related to physical processes and coastal structures, Loutraki, Greece

van Bendegom, L. (1949), Considerations of the fundamentals of coastal defence (in Dutch), R.W.S., Dir. ben.riv.

Bruun, P. (1983), Review of conditions for uses of the Bruun rule of erosion, Coastal engineering 7, pp 77-89

Edelman, T. (1961), Erosion and accretion of the coast between Den Helder and Hoek van Holland (in Dutch), R.W.S., WWK61-1

Edelman, T., Eggink, D.N. (1963), Some characteristics of the Dutch coast, Proc.8th Conf.on Coastal Eng.,Mexico,ch41 pp 756-764

Edelman, T. (1974), Historical geography of the Dutch coastal area, R.W.S., Dept. of Water Management and Hydr. Res., publication 14

Gelb, A. (1974), Applied optimal estimation, T.A.S.C., Massachusetts Institute of Technology

Jelgersma, S. (1979), Sea-level changes in the North Sea basin, The quaternary history of the North Sea, pp 233-248

Jelgersma, S., Beets, D., Schuttenhelm, R. (1985), A Geological view on the coast (in Dutch), kustmorfologie in ontwikkeling, pt/Civ.Techn. nr 11, pp 3-8

Kalman, R.E. (1969), A new approach to linear filtering and prediction problems, J. Basic Eng., pp. 35-46

Ligtendag, W.A. (1987), From Yzer to Jade (reconstruction of the Southern North Sea coast in the years 1600 and 1750), Univ. of Amsterdam, Hist. Geograph. Seminarium

Niederoda, A.W., Swift, D.J.P., Figueredo (jr), A.G., Freeland, G.L. (1985), Barrier island evolution, middle Atlantic shelf, U.S.A., part 2: evidence from the shelf floor, Marine Geology, 63, pp 363-396

Pelnard-Considère, R. (1954), Essai de théorie à l'evolution des formes de rivages en plages de sable et de galets, Quatrième Journées de l'Hydraulique, Paris, Les energies de la mer, question III

Roep, T.B. (1984), Progradation, erosion and changing coastal gradient in the coastal barrier deposits of the Western Netherlands, Geol. en Mijnb. nr 3, vol.63, pp 225-336

Smit, E.S.P. (1987), The role of cross-shore transport in the development of the dutch coast (in dutch), Delft univ. of Techn., Hydraulic Engineering Group

van Straaten, L.M.J.U. (1965), Coastal barrier deposits in South- and North-Holland, Mededelingen Geologische Stichting, nr 17, pp 41-75

Swart, D.H. (1974), Offshore sediment transport and equilibrium beach profiles, Ph.D. thesis

Vellinga, P., Zitman, T.J. (1988), Coastal zone response to an increasing sea level rise, Proc. 21st Conf. on Coastal Eng., Malaga, Spain

Wright, L.D., May, S.K., Short, A.D., Green, M.O. (1984), Beach and surf zone equilibria and response times, Coastal Engineering

Zagwijn, W.H. (1974), The paleo-geographic evolution of the Netherlands during the Quaternary, Geologie en mijnbouw 53, pp 369-385

CHAPTER 211

ARTIFICIAL BEACH UNITS ON LAKE MICHIGAN

K.J., MacIntosh,* Associate Member, C.D. Anglin,* Associate Member ASCE

ABSTRACT

High water levels on Lake Michigan during 1985 and 1986 created substantial erosion of the shoreline and reduced the size and recreational potential of many of the parks and beaches. To prevent further erosion, protect existing properties and structures, and to create and improve recreational areas along the Lake Michigan shoreline, four coastal engineering projects were designed and constructed during this time. Artificial beach units stabilized by offshore breakwaters were used as the main component of each project.

Physical hydraulic model studies were used to determine the orientation, size, and spacing of breakwaters and artificial beach units. Model predictions of beach profiles and plan shapes compare closely with prototype surveys. Surveys completed since construction demonstrate the stability of the beaches and support their use as effective, low maintenance shoreline protection.

Prototype experience has shown that these projects are extremely successful both in their ability to withstand storms on the Great Lakes and to attract people for recreational activities.

INTRODUCTION

Four coastal projects were constructed on the west shore of Lake Michigan during 1985 and 1986 to prevent excessive erosion during a period of record high water levels. Of the four projects discussed in this paper, two are located in Lake Forest, Illinois and the other two are located in Milwaukee, Wisconsin.

Each project has a combination of offshore breakwaters and artificial beaches as the basic component providing both shoreline protection and new recreational areas. In assessing the attributes of various shoreline protection schemes it was evident that the benefits of creating new recreational areas outweighed the additional costs in comparison to the costs associated with simply providing basic shoreline protection.

Physical hydraulic models were used in all four projects to design the shoreline protection, determine optimum beach fill profiles, and to assess the impact of design storms on these projects.

---

* Senior Engineers, W.F.Baird and Associates, Coastal Engineers, 38 Antares Drive, S-150, Ottawa, Ontario, Canada, K2E7V2

The final designs are unique to each project depending on the particular recreational requirements and environmental conditions specific to each site. Varied combinations of breakwaters, groins, shoreline revetments, and sand fills have been used. Specific items which were varied between projects are discussed in further detail below:

- The type of beach fill material varies from a coarse sand (D50=2.6mm) to waste rock from a tunneling project. Criteria for selection of beach fill materials included:
  1) beach stability -grain size must be sufficiently large to provide stable, low maintenance beaches.
  2) aesthetics and recreation -the material had to be acceptable for pedestrian traffic.
  3) cost -the total cost of material per unit length of shoreline had to be reasonable relative to providing other forms of basic shoreline protection.

- The length, height , and orientation of steel sheet pile groins between the individual beach cells was varied in all four projects. The groins were used to control the amount of alongshore sediment transport behind the breakwaters. In two of the projects the groins were tapered to match the stable beach profiles estimated from hvdraulic model studies. This allows native sand to bypass the projects by moving between the breakwaters and artificial beach. Alongshore currents and waves overtopping the breakwaters move the sand through the projects to the downdrift shoreline.

- The constructed plan and profile shapes of the artificial beaches varied significantly between projects. Depending on the cost of the beach fill and room available for beach profile development, physical models were used to determine the most effective elevations and shapes of the beaches, breakwaters and revetments.

- Breakwater design details including both layout and cross-section (for example, the distance that the breakwaters were constructed offshore, breakwater crest heights, and breakwater gap widths) were varied depending on the bathymetry at the site, size of recreational area required, and environmental conditions.

Literature Review

The large grain size of beach fill materials used in the projects discussed in this paper create stable beaches with unique profiles and plan shapes and make direct comparisons to other projects discussed in the literature generally inappropriate. However, a number of papers dealing with related topics have been published. Most of the published information deals with the construction of offshore breakwaters along coastlines with significant littoral drift. In the majority of cases described in the literature the breakwaters are used to trap sand.

For example, Toyoshima (1974 and 1982), and Nir (1982) discuss offshore breakwaters and their effect on coastlines in Japan, and Israel and Sinai respectively. Both authors report rapid formation of large tombolos behind the structures; the tombolo shape tends to stabilize after several years, following which only seasonal fluctuations occur.

Chew et al (1974) and Silvester et al (1980) discuss the use of offshore breakwaters to provide artificial headland control along beaches with substantial littoral drift. In these cases, construction of the structures is followed by the development of pocket or log-spiral beaches, also known as zeta bays, between the offshore breakwaters.

Theoretical aspects of sedimentation behind offshore breakwaters is discussed by Dean (1978). Dean presents a procedure for calculating beach planforms behind closely spaced offshore breakwaters. Perez and Fernandez (1988) also provide a method of predicting beach planforms based on the analysis of data collected from 33 beaches on the Mediterranean coast of Spain.

Model studies assessing the sedimentological influences of offshore breakwaters were discussed in a number of papers, including Shinohara and Tsubaki (1966), Rosen (1982) and Noda (1984).

Curren and Chatham (1977, 1980) describe model investigations undertaken to assess the performance of various schemes of protecting eroding beaches at Imperial Beach, California, and Oceanside Harbor, California. Both offshore breakwater and groin systems were tested.

The most recent information available on the use of detached breakwaters for shore protection is found in Dally and Pope, 1986. The authors review existing systems in the United States, including four single detached breakwaters and six segmented breakwater systems.

The papers described above generally discuss depositional effects behind an offshore breakwater, assuming a suitable source of sand, such as net alongshore drift. Several other papers discussing the use of offshore breakwaters to protect an artificial beach were also reviewed. These included Tourman (1968) (Larvotto Bay, Monte Carlo), Sato and Tanaka (1980) (Suma and Ito Beaches, Japan) and to a limited extent, Walker et al (1980) (Lakeview Park, Ohio). In addition, information on an artificial beach system, constructed near Cannes, France was obtained from the engineering firm Sogreah in France.

ENVIRONMENTAL CONDITIONS

Environmental conditions specific to each site were determined prior to the preparation of designs. The data has been summarized below for the projects discussed in this paper to allow a general comparison of the environmental conditions to other locations.

Water Levels

The maximum and minimum water levels recorded by the U.S. Army Corps of Engineers for Lake Michigan are summarized in Table 1:

Table 1

Lake Michigan	International Great Lakes Datum (IGLD)	Low Water Datum (LWD)
Record high water level	177.30m (October 4, 1986)	+1.5m
Record low water level	175.18m	-0.63 m

Lake Michigan low water datum (LWD) = 175.8m IGLD

The projects described in this paper were designed for maximum water levels of +1.4 m to +2.5 m. These values include an allowance for storm surge, which may increase water levels up to approximately 1m along the west shore of Lake Michigan. For example, a storm surge of 0.88 m occurred at Milwaukee on February 8, 1987 during a severe winter storm. However, it is unlikely that the maximum storm surge levels will ever occur during the summer months which generally correspond with the period of highest water levels on Lake Michigan.

In summary , it was appropriate to design the structures and land based facilities so that severe damage will not occur with water levels in the order of +1.4 to+2.5 m above LWD. These water levels account for extreme water levels plus storm surge. In addition, the facilities are designed to be functional for recreational purposes with both high and low water levels.

Wave Climate

Table 2 provides the range of maximum offshore significant wave conditions estimated using a 20 year parametric wind-wave hindcast. These results represent typical maximum offshore wave climates for the projects described in this paper.

Table 2 Summary of Typical Offshore Wave Climate

Direction	Maximum Deepwater Significant Wave Height (m)	Peak Period Tp (s)
N	4.0 to 5.0	8 to 10
NE	4.0 to 4.6	8 to 10
E	3.0 to 3.3	6 to 8
SE	3.0 to 3.3	7 to 8
S	2.7 to 3.0	6 to 7

In all of the projects discussed in this paper the breakwaters are constructed in depths ranging from -1.5 m to -3.5 m LWD. Therefore, the wave climate incident on the breakwaters and beaches is depth limited during storms. For each project the inshore refracted wave climate was estimated and model tests were completed for the range of incident wave directions and water levels. This procedure provided sufficient information to estimate the possible range of beach profiles and plan forms. It also provided data on the effects of storms from different directions.

Beach Materials

Two types of primary beach fill materials were used in the projects discussed in this paper. Waste material from a tunnelling project was used as the main source of material for beach fills in the two Milwaukee projects and a coarse sand was used as the primary fill material in the two Lake Forest projects. Figure 1 provides a gradation curve for the tunnel spoil material used in the Milwaukee projects and the coarse sand fill used in all the projects.

The waste material used in the Milwaukee projects consisted of limestone fragments ranging in size from 0.1mm to approximately 80mm in the largest dimension. The D_{50} was approximately 3mm. The material is very angular as shown in Figure 2 but stone quality tests and prototype experience since construction indicate that the majority of sharp edges will be removed as wave action continues to move the beach materials in the swash zone. The areas above the swash zone have compacted since construction was completed and can be comfortably walked on by pedestrians.

The two Lake Forest projects used a coarse sand fill referred to as Bird's Eye sand as the primary source of beach fill material. This sand consists of a well sorted natural source of sand with a D_{50} of 2.6mm. Other finer materials were used in the initial stages of filling the beach cells in areas where it was predicted that the coarse sand would not be removed from the outer face of the beach. Several other types of sand were considered however, model study results and cost/benefit analyses indicated that the Bird's Eye sand provided the optimum solution. The Bird's Eye sand provided a stable beach profile which allowed maximum development of land based facilities without having to construct the offshore breakwaters in deep water.

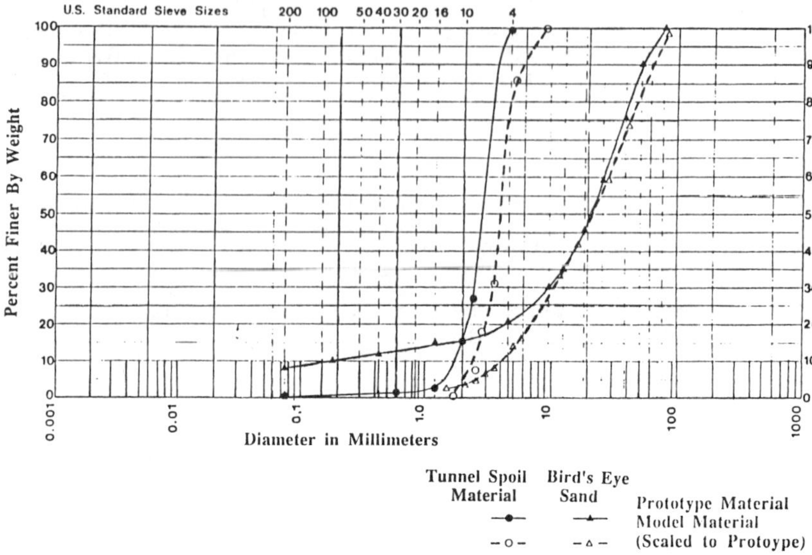

Figure 1

BEACH MATERIAL GRADATION CURVES

Figure 2 TUNNEL SPOIL BEACH FILL MATERIAL

PROJECT SUMMARIES

Of the projects discussed below one was completed in 1986, one in 1987 and the other two in 1988. Information available at this time is provided below. This includes: photographs of prototype structures; survey results of the 'as-built' projects and surveys completed since construction; and photographs showing the public's perspective.

Forest Park Beach Development, Lake Forest, Illinois

The shoreline development at Forest Park extends approximately 1067 m along the Lake Michigan shoreline, and consists of a series of four individual beach cells contained by detached offshore rubblemound breakwaters and shore connected steel sheet pile groins. The project also includes a protected boat launch basin, a shoreline revetment, and considerable greenspace and land based development as shown in Figure 3. Construction of the project started in the spring of 1986, and was completed approximately one year later.

The Lake Forest project was designed using a three dimensional hydraulic model study at a geometric scale of 1:20. Figure 4 shows the Forest Park model study in progress. During the hydraulic model tests it was demonstrated that it would be necessary to separate the beach cells with groins and to use a coarse sand as the artificial beach fill. The selected beach fill consists of sand with a D_{50} of approximately 2.6 mm. The coarse sand was necessary to reduce the amount of material transported between beach cells during storms, and to provide stable beaches within the limited area available.

An extensive review of the history of the shoreline and coastal processes was completed at this site prior to construction. It concluded that the project would have no negative impacts on the shoreline, since the alongshore sediment transport is negligible. A comparison of surveys taken in late April and August 1988 supports this conclusion. Cross sections were surveyed every 15m (50 ft) along the beach and indicate a total net accretion of $178m^3$ for the project. Considering the accuracy of the surveys, the net accumulation of sand within the project is considered to be negligible. Figure 5 shows the project in plan view including the locations of the breakwaters, design and August 1988 plan forms, and locations where beach cross sections were completed.

Beach cross sections are presented in Figures 6 and 7 for the following conditions: original beach profile, the as-built profile, April 1988 survey, and August 1988 survey. Profiles 2+00 to 3+00 indicate a net accumulation of sand whereas profiles 12+00 through 14+00 show a net loss between the as-built and August surveys. The elevation of the groins, +2.4 m LWD, is insufficient to prevent movement to the south as evidenced by the net accumulation of sand in beach cell 1.

The project was subjected to several severe storms during the winters of 1986-87 and 1987-88. It was noted during the model studies and during the first winter following construction that the storm profiles included development of a sand berm up to elevation +3.7m LFD. The comparison between model and actual beach plan and profile development under storm conditions was excellent as shown in Figure 8.

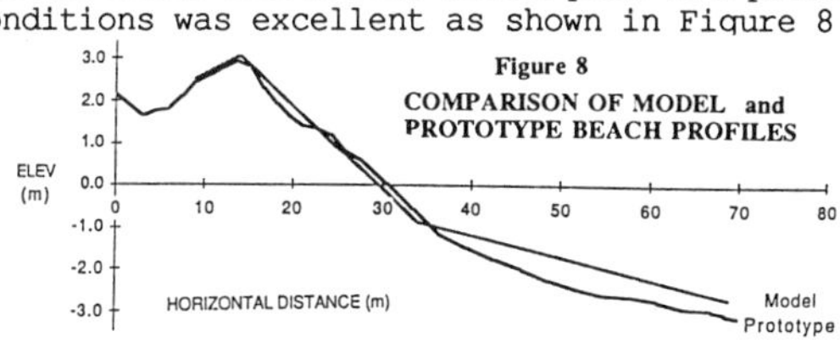

Figure 8
COMPARISON OF MODEL and PROTOTYPE BEACH PROFILES

Figure 3 COMPLETED FOREST PARK PROJECT

Figure 4 FOREST PARK MODEL STUDIES IN PROGRESS

FIGURE 5 FOREST PARK PLAN SURVEYS and LOCATION OF PROFILES

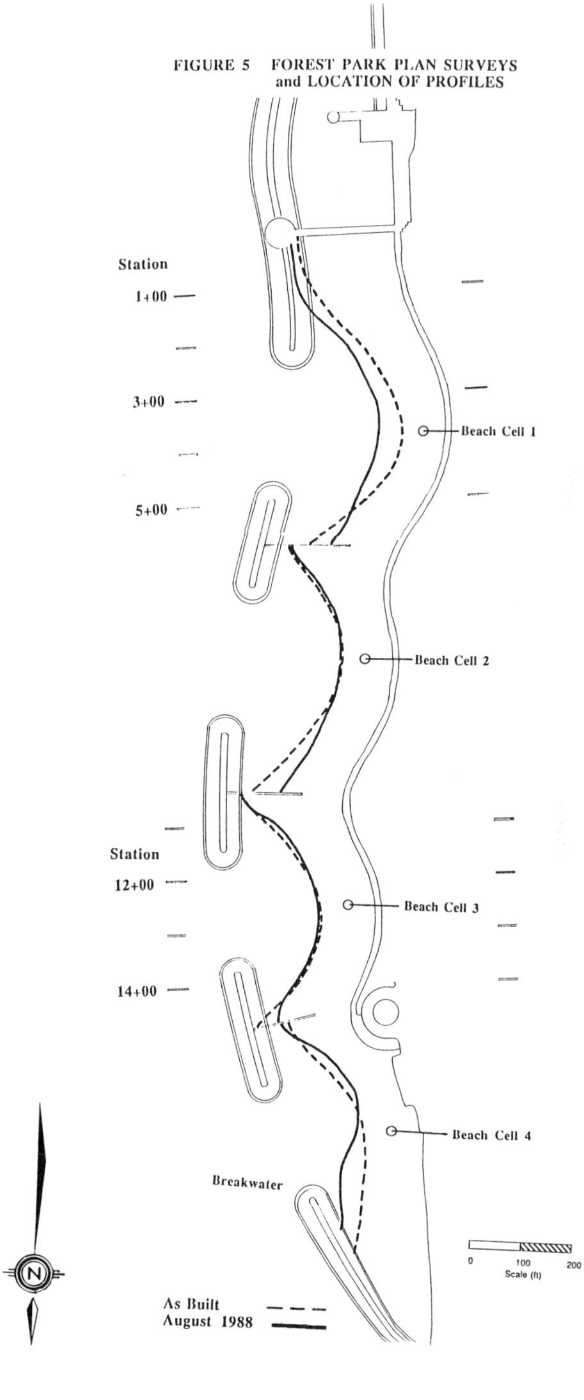

Figure 6 FOREST PARK BEACH PROFILES

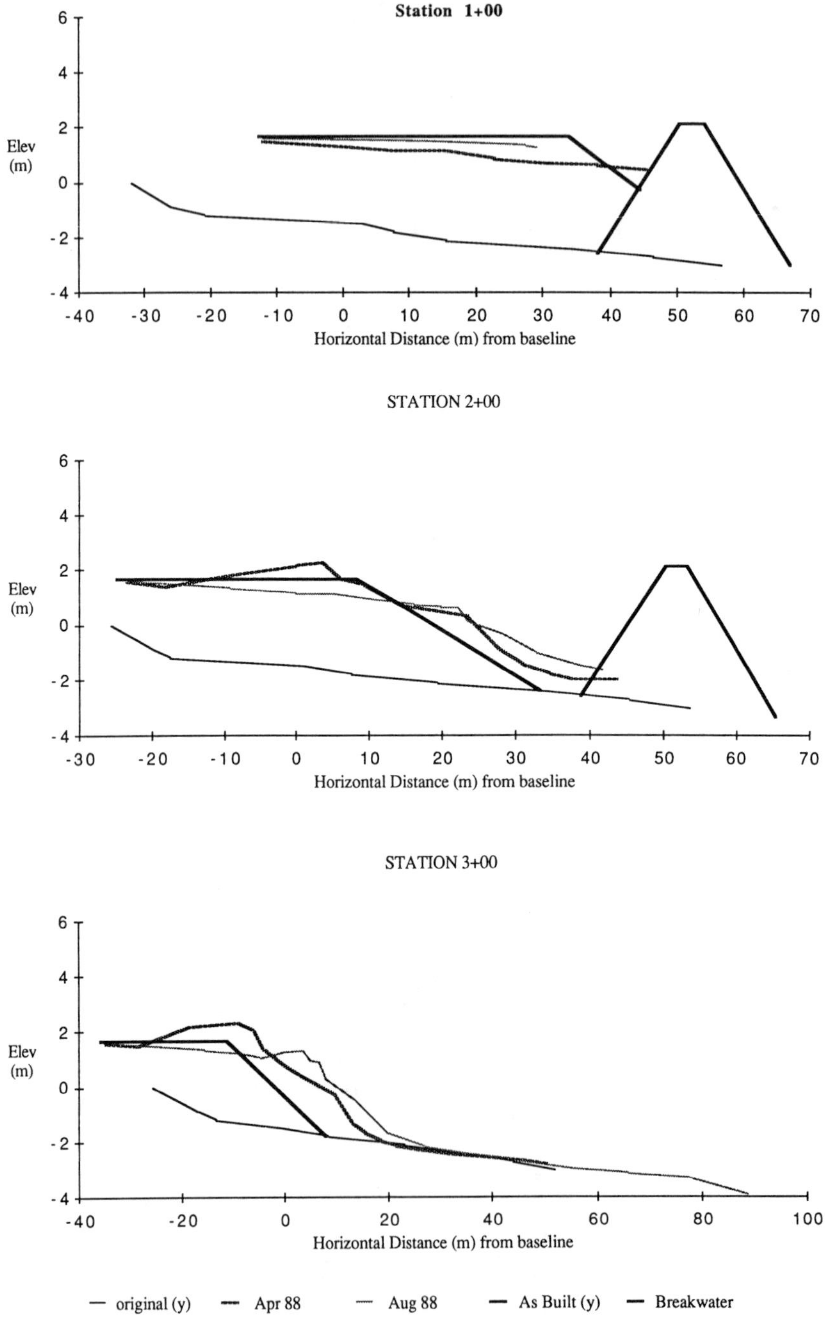

Figure 7 FOREST PARK BEACH PROFILES

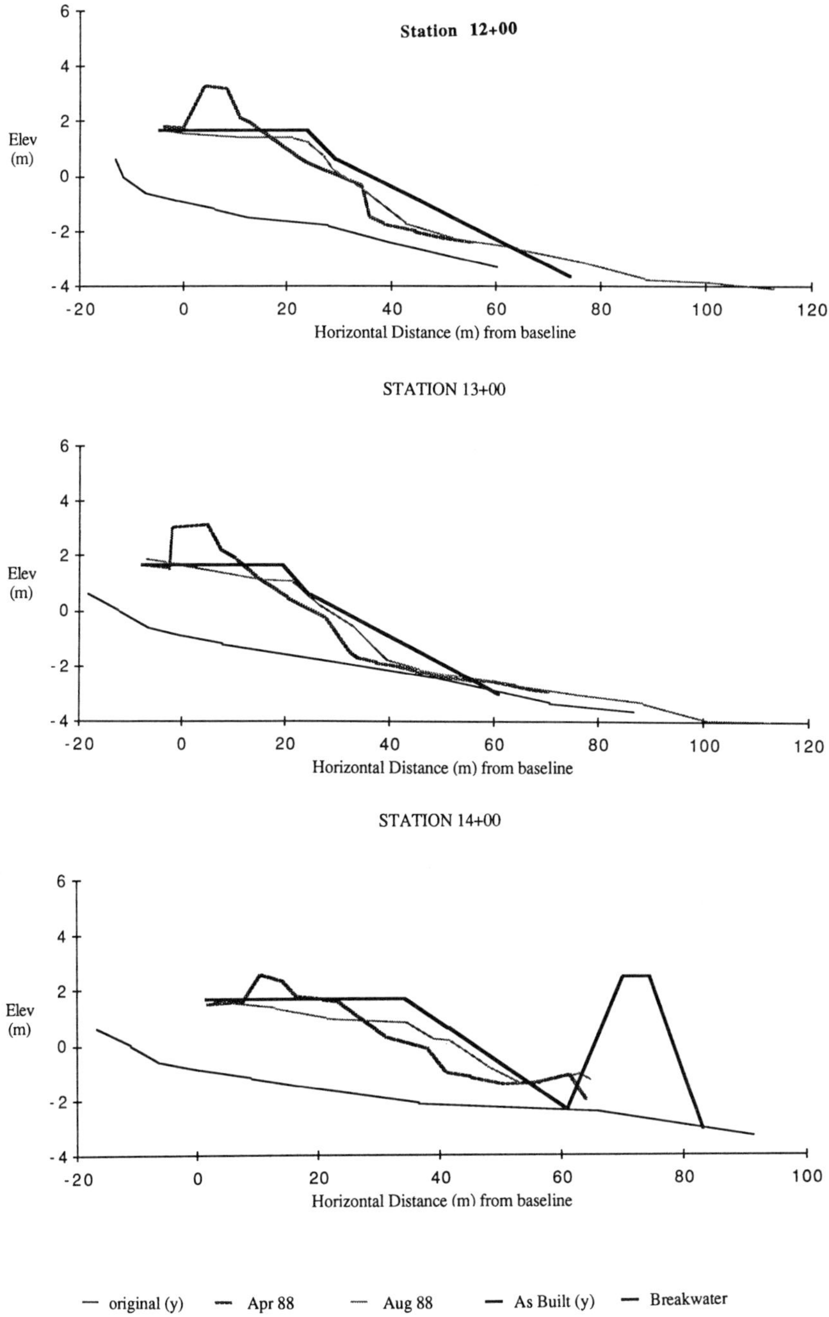

The prototype berm profile is also evident in Figures 6 and 7 in profiles 3+00,12+00 and 13+00 which are located primarily in the center of the beach cells. The berm is now constructed each fall by pushing the existing beach sand up into the storm profile. This protects the land based facilities behind the beach during fall and winter storms. and each spring the berm is levelled to provide a clear view of lake from the walkways for the summer season.

The Forest Park project has been extremely successful since its completion. Thousands of people visit the beach each day during the summer and the facility is also used for cross country skiing in the winter.

McKinley Park Shoreline Restoration, Milwaukee, Wisconsin

A design to protect a 762m long section of shoreline at McKinley Park was developed using a combination of desk and model studies. The project is located along a main commuter roadway which was frequently flooded during major storms.

The project consists of a series of rubblemound groins, revetments, and offshore breakwaters to contain landfill, a stone beach, and a coarse sand beach. Alongshore sediment transport was not a concern at this site, as the project is immediately north of Milwaukee harbor, which extends much further out into Lake Michigan.

The landfill and stone beach fill consist of crushed limestone produced by a large sewer tunnelling project underway in Milwaukee during construction of the McKinley Park project. This material is very angular, and ranges in size from 0.1 to 80 mm, as mentioned previously. The material was delivered to the site at no cost making it an extremely attractive product to be used as beach fill.

Figure 9 provides a photograph of a 180m wide beach cell which was filled with tunnel material. The photograph was taken in the fall of 1987 immediately following construction. A vertical edge is visible near the top of the beach which was removed during winter storms leaving a smooth beach face as shown in Figure 2.

Figure 9 McKinley Park Beach

There have been no obvious changes in the plan form shape of the tunnel spoil beach since construction. As evidenced in Figure 9 the beach has a very flat plan profile along the majority of its length.

Since construction was completed the McKinley project has prevented flooding and damage to the roadway in addition to adding a significant recreation area for the public's enjoyment. By taking advantage of the free tunnel material it was possible to construct this project for a total sum estimated to be similar to the cost of a conventional revetment using quarried armour stone.

Klode Park Beach Development, Whitefish Bay, Wisconsin

A three dimensional hydraulic model study conducted at a geometric scale of 1:20 was completed to develop a design for a shoreline restoration project to protect the Klode Park shoreline (approximately 200 m long), and to provide increased recreational space. The shoreline at Klode Park has a bluff with a height of approximately 20m. At the time the project was initiated the bluffs and shoreline had been steadily eroding to the point that slope failures were occurring which were threatening a water filtration plant located within the park.

Due to the existence of alongshore sand transport in this area it was necessary to limit the extent to which the project extended out into the lake and to account for sand bypassing the project. This was accomplished by constructing the offshore breakwaters within 60m of the original shoreline and by tapering the steel sheet pile groins which separated the beach cells. The groins were tapered to match the beach profiles to minimize the loss of beach fill during severe storms. The tapered groins also allowed native sediments to bypass the project behind the breakwaters. Because the distance from the offshore breakwaters to the permanent land based facilities is substantially shorter in this park as compared to Forest Park it was necessary to significantly modify the design of the beaches and other shoreline protection measures.

Variations over the Forest Park project include significantly reducing the breakwater gap width, tapering the groins, increasing the breakwater crest height, and adding stone revetments to the upper beach area. The revetments are buried and not noticeable to the public, but become exposed during severe storms at extreme water levels providing protection to the land based facilities and water filtration plant.

The final design, shown in Figure 10, consists of three offshore rubblemound breakwaters, which contain artificial beaches. The beaches consist of either all tunnel material similar to the McKinley project or tunnel material covered with coarse sand fill. The coarse sand has a D50 of approximately 2.6 mm placed over a general fill consisting of tunnel material similar to the McKinley project.

FIGURE 10 KLODE PARK FINAL DESIGN

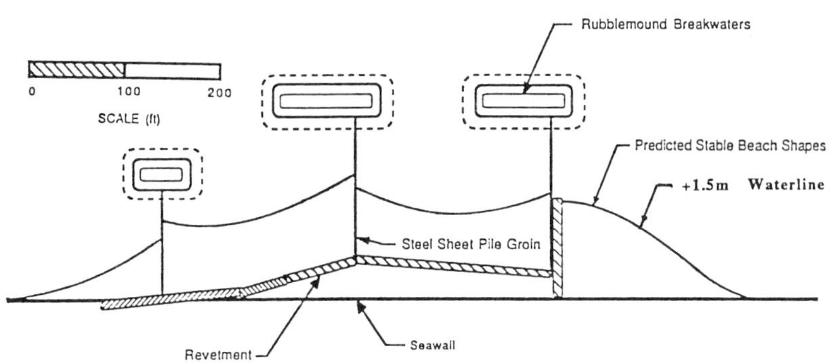

Public's perspective from the park down onto the beach and from the beach looking down the adjacent shoreline are provided in Figure 11.

FIGURE 11 PUBLIC PERSPECTIVES

Pesch Property Shoreline Protection, Lake Forest, Illinois

Similar to the other projects the high Lake levels in 1985 and 1986 had caused considerable shoreline erosion and bluff retreat along this section of the Lake Michigan shoreline. The project was required to prevent further erosion of a 130 m long privately owned section of shoreline where erosion was threatening the stability of the bluff and the property owner's house.

The project was completed in the winter of 1986 and consists of two offshore rubblemound breakwaters and two steel sheet pile groins which contain small pocket beaches filled with a coarse sand as shown in Figure 12. Rubblemound revetments provide additional protection along the shoreline behind the breakwater gaps where the beach fill is not sufficient to protect the bluff during storms. A physical model study was used to design the combination of offshore breakwaters, steel sheet pile groins and sand beaches.

This project was subjected to several severe storms during the winters of 1986-87 and 1987-88 and similar to the Forest Park project performed as predicted by the model investigation.

FIGURE 12 PESCH PROPERTY ARTIFICIAL BEACH

SUMMARY

Shoreline protection was the basic goal of each of the projects described above. In assessing the attributes of various shoreline protection schemes it was evident that the benefits of incorporating recreational areas into the projects outweighed the additional costs associated with creating them. Prototype experience has shown that these projects are extremely successful both in their ability to withstand storms on the Great Lakes and to attract people for recreational activities. Therefore each of these projects meets the basic criteria of providing effective shoreline protection while also providing excellent recreational areas, and demonstrates the numerous advantages associated with choosing artificial beach units versus standard protection measures.

REFERENCES

Bascom, W.N., (1951) from King, C.A.M., Beaches and Coasts, published by Edward Arnold Ltd., London, 1972.

Chew, S.Y., Wong, P.P., and Chin, K.K., "Beach Development Between Headland Breakwaters", Proceedings of the 14th Coastal Engineering Conference, Copenhagen, June 1974, pp. 1300-1418.

Clark, G., Personal Communication, January 1986.

Curren, C.R. and Chatham, C.E., Jr., "Imperial Beach, California, Design of Structures for Beach Erosion Control", TR H-77-15, U.S. Army Engineer Waterways Experiment Station, Vicksburg, Mississippi, August 1977.

Curren, C.R. and Chatham, C.E., Jr., "Oceanside Harbour and Beach, California, Design of Structures for Harbour Improvement and Beach

Erosion Control", TR HL-80-10, U.S. Army Engineer Waterways Experiment Station, Vicksburg, Mississippi, June 1980.

Dally, W.R. and Pope, J., "Detached Breakwaters for Shore Protection", draft report TR CERC-86-1, U.S. Army Engineer Waterways Experiment Station, Vicksburg, Mississippi, January 1986.

Dean, R.G., "Diffraction Calculations of Shoreline Planforms", Proceedings of the 16th Coastal Engineering Conference, Hamburg, August 1978, pp. 1903-1917.

Noda, H., "Depositional Effects of Offshore Breakwater Due to Onshore-Offshore Sediment Movement", Proceedings of the 19th Coastal Engineering Conference, Houston, September 1984, pp. 2009-2025.

Perez, J.M.B, and Fernandez, J.E., "Design of Pocket Beaches. The Spanish Case", Proceedings of the 21st Coastal Engineering Conference, Torremolinos, Spain, June 1988, pp. 463-464, Abstracts.

Rosen, D.S., and Vajda, M., "Sedimentological Influences of Detached Breakwaters", Proceedings of the 18th Coastal Engineering Conference, Cape Town, November 1982, pp. 1930-1949.

Sato, S. and Tanaka, N., "Artificial Resort Beach Protected by Offshore Breakwaters and Groins", Proceedings of the 17th Coastal Engineering Conference, Sydney, March 1980, pp. 2003-2022.

Seeling, W.N. and Walton, T.L., Jr., "Estimation of Flow Through Offshore Breakwater Gaps Generated by Wave Overtopping", CETA 80-8 U.S. Army Corps of Engineers, Coastal Engineering Research Center, Fort Belvoir, Virginia, December 1980.

Shinohara, K. and Tsubaki, T., "Model Study of the Change of Shoreline of Sandy Beach by the Offshore Breakwater", Proceedings of the 10th Coastal Engineering Conference, Tokyo, September 1966, pp. 550-563.

Silvester, R., Tsuchiya, Y. and Shibano, Y., "Zeta Bays, Pocket Beaches, and Headland Control", Proceedings of the 17th Coastal Engineering Conference, Sydney, March 1980, pp. 1306-1319.

Tourman, L., "The Creation of an Artificial Beach in Larvotto Bay - Monte Carlo Principality of Monaco", Proceedings of the 11th Coastal Engineering Conference, London, September 1968, pp. 558-569.

Toyoshima, O., "Design of a Detached Breakwater System", Proceedings of the 14th Coastal Engineering Conference, Copenhagen, June 1974, pp. 1419-1431.

Toyoshima, O., "Variation of Foreshore Due to Detached Breakwaters", Proceedings of the 18th Coastal Engineering Conference, Cape Town, November 1982, pp. 1873-1892.

Walker, J.R., Clark, D. and Pope, J., "A Detached Breakwater System for Beach Protection", Proceedings of the 17th Coastal Engineering Conference, Sydney, March 1980, pp. 1968-1987.

CHAPTER 212

THE BEACH REHABILITATION PROJECT AT OSTEND-BELGIUM

ir. R. SIMOEN*
ir. H. VERSLYPE**
ir. D. VANDENBOSSCHE***

Abstract

Steady erosion has occured on the beaches at Ostend, Belgium, especially on the section between the Casino and the Harbour Entrance. The gravel beach supported by a sandfill core has been proven to be a feasible and cost effective flood protection alternative, the first objective of the long term beach enhancement programme. Secondary goals are also met such as : new beach recreational facilities in safe conditions for the public ; the economic incentive for the needed redevelopment of this part of down-town Ostend and the beach front ; the conservation of the present sea-view from the promenade.

1. Introduction

Since the end of the 19th century the beaches at Ostend, Belgium have been popular holiday resorts. However, steady erosion has occured, especially on the beach section between the Casino-Kursaal and the Harbour entrance (see fig. 1). Indeed, the beach front promenade was constructed along the line of the 17th century fortifications of the city, which protrude into the sea beyond the natural coastline over ca. 150 m. A system of huge perpendicular groynes, ca. 200 m long could only slow down the erosion process.

----

* Inspector General, Ministry of Public Works, Coastal Service, Administratief Centrum, Vrijhavenstraat 3, B-8400 Oostende, Belgium

** Chief Engineer - Director, Ministry of Public Works, Coastal Service, Administratief Centrum, Vrijhavenstraat 3, B-8400 Oostende.

*** Technical Director, HAECON (Harbour and Engineering Consultants) N.V., Deinsesteenweg 110, B-9810 Gent.

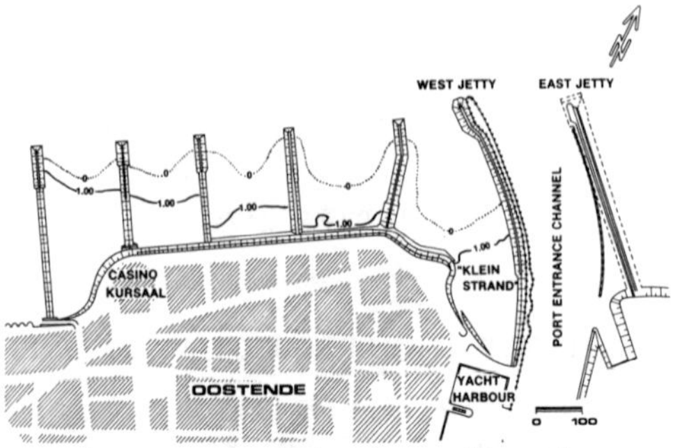

Figure 1. Existing situation

At that time, the coastal protection consisted of :

- a seawall up to Z + 9.80 at the promenade level, with a slope revetment of natural stone masonry on 10/4 slope ;

- a sandy beach with a gentle slope of 1:60, the level at the seawall toe is around Z + 2.80 to 3.00 m, almost the Mean Sea Level (MSL). As the beach had reached this rather low level, the main functions of the coastal defence, as depicted on cross-section (fig. 2) became progressively inadequate e.g. :

* flood protection : during storm surges Ostend's lower downtown became more frequently flooded by waves overtopping the seawall. During extreme storm surges as in February 1953, the seawall was even partially destroyed.

* recreation : remained only viable on the "Klein Strand" beach section (fig. 1). The remainder being flooded at mean sea level up to high water. Practical use of the beach is limited to only a brief period at low-water. Compared with other similar beach fronts at Ostend and elsewhere along the Belgian Coast the real estate properties are consequently less developed to modern age standards.

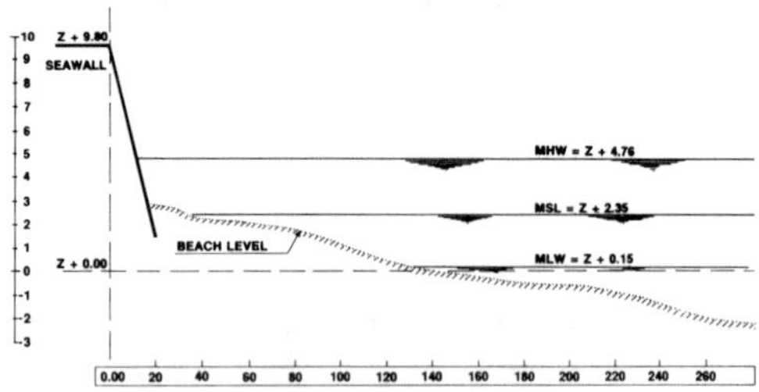

Figure 2. Typical cross-section existing beach

A flood protection enhancement programme, realized after the dramatic storm surge of February 1953 and extended after the winter storms of January 1976, resulted in a small storm surge barrier wall on top of the seawall up to level

Z+10.20 and associated storm surge sewer system. This programme gives results but only in the short term.

2. The beach rehabilitation project

The National Ministry of Public Works, Coastal Service entrusted HAECON, Consulting Engineers of Ghent, to elaborate a beach enhancement programme in order to provide in the long term :

- flood protection by preventing storm surge waves overtopping the seawall ;
- additionally the realization of a recreational beach potential by both a dry sun beach, and to a lesser extent a bathing beach.

3. Physical design data

3.1. Beach and shore morphology has been analyzed from beach profiles and bathymetry soundings from '71 up to '82.

The beach in the project area although at a low level is compared to other sections stabilized on a typical average slope of 1:60. The beach foot is situated at the Z-5.00 contour. From this to the LW-line the depth contours are almost parallel to the coast. Off-shore to deep water the wave attenuation is influenced by the system of banks (ca. Z-6.00) and channels (Z-8.00 to Z-10.00) of the Flemish Bank area.

3.2. Hydrographic data

A comprehensive set of data was available concerning : tides, currents, surges and waves.

3.2.1. Tides : the high and low water characteristics at spring tide are : MHWS = Z+4.75 ; MLWS = Z+0.15. The period of the semi-diurnal tide is : 12 h 17 min.

3.2.2. Surges : the exceedance frequency of high waters were established by linear regression of the tide observations of the Ministry of Public Works' Hydrographic Service from 1952 to 1967 (see fig. 4). From this the decennial surge level has been estimated at Z + 6.00 m ; the centennial level at 6.40 m.

3.2.3. Waves : from deepwater wave data, measured from 1977 to 1980 at the lightship Westhinder. The near-shore exceedance frequency of significant waves was established with mathematical refraction analysis. The annual significant wave height is estimated at Hs = 4.00 m ; the centennial significant wave height 4.80 m is depth limited. From this data the frequency of occurrence of wave periods was also established. The peak period of the wave energy spectrum is around 6 to 6.5 sec.

3.3. Beach and beach fill materials

The existing beach sand is a rather fine material, on average characterized by : D_{50} = 173 μ and D_{90} = 688 μ.

The typical grain size distribution of the beach fill sand from the Kwintebank and the gravel to be found off Dieppe, Wisant and Wight, Thames, is given on fig. 3. One should note that the sand quantity in the gravel load can reach 30 % by weight.

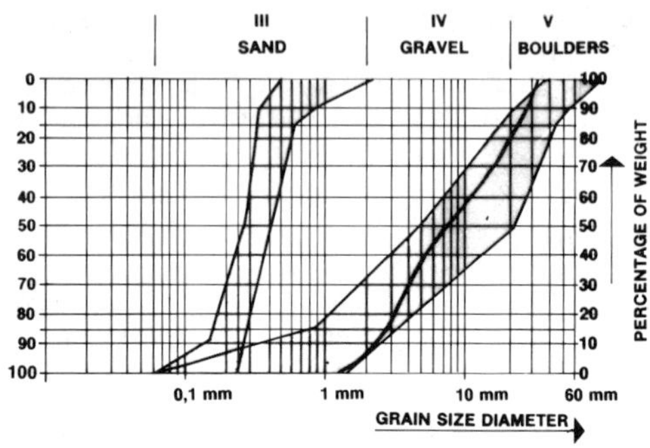

Figure 3. Grain size distribution tolerances of beach fill sand and gravel

4. Planning and design

4.1. For the beach rehabilitation, a straight sand fill solution was technically unfeasible because of the adjoining access-channel to the Port of Ostend. Indeed, in order to prevent higher sedimentation in this channel, the beach replenishment should be limited within the confines of the existing groyne system and the western harbour jetty.

4.2. The main conceptual design alternatives were assessed :

- Figure 4 - the spending beach concept : consisting of a beachfill from HW-level (Z+4.75) at the seawall down to LW-level whereas the beach foot is supported by a low crested rubble mound at level Z+0,75 parallel to the coast. The slope of the replenished beach is designed at 1:50.

- Figure 5 : the concept of a gravel beach supported by a 50 m wide sandfill core with the gravel beach crest at Z+6.00, the sand level at the seawall up to Z+7.00. Based on literature data the gravel slope has been designed at 1:7 and the required gravel layer thickness estimated around 3.00 m, for gravel with size distribution D_{50} = 12 mm / D_{90} = 30 mm, Ho = 4.00 m and T = 6 sec.

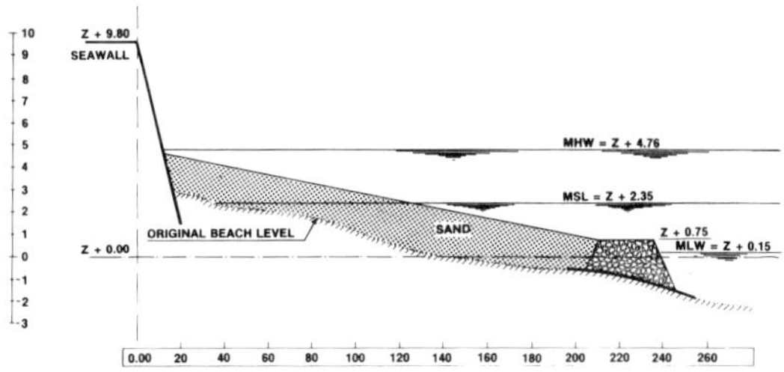

Figure 4. The spending beach concept

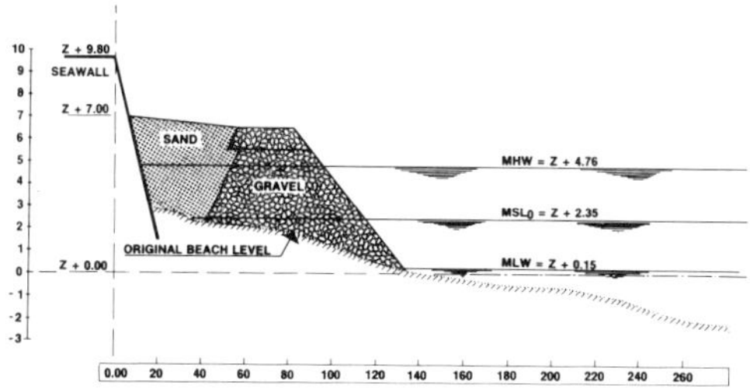

Figure 5. The gravel beach concept

Both the designs, together with the existing beach seawall configuration were evaluated for the occurrences of wave overtopping, transport capacity of beach material, accomplishment of design functions and other related aspects.

4.3. The results of the wave overtopping calculation are given in the table below.

	Average occurrence of wave overtopping (N/year)
* Existing situation (fig. 2)	21
* Spending beach	
- at designed level (fig. 4)	1.8×10^{-2}
- at assumed eroded level (-1.00 m)	2
* Gravel beach (fig. 5)	1.24×10^{-5}

TABLE : Estimated wave overtopping rate.

The results in the existing situation fit well with those experienced on site.

4.4. The morphological aspects have been assessed mainly by evaluating the longshore transport capacity especially the eastbound component towards the Port Entrance Channel.

4.4.1. Present situation : Longshore transport is to a large extent blocked by the groynes. The eastbound transport capacity for D_{50} = 173 μ / D_{90} = 242 μ can be estimated at 400,000 m^3/year. The resultant longshore transport is up to 180,000 m^3/year eastbound.

4.4.2. Spending beach : The longshore transport capacity for a beach fill of D_{50} = 371 μ /D_{90} = 600 μ has been estimated on 140,000 m^3/yr (eastbound component) and 20,000 m^3/year (resultant eastbound). The resultant longshore transport may be caught in the "Klein Strand" beach section near the West Harbour jetty. Consequently, a higher sedimentation risk in the access-channel is to be excluded.

Despite the decreased transport capacities, compared with the present situation, increased losses due to the higher turbulence caused by wave reflection on the seawall and breaking waves at the rubble mound toe, giving rise to concern about the long-term flood protection efficiency of the concept. Indeed, a lowering of only 1.00 m brings the wave overtopping almost to the same order of magnitude as the present situation ! (See table : from 1.8 x 10^{-2} tides/yr designed to 2 tides/yr, cfr. the present 21 tides/year).

The rubble mound structure prevents the normal exchange of beach material between the replenished upper beach and the foreshore. A prerequisite for a profile in dynamic equilibrium is therefore not met.

As all waves will break on the rubble mound the erosion pits near the structure would create a dangerous situation for the public. Bathing and swimming should not be tolerated.

4.4.3. Gravel beach : The longshore transport capacity will decrease as the beach material is increased by grain size. The resultant is lower than 200 m^3/yr and the eastbound component less than 8,000 m^3/yr.

The gravel beach and the foreshore represents a natural beach system both in morphology and in grain size distribution. A dynamic equilibrium can be expected with greatly reduced cross-sectional transport losses.

Compared with a sandy beach the gravel beach would not be the preferred choice for a bathing beach, however, by no means create dangerous situations for the public.

5. Detailed design

5.1. The retained gravel beach concept (fig. 5) has been designed in detail based on extensive physical scale model testing in the wave flume and wave tank facilities of the Ministry of Public Works' Hydraulic Research Station at Antwerp.

5.2. The aims of the test programme were :

1) the definition of the equilibrium slope and active layer thickness as a function of the waves ;
2) the study of wave overtopping by defining the water quantities overwashing the seawall ;
3) the study of some construction phases of the gravel mound in order to establish guidelines for construction.

5.3. The studies on items 1) and 2) were realized on the wave flume in a 2-dimensional model. The 3-dimensional model has been used for item 3).

The wave flume measures 63.5 x 4.08 x 1.45 m^3 and is e-quipped with a Arenfield wave generator and a hydraulic system to simulate the tidal changes of the water levels. Maximal tidal difference can be set up to 1.20 m. The geometric scale of the models was 1/12, time scale 1/3.464.

The model tides schematization was calibrated around a spring tide (LW = Z-0.10, HW = Z+4.00 and T = 12 h 30).

The irregular model waves correspond to the Pierson Moskowitz Spectrum, the theoretical spectrum best fitting with the shallow water wave conditions off the Belgian Coast.

The grain size distribution of the model gravel is given on fig. 3.

5.4. Equilibrium slope and required layer thickness

5.4.1. The results of the tests regarding equilibrium slopes as a function of Hs and active layer thickness are summarized in fig. 6.

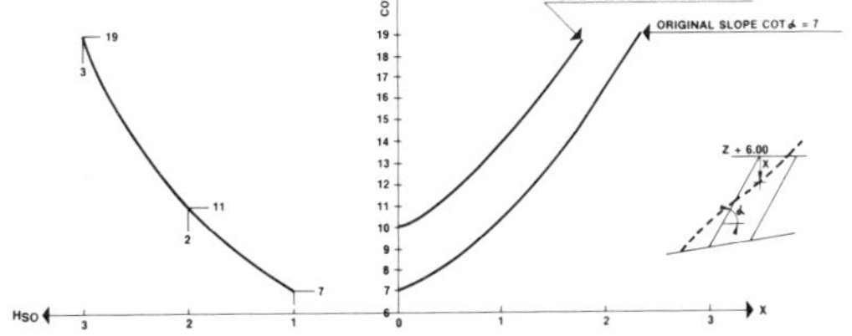

Figure 6. Equilibrium slopes (cotg α) and active layer thickness (x) in function of significant design wave (Hs)

The behaviour of the gravel beach under a schematic annual wave climate is reported on fig. 7.

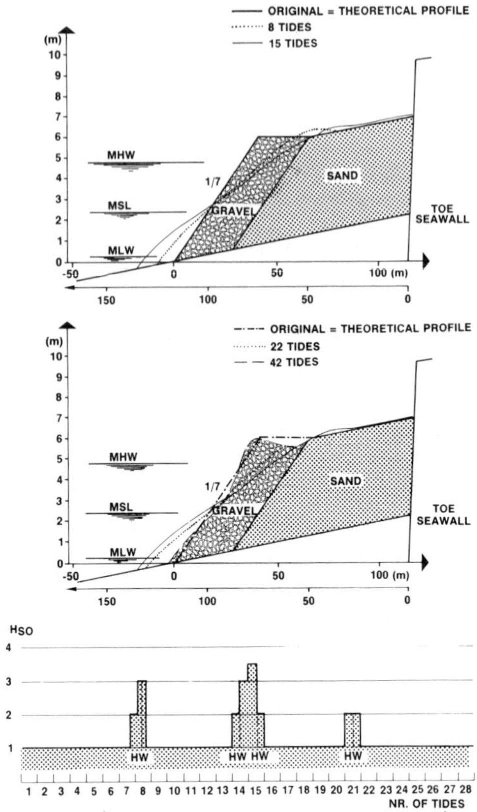

Figure 7. Behaviour of gravel beach profile line under schematic annual wave action

5.4.2. From the test series one determines that :

- the equilibrium slopes range from 1:7 over 1:11 to 1:19 for wave classes Hso = 1.00, 2.00 and 3.00 m respectively ;
- the equilibrium slopes are found roughly by a rotation of the slope around MSL.

5.4.3. For the design of the gravel beach following conclusions could be retained :

1) the theoretical design slope should be 1:8, because :

   * this slope corresponds with a wave of Hs = 1.25 which is exceeded 10% of the tides ;
   * with the slope of 1:7 the gravel beach crest at Z+6.00 could be eroded over the whole width of 25.00 m during one storm with Hs = 3.00 to 3.50 m (ca. 2 times/year) ;
   * for the schematic annual wave climat test the final slopes were : 1:7.50 to 1:9.

2) the theoretical design crest width may be 25.00 m with a slope of 1:8. Two consecutive tides of Hs = 3 to 3.50 m are needed to erode this width ;

3) the active layer thickness with a slope of 1:8 for H = 3 to 3.50 m is at equilibrium 2.00 m. A layer thickness of 2.50 to 3.00 m seems adequate, depending on the safety margin needed.

5.5. Wave overtopping

5.5.1. The test series on wave overtopping have been conducted starting from the equilibrium profile resulting from the annual wave climate test serie. With consecutive constant water levels of Z+4.75 m ; Z+5.75, Z+6.25 and 6.75 m the water quantities have been measured from irregular overtopping wave fields with Hs = 3.00, Hs = 3.50 and Hs = 4.00 m.

5.5.2. In the existing situation wave overtopping starts when HW = 4.75 m (ca. 100 times/year) and Hs = 2.00 m (ca. 25 times/year) or 3.4 tides per year. The same quantity of water (ca. 60 l/m'/min.) will pass by the gravel beach concept when HW = 6.25 m (1 time/25 years) and Hs = 4.00 m (1 time/year) or 5.5×10^{-5} tides per year.

As the beach level of the gravel beach design is lowered by between 0.5 and 1.00 m the quantities of overtopping water will only be increased by ca. 10 times. The long term flooding protection efficiency is thus well demonstrated.

5.5.3. Upon completion of all the wave overtopping test series the remaining profile is measured. The test record is given on fig. 8.

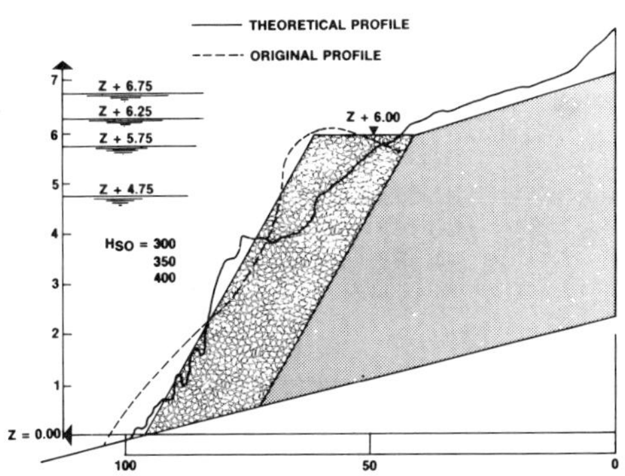

Figure 8. Gravel beach slope after wave overtopping test series ("crach-storm" surges)

The wave overtopping test series can be also used as "crash tests" for the design of wave stability. For the very unlikely event of combined surge waterlevels, wave action and duration, we observe that under no circumstances is the sandlayer touched. The beach, at worst, will only be covered by a gravel layer. After such a "crash-storm" the beach needs only be cleaned of the overwashed gravel.

5.6. Construction phases

5.6.1. In the wave flume two provisional round heads - one at level Z+5.00, the other at level Z+3.50 - have been tested under orthogonal irregular wave action. The test cycles consisted each time of 7 tides with Hs = 1.00 m, 1 tide with Hs = 2.00 and 1 tide with Hs = 3.00 m.

As shown on fig. 9, using a cross section taken 10.00 m out from the round head, some material will be transported towards the sea-wall mainly from Hs = 1.00 m, exceedance frequency ca. 10 % of the tides.

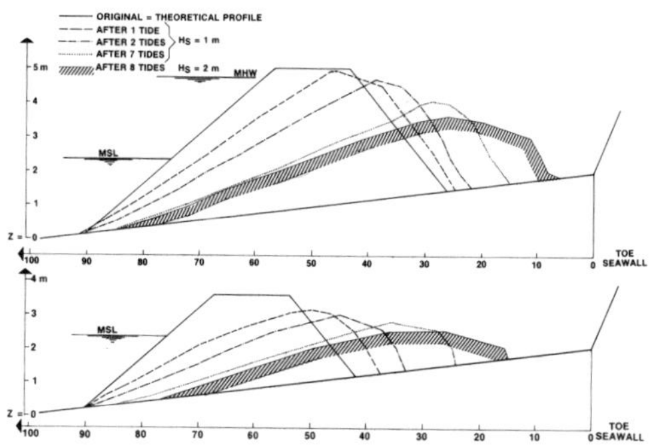

Figure 9. Construction phases - test series

5.6.2. The following guidelines for the construction could be formulated from these tests .

- the gravel beach structure should be constructed (at a maximum) in a calm wave season to avoid waves exceeding Hs = 1.00 m ;
- the gravel mound should be first constructed to a level above HW (Z+4.76) in order to decrease the amount of material out of the gravel profile and so as to redress the incurred damages during construction from a dry work level ;
- a gravel mound running section of that nature will probably resist waves of between Hs = 1.00 m and 1.50 m. At the round head some damage will occur which can be repaired.

5.6.3. Finalizing the detailed design : taking into account the construction constraints and guidelines several execution methods have been identified and analyzed. The final choice will be that of the succesful contractor in accordance with the job-specifications.

5.6.4. The detailed design has been finalized as shown on fig. 10 and 11. The construction phases anticipate the possible gravel losses towards the sandfill section. The project covers ca. 850 m of beach front. At both ends some minor works had to be conceived to finish the gravel beach. The volumes of material are : sea sand : 300,000 m^3 ; sea gravel : 180,000 m^3. The works have been estimated at 260 M BF or some 7,5 M US$. A construction time of 3 months inclusive down-time is thought feasible.

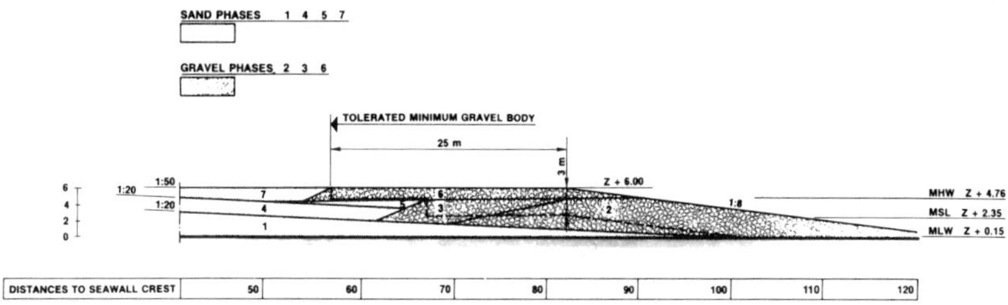

Figure 10. Final design cross section gravel mound and sandfill beach construction phases

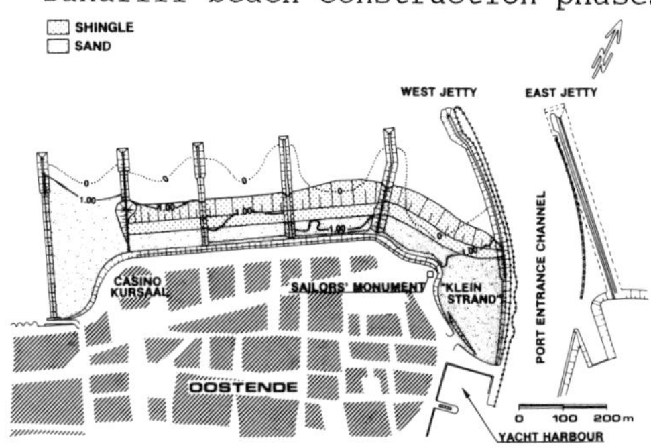

Figure 11. Final layout beach rehabilitation project depicted on the existing bathymetry, the groyne system and jetties are existing and will remain

6. Conclusion

6.1. The gravel beach supported by a sandfill core has been proven to be a feasible and cost effective flood protection alternative, the first objective of the scheme.

6.2. The following secondary goals have also been met :

- new beach recreational facilities in safe conditions for the public ;
- the economic incentive for the much-needed redevelopment of this part of down-town Ostend and the beach front ;
- the conservation to a large extent of the present sea-view from the sea promenade.

REFERENCES

- HAECON Gent, various planning and design reports (in dutch).

- Hydraulic Research Station Antwerp : Reports on the physical model testseries of the gravel beach concept (in dutch).

CHAPTER 213

AN ENGINEERING STUDY OF OCEAN CITY'S BEACHES, NEW JERSEY, USA

J. Richard Weggel, F ASCE (1)
Scott L. Douglass, AM ASCE (2)
Robert M. Sorensen, F ASCE (3)

INTRODUCTION

Ocean City, New Jersey is a major coastal resort in the heavily populated northeastern part of the United States. It is located on a 13 km-long barrier island about 13 km south of Atlantic City and 48 km north of Cape May, the southernmost point in New Jersey. See Figure 1. The barrier island, called Peck's Beach, is bounded on the north by Great Egg Harbor Inlet and on the south by Corsons Inlet.

Because wide recreational beaches are important to Ocean City's economy, an engineering study of the beaches was undertaken: a) to quantify natural and man-made shoreline changes, b) to quantify tides, sea level changes, waves, longshore sand transport rates, seasonal variations in beach width and Great Egg Harbor Inlet processes and incorporate them into a sediment budget for Ocean City, c) to determine why past attempts to maintain wide recreational beaches were only partially successful, and d) to recommend a plan to establish and maintain wide beaches.

EXISTING CONDITIONS

The northerly end of Ocean City is the most densely developed part of the city. Here the barrier island is at its widest at about 1.4 km. Heavily used commercial development and tourist facilities are also located in this area. Further southward the island narrows to about 0.9 km and development consists primarily of single and multiple family dwellings. The streets in Ocean City are numbered from north to south with 9th Street being about the commercial and tourist center. A boardwalk extends from First Street in the north, southward to 23rd Street. Timber sheet-pile bulkheading parallels the shoreline behind much of Ocean City's beach. In some areas, e.g. south of 47th

---

(1) Professor & Head, Department of Civil & Architectural Engineering, Drexel University, Philadelphia, PA, USA.
(2) Senior Project Engineer, PARE Engineering, Inc., Lincoln, RI, USA.
(3) Professor, Department of Civil Engineering, Lehigh University, Bethlehem, PA, USA.

Street, the bulkhead is fronted by rubble toe protection and the high water shoreline is up against the bulkhead. At the heavily used northern end of Peck's Beach the shoreline has been stabilized by a number of long, high, impermeable rubble-mound groins. Groin crest elevations vary but average about 2.5 m above the local mean low water datum which is 0.58 m below the mean sea level of 1929 (National Geodetic Vertical Datum, NGVD). There are 18 groins between the terminal groin at Great Egg Harbor Inlet and 21st Street. Groin spacing ranges from 225 m to 370 m and averages about 300 m. In addition, there is a terminal groin at 59th Street, the southern end of the developed part of the island. There are no groins between 21st Street and 59th Street. South of 59th Street the shore is characterized by natural sand dunes on the sand spit which comprises the northern boundary of Corson's Inlet.

HISTORICAL SHORELINE CHANGES

Historical shoreline changes at Ocean City were studied by analyzing 21 sets of controlled, vertical, aerial photographs taken between April 1952 and September 1984. Most photos had a scale of 1:4800. The "wet line" or the interface between the wet and dry beach was identified at 46 stations along Ocean City's shoreline. This line is easily identifiable and is close to the limit of wave runup during the preceding high tide and to the berm line. The location of the wet line relative to a local baseline, usually the centerline of a shore-parallel street, was noted. Most of the 46 stations were located close together along the more northerly beaches to define shoreline changes within the groin compartments there. Stations were spaced further apart along the southerly beaches where spatial variations were smaller. Linear regression lines were fitted to the plots of wet line location as a function of time.

Interestingly, with only minor exceptions, the analysis found that the shoreline in 1984 was in almost the same location as it had been 32 years earlier in 1952. In addition, the regression lines showed that the time-averaged erosion rate was small or nearly zero even though there had been significant fluctuations of the shoreline about the average. In at least two locations, narrowing of the beach was due to encroachment from the landward side rather than by beach erosion.

The air photo analysis incorporated into it the effect of beach nourishment activities carried out by the Corps of Engineers and the City of Ocean City between 1952 and 1984. During that time, 6.4 million cu m (8.4 million cu yd) of sand were placed on Ocean City's beaches in three major beach fills in 1952, 1959 and 1982 and in numerous smaller fills between 1970 and 1981. Watts (1956) documented the performance of the relatively successful 1952 fill, one of the earliest beach fills ever undertaken, in which 1.95 million cu m (2.6 million cu yd) of sand was placed on the beach. During the 11 year period from 1970 to 1981, the City of Ocean City operated a dredge behind the barrier island and pumped an average of 166,000 cu m (217,000 cu yd) of relatively fine sand onto the beaches each year. Thus the average rate of beach filling from all sources between 1952 and 1984, the period covered by the air photo analysis, was 198,000 cu m/yr (259,000 cu yd). While the 1952 fill was relatively successful, the 1982 fill, placed during September and October, was short-lived and most of it was lost during that first winter. The winter of 1982-83 was marked by 18 storms out of the northeast which contributed to the rapid loss of fill (Farrell & Inglin, 1988).

Ocean City's beaches are relatively stable, but only at the expense of placing an average of 200,000 cu m (262,000 cu yd) of fill per year on them. However, the problem is beaches that are too narrow rather than beaches that are experiencing excessive erosion. The solution is to initially widen Ocean City's recreational beaches and then to stabilize and maintain these wider beaches.

SEASONAL BEACH PROFILE CHANGES

Quarterly surveys of 17 beach profile lines were made in September 1985, December 1985, May 1986 and June 1986 to determine what seasonal changes might be typical of Ocean City's beaches. Horizontal movements of 25 m at the mean sea level shoreline were not unusual during this period. In fact, changes of about 35 m occurred near the center of the island between 14th and 22nd Streets. See Figure 2. Most changes occurred between the September survey and the December survey with a further, less dramatic, shoreline recession between December and May. In general, there was little change between May 1986 and June 1986. A dramatic loss of sand occurred along the Great Egg Harbor Inlet shoreline between September and December 1985. The wide beaches and large sand dunes in the inlet area were eroded during two storms that occurred in late September and October. Shortly after the September survey, hurricane "Gloria" threatened the New Jersey coast. While most of Ocean City's beaches were spared major damage because the hurricane was too far offshore, the Great Egg Harbor Inlet beaches of Ocean City were severely eroded. Large volumes of sand, stockpiled in dunes along the inlet shoreline, dissappeared and two groins, which had been constructed in 1979 and buried beneath the sand, were exposed. See Figure 3.

WAVE AND POTENTIAL LONGSHORE TRANSPORT ENVIRONMENT

The wave and longshore sand transport environment in the vicinity of Ocean City were investigated using the hindcast wave data developed under the Corps of Engineers' Wave Information Study (WIS) (Jensen, 1983). WIS Station 62, Peck Beach, is located approximately offshore of 40th Street and best represents the average wave conditions in Ocean City. The next station to the north (Station 61) is at Atlantic City while the next station to the south (Station 63) is at Seven Mile Beach. Maximum offshore wave heights in 10 meters of water for Station 62 are given in Figure 4 along with an estimate of their return period.

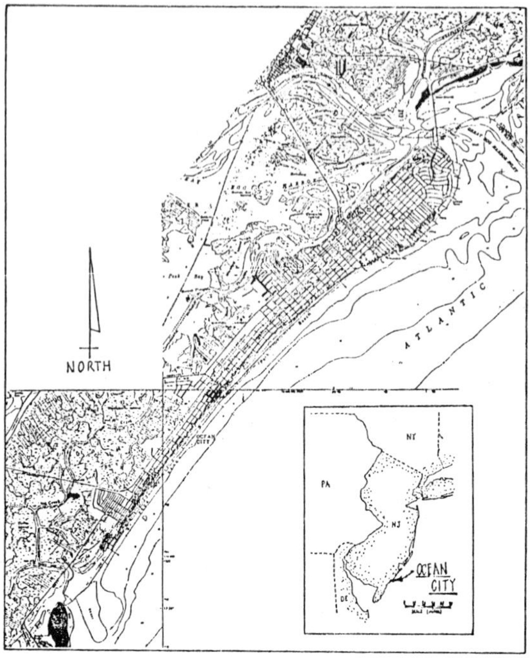

Figure 1 Location Map, State of New Jersey and Ocean City, New Jersey

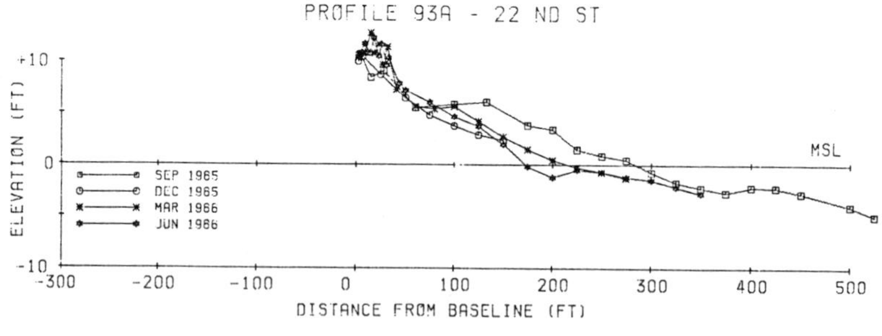

Figure 2 Beach Profiles at 22nd Street Showing Seasonal Shoreline Movement

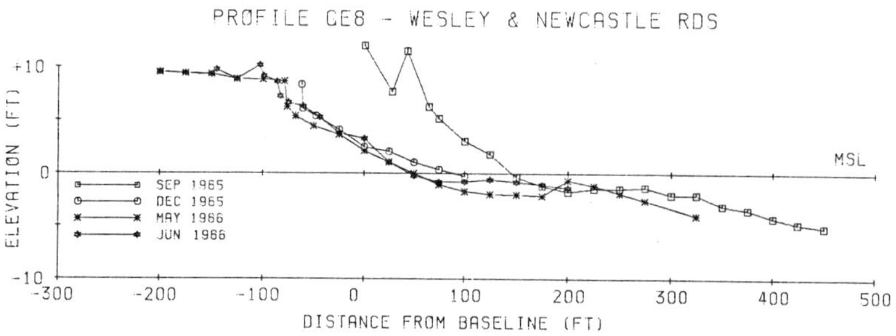

Figure 3 Beach Profiles at Great Egg Harbor Inlet Shoreline Showing Sand Losses between September and December attributable to Hurricane "Gloria"

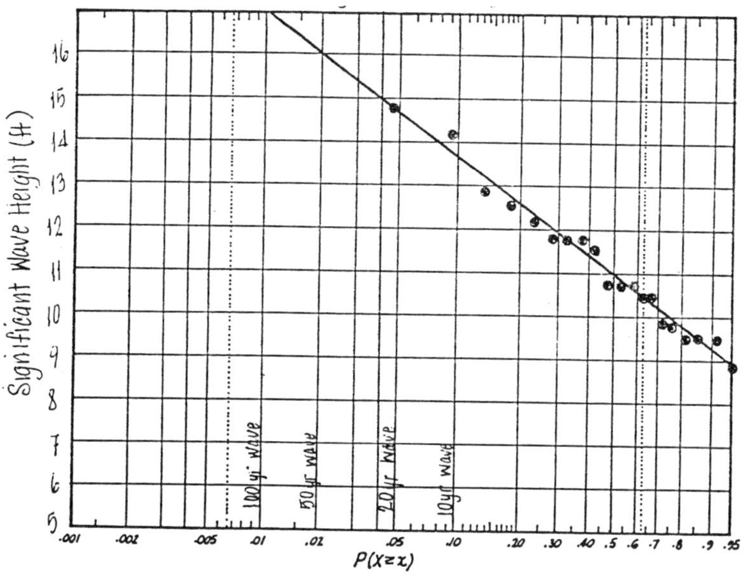

Figure 4 Frequency of Occurrence of High Waves in 10 Meters of Water at Station 62, Peck Beach (WIS Hindcast Data)

The WIS tabulated wave heights, periods and directions were used with the CERC formula to calculate potential longshore sand transport rates. In most cases calculated transport rates represent only potential transport because of: a) the effect of groins on transport, b) local reorientation of the shoreline, and c) the limited amount of sand available for transport in some areas. Transport rates calculated from the Station 62 WIS data were analysed statistically to determine how often given levels of transport are exceeded. The calculated transport rates were separated into two populations: positive transport (transport to the right for an observer looking seaward) and negative transport. Each data set was then listed in rank from the highest transport rate to the lowest. The number of hours that each transport rate prevailed was accumulated to obtain two freqency plots, one for positive transport and one for negative transport. The resulting longshore transport frequency data was plotted on log-normal probability paper to obtain the plots shown in Figure 5. Thus the longshore transport environment can be summarized by six parameters, the mean and variance of each of the two log-normal distributions (a positive transport distribution and a negative transport distribution) and the fractions of time the transport is positive, negative or calm (Weggel & Perlin, 1988; Weggel, Douglass & Tunnell, 1988). Driven by waves out of the northeast, sand moves southwestward only about 38% of the time at Station 62, Ocean City. Sand moves northeastward about 60% of the time. While northeastward transport occurs over a greater fraction of time, it is at a lower rate since it is driven by the relatively smaller waves out of the southeast during the summer months. Episodes of southwestward transport (northeasters), while less frequent, are more intense and carry more sand in the course of a year. This results in a net southwestward longshore transport. The potential net southwestward transport is estimated to be 56,000 cu m/yr (73,000 cu yd/yr). The northeastward transport is estimated at 540,000 cu m/yr (706,000 cu yd/yr) and the southwestward transport is 596,000 cu m/yr (779,000 cu yd/yr) for an annual gross transport rate of 1,136,000 cu m/yr (1,485,000 cu yd/yr). Obviously, these net and gross transport rates vary from year to year. The numbers given here are estimates of the 20-year average conditions summarized in the WIS hindcast data.

Similar analyses were performed for WIS Stations 61 and 63. Analysis of the Station 61 data, Atlantic City, predicts a northward net longshore transport on Absecon Island. This is contrary to evidence obtained from observations of sediment accumulation at structures on Absecon Island, particularly the terminal groin on the Longport side of Great Egg Harbor Inlet; consequently, Station 61 data was not used in the present analysis. This discrepancy between predicted and observed net transport rates shows that care must be exercised in interpreting transport rates computed from the WIS data and that other evidence to confirm or deny transport predictions should be sought. Quite often, as is the present case, net transport rates are the difference between large upcoast and large downcoast transport rates.

Small errors in estimating either the upcoast or downcoast rate can result in a large error in the net rate.

GREAT EGG HARBOR INLET PROCESSES

Great Egg Harbor Inlet has a significant effect on Ocean City's beaches, particularly those beaches adjacent to the inlet at the north end. However, even the beaches exposed to the ocean in the vicinity of the inlet are influenced by the effect of the inlet's ebb tidal shoal on incident waves. The prevailing southerly longshore sediment transport in the area causes the ebb tidal shoal to elongate so that it frequently extends southward from the inlet in the form of a shore-parallel offshore bar. This bar affords some protection from high waves to these beaches and thus influences the magnitude and direction of longshore transport near the inlet. In addition, Great Egg Harbor Inlet serves as a sediment trap for sand carried into it from both Longport's beaches to the north and Ocean City's beaches to the south. Historically, before construction of the stabilizing terminal groin/jetty on the Longport side of the inlet, the inlet was migrating northward. See Figure 6. Following construction of the jetty, the inlet's location stabilized but Ocean City's inlet shoreline has since been subjected to alternating periods of accretion and erosion as the channel thalweg moves back and forth in response to the influx and trapping of sand.

An analysis was made of the sedimentation patterns and of the volume of sediment trapped within the inlet. Extensive hydrographic surveys of the inlet had been made by the U.S. Army Corps of Engineers in 1965 and 1984. These two surveys coincided with the time period for which a sediment budget of the area was developed. Contours of the differences in depth between the two surveys were constructed as shown in Figure 7. The resulting contours show areas within the inlet where accretion occurred (solid lines) and areas where scour occurred (dashed lines). The figure demonstrates the southward movement of the channel thalweg since the accretion contours on the northeasterly side show filling of the old channel thalweg and the scour contours on the southwesterly side show the scouring of a new channel thalweg. The southward migration of the thalweg is also shown in Figure 8 which is a profile through the inlet's throat.

In addition to the southward migration of the inlet channel, there was an net increase of the volume of sand within the inlet. Between 1965 and 1984 the inlet accumulated approximately 3.6 million cubic meters (4.7 million cubic yards) of sand, presumably stripped from the Longport beaches to the north and Ocean City beaches to the south. This corresponds to an annual rate of trapping of 179,000 cu m/yr (234,000 cu yd/yr).

SEDIMENT BUDGET

A sediment budget for Ocean City's beaches was developed for the 19 year, 9 month period between 1964 and 1983. These

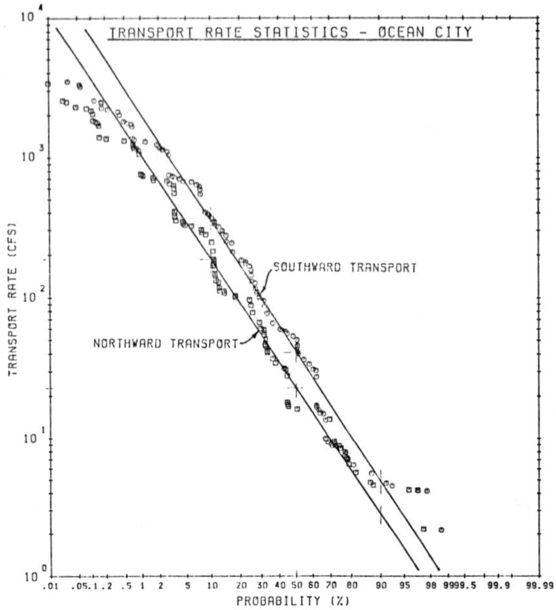

Figure 5 Log-Normal Probability Distributions for Positive and Negative Potential Longshore Sand Transport Rates at Station 62, Peck Beach (Computed from WIS Hindcast Data)

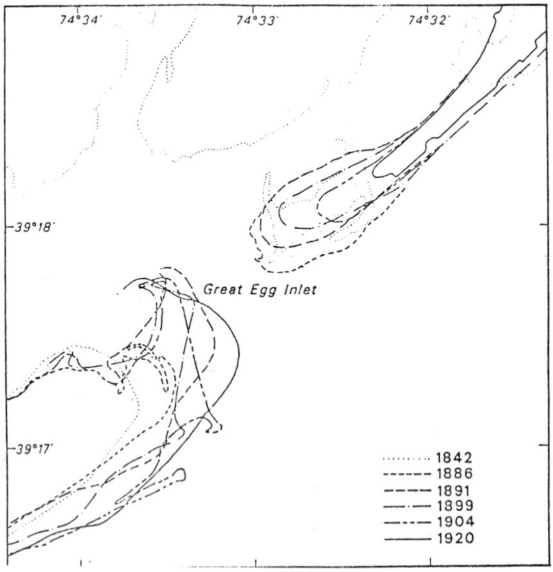

Figure 6 Historical Locations of Great Egg Harbor Inlet Showing Northward Migration Prior to Construction of Terminal Groin at Longport

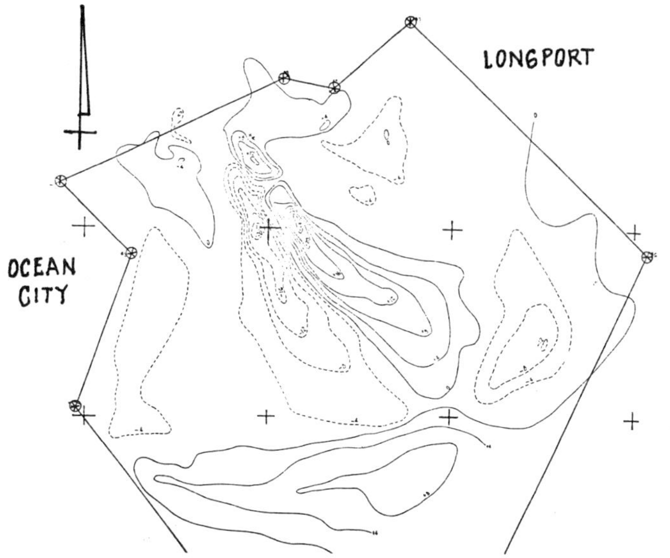

Figure 7 Bathymetric Changes Occurring within Great Egg Harbor Inlet Between 1965 and 1984

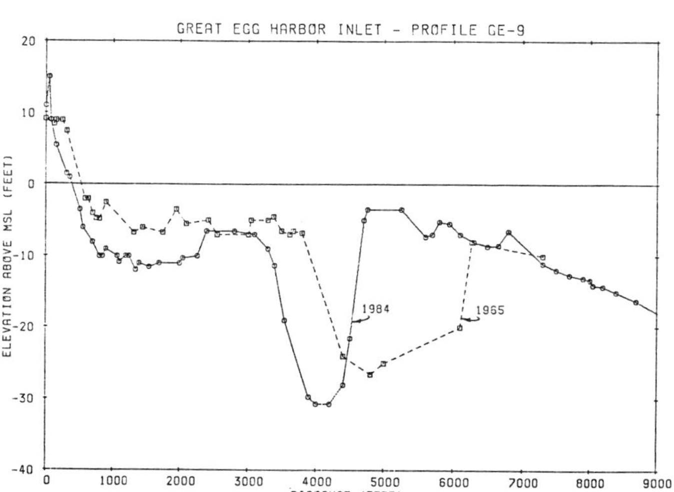

Figure 8 Profiles Through Throat of Great Egg Harbor Inlet Showing Southward Migration of Channel Thalweg

dates were dictated by the times for which both inlet bathymetric data and beach profile data were available. The area was divided into three cells and the transport of sediment into and out of each cell was balanced against the change in sediment volume within the cell. The three cells were: Ocean City's beaches from the inlet south to about 40th Street, Great Egg Harbor Inlet, and Longport's beaches on the southern end of Absecon Island just north of the inlet. See Figure 9. Sand volume changes on both Longport's beaches and Ocean City's beaches were determined from beach profile surveys obtained by the Philadelphia District of the Corps of Engineers. Sand volume changes in the inlet were obtained from two bathymetric surveys obtained in 1964 and 1983 by the Corps. Offshore losses were assumed to be caused by the long term increase in relative sea level as measured at Atlantic City. This rate is about 0.01 foot/year and is believed to be one-half due to sea level rise and one-half to local subsidence. Offshore losses were determined using the Bruun rule (Bruun, 1962) with the closure depth determined by the semi-logarithmic profile methodology outlined in Weggel (1979). The closure depth was estimated at about 7.9 to 8.5 meters (26 to 28 feet) below mean low water and erosion rates were estimated at between 7.0 and 9.3 cu m/m/yr (75 to 100 cu ft/ft/yr) yielding a shoreline recession rate of about 0.96 m/yr (2.5 ft/yr).

The results of a simplified sediment budget are shown in Figure 9. The three cells are shown schematically along with the contributions to and losses from each cell. Units on the figure are in thousands of cubic meters per year. The 7,000 m (23,000 ft) long Longport cell lost 47,400 cu m/yr (62,000 cu yd/yr) offshore, gained an unknown amount of sand from Atlantic City's beaches to the north and lost an unknown amount of sand to Great Egg Harbor Inlet to the south. The annual rate of increase in sand volume on Longport's beaches was 40,500 cu m/yr (53,000 cu yd/yr). The inlet lost an estimated 17,600 cu m/yr (23,000 cu yd/yr) offshore and gained unknown amounts of sand from Longport's and Ocean City's beaches. The net accumulation of sand in the inlet was at a rate of 178,900 cu m/yr (234,000 cu yd/yr). The Ocean City cell lost an unknown amount of sand to the inlet, 79,500 cu m/yr (104,000 cu yd/yr) offshore and 55,800 cu m/yr (73,000 cu yd/yr) to the beaches at the south end and the spit at Corsons Inlet. Ocean City gained 141,100 cu m/yr (195,000 cu yd/yr) through beach nourishment; however, there remained a net loss of 100,900 cu m/yr (132,000 cu yd/yr) from the beaches. Solving the three simultaneous equations that result, yields values for the three unknown transport rates: the transport into the Longport cell from Atlantic City is 169,700 cu m/yr (222,000 cu yd/yr), transport into the inlet from Longport is 81,800 cu m/yr (107,000 cu yd/yr), and transport into the inlet from Ocean City is 114,700 cu m/yr (150,000 cu yd/yr). Obviously, these results are sensitive to the quality of the data used to determine them. The longshore transport rates are most suspect; in fact, longshore transport into the Longport cell from Atlantic City should be known from the WIS wave hindcast longshore sand transport analysis. However, because the magnitude and direction of this longshore rate

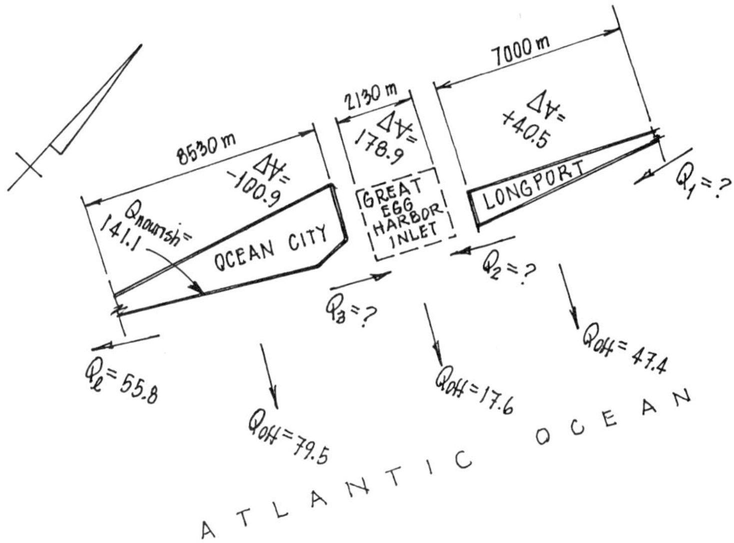

Figure 9 Schematic of Sediment Budget Analysis Showing Longport Cell, Great Egg Harbor Inlet Cell, and Ocean City Cell

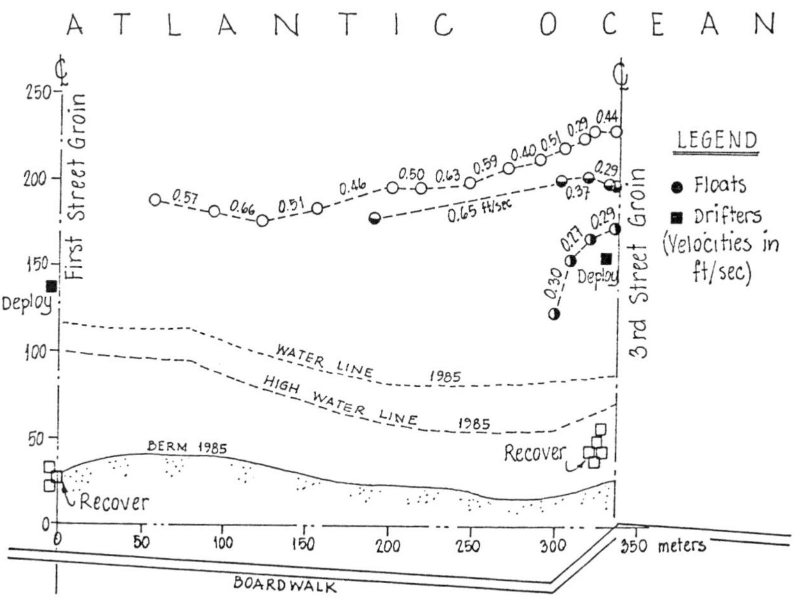

Figure 10 Current Patterns in Compartment Between Groins at First and Third Streets, Ocean City, New Jersey, 2 November 1986

is questionable, it was taken to be an unknown in the analysis. Actually, the rate of sand loss to the beaches south of Ocean City might also differ from the 55,800 cu m/yr (73,000 cu yd/yr) assumed here. Larger losses to downdrift beaches would result in larger transport rates into the Longport cell at the northern end of the analysis; smaller losses would result in smaller transport rates into the Longport cell.

CIRCULATION STUDY IN GROIN COMPARTMENT BETWEEN FIRST AND THIRD STREETS

The groin compartments along the northerly beaches of Ocean City are believed to induce circulation cells that exacerbate the loss of beach fill during storms. Even during high storm water levels, the high rubble mound groins block longshore transport and divert sand offshore in rip currents that form adjacent to the groins. A field study was undertaken offshore of the beach compartment formed by the groins at First and Third Streets to evaluate the circulation patterns that might arise within the groin compartments. Surface currents were measured by tracking plywood surface floats equipped with a small flag. (In subsequent tests, plastic milk bottles partially filled with sand and water were used instead of plywood floats. These proved easier to deploy and always floated with the flag upward since the body of the bottle was submerged.) The floats were released from the groins and followed using two surveyor's transits set up a known distance apart along a shore-parallel baseline. Simultaneous recording of the angle between the float's location and the baseline by the two transits along with the time of the observation allowed the pathlines of the floats to be plotted and their average velocity between observations calculated.

The results of a field test conducted on 2 November 1986 are shown in Figure 10. During that test, transport was northward with current velocities averaging about 0.15 m/sec (0.5 ft/sec). The wave period was about 8.5 sec and breaker heights averaged about 0.55 m (1.8 feet). Longshore current speeds, measured by throwing dye into the surf, were about 0.2 m/sec (0.66 ft/sec). Figure 10 shows no evidence of rip current development along the First Street groin; however, initial observations made with floats released from the First Street groin showed offshore movement. Unfortunately, the floats often became trapped within the rocks of the groin. The floats also capsized and could not be seen with the transits.

FINDINGS AND RECOMMENDATIONS

Ocean City's beaches have been relatively stable, at least over the period of observation covered by this study. In 1984 the beaches were in essentially the same location as they were in 1953. This is believed in part to be due to the high, impermeable, rubble-mound groins constructed to stabilize the northerly beaches, and in part to the occasional renourishment of the beaches. Over the 32 year period between 1952 and 1984, more than 6.4 million cu m

(8.4 million cu yd) of sand were placed on the beaches. Some of these beach nourishment projects were relatively successful, e.g. the 1952 fill, while others, such as the 1982 fill, were very short-lived. Between 50% and 90% of the beach fill placed late in 1982 was lost before the following summer (Farrell & Inglin, 1988). The existing beaches are not wide enough to accomodate the many beach users during the summer months. Also, in the commercially developed areas along Ocean City's narrow northerly beaches, the beach affords little protection to the area behind it. The problem is to widen the beach and to stabilize it in a new, equilibrium position that is farther seaward than the present equilibrium position.

The long, high, impermeable groins contribute a level of stability to the beaches by sheltering a length of the shoreline from northeast storms; however, they probably also contribute to the rapid loss of sand during storms by inducing the formation of rip currents. Following the placement of a beach fill, groin compartments are full of sand and the more-seaward, less-protected shoreline is exposed and vulnerable to wave attack during storms. During storm periods with high water levels and swift longshore currents, the longshore currents are deflected seaward as rip currents form along the groins. The rip currents quickly transport beach fill sand offshore. After the shoreline has eroded back to approximately its pre-fill location, it is again sheltered by the groins and a dynamic, near-equilibrium condition is again established.

Lowering and lengthening the existing groins was recommended to increase the residence time of future beach fills. Lower groins would allow sand to spill over them during high storm water levels rather than have the sand diverted seaward by rip currents. While the sand might not remain in the groin compartment in which it was placed, it would at least remain on the beach rather than be diverted offshore. The modified groin profiles would approximate the natural beach profile from about the +2.4 m (+8 foot) berm elevation out to the mean high water line; from here, the groin crest would extend seaward horizontally. See Figure 11. Stone removed by lowering the groins will be used to lengthen and extend them seaward. Also, several existing groin spurs (rock extensions to the groins that extend perpendicular to the groin axis) would be lowered or removed and the stone used to lengthen the groins.

Approximately 1.1 million cu m (1.4 million cu yd) of beach fill is required to provide at least a 30.5 m (100 ft) wide berm having an elevation of +3.0 m (+10 ft) between the northern end of the barrier island and about 20th Street. The Corps of Engineers has recommended a more extensive beach fill extending from the inlet southward to 59th Street. The fill proposed by the Corps will require about 3.8 million cu m (5 million cu yd) of sand. Construction of the groin modifications is scheduled to begin in late 1988 or early 1989 and the Corps of Engineers beach fill project is scheduled for 1990.

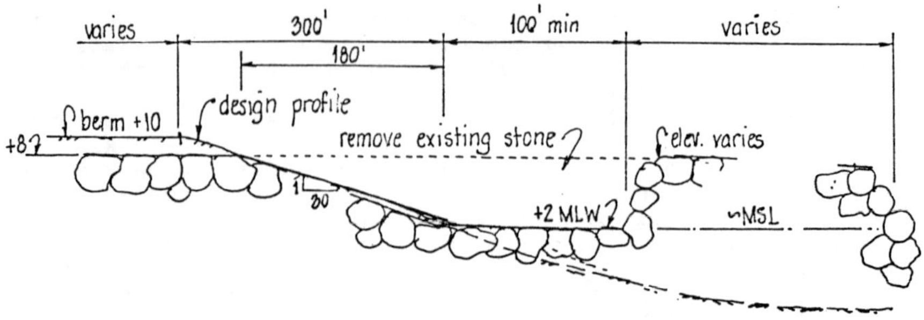

Figure 11 Proposed Modified Groin Profile

ACKNOWLEDGEMENTS

The authors wish to acknowledge the contributions to this paper made by Monica Ippolito, Michael Strzelecki, Craig Young, Sherrerd Steel, Steve Kemp, David Weggel, Ted Wilson, Juan Carlos Escajadillo, and Dave Givler, all of whom, at one time or another were involved in data collection and/or analysis. In addition, John Carey, Joseph Kane, John Gideonse, Jack Schreiner, Jr. of Ocean City; Bernard Moore and John Garofolo of the Bureau of Coastal Engineering, and Jeff Gebert of the Philadelphia District of the Corps of Engineers were all helpful in providing information used in the course of this study.

LITERATURE CITED

BRUUN, P. (1962) "Sea Level Rise as a Cause of Shore Erosion," Journal of the Waterways and Harbors Division, ASCE, Vol. 88, No. 1, pp 117-130

FARRELL, S.C. & D.C. INGLIN (1988) "A Summary of Beach Survey Data from the 1982-83 Ocean City, NJ Beach Fill Administered by the NJ Department of Environmental Protection," prepared for the NJ Division of Coastal Resources, Coastal Research Center, Stockton State College, Pomona, NJ, September 1988.

JENSEN, R.E. (1983) "Atlantic Coast Hindcast, Shallow Water, Significant Wave Information," WIS Report 9, U.S. Army Waterways Experiment Station, Vicksburg, MS 39180.

WATTS, G.M. (1956) "Behavior of Beach Fill at Ocean City, New Jersey," U.S. Army Corps of Engineers, Beach Erosion Board, Technical Memorandum 77, February 1956.

WEGGEL, J.R. (1979) "A Method for Estimating Long Term Erosion Rates from a Long Term Rise in Water Level," CETA 79-2, U.S. Army Coastal Engineering Research Center, Fort Belvoir, VA 1979.

WEGGEL, J.R. & M. PERLIN (1988) "Statistical Description of Longshore Transport Environment," Journal of Waterway, Port, Coastal and Ocean Engineering, ASCE, Vol. 114, No. 2, March 1988.

WEGGEL, J.R., S.L. DOUGLASS & J. TUNNELL (1988) "Sand Bypassing Simulation Using Synthetic Longshore Transport Data," Journal of Waterway, Port, Coastal & Ocean Engineering, ASCE, Vol. 114, No. 2, March 1988.

CHAPTER 214

SHORE PROTECTION OF KAOHSIUNG HARBOR OUTER BREAKWATER

Kuo, Shih-Duenn *

ABSTRACT

As shown in Fig.1 the main fairway of Port of Kaohsiung is running from north to south with a length of 12 kilometers. To the west, there is a sand bar served as a natural breakwater, the so called outer Breakwater, to provide the port and resident's property the necessary safety and security. Recently, due to rapid growth of container traffic and bulk cargo transportation, Port of Kaohsiung has developed some parts of the outer breakwater near the 2nd Harbor Entrance as Container Terminal No.4 with seven 14-meter-deep, 320-meter-long container berths and about 100 hectares container yard on northern side, and coal Terminal with 16-meter-deep, 320-meter-long berth and a huge storage yard on southern side. Therefore, the shore protection along the area become more and more important year by year. So far, some 6-kilometer seawall and 20 groynes have built successively, and already played a very important role for the shore protection there.

INTRODUCTION

Port of Kaohsiung is the largest port in Taiwan, R.O.C. and handles more than two-thirds of the Island's total imports and exports, over 100 million tons of cargo annually. It is expected to become one of the top international seaports in the world in terms of freight handled and physical size resulting from the huge expansion projects which have completed, now under construction or in planning stage.

Starting from 1967, Port of Kaohsiung come to a new era due to the construction of 2nd Harbor Entrance to navigate ships under 100,000 D.W.T. and to bring city of Kaohsiung to a international industrial metropolis. Since then, the outer breakwater between 1st Harbor Entrance and 2nd Harbor Entrance has been isolated. It not only influenced the development of the bar area, but also affected the stability of the shore line. Anyway, no protection is done until a bule-print for developing that area was made in 1970's by the Harbor Bureau to meet the overall expansion project of the whole harbor area. Further, following the completion of Cross Harbor Tunnel in 1984 to pave the way for the transportation between city of Kaohsiung and the island, the pace of development of the bar area, including construction of Container Terminal No.4 and seawall or groynes, is accelerated. So far, all projects go on well and in good performance, As a result, its container

* *Senior Civil Engineer & Section Chief of Engineering Affairs of Engineering Dept., Kaohsiung Harbor Bureau, Taiwan, R. O. C.*

throughput of about 2.8 million TEUs in 1987 has put Port of Kaohsiung in third place in the world. This tremendous achievement is, no doubt, attributed to the perfect shore protection of the outer breakwater.

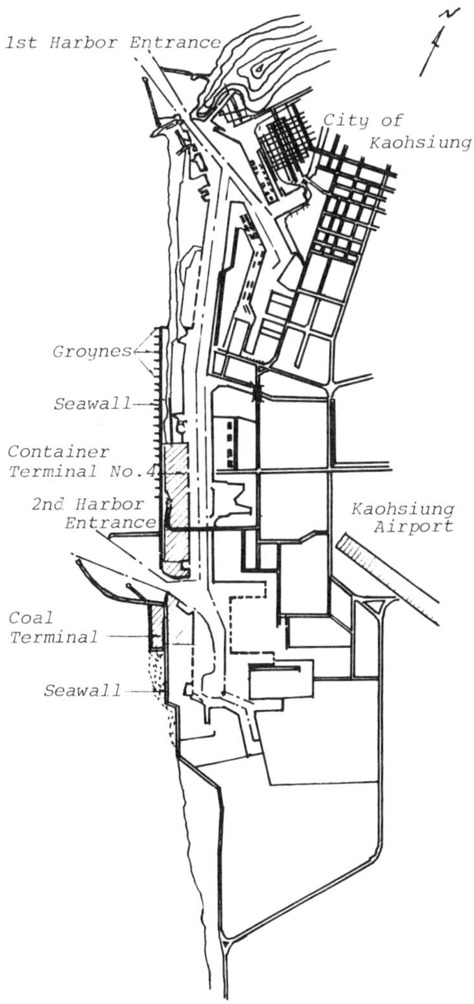

Fig. 1 Port of Kaohsiung

NATURAL CONDITIONS

(1) Sea bed condition:

The sea bed condition off the sand bar is very flat and homogeneous. Most of the 12 kilometers coastal area is distributed with fine sand of Dm=0.15m/m, and in a slope ranging from 1:30 to 1:100, as shown in Fig.2

except section F5 which located near 2nd Harbor Entrance has Dm value of about 0.30-0.35m/m at depth of 15m-20m, (see Fig.3).

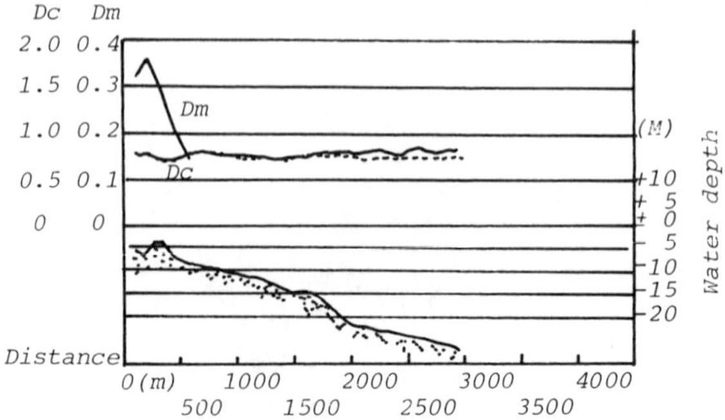

Fig. 2 Typical section of sand size VS sea bed gradient

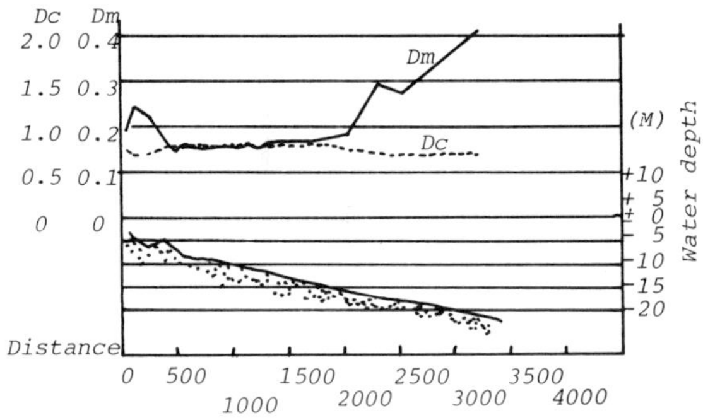

Fig.3 F5 section

(2) Wind condition:

Due to different paths of typhoon and different seasons of monsoon, the transport of sediment along the sea shore will be changed for different periods.

a) Every year, typhoon (cyclone) from May through October occasionally packs wind of 11-65 m/sec to hit the area and to cause a lot of sediment transport along the coastal area. Among the paths which are probable to pass the area, path D is the main one to influence the coast area as shown in Fig.4 due to its big wave hight & high wind velocity. Fig.5 shows the actural example of typhoon path D and its corresponding wave height and wind velocity.

b) Winter monsoon, appearing mostly in NNE, N, NNW directions with wind speed of about 0.3-10 m/sec,is longer in duration, but less in unit sediment transport, as shown in Fig.6.

c) Summer monsoon, prevailing in SE, SSE, S directions with 5.5-13.8 m/sec wind speed, performed in a different way as winter monsoon does (see the same figure as b.)

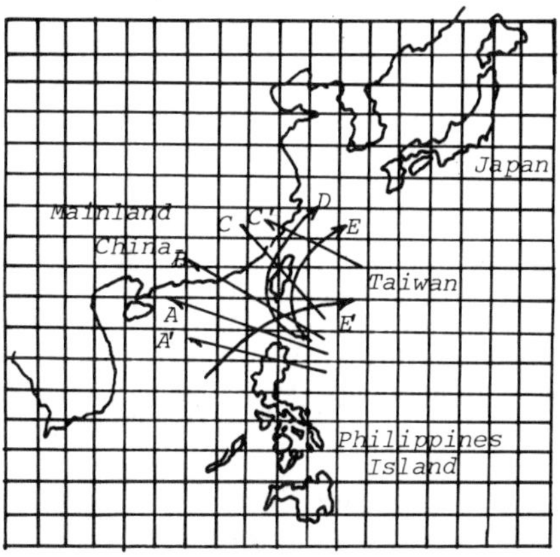

Fig. 4. Typical typhoon paths hitting taiwan area.

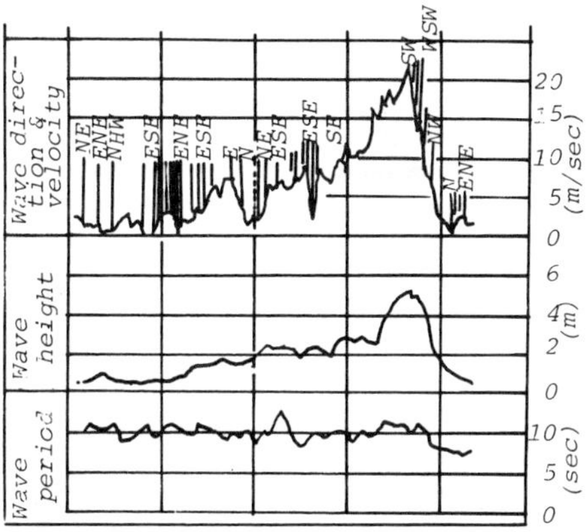

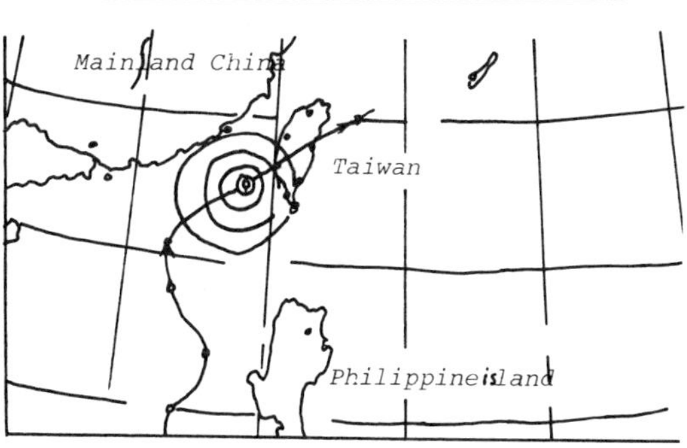

Fig. 5. Winds & waves for judy typhoon in May 30, 1966.

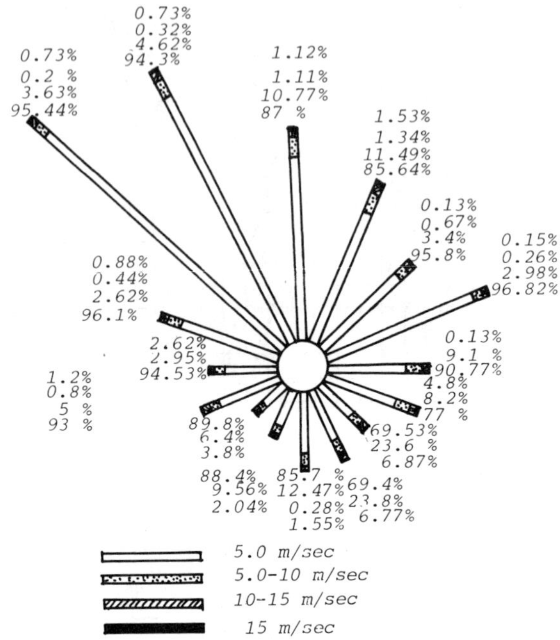

Fig. 6. Typical wind rose for Kaohsiung Harbor outer breakwater area

(3) Wave condition:

The wave condition in the area will depend largely on the magnitude and duration of typhoon and monsoon. A statistical wave data is shown in Table. 1. The max. wave height is some thing like 5-6 meters in 12 years from 1966-1977. while the corresponding max. significant wave height is about 4 meter.

	Year	1966	1967	1968	1969	1970	1971	1972	1973	1974	1975	1976	1977
Wave direction (S.W.)		63	56	53	50	48	49	47	49	50	52	55	57
Max. wave	Wave height (M)	7.00	3.13	6.96	5.30	4.20	4.60	4.31	9.09	6.00	8.50	5.10	6.80
	Mean (M)	2.54	1.52	2.90	2.64	2.64	3.05	2.95	3.63	3.11	2.70	1.60	2.90
	Period (sec)	7.4	8.1	11.0	9.5	7.7	9.4	10.2	8.1	9.0	8.6	8.11	11.0
Sig. wave	Wave height (M)	5.24	1.67	4.95	3.70	2.04	3.85	2.96	6.29	4.01	6.40	3.10	4.10
	Mean (M)	1.71	0.99	1.75	3.09	1.51	1.98	2.95	2.78	3.90	2.98	1.90	1.71
	Period (sec)	7.7	8.5	9.3	10.8	7.8	5.7	10.5	8.8	10.1	8.3	8.4	10.0

Table. 1. Max. wave height and significant W. H.

(4) Others:

Such as tide, current etc. will also play an important factor to the shore protection. Tide in the area is about 0.7 meter in range as shown in Table. 2 and current is about 0.25 m/sec or 0.5 knots (nautical mile). But it will be changed depending on the wave height.

TIDE \ YEAR	1966	1967	1968	1969	1970	1971	1972	1973	1974	1975	1976	1977	M.V.
H.H.W.L.	1.60	1.55	1.75	2.60	1.53	1.60	1.63	1.62	1.60	1.51	1.56	1.56	1.60
L.L.W.L.	2.02	0.04	0.03	0.02	0.06	0.05	0.16	0.10	0.15	0.10	0.13	0.02	0.03
M.H.W.L.	1.09	1.14	1.10	1.11	1.16	1.11	1.12	1.13	1.14	1.15	1.14	1.16	1.15
M.L.W.L.	0.37	0.42	0.39	0.41	0.42	0.41	0.42	0.42	0.44	0.46	0.49	0.53	0.49
M.S.L.	0.69	0.74	0.71	0.71	0.76	0.73	0.75	0.74	0.76	0.77	0.78	0.78	0.76
H.W.O.S.T.	1.20	1.25	1.19	1.21	1.24	1.20	1.22	1.22	1.21	1.24	1.24	1.25	1.24
L.W.O.S.T.	0.32	0.40	0.32	0.35	0.37	0.36	0.38	0.37	0.39	0.41	0.47	0.49	0.43
H.W.O.N.T.	0.94	0.99	0.99	0.98	1.05	1.02	1.02	1.04	1.05	1.05	1.03	1.00	1.01
L.W.O.N.T.	0.42	0.44	0.44	0.44	0.47	0.47	0.48	0.48	0.49	0.51	0.54	0.56	0.51
Sprang Tide Range	0.89	0.85	0.88	0.86	0.87	0.84	0.84	0.85	0.82	0.81	0.75	0.78	0.81
Neap Tide Range	0.54	0.55	0.55	0.55	0.58	0.55	0.54	0.56	0.55	0.52	0.51	0.46	0.50
Mean Tide Range	0.72	0.71	0.70	0.70	0.73	0.70	0.70	0.71	0.70	0.69	0.67	0.63	0.66

Table. 2. Tide levels surveyed from 1966-1977

SEDIMENT TRANSPORT AND SHORE CHANGE

Through long term observation, the change of shore profile near 2nd Harbor Entrance is quite similar with the results of model test which carried out during construction of the 2nd Harbor Entrance.

(1) Transport of sediment before construction of 2nd Harbor Entrance

a) *In winter season, sediment transport, from north to south, was governed by winter monsoon NNW, N, NNE. etc. and shore line appered in an accretion condition along the whole area. While in summer, typhoon and summer monsoon SE, S. SSE etc. dominated the process of sediment transport, and shore line was eroded by the north-bound wave force (current). But finally the processes of accretion and erosion would come to balance, and return to it original state unless in abnormal condition such as being hit by super typhoon, and if so, the shore line would be scoured until next cycle. The contour line of water depth in the coastal area before 2nd Harbor Entrance being constructed is shown in Fig. 7.*

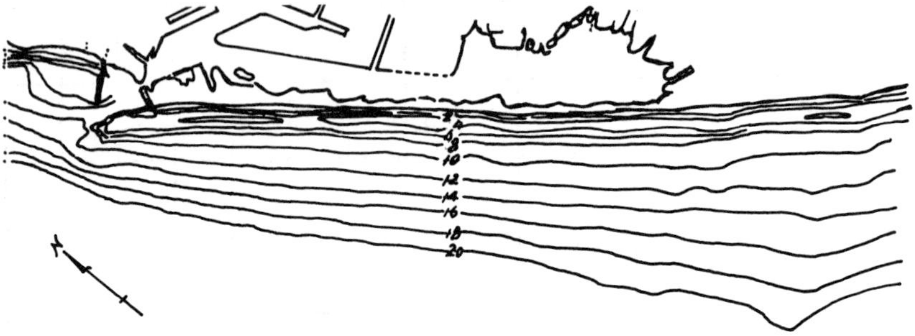

Fig. 7. Shore contours before construction of 2nd Harbor Entrance

b) *Due to the southern breakwater of 1st Harbor Entrance being constructed almost to parallel the shore line, the channel near the Entrance will be silted in summer time, and will be scoured in winter season, as shown in Fig. 8 and Fig. 9.*

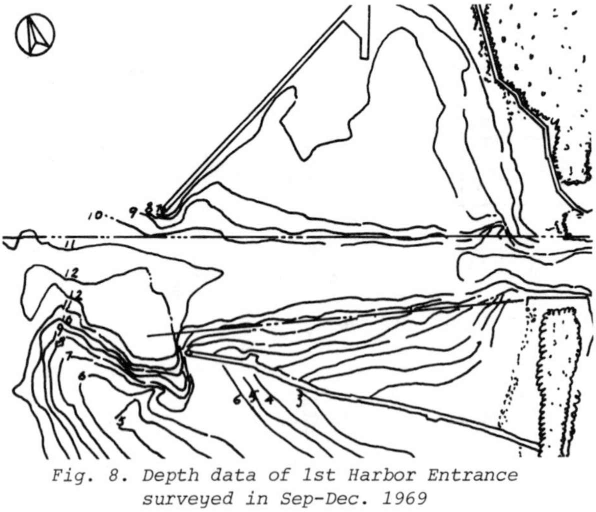

Fig. 8. Depth data of 1st Harbor Entrance surveyed in Sep-Dec. 1969

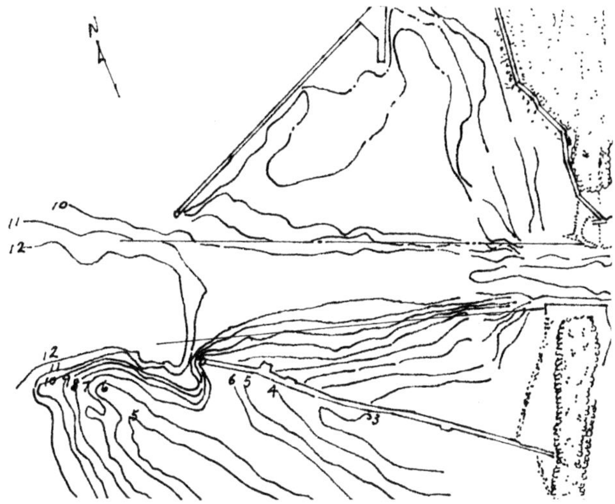

Fig. 9. Depth data of 1st Harbor Entrance surveyed in April-July 1970

(2) Transport of sediment after construction of 2nd Harbor Entrance:

Due to the prevention of the breakwater of 2nd Harbor Entrance, the change of shore line on both sides of the entrance has different result under the same conditions as described above, that is:

a) In winter season, the northern shore of the entrance is a creative beach, while the southern shore is an erosive one, and vice versa in summer time. The contour line of water depth in the coastal area after 2nd Harbor Entrance being constructed is shown in Fig. 10.

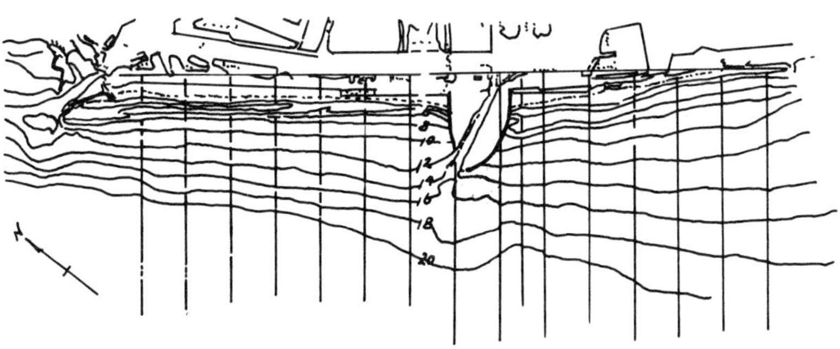

Fig. 10.Shore contoures after construction of 2nd Harbor Entrance

b) Due to the source of littoral drift being cut by the breakwater of 2nd Harbor Entrance, the shore line between 1st Entrance and 2nd Entrance may draw a sharp change toward the land area during typhoon season.

c) Although, the channel of 1st Harbor Entrance is unable to totally free from silting or scouring, but it already has a great improvement to compare with the condition of which before construction of 2nd Harbor Entrance. It said from 50,000 m^3/year dereasing to less than 10,000 m^3/year.

PROTECTIVE METHOD

In addition to providing the port and residents the protective function, the outer breakwater of Kaohsiung Harbor is also to be considered as an integral part of port facility. Therefore, the protective facilities are designed as permanent structure as following.

(1) Parallel seawall:
*Parallel seawall was built on the land area about 30 meters far from the shore line to parallel to the quay wall of container berth as a result of making up the required space of about 450m *320m of back-up area in each berth to cope with the development of Container Terminal No.4 in northern side of the 2nd Harbor Entrance as shown in Fig. 1. In the south, the seawall was located about 300-meter off the shore line to constitute a big coal storage yard for the Coal Terminal, also shown in Fig. 1. So far, totally, some 6 kilometers of this seawall has been built to form the main structure for shore protection of the outer breakwater of Kaohsiung Harbor. For the seawall, besides the concrete parapet and wall itself, it also includes a earth dike planting with tropic trees to serve as a wind shield for the port area, and some side ditches to collect rainfall and overtopping salt water, if any, as shown in Fig. 11.*

(2) Perpendicular groynes:
Perpendicular groynes, being constructed on every other 200 to 300 meter interval with each a length of about 90 meters, are perpendicular to the parallel seawall with a view to stabilizing the shore line and keeping the approach channel of the entrance from silting. In order to meet the attack of 6-meter height wave force, the groynes are designed as rubble mound type structure with 2 tons of modified Tetrapod for the armored layer as the conventional breakwater does, as shown in Fig. 12. From an engineer point of view, it sounds to cost too much, but for the point view of safety and security, it's worth-while to do so and perhaps it's the best choice for the long-range-project consideration.

(3) Artificial beach nourishment
In addition to the protective facilities mentioned above, the erosive beach between 1st and 2nd Entrances has been nourished with good material occasionally to prevent the shore from further scouring. From 1968 to 1982, the total spoil disposed to the area was about 34 million m^3, But no headland is formed due to typhoon reason.

RESULT

Referring to the purpose as many other protective works do, the design of shore protective facility should take both direct function and indirect function into account, the former is aimed at protecting the shore and facility itself from eroding and damaging under the normal

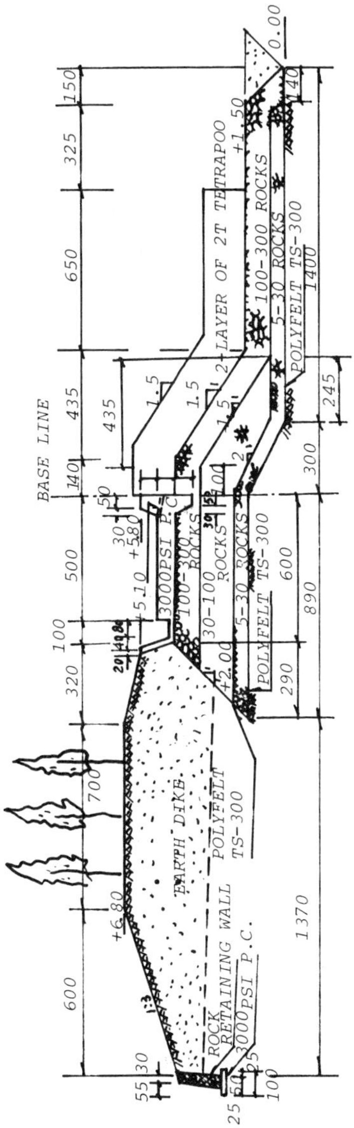

Fig. 11. Typical cross section of seawall

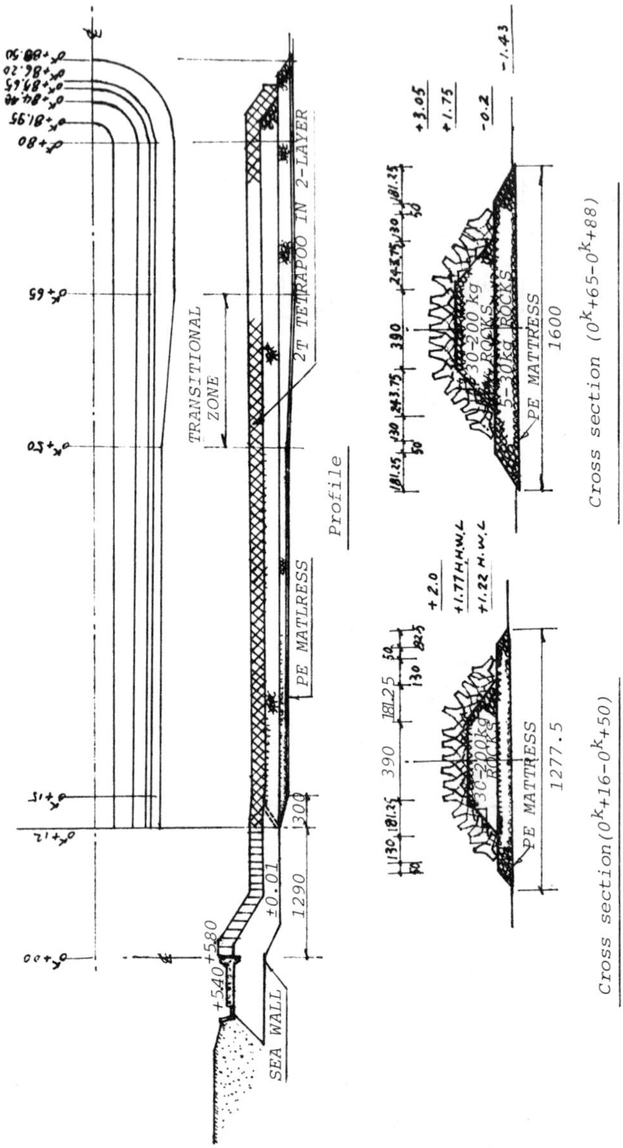

Fig. 12. Typical cross section & profile of groyne

condition. While the latter is emphasized on the function of durability from which the structure can still provide a retard action to decrease the loss to a minimum, should the facility once has been damaged in extraordinary conditions.

Basically, the protective facilities in Kaohsiung Harbor have followed the above-said guiding principle in design and construction. Therefore the shore protection of the outer breakwater has performed well up to now since its completion for about 13 years during which a number of big typhoons have experienced. Fig. 13. shows the field results of seawall & groynes and beach erosive conditions after being attacked by a great of big typhoons. Though, the seawall and groynes keep no damage at all, while the beach in between the groynes presents some different extend erosion due to different width(w) between the groynes and different length(L) of the groynes. Normally, for normal wave w=4L is the best solution and for inclined wave, 35°-50° against shore line, the best solution will be w=3L.

Fig. 13. Field results of shore protection of Kaohsiung Harbor outer breakwater

CONCLUSION

(1) For the protective structure, the rubblemound groyne (jetty) may be the most effective and reliable means in using against big wave force, should the layout and design method be right in principle. The case in Kaohsiung is a good example as the goynes built there have brough function into fully play up to now.

(2) The width between every two groynes is a main factor to beach control, it also means best result comes to perfect layout, from the field results got in Kaohsiung Harbor outer breakwater, we found that the width between every two groynes designing as 3 times and 4 times of the groynes length (length from shore line to head) for inclined wave (angle from 35°-50° against shore line) and normal wave respectively, will be the solution for shore protection.

(3) Taking eroded beach into account, artificial beach nourishment is a very important approach to the shore protection. For instance, the northern shore of the 2nd entrance of Kaohsiung Harbor is something like an erosive beach for long term consideration, due to the source of sediment transport being cut by the breakwater. But, through artificial beach nourishment management, occasionally by disposing the good material of sand from channel dredging, the shore, however, has maintained in a stable condition year by year. It means that to the erosive shore line the best salution is to do beach nourishment and protective work simultaneously, if possible.

(4) Due to the breakwater of the 1st Entrance of Kaohsiung Harbor being laid in a way, almost, to parallel to the shore line, the channel near the Entrance was, frequently, to be silted by sand drift from the shore area between 1st and 2nd Entrance before shore protection in the area was done. While conditions have been improved since the completion of parallel seawall and perpendicular groynes. It means that shore protective management can play a multifunction role depending on what the case needs. The successful method in Kaohsiung Harbor, no doubt, has set a good example in this field.

(5) No headland or other simple methods can be used for shore protection like Kaohsiung Harbor outer Breakwater since it will resist the big wave up to 6-meter height and it left no tracks after disposing about 34 million M^3 spoil in the area in last 14 years from 1968-1982.

REFERENCES

(1) Report on study of construction planning of Kaohsiung 2nd Harbor Entrance (1968), Kaohsiung 2nd Harbor Entrance Construction Department, Kaohsiung Harbor Bureau.

(2) Gong. C. I. et all (1976), Report on Improvement study of Kaohsiung 1st Harbor Entrance.

Listing of key words to index the paper entitled SHORE PROTECTION OF OF KAOHSIUNG HARBOR OUTER BREAKWATER is:

Kuo, Shih-Duenn,

CHAPTER 215

RIA DE FOZ: SOLUTION FOR A HARBOUR AND COASTAL PROBLEM

Jose Javier Díez Gónzález*
Víctor A. Escobar Paredes
José Aguilar Herrando

SITUATION.-

The ría of Foz is located in the Northwestern part of the Ibérica - Península, inside the Rías Altas group, in the Eastern extrem of Lugo/ province. It is placed on the natural form known as Golfo of Masma, - topographical accident that takes its name from the river that flows - into the ría.

From the point of view of the materials that constitute the environ ment, they are mainly composed by slaters and quarzites, with some le- vels of carbonates from the Low Paleozoical period. To the West, the - granitic materials are predominant.

The ría is limited in its margens by some little slope and relative ly low cliffs. Its bounds have a little depth, due to the filling pro- cess, natural in the Rías Altas. The materials that refill the ría, - come from the sea erosion of the cliffs (1), and the refilling is due/ to the sea waves action that produces in the tidal flow and ebb an - - asymetrical capacity of erosion and transport (2). That is a natural - situation, but the human action, building some channeling constructions has accelerated this refilling process.

The traditional economic development on this zone, has been based - on some principal activities: fishing and agriculture, but nowadays - turistic use rounds off the others. This situation and the evolution - of sailing neccesities on the ría, has been a conflict reason, which - is a consequence of the competition between these uses.

In these terms, the first problem to the possible feasibility of - Foz harbour was its accesibility, and that depended on the stability - of the main tidal channel. Therefore, the harbour authorities developped a planning design to fix the position of the entry channel, by the - - construction of a couple of jetties; however only one of them was built up to the date. In a certain phase of the construction, when the inlet gorge was affected, the sandy barrier was destroyed. Since then, seve- ral erosion problems have affected to the neighbour littoral and bea-- ches. And the aim of this study is to regenerate the coast without - - decreasing the harbour availability. We might not forget that the natu ral tendence of the ría is to refill itself; and that means a hight - variability of channel, trying to adapt themselves to the new condi -- tions and distribution of materials that refill the ría.

The light was built because of the risks derived of the existence -

*Universidad Politecnica de Madrid, Ciudad Universitaria, Madrid 3, SPAIN

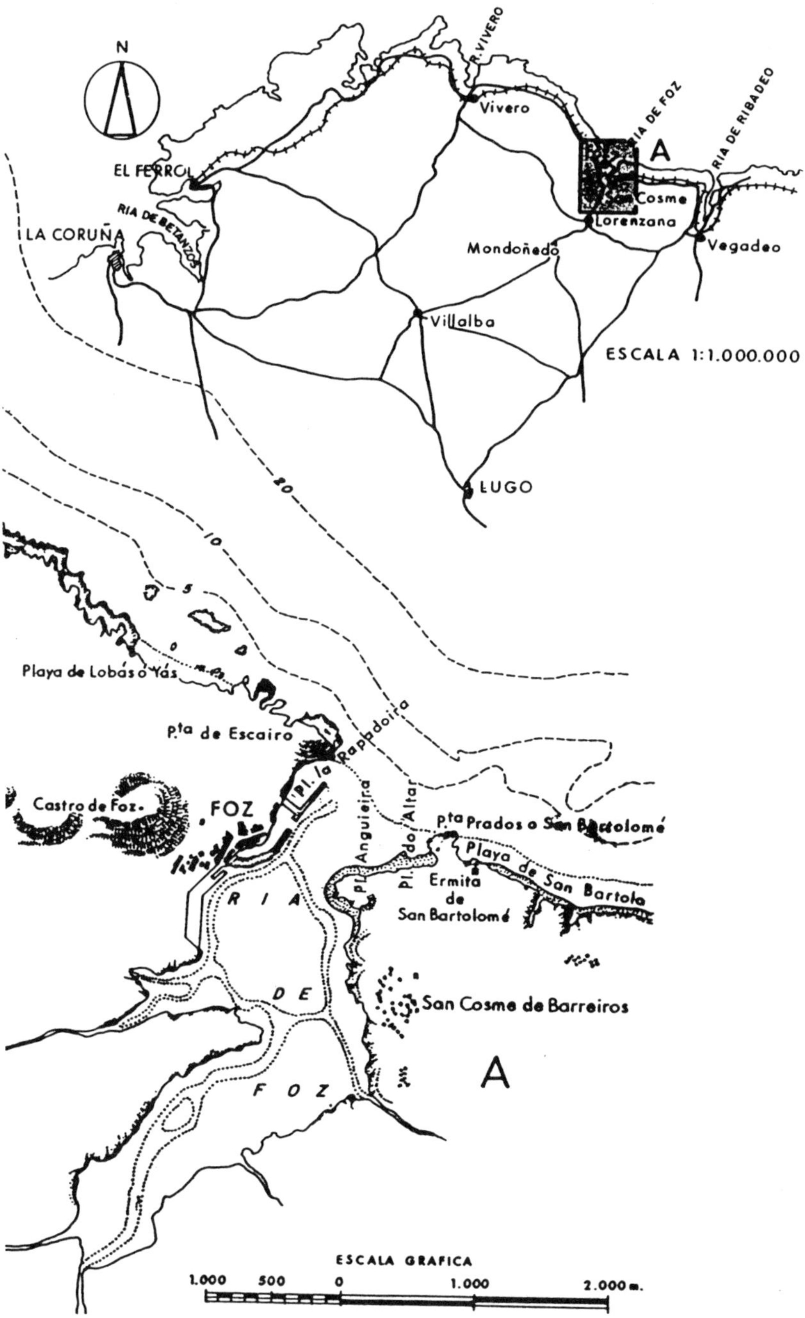

FIG. 1 Situation

of a dangerous rocky system (shoals and rocks of the Rapadoira). This/ light caused degeneration of a little beach on the left hand of the - mouth and stabilized in a certain way the barrier on the inlet. Nevertheless, in this brief note, we have explained the situation we - have to face up, which is derived from the general development of the/ fishing activity that requests every day bigger depths and safety.

So the objective was to define the neccesary works to protect the - coast considering the natural conditions of littoral dynamics, in such a way that the conditions of the harbour entrance and explotation do - not suffer additional problems.

ELEMENTS OF LITTORAL DYNAMICS.-

Relating to the processes that affect to the mouth, the determinant elements are tides and sea waves, and, in a minor order, winds and - - fluvial contribution.

Tides: They introduce some flow and ebb currents that carry the - - solid materials. The net component of this transport is to the basin./ The main high water appears in equinox (but it is little out of phase). The M.H.W. has an amplitude of 4'50 m. aproximately. The tidal prism - considerated in the project is between 1'7 and 7 hm^3.

Waves: Due to the special morphology and topography of the exterior part of the ría, the littoral currents induced by wind waves have a - quasy permanent sense from E. to W. in Anguieira and Altar beaches; on the opposite, out of the ría mouth and along all Masma Gulf, currents/ have a variable direction, related to the wind wave direction, but the dominant currents and littoral transport are there from W. to E..

The special direction of currents and drifts in the beaches is the/ most important reason of the shaping of the lock barrier as a spit, - what may be checked by observing the distribution of materials in the/ tombolos produced by the Peña do Altar.

Winds: They are responsible of the formation of dunes on the lock - barrier before it were demolished. They are also responsible of the - filling of some marginal zones. Winds effect can be considered as a - second order action.

Fluvial waters: Their effect can be compared to the tides only in - extraordinary floods. But they are not so important in normal condi -- tions. In the project it is considered a flow between 10 and 110 m^3/s. and 170 m^3/s. in extraordinary floods.

MATERIALS.-

From the point of view of the materials, Masma Gulf can be conside- red as a morphodynamic unity, with some located differences, but the - materials are generally very similar.

Several samples have been obtained and analized in their grain size distribution and mineralogy. A group of the samples were obtained from the different sandy formations of the ría, but other group corresponds to different beaches along Masma Gulf. The analysis of sediments shows that the anterior afirmation is all right, and that is proved with a - high certainty between Foz and Ribadeo.

On the other hand, the mineralogical analysis show that in broad - outline, materials in this zone proceed from the erosion of the cliffs principally, and that the sands from the beaches around ría mouth and/ from the ría sandbanks are completely similar.

The similarity of the sediments showing in the analysis of sands - from Anguieira and Altar beaches and from the sandbanks in the interior of the ría allows the use of the last for refilling of beaches with - them.

PRELIMINARY CONCLUSIONS.-

Some historic and recent cartographic and topographic documentation has been analized and it is possible to propose some reflexions that - can help to puzzle out the present situation of the inlet-outlet system:

-Based on the available documentation before the building of the - channeling jetty, the location of the exterior tidal channel was very/ variable. This situation was foreseen to reappear (2) and is now - - reappearing, but in the extrem of the jetty; therefore the entry pro-- blems have not been definitively solved.

-Though the spit seemed to be stabilized, its morphology was adap-- ting itself to the new hydrodynamic situation, changing its orientation Northwestward and modifying its relative dimensions, its anterior - - cyclic evolution, however seems to have been broken, after the cons -- truction of the light and before the building of the channeling jetty, dunes and vegetables colonization grew up; this situation went down - during the construction of the channeling jetty, but in 1.978 the lock barrier disappeared, just when the interior stretch of the jetty was - constructed.

-The accretion in Rapadoira beach was following on the growing of - the jetty.

-After destruction of the barrier, the erosion on the right hand - beaches began to occur and with some little delay the erosion of the - beaches Eastern towards Ribadeo began too.

-Since then, the inner channel suffers quicker changes in a dialec- tic relationship with the cyclic regeneration and demolishing of the - lock barrier. The successive growths of the spit has been permited by/ the erosion of Anguieira and Altar beaches and by the mentioned perma- nent Western littoral drift along them. Finally the successive destruc tions of the spit became quicker. These variations depends on the sea- sonal changes of the conditions of the sea dynamics; it will mean that the variations of the channel are now very quick in this zone.

-The monthly variations of the tidal prism (due to the tide compo-- nents in this area) produce a very high mutability of the channel - - depths and breadths. And the increased filling process of the ría - - after the first demolishing of the spit increases the sailing difficul ties.

SOLID TRANSPORT ON THE OUTLET.-

The values of the littoral transport contained in a first numeri-- cal study based on hindcasting methodology are very important, for the net as for the gross transports. However these values correspond to a/

potential transport and do not mean the same importance in the littoral drift, which is limited by the existence of materials. In any case, the littoral drift is important enough to impose a very high variability of the tidal channel, even seasonally, following the motions of the transported solids.

In the exterior mouth of the ría, like in other cases, the natural mechanism of by-passing of materials is stablished. In this way, the bars that exist in front of the ría, with their net movements to the SE., catch up their aim of solid transport when they arrive to the sandbanks next to the Punta de St. Bartolomé; in this moment the channel opens again suddenly in the Northwest end. The changes of the position of the outlet are so progressive towards the East up to the sudden openings.

In terms of the transport of solids to the inner basin, it becomes smaller when the effect of the lock barrier is efficient; in this way, before 1.978, when the lock barrier was demolished, the net transport was less than it is possible to observe nowadays.

In the exterior beaches, the transport has a special direction, due to the intrinsic characteristics of this zone. The bay which is open between Escairo and St. Bartolomé headlands, stays widely emerged in low tide; therefore the transport only occurs in mid and high tide there.

If we observe the materials in the shelter formed in the Peña do Altar, and we accept that the disposition of materials should respect the orientation of predominant waves in term of their main annual resultant (recognizing the seasonal variations),we can conclude that the disposition and growth of the lock barrier is logical, in spite of the fact that out of the mouth of the ría the transport has frequently the opposite direction. We should not forget that in the bay of the ría it is neccesary to wait for high water to observe solid transport and in this moment the refraction, the diffraction in the Escairo headland and even the reflexion on the St. Bartolomé headland and Altar Rock are important. The tide currents help this current direction too. We have proved that the solid transport is from E. to W. along the bay beaches; so the genesis and orientation of the lock barrier may be perfectly explained as a spit.

CONSTRUCTIONS: EVOLUTION.-

The morphologic changes, as an evolution of the materials distribution in relationship with the construction wich are made along the time, are analized as follows:

-Between 1.931-1.953 it is built a little transversal jetty, for giving access to the light in the Rapadoira Rocky shoals. Leaned and supported on this jetty a little beach was formed; that system seems to have operated as a stabilizer of the lock barrier which moved innerward and became wider allowing an increasing colonization of dunes.

-In 1.969 it began the construction of the jetty which starting from the light and parallel to the left margen, tried to stabilize the position of the main tidal channel. The response of the lock barrier materials was fast; the spit began to take a progressive orientation Northward with evident lengthening and lost of thickness/

in its extrem head due to the following processes:
. The stabilization of the tidal channel caused a reduction of its - - curvature in the interior zone, and so, it began to erode the lock - - barrier from the internal side.
. In the area extended between the jetty and the original shoreline - the Rapadoira beach became progressively wider what supposed a bleeding of a part of the materials that constituted the feed of the spit, Al-- tar beach and beaches further to the East.

-Between 1.975-1.977 the construction of the last exterior of the - channeling jetty was ended. These buildings supposed the outer prolongation of the jetty in the left margen to the present position. Its - inner prolongation began then producing a subsequent reduction of the/ gorge.

-The effects were inmediate and spectacular; in March 1.978 the - - lock barrier had disappeared. It is possible to observe a faster - - filling of the ría and an increasing of the exterior bars since then,/ and also Altar beach began to be eroded, trying the materials to feed/ the lock barrier that was demolished every time by the littoral dyna-- mics. But this process was not inmediately evident in the regression - of the shoreline up to three years later; initially the erosion only - affected to the intertidal beach (depth erosion).

-In 1.981-1.982, we could confirm a very important backward move -- ment of the coast-line due to the processes we have explained.

-Since 1.983, we could observe a relative important erosion (changes in the transversal profiles) of the beaches East from ría of Foz.

OBJECTIVES.-

The most adjusted solution to the problem we have exposed, might - accomplish therefore the following objective:

1. Regeneration and stabilization of the shore between Anguieira - and St. Bartolomé headlands. This is the principal objective; the - - follows are secundary and act as limitant conditions.

2. Preservation and improvement of Foz harbour facilities, so that/ particularly the sailing conditions in the access channel do not worsen.

3. To avoid negative impacts in the ría and its environment.

4. To respect the natural conditions of littoral drift Eastward - - from the ría.

Securing theses objectives it will be possible to obtain an inte -- gral solution to Foz coastal and harbour problem. And a discussion of/ the possible element of the solution will be done previous to the - - analysis and definition of the complete solution.

A. Artificial filling with sand.-

The level of the regression process of the Altar beach is so important and the risk of damages so high that it seemed neccesary an artificial refilling on this beach. But that decision planned two suplementary problems: the source of sands and their stabilization.

- In relation to the sand source the sedimentological analysis have shown that the sands from the bay and from the channel as the ones - - from the banks of the ría have the same characteristics than the sand/ of the beaches; the use of them is, in consequence, allowed. We have - also test the materials of Rapadoira beach; these sediments have not - the same characteristics but they can be used too. Nevertheless the Ra padoira beach has become too wide and now it needs a bleeding because/ the sediments are in the way to overstep the jetty, bringing some nave gability problems in the exterior end of the channel.

-Nowadays, due to the natural processes that the littoraz dynamics impose in Altar and Anguieira beaches, the balance of sediments is - - negative; that means an evident erosion of them. This erosion has some inmediate consequences: the acceleration of filling speed in the ría,/ and the lost of the materials that feeded the beaches, further to the/ East from the ría. Therefore, in relation with the stabilization of - materials, it would be neccesary some kind of artificial shelter, - - barrier or protection. Taking into account both problems. Two different types of solutions may be considered:

. A continuous feeding after an adecuatelly great first reclamation - what should interrupt the actual rythm of filling in the inner basin./ The simple reclamation surely would be not enough; it is also neccesa- ry to feed periodically with, to counteract the deficit that imposes - the natural eroding tendence.

. Construction of a kind of element that allows the stabilization of - sands, by support, shelter or protection. These elements are analized/ in the following discussion.

B. Transversal jetties.-

It is a parcial solution that should allow the defence of shore - and buildings (apartments and hotels) in the zone of Altar beach, - - where the erosion has been stronger, attacking the quaternary cliffs/ over what these constructions are located. They behave principally as barriers or supports. Nevertheless this solution has some important - disadvantages:

-On the one hand, it does not resolve all the problem because it - does not give a really solution to the accessibility to Foz harbour.

-On the other, the traspassing of materials to the beaches Eastern from Foz is not favoured and possibly, if the barrier was long enough, it would be even not allowed; so there the erosion process should - - increased.

-In addition, this kind of jetties would produce indesirable visual impacts.

C. Free jetties.-

Though they behave a shelter works, like in the previous case, it - is a parcial solution if they are not complemented with other channe-- ling constructions and by-passing disposal.

D. Channeling constructions.-

The only way to complay with the demands of Foz harbour is to sta-- blish a channeling element on the right margen that completes the old/ jetty. The old channeling project defined two parallel jetties already.

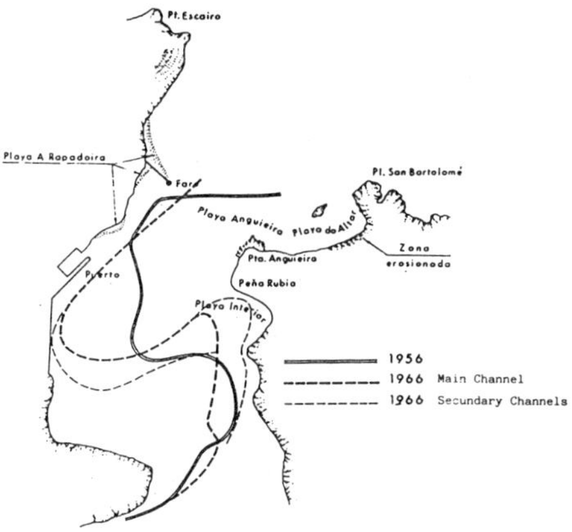

FIG. 2
Evolution of the inner channel-1956-1966

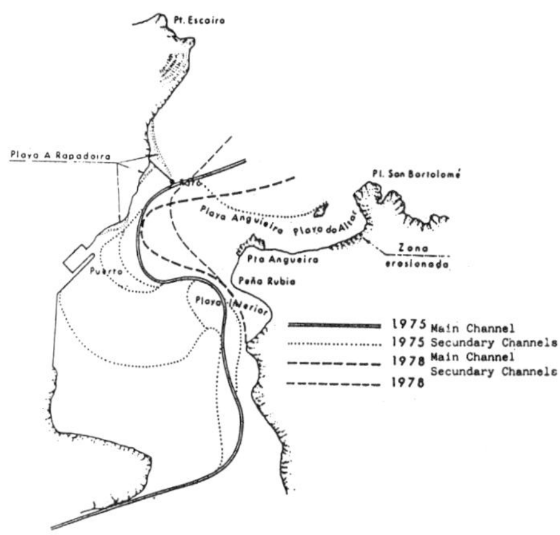

FIG. 3
Evolution of the inner channel-1975-1978

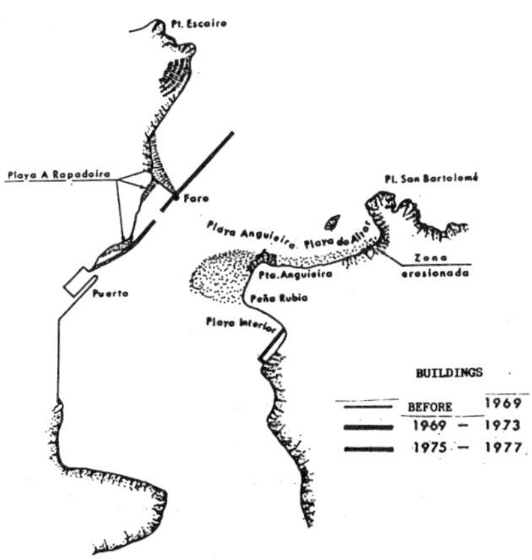

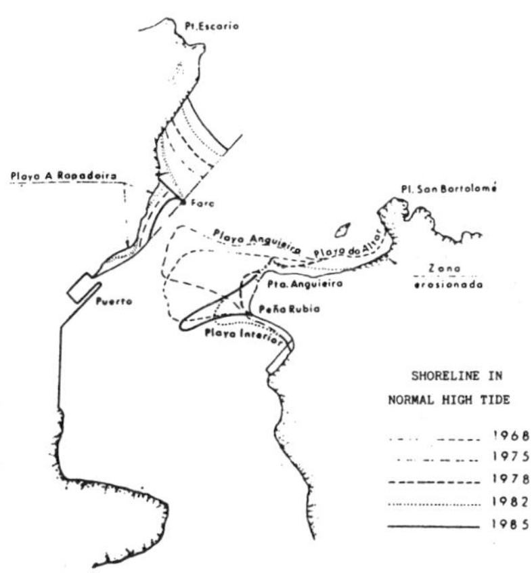

FIG. 4
Schedule of the channeling constuctions

FIG. 5
Changes in shoreline

But only the one of the left margen was built. That was other reason - of the posterior erosion happened. At present the second jetty is also necessary for channeling, however nowadays it is necessary to regenerate the lock barrier too.

There is a difficulty derived from the shortage of complete studies about bathymetric and hydrodynamic conditions, and in consequence, - - there are not dates enough to assure the success of the solution, - - because the definition of the exact cross section of the tidal channel, that would be necessary to improve the accessibility conditions of Foz harbour needed a wider research.

The channeling works on the right margen made behave as bracket and protection against the waves and tidal erosion of the spit that may be regenerated in base of an artificial filling. This is, in consequence, the most complete and economic solution in relationship with all the - planned objectives. It has the advantage of beeing the less detectable visually because the jetty stays sumerged under the lock barrier, - - except on the tide channel side.

ADOPTED SOLUTION.-

The adopted solution is composed of the following elements:

1. Artificial filling with sand.

2. Extraction of sand from the ría increasing the tidal prism and - reconstituting the primitive condition. It is possible to take out the sands from Rapadoira beach too. We have explained that Rapadoira beach needs a bleeding to avoid both posterior problems of accessibility to/ Foz harbour and risks in the own beach.

3. Construction of a bracket-channeling jetty that starting from - Anguieira headland approaches enough to Foz light to allow a breadth - compatible with the final tidal prism. This jetty should stabilize the tidal channel on one hand, and avoid the breaking and eroding of the - spit due to the tidal and wave actions on the other hand. The following conditions must be observed:

-The jetty would be short enough to be sure that the distance from/ its head to the Rapadoira jetty exceeds the theoric channel breadth, - to make easier the foreseen construction of its extension as the right margen channeling jetty; but at the same time it must be as long as - possible to favour the natural inlet-outlet stability.

-It might adopt in the ground plane a compatible form with the - - tidal ebb channel eroding action.

In abstract, the solution can be defined as the artificial settle-- ment of a lock barrier, stabilized by a channeling and bracket jetty,/ in such a way that this may constitute the first stretch of the complete right side channeling jetty.

THE JETTY.-

1. The ground plane will be conditioned by two kind of considera--- tions: fluent circulation and constructive facility.

They have been taking into account three different layouts which -

FIG. 6 View of the spit in 1975

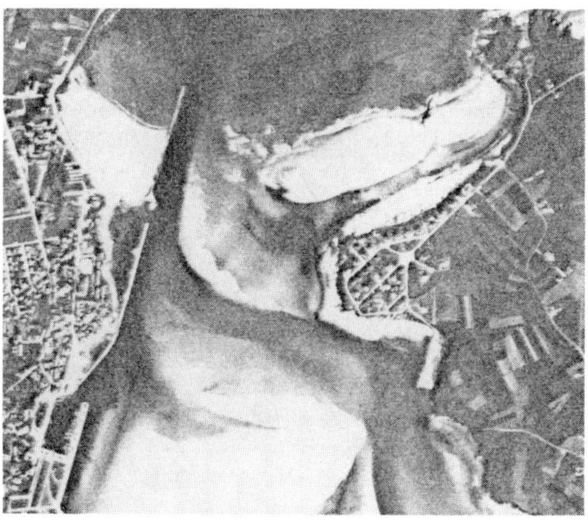

FIG. 7 View of the spit in 1978

FIG. 8 General view 1975

FIG. 9 Alternatives for the jetty to tesig.

are shown in figure 3. The following discussion was carried on:

The hydrodynamical conditions improved for number 1 to number 3; - number 3 allows better the evacuation of water in the ebb, and the - - changes of direction are less than in number 1; the curvature is minor in number 3 too. At the same time, solution number 1 corresponds to - the maximum increasing of the lock barrier. Therefore, all these consi derations select layout number 3. However the easiness decreases from/ number 1 to number 3; while the number 3 layout suposes a construction through the presently flattened lock barrier sandbanks, demanding to - pump out a great quantity of sand, number 1 layout demands a minor - - previous sandy pumping. This need for pumping sandy out is imposed by/ the necessity of getting an adecuate level for jetty fundation. Nevertheless on the other hand, solution number 3 has the advantage of - - beeing shorter than the others; so, the anterior disadvantages are - - parcially balanced. Therefore, solution number 3 has been selected, - taking into account other motives:

-The present changes of the tidal channel do not allow a clear defi nition of the jetty based on the present morphology and other condi -- tions in the moment of the design, which undoubtedly will be some - - different from the ones of the unforeseeable moment of the construc -- tion.

-The jetty number 3 has the advantage of starting from the rocky - massif known as Peña Rubia, what means a good stability in the star-- ting point.

-This solution guarantees the best stability of depths in the har-- bour access channel.

2. Refering to the jetty cross section, its double function as a - channeling and bracket jetty may allow the construction of an asymetri cal section. The channel zone demands bigger slopes to decrease the - roughness and in consequence to improve the evacuation of waters in - ebb tide. On the other hand, the right slope is going to be covered by the sands of the supported lock barrier.

OTHER ACTIONS.-

As it has been pointed out before in this paper, it is possible to/ observe some changes in the natural transversal profiles of the bea -- ches Eastern from the ría of Foz. These changes forseeably mean some - erosion of the shore and therefore it is recomended to refill the - - shortage of sands of those beaches. It is also proposed to carry the - sands out of the inner ría for this reclamation in order to increase - its tidal prism.

ECOLOGICAL ASPECTS.-

The proposed sandy extraction from the ría may affect the ecosystem, but this affection may be considered positive, because it is just the/ acceleration of the speed of filling of the ría and the variability in the position of the internal sandbanks (and channels), the factors - - that more ecological impact have represented in the past.

On the other hand, the bentonic fauna and flora needs a certain - - stability of bounds and this project foreseeably will favour this stabi lity, and the ecological impact for sandy extraction is not so important as the stabilization of bounds.

SOME INTERESTING RESULTS.-

The construction of the jetty has begun on the beginning of 1.987 - and as early in summer of that year was still possible to observe as - an interesting result, the advance from the near shore of an important sandbank towards the beach all along the Altar and Anguieira beaches./ Besides it was the first summer without erosions in Altar beach since/ 1.978. By then, the artificial refilling was still scarce. It is a - - good prove of the correct behaviour of the jetty as a support for the/ lock barrier. The work will foreseeably be finished in summer of 1.988.

POST CONCLUSIONS.-

Since 1.987, based on the results of the measurements of currents - speed and tidal levels, we are working in the sense of determining the hydrodynamics of the ría. Periodically we make some bathymetrics cam-- paigns, to allow us to discern the changes of the volume of materials/ that come into the ría in relation to the volume that entered before - the construction of the jetty.

On the other hand it is foresight to make a pursuit for the project to verify the grade of adjusting of it to the foreseen behaviour.

In other term, we are measuring the relationship between the tidal/ prism and the cross section of the gorge. We have not forget that soon in the future it will be necessary to determine the breadths and the - depths to the definitive stabilization of the harbour entrance and it/ has been previously exposed.

All these works in the future can serve to elaborate some conclu -- sions about ría of Foz hydrodynamics to extrapolate them to other - - analogous outlets.

BIBLIOGRAPHY.-

(1) Asensio Amor

(2) Diez J. Introducción al Estudio Geomorfológico y de los procesos Litorales en la Ria de Foz. Rev. O. P. 1.980. Madrid.

(3) P. Bruun. Stability in tidal inlets.

Puerto de la Coruna

PART VI

SHIP MOTIONS

"Camino Il," —puerto de Cadiz

CHAPTER 216

WAVE-INDUCED SHIP MOTIONS IN HARBOUR APPROACH CHANNELS

A C VAN WYK* and J A ZWAMBORN**

ABSTRACT

The results of scale-model simulation of wave-induced hull motions of large bulk carriers under conditions representative of those for coastal ports are presented and discussed to illustrate their usefulness in studies aimed at evaluating the optimum depth requirements for port approach channels.

1. INTRODUCTION

Port accessibility, particularly with regard to safe underkeel clearance for deep-draught vessels, is receiving close attention at South Africa's major export harbours. Saldanha Bay harbour on the west coast provides for ships up to about 270 kt deadweight and Richards Bay harbour, on the east coast, for ships up to 180 kt deadweight. Model studies are at present being carried out to determine the conditions under which ships of 270 kt deadweight could be considered in the future.

The CSIR has been engaged in an extensive programme of research including field measurements and physical and mathematical model simulations to study the behaviour of these deep-draught ships in port approaches of limited depth and exposed to waves, particularly long-period swell.

This paper deals mainly with the ship responses as function of the wave direction and water depth.

2. APPROACH

Fundamental model simulation studies were conducted to determine the responses of two models, the M150 (draught = 17 m) and the M270 (draught = 21 m), representing typical 150 and 270 kt deadweight bulk carriers in terms of their

* Research Engineer, Maritime and Hydraulic Structures, CSIR, P O Box 320, Stellenbosch, 7600, South Africa.
**Programme Manager: Maritime and Hydraulic Structures, CSIR, P O Box 320, Stellenbosch, 7600, South Africa.

vertical hull motions resulting from heaving, pitching and rolling in waves from various directions and in different depths of water. The primary test conditions were selected to represent, in broad outline, the range of conditions most likely to influence ship behaviour at the major South African ports.

The ship model responses to these conditions were determined by physical modelling to a 1:100 scale as well as by mathematical simulation using a 3-D source method.

Subsequent to these fundamental studies the physical model tests were extended to cover the operational conditions characteristic of the approach channel to Richards Bay harbour. During these tests due account was taken of the spatial variations in the wave conditions due to the local wave refraction for different deep-sea swell directions. Prior to and also during the course of these model studies, a comprehensive field monitoring study was conducted at Richards Bay harbour to record the behaviour of prototype vessels presently using the entrance channel to this port. The results of these field measurements primarily served to validate the results obtained with the model studies for the 150 kt deadweight class ship.

3. TECHNIQUES AND FACILITIES

The techniques and facilities used for both the field monitoring and scale-model studies were described in detail by Van Wyk (1982) and Van Wyk and Zwamborn (1984).

The field monitoring basically comprised the photographic recording of the behaviour of ships while transiting the approach channel to Richards Bay harbour during both entry and departure. The photographic records were processed to derive the time series of the wave-induced hull motions of the ships during the selected channel transits and wave conditions.

The scale model tests were conducted in a 23 x 28 m flat-bottomed basin with variable water-depth and in a scaled model of the Richards Bay harbour entrance channel. A bank of programmable wave generators, of the modular paddle type, was used for simulating typical swell conditions.

The hull geometries for the two 1-in-100 scale model ships were determined from line drawings of existing prototype vessels. The propulsion and steering systems were designed in accordance with the scale model laws to ensure realistic steerability. The required loading conditions were obtained by the proper distribution of lead weights, both horizontally and vertically, to ensure correct metacentric heights and radii of gyration for both rolling and pitching. These determine the natural periods for the two rotational modes of movement.

Four ultrasonic sensors fitted at relevant positions on the hulls of these models served to measure the instantaneous distances between these hull points and the basin floor (underkeel clearances). These sensors had a measuring resolution of about 2 mm and could sample at 0,2 s intervals. The models' horizontal excursions were monitored by two remote-controlled overhead cameras. Three flashing lights on the decks, triggered at 3 s intervals, were time-exposed on film and served to position the models relative to their intended course.

Both the propeller revolutions and rudder movement were remote radio-controlled. The propeller speed, rudder angle and underkeel clearances were sampled at 0,2 s intervals and were transmitted by six-channel telemetry equipment to a data acquisition system linked to an in-house mini computer. Three additional channels were used for recording wave conditions in the test basin.

During testing each model was run along a straight course at the appropriate angle to the waves, covering a distance of about 20 m (2 km prototype). Each test condition was repeated between 20 and 30 times to obtain sufficient data for reliable statistical analysis.

During a time-domain analysis the recorded motion time series were used to derive time series of roll, pitch and heave and these principal motions were then used to calculate the vertical motion time series at the perpendiculars, shoulders and quarters of the models for each individual run.

These time series together with the recorded waves were spectrally analyzed individually and averaged over the number of runs during the test. The averaged spectra were then used to calculate the significant wave height and significant motion amplitudes (defined as twice the standard deviation of the motion time series) for roll, pitch and heave and for the vertical motions at the six hull points (i.e. the perpendiculars, shoulders and quarters). The data were also reduced to amplitude response functions for the said principal and hull motions using the encountered wave spectra derived from the incident wave spectra, average angle of wave incidence and average ship speed during the test.

4. SCALE-MODEL TEST PROGRAMME

A test programme was devised to enable coverage of as wide a range of test variables as possible in the available time. The test conditions were selected to represent, in broad outline, the range of conditions most likely to influence ship behaviour at the major South African ports.

In this respect only swells ranging from 1 to 4 m in height and with spectral periods of 12 to 16 s and longer would have to be considered. The angles of wave incidence relative to the models could, depending on their sailing directions, vary from following waves, through beam waves, to heading waves.

Overdepths could range from as little as 10 per cent to as much as 50 (to 75) per cent of the draughts of the model ships.

Ship speeds in and around ports would probably not exceed about 6 m/s (12 kn).

In planning a justifiable test programme, the following assumptions, based on earlier literature surveys, were first made:

(i) The irregular nature of both waves and wave-induced ship motions could be described by a linear superpositioning of sinusoidal components of different frequencies, these frequency compositions being fully described by one-dimensional energy density spectra.

(ii) In a given water-depth and for a given ship speed, the wave-induced ship motion amplitudes were assumed to be linearly related to wave amplitudes for each frequency of wave encounter with the result that the model's motion responses could be expressed in terms of frequency response functions for each angle of wave incidence.

Based on the above assumptions, the initial tests were done with one spectral shape (fixed wave height and spectral peak period) and one ship speed. Thus, only the angle of wave incidence and the relative water-depth were varied which were expected to be the most important variables.

A typical swell spectral shape, based on field measurements, was used to reconstruct a swell spectrum with a significant wave height of 3 m and a spectral peak period of 14 s. An appropriate ship speed of advance of 4 m/s (8 kn) was chosen to run the models in the specified swell conditions with the angles of wave incidence relative to the models to be incremented in steps of 15°. These tests were all done in a relative water-depth (depth-to-draught ratio) of 1,3.

The influence of different relative water depths, ranging from 1,1 to 1,5 in steps of 0,1 was studied for the primary angles of wave incidence, that is, following (α = 0°), heading (α = 180°), beam (α = 270°) and stern- and bow quartering waves (α = 315° and 135°).

In addition to the primary tests, secondary test programmes were devised to serve both for checking the validity of the above assumptions and for studying the influence of additional factors such as the absence/presence of bilge keels on the models, different loading conditions and different spectral shapes.

On completion of the fundamental model tests, a series of tests were planned to simulate typical port entries and departures with both models in a scale model of the Richards Bay harbour approach channel for comparison with data collected during the field monitoring study as well as for determining port accessibility for the deeper draught (21 m draught) model ship.

5. TEST RESULTS

5.1 Wave-induced Hull Motions Versus Angle Of Wave Incidence And Depth Of Water

The significant motion responses, A_{moz}/H_{mo}, versus angle of wave incidence, α, are shown for the perpendiculars, shoulders and quarters of the two models for the case with a relative water-depth, $d/_D$, of 1,3 in Figures 1a and b.

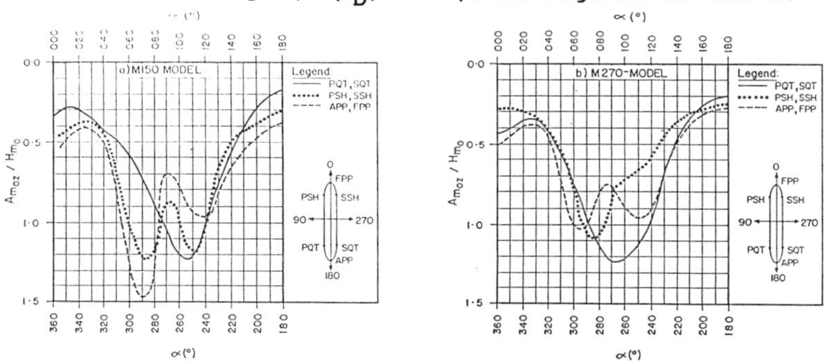

Figure 1a & b. Significant motion response at critical hull points versus angle of wave incidence for $d/_D$ = 1,3 and H_{mo} = 3m, T_p = 14s

The **maximum** significant motion responses are given by the lower limits of the envelopes described by these trend curves.

The effect of the relative water-depth (depth-to-draught ratio, $d/_D$) on these maximum significant hull motions was studied for following (α = 0°), heading (α =180°), beam (α = 270°) and bow and stern quartering (α = 135° and 315°) waves. The results for the M270 model are shown in Figure 2.

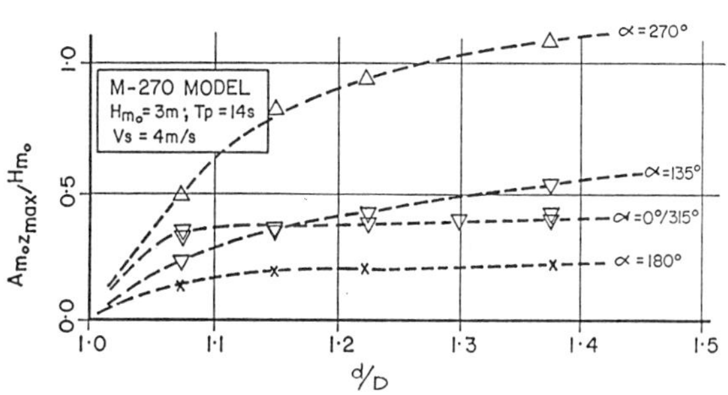

Figure 2. Maximum significant motion amplitude versus water depth

The dependence on water-depth is most noticeable (for both models) in beam and quartering waves indicating a gradual reduction in motion with decrease in underkeel clearance.

The maximum significant motion responses versus angle of wave incidence and relative water-depth, based on the above data, are shown for the two models in Figures 3a and b.

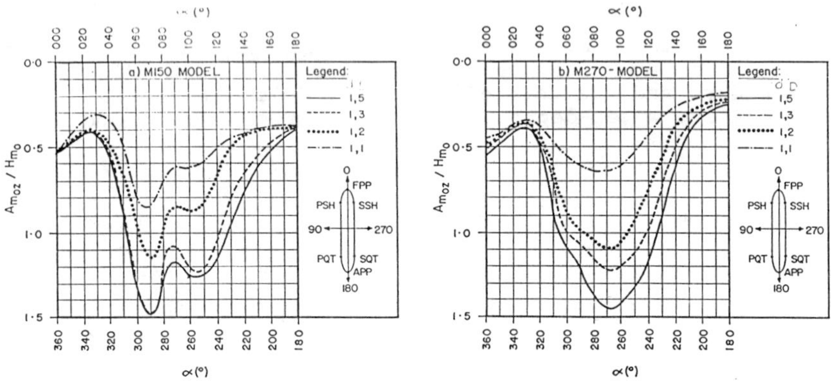

Figure 3a & b. Maximum significant motion response versus angle of wave incidence and relative water depth for H_{mo} = 3m, T_p = 14 s

Figures 1 and 3a and b show significantly different responses for the two models in equal wave conditions and relative water-depths. The most severe motions, for the M150 model, occurred at the perpendiculars for values of α around 70° (or 290°) and were caused mainly by pitching. Fairly severe motions were, however, also recorded at the quarters and shoulders for values of α between 90° and 120° (or 270° and 240°).

In the case of the M270 model the motions were largest at the quarters during rolling in beam waves (α = 90° or 270°). Motions were also quite severe at the shoulders for values of α around 75° (or 285°) and at the perpendiculars around 65° and 110° (or 295° and 250°).

The least vertical motions, for both models, occurred during heading waves (α = 180°) and were usually largest at the perpendiculars.

Motions in following waves (α = 0°) were generally larger than those in heading waves particularly for the M270 model where these were almost twice the magnitude of those in heading waves.

The differences in the motion responses for the two models can be ascribed to the differences in their principal dimensions and loading conditions as well as to the differences in absolute underkeel clearances for equal relative water-depths (e.g. with $d/_D$ = 1,3 the underkeel clearance for the M150 model (D = 17 m) was 5,1 m while that for the M270 model (D = 21 m) was 6,3 m).

Recorded maximum significant motion responses for both models are compared with mathematically predicted responses (using the VESDYN model) in Figures 4a and b.

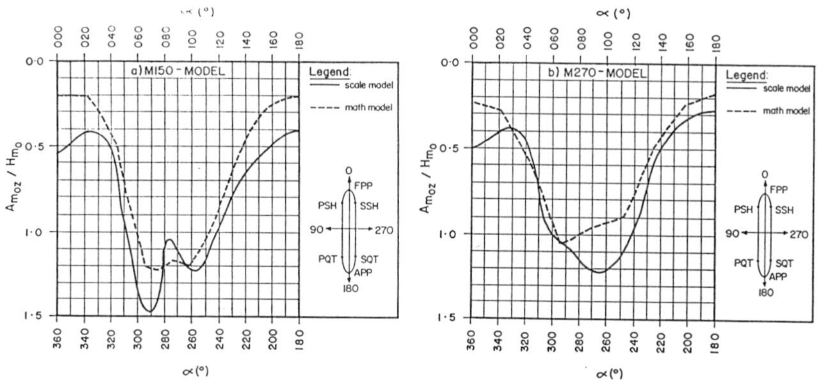

Figure 4a & b. Measured versus predicted significant motion response for $d/_D$ = 1,3 and H_{mo} = 3 m, T_p = 14 s

Measured motions were generally larger than those predicted particularly in following and heading waves and for angles of wave incidence where extreme motions were recorded.

5.2 <u>The Effect Of Wave Height</u>

The effect of wave height on the vertical motion response of the model ships was studied for the primary angles of wave incidence, namely heading (α = 180°), following (α = 0°), beam (α = 270°), bow quartering (α = 135°) and stern quartering (α = 315°) waves. These tests were all done in a relative water-depth of 1,3 and in a 14 s peak period swell with the models advancing at 4 m/s. Wave heights were varied between about 1 m and 5 m.

Figure 5 illustrates the effect of wave height on the maximum significant motion amplitudes for the M270 model for the different angles of wave incidence. Similar trends were also obtained for the M150 model.

The significant motion amplitudes showed an almost linear increase with increase in wave height for all test conditions except those in following waves (α = 0°). During this latter conditions the maximum motions were recorded at the aft perpendiculars of the models and would therefore have resulted from a combination of heaving and pitching.

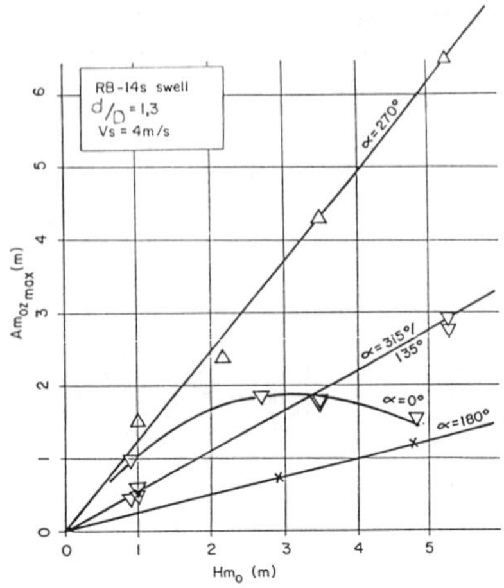

Figure 5. Maximum significant motion amplitude versus wave height for M270 Model

5.3 <u>Sensitivity To Wave Spectral Shape, Peak Period And Ship Speed</u>

Sensitivity tests were conducted to study the influence on wave-induced motion response for different wave spectral shapes, spectral peak wave periods and ship speeds.

The tests confirmed the assumption of linearity in response with respect to the above variables in the sense that hull motion **response amplitude operators** (RAOs) show only insignificantly small changes for the range of conditions tested.

As would be expected significantly large differences were, however, recorded in the resulting significant hull motions for different spectral shapes, peak periods and ship speeds for different angles of wave incidence. The differences primarily result from the differences in the distribution of wave energy with frequency of wave encounter and emphasize the need for using measured (or calculated) RAOs (response amplitude operators) as functions of angle of wave incidence and relative water-depth together with the operating condition, determined by the ship speed, angle of wave incidence and wave spectrum, to determine the resulting hull motion response. Sample ROAs as function of angle of wave incidence for the M270 model in a relative water-depth of 1,3 are shown in Figure 6.

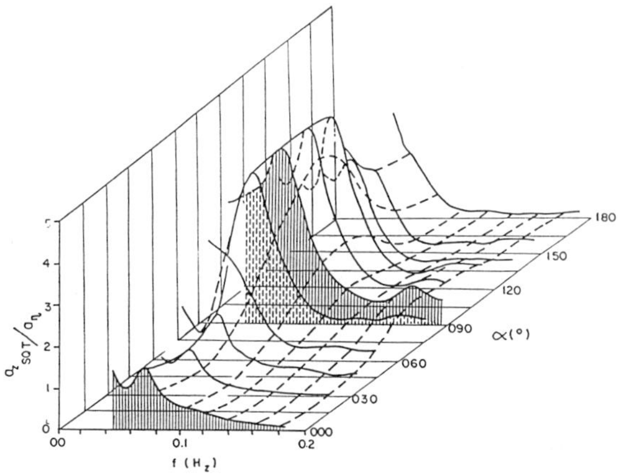

Figure 6. Motion amplitude response function for M270 Model

5.4 Maximum Versus Significant Vertical Hull Motions

The **most probable** maximum motion amplitude can be approximated by the mean of maxima obtained for the number of repeat runs for each test condition. This mean value, A_{max}, is shown versus A_{moz} for a representative number of tests with the M150 model in different relative water-depths and with different significant wave heights and angles of wave incidence in Figure 7.

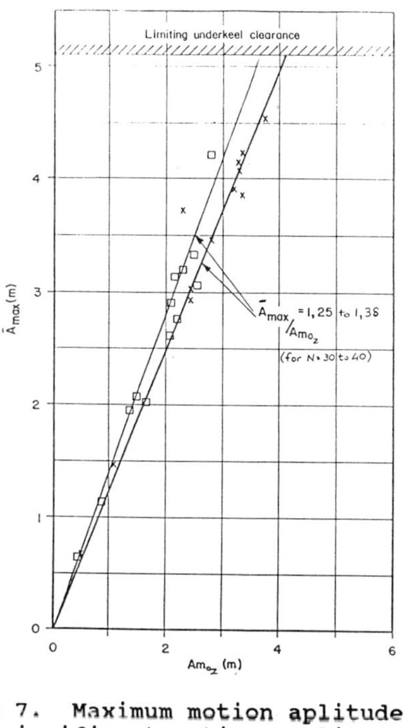

Figure 7. Maximum motion aplitude versus significant motion amplitude

The data indicate ratios of A_{max}/A_{moz} of between 1,25 and 1,38 for motion time series resulting in, on average, 30 to 40 motion oscillations. The apparent linearity of this relationship implies that extreme hull motions are also not affected by the limited underkeel clearances. The data of Figure 7 also appear to confirm the assumption that ship motions are approximately Rayleigh distributed.

The data of Figure 7 also appear to confirm the assumption that ship motions are approximately Rayleigh distributed. This assumption implies that the most probable maximum motion amplitude, μ (A_{max}) can be found from the relationship

$$\mu\ (A_{max}) = \frac{1}{\sqrt{2}} (\ln N)^{0,5} \cdot A_{moz}$$

where N is the average number of oscillations of the ship for the duration of the tests. For values of N of 30 to 40 the theoretical most probable maxima show good agreement with the means of maxima obtained from the model data (Figure 7).

5.5 Ship Motions in the Richards Bay Harbour Entrance Channel

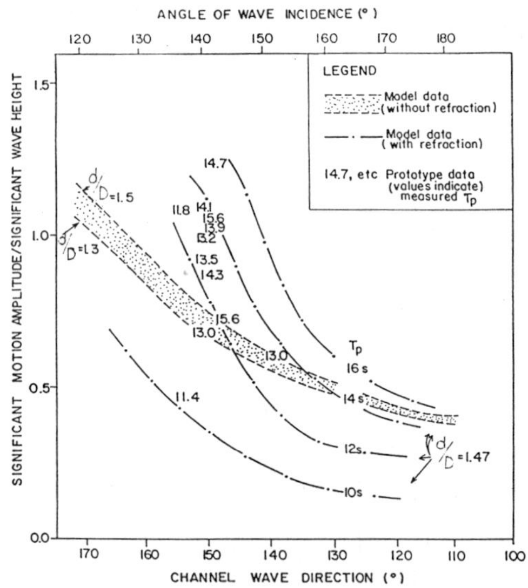

Figure 8. 150kt deadweight (17 m draught) ship: model and prototype response in Richards Bay harbour entrance channel and comparison with model response in undirectional waves

Figure 8 presents the results of the model tests with the M150 model during departure in the Richards Bay harbour approach channel together with those obtained from the field measurements and the model tests in the flat-bottomed basin.

The significant hull motions of the model in the channel were significantly larger than those obtained from the tests in the flat-bottomed basin. This phenomenon is explained by the fact that the wave energy, having accounted for wave directionality due to refraction for any given incident wave spectrum, was much more concentrated at frequencies of wave encounter close to the natural frequencies of motion of the model.

Good agreement was also found between the M150 model results and the prototype data and thus gave confidence in the model study approach and results also for the larger ship model.

CONCLUSIONS

The tests with the two model ships showed that there are significant differences in their vertical motion responses even with identical wave conditions and identical relative water-depths.

For both models the water-depth appeared to have effected the vertical motions for relative depths less than about 20 to 30 per cent of their draughts. This resulted in a non-linear reduction in motion with decrease in water-depth.

Vertical motions at any given depth of water appeared to have increased almost linearly with wave height except during following waves in which case the hull motions appeared to be relatively insensitive to wave height.

The amplitude response functions were found to be insensitive to ship speeds ranging from 0 to about 5 m/s. The amplitude response functions also were not affected by the spectral shape (peak period) although the significant motion amplitudes as such could differ appreciably.

In this respect use should rather be made of the actual amplitude response functions versus angle of wave incidence for given ships and of the wave spectra (accounting for directional spreading and refraction) to be encountered by these ships to determine their significant motion response for operating conditions other than those used during the tests.

The study further showed that local wave conditions in a harbour entrance can cause large differences in ship behavior. In the case of Richards Bay, vertical motion predictions based on the fundamental study results (flat-bottomed basin tests) were considerably less than those recorded by prototype ship monitoring at Richards Bay. Only after carrying out detailed tests in an accurately scaled model of the entrance area of Richards Bay and by reproducing correctly refracted directional swells could the recorded prototype motions be adequately reproduced.

The results of these studies, in general, are believed to offer a significant contribution to the state-of-the-art knowledge of wave-induced ship behaviour in limited depths of water. Since the model results could be verified by prototype measurements, these can be applied with more confidence in the determination of optimum depths for safe harbour approaches.

ACKNOWLEDGEMENT

The permission of South African Transport Services to publish this paper is gratefully acknowledged.

REFERENCES

VAN WYK, AC (1982). Wave-induced ship motions in harbour entrances. Proceedings of 18th ICCE, Cape Town.

VAN WYK, AC and ZWAMBORN, JA (1984). The effect of wave direction on ship motions in harbour entrance channels - model study approach. Proceedings of 19th ICCE, Houston.

CHAPTER 217

DYNAMIC BEHAVIOR OF A MOORING BUOY INSTALLED IN OPEN SHORE

TOSHIYUKI SHIGEMURA*, M.ASCE
AND
KENJIRO HAYASHI**

Abstract

To investigate the dynamic behavior of a mooring buoy installed in an open shore, a simplified simulation model is developed, which consists of the surging and heaving motions of a mooring buoy.

A series of the model tests is performed to examine the reliability of the developed model. The model tests verify that the developed model can predict the dynamic motion of a mooring buoy satisfactorily.

Finally, simulation is made for the buoy installed in the west shore of Iwo-jima which failed during a stormy weather. This simulation indicates that the external loads enough to cause the failure has been loaded on the buoy at the moment of the failure.

Introduction

A mooring buoy is normally installed inside a port or a bay where it is well protected against severe wind and wave loads. However, on some shores where a port can't be built for some reason, it has to be installed in the open shore where it is fully exposed to severe environmental condition. Iwo-jima is an example of this.

Iwo-jima is an isolated small volcanic island located in the Pacific Ocean about 1250 kilometers south of Tokyo. Because of the continuous upheaval of the island, no port has been built there yet. Instead, four single point mooring buoys have been installed off the west shore of the island, to moor the tankers or cargo ships which occasionally come to the island. The problem experienced by these mooring buoys is the frequent failures of the buoy systems during a period of stormy weather although no ship is moored to the buoys. In fact, the mooring buoys installed in Iwo-jima often fail during the typhoon season, which causes a lot of trouble for the supplying operation to the island.

Thus, to insure efficient operation of the mooring buoys installed in the open sea, it is important to clarify the

---

*Professor, Department of Civil Eng., National Defense Academy, 1-10-20 Hashirimizu, Yokosuka, Kanagawa 239, Japan.
**Assistant Professor, Department of Civil Eng., National Defense Academy.

dynamic behavior of the mooring buoys to which no ship is moored. This paper intends to investigate the dynamic behavior of a single point mooring buoy when no ship is moored to it. The investigation will be performed by putting the emphasis on the research of tensile stress induced on a mooring line as follows:
(1)Development of a simplified simulation model of a single point mooring buoy subject to severe environmental loads.
(2) Examination of the adequacy of the developed model through the model tests.
(3) Investigation of the possible causes of a failure which occurred in Iwo-jima through the simulation done by using the developed model.

Motion Equations of a Mooring Buoy

When a ship is not moored to a buoy, the buoy may experience the translational motions called surging, swaying and heaving and the rotatory motions called rolling, pitching and yawing due to the environmental loads. Thus, six equations should be developed and solved simultaneously to clarify the dynamic behaviors of the mooring buoy. However, it is difficult at the present stage to get reliable equations which represent each of these six motions. Further, one of the aims of this paper is to develop a simplified simulation model which can predict the dynamic behaviors of the tensile stress induced on a mooring line satisfactorily. Thus, the following assumptions were made to derive the motion equations:
(1) A mooring buoy doesn't take any rotational motions but takes the translational motions of surging and heaving.
(2) Dynamic motion of the mooring line is not taken into account in the derivation of the simulation model.
(3) A buoy is moored to the sea floor by an open link chain of uniform size which is elastically extensible.
(4) Only the wind and wave loads act on a buoy as the external loads.
(5) Tensile stress induced on the line by external loads can be substituted for the static tensile stress evaluated by an available formula.

Assumption (1) means that the motion of a mooring buoy can be treated as a two dimensional motion. Figure 1 shows the coordinate system used for deriving the motion equations. Here, each notation in figure 1 indicates the following:
a=draft of the buoy
b=freeboard of the buoy
F_{wd}=wind load acting on the buoy
T=tensile stress induced on the mooring chain
T_x=horizontal component of T
T_z=vertical component of T
F_x=horizontal component of the wave load acting on the buoy
F_z=vertical component of the wave load acting on the buoy
η=elevation of wave surface

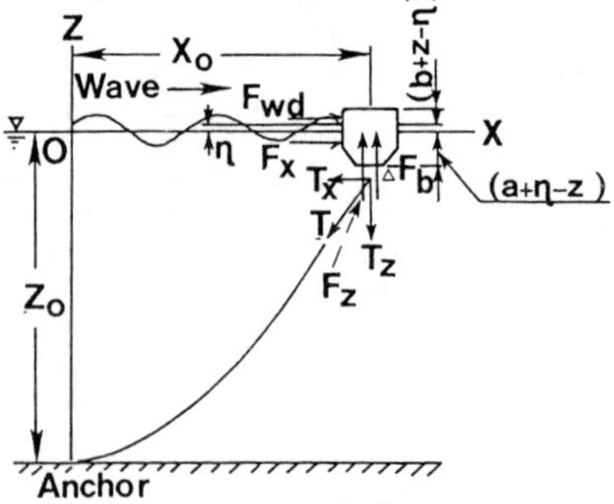

Figure 1. Coordinate system used for the analysis.

X_o=original position of the model buoy on X axis
Z_o=original position of the model buoy on Z axis

Suppose the model buoy moved to (X_o+x, Z_o+z) from its initial position (X_o, Z_o) for time t due to the wind and wave load. Then, basing on the assumptions together with the coordinate system provided, the following two equations can be derived as the motion equations of a mooring buoy:

$$(m+m'_x)\ddot{x}=F_{wd}+F_x-T_x-C_x\dot{x} \quad ....(1)(\text{for surging})$$

$$(m+m'_z)\ddot{z}=F_z-T_z+\Delta F_b-C_z\dot{z} \quad ....(2)(\text{for heaving})$$

where
m=mass of a buoy
$m'_{x,z}$=horizontal and vertical components of virtual mass of model buoy
$C_{x,z}$=horizontal and vertical components of damping coefficient of a buoy system
ΔF_b=dynamic buoyance of model buoy
X=horizontal displacement of a buoy from the origin
x=horizontal displacement of a buoy from its initial position
Z=vertical displacement of a buoy from the origin
z=vertical displacement of a buoy from its initial position

The former equation includes the horizontal components of wind load F_{wd} and wave load F_x, both acting on a buoy, tensile force induced on a chain T_x, and a damping force of the buoy system $C_x dx/dt$. The latter includes the vertical components of the wave load F_z, tensile stress T_z, damping force of the buoy system $C_z dz/dt$, and the dynamic buoyance of the buoy ΔF_b. However, values of these components together with the coefficient values are all unknown and should be somehow determined at an arbitrary time to solve eqs.(1) and (2) simultaneously.

Evaluation of Tensile Stress and External Loads

The catenary theory is generally used to calculate the static tensile stress induced on a mooring line. However, the value of the tensile stress T varies due to the hanging state of a mooring line under the water. The hanging state is usually classified into three states as follows:
(1) State I in which a part of the mooring line is hanging down vertically from a buoy to the sea floor and the rest of it is resting on the sea floor.
(2) State II in which the most part of the mooring line is hanging in the water forming a catenary curve although the rest of the chain is resting on the sea bed near the anchor.
(3) State III in which all of the mooring line is suspended in the water and no part of the line is resting on the sea bed near the anchor.

Figure 2 shows these states of the hanging of a mooring line. The notations in this figure indicates the following:
X_1=limiting displacement of a buoy from its origin in which the hanging state of the mooring line stays in state I.
X_2=limiting displacement of a buoy from its origin in which the hanging state of the mooring line stays in state II.
X_3=limiting displacement of a bouy from its origin in which the hanging state of the mooring line stays in state III.

a_r=length of the mooring line resting on the sea floor.
θ_a=directional angle of the tangential line to the mooring line at the anchoring point measured from the sea floor.

Calculation of the tensile stress is then made for each of these three states separately. Shoji(1975) introduced the effect of the elasticity of a mooring line to the catenary theory and derived three approximate polynomials expanded into a power series of T_x. However, there is some uncertainty in his equation for the line in state II. Further, there is some difficulty in finding the real value of T_x among the solutions of the polynomials. Thus, the authors developed new equations for each of the three states (1987), following Shoji's deriving method. Table 1 summarizes these equations developed for each state of hanging.

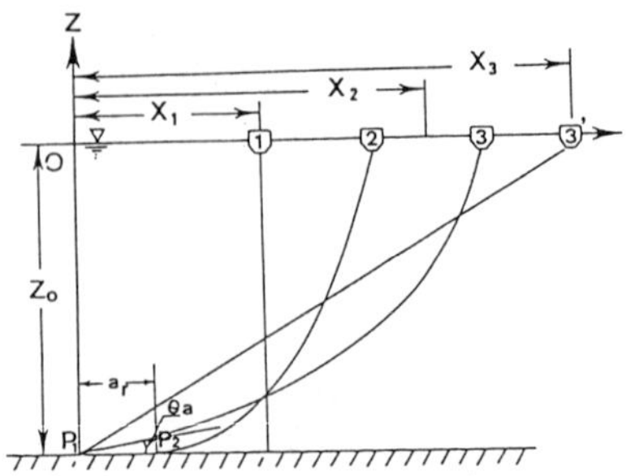

Figure 2. Hanging states of a mooring lines.

Table 1. Summary of the equations for evaluating the tensile stress induced on a mooring line and other related terms

Equations for a mooring line in State I

$$T_x=0 \quad \ldots\ldots (3)$$

$$T_z=W\,Z_o-WZ_o^{\,2}/(2EA) \quad \ldots\ldots (4)$$

$$X_1=S-Z_o+WZ_o^{\,2}/(2EA) \quad \ldots\ldots (5)$$

Equations for a mooring line in State II

$$X=S(1+T_x/EA)-(S-a_r') +T_x/W\left\{\ln W/T_x(S-a_r')+\sqrt{[W(S-a_r')/T_x]^2+1}\right\} \quad \ldots\ldots (6)$$

$$Z_o=\sqrt{(T_x/W)^2+(S-a_r')^2}-T_x/W+W(S-a_r')^2/(2EA) \quad \ldots\ldots (7)$$

$$T_z=W(S-a_r') \quad \ldots\ldots (8)$$

$$X_2=T_xS/EA+T_x/W\left\{\ln WS/T_x+\sqrt{(WS/T_x)^2+1}\right\} \quad \ldots\ldots (9)$$

Equations for a mooring line in State III

$$X=T_xS/EA+T_x/W\left\{\ln\left[\tan\theta_a+WS/T_x+\sqrt{(\tan\theta_a+WS/T_x)^2+1}\right] -\ln\left[\tan\theta_a+\sqrt{\tan^2\theta_a+1}\right]\right\} \quad \ldots\ldots (10)$$

$$Z_o=T_x/W\left\{\sqrt{(\tan\theta_a+WS/T_x)^2+1}-\sqrt{\tan^2\theta_a+1} +WS(\tan\theta_a+WS/2T_x)/EA\right\} \quad \ldots\ldots (11)$$

$$T_z=T_x\tan\theta_a+WS \quad \ldots\ldots (12)$$

$$X_3=\sqrt{S^2-Z_o^{\,2}} \quad \ldots\ldots (13)$$

In this table, several new notations appear. These notations indicate the following:

S=total length of a mooring line before getting deformation

W=weight of a mooring line per unit length under the water

A=sectional area of a mooring line

a_r'=original length of a_r before getting a deformation

Using these equations developed above, we can determine the static tensile stress and its components induced on a mooring line for any location of buoy if the physical properties of the mooring line are given. Namely, the values of T_z and X_1 are determined immediately from eqs.(4) and (5) because every variable in the equations is known. In case of State II, values of T_x and T_z can be determined from eqs.(7) and (8) if a_r' is given. Once the value of T_x is determined, the corresponding X_2 is also determined from eq.(6). In case of State III, the values of T_x and T_z are determined similarly from eqs.(11) and (12) if the value of $\tan\theta_a$ is given. The corresponding value of X is then determined from eq.(10) easily.

Next, let's derive the equations for wind and wave loads, referring to figure 1. It was decided to evaluate the wind load F_{wd} by a conventional type formula as shown below.

$$F_{wd}=1/2\rho_a C_d D(b+Z-\eta)V^2 \quad \text{........................(14)}$$

where ρ_a is air density, C_d is drag coefficient of a buoy, D is diameter of a buoy, b is freeboard of a buoy, η is the elevation of wave surface, z is the elevation of the gravity center of the buoy and V is the mean velocity of wind. Further, it was assumed that wave load can be evaluated by the following formulas derived from the Morison equation:

$$F_x=1/2\rho_w C_{dx} D(a+\eta-z)u|u| +1/4\rho_w C_{mx} D^2(a+\eta-z)\dot{u} \quad \text{....(15)}$$

$$F_z=1/8\rho_w C_{dz} D^2 v|v| +1/4\rho_w C_{mz} D^2(a+\eta-z)\dot{v} \quad \text{..........(16)}$$

where

ρ_w=water density

u=horizontal component of water particle's velocity at still water level

$\dot{u}$=horizontal component of water particle's acceleration at still water level

C_{dx}=drag coefficient of a buoy for horizontal flow

C_{mx}=mass coefficient of a buoy for horizontal flow

v=vertical component of water particle's velocity at still water level

$\dot{v}$=vertical component of water particle's acceleration at still water level

C_{dz}=drag coefficient of a buoy for vertical flow

C_{mz}=mass coefficient of a buoy for vertical flow

Finally, it was decided to evaluate the dynamic buoyance ΔF_b by the following equation:

$$\Delta F_b=1/4\,\rho_w\,D^2(a+\eta-z) \quad \text{........................(17)}$$

Using the equations developed above, respective terms in the righthand sides of eqs.(1) and (2) could be evaluated once we get the values of respective coefficients, flow field induced by the wave, physical properties of the buoy and mooring line, and other values of the related terms.

Model Tests of a Single Point Mooring Buoy

As described previously, a mooring buoy installed off the west shore of Iwo-jima often fail during the typhoon season. Photo 1 shows the failure which occurred at Iwo-jima on August 4, 1986. In this case, the mooring chain was parted due to the severe environmental loads caused by the typhoon and the buoy was cast on the west coast of the island.

Photograph 1. Buoy cast on the west coast of Iwo-jima.

Thus, it was decided to perform a series of model tests in light of the buoy systems and their surrounding environmental conditions at the west shore of Iwo-jima. The following basic principles were established for performing the model tests:

(1) Model tests are to be performed in a two dimensional wave tank with a constant depth of water.

(2) Regular waves of various size will be used as the waves acting on the buoy.

(3)A constant weight will be loaded on the buoy horizontally instead of blowing wind at the bouy.

(4)An open link chains will be used as the mooring line because of its availability, although stud link chains are used in Iwo-jima.

The geometric scale of the test model was set to be 1/35 of the prototype and the time scale was determined by referring to the Froude law of similarity. Table 2 summarizes the detail of the test conditions.

Table 2. Details of the model tests

Dimension of the model buoy	
Diameter	7.9 cm(2.76 m)
Height	5.2 cm(1.83 m)
Draft	2.4 cm(1.0 m)
Weight in the air	70.0 grf(3.0 tonf)
Dimension of the model chain	
Diameter	1.9 mm(68.0 mm)
Weight under the water	0.7 grf/cm(86.5 kgf/m)
Length	57.1 cm(20.0 m), 68.6 cm(24.0 m), 91.4 cm(32.0 m), 114.3 cm(40.0 m)
Type	Open link(Stud link)
Range of the test conditions	
Water depth	45.7 cm(16.0 m)
Wave height	3.0--16.5 cm(1.0--5.8 m)
Wave period	0.7 --2.0 s(4.1--11.8 s)
Steepness	0.01--0.10
Wind load	12.9 grf(55.0 m/s)
Scope of the mooring line	1.25, 1.50, 2.0 and 2.5

In this table, the numerical figures in parentheses indicate the corresponding values for the prototype in Iwo-jima. The model buoy was made of aluminum. Chains of four different length were prepared to use as mooring line. The length of these chains was determined so that the scope of the line may become the respective values shown in table 2. Here, the scope is a parameter defined as the ratio of line length to the depth of water.

A 39.6m long, 0.6m wide and 0.8m high wave channel was used for the model tests. Around the middle portion of this channel, a model buoy was moored to the channel bottom by one of the four chains shown in table 2. Figure 3 shows the setup of the model test. A waterproof tension meter was placed on the line at the position just below the bottom of the model buoy, and string-pully system was provided upon the upper structure of wave channel so that constant weight may be loaded to the buoy horizontally through that system.

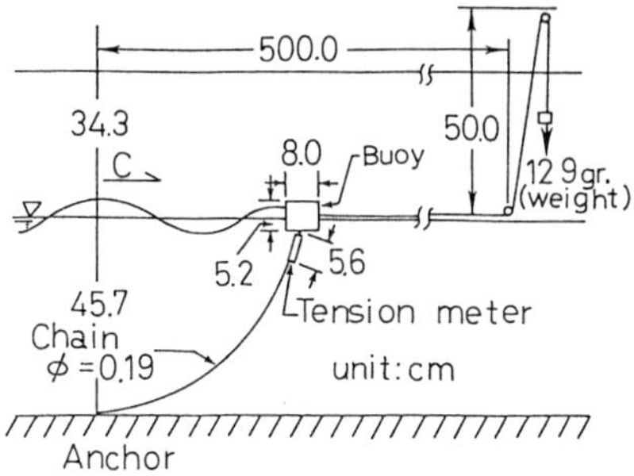

Figure 3. Setup of the model test

Beside the initial position of the buoy a capacitance type wave meter was installed and both the tension and wave meters were connected to a data recorder. Further, a video camera was placed in front of the buoy 2m from the glass wall to record the motion of buoy.

Regular waves of 24 kinds were chosen as the external waves acting on the buoy. Their characteristic values are shown in table 2. In addition, it was decided to load a constant weight of 12.9grf to the buoy horizontally as the substitute for the wind load. This weight of 12.9grf corresponds to the wind load which might be generated on the buoy by a wind having a speed of 55m/s. Here, the wind speed of 55m/s is the maximum one which was recorded in Iwo-jima for the past 20 years.

Prior to starting the model tests, performance of tension meter was examined statically. Tests were made in another small channel, changing both water depth and size of weight loaded on the model buoy horizontally through the string-pully system. The position of the model buoy, the size of the loaded weight, the hanging state of the mooring chain, and the tensile stress detected by the tension meter were all recorded. These tests were repeated a couple of times for three of the four chains prepared. The main tests were conducted for the model buoy moored by each of the four chains mentioned previously. This was done by exerting regular waves of 24 kinds one after another. In each run, both waves and tensile stress were recorded on the data recorder and the motion of the model buoy was recorded by video recorder.

Results of the Model Tests

To check the reliability of equations developed for evaluating the static tensile stress induced on the mooring line, measured values of the horizontal components of tensile stress T_x were compared with the corresponding values calculated by the equations shown in table 1, using the physical properties of moorimg line shown in table 2. Further, it was assumed in the calculation that modulus of elasticity E of the mooring line is 2.1×10^6 kgf/cm^2.

Figure 4 shows the results of this comparison. In this figure, CL means the length of the mooring line and D means the water depth at which the tests were made. As can be seen from this figure, calculated values of tensile stress agree quite satisfactorily with the measured ones within the range of the expected position of model buoy. This fact clearly verified that the developed equations would be quite reliable for evaluating the statical tensile stress induced on the mooring line.

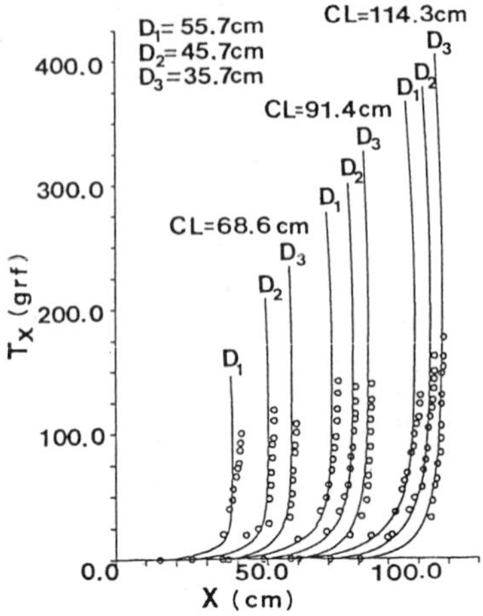

Figure 4. Reliability of the equations for evaluating the static tensile stress.

It was assumed previously that tensile stress induced on the mooring line dynamically could be substituted for the static tensile stress evaluated by some available equations. Thus, tensile stress T and its components T_x and T_z were all calculated similarly by the equations shown in table 1 by varying the position of the buoy every one centimeter within the expected range of the buoy's movement. This calculation was made for each of the four chains and the results were stored on floppy disks. Next, results of the model tests were analyzed. From the recorded data of tensile stress T, the extreme values of it were extracted from the record over 20 cycles, and their mean value T_m was determined by simply averaging them.

Photograph 2 is a picture of the orbital motion of model buoy. Similar pictures were taken at every run shading the laboratory from the sun light. At the same time, motion of the model buoy was recorded on a video recorder. Based on these data, the authors read the data such as horizontal and vertical displacements of model buoy x_m and z_m, the length of both major and minor axes of the inclined orbit of model buoy a_m and b_m, inclination of the major axis against the horizontal axis θ_{am} and the inclination of the mooring chain at the bottom of model buoy θ_{cm}. This data was analyzed using the steepness H_o'/L_o, ratio of wave height to water depth H/h of the

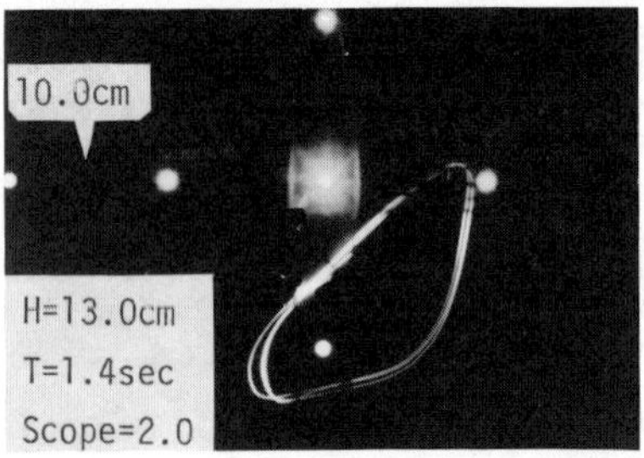

Photograph 2. Orbital motion of a buoy moored by a chain.

incident wave and scope of the mooring line as the controlling variables.

Figure 5 shows the relationship between T_m/W and H_o'/L_o plotted introducing the scope as a parameter. Here, W is the dead weight of model buoy. As can be seen from this figure, quite a high correlation exists between both variables. Further, it can be seen that the gradient between two variables decreases almost linearly as the value of scope increases.

Next, it was determined how the displacement of the buoy varies with the variation of external loads and mooring conditions. Figures 6 shows the relationship between x_m/L_o and H_o'/L_o. In this figure scope of the mooring line is taken as a parameter. From this figure, it can be seen that both x_m/L_o increases almost linearly as H_o'/Lo increases although the effect of the scope to the distribution is not clear. Similar distribution was also found between z_m/L_o and H'_o/L_o. Further, figure 7 shows the relationship between z_m/x_m and H_o'/L_o taking the scope as a parameter. Data scatters considerably. However, we can see that the ratio of both displacements tends to increase as H_o'/L_o increases in the range of H_o'/L_o less than approximately 0.3 and that it tends to have a constant value surrounding 1.0 in the range of H_o'/L_o greater than nearly 0.3. Effect of the scope is also not clear in this case.

Similar analyses were also made for the axial length of the inclined orbit of the buoy. Figure 8 shows the relationship between a_m/L_o and H_o'/L_o. From this figure, it can be seen that there exists a firm correlation between a_m/L_o and H_o'/L_o although the effect of the scope to this relationship is not seen clearly. Relationship between b_m/L_o and H'_o/L_o was also checked similarly. However, cor-

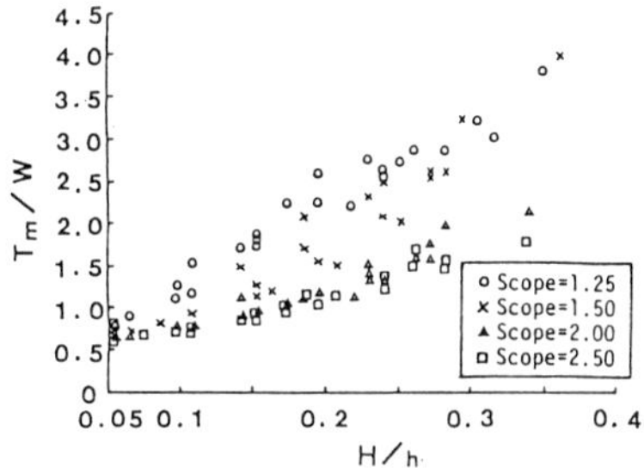

Figure 5. Relationship between T_m/W and H_o'/L_o.

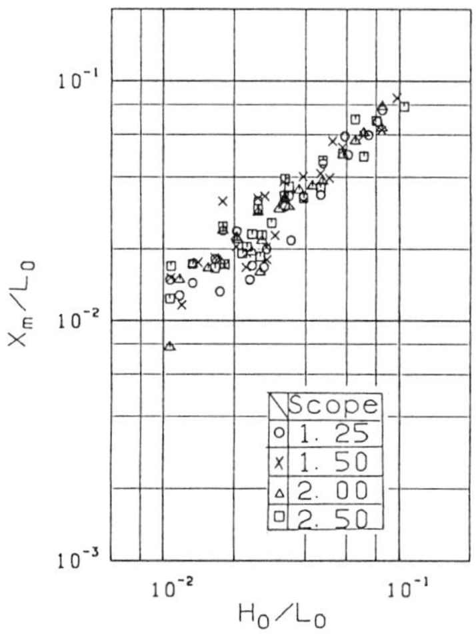

Figure 6. Relationship between x_m/L_o and H_o'/L_o.

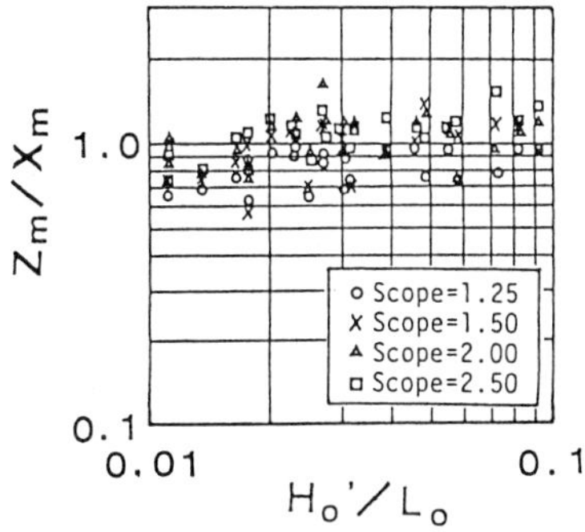

Figure 7. Relationship between z_m/x_m and H_o'/L_o.

relation between b_m/L_o and H_o'/L_o was not as high as that found for a_m/L_o. Especially, in the range of H_o'/L_o less than 0.03, data of b_m/L_o scattered considerably against the corresponding values of H_o'/L_o.

Figure 9 shows the relationship between the ratio of both axial length a_m/b_m and H_o'/L_o. This figure shows that b_m/a_m tends to increase as H_o'/L_o increases in the range of H_o/L_o less than nearly 0.4 although data scatters considerably. Further, this ratio tends to converge to a constant value of nearly 0.05 in the range where H_o'/L_o is greater than approximately 0.05.

Finally, let's review the inclination of the mooring line at the bottom of model buoy. Figure 10 shows the relationship between θ_{cm} and H/h. In this case, scope of the mooring buoy is also taken as a parameter. As can be seen from this figure θ_{cm} tends to decrease almost linearly as the value of H/h increases in the range of H/h less than 0.25. This figure further indicates that θ_{cm} tends to keep a constant value of approximately 45 degrees in the range of H/h greater than 0.25.

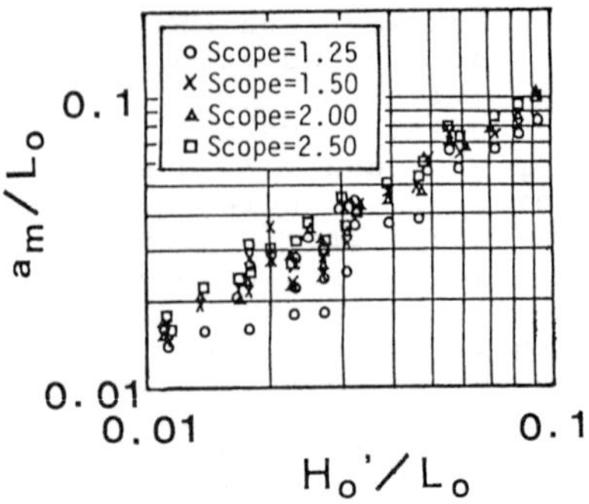

Figure 8 Relationship between a_m/L_o and H_o'/L_o.

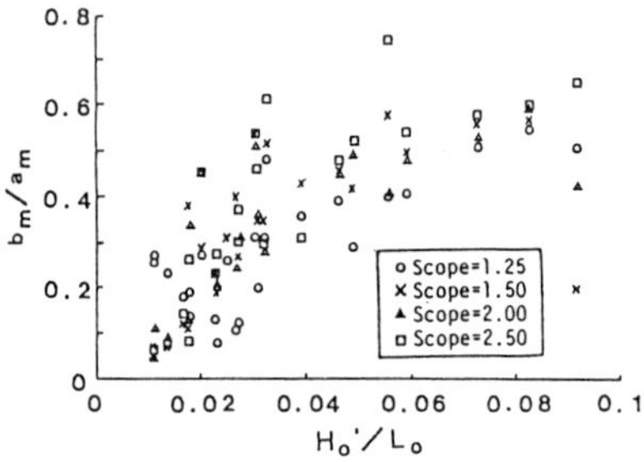

Figure 9. Relationship between a_m/b_m and H_o'/L_o.

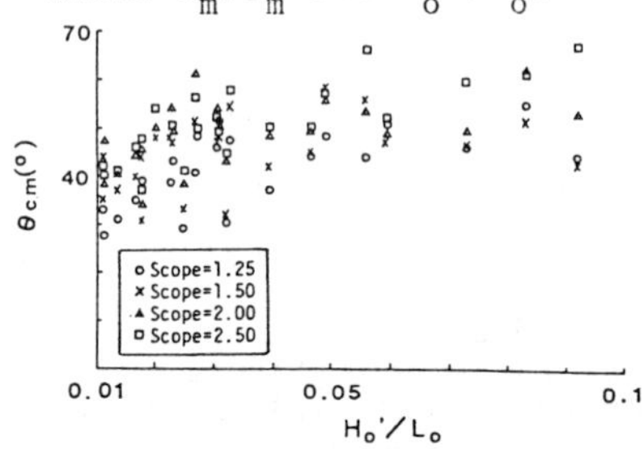

Figure 10. Relationship between θ_{cm} and H_o'/L_o.

Adequacy of the Simulation Model

The following assumptions were made prior to starting the numerical analysis:

(1) virtual mass of the model buoy m'_x and m'_z in eqs.(1) and (2) are both equal to the mass of water displaced by the model buoy.

(2) drag coefficients of model buoy C_{dx} and C_{dz} in eqs.(15) and (16) are both equal to 1.0.

(3) mass coefficients of model buoy C_{mx} and C_{mz} in eqs.(15) and (16) are both equal to 2.0.

(4) velocity and acceleration fields of the wave induced flow can be evaluated by the theory of small amplitude wave.

(5) damping coefficients of the buoy system C_x and C_z can be determined by trial and error method in the process of numerical analysis although the coefficient values determined

by Shoji(1975) through his tests of similar size are used as the initial values of them.

Based on these assumptions, we can now calculate the wave load and dynamic buoyance included in eqs.(1) and (2), for any time and location of model buoy. Further, wind load has been set to be 12.9grf and tensile stress and its components induced on the mooring line have been calculated for any location of model buoy. Thus, we are ready to solve eqs.(1) and (2) simultaneously. Numerical analyses were made for every case of the model tests by using the Runge-Kutter-Gill method. In the actual analyses, time interval was set to be 0.01 second and the analyses were proceeded until the calculated orbit of the model buoy reached almost stable state.

Figure 11 shows some examples of the numerical analyses. Namely, tensile stress T_c induced on the line and orbit of the model buoy both obtained from the analyses are compared in this figure with the corresponding ones measured in the model tests. The solid lines indicate the results obtained from the analyses and the dotted lines indicate the results measured in the model tests. From this figure, we can see that eqs(1) and (2) can simulate the dynamic motion of model buoy fairly well although the developed model tends to give a slight overestimation.

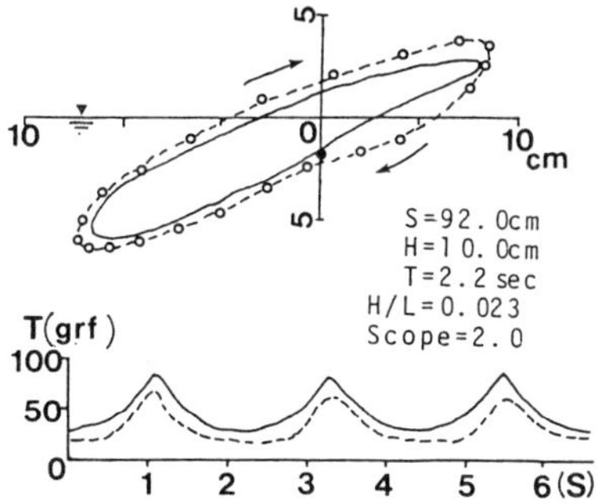

Figure 11. Example of the simulation of buoy's motion.

To see the adequacy of the simulation model more quantitatively, the maximum values of the calculated tensile stress T_c, horizontal and vertical displacements of model buoy x_c and z_c, and inclination of the mooring chain θ_{cc} were compared with the corresponding values measured through the model tests.

Figures 12 shows the distribution of T_c/T_m plotted against H_o'/L_o for the scope greater than 2.0. From this figure, we can see that developed model can predict the dynamic tensile stress induced on the mooring line quite satisfactorily when the scope value is greater than 2.0 and steepness of the acting wave is less than approximately 0.04. Further, the value of T_c/T_m scatters considerably when H_o'/L_o is greater than approximately 0.04. Similar analysis was made for the case whose scope is less than 1.5. In such a case, the developed model has a tendency to considerably overestimate the ten-

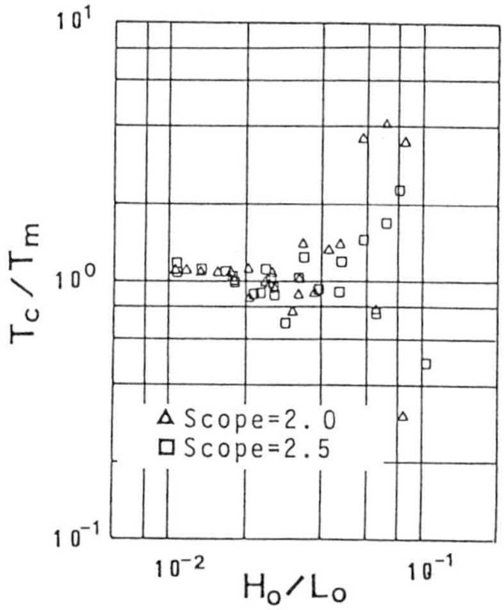

Figure 12. Relationship between T_c/T_m and H_o'/L_o.

sile stress.

Figures 13 shows the distribution of z_c/z_m plotted against H_o'/L_o. This figure indicates that Z_c agrees quite satisfactorily with z_m in the whole range of H_o'/L_o used in this model test. However, the model still has a tendency to overestimate the respective displacements of the model buoy when the scope of the line is 1.25. Analysis done for x_c/x_m also indicates a similar distribution for the values of H_o/L_o.

Figure 14 shows the distribution of θ_{cc}/θ_{cm} plotted against H_o'/L_o. From this figure, we can see that θ_{cc}/θ_{cm} scatters around the value of 1.2 in the whole range of H_o'/L_o although the model has a tendency to overestimate θ_c.

These comparisons indicate that the developed model can trace the dynamic behaviors of the mooring buoy fairly well when scope of the mooring line is greater than 2.0 and the steepness of the acting waves is less than nearly 0.04. Further, these comparisons may indicate the possibility that the developed model can be improved more when some modifications are made on the values of respective coefficients and the formulae to evaluate wind and wave load.

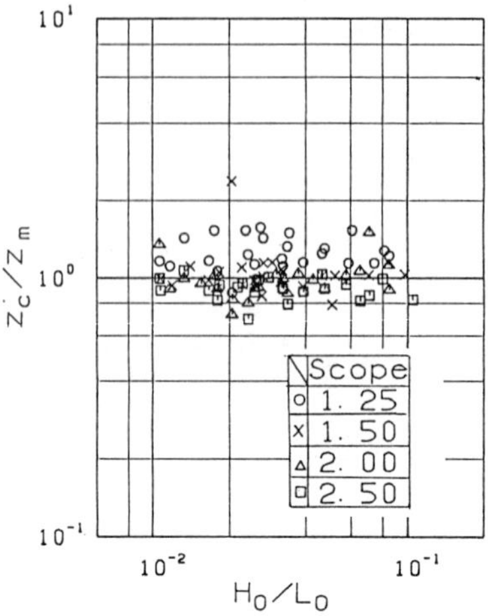

Figure 13. Relationship between z_c/z_m and H_o'/L_o.

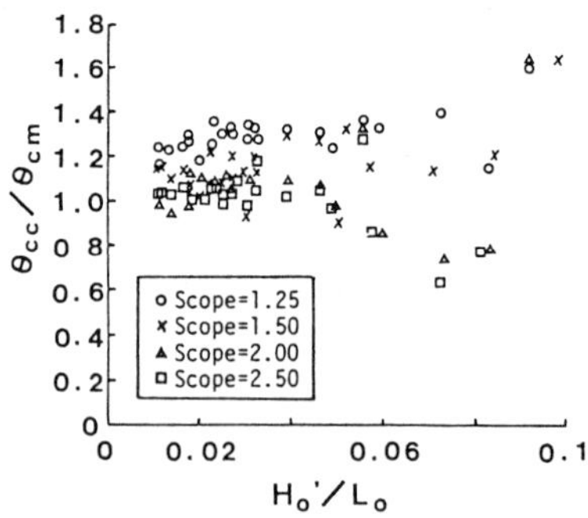

Figure 14. Relationship between θ_{cc}/θ_{cm} and H_o'/L_o.

Application of the Developed Model to a Real Buoy

To determine whether the developed model can be applied to simulate the dynamic behaviors of a real buoy, and further to see whether the simulation done by this model can serve to find any possible causes for the failures of buoy systems which have often occurred in Iwo-jima, it was decided to choose a failure incident which occurred in Iwo-jima on May 23, 1982, as an object of the simulation, and to simulate the dynamic behaviors of the buoy just before its failure.

This incident occurred during a stormy weather which was caused by the typhoon No.4 which had been hitting the island since the day before the failure. In this incident, the harp shackle was distorted seriously due to a huge tensile force induced on it by the severe environmental loads, and was pulled out through the chain hole of the buoy. As a result, the buoy was cast away on the west coast of the island. Figure 15 is the side view of a harp shackle of the same size as that used at the island. As can be seen from this figure, it is clear that a huge tensile force strong enough to squash the shackle's width from 420 mm into 352 mm was loaded on it at the moment of the failure.

Soon after the incident, the harp shackle was recovered by members of the Japan Maritime Self Defense Force (JMSDF) stationed there, and it was reported later that the width of the harp shackle was 364 mm and that several scratches were found on the inner wall of the chain hole of the recovered buoy. Further, the Engineering Office of JMSDF made a series of static pulling test of harp shackles in a laboratory and concluded that external loads of approximately 110 tonf were loaded on the harp shackle at the moment of the failure(1982). This buoy was one of four buoys installed off the west coast of Iwo-jima, and it had been just swapped two months before the incident. The depth of installation was 16m and, and the buoy had been moored by a 20 m long mooring chain with a diameter of 68mm. Thus, the scope of mooring line was 1.25. Just two weeks before the failure incident, the authors had installed a supersonic wave meter about 450 m off the west coast of Iwo-jima at a depth of 12 m(1984). This site is about 500 m south of the installation site of the four mooring buoys. Fortunately, the wave meter recorded the wave data for ten minutes every two hours during this typhoon although the wave meter was washed away by typhoon No.17 which hit the island in November, 1983.

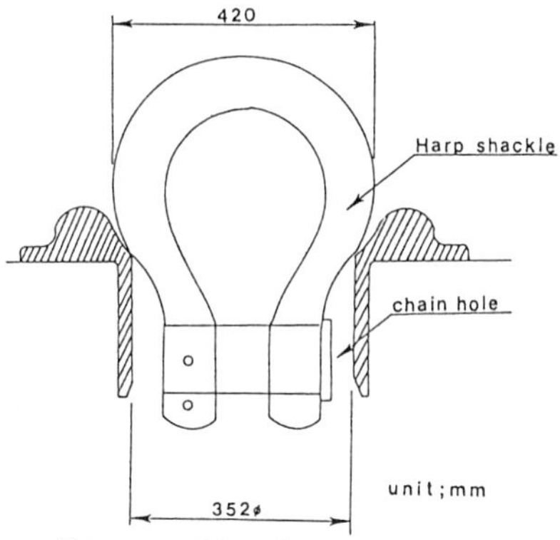

Figure 15. Side view of a harp shackle used in Iwo-jima in 1982.

Through the analysis of this record, it was found that the maximum wave height was nearly 7 m and the significant wave period was approximately 10 second. The JMSDF on Iwo-jima has a meteorological station which is built near the center of the island and whose elevation is approximately 110 m above the mean sea level. At 6 a.m. of May 23, 1982, the station recorded the instantaneous maximum wind speed of 26.0 m/s. However, wind data around the installation site of the buoy is unknown. Thus, it was decided to use the wind of 55m/s which is the maximum wind speed recorded on the island for the

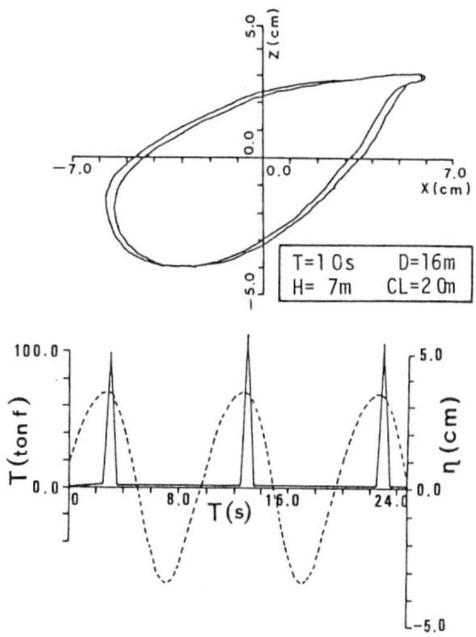

Figure 16. Results of the numerical analysis done for the buoy failed in Iwo-jima.

past 20 years. Further, it was decided to use the value of 1200 kgf.s/m for the damping coefficient C_x and the value of 560 kgf.s/m for the damping coefficient C_z, respectively. These values were determined by applying the Froude law of similarity to the corresponding coefficient values obtained through the model tests. Numerical analysis was made for the conditions stated above.

Figure 16 shows the analytical results. Namely, the upper figure shows the orbit of the buoy, and the lower one shows the time variations of both wave profile and tensile stress T. This figure clearly indicates that tensile stress exceeding 110 tonf may possibly be induced on the mooring line if the buoy is subject to as large external loads as those used for the numerical analysis.

Concluding Remarks

A rather simplified model was developed for simulating the dynamic behavior of a simple point mooring buoy which is installed in the open sea and the adequacy of the model was examined through the model tests. Further, using the developed model, numerical analysis was made for the dynamic motion of a real buoy in Iwo-jima whose system failed during a stormy weather caused by a typhoon. As a result, it was verified that the developed model can satisfactorily predict the dynamic behavior of a mooring buoy in the open sea.

References

1. Engineering Office of JMSDF,"Investigation report on the failure of No.1 buoy in Iwo-jima", p.13, June 1982.
2. Shigemura, T. and Komiya, Y.,"Formation of tombolo at the west coast of Iwo-jima",Proc. of the 19th Conf. on Coastal Eng.,Chap.96, pp.1403-1419, 1984.
3. Shigemura, T.,Hayashi, K and Kouzaki, T.,"Dynamic behavior of a single point mooring buoy installed in open shore", Proc. of the 34th Japanese Conf. on Coastal Eng., pp.621-625, 1987.
4. Shoji, K.,"Study on the motion of a moored buoy and tension of mooring lines", Journ. of the Society of Naval Architects of Japan, Vol.138, pp.233-246, 1975.

CHAPTER 218

The Fundamental Study to Reduce The Settled Area of The Fish Aggregation Devices on The Sea Floor Thrown from A Ship

by Yoshiharu Matsumi* and Akira Seyama**

ABSTRACT

This study aims at clarifying the effective method to set up the fish aggregation device (FAD) thrown down from a ship on the designed position most accurately. The numerical simulation technique is developed to analyzed the behavior of the settling FAD in consideration of the effect of vortices generated behind it. From the systematic calculations for the change of the initial condition on the posture of the FAD at throwing, the effective initial posture of the FAD which reduces the settled area on the sea floor most is investigated. Furthermore, the added-mass coefficient of the settling FAD is numerically analyzed from the kinetic energy of the fluid induced by the settling FAD. The FAD treated in this study is the structure to gather fishes.

1. INTRODUCTION

The fish aggregation device is usually placed on the sea floor of a depth from about 30m to 100m to gather fishes. The most common shape of the FAD is the cubic type of 2mx2mx2m with a hollow inside and a gap in individual surfaces as shown Fig.1. Although the reason why and how the FAD can gather fishes has not been enough clarified yet, vortices formed at the FAD and their shedding behind the FAD have been pointed out to be the major reason to gather fishes around it. The FAD supplies for fishes not only a hiding place but also nourishing materials which are flung up from the sea bed by the vortex induced. The multi-FAD system in which the FADs are arrayed with a constant interval along a dominant flow in planned sea area has been adopted to activate the mentioned properties of the FAD. The multi-FAD system is mainly constructed by throwing them down from a ship to lay out on the fishing ground. It is difficult to set up the FAD in the fluid is consisted by following three complex motions (i.e. an oscillatory motion in horizontal direction, a rotational motion and vertical

* Research Associate Dept. of Ocean Civil Engg. Faculty of Engineering, Tottori University, Tottori, Japan.
** Prof. Dept. of Ocean Civil Engg. Faculty of Engineering, Tottori University, Tottori, Japan.

Fig.1 Illustration of fish aggregation device

drop). As a result of the position errors induced in the set up, the work of the system on gathering fishes may not be demonstrated sufficiently. Hence, to make it clear the settling behaviors of the FAD becomes urgent to carry out their accurate settling on the designed position.
On the other hand, from the strength point of view, the impulsive force exerted on the FAD at landing on the sea bottom must be evaluated accurately. The impulsive force F is given by

$$\int_{\Delta t} F dt = (M + C_{MA} M') v_0 (1+e) \qquad (1)$$

where M and M' are the mass of the FAD and the liquid displaced by FAD respectively, C_{MA} is the coefficient of the added-mass coefficient, v_0 is the speed of the settling FAD at immediate time before landing, e indicates the repulsion factor between the FAD and the sea bottom. In general, the added-mass coefficient of the settling body is affected by the distance between the body and the bottom. Therefore, in order to evaluate the stability of the FAD against the impulsive force at landing efficiently, the affection of the sea bottom on the added-mass coefficient should be taken into account when we determine the design impulsive force on the FAD.
This study aims to clarify the following points; (1) the numerical simulation technique to analyze the behaviors of the settling FAD considering the effect of vortices generated behind it, (2) the initial condition of posture of the FAD at throwing to set up the FAD thrown down from a ship on the designed position most accurately, (3) the change of the added-mass coefficient of the settling FAD.

2. NUMERICAL SIMULATION TECHNIQUE OF THE SETTLING FAD

The oscillatory motion of the settling FAD in a periodic wave is induced by the instantaneous fluctuations of the pressure distribution around it, which is induced by the vortices generated from it and the wave motion. Hence, in the present numerical simulation technique, the fluid resistance exerted on the settling FAD is estimated firstly by integrating the pressure distribution around its surface. This surrounding pressure distribution is numerically analyzed with both the discrete vortex approximation method (see Sarpkaya 1968, Clements 1973) for the simulation of the vortex formation behind the FAD and the source-sink points method (see Chakrabarti 1987) for the formulation of the boundary condition on the surface of it. Secondly, the oscillatory motion of the settling FAD at every moment is numerically calculated from the equation of horizontal, vertical and rotational motion for the settling FAD.

2-1 DESCRIPTION OF THE FLUID FIELD AROUND THE SETTLING FAD

In the case of two-dimensional wave field as shown in Fig.2, applying the small amplitude wave theory to the incident wave, the free surface in the wave field can be approximately replaced by the fixed rigid surface. Then, the appropriate complex velocity potential for the flow around the settling FAD can be determined by using the Schwartz-Christoffel transformation to project the interior region

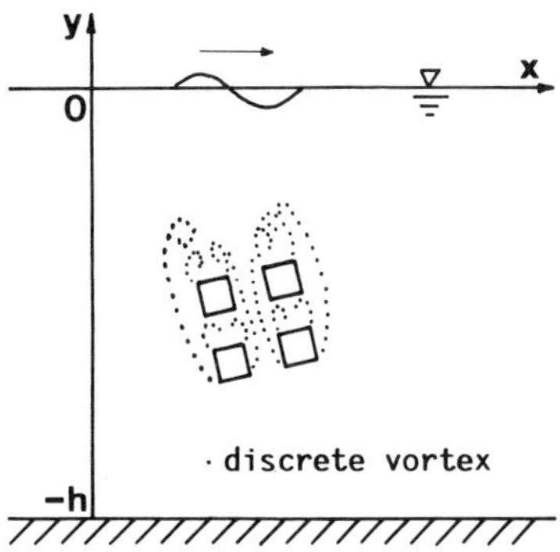

Fig.2 Schematic figure of flow field around the settling FAD

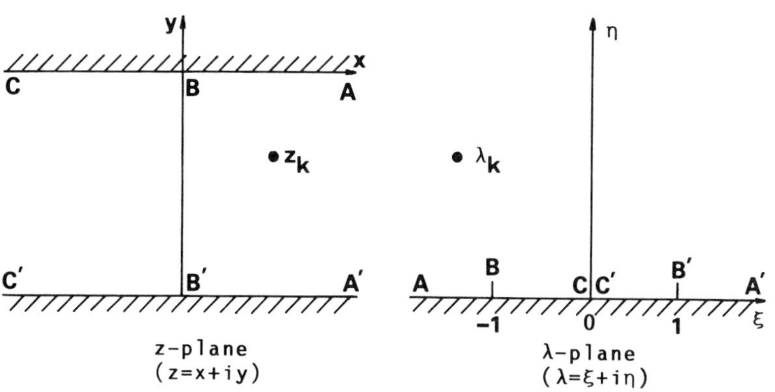

Fig.3 (a) physical plane z ; (b) transformed plane λ

between the boundaries in the z-plane into an upper half of the λ-plane with the boundary along the real axis (see Fig.3). The λ-plane is transformed into the physical z-plane by the function

$$z = x+iy = K\log(-\lambda) + C \tag{2}$$

where K is the constant determined by the water depth, and C is a integral constant. Since $\lambda=-1$ at $z=0$ and $\lambda=1$ at $z=-ih$ in which $i=\sqrt{-1}$ and h is the water depth, Eq.(2) is rearranged in term of z as,

$$\lambda = -\exp(C_0 z) \quad (C_0=\pi/h) \tag{3}$$

The complex velocity potential ($\omega_{V\lambda k}$) of the discrete vortex (λ_k) in λ-plane (see Fig.3) is given by the following equation in terms of the imaginary discrete vortex which is necessary to maintain the boundary condition of zero flow across the real axis in λ-plane.

$$\omega_{V\lambda k} = \frac{i\Gamma_k}{2\pi}[\log(\lambda-\lambda_k) - \log(\lambda-\overline{\lambda}_k)] \tag{4}$$

where Γ_k and λ_k are the circulation and the complex coordinate of the vortex respectively, the circulation is defined as positive being clockwise, and the over bar denotes the complex conjugate. Substituting Eq.(3) into Eq.(4), the complex velocity potential (ω_{Vzk}) in z-plane is introduced as

$$\omega_{Vzk} = \frac{i\Gamma_k}{2\pi}[\log(e^{C_0 z_k}-e^{C_0 z}) - \log(e^{C_0 \overline{z}_k}-e^{C_0 z})] \tag{5}$$

where z_k is the complex coordinate of the vortex in z-plane. With the same way that used for the discrete vortex (Eq.(5)), the complex velocity potential (ω_{Rzc}) of the source point in z-plane is given by

$$\omega_{Rzk} = \frac{D(z_c)}{2\pi}[\log(e^{C_0 z_c}-e^{C_0 z}) + \log(e^{C_0 \overline{z}_c}-e^{C_0 z})] \tag{6}$$

where z_c and $D(z_c)$ are a location and a strength of the source point respectively.
When the flow around the settling FAD in z-plane consists of a periodic wave, N discrete vortices with some circulations generated from P separation points and the flow from the source points on the FAD surface, the complex velocity potential ω_z at the point z in z-plane is given by

$$\omega_z = \phi_z + i\psi_z = \omega_W + \omega_V + \omega_R \tag{7}$$

where ϕ_z and ψ_z are the velocity potential and the stream function for the total flow in the z-plane respectively, ω_W, ω_V and ω_R are the complex velocity potential of the incident wave, the discrete vortices and the source points, these are derived as follows with Eq.(5) and Eq.(6).

$$\omega_W = \frac{\sigma H}{2k\sinh kh}\sin[k(ih+z)-\sigma t] \tag{8}$$

$$\omega_V = \frac{i}{2\pi}\sum_{j=1}^{P}\sum_{k=1}^{N}\Gamma_{jk}[\log(e^{C_0 z_{jk}}-e^{C_0 z})-\log(e^{C_0 \bar{z}_{jk}}-e^{C_0 z})] \tag{9}$$

$$\omega_R = \frac{1}{2\pi}\oint_c D(z_c)[\log(e^{C_0 z_c}-e^{C_0 z})+\log(e^{C_0 \bar{z}_c}-e^{C_0 z})]dc \tag{10}$$

Here H, σ and k are a wave height, angular frequency and wave number respectively, $\oint_c$ represents the contour integral along the surface c of the settling FAD. In the numerical calculation of Eq.(10), the surface c of the FAD is divided into M sections of length ΔC_m ($m = 1 \sim M$) and the source point is set on the center of the each section. Then the corner of the FAD is approximated with a circular arc of a radius 0.02a, in which a is the side length of the FAD. The integration in Eq.(10) is evaluated as

$$\omega_R = \frac{1}{2\pi}\sum_{m=1}^{M} D(z_m)[\log(e^{C_0 z_m}-e^{C_0 z})+\log(e^{C_0 \bar{z}_m}-e^{C_0 z})]\Delta m \tag{11}$$

The strength of the individual source point in Eq.(11) is determined by the boundary condition on the FAD surface (i.e. the fluid velocity normal to the FAD surface is equal to the settling velocity normal to the FAD surface). When the FAD is settling with speed (u_G, v_G) and with an angular velocity ω_r as shown in Fig.4, the identical equation of the strength $D(z_m)$ of the source point becomes as

$$R[\frac{\partial \omega_W}{\partial n}] + R[\frac{\partial \omega_V}{\partial n}] + R[\frac{\partial \omega_R}{\partial n}] = (u_G+u_r)n_x + (v_G+v_r)n_y \tag{12}$$

(on the surface of the FAD)

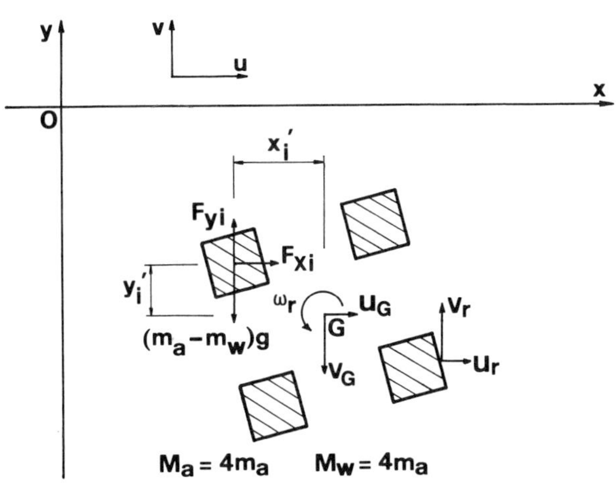

Fig.4 Fluid resistance on the settling FAD

where $\partial/\partial n$ denotes the derivative taken normal to the surface of the FAD, n_x and n_y are the x and y components of the unit vector taken outward normal to the FAD surface respectively, u_r and v_r are the x, y components of the velocity induced by the FAD rotation with angular velocity ω_r, **R**[ ] denotes the real part.
In Eq.(9), the velocities of the vortices are decided from the kinematic condition for the discrete vortices (i.e. a marked vortex is affected only from the incident wave, the other vortices and the source points on the FAD surface). Therefore, the velocity components u_{jk}, v_{jk} of the j-th vortex generated from the k-th separation point are expressed as

$$u_{jk} - iv_{jk} = \frac{d}{dz}\left[\omega_z - \frac{i\Gamma_{jk}}{2\pi}\log(e^{C_0 z_{jk}} - e^{C_0 z})\right]\Bigg|_{z=z_{jk}} \tag{13}$$

If there are PN pieces of discrete vortices simultaneously around the settling FAD, a set of PN ordinary differential equations are derived as equations of the kinematic conditions for the discrete vortices.

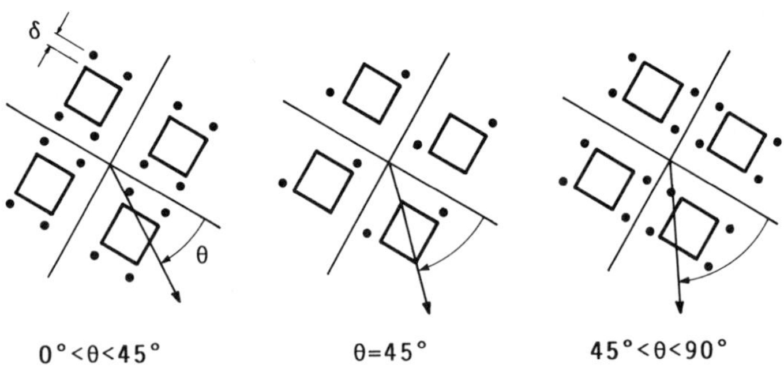

Fig.5 Initial positions of nascent vortices

In Eq.(13), the unknown initial position of the nascent vortex and the circulation of the vortices are included. There have been no established method, however, to calculate the initial position of a nascent vortex, its position has been determined with (A) the Kutta condition at the edge of the body or (B) the boundary layer thickness. In this study, (B) is adopted, and the initial positions are located at the distance of δ (boundary layer thickness which is determined by $\delta=\sqrt{\nu T/\pi}$ in which ν is the kinematic viscosity coefficient and T is the period of the wave) from the edges of the settling FAD as shown in Fig.5. Each initial position of the nascent vortex is changed with the relation between the angle of the FAD and the settling direction. The three separation patterns are treated in this study as shown in Fig.5. Secondly, the circulation of the nascent vortex is calculated with the Roshko's approximate equation (Roshko 1954). Taking the direction of the rotation of vortices into consideration, the circulation of the nascent vortex is given by

$$\partial\Gamma/\partial t = U_s|U_s|/2 \tag{14}$$

where U_s is a fluid velocity at the initial position of the nascent vortex. Although some vortices induce large velocities when they approach each other because we neglect the viscosity, this phenomenon is avoided by replacing the distributed discrete vortices with the Rankine vortex. The core radius of the Rankine vortex is given by $|\Gamma_{jk}|/2\pi U_s$ (Stansby 1977) in this study.

2-2 FLUID RESISTANCE ON THE SETTLING FAD

The fluid resistance exerted on the settling FAD can be obtained by integrating the pressure distributions around it, which is given in terms of $\partial\phi_z/\partial t$, u and v on the FAD surface ($z=z_p$) as

$$P(z_p) = -\rho\left[\frac{\partial\phi_z}{\partial t} + \frac{1}{2}(u^2+v^2)\right]\Bigg|_{z=z_p} \tag{15}$$

where ρ is the fluid density. Since the position z_{jk} of the discrete vortices, the strength $D(z_m)$ and the location z_m of the source points; and the position z_p on the FAD surface are functions of time, the first term of Eq.(15) is given by

$$\begin{aligned}\left.\frac{\partial\phi_z}{\partial t}\right|_{z=z_p} = &\Bigg[\mathbf{R}\Big[\frac{\partial\omega_W}{\partial t}\Big]+\mathbf{R}\Big[\frac{\partial\omega_W}{\partial z}\frac{dz}{dt}\Big]+\mathbf{R}\Big[\frac{\partial\omega_V}{\partial z_{jk}}\frac{dz_{jk}}{dt}\Big]+\mathbf{R}\Big[\frac{\partial\omega_V}{\partial z_{jk}}\frac{dz_{jk}}{dt}\Big]\\ &+\mathbf{R}\Big[\frac{\partial\omega_V}{\partial z}\frac{dz}{dt}\Big]+\mathbf{R}\Big[\frac{\partial\omega_R}{\partial z_m}\frac{dz_m}{dt}\Big]+\mathbf{R}\Big[\frac{\partial\omega_R}{\partial z_m}\frac{dz_m}{dt}\Big]+\mathbf{R}\Big[\frac{\partial\omega_R}{\partial z}\frac{dz}{dt}\Big]+\mathbf{R}\Big[\frac{\partial\omega_R}{\partial D}\frac{dD}{dt}\Big]\Bigg]\Bigg|_{z=z_p}\end{aligned} \tag{16}$$

In Eq.(16), dD/dt can be calculated by differentiating Eq.(12) with respect to time t, the differentiated equation is analyzed in the same manner as the above-mentioned method for the strength of the source point.
The horizontal and the vertical fluid resistance F_x, F_y exerted on the settling FAD are given as follows lastly

$$F_x = -\sum_{p=1}^{M} P(z_p)n_{xp}\Delta C_p \quad , \quad F_y = -\sum_{p=1}^{M} P(z_p)n_{yp}\Delta C_p \tag{17}$$

2-3 THE EQUATIONS OF MOTION OF THE SETTLING FAD

Referring to Fig.4, the equations of horizontal, vertical and rotational motion for the settling FAD are give as

$$M_a\frac{du_G}{dt} = \sum_{i=1}^{4} F_{xi} \quad , \quad M_a\frac{dv_G}{dt} = \sum_{i=1}^{4} F_{yi} - (M_a - M_w)g$$

$$I_r\frac{d\omega_r}{dt} = \sum_{i=1}^{4}(x_i'F_{yi} - y_i'F_{xi}) \tag{18}$$

where M_a is a mass of the FAD per unit thickness, M_w is the liquid (per unit thickness) displaced by the FAD, g is the gravity acceleration, and I_r is an inertia moment of the FAD around the gravitational center axis of the FAD. x_i' and y_i' are the distance from the center of the FAD cross section to the center of an i-th piece in the x and y direction as illustrated in Fig.4.

3. APPLICATION OF THE NUMERICAL SIMULATION TECHNIQUE

The application of the numerical simulation technique for the settling behavior of the FAD was investigated by comparing the calculated and experimented results. The experiments were carried out by using the wave tank of 30m long, 60cm wide and 1m depth to observe the behaviors of the settling FAD. The FAD model used in the experiments was consisted of four pieces of slender rectangular (1cmx1cm in the cross section) bodies. This study restricted the FAD motion on the vertical plane. Therefore, to prevent the FAD from 3-dimensional motion, a long model which had the same cross section was used in the experiments. Size of the model section is 3cmx3cm, length is 56cm, the void ration per unit thickness is 55.6%. The photographs of the settling figures of the FAD model were taken on a same film by using a strobe flash, the interval time of the flash is 0.2s. 2-dimensional motion is strictly inspected through the experiments.
Since the behaviors of the settling FAD model were very irregular in the experiments, it was difficult to compare the simulated FAD behaviors with the measured behaviors. Therefore, we compared the simulated and measured horizontal fluctuation of FAD and the angular velocity of the rotational motion of the settling FAD.

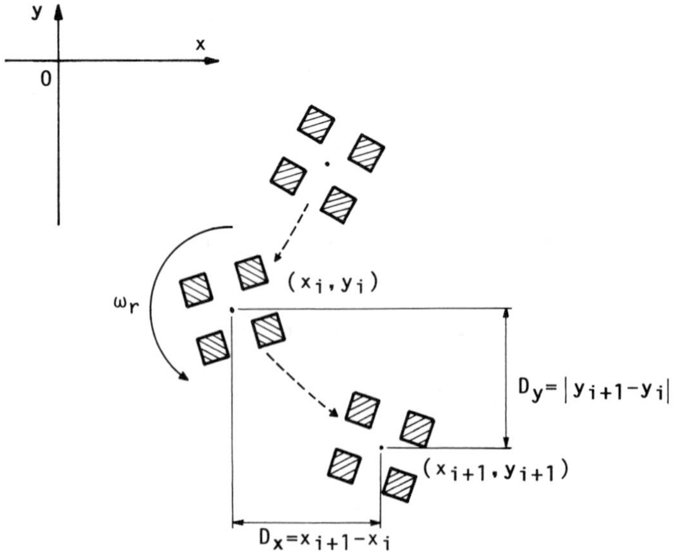

Fig.6 Definition of the fluctuation of the settling FAD

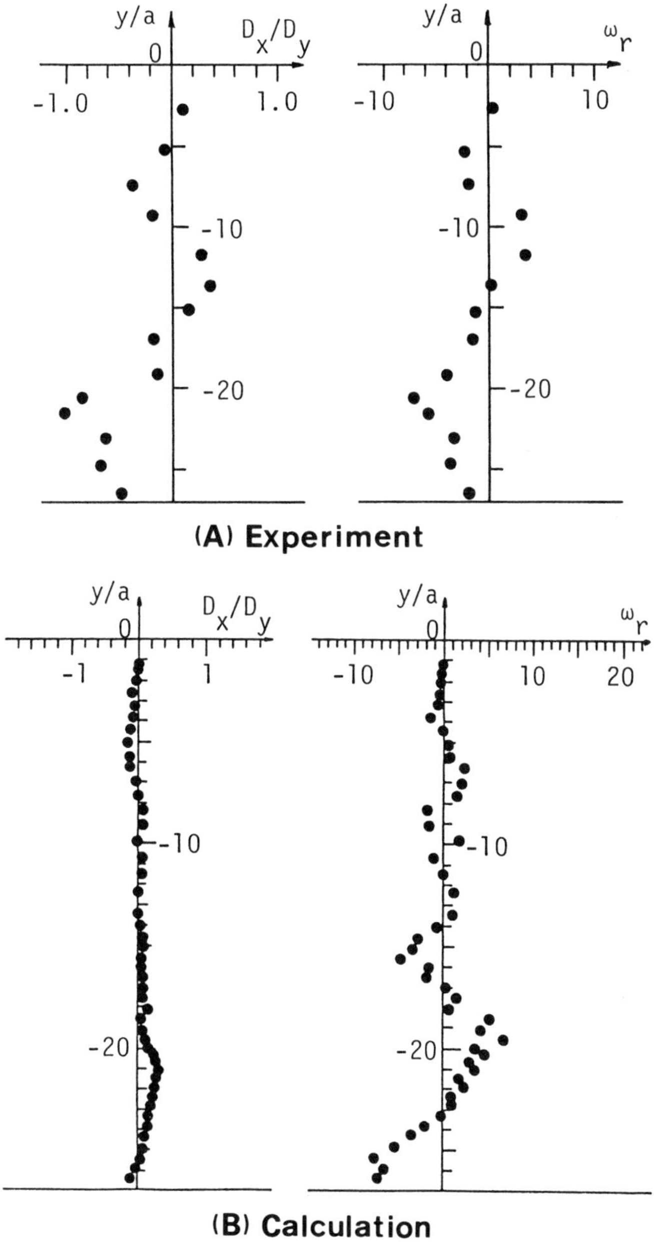

Fig.7 Measured and calculated oscillatory motion of the settling FAD

Fig.6 shows the definitions of the fluctuation of the settling FAD, D_x and D_y are the horizontal fluctuation and the settling distance from the simulated position. The angular velocity ω_r is defined as positive being anti-clockwise. When ω_r is positive, since the FAD moves in the positive direction of horizontal axis in Fig.6, the relative horizontal fluctuation to the settling distance D_x/D_y becomes positive.
Fig.7 shows examples of the experimented and calculated oscillatory motions of the settling FAD, x/a and y/a are the normalized x, y positions of the center of the FAD cross-section at each measured and calculated position. From the experiment (Fig.7(A)), the dominant behaviors of the settling FAD are the fluctuation in the horizontal direction which corresponds to a change in the rotational direction and the time lag between the inflection points of D_x/D_y and ω_r. This time lag may come from the change in the direction of the horizontal fluctuation which can not instantly correspond to the change in the rotational direction, because a motion of the settling FAD is the inertia motion. On the other hand, in the calculated results (Fig.7(B)), the present simulation technique can simulate also the dominant settling behaviors which are recognized in the experiments, i.e., the fluctuation of the settling FAD in horizontal direction and the change of its direction of rotation, the points of inflection. From these investigations, we may conclude that the settling behavior of the FAD can be evaluated with this numerical simulation technique.

4. RELATION BETWEEN INITIAL ANGLE OF FAD AT THROWING AND SETTLING BEHAVIOR

To investigate the effective initial posture of the FAD at throwing to reduce the settled area on the sea floor most, the systematic calculations for the change of the initial angle from the still water surface at throwing were carried out by using the present simulation technique.
Fig.8 shows the calculated behaviors of the settling FAD for three kinds of the initial angles, $\theta_0=0°$, $22.5°$, $45°$. In these cases, wave period of T=1s is applied. From the settling figures, it is found that the fluctuation of the settling FAD in the horizontal direction has close relation to the initial angle of the FAD at throwing. Therefore, the initial angle is the important factor to dominate the scatter range of the landing FAD on the sea floor. In Fig.8(A) and (B), the ω_r takes the positive value for the major part of water depth, D_x/D_y also takes the positive values. From these results, it can be seen that the FAD settles in the positive direction of the horizontal axis. When θ_0 is 45°, the period of the change in the rotational direction is shortest among these three cases, and the horizontal fluctuation becomes very small. This situation for the settling FAD becomes the most stable, i.e., less oscillation may take place. From this result, it may be estimated that this situation reduces the settled area on the sea floor most. Furthermore, Fig.9 shows the calculated behaviors of the settling FAD in the case of T=1.5s. When θ_0 is 45° (Fig.9(C)), the period of the change in ω_r becomes the shortest, and the most stable situation for the settling

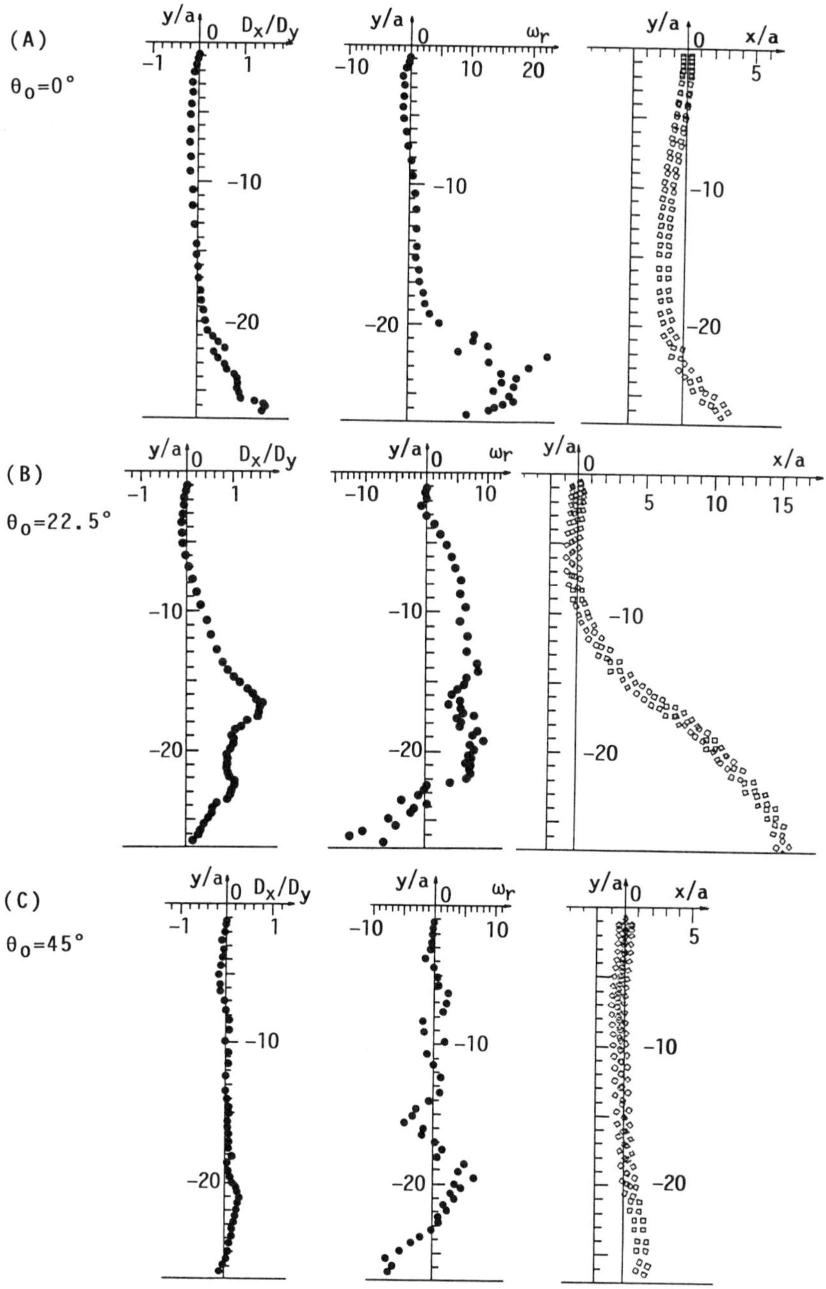

Fig.8 D_x/D_y, ω_r and behaviors the settling FAD with respect to the initial angle θ_0 (T=1s)

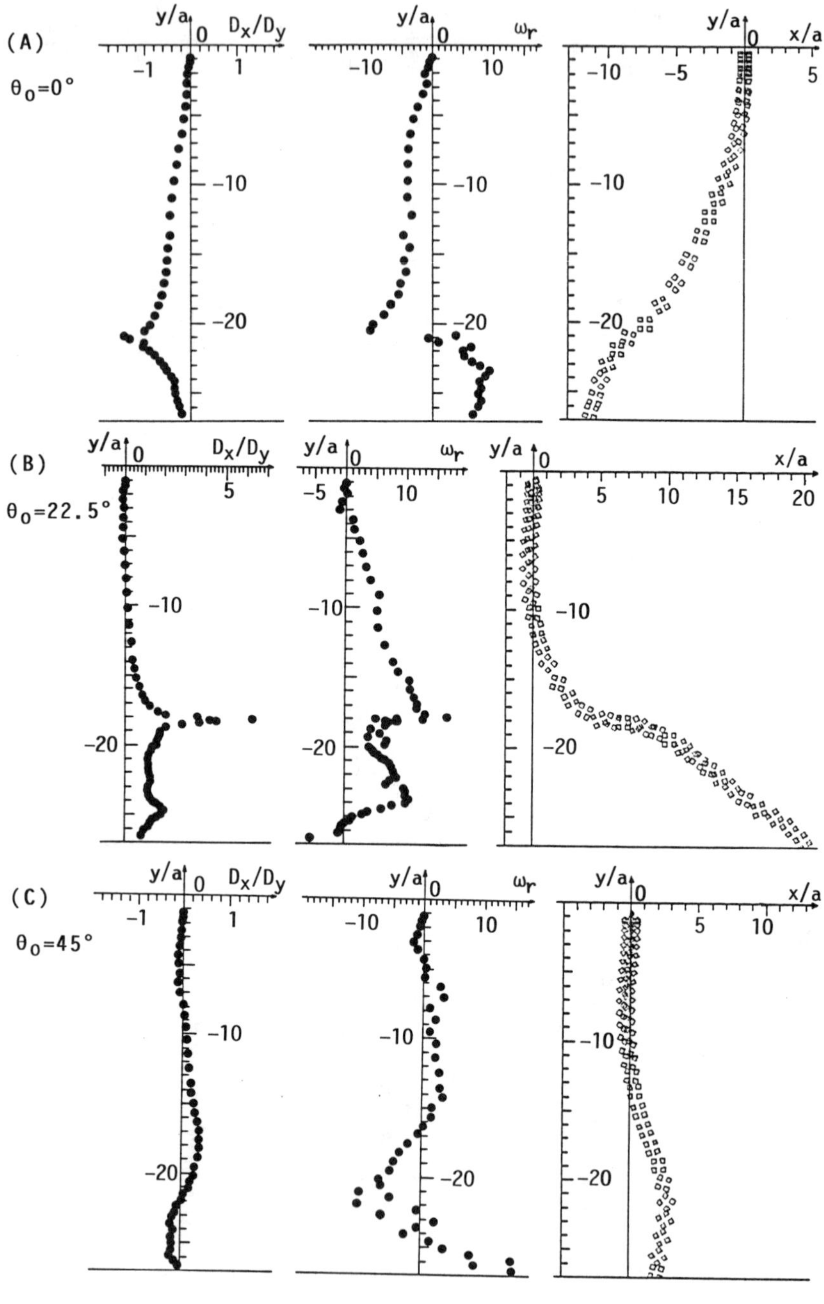

Fig.9 D_x/D_y, ω_r and behaviors the settling FAD with respect to the initial angle θ_0 (T=1.5s)

FAD can be observed again. From these investigations, it can be concluded that the initial posture of the FAD at throwing with the angle of 45° from the still water surface reduces the settled area on the sea floor most.

5. ADDED-MASS COEFFICIENT OF THE SETTLING FAD

The added-mass coefficient of the settling FAD is evaluated from the kinetic energy of the fluid induced by the settling FAD (see Milne-Thomson 1968). The kinetic energy T_w of the flow induced by the settling FAD is given by

$$T_w = -\frac{1}{2}\rho \oint_c \phi \frac{\partial \phi}{\partial n} dc \tag{19}$$

where ϕ is the velocity potential. If V is the volume (per unit thickness) of the FAD, the kinetic energy E_w of the fluid replaced by the settling FAD is

$$E_w = \frac{1}{2} C_{MA} \rho V v_0{}^2 \tag{20}$$

Since the added-mass $C_{MA}V$ is defined as the mass such that the kinetic energy E_w becomes equivalent to the kinetic energy T_w, the added-mass coefficient C_{MA} becomes from Eq.(19) and Eq.(20) as

$$C_{MA} = -\oint_c \phi \frac{\partial \phi}{\partial n} dc \,/(V v_0{}^2) \tag{21}$$

In the numerical calculation technique of the added-mass coefficient, the source-sink points method is adopted for the formulation of the boundary condition on the surface of the settling FAD. The velocity potential of flow around the settling FAD is given by

$$\phi = \mathbf{R}[\frac{1}{2\pi}\oint_c D(z_c)[\log(e^{C_0 z_c} - e^{C_0 z}) + \log(e^{C_0 \bar{z}_c} - e^{C_0 z})]dc] \tag{22}$$

The second term of Eq.(22) is the imaginary part which is necessary to maintain the bottom boundary condition of zero flow across the sea floor. The calculations are carried out for the settling FAD in a still water, the horizontal and the rotational motion for the settling FAD are neglected for simplicity.

Fig.10 shows the change of the added-mass coefficient of the settling FAD with respect to the relative distance h_q/a from the sea bottom. When $h_q/a>4$, C_{MA} is less affected by h_q/a. When $h_q/a<4$, however, C_{MA} is affected by the "bottom" effect, and in the region of $h_q/a<1$ the shorter the relative distance becomes, C_{MA} starts to increase very quickly in spite of the void ratio of the FAD. Furthermore, from Table-1 which shows the C_{MA} at landing on the sea bottom ($h_q/a=0.5$) and that in the infinite fluid field ($h_q/a=\infty$), it

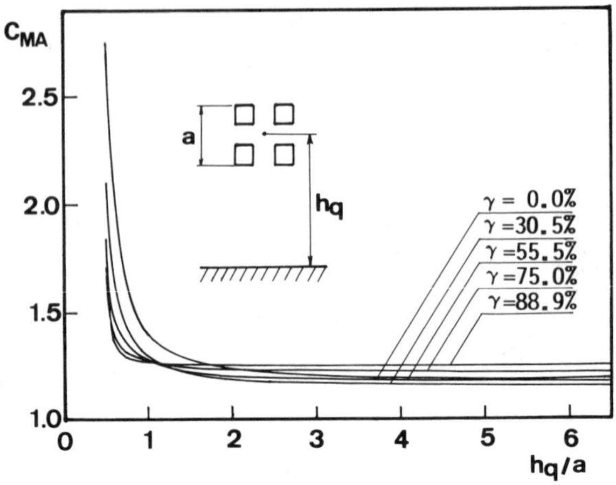

Fig.10 Change of the added-mass coefficient of the FAD with settling

Table-1 C_{MA} at landing and C_{MA} in the infinite field

γ (%)	0	30.5	55.5	75.0	88.9
C_{MA} at hq/a= ∞	1.19	1.15	1.18	1.22	1.24
C_{MA} at hq/a=0.5	2.74	2.08	1.84	1.82	1.78

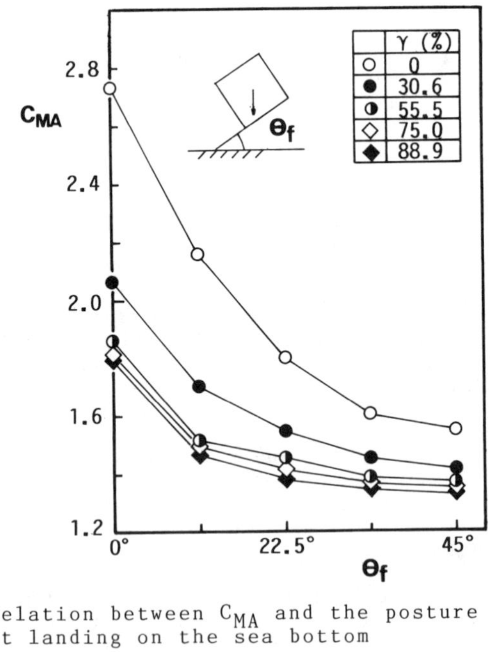

Fig.11 Relation between C_{MA} and the posture of the FAD at landing on the sea bottom

can be found that C_{MA} at landing on the sea bottom reaches about 1.4 to 2.3 times of that in the infinite fluid field. Fig.11 shows the relation between the C_{MA} and the posture of the FAD at landing on the sea bottom, θ_f is the angle between the FAD surface and the bottom, C_{MA} is the value at landing on the sea bottom. From this figure, it is seen that the added-mass coefficient of the settling FAD is affected by θ_f and C_{MA} in the case of $\theta_f=0^{\circ}$ becomes the largest among these five angles.

6. CONCLUSION

Concluding remarks are as follows.
(1) The present simulation technique for the settling behaviors of the FAD can successfully simulate the dominant behaviors of the settling FAD in the experiments (i.e. the fluctuation of the settling FAD in the horizontal direction corresponding to the change of its rotational direction, the time lag between the inflection points in the relative horizontal fluctuation to the settling distance and those in the angular velocity in the rotational motion).
(2) The fluctuations of the settling FAD in the horizontal direction are closely related to the initial angle of the FAD at throwing. The most stable situation for the settling FAD is achieved when the initial posture of the FAD is thrown with the angle of 45° from the still water surface. This situation may reduce the settled area on the sea floor most.
(3) The added-mass coefficient of the settling FAD may be evaluated through the calculation of the kinetic energy of the flow around the settling FAD using the potential theory with a sufficient accuracy. The added-mass coefficient at landing on the sea bottom reaches about 1.4 to 2.3 times of that in the infinite field. When the settling FAD lands on the sea bottom with zero angle, the added-mass coefficient becomes the largest.

REFERENCES

Chakrabarti S.K.(1987): Hydrodynamics of Offshore Structures, Springer-Verlag, pp.286-324.
Clements R.R.(1973): An Inviscid Model of Two-Dimensional Vortex Shedding, J. F. Mech. Vol.57, Part 2, pp.321-336.
Milne-Thomson L.M.(1968): Theoretical Hydrodynamics, Macmillan Education LTD, pp.240-268.
Roshko A.(1952): On The Development of Turbulent Wakes from Vortex Street, NACA TN 2913.
Sarpkaya T.(1968): An Analytical Study of Separated Flow about Circular Cylinders, J. Basic Engg., TRANS., ASME, Series D, Vol.90, pp.511-518.
Stansby P.K.(1977): An Inviscid Model of Vortex Shedding from A Circular Cylinder in Steady and Oscillatory Far Flow, Proc. Instn Civ. Engrs, Part 2, pp.865-880.

CHAPTER 219

THE TIME DOMAIN ANALYSIS ON MOORED SHIP MOTIONS

Masayoshi KUBO*
M.JSCE, Dr.Eng., Associate Professor

Naokatsu SHIMODA**
M.JSCE, B.Eng., Vice-Head

Shunsaku OKAMOTO***
B.Sc.

ABSTRACT

The ship refuge inside a harbor in storm requires the analysis of moored ship motions along a quay wall. In this case the time domain analysis with the convolution integral method becomes effective. But the calculation accuracy is not enough and must be improved to analyze actual moored ship motions. In this paper some methods of the improvement are proposed and their efficiency is verified by comparing the calculation results with the experimental ones.

1.INTRODUCTION

Recently, many developments in the coastal zone are planed in Japan, for example the construction of artificial islands and so on. Since the reclamations of the sea narrow the anchoring areas outside a harbor in storm, the ship refuge inside a harbor in storm has been investigated by the Ministry of Transport.

When moored ship motions along a quay wall are considered, the time domain analysis with the convolution integral method becomes effective. But the papers with this method are very few1)2) and moreover the calculation accuracy for actual moored ship motions does not seem to be satisfactory. In this paper, some calculation methods are proposed to improve the accuracy and their efficiency is verified by comparing the calculation results with the experimental ones.

---

* Associate Professor, Kobe University of Mercantile Marine 1-1, 5-Chome, Fukae-Minami-Machi, Higashinada-Ku, Kobe 658, Japan
** Vice-Head, Technical Laboratory, Giken Kogyo Co.,LTD. 1540, Kawaguchi-Cho, Hachiohji 193, Japan
*** Technical Laboratory, Giken Kogyo Co.,LTD. 1540, Kawaguchi-Cho, Hachiohji 193, Japan

2. ON HYDRODYNAMIC COEFFICIENTS

2.1 Calculation Method

a) Equations of Ship Motions in the Time Domain

The calculations and the experiments are carried out by using the model ship shown in Fig.-1. The model is a rectangular floating body moored symmetrically along a quay wall. The incident wave angle θ is 60 (deg). The equation of ship motions in the time domain and the coefficients are given by,

$$\sum_{i=1}^{6}\{M_{ij}+m_{ij}(\infty)\}\ddot{x}_i(t) + \sum_{i=1}^{6}\int_{-\infty}^{t}\dot{x}_i(\tau)L_{ij}(t-\tau)d\tau + \sum_{i=1}^{6}(C_{ij}+G_{ij})x_i(t)=F_j, \quad (1)$$

$$(j=1,2,\cdots,6)$$

$$L_{ij}(t)=\frac{2}{\pi}\int_0^{\infty}B_{ij}(\omega)\cos\omega t\,d\omega, \quad (2)$$

$$m_{ij}(\infty)=A_{ij}(\omega)+\frac{1}{\omega}\int_0^{\infty}L_{ij}(t)\sin\omega t\,dt, \quad (3)$$

where $x_i(t)$=displacement, M_{ij}=mass or moment of inertia, $m_{ij}(\infty)$=constant added mass or constant added moment of inertia, $L_{ij}(t)$=retardation function, C_{ij}=hydrostatic restoring force coefficient, G_{ij}=mooring force coefficient, F_j=external force, t=time, τ=integral variable, $A_{ij}(\omega)$=added mass or added moment of inertia (which is called added mass hereafter) and $B_{ij}(\omega)$=damping coefficient which correspond to the angular frequency ω. The subscripts i and j describe the mode of ship motions, namely, 1:Sway, 2:Surge, 3:Heave, 4:Pitch, 5:Roll and 6:Yaw. The coefficients with the subscript ij show the effect on the i-th mode from the motion of the j-th mode.

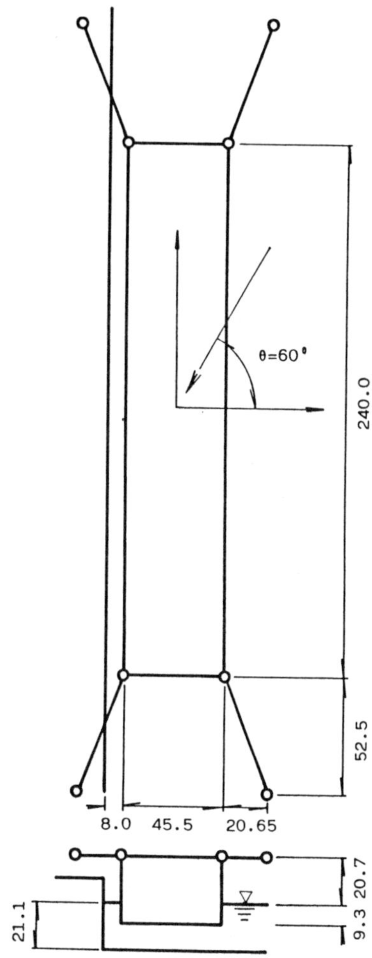

Fig.-1 The model ship (unit:cm)

As is obvious from the above, the hydrodynamic coefficients in Eq.(1) are decided by calculating the

frequency dependent added mass and the damping coefficient.

b) Calculation of Added Mass and Damping Coefficient

The added mass and the damping coefficient are calculated by the method of Green's function. Then the length of the mesh must be divided into less than a quarter of the wave length, so that the meshes represent the wave profile. In the case shown in Fig.-1, there are 18 coupling terms of $A_{ij}(\omega)$ and $B_{ij}(\omega)$ as shown in Table-1 because of the influence of the reflected waves. Further, the number of coupling terms is decreased to 12 by confirming the symmetry of the hydrodynamic coefficients, $A_{ij}(\omega)=A_{ji}(\omega)$ and $B_{ij}(\omega)=B_{ji}(\omega)$.

c) Calculation of Retardation Function

Since the damping coefficients in the whole angular frequency are required to calculate the retardation functions by Eq.(2), the damping coefficients must be approximated in the higher frequency than the maximum calculation angular frequency. In the two dimensional case the approximating formula for the higher frequency is given by Takagi and Saito[3] as follows,

$$B_{ij}(\omega)=k_{ij}\,\omega^{-n}, \tag{4}$$

where k_{ij} is a coefficient for joining calculated values to approximated ones and n is a multiplier. It's considered that the same approximation can be possible in the three dimensional case, then the values of n are given as shown in Table-2.

As mentioned above, the retardation functions are given by substituting Eq.(4) into Eq.(2).

Table-1 The coupling between each motion mode

j \ i		1 Sway	2 Surge	3 Heave	4 Pitch	5 Roll	6 Yaw
1	Sway	o	×	o	×	o	×
2	Surge	×	o	×	o	×	o
3	Heave	o	×	o	×	o	×
4	Pitch	×	o	×	o	×	o
5	Roll	o	×	o	×	o	×
6	Yaw	×	o	×	o	×	o

o = coupling term
× = no coupling term

Table-2 The values of n in Eq.(4)

j \ i		1 Sway	2 Surge	3 Heave	4 Pitch	5 Roll	6 Yaw
1	Sway	3	-	5	-	5	-
2	Surge	-	3	-	5	-	3
3	Heave	5	-	7	-	7	-
4	Pitch	-	5	-	7	-	5
5	Roll	5	-	7	-	7	-
6	Yaw	-	3	-	5	-	3

$$L_{ij}(t)=\frac{2}{\pi}\{\int_0^{\omega_e}B_{ij}(\omega)\cos\omega t d\omega+k_{ij}\int_{\omega_e}^{\infty}\omega^{-n}\cos\omega t d\omega\}, \tag{5}$$

where ω_e is the maximum calculation angular frequency. The analytical integral in the second term of the right hand side can be carried out by using the integration by parts and cosine integral.

d) Calculation of Constant Added Mass

The retardation functions are also required in the whole time domain to calculate the constant added masses in Eq.(3). On the other hand, the retardation functions show the decreasing osicillatory curve, so they are approximated by the next Eq.,

$$L_{ij}(t)=re^{-pt}\cos(qt+\varepsilon), \tag{6}$$

where r, p, q and ε are arbitrary constants. And the damping coefficients are given by applying the inversion formula of Fourier transform to Eq.(2),

$$B_{ij}(\omega)=\int_0^{\infty}L_{ij}(t)\cos\omega t dt. \tag{7}$$

Substituting Eq.(6) into Eq.(7) and carrying out the integrations by parts in the right hand side, then Eq.(8) is obtained,

$$B_{ij}(\omega)=\frac{r}{2}\{\frac{p\cos\varepsilon-(q+\omega)\sin\varepsilon}{p^2+(q+\omega)^2}+\frac{p\cos\varepsilon-(q-\omega)\sin\varepsilon}{p^2+(q-\omega)^2}\}, \tag{8}$$

where the constants r, p, q and ε are decided by comparing the values obtained from Eq.(6) and (8) with ones from the numerical integral calculation. The truncation error S_p which is occurred from the halfway end of the infinite integral Eq.(7) is defined by using these constants and the end time of the integral t_e, namely,

$$Sp=\frac{\frac{r}{\omega}\int_{t_e}^{\infty} e^{-pt}\cos(qt+\varepsilon)\sin\omega t dt}{A_{ij}(\omega)+\frac{1}{\omega}\int_0^{t_e} L_{ij}(t)\sin\omega t dt}. \qquad (9)$$

Since both the numerator and the denominator in Eq.(9) can be calculated analytically, t_e for the enough accuracy can be decided by Eq.(9). Eventually, the infinite integral in Eq.(3) is changed to the finite integral from 0 to t_e.

2.3 Calculation Results

a) Added Mass and Damping Coefficient

The length of the mesh is decided as the value which makes the maximum calculation angular frequency ω_e 10.0 (2π/sec). To confirm the symmetry of the hydrodynamic coefficients, the added masses and the damping coefficients in the case of i=1, j=5 and i=5, j=1 are shown in Fig.-2. In the figure the dotted line shows the case of i=1, j=5 , the broken line shows the case of i=5, j=1 and the solid line shows the average values of them. Good agreements between the results of the two cases are obtained but there are a little difference. So the average values are used in the cases of $i \neq j$ hereafter.

In Fig.-2, there is a discontinuous value at ω=7.9 (2π/sec). In particular, the value of the damping coefficient is negative at this frequency, therefore, the calculation of ship motions diverges. So we use the smoothed value at the angular frequency as shown in Fig.-3.

b) Retardation Function

The damping coefficients are shown in Fig.-4. The solid line shows the calculation results and the dotted line shows the approximate curve by Eq.(4). The results in the range from 10.0 to 14.0 (2π/sec) become valid by shortening the mesh length to a half and the converging tendency proves the validity of the approximation. The converging tendency of all damping coefficients in the high angular frequency is completed by the results of B_{22}.

The retardation functions by Eq.(5) are shown in Fig.-5. The abscissas show the elapsed time.

c) Constant Added Mass

As an example, the value of t_e will be shown in the case of i=5, j=5. The approximate results of the retardation functions and the damping coefficients by Eq.(6) and (8) are shown as the solid line in Fig.-6. The dotted line shows the damping coefficients by the numerical calculation, when $r=1.6\times10^8$ ($g\cdot cm^2/sec^2$), p=0.416 (1/sec), q=5.153 (1/sec) and ε=0. The relation between t_e and S_p is shown in Fig.-7. Since S_p is less than 0.01 at t_e=20 (sec), t_e is decided 60 (sec) to secure the calculation accuracy.

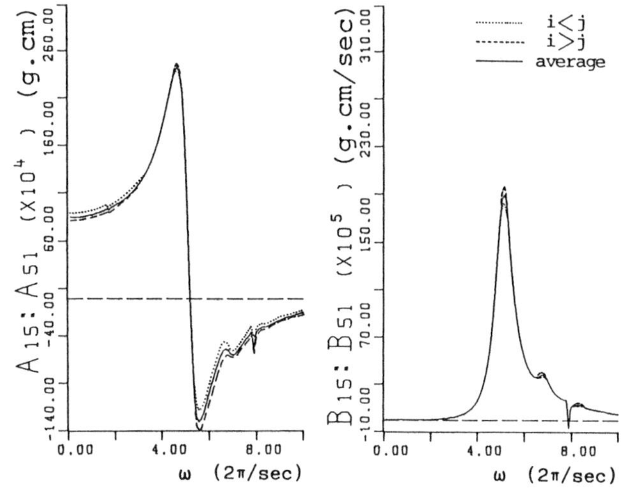

Fig.-2 The added mass and the damping coefficient

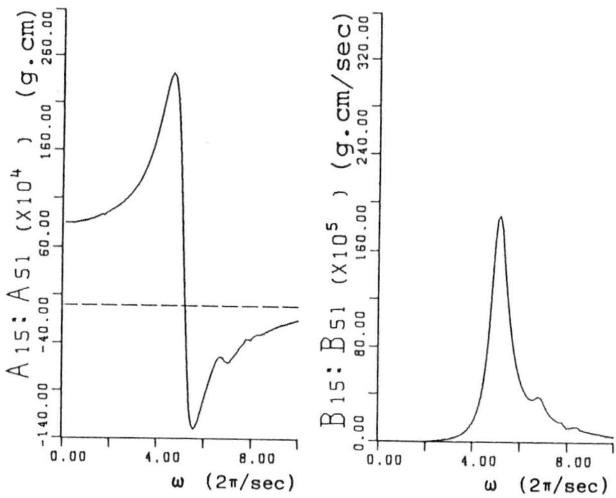

Fig.-3 The smoothed added mass and the damping coefficient

Next, the constant added masses calculated by using the te and Eq.(3) are shown as the thick lines in Fig.-8. The thin lines show the average values of them and the two-dot-dash-lines show the added masses which correspond to the angular frequency ω. It can be considered that the constant added masses show the constancy on the angular frequency.

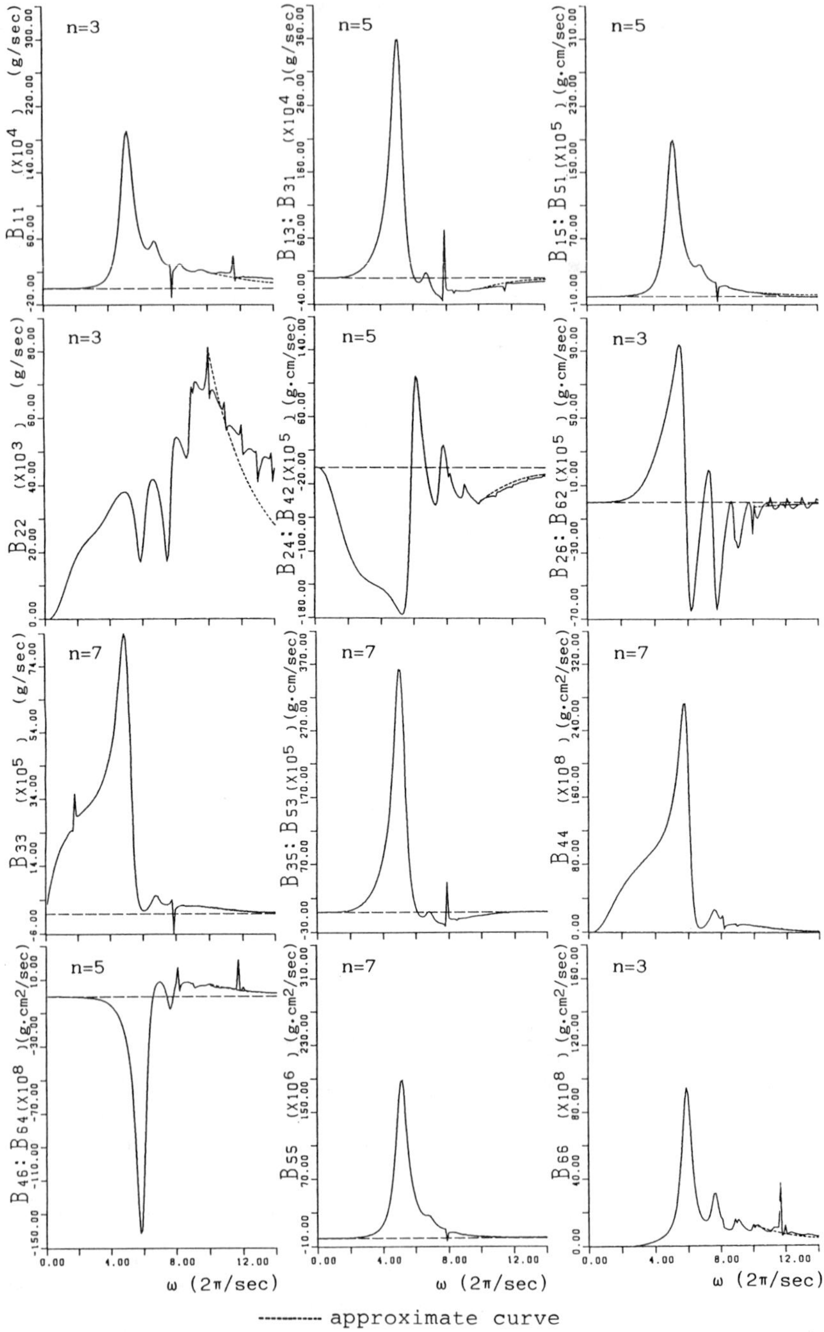

--------- approximate curve

Fig.-4 The approximate curves of the damping coefficients

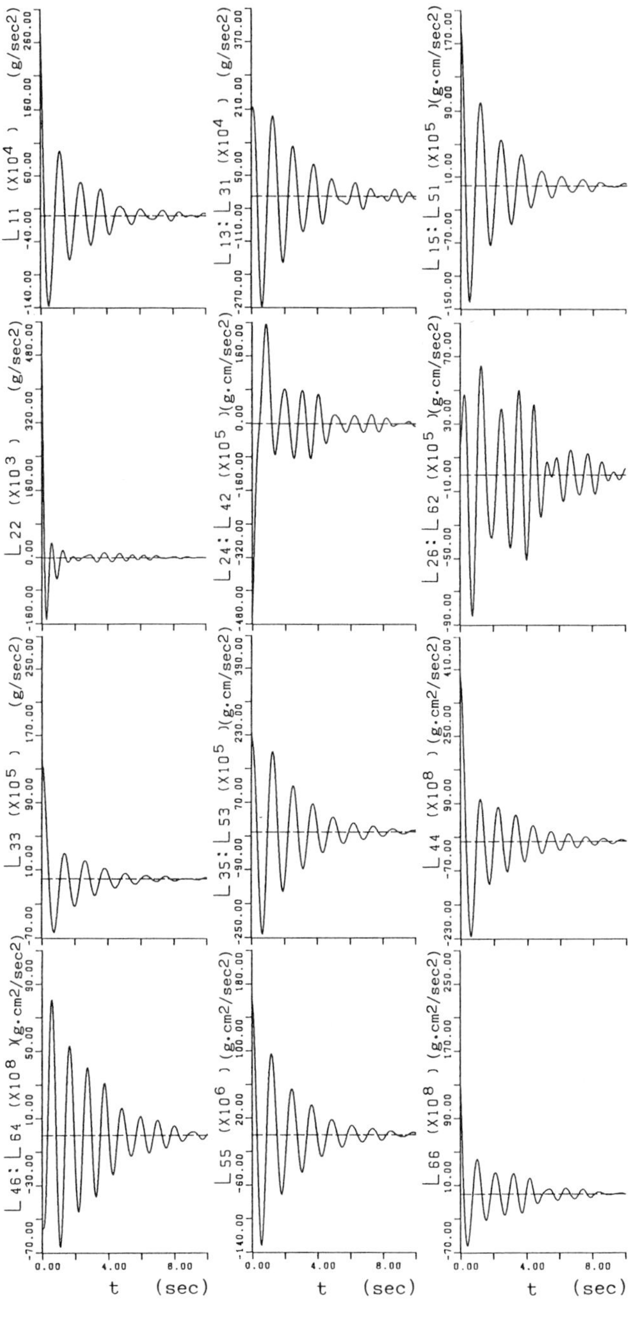

Fig.-5 The retardation functions

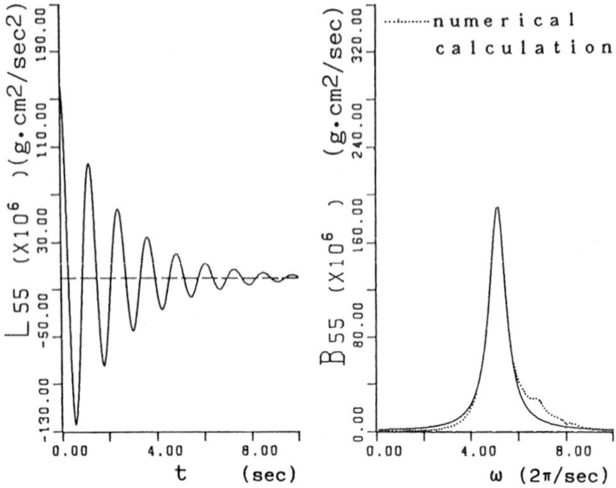

Fig.-6 The approximate results of the retardation function and the damping coefficient

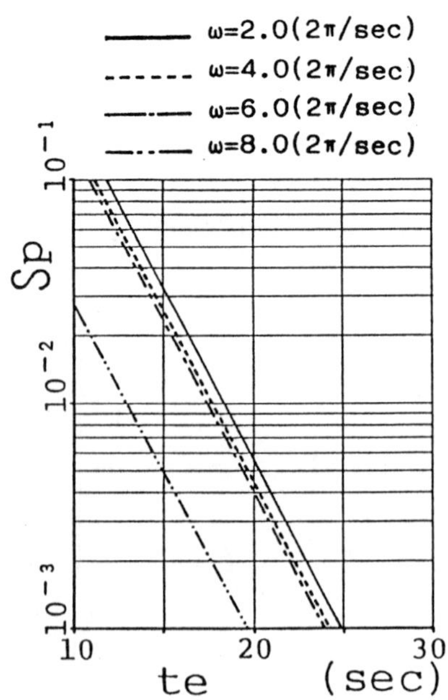

Fig.-7 The relation between te and Sp

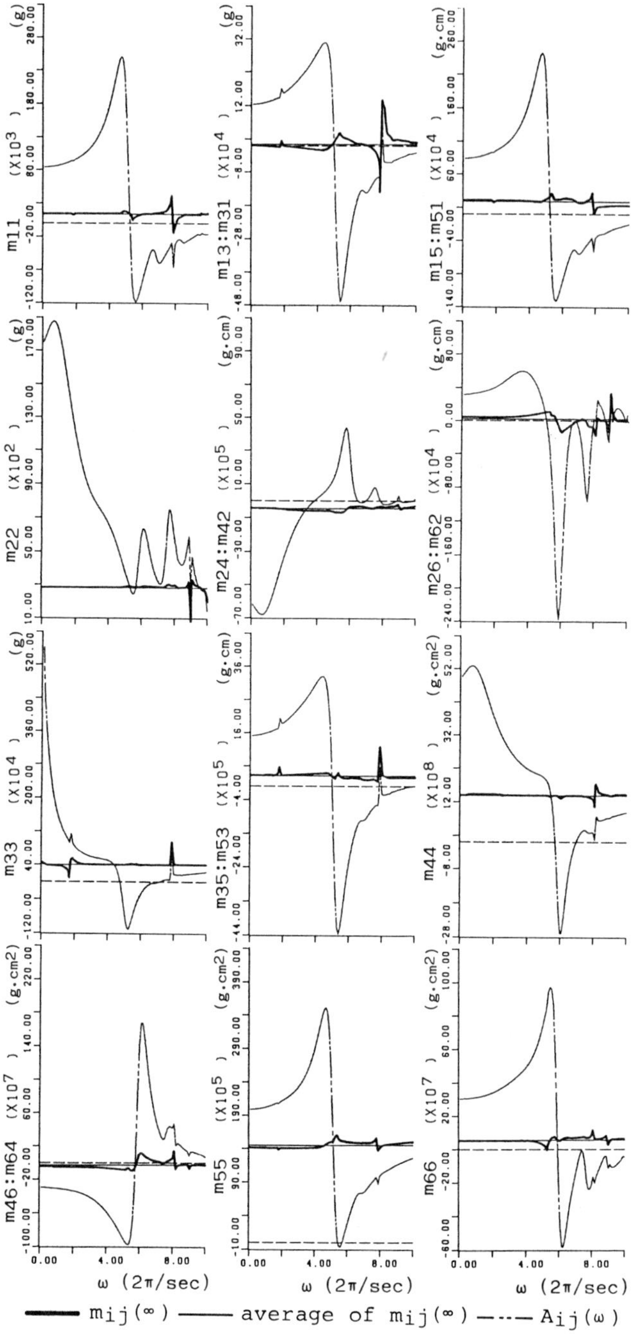

Fig.-8 The constant added masses

3. TIME DOMAIN ANALYSIS

3.1 Calculation Method

A time domain analysis can be carried out by calculating Eq.(1) with the obtained results. But the attention should be paid to the following matters.

a) Natural Period

Since the potential theory dose not give the viscosity term, the amplitude of ship motions is overestimated at the natural period because of the lack of damping force. Therefore, the existence of the natural period must be investigated within the range of used period. The natural period is obtained from the calculation of the free oscillations by giving an initial displacement in Eq.(1).

b) Introduction of Viscosity Term

The calculations are corrected by introducing the viscosity term $N_i \cdot x_i(t)$ to the second term of the left hand side in Eq.(1). The coefficient N_i is decided so that the calculated amplification factors of ship motions coincide with the experimental ones at the natural periods.

3.2 Calculation Results

a) Natural Period

The calculation results of the free oscillations are shown in Fig.-9. Each natural period is 10.1 (sec) in Sway, 3.8 (sec) in Surge, 0.8 (sec) in Heave, 0.8 (sec) in Pitch, 1.4 (sec) in Roll and 9.6 (sec) in Yaw.

b) Ship Motions in Beam Seas

The calculations are carried out in the range of the incident wave period from 0.6 to 2.4 (sec) and the results are compared with the experimental ones. The viscosity terms are introduced in Heave and Roll which have the natural periods in the above wave period range. The viscosity damping are evaluated by the logarithmic damping coefficient λ. The calculated free oscillations and the experimental ones are shown in Fig.-10. The ordinates, α, show the ratio of the displacement to the initial one. λ of Heave is varied widely because of the considerable change of the neighboring wave amplitude ratio. In Roll, 0.58 of λ is suitable for the calculations of ship motions, but 0.26 of λ is adequate for the above free oscillations. That is to say, the viscosity coefficient in ship motions is bigger than the one in the free oscillations. The difference may be modified by considering that the viscosity force is in proportional to the velocity squared.

Both results of the experiments and the calculations with and without the viscosity terms are shown in Fig.-11.

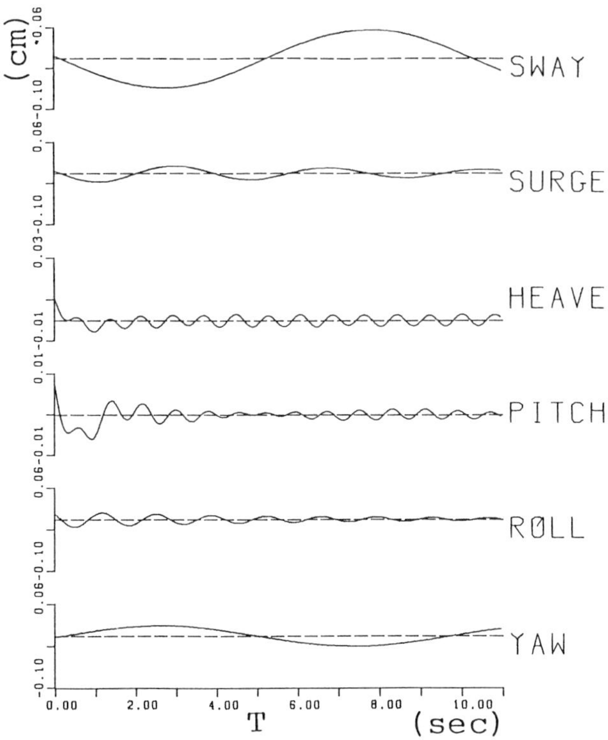

Fig.-9 The free oscillations

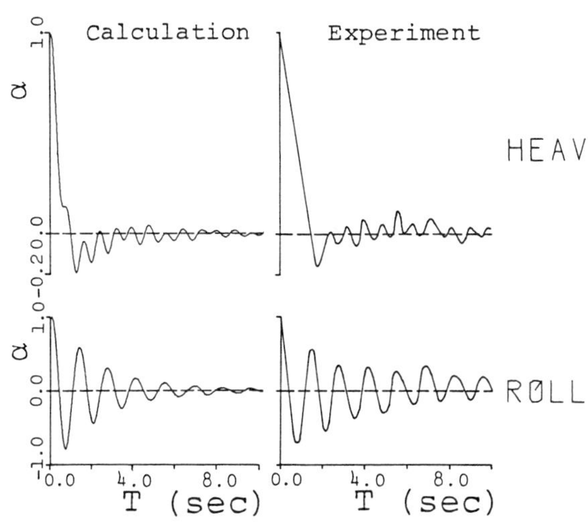

Fig.-10 The experiments and the calculated free oscillations with the viscosity terms

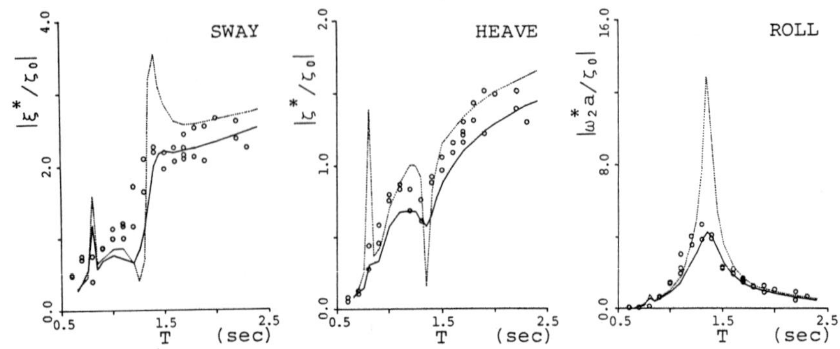

o exp. ——— cal.(with viscosity terms)
-------- cal.(without viscosity terms)

Fig.-11 The calculations and the experiments in the beam sea

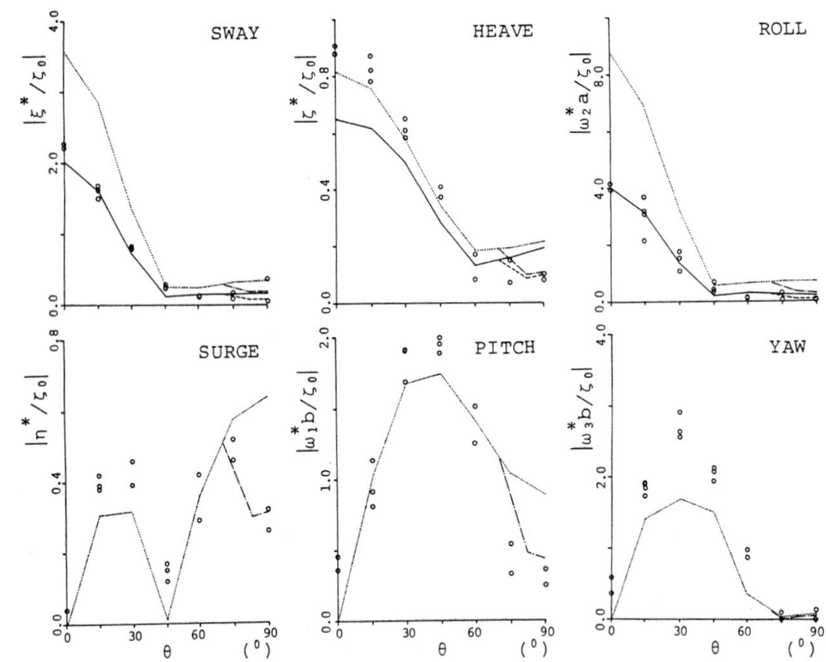

o exp. ——— cal.(with viscosity terms)
-------- cal.(Without viscosity terms)

Fig.-12 The calculations and the experiments when the incident wave period is fixed at 1.4 (sec) and the incident wave angle is changed.

The ordinates show the amplification factor of ship motions. ξ^*, ζ^* and ω_2^* correspond to the complex amplitudes of Sway, Heave and Roll. ζ_0 is the incident wave amplitude and a is a half ship width. On Sway and Roll, the correction by the viscosity term is effective, but on Heave the correction tends to be overestimated except in the vicinity of the natural period.

c) Ship Motions in Oblique Waves

When the incident wave period is fixed at 1.4 (sec) and the incident wave angle θ is changed, the results are shown in Fig.-12. The ordinates show the amplification factor of ship motions. The abscissas θ show the incident wave angle and $\theta=0$ (deg) corresponds to the beam sea. η^*, ω_1^* and ω_3^* correspond to the complex amplitudes of Surge, Pitch and Yaw. b is a half ship length and the others are the same in Fig.-11. The viscosity terms are not introduced into Surge, Pitch and Yaw because of the disagreement between the wave period and the natural periods. The dotted line and the one-dot-dash-line show the results corrected by considering the effect of the finite length of a quay wall when θ is bigger than 75 (deg). It can be seen from Fig.-12 that there is a little error in the modes except Sway and Roll.

4.CONCLUSION

It is confirmed that the calculation accuracy is advanced by the improvement of the calculation methods on the retardation functions, the constant added masses and the introduction of the viscosity terms into equation of ship motions. But the introduction of the viscosity terms into the other modes where the viscosity forces are not considered in this paper must be discussed hereafter. They will be a main problem when the long period ship motions are calculated.

REFERENCES

1) Swaragi, T., M. Kubo and S. Aoki: New mooring system to reduce ship motions and berthing energy, Coastal Engineering in Japan, pp.303-313, Vol.27, 1984.
2) Oortmerssen, G. Van: The motions of a moored ship in waves, Publication No.510, Netherlands Ship Model Basin, Wageningen, The Netherlands, 1976.
3) Takagi, M. and K. Saito: On the description of non-harmonic wave problems in the frequency domain (5-th report), Journal of the Kansai Society of Naval Architects, Japan, pp.51-59, No.191, 1983 (in Japanese).

CHAPTER 220

FENDER SELECTION CRITERIA

F. Vasco Costa*

F. ASCE

ABSTRACT

Fenders are to be selected taking into account two quite distinct functions they have to fulfill:

- During berthing manoeuvres fenders have to avoid damage to the ship and to the berthing structure;

- After ships are already berthed and moored, fenders have to keep them quiet during loading and unloading operations.

To fulfill the first function fenders have to be able, while deflecing, to absorb a very large amount of energy.

To fulfill the second function fenders need to be able, while recovering their form, to dissipate a large amount of energy. Besides, they have to contribute to reduce or to increase the natural period of oscillation of already moored ships, depending on their tendency to oscillate in phase or out of phase with waves reaching the berths.

INTRODUCTION

The dimensions of ships have increased very rapidly in the last thirty years. The power of tugboats and the strength of ropes needed to assist the ships during berthing manoeuvres did not increase as fast as the dimensions of the ships. Besides, because they are larger, ships have now, at least in some harbours, to be moored in where ships will possibly be brought to move in ressonance with very stable long waves reaching such berths.

For these reasons fenders need to be selected taking into account that they will have not only to be able to absorb large amounts of energy during risky berthing operations but, as well, to be able to oppose, working as dash-pots, the tendency of moored ships to oscillate in ressonance with waves reaching their berths.

*Rua Borges Carneiro 6, 2o, Lisboa 1200, PORTUGAL

Just by passing ashore several strong ropes and keeping them tight, it is possible to keep small ships quiet at their berths, independently of the periods of the waves reaching the berths. But to prevent large ships from starting to move in ressonance with long waves it is preferable, in most cases, to just use a few soft long mooring ropes passed not very tightly, so as to bring the natural periods of oscillation of the moored ships well above the periods of the waves reaching the berths.

THE FIRST IMPACT DURING A BERTHING MANOEUVRE

Berthing manoeuvres are very risky operations, especially when the ships are to be berthed in locations exposed to the action of waves, of currents and of wind.

A sudden change in the magnitude or in the direction of the wind or current, a rope that breaks, a deficient evaluation of the velocity of the point of the ship about to contact the first fender, an order to a tugboat which is not well understood or not promptly obeyed, and the ship will hit the fender moving at a too high velocity (Fontijn 1987, Svensen 1970 and Tryde 1987).

The cost of a fender is just a small fraction of the cost of a repair in the ship's hull or in the berthing structure. Preference is therefore to be given to fenders that are not only able to absorb a large amount of energy without being damaged but that, in case of something going wrong during the berthing manoeuvre, will keep absorbing an extra amount of energy while being damaged and having afterwards to be substituted.

Fenders that just "bust" in case of being overloaded, are to be avoided.

Berthing manoeuvres, which must always be planned and performed with the greatest care, are usually accomplished in the following phases:

- Placing the ship in front of the berth in a position parallel to the berth;

- Bringing the ship by a rotation, or a translation at a small angle, into contact with a first fender;

- Rotation of the ship around the first point of contact until she touches a second fender at the other end, becoming again parallel to the berthing surface.

In case of the velocity and the direction of the translation being known, the amount of the energy to be absorbed during the impact of the ship with the first contacted fender can be evaluated by the expression (see fig. 1)

$$E = \frac{mu^2}{2} \times \frac{k^2+r^2\cos^2\gamma}{k^2+r^2} \tag{1}$$

where

m is the mass of the ship plus the added mass;

u is the ship's velocity of translation;

r is the ship's radius of gyration;

k is the distance from the center of mass of the ship to the point of contact;

γ is the angle of the velocity direction with the line connecting the center of mass of the ship with her point of contact with the fender.

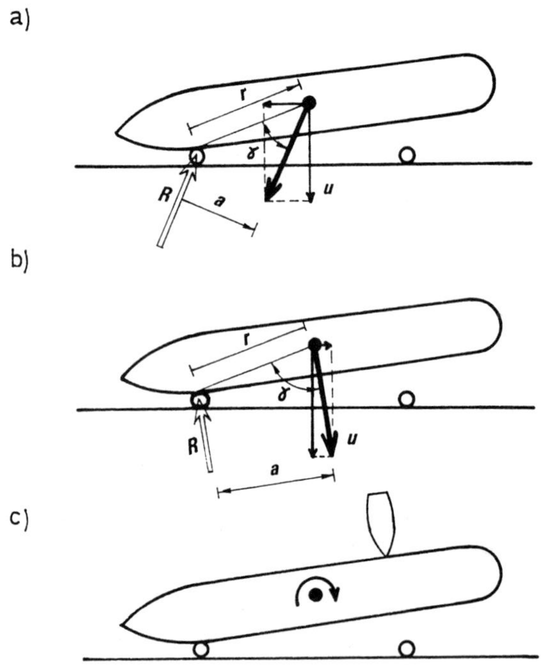

Fig. 1 - The berthing manoeuvre

In order to render the first impact of the ship as smooth as possible, the ship has to be moved in such a way as to render both the distance r and the angle γ as large as possible (Vasco Costa, 1964).

In case of r/k = 1,6, the making of γ = 90° instead of γ = = 70° will permit the amount of energy to be absorbed by the fender during the first impact to be halved.

The bringing of the ship not in a perpendicular but in an oblique direction towards the berth will imply the risk of her hull sliding on the fender surface. If this happens, part of the ship's kinetic energy will be dissipated overcoming the friction between hull and fender. If the fender is not damaged in the process, nor the neighbouring ships affected, the sliding contributes to the reduction of the amount of energy to be absorbed by the deflection of the fender.

What is to be avoided is the bringing of the ship to berth moving in a direction with a tangential component oriented towards the point of contact, as indicated in the British Standard Code of Practice for Maritime Structure BS 6349, Part 4, and in fig. 1a. The bringing the ship as represented in fig. 1b, will not only imply a reduction on the amount of energy to be absorbed by the fender but, as well, a reduction on the magnitude of the fender reaction on the ship's hull. The only inconvenient of the recomended procedure is the increase on the amount of energy to be absorbed later, when a second fender is contacted at the other end of the ship.

If the ship enters in contact with the fender while moving in a simple rotation, instead of a translation, the amount of energy to be absorbed can still be evaluated by expression (1) by making γ = 0. The rotation permits, therefore, the minimization of the energy to be absorbed for a given velocity of the point of the ship's hull entering into contact with the first fender.

THE SECOND IMPACT

When a ship about to contact a first fender is moving at an excessive velocity, there is a tendency to apply on her, through ropes or tugboats, a force at the other end of the ship, in order to reduce the velocity of the point about to hit the first fender (see fig. 1c). In this way the risk of an accident during the first impact is, no doubt, reduced, but at the cost of increasing the amount of energy to be absorbed at the other end of the ship during the second impact (see fig. 1c). This must be the reason why second impacts are usually harder than the first (Jamm and Russman 1966). As second impacts take place after the motion of the ship is already under control, serious accidents never happen during the second impact.

The bringing of the ship to berth moving in a direction with a tangent oriented as represented in fig. 1b, and applying on a berthing ship torques that will reduce the velocity of the point of the hull about to contact the first fender, are practices to be recomended. Although implying an increase on the amount of energy to be absorbed during the second impact, they will permit to reduce the risk of accidents during the first impact.

It is customary practice to assume, in the evaluation of the energy to be absorbed by the fender, that a certain amount of water, to be denoted as added or hydrodynamic mass, is moving solidary with the ship, contributing to increase the amount of energy to be absorbed by the fender. In fact most of the water in the vicinity of the ship has to move in a direction opposed to that of the ship, not in the direction the ship is moving, in order to give room for the ship to move.

Much has been written about the elusive concept of added mass. For the purpose of fender selection the suggestion is advanced to evaluate the mass to be decelerated just by multiplying the mass of the ship by a coefficient

$$1 + 2\ D/B$$

where D is the draught and B the beam of the ship (Vasco Costa, 1964).

To reduce the influence of the added mass, and therefore the amount of energy the fenders will be required to absorb, the suggestion is advanced to stop the ship, or at least to reduce her velocity, just before the ship contacts the fender. In this way the inertia of the water will keep for a while the water moving faster than the ship. When the ship is brought again into motion, the water that has overtaken the ship will contribute to reduce her velocity during the impact with the fender.

MOORED SHIPS AS OSCILLATING SYSTEMS

It has been found in some harbours, by trial and error, that the most convenient way to keep moored ships quiet at some berths is by reducing the number of mooring lines and making them longer and slacker, while at some other berths it is better to make them shorter and to keep them very tight (Khanna and Bert 1977, Kilner 1960, Vasco Costa 1983 and Wilson 1970).

Why does it so happen?

Just because some waves, of so small height as to pass unnoticed, but of such great length as to render them very stable, succeed in reaching berths far inside harbours and in bringing the ship to oscillate in ressonance with the periodic forces they exert on the ship.

These long waves, with periods of 30 seconds and longer, may have been generated by long/lasting violent storms in far-off places, or be the result of the interference of distinct trains of shorter waves.

If the basins they reach inside a harbour happen to have good reflective walls, and natural modes of oscillation coincident with any of the periods of the incoming waves, stationary waves will be formed inside the basins, their height possibly becoming larger than that of the waves in the open seas (Martinez and Naverac 1988, Oortmerssen 1987, and Portela 1970).

The periods of the distinct modes of oscillation of a closed rectangular basin can be evluated by the expression

$$T = \frac{2\ \ell}{n\ \sqrt{gd}} \qquad (2)$$

where

l is the length of the basin;
d is the depth of the water;
n is an integer defining the mode of oscillation.

In table 1 are given the natural periods of oscillation of closed basins with rectangular form. For the evaluation of periods of oscillation of basins with other simple geometrical forms see Wilson (1970).

The natural period of oscillation of a moored ship of mass m being held at berth, without clearances, by mooring lines and fenders that deflect proportionally to the forces to which they are submitted, is to be evaluated by the expression

$$T = 2\ \pi\ \sqrt{m/k} \qquad (3)$$

where

K is the spring factor of the moorings (see table 2).

The smaller the mass of a ship, the stiffer the fenders and shorter her mooring lines, the shorter will be her natural period of oscillation. The larger the mass of a ship the softer the fenders and the longer her mooring lines, the longer will be the natural period of oscillation of the moored ship.

Small ships, namely pleasure and fishing boats, have natural periods of oscillation that are in general much shorter than the natural periods of oscillation of the basins in which they are moored. For that reason they oscillate in phase with the periodic forces exerted on their hulls by the waves. To reduce the amplitude of their motions after being moored, the spring factors of

TABLE 1

Periods of oscillation of closed basins

$$T = \frac{2l}{n\sqrt{gh}}$$

Dimensons of basin		Periods of oscillation				
Length	Depth	Uninodal	Binodal	Trinodal	Quadri-nodal	Quinqui-nodal
500 m	5 m	2min23s	1min12s	48s	36s	29s
	8 m	1min53s	56s	38s	28s	23s
1 000 m	8 m	3min46s	1min53s	1min15s	56s	45s
	15 m	2min45s	1min22s	55s	41s	33s

TABLE 2

Periods of oscillation of ships moored without clearences

$$T = 2\pi\sqrt{m/K}$$

Mass of ship m	Spring factor of moorings K				
	200 kN/m	500 kN/m	1000 kN/m	2000 kN/m	5000 kN/m
100 t	4.4s	2.8s	2.0s	1.4s	0.9s
500 t	9.9s	6.3s	4.4s	3.1s	2.0s
1 000 t	14.0s	8.9s	6.3s	4.4s	2.8s
5 000 t	31.4s	19.9s	14.0s	9.9s	6.3s
10 000 t	44.4s	28.1s	19.9s	14 s	8.9s
50 000 t	1min40s	1min3s	44.4	31.4s	19.9s
100 000 t	2min20s	1min29s	1min3s	44.4s	28.1s
500 000 t	5min14s	3min19s	2min20s	1min40s	1min3s

their moorings are to be increased by resort to a large number of shorter, stiffer and tighter ropes, as well to stiffer fenders.

Large ships, especially if moored in basins of relatively small dimensions, present natural periods of oscillation that are larger than the natural periods of oscillation of such basins. As a consequence, they will tend to oscillate out of phase with the periodic excitation forces exerted in them by the waves. To reduce the amplitude of their movements after being moored it is convenient to increase their natural periods of oscillation by holding them with just a few very long ropes, to be kept not too tight, and by resort to soft fenders.

Unfortunately, large moored ships never have well-defined natural periods of oscillation. This is not only because surge, sway, yaw, heave, pitch and role motions interfere with each other but, besides, because mooring ropes, instead of having elongations proportional to the forces to which they are submitted, become stiffer when the forces applied on them increase in magnitude, the same happening to some types of fenders. When mooring ropes and fenders become stiffer, the natural period of oscillation of the moored ship decreases, this implying a greater risk of her starting to move in ressonance with shorter waves.

The places where the consequences of moored ships starting to move in ressonance with stationary waves in a basin can be more serious are the nodes of such waves. At the anti-nodes a moored ship will only be made to move up and down a few centimeters, while at the nodes the vertical movements will be almost non-existent, but the slope of the water surface will keep changing all the time. The larger the slope of the water surface the larger will be the force tending to bring the ship to slide down the water surface.

A stationary wave with a length of 1 500 m and a height of 0.5 m will cause the water slope to reach every half period a value of

$$\pi \frac{H}{L} = \pi \frac{0{,}5\ \text{m}}{1500\ \text{m}} \cong \frac{1}{1000} \qquad (4)$$

Such slope will pass unnoticed to visual observation. Notwithstanding, a ship with a displacement of 300 000 t will be submitted to an horizontal force of 300 t, which every half period will change direction (see fig. 2).

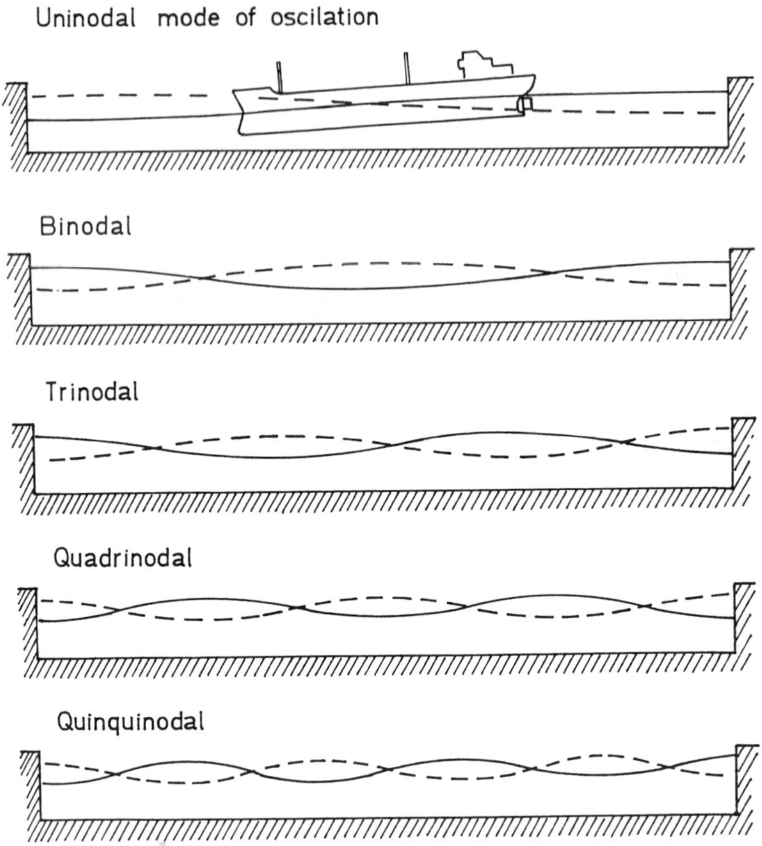

Fig. 2 - Modes of oscillation of a closed basin

Such periodic force will bring the moored ship to have oscillatory motions whose amplitude will increase from motion to motion if the natural period of oscillation of the moored ship happens to be coincident, or near coincident, with that of the wave. In such case not only loading and unloading operations will have to be suspended but the ship will possibly break her mooring ropes or damage the fenders.

What can be done to avoid such situations?

Confronted with the difficulty of keeping ships quiet at their berths, some entities just prefer to ignore the nature of problem. This is the reason why, in the book "Guidelines and Recomendations for the Safe Mooring of Large Ships at Pier Piers and Sea Islands", published by the Oil Companies International Marine Forum, is stated on page 3: The design mooring conditions do not include wave conditions able to cause significant dynamic loadings in mooring.

Let us present some considerations on how to deal with the oscillation of moored ships in ressonance with long waves reaching their berths.

First of all the possible modes of oscillation of the water in front of the berth must be identified. Just by throwing a few pieces of cork along a quay berth it will be possible to find out if stationary waves are present, which are their periods, and where are located their nodes and anti-nodes. But equipment is already available to study the motion of the water in front of a berth. In case of basins with simple geometrical forms, it will be possible to find the main forms of oscillation of a basin just by resort to simple analytical expressions (Kubo et al, 1988).

Once the periods of stationary waves in the basin and the natural periods of oscillation of the moored ship have been found, an option has to be made between two alternative ways of reducing the risk of the ship starting to move in ressonance with the waves. Are the stiffnesses of mooring ropes and of fenders to be increased in order to reduce the natural period of oscillation of the moored ship, or are their stiffnesses to be decreased in order to increase her natural period of oscillation?

In general, as referred, small ships are to be held by stiff mooring cables and stiff fenders while large ships, especially when berthed in small basins, are to be held by just a few, soft, long mooring cables and relatively soft fenders.

It must be pointed out that the avoidance of ressonance by increasing the natural period of oscillation of a moored ship implies the risk of the ship starting to move with undetected longer waves, that can be the occasional result of the superposition of trains of long waves with different lengths. For such reason, resort to soft moorings and soft fenders should only be adopted when the occurence of very long waves is not to be feared.

When selecting fenders having in view the keeping of moored ships quiet at berth, one has to take into account that fenders, besides contributing to increase or decrease their natural periods of oscillation, will also have to act as dash-pots, dissipating as much energy as possible while recovering their form after deflection (see fig. 3).

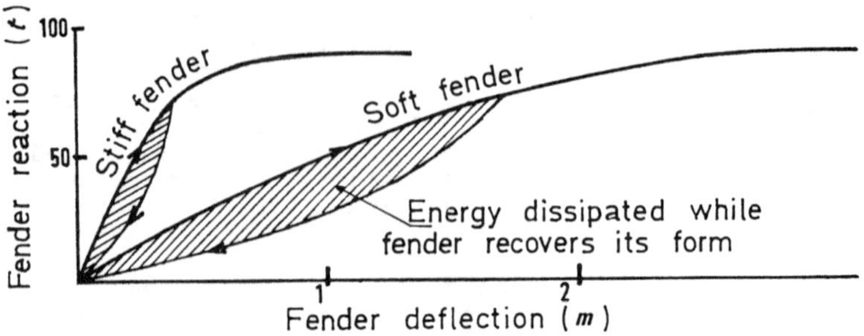

Fig. 3 - Stiff versus soft fenders

RECOMENDATIONS

In order to reduce not only the frequency of accidents during berthing manoeuvres but also their consequences, fenders need to be able to absorb very large amounts of energy when something goes wrong during such operations. Preference is to be given to fenders that, instead of just "busting" when overloaded, will keep absorbing energy while being damaged.

In order to reduce "down time" during loading and unloading operations, the fact has to be faced that moored ships are oscillating systems, and that the problems they pose are to be studied and solved, just as similar systems have been studied and solved, namelly the suspension of automobiles. For such purpose the modes and periods of oscillation of the water inside harbour basins and the location of nodes and anti-nodes of stationary waves have to be identified. Spring factors of mooring ropes and fenders and of their combinations are to be evaluated, in order to find out if the natural periods of oscillation of the moored ships are to be increased or decreased.

When selecting fenders having in view the quitness of moored ships, preference has to be given to fenders which present deflections proportional to the forces they are being submitted to, because such fenders permit moored ships to present well defined natural periods of oscillation. Preference has also to be given to fenders that dissipate a large amount of energy while recovering their original form after being deflected.

Great progress is the study of modes of oscillation of water basins and of moored ships has been made in the last twenty years.

In the Procedings of the NATO-ASI on "Advances in Berthing and Mooring of Ships and Offshore Structures", held in 1987 in Trondheim, reference is made to most recent publications on the subject.

ACKNOWLEDGEMENTS

Tanks are due for valuable comments by Eugene H. Harlow and by Bela Komman, both of Soros Associates, on the original version of this contribution.

REFERENCES

Bruun, Per [1986], "Can we save on breakwaters by providing better mooring-fendering?" PIANC Bolletin Nº 53, Brussels.

Fontijn, H. L. [1987], "On the prediction of fender forces on berthing structures" NATO-ASI on Advances in Berthing and Mooring of Ships, Trondheim, Norway.

Jamm and W. Russman [1965], "Measurement of tankers berthing manoeuvres" Hamburg, Report submitted to the Int. Oil Tankers Commission of PIANC.

Khanna, J. and C. Brit [1977], "Soft mooring systems at exposed terminals", Ports 77.

Kilner, F. A. [1960], "Model tests on the motion of moored ships placed on long waves". Proc. VII Conf. on Coastal Eng., Den Haag.

Kubo, M. N. Shinoda and S. Okamoto [1988], "The time domain analysis on moored ship motions", 21st Int. Conf. Coastal Eng., Torremolinos.

Martinez, F. and V. S. Naverao [1988], "An experimental study of harbours ressonance phenomena" 21st Int. Conf. Coastal Eng., Torremolinos.

Oortmerssen, G. Van [1987], "Forces related to motions of moored ships - Analytical methods of moored ship motions" NATO-ASI on Advances in Berthing and Mooring Ships, Trondheim, Norway.

Oumeraci, H., O. Burkhardt and S. Kohlhase [1987], "Influence of ship motions on the optimum design of ports layouts" 2nd Conf. Coastal and Port Eng. in Developing Countries. Beijing, China.

Scheffer, Hans-Joachim, and Sosen Kohlhase, "On the vertical motion of ships at berth" 2nd Conf. Coastal and Port Eng. in Developing Countries, Beijing, China, 1987.

Svensen, I. A. [1970], "Measurement of impact energies on

fenders" The Dock and Harbour Authority, Oct. 1970.

Tryde, P. [1987], "Use of statistical data and methods in impact evaluation of fender design" NATO-ASI on Advances in Berthing and Mooring of Ships, Trondheim, Norway.

Ueda, Shigeru, Satorn Shiraishi and Mazakasu Yoshimi, "Estimation of Port Operation Rate by use of numerical simulation method of moored ship motions", 2nd Conf. Coastal and Port Eng. in Developing Countries, Beijing. China, 1987.

Vasco Costa, F. [1964], "The berthing ship" The Dock and Harbour Auth. May, June & July 1964, also published in booklet form by Foxlow Publications, London. U.K.

Vasco Costa, F. [1973], "Selection of fender performance", Proc. NATO-ASI on Analytical Treatment of Problems in the Berthing and Mooring of Ships, Wallingford, England.

Vasco Costa, F. [1983], "Moored ships as oscillating Sustems" The Dock & Harbours Auth. Feb.

Wilson, Basil W. [1968], "The threshold of surge damage for moored ships" Proc. Inst. Civil Eng. Vol. 40, London.

Wilson, Basil W. [1965], "Seiches and other causes of the motion of ships already moored" Proc. of a NATO-ASI on Analytical Treatment of Problems of Berthing and Mooring Ships, Lisbon, published by the ASCE 1970.

Wyk, A. C. Van, and J. A. Zwanborn [1988], "Wave-induced ship motions in harbours approach channels", 21st Int. Conf. Coastal Eng., Torremolinos.

CHAPTER 221

UTILIZATION OF MOORED VESSELS IN HYDRAULIC MODELS OF HARBORS.

Dr. Ing. Rafael Guarga (IAHR member)*
Ing. Eduardo Goldsztejn**
Ing. Fernando Puntigliano**

Abstract

The behavior of the model of a moored vessel in a harbor's hydraulic model (Froude's model) is studied, taking into account the difficulties arising from the fact that modelling a moored vessel requires to incorporate the Reynolds number. Experiments made with the model of a large grain carrier at a 1/175 scale are shown. The horizontal motions of this ship due to the waves were measured when it is moored in a zone in which the length scale along the vertical axis was modified (vertical distortion), and the influence of different depths on these motions was studied.

This study was performed in order to be able to use this ship model for the evaluation of different alternatives of harbor design, in the building of new facilities at a port placed in the Atlantic coast of Uruguay.

*Director, Institute of Fluid Mechanics and Environmental Engineering, J. Herrera y Reissig 565, Montevideo, Uruguay

**Assistant Professors, Institute of Fluid Mechanics and Environmental Engineering

Introduction

The amount of agitation in the hydraulic model of a harbor is a main fact to be taken into account when decisions about harbor design are to be made. The traditional way for the evaluation of agitation is directly measuring wave heights in those places of the berth which are considered important for its performance. Comparison between two harbor designs is done by comparing the wave heights associated to each design, and frequently, design criterium is not reaching certain values of wave height which are accepted as maximum for a safe and economical developement of the port's activities. Nevertheless, this method, though useful to compare two harbor designs, does not provide an absolute criterium for the evaluation of the port. The port's ability to fulfill performance requirements must be based on experience about the influence of the action of waves. An absolute criterium for the evaluation of a harbor design must provide the prediction, by means of the model, of the motions to be experienced by the ships which will be moored to the prototype's docks. A good design will be that in which these motions allow cargo handling to be safe and economical. These motions shall also be safe to all the port's facilities. References related to the behavior of vessels' physical models and their utilization to harbor design is found in bibliography about harbor design |2,3,4,5|.

Stating the problem

When harbor's modelling is performed to study the behavior of a moored vessel, a similitude problem arises. The harbor's hydraulic models are Froude's models with length scales e_L roughly varying from 1/100 to 1/200, and the modelling of a moored vessel makes it necessary to introduce the Reynolds number. Simulation of the motion of a body subjected to external forces produced by the fluid, the mooring lines and the fenders system must be performed. Oscillatory flows between the ship's hull, the sea bottom and the docks take place and in such flows, power dissipation per unit of surface area is a function of the Reynolds number of the flow. Reynolds number scale (e_{Re}) in a Froude's model is $e_{Re} = e_L^{3/2}$. Considering, for example, e_L =1/100, it results e_{Re} = 1/1000. Reynolds number reduction from prototype to model is then very important, and therefore there will be no similitude between the model of a moored ship in an agitation hydraulic model, and its prototype.

As an exact theoretical solution for the complex dissipative process which takes place in the turbulent boundary layer placed between the ship's hull, the sea bottom and the docks, is still not available, the problem must be tackled experimentally. This will be performed taking into account recent results related to the turbulent boundary layer created by a wave. Following Blondeaux |1|, it is seen that the dimensionless mean power D dissipated by the boundary layer generated by the wave, per unit of surface area, usually decreases with the Reynolds number. D is defined as $D = D^*/\rho U_b{}^{*3}$, D* being the mean power dissipated by the boundary layer per unit of surface area, and ρ is the water density. The Reynolds number is defined as $U_b{}^* a_b/\nu$, being $U_b{}^*$ the maximum velocity of the irrotational wave in the bottom, $a_b = U_b{}^* t/2\pi$, t being the period of the wave, and ν the kinematic viscosity of water. In the referred paper it is stated that in hydraulically smooth regime, D can be calculated by means of the relationship: $D = 0.14\ Re^{-0.23}$. Considering now the average total power per unit of surface area in a simple harmonic progressive wave, E*, this power can be expressed as $E^* = H^2\gamma/8t$, where H is the wave amplitude, and t the wave period. Therefore $D^*/E^* = 8U_b{}^{*3}\ D/gH^2T$. If model's and prototype's D*/E* values are compared and a D*/E* scale is defined it is obtained

$$e_{D^*/E^*} = (D^*/E^*)m/(D^*/E^*)p = e_v{}^3\ e_t\ e_L{}^{-2}\ e_{Re}{}^{-0.23} = e_L{}^{-0.345},$$

being e_t the time scale, and e_v the velocity scale. Then

$$d\,(e_{D^*/E^*})/d\,(e_L) < 0$$

If this result is applied to a ship moored in a harbor, the conclusion is that the D*/E* value in the model is higher than in the prototype. The precedent reasoning shows that if dissipation in the model is not reduced, results obtained about angular and linear desplacements of a moored vessel in a hydraulic model of a harbor will be smaller, at prototype's scale, than real desplacements.

Experimental solution of the problem.

The installations:

The experiments described in the paper were performed in a hydraulic agitation model*. The ship which was modelled is the largest grain carrier that will be able in the future to operate in the new facilities to be built at this port. The model is shown in figure 1, in working position, with the mooring lines and fenders system described later in this paper. The length scale was e_L =1/175. The main dimensions of the ship's model are:

*built in the Institute of Fluid Mechanics and Environmental Engineering, of the Faculty of Engineering of Uruguay.

Figure 1. Ship's model moored in harbor's hydraulic model.

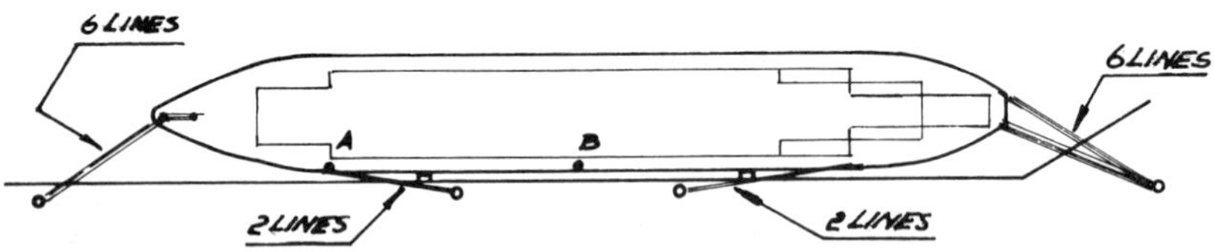

Figure 2. Arrangement of mooring lines in ship's model.

Length between perpendiculars:......Lppm =1.41m
Maximum moulded breadth:............Bm =0.184m
Displacement (maximum):.............Dm =146N
Draught (maximum):..................Tm =0.071m
Displacement (ballast condition):...D'm =75.2N
Draught (ballast condition):........T'm =0.039m
Block coefficient:..................Cb=0.84

In addition to the hull's shape, longitudinal distribution of the weight (longitudinal position of the center of gravity, LCG) was simulated. The model's behavior was studied in ballast condition, and for this case, transverse distribution of weight, and metacentric radius were simulated. Being the time scale $e_t = e_L^{1/2}$, as it happens in Froude's models, the natural period of roll, which is of 8s in the prototype, resulted of 0.605s in the model.

This model was built to study the new breakwaters location in the port adjacent to the city of La Paloma, in the Atlantic coast of the country.

Mooring of the model and fenders system

Figure 2 shows the mooring of the vessel. The lines in the bow and in the stern simulate six synthetic-fibres lines of 80 mm diameter in each case, as required for this kind of ships. The lines on the sides simulate the behavior of two of these 80mm diameter lines, for each mooring. In order to measure the motions of the ship's model, the moorings of bow and stern were pretensioned with a force of 0.137 N, correspondent to 1.23×10^5 N for each line in the prototype. Forces on the lines were measured in the model by joining them to two force transducers which operate with strain-gages. In these transducers, a varible current is obtained when a horizontal force, perpendicular to them, is exerted. When forces vary from 0 to 0.98 N , the intensity of this current vary from 4 to 20 mA, and the variation is linear. Pretensioning of the lines meant therefore an increase of 2.24 mA in the intensity of the current generated at each transducer. The simulation of the mooring lines was performed using textil fibres which have a force-deformation law reproducing, at model's scale, the behavior of real mooring lines. The force-deformation law of several 80 mm dimeter lines are shown in figure 3, as function of a reference force Fo of 0.98×10^5 N in the prototype. In this figure, δ is a relative deformation, expressed in percentage ($100 \times \Delta L/L$), and δ_o is the relative deformation corresponding to Fo. With regard to the fenders system, simulation was achieved by using two force transducers of the same kind used to measure the forces on the mooring lines. Two springs were added in parallel to these transducers (figure4).The elasticity constant of the device formed by each transducer and it's spring is 1.23×10^7 N/m, so that both of them, when placed in the model's docks simulate the behavior of a fenders system adequate for the reception of ships of the size of the one which was modelled.

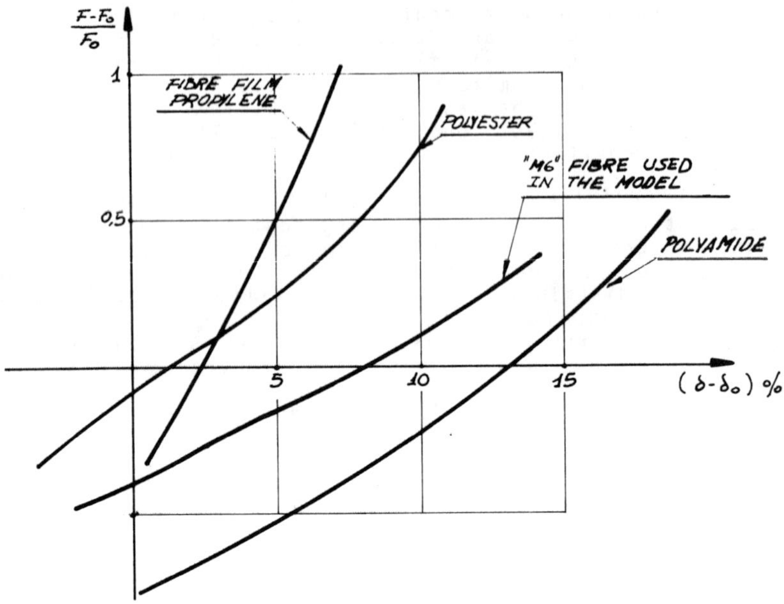

Figure 3. Force-deformation laws of 80 mm diameter lines, and of line used in the experiments.

Figure 4. Device used for fenders simulation.

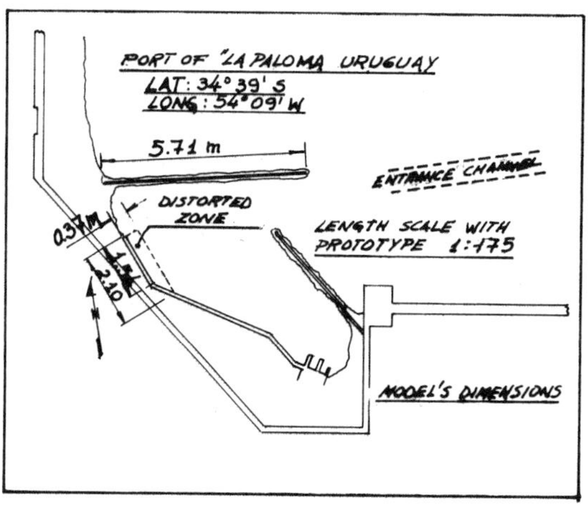

Figure 5. Location of the distorted zone.

Horizontal motions of points A and B (figure 2), were measured visually with the aid of graduated scales

With the aim of reducing power dissipation in the model, and taking into account that power transfer from the mean flow to turbulent fluctuations is proportional to the magnitude of the components of the mean deformation velocities tensor |6|, the harbor's model depth in the ship's neighborhood was distorted. The distortion was made in a zone in which reception of large vessels is planned. The area of this distorted zone is 3.72% of the total area of the port, and with regard to wave heights, no change in the behavior of the hydraulic model of the port was detected, either inside or outside the distorted zone, when measuring these wave heights with resistive devices.

In figure 5, the location of this zone is shown, and in figure 6 an overall view of it is provided. In figure 7, wave heights in five points named 1,2,3,4 and 5, in the distorted region are shown. Wave heights are referred to the amplitude of the wave in the generation zone, and they were measured with the normal depth of the port design (without any distortion), and with a depth of 0.23 m in the model, correspondent to the maximum distortion made. In both cases, wave periods in the model were 1.06s, and 0.378s. In figure 8, location of the points above mentioned in the distorted zone, is indicated.

The vertical distortion proportionally reduces the vertical gradients of mean velocities, which are the most important in this case, for turbulent dissipation. The effect of different depths in the distorted zone, on the motions of points A and B of the ship's model was studied. The depths used were d =0.051m, d =0.15m, and d =.23m, and referring these depths to the model's draught in ballast condition, values of d/T' were 1.32, 3.9 and 5.88. In figure 9 (a,b), results of the measurement of horizontal motions' magnitudes at points A and B (denoted as HA and HB) are summarized. In order to make comparison possible, HA and HB were divided by the height Hi of the generated wave in the hydraulic model of the harbor. Wave periods utilized were t_1 =1.06s, t_2 =0.907s, t_3 =0.756s, t_4 =0.605s and t_5 =0.454s, which are correspondent to periods of 14s, 12s 10s, 8s and 6s in the prototype. In figure 10 (a,b), the ratios HA/Hi and HB/Hi are plotted against increasing values of Lw/Lppm, Lw being the wave longitude in the generation zone, and, as already stated, Lppm is the length between perpendiculars of the ship's model.

It is found that when depth increases in the distorted zone, the amplitudes of oscillations of the ship's model also increase. It's found also that the effect of depth on these amplitudes diminishes when distortion is increased, and that a condition is reached in which, beneath the appreciation of the measurements performed, no more effects on these motions are observed when depth is increased. the same behavior was observed by Giraudet |4|, when studyng berthing speeds in vessels' models, and energy transfer from ships to the fenders system.

These results are interpreted as a limit situation in which power transfer from mean flow to turbulent

Figure 6. View of the distorted zone.

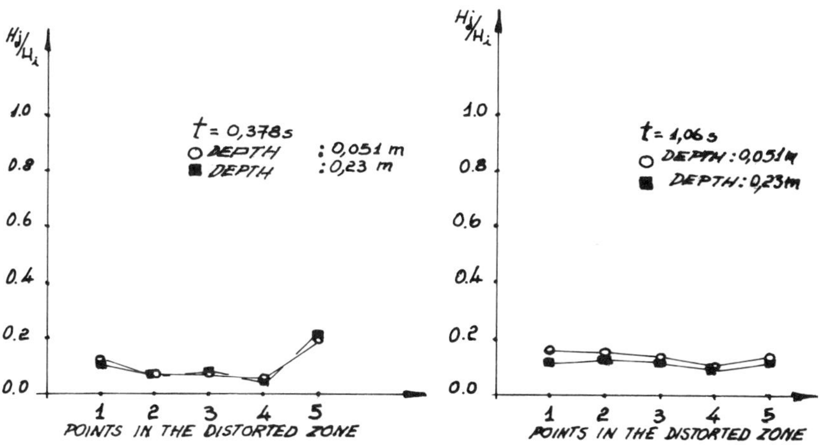

Figure 7. Measurement of wave heights in the distorted for different depths.

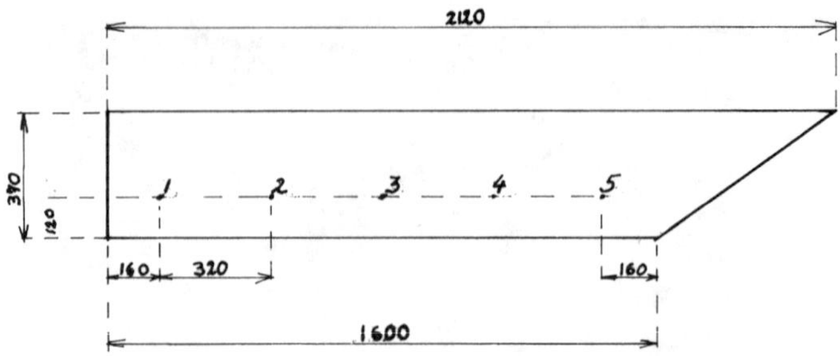

Figure 8. Location of points 1 to 5 in the distorted zone.

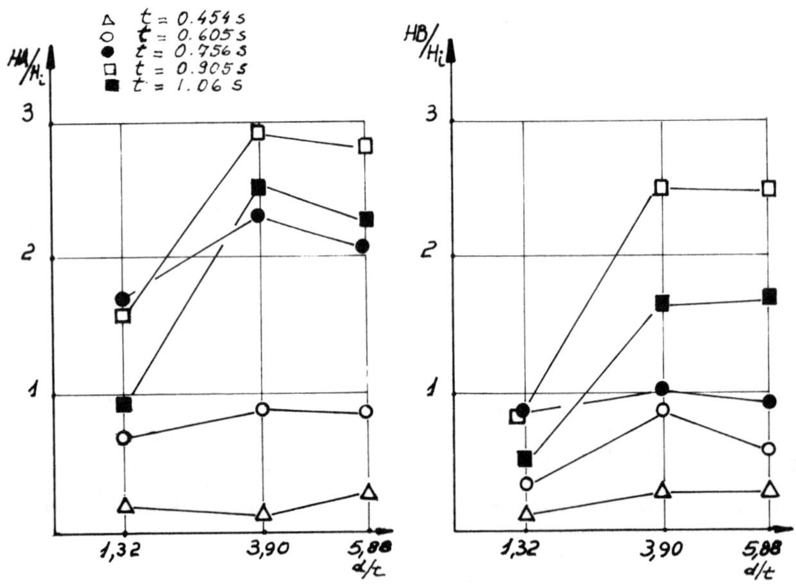

Figure 9(a,b). Displacement of the ship's model for different depths.

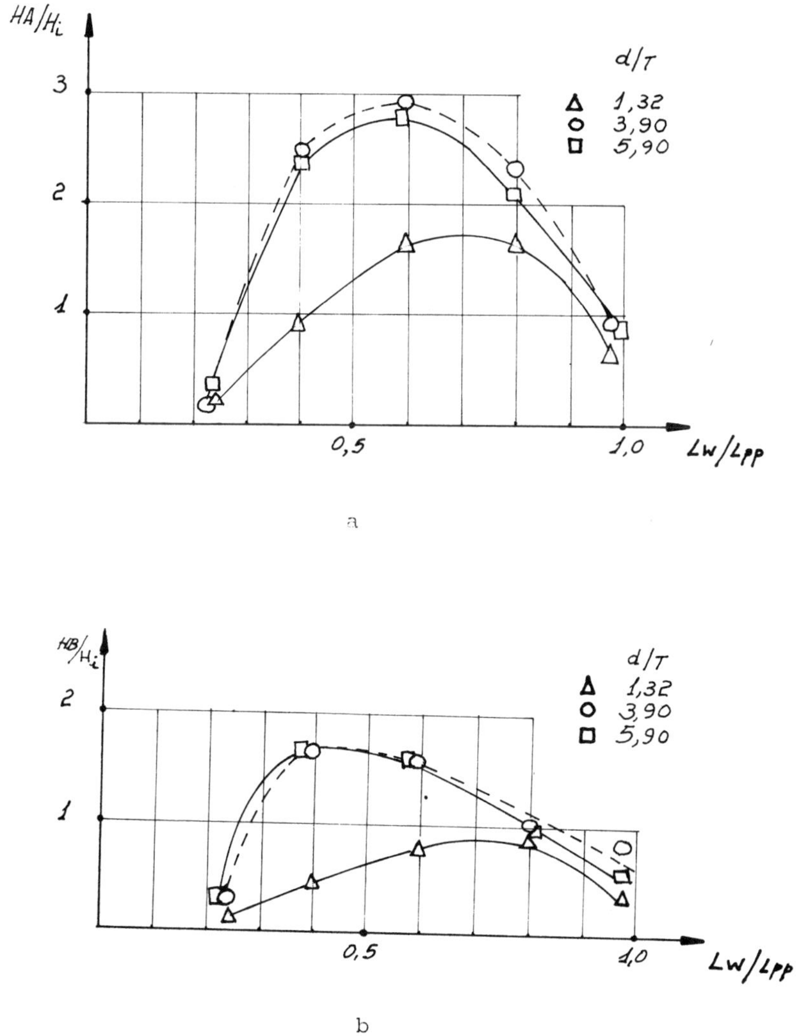

Figure 10(a,b). Displacement of ship's model as function of incident wave length.

fluctuations does not depend now on geometry, and, thus on the Reynolds number. Therefore the model operates in an "automodelling condition" with regard to Reynolds number. As the magnitude of the motions increases with depth, this condition places the model in a situation in which these motions will not be of less magnitude, at model's scale than those of the prototype, and results related to harbor's agitation by means of these ship's motions are on the safety side.

Conclusions

A method for evaluating agitation in harbors' hydraulic models is studied, which is more representative of port operation than measuring only wave heights in a mesh of points of the berth. It is shown that by using this method, safety factors are introduced if the bathimetry of the model is modified in a reduced area, which in view of it's small size, does not change the behavior of the port.

BIBLIOGRAPHY

1 - Blondeaux, P., "Turbulent layer at the bottom of gravity waves", Journal of Hydraulic Research, Nº 4, vol 25, 1987 p. 447-464.

2 - Giraudet, P., "Recherches expérimentalles sur l'energie d'accostage des navires", Annales des Ponts et Chaussées, Mars-Avril 1966, Nº II p. 103-127.

3 - Keith, J. and Murphy, E., "Harbor Study for San Nicolás Bay, Perú", Journal of the Waterways and Harbors Division, vol 2, May 1970 p. 251-273.

4 - Kim, C.H., "The influence of water depth on ship motions in oblique waves", Civil Engineering in the Oceans, vol.1, 1975, p.653-667.

5 - Schaeffer, H.and Kolhase, S., "Functional harbor design under consideration of ship motions", International Conference on Numerical and Hydraulic Modelling of Ports and Harbors, BHRA Birmingham, England, April 1985, p.221-229.

6 - Tennekes, H.and Lumley, J., "A first course in turbulence", MIT Press, 10th edition, 1985, Chapter 3.

7 - Van Wart, T. "Model testing a LNG terminal at Bintulu, Malasia", Civil Engineering in the Oceans, vol.1, 1979.

CHAPTER 222

A NUMERICAL WAVE PREDICTION MODEL FOR PERSONAL COMPUTERS[1]

David J. Schwab[2]

ABSTRACT

A two-dimensional wave prediction model suitable for use on personal computers is described. The model requires the two-dimensional time-dependent wind field as input. Output consists of wave height, wave period, and wave direction estimates at all grid points on a computational grid representing an enclosed or semi-closed basin. Model predictions compare favorably with observations from a wave research tower in Lake Erie. A formula is provided to estimate how long a model simulation would take on a personal computer given the surface area of the computational domain, the grid size, and the computer clock speed.

1. Introduction

Numerical wave prediction models have been categorized in various ways according to their development as one-dimensional or two-dimensional, parametric or spectral, discrete or hybrid, and first, second, or third generation. The SWAMP Group, 1985, provides a summary and intercomparison of the various model types. The simple empirical formulas for wave height and period estimation such as the SMB formulas (Bretschneider, 1970) or JONSWAP relations (Hasselmann et al., 1973) are basically one-dimensional in that they depend only on local wind speed and upwind fetch distance. Two-dimensional models take account of advective affects on the wave field by predicting the spatial distribution of wave energy over an entire two-dimensional domain. Until recently, the two-dimensional prediction models have only been implemented on multi-user mainframe or minicomputers while microcomputer (PC) users were limited to using empirical, one-dimensional wave prediction formulas for wave estimation. With improvements in the capabilities of the personal computers, it is now

---

1 GLERL Contribution number 632.

2 National Oceanic and Atmospheric Administration
Great Lakes Environmental Research Laboratory
2205 Commonwealth Blvd.
Ann Arbor, Michigan 48105 USA

possible to make practical use of some two-dimensional models with relatively inexpensive computing machines.

This paper describes the formulation and implementation of a particular two-dimensional wind wave prediction model (Donelan, 1977, Schwab et al., 1984a and Liu et al., 1984) for use on personal computers. The model is a parametric type that numerically solves a local momentum balance equation on a computational grid covering a closed or semi-closed basin. Only actively generated wind waves are predicted, swell is not included. The model is driven by the two-dimensional, time-dependent wind field. Output includes estimates of significant wave height, peak energy period, and wave propagation direction at each grid point. The fact that the model is a parametric model and only needs to retain a few parameters describing the wave spectrum at each grid point instead of a complete spectral description makes it practical for use on personal computers.

2. Techniques

The basic model equations relate the time rate of change of wave momentum and the divergence of the wave momentum flux to wind forcing as follows:

$$\frac{\partial M_x}{\partial t} + \frac{\partial T_{xx}}{\partial x} + \frac{\partial T_{xy}}{\partial y} = \frac{\tau_w^x}{\rho_w}$$

$$\frac{\partial M_y}{\partial t} + \frac{\partial T_{yx}}{\partial x} + \frac{\partial T_{yy}}{\partial y} = \frac{\tau_w^y}{\rho_w} \tag{1}$$

The x and y momentum components are defined as

$$M_x = g \int_0^\infty \int_0^{2\pi} \frac{F(f,\theta)}{c(f)} \cos\theta \, d\theta \, df$$

$$M_y = g \int_0^\infty \int_0^{2\pi} \frac{F(f,\theta)}{c(f)} \sin\theta \, d\theta \, df. \tag{2}$$

$F(f,\theta)$ is the wave energy spectrum as a function of frequency, f, and direction, θ. $c(f)$ is the phase speed. T_{xx}, T_{xy}, T_{yx}, and T_{yy} are the components of the momentum flux tensor defined as:

$$T_{xx} = \frac{g}{2} \int_0^\infty \int_0^{2\pi} F(f,\theta) \cos^2\theta \, d\theta \, df$$

$$T_{xy} = T_{yx} = \frac{g}{2} \int_0^\infty \int_0^{2\pi} F(f,\theta) \cos\theta \sin\theta \, d\theta \, df \qquad (3)$$

$$T_{yy} = \frac{g}{2} \int_0^\infty \int_0^{2\pi} F(f,\theta) \sin^2\theta \, d\theta \, df$$

If we assume that deep water linear theory applies and that wave energy is distributed about the mean wave direction as cosine squared, i.e.,

$$F(f,\theta) = \frac{2}{\pi} E(f)\cos^2(\theta-\theta_o) \qquad \text{for } |\theta-\theta_o| \le \frac{\pi}{2} \qquad (4)$$

then the momentum fluxes can be expressed as (Schwab et al., 1984a)

$$T_{xx} = \frac{g\sigma^2}{4} \left[\cos^2\theta_o + \frac{1}{2}\right]$$

$$T_{xy} = T_{yx} = \frac{g\sigma^2}{4} \left[\cos\theta_o \sin\theta_o\right] \qquad (5)$$

$$T_{yy} = \frac{g\sigma^2}{4} \left[\sin^2\theta_o + \frac{1}{2}\right]$$

where σ^2 is the variance

$$\sigma^2 = \int_0^\infty \int_0^{2\pi} F(f,\theta) \, d\theta \, df. \qquad (6)$$

In the numerical model, only the two components of the wave momentum vector (M_x, M_y), the wave phase speed at the peak of the spectrum (c_p), and the total energy in the spectrum (σ^2) are calculated at each grid point. To calculate the wave momentum vector components, the time derivatives in (1) are represented by central differences. An upwind difference scheme is used for the momentum flux advection terms. The time step is determined dynamically as

$$\Delta t = \Delta s/\sqrt{2}c_m \tag{7}$$

where c_m is either the maximum wind speed during that time step or the maximum wave phase velocity calculated at the previous time step, whichever is greater and Δs is the grid interval. The wave momentum fluxes in (4) are calculated using the approximate relation

$$\sigma^2 = \frac{c_p|M|}{g} \quad . \tag{8}$$

Although this relation between variance (or energy), peak energy frequency ($f_p = g/2\pi c_p$), and integrated wave momentum is basically empirical, it applies quite well over a wide range of conditions.

The numerical model is based on a wave momentum conservation equation rather than wave energy conservation so that the following formula can be used for the wind input source function:

$$\frac{\tau_w}{\rho_w} = 0.028\ c_d|\vec{u} - 0.83\ \vec{c}_p|(\vec{u} - 0.83\ \vec{c}_p) \tag{9}$$

where u is the 10 m wind vector and c_d is the form drag coefficient defined as

$$c_d = [0.4/\ln(50/\sigma)]^2 \tag{10}$$

with σ in meters. This formula for the drag coefficient is based on a logarithmic atmospheric boundary layer profile with a surface roughness height of $\sigma/5$. Note that when the wind and waves are travelling in the same direction, the momentum input function goes to zero when $u = 0.83\ c_p$ or $c_p = 1.2\ u$ (full development). This form of wind input is useful in describing the relative amounts of atmospheric momentum being expended on wave generation and the amount going into the water column to generate currents (see Donelan, 1977).

In the numerical scheme, after the new wave momentum vector is calculated from (1), peak energy frequency is found from the empirical relation

$$\frac{g^2\sigma^2}{u^4} = 6.23\text{x}10^{-6} \left(\frac{f_p u}{g}\right)^{-10/3} . \qquad (11)$$

By using (8) this relation can also be written as

$$f_p = 0.0179 \left(\frac{u^2}{|M/g|^3}\right)^{1/7} . \qquad (12)$$

It should also be pointed out, however, that other values of the coefficient and exponent in the empirical relation (11) between nondimensional energy and nondimensional peak energy frequency could be used and, in fact, are currently being evaluated in the model.

3. Results

The numerical model has been implemented on an IBM-XT type personal computer with an 8087 numeric coprocessor running at a clock rate of 4.77 MHz. With this configuration, a thirty day hindcast for a 10 km grid covering Lake Erie (253 grid squares) takes about one hour. Because the time step (7) in the numerical model is proportional to the grid interval, the total computation time is proportional to the cube of the grid interval, i.e., doubling the resolution of the computational grid (halving the grid interval) increases computation time by a factor of eight. Model tests on several different types of mainframe and microcomputers using various lake grids led to the development of a rough formula for the expected computation time. As shown in (7), the total computation time in seconds for one day of simulation, T, is going to depend on the maximum phase speed of the waves predicted during a run, but for typical wind speeds of 5-10 m s^{-1}, T can be estimated for a lake of surface area A with a computational grid of grid interval Δs and a computer running at a CPU clock rate of R Mhz as

$$T = \frac{1.7 \text{ x } 10^4 \text{ A}}{(\Delta s)^3} \left(\frac{4.77}{R}\right) \frac{\text{m-sec}}{\text{day}} . \qquad (13)$$

For Lake Erie with $A=2.54\text{x}10^{10}$ m^2, Table 1 shows some typical expected run times on a 4.77 Mhz machine with various grid intervals.

Table 1. Typical PC run times for various grid sizes on Lake Erie

Δs (km)	T (sec/day)
15	128
10	432
5	3456

In September and October of 1981, a solar-powered wave research tower was deployed in the eastern part of the central basin of Lake Erie (see Schwab et al., 1984b for the details). The tower was 6 km offshore in 14 m of water. Instruments on the tower measured wave height, wave period, and wave direction at hourly intervals. The numerical model was run for this entire period using the wind measured at the tower for the forcing function (9). The results from the model are compared to observed wave parameters in Fig. 1.

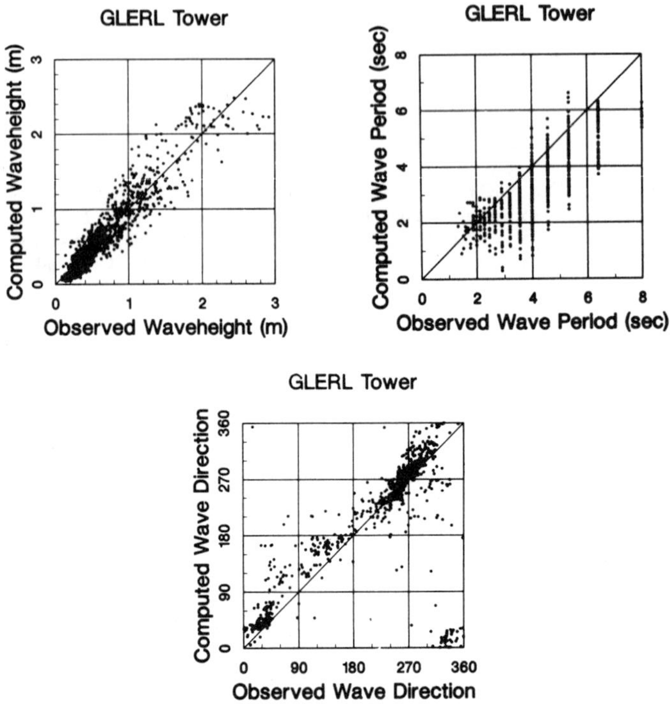

Figure 1. Comparison of observed and computed wave height, period, and direction in Lake Erie for September and October, 1981.

The agreement for wave height is quite good, with a root mean square error of 0.20 m and a correlation coefficient of 0.93 between computed and observed values. The computed wave periods, however, tend to underestimate the observed wave period. This may be due to using (12) to estimate wave period. Other formulas of the general form of (14) are currently being tested. Wave direction is modeled quite well, except for waves travelling in the 90-180° range. These are small waves generated from winds coming directly offshore (6 km fetch) so that it is very difficult to measure dominant wave direction accurately.

The model has also been extensively tested against other field data from the Great Lakes (typical fetch distances 100-300 km) and has been applied to Puget Sound, Hudson's Bay and the Beaufort Sea. Calibration and verification runs from deep and intermediate water depths show typical accuracies of 10-20% for wave height and 20-30% for wave period. The model is currently being used operationally by the National Weather Service for routine wave forecasting for the Great Lakes.

4. Conclusions and Recommendations

Although this type of model applies only to active wave generation, it can provide practical estimates of design waves for lakes, bays, estuaries, and semi-enclosed coastal areas. The main limitations of the model are that the wave spectrum is assumed to be single-peaked (no swell or multiple wind seas) and that shallow water effects are ignored. It is possible to extend the parametric formulation of the model to finite depth water by incorporating depth-dependent phase speed and depth-dependent spectral shape functions in the model equations along with a bottom dissipation term. Experiments with a shallow water model and comparisons of the results with field measurements are currently underway.

REFERENCES

Bretschneider, C. L., 1970: Wave forecasting relations for wave generation. Look Lab, Hawaii, Vol. 1, No. 3.

Donelan, M.A., 1977: A simple numerical model for wave and wind stress prediction. Report, Nat. Water Res. Inst., Burlington, Ontario.

Hasselmann, K.D., T.P. Barnett, E. Bouws, H. Carlson, D.E. Cartwright, K. Enke, J.A. Ewing, H. Gienapp, D.E. Hasselman, P. Kruseman, A. Merrburg, P. Muler, D.J. Olbers, K. Richter, W. Sell, and H. Walden, 1973: Measurements of wind-wave growth and swell decay during the joint North Sea wave project (JONSWAP). Deutsh. Hydrogr. Zeit., Sup. A, Vol. A, No. 12, 95pp.

Liu, P.C., D.J. Schwab and J.R. Bennett, 1984: Comparison of a two-dimensional wave prediction model with synoptic measurements in Lake Michigan. J. Phys. Oceanogr., 14(9), 1514-1518.

Schwab, D.J., J.R. Bennett, P.C. Liu and M.A. Donelan, 1984a: Application of a simple numerical wave prediction model to Lake Erie. J. Geophys. Res., 89(C3), 3586-3592.

Schwab, D.J., G.A. Meadows, J.R. Bennett, H. Schultz, P.C. Liu, J.E. Campbell, and H.H. Dannelongue, 1984b: The response of the coastal boundary layer to wind and waves: analysis of an experiment in Lake Erie. J. Geophys. Res., 89(C5), 8043-8053.

The SWAMP Group, 1985: Ocean Wave Modeling. Plenum Press, New York, 256 pp.

SUBJECT INDEX

Page number refers to first page of paper.

AUTHOR INDEX

Page number refers to first page of paper.